Lecture Notes in Civil Engineering 328

Lecture Notes in Civil Engineering (LNCE) publishes the latest developments in Civil Engineering—quickly, informally and in top quality. Though original research reported in proceedings and post-proceedings represents the core of LNCE, edited volumes of exceptionally high quality and interest may also be considered for publication. Volumes published in LNCE embrace all aspects and subfields of, as well as new challenges in, Civil Engineering. Topics in the series include:

- Construction and Structural Mechanics
- Building Materials
- Concrete, Steel and Timber Structures
- Geotechnical Engineering
- Earthquake Engineering
- Coastal Engineering
- Ocean and Offshore Engineering; Ships and Floating Structures
- Hydraulics, Hydrology and Water Resources Engineering
- Environmental Engineering and Sustainability
- Structural Health and Monitoring
- Surveying and Geographical Information Systems
- Indoor Environments
- Transportation and Traffic
- Risk Analysis
- Safety and Security

To submit a proposal or request further information, please contact the appropriate Springer Editor:

– Pierpaolo Riva at pierpaolo.riva@springer.com (Europe and Americas);
– Swati Meherishi at swati.meherishi@springer.com (Asia—except China, Australia, and New Zealand);
– Wayne Hu at wayne.hu@springer.com (China).

All books in the series now indexed by Scopus and EI Compendex database!

Wei Guo · Kai Qian · Honggang Tang · Lei Gong
Editors

Proceedings of the 2023 International Conference on Green Building, Civil Engineering and Smart City

GBCESC

Springer

Editors
Wei Guo
Central South University
Changsha, China

Kai Qian
Guilin University of Technology
Guilin, China

Honggang Tang
Guizhou University
Guiyang, Guizhou, China

Lei Gong
Guizhou University
Guiyang, Guizhou, China

ISSN 2366-2557 ISSN 2366-2565 (electronic)
Lecture Notes in Civil Engineering
ISBN 978-981-99-9946-0 ISBN 978-981-99-9947-7 (eBook)
https://doi.org/10.1007/978-981-99-9947-7

This Springer imprint is published by the registered company Springer Nature Singapore Pte Ltd.
The registered company address is: 152 Beach Road, #21-01/04 Gateway East, Singapore 189721, Singapore

Paper in this product is recyclable.

Preface

Civil engineering plays a very important role in urbanization. As urban areas continue to expand and the global population increases, there is an urgent need to adopt sustainable development measures and incorporate smart technologies to minimize the impact on the environment in order to increase resource efficiency, improve the quality of life of the inhabitants, and provide good working conditions for people. The 2nd International Conference on Green Building, Civil Engineering, and Smart City is dedicated to addressing these challenges head-on, with a specific focus on the role of civil engineering in achieving Sustainable Development Goals. This conference acts as a platform for researchers, professionals, and industry experts to come together and share their knowledge and experiences in this crucial field.

This proceedings is a collection of papers presented at the 2nd International Conference on Green Building, Civil Engineering, and Smart City (GBCESC) that deals with the issues stated above. After a rigorous peer-review process, 99 papers were selected for presentation at the conference, covering a wide range of topics related to civil engineering and smart cities. The selected papers represent the latest research and developments in these areas and showcase the diverse perspectives and approaches adopted by researchers and practitioners. The accepted papers were organized into thematic sessions covering topics such as geological engineering, intelligent construction, innovative materials and technologies, and smart city applications. Each session was chaired by an expert in the relevant field, who provided insights and led discussions on the papers presented.

Several outstanding keynote speakers presented the state-of-the-art findings. Our esteemed speakers are Prof. Kejian Ma, Prof. Zhishen Wu, Prof. Kai Qian, Prof. Wei Guo, Prof. Liu Jin, Prof. Kai Wei, Prof. Zhonghua Gou, and Prof. Linchuan Yang. In addition, there are two invited presentations delivered by Prof. Junxian Zhao and Dr. Filipe Afonso.

The editors would like to thank all members of the scientific and organizing committees as well as the reviewers for their important contributions to the organization of this conference and for their support in the review process. We would like to thank all the keynote speakers, authors, and attendees for the time and effort they put into their presentations. We would like to acknowledge the major sponsor, Guizhou University, and other co-sponsors who helped organized the GBCESC 2023 event, including Guilin University of Technology, Central South University, Soochow University, High-Speed Railway of Construction Technology of National Engineering Research Center of China, China-Portugal Joint Laboratory of Cultural Heritage Conservation Science Supported—The Belt and Road Initiative, Xi'an University of Architecture and Technology, Shaanxi Key Laboratory of Geotechnical and Underground Space Engineering, Key Lab for Intelligent Infrastructure and Monitoring of Fujian Province, China Association of Building Energy Efficiency, and Committee of Earthquake Prevention and Disaster Reduction in Infrastructure of Seismological Society of China. Additionally, we would also like to express our sincere appreciation to Springer for their assistance in making the conference a success.

We hope that the proceedings of the 2nd International Conference on Green Building, Civil Engineering, and Smart City will significantly contribute to the advancement of these fields and will inspire future research and innovation. We hope that readers will find the book useful for their current projects and a source of inspiration for future ones.

Kai Qian
Wei Guo
Honggang Tang
Lei Gong

Organization

General-Chairs

Kai Qian	Guilin University of Technology, China
Wei Guo	Central South University, China
Honggang Tang	Guizhou University, China
Liu Jin	Beijing University of Technology, China

Co-chairs

Yu Zhao	Guizhou University, China
Yao Wu	Soochow University, China
Huagang Zhang	Guizhou University, China

Honorary Advisor

Stephen Morgan	University of Saint Joseph, Macao

Technical Chairs

Lei Gong	Guizhou University, China
Zhen Guo	Zhejiang University, China
Li Zhou	Guizhou University, China
Dillip Kumar Das	University of KwaZulu-Natal, South Africa

Technical Program Committees

Feng Chen	Tongji University, China
Jianlin Liu	China University of Petroleum, China
Yonghui Li	University of Sydney, Australia
Lin Zhao	Tongji University, China
Jun Wang	Wenzhou University, China
Lizhong Jiang	Central South University, China
Jiangfeng Dong	Sichuan University, China
Bing Fu	Jinan University, China
Piguang Wang	Beijing University of Technology, China
Jiang He	Gunagxi University, China
Xuhui He	Central South University, China

Tony Yang	University of British Columbia, UK
Yan Han	Changsha University of Science and Technology, China
Yifei Hao	Hebei University of Technology, China
Xiyin Zhang	Lanzhou Jiaotong University, China
Wei Wang	Hefei University of Technology, China
Junjie Zeng	Guangdong University of Technology, China
Yulin Zhan	Southwest Jiaotong University, China
Jin Jiang	Guangzhou University, China
Mengfu Wang	Hunan University, China
Hongniao Chen	Guizhou University, China
Marco Cimillo	Xi'an Jiaotong-Liverpool University, China
Hazem Samih Mohamed	Fujian Agriculture and Forestry University, China
Gao Ma	Hunan University, China
Irineu da Silva	The University of São Paulo, Brazil
Nur Izzi Md Yusoff	Universiti Kebangsaan Malaysia, Malaysia
Junsheng Su	Tianjin University, China
Ali M. Shehadeh	Yarmouk University, Jordan
Ghassan Almasabha	The Hashemite University, Jordan
Syuhaida Ismail	Universiti Teknologi Malaysia, Malaysia
Zhenhua Duan	Tongji University, China
Peng Wei	South China University of Technology, China
Guan Lin	Southern University of Science and Technology, China
Mohammad Arif Kamal	Aligarh Muslim University, India
Zuolong, Li	China University of Petroleum (East China), China
Jian Zhong	Hefei University of Technology, China
Pleasa Serin Abraham	Bangalore University, India
Dat Doan Auckland	University of Technology, New Zealand
Oluwole A. Odetunmibi	Covenant University, Nigeria
Hilma Tamiami F.	University of Sumatera Utara, Indonesia
Yu Cao	University of Malaya, Malaysia
Azhan Abdul Aziz	Universiti Teknologi MARA, Malaysia
Mohd Norazam Bin Yasin	Universiti Tun Hussein Onn Malaysia, Malaysia
Nazirah bt Mat Russ	International Islamic University, Malaysia
Yeong Huei Lee	Curtin University, Malaysia
Lisaia Daria	Vanke Urban Research, China
Xiaolin Qi	China University of Petroleum (East China), China

Contents

Green Building

Civil Engineering

Experimental Investigation on Shear Connectors for Glulam-UHPC Composite Structures

Wanru Huang, Pengcheng Li, and Xiaoyue Zhang(✉)

Chongqing University, Chongqing 400045, China
xiaoyuezhang@cqu.edu.cn

Abstract. The combination of glulam timber and concrete has shown to have significant potential for sustainable building design, exploiting the compressive strength of concrete and the tensile strength of timber. To improve the structural performance of timber-concrete composite beam (TCC), this paper proposed a glulam-UHPC composite beam structure that incorporated Ultra-High Performance Concrete (UHPC) material, which possessed high strength and durability. This new composite system aimed to reduce dead weight, minimize long-term deformation, and enhanced the span capacity of the structure. However, useful shear connection between the glulam beams and concrete slabs was crucial to ensure the structural feasibility of this system. To investigate the mechanical behavior of the Timber-UHPC connectors in composite structures, various connection systems were analyzed and compared through push-out tests. Results revealed that the notch-screw connection exhibited superior bearing capacity, anti-sliding stiffness, and ductility compared to pure screw and pure notch connectors. This study provided valuable insights into the design and construction of composite structures using TCC and UHPC materials, with potential applications in a range of structural systems.

Keywords: Timber-UHPC composite structures · Shear connector · Pust-out tests

1 Introduction

The Timber-Concrete Composite (TCC) structure has emerged as a viable construction solution, and has found widespread use in new building construction, small-span bridges, and the upgrading and maintenance of existing structures [1]. Nonetheless, the application of TCC structures has been marred by certain issues, the concrete shrinks as it dries, and if not controlled properly, this can lead to cracking. Furthermore, concrete is a dense and heavy material, which can limit its use in certain applications where weight restrictions are essential, such as lightweight timber bridges. Finally, the normal concrete slab has poor durability and bonding failure between timber-concrete interfaces [2]. To surmount these challenges, a novel composite structure known as glulam-UHPC Composite (GUCC) has been proposed. A bridge deck made of UHPC adds much less dead weight to the timber structure compared with a conventional standard-strength concrete bridge deck. When using a precast UHPC bridge deck, the effects of creep and shrinkage of concrete are significantly reduced [3].

W. Guo et al. (Eds.): GBCESC 2023, LNCE 328, pp. 3–11, 2024.
https://doi.org/10.1007/978-981-99-9947-7_1

Glulam-UHPC Composite structures offer tremendous potential for sustainable building design, combining the natural advantages of timber with the exceptional strength and a low creep coefficient of UHPC. One critical element in such structures is the shear connector - a device used to transfer shear forces between the timber and concrete components. Over the past few decades, several connectors have been developed, including metal fasteners like screws, studs, notches cut in the timber and filled with concrete, glued notches, and glued steel plates [4]. The shear performance of the connection is typically evaluated through push-out tests. However, limited studies are currently available on the connectors for glulam-UHPC composite structures. Holý and Vr´ablík [5] proposed a novel type of composite connection between timber and ultra-high-performance concrete (UHPC) elements in structural engineering. This paper described the experimental evaluation of the unique connection system, which consisted of a UHPC dowel cast into a precast element that connected with the timber member through cutting teeth and gluing. The results demonstrated the high strength and ductility of the connections and their potential for use in a wide range of structures requiring hybrid timber-concrete solutions.

In this paper, three different connection methods which include pure screw shear connectors, pure notch shear connectors, and hybrid notch-screw shear connectors were considered and in order to examine the best reliable connectors forms for the Timber-UHPC interface in composite structures.

2 Experimental Tests

2.1 Material Properties

The timber elements used in this investigation were made of Douglas fir glued laminated timber (glulam). The glulam specimens were sized to 20 mm × 20 mm × 20 mm, and their average moisture content was determined by testing eight specimens, yielding a value of 11.82%. To ascertain the compressive strength and elasticity modulus of the timber parallel to the grain, the specimens were subjected to testing in compliance with EN standard 408 [6]. The glulam demonstrated an average elasticity modulus of 12363 MPa and a compressive strength of 36.5 MPa, with the outcomes tabulated in Table 1.

To connect the timber element, self-tapping screws (Konstruc X) were used, which were specifically designed for this purpose. The self-tapping screws had a nominal diameter of 10 mm and a length of 195 mm, with a shank and tip diameter of 6.1 mm and 5.7 mm, respectively. The pre-drilling hole diameter was 7 mm, and the characteristic yield moment of the self-tapping screw was 40000 MPa, as provided by the manufacturer (Euro Tec Inc).

The UHPC used in this study is UDC(II)-150 supplied by Jiangsu Subot New Materials Co., Ltd., commercially available and contains 7% volume content of steel microfibers with dimensions of 12.7 mm length and 0.2 mm diameter. To determine the mechanical properties of the concrete, tests were conducted in accordance with T CBMF 37–2018 [7]. Table 1 presents the mean values and standard deviations of the concrete properties characterized through laboratory testing.

Table 1. Mechanical properties of glulam and UHPC used in the test (MPa).

Component	Mechanical properties			
	Modulus of elasticity	Compressive strength	Bending strength	Shear strength
Glulam	12363	36.5	54.97	6.2
UHPC	40100	120	–	–

2.2 Push-Out Tests

Test Specimens

Three test series were considered, each test series has two replicates and a total of 6 specimens were fabricated for the push-out tests, as shown in Fig. 1. The connection details are illustrated in Fig. 2. Table 2 presents the parameters of test specimens. The number of self-tapping screws that embedded into series S-4 and series SN-2 were 4 and 2, respectively.

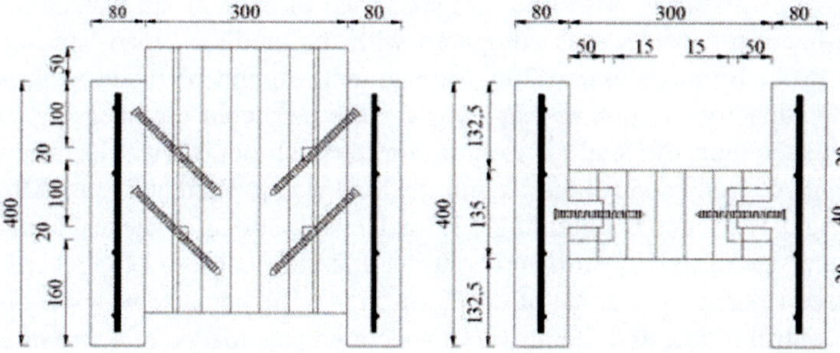

(a) (b)

Fig. 1. Schematics of a push-out specimen (unit: mm): (a) elevation view, (b) top view.

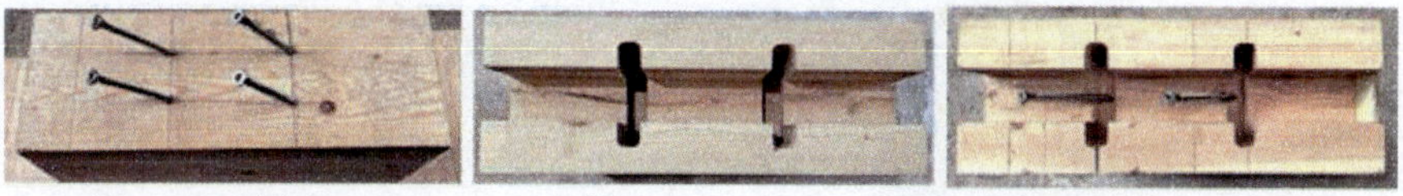

(a) (b) (c)

Fig. 2. Connection details for test series: (a) series S-4, (b) series N-2, (c) series SN-4.

Test Setup and Loading Procedure.

For the push-out tests, a 1000 kN universal testing machine with measurement systems were used. The load was applied directly to the specimen and four linear variable different transformers (LVDT), were used to measure the relative slip between the UHPC slab

Table 2. Notched connection specimen design.

Test Series	Replicates	Screw number	embedded angle	Notch dimension (mm)
				length × width × depth
S-4	2	4	45°	–
N-2	2	0	–	20 × 100 × 65 + 400 × 40 × 50
SN-2	2	2	45°	20 × 100 × 65 + 400 × 40 × 50

and the glulam member. The LVDTs were symmetrically arranged at both sides of the specimens. To prevent lateral movement of the concrete slabs during loading, angle irons were used at the bottom of the push-out specimen to provide proper constraint, with the limit being that the angle irons did not slide off. Please refer to Fig. 3(a) for a detailed depiction of the boundary conditions and the measurement devices used in the test.

According to European standard EN26891 [8], the loading system for push-out testing of shear connectors is shown in Fig. 3(b). Before adjusting the loading system, it was necessary to estimate the bearing capacity of the push-out specimens. The predicted load-bearing capacity of a push-out specimen could be determined by referencing literature or previous research. After the first specimen of each group was completed, the actual load-bearing capacity was compared with the predicted load-bearing capacity, and if they differ by more than ±20%, adjustments should be made to the estimated load-bearing capacity. Assuming that F_{est} was the predicted load-bearing capacity of the push-out specimen, the loading program for the push-out test was as follows: 1) The actuator adopted a load control mode, and the load was applied at a rate of $0.2F_{est}$/min until reaching $0.4F_{est}$, and maintaining this load level for 30 s; 2) The test load decreased to $0.1F_{est}$ at a rate of $0.2F_{est}$/min and continued at that load level for 30 s; 3) The test load increased to $0.7F_{est}$ at a rate of $0.2F_{est}$/min; 4) The actuator was switched to displacement control mode, and the push-out specimen was loaded at a rate of 2 mm/min until failure.

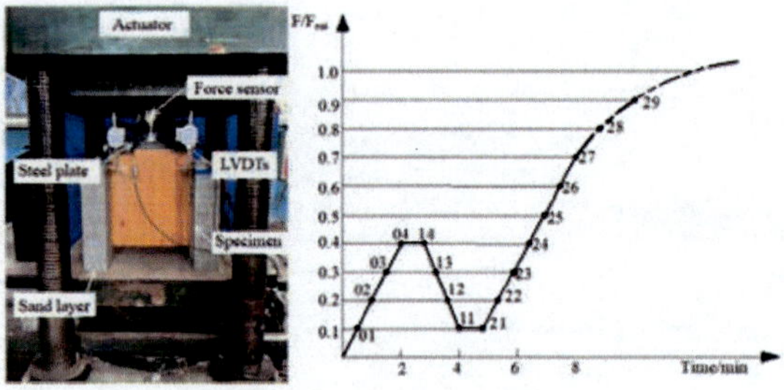

(a) (b)

Fig. 3. Test setup and measurements: (a) test setup; (b) the procedure for loading the specimen.

3 Tests Results and Discussion

3.1 Failure Mode

The results of the push-out tests are depicted in Fig. 4, where the failure modes of the different specimens are presented. The series S-4 demonstrated primary failure in the form of compression deformation of the glulam member and bending failure of the screws, as illustrated in Fig. 4(a). Conversely, the series N-2, which featured a notch cut into the timber and filled with UHPC, exhibited shear failure of the UHPC tenon accompanied by timber shear and splitting failure, as shown in Fig. 4(b). Similarly, the series SN-2 displayed timber splitting failure after embedding self-tapping screws into the timber notch, as seen in Fig. 4(c). In addition, all self-tapping screws experienced shear failure, alongside the shear failure of the UHPC tenon. However, the series SN-2 demonstrated a significant improvement in preventing the concrete tenon from slipping out of the timber notch. The self-tapping screws embedded into the timber notch effectively restrained the slipped concrete tenon after the timber splitting failure, resulting in excellent post-peak bearing performance.

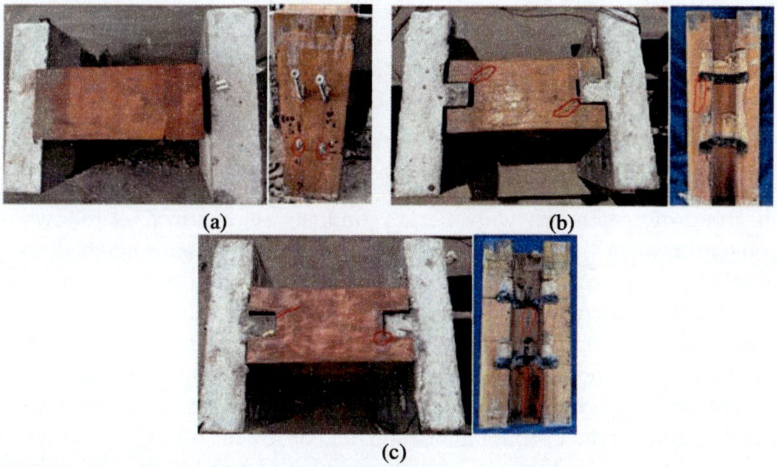

Fig. 4. Failure diagram of the shear connector: (a) bending failure of the screw; (b) UHPC tenon shear failure; (c) timber splitting failure.

3.2 Shear Capacity

Table 3 summarizes the testing results. The peak load (F_{max}) was determined as the failure load of the tested specimens; the slip value at the failure (s_u) and the maximum slip (s_{max}) were determined in accordance with EN standard 12512 [9]; the serviceability stiffness of the connection (Ks) was determined according to EN standard 26891 [8].

The notch-screw connection (series SN-2) exhibited a significantly improved bearing capacity when compared to the pure screw connection (series S-4). The pure screw push-out specimens had a s_u of 13.7 mm, which indicated significant deformation within the bearing capacity limit range and a limited anti-slip stiffness. Conversely, the series SN-2 demonstrated a clear increase in the ultimate bearing capacity compared to the series N-2, with a 47% increase resulting from the increase in screw number. This suggests that the use of notches cut into the timber and reinforced with screws effectively restrains the deformation of concrete tenons and prevents premature shear damage to the timber.

Table 3. Push-out test results.

Specimen	F_{max} (kN)	F_u (kN)		s_u (mm)		s_{max} (mm)		K_s (kN/mm)	
	Tv	Tv	Av	Tv	Av	Tv	Av	Tv	Av
S-4-a	280.4	140.2	144 (2.6%)	2.8	3.0 (6.7%)	12.3	13.7 (8.8%)	208.7	205 (2.0%)
S-4-b	295.6	147.8		3.2		14.8		201.3	
N-2-a	292.3	146.2	151.4 (3.4%)	1.6	1.5 (10.5%)	15	15 (0%)	310.7	335.1 (7.3%)
N-2-b	312.9	156.5		1.3		15		359.5	
SN-2-a	459.3	229.7	222.8 (3.1%)	3.1	2.9 (8.8%)	15	15 (0%)	327.8	346.6 (5.4%)
SN-2-b	431.6	215.8		2.6		15		365.4	

3.3 Load-Slip Curve

Figure 5 displays the test results which includes maximum loads and relative slips for each group of specimens, with the key findings summarized as follows: Firstly, a comparison between the series S-4 and N-2 revealed that pure notched connectors utilizing self-tapping screws had low shear capacity, poor ductility, and exhibited brittle failure. Secondly, an increase in screw number in the notch of the series N-2 and SN-2 resulted in significant improvements in strength of the notched connectors. Thirdly, the adoption of an internal notch in series SN-2 led to a significant improvement in bearing capacity compared to series S-4. This improvement was attributed to an increase in the shear area of timber at the opening tenon and the constraints of timber on both sides of the notch, as evident from Fig. 5(c). Additionally, the screws in the concrete of series SN-2 continued to play a role, exhibiting better post-yield shear performance. Further enhancement of the shear capacity of the concrete tenon could be achieved by placing structural reinforcement bars in the tenon.

3.4 Slip Modulus

The slip modulus of the specimens was determined at different stress stages according to the EN 26891 standard [8]. This can be obtained as follows:

$$K_{0.4}=\frac{0.4F_u}{\frac{4}{3}(s_{0.4}-s_{0.1})} \quad (1)$$

$$K_{0.6}=\frac{0.6F_u}{s_{0.6}-s_{24}+\frac{4}{3}(s_{0.4}-s_{0.1})} \quad (2)$$

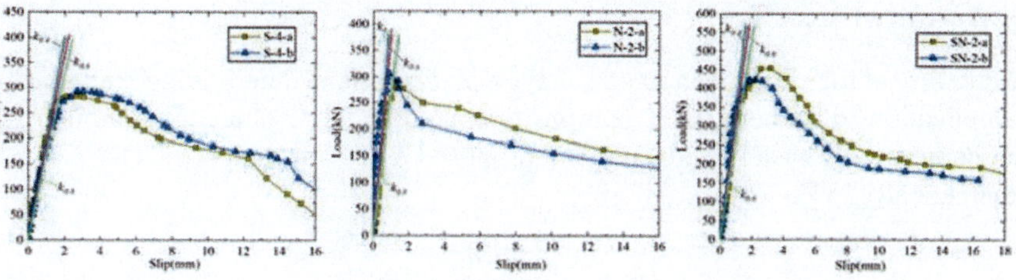

(a) (b) (c)

Fig. 5. Load-slip curve of six groups: (a) series S-4; (b) series N-2; (c) series SN-2.

$$K_{0.8}=\frac{0.8F_u}{s_{0.8}-s_{24}+\frac{4}{3}(s_{0.4}-s_{0.1})} \tag{3}$$

where F_u is the maximum test load of the specimen. The physical meaning of each slip displacement (s) can be referenced in Fig. 3 (b), where $s_{0.1}$ and $s_{0.4}$ are the relative interfacial slips (mm) at $0.1F_{est}$ and $0.4F_{est}$ during the first loading cycle, respectively; s_{24} is the relative interfacial slip (mm) of the push-out specimen at $0.4F_{est}$ during the second loading stage; $s_{0.6}$ and $s_{0.8}$ represent the relative interfacial slips (mm) when the test loads reach $0.6F_{max}$ and $0.8F_{max}$, respectively.

Table 3 and Fig. 5 illustrate the calculation of $K_{0.4}$, $K_{0.6}$, and $K_{0.8}$ of the specimens, and Fig. 6 illustrates the comparison of slip stiffness at different stress stages. The variations of series S-4 were smoother than those of series N-2 and SN-4. Compared to series S-4 (pure screw connector), the slip modulus of notch-screw connectors had been significantly improved. Comparing the series N-2 (pure notch connector) with series SN-2 (hybrid notch-screw connector), it could be found that increasing the number of screws has an improvement in the slip stiffness of the specimens at various stages. The $K_{0.4}$ of the SN-2 specimens increased by 3.43% compared to the N-2 specimens, while $K_{0.6}$ and $K_{0.8}$ increased by 5.13% and 14.55% respectively compared to the N-2 specimens.

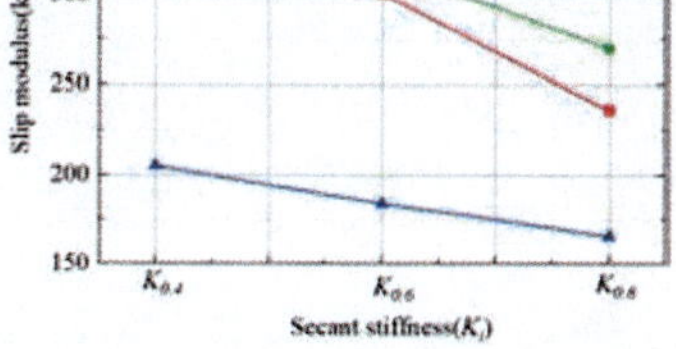

Fig. 6. Comparison of slip stiffness at different stress stages.

3.5 Ductility Factor

The ductility of the shear connections plays a pivotal role in determining their feasibility for application in timber-UHPC composite structures. To evaluate the ductility of the shear connections, we adopted a method proposed by Deam et al. [10]. This is captured mathematically in Eq. (4):

$$D_1 = \frac{F_u - F_{10}}{F_u} \times 100\% \tag{4}$$

where F_u is the maximum test load of the specimen, and F_{10} represents the load corresponding to a slip of 10mm. When D_1 is no greater than 20%, this connection can be defined as a ductile connection.

Table 4 summarizes the ductility results along with the maximum interface relative slips, which provide a measure of the shear connections deformation ability. Among the tested series, the pure screw connections in series S-4 displayed a completely ductile failure mode with excellent deformability. On the other hand, the specimens in series N-2 exhibited brittle failure and poor deformability, indicating a non-ductile behavior. The internal rectangular notched connections with screws showed an improvement in ductility performance with a ratio of 30.5%, a significant increase of almost 26.5% compared to series N-2. This improved behavior can be attributed to the reinforcement of self-tapping screws, which resulted in better deformation ability and ductility performance. Overall, the results suggest that the self-tapping screw reinforcement can effectively enhance the ductility behavior of shear connections, thus improving their performance and reliability.

Table 4. Determination of ductility of specimens.

Specimen	D1 (%)	Average (%)	Ductile/brittle	Maximum slip (mm)
S-4-a	11.5%	12.9%	Ductile	12.3
S-4-b	14.2%			14.8
N-2-a	37.5%	41.5%	Brittle	15.0
N-2-b	45.5%			15.0
SN-2-a	27.6%	30.5%	Fairly ductile	15.0
SN-2-b	33.4%			15.0

4 Conclusion

This paper investigated the mechanical properties of the timber-UHPC shear connector in composite structures, based on the six groups experimental results and discussion, the following conclusions can be drawn:

Firstly, the notch-screw connection exhibited superior mechanical properties compared to the pure screw and pure notch connections. The load-slip relationship showed higher bearing capacity, anti-sliding stiffness, and higher ductility.

Secondly, the shear capacity ranged from 280.4 to 459.3 kN, with the primary failure mode being the shear failure of the concrete tenons and the bending failure of self-tapping screws. These findings provide important insights into the failure modes and mechanisms of shear connections in glulam-UHPC composite beams.

Finally, increasing the number of screws in the notch improved the strength and ductility of the notched connectors. This suggests that the notch-screw connection can be further optimized by increasing the number of screws to achieve better mechanical properties.

Overall, this study provides valuable information on the mechanical properties of shear connectors in glulam-UHPC composite beams and highlights the importance of using the notch-screw connection for enhancing the performance of these structures.

Acknowledgement. The authors would like to thank the National Key R&D Program of China (Grant Nos. 2021YFF0500804); The National Natural Science Foundation of China (Grant Nos. 52108192); The Chongqing Technology Innovation and Application Development Project (Grant Nos. CSTB2022TIAD-KPX0138); General funded project of China Postdoctoral Science Foundation (Grant Nos. 2022M710528); the OEICDI fund (Grant Nos. B13041); Open Research Fund for Key Laboratory of Building Structure Reinforcement and Underground Space Engineering, Ministry of Education (Grant Nos. MEKL202204); the Entrepreneurship and Innovation Support Program for Overseas-educated student in Chongqing China (Grant Nos. CX2021085).

References

1. Wang, H., Li, W., Liu, X., Wang, X., Chen, A., Qin, H.: Behavior of horizontal steel plate + studs connectors in a glulam-UHPC composite system: experiment and analysis. Eng. Struct. **275**, 115187 (2023)
2. Ferrier, E., Agbossou, A., Michel, L.: Mechanical behaviour of ultra-high-performance fibrous-concrete wood panels reinforced by FRP bars. Compos. B Eng. **60**, 663–672 (2014)
3. Holý, M., Čítek, D., Tej, P., Vráblík, L.: The experimental timber–UHPC composite bridge. Sustainability **13**(9), 4895 (2021)
4. Ceccotti, A.: Timber-concrete composite structures. Timber Eng. Step **2**(1), E13 (1995)
5. Holý, M., Vráblík, L: Push-out shear tests for timber-UHPC composite footbridge. In Proceedings of the 12th International fib PhD Symposium in Civil Engineering, pp. 195–202. FIB - Féd. Int. du Béton, Prague (2018)
6. EN 408: Timber structure-Structural timber and glued laminated timber-Determination of some physical and mechanical properties. European Committee for Standardization, Brussels, Belgium (2009)
7. T CBMF37–2018: Fundamental characteristics and test methods of ultra-high performance concrete. China Building Material Council, Beijing, China (2018)
8. EN 26891: Timber structure-Joint made with mechanical fastener-General principles for the determination of strength and deformation characteristic. European Committee for Standardization, Brussels, Belgium (1991)
9. EN 12512: Timber structure-Test methods-Cyclic testing of joints made with mechanical fasteners. European Committee for Standardization, Brussels, Belgium (2001)
10. Deam, B.L., Fragiacomo, M., Buchanan, A.H.: Connections for composite concrete slab and LVL flooring systems. Mater. Struct. **41**, 495–507 (2008)

A Dynamic Detection Method for Railway Track Irregularities Combining Line-Structured Lasers and GNSS/IMU

Tong Wang[1], Haoxuan Xu[1], Qingzhou Mao[1,2](✉), Yuanbo Mu[1], and Guangqi Wang[1]

[1] School of Remote Sensing and Information Engineering, Wuhan University, Hubei 430072, China
qzhmao@whu.edu.cn
[2] Hubei Luojia Laboratory, Hubei 430079, China

Abstract. A method combining line-structured lasers and GNSS/IMU sensor groups was proposed for dynamic detection of geometric irregularity parameters of railway tracks. 4 sets of line-structured lasers and a set of GNSS/IMU modules are mounted on a small track inspection vehicle and used to collect 3D point clouds of the track area, then through hard synchronization, the point cloud of the left and right track areas was fused with the sensor space pose parameters to obtain the point cloud model of the track area in absolute coordinates system. Use the combination of straight-through filter and cloth filter to extract the center line of the top surface of the rail and the gauge measurement reference line, then calculated the track irregularity parameters based on the filtering results. The method in our paper was verified on a test railway section. The results show that the proposed method can accurately obtain the center line of the rail top and the reference line of the gauge measurement, and can quickly and dynamically implement the five irregular geometric parameters of height, level, twist, rail direction, and gauge, and the accuracy can reach mm level measurement.

Keywords: Track irregularity · Line structured light camera · 3D point cloud · Cloth simulation filtering · Parameter measurement

1 Introduction

In the process of railway operation, the dynamic load of the train, subgrade settlement, natural environment and other reasons will cause the railway track to deviate from the design, which is called track irregularity [1]. Track irregularities will affect ride comfort and line safety operation by stimulating train vibration during travel [2], and regular inspection and maintenance are required to ensure that the line meets the safe operation standards [3]. With the continuous increase of railway mileage, how to implement efficient line inspection and repair in a short maintenance skylight is an urgent problem to be solved.

W. Guo et al. (Eds.): GBCESC 2023, LNCE 328, pp. 12–21, 2024.
https://doi.org/10.1007/978-981-99-9947-7_2

The existing track irregularity detection technology can be divided into static and dynamic detection [4]. Static detection mainly relies on instruments such as track ruler, track measuring instrument, laser long string inspection instrument, etc., and track geometric parameters are measured by static string measurement method [5] or vector distance difference method [6], but the method mentioned above has low detection efficiency, poor safety, and is affected by the dynamic load of the train, which is different from the dynamic detection results in terms of high and low and uneven rail direction [7]. In terms of dynamic detection, China's rail inspection vehicles mainly rely on on-board inertial navigation and acceleration, and calculate the uneven geometric parameters of the track through the inertial reference method [8]. However, this method has relatively low signal-to-noise and high requirements for vehicle speed during the measurement process, which is easy to produce error accumulation, and it is difficult to accurately measure long-wave unevenness [9]. Qijin Chen etc. [10] use the GNSS/INS integrated system to dynamically measure the unevenness of the track, Its relative positioning accuracy can reach 1mm, but it depends on data post-processing. Zhang Yuxuan et al. [11] combined the ten-degree-of-freedom rail vehicle model and the Kalman filter algorithm to study the unevenness detection effect of the on-board vibration observation scheme under different sensor combinations. Although the above scheme can achieve dynamic detection, it only considers one-dimensional vectors such as displacement and acceleration during vehicle driving.

3D reconstruction technology can obtain precise orbital geometry [12], which can be used to detect orbital irregularities at any wavelength. Compared with lidar, the linear structured light sensor can directly realize the three-dimensional scene reconstruction within a defined angle, which is more suitable for the track unevenness detection task that only models the rail structure. Xu Wanyang et al. [13] designed a track measurement system based on industrial camera and linear structure light source, the maximum gauge measurement error of the system is 0.96 mm, but the system cannot obtain the displacement and attitude parameters of the measuring trolley, so the measurement content is limited, In the inertial reference method optimized by Chen Shiming et al. [14], the linear structured light sensor is only used to provide ranging information and direct relative displacement of the bogie.

In this paper, a non-contact track unevenness dynamic detection method based on line structured light sensor is proposed. The track area was modeled in 3D using four structured light cameras and a GNSS/IMU sensor set mounted on a light rail inspection vehicle. Then, the orbital point cloud data and the sensor pose data are fused by hard synchronization to obtain the point cloud data under the absolute coordinate system. Through the point cloud processing of pass-through filtering and cloth filtering, the top plane and gauge measurement reference lines of the two rails were extracted, and with this result, the measurement of high and low, distortion, level, orbital direction, track gauge and five types of uneven parameters is realized.

2 Track Geometric Parameters Calculation Based on Line Structured Light Point Cloud

2.1 Data Collection

The measurement system used for dynamic three-dimensional modeling for tracks and other related structures, showed in Fig. 1, is mounted on a light rail inspection vehicle. Four sets of line structured light sensors installed at both ends of the vehicle, as well as GNSS/IMU positioning and attitude determination modules constitute this system. The complete section contours on both sides of the track are dynamically collected through the line structured light sensors, and the point cloud of each frame is fused with the position and attitude parameters measured by GNSS/IMU in a hard synchronous manner through the synchronization control board to obtain the three-dimensional point cloud model of the tracks in the absolute coordinate system. After data fusion, the point cloud data is cut according to the time frame. The former N-1 files are of the same size, for instance, 1 GB, while the Nth file contains the remaining. As the vehicle runs at a speed of 20 km/h, the coverage range of each cut files is around 15 m, which can be considered as a straight line approximately to achieve the data processing automatically. The range of original point cloud $\{P_{origin}\}$ for each individual file is represented by formula (1), where parameters X_{min}, Y_{min}, Z_{min} are offsets to decrease the coordinate values and to increase the efficiency.

$$\{P_{origin}\} \subset \begin{cases} 0 \leq x - X_{min} \leq X_{max} - X_{min} \\ 0 \leq y - Y_{min} \leq Y_{max} - Y_{min} \\ 0 \leq z - Z_{min} \leq Z_{max} - Z_{min} \end{cases} \tag{1}$$

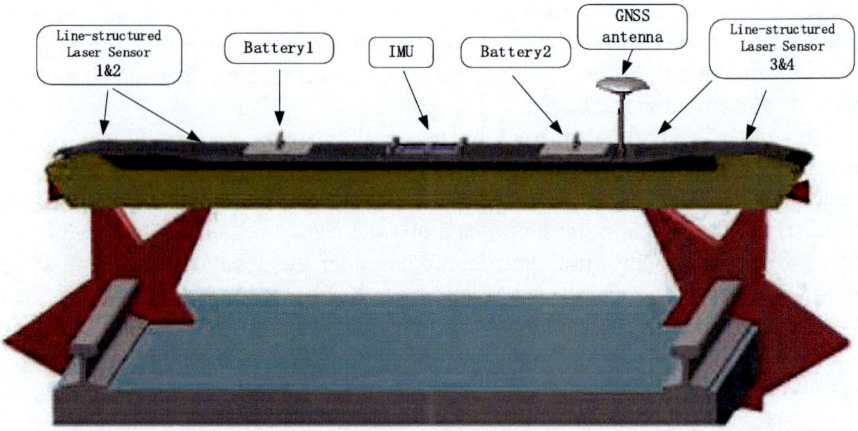

Fig. 1. Schematic diagram of track comprehensive inspection vehicle

2.2 Extraction of Key Point Sets for Comfort Parameters Calculation

The top centerline points of tracks and the reference points located 16 mm below the top centerline for track gauge measurement are most significant for calculating the geometric parameters of tracks. As the point cloud data acquired by line structured light sensors in a fixed angle contains the non-track part, it is of great necessity to get tracks filtered. Pass through filter along the elevation direction is the most intuitive method, treating points exceeding a certain height as track points, but causing truncation error due to the fluctuation of track. Cloth simulation filter can obtain ground points and off-ground points quickly from large scale point cloud data, but results in missing of tracks depending on calculation parameters (scene, cloth resolution, max iteration and classification threshold). A combined filter based on pass through and cloth simulation methods to extract track points accurately is applied. Approximate bounding box of track point set $\{P_{track}^{CSF}\}$ shown in formula (2) is obtained by cloth simulation filter with appropriate parameters.

$$\left\{P_{track}^{CSF}\right\} \subset \left\{\begin{array}{l} X_{min}^{CSF} \leq x \leq X_{max}^{CSF} \\ Y_{min}^{CSF} \leq y \leq Y_{max}^{CSF} \\ Z_{min}^{CSF} \leq z \leq Z_{max}^{CSF} \end{array}\right. \tag{2}$$

Based on the cloth simulation filter bounding box, a threshold in Z direction is set to extract exact track point cloud set $\{P_{track}^{CSF}\}$, which is shown as formula (3).

$$\{P_{track}\} \subset \left\{\begin{array}{c} 0 \leq x - X_{min} \leq X_{max} - X_{min} \\ 0 \leq y - Y_{min} \leq Y_{max} - Y_{min} \\ Z_{min}^{CSF} - z_{Thres} \leq z \leq Z_{max} - Z_{min} \end{array}\right. = \left\{\begin{array}{l} X_{min}^{track} \leq x \leq X_{max}^{track} \\ Y_{min}^{track} \leq y \leq Y_{max}^{track} \\ Z_{min}^{track} \leq z \leq Z_{max}^{track} \end{array}\right. \tag{3}$$

Figure 3 shows the different effects of different filters. Figure 2(a) is the original point cloud. Figure 2(b) is the result of pass through filter, containing data missing (b1) and noises of non-track points (b2), while Fig. 2(c) is got by using cloth simulation filter directly, including a lot non-track points (c1) and data missing (c2), (c3)as well. These two figures are both the optimal images selected from different combinations of parameters, failing to extract complete and correct track points. Figure 2(d) shows the result by combined filter and gets the track data completely without obvious noise.

In order to extract the top centerline point sets $\left\{P_{track}^{left}\right\}$ and $\left\{P_{track}^{right}\right\}$ of left and right tracks respectively, segmentation of the point cloud after filtering is necessary. Bounding box of the filtered point cloud can be rotated to be orthogonal to the coordinate axis according to formula (4) and (5) because the point cloud of one file could be considered to linear.

$$\alpha = arctan\left(\frac{Y_{max} - Y_{min}}{X_{max} - X_{min}}\right) \tag{4}$$

$$\begin{bmatrix} x_i^r \\ y_i^r \\ z_i^r \end{bmatrix} = \begin{bmatrix} cos\alpha & -sin\alpha & 0 \\ sin\alpha & cos\alpha & 0 \\ 0 & 0 & 1 \end{bmatrix} \begin{bmatrix} x_i \\ y_i \\ z_i \end{bmatrix} \tag{5}$$

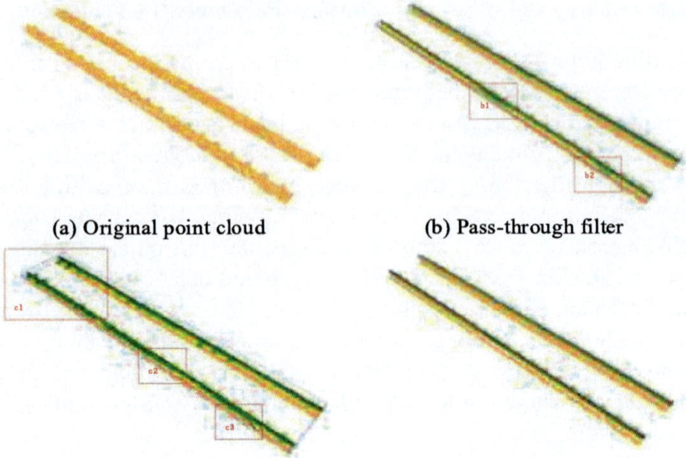

(a) Original point cloud

(b) Pass-through filter

(c) Cloth Simulation Filter

(d) combined filter

Fig. 2. Results of track extraction in different filter

Points on left or right track should be divided based on special relationship as the need of geometric parameters calculation. If the shorted side of bounding box is close to standard track gauge, 1435 mm in China, the rotated point cloud will be divided by the line connecting the central points of two short sides (Fig. 3(a)). Conversely, if the longer side of bounding box is close to 1435 mm, the dividing points will on the longer sides (Fig. 3(b)). Besides, if the length of two sides of bounding box is close, the limitation line would not pass through any profile of tracks (Fig. 3(c)).

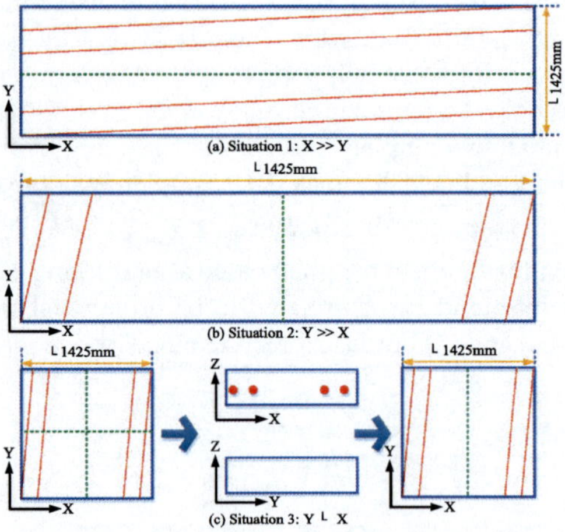

Fig. 3. Diagram of segmenting the left and the right track

Track gauge is calculated by the points 16mm below the top centerline of track. In order to extract these key points, the point cloud is divided into *m* strips along the direction of the track in the same length. Theoretically, the top centerline points could be identified by searching the highest point in each strip. However, due to the external errors such as noise, there are misjudged top centerline points by directly determination. To overcome the problem, taking the left track as an example, the extraction method of top centerline point set $\left\{P_{tc}^{left}\right\}$ is introduced. Firstly, random sampling consensus algorithm is used to eliminate the influence of outer points on XOY plane and get robust calculation point sets $\{P_m^{inner}\}$ and $\{P_m^{outer}\}$ based on the innermost and outermost point sets within *m* strips. Then, the midpoint $P(x, y, z)_m^{ct\_ori}$ of the track is computed by formula (6), where n_1 and n_2 are the number of points of $\{P_m^{inner}\}$ and $\{P_m^{outer}\}$.

$$P(x, y, z)_m^{ct\_ori} = \frac{1}{2}\left[\sum_{n_1=0}^{N_{n_1}} P(x, y, z)_{m,n_1}^{inner} + \sum_{n_2=0}^{N_{n_2}} P(x, y, z)_{m,n_2}^{outer}\right] \tag{6}$$

After the determination of approximate range of the top centerline point, the points with the standard deviation greater than 3σ in Z direction within the circular neighborhood Φ_m of the rough top centerline point are excluded. According to formula (7), the center of gravity of the left points is considered to be the top centerline point, where $P(x, y, z)_j \in \Phi_m$ and j is the number of points of the circular neighborhood after the elimination of noise.

$$P_{tc}^{left} = P(x, y, z)_m^{ct} = \frac{1}{N_m}\sum_{j=0}^{N_m} P(x, y, z)_j \tag{7}$$

Point set $\left\{P_{tc-16mm}^{left}\right\}$ to compute the track gauge is composed of the points 16 mm below $\left\{P_{tc}^{left}\right\}$. Figure 4 shows the key point sets to calculation the comfort parameters of the two tracks.

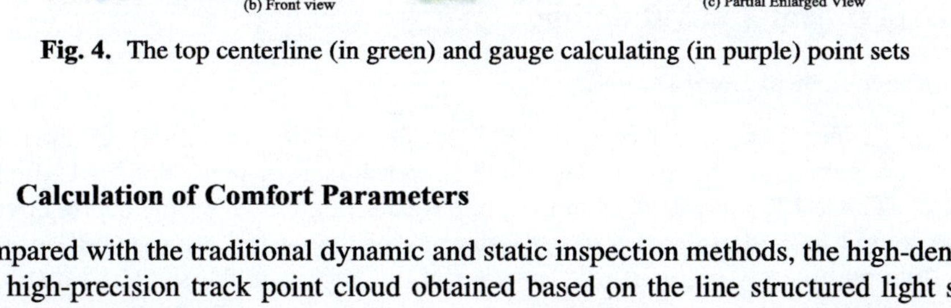

Fig. 4. The top centerline (in green) and gauge calculating (in purple) point sets

2.3 Calculation of Comfort Parameters

Compared with the traditional dynamic and static inspection methods, the high-density and high-precision track point cloud obtained based on the line structured light sensors and the GNSS/IMU integrated navigation modules can calculate the five typical types comfort parameters, i.e., vertical irregularity, superelevation, twist, irregularity and gauge of track, in any length of wave quickly, greatly improving the flexibility and efficiency of the inspection. The specific calculation method is as follows.

The irregularity of the track indicates the horizontal fluctuation of the inner measuring surface along the track direction. Generally, the curvature of the track changes smoothly along the forward direction, and a higher-order equation can be used to fit the smooth model of the track. Cubic curves Γ_{left} and Γ_{right} of $\left\{P_{tc}^{left}\right\}$ and $\left\{P_{tc}^{right}\right\}$ can be fitted on XOY projection plane using the RANSAC method respectively. Then, the track irregularity can be calculated by formula (8), where * presents the accord left or right track.

$$GX_{m,*} = y_{m,*}^{ct} - \Gamma_*(m) \tag{8}$$

Vertical irregularity of track refers to the vertical undulation of the tracks along the forward direction. Similar to track irregularity, H_{left} and H_{right} can be computed respectively in XOZ projection plane, and the parameter can be obtained by formula (9).

$$GD_{m,*} = z_{m,*}^{ct} - \mathrm{H}_*(m) \tag{9}$$

Gauge of track indicates the difference between the measured gauge and the standard gauge (1435 mm). Taking the left reference point sets $\left\{P_{tc-16\,\mathrm{mm}}^{left}\right\}$ as the basis, the nearest point on the right set $\left\{P_{tc-16\mathrm{mm}}^{right}\right\}$ in the mileage $x = m$ can be calculated according to formula (10).

$$GJ_m = \|P(x, y, z)_{m,tc-16\mathrm{mm}}^{left} P(x, y, z)_{nearest}^{right}\| \tag{10}$$

Superelevation of track refers to the height difference between the left and right tracks on a vertical profile. It can be computed by searching the nearest point like the gauge by formula (11).

$$CG_m = z_{m,tc-16\mathrm{mm}}^{left} - z_{nearest}^{right} \tag{11}$$

Twist of track indicates the inclination of the left and right tracks corresponding to the track plane. For wavelength λ, the twist can be calculated by formula (12).

$$SJK_{m,m+\lambda} = \left(z_{m,tc}^{left} - z_{m,nearest}^{right}\right) - \left(z_{m+\lambda,tc}^{left} - z_{m+\lambda,nearest}^{right}\right) \tag{12}$$

3 Experiment and Validation

3.1 Experimental Environment

Field experiments began in November 2022 at a test section of the railway area. The test scene is open and unobstructed, which is convenient for the GNSS/IMU module to work. The test line is about 1 km long, the right straight line, easing curve, circular curve composition, during the test, the tester pushes the trolley at a speed of no more than 20 km/h to carry out dynamic detection of the track, and the test site situation is shown in Fig. 5.

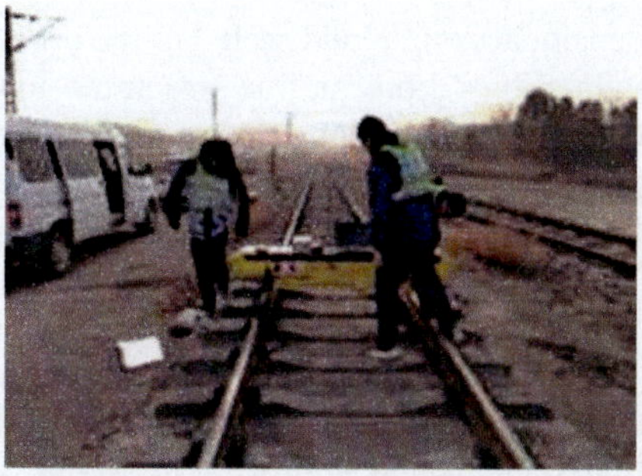

Fig. 5. Schematic diagram of the on-site operation

3.2 Analysis of Experimental Results

Based on the measured data, the track uneven comfort parameters of the test section are solved. On the same road section, two repeated experiments are carried out to verify the accuracy stability of the proposed method by taking the results of gauge measurement as an example. As shown in Fig. 6(a), the results of two gauge measurement experiments. According to the results, the maximum error of the two measurements is 0.08 mm, MAE is 0.055 mm, RMSE is 0.0577 mm, and the measurement accuracy reaches the millimeter level, which can meet the detection needs. The remaining four uneven measurements are also shown in Figs. 6(b)–6(e).

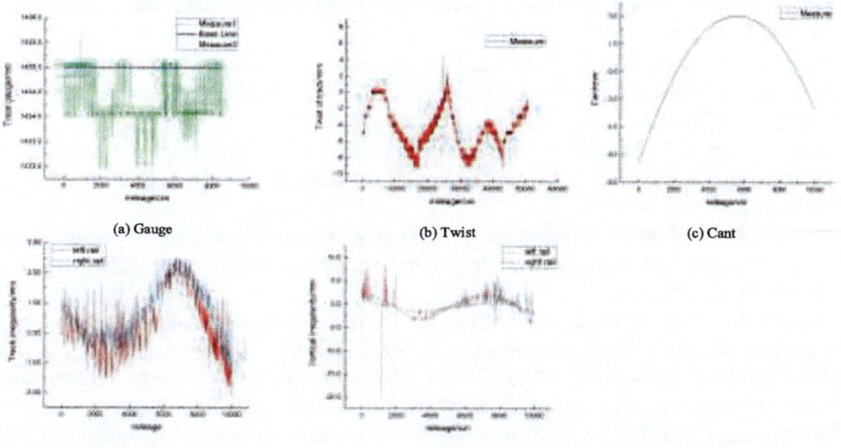

(a) Gauge (b) Twist (c) Cant (d) Irregularity (e) Vertical

Fig. 6. Schematic diagram of the results of five types of unevenness detection

4 Conclusion

In this paper, a three-dimensional point cloud model of the orbital region under absolute coordinate system is obtained based on the line structured light-GNSS/IMU module. Then, the combination of pass-through-cloth analog filter is used to accurately extract the center line of the top surface of the rail and the gauge measurement reference line;

Based on the above extraction results, the geometric parameters of five types of orbital unevenness are accurately measured. The experimental results show that the calculation of the geometric parameters of five types of orbital unevenness can be quickly implemented according to the method proposed in this paper, and the detection accuracy of millimeters can be achieved. In the next step, the model will be verified in more environments, and different dynamic and static orbit unevenness detection methods will be combined to verify the absolute accuracy of the model.

Acknowledgement. This work was funded by the National Natural Science Foundation of China (NSFC) under the Grants No.41971413, National Natural Science Foundation of China (NSFC) under the Grants No.U1934215 and research project of China Academy of Railway Sciences Corporation Limited, and the project number is 2021YJ047, The Second Batch of Curriculum Assessment Reform Pilot Project of University of Sanya (SYJGKH2023112).

References

1. Tian, G.Y., Gao, J.M., Zhai, W.M.: Comparative analysis of track irregularity management standards for high-speed railways. J. China Railw. Soc. **37**(03), 64–71 (2015)
2. Zhao, X., Zhang, P., Wen, Z.: On the coupling of the vertical, lateral and longitudinal wheel-rail interactions at high frequencies and the resulting irregular wear. Wear **430**, 317–326 (2019)
3. Wang, Z., Song, Y., Yin, Z., et al.: Random response analysis of axle-box bearing of a high-speed train excited by crosswinds and track irregularities. IEEE Trans. Veh. Technol. **68**(11), 10607–10617 (2019)
4. Yang, F., Sun, X.F., Tan, S.H., et al.: Evaluation difference of dynamic and static track irregularity and characteristics of dynamic chord measurement method. J. Sout. Jiaoton. Univ. **57**(06), 1239–1249 (2022)
5. Yang, F., Zhao, W.B., Gao, M.M., et al.: Static measurement and control standard for long-wave irregularity of high-speed railway track during operation period. China Railw. Sci. **41**(03), 41–49 (2020)
6. Tsunashima, H., Naganuma, Y., Kobayashi, T.: Track geometry estimation from car-body vibration. Veh. Syst. DYN. **52**(sup1), 207–219 (2014)
7. Aceituno, J.F., Chamorro, R., Muñoz, S., et al.: An alternative procedure to measure railroad track irregularities. Appl. Scaled Track Meas. **137**, 417–427 (2019)
8. Wang, X.K., Cai, D.G., Wang, P.: Research on the present situation and improvement method for precise adjustment of railway tracks. J. Railw. Eng. Soc. **39**(11), 33–37 (2022)
9. Lee, J.S., Choi, S., Kim, S.S., et al.: A mixed filtering approach for track condition monitoring using accelerometers on the axle box and bogie. Ieee. T. Instrum. Meas. **61**(3), 749–758 (2011)
10. Chen, Q., Niu, X., Zhang, Q., et al.: Railway track irregularity measuring by GNSS/INS integration. Nav. J. Inst. Nav. **62**(1), 83–93 (2015)

11. Zhang, Y.X., Li, Q., Wu, Y., et al.: Investigation on the influence of observation scheme of vehicle vibration on track irregularity estimation. J. Rail. Sci. Eng. 1–12 (2023). https://doi.org/10.19713/j.cnki.43-1423/u.T20221690
12. Zhangyu, W., Guizhen, Y., Xinkai, W., et al.: A camera and LiDAR data fusion method for railway object detection. IEEE Sens. J. **21**(12), 13442–13454 (2021)
13. Xu, W.Y., Liu, Y.J., Li, P., Wang, S., et al.: Application of vision - linear structured light's combination in track detection. Laser. Infra. **48**(10), 1218–1222 (2018)
14. Chen, S.M., Wei, S.B., Li, Y., et al.: Dynamic measurement method of track irregularity based on complementary filter **43**(01), 52–62 (2022)
15. Zhang, W., Qi, J., Wan, P., et al.: An easy-to-use airborne LiDAR data filtering method based on cloth simulation. Remote Sens. **8**(6), 501(2016)

Seismic Performance of Bridge Piers with Pile Foundations Under Frozen Soil Conditions

Wanping Wang, Xiyin Zhang(✉), Shengsheng Yu, and Jiada Guan

School of Civil Engineering, Lanzhou Jiaotong University, Lanzhou 730070, China
zhangxiyin@mail.lzjtu.cn

Abstract. In order to study the effect of permafrost on the seismic performance of bridge piers with pile foundations, a three-dimensional finite element model considering the effect of frozen soil was firstly established. Then the effects of unfrozen soil, seasonal frozen soil and permafrost on the seismic performance of bridge piers with pile foundations were compared and analyzed. The results show that the variation of frozen soil conditions has a great influence on the seismic performance of bridge piers with pile foundations. The presence of frozen soil can greatly enhance the lateral bearing capacity, energy dissipation and initial stiffness of piers with pile foundation, and can improve the deformation capacity of pile-soil system to a large extent. Frozen soil has a great influence on the distribution of pile foundation displacement. It is found that permafrost can significantly reduce the displacement of pile base, while seasonal frozen soil can significantly reduce the displacement of pile head.

Keywords: Bridge engineering · Seismic performance of bridges · Finite element analysis · Frozen soil effect

1 Introduction

China is the third-largest frozen soil distribution country in the world. Statistics show that the area of frozen soil in China is close to 70% of the total area [1], and it is found in the eastern monsoon region, the arid region of northwest China and the Qinghai-Tibet plateau [2]. In recent years, with the implementation of the "Belt and Road" and "China Western Development" national strategies, the construction of infrastructure including roads and railways in Qinghai-Tibet region in china has made rapid progress. To cross the ravine and minimize the disturbance of construction to the frozen soil environment, a large number of "bridge instead of road" construction measures have been adopted in line engineering in the frozen soil area [3]. And given the powerful seismic performance and thermal stability of frozen soil, pile foundation is widely used in bridge construction in frozen soil area. In addition, the Qinghai-Tibet region in china is an earthquake-prone area, with active geological structures and frequent seismic activities. Under the action of earthquake, the movement and deformation of each component of frozen soil-pile-structure system restrict each other, and the mechanical characteristics of frozen soil are easily disturbed by many external factors. The change of mechanical

W. Guo et al. (Eds.): GBCESC 2023, LNCE 328, pp. 22–30, 2024.
https://doi.org/10.1007/978-981-99-9947-7_3

characteristics will affect the mechanical behavior of pile foundation and superstructure. So, it brings difficulties to the comprehensive evaluation of seismic fortification and seismic performance of bridges in frozen regions [3].

Many scholars have studied the seismic performance of pile-based bridges in frozen soil regions. For example, Xu Xueyan et al. [4] studied the seismic acceleration response spectrum of frozen soil by numerical method and considered the soil nonlinearity based on the wave theory, the result shows that the negative temperature and frozen soil had a significant impact on the acceleration response spectrum. Chen Xingchong et al. [5] studied the influence of frozen soil layer change on the seismic response of the bridge, and analyzed the internal force's rules of the bridge pier with different thicknesses of frozen soil and different pier heights under the earthquake action. Plotnikova et al. [6] showed that seasonal soil freezing-thawing had a significant impact on the vibration frequency and vibration modes of the bridge, and the change of dynamic characteristics and boundary conditions of the bridge caused by seasonal soil freezing-thawing would further affect its seismic performance.

It can be seen that the presence of frozen soil has a great influence on the seismic response of bridge pile foundations. However, the existing studies do not mention the differences in seismic performance of the same bridge pile foundation under different frozen soil conditions. This is detrimental to the accurate evaluation of the seismic performance of pile-founded bridges in areas where frozen soil is present and to the determination of reasonable seismic design methods. Therefore, in this paper, a 3D finite element model of pile-soil foundation bridge pier that wildly used in frozen soil area was established. Based on the finite element analysis, the influence rules of the change of the bridge pile foundation soil conditions (unfrozen soil, seasonal frozen soil, permafrost) on lateral bearing capacity, stiffness characteristics of pile foundation, the deformation and displacement of pile body structure is analyzed, which provides reference and basis for the seismic design and comprehensive evaluation of the seismic performance of pile foundation bridges in frozen soil area.

2 The Establishment of Finite Element Model

It is difficult to achieve accurate modeling, because lack of the geological conditions of the bridge site and related parameters of the original soil. In addition, to achieve rapidness and precise modeling and ensure the accuracy of the results to greatest extent. Only by changing some parameters in the model to explore the degree of influence of the parameter changes on the seismic performance of bridge pile foundations. So, the finite element model of prototype structure was not adopted in the modeling process, instead, the 1/8 scale model is used to simulate and analyze the influence of unfrozen soil, seasonal frozen soil and permafrost on the seismic performance of bridge pile foundation.

According to similarity ratio, the height of the bridge pier was 1 m, the height of the cap was 0.3 m and the length of the pile foundation was 1.6 m in the modeling process. The width of the soil was 5 m and the height of the soil was 2 m to consider the influence of the boundary effect. The model adopts high pile cap foundation, the bridge pier, cap and pile foundation are all made of C30 concrete. The model size and reinforcement of pile foundation are shown in Fig. 1. Considering the dead load of the model and the sum of all kinds of load, the force that applied to the top of the model pier consists of two parts, one is the self-weight of the superstructure is 97.4 kN, and the other is the counterweight is 91.1 kN that considering the insufficient bulk density of the scaled model according to the similarity scaling. The calculation applying low cycle horizontal reciprocating load, and adopt displacement control loading system in the model. In terms of loading displacement using, adopt 2 mm amplitude before 20 mm displacement, over 20 mm, 5 mm amplitude is used for loading.

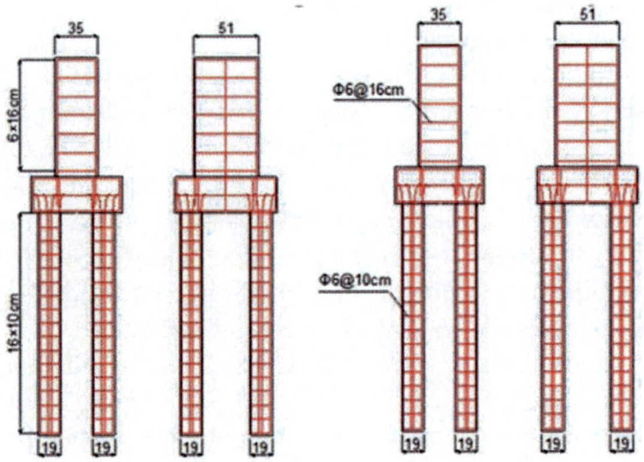

Fig. 1. Model size and reinforcement diagram

For model units division, soil and the concrete adopts three-dimensional entity unit (C3D8R), steel adopts truss element (T3D2). As for contact, the soil and pile foundation adopts "hard" contact in the outside that can unlimited pass pressure between two contact surfaces, when the two contact surfaces pressure nearly to zero or negative, the interface start to separate, at this time, it can better simulate the separation between pile and soil. And the tangential adopted "penalty" contact, which is similar to the setting of friction coefficient. In terms of the constitutive of the model, concrete adopts damage plastic model [7], reinforcement adopts bilinear hysteretic model and soil adopts Mohr-Coulomb model. Figure 2 shows the schematic diagram of the finite element model. Figure 3 illustrates the constitutive behavior of concrete under uniaxial compression. Figure 4 illustrates the constitutive behavior of steel under stretch.

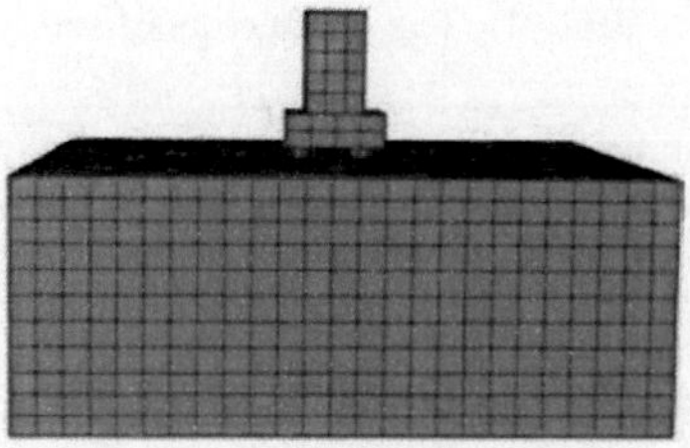

Fig. 2. 3D finite element model

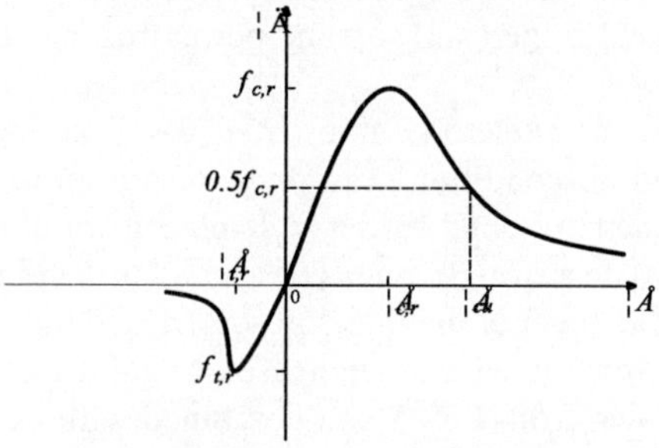

Fig. 3. Stress-strain curve of concrete

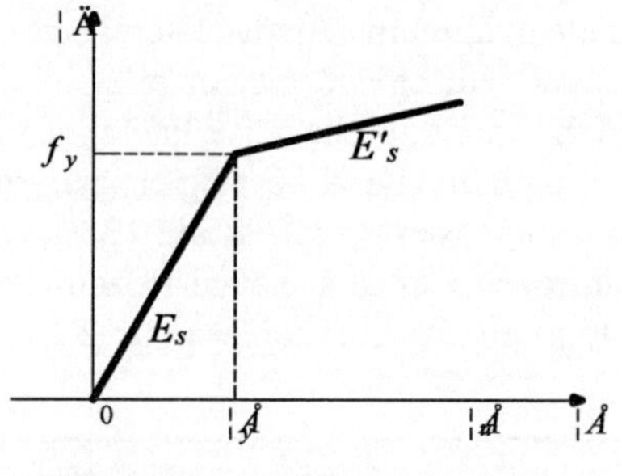

Fig. 4. Stress-strain curve of steel

3 Parametric Analysis of Influence for Frozen Soil on Bridge Pile Foundation

To further study the influence of the presence of permafrost layer on the seismic performance of bridge pile foundations. In this paper, the influence law of the variation of bridge pile foundation soil (unfrozen soil, seasonal frozen soil and permafrost) on the seismic performance of pile foundation is analyzed by numerical simulation. The parameters used in the model were obtained by experiments, Table 1 shows the parameters used in the model.

Skeleton Curve. Skeleton curve is an important basis for judging the excellent seismic performance of structures. It is the outer complex line of hysteresis curve and

Table 1. The parameters of soil

Test conditions	Temperature/°C	Elastic modulus/kPa	Cohesive force/kPa	Internal friction angle/°
1	2	5433.04	15.57	19.52
2	−9	158985.18	798.41	30.49

reflects the deformation stress characteristics of the structure at different stages. In this paper, extracted skeleton curves of the pile-soil system for bridge pile foundations under unfrozen soil site, seasonal frozen soil site and permafrost site conditions are shown in Fig. 5.

As shown in Fig. 5, the skeleton curves of pile-soil system show S shape under different conditions, and all conditions can divided into three phases. The first is the elastic stage, in which the horizontal load and displacement of pier top were positively correlated, and approximate a straight line. The second is the plastic phase, in which the curve between the displacement of the pile-soil system and the lateral load on top of the pile is no longer linear. Finally, with the increase of load displacement, the structure is destroyed. It can be known from Fig. 5 and combined with extracted data. The maximum positive lateral bearing capacity of pile-soil system is 21.82 kN under unfrozen soil conditions, and the maximum negative lateral bearing capacity is −21.58 kN. The maximum positive lateral bearing capacity of pile-soil system is 49.6 kN under seasonal frozen soil conditions, and the maximum negative lateral bearing capacity is −48.04 kN. The maximum positive lateral bearing capacity of pile-soil system is 59.65 kN under permafrost conditions, and the maximum negative lateral bearing capacity is -59.97 kN. According to above date, maximum lateral bearing capacity of pile-soil system in positive and negative directions increased by 127% and 122%, respectively, when the site conditions change from unfrozen soil to seasonal frozen soil. When the site changes from seasonal frozen soil to permafrost, it increased by 20.3 and 24.8% respectively.

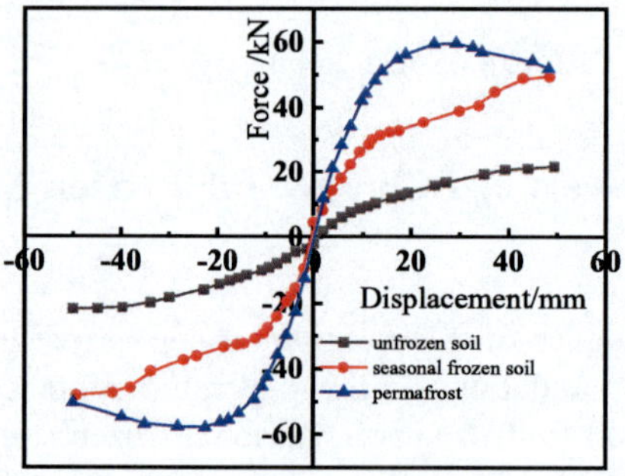

Fig. 5. Skeleton curves of the pile-soil system under different conditions

Energy Dissipation. For arbitrary structures, the hysteretic energy dissipation is the sum of the areas surrounded by loading curves and unloading curves. The energy dissipation curves extracted in this paper for the pile-soil system under unfrozen soil site, seasonal frozen soil site and permafrost site conditions are shown in Fig. 6.

It can be seen from Fig. 6 that the presence of frozen soil has a great influence on the energy dissipation of the pile-soil system. Compare the energy consumption curves of different conditions, the energy dissipation of pile-soil system in frozen conditions (seasonal frozen soil and permafrost) is greater than that in unfrozen condition, while the energy dissipation of permafrost is the largest.

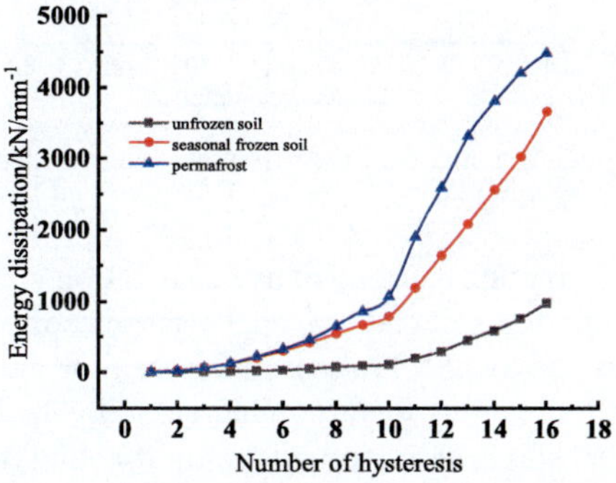

Fig. 6. Energy dissipation curves of the pile-soil system under different conditions

Stiffness Property. Stiffness is the ability of material or structure to resist deformation under stress. The stiffness of the structure will gradually decrease and the stress state of the structure will change under the action of reciprocating load, which have an adverse effect on the structure. Therefore, the stiffness degradation of the structure should be considered in the seismic design. The stiffness degradation of a structure is generally expressed in terms of equivalent stiffness.

It can be seen from Fig. 7, on the whole, the stiffness degradation rule are similar under different conditions of the pile-soil system. It was shown that the stiffness decreased with the increase of load drops rapidly before reaching the yield displacement. And the stiffness attenuation rate becomes slowly when reaches the yield displacement. When the loading displacement is about 40 mm, the stiffness is trend steady almost unchanged. It can be known from Fig. 5 and combined with extracted data, the stiffness of the pile-soil system trend to increase with the freeing of the site. Taking the positive stiffness degradation as an example, the initial stiffness of the pile-soil system increases rapidly to 194% from unfrozen soil to seasonal frozen soil, while the stiffness increases slowly to 39% from seasonal frozen soil to permafrost.

Deformability. Generally, displacement ductility coefficient is used to describe the deformation of a local component or of the whole structure in seismic design analysis. The definition of ductility coefficient is the ratio of deformation displacement, when the structure is damaged and yield. The larger the ductility coefficient is, the structure can bear large plastic deformation without collapse under the action of strong earthquake, and the seismic effect can be reduced.

The displacement ductility coefficients of different site conditions extracted in this paper are shown in Fig. 8. The ductility coefficient was calculated by the skeleton curve,

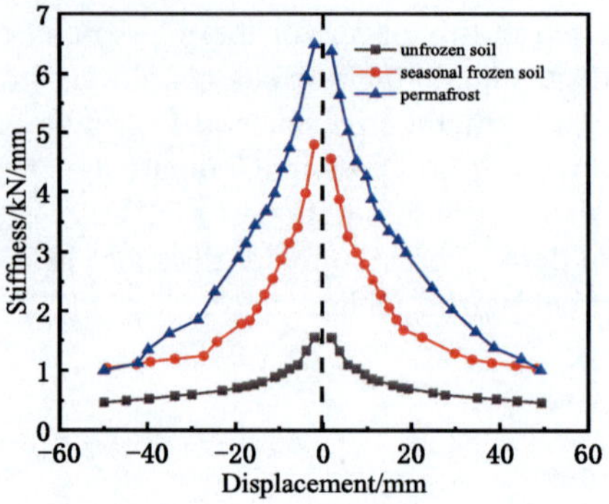

Fig. 7. Stiffness degradation curves of the pile-soil system under different conditions

it reflect the common deformation capacity of pile foundation-soil system. It can be know from Fig. 8. Contrast unfrozen soil and seasonally frozen soil, the ductility coefficient increased by 69.3% from 1.50 to 2.54. For seasonal frozen soil and permafrost conditions, the ductility coefficient of the pile-soil system increased by 38.2% from 2.54 to 3.51. It can be concluded that that soil is an important factor affecting ductility, and pile-frozen soil system has better deformation ability.

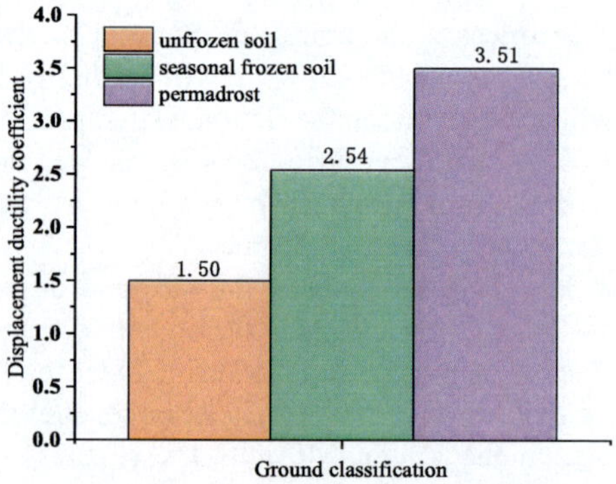

Fig. 8. Ductility coefficient of the pile-soil system under different conditions

Displacement of Pile. To explore the influence of frozen soil on seismic performance of bridge pile foundation, extracted the curve of the displacement along the pile of the model as shown in Fig. 9.

We can know from Fig. 9 that the displacement of pile head is larger in permafrost condition, but the displacement of pile bottom in permafrost condition almost didn't change. As for seasonal frozen soil condition, the displacement of pile head place small because of the limitation of frozen soil on the surface. As for unfrozen soil condition, the deformation of pile is almost linear distribution under the site, and the pile head and pile bottom both produced large displacement. Therefore, the presence of permafrost

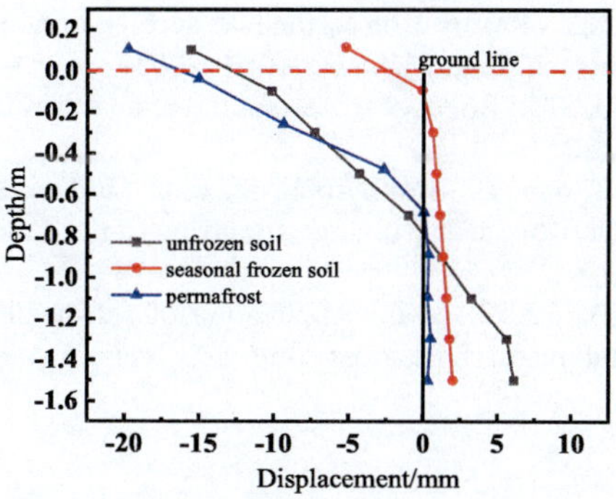

Fig. 9. The curve of pile displacement under different conditions

condition can significantly reduce the displacement of pile bottom, while the seasonal frozen soil condition can significantly reduce the displacement of pile head.

4 Conclusions

In this paper, a 3D finite element model considering the effect of frozen soil is established to discuss the influence of frozen soil on seismic performance of the bridge pier with pile foundation, and some preliminary conclusions are drawn:

1) The existence of frozen soil can greatly improve the lateral bearing capacity, energy dissipation and initial stiffness of the bridge pier with pile foundation;
2) The existence of frozen soil can increase the deformation capacity of the pile-soil system, so soil is an important factor affecting ductility, and pile-frozen soil system has better deformation ability;
3) The existence of frozen soil can change the distribution of pile displacement. It is found that the existence of permafrost can obviously reduce the displacement of pile bottom. While, the existence of seasonal frozen soil can reduce the displacement of pile head.

Acknowledgement. This research is supported by the National Natural Science Foundation of China (No. 51808273 and 52068045), Science and Technology Program of Gansu Province for Distinguished Young Scholars (No. 20JR5RA430).

References

1. Gao, F., Chen, X.C., Yan, S.H.: Influence of permafrost and seasonally frozen soil on seismic response of sites. Chin. J. Rock Mech. Eng. **25**(8), 1639–1644 (2006)
2. Zhou, Y.W.: China's Frozen Soil. Science press, Beijing (2000)
3. Wang, W.P., Zhang, X.Y., Chen, X.C., et al.: Study on dynamic internaction between bridge pile and soil with permafrost effect: review and research trend analysis. J. Glaciol. Geocryol. **42**(4), 1–7 (2020)

4. Xu, X.Y., Xu, C.H., Li, X.Z.: Research on earthquake acceleration response spectrum of frozen soil ground. Chin. J. Geotech. Eng. **25**(6), 680–683 (2003)
5. Chen, X.C., Gao, F., Wu, S.H.: Effect of frozen soil layer on seismic response of bridges. Eng. Mech. **24**(3), 120–125 (2007)
6. Plotnikova, A., Wotherspoon, L., Beskhyroun, S., et al.: Influence of seasonal freezing on dynamic bridge characteristics using in-situ monitoring data. Cold Reg. Sci. Technol. **160**, 184–193 (2019)
7. Zhu, X.Y., Lu, C.H., Dai, Z.W., et al.: Application of self-healing engineering materials: mechanical problems and research progress. Bull. Sci. Technol. **66**(22), 2802–2819 (2021)

Application of Big Data Analysis in Bridge Monitoring System

Xian Xiao(✉)

School of Engineering, China University of Geosciences, Wuhan, China
xiaoxian@cug.edu.cn

Abstract. In this paper, bridge health monitoring and damage detection are studied. The development status of the bridge health monitoring system is investigated, especially the application of big data analysis in this field. Firstly, the steps and methods of big data analysis are introduced in detail. Then the Ganjiang River high-speed railway cable-stayed bridge is selected as the monitoring object, and the possibility of the real-time online wind field, fatigue degree, statistics, and safety evaluation are analyzed. Finally, the problems existing in the bridge health monitoring and safety evaluation system are summarized.

Keywords: Monitoring system · Structural health monitoring · Railway bridge · Bridge safety assessment

1 Introduce

With the development of the social economy and the rise of people's living quality, and the population surge in large and medium-sized cities, railway traffic is promoted vigorously because of its effective advantage of solving the traffic pressure. The railway traffic may be damaged by the environment, operation, structure itself, and other factors, thus threatening the safety of life and property [1]. It is necessary to conduct health detection and damage identification for rail transit to reduce the likelihood of accidents.

The definition of bridge health monitoring is to provide basis and guidance for bridge maintenance and management decisions when the bridge is in special climate, traffic conditions, or serious abnormal bridge operation through the monitoring and evaluation of bridge structural status [2]. Changes in the material or geometric characteristics of the bridge structural system are defined as bridge damage, and changes in boundary conditions or system connectivity are also included, which will adversely affect the safety and stability of the bridge [3].

Previously, bridge health detection technology is mainly applied in the field of engineering technology. However, with the increase of analysis data, traditional analysis means are no longer sufficient to complete the analysis process with high quality. As a new technology, big data has attracted wide attention.

W. Guo et al. (Eds.): GBCESC 2023, LNCE 328, pp. 31–37, 2024.
https://doi.org/10.1007/978-981-99-9947-7_4

2 Bridge Structure Health Monitoring Mass Data Analysis

At the early stage of the bridge health monitoring system, data acquisition, data management, and bridge early warning are mainly used in the Functional monitoring system [4]. Due to the lack of perfect data analysis and safety assessment modules, the requirements for complex bridges could be met. Especially the health monitoring of high-speed railway cable-stayed bridges. Therefore, the data analysis function and security improvement of the evaluation function are very important. In this way, a monitoring system is necessary. A structural health monitoring system consists health monitoring system, which consists of four parts, they are sensor subsystem, data acquisition and transmission subsystem, data management subsystem, security assessment, and early warning subsystem [5].

In recent years, due to the development of big data technology, bridge structure health monitoring is also advancing with the time series. Big data analysis of structural health monitoring is a new topic in the field of structural health monitoring in recent years [6].

2.1 Data Gathering

The data gathering should be completed before establishing a bridge monitoring system to identify bridge damage. For different bridges and different properties, the type and amount of data to be collected varies. The subsequent analysis of this paper is based on the Ganjiang bridge, so the data gathering refers to the Ganjiang bridge.

The sensor of the monitoring system is installed during the construction of the Ganjiang railway bridge. Sensor type includes thermometer, anemoclinograph, GPS, strainmeter, flexometer, displacement meter, accelerometer, hygrometer, stress gauge, and rope dynamometer. The number of various types of sensors is shown in Table 1 (Fig. 1).

Table 1. The number of sensors.

Sensor type	Rope dynamometer	Anemoclinograph	GPS	Strainmeter	Flexometer
Numbers	96	2	2	91	13
Sensor type	Displacement meter	Accelerometer	Hygrometer	Stress gauge	Thermometer
Numbers	30	8	11	18	99

2.2 Data Preprocessing

Data preprocessing is a very important preparation before data mining, aiming to handle all the redundant data, missing data, and noisy data. Redundant data refers to repeatedly recorded data, and sorting and similarity calculation are generally used to remove redundant data. Especially in the analysis of the bridge, the influence of noises should be fully considered. Because various noises are also important factors affecting bridge life, it is not possible to remove all noise data.

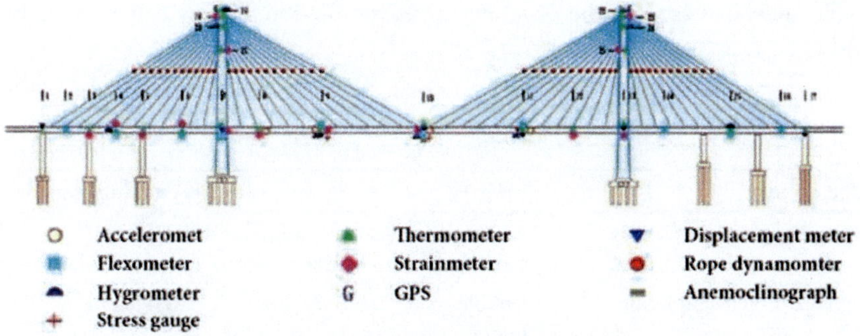

Fig. 1. The position distribution diagram of Sensor.

2.3 Massive Monitoring of Big Data Analysis

The monitoring data of the bridge structure includes vehicle load, strain, cable force, wind field, vibration acceleration, etc. Through the deep mining of the information contained in the monitoring data, it may be helpful to grasp the working state and environment of the bridge [7].

Currently, the common methods for big data analysis of structural monitoring at home and abroad include analysis of time series, fuzzy clustering analysis, analytic hierarchy process, and so on [8].

Time series analysis is an important branch of mathematical statistics, which can be used to mine the important information contained in monitoring data [9]. Based on the time series, we can identify the damage that may occur in a certain period of the bridge, including dynamic response calibration in the healthy condition of the bridge, and residual error minimization using the model coefficient [10].

Fuzzy clustering analysis is a clustering analysis method, which establishes fuzzy similarity relation based on the sample characteristics, degree of affinity, and similarity between objective things. Fuzzy clustering is a classification method of unsupervised learning, which can complete the classification according to the characteristics of samples. Since there are no specific sample standards before classification, the classification standard is inconsistent, so the uncertain description of samples can be established, which can more accurately reflect the actual condition [7]. The fuzzy clustering method could be used for the construction of a state system for bridge [11]. The fuzzy clustering in this paper is mainly used to identify train speed.

At present, many countries all over the world are carrying out comprehensive assessment of bridge safety, the Analytic Hierarchy Process (AHP) is generally used in the safety assessment of large and important bridge structures. AHP divides the factors affecting the safe use function of the bridge, and determines the influence indicators under each level [12]. The index weight is determined by the influence degree of different indexes on the structure, and the status of other indicators is calculated.

There are numerous big data analysis methods for structural monitoring, but due to the immature development of bridge monitoring system technology, many efficient and accurate methods cannot be realized in the short term. So the following part, I only do a simple analysis based on the Ganjiang River high-speed railway cable-stayed bridge (Table 2).

Table 2. List of safety assessment methods for some large bridges in China.

Bridge name	Brige type	Assessment methods
Nanjing Yangtze River Bridge	Railway multi-span continuous steel truss beam bridge	AHP
Donghai Bridge Main Bridge	Double tower cable-stayed bridge	AHP and FIS
Jiangsu Runyang Yangtze River Bridge	Cable-stayed bridge and Single span suspension bridge	Reliability and FAHP
Zhanjiang Bay Bridge Main Bridge	Cable-stayed bridge	AHP and FIS
Hangzhou Wenhui Bridge	Cable-stayed bridge	AHP, GRA and VWS
Xupu Bridge in Shanghai	Cable-stayed bridge	AHP, GRA and VWS
The Second Yangtze River Bridge in Wuhan	Cable-stayed bridge	AHP and FIS
The third Yangtze River Bridge in Wuhan	Arch bridge	AHP and FCE
Wuhu Yangtze River Bridge	Double tower cable-stayed bridge	AHP and FNN
Zhengzhou Yellow River Highway Bridge	Railway multi-span continuous steel truss beam bridge	AHP and FNN

3 The Process of System Design

During the operation of the railway bridge, the damage to various structures and components has an impact on the safe operation of the bridge. Even if one of the components fails, the bridge may lose its safety completely. To facilitate the staff intuitively and quickly carry on the bridge health monitoring and damage detection, health monitoring and damage detection for the bridge are necessary. Ganjiang Railway bridge is a high-speed railway cable-stayed bridge, it is necessary to consider the influence of cable-stayed cable, ballastless ballast bed, steel rail, fasteners, and temperature regulator on the safety of bridge operation. Wind parameter analysis, fatigue analysis, and high-speed train speed identification need to be considered. In this way, this paper introduces the design process of the Ganjiang Railway bridge structural health monitoring and safety evaluation system.

Firstly, considering the effect of the wind field characteristic on the bridge, average wind and pulsating wind analysis, turbulence, turbulence integration scale, gust coefficient, pulsating wind power spectrum, and other wind field analyses need to be focused. Based on the definition and analysis theory of wind field characteristics, an algorithm for calculating wind field parameters can be developed to provide program support for wind speed and direction analysis of bridge monitoring systems. Thus, the wind field information of the Ganjiang bridge during the monitoring period is calculated in real-time.

Secondly, it is necessary to analyze the fatigue of the steel structure of the bridge. Based on fatigue analysis theory, a fatigue analysis algorithm can be designed, including accessing the data of the InfluxDB database and MySQL database, calculating the actual stress cycle times using the rain flow counting method, the influence of Goodman straight line calculation, calculating the fatigue life of miner linear fatigue cumulative damage theory and calculating the cycle times of fatigue life curve. Based on the algorithm, we can take the measured strain sample data of a certain day for analysis and calculate the fatigue life of the bridge. If the fatigue life meets the requirements of the service life of the bridge, then the bridge can be considered qualified.

Thirdly, the speed identification method of high-speed trains can be established, based on time series analysis and fuzzy clustering. After extracting the characteristics of the response data of the train crossing the bridge under the condition of known speed or unknown speed, the fuzzy clustering model of the characteristic parameters could be established, which can identify the high-speed train speed online in real-time. Choosing the train on the bridge as the research object, the Auto-Regressive (AR) model could be established for the acceleration time series at different speeds, the order of the AR model could be determined, the AR coefficient could be calculated, and the AR coefficient under each working condition could be analyzed by fuzzy clustering.

Fourthly, according to the bridge safety evaluation theory and railway bridge standards, bridge safety can be comprehensively analyzed. Specific methods can be used to establish a safety comprehensive assessment model for the Ganjiang railway bridge by AHP, variable weight synthesis method, and grey relational analysis (GRA) method. To distinguish the difference in evaluation index, the safety comprehensive evaluation model of Ganjiang Railway bridge can be divided into two parts: quantitative index evaluation model and qualitative index evaluation model. The bridge scores of the two evaluation models are calculated respectively, and then the overall safety comprehensive evaluation scores of Ganjiang Railway bridges are calculated by the variable weight synthesis principle. According to the score, the current safety of the railway can be judged.

Fifthly, to achieve real-time monitoring, the Spring MVC framework can be used to establish a bridge monitoring system software platform. Through the platform to achieve the integrated data display, data analysis, safety assessment and structural warning, and other functional modules of the online visualization platform. It means that the comprehensive embodiment of the above functions could be achieved.

Through the above steps, we can complete the health monitoring and safety assessment of Ganjiang bridge (Fig. 2).

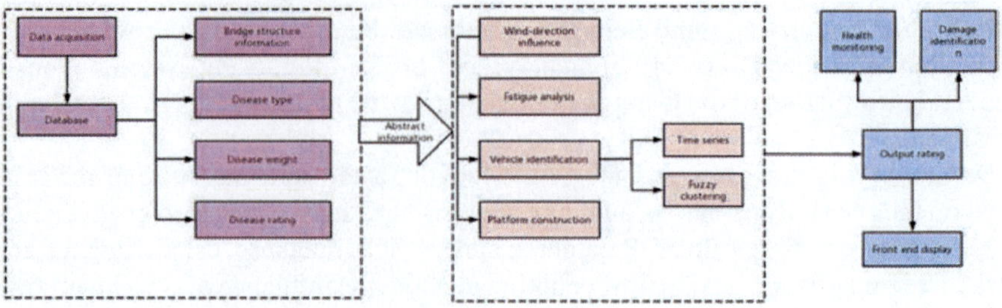

Fig. 2. Flow diagram.

4 Conclusion

This paper studies the application of big data analysis technology in bridge health monitoring and damage identification. And we complete the description of Ganjiang bridge health monitoring system establishment and damage identification analysis process. This is a new field, represents the combination of two industries with a good promising development. However, big data technology is not mature enough, and there is still a lot of room for improvement and application in bridge detection. Considering the potential application of big data technology in this field, big data analysis can also be used to predict the safety of bridges under different conditions. The research and development of the health monitoring and safety assessment system of high-speed railway bridge structure not only involves the knowledge of civil engineering, such as bridge structure, structural dynamic and dynamic response analysis methods, but also involved in high-speed railway design, computer programming and algorithm design. The analysis process includes massive data processing, which is an extremely complex work for manual inspection. Therefore, the intervention of big data as a new technology is very valuable. In the future, we will continue to do research on Ganjiang bridge and the potential applications of big data analysis on bridge detection.

Funding Information. The 2023 Innovation and Entrepreneurship Plan for College Students of China University of Geosciences, Wuhan, China.

References

1. Montenegro, P.A., Carvalho, H., Ribeiro, D., Calçada, R., Tokunaga, M., Tanabe, M., Zhai, W.M.: Assessment of train running safety on bridges: a literature review. Eng. Struct. **241**, 112425 (2021)
2. Zhang, Y.-M., Wang, H., Bai, Y., Mao, J.-X., Xu, Y.-C.: Bayesian dynamic regression for reconstructing missing data in structural health monitoring. Struct. Health Monit. Int. J. **21**(5), 2097–2115 (2022)
3. Bien, J., Kuzawa, M.: Dynamic tests in bridge health monitoring. Studia Geotechnica Et Mechanica **42**(4), 291–296 (2020)
4. Wang, Z., Yi, T.H., Yang, D.H., Li, H.N., Liu, H.: Multiorder frequency-based integral performance warning of bridges considering multiple environmental effects. Pract. Period. Struct. Des. Constr. **28**(2), 04023011 (2023)

5. Fan, X.: Research on bridge health and safety monitoring system. Appl. Mech. Mater. **39**, 131–135 (2011)
6. Cremona, C., Santos, J.: Structural health monitoring as a big-data problem. Struct. Eng. Int. **28**(3), 243–254 (2018)
7. Sun, L., Shang, Z., Xia, Y., Bhowmick, S., Nagarajaiah, S.: Review of bridge structural health monitoring aided by big data and artificial intelligence: From condition assessment to damage detection. J. Struct. Eng. **146**(5), 04020073 (2020)
8. Wang, M., Ding, Y., Wan, C., Zhao, H.: Big data platform for health monitoring systems of multiple bridges. Struct. Monit. Maint. **7**(4), 345–365 (2020)
9. Kostić, B., Gül, M.: Vibration-based damage detection of bridges under varying temperature effects using time-series analysis and artificial neural networks. J. Bridge Eng. **22**(10), 04017065 (2017)
10. Farahani, R.V., Penumadu, D.: Damage identification of a full-scale five-girder bridge using time-series analysis of vibration data. Eng. Struct. **115**, 129–139 (2016)
11. Jiao, Y., Liu, H., Zhang, P., Wang, X., Wei, H.: Unsupervised performance evaluation strategy for bridge superstructure based on fuzzy clustering and field data. Sci. World J. (2013)
12. Ikpong, A., Chandra, A., Bagchi, A.: Alternative to ahp approach to criteria weight estimation in highway bridge management. Can. J. Civil Eng. **48**(9), 1181–1191 (2021)

Check for updates

Design of the Stiffener Layout for Dome Structures Based on Topology Optimization

Yougang Wang[1], Dingkun Chen[2], Yunlun Sun[1], Zitong Bao[1], Junhong Zhang[1], Weipeng Xu[2], Liang Hong[2], and Peng Wei[2(✉)]

[1] Chinergy Co., Ltd., Beijing 100193, China
[2] School of Civil Engineering and Transportation, State Key Laboratory of Subtropical Building and Urban Science, South China University of Technology, Guangzhou 510640, Guangdong, China

ctpwei@scut.edu.cn

Abstract. The inner steel plate of the containment dome in the nuclear power plant is commonly used as formwork for concrete pouring. After construction, the steel plate also becomes a part of the load-bearing system of the dome. To prevent excessive deflection and out-of-plane instability during construction and operation, stiffeners need to be added to the surface of the inner steel plate. In this paper, we combine the parameterization scheme based on radial basis functions and topology optimization to conduct the optimization design for the stiffener layout on the inner steel plate of the containment dome. Various loading conditions and symmetry models are employed to optimize the stiffener layout. The secondary design of the dome with stiffeners is then modeled based on the optimal results. Finally, the finite element analysis of the secondary design is performed using ANSYS to evaluate structural performance. The results of the static analysis, eigenvalue buckling analysis, and ultimate bearing capacity demonstrate the effectiveness of the proposed optimization method in the optimal design of the stiffener layout.

Keywords: Topology Optimization · Stiffener · Containment Dome

1 Introduction

Stiffener, as a lightweight reinforcement form for thin-walled structures, effectively improves structural mechanical properties such as stiffness and strength. It is widely used in various engineering structures, including steel box girders, steel plate walls, and steel dome structures. The shape, layout, and other parameters of stiffeners directly affect the structural performance. Therefore, how to design the layout of stiffeners in thin-walled structures to achieve the best structural performance has always been a prominent concern in the field of structural optimization.

Topology optimization, as a rapid form-finding optimization method for conceptual design, has attracted significant attention from researchers and has been rapidly applied to the design of the stiffener layout. In the early stages, Cheng and Olhoff [1, 2], investigated

W. Guo et al. (Eds.): GBCESC 2023, LNCE 328, pp. 38–46, 2024.
https://doi.org/10.1007/978-981-99-9947-7_5

the optimal distribution of stiffeners by optimizing the thickness of plate elements. Gea and Luo [3] proposed a stiffener layout design method that considers the positioning and orientation of stiffeners. Belblidia and Afonso [4] introduced an integrated approach for optimizing stiffeners in shell structures, where topology optimization was used to obtain the stiffener layout, followed by size optimization to determine the stiffener dimensions. Ding and Yamazaki [5] proposed a plant growth method to obtain the optimal stiffener layout design. Wei *et al.* [6] proposed the stiffness spreading method to incorporate the stiffness of discrete bar elements into the base plate structure. Guo *et al.* [7] proposed the MMC (Moving Morphable Components) method based on the stiffness spreading method and Sun *et al.* [8] applied it to the optimization design of the stiffener layout. Additionally, Liu *et al.* [9] employed a density-based method with casting constraints to optimize the stiffener layout in telescope structures. Feng [9], and Zhang [10] proposed a stiffener layout optimization method for irregular surfaces based on B-spline parametrization and mesh parametrization.

Considering the efficiency of topology optimization, this paper combines the radial basis functions (RBFs) based parametrization scheme and topology optimization method to optimize the layout of stiffeners in the containment dome of a nuclear power plant. The secondary design of the dome with stiffeners is modeled based on the optimal results. Finally, the finite element analysis (FEA) including the static analysis, eigenvalue buckling analysis, and ultimate bearing capacity analysis is performed to validate the effectiveness of the proposed method.

2 Stiffener Layout Optimization Method Based on RBFs

2.1 Optimization Method

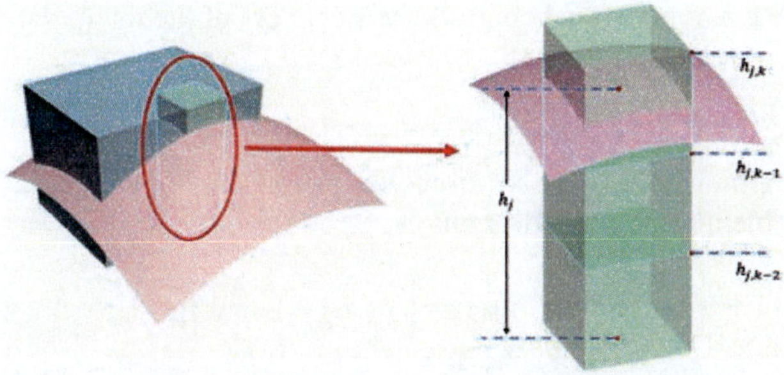

Fig. 1. Illustration of the pseudo-density interpolation scheme.

In this paper, we employ a topology optimization method based on radial basis functions for parametrically optimizing the layout of stiffeners in structures. The fundamental idea of this method is derived from the parametrized level set approach, utilizing a radial basis function-based parametrized surface to explicitly represent the outer surface of stiffeners.

After partitioning a thin-walled structure with a certain thickness using a parameterized surface, the portion above the surface represents the void region filled with weak material, while the region below the surface represents the solid material region. The parameterized surface is used to represent the outer surface of the stiffener structure. This approach effectively avoids the formation of voids in the middle part of the structure, providing a similar effect to the casting constraint in density-based methods. The expression for this parameterized surface can be written in the following form:

$$h(\boldsymbol{x}) = \sum\nolimits_{i=1}^{n} \alpha_i g_i(\boldsymbol{x}) \tag{1}$$

where $\boldsymbol{x}$ represents the coordinates, $h(\boldsymbol{x})$ represents the height of the stiffener at point $\boldsymbol{x}$, α_i denotes the expansion coefficients of the i th basis function, and g_i represents the i th radial basis function.

Based on the defined parameterized surface, the height h of the surface at the centroid of each column element is obtained. Subsequently, interpolation functions are used to obtain the density of each element.

The pseudo-density interpolation scheme for each element is shown in Fig. 1. The relationship between the pseudo-density of each element and the surface height at the centroid of that element is as follows:

$$\rho_{j,k} = H\big(h_j - \big(h_{j,k} + h_{j,k-1}\big)/2\big) = \begin{cases} 0 \; h_j < \big(h_{j,k} + h_{j,k-1}\big)/2 \\ 1 \; h_j \geq \big(h_{j,k} + h_{j,k-1}\big)/2 \end{cases} \tag{2}$$

where $H(x)$ is a Heaviside function that ensures continuity during the interpolation process [9]. The expression of $H(x)$ is given by:

$$H(x) = \frac{e^{\beta x}}{1 + e^{\beta x}} \tag{3}$$

where, $\beta > 0$ is a parameter that controls the steepness of the interpolation function.

2.2 Problem Formulation

Based on the optimization method discussed in the previous section, the minimum compliance problem can be stated as follows:

$$\begin{aligned} & F\,ind\ h(\boldsymbol{\alpha}) = \begin{bmatrix} \alpha_1 \; \alpha_2 \; \cdots \; \alpha_n \end{bmatrix} \\ & Min\ J(\boldsymbol{u}, h) = \boldsymbol{u}^T \boldsymbol{f} \\ & s \cdot t \cdot \begin{cases} \boldsymbol{Ku} = \boldsymbol{f} \\ \sum_{k=1}^{n_e} \rho_k V_k \leq V_{frac} \\ 0 \leq \rho_k \leq 1 \end{cases} \end{aligned} \tag{4}$$

where J is the structural compliance, $\boldsymbol{u}$ is the displacement field satisfying the physical equations, $\boldsymbol{f}$ is the load vector, ρ_k is the pseudo-density of the k th element, $\boldsymbol{K}$ denotes the global stiffness matrix of the structure. V_{frac} is the maximum allowable material volume fraction.

3 Stiffener Layout Design Using Topology Optimization

The outer containment dome model of a nuclear power plant is shown in Fig. 2(a) and (b), with a height of 19 m, a diameter of 60 m, and the bottom side fixed. The thickness of the inner steel plate is 0.02 m. The thickness of concrete poured on the inner steel plate is 1 m. The maximum allowable height of the stiffener is 0.8 m.

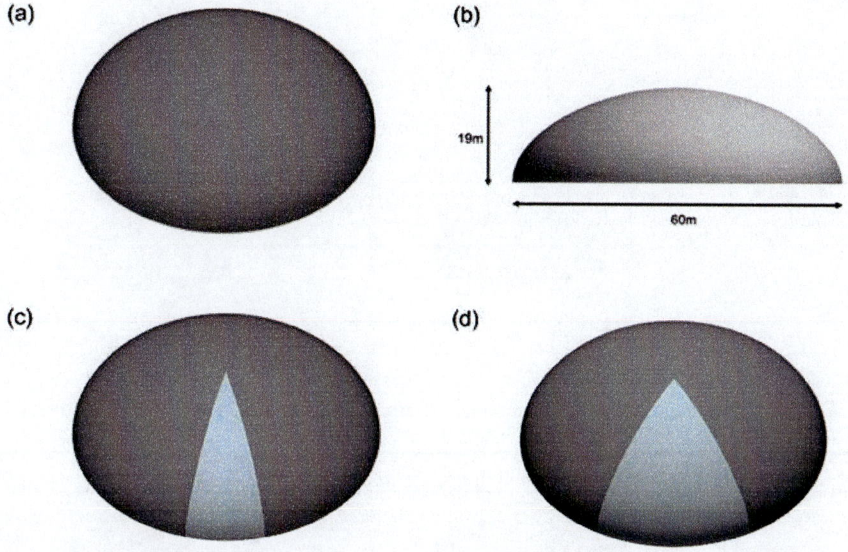

Fig. 2. Outer containment dome model of a nuclear power plant. (a) Vertical view. (b) Side view. (c) One-twelfth part. (d) One-sixth part

To reduce the computation efforts of the finite element analysis, we choose one-twelfth and one-sixth parts of the structure according to the symmetry for modeling and optimization design, as shown in Fig. 2(c) and (d), respectively. The one-twelfth model is discretized into 25128 quadrilateral elements and 25128 $\times$ 4 hexahedral elements, and the one-sixth model is discretized into 32137 quadrilateral elements and 32137 $\times$ 4 hexahedral elements. The bottom layer elements represent the steel plate and are treated as non-designable domains.

The elastic modulus and Poisson's ratio of the steel plate are set to 2.1×10^{11} Pa and 0.3 respectively. The elastic modulus of the weak material E_{min} in the optimization is set to 2.1×10^5 Pa. The density of steel and concrete is set to 7850 kg/m^3 and 2500 kg/m^3, respectively.

Since the bottom steel plate is used as a formwork for pouring concrete during the construction stage, the structure is subjected to the load from the concrete pouring and the self-weight load of the steel plate. Without considering the layered pouring of concrete, the total external load of the model is set to 26.5 KN/m^2. Referring to the layout of the lattice shell structure, we consider two methods of converting the uniform distributed loadings into the nodal concentrated loadings as shown in Table 1.

Table 1. Two types of load application method

	Load type 1	Load type 2
One-twelfth model	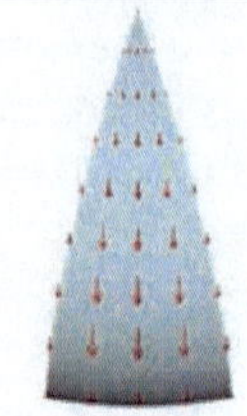	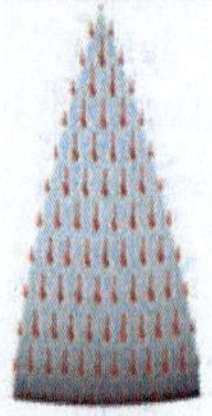
One-sixth model	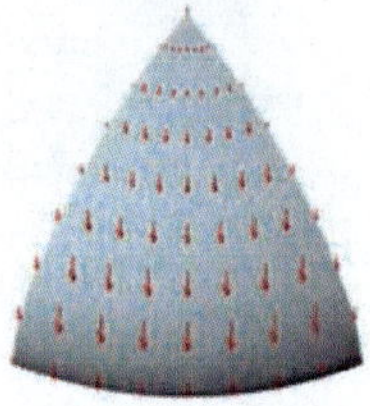	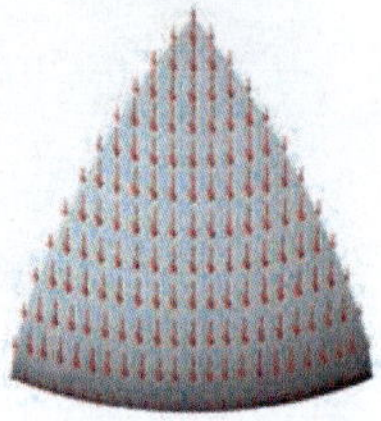

Table 2. Optimal results for different symmetry models and load application methods.

	Load type 1	Load type 2
One-twelfth model	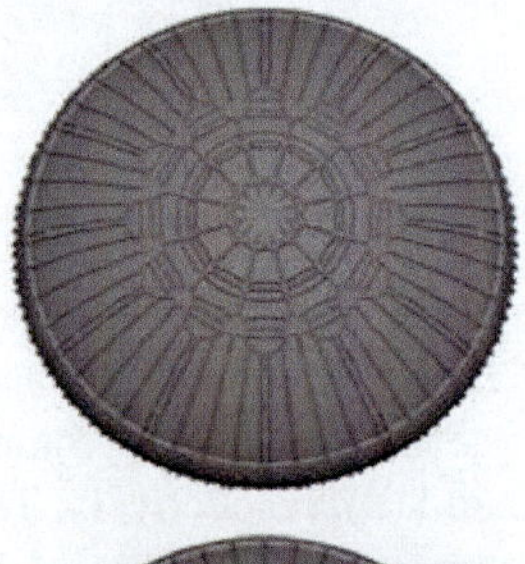	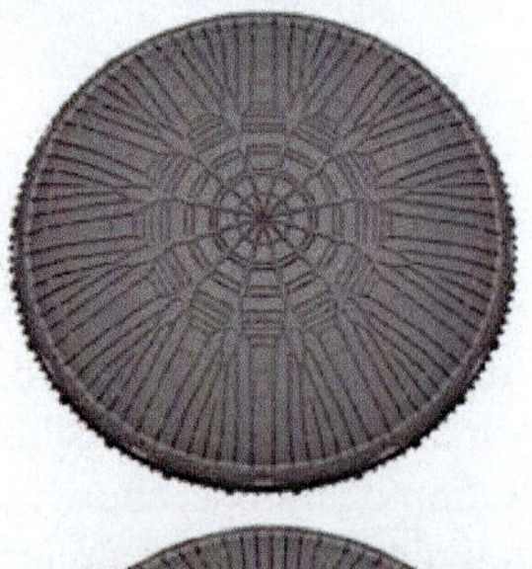
One-sixth model	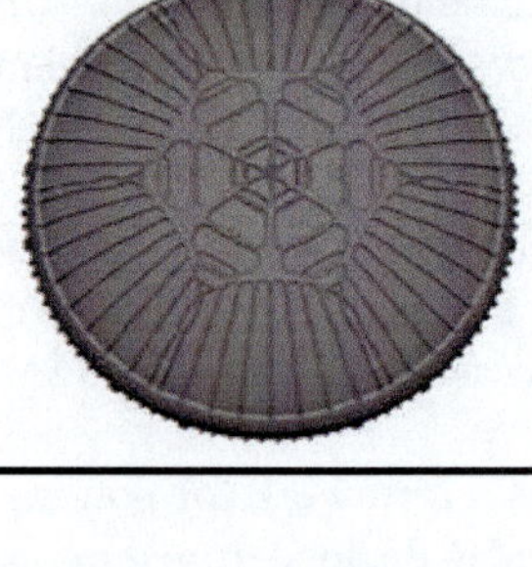	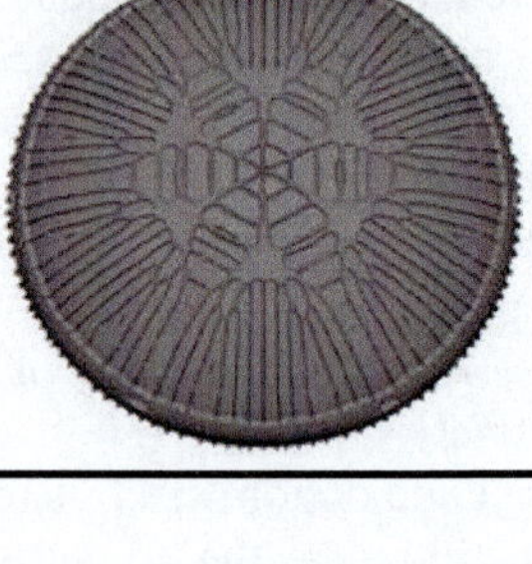

The corresponding optimization results assembled into the overall structure are shown in Table 2. Similar results are obtained under different loading conditions and different symmetry models. It can be observed that the stiffeners away from the top are mainly distributed along the longitudinal direction, and there are thick circular stiffeners distributed near the bottom. The stiffeners near the top are mainly distributed in the circular direction. Thick longitudinal stiffeners and small circular stiffeners are distributed at the symmetry boundary.

Besides, a certain amount of material gathers in the intermediate area between the top and the bottom of the dome. The reason for this phenomenon is the presence of both longitudinal and circumferential stresses in this region. The complex stresses in this region lead to an unclear distribution of stiffeners even when the optimization process converges.

4 Secondary Design of the Stiffener Layout

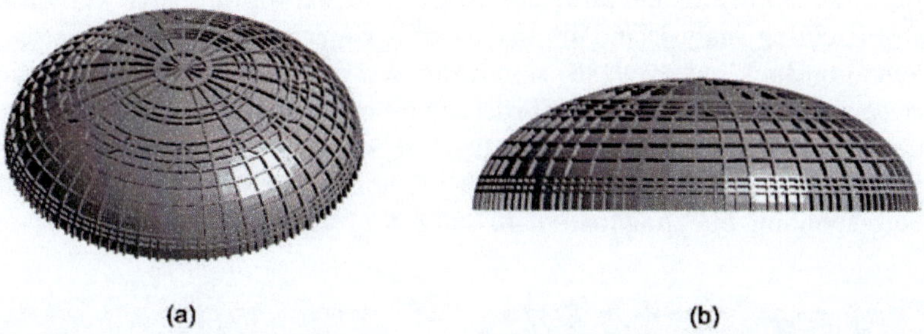

Fig. 3. Secondary design model of the dome with stiffeners. (a) Vertical view. (b) Side view

Based on the optimal results obtained in the previous section, we carry out the secondary design of the stiffener layout. The secondary design of the dome with stiffeners is shown in Fig. 3. In this section, the finite element analysis is performed using ANSYS to evaluate the structural response.

4.1 Static Analysis

As mentioned above, the structure is subjected to the load from the concrete pouring and the self-weight load of the steel plate. We first perform the linear elastic static analysis on the structure with the above loads. The distribution of the displacement and the Von Mises stress is shown Fig. 4. The maximum displacement and Von Mises stress of the secondary design are 6.8 mm and 33.1 MPa, respectively.

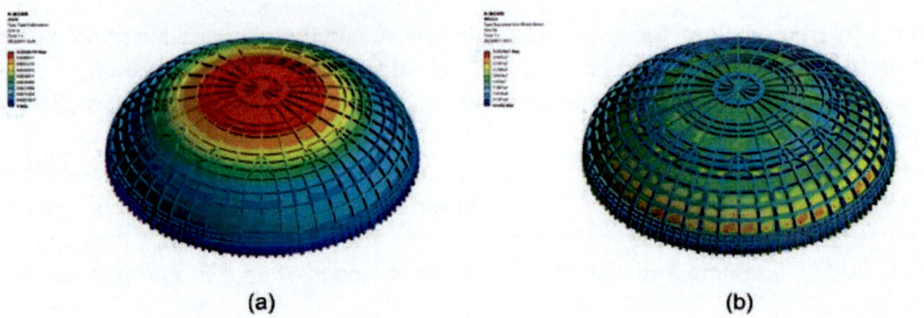

Fig. 4. Structural response of the secondary design. (a) Displacement. (b) Von Mises stress

4.2 Structural Stability Analysis

To ensure the stability of the structure during the pouring process, we conduct the eigenvalue buckling analysis and ultimate bearing capacity analysis on the structure.

Eigenvalue Buckling Analysis. A uniform load of 1 KN/m^2 is applied to the structure surface. The calculated first four order modes are shown in Fig. 5. Table 3 presents the load multipliers corresponding to the first ten order modes. The results of the eigenvalue buckling analysis indicate that the load factor for the first order mode is 407.85, with a corresponding buckling load of 407.85 KN/m^2.

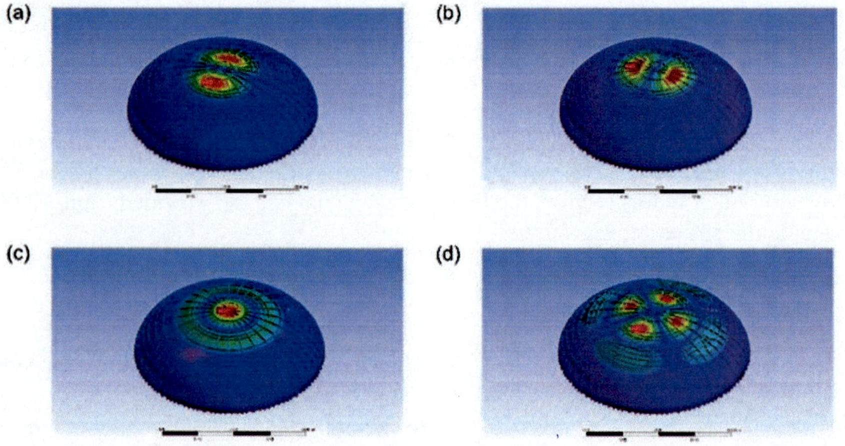

Fig. 5. The calculated first four order modes. (a) Order 1. (b) Order 2. (c) Order 3. (d) Order 4

Ultimate Bearing Capacity Analysis. Since structural buckling only occurs under ideal conditions, the nonlinear stability analysis is required to obtain the ultimate bearing capacity. Considering that the previously calculated buckling load for the first order mode is 407.85 KN/m^2, a uniform load of 300 KN/m^2 in the same direction is applied to the structure for the ultimate bearing capacity analysis. From Fig. 6, it can be observed

Table 3. The load multiplier for different order modes.

Modal order	1	2	3	4	5
load multiplier	407.85	408.01	511.75	534.73	535.06
Modal order	6	7	8	9	10
load multiplier	562.93	569.44	576.09	584.66	587.20

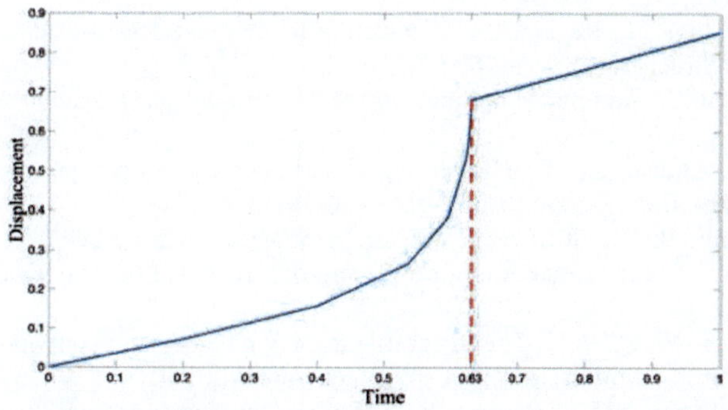

Fig. 6. The time-dependent variation curve of the calculated maximum displacement of the structure.

that the structure experiences instability when the load is applied at 0.63 s, indicating that the ultimate load-bearing capacity of the structure is 189 KN/m^2. According to relevant standards, the allowable load-bearing capacity for the stability of spatial lattice shell structure (with loads taken as the standard value) should be equal to the ultimate load-bearing capacity divided by the safety factor K. When considering elastic-plastic behavior, the safety factor K can be taken as 2.0. Therefore, the stability allowable load-bearing capacity for this structure is 94.5 KN/m^2, indicating that the structure meets the design requirements.

5 Conclusion

This paper utilizes a topology optimization method for the design of the stiffener layout. Firstly, the stiffener layout of the containment dome in a nuclear power plant is optimized using a parameterization approach based on radial basis functions. Subsequently, a secondary design of the dome with stiffeners is modeled based on the optimization results. The results of the finite element analysis of the secondary design, including the static analysis, eigenvalue buckling analysis, and ultimate bearing capacity analysis, demonstrate the effectiveness of the proposed method in the optimization design of the stiffener layout.

Acknowledgement. This research was supported by the National Natural Science Foundation of China (Grant No. 12072114), the National Key Research and Development Plan (Grant

No. 2020YFB1709401), Guangdong Basic and Applied Basic Research Foundation (Grant No.: 2023A1515012830), and the Guangdong Provincial Key Laboratory of Modern Civil Engineering Technology (2021B1212040003).

References

1. Cheng, K., Olhof, N.: An investigation concerning optimal design of solid elastic plates. Int. J. Solids Struct. **17**(3), 305–323 (1981)
2. Cheng, K., Olhof, N.: Regularized formulation for optimal design of axisymmetric plates. Int. J. Solids Struct. **18**(2), 153–169 (1982)
3. Gea, H.C., Luo, J.: Automated optimal stiffener pattern design. J. Struct. Mech. **27**(3), 275–292 (1999)
4. Belblidia, F., Afonso, S.M.B., Hinton, E., et al.: Integrated design optimization of stiffened plate structures. Eng. Comput. **16**(8), 934–951 (1999)
5. Ding, X., Yamazaki, K.: Stiffener layout design for plate structures by growing and branching tree model (application to vibration-proof design). Struct. Multidiscip. Optim. **26**(1–2), 99–110 (2004)
6. Wei, P., Ma, H., Wang, M.Y.: The stiffness spreading method for layout optimization of truss structures. Struct. Multidiscip. Optim. **49**(2), 667–682 (2014)
7. Guo, X., Zhang, W.S., Zhong, W.L.: Doing topology optimization explicitly and geometrically-a new moving morphable components based framework. J. Appl. Mech. Trans. ASME **81**(8), 081009 (2014)
8. Sun, Z., Cui, R., Cui, T., et al.: An optimization approach for stiffener layout of composite stiffened panels based on moving morphable components (MMCs). Acta Mech. Solida Sin. **33**(5), 605–662 (2020)
9. Liu, S., Li, Q., Chen, W., et al.: H-DGTP-a Heaviside-function based directional growth topology parameterization for design optimization of stiffener layout and height of thin-walled structures. Struct. Multidisc. Optim. **52**(5), 903v13 (2015)
10. Feng, S., Zhang, W., Meng, L., et al.: Stifener layout optimization of shell structures with B-spline parameterization method. Struct. Multidiscip. Optim. **63**, 2637–2651 (2021)
11. Zhang, W., Feng, S.: Combined parameterization of material distribution and surface mesh for stiffener layout optimization of complex surfaces. Struct. Multidisc. Optim. **65**(3) (2022)

Feasibility Study of Optimization of Ultrasonic Tomography Algorithm in Concrete

Lu Zhang, Chong Qiao, Shangda Jia, and Hongyu Li(✉)

College of Civil Engineering and Architecture, Guilin University of Technology, Guilin 541004, Guangxi, China
lihongyu@glut.edu.cn

Abstract. Concrete, as the main material in modern engineering construction, is one of the most determinant factors of the mechanical performance in structure. To rapid evaluation of its status is essential for the structural safety. Ultrasonic tomography is a non-destructive technique that can visualize the internal defects in the concrete by reconstructing the velocity distribution. In this paper, we proposed some feasible implementations to optimize the ray-trace based tomography algorithm. The influential factors on the accuracy of ultrasonic tomography were studied using numerical models. The errors source due to the nature of ray-trace method was discussed. The corresponding be reduced in terms of solution-correction based on the benchmark for SART. Then the influential parameters (e.g. flaw size, transducer layout, and frequency) were discussed. Finally, the feasible optimization measures have been proposed and evaluated.

Keywords: ultrasonic tomography · ray-trace method · concrete · flaw

1 Introduction

Concrete is a very important material for the construction industry, which is used to build various structural systems such as roads, bridges, tunnels, dams and power plants [1]. However, concrete structures may be confronted with some effects that undermine their structural integrity. Especially for the aging and environmental factors, material degradation and deterioration in concrete are concerned. Therefore, it is urgent to find the efficient and reliable status assessment method. Though many destructive testing explores failure mechanisms to determine the mechanical properties of materials, such as yield strength, punching-shear capacity, and flexural behavior [2], it is not applicable for on-site test and usually involves complex experimental configurations. Thus, non-destructive testing has many merits over the conventional destructive tests, which can evaluate and characterize the internal and subsurface damages rapidly without affecting the performance of the object [3].

In the field of concrete structures, non-destructive testing is playing an increasingly important role in identifying and detecting internal flaws and cracks [4, 5]. Herein, ultrasonic testing is widely used because of its advantages of easy use, speed of scan high resolution and accuracy, portability, and versatility. Lee et al. [6] found that the

W. Guo et al. (Eds.): GBCESC 2023, LNCE 328, pp. 47–58, 2024.
https://doi.org/10.1007/978-981-99-9947-7_6

ultrasonic pulse velocity varies with decreasing W/C ratio of concrete, and the velocity is slightly higher in concrete than in mortar. E. et al. [7] determined methods for estimating the porosity of pervious concrete using the measured ultrasonic wave velocity. However, conventional UT fails in localizing the flaw. Combined with the computed tomography algorithm, ultrasonic tomography can provide the visual evaluation of the interior of objects, which has been applied to steel, rock and concrete [8–10]. For evaluation of the engineering materials, the ray-based tomography is for the mainstream, which is a perfect tradeoff among the computational cost, efficiency and the accuracy. For example, Bond et al. [11, 12] demonstrated that ray-based tomography may provide cross-sectional images of structures that can be used to locate cracks, identify areas of structural damage, and other anomalies at depth in large-scale concrete structures.

But this method has a natural drawback in the path assumption, to be more specific, the reflection and diffraction were ignored, the path of ray-trace method is assumed as a straight line, which definitely cause errors. In this paper, the ray-tracing method with straight ray path assumption is applied. The influential factors on the accuracy of the tomography algorithm were investigated in terms of the location and size of the inclusion, transducer layout, and frequency. First, the error and noise of the tomography algorithm are discussed from the nature of the method. The effect of inclusions location and size on tomography was investigated. Additionally, numerical results of multiple sets of frequencies and transducer layouts were conducted to understand the relationship between transducer layout and frequency. Finally, the feasible measures have been proposed to optimize the ray-trace based tomography algorithm.

2 Ray-Trace Based Ultrasonic Tomography Theory

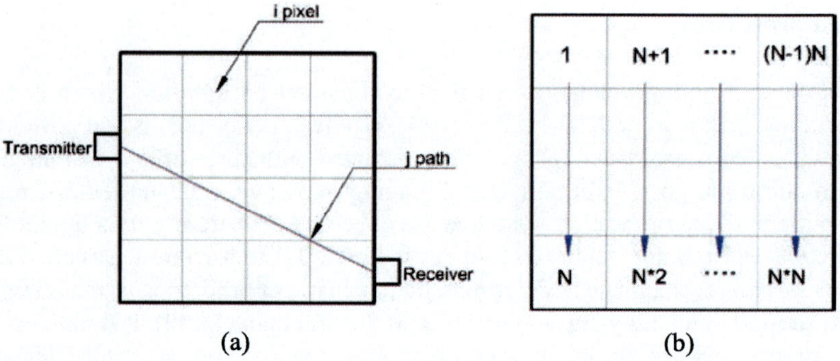

Fig. 1. (a) Schematic of one single ray-path and (b) regulation of pixel index.

The major input to the algorithm of any single ray-path is the *time of flight* (*TOF*) between transmitter and receiver, which is converted into the sum of TOF_i in each mesh, see Fig. 1, as shown in Eq. (1). The TOF_i in one mesh can be calculated by dividing the length by the velocity, as shown in Eq. (2). The velocity is transformed into slowness and substituted into Eq. (1), as shown in Eq. (3). It is assumed that the slowness or velocity of wave propagation within one mesh is constant and unique. Multiple measurements can yield multiple paths to generate a system of linear equations, as shown in Eq. (4).

$$TOF = \sum_{i=1}^{n} TOF_i \quad (1)$$

$$TOF_i = \frac{l_i}{v_i} \rightarrow TOF_i = l_i * s_i \quad (2)$$

$$TOF = \sum_{i=1}^{n} l_i * s_i \quad (3)$$

$$\{TOF_j\} = \begin{Bmatrix} TOF_1 \\ TOF_j \\ \vdots \\ TOF_m \end{Bmatrix} = \begin{bmatrix} l_{11} & l_{12} & \cdots & l_{1n} \\ l_{21} & l_{22} & & l_{2n} \\ \vdots & & \ddots & \vdots \\ l_{m1} & l_{m2} & \cdots & l_{mn} \end{bmatrix} \begin{Bmatrix} s_1 \\ s_2 \\ \vdots \\ s_n \end{Bmatrix} = [l_{ij}]\{s_i\} \quad (4)$$

where, i represents any mesh; j represents any ray-path; $s_i = 1/v_i$, s_i is the slowness, s/m, which is the inverse of velocity v_i, m/s; l_{ij} is the length of ray-path j in the pixel i. The [L] matrix is determined by layout of transducers. The *TOF* is the only input for Eq. 4, which is critical to the quality of tomography. So, in order to increase the accuracy of *TOF*, the arrival time is extracted using the based on identifying the outlier method in this paper [13]. This method features simple algorithm design and high accuracy. The solution of Eq. (4) is vector of slowness/velocity, which can be used for reconstructing the tomographic image.

To solve this system of linear equations, the numerical calculation with iterative procedures are employed. Many iterative methods can be used, while the algebraic reconstruction technique (ART) is the most widely used. For selecting the proper algorithm for concrete, the ART method has some issues in efficiency and accuracy since it can only treat one equation at a time during the iterations; while the Simultaneous Algebraic Reconstruction Technique (SART) [14] method handles all the equations at the same time in one iteration, which reduces the influences of noise. Especially for concrete, the heterogenous material can produce more noises for the propagation of ultrasound. Hence, in this study, the SART is adopted to solve Eq. (4). The open source tool (AIR tool) associated with MATLAB is used [15].

3 Numerical Test

For verifying the performance of the optimized algorithm, the numerical test was conducted by models using COMSOL Multiphysics software. The prototype object in this study is cube, as shown in the Fig. 2 (a). The sample size is 300 mm × 300 mm × 300 mm. In order to reduce the computational time, 2D plane strain model is selected, which represents the cross section of cube sample, see in Fig. 2 (a). The sample without inclusion are used as a benchmark. And three models with different types of inclusions are designed to evaluate the performance of tomography algorithm, see in Fig. 2 (b), (c), (d) and (e). For simplification, the transducers are modeled as point and distributed evenly along the side length, see in Fig. 3, two layouts of transducers were compared and stated in details in the following section. In the numerical models, the speeds of pressure-wave and shear-wave were specified. The material properties used in the numerical models are summarized in Table 1. A free triangular mesh was used with mesh size as one-sixth of the wavelength. The excitation signal is five-cycles tone burst signal with Hanning window, as shown in Fig. 4, and applied as point load on the edge. Three frequencies (60 kHz, 100 kHz and 150 kHz) were conducted to understand the correlation between frequency and transducer layout for further discussion. The number of pixels in one direction is chosen to be the same as the number of transducers. For example, a 10 × 10 pixels image is created when ten transducers are attached on one direction.

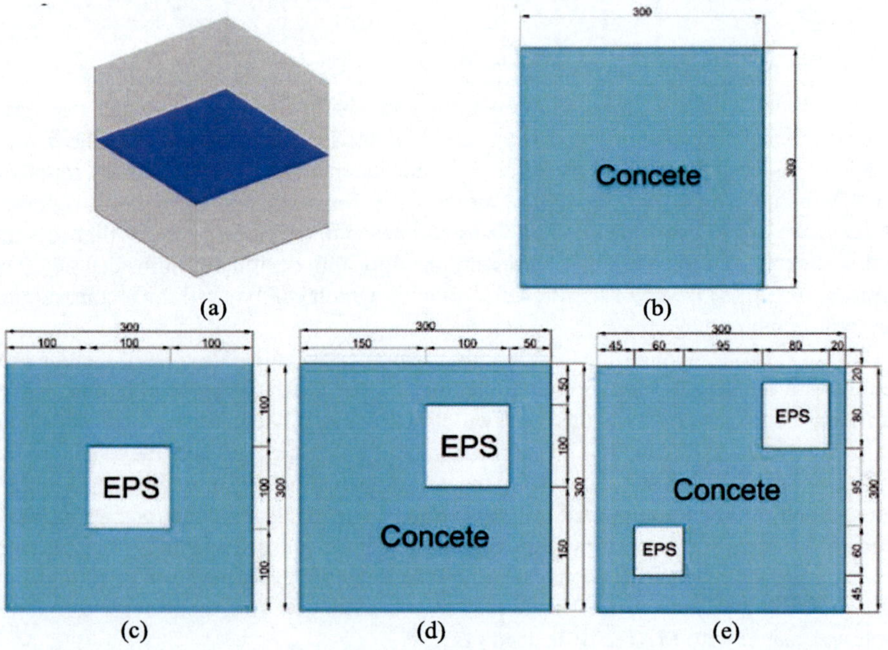

Fig. 2. The numerical model: (a) the prototype sample, (b) sample without inclusions (c) sample with inclusion type 1, (d) sample with inclusion type 2, and (e) sample with inclusion type 3

Table 1. Material properties used in numerical models

Properties	Base material	Inclusion (EPS foam)
Density (kg/m^3)	2360	12
Pressure Wave Velocity (m/s)	4259	711
Shear-wave speed (m/s)	2608	526

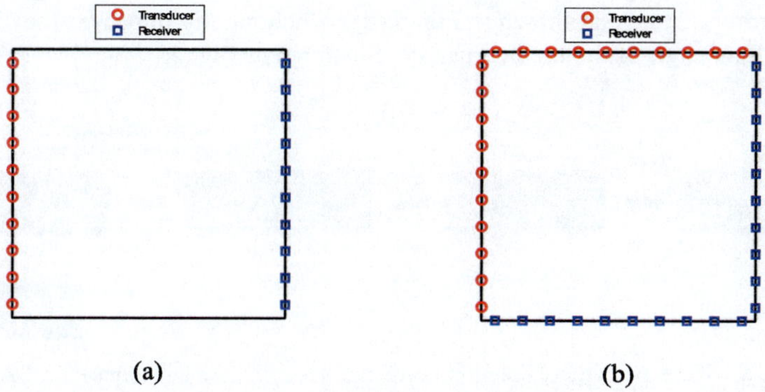

Fig. 3. The transducer layout, (a) Only one opposite side arrangement, and (b) Both opposite sides are arranged

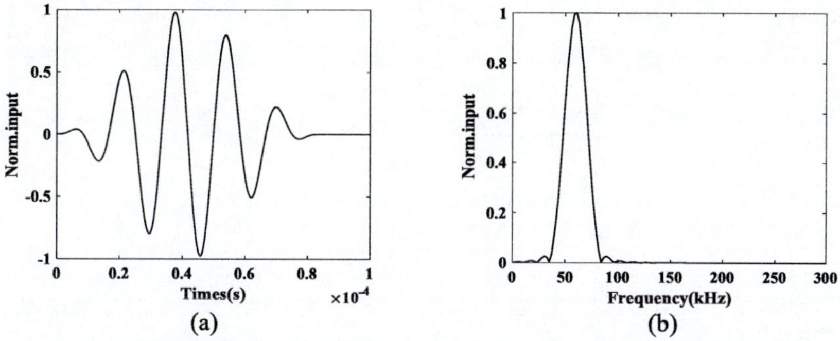

Fig. 4. Excitation signal, (a) time domain signal, and (b) frequency spectrum.

4 Optimization of Ultrasonic Tomography: Solution-Correction Based on the Benchmark for SART

Combining the ray-trace method with SART, if we want to increase the accuracy and resolution, the pixel and transducers should be increased, accordingly. Meanwhile, the obvious error of "X" shaped low velocity zone will happen see in Fig. 6 (a). It is inevitable due to the nature of ray-trace method. A solution-correction method has been proposed: solution-correction based on the benchmark for SART. As shown in Eq. (4), the *TOF* is the only input variable for most of the ray-trace based tomography methods. Therefore, it can cause the significant impact on the accuracy of proposed tomography. However, the length of each ray-path is different. It is not possible to analyze the source of error directly from the *TOF*. So, the average velocity is introduced to find the source of error. Using the transducer layout shown in Fig. 6 (b), the *TOF* is extracted and the average velocity is calculated. As shown in Fig. 6 (c), When the absolute value of the slope of the ray-path is too large, the wave speed decreases significantly. This is because the ray-trace algorithm with straight ray-path assumption introduces path errors, which approximates the ray-path as a straight line. Thus, the paths with slope greater than 0.5 are removed. The tomographic image is shown in Fig. 6 (d), which greatly improves the "X" shaped error. The flow chart of the method can be illustrated as Fig. 5.

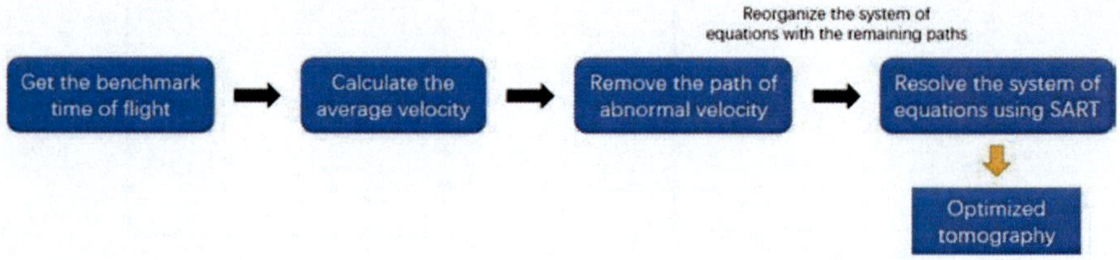

Fig. 5. The schematic of solution-correction based on the benchmark for SART

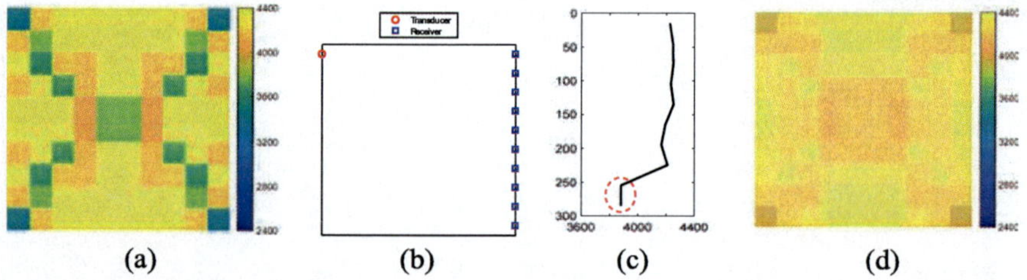

Fig. 6. (a) Tomography results without processing, (b) transducer layout, (c) average wave velocity and (d) tomography results with processing.

In addition, to further improve the accuracy of the algorithm, the tomographic images of two transducer layouts were compared. Figure 7 (b) is the tomographic image of sample with inclusion type 1 using the transducer layout shown in Fig. 3 (a). There are noises as marked by black circle in Fig. 7 (a). Increasing the number of transducers and arranging them as in Fig. 3 (b). The tomography result is shown in Fig. 7 (b), which shows that the noise is reduced and the images are more accurate. It is obvious that the effect of arranging the transducer in one direction is not as well as in two directions. In this way it is also feasible to optimize ultrasonic tomography. Therefore, this transducer layout is adopted in this paper.

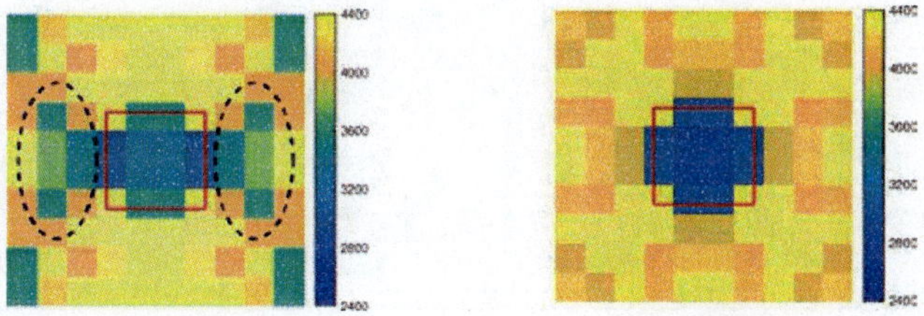

(a) (b)

Fig. 7. Tomography results, (a) transducer arrangement in one direction only, and (b) transducer arrangement in both directions.

5 Parametric Study and Results

5.1 Influence of Inclusion Size and Position

Using 60 kHz excitation frequency and 10 × 10 transducer layout, tomographic images of all samples are shown in Fig. 8. The scatter plot of the velocities is shown in the Fig. 9, with the horizontal axis representing pixel index and the vertical axis representing velocity. Using the sample without inclusion as the baseline, the mean and standard deviation of the velocities generated by the algorithm are calculated. From Fig. 6, the velocities inversed by tomography are mostly in the interval of $\mu \pm 6 * \sigma$. Obtained velocities outside of this interval are defined as outlier. Since the wave velocity of the inclusion is set to be lower than that of the concrete, we are only interested in the portions below $\mu - 6 * \sigma$. When the velocities obtained by the tomography algorithm are lower than $u - 6 * \sigma$, these pixels are considered to be defective ones. The comparison between the area of the defective region obtained by the algorithm and that of the actual inclusions is shown in Table 2. For samples without inclusion, as shown in Fig. 8 (a), the images are uniform due to the absence of inclusions interfering with the propagation of ultrasound. The presence of inclusions causes refraction and reflection of waves. The samples with inclusion type 2 and type3 being near the boundary, the waves produce complex reflections during propagation, which can bring errors to the tomography results, as shown in Fig. 8 (c) and Fig. 8 (d). On the contrary, the results of tomography with inclusions far from the boundaries enable better identifying and locating defects because less uncertainty is introduced, as shown in Fig. 8 (b) and Fig. 8 (c).

5.2 Influence of Frequency and Transducer Layout

Samples with inclusion type 1 and type2 can be well detected using 60 kHz excitation frequency and 10 × 10 transducer array. But the tomographic image of sample with inclusion type 3 is not satisfactory. In order to obtain better tomography result of sample with inclusion type 3, Three frequencies (60 kHz, 100 kHz and 150 kHz) and three transducer layouts (6 × 6, 8 × 8 and 15 × 15 transducer array) are employed in this section. The tomography results of excitation frequency as 60 kHz, 100 kHz and 150 kHz are shown in Fig. 10 (b), (c) and (d), respectively. It can be seen from the figure that even though the excitation frequency and transducer arrangement are changed, there are

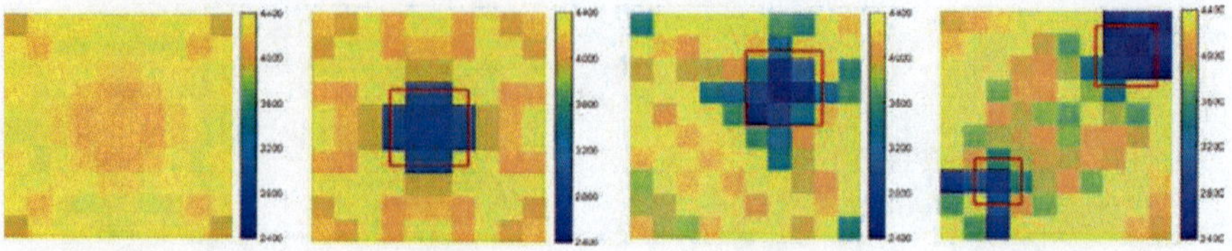

(a) (b) (c) (d)

Fig. 8. Tomography results with 60 kHz excitation frequency and 10 × 10 transducer layout, (a) sample without inclusions (b) sample with inclusion type 1, (c) sample with inclusion type 2, and (d) sample with inclusion type 3.

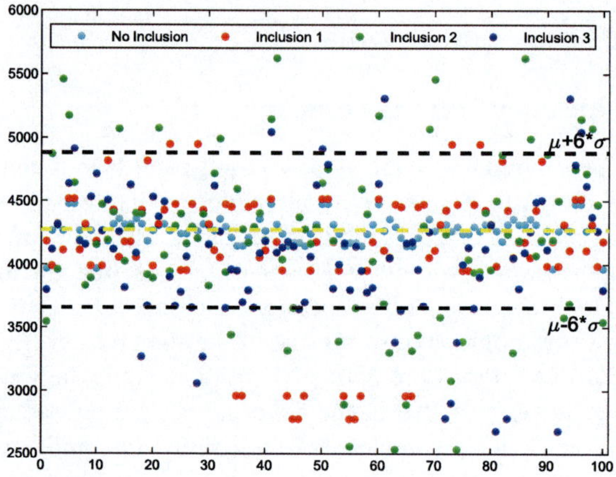

Fig. 9. The velocities scatter plot of all samples.

Table 2. Area of the defective region obtained by the algorithm and that of the actual inclusions

Sample	Actual inclusions area (mm^3)	Area obtained by the algorithm (mm^3)
Sample with inclusion type 1	10000	10800
Sample with inclusion type 2	10000	18900
Sample with inclusion type 3	10000	16200

still many errors in tomographic image of sample with inclusion type 3. This is an error caused by the nature of the methodology, which cannot be eliminated by changing the excitation frequency and the number of transducers. However, using the same transducer layout, different excitation frequencies have different imaging effects. The relationship between frequency and the number of transducers can be explored. This requires an evaluation of tomographic image in Fig. 10.

The method for evaluating the performance of the tomography algorithm is mentioned above. However, it is just a rough evaluation. Much information is ignored. In

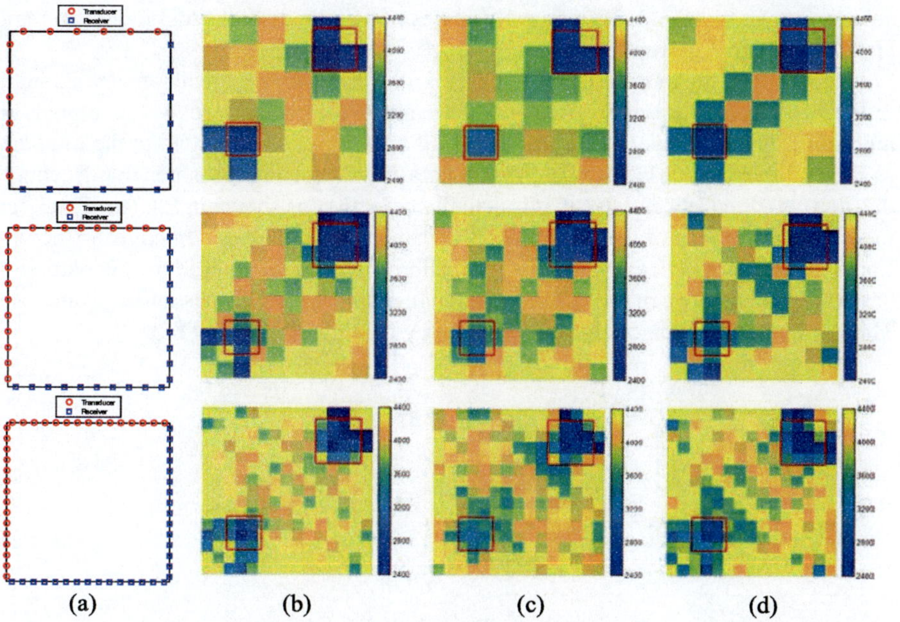

Fig. 10. Tomography results of sample with inclusion type 3, (a) Transducer arrangement (b) 60 kHz excitation frequency, (c) 100 kHz excitation frequency, (d) 150 kHz excitation frequency.

order to evaluate the performance of the tomography algorithm in a more refined way, new evaluation indicators are proposed in this paper to evaluate the algorithm. It is on the basis of evaluation indicators commonly used in the field of Machine Learning [16, 17]. Before introducing the evaluation indicators, it should be taken into account that the result of tomography provides four possible outcomes as shown in the Fig. 11.

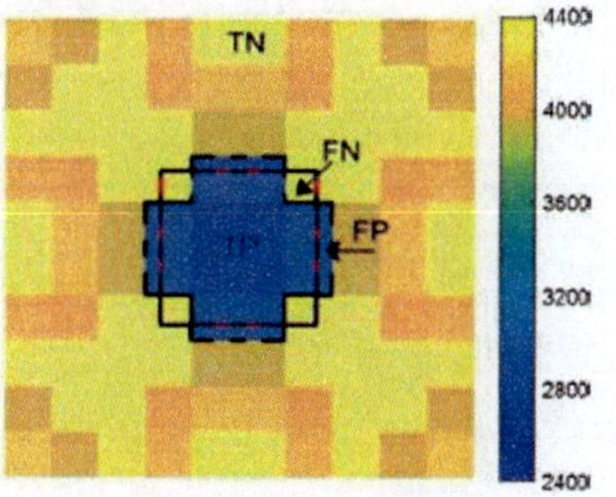

Fig. 11. Four possible outcomes of tomography: True Positive (TP), False Positive (FP), True Negative (TN) and False Negative (FN).

TP: True Positives, area within the actual inclusion region and with velocity below threshold; FP: False Positives, area outside the actual inclusion region and with velocity below threshold; FN: False Negatives, area within the actual inclusion region and with velocity above threshold; TN: True Negatives, area outside the actual inclusion region and with velocity above threshold.

In this paper, the threshold is $\mu - 6 * \sigma$ as above. Accuracy, Precision, Recall and F1 score are applied as evaluation indicators to evaluate the performance of the tomography algorithm. The Accuracy is the percentage of area correctly classified to the total area, as shown in Eq. (5). It is a statistic for all data. The Precision measures that fraction of the area classified as defective that is actually defective, as shown in Eq. (6). The Recall measures the fraction of the actual defect area that is classified as defective, as shown in Eq. (7). F1 score (F1) combines precision and recall into one metric by calculating the harmonic mean between those two, as shown in Eq. (8). It reaches its best value at 1. These evaluation indicators for all tomography result are shown in Table 3.

$$\text{Accuracy} = \frac{TP + TN}{TP + FP + TN + FN} \tag{5}$$

$$\text{Precision} = \frac{TP}{TP + FP} \tag{6}$$

$$\text{Recall} = \frac{TP}{TP + FN} \tag{7}$$

$$F1 = 2\frac{\text{Precision} \cdot \text{Recall}}{\text{Precision} + \text{Recall}} \tag{8}$$

Table 3. Evaluation indicators for all tomography results of sample with inclusion type 3.

excitation frequency and transducer array	Accuracy	Precision	Recall	F1 score
60 kHz, 6 × 6 transducer array	0.91	0.57	0.85	0.68
60 kHz, 10 × 10 transducer array	0.89	0.51	0.82	0.63
60 kHz, 15 × 15 transducer array	0.91	0.56	0.76	0.64
100 kHz, 6 × 6 transducer array	0.95	0.80	0.80	0.80
100 kHz, 10 × 10 transducer array	0.88	0.48	0.82	0.61
100 kHz, 15 × 15 transducer array	0.86	0.43	0.78	0.56
150 kHz, 6 × 6 transducer array	0.75	0.30	0.90	0.45
150 kHz, 10 × 10 transducer array	0.78	0.31	0.81	0.45
150 kHz, 15 × 15 transducer array	0.85	0.42	0.88	0.57

From the data in Table 3, the excitation frequency of 150 kHz gives the best tomography results for the 15 × 15 transducer arrangement compared to the other two transducer layouts. Using 100 kHz excitation frequency, the 10 × 10 transducer layout is the best among the three transducer arrangements. This indicates that there is an optimal relationship between the number of transducers and the excitation frequency.

6 Conclusion

In this paper, the optimization of ultrasonic tomography algorithm has been proposed in terms of solution-correction based on the benchmark for SART. And the performance of the optimization has been proved to be feasible using numerical models. And also, the effects of different types of defects and the effects of frequency and transducer layout on ultrasound tomography are studied. It was found that for ray-trace based method, errors introduced by the natural principle of algorithms or real paths of wave propagation must be considered in order to reach better imaging results in tomography. Meanwhile, the reflection of waves by the boundaries and inclusions boundaries can also cause the negative impact on the accuracy of the tomography. The resolution of image highly depends on the excitation frequencies and leading to different imaging effects, and there is an optimal frequency for one transducer arrangement. In future work, curved ray methods can be used to reduce errors in the nature of the methodology, and three-dimensional tomography of concrete also needs to be studied.

Acknowledgement. This work was supported by National Natural Science Foundation of China (Grant Nos. 52068015 and 52168068), and Guangxi Science and Technology Base and Special Fund for Talents Program (Grant No. Guike AD20159011). Any opinions, findings and conclusions expressed in this paper are those of the writers and do not necessarily reflect the view of Natural Science Foundation of China and Guangxi Science and Technology Base and Special Fund for Talents Program.

References

1. van Damme, H.: Concrete material science: past, present, and future innovations. Cem. Concr. Res. **112**, 5–24 (2018)
2. Miller, R., Aktan, A., Shahrooz, B.: Destructive testing of decommissioned concrete slab bridge. J. Struct. Eng. **120**(7), 2176–2198 (1994)
3. Gholizadeh, S.: A review of non-destructive testing methods of composite materials. Procedia Struct. Integrity **1**, 50–57 (2016)
4. Wankhade, R.L., Landage, A.B.: Non-destructive testing of concrete structures in Karad region. Procedia Eng. **51**, 8–18 (2013)
5. Goueygou, M., Abraham, O., Lataste, J.-F.: A comparative study of two non-destructive testing methods to assess near-surface mechanical damage in concrete structures. NDT & E Int. **41**(6), 448–56 (2008)
6. Lee, T., Lee, J.: Setting time and compressive strength prediction model of concrete by nondestructive ultrasonic pulse velocity testing at early age. Constr. Build. Mater. **252**, 119027 (2020)
7. Ridengaoqier, E., Hatanaka, S., Palamy, P., et al.: Experimental study on the porosity evaluation of pervious concrete by using ultrasonic wave testing on surfaces. Constr. Build. Mater. **300**, 123959 (2021)
8. Samokrutov, A., Shevaldykin, V.: Ultrasonic tomography of metal structures using the digitally focused antenna array method. Russ. J. Nondestr. Test. **47**(1), 16–29 (2011)
9. Shiotani, T., Kobayashi, Y., Chang, K.-C.: Hybrid elastic-wave CT with impact acoustics for single-side measurement in concrete structures. Const. Build. Mater. **112**, 907–14 (2016)

10. Li, X., Lei, X., Li, Q., et al.: Experimental investigation of Sinian shale rock under triaxial stress monitored by ultrasonic transmission and acoustic emission. J. Nat. Gas Sci. Eng. **43**, 110–123 (2017)
11. Bond, L.J., Kepler, W.F., Frangopol, D.M.: Improved assessment of mass concrete dams using acoustic travel time tomography. Part I—Theory. Const. Build. Mater. **14**(3), 133–146 (2000)
12. Kepler, W.F., Bond, L.J., Frangopol, D.M.: Improved assessment of mass concrete dams using acoustic travel time tomography. Part II—Application. Const. Build. Mater. **14**(3), 147–56 (2000)
13. Zhang, L., Dong, J., Godinez-Azcuaga, V., et al.: The identification of accurate and computationally efficient arrival time pick-up method for acoustic tomography. In: Proceedings of the Nondestructive Characterization and Monitoring of Advanced Materials, Aerospace, Civil Infrastructure, and Transportation XIII. SPIE (2019)
14. Andersen, A.H., Kak, A.C.: Simultaneous algebraic reconstruction technique (SART): a superior implementation of the ART algorithm. Ultrason. Imaging **6**(1), 81–94 (1984)
15. Hansen, P.C., Saxild-Hansen, M.: AIR tools—a MATLAB package of algebraic iterative reconstruction methods. J. Comput. Appl. Math. **236**(8), 2167–2178 (2012)
16. Davis, J., Goadrich, M.: The relationship between Precision-Recall and ROC curves. In: Proceedings of the 23rd International Conference on Machine Learning (2006)
17. Giglioni, V., García-Macías, E., Venanzi, I., et al.: The use of receiver operating characteristic curves and precision-versus-recall curves as performance metrics in unsupervised structural damage classification under changing environment. Eng. Struct. **246**, 113029 (2021)

Static and Dynamic Analysis of Construction Catwalk of Long-Span Suspension Bridge

Jinguo Jiang(✉), Jihua Xiong, and Feng Wang

Sichuan Highway and Bridge Construction Group Co, Ltd. Bridge Engineering Branch, Chengdu 610041, China
Jingguo_Jiang@126.com

Abstract. With the increasing span of the suspension bridge, the wind resistance design of the catwalk has gradually become the dominant factor controlling its design. Therefore, it is necessary to analyze its static and dynamic characteristics and the factors that affect its dynamic characteristics. Taking the catwalk of a long-span suspension bridge as an example, the static and dynamic analysis of the catwalk is carried out based on Midas/Civil software and cable vibration theory. The results show that the dynamic characteristics of the catwalk obtained by the finite element model are consistent with the results of theoretical analysis. The stress of various components in the catwalk meets the safety factor requirements under extreme loads. However, there are obvious low-order vibrations in the local nodes of the catwalk structure, which will generate significant wind-induced responses at high wind speeds and significantly impact construction. The research can provide a further reference for the wind resistance design of catwalks of the same type.

Keywords: Long Span · Suspension Bridge · Construction · Catwalk · Static and Dynamic Analysis · Finite Element Analysis

1 Introduction

The catwalk is a temporary structure used as a working platform and channel for construction personnel during the construction of the main cable of the suspension bridge. The catwalk structure is the operating platform for the erection of the main cable of the suspension bridge [1, 2]. It plays an essential role in construction stages such as cable strand traction, cable adjustment, shaping into the saddle, tightening the cable, cable clamp and sling installation, main cable winding, protective coating, etc. Its structure mainly comprises catwalk load-bearing cable, catwalk surface layer, railings and handrails, wind-resistant system, frame system, horizontal overpass, and various anchorage connections [3, 4]. During construction, the temporary structure catwalk of the long-span suspension bridge has low stiffness and small damping, so it is sensitive to wind load. At the same time, the construction period of the suspension bridge is up to several years, and the wind environment is complex [5, 6]. As the operating platform of the superstructure, the catwalk is very important for constructing a suspension bridge.

W. Guo et al. (Eds.): GBCESC 2023, LNCE 328, pp. 59–71, 2024.
https://doi.org/10.1007/978-981-99-9947-7_7

In the construction of long-span suspension bridges at home and abroad in the past, the measures of setting wind-resistant cables were adopted to improve the wind resistance of the catwalk. When analyzing the vibration characteristics of the catwalk, it is generally considered that the tower is rigid relative to the catwalk, and only the catwalk collapsed while ignoring the influence of side spans and towers [7, 8]. Some scholars studied the dynamic characteristics of the abovementioned construction catwalks with wind-resistant cable measures and gave the theoretical calculation and approximate empirical formula of the fundamental frequency [9, 10]. The theoretical calculation results were compared with the finite element analysis results through examples. The influence of tension, rise-span ratio, and wind load on the dynamic characteristics of the catwalk was also analyzed [11]. At present, with the construction of long-span suspension bridges on the river and sea, in order to speed up the construction progress, reduce the cost and ensure the standard navigation requirements, the construction catwalks without wind-resistant cables have been used, including Akashi Strait Bridge and the domestic Runyang Bridge South Branch Suspension Bridge, which puts forward higher requirements for the wind-resistant safety of such catwalks [12, 13]. To ensure the wind-resistant safety of catwalks without wind-resistant cables, much literature has studied the wind-resistant performance and wind-resistant measures of such catwalks using the finite element method. However, to thoroughly understand the wind-resistant performance of such catwalks and find a theoretical basis for better wind-resistant measures, it is vital to analyze their dynamic characteristics and the factors affecting them.

In theory, flutter, galloping, buffeting, vortex-induced vibration, and static instability may occur in catwalks without wind cable under wind load. Researchers mainly study the static stability and flutter of the catwalk [14, 15]. Some literature studied the static stability of the catwalk of a long-span suspension bridge under the action of orthogonal wind and non-orthogonal wind by wind tunnel test and numerical simulation and studied the buffeting performance of the catwalk under the action of turbulent wind. The turbulent wind has little effect on the wind resistance stability of the catwalk, and the influence of turbulent wind can not be considered in the catwalk design [10, 16]. The critical wind speed of aerostatic instability of the catwalk of Xihoumen Bridge was studied by wind tunnel test at Southwest Jiaotong University [17, 18]. Many kinds of literature show that the research method of bridge static wind stability has developed from full model wind tunnel test to segment model wind tunnel test and numerical calculation. The accuracy of the research has been developed from linear to geometric and material nonlinearity to considering geometric nonlinearity, material nonlinearity, and aerodynamic nonlinearity simultaneously. Most of the research methods are calculated by compiling a finite element program [8, 19]. The popularization of sizeable commercial software such as ANSYS is a good research measure to carry out secondary development to calculate aerostatic instability. Due to the convenience of ANSYS, the original programming calculation steps can be omitted, significantly simplifying the calculation difficulty [20]. With the increasing span of the suspension bridge, the problem of aerostatic instability of the catwalk has gradually become the dominant factor controlling its design. The influence of the components of the catwalk on the stability of the catwalk is bound to attract more and more attention.

With the arrival of the construction boom of the cross-sea bridge, the span of the catwalk is also getting larger and larger, which puts forward higher requirements for the wind resistance of the catwalk without wind cable [21]. The buffeting of the catwalk under the action of turbulent wind significantly impacts the comfort and safety of the construction personnel. Whether other vibration forms of the wind-induced vibration of the catwalk can appear remains to be studied. Therefore, to ensure the wind resistance safety of the catwalk during the construction of the leading cable and provide a basis for the study of wind resistance performance, it is necessary to analyze its vibration characteristics and the influence of relevant design parameters on its dynamic characteristics.

2 Project Overview

2.1 Engineering background

Heishuihe Bridge is a vital control project of the Ningnan Panzhihua Expressway Project on the G4216 line of China's expressway network. The main span of the bridge is a 550 m steel-truss-girder suspension bridge, with a span of 138.0 m + 550 m + 131.5 m, a rise span ratio of 1/10 in the middle span, a center spacing of 30 m in the transverse direction of the main cable, a standard spacing of 13.0 m in the longitudinal direction of the slings, and a distance of 15.0 m from the center line of the tower at the place near the tower. The bridge site is located at Shanhoutou, Makou Town, Ningnan County, Liangshan Yi Autonomous Prefecture, spanning both banks of the Heishui River and on two active fault zones. The anchor on the south bank of Qining is a gravity anchor, while the anchor on the bank of Panzhihua is a tunnel anchor. The effect diagram of the complete plan for the Heishui River Grand Bridge is shown in Fig. 1.

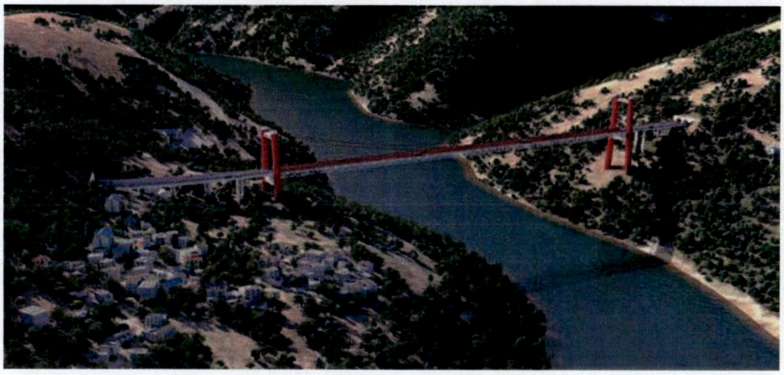

Fig. 1. Bridge rendering.

2.2 Overview of the Catwalk

As shown in Fig. 2, the bridge's catwalk adopts a three-span continuous form with a maximum rise span ratio of 1/10. The catwalk system consists of load-bearing cables,

a catwalk surface layer, a catwalk portal frame, a transverse channel, a load-bearing cable anchoring and turning device, a displacement steel frame, a pull-down device, an operation platform, etc. There are five horizontal passages, three mid spans, a spacing of 137.5 m, and one side span on both sides of the bridge. Fourteen catwalk load-bearing frame frames are arranged, with a spacing of 50 m. Each catwalk is arranged with a total of 6 pieces ϕ 48 mm load-bearing rope, two pieces ϕ 48 mm frame rope, four pieces arranged on both sides ϕ 22 mm protective rope, two pieces ϕ 24 mm handrail rope, connected to the frame through M12U bolts; The catwalk adopts double-layer steel wire mesh as the walking panel, with large beams spaced at 6 m, small beams spaced at 3 m, and anti-slip wooden strips spaced at 50 cm at the bottom. According to the construction organization design of the main cable, the catwalk structure adopts a continuous type,

The span from the west anchored frame to the bridge tower top cable saddle is 134.635 m, while the span from the east anchored frame to the bridge tower top cable saddle is 134.635 m. The span of the mid-span catwalk is 550 m (from the center of the west tower top to the center of the east tower top). The plane and cross-section of the catwalk structure are shown in Fig. 2 (b) (c), and the details of the frame and transverse overpass are shown in Fig. 2 (d) (e).

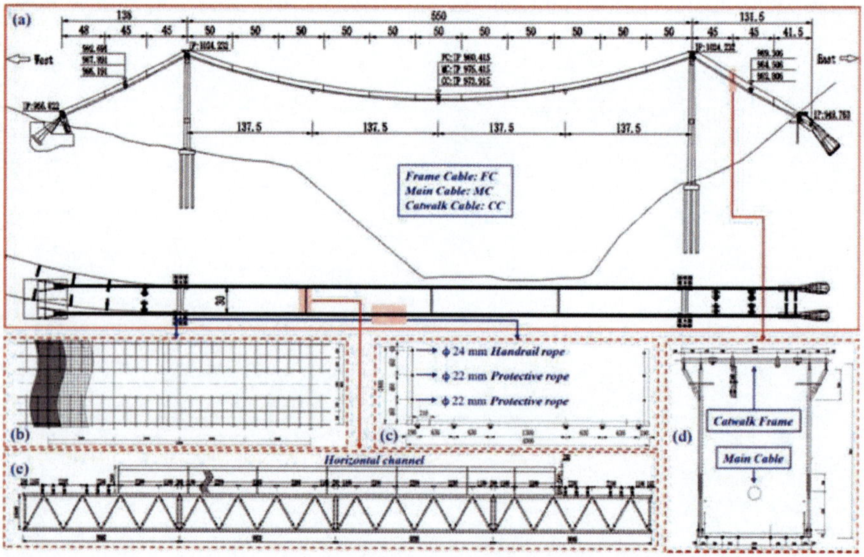

Fig. 2. Overall Layout of Catwalk (Unit: mm).

3 Static Bearing Capacity Analysis of Catwalk Structures

3.1 Component Cross-Section, Material Characteristics, and Catwalk Design Loads

Under the action of dead load, main cable strand weight, a live load such as construction personnel or machinery, and wind load, construction catwalks, like bridge structures, will also undergo static deformation and dynamic response in the transverse, vertical, axial, and torsional directions of the bridge. To investigate its static and dynamic effects in detail, it is necessary to establish a three-dimensional finite element model that can reflect the characteristics of the stress and deformation effects of the catwalk structure. To simplify the modeling, without considering the role of steel wire mesh, only load-bearing cables, handrail cables, frame support cables, frame structures, transverse overpass structures, and handrail railing structures are established in the model. The specifications and physical characteristics of the main components of the catwalk are shown in Table 1 and Table 2, respectively, and the design loads of the catwalk structure are shown in Table 3.

Table 1. Specifications of various components of catwalk structure.

Structural components	Specification
Main tower frame chord	H400 × 400
Main tower frame diagonal rod	I 40
Main tower frame construction ramp diagonal rod	][14b
Main tower frame construction chord	][20b
Catwalk surface mesh crossbeam	□80 × 80 × 4
Catwalk surface mesh small crossbeam	□50 × 50 × 2.5
Catwalk handrail column	∠75 × 75 × 5
Upper crossbeam of catwalk frame	□160 × 160 × 4
Catwalk portal slant support	□140 × 140 × 4
Upper pillar of catwalk frame	□140 × 140 × 4
Lower pillar of catwalk portal frame	□120 × 120 × 5
Catwalk frame column support	□140 × 140 × 4
Transverse channel chord	□120 × 120 × 4
Transverse channel web member	□80 × 80 × 3
Catwalk load-bearing rope	ϕ 48
Catwalk handrail rope	ϕ 24
Catwalk protective rope	ϕ 22
Frame load-bearing cable	ϕ 48

Table 2. Material characteristics of various components of catwalk structure.

Structural components	Material	E (Gpa)	Poisson's ratio	Density (kg/m3)
Catwalk load-bearing rope	Strand1960	1050	0.3	16447
Catwalk handrail rope	Strand1960	1050	0.3	8851
Catwalk protective rope	Strand1960	1050	0.3	8870
Frame load-bearing cable	Strand1960	1050	0.3	9157
Large crossbeam	Q235	2060	0.3	8700
Small crossbeam	Q235	2060	0.3	8200
Horizontal channel	Q235	2060	0.3	10334
Catwalk handrail column	Q235	2060	0.3	7850
Catwalk frame	Q235	2060	0.3	9000
Tower frame column	Q345	2060	0.3	7850
Other components of tower frame	Q235	2060	0.3	7850

Table 3. The design load of a single catwalk.

Load form	Composition		Count
Dead load (kg/m)	--	--	131.17
Live load (kg/m)	--	--	23.89
Wind load	Lift force (kN/m)	0.191	0.191
	Drag force (kN/m)	0.755	0.755
	Torsional moment (kN·m/m)	0.163	0.163
Temperature (°C)	Reference temperature	15.00	15.00
	System heating	25.00	25.00
	Cooling	20.00	20.00

3.2 Finite element model

The model was established using large-scale commercial software MIDAS/CIVIL, and each component's material properties and unit types are shown in Table 4. At the same time, only load-bearing truss elements were used to simulate the catwalk load-bearing cable, handrail rope, protective rope, and load-bearing frame cable, while other components were manufactured using beam elements. The finite element model is shown in Fig. 3.

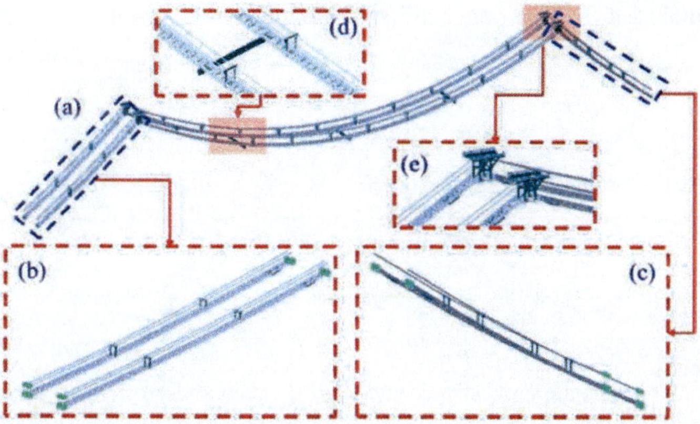

Fig. 3. The finite element model of the catwalk: (a) overall model diagram of the catwalk, (b) Ningnan side catwalk, (c) Panzhihua side catwalk, (d) catwalk frame and transverse passage, (e) tower top frame.

3.3 Strength calculation of catwalk structure

The typical states where the strength of the construction catwalk is relatively unfavorable are the catwalk erection stage and the traction cable strand stage. For the catwalk erection stage, the load-bearing catwalk rope, bottom net, side net, etc. have been erected, but they are not subjected to joint forces with the catwalk frame and frame load-bearing rope; For the traction cable strand stage, all catwalks are erected, and the main cable construction is carried out, with one cable strand on each drum of each catwalk. The more unfavorable load combination in the two states should be combination one: catwalk weight & temperature change & crowd load; Combination 2: catwalk self-weight & extreme wind load. Based on the above two different load combinations and unfavorable states, MIDAS is used to model and perform nonlinear analysis and calculation on the catwalk, and each state's structural internal forces and displacements can be obtained.

There are two main types of combinations for the weight of the catwalk, temperature change and crowd load. One is to consider system heating (under S1 working conditions), and the other is to evaluate system cooling (under S2 working conditions); Similarly, the combination of catwalk self-weight and extreme wind load can also be divided into two types: one is to consider system heating (under S3 working condition), and the other is to evaluate system cooling (under S4 working condition). The stress and internal force diagrams corresponding to each load-bearing cable and catwalk frame are shown in Figs. 4, 5, 6and 7. It can be seen from the figure that for combination condition 1, the maximum stress of the catwalk load-bearing cable is 618.1 MPa (safety coefficient of $1960.3/618.1 = 3.17 > 3$), and for combination condition 2, the maximum pressure is 618.6 MPa (safety coefficient of $1960.3/423.1 = 4.63 > 2.5$). Both catwalk load-bearing cables meet the requirements of safety coefficients 3 and 2.5; For the catwalk frame, the maximum tensile stress corresponding to combination condition one is 29.5 MPa, and the maximum compressive stress is 57.0 MPa; The maximum pressure corresponding to combination condition 2 is 63.5 MPa, and the maximum compressive stress is 25.5 MPa.

It can be concluded the that under the extreme wind load, the pressure of the catwalk frame meets the requirements.

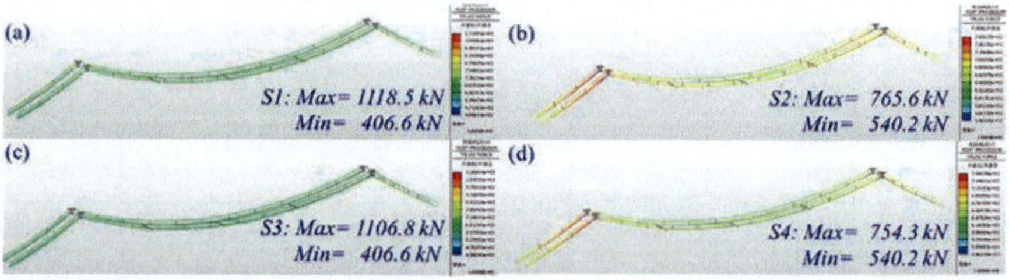

Fig. 4. Internal force diagrams of catwalk load-bearing cables under different working conditions: (a) S1 working condition, (b) S2 working condition, (c) S3 working condition, (d) S4 working condition.

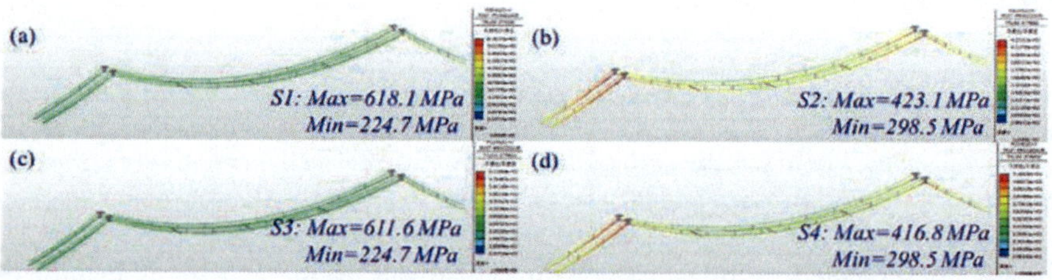

Fig. 5. Stress diagrams of catwalk load-bearing cables under different working conditions: (a) S1 working condition, (b) S2 working condition, (c) S3 working condition, (d) S4 working condition.

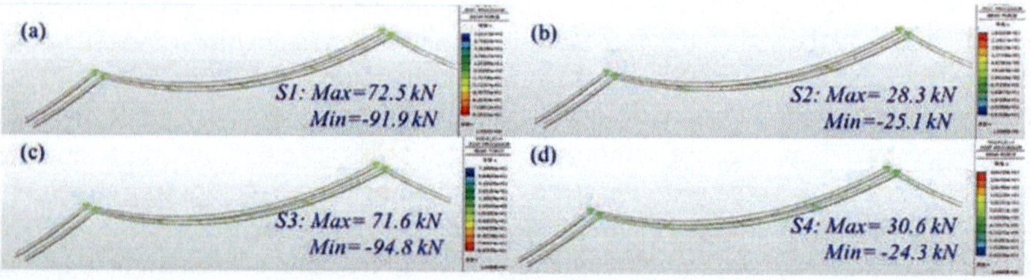

Fig. 6. Internal force diagrams of catwalk frame under different working conditions: (a) S1 working condition, (b) S2 working condition, (c) S3 working condition, (d) S4 working condition.

4 Dynamic characteristics analysis of catwalk

4.1 Theoretical analysis of dynamic characteristics of catwalks

The catwalk is a typical cable truss structure. If the bending stiffness of the catwalk is ignored, based on the analysis theory of cable structures and the characteristics of non-wind resistant cable catwalk structures, it can be simplified as a cable truss structure

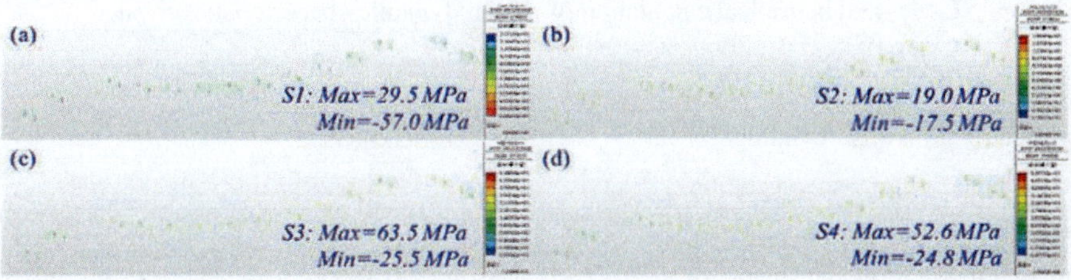

Fig. 7. Stress diagrams of catwalk frame under different working conditions: (a) S1 working condition, (b) S2 working condition, (c) S3 working condition, (d) S4 working condition.

with pretension, and only the mid-span analysis of the catwalk is taken. At this time, full load effects such as frame and transverse passage are not considered, and the mass of the catwalk is assumed to be uniformly distributed along the length of the span. Based on the above assumptions and the literature [9, 20], the frequencies of the positive symmetric and antisymmetric transverse, antisymmetric vertical, and antisymmetric torsion of the catwalk can be calculated using Eq. (1), Eq. (2), and Eq. (3), respectively.

$$f_n = \frac{\omega_n}{2\pi} = \frac{n}{2l}\sqrt{\frac{H_d}{m}} \quad (n = 1, 2, 3 \cdots). \tag{1}$$

$$f_k = \frac{\omega_k}{2\pi} = \frac{k}{2l}\sqrt{\frac{H_d}{m}}(k = 2, 4, 6 \cdots). \tag{2}$$

$$f_r = \frac{\omega_r}{2\pi} = \frac{k}{2l}\sqrt{\frac{H_d}{m}}(r = 2, 4, 6 \cdots). \tag{3}$$

When using theoretical formulas to calculate the dynamic characteristics of the catwalk, the specific design parameters of the catwalk are the calculated span $l = 543.1$ m; The mass per linear meter along the span direction of the catwalk is m = 326.26 kg/m; Load bearing cable elastic modulus E = 1.05×10^5 MPa; Total area A = 1.2624×10^{-2} m^2; Total horizontal tension H_d = 2532.604 kN; The height of the load-bearing rope is 47.497 m (in the empty cable state). Substitute the above parameters into Eq. (1), Eq. (2), and Eq. (3), respectively, to obtain the required natural frequency of the catwalk. The theoretical calculation results of the dynamic characteristics of the catwalk are shown in Table 5.

4.2 Finite element analysis of dynamic characteristics of catwalks

Due to the consideration of large deformation and prestressing effects in the cable element during modeling, it is necessary to use prestressing modal analysis. Before conducting prestressed modal analysis, the static analysis must be conducted first, and only based on the static analysis results can prestressed modal analysis be conducted. Figure 8 shows the modes of the first four sections. Table 6 shows that the error between the theoretical method results and the finite element method results does not exceed 5%, which verifies the high reliability of this study's theoretical method and finite element model.

Table 4. Theoretical calculation of partial dynamic characteristics results.

Vibration mode	Frequency (Hz)
symmetric transverse bending-1	0.0811
antisymmetric transverse bending-1	0.1622
antisymmetric vertical bending-1	0.1622
antisymmetric torsion-1	0.1622
symmetric transverse bending-2	0.2433
antisymmetric transverse bending-2	0.3244
antisymmetric vertical bending-2	0.3244
antisymmetric torsion-2	0.3244

In addition, there is a significant local oscillation between the transverse channels of the catwalk. Under wind load, in addition to the possibility of overall vibration and instability of the catwalk, small spans between the transverse channels may also experience local vibration and instability. Therefore, further analysis is needed on small local spans between horizontal channels.

Table 5. Theoretical calculation of partial dynamic characteristics results.

Vibration mode	Midas (Hz)	Theoretical (Hz)	Relative error
symmetric transverse bending-1	0.0809	0.0811	0.25%
antisymmetric transverse bending-1	0.1552	0.1622	4.32%
antisymmetric vertical bending-1	0.1583	0.1622	2.40%
antisymmetric torsion-1	0.1621	0.1622	0.06%
symmetric transverse bending-2	0.2395	0.2433	1.56%
antisymmetric transverse bending-2	0.3238	0.3244	0.18%
antisymmetric vertical bending-2	0.3241	0.3244	0.09%
antisymmetric torsion-2	0.3263	0.3244	0.59%

4.3 Local vibration analysis of the catwalk

Considering that the mass and stiffness of the transverse channel are much larger than those of the catwalk when considering the local vibration of the catwalk, the transverse channel can be considered as a fixed-end constraint of the catwalk. Taking the main tower frame to the first transverse channel of the mid-span and the small span from the first transverse channel of the mid-span to the transverse channel of the mid-span as examples, the finite element model is shown in Fig. 9 and Fig. 10, and the local vibration

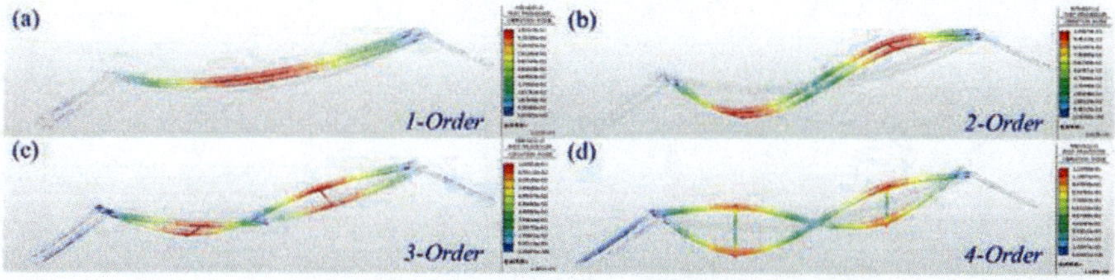

Fig. 8. First 4 vibration modes.

modes are shown in Figs. 9 (a) ~ (c) and 10 (a) ~ (c). Due to symmetry, only half span is considered, and the local vibration natural frequencies between the transverse channels of the catwalk are shown in Table 7. It can be seen that there are apparent low-order vibrations in the local nodes of the catwalk structure, which will generate significant wind-induced responses at high wind speeds and have a significant impact on construction.

Table 6. Theoretical calculation of partial dynamic characteristics results.

Vibration mode	Local internode-1 (Hz)	Local internode-2 (Hz)
symmetric transverse bending-1	0.336	0.323
symmetric vertical bending-1	0.530	0.552
symmetric torsion-1	0.644	0.615

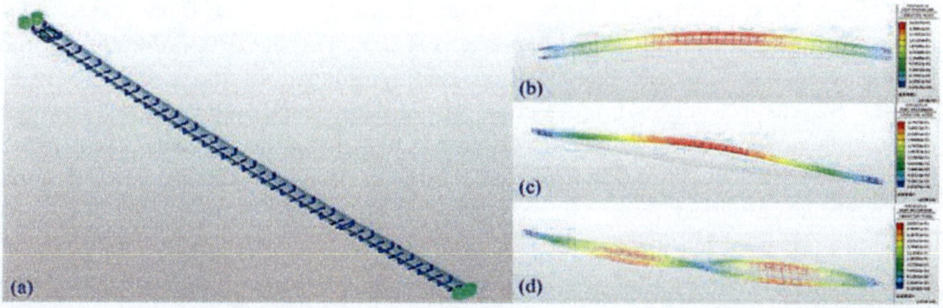

Fig. 9. The first-order symmetric transverse bending, first-order symmetric vertical bending, and first-order symmetric torsion corresponding to the first transverse channel from the main tower frame to the mid-span.

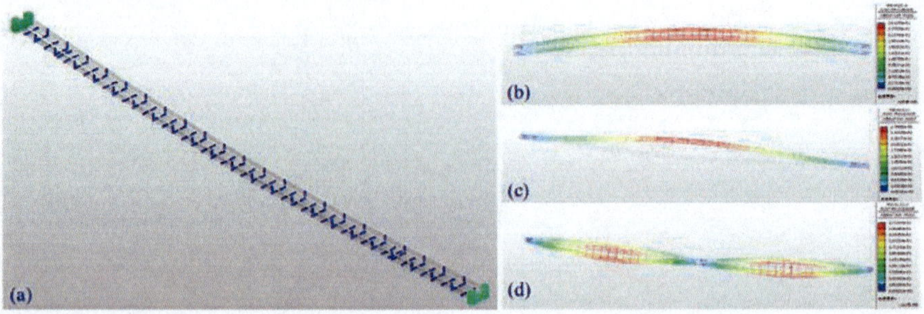

Fig. 10. The first transverse channel corresponds to the first-order symmetric transverse bending, first-order symmetric vertical bending, and first-order symmetric torsion of the midspan transverse channel

5 Conclusion

Based on finite element and theoretical analysis, static and dynamic analyses were conducted on the catwalk during the construction period, and the following conclusions can be drawn:

- The dynamic characteristics of the catwalk obtained through the finite element model are consistent with the results of theoretical analysis, which can provide a further reference for the wind resistance design of other catwalks.
- Under various working conditions, the maximum stress of the catwalk load-bearing cable is 618.1 Mpa (safety factor of $1960.3/618.1 = 3.17 > 3$), and the maximum stress of the catwalk portal frame is 63.5 Mpa. The stress of various components meets the safety factor requirements.
- The first-order symmetrical transverse bending of the catwalk is 0.0809 Hz, and the first-order antisymmetric vertical bending is 0.1552 Hz, the first-order antisymmetric transverse bending and the first-order antisymmetric torsion is 0.1583 Hz and 0.1621 Hz, respectively, which is close to the peak frequency of 0.142 Hz in the fluctuating wind von Karman spectrum and 0.106 Hz in the transverse and vertical directions. Therefore, attention should be paid to the windbreak of the catwalk during the construction period.
- There are apparent low-order vibrations in the local nodes of the catwalk structure, which will generate significant wind-induced responses at high wind speeds and significantly impact construction.

References

1. Kwon, S.-D., Lee, H., Lee, S., Kim, J.: Mitigating the Effects of Wind on Suspension Bridge Catwalks. J. Bridg. Eng. **18**, 624–632 (2013)
2. Zhang, J., Zhang, M., Li, Y., Jiang, F., Wu, L., Guo, D.: Comparison of wind characteristics in different directions of deep-cut gorges based on field measurements. J Wind Eng. Ind. Aerodyn. 212, 104595 (2021)

3. Piesla, M.J., et al.: Abnormal gait, due to inflammation but not nerve injury, reflects enhanced nociception in preclinical pain models. Brain Res. **1295**, 89–98 (2009)
4. Alavinezhad, M., Hassanabad, M.G., Ketabdari, M.J., Nekooei, M.: Sensitivity-based modal strain energy damage identification in an offshore catwalk to prevent environmental hazards. Int. J. Environ. Sci. Technol. **19**, 6639–6654 (2022)
5. KimHoKyung: Evaluation of Torsional Behaviour for the Catwalk System on A Suspension Bridge by Cross Bridge Interval. Journal of Korean Society of Steel Construction. 27, 371–376 (2015)
6. Shengli, L., Chaoqun, W., Fuyou, X., Dongwei, W.: Aerodynamic influence of catwalk on main cables of suspension bridges based on wind tunnel test. Chin. Civil Eng. J. **50**, 79–84 (2017)
7. Ning, J.I.A., Jianxin, L.I.U., Wanfeng, L.I.U.: Analysis of wind-resistant stability on the catwalk for suspension bridge. J. Highw. Transp. Res. Dev. **25**, 99–102 (2008)
8. Guo, P., Li, S., Wang, C., Hu, Y., Wang, D.: Influence of catwalk design parameters on the galloping of constructing main cables in long-span suspension bridges. J. Vibroengineering. **19**, 4671–4684 (2017)
9. Jianxin, L.I.U., Ning, J.I.A.: Vibration control of catwalks for suspension bridges. J. Chang'An Univ. Nat. Sci. Ed. **26**, 54–57 (2006)
10. Wan, J., Wang, Q., Liao, H., Li, M.: Study on aerodynamic coefficients and responses of the integrated catwalk of Halogaland Bridge. Wind Struct. **25**, 215–232 (2017)
11. Bozkurt, A., et al.: Aspects of static and dynamic motor function in peripheral nerve regeneration: SSI and CatWalk gait analysis. Behav. Brain Res. 219, 55–62
12. Wang, H., Li, A., Jiao, C., Li, X.: Characteristics of strong winds at the Runyang Suspension Bridge based on field tests from 2005 to 2008. J. Zhejiang Univ. Sci. A. **11**, 465–476 (2010)
13. Deng, Y., Ding, Y., Li, A., Zhou, G.: Prediction of extreme wind velocity at the site of the Runyang Suspension Bridge. J. Zhejiang Univ. Sci. A. **12**, 605–615 (2011). https://doi.org/10.1631/jzus.A1000446
14. Jie, W., Yu, L., Jianxin, L., Jiawu, L., Junqing, L.: Nonlinear Aerostatic Response of Catwalk of Suspension Bridge. J. Highw. Transp. Res. Dev. 29, 41–45, 54 (2012)
15. Li, Y., Wang, D., Wu, C., Chen, X.: Aerostatic and buffeting response characteristics of catwalk in a long-span suspension bridge. Wind Struct. **19**, 665–686 (2014)
16. Li, Z.-G., Chen, F., Pei, C., Zhang, J.-M., Chen, X.: Comfort Evaluation of double-sided catwalk for suspension bridge due to wind-induced vibration. Math. Probl. Eng. **2021**, 6673816 (2021)
17. Zheng, S., Liao, H., Li, Y.: Stability of suspension bridge catwalks under a wind load. Wind Struct. **10**, 367–382 (2007)
18. Ma, C., Duan, Q., Li, Q., Liao, H., Tao, Q.: Aerodynamic characteristics of a long-span cable-stayed bridge under construction. Eng. Struct. **184**, 232–246 (2019)
19. Shengli, L., Jinping, O.: Galloping vibration control for transient main cables of a long-span suspension bridge at construction stage. J. Vibr. Shock. 29, 137–143,215 (2010)
20. Shengli, L.I., Jinping, O.U.: Nonlinear aerostatic stability analysis of the construction catwalks for long-span suspension bridge. China Railway Sci. **30**, 19–26 (2009)
21. Blasko, J., Szekiova, E., Slovinska, L., Kafka, J., Cizkova, D.: Axonal outgrowth stimulation after alginate/mesenchymal stem cell therapy in injured rat spinal cord. Acta Neurobiol. Exp. **77**, 337–350 (2017)

Feasibility Study on Angle Integral Deformation Measurement Method of Inclination Sensor in Existing Railway Deformation Monitoring

Yufeng Xu[1,2](✉), Yongmao Tang[3,4], Gui Li[4], Fentao Guo[5], and Zhuobin Huang[5]

[1] South China University of Technology, Guangzhou, China
xuyf@scut.edu.cn
[2] Guangdong Hua Jiao Ke Engineering Technology Co., Ltd., Shenzhen, China
[3] Guangzhou Metro Construction Management Co., Ltd., Guangzhou, China
[4] Guangdong Shunguang Rail Transit Co. Ltd., Shunde, China
[5] Guangdong Huitao Engineering Technology Co., Ltd., Shunde, China

Abstract. In order to monitor the influence of the actual project construction on the existing subway structure, it is necessary to monitor the deflection of the existing subway structure in the construction process. At present, the total station is used to measure the deformation of the project. Practice has proved that the total station measurement method has a high cost of instruments and personnel, the layout of measuring points is difficult and easy to be damaged, and the monitoring conditions are relatively harsh. The angle integral deformation measurement method of inclination sensor is adopted for the deformation monitoring of existing railway. The method mainly uses the high-precision inclination sensor to measure the sectional angle of the existing subway structure, and then calculates the deformation of the subway structure by subsection integral deformation. This method has a mature application in the horizontal displacement measurement of deep foundation pits, but it is an innovative application in the deformation measurement of subway.

Keywords: Inclination sensor · Metro deformation · error analysis

1 Introduction

With the continuous development of flexural deformation monitoring schemes, numerous flexural deformation monitoring schemes have emerged both domestically and internationally. By analyzing and studying the operational environment, material structure, and measurement locations of the tested structure, different flexural measurement schemes are selected to meet the characteristics of different structures. Early methods included the use of total stations, dial gauges, leveling methods, and theodolite methods. Existing methods include electronic micrometer flexural measurement, GPS flexural monitoring, communication pipe horizontal flexural measurement, optical-based flexural measurement, and angle-based flexural measurement.

W. Guo et al. (Eds.): GBCESC 2023, LNCE 328, pp. 72–82, 2024.
https://doi.org/10.1007/978-981-99-9947-7_8

The tunnel deformation monitoring adopts the angle integration deformation measurement method using inclinometers. This method primarily utilizes high-precision inclinometers to measure the angles of tunnel sections and then calculates the deformation integration of the tunnel section. This method has been successfully applied in horizontal displacement measurement of deep excavations but represents an innovative application in tunnel deformation measurement. This study aims to verify the feasibility of the angle integration deformation measurement method using inclinometers in tunnel deformation monitoring.

2 Inclination Sensor Measuring Deformation Principle

Inclination sensor is a kind of acceleration sensor based on Newton's second law. It is a kind of fixed inclination measuring instrument which uses biaxial inclination sensor developed and produced by a microelectromechanical system as sensitive element and combined with intelligent chip technology [1]. It is used to observe the biaxial tilt Angle of bridge, building, railway and other structures relative to the horizontal. It is suitable for the deformation of hidden parts that are difficult to be observed by conventional geodetic survey method. It can be used for long-term testing with automatic system. The size of the inclination sensor is 120 mm × 150 mm × 40 mm, and the inclination sensor used is shown in Fig. 1.

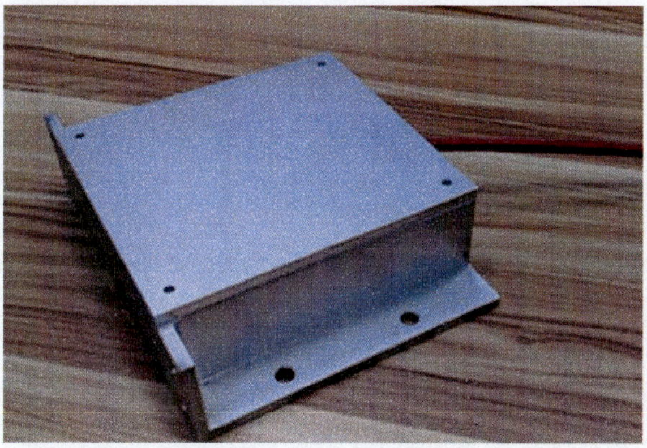

Fig. 1. Inclination sensor.

The method of converting deflection by inclination includes obtaining a set of optimal solutions by using the least square method and directly integrating the inclination function to obtain the deflection value, but these methods all involve more complicated mathematical calculation. Therefore, the simplest arrangement measuring point method and deflection calculation method in the conversion process are selected for research: that is, n positions on the structure are selected to place the inclination sensor, as shown in Fig. 2, and the structural deformation is assumed to be within the linear range. By loading the structure, the change value of the inclination before and after loading can be obtained. The tangent value of the inclination Angle can be multiplied by the distance of the segment to obtain the deflection value of the segment [2–5].

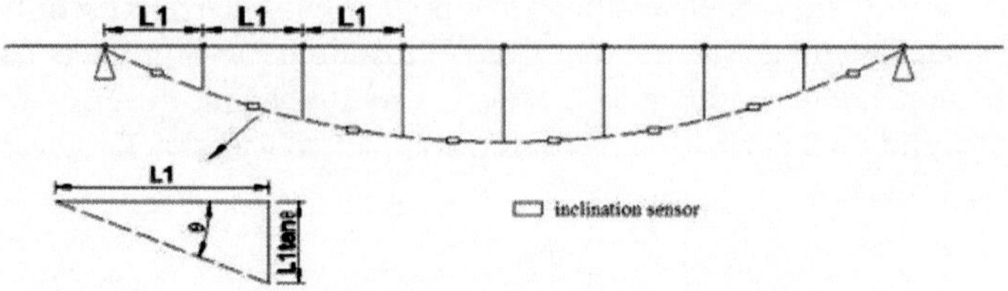

Fig. 2. Schematic diagram of deflection calculation method.

According to the knowledge of material mechanics, we know that the approximate differential equation of beam deflection is:

$$\omega'' = -\frac{M(x)}{EI} \tag{1}$$

If it is a straight beam with constant section, its bending stiffness EI is a constant, and the above equation can be rewritten as:

$$EI\omega'' = -M(x) \tag{2}$$

The Angle equation of the beam can be obtained by integrating the above equation once, which is:

$$\mathrm{EI}\omega' = -\int M(x)dx + C_1 \tag{3}$$

If Eq. (3) is integrated again, the deflection equation of the beam can be obtained.

To sum up, the Angle of any section of the structure is equal to the Angle of the flexural line at this point, that is, the Angle between the tangent line of the flexural line at this point and the X-axis. There is an integral relationship between the deflection of the beam and the Angle. From this we can obtain the deflection of the structure by measuring the Angle of some points when the structure is bent.

There is a beam of length L, and the beam is divided into n segments, the length of each segment is $L_1 = \frac{L}{n}$. An inclination sensor is placed at the midpoint of each section. When a beam is loaded at some point, the beam flexes. The deflection increment of each segment is:

$$\Delta\omega_i = L_1 tan\theta_i \tag{4}$$

Then, the deflection at the end of segment i is the accumulation of all segment deflections of the preceding segment i-1, which is:

$$\omega_i = \sum\nolimits_1^{i-1} L_1 \tan\theta_i \tag{5}$$

where, L_1 is the length of each segment; θ_i is the change value of inclination at the middle point of segment i, namely the measurement value of the segment inclination sensor; $\Delta\omega_i$ is the difference of deflection at the front and rear ends of segment i; ω_i is the deflection value at the end of segment i. Calculating the deflection by measuring the inclination is an indirect method, and calculating the deflection by different mathematical models is an approximate result rather than an accurate deflection value. In this model, the calculation accuracy increases with the increase of the number of segments. Generally, when the number of segments is the same as the number of curves, the relative error between the calculated value and the actual value can reach a range of less than 5%.

3 Laboratory Model Bridge Validation Analysis

3.1 Experimental Design

The theoretical deflection of the span of the simply supported beam was calculated by building a model of the beam with finite element software, and then the deflection was measured by applying a load to the beam by arranging inclination sensors uniformly on the simply supported beam and a micrometer in the span. The deflection measured by the micrometer is used as the actual deflection value, and the deflection in the span measured by the inclination sensor is compared with the deflection measured by the micrometer, and the error probability interval of multiple measurements is calculated.

The actual simply supported beam experiments were carried out using a double-split channel steel simply supported beam with a calculated span of 2m and a hinged boundary. The loading process was carried out using a heavy load on the beam. Two loading conditions were designed for this experiment: (1) loading in the middle of the span of the weight; (2) loading at a quarter of the span of the weight. Firstly, the deflection curves were calculated for the first and second loading conditions. It can be determined that the maximum deflection in the span is 1.716 in Case 1 and 0.69 in Case 2.

If the method of arranging four inclination sensors with an absolute accuracy of 0.0003° is chosen, using the Monte Carlo sampling assessment method [6–9], the errors in the deflection measurements of the method are shown in Table 1.

Table 1. Error in loading deflection values across the span (in mm).

Case	Average value	Standard deviation	95% probability Error interval	95% probability relative error interval
Case1	1.727	0.026	(−0.056, +0.068)	(−2.41, 2.93)%
Case2	0.695	0.026	(−0.047, +0.057)	(−6.81, 8.26)%

In case 1, the deflection value of the simply supported beam is larger, so the 95% probability error interval of the calculated deflection value is smaller; while in case 2, the deflection of the simply supported beam is smaller, and the 95% probability error interval of the calculated deflection by inclination is larger. Under different working conditions, the mean value of the calculated deflection of the simply supported beam in working condition two is about one-third of that in working condition one, but the standard deviation of the calculated deflection value is almost the same, which indicates that it is the sensor itself that has a greater influence on the measurement accuracy, making the probability error interval larger.

3.2 Experimental Procedure

Four inclination sensors, numbered 1 to 4, are arranged equally spaced within the half span of the simply supported beam. The layout is shown in Fig. 5 shown in Fig. 3.

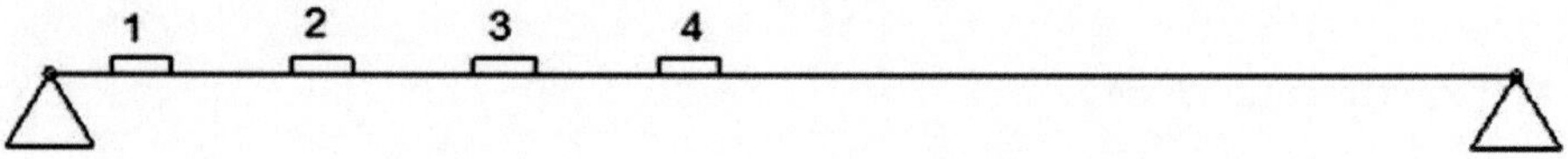

Fig. 3. Inclination sensor arrangement diagram.

The experiment starts with the installation of the sensor and the initial value of the sensor reading is recorded after 5 min of standing. Then the loading of the different working conditions is carried out in turn. When loading, care should be taken to apply the weight slowly so as not to cause the simple beam to sway. The total weight of the load is 30 kg for working condition 1 and 22 kg for working condition 2. Each time the load is stabilised for 10 min, the reading is then taken for the next working condition.

3.3 Test Results

During the test, the measured values of the inclination sensors are recorded and calculated for several moments, and the same moments are used to calculate the span deflection, then the mean, standard deviation and 95% probability confidence interval can be obtained from the data of several moments. The results are shown in Tables 2 and 3.

Table 2. Mean and standard deviation of measured deflection values at inclination.

Case	Average value	Standard deviation	Micrometer measurements (mm)	Relative error of means
Case1	1.727	0.026	1.584	0.95%
Case2	0.695	0.026	0.672	−1.05%

Table 3. 95% probability confidence interval for the measured deflection value of the inclination angle.

Case	Measured value interval	Absolute error interval	Relative error interval
Case1	1.727	(−0.024, 0.08)	(−1.52, 5.05) %
Case2	0.695	(−0.064, 0.036)	(−9.52, 5.36) %

As can be seen from the above table: the micrometer measurements are in error with the theoretical model calculations, while the mean value of the deflection values calculated from the tilt angle values obtained from the tests is less different from the micrometer measurements, so it can be said that the micrometer measurements are reliable as the actual deflection values. In addition, the standard deviation calculated from the test data in the table is less than the difference in the table, which verifies the results in the previous example in this paper.

Through the experiment, data analysis can be carried out to draw the following conclusions:

1. The use of tilt sensors to measure the bridge deflection of the single measurement results obey a normal distribution, and the distribution of the mean and standard deviation and the number of selected sensor arrangements and the accuracy of the sensor itself, can be predicted through the pre-calculation.
2. The measured deflection values measured in the experiment are smaller than the theoretical values calculated by the finite element model, so the actual error of the measurement method needs to be re-analysed based on the measured data. However, the standard deviation and relative error intervals are still predictive and instructive.
3. The smaller the beam deflection or turning angle, the greater the influence of the sensor's own accuracy on the deflection measurement accuracy and the greater the 95% probability relative error interval.
4. By first carrying out theoretical calculations of the bridge deflection, and then designing the sensor accuracy and number of arrangements, the accuracy of each deflection measurement can be better controlled, relatively within ± 5%.

4 Selection and Deployment of Inclination Sensors for Affected Tunnel Structures

The finite element simulation was carried out using the bridge finite element software MIDAS/GTS, developed by MIDAS IT Software, which was able to meet the requirements for the numerical simulation of the existing railway. An underground tunnel model with a length of 85.4 m was established. The calculation model is shown in Fig. 4.

4.1 Solution for the Inclination Sensor Arrangement in the Affected Tunnel

Based on the above model, 25 calculated theoretical deformation values for existing metro structures were taken, as shown in Table 4:

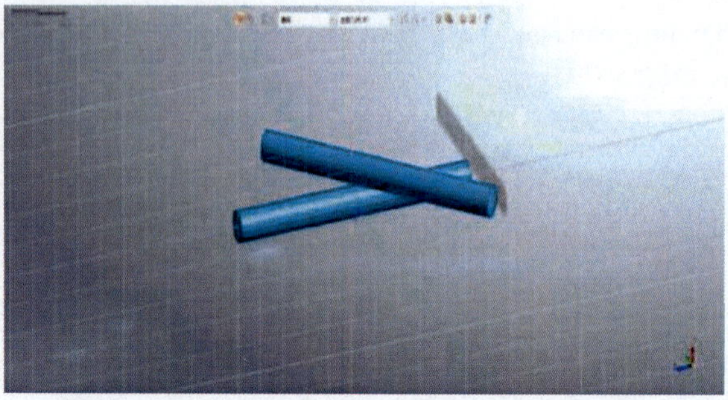

Fig. 4. Finite element calculation model diagram.

Table 4. Model theoretical deformation value.

Distance to zero (m)	Deformation values (mm)	Distance to zero (m)	Deformation values (mm)
0	−1.76	44.9	−8.23
4.8	−3.58	47.4	−7.89
9.8	−1.85	49.9	−7.03
14.6	−0.48	52.4	−5.91
19.4	0.44	54.9	−4.21
24.3	−0.96	57.4	−1.68
26.8	−2.35	59.9	0.3
29.3	−3.23	62.3	1.21
31.8	−4.86	64.7	2.35
34.3	−5.81	69.5	3.4
36.8	−6.67	74.3	3.98
39.3	−7.46	79.1	4.96
42.5	−8.12	0	−1.76

As we cannot obtain the deflection curve of the metro directly from the calculation results of the finite element calculation software, we can only obtain the deflection values of certain segmental nodes. In order to facilitate our calculation of the corner curve, the deflection curve of the metro needs to be fitted by the deflection of each node. Using the distance of the nodes from the zero point as the x-coordinate and the deflection value as the y-coordinate, the deflection curve is plotted using Matlab and the deflection curve function is fitted using Curve Fitting. The function image is shown in Fig. 5.

The polynomial function was used to fit the deflection curve function, and when adjusting the Degree (number of terms) of the polynomial function, it was found that at a Degree value of 7, the R-square of the fitted curve and the nodal deflection coordinates was approximately equal to 1. When fitting a model to the data, due to the discontinuity

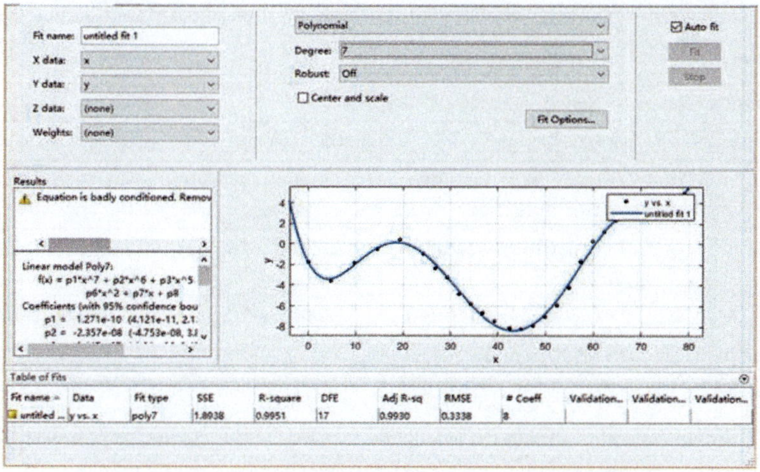

Fig. 5. Theoretical metro deformation curve fitting diagram.

of the data, the model is subject to error. The degree of approximation of the fit of the regression equation to the observations is called the goodness of fit, or R-square in this case, and is a statistical measure of goodness of fit, which is called the coefficient of resolvability. The closer the coefficient of determination is to 1, the better the fit approximation is and the closer the fitted curve is to the actual curve.

From Tables 5, 6, 7, and Table 8, it can be seen that the absolute accuracy of the sensor is different when the discrete difference in the deflection calculation results is small. Using absolute accuracy of 0.01°, 0.005° 0.001° of the inclination sensor, the error of deflection are small, such error range in engineering applications is acceptable. In the case of comprehensive consideration of measurement accuracy and economy, the existing underground is affected by the range of seven absolute accuracy of 0.01° of the tilt sensor scheme can meet the needs.

Table 5. Deflection simulation calculates the mean value of the y distribution (unit: mm).

Number of segments n	Sensor accuracy δ			
	0.01°	0.005°	0.001°	Theoretical value
7	−7.5366	−7.5365	−7.5365	−7.8765
8	−7.6647	−7.6647	−7.6647	−7.8765
9	−7.7806	−7.7804	−7.7805	−7.8765

4.2 Monitoring Site Design

According to a project tunnel deflection deformation measurement accuracy analysis, taking into account the actual out of the section line part of the impact of the larger, so in

Table 6. Deflection simulation calculates the standard deviation of the y-distribution value (unit: mm).

Number of segments n	Sensor accuracy δ		
	0.01°	0.005°	0.001°
7	(−4.2534, −4.3807) %	(−4.2840, −4.3479) %	(−4.3096, −4.3222) %
8	(−2.6149, −2.7635) %	(−2.6508, −2.7261) %	(−2.6820, −2.6969) %
9	(−1.1524, −1.2839) %	(−1.1846, −1.2511) %	(−1.2118, −1.2253) %

Table 7. Single deflection simulation calculates 95% probability confidence interval (in mm).

Number of segments n	Sensor accuracy δ		
	0.01°	0.005°	0.001°
7	0.0025	0.0012	0.0025
8	0.0029	0.0015	0.0029
9	0.0025	0.0013	0.0026

Table 8. Single deflection simulation calculates 95% probability error interval (in mm).

Number of segments n	Sensor accuracy δ		
	0.01°	0.005°	0.001°
7	(−7.5415, −7.5314)	(−7.5391, −7.5340)	(−7.5370, −7.5361)
8	(−7.6705, −7.6588)	(−7.6677, −7.6616)	(−7.6652, −7.6641)
9	(−7.7857, −7.7754)	(−7.7832, −7.7779)	(−7.7810, −7.7800)

the out of the section line part of the encrypted layout of 6 inclination sensors, the final actual in the existing metro construction of the area affected by the layout of 13 absolute accuracy of 20″ of the inclination sensor. The actual site layout of 13 absolute accuracy of 20″ of the inclination sensor. The range of monitoring points is shown in Fig. 6.

There are 13 inclination sensors on site and 2 data collection boxes. Q1-Q5 measurement points are collected in box 1 (400 × 300 × 150 mm) and Q6-Q10 measurement points are collected in box 2. Conclusion and outlook.

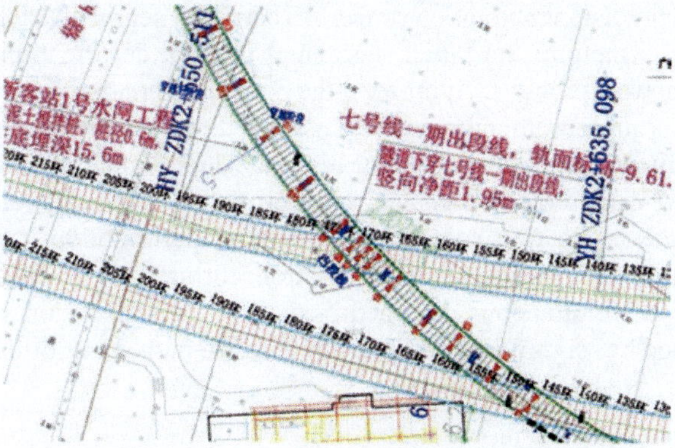

Fig. 6. Monitor the floor plan of the measurement point.

4.3 Conclusion

This paper presents a study on the feasibility of inclination sensors to measure the deformation of existing metro structures for a project engineering construction, providing a new monitoring method reference for the same type of project.

The inclination sensor is an instrument that measures the horizontal angle of a structure and is widely used in bridge erection and other areas. However, its application in metro tunnel deformation monitoring is still relatively rare. Compared to other deflection measurement methods, the inclination sensor has its unique advantages.

(1) Briefly describe the relevant deflection measurement methods and compare them to show the innovation of inclination sensors in tunnel applications, and derive the feasibility of using inclination sensors to measure tunnel deformation.
(2) Analysis of inclination sensor errors: segmentation error and goniometric error. Considering the characteristics of the sensor measurements conforming to a normal distribution, the uncertainty of the inclination sensor measurements is assessed using the Monte Carlo method. Through laboratory experimental design and practical case analysis, it is concluded that the number of inclination sensor arrangements and the choice of accuracy need to take into account the segmentation error, the angular error as well as the actual economy.
(3) The theoretical deformation values of the model are obtained by establishing a model and carrying out finite element simulation analysis. According to the selected mathematical model as well as the measurement method, the factors that mainly affect the accuracy of the inclination sensor measuring bridge deflection deformation are analysed, and through calculation, it is shown that such an error range is acceptable in engineering applications. In the case of comprehensive consideration of measurement accuracy and economy, the affected tunnel structure inclination sensor selection and deployment plan is derived.

4.4 Outlook

The deformation monitoring of existing railways described in this paper can be carried out using the inclination sensor angle integral deformation measurement method, which mainly uses high precision inclination sensors to carry out segmental angle measurement of existing metro structures, and then segmental integral deformation to calculate the deformation of metro structures. At present, the main deformation measurement is to use the total station for measurement, practice has proved that the total station measurement method of instruments, personnel costs are high, the deployment of measurement points more difficult and vulnerable to damage, monitoring conditions are relatively harsh.

The use of the inclination sensor angle integral deformation measurement, high accuracy, adaptability, high degree of digitisation, through the hardware and software method to achieve simultaneous data acquisition of the inclination sensor. This new monitoring method has significant guiding significance in the promotion and application of future projects.

References

1. Design and Research of MEMS Acceleration-Inclination Sensor for Bridge Deformation Monitoring[D]. South China University of Technology (2020)
2. Yang, X.S., Hou, X.M., Liao, Z.P., Huang, Z.P.: A new method for bridge deflection measurement. J. Civ. Eng. **02**, 92–96 (2002)
3. Xu, J.F.: Theoretical analysis and research on the method of measuring bridge deflection using a new inclinometer. Lanzhou Jiaotong University (2009)
4. Xie, H.Y., Yu, Y., Ou, J.-P.: Research on bridge deflection measurement based on wireless inclination sensors. J. Disaster Prevention Mitigation Eng. **30**(s1):31–35 (2010)
5. Deng, C.J.: Research on the key technology of the application of inclination sensor in bridge deflection measurement. South China University of Technology (2018)
6. JJF1059–2012 Assessment and representation of measurement uncertainty (2012)
7. Chen, H.Y., Cao, Y., Han, J.: Measurement uncertainty assessment based on Monte Carlo method. J. Electron. Measur. Instrument. **25**(04), 301–308 (2011)
8. Li, Z.M.: Research on the Confidence of Structural Reliability Indicators Considering Multiple Errors. Shanghai Jiaotong University (2012)
9. Yang, Y.G.: Research on deflection measurement method of tunnel structure. Chang'an University (2009)

Seismic Response of Bridge Pile Foundation in Permafrost Incorporating Advanced Pile-Soil Dynamic Interaction Model

Shengsheng Yu, Xiyin Zhang(✉), Wanping Wang, and Jiada Guan

School of Civil Engineering, Lanzhou Jiaotong University, Lanzhou 730070, China
zhangxiyin@mail.lzjtu.cn

Abstract. High-rise pile cap foundations play a crucial role in bridge engineering within permafrost regions. This study aims to investigate the seismic response of bridge pile foundations with high-rise caps during both cold and warm seasons. Initially, the existing dynamic analysis model of pile-soil interaction is enhanced. Subsequently, a dynamic analysis finite element model incorporating pile-soil interaction is established, focusing on a bridge pile foundation with a high-rise cap along the Qinghai-Tibet Railway. The seismic response of the bridge pile foundation in the permafrost region under varying seasonal conditions is analyzed. Compared with thawed active layer condition, the acceleration response at the pier top is larger, the maximum shear force and moment of the pile foundation are larger under the frozen active layer condition. After the thawing of the active layer, the pile foundation and the pier top lateral displacement of increases significantly, and the residual gap between the pile-soil increases during earthquake. The pile plastic zone length is small and mainly concentrated near the soil surface when the active layer is frozen, while the pile plastic zone length increases after the thawing of the active layer, and the severe damages occur near the freezing and thawing interface of the soil layer.

Keywords: Seasonal frozen soil · Permafrost · Pile foundation bridge · Pile-soil interaction · Seismic response · Nonlinear

1 Introduction

In order to protect the stability of permafrost and reduce the damage of permafrost degradation caused by seasonal changes and climate warming to railway, high-pile cap foundation bridges are widely used in Qinghai-Tibet Railway. Studies have shown that the existence of frozen soil layer will change the dynamic characteristics of the site, and have an important impact on the pile foundation bridge seismic response and damage degree on the site. Vaziri and Sato et al. studied the influence of foundation soil freezing on the pile foundation dynamic interaction, and found that even the frozen soil layer with small thickness would have a great influence on the dynamic response of pile foundation [1, 2]. Sritharan and Suleiman et al. studied the response of pile-soil system under lateral load, and found that the existence of frozen soil layer would improve the elastic stiffness

W. Guo et al. (Eds.): GBCESC 2023, LNCE 328, pp. 83–91, 2024.
https://doi.org/10.1007/978-981-99-9947-7_9

and shear strength of pile-soil system to varying degrees, reduce its lateral deformation ability and the length of pile foundation plastic zone, and change the position of pile max moment [3–5]. In addition, Fei et al. studied the influence of soil freezing on the pile-soil system seismic performance by numerical analysis and quasi-static test method. The results show that soil freezing can make the pile-soil system change from ductility to brittleness [6]. Therefore, the pile-soil interaction effect should be fully considered in seismic response analysis of pile foundation bridges in frozen regions.

However, the dynamic interaction of pile-soil is very complex. The existing pile-soil dynamic interaction models include concentrated spring (SR) model, Penzien model and various improved models based on Winkler model [7]. Although there are many pile-soil interaction models, which can be used in frozen soil is very limited. Based on the nonlinear BNWF model, Li Yongbo proposed an improved nonlinear analysis model of pile-soil interaction and carried out finite element analysis. The pile top hysteresis curve under dynamic displacement load and the numerical analysis results of moment dynamic response at pile foundation different depths were extracted. Compared with the corresponding model test results, it is found that the two have good fitting degree [8]. However, this model does not consider the nonlinear characteristics of pile concrete materials and the soil inertial effect on the pile under earthquake. Therefore, based on the BNWF model improved by Li Yongbo, this paper proposes a dynamic analysis model that can consider the nonlinear of piles, and uses this model to analyze the influence of seasonal freezing and thawing effect of surface soil on the pile foundation bridges seismic failure characteristics and seismic response in permafrost regions.

2 Dynamic Analysis Model of Pile-Soil Interaction

2.1 Dynamic Analysis Model of Frozen Soil-Pile-Pier Interaction

Based on the nonlinear BNWF model, an improved nonlinear analysis model of pile-soil interaction is proposed by Li Yongbo. However, this model does not consider the nonlinearity of pile concrete material and the inertia effect of soil on pile under seismic action. Under strong earthquakes, the pile is also easy to enter a nonlinear state or even cause serious damage, and the failure characteristics of pile are significantly affected by the permafrost conditions. The present paper introduces a novel nonlinear dynamic analysis model for pile foundations, which employs fiber beam elements to accurately simulate pile elastoplastic behavior. This model not only takes into account the impact of seismic loads on pile elastoplasticity resulting from changes in axial forces, but also incorporates an automated algorithm to identify the precise location of the pile's plastic hinge and determine the length of the plastic zone. In addition, this model considers the inertial effect of soil layer on the pile by applying concentrated mass at the pile joint. The improved model is shown in Fig. 1.

2.2 Nonlinear Spring Parameters of Soil

The formula for calculating the relationship between resistance and displacement of foundation soil around pile and pile bottom can be found in reference [9]. The calculation

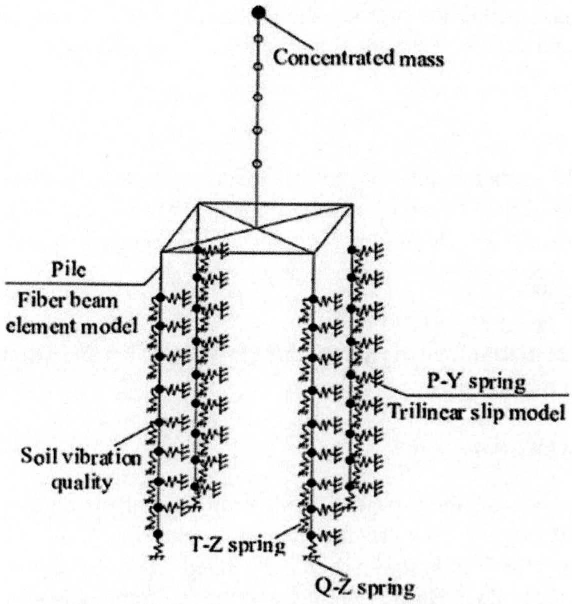

Fig. 1. Dynamic nonlinear model of frozen soil-pile

of non-frozen soil P-Y curve can be found in API specification [10]. The frozen soil P-Y curve is determined according to the weak rock P-Y curve modified by Li Yongbo [8]. The calculation method is as follows:

$$\begin{cases} p = \frac{p_u}{2}\left(\frac{y}{y_m}\right)^{1/3} & y \le y_u \\ p = p_u & y > y_u \end{cases} \tag{1}$$

p_u is the ultimate resistance of frozen soil; y_m is the pile deformation when the load pressure on frozen soil is half of p_u value.

$$y_m = k_m d \tag{2}$$

$$\begin{cases} p_u = q_u d\left(1.5 + 0.25\frac{x}{d}\right) & 0 \le x \le 12d \\ p_u = 4.5 q_u d & x > 12d \end{cases} \tag{3}$$

The q_u is the compression strength of frozen soil; d is pile diameter or width; x is the depth of frozen soil below the surface; k_m is the strain corresponding to the ultimate strength of frozen soil.

2.3 Soil Layer Vibration Quality

When calculating the seismic response of the bridge considering the pile-soil interaction, it is assumed that the soil around the pile vibrates with the pile, so the soil spring damping system must be given a certain mass in the modeling. The mass of the soil spring that vibrates with the pile can be calculated by the following equation.

$$m_i = \rho_i b_p^2 H_i/2 + \rho_{i-1} b_p^2 \frac{H_{i-1}}{2} \tag{4}$$

In the formula, m_i is the mass of the soil spring; b_p is the calculated width of pile (for circular pile, when $d \leq 1$ m, $b_p = 0.9$ (1.5 d + 0.5), when $d > 1$m, $b_p = 0.9$ (d + 1)); ρ_i and ρ_{i-1} are the density of upper and lower soil respectively; H_i and H_{i-1} are the thickness of upper and lower soil respectively.

3 Seismic Response Analysis of High-Pile Cap Foundation Pier in Permafrost Region

3.1 Summary of Bridge

Taking a gravity pier of Qinghai-Tibet Railway in permafrost regions as the research object, the pier adopts high-pile cap foundation. The pier, cap and pile foundation are C30 concrete. The pile foundation is evenly arranged with 24φ20 main reinforcement. The reinforcement of pile is HRB335 rebar, and the reinforcement ratio is 0.427%. The stirrup is HPB300 rebar with diameter of 8 mm, and the stirrup spacing is 0.15 m. The foundation soil type is silt. The upper structure adopts 32 m prestressed concrete T-beam. The dimensions of pier and pile foundation are shown in Fig. 2. The basic seismic intensity of bridge site is VIII degree and type II site, and the peak ground acceleration (PGA) is 0.20 g.

3.2 Bridge Finite Element Model

The finite element model is divided into two groups. Model1 is the complete thawing state of frozen soil in warm season, and the thickness of surface thawing soil is 3.5 m. Model2 is the complete frozen state of frozen soil layer in cold season, that is, the thickness of surface thawing soil layer is 0 m. The free length of pile body of high-pile cap foundation is 0.5 m. The bridge pier and pile cap are connected by common nodes, and the pile cap and pile foundation are connected by rigid connection to realize the interaction between pile and pile cap. The pile-soil interaction is simulated by the dynamic nonlinear analysis model proposed in this paper.

3.3 Seismic Input

Three seismic waves conforming to the site characteristics of the bridge site were selected from the database of the Pacific Earthquake Research Center of the United States and adjusted to 0.20 g. The acceleration time histories of the first 40 s of the three seismic waves were input transversely to the pile foundation pier. It was found that the basic laws of the seismic response of the pile foundation pier were similar. Due to space limitation, only the analysis results under the action of one seismic wave were listed. The acceleration history curve of the seismic wave is shown in Fig. 3.

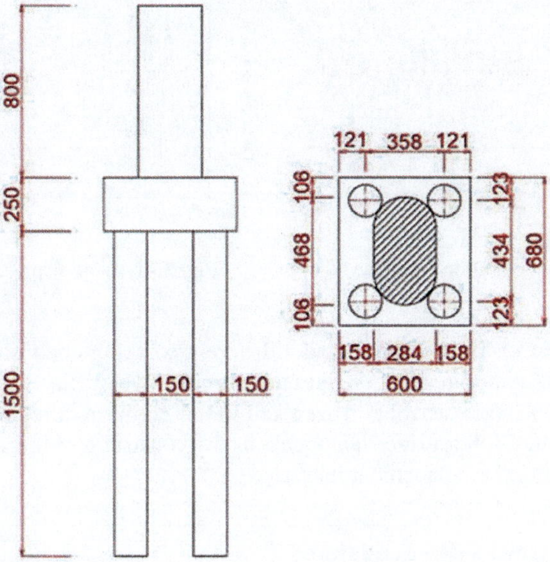

Fig. 2. Pile foundation pier size structure diagram (unit: cm)

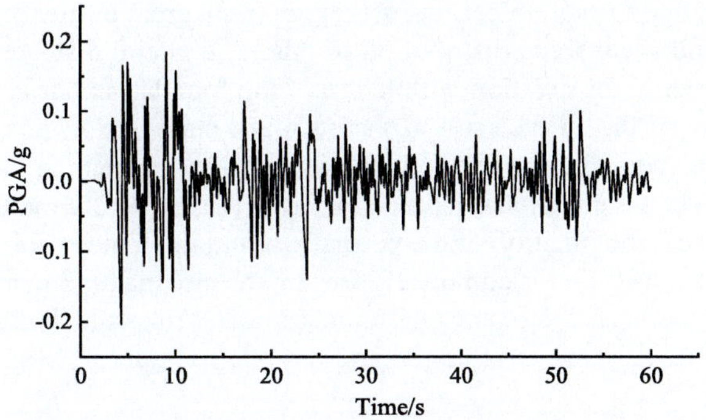

Fig. 3. Acceleration time history of bedrock ground motion

4 Analysis of Calculation Results

4.1 Dynamic Time History Response of Bridge Piers

Figure 4 shows the acceleration time history curves of the top of the pier under two working conditions. Generally speaking, the pile foundation pier in permafrost region has amplification effect on the bedrock ground motion. When the surface seasonal frozen soil is completely frozen, the seismic response of the pile foundation pier is enhanced, which is 25% higher than that of the seasonal frozen soil. In addition, the seasonal freeze-thaw layer has obvious influence on the spectral characteristics of the acceleration time history of the pile foundation pier top. Figure 5 shows the displacement time histories

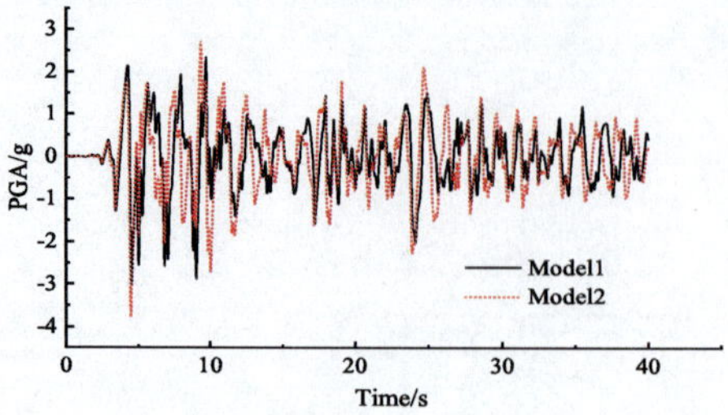

Fig. 4. Acceleration response

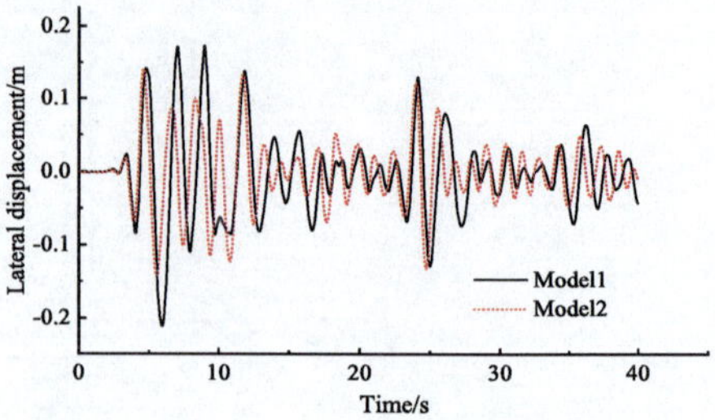

Fig. 5. Lateral displacement response

of the pier top relative to the ground under different working conditions under seismic load. From the calculation results, the pier top max relative displacement increases with the thawing of the surface seasonal frozen soil in the permafrost region. This is mainly because the thawing of the surface soil weakens the constraint of the soil on the pile, so that the pier top lateral displacement increases.

4.2 Max Response of Pile Foundation

Figure 6 shows the time history response envelope diagram of high-pile cap foundation under different working conditions. It can be seen that the seasonal freezing and thawing effect of surface soil in permafrost region has a great influence on the displacement, moment and shear force distribution of pile. The lateral displacement of the pile foundation is larger at the position within 4 m of the foundation burial depth, while the pile displacement of the surface soil in the unfrozen state is significantly greater than that in the frozen state, and the pile top max lateral displacement is more than 2 times the difference. When the seasonal frozen soil layer is completely melted, the pile max moment occurs near the junction of the permafrost and the thawed soil layer, and when the seasonal frozen soil layer is completely frozen, the pile max moment is at the surface position. Taking the buried depth of pile foundation at 2 m as the dividing line, the pile moment in Model2 above 2 m is larger, and the pile moment in Model1 below 2m is larger.

4.3 Nonlinear Characteristics of Pile Foundation

Hwang et al. defined the curvature of the steel bar at the first yield as the threshold for slight failure in the pile section, while the equivalent yield curvature of the steel bar served as the threshold for medium failure in the pile section. The values of the first yield curvature and equivalent yield curvature of the pile foundation section were determined to be 0.00169 m^{-1} and 0.00210 m^{-1}, respectively, as shown in Fig. 7. Furthermore, Fig. 8 demonstrated the notable influence of the seasonal freeze-thaw layer on the failure characteristics of the high pile cap foundation. When the local surface seasonal frozen soil layer is completely melted, the pile max curvature occurs near the junction of the freeze-thaw layer and reaches medium damage. When the seasonal frozen soil layer is fully frozen, the pile's maximum curvature moves upward, reaching the surface and

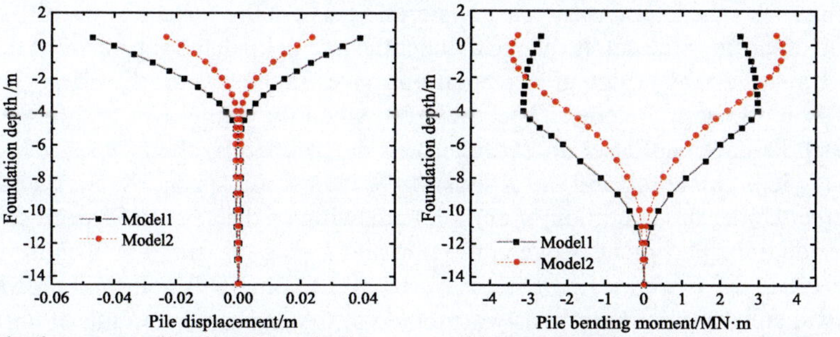

(a) Displacement distribution of piles (b) Moment distribution of piles

Fig. 6. Time history response envelope diagram of pile

entering a state of moderate failure. However, it is noteworthy that the length of the pile's plastic zone is smaller in comparison to when the seasonal frozen soil layer is thawed.

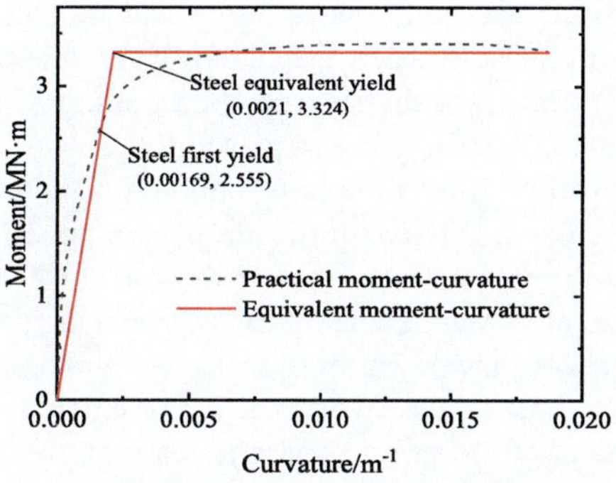

Fig. 7. The moment-curvature relationship

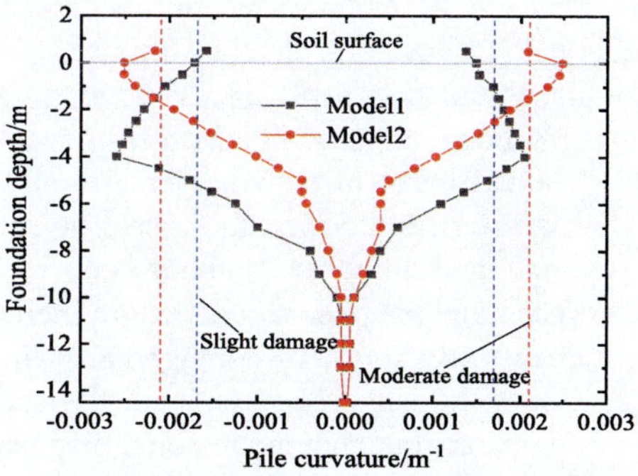

Fig. 8. Curvature distribution of pile

4.4 Nonlinear Characteristics of Pile Side Soil

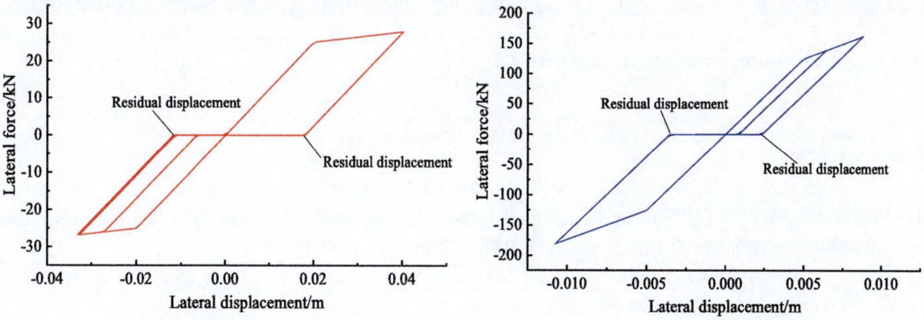

(a)Model1 (b)Model2

Fig. 9. Hysteresis curve of surface lateral soil spring

Figure 9 is the lateral soil spring hysteresis curve at the surface of the pile side. After the soil of the pile side enters the yield under the lateral load, the gap between the pile-soil will occur when the bridge pile foundation moves in the reverse direction. The support force from the spring element becomes zero when the pile moves within the gap. The residual displacement after an earthquake is determined by the intersection of the last contact yield unloading and the X-axis of the lateral soil spring hysteresis curve at the pile side. Under the condition of complete thawing of the seasonal frozen soil surface, the residual displacement between the pile and soil is measured at 17.7 mm, whereas under complete frozen state, it is measured at 3.6 mm. This substantial difference of approximately 5 times is primarily attributed to the low yield strength of unfrozen soil and significant plastic deformation.

5 Conclusions

(1) The seismic response of bridge pile foundations with high-rise caps in permafrost regions exhibits notable variations between cold and warm seasons. In the cold season, when the soil layer is completely frozen, the acceleration response at the pier top is magnified, resulting in increased maximum shear force and moment on the piles. Conversely, during the warm season, when the active layer thaws, the foundation experiences reduced constraint from the surface soil, leading to a significant increase in lateral displacement at both the pile and pier top.
(2) The existence of active layer also significantly affects the pile foundation failure characteristics. When the active layer is frozen, the length of pile plastic zone is small, and the severe damage occurred at the pile foundation near the soil surface. When the pile plastic zone length increases after the thawing of the active layer, the pile is most seriously damaged near the freeze-thaw layer interface.
(3) Under the same seismic load, the surface soil of the two models has entered a nonlinear state, but the residual displacement of the surface soil after earthquake in the unfrozen state is larger, which is about 5 times that in the frozen state. The lateral support force of the surface soil to the pile foundation in the frozen state is larger than that in the unfrozen state, which is about 6.5 times that in the unfrozen state.

Acknowledgements. This research is supported by the National Natural Science Foundation of China (No. 51808273 and 52068045), Science and Technology Program of Gansu Province for Distinguished Young Scholars (No. 20JR5RA430).

References

1. Vaziri, H., Han, Y.C.: Full-scale field studies of the dynamic response of piles embedded in partially frozen soils. Can. Geotech. J. **28**(5), 708–718 (1991)
2. Sato, T., Konagai, K., Ikeda, T., et al.: Effect of surface layer freeze to soil-pile interaction. In: MATEC Web Conference (2019)
3. Sritharan, S., White, D.J., Suleiman, M.T.: Bridge column foundation-soil interaction under earthquake loads in frozen conditions. In: Proceedings of the 13th World Conference on Earthquake Engineering (2004)

4. Suleiman, M.T., Sritharan, S., White, D.J.: Cyclic lateral load response of bridge column-foundation-soil systems in freezing conditions. J. Struct. Eng. **132**(11), 1745–1754 (2006)
5. Sritharan, S., Suleiman, M.T., White, D.J.: Effects of seasonal freezing on bridge column–foundation–soil interaction and their implications. Earthq. Spectra **23**(1), 199–222 (2007)
6. Fei, W., Yang, Z.J., Sun, T.: Ground freezing impact on laterally loaded pile foundations considering strain rate effect. Cold Reg. Sci. Technol. **157**, 53–63 (2019)
7. Wang, W., Zhang, X., Chen, X., et al.: Study on dynamic interaction between bridge pile and soil with permafrost effect: status and review. J. Glaciol. Geocryol. **42**(4), 1213–1219 (2020)
8. Li, Y., Zhang, H.: Study on frozen soil pile interaction model based on dynamic beam on nonlinear winkler foundation method. China Earthq. Eng. J. **37**(2), 453–459 (2015)
9. Japan Railway Technical Research Institute: Seismic design standards and explanations for railway structures. Tokyo: MARUZEN Co. Publishing Division (1999).
10. Geotechnical and foundation design considerations. Washington D.C.: American Petroleum Institute (2011)

Study on the Calculation of Bending Capacity Based on UHPC Design Codes

Lei Sun[1](✉) and Jianluan Li[2]

[1] Anhui Transportation Holding Group Co., Ltd, Hefei 230000, China
1968457282@qq.com

[2] Anhui Transport Consulting and Design Institute Co., Ltd, Hefei 230000, China

Abstract. Based on ultra-high performance concrete (UHPC) codes of China, France and Switzerland, the comparative calculation and analysis were carried out. Besides, finite element analysis (FEA) is conducted based on MIDAS software. Results of FEA indicated that the maximum bending moment of the midspan section is 25084.11 kN-m. moreover, the calculation results obtained from the above codes meet the requirements of bending moment. Addiction, analysis results of codes indicate that there is little difference in the calculation of bending capacity in those codes. The bending capacity calculated by Chinese, French and Switzerland codes are 39613.3 kN·m, 38423.61 kN·m, and 41067.36 kN·m, respectively.

Keywords: ultra-high performance concrete (UHPC) · UHPC bending capacity · UHPC-RC composite beam bridge · Design codes

1 Introduction

At present, there is no code for design and calculation of UHPC bridges in China, only the material code GB/T31387-2015 *Reactive Powder Concrete*. However, foreign design codes are relatively mature. At present, there are some codes and standards for design and calculation of UHPC bridges, such as French standards and Switzerland codes.

As UHPC is a new kind of concrete material, the calculation of bending capacity of UHPC bridge in various countries is different. Three codes are adopted to calculate the bending capacity of the section of UHPC-RC composite beam bridge, include *Specification for Application of Highway Ultra High Performance Concrete Bridge and Culverts (China)* [1], *Recommendation: Ultra-High Performance Fiber-Reinforced Concrete (France)* and *Recommendation* [4], *Ultra-High Performance Fibre Reinforced Cement-based composites (UHPFRC) (Switzerland)* [2]. For convenience of expression, the above codes are referred to as Chinese code, French code and Japanese code.

2 Design Comparison of Shear Calculation of UHPC-RC Composite Girder Bridge

This study is based on the design example of a box composite beam bridge in Anhui, China. The bridge deck is C40 concrete, the material label of high-performance concrete "U" beam is UC150, there are no longitudinal bars and stirrups in the "U" beam, and

W. Guo et al. (Eds.): GBCESC 2023, LNCE 328, pp. 92–97, 2024.
https://doi.org/10.1007/978-981-99-9947-7_10

the prestressed reinforcement is 1 × 7 steel strand. See Table 1 for mechanical property indexes of materials. The section of the composite beam bridge is a box section (see Fig. 1a). And its equivalent section is shown in Fig. 1b.

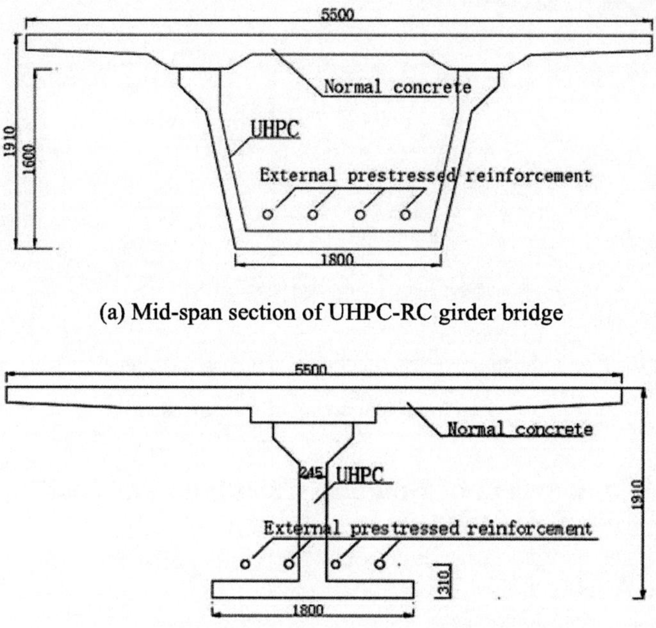

(b) Schematic diagram of equivalent section

Fig. 1. Schematic diagram

Table 1. Mechanical property index of material(MPa)

Material	f_{ck}	f_{tk}	f_{cd}	f_{td}	E
UC150	135	11	93.1	7	48000
C40	26.8	2.4	18.4	1.65	32500
steel strand	/	1860	330	1260	195000

f_{ck} is characteristic value of compressive strength of material; f_{tk} is characteristic value of tensile strength of material; f_{cd} is design value of the compressive strength of material; f_{td} is design value of the tensile strength of material; E is Elastic modulus.

The finite element model (FEM) of UHPC-RC composite bridge is conducted based on MIDAS software. The results of FEM indicated that the maximum bending moment in the mid-span section is 25599 kN·m under the basic load combination (see Fig. 2 b).

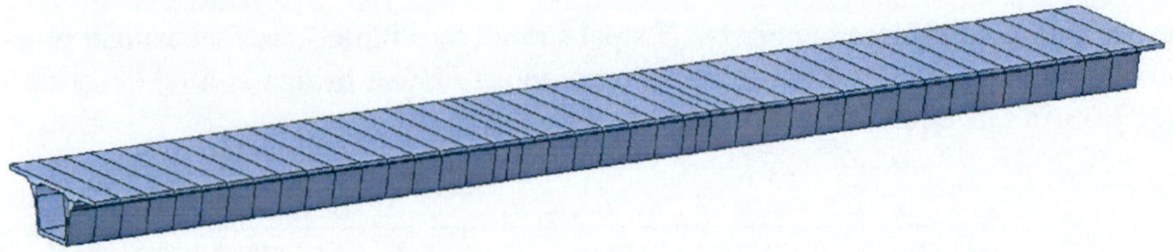

(a) finite element model

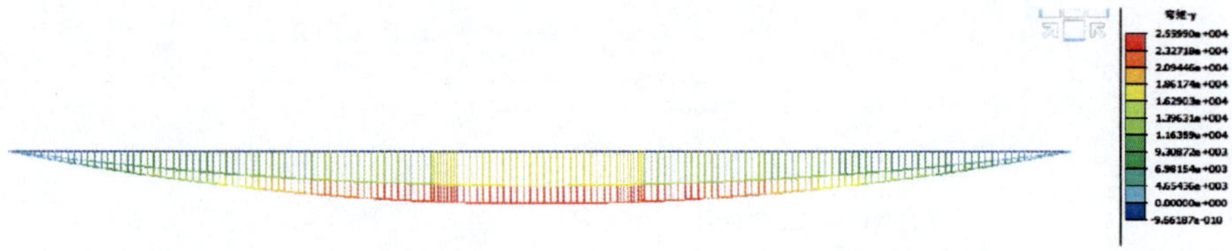

(b) Bending moment envelope diagram under basic combination (kN · m)

Fig. 2. 3D rendering of specimens

3 Design Comparison of Bending Calculation of UHPC-RC Composite Girder Bridge

3.1 Chinese Specification

According to the codes of Application of Highway Ultra High Performance Concrete Bridge and Culverts (Exposure Draft), box-section flexural members can be calculated in accordance with section 6.3.10 for T-section UHPC flexural members whose flange plate is located in the pressure zone, and their positive section flexural bearing capacity is calculated in accordance with the following method:

$$f_{\mathrm{sd}}A_{\mathrm{s}} + f_{\mathrm{pd,i}}A_{\mathrm{p,i}} + f_{\mathrm{pd,e}}A_{\mathrm{p,e}} + 0.5f_{\mathrm{td}}b(h - h'_{\mathrm{f}}/\beta) \geq f_{\mathrm{cd}}b'_{\mathrm{f}}h'_{\mathrm{f}} + f'_{\mathrm{sd}}A'_{\mathrm{s}} + (f'_{\mathrm{pd,i}} - \sigma'_{\mathrm{p,i0}})A'_{\mathrm{p,i}} \tag{1}$$

The calculation satisfies the formula (1), that is, the compressive zone should take the width of the rectangular section, as shown in Figure x. At the same time, the tensile effect of the web UHPC in the tension zone should also be considered, and its positive section bending capacity should be calculated according to Eq. 2.

$$M_{\mathrm{u}} = f_{\mathrm{cd}}b'_{\mathrm{f}}x(h_0 - \frac{x}{2}) + f'_{\mathrm{sd}}A'_{\mathrm{s}}(h_0 - a'_{\mathrm{s}}) + (f'_{\mathrm{pd,i}} - \sigma'_{\mathrm{p,i0}})A'_{\mathrm{p,i}}(h_0 - a'_{\mathrm{p,i}}) - 0.5f_{\mathrm{td}}bx_t(\frac{x_t}{2} - a) \tag{2}$$

In which, f_{sd} and $f_{pd,j}$ denote the design tensile strength of longitude reinforcement and prestress steel bar, respectively, MPa. As and $A_{p,j}$ are sectional area of reinforcement and prestress steel bar, respectively, mm2. The design value $f_{pd,e}$ of the ultimate stress of the longitudinal external prestressed reinforcement is based on the effective prestress value during the use stage, MPa. $A_{p,e}$ is the cross-sectional area of longitudinal external prestressed reinforcement in the tensile zone. F_{td} is design value of UHPC axial tensile strength, which is 7 MPa in this study. b is web width of T-shaped section, mm. β is ratio of the height of rectangular blocks in the compression zone of flexural members to the neutral axis height (actual compression zone height). is The stress of the prestressed reinforcement at the resultant point of the prestressed reinforcement in the compression zone when the UHPC normal stress is equal to zero, MPa. is The cross-sectional area of the longitudinal internal prestressed reinforcement in the compression zone. a is the distance from the combined force point of the ordinary reinforcement and prestressed reinforcement (internal and external prestressed reinforcement) in the tensile zone of the section to the edge of the tensile zone, mm.

It is calculated that the height edge of the compressive zone is approximately the junction of the combined beam, so the moment is taken at the joint action point of the extracorporeal prestressing steel bars: The calculated height of the compressive zone is $x = 338.58\,\text{mm} > 310\,\text{mm}$. Thus, the resistance of the cross-section in the midspan is 39613.31 kN-m.

$$\begin{aligned} M_u &= 18.4 \times 1048000 \times (205.15 + 1290) + 330 \times 18501.2\times \\ &(1910 - 90 - 310) + 93 \times 720 \times 28.58 \times (1290 - 14.29) \\ &- 0.5 \times 7 \times 744618.18 \times (647.07 - 310) = 39613.31\,\text{kN} \cdot \text{m} \end{aligned} \tag{3}$$

3.2 French Codes

For the pressure zone is located in the T-shaped section flange plate, e. g.,

$$f_{pd} \cdot A_p + 0.5 f_{ctm} \cdot A_{Ut} + f_{sd} \cdot A_s = \eta f_{cd} \cdot \lambda x \cdot b + f'_{sd} \cdot A'_s \tag{4}$$

$$M_{\text{u}} = \eta \cdot \lambda \cdot f_{\text{cd}} b_{\text{f}'} x (h_0 - \frac{x}{2}) + f_{\text{sd}'} A_{\text{s}'} (h_0 - a_{\text{s}'}) - 0.5 f_{ctm} b x_t (\frac{x_t}{2} - a) \tag{5}$$

The process of calculation, the calculated height of the compressive zone is $x = 455.99\,\text{mm} > 310\,\text{mm}$.

$$\begin{aligned} M_{\text{u}} &= 0.7 \times 0.8 \times 720 \times 145.99 \times 93 \times 1217 + 0.8 \times 18.4 \\ &\times 1048000 \times (1290 + 205.15) + 18501.2 \times 330 \times 1510 - 3.5 \\ &\times 660118 \times (536.26 - 310) = 38423.61\,\text{kN} \cdot \text{m} \end{aligned} \tag{6}$$

Therefore, the French UHPC code calculated the flexural bearing capacity of 38423.61 kN-m.

3.3 Switzerland Code

A schematic diagram of the flexural calculation when UHPC is in tension is provided in Switzerland code (see Fig. 3). When the cross-sectional design satisfies Eq. 7, the flexural load capacity of the rectangular section is calculated jointly by Eq. 8 and Eq. 9.

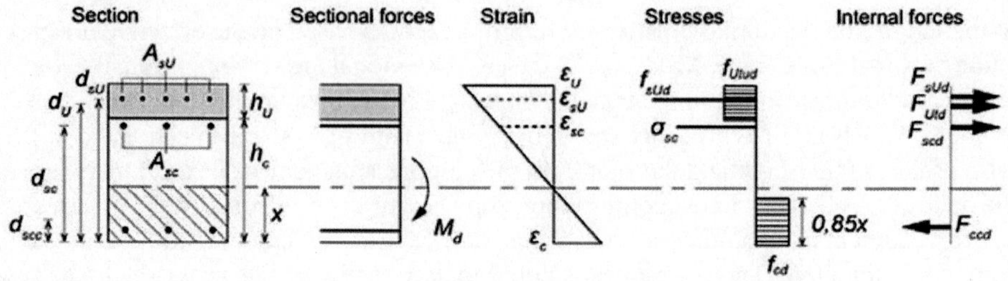

Fig. 3. Switzerland code

$$0.85 \cdot h_c \cdot b \cdot f_{cd} + f_{sd'} \cdot A_{s'} < f_{pd} \cdot A_P + f_{td} \cdot b \cdot h_U \tag{7}$$

$$0.85 \cdot b \cdot h_c \cdot f_{cd} + f'_{sd} \cdot A'_s + 0.85 \cdot (x - h_c) \cdot b \cdot f_{cd} + f'_{sd} \cdot A'_s = f_{\text{pd}} \cdot A_{\text{p}} + f_{td} \cdot b \cdot (h - x) \tag{8}$$

$$M_{\text{u}} = 0.85 \cdot b \cdot h_c \cdot f_{\text{cd}} \cdot (d_{sU} - \frac{h_c}{2}) + 0.85 \cdot b \cdot (x - h_c) \cdot f_{Ucd} \cdot (d_{sU} - h_c - \frac{x - h_c}{2}) + f'_{\text{sd}} \cdot A'_{\text{s}} \cdot (d_{sU} - d_{\text{scc}}) - f_{\text{td}} \cdot b \cdot (h_U - \frac{x - h_c}{2}) \cdot [(h_U - \frac{x - h_c}{2})/2 - h - d_{sU})] \tag{9}$$

The calculated height of the compressive zone is $x = 431.9\,\text{mm} > 310\,\text{mm}$.

$$M_{\text{u}} = 0.85 \times 720 \times 121.9 \times 93 \times 1229.05 + 0.85 \times 18.4 \times 1048000 \times (1290 + 205.15) + 18501.2 \times 330 \times 1510 - 7 \times 677427 \times (560.02 - 310) = 41067.36\,\text{kN} \cdot \text{m} \tag{10}$$

The calculation shows that the neutral axis is located in the UHPC, that is, part of the UHPC is under compression and part of the UHPC is under tension, because the Swiss code does not discount the design value of the tensile strength of the UHPC, so the Swiss UHPC code calculation result 41067.36 kN-m is a little larger than other codes.

3.4 Discussion

The main difference of flexural calculations between Chinese and French. The positive section flexural bearing capacity calculation of UHPC girder bridge, China codes basically follows the normal reinforced concrete structure's positive section flexural bearing capacity calculation method, without considering the tensile strength of UHPC in the tensile zone, compared with the French UHPFRC standard, which stipulates that the tensile strength of UHPC in the tensile zone should be considered.

Based on the assumption of flat section and the intrinsic relationship of UHPC concrete, the equivalent stress-strain distribution of T-beam in the limit state can be obtained. For the convenience of calculation, both Chinese and French standards consider the equivalent conversion of curvilinear stress distribution of UHPC concrete into rectangular fractional form, only the conversion coefficients are slightly different among these countries.

When UHPC concrete stress diagram, the Chinese standard introduces two dimensionless coefficients α, β to discount the design value of UHPC axial compressive strength and neutral axis height, respectively, and gives the recommended values of the two coefficients directly. The French UHPFRC standard also introduces two dimensionless coefficients η, λ (Table 2).

Table 2. Sectional bending capacity

Codes	Bending capacity (kN·m)
Specification for Application of Highway Ultra High Performance Concrete Bridge and Culverts (Exposure Draft, China)	39613.31
Recommendation: Ultra High Performance Fiber-Reinforced Concrete (2015, Frence)	38423.61
Recommendation: Ultra-High Performance Fibre Reinforced Cement-based composites (UHPFRC) (Switzerland)	41067.36

4 Conclusion

Based on the bending capacity calculation of 40m span UHPC-RC composite beam bridge, this study compares the bending moment calculation of three countries's codes, namely, China, France and Switzerland. The following conclusions are obtained:

(1) The bending capacity calculated by Chinese, French and Switzerland codes are 39613.3 kN·m, 38423.61 kN·m, and 41067.36 kN·m, respectively.
(2) Based on FEA analysis, under the basic combination of loads, the maximum bending moment of the midspan section is 25084.11 kN-m. The resistance results obtained from the three codes meet the requirements of bending necessary, indicating that there is little difference in the calculation of bending capacity in each code.

References

1. Specification for Application of Highway Ultra High Performance Concrete Bridge and Culverts (Draft for comments) [EB/OL] (2018)
2. National addition to Eurocode 2 - Design of Concrete Structures: Specific Rules for Ultra-High Performance Fiber-Reinforced Concrete (UHPFRC) (2016)
3. Recommendations for Design and Construction of High Performance Fiber Reinforced Cement Composites with Multiple Fine Cracks (HPFRCC) (2008)
4. Association Francaise de Génie Civil (AFGC). Ultra High Performance Fibre-Reinforced Concretes—Interim Recommendations. AFGC Scientific and Technical Documents, Paris (2002)

Choice of Soil Constitutive Models in Numerical Analysis of Foundation Pit Excavation Based on FLAC3D

Shang Xiao[1](✉), Ming Xu[1], and Riyan Lan[2]

[1] School of Civil Engineering, Chongqing University, Chongqing 400045, China
202016021057@stu.cqu.edu.cn

[2] Guangxi Xinfazhan Communications Group Co., Ltd., Nanning 530029, China

Abstract. The selection of soil constitutive model is very important in numerical analysis of foundation pit excavation. In the process of foundation pit excavation, the stress path of soil in different areas is different. And the stress path has a significant influence on the mechanical parameters of soil. Therefore, the selected constitutive model should be able to reflect the influence of stress path. FLAC^{3D} 6.0 embeds directly usable hardened soil model and small strain hardened soil model, which expands the application range of the numerical method in geotechnical engineering. Based on FLAC^{3D}, the applicability of common soil constitutive models including linear elastic model, Mohr-Coulomb (MC) model, hardened soil (HS) model and small strain hardened soil (HSS) model for foundation pit engineering are studied. It is founded that the linear elastic model is not suitable for the analysis of foundation pit excavation. MC model has poor applicability for shallow foundation pit excavation, but the increase of excavation depth improves its applicability in some extent. HS model and HSS model can reasonably describe the deformation of surrounding soil and supporting structure in the process of excavation, and HSS model can consider the elastic nonlinearity in the small strain range, hence the numerical results are closest to the measured results. The results can provide a theoretical basis for selecting a reasonable constitutive model for foundation pit engineering numerical analysis based on FLAC^{3D} in the future.

Keywords: Foundation pit · FLAC^{3D} · Constitutive model · Numerical simulation

1 Introduction

The excavation of the foundation pit is essentially a process of soil unloading, the unloading of the soil in the foundation pit will inevitably lead to the change of the support structure and the stress field and displacement field of the soil around the foundation pit, which will induce the deformation of the surrounding buildings and underground pipelines and even endanger its safe use [1, 2]. With the continuous expansion of urban underground space, the number of foundation pit projects is increasing, the scale of

W. Guo et al. (Eds.): GBCESC 2023, LNCE 328, pp. 98–110, 2024.
https://doi.org/10.1007/978-981-99-9947-7_11

excavation is expanding, and the geological conditions and surrounding environmental conditions are becoming more and more complex, which poses a great challenge to the design and construction of supporting structure [3]. Reasonable supporting structure design should not only ensure the stability of its own system, but also ensure that the deformation of the soil around the foundation pit is controlled [4]. Therefore, it is important to accurately predict the deformation of the foundation pit side wall and the surrounding soil.

At present, the prediction method for deformation during foundation pit excavation mainly includes field test [5], indoor test [6] and numerical simulations [7]. In numerical simulation, for foundation pit engineering with complex stress path, especially deep large foundation pit engineering, the selection of constitutive model is very important for the reliability of numerical results.

Liu et al. [8] use FLAC3D to predict the deformation of envelope according to the dynamic monitoring data during the excavation of the deep foundation pit. Yang [9] use numerical simulation based on FLAC3D and the data of field dynamic monitoring to analyze the rules of the soil pressure, anchor tension, pile deformation and the bending moment with foundation excavation. But due to the FLAC3D has not embedded HS model [10] and HSS model [11] in early version, and these numerical results are all obtained based on the MC constitutive model, resulting in a large deviation between the numerical results and the measured results.

It is in this context that this paper adopts different soil constitutive models embedded in FLAC3D6.0, such as linear elasticity model, MC model, HS model and HSS model, are used to analyze the process of foundation pit excavation, and the applicability of these constitutive models in foundation pit engineering is discussed qualitatively or quantitatively through the comparative analysis of calculation examples and engineering examples. The findings are expected to provide a reference for engineers when use the numerical simulation for foundation pit excavation by FLAC3D.

2 Unloading Stress Path for Foundation Pit Excavation

The mechanical properties of the soil depend not only on the current stress state, but also on the stress history and the stress path. Generally, the stress path of the soil during the excavation of the foundation pit can be simplified to three partition types, as shown in Fig. 1. Active area stress path a: with the excavation of the foundation pit, the lateral stress of the soil in the active area decreases and the vertical pressure remains unchanged, lead to the average main stress p decreases and the generalized shear stress q increases, and the corresponding stress path is shown in PB in Fig. 2. The path of shallow passive area b: the excavation surface shallow soil by excavation unloading and the support structure extrusion, the overlying pressure reduction and the lateral stress increases, $\Delta\sigma_h > 0$, $\Delta\sigma_v < 0$, the maximum main stress and the minimum main stress direction change, the average main stress p decrease, the generalized shear stress q reduced after increase, the corresponding stress path is shown in the PCD in Fig. 2. Stress path in the deep passive area c: the overlying pressure of the soil in the deep area below the excavation surface decreases, and the lateral pressure remains basically unchanged that is $\Delta\sigma_h \approx 0$, $\Delta\sigma_v < 0$, but the maximum main stress and the minimum main stress direction do not change, resulting in the average main stress p decreases and the generalized shear stress q decreases, and the corresponding stress path is shown in PC in Fig. 2.

A large number of experimental studies show that the modulus of soil is clearly different under different stress path, which is higher than that under the conventional tri-axial loading path 1–2 times and 2–5 times under the active stress path a and the stress path b respectively. Under the action of the stress path c in the passive region, the soil modulus is higher than that under the conventional tri-axial loading path at a larger unloading ratio. These significant effects demonstrate the great need for selecting constitutive models that can reflect the effects of stress paths in numerical simulations of foundation pit excavation.

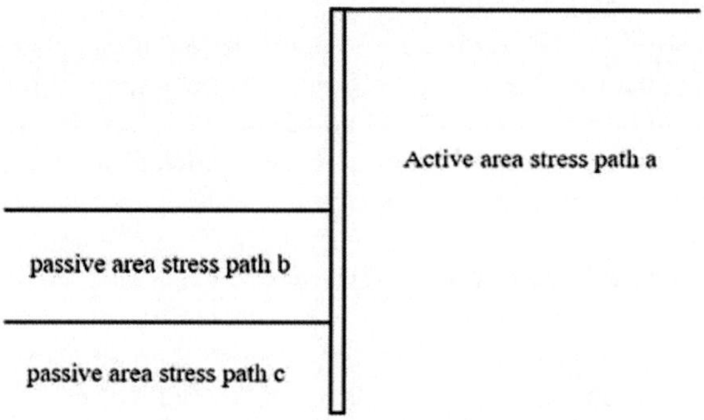

Fig. 1. Distribution of stress path

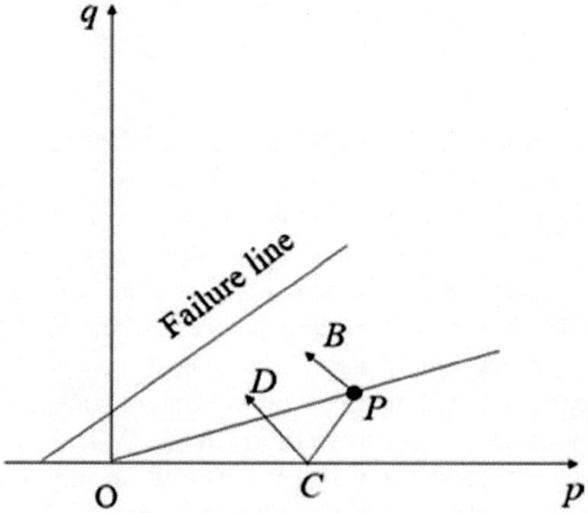

Fig. 2. Schematics of various stress paths in the p-q plane

3 The Constitutive Model of the Soil in FLAC3D

3.1 The Linear-Elastic Model

The linear-elastic model in the FLAC3D is the simplest constitutive model for describing the mechanical behavior of soil, which is suitable for continuous materials describing isotropic, uniform and linear stress-strain relationship, with parameters including Young's modulus E and Poisson's ratio v. However, the strain-strain relationship of soil is usually nonlinear, rigid and stress path correlation, so the linear elasticity model is too simplified for rock and soil mass.

3.2 The Mohr-Coulomb (MC) Model

MC model is essentially an ideal elastic-plastic constitutive model that is composed of linear elastic model and MC yield criterion, which is widely used to describe the shear failure of rock and soil mass. Before the MC model enters the plastic stage, the stress-strain relationship of soil is still assumed to be linear elasticity. Therefore, just like the linear elastic model, it cannot reflect the nonlinear stress-strain relationship, nor can consider the influence of stress path. In view of this, MC model is not applicable to the problem of foundation pit excavation.

3.3 Hardened Soil(HS) Model

HS model in FLAC3D6.0 as an elastic-plastic model considering both shear hardening and compression hardening based on the MC failure criterion. As shown in Fig. 3, the elastic phase $q > q_f$ HS model assumes the partial stress q and axial strain ε_1 in the tri-axial drainage test, the relationship between them is approximated by a hyperbolic relationship as follows:

$$\varepsilon_1 = \frac{1}{E_i}\frac{q_a q}{(q_a - q)} \tag{1}$$

where q is the partial stress, $\varepsilon 1$ is the axial strain and $q_a = \frac{q_f}{R_f}$ is the progressive value of the shear strength, where R_f as the failure ratio is usually 0.9, and q_f is the final partial stress, which can be calculated by Eq. (2):

$$q_f = \frac{2\sin\varphi(c\cot\varphi - \sigma_3)}{1 - \sin\varphi} \tag{2}$$

where σ_3 is the minimum principal stress.

$$E_i = \frac{2E_{50}}{2 - R_f} \tag{3}$$

where E_i is the initial stiffness, E_{50} is the secant modulus corresponding to the 50% peak shear stress under the main loading condition.

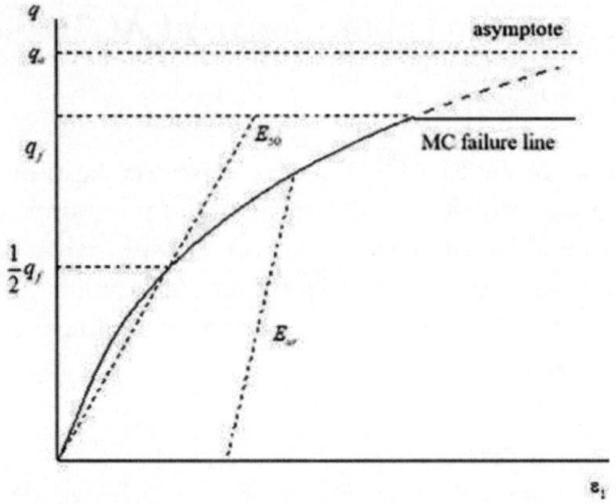

Fig. 3. Hyperbolic stress-strain relation in primary shear loading

Defines the initial modulus of the hyperbolic stress-strain relationship as follows:

$$E_{50} = E_{50}^{ref}\left(\frac{c\cot\varphi - \sigma_3}{c\cot\varphi + p^{ref}}\right)^m \tag{4}$$

where p^{ref} is the reference pressure is usually 100 kPa, m is the power index, which is usually 1 for clay, and 0.4–0.9 for sand. Where E_{50}^{ref} is the loading modulus corresponding to the reference pressure can usually be obtained by the tri-axial test.

Using the unloading-reloading modulus E_{ur} to update the volume modulus K and shear modulus G, E_{ur} expressed as:

$$E_{ur} = E_{ur}^{ref}\left(\frac{c\cot\varphi - \sigma_3}{c\cot\varphi + p^{ref}}\right)^m \tag{5}$$

where E_{ur}^{ref} is the unload-reload modulus corresponding to the reference pressure.

When $q > q_f$, the soil enters the plastic stage, producing plastic deformation, with the change of hardening parameters γ^p, the yield surface of HS is also changing. The shear yield function of the initiation and evolution of shear hardening is determined as:

$$f_s = \frac{q_a}{E_{50}}\frac{q}{q_a - q} - \frac{2q}{E_{ur}} - \gamma^p \tag{6}$$

where γ^p is an internal variable, which can be calculated by the following equation:

$$\gamma^p = \varepsilon_1^p - \varepsilon_2^p - \varepsilon_3^p \tag{7}$$

The Volume yield function is defined as:

$$fv = \frac{\tilde{q}^2}{\alpha^2} + p^2 - p_c^2 \tag{8}$$

where α is the constant related to the pressure coefficient of the static soil, which determines the flattening degree of the volume yield surface, $\tilde{q}$ is a shear stress measure and p_c is the early consolidation pressure.

3.4 Small Strain Hardened Soil Model

In order to describe the modulus of soil increase with the increase of strain in the small strain range, the HSS model is obtained on the basis of the HS model:

$$\frac{G}{G_0} = \frac{1}{1 + 0.385\gamma/\gamma_{0.7}} \tag{9}$$

where G_0 is the initial shear modulus is obtained from the initial modulus E_0^{ref}, $\gamma_{0.7}$ is the corresponding reference shear strain of $0.7G_0$. Other parameters of the HSS model are exactly the same as those of the HS model.

4 Study on the Suitability of Soil Constitutive Model

The purpose of this section is to study the applicability of the constitutive model in FLAC3D for the analysis of foundation pit engineering excavation process, which is evaluated by the qualitative comparative analysis of the rebound of the pit bottom, retaining wall horizontal displacement and the surface subsidence behind the wall.

4.1 Model Introduction and Parameters

The selected computational standard tests are shown in Fig. 4. The study profile consists of two horizontal sand layers from top to bottom. The thickness of the first layer of sand is 20.0 m, and the thickness of the second layer is 80.0 m. The excavation width is 60.0 m and the excavation depth is 17.0 m. The initial groundwater level is −3.0 m, and the final depth of precipitation in the pit before excavation is −18.0m. The retaining wall has a depth of 30 m and a width of 1m. The plane strain assumption is adopted, the model thickness is 2.7 m, and only the right side of the foundation pit is selected as the research object to simplify the calculation.

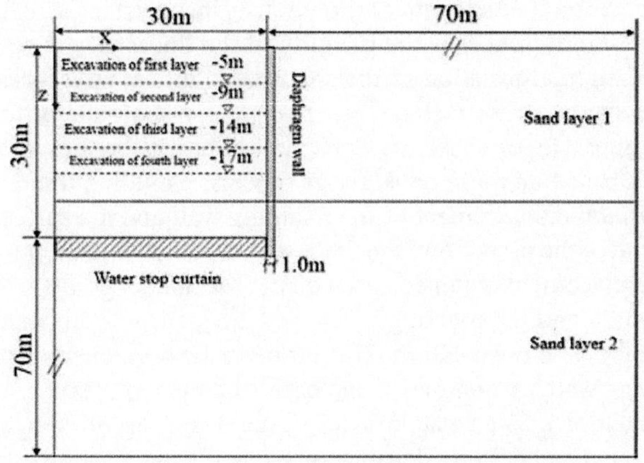

Fig. 4. Schematic of the benchmark exercise

The dry density of the soil above the groundwater level is 1900 kg/m^3, and the wet density of the soil below the groundwater surface is 2000 kg/m^3. The retaining wall adopts a linear-elastic model, and the density is 2400 kg/m^3, Young's modulus is 30 GPa and Poisson's ratio is 0.15, which were simulated using the inner grid cell of the adjacent foundation pit.

Linear elastic model, MC model, HS model and HSS constitutive models are used to simulate the soil respectively, and the corresponding parameters are shown in Table 1.

Table 1. Mechanical parameters of different constitutive models

linear elastic		MC		HS		HSS	
E(MPa)	180/300	c(kPa)	1/1	E_{50}^{ref}(MPa)	450/750	E_0^{ref}(MPa)	540/900
v	0.2/0.2	φ(°)	35/38	E_{ur}^{ref}(MPa)	180/300	$\gamma_{0.7}$	1.6×10-4/2.3×10-4
		ψ(°)	5/6	m	0.55/0.55		
				Rf	0.9/0.9		

* The left side of the "/" symbol represents the first soil layer, and the right side of the "/" symbol represents the second soil layer.

4.2 Comparative Analysis of Constitutive Models

Figure 5 shows the deformation form of the foundation pit after the first soil excavation obtained by different constitutive models, where the deformation coefficient is set to 200 to clearly distinguish the deformation characteristics. As shown in Fig. 5(a), after the excavation of the foundation pit, rebound of the pit bottom, the supporting retaining wall slopes along the direction of the foundation pit and the surface behind the wall is lifted, these deformation characteristics are obviously inconsistent with field observation. The reason for these differences is the inability of the linear elastic constitutive model to describe the plastic deformation of the soil, the nonlinear stress-strain relationship, and the influence of the stress path. In order to clearly show the differences between these models. Figure 6(a–c) show the deformation characteristics of the soil around the foundation pit based on these constitutive models, including the rebound of the pit bottom, the horizontal displacement of the retaining wall and the surface subsidence.

For the rebound of the pit bottom, the linear elastic model is similar to the MC model. The horizontal displacement obtained based on the MC model points into the pit, but the deformation is small, and the maximum value occurs at 1/3 of the wall body, which is inconsistent with the field observation. This is due to the horizontal displacement of the MC model retaining wall is a common result of the rebound of the pit bottom and the soil of part of the foundation pit side wall into the plastic stage. The horizontal displacements obtained by HS model and HSS model all point to the pit, and the maximum value occurs at the surface, among which the horizontal displacements obtained by HSS model are

(a) The linear-elastic model (b) MC model (c) HS model (d) HSS model

Fig. 5. Deformed pattern of foundation trench based on various constitutive models

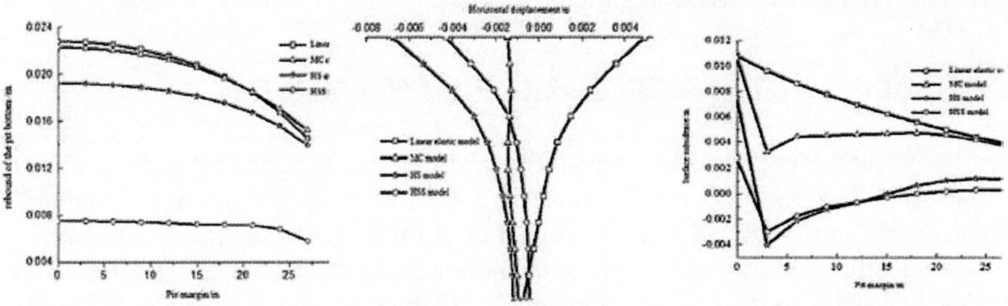

(a) The rebound of the pit bottom (b) the horizontal displacement of the retaining wall (c) Surface subsidence

Fig. 6. Deformed characteristics of soil surrounding foundation trench after the excavation of the first soil layer based on various constitutive models

small. In short, for the excavation of the first layer of soil, the prediction results of the linear-elastic model and the MC model are obviously inconsistent with the field observation. The HS model and the HSS model provide a reasonable numerical solution.

4.3 Depth Effect of Foundation Pit Excavation

Depth effect of foundation pit excavation means that the adaptability of different constitutive model change with the foundation pit excavation depth. Figure 7(a–c) respectively shows after the excavation of the second layer of soil by different model predict rebound of the pit bottom, horizontal displacement of retaining wall and surface subsidence curve.

With the increase of the excavation depth, the rebound of the pit bottom, the horizontal displacement of the retaining wall and the surface subsidence all increase. For the rebound of the pit bottom predicted by the linear elastic model and the MC model is still the largest, while the predicted value of the HS model and the HSS model is still small.

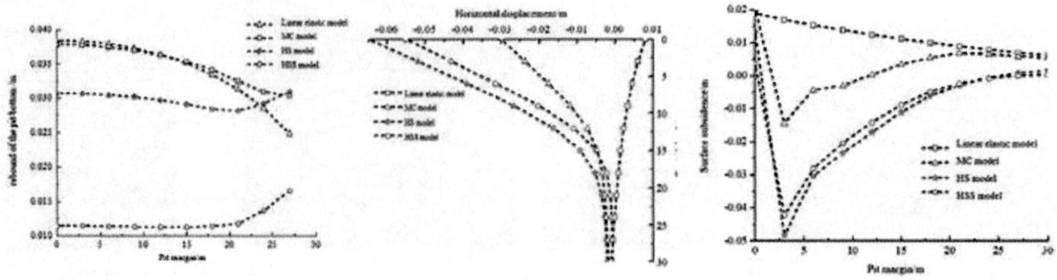

(a) The rebound of the pit bottom

(b) the horizontal displacement of the retaining wall

(c) Surface subsidence

Fig. 7. Deformed characteristics of soil surrounding foundation trench after the excavation of the second soil layer based on various constitutive models

5 Example Analysis of Foundation Pit Engineering

In order to further explore the applicability of different soil constitutive models for the foundation pit excavation process, different constitutive models are used to numerical simulate the same foundation pit engineering example, and the numerical results are compared with the measured results to quantitatively evaluate the applicability of each model. According to the discussion in Sect. 4, the foundation pit deformation predicted by the linear elastic constitutive model is obviously inconsistent with the actual situation, so only the numerical results based on the MC model, HS model and HSS model are compared and analyzed in this section.

The foundation pit engineering examples studied in this section are selected from reference [12], the foundation pit project is composed of soil and rock, which is located in the south of Liaoyang Road and west of Fuzhou Road. The area has an 18-story and a 3-story existing building, and the two existing buildings have two-story basement. The side adjacent to the 3-storey building is selected as the calculation area, as shown in Fig. 8. The excavation width of the foundation pit is 8m and the depth is 15.5 m. In order to reduce the influence of boundary conditions on the soil deformation in the excavation affected area, the depth of the calculated area is 27 m and the width is 30 m. The plane strain assumption was used, and the thickness of the calculated area was set to 2 m. The existing building is 2.7 m away from the excavation edge of the foundation pit, and the buried depth of the basement is 4m. The calculation area is composed of silty clay layer of 4.4 m, strongly weathered granite of 11.7 m and slightly weathered granite of 10.9 m. Both strongly weathered granite and slightly weathered granite adopt MC model, and the mechanical parameters are shown in Table 2. The silty clay layer was simulated using MS model, HS model and HSS model, respectively, and the corresponding parameters are shown in Table 3. In order to ensure the construction safety, the micro steel pipe pile supplemented by soil nails and pre-stressed anchor rod support. The diameter of micro pile is 0.273 m, the elastic modulus is 220 GPa, using Pile unit simulation, the diameter of soil nail and pre-stressed anchor is 0.032 m, and the elastic modulus is 220 GPa. Using Cable unit simulation, the bond stiffness of the pre-stressed anchor is 0.56 GPa, and the bonding friction Angle is 25°. The pre-stressed anchors MG1, MG2 and MG3 are 281.25, 281.25 and 320.15 kN, respectively.

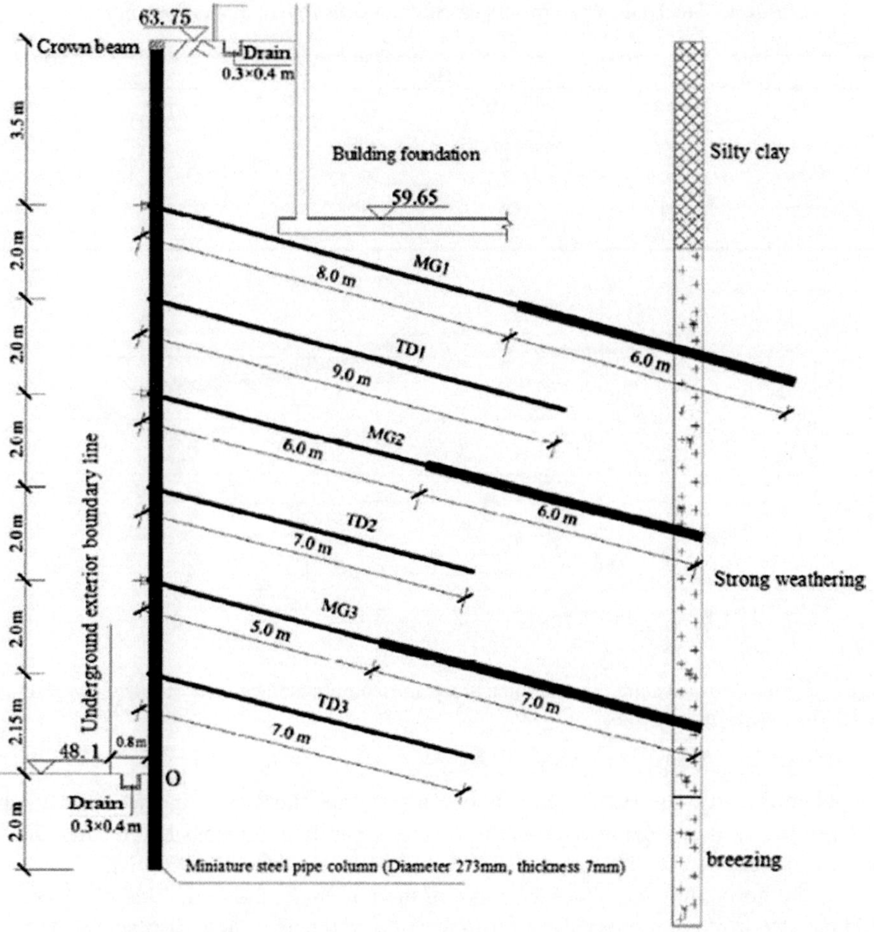

Fig. 8. Schematic of the section the foundation pit example and supporting structure

Table 2. Mechanical parameters of MC constitutive model of rock layer

Rock layer	γ (kN/m3)	*E*(MPa)	ν	*c*(kPa)	φ(°)	ψ(°)
Strong weathering	21	207.7	0.3	25	35	5
Slightly weathered	25	25000	0.28	50	65	35

Figure 9 show the numerical and measured results of the surface subsidence behind the wall and pile horizontal displacement predicted by different constitutive models after the excavation of the foundation pit. For the surface subsidence behind the wall, the numerical prediction value of the MC model is smaller than the measured value, because the foundation pit deformation induced by the plastic deformation of the soil

Table 3. Mechanical parameters of different constitutive models of soil layer

MC		HS		HSS	
E(MPa)	98.8	E_{50}^{ref}(MPa)	220	E_0^{ref}(MPa)	900
v	0.35	E_{ur}^{ref}(MPa)	88	$\gamma_{0.7}$	1.6×10-4
c(kPa)	18	m	1.0		
φ(°)	19	Rf	0.9		
ψ(°)	0				

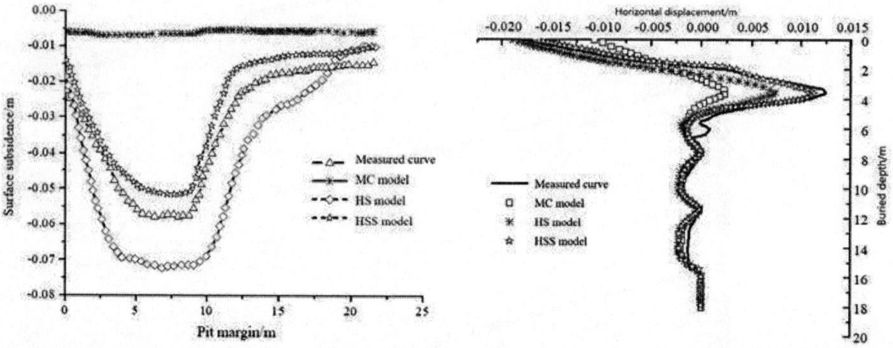

Fig. 9. Ground surface settlement predicted and horizontal displacement of pile shaft predicted by various constitutive models

is smaller. Because the HS model cannot consider the elastic nonlinearity in the small strain stage, its numerical prediction results are larger than the measured results, but the numerical results of the HSS model are the closest to the measured results.

For the horizontal displacement curve of the pile body, the horizontal displacement predicted by different models all points to the pit, and the maximum displacement occurs at the pile top. Because the granite layer adopts MC model, the horizontal displacement of the pile below the rock layer is roughly the same, and the horizontal deformation of the pile above the rock layer obviously depends on the choice of constitutive model. Due to the role of the pre-stressed anchor MG1, the horizontal displacement predicted by the different models is all pointed outside the pit at MG1. The maximum horizontal displacement predicted by the MC model and the horizontal displacement at MG1 are 59.5% and 18% of the measured values respectively, which differ greatly from the measured results. The maximum horizontal displacement predicted by the HS model and the horizontal displacement at MG1 are 104.9% and 59.3% of the measured values respectively, and the predicted values of the HSS model are 92% and 90% of the measured values respectively, indicating that the predicted results of the HSS model match the measured results.

6 Conclusion

This paper first discusses the stress path of soil during foundation pit excavation, and then based on FLAC3D qualitatively studies the applicability of common soil constitutive model including linear elasticity model, MC model, HS model and HSS model for numerical simulation of foundation pit excavation. Finally, we study the applicability of different constitutive models with engineering examples quantitatively and draw the following conclusions:

(1) In the process of foundation pit excavation, the stress path can be simplified to lateral unloading in active area, axial unloading in shallow passive area and axial unloading stress path in deep passive area. The stress path has a significant impact on the mechanical parameters of soil.
(2) The stress-strain relationship of the linear elastic model and the MC model are linear and cannot consider the difference in the elastic modulus of loading and unloading. The linear elastic model cannot describe the plastic deformation of the soil, so the linear elastic model is not suitable for the numerical simulation of foundation pit excavation. MC model can describe the plastic deformation of soil, which has poor applicability for shallow foundation pit and improved applicability for deep foundation pit.
(3) The HS model and the HSS model can consider the difference stiffness during the process of loading and unloading, and consider the shear hardening and compression hardening of the soil, which can reasonably reflect the deformation during the foundation pit excavation, and the HSS model considers the elastic nonlinearity of the small strain stage, so the model is very consistent with the engineering application, and the numerical prediction results match well with the measured results.

Acknowledgments. This research was funded by the National Natural Science Foundation of China, grant numbers, 52279094 and 51478065, and the Key Research and Development Program of Guangxi, grant numbers AB20238036.

References

1. Feng, X.L., Xiong, Z.H., Mo, Y., Pang, J.C.: Numerical simulation and analysis of surrounding environment deformation influenced by excavation of foundation pits under complex conditions, **36**(2), 330–336 (2014)
2. Liu, C., Ji, F.F., Zheng, G., Liu, T., Liu, Y.C.: Measurement and mechanism of influences of rainfall on supporting structures of foundation pits in soft soils, **42**(3), 4476–456 (2020)
3. Wang, S.G.: Deformation control of excavation engineering with complicated surroundings, **35**(S1), 474–477 (2013)
4. Sheng, Z.Q., Teng, Y.J., Li, P.: Discussion on several problems in design of retaining structures of deep excavation, **43**(1), 94–101 (2021)
5. Li, Z., Liu, S.Y., Wu, K., Cai, G.J., Tong, L.Y., Liu, W.L.: Determination of the disturbance depth due to excavations using multifunctional CPTU tests. **43**(01), 181–187 (2021)
6. Kong, D.S., Zhang, J., Wang, S.Q., Liu, Y.: Experimental study on stress and deformation characteristics of cantilever inclined retaining pile for foundation pit support, **16**(1), 160–168 (2020)

7. Gu, D.P., Ling, T.H., Yin, Z.R., Zheng, H.J., Yang, J.K.: Analysis of deformation characteristics of diaphragm wall in deep foundation pit under load of temporary bridge, **16**(6), 1781–1791 (2020)
8. Liu, Y., Feng, Z., Huang, G.C., Feng, X.D.: The study in predicting the deformation of supporting structure for deep foundation pit, **5**(2), 329–335 (2009)
9. Yang, Z.P., Liu, Y.Q., Liu, X.R., Xie, Y.K.: Excavation response of the external loads and characteristics of internal force and deformation of anchorage piles in foundation pit, **12**(2), 503–510 (2016)
10. Wang, C.B., Ding, W.Q., Qiao, Y.F.: Development and application of hardening soil constitutive model in FLAC3D, **33**(01), 199–208 (2014)
11. Tang, D.F., Wang, C.H.: Development and engineering application of hardening soil model with small strain stiffness in FLAC3D, **13**(01), 1–13 (2023)
12. Hu, R.G., Liu, H.J., Wang, Z.Y., Xi, Y.T.: Deformation analysis of supporting structure for soil rock combination foundation pit with adjacent building, **28**(6), 1368–1377 (2020)

Application of Endurance Time Method in the Seismic Responses Analysis of Free-Field Site

Wenting Li(✉) and Haozhe Xu

Shanghai Normal University, Shanghai 200234, China
lwt@shnu.edu.cn

Abstract. The Endurance Time Method is a promising seismic analysis technique due to its gradual increase in seismic excitation amplitude over time. It involves the prediction of a structure's gradual transition from an elastic state to a nonlinear damage state during a single time-history calculation. The application of Endurance Time Method in soil-structure problems can significantly reduce computing time. In this paper, the Endurance Time Method is applied on the seismic prediction of free-field site. The results of Endurance Time Method are compared with that of incremental dynamic analysis. For the problem in this paper, Endurance Time Method is well applied in the seismic responses analysis of free-field site, particularly when the induced seismic intensity is small.

Keywords: Endurance Time Method · Seismic Analysis · Finite Element Analysis · Free-Field Site · Time-History Analysis

1 Introduction

Earthquake is a natural disaster with high suddenness and destructiveness. During earthquake, the damage of underground rail transit brings significant casualties, economic losses and social impacts [1, 2]. It is therefore critical that we design and construct buildings and infrastructure that can withstand such events. Underground structures, including tunnels, underground parking lots, and subways, are often subjected to intense ground shaking during an earthquake, as well as secondary effects such as soil liquefaction. Analyzing underground structures for seismic activity is a complex issue in which the behavior of site is vital [3–5]. Traditionally, a large number of cases has to been simulated in the seismic fragility analysis [6, 7], which makes the research process tedious and time-consuming. Therefore, a method will be introduced to simplify the process of seismic fragility analysis to the greatest extent possible, in order to achieve the goal of saving manpower and material resources.

W. Guo et al. (Eds.): GBCESC 2023, LNCE 328, pp. 111–116, 2024.
https://doi.org/10.1007/978-981-99-9947-7_12

Endurance Time Method (ETM) is a method to conduct structural seismic response analysis [8]. By ETM, the structural damage evolution under different earthquake intensities can be obtain through only one seismic time history analysis. Researchers [9–12] have applied ETM in the analyzing of seismic responses of steel frame structures and soil-foundation-reinforced concrete frame structure, concluding that ETM well predicts the seismic responses.

In this paper, the seismic responses results of free-field calculated by ETM is compared with that by incremental dynamic analysis (IDA). Both procedure ABAQUS and SHAKE91 are employed.

2 Basic Concepts of ET Analysis

In ETM, ET excitation causes increasing amplitude over time, resulting in structures that will gradually transition from elastic to nonlinear damage states before ultimately collapsing. The quality of endurance time excitation plays a decisive role in the success of implementing the ET method. A straightforward way to represent the objective function for simulating endurance time excitation can be described in Eq. (1):

$$F\left(a_g\right) = \int_{T_{\min}}^{T_{\max}} \int_0^{t_{\max}} \left\{[S_a(T,t) - S_{aT}(T,t)]^2 + \alpha[S_u(T,t) - S_{uT}(T,t)]^2\right\} dTdt \quad (1)$$

where $a_g(t)$ is the acceleration time history. T_{min} and T_{max} are the minimum and maximum considered periods. t_{max} is the duration of excitations. $S_a(T,t)$ and $S_u(T,t)$ is acceleration and displacement spectra at a given period T and time t.

3 Problem Definition

The soil surrounding the structure is modeled with 7 horizontal layers, and their corresponding profile characteristics are provided in Table 1. To account for nonlinear soil behavior during an earthquake, the equivalent linearization method is employed. The soil's nonlinear phase refers to the soil response under large seismic amplitudes when nonlinear deformation occurs. During this phase, The stiffness and damping characteristics of the soil undergo substantial changes during this phase, attributed to variations in both its modulus and hysteresis loop. Therefore, it is necessary to approximate the nonlinear phase of soil by using an equivalent linear system with the same fundamental period and damping ratio, i.e. equivalent linearization method, to analyze the soil's response to seismic motion. The data for the shear-strain-dependent shear modulus and damping properties of related material are presented in Table 1, while the graphical representation of the data is shown in Fig. 1.

Table 1. Site properties

No	Type	Thickness/m	Density/(kg/m^3)	Gmax (MPa)	Poisson's ratio υ
1	Artificial filled soil	1.0	1,900	38.00	0.333
2	Sand in Pleistocene	4.1	1,900	38.00	0.32
3	Sand in Pleistocene	3.2	1,900	56.03	0.32
4	Clay in Pleistocene	3.1	1,900	69.99	0.40
5	Clay in Pleistocene	5.8	1,900	111.67	0.30
6	Sand in Pleistocene	5.0	2,000	222.24	0.26
7	Sand in Pleistocene	11.8	2,000	2,000	0.26

In the ETM analysis, the finite element simulation was developed by ABAQUS. The endurance time excitation whose acceleration time history plotted in Fig. 2 is employed. The bottom of soil column is fixed. The ET excitation is applied to the ETM's bottom boundary, i.e., bedrock level.

In the IDA, the procedure SHAKE91 is employed. SHAKE91 can be used to analyzes seismic response in horizontally layered soil deposits using equivalent linear methods. 20 seismic records are selected as input shakings whose peak acceleration are scaled to increase from 0.1 g to 0.6 g, with an increment of 0.1 g each time, respectively. Among the 20 seismic records, there are 10 measured records and 10 artificial records.

4 Results

The peak ground acceleration calculated by incremental dynamic analysis and ETM are compared in Fig. 3. As the intensity of input shaking increased for 0.1 g to 0.6 g, the peak ground acceleration responses of soil column increase significantly. The results by ETM is close to that of incremental dynamic analysis, particularly when the seismic intensity is small. When the input peak acceleration is 0.1 g, the difference ratio of peak ground acceleration of ETM results and the average value of IDA results is 2.4%. As the intensity of input shaking increased to 0.6 g, the difference ratio increases to 4.2%. Therefore, it is deduced that ETM is well applied in the seismic responses analysis of free-field site, particularly when the induced seismic intensity is small. Even when the induced seismic intensity is relatively high, ETM still maintains a high effectiveness and can be used for subsequent research and analysis.

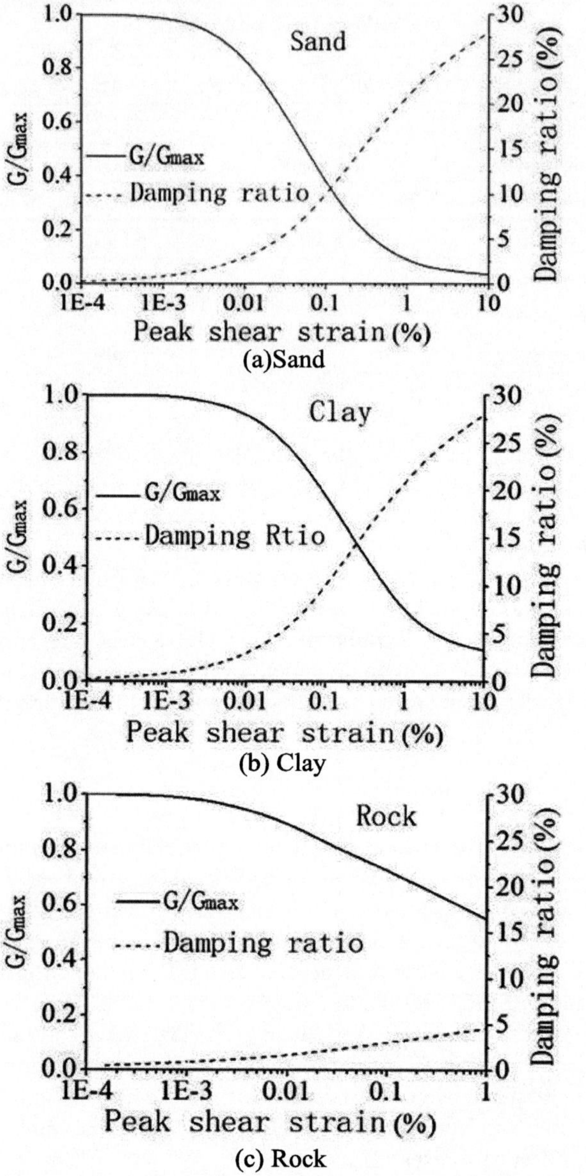

Fig. 1. Curves of $G/G_{max} \sim \gamma$ and $D \sim \gamma$ for sand, clay and rock

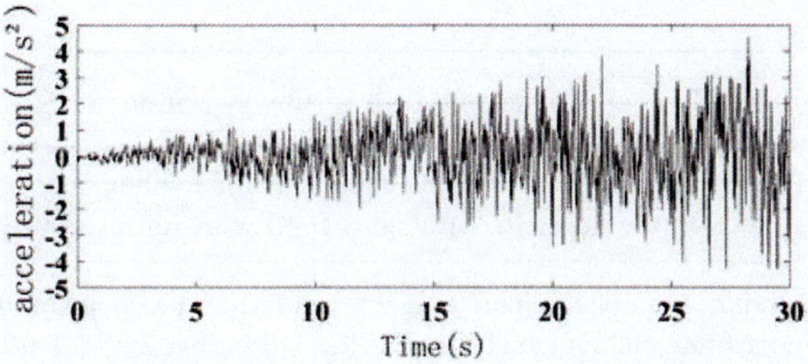

Fig. 2. Acceleration time history of endurance time excitation

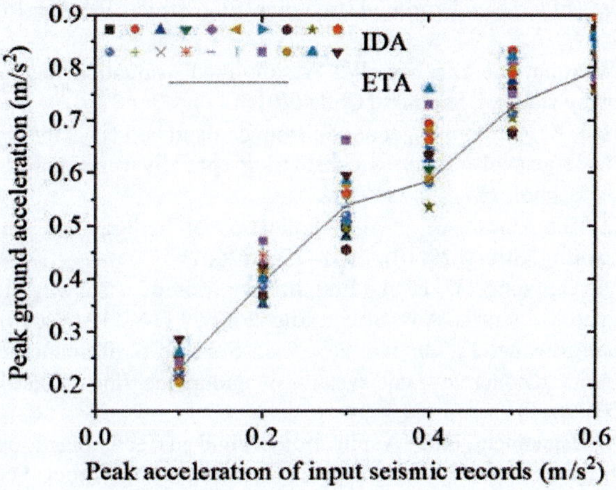

Fig. 3. Comparison of the acceleration responses calculated by incremental dynamic analysis and ETM

5 Conclusions

This study compares the numerical results obtained through the ETM and the IDA Method in analyzing the seismic responses of horizontal layered free-fields. The difference ratio of peak ground acceleration between Endurance Time Method results and the average values obtained via IDA is found to be 2.4% at an input peak acceleration of 0.1 g, which further increases to 4.2% as the input shaking intensity reaches 0.6 g. Overall, the results of this study suggest that the ETM can be effectively used for the seismic response analysis of free-field sites, especially when the induced seismic intensity is relatively low.

Acknowledgement. This work is funded by the National Natural Science Foundation of China (52008248).

References

1. Hiroto, N., Yoshida, N., et al.: Damage to Daikai subway station. Soils Found. **36**, 283–300 (1996)
2. Lanzano, G., Bilotta, E., Russo, G.: Tunnels under seismic loading: a review of damage case histories and protection methods. In: Strategies for Reduction of the Seismic Risk, p. 65 (2008)
3. Wenting, L., Qingjun, C.: Seismic performance and failure mechanism of a subway station based on nonlinear finite element analysis. KSCE J. Civ. Eng. **22**(7), 1–12 (2017)
4. Qingjun, C., Wenting, L.: Effects of a group of high-rise structures on ground motions under seismic excitation. Shock. Vib. **2015**(3), 1–25 (2015)
5. Qingjun, C., Wenting, L.: Effect of vertical ground motions and overburden depth on the seismic responses of large underground structures. Eng. Struct. **205**(Feb.15), 110071–110073 (2020)
6. Qingjun, C., Wenting, L., Yanchao, W.: A simplified evaluation method for the seismic fragility of subway stations, Santiago, Chile (2016)
7. Zhiming, H., Hao, X., Gardoni, P., et al.: Seismic demand and capacity models, and fragility estimates for underground structures considering spatially varying soil properties. Tunn. Undergr. Space Technol. **119**, 104231 (2022)
8. Estekanchi, H., Valamanesh, E., et al.: Application of Endurance Time method in linear seismic analysis. Eng. Struct. **29**(10), 2551–2562 (2007)
9. Tong, L., Zhiyi, C., Yong, Y., et al.: Fragility analysis of a subway station structure by incremental dynamic analysis. Adv. Struct. Eng. **20**(7), 1111–1124 (2016)
10. Jinlin, B., Shuangshuang, J., Junxian, Z., et al.: Seismic performance evaluation of soil-foundation-reinforced concrete frame systems by endurance time method. Soil Dyn. Earthq. Eng. **118**, 47–51 (2019)
11. Foyouzat, M.A., Estekanchi, H.E.: Application of rigid-perfectly plastic spectra in improved seismic response assessment by Endurance Time method. Eng. Struct. **111**, 24–35 (2016)
12. Shirkhani, A., Mualla, I.H., Shabakhty, N., et al.: Behavior of steel frames with rotational friction dampers by endurance time method. J. Constr. Steel Res. **107**, 211–222 (2015)

Compressive Stress-Strain Relationships of Wall Sheathings Used in Cold-Formed Thin-Walled Steel Shear Walls

Song Hu[1,2], Li Zhou[3,4](✉), Yong Huang[2], Chao Yin[3], Qingyu Zou[3], and Yifeng Xu[3]

[1] College of Agriculture and Forestry Engineering and Planning, Tongren University, Tongren 554300, China

[2] Research Center of Space Structures, Guizhou University, Guiyang 550025, China

[3] College of Architecture and Urban Planning, Guizhou University, Guiyang 550025, China
lzhou5@gzu.edu.cn

[4] Intelligent and Green Mountain Residence Engineering Research Center of Guizhou Province, Guiyang 550025, China

Abstract. Structural wall sheathings are commonly used in cold-formed thin-walled steel (CFS) shear walls. However, there are few reports on the compressive stress-strain relationships of wall sheathing. To investigate the compressive stress-strain relationships of the wall sheathing, 54 rectangular samples of gypsum wallboard (GWB), oriented strand board (OSB), and fiber cement board (FCB) were designed and manufactured. Compressive loading tests were also conducted, and failure phenomenon, stress, strain, elastic modulus, and Poisson's ratio of samples were investigated. The results indicate that the failure mode of GWB is a compressive failure at the end of the sample while that of OSB and FCB is an oblique shear failure at the middle of the sample. Finally, a three-stage stress-strain model was proposed. The comparison between the test and calculated curves shows that the proposed model has good accuracy. The results of this paper can be utilized as a reference for engineering applications and theoretical research of CFS shear walls.

Keywords: Stress-strain curve · Gypsum wallboard · Oriented strand board · Fiber cement board

1 Introduction

Wall sheathings, such as gypsum wallboard (GWB) [1], oriented strand board (OSB) [2], and fiber cement board (FCB) [3], are commonly used in cold-formed thin-walled steel (CFS) shear walls and listed in CFS specifications [4, 5]. As one of the main load-bearing components in CFS walls, wall sheathings are connected to the CFS frame by self-drilling screw connections and develop a stressed skin action to resist lateral deformation along with the CFS frame when the wall is subjected to wind or seismic loads [6]. CFS structures are widely used in North America, Europe, Japan, Australia, China, and other countries and regions [7], which has led to an increased interest in studying the structural behavior of wall sheathings.

W. Guo et al. (Eds.): GBCESC 2023, LNCE 328, pp. 117–127, 2024.
https://doi.org/10.1007/978-981-99-9947-7_13

It is noteworthy that the mechanical indexes of wall sheathings, including strength, elastic modulus, Poisson's ratio, etc., are crucial parameters for designing CFS walls [8]. However, there is limited research on the mechanical properties of wall sheathings that prevent the engineering application of CFS shear walls. In addition, to conduct a numerical simulation analysis of CFS walls, it is necessary to first understand the stress-strain model of wall sheathing. For simplification, researchers often use a bilinear model of wall sheathing, assuming that the material is fully plastic when damaged [9]. However, this model does not accurately reflect the true state of the material under compression. Therefore, studying the constitutive model of wall sheathing under compression is of great scientific significance for the engineering application and numerical simulation of CFS walls.

In this study, 54 rectangular samples of various wall sheathings, including GWBs, OSBs, and FCBs, were fabricated and tested under uniaxial compression. The failure characteristic, stress, strain, elastic modulus, Poisson's ratio, and stress-strain curve were investigated. Moreover, a simplified stress-strain model was proposed. The results of this paper can be utilized as a reference for the engineering application and numerical simulation of CFS walls.

2 Samples and Test Methods

The European specification BS EN 789 [10], Chinese specifications LY/T 1580-2010 [11] and GB/T 17657-2013 [12] suggest that the compression sample of wall sheathing should be a rectangular specimen. While almost all wall sheathing products in the market are thin and their plane dimension is generally 2440 mm × 1220 mm. Therefore, the rectangular plate of wall sheathing must be taken from the wall sheathing product according to specification requirements. As shown in Fig. 1, firstly, square plates 500 mm × 500 mm in dimensions were cut from the sheathing products (see Fig. 1(a)), and secondly, rectangular plates of 50 mm × 240 mm dimensions were cut from the square plate (see Fig. 1(b)). In the figure, the symbol ① indicates that the length direction of the rectangular plate is parallel to the horizontal direction, while the symbol ② represents that the length direction is perpendicular to the horizontal direction. Due to the thinness of wall sheathings, with the thickness of GWB, OSB, and FCB being 12 mm, 9 mm, and 8 mm, respectively, they could not meet the shape and size requirements of the compression test method specified in BS EN 789 [10]. Hence, structural adhesive was used to bond the thin-walled plates into a thick sample, as shown in Fig. 2. GWB, OSB, and FCB samples were bonded by 4, 5, and 6 plates, respectively, and a total of 18 rectangular samples were created for each wall sheathing, with 9 samples for ① numbered 4–12, and 9 samples for ② numbered 22–30. To ensure the randomness of sampling, rectangular samples were taken from different positions of different sheathing products.

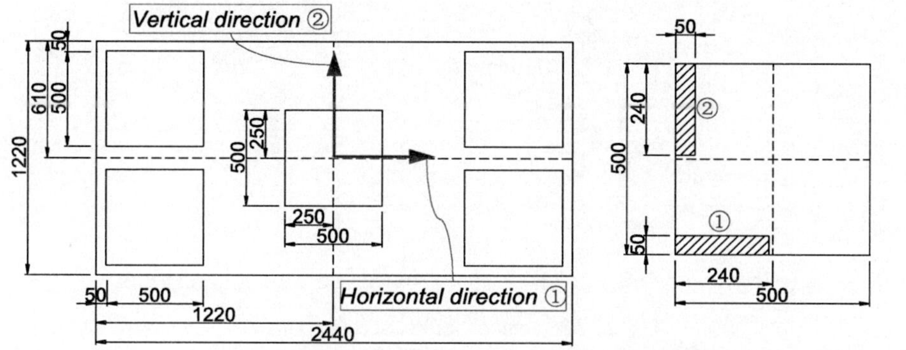

(a) Cutting square plates (b) Cutting rectangular plates

Fig. 1. Cutting scheme of the wall sheathing (unit: mm).

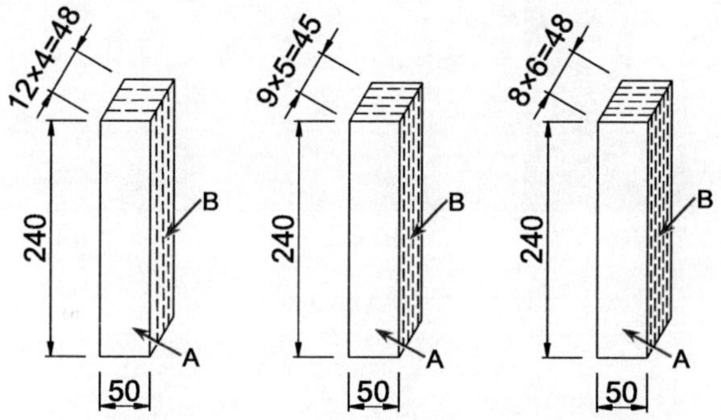

(a) GWB (b) OSB (c) FCB

Fig. 2. Sizes of rectangular samples (unit: mm).

As shown in Fig. 2, the rectangular sample had a smooth surface designated as A, while the adjacent surface was marked as B. Prior to loading, a vertical and a transverse strain gauge were affixed to the middle of surfaces A and B, respectively. As shown in Fig. 3, six samples had vertical strain gauges attached to surface A (numbers 4–9, 22–27), while three samples had transverse strain gauges attached to surface A (numbers 10–12, 28–30). According to BS EN 789 [10], a displacement-controlled loading procedure was proposed. When the applied load was below 60% peak load, the loading speed was set as 2 mm/min, which was then decreased to 1 mm/min to observe the sample failure in detail. All samples were tested on the WAW-100B testing machine. Figure 4 shows the axial compression test.

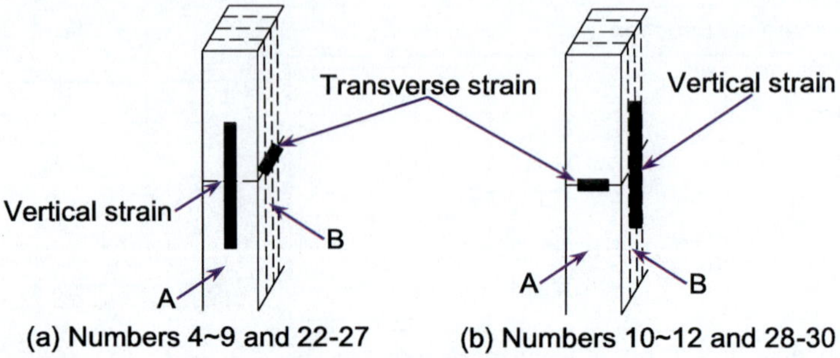

Fig. 3. Arrangement of measuring points.

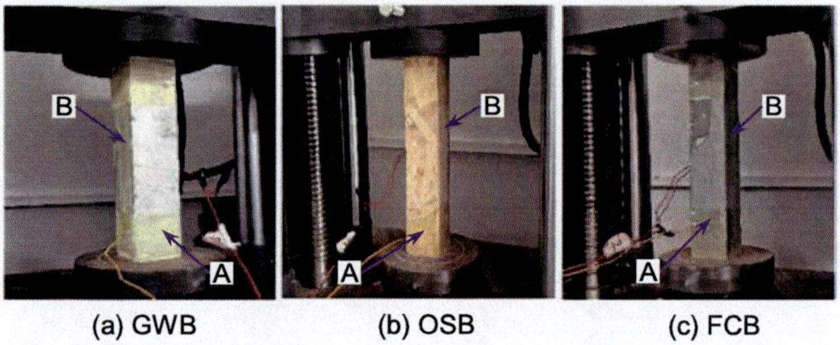

Fig. 4. Axial compression test.

3 Results and Discussion

3.1 General Observations

Figure 5 shows the final failure phenomena of samples. The observations indicated that the final failure phenomena were similar for both sets of samples, i.e., samples ① and samples ②. Therefore, Fig. 5 only displays the final failure phenomena of samples GWB-22, OSB-10, and FCB-10.

The failure process of all samples includes three stages. Firstly, in the elastic stage, the samples were in the elastic working stage, and the vertical compression deformation increased with the increase of load. Thin and short vertical cracks emerged in the middle of the sample indicating that the sample had reached its proportional limit of elasticity, which was consistent with the test observation in reference [9]. Secondly, in the elastic-plastic stage, oblique cracks appeared in the middle (see Figs. 5(a), (c), (e)), and they extended towards both ends as the loading increased. At peak load, the middle of the samples protruded outward and wrinkles appeared (see Figs. 5(b), (d), (f)). In the failure stage, the failure intensified until complete failure occurred.

The end-bottom of GWB samples was crushed due to their low material strength, as shown in Fig. 5(a). On the other hand, OSB samples did not crush and the final damage was relatively light because of their good deformation capacity. Both of these examples demonstrate a ductile failure characteristic. However, the failure phenomena of FCB samples were severe and exhibited a brittle failure characteristic.

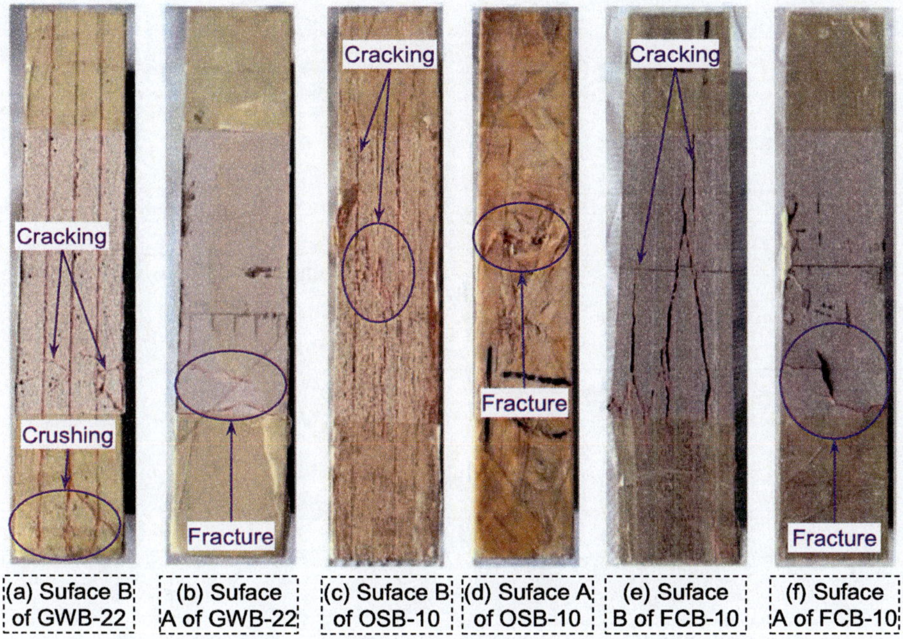

Fig. 5. Final failure phenomena of samples.

3.2 Stress-Strain Curves

The stress and strain data of samples were obtained by test machine sensors and strain gauges, and stress-strain curves of samples are shown in Fig. 6. However, it can be observed from Figs. 6(c), (d), and (f) that the strain values of samples OSB-8, OSB-24, OSB-26, FCB-25, and FCB-30 may be distorted. Despite this, the stress-strain curves of samples can be divided as elastic stage, elastic-plastic stage, and descending stage.

3.3 Characteristic Values

Stress and strain characteristic values of samples are elastic stress σ_e, peak stress σ_c, and ultimate stress σ_u, and the corresponding elastic strain ε_e, peak strain ε_c, and ultimate strain ε_u. As shown in Fig. 7, peak point D is located at the vertex of the stress-strain curve, stress and strain corresponding to point D are peak stress σ_c and peak strain ε_c, respectively. According to reference [9], elastic stress σ_e is suggested to be $0.6\sigma_c$, and strain corresponding to $0.6\sigma_c$ on the ascending curve is elastic strain ε_e. Ultimate stress σ_u is suggested to be $0.8\sigma_c$, and strain corresponding to $0.8\sigma_c$ on the descending curve is ultimate strain ε_u. Chinese standard GB/T 50081-2002 [13] suggests that all indexes

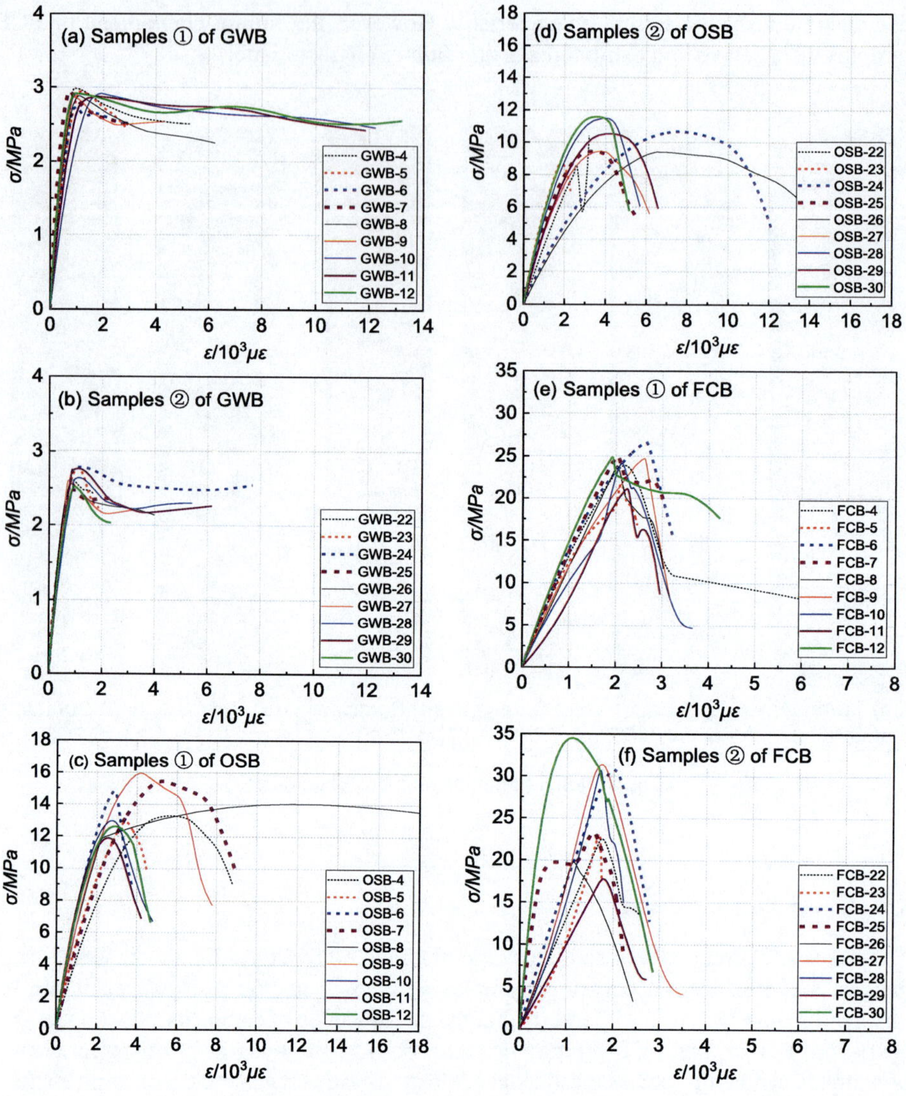

Fig. 6. Stress-strain curves of samples.

should be taken as the average of test values. If the subtractive value of a test value and the average value is greater than 20% average value, the test value would be discarded and a new average value would be calculated until it meets the requirements. Peak stress, strain, elastic modulus, and Poisson's ratio of each sample are presented in Table 1, while elastic stress and strain values, and ultimate stress and strain of samples are shown in Table 2.

From Tables 1 and 2, it can be observed that the compressive strength of GWB, OSB, and FCB samples are 2.76 MPa, 11.75 MPa, and 23.32 MPa, respectively; the elastic modulus values are 4890 MPa, 6392 MPa, and 14198 MPa, respectively; the Poisson's ratios are 0.207, 0.216, and 0.258, respectively. It can be concluded from the tables that three boards are anisotropic materials. The peak stress of GWB, OSB, and FCB samples along the horizontal direction is higher than that along the vertical direction by 9.89%, 30.97%, and 6.33%, respectively. Moreover, among the three types of wall sheathings, the OSB sample exhibits the highest peak strain and the best deformation performance.

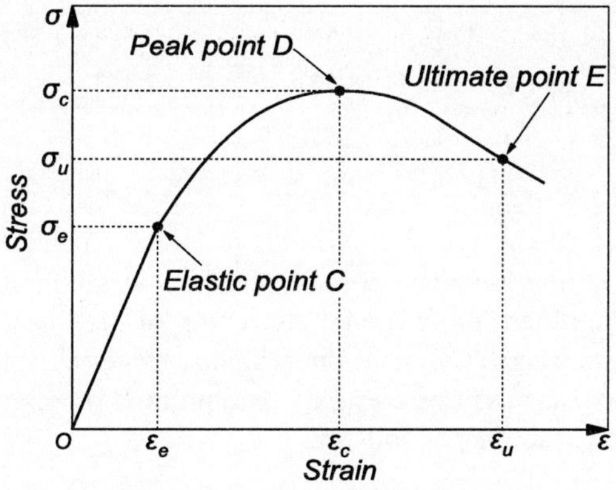

Fig. 7. Characteristic points.

Table 1. Peak stress and strain, elastic modulus, and Poisson' ratio of samples.

Material type	Loading direction	Peak strain $\varepsilon_c/10^3\,\mu\varepsilon$		Peak stress σ_c/Mpa		Elastic modulus E/Mpa		Poisson' ratio μ	
		Test	Average	Test	Average	Test	Average	Test	Average
GWB	Horizontal	1.147	1.015	2.89	2.76	4852	4890	0.193	0.207
	Vertical	0.883		2.63		4927		0.220	
OSB	Horizontal	4.058	3.991	13.32	11.75	7491	6392	0.226	0.216
	Vertical	3.923		10.17		5292		0.205	
FCB	Horizontal	2.069	1.906	24.03	23.32	15820	14198	0.255	0.258
	Vertical	1.743		22.60		12575		0.261	

Table 2. Elastic stress and strain, and ultimate stress and train of samples.

Material type	Loading direction	Elastic strain $\varepsilon_e/10^3\mu\varepsilon$		Elastic stress σ_e/Mpa		Ultimate strain $\varepsilon_u/10^3\mu\varepsilon$		Ultimate stress σ_u/Mpa	
		Test	Average	Test	Average	Test	Average	Test	Average
GWB	Horizontal	0.432	0.411	1.73	1.66	6.846	6.582	2.31	2.21
	Vertical	0.389		1.58		6.317		2.10	
OSB	Horizontal	2.047	1.785	8.10	7.11	4.261	4.873	10.80	9.48
	Vertical	1.522		6.12		5.484		8.15	
FCB	Horizontal	1.184	1.114	14.54	14.06	2.621	2.544	19.38	18.74
	Vertical	1.044		13.57		2.467		18.10	

4 Stress-Strain Model

As illustrated in Fig. 8, the uniaxial compressive stress-strain models of the three wall sheathings can be simplified into a model consisting of the linear elastic stage (*OC*), nonlinear elastic-plastic stage (*CD*), and linear failure stage (*DE*). Point *C* on the curve represents the proportional limit of elasticity, and point *E* represents the failure point. The expression can be described as follows.

$$y = \begin{cases} ax & (0 \le x \prec x_e) \\ bx^2 + cx + d & (x_e \le x \prec x_c) \\ ex + f & (x \ge x_c) \end{cases} \tag{1}$$

where $x = \varepsilon/\varepsilon_c$, $y = \sigma/\sigma_c$, $x_e = \varepsilon_e/\varepsilon_c$, $x_c = \varepsilon_c/\varepsilon_c = 1.0$, and *a*–*f* are constant coefficients satisfying the boundary conditions at points *C*, *D*, and *E*.

Submitting average values of stress and strain at points *C*, *D*, and *E* (see Tables 1 and 2) to Eq. (1), respectively, follows can be obtained.

$$\begin{cases} y_e = ax_e \\ y_e = bx_e^2 + cx_e + d \\ y_c = bx_c^2 + cx_c + d \\ y_c = ex_c + f \\ y_u = ex_u + f \end{cases} \tag{2}$$

where $x_u = \varepsilon_u/\varepsilon_c$, $y_e = \sigma_e/\sigma_c$, $y_c = \sigma_c/\sigma_c = 1.0$, $y_u = \sigma_u/\sigma_c$.

Line *OC* and curve *CD* must have a smooth transition at point *C*. Hence Eq. (3) can be obtained.

$$\frac{dy}{dx}\Big|_{x=x_e} = 2bx_e + c = a \tag{3}$$

Coefficients *a*–*f* can be obtained by associating Eqs. (1)–(3). Figure 7 shows the calculated curves and test values. It can be seen from the figures that the models correctly reflect the overall trend of the test values, indicating that the proposed model is reasonable.

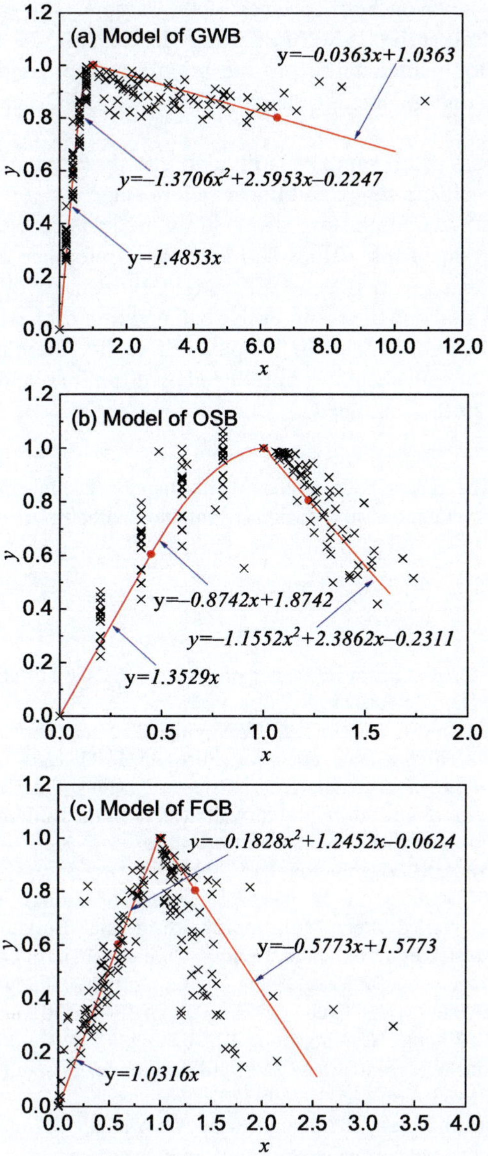

Fig. 8. Calculated curves and test values.

5 Conclusions

In this study, stress-strain curves of wall sheathings were investigated. Additionally, a stress-strain model of wall sheathing was proposed. Following can be drawn:

(1) The GWB exhibits better-cracking resistance and the least number of cracks due to the presence of paper on both surfaces of the sample. The failure mode of GWB is gypsum crushing at the end of the sample. In contrast, the failure mode of OSB and FCB is oblique shear failure in the middle of the sample. Moreover, GWB and OSB exhibit ductile damage characteristics, while FCB exhibits brittle damage characteristics.
(2) The failure process of all samples is divided into three stages: linear elastic stage, nonlinear elastic-plastic stage, and linear failure stage.
(3) A simplified uniaxial compressive stress-strain model is proposed, which can better reflect the evolution of test values and serve as a reference for the finite element analysis of CFS walls with GWB, OSB, and FCB, respectively.
(4) An experimental study on the compressive behaviour of each wall sheathing from the same group has been conducted in this paper. In reality, there may be differences in the performance of wall sheathings produced by different manufacturers. Therefore, further in-depth studies are needed.

Acknowledgments. The research described in this paper was financially supported by the National Natural Science Foundation of China (Grant: No. 52068007).

References

1. Reynaud, S., Kehinde, O.: Shear resistance of gypsum-sheathed light-gauge steel stud walls. J. Struct. Eng. **122**(4), 383–389 (1996)
2. Yan, Z., Cheng, Y., Shicai, C., Ziqin, J., Wengying, Z.: Shear performance of cold-formed steel shear walls with high-aspect-ratios. Structures **33**, 1193–1206 (2021)
3. Chao, Y., Li, Z., Qingyu, Z., Yifeng, X.: Effect of filling phosphogypsum on the axial compression behavior of cold-formed thin-walled steel walls. Buildings **12**, 1325 (2022)
4. AISI (American Iron and Steel Institute): Standard for Cold-Formed Steel Framing-Lateral design (AISI S213). AISI, Washington, DC (2007)
5. PSPRC (Profession standard of the people's republic of China): Technical Standard for Cold-Formed Thin-Walled Steel Multi-Storey Residential Buildings (JGJ/T 421-2012). Architecture & Industry Press of China, Beijing, China (2018). (in Chinese)
6. Xuhong, Z., Yu, S., Tianhua, Z., Zhengning, Y.: Experimental study of the shear resistance of cold-formed steel stud walls. China Civ. Eng. J. **43**, 38–44 (2010). (in Chinese)
7. Yi, H., Liqiang, J., Jihong, Y., Xingshuo, Z., Lizhong, J.: Seismic responses and damage assessment of a mid-rise cold-formed steel building under far-fault and near-fault ground motions. Thin-Walled Struct. **163**, 107690 (2021)
8. Wenying, Z., Xiangzhi, X., Yu, Z., Shuangshuang, W., Yuanqi, L.: Influencing factors analysis on shear capacity of cold-formed steel light frame shear walls. Structures **33**, 3588–3604 (2021)
9. Guo, C., Bin, H.: Experimental study on constitutive relation of oriented strand board under uniaxial compression. Build. Struct. **48**(10), 64–101 (2018). (in Chinese)

10. BS (British Standard): Timber structures-Test methods-Determination of Mechanical Properties of Wood Based Panels (EN 789:2004), London, UK (2004)
11. NFGA (National Forestry and Grassland Administration): Oriented strand board (LY/T 1580-2010). Standards Press of China, Beijing, China (2010). (in Chinese)
12. GAQSIQ (General Administration of Quality Supervision, Inspection and Quarantine of the People's Republic of China): Test methods of evaluating the properties of wood-based panels and surface decorated wood-based panels (GB/T 17657-2013). Standards Press of China, Beijing, China (2013). (in Chinese)
13. MHURD (Ministry of Housing and Urban-Rural Development of the People's Republic of China): Standard for test method of mechanical properties on ordinary concrete (GB/T 50081-2002). Architecture & Industry Press of China, Beijing, China (2002). (in Chinese)

Research on Impact-Abrasion Resistance of High-Strength Concrete with Recycled Rubber

Yuancong Liu[1], Jiangfeng Dong[1,2](✉), Yi Xu[2], Qingyuan Wang[2], and Dekun Peng[3]

[1] MOE Key Laboratory of Deep Earth Science and Engineering, School of Architecture and Environment, Sichuan University, Chengdu 610065, China
apiver@sohu.com
[2] Failure Mechanics and Engineering Disaster Prevention and Mitigation Key Laboratory of Sichuan Province, Sichuan University, Chengdu 610065, China
[3] China MCC5 Group Corp. Ltd., Chengdu 610063, China

Abstract. To improve the impact and abrasion resistance of concrete while minimizing the detrimental effects on compressive strength caused by the addition of rubber, a novel high-density rubber concrete was designed based on the closest packing model. The influence of different rubber content on the impact and abrasion performance, as well as the mechanical properties, was analyzed. The underlying mechanism behind the impact and abrasion resistance of high-density rubber concrete was elucidated. The results revealed that the high-density rubber concrete exhibited excellent resistance to impact and abrasion. Compared to the high-density reference concrete, the impact and abrasion strength of rubber concrete initially increased and then decreased as the rubber content gradually increased from 10% to 20% and 30%. Considering both the degradation of compressive strength and the enhancement of impact and abrasion strength, the optimal rubber particle content was determined to be 10% for high-density rubber concrete.

Keywords: Rubber Concrete · High Strength · Impact · Abrasion Resistance · Compressive Strength · Surface Morphology

1 Introduction

Rubber concrete has attracted significant attention in recent years due to its exceptional elasticity and impact energy absorption capabilities, allowing it to be widely employed in a variety of sectors such as hydraulic structures, airport runways, and docks, with promising future uses. According to statistics, over 70% of hydraulic concrete structures in China are prone to erosion and abrasion issues [1]. The high-speed erosion caused by sediment-laden water flow leads to varying degrees of concrete surface spalling and exposed reinforcement damage [2–4], seriously compromising the safety of hydroelectric power stations and resulting in substantial economic losses when repairing the damaged areas.

W. Guo et al. (Eds.): GBCESC 2023, LNCE 328, pp. 128–135, 2024.
https://doi.org/10.1007/978-981-99-9947-7_14

There are extensive research on the impact and abrasion resistance of rubber concrete and influenced factors using numerical simulations and experimental tests. The results has indicated that the addition of rubber powder as a replacement of fine aggregates in concrete could increase the impact and abrasion strength greatly. And the water-cement ratio and rubber content were the key factors affecting the impact and abrasion resistance of rubber concrete, while the particle size of rubber particles gave a minor impact [5–7]. However, the workability of rubber concrete decreased when the rubber particle content exceeded 10% or the direct addition exceeded 20 kg/m^3 [8–10]. In addition, with the decreasing of the rubber particle size, the improvement in impact and abrasion resistance became less significant [11], even if an reduction occurred when the rubber particles reduced to 0.6 mm [12, 13]. This was mainly attributed to the fact that smaller rubber particle sizes led to a higher concentration of weak interfaces between the rubber and cement mortar, which will reduce the density of the concrete significantly. In general, the introduction of rubber particles of any size or shape reduced the strength of compression of rubber concrete. The reduction on compressive strength will reach upto 52.6% when the 20% fine aggregates was replaced by rubber particles, which will limits the applicability of rubber concrete in the specific engineering fields [14, 15]. Therefore, it is of great engineering significance to increase the impact and abrasion resistance, and compressive strength of rubber concrete simultaneously and expand the range of applications in water conservancy and hydropower engineering.

This study aims to explore a new type of high-density rubber concrete based on the closest packing model to enhance the impact and abrasion strength of the concrete while mitigating the adverse effects on compressive strength caused by the addition of rubber. The impact and abrasion tests on high-density rubber concrete will be conducted using the underwater steel ball method, and the influence of different rubber content on its impact and abrasion resistance and mechanical properties will be analyzed. Additionally, the surface morphology of the concrete specimens after erosion and abrasion will be observed and summarized to reveal the underlying mechanism of the impact and abrasion resistance of high-density rubber concrete.

2 Experimental Overview

2.1 Materials

Following the requirements specified in the standard JGJ 52-2006, as the coarse aggregate, a constant gradation of basalt crushed stone with particle diameters ranging from 4.75 to 16 mm was used. Mechanism sand in Zone II was chosen as the fine aggregate. The cement material used was PC.42.5. To enhance the density and compressive strength of rubber concrete, Grade II ultrafine cement, Grade I flyash, and Grade II silica fume were selected as powder materials. The rubber particles were processed from discarded rubber tires, with a particle size of 1–3 mm. A high-performance polycarboxylate-based water reducer was used as the admixture. Water from the Chengdu region's standard tap was used in the trials.

2.2 Concrete Mix Proportion

The high-density reference concrete was designed based on the Dinger-Funk equation of the closest packing model, which calculates the proportions of solid materials with different particle size distributions. The target strength grade was set as C60. The initial design mix ratio for high-density concrete was cement: sand: aggregate: water reducer = 1:2.17:3.11:0.008, with a designed water-cement ratio of 0.55. Subsequently, the fine aggregate was replaced by rubber particles using a volume replacement technique. The replacement rates were set at 10%, 20%, and 30%. The mix proportions of the tested concrete are presented in Table 1, where "PC" represents the high-density reference concrete, "RC" represents the rubber concrete, and the numbers in the identification indicate the replacement rates of rubber particles.

Table 1. Concrete test mix ratio (unit: kg/m^3)

NO	Amount of concrete material							water reducer	rubber
	cement	Silica fume	superfine cement	Fly-ash	sand	pebble	water		
PC	354	52	87	87	580	1156	140	8.7	–
RC-10	354	52	87	87	580	1156	128.6	8.7	15.1
RC-20	354	52	87	87	464	1156	128.6	8.7	30.2
RC-30	354	52	87	87	406	1156	128.6	8.7	45.3

2.3 Test Setup

The abrasion resistance test of high-density rubberized concrete was conducted in an underwater steel ball apparatus set up in the laboratory. To prevent the steel ball from abrading surfaces other than the working surface of the concrete during the experiment, four concrete specimens were embedded in assembly blocks of the same diameter as the apparatus, as shown in Fig. 1. In addition, the test parameters were set according to the specifications of SL 352-2006. The rotational speed for the abrasion test was set at 1200 r/min. By measuring the mass loss of the concrete per unit area within a specified time, the abrasion resistance strength of the concrete was calculated. The abrasion resistance strength was used as an indicator of the concrete's abrasion resistance.

Cubic blocks with dimensions of 100 mm × 100 mm × 100 mm were used to evaluate the strength in compression of the specimens. Each group consisted of three parallel specimens. Due to the non-standard shape of the specimens, a correction factor of 0.95 was applied to the measured compressive strength. A HUT-106A computer-controlled hydraulic servo universal testing machine was used to load the specimens. The loading was done continuously and evenly, at a rate of 0.5 MPa/s.

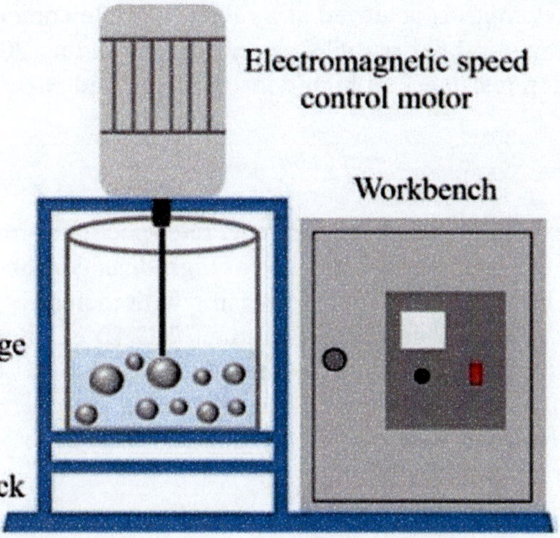

Fig. 1. Concrete impact abrasion resistance test

3 Results and Analysis

3.1 Abrasion Resistance Strength

The specimens' abrasion resistance strength may be represented as [16]:

$$f = \frac{\Delta m}{TA} \tag{1}$$

where f is the abrasion resistance strength (h/(g/cm^2)), Δm is the cumulative mass loss of the specimen after T hours of abrasion (g), T is the abrasion time (h), and A is the abrasion area of the specimen (cm^2).

Table 2. Highly dense rubber concrete impact abrasion resistance parameters

Specimens	Mass wear rate/%	Abrasion strength/(h/(g/cm^2))	Relative abrasion strength
PC	0.215	42.86	1
RC-10	0.166	57.14	1.33
RC-20	0.126	75.00	1.75
RC-30	0.351	27.59	0.64

The impact abrasion strength of high-density rubberized concrete is shown in Table 2 and Fig. 2a. Compared to the reference concrete, the impact abrasion strength increased by 33.3%, 75.0%, and −35.6% with rubber particle contents of 10%, 20%, and 30%, respectively. It can be observed that the impact abrasion strength of the concrete initially increases and then decreases as the rubber content gradually increases. The maximum impact abrasion strength is achieved at a rubber particle content of 20%. Therefore, it can be inferred that a rubber particle content not exceeding 20% effectively enhances the impact abrasion resistance of high-density rubberized concrete.

3.2 Compressive Strength

The compressive strength results of each concrete specimen are shown in Fig. 2b. It can be observed that the compressive strength of high-density rubberized concrete shows a trend of initially increasing and then decreasing with increasing rubber particle content compared to the reference concrete. Specimen RC-10 exhibited the highest increase in compressive strength, with a maximum enhancement of 11.0%. On the other hand, specimens RC-20 and RC-30 showed a decrease in compressive strength by 3.6% and 5.2%, respectively.

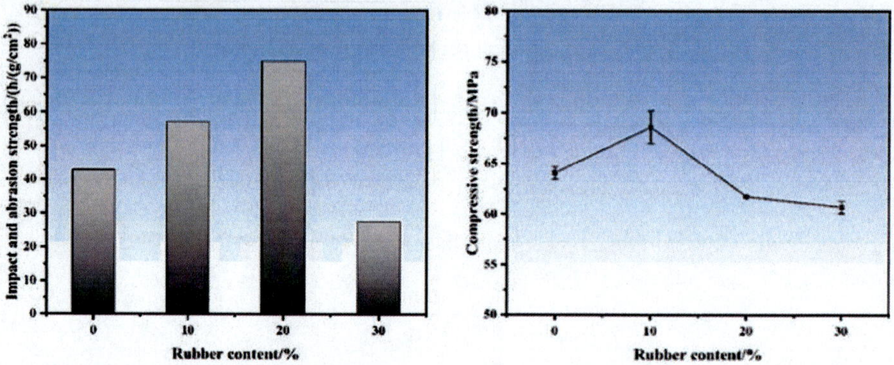

(a) Impact abrasion strength (b) Compressive strength

Fig. 2. Test results of high-density rubber concrete

The reason for this variation could be that when the rubber particle content is below 10%, the rubber particles are uniformly distributed in the mixture. The dense powder particles fill the small gaps between the rubber particles and the cement matrix more effectively, improving the bonding performance of rubberized concrete and thus increasing the compressive strength. However, when the rubber particle content exceeds 10%, the rubber particles cannot be uniformly distributed in the less flowable dense mixture. This leads to significant agglomeration of the rubber particles, and the cracks between the rubber particle agglomerates are larger than the filling effect of the dense powder material. As a result, the compressive strength decreases. Therefore, if both impact abrasion resistance and compressive strength are considered, it is recommended to select a rubber particle content of 10% for high-density rubberized concrete.

3.3 Surface Morphology of Specimens

For Reference Specimens. Before abrasion testing, the specimens exhibited some rough spots on the surface, which were attributed to the multiple uses of the molds during casting and slight surface roughness. However, the performance of the specimens as a whole was unaffected by these spots.. As shown in Fig. 3, none of the specimens exhibited extensive cement mortar matrix detachment after the completion of the experiments, indicating the impact abrasion resistance of high-density concrete. Therefore, the reference concrete showed multiple cracks on the surface after abrasion, and shallow abrasion pits appeared at the four corners of the specimens.

Rubber Concrete Specimens. As shown in Fig. 4, It showed that specimen of RC-10 exhibited wear scratches on the edges and corners after abrasion, but no significant abrasion pits were observed. Specimen RC-20 showed shortened wear scratch lengths at the edges and corners, with only two shallow abrasion pits near the center. Specimen RC-30 exhibited significant exposure of coarse aggregates after abrasion, and extensive agglomeration and uneven distribution of rubber particles were observed. In comparison to the reference concrete, the rubberized concrete surfaces did not show any cracks. This can be attributed to the rubber particles acting as elastic centers, absorbing the energy applied externally and effectively delaying damage.

In conclusion, basing on the surface morphology results, it can be concluded that high-density concrete, after abrasion, exhibits a trend of decreasing surface damage with increasing rubber content, followed by an increase. The impact abrasion resistance initially improves and then decreases. Therefore, the rubber particle content should not exceed 20%.

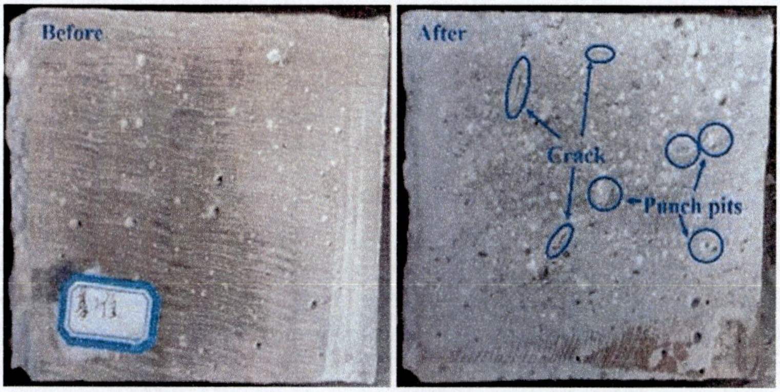

Fig. 3. Surface morphology of reference specimens

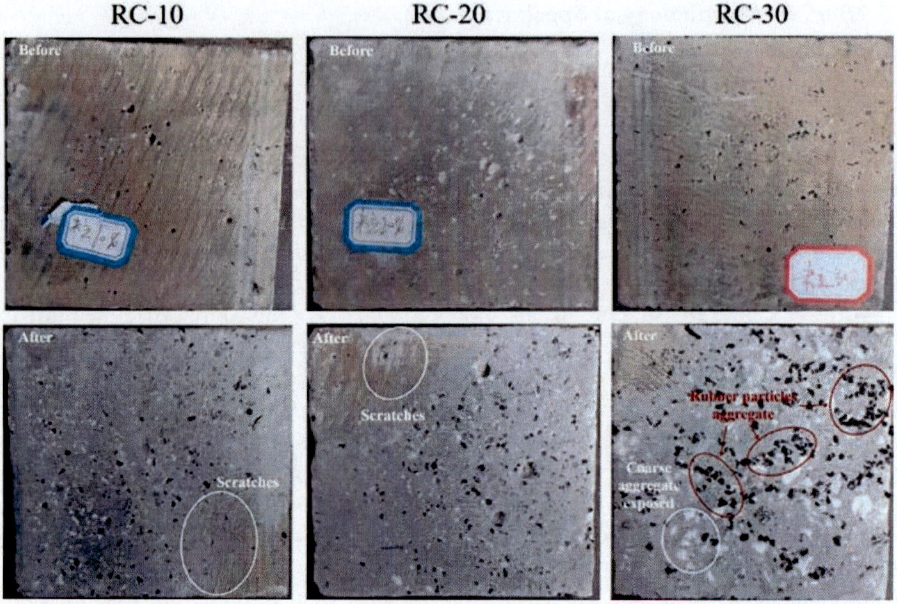

Fig. 4. Surface morphology of rubber concrete after punching and grinding

4 Conclusions

In this study, a high-density rubberized concrete mixture was designed, and experimental investigations were conducted to evaluate its impact on abrasion resistance and mechanical properties. The key findings include the following:

(1) High-density rubberized concrete exhibits excellent impact abrasion resistance. The impact abrasion strength showed a trend of initially increasing and then decreasing with the increase in rubber content, ranging from 10% to 30%. Compared to the high-density reference concrete, the maximum increase in impact abrasion strength was 75.0%.
(2) The compressive strength of high-density rubberized concrete decreased with increasing rubber content. Taking into account the compressive and impact abrasion strength, the optimal rubber particle content for high-density concrete was found to be 10%.
(3) The mechanism behind the excellent resistance to abrasion of high-density rubberized concrete can be summarized as follows: The dense powder material effectively improves the initial cracks caused by rubber, and the rubber particles inhibit the initiation and propagation of damage cracks. However, when the rubber content exceeds 10%, the influence of cracks between rubber particles becomes dominant in terms of impact abrasion resistance.

In conclusion, the developed high-density rubberized concrete exhibited remarkable impact abrasion resistance. It is recommended to use a rubber particle content of 10%

in the design of high-density concrete, considering both impact abrasion resistance and compressive strength. The findings of this study contribute to the understanding and application of rubberized concrete in areas that require enhanced durability and resistance to impact abrasion.

Acknowledgment. The research was supported by the Open Fund of Sichuan Provincial Engineering Research Center of City Solid Waste Energy and Building Materials Conversion and Utilization Technology (No. GF2022ZD001). The scholarship granted and financial support by Sichuan University are greatly appreciated. Special thanks to Mr. Shucheng Yuan and Mr Kuipeng Li for their assistance in the laboratory work.

References

1. Chen, X.: Research status of mechanical and impact wear resistance of modified rubber concrete. Henan Sci. Technol. **41**(02), 65–68 (2022)
2. Dong, J.F.: High temperature behaviour of basalt fibre-steel tube reinforced concrete columns with recycled aggregates under monotonous and fatigue loading. Constr. Build. Mater. **389**, 131737 (2022)
3. Bayazıt, Y.: Investigation of erosive wear effect on concrete water structures: the case of Porsuk Dam, Turkey. Mag. Concr. Res. **75**(10), 529–540 (2023)
4. Dong, J.F.: High-temperature behaviour of basalt fibre reinforced concrete made with recycled aggregates from earthquake waste. J. Build. Eng. **48**, 103895 (2022)
5. Bu, C.: Research progress on rubber concrete properties: a review. J. Rubber Res. **25**, 105–125 (2022)
6. Yu, Y.: Effect of rubber particles on impact resistance of concrete at a temperature of −20 °C. Arch. Civil Mech. Eng. **21**, 107 (2021)
7. Dong, J.F.: Freeze-thaw behaviour of basalt fibre reinforced recycled aggregate concrete filled CFRP tube specimens. Eng. Struct. **273**, 115088 (2022)
8. Zhu, X.: Reuse of waste rubber in pervious concrete: experiment and DEM simulation. J. Build. Eng. **71**(15), 106452 (2023)
9. Karimi, H.R.: Strength and cracking resistance of concrete containing different percentages and sizes of recycled tire rubber granules. J. Build. Eng. **67**(15), 106033 (2023)
10. Ghaleh, M.B.: Enhancing mechanical performance of waste tire concrete with surface double pre-coating by resin and micro-silica. J. Build. Eng. **50**, 104084 (2022)
11. Eltayeb, E.: Dynamic performance of rubberized concrete and its structural applications – an overview. Eng. Struct. **234**, 111990 (2021)
12. Xu, J.: Research on crumb rubber concrete: From a multi-scale review. Constr. Build. Mater. **232**, 117282 (2020)
13. Jin, H.: Mesoscale research on electric potential of rubberized concrete affected by rubber geometry. Constr. Build. Mater. **340**, 127851 (2022)
14. Chen, A.: Analytical evaluation of compressive strength for concrete with rubber fine aggregates and the predictive model. Constr. Build. Mater. **345**, 128359 (2022)
15. Gupta, T.: Assessment of mechanical and durability properties of concrete containing waste rubber tire as fine aggregate. Constr. Build. Mater. **73**, 562–574 (2014)
16. Wang, L.: Influences of MgO and PVA fiber on the abrasion and cracking resistance, pore structure and fractal features of hydraulic concrete. Fractal Fractional **6**(11), 674 (2022)

Structural Force Analysis and Service Condition Monitoring of a Port Door Machine

Wei Sun, YaYa Gao(✉), and PeiXuan Yan

Shenyang Jianzhu University, Shenyang 110168, China
gaoyaya2022@126.com

Abstract. Based on the service state of the port door machine, the structural force characteristics are analysed from multiple angles, the optimal online real-time monitoring solution is designed and the real service state of the door machine is given based on the monitoring data. According to the field working condition of the door machine, six typical service conditions of the door machine are summarised according to the wind load, boom extension and lifting weight. The finite element method is used to establish a refined model of the door machine, and the structural stress performance of the door machine under the six operating conditions is analysed comprehensively to find out the weak points of the structure so as to determine the optimal sensor arrangement, and then realise online real-time monitoring of the health information of the working stress and strain of the door machine. The degree of influence of different loads on the stresses of the door machine was obtained. The results show that the most influential load is the self-weight load of the door machine, followed by the crane load of the door machine, while the wind load has less influence on the strain, and the dangerous cross-section of each member of the door machine is located at the member joints. The strain values of the members of the door machine did not exceed the permissible strain of the material according to the actual measurements. The strains in the actual monitoring corresponded to the simulation results, and the monitoring data of all parts of the door machine fluctuated steadily, with no abnormalities in the structural health status, effectively ensuring the safe and stable operation of the door machine.

Keywords: Structural health monitoring · Steel structure · Finite element analysis · Gantry crane · Service condition

1 Introduction

Gate machine, known as gantry crane, is an important tool for modern port handling materials, port work environment is relatively complex such as high winds, corrosion, high temperature, etc., in the longer cycle of full load carrying process, as well as the role of the harsh environment, which then leads to different degrees of crane damage, there are three main reasons fatigue crack expansion, local instability leading to the destruction of the whole machine, rib corrosion [1], and even On February 3, 2016, a crane collapsed at a wharf in Jingjiang, Jiangsu Province, resulting in the death of one person and damage

W. Guo et al. (Eds.): GBCESC 2023, LNCE 328, pp. 136–144, 2024.
https://doi.org/10.1007/978-981-99-9947-7_15

to the ship. In June 2019, Jiangsu Yangzi Xinfu shipyard a gantry crane overturning collapse accident resulting in one death and one injury. Therefore, the use of effective online monitoring of large and complex structures is particularly important, the extent of damage to the structure does not exceed the safety tolerance value of the structure before the structure is monitored, thereby ensuring the safe operation of the structure in service.

Gantry health monitoring is mainly for the main boom, nose beam, large tie rod, balance beam, small tie rod, herringbone frame and other key structural weaknesses of the arrangement of sensors, the use of advanced sensing technology to sense the dynamic response of the structure, and further combined with the signal analysis of structural health. At present, scholars at home and abroad conduct a series of studies on crane structural health monitoring. Yang Yang et al. [2] studied the horizontal boom of tower crane, and based on the relevant characteristics of the crane in service state lifting arm to carry out research. Caglayan et al. [3] studied the vulnerable point on the tower arm of the tower crane, and based on the combination of numerical analysis and experimental method, the structure vulnerable point for structural health monitoring. Meng Wenjun et al. [4] established a portal crane health monitoring safety assessment system, real-time monitoring and condition assessment of large machinery and equipment, concluding that online real-time monitoring is conducive to the detection of minor faults, and plays an important role in the design, maintenance and extension of the service life of the complete set of equipment. Niu Qingliang et al. [5] designed an online health monitoring system for ship-type cranes, and optimized the system according to the experimental results, through which Ozden Caglayan et al. [6] modified the finite element model based on ANSYS analysis by testing the crane on site and based on the actual data collected, and finally used the modified model to assess the crane health status.

Li Xiangdong et al. [7] used the simulation software Adams was used to construct a virtual model to realize the motion simulation of the crane working process and output the load spectrum under various working conditions. The maximum stress location was derived to verify that the strength and stiffness of the gantry crane main girder meet the requirements of various operating conditions. Mosbeh R. Kaloop et al. [8] analyzed and evaluated the motion of a container crane based on structural health monitoring (SHM). And concluded that the average relative dynamic displacement can reflect the relative static displacement of the structure under the action of environmental loads, environmental load conditions significantly affect the crane deformation under different load cases, the crane deformation under different loads are within the safe range. Li Na et al. [9] Using strain and vibration sensing means, a port crane operating condition monitoring system has been developed for the typical working conditions of port cranes, with data flow as the guide to achieve the functions of data integration, storage, processing and remote transmission in turn. Zhang Chong et al. [10] used the tower crane in service at Shenzhen Lepu Building as the monitoring object and concluded that the static monitoring measurement points of the crane are generally distributed at the bottom of the transition section, the connection interface between the sections, and the top of the tower body. The dynamic monitoring measurement points are generally distributed in the area between the tower body and the rotary connection.

Although crane structural health monitoring has also been a lot of research, but the research system is not perfect.

2 Structural Force Analysis of Portal Machine

As the structure form inside the door machine is more complex, the complex structure will lead to slow model operation or even non-convergence during the finite element analysis, the door machine model is simplified according to the St. Saint Venant's Principle. The cell types of each component of the door machine are selected as shown in Table 1.

Table 1. Simplification principle of portal crane components

Component name	Simplification principle
Booms	The boom structure is divided into two forms, as box type and truss type. The boom structure of the door machine studied in this paper is a box type structure, so it is simplified to a shell structure
Herringbone frame	The herringbone frame is a solid structure composed of different plate structures, and its connecting parts are simplified to square solid members
Elephant trunk beam	The elephant trunk beam is two sets of truss structure connected together, so it is simplified to a set of truss members in tension
Balancing beam	The balance beam is a combined member consisting of counterweight and two plates, so it is simplified to a solid member
Small Tie Rods	The small tie rod is a rod system member that connects the boom to the balance beam, reducing it to a solid member
Large Tie Rods	The big ties are rod system members with holes, which are simplified to a rod without holes because of the uneven force at the holes

2.1 Simplification of Force Analysis Working Condition of Door Machine

According to the 40t-40m door machine design drawing, ABAQUS is used to establish the overall model, as the overall force form of the door machine is more complex, and more working conditions, so its working conditions are numerous. The operating load of the door machine is mainly the self-weight of the door machine, crane load and wind load, crane weight on its influence is mainly full load crane weight and no crane weight, wind load mainly has a normal state wind load and extraordinary state wind load (generally refers to 8 levels of wind and above), the author will be these kinds of load combination into 6 kinds of working conditions as shown in Table 2.

2.2 Overall Force Analysis of the Door Machine

It can be seen from Fig. 1 that the main components and locations of deformation of the door machine with wind load are the same as without wind load, the strain of each

Table 2. Working condition type of portal crane

Working conditions	Boom extensions	Wind loads	Hoist weight
Working condition 1	Maximum magnitude	Windless	Fully loaded
Working condition 2	Maximum magnitude	Normal wind	Fully loaded
Working condition 3	Minimum range	Windless	unloaded
Working condition 4	Minimum range	Normal wind	unloaded
Working condition 5	Maximum magnitude	Normal wind	Fully loaded
Working condition 6	Maximum magnitude	Extraordinary wind	Fully loaded

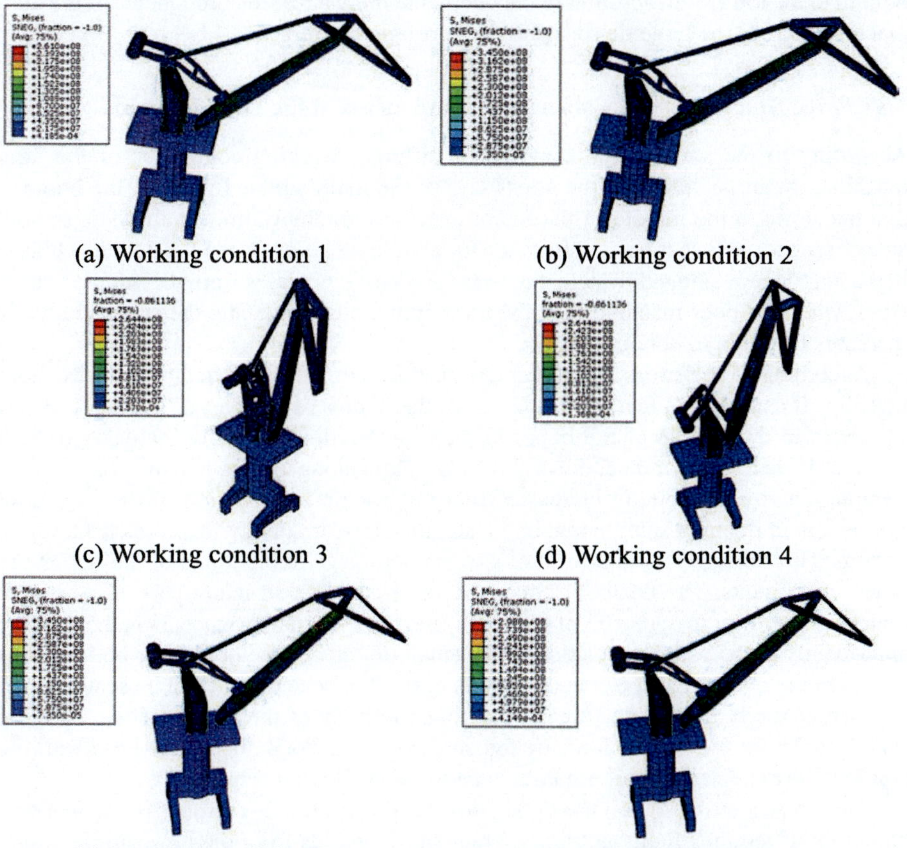

(a) Working condition 1 (b) Working condition 2

(c) Working condition 3 (d) Working condition 4

(e) Working condition 5 (f) Working condition 6

Fig. 1. Stress nephogram of portal crane under 6 working conditions

component of the door machine increases when there is wind load, but the increase is not large, so it can be seen that the wind load has a small effect on the working condition of the door machine. The maximum stress is 261.1 MPa at the bottom to the top 2/3 of the arm when the machine is in its maximum working condition and 264.4 MPa at the weld between the arm and the small tie rod when the machine is in its minimum working condition, so the working condition of the machine has a great influence on its strain. From Fig. 1a and Fig. 1e, the strain difference between the door machine at full load and no load is smaller than the strain difference between the door machine at maximum and minimum working amplitude, thus it is concluded that the load that has the greatest influence on the stress is the self-weight load of the door machine, followed by the crane load of the door machine; Fig. 1f shows the stress cloud before the door machine is destabilised, from the figure it can be seen that the maximum working stress of the door machine is 299.8 MPa at the bottom to 2/3 of the boom, and its The damage is caused by buckling of the outer steel plate on the upper side of the boom from the bottom to the top 2/3. In addition to the boom, the more stressed components of the door machine include the large tie rods and the herringbone frame.

2.3 Force Analysis of the Vulnerable Components of the Door Machine

According to the stress cloud diagram of different working conditions of the door machine, it can be seen that the stress law of the main stressed parts of the boom is that the strain in the middle of the boom increases gradually from the middle to both sides from the cross section, and from the longitudinal section, when the door machine is in the maxi1mum elongation state, the stress gradually increases from the middle to both sides; when the door machine is in the minimum elongation state, the stress gradually increases from top to bottom (Fig. 2).

According to the stress cloud diagram of different working conditions of the door machine, it can be seen that the stress law of the main stressed part of the boom is that the strain in the middle of the boom in the cross section gradually increases to both sides, and when the door machine is in the maximum elongation state in the longitudinal section, the stress gradually increases from the middle to both sides. When the door machine is in the minimum elongation state, the stress gradually increases from top to bottom. The damage is in the form of extrusion damage at the joints of the boom and other components. The flexural damage occurs in the box structure part of the boom, mainly in the lower to upper 2/3 of the boom, and the shear damage occurs in the welding part, mainly in the welding position of the small tie rod connector and the boom.

From the overall finite element analysis of the door machine, it can be seen that the location of the largest strain in each working condition of the door machine is in the middle and both sides of the large tie rod, and the stress cloud diagram of each working condition is extracted from these three locations, as shown in Fig. 3.

The stress distribution on the upper surface of the large tie member in the working condition of the maximum amplitude gradually increases from the lower to the upper 4/5 to both sides, and the stress distribution on the lower surface gradually increases from the lower to the upper 1/6 to both sides, and the stress on the windward side is large; the stress distribution of the large tie member in the working condition of the minimum amplitude is that the stress gradually decreases from the lower to the upper stress. Therefore, the large tie rod is prone to extrusion damage at the connection with other members.

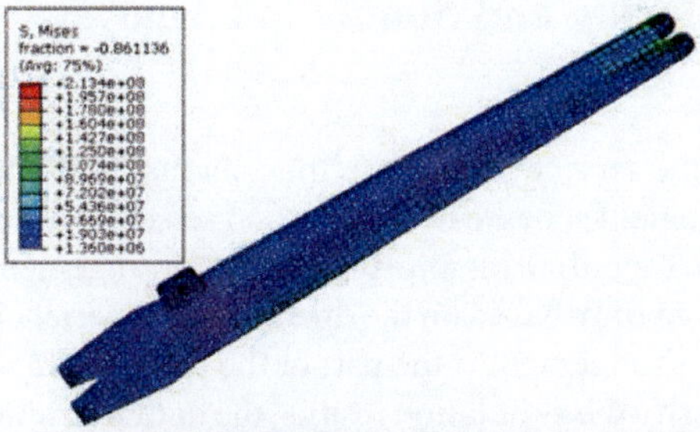

(a) Full maximum load

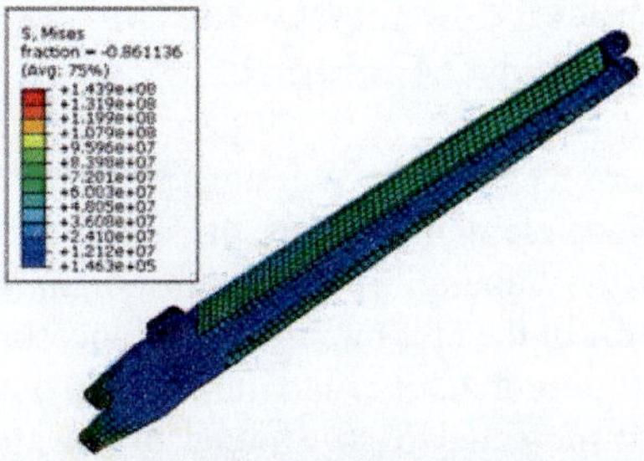

(b) Full load minimum range

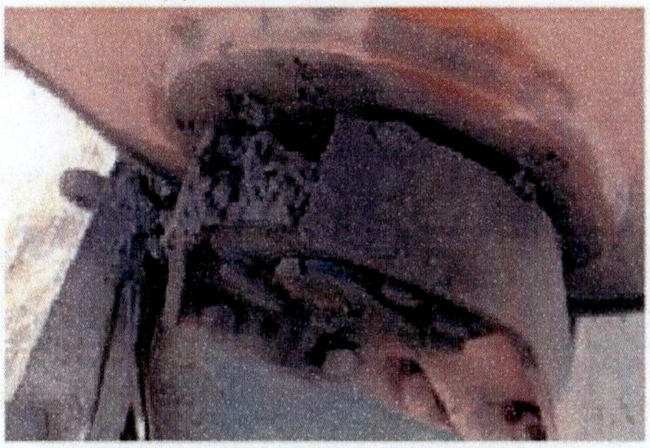

(c) Broken lower hinge point of arm

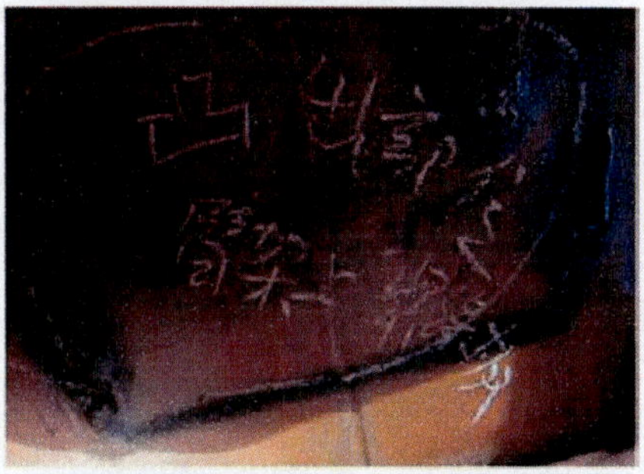

(d) Arm upper flange plate flexion

Fig. 2. Mises stress nephogram of boom structure and its partial failure location

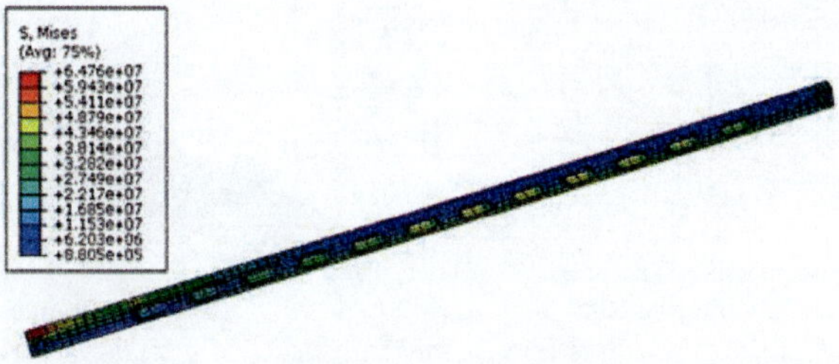

(a) Full maximum load

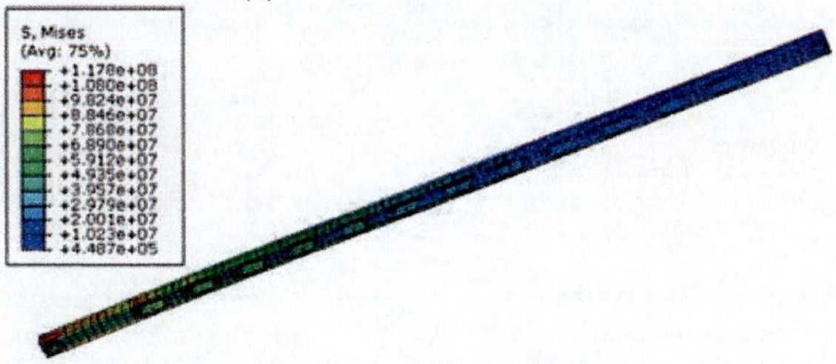

(b) Full load minimum

Fig. 3. Mises stress nephogram and partial failure location of large tie rod structure

3 Service Safety Condition Monitoring and Simulation Analysis of Door Machines

According to the crane design specification, the service situation of the door machine and the finite element analysis, the sensor arrangement scheme is derived, and sensors are set up in the vulnerable parts of the door machine for online monitoring, and the arrangement scheme of the measurement point location is mainly based on the operational experience of the door machine, and the location of the arrangement is the part of the door machine that is subjected to large forces. In order to verify the simulation results, the door machine was monitored for a period of 8 months from 26.9.2019 to 26.5.2020.

3.1 Monitoring and Simulation Analysis at the Boom

From the curves of the individual sensors extracted from the monitoring platform it can be seen that the strain is greatest at the boom and the trend of strain is the same as the trend of temperature. yb-5 is located at the lower part of the boom boom monitoring the main boom from the bottom to the top 1/3 of the way up, yb-6 is at the upper part of the boom monitoring the main boom from the bottom to the top 2/3 of the way up and yb13-2 is located at the base of the main boom. The sensor with the greatest strain variation is YB13-2, at the point where the boom is connected to the other components. The boom monitoring data and simulation results are shown in Fig. 4.

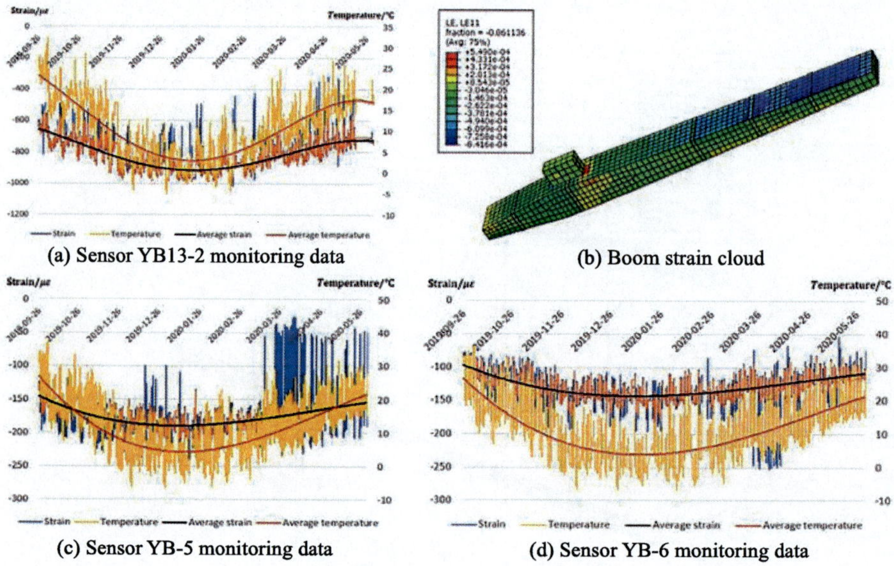

(a) Sensor YB13-2 monitoring data

(b) Boom strain cloud

(c) Sensor YB-5 monitoring data

(d) Sensor YB-6 monitoring data

Fig. 4. Strain data at the boom

The measured and simulated data show that the boom has a large variation at the base of the YB13-2 main boom (near the door end of the engine room), where the maximum strain at the bottom to top 1/3 of the main boom is −229.8 με. The strain variation is greater than the maximum strain at the bottom to top 2/3 of the main boom, which is −250.6 με, the further down the boom the greater the strain. The maximum strain was −917 με in the measured data and −841.6 με in the simulated data, both of which did not exceed the allowable material strain of 950 με. There are no stress concentrations and the overall condition is safe.

3.2 Monitoring and Simulation Analysis at Large Tie Rods

The strain transducer YB7-1 at the large tie rod is located at the upper end of the tie rod where it is connected to the herringbone frame. YB8-1 is located at the lower end of the large tie rod where it is hinged to the secondary tie rod and YB9-1 is located at the lower end of the large tie rod where it is hinged to the base. The measured data and simulation results are output as shown in Fig. 5.

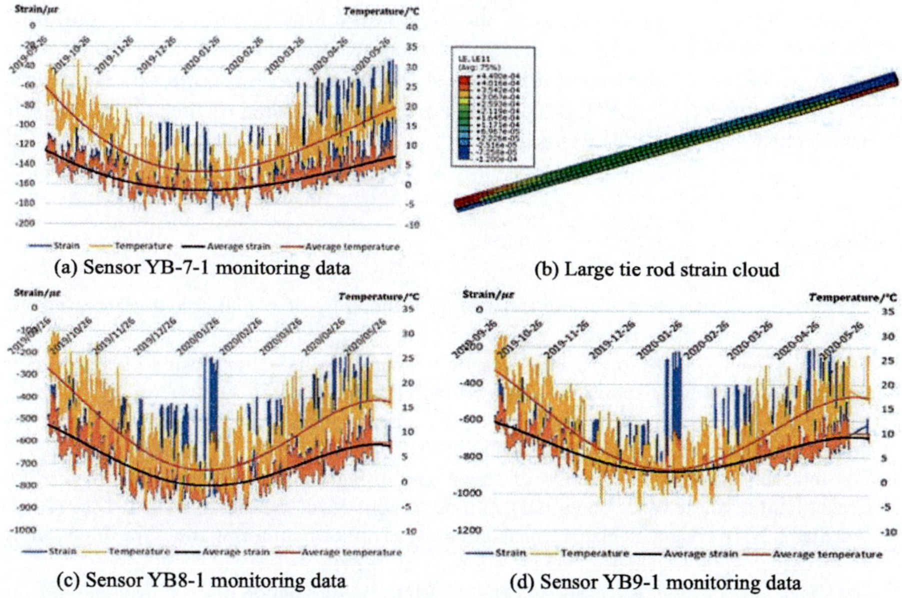

(a) Sensor YB-7-1 monitoring data

(b) Large tie rod strain cloud

(c) Sensor YB8-1 monitoring data

(d) Sensor YB9-1 monitoring data

Fig. 5. Sensor strain trend at large tie rod

The simulated and measured data show that the simulated and measured data have the same strain trend, which means that the simulated conditions are correct. YB8-1 and YB9-2 have higher strains than YB7-1, which means that the strain at the lower end of the large tie rod is greater than the strain at the upper end. This is because the self-weight of the large tie rods themselves counteracts some of the strain. The maximum strain of −896.4 με is less than the permissible strain of 950 με, so the big tie rod is in good working condition.

4 Conclusions

(1) Analysis and comparison of the stress conditions of the door machine under six different working conditions, to derive the degree of influence of different loads on the stress of the door machine, where the load that affects the stress change of the door machine most is the self-weight load of the door machine, followed by the lifting load of the door machine, the wind load has less influence on its strain.
(2) The analysis of the vulnerable parts of the door machine identifies the locations of the hazardous sections of the vulnerable parts of the door machine, all of which are at their connections with other components. The sensors were installed according to the optimal sensor arrangement derived from the finite element simulation. The monitoring data was analysed and compared with the finite element data to obtain the strain distribution pattern of the door machine, and the strain trend was consistent with the temperature trend.
(3) The finite element simulation results and real-time monitoring results show that the most dangerous component of the door machine is the boom, followed by the large tie rod. In the actual monitoring of the project, the focus should be on monitoring these two components. The monitoring data shows that the maximum strain of the boom $-917\ \mu\varepsilon$ and the maximum strain of the big tie rod $-896.4\ \mu\varepsilon$ do not exceed the permissible strain of $950\ \mu\varepsilon$, so the door machine studied by the author is in safe service under all working conditions.

References

1. Daoman, Q.: Research on remote monitoring technology of metallurgical crane structure stress based on wireless network. Wuhan University of Technology (2011)
2. Yang, Y., Datong, Q., Tao, Y.: Research on structural optimization design of tower crane double lifting point horizontal boom. Eng. Mach. 16–19 (2003)
3. Caglayan, O., Ozakgul, K., Tezer, O., et al.: Fatigue life prediction of existing crane runway girders. J. Constr. Steel Res. **66**, 1164–1173 (2010)
4. WenJun, M., Tianwei, H.: Design of stress monitoring and safety assessment system for CraneMetal structure based on virtual instrument. Appl. Mech. Mater. **152**, 1492–1497 (2012)
5. Qingliang, N., Weixiao, T., Nan, S., et al.: Stability of online monitoring stress for shipbuilding portal crane. Appl. Mech. Mater. **541**, 4124–1428 (2014)
6. Caglayan, O., Ozakgul, K., Tezer, O., et al.: Fatigue life prediction of existing crane runway girders. J. Constr. Steel Res. **66**(10), 1164–1173 (2010)
7. Xiangdong, L., Xu, Y., Fei, S., et al.: Study on estimating fatigue life of main girder of portal crane. In: Prognostics and System Health Management Conference, pp. 1133–1137 (2018)
8. Kaloop, M.R., Kim, E., et al.: Movement identification model of a steel structure based on structural health monitoring system. Adv. Struct. Eng. Mech. (ASEM13) **8**, 2878–2888 (2013)
9. Na, L., Guansi, L., Zhijie, W.: Development and application of integrated analysis system for port crane operating condition monitoring data. China Spec. Equip. Saf. **38**(9), 1–5 (2022)
10. Chong, Z., Tingsheng, Z., Ling, J., et al.: Selection of safety monitoring parameters and layout of measuring points for tower crane. China Saf. Sci. J. **31**(8), 112–118 (2021)

A Novel Self-Recovery Tri-stable Damper: Design and Analysis of the Energy Dissipation Performance

Hongyu Li, Xiangxing Zeng, Liling Xie, and Lu Zhang(✉)

College of Civil Engineering and Architecture, Guilin University of Technology, Guilin 541004, Guangxi, China
zhanglu@glut.edu.cn

Abstract. Mechanical metamaterials with the natural material engineered can give them properties and functions, which are not available in any existing materials. Recently, the mechanical metamaterials have already become a cutting-edge research topic and are widely used in many fields such as electronics, biology, medicine etc. In this paper, we introduced the concept of the mechanical metamaterials to the damper design for seismic resistance in civil engineering. A novel self-recovery tri-stable metal damper (SRTMD) was proposed, which takes fully advantage of the elastic deformation to realize the ability of self-recovery. Tri-stablility consists of two bistable elements (BE). The detailed design of BE and SRTMD was described in the paper. The mechanism was illustrated theoretically. Further, the energy dissipation performance of SRTMD was evaluated numerically. The results show that the proposed damper has an excellent capability in energy dissipation and post-earthquake restorability.

Keywords: Mechanical metamaterials · tri-stability · metal dampers

1 Introduction

Mechanical metamaterials are artificial structures with a prescribed mechanical behaviour by reconstructing the structures of the existing material other than their chemical composition. Recently, in many fields, the applications of mechanical metamaterials have been reported. By making use of its own structure design, the metamaterial can perform different mechanical behaviours (movement, deformation, folding, etc.) under the external loads, and those characteristics and functions cannot be observed in any conventional and natural materials. Moreover, mechanical metamaterials are materials with desirable properties and adjustable performance, its unique properties have led to a wide range of applications in electronics, energy storage and biomedicine [1–3]. However, in the field of civil engineering, there are few applications of metamaterial principles reported in literature. In this paper, we try to propose a mechanical metamaterial design which is combined with the metal dampers to enhance the seismic performance.

A structural damper is a device that can be used to absorb or dissipate energy. In civil engineering, the principle of damper system is to reduce the seismic energy. Many types

W. Guo et al. (Eds.): GBCESC 2023, LNCE 328, pp. 145–155, 2024.
https://doi.org/10.1007/978-981-99-9947-7_16

of damper are currently used: velocity related, displacement related and other types. To be more specific, the velocity dependence is further divided into viscous dampers [4], viscoelastic dampers [5], etc. The displacement type dampers are further divided into the metal dampers [6], friction dampers [7], etc. Herein, since the first metal dampers were proposed by Kelly et al. in 1972 [8], it has been widely used because of high energy consumption, superior performance, low price and easy installation and replacement. Metal dampers are generally made of soft steel or alloy. Its basic principle is to dissipate seismic energy by the inelastic characteristics of the energy dissipating material after yielding, its energy dissipation performance is limited by the material, and its damping effect is limited in structural applications.

Traditional metal dampers mainly use materials to enter the inelastic phase to achieve the energy dissipation effect, Therefore, it is limited by the plastic deformation force of the material, which makes it difficult to obtain efficient energy dissipation performance. Though many researchers have made various optimizations on the original metal dampers [9–13], and previous studies on the optimization of metal dampers have focused on the size, shape, and design deformation of the energy dissipating materials and materials of the dampers, it cannot change its essential drawbacks: performance degradation due to the plastic deformation and limitation in its capability of energy dissipation In this paper, the concept of mechanical metamaterials is added to the design of metal dampers, a novel design was proposed to utilize multi-stability and self-recovery of the damper to address the drawbacks of the traditional metal damper. By means of multi-stability, no plastic deformation happened in any components of the damper and superb capability of the energy dissipation can be realized. Furthermore, with the proposed design, the self-recoverability can be obtained: the structure can return to its initial stable state after unloading. This paper illustrated the design of a self-recovery tri-stable metal damper consisting of two bistable elements. By theoretical analysis of a single bistable element, to obtain bistable element's mechanical properties and influencing factors. Further analysis obtained self-recovery tri-stable dampers' energy dissipation performance and mechanical properties.

2 Bistable Element: Design and Analysis

2.1 Damper Design

Bistable element a structure that can occur two steady state switching under external load, as that in Fig. 1, the structure includes: 1. Sliding components consists of slider and Elastic rod, of which the elastic rod is a key component to achieve switching between two steady states; 2, External sleeve consists of built-in sleeve and rigid Fan-shaped recess component, the role of the built-in sleeve is to limit the slider displacement direction, the role of the rigid Fan-shaped recess component is to make the elastic rod bending deformation; 3. Compression Spring, the two ends are connected to the slider and the sleeve respectively; 4. External connectors, which serve to facilitate joining to the main structure; 5. Rolling bearing, which is set at the end of the elastic rod and is in contact with the fan-shaped recess, The function is to prevent the elastic rod to occur friction damage.

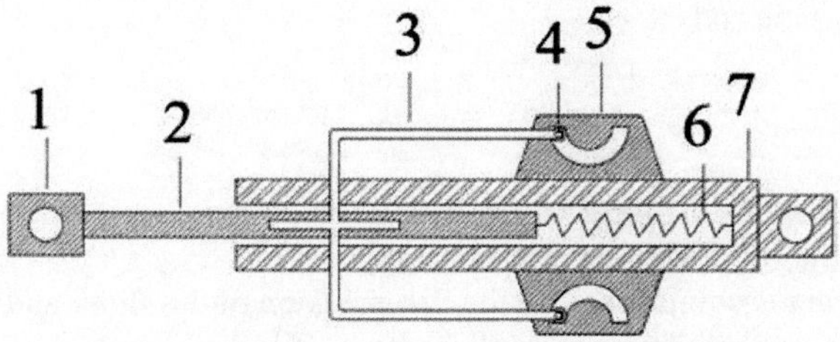

Fig. 1. Structure diagram of bistable dampers

2.2 The Mechanism of Bistable Elements

The working principle of the bistable element can be described from two aspects: energy dissipation and reset. When the bistable element is subjected to external pressure at both ends, the slider moves relatively to the outer sleeve. The elastic rod bends by interacting with the rigid Fan-shaped recess component, and the spring compresses by squeezing the slider and the outer sleeve, as shown in Fig. 2. These two relative motion together makes the entire system jumps from one stable stage to another as long as the elastic rod pass through the vertex of the half slot, which can provide the energy consuming effect. When the element is unloaded, the spring provides the reset load at this point because the spring is in compression.

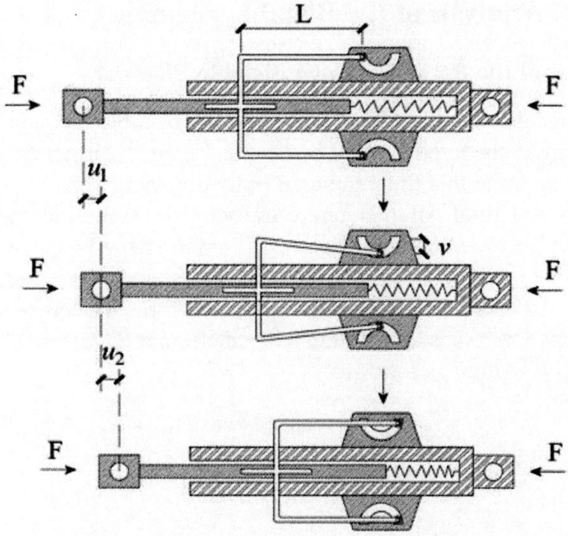

Fig. 2. Bistable element deformation process

When the bistable element is subjected to external loads at both ends. We assume that the bending displacement at the rod end of the elastic rod is v. Potential energy equation of the elastic rod:

$$\Pi_1 = \frac{1}{2} \cdot k_b \cdot v^2 \tag{1}$$

where, k_b is the bending stiffness of the elastic rod, $k_b = 2 \cdot \frac{1}{L^3} \cdot 3\mathrm{EI}$; v is the vertical displacement of the rod end of the elastic rod.

When the spring is compressed by the compression of the slider and the outer sleeve, assuming that the spring compression displacement is u, the potential energy equation of the spring:

$$\Pi_2 = \frac{1}{2} \cdot k_s \cdot u^2 \tag{2}$$

where, $\mathrm{k_s}$ is the spring stiffness; u is the slider displacement with respect to the outer sleeve.

The total potential energy of the bistable element consists of the elastic rod, the spring and the external load, so that a bistable element has the following total potential energy equation:

$$\Pi = \frac{1}{2} \cdot k_s \cdot u^2 + \frac{1}{2} \cdot k_b \cdot v^2 - F \cdot u\left(0 < u < \frac{1}{2} \cdot L_1\right) \tag{3}$$

where, F is the external load.

3 Theoretical Analysis of the Bistable Element

3.1 The Analysis of the Response of the Bistable Element

Assuming that only the elastic rod and Spring in the bistable element can deform, the rest components are considered as rigid bodies, and their deformation is neglected. So, for the bistable element at this time, the total potential energy can be formulated as the Eq. (3). Now two structural parameters are introduced: (i) the inclination angles θ of the connection line between the circle centre of the fan-shaped recesses, and the elastic rod ends with respect to the horizontal line; (ii) Fan-shaped Recesses centre of circle and rigid rod end distance L_1, as shown in Fig. 3. Thus, the trajectory of the elastic rod end in the fan-shaped recess is along a fan-shaped circular edge, and the corresponding displacement can be written as follows:

$$u = L_1(\cos\theta_0 - \cos\theta) \tag{4}$$

$$v = L_1(\sin\theta - \sin\theta_0) \tag{5}$$

where, θ_0 is the initial tilt angle; θ is the tilt angle when the elastic rod end is moving.

Combined with the multiple structural parameters $\{\theta, \theta_0, k_s, k_b, \mathrm{F}, L_1\}$ in Eq. (3), and a $k_b \cdot {L_1}^2$ is now introduced into it to re-write. Obtain the dimensionless potential energy equation expressed only in terms of four dimensionless parameters $\{\theta, \theta_0, f, k\}$:

$$U\{\theta, f, k, \theta_0\} = \frac{1}{2} \cdot k \cdot x^2 + \frac{1}{2} \cdot y^2 - f \cdot x \tag{6}$$

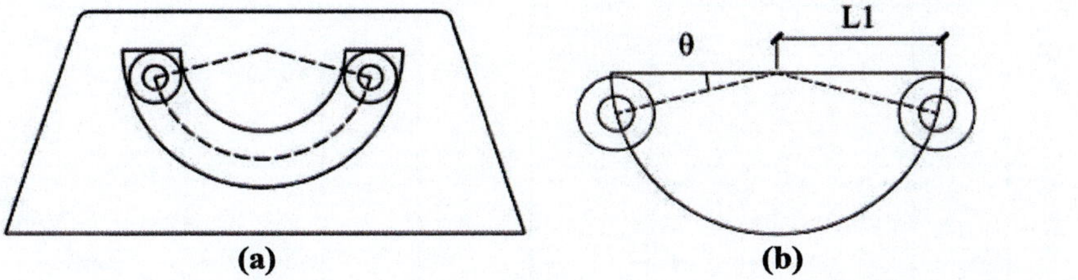

Fig. 3. Fan-shaped recesses component (a) Bearing trajectory; (b) Structure parameter schematic

$$k = \frac{k_s}{k_b}; f = \frac{F}{k_b \cdot L_1}; x = \frac{u}{L_1} = (\cos\theta_0 - \cos\theta); y = \frac{v}{L_1} = (\sin\theta - \sin\theta_0) \quad (7)$$

where, k is the ratio of the spring stiffness to the elastic rod stiffness; f is the ratio of the external load to the elastic rod stiffness and the radius of the fan-shaped Recesses.

The hysteresis curve equation is expressed as a first order differential equation for the total potential energy equation, so the bistable element's hysteresis curve equation is shown below:

$$\frac{\partial U}{\partial l}\{\theta, \theta_0, f, k\} = (1-k) \cdot cos\theta + k \cdot cos\theta_0 - cot\theta \cdot sin\theta_0 - f = 0 \quad (8)$$

combining Eq. (8) with Eq. (7), the bistable element hysteresis curve is analytically analyzed by MATLAB. The prescribed values of structural parameters are: $k = 0.15$, $\theta_0 = 45°$ and obtain the response of the bistable element with the relationship between displacement and load (f - x curve), see in Fig. 4. The results show that the bistable element's f - x curve is an S-shaped curve. According to Hooke's law, when a load acts on the structure, the applied load continuously increase as the displacement increases; but for the bistable element, it has two changing points in the increment of the load throughout the increase in displacement. For these two changes, the bistable element undergoes a transformation between stable states and the slider has an unstable displacement (The moment the bistable element switches between steady states, the slider undergoes a large displacement.), as shown in Fig. 4(a), When the external load increases to about 0.14, an unstable displacement X_{se1} occurs and the slider jumps from displacement x_2 to x_4. Similarly, when the external load is released, and the displacement of the slider drops to x_3; at the same time, the bistable element undergoes an inter-stable transformation; While, the slider undergoes an unstable displacement X_{se2}, and the slider displacement drops from x_3 to x_1. It can be seen that the bistable element does undergo a steady state switch during both loading and unloading in the process. This shows the actual f - x curve of the bistable element follows the solid line part as shown in Fig. 4(b). The results show that in the loading and unloading process of bistable element, when the external load reaches f_c (critical load for steady state transformation), the occurrence of unstable displacement X_{se} and the transformation to steady state take place simultaneously. Since the two critical loads occur at different displacements, the resulting hysteresis curve closure area is fuller compared to the conventional damper hysteresis curve closure area, indicating a better energy dissipation performance.

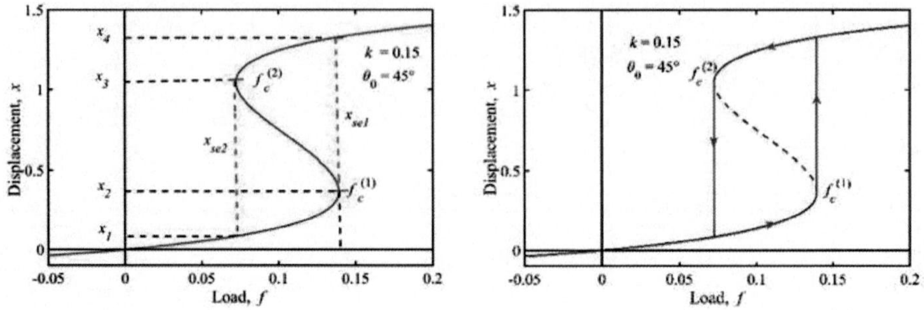

(a) (b)

Fig. 4. Bistable element response curve (a) Response curve equation trajectory; (b) Response curve real trajectory

3.2 Parametric Analysis of Bistable Element

In this section, the influential factors of the bistable element's hysteresis curve are analysed by parametric study with the values of the structural parameters k andθ_0. The hysteresis curves of the bistable with different parameters were shown in Fig. 5. With a certain stiffness ratio$k$, the bistable element's hysteresis curves at different initial angles θ_0 were compared in Fig. 5(a) with (c). The corresponding hysteresis curves can be classified into three types. (i) An irreversible bistable hysteresis response: the bistable element of this response cannot be reset even when the external load is removed to zero; while only when the load is applied in reverse, the structure can be reset, see in Fig. 5(a), which is called a super-plastic bistable response curve; (ii) A reversible bistable hysteresis response: this response behaviour is under external loading to develop to the second steady state. When unloaded can be reset, see in Fig. 5(b). And this response is called a super-elastic bistable response. (iii) No bistable change during loading and unloading, as shown in Fig. 5(c), and this response is called as the monostable response. This study of self-recovery tri-stable metal dampers belongs to response (ii). Comparing Fig. 5(d) to (f), when the initial angleθ_0, is set as constant, by adjusting the value of the stiffness ratio k, three types of response of the bistable element can be obtained. It was finally determined that the bistable element response type is a joint effect of the structural design parameters$\{k, \theta_0\}$.

The area closed by the unloading curve and the loading curve of the hysteresis curve indicates the dissipated energy. The comparison of the hysteresis curves shows that parameter $\{k, \theta_0\}$ affects the energy dissipation performance of the structure. Comparing Fig. 5(a) to (c), when the stiffness ratio k is set to a constant, the structure's energy dissipation performance decreases as the initial angle θ_0 increases. In addition. Comparing Fig. 5(d) to (f), when the initial angle θ_0 is set to a constant, the structure's energy dissipation performance decreases as the stiffness ratio k increases. It can be seen that only with appropriate structural parameters$\{k, \theta_0\}$, the bistable structure has super elastic response and efficient energy dissipation performance.

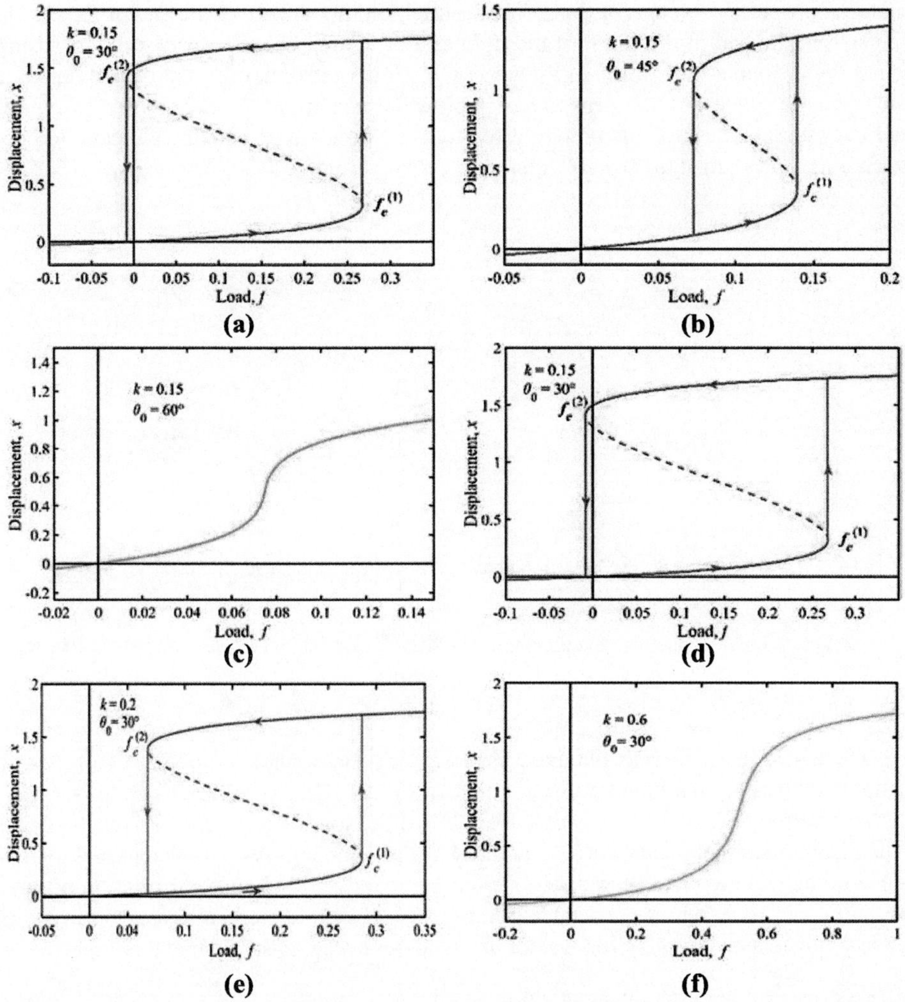

Fig. 5. Response curves under different parameters (a) Structure parameters: $k = 0.15, \theta_0 = 30°$; (b) Structure parameters: $k = 0.15, \theta_0 = 40°$; (c) Structure parameters: $k = 0.15, \theta_0 = 60°$; (d) Structure parameters: $k = 0.15, \theta_0 = 30°$; (e) Structure parameters: $k = 0.2, \theta_0 = 30°$; (f) Structure parameters: $k = 0.6, \theta_0 = 30°$

4 Tri-stable Metal Dampers Analysis

4.1 Tri-stable Metal Damper Construction Design

The tri-stable damper structure can be created by the design connection of two bistable elements. From the analysis in Sect. 2, Obviously, the bistable element is only under pressure, the slider and the outer sleeve can make a relative motion between them, and the switch between stable states occurs. Now consider a bi-stable element with the same material properties except for the spring's initial length, and move the end of the elastic rod so that the end of the rod is at the right end of the fan-shaped Recesses (The spring is not under load at this time), as that in Fig. 6. Under tension, this bistable element has the opposite process to the pressure-type bistable element in Sect. 2, as that in Fig. 7. For easy understanding, it is now referred to as a tension-type bistable element and the former as a pressure-type bistable element.

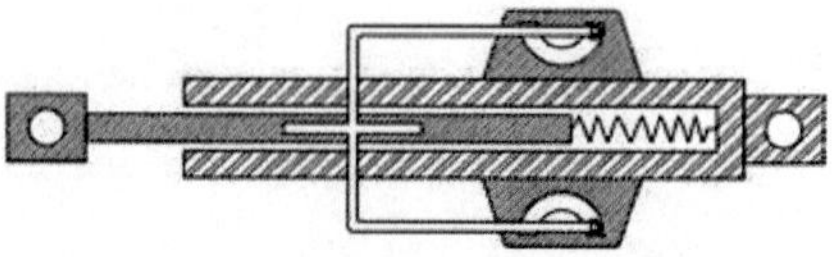

Fig. 6. Tension-type bistable element

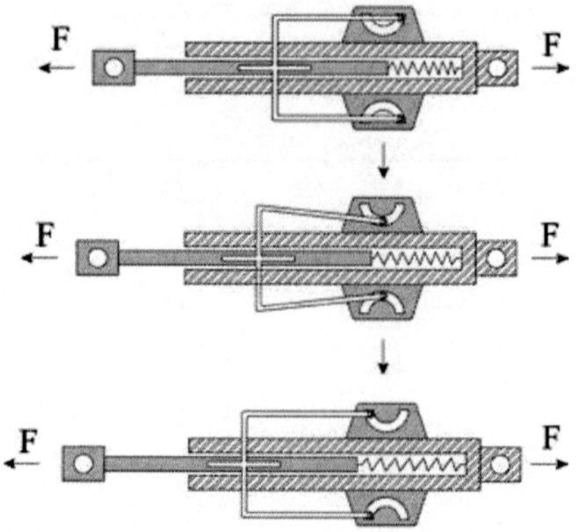

Fig. 7. Tensile type bistable element loading history

The tri-stable damper is obtained by designing the connection of the two types of bistable elements mentioned above, as that in Fig. 8. The working principle is: when the structure is subjected to external load, when the load is pressure, the left pressure type bistable element bears the load and deformation occurs; When the external load is tension, the right tension type bistable element bears the load and deformation occurs, as that in Fig. 9. When the tri-stable damper's two bistable elements' response type is set to hyper-elastic, at this time we obtain a self-recovery tri-stable metal damper.

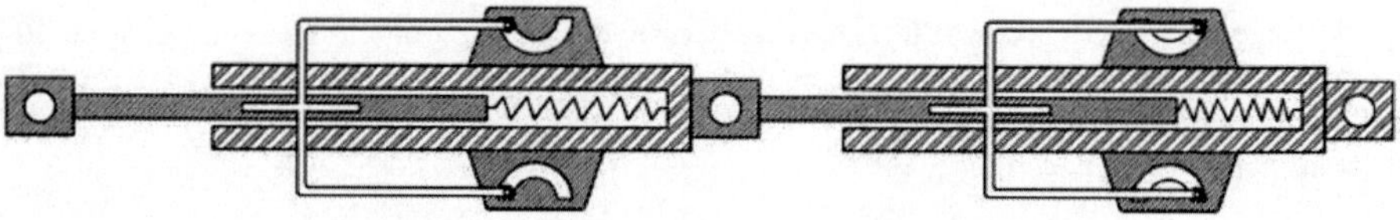

Fig. 8. Tri-stable structure construction design

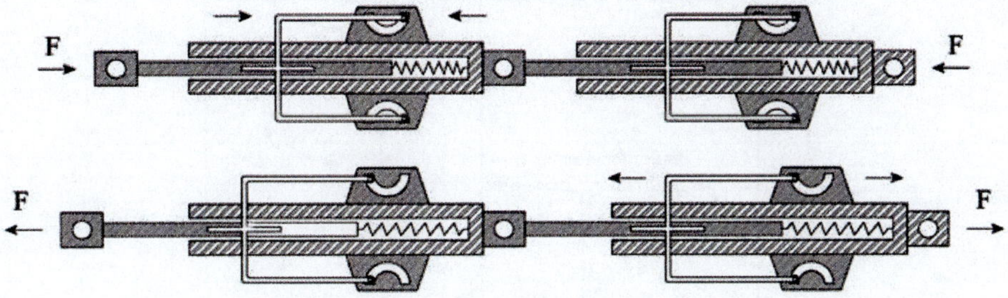

Fig. 9. Tri-stable structure loading deformation

4.2 Tri-stable Metal Damper's Performance Analysis

The tri-stable damper element consists of two different types of bistable elements. Based on Eq. (3), obtained the tri-stable structures with dimensionless potential (Assume that the load pressure is positive):

$$U_2\{\theta, \theta_0, f, k\} = \frac{1}{2} \cdot k \cdot (x_1 + x_2)^2 + \frac{1}{2} \cdot (y_1 + y_2)^2 - f \cdot (x_1 + x_2) \quad (9)$$

$$x_1 = \frac{u_1}{L_1} = cos\theta_0 - cos\theta_1;\ y_1 = \frac{v_1}{L_1} = sin\theta_1 - sin\theta_0 \quad (10)$$

$$x_2 = \frac{u_2}{L_1} = cos(\pi - \theta_0) - cos\theta_2;\ y_2 = \frac{v_2}{L_1} = sin\theta_2 - sin(\pi - \theta_0) \quad (11)$$

where, u_1 is the slider displacement of the pressure-type bistable element; u_2 is the slider displacement of the tension-type bistable element; v_1 the elastic rod bending displacement of the pressure-type bistable element; v_2 the elastic rod bending displacement of the tension-type bistable element; θ_1 is the changing tilt angle of the pressure-type bistable element; θ_2 is the changing tilt angle of the tension-type bistable element.

Self-recovery tri-stable damper element response curve equation as follows:

$$\frac{\partial U_2}{\partial l}\{\theta, \theta_0, f, k\} = (1 - k) \cdot \cos(\theta_1 + \theta_2) + k \cdot \cos\theta_0 - \cot(\theta_1 + \theta_2) \cdot \sin\theta_0 - f \quad (12)$$

Combining the Eqs. (10)–(12), the tri-stable structure hysteresis curves were numerically analysed by MATLAB. The structural parameters are pre-set at $k = 0.15$ and $\theta_0 = 45°$ to obtain the tri-stable structure's hysteresis curve, as that in Fig. 10. The results show that the super-elastic tri-stable structure exhibits good energy dissipation performance either in tension or pressure. When the external load is unloaded, the structure can be restored.

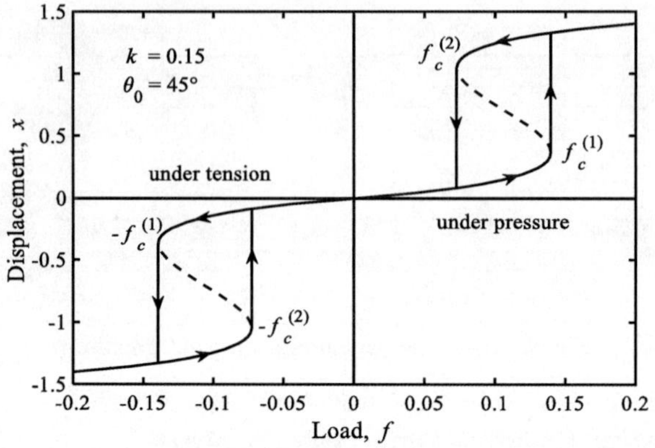

Fig. 10. The hysteresis curve of the super-elastic tri-stable structure

5 Conclusion

In this paper, we combine the concept of mechanical metamaterials to design a self-recovery tri-stable metal damper. Compression-type bistable element are theoretically analyzed to obtain the response curves under compression and analyse its energy dissipation performance and influencing factors. Further analysis of energy dissipation performance and structural feasibility of the self-recovery triple-stabilized metal dampers was performed. Three conclusions are as follows:

(1) The bistable structure has three types of response (super-elastic response, super-plastic response, and mono-stable response), and setting the appropriate structural parameters $\{k, \theta_0\}$ can obtain the structural super-elastic response.
(2) Compared with the hysteresis curve of conventional dampers, the super-elastic bistatic structure has a larger area of closed loop, and by setting the structural parameter$\{k, \theta_0\}$, it can obtain higher energy dissipation performance.
(3) Self-recovery tri-stable metal dampers exhibits the good energy dissipation performance, whether under tension or compression. And the structure can be reset after unloading. It can improve structural seismic performance and reduce post-earthquake structural residual deformation.

In this paper, the research on tri-stable dampers is still at the theoretical analysis stage, and the current research stage is to investigate the compression type bistable element in the damper structure, by creating scaled-down specimens to verify the response curve and mechanical properties of the element. Follow-up will produce scaled-down specimens of the tri-stable damper structure to verify the mechanical and energy dissipation properties of the structure.

Acknowledgement. This work was supported by National Natural Science Foundation of China (Grant Nos. 52168068 and 52068015), and Guangxi Science and Technology Base and Special Fund for Talents Program (Grant No. Guike AD20159011). Any opinions, findings and conclusions expressed in this paper are those of the writers and do not necessarily reflect the view of Natural

Science Foundation of China and Guangxi Science and Technology Base and Special Fund for Talents Program.

References

1. Mirzaali, M.J., Janbaz, S., Strano, M., et al.: Shape-matching soft mechanical metamaterials. Sci. Rep. **8**(1) (2018)
2. Kelkar, P.U., Kim, H.S., Cho, K.H., et al.: Cellular auxetic structures for mechanical metamaterials: a review. Sensors **20**(11), 3132 (2020)
3. Jiang, Y., Liu, Z., Matsuhisa, N., et al.: Auxetic mechanical metamaterials to enhance sensitivity of stretchable strain sensors. Adv. Mater. **30**(12) (2018)
4. Lee, D., Taylor, D.P.: Viscous damper development and future trends. Struct. Design Tall Build. **10**(5), 311–320 (2001)
5. Xu, Z.D., Wang, D.X., Shi, C.F.: Model, tests and application design for viscoelastic dampers. J. Vib. Control **17**(9), 1359–1370 (2011)
6. Javanmardi, A., Ibrahim, Z., Ghaedi, K., et al.: State-of-the-art review of metallic dampers: testing, development and implementation. Archiv. Comput. Methods Eng. **27**, 455–478 (2020)
7. Jaisee, S., Yue, F., Ooi, Y.H.: A state-of-the-art review on passive friction dampers and their applications. Eng. Struct. **235**, 112022 (2021)
8. Kelly, J.M., Skinner, R.I., Heine, A.J.: Mechanisms of energy absorption in special devices for use in earthquake resistant structures. Bull. New Zealand Soc. Earthq. Eng. **5**, 63–88 (1972)
9. Ghabraie, K., Chan, R., Huang, X., et al.: Shape optimization of metallic yielding devices for passive mitigation of seismic energy. Eng. Struct. **32**(8), 2258–2267 (2010)
10. Lin, Y., Guo, Z., Yang, S., et al.: Development of duplex assembled I-shaped steel panel dampers. J. Construct. Steel Res. **174**, 106267 (2020)
11. Wang, W., Liu, Y.: Concept and performance testing of an all-steel miniature dual stiffness damper. J. Construct. Steel Res. **183**, 106772 (2021)
12. Zheng, J., Zhang, C., Li, A.: Experimental investigation on the mechanical properties of curved metallic plate dampers. Appl. Sci. **10**(1), 269 (2019)
13. Zhang, C., Zhu, J., Wu, M., et al.: The lightweight design of a seismic low-yield-strength steel shear panel damper. Materials **9**(6), 424 (2016)

Effect on Autogenous Volume Deformation of Concrete Mixed with Magnesium Oxide and Polyethylene Fiber

Shaolian Yan[1], Weiwei Li[2](✉), Tijiang Fu[1], Ziyu Song[2], Xue Luo[1], and Guigang Jin[1]

[1] Guizhou Institute of Water Resources Science, Guiyang 550002, China
[2] School of Materials and Architectural Engineering, Guizhou Normal University, Guiyang 550025, China
ghtliweiwei@163.com

Abstract. In order to explore the effect on autogenous volume deformation of concrete mixed with magnesium oxide (MgO) and polyethylene fiber, this paper measured the autogenous volume deformation of concrete different admixtures of magnesium oxide and polypropylene fiber revealing the mechanism of compounding in concrete. The results showed that the compounding of MgO and polypropylene fiber has a significant effect on the autogenous volume deformation of concrete, and the more obvious with the increase of mixing amount. It was revealed from the experimental point of view that the cracking effect of concrete compounded with MgO and polypropylene fiber can be effectively improved, which promote the application of panel rockfill dam.

Keywords: magnesium oxide concrete · polyethylene fiber · autogenous volume deformation · concrete

1 Introduction

Panel rockfill dams are indispensable for the construction of hydropower projects [1, 2]. Especially in the economically underdeveloped and traffic-closed mountainous areas of the canyon, which are constrained by topographical and geological conditions, dam construction materials and other factors, panel rockfill dams have become the most competitive dam type [3, 4]. As a special hydraulic structure, the panels of panel rockfill dams are in long-term contact with water or in an exposed state, and are subject to environmental conditions (water pressure, water quality, etc.), boundary conditions (dam settlement and deformation, foundation restraint, block movement), meteorological conditions (temperature, humidity, wind speed, sun and rain) and many other factors, which can lead to cracks to varying degrees during long-term service, resulting in a reduction in the crack resistance of concrete projects [5]. The capacity of concrete projects is reduced [5]. Some penetrating cracks can even endanger the safety of the whole project [6].A large number of engineering practices in China have also proved this point [7]. However,

W. Guo et al. (Eds.): GBCESC 2023, LNCE 328, pp. 156–166, 2024.
https://doi.org/10.1007/978-981-99-9947-7_17

so far, domestic and foreign can only curb the chances of cracks occurring in panel rockfill dams, but cannot fundamentally eliminate the generation of cracks, especially there are few studies related to targeted basic information on the effect of compounded magnesium oxide (MgO) and polypropylene fibres on the autogenous volume deformation of panel concrete.

Therefore, how to fundamentally prevent the occurrence of panel cracks, the extension of control panel cracks, or minimise the generation of panel cracks has become a pressing problem in the engineering community. In this paper, the effect of compounding MgO and polypropylene fibres on the autogenous volume deformation of concrete is investigated to reveal the mechanism of the co-mixing of MgO and polypropylene fibres in concrete, and to study whether the expansion deformation of compounded concrete can compensate for the temperature drop shrinkage of concrete, thus improving the crack resistance of concrete, simplifying the temperature control measures, shortening the construction period, and thus providing a better solution for the crack resistance of concrete in rockfill dam panels.

2 Experimental Programs

2.1 Raw Materials

Cement: The cement used is Lafarge P.O 42.5 ordinary Portland cement produced in Shuicheng, Guizhou. Density 3.06g/cm^3, specific surface area 347m^2/kg, standard consistency water consumption 25.6%, stability qualified, the main chemical composition of cement is shown in Table 1.

Magnesium oxide: The MgO used in the test is produced by Wuhan Sanyuan Special Building Materials Co., Ltd, with a density of 3.34g/cm^3, activity index of 215s, fineness of 200 mesh, and quality in accordance with the technical indexes of the Technical Requirements for Quality of Lightly Burnt Magnesium Oxide Materials for Water Conservancy and Hydropower Projects (for Trial Implementation).MgO is added to the cement for mixing by dry blending in this project. Its main chemical composition is shown in Table 2.

Table 1. Chemical composition of Shuicheng P.O 42.5 cement (mass fraction)/%.

CaO	S_iO_2	Al_2O_3	Fe_2O_3	MgO	SO_3	Na_2O	K_2O	I.L
60.17	20.56	5.55	2.83	2.48	2.18	0.17	0.47	4.03

Table 2. Chemical composition (mass fraction)/% of MgO.

MgO	SiO_2	CaO	Fe_2O_3	Al_2O_3	SO_3	I.L
93.36	2.05	0.77	0.62	0.36	0.12	1.74

Polypropylene fibres: The polypropylene fibres used in the tests were produced by Changsha Zhengde Building Materials Technology Co. Polypropylene fiber in this project is added to the water by means of wet mixing for agitation.

Aggregate: The coarse and fine aggregates used for the test were crushed stone and artificial sand processed from the chert at a hydropower station site in Guizhou. The coarse aggregate was processed into small stones (5–20 mm diameter) and medium stones (20–40 mm diameter) at the site. The fineness modulus of the artificial sand is 2.80, which satisfies the requirements of the Specification for Hydraulic Concrete Construction (SL677-2014) and belongs to medium sand, with particle gradation belonging to zone II and particle gradation qualified.

Additives: Naphthalene high efficiency water reducing agent produced by Chongqing Sangsheng Company, the quality of which meets the current standard.

2.2 Mix Proportions

The concrete proportions for the laboratory mixes are shown in Table 3. The dosages of raw materials used in the mix is designed according to MgO content (0, 6%, 8%, 10%), polypropylene fiber content (0, 0.9 kg/m^3, 1.2 kg/m^3, 1.5 kg/m^3) water-cement ratio (0.4), sand rate of 38%, aggregate gradation (secondary grading), admixture content (0.8%).The concrete mixes are again classified according to the purpose and content of the tests required for the different studies,as shown in Table 4. Where the MgO admixture was calculated as a percentage by mass of the total amount of cementitious material. During the mixing of the concrete, the mixture was controlled to achieve the same slump (30 to 70 mm) by adjusting the admixture dosage.

Table 3. Amount of raw materials for concrete (kg/m^3).

Group ID	Water	Cement	Admixture	Sand	Small stone	Medium stone	MgO	Polypropylene fibers
KTI	132.00	330.00	2.64	755.44	493.02	739.54	0.00	0.00
KT2	132.00	330.00	2.64	755.44	493.02	739.54	0.00	0.90
KT3	132.00	330.00	2.64	755.44	493.02	739.54	19.80	0.00
KT4	132.00	330.00	2.64	755.44	493.02	739.54	19.80	0.90
KT5	132.00	330.00	2.64	755.44	493.02	739.54	26.40	0.90
KT6	132.00	330.00	2.64	755.44	493.02	739.54	33.00	0.90
KT7	132.00	330.00	2.64	755.44	493.02	739.54	19.80	1.20
KT8	132.00	330.00	2.64	755.44	493.02	739.54	26.40	1.20
KT9	132.00	330.00	2.64	755.44	493.02	739.54	33.00	1.20
KT10	132.00	330.00	2.64	755.44	493.02	739.54	33.00	1.50

Table 4. Concrete proportioning parameters for different MgO and polypropylene fibre admixtures.

Group ID	MgO/%	polyethylene fiber /kg/m^3	water-cement ratio	aggregate gradation
KT1, KT3	0, 6	0	0.4	secondary grading
KT2, KT4, KT5, KT6	0, 6, 8, 10	0.9	0.4	secondary grading
KT7, KT8, KT9	6, 8, 10	1.2	0.4	secondary grading
KT10	10	1.5	0.4	secondary grading

2.3 Test Conditions and Methods

Using the same raw materials, fixed water-cement ratio (0.4), fixed sand rate (38%), varying the MgO admixture (0, 6%, 8%, 10%), varying the polypropylene fibre admixture (0, 0.9 kg/m^3, 1.2 kg/m^3, 1.5 kg/m^3) and forming standard specimens of autogenous volume deformation of secondary aggregate grading concrete in accordance with the current test protocols (200 mm diameter, 500 mm height).

(1) Preparation before forming: Weigh the raw materials according to the concrete mix ratio, the amount of materials by mass, there are cement, magnesium oxide, polypropylene fibre, admixture, sand, small stone and medium stone. Strictly check the sealed specimen barrel, No water penetration and no air permeability.
(2) Forming process: The specimen is formed in accordance with the current "Hydraulic Concrete Test Procedure (SL352-2020)", Pour the admixture into the water and stir evenly with a fine wire, and all the weighed raw materials except the admixture are poured into the concrete mixer for mixing. After mixing for 90s, pour the water with admixture into the mixer again, and finally mix for 150s, and then manually mix for 2–3 times to make it even, to complete the mixing of concrete. The freshly mixed concrete is loaded into the test barrel in three layers as quickly as possible, then the DI-15 strain gauge is buried vertically in the centre of the barrel (taking care not to damage the strain gauge) and the barrel is then placed on the vibrating table to vibrate the concrete densely. After forming the specimen, the cover of the barrel should be covered as soon as possible and the perimeter of the barrel and the outlet of the strain gauge cable should be sealed with glass glue to avoid loss of moisture and to finalise the forming of the specimen (two groups of concrete specimens of the same ratio).The test procedure is shown in Fig. 1 and Fig. 2.
(3) Specimen curing: Number the top of the above formed autogenous volume deformation concrete specimen barrel and place the specimen barrel in a constant temperature (20 ± 2 °C) and moisture-proof environment for curing, avoiding vibration of the concrete specimen during the curing process.
(4) Calculation of autogenous volume deformation: The autogenous volume deformation concrete test is measured according to the frequency specified in the test procedure "Hydraulic Concrete Test Procedure (SL352-2006)": 2 h, 6 h, 12 h and 24 h after forming with SQ-5 digital proportional bridge to measure the resistance (R)

and resistance ratio (Z) of the specimen, of which the measured value at 24 h is the reference value. The measurements were taken once a day for the next two weeks, then once or twice a week, and once or twice a month after six months. The results of the autogenous volume deformation values of concrete compounded with MgO and polypropylene fibres were calculated according to the formula (1–2) for autogenous volume deformation of concrete in the specification "Test Procedure for Hydraulic Concrete (SL352-2006)" (the results are accurate to 1×10^{-6}), with the calculated value of autogenous volume deformation of each concrete being the average of two sets of concrete specimens with the same mix proportion.

$$Gt = f\,(Z - Z0) + (b - a)(T - T0) \tag{1}$$

$$T = a'(R - R0) \tag{2}$$

where:

Gt - autogenous volume deformation of concrete, 10^{-6}.

f - strain gauge strain sensitivity, $10^{-6}/(0.01\%)$;

Z - resistance ratio of the autogenous volume deformation measurement, 0.01%;

Z0 - reference value of the resistance ratio for autogenous volume deformation measurement, 0.01%;

b - strain gauge temperature compensation coefficient, $10^{-6}/°C$.

a - coefficient of linear expansion of concrete, $10^{-6}/°C$;

T - measured temperature, °C;

T0 - temperature reference value, °C;

a′ - strain gauge temperature sensitivity coefficient, °C/Ω;

R - observed resistance, Ω;

R0 - 0 °C resistance, Ω.

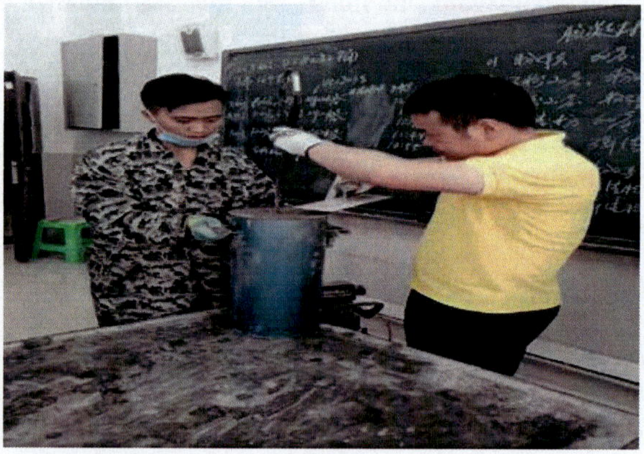

Fig. 1. Experimental procedure for the autogenousvolume deformation of concrete.

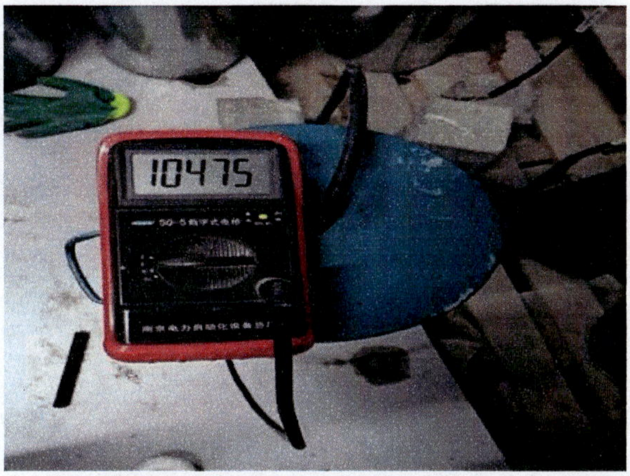

Fig. 2. Measure the data of autogenous volume deformation of concrete.

3 Results and Discussion

3.1 Experimental Result

The results of the autogenous volume deformation test of the compound concrete are shown in Figs. 3, 4, 5, 6, 7, 8, 9, 10, 11, and 12 and Table 5. Figures 3, 4,, 5, 6, 7, 8, 9, 10, 11, and 12 represent the autogenous volume deformation curves of the concrete, Table 5 represents the autogenous volume deformation values at typical ages and the observed ages of the autogenous volume deformation test are 1 d, 3 d, 7 d, 14 d, 28 d, 90 d and 180 d. Figure 3 shows the autogenous volume deformation curves for concrete without MgO and polypropylene fibres (base concrete), Figs. 4 and 5 show the autogenous volume deformation curves for concrete with MgO alone and polypropylene fibres alone, and Figs. 6, 7, and 8 show the autogenous volume deformation curves for compound concrete with different MgO admixtures at 0.9 kg/m^3 of polypropylene fibres. Figures 9, 10, and 11 show the autogenous volume deformation curves for compound concrete with different MgO admixtures at 1.2 kg/m^3 of polypropylene fibre and Fig. 12 shows the autogenous volume deformation curves for compound concrete with 1.5 kg/m^3 of polypropylene fibre and 10% of MgO.

3.2 Analysis of Test Results

From Fig. 3 and Table 5, it can be seen that the autogenous volume deformation of the base concrete without MgO is negative, i.e. the concrete is shrinking. The autogenous volume deformation values of concrete mixed with MgO alone or with MgO and polypropylene fibres minus the autogenous volume deformation values of concrete not mixed with MgO are positive, and the difference is the autogenous volume expansion caused by MgO in concrete. This indicates that the delayed micro-expansion of MgO concrete is objective [8, 9].

Table 5. Results of autogenous volume deformation tests (1×10^{-6}).

Group ID	1 d	3 d	7 d	14 d	28 d	90 d	180 d
KT1	0.01	−20.69	−20.60	−33.47	−42.61	−49.93	−70.38
KT2	0.00	−9.06	−26.94	−35.94	−50.22	−50.69	−66.91
KT3	−0.01	−7.37	−5.10	−5.39	−3.85	10.98	10.59
KT4	0.00	−10.29	−7.77	−7.70	−6.17	−0.63	0.10
KT5	0.00	−5.52	4.53	7.36	12.79	36.41	43.12
KT6	−0.01	−2.68	6.45	13.03	17.29	45.26	60.60
KT7	0.00	−8.38	−10.15	−13.02	−11.82	7.03	10.5
KT8	0.00	−4.07	−1.06	1.74	3.43	25.32	33.27
KT9	0.00	−4.32	4.87	11.49	20.92	46.08	60.33
KT10	−0.02	−3.22	4.63	11.01	14.02	46.19	59.62

(1) The incorporation of MgO has a great influence on the autogenous volume deformation of concrete. That is to say, the concrete mixed with MgO can increase the expansion deformation of concrete, as shown by the value of the autogenous volume deformation of concrete mixed with MgO is larger than that of concrete not mixed with MgO. As Fig. 3 ((1), (3)–(6)) in the age of 180 d, The autogenous volume deformation of concrete without MgO is -70.38×10^{-6}, and increases to 10.59×10^{-6}, 43.12×10^{-6} and 60.60×10^{-6} for concrete with 0.9 kg/m^3 polypropylene fibres after incorporation of 6%, 8% and 10% MgO, an increment of 80.97×10^{-6}, 113.5×10^{-6}, 13.98×10^{-6}.

This is because concrete mixed with MgO has swelling properties, which are mainly derived from the swelling properties of MgO itself [10–12]. MgO swelling is mainly caused by the hydration products formed after its hydration, which are magnesium hydroxide ($Mg(OH)_2$) crystals. The driving force for its swelling is the swelling force of the $Mg(OH)_2$ crystals and the growth pressure of the crystals, the former being the dominant factor for early hydration and the latter for later growth. The appearance of this crystal and its growth with the prolongation of the curing time led to the further development of the expansion pressure of the crystal, which caused the expansion effect of the concrete in the corresponding area. In this test, MgO plays a role to some extent in absorbing the shrinkage deformation of the concrete, i.e. the swelling properties of MgO compensate for the shrinkage of the concrete, i.e. MgO increases the autogenous volume deformation of the concrete.

(2) The addition of polypropylene fibres alone had a small effect on the autogenous volume deformation of the concrete. From Fig. 3 ((1)–(3)) and Table 5, it can be

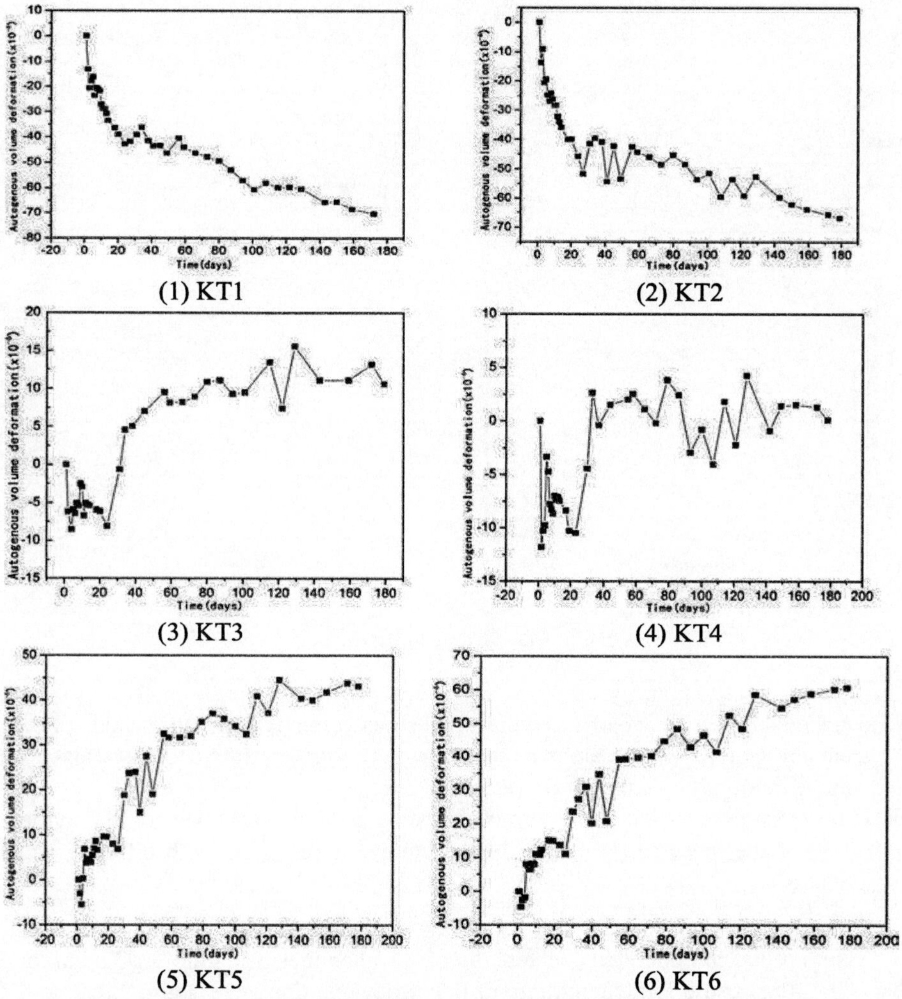

Fig. 3. Autogenous volume deformation curve of concrete for KT1-KT10.

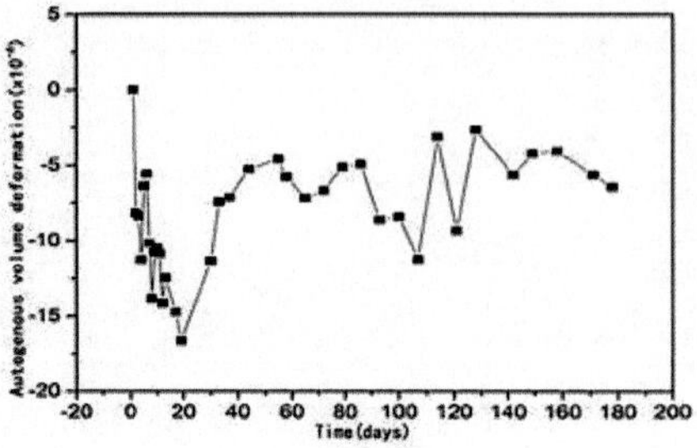

(7) KT7

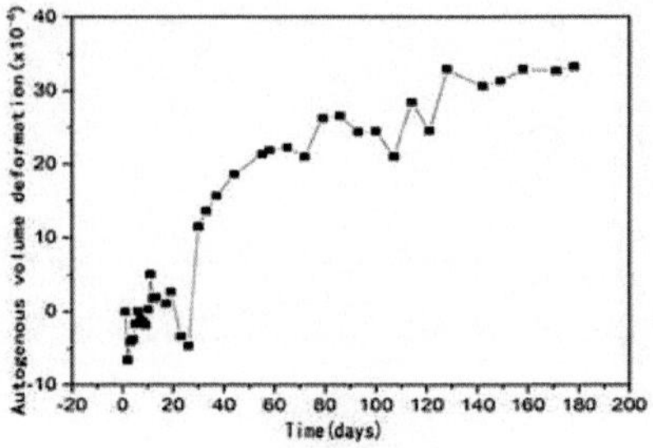

(8) KT8

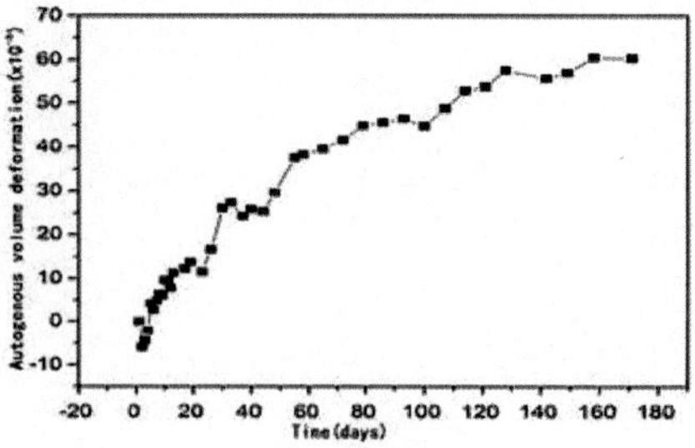

(9) KT9

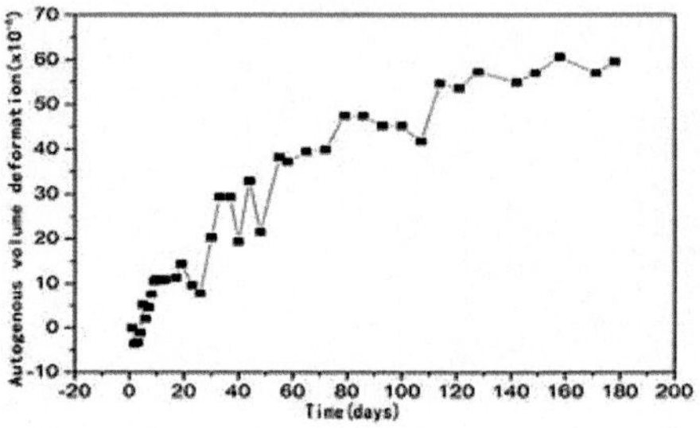

(10) KT10

Fig. 3. *(continued)*

concluded that at an age of 180 d the autogenous volume deformation value of concrete with polypropylene alone increases by 3.47×10^{-6} relative to the autogenous volume deformation value of the base concrete.

(3) Different polypropylene fibre admixtures have a significant effect on the autogenous volume deformation values of the compound concrete with different MgO admixtures.

From Fig. 3 ((7)–(9)) and Table 5, it can be seen that the autogenous volume deformation values of the compounded concrete increased when increasing the MgO admixture (6%, 8%, 10%) at the same admixture of polypropylene fibers (1.2 kg/m^3), e.g. at the age of 90 d, the autogenous volume deformation values of the compounded concrete were 7.03×10^{-6}, 25.32×10^{-6}, 46.08×10^{-6}. At age 180 d, the autogenous volume deformation values of the compound concrete were 10.5×10^{-6}, 33.27×10^{-6} and 60.33×10^{-6}, with increments of 22.77×10^{-6} and 49.83×10^{-6} respectively.

It can be seen from Fig. 3 ((6), (9), (10)) and Table 5 that the autogenous volume deformation values of the compounded concrete increase when the amount of polypropylene fibre is varied at the same MgO dose (10%). For example, at the age of 90 d, the autogenous volume deformation value of compound concrete increases with the increase of polypropylene fibre admixture (0.9 kg/m^3, 1.2 kg/m^3, 1.5 kg/m^3) to 45.26×10^{-6}, 46.08×10^{-6}, 46.19×10^{-6}, with an increment of 0.82×10^{-6}, 0.93×10^{-6} respectively; That is, while keeping the MgO admixture constant, increasing the admixture of polypropylene fibres in the compound concrete can still improve the autogenous volume deformation value of the compound concrete to a certain extent, by 1.81% and 2.1%.

Polypropylene fibres are effective in preventing and reducing cracking by the mechanism that when tens of millions of fibres are added to each m^3 of concrete, the fibres form a disordered support system within the concrete, producing an effective multi-directional secondary strengthening effect, thereby increasing the tensile strength of the cementitious gel [13, 14]; According to the theory of average fibre spacing, the more fibres in the concrete per unit volume, the better the anti-cracking performance of fibre concrete, Dispersing the shrinkage energy of the concrete onto fibre monofilaments with high tensile strength and relatively low modulus of elasticity, preventing the expansion of existing defects (microcracks) in the concrete and retarding the appearance of new cracks, preventing the expansion of the original defects in the concrete (microcracks) and delaying the appearance of new cracks; reducing the capillary channels of the concrete, reducing the rate of loss of moisture from the exposed surface of the concrete. At the same time, the support system formed by the numerous fibre filaments effectively ensures uniform bleeding and prevents settlement from occurring.

4 Conclusions

(1) The delayed micro-expansion characteristics of MgO concrete are objective. MgO plays a role in dissipating the shrinkage deformation of concrete to a certain extent, increasing the autogenous volume deformation of concrete.
(2) MgO or polypropylene fibre alone can increase the autogenous volume deformation value of concrete to a small extent, and the increase in autogenous volume deformation of concrete will be greater with MgO alone than with polypropylene fibres alone.
(3) The compounding of MgO and polypropylene fibres has a significant effect on the autogenous volume deformation value of concrete, and the more obvious with the increase in the amount of both compounded MgO and polypropylene fibres, further revealing that the anti-cracking effect of concrete compounded with MgO and polypropylene fibres can be effectively improved from an experimental point of view.

Acknowledgement. Guizhou Provincial Science and Technology Program Project Grant (Qiankehe Foundation [2020]1Y251); Guizhou Provincial Science and Technology Program Project Grant (Qiankehe service enterprise [2021] No. 4); Guizhou Provincial Water Resources Science and Technology Funding Project (KT202008, KT202207); Guizhou Water Resources Research Institute Scientific Research Project (ZC202005).

References

1. Mehta, P.K., Pirtz, D.: Magnesium oxide additive for producing selfstress in mass concrete. In: Proceedings of the 7th International Congress on the Chemistry of Cement, Paris, vol. III, pp. 6–9 (1980)
2. Li, C.M., Yuan, M.D.: A review of the application of MgO-doped micro-expansion concrete for dam construction technology. Adv. Water Resour. Hydropower Sci. Technol. **23**(6), 57–63 (2003)

3. Jiang, G.C., Zhao, Z.K., Sun, Y.: 20 Years of Concrete Panel Rockfill Dams in China. China Water Conservancy and Hydropower Press, Beijing (2005)
4. Yang, Z.Y.: Causes of cracks and countermeasures in concrete panel rockfill dam projects. China New Technol. New Prod. **12**, 58 (2013)
5. Chen, C.L., Tang, C.S.: Application of magnesium oxide concrete in the foundation of Dongfeng arch dam in Guizhou and analysis of long-term observation results. J. Hydropower Gen. **4**, 102–107 (2006)
6. Xie, Y.J., Zhang, G.B., Feng, J.D.: The causes and countermeasures of cracks in concrete panel rockfill dams. Sichuan Hydropower **26**(5), 76–77 (2007)
7. Zhang, H., Liu, W.R.: Analysis and treatment of concrete panel cracks in Xiaoshankou hydropower Station, Kaidu River, Xinjiang. Build. Mater. Decorat. **37**, 282–283 (2016)
8. Fang, K.H.: Hydration of overburned magnesium oxide and its effect on autogenous volume deformation of concrete. J. Hydropower **23**(4), 46–49 (2004)
9. Chen, C.L.: Influence of the main parameters of the mix ratio on the autogenous volume deformation of exogenous MgO-doped concrete. J. Hydropower **32**(2), 253–256 (2013)
10. Liu, P., Zhao, M.J., Jiang, B.W.: Application study on rigidity correlation method for density inversion of rockfill. Engineering **7**, 331–336 (2015)
11. Zhu, B.F.: Temperature Stresses and Temperature Control of Mass Concrete. China Electric Power Industry Press, Beijing (2000)
12. Gao, P.W., Lu, X.L.: Production of MgO-type expansive agent in dam concrete by use of industrial by-products. Build. Environ. **43**, 453–457 (2008)
13. Lv, Z.D., Wang, J.: Effect of polypropylene fibers on the properties of concrete. Shanxi Water Conserv. Sci. Technol. **2**, 107–109 (2016)
14. Zhang, H.D.: Research on the application of polypropylene fiber concrete in concrete panel rockfill dams. Henan Water Conserv. South-North Water Divers. **2**, 53–54 (2014)

Research on Critical Technology of Cable Hoisting Construction of Large-Span Bridge

Jihua Xiong(✉), Jinguo Jiang, Xu Liu, and Pengcheng Li

Sichuan Road & Bridge (Group) Co., Ltd. (SRBG), Bridge Engineering Branch, Chengdu 610041, China
jihuaxiong@126.com

Abstract. Bridges are important nodes in the transportation network, and the rationality of design and the safety of construction should be paid more attention to. Erecting the main cable strand and hoisting the main girder are important links in the construction of suspension bridges, which will affect the quality and efficiency of the whole construction. To research the critical technology of cable hoisting construction, a truss-steel suspension bridge is investigated. The main cable of the research object is carried out by the Prefabricated Parrel Wire Strand (PPWS) method. And, the erected catwalk system then undertakes the main girder hoisting process. The lineshape and cable length are calculated based on the parabolic theory, and then the tractive force required for the main cable erection is also obtained. The force analysis of carrier cable, hoisting cable, and haulage cable of the catwalk cable hoisting system has been conducted, which ensures the safety of the system. The results show the tractive force of the main cable construction is mainly affected by the reverse tension and gravity. The mid-span sag between the portals will affect the tractive force of winches. Reasonable design can ensure the bearing capacity of the cable when hoisting the main beam. In the discussed study case, the successful application of the cable hoisting system can prove a method and reference to similar construction.

Keywords: PPWS · Parabolic theory · Traction system · Cable hoisting system · Construction and design

1 Introduction

With the development of the economy and society, transportation network plays an irreplaceable role today. As important nodes of the transportation network, the research on the design and construction of bridges has received more and more attention. Suspension bridges can span a long distance, and the rationality of design and the safety of construction is very important [1, 2].

The main cable construction of suspension bridges is the main control step of bridge construction. The main cable construction mainly includes the Air Spinning method (AS) [3] and the Prefabricated Parallel Wire Strand method (PPWS) [4]. The AS method has low requirements for the transportation and erection of traction equipment, which can

W. Guo et al. (Eds.): GBCESC 2023, LNCE 328, pp. 167–183, 2024.
https://doi.org/10.1007/978-981-99-9947-7_18

meet the construction needs of super-long-span suspension bridges with poor transportation conditions [5]. The AS method is used in the construction of the famous Great Belt East Bridge in Denmark [6]. PPWS method can ensure that the steel wire is parallel, with no distortion, and no cross, which can improve the density of the main cable [7]. The PPWS method has been widely used in the construction of long-span suspension bridges and has been continuously developed [8]. Yoo et al. describes the mechanism of the bonding between the Zn-Al-coated wires and Zn-Cu alloy casting medium of the socket terminations of the PPWS method [9]. Seo et al. ultimates load factors of PPWS sockets in main cables of suspension bridges are studied concerning the tapered angles of the inner surface of sockets [10]. Jiang et al. adopts the PPWS method to build the Yangxi main bridge and discuss the process of the construction [11]. Although the PPWS method has been relatively perfect, its mechanism and force analysis still has the value of discussion. Catwalk is an important structure in main cable construction, therefore, it is necessary to carry out corresponding research on the catwalk. A nonlinear numerical method was developed to assess the stability of suspension bridge catwalks under a wind load [12]. Tarui et al. studied steel wire materials for catwalks [13]. The buffeting response of a double-sided catwalk designed for Maputo Bridge was investigated considering wind load nonlinearity, geometric nonlinearity, and self-excited forces [14].

The main girder erection construction of suspension bridges mainly includes the jacking method, scaffolding method, and hoisting method. The suspension bridge girder method can be divided into two types according to the propulsion method of stiffening girder erection: the first is to symmetrically advance from the mid-span position to the direction of both sides of the bridge tower. The second is symmetrical hoisting from the bottom of the main tower to the mid-span. Deng et al. simulates the hoisting process of the main girder using the ANSYS model and studies the steel girder hoisting construction of Yangsi Port YangtzeRiver Bridge, the bearing capacity, applicability, safety, and operation efficiency of 9000 KN hydraulic hoisting cable-borne crane [15]. Zhang et al. studied the hoisting equipment to ensure the safety of the steel box girder structure during large-scale hoisting [16]. Xia et al. put forward a suitable FRP truss bridge construction hoisting technology [17].

This paper investigates the construction of main girders and cables for large-span suspension bridges, one of which is a two-tower, single-span suspension bridge. The main span of the suspension bridge is 550 m long and the main cables are constructed using the PPWS method and the main girders are constructed using the hoisting method. The lineshape of the main cables was calculated using parabolic theory and the traction forces for the construction of the main cables are also calculated. In addition, the carrier cable, hoisting cable, and haulage cable for the erection of the main girders have also been verified. This study can be used as a reference for similar projects. The main contents of the article are, Sect. 2 presents an overview of the project. Section 3 introduces the critical parameters in the construction of the main cables. Section 4 presents the mechanical analysis of the carrier cable, hoisting cable, and haulage cable during the construction of the main girder. Finally, some conclusions are given in Sect. 5.

2 Project Description

To study the technology of construction, a steel-truss girder suspension bridge is taken as the research object, which is across a V-shaped valley. The design elevation of this bridge between the bridge deck slab and the impounded level is large, with 144.232 m. The bridge is a double-tower single-span. The length is 500 m, while the span arrangement is 138.0 + 550 + 131.5 m. The rise-span ratio of the mid-span is 1/10. Besides, the height of the towers is 104.6 m and 141.6 m, respectively. The Prefabricated Parrel Wire Strand (PPWS) is used to install the main cable, erection construction adopts the traction system of the Portal haulage method and the two-wire reciprocation method. The effect picture of the completed bridge is shown in Fig. 1.

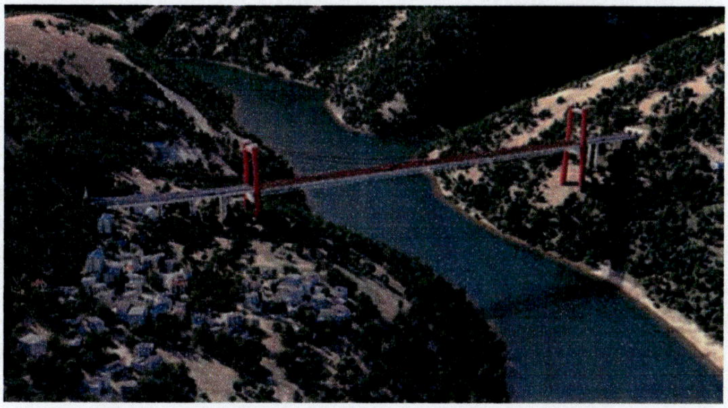

Fig. 1. The effect picture of the completed bridge

The steps of the main cable and girder construction using the cable hoisting system are shown as follows. The first stage is adopting the unmanned aerial vehicle to tractive pilot cable across the river. Then, through the winches at the west bank pull the larger haulage cable to the east bank, and cycle this process until the catwalk cable hoisting system has been completed. The completed system is used to install the catwalk carrier cable in the next stage. After that, the main cable can be erected through haulage cable and catwalk carrier cable. Finally, the main girder can be installed through the erected main cable, hoisting cable, and carrier cable. The layout diagram of the traction system is shown in Fig. 2.

It's noticeable that there are many system conversions process in the construction. Besides, due to the complex terrain conditions in the bridge site area, construction is difficult and involves many construction problems. The construction of the main cable and main girder is an important step in the whole bridge construction step and is also a critical research problem in bridge design. The bridge is therefore of great research value and can be used as a reference for related projects.

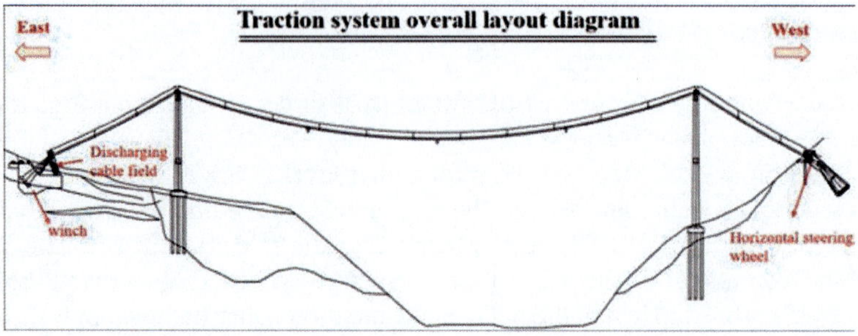

Fig. 2. The layout diagram of the traction system

3 Calculation and Analysis of the Main Cable Traction System

3.1 The Lineshape of the Main Cable

The material of the main cable is the high tensile steel wire rope, which has the typical flexible cable mechanics characteristics. Assuming that the gravity load is uniformly distributed along the chord of the cable, the parabolic theory can be used to calculate the shape of the flexible cable. Besides, the parabolic theory assumes that the gravity load acting on the suspension cable is distributed along the span direction of the suspension cable. The suspension cable is subjected to vertical uniform load $q(x)$, and the geometric position curve of the cable can be expressed by the function $y = y(x)$. The horizontal component of tension at any point is H. Equation (1) can be obtained by taking any suspension segment $d(x)$ as the isolator.

$$H\frac{d^2y}{dx^2} = -q(x) \tag{1}$$

According to the boundary condition, $x = 0, y = 0; x = l, y = c$, the Eq. (1) can be change as,

$$y = \frac{q}{2Hcos\alpha}x(l - x) + \frac{c}{l}x \tag{2}$$

Substituting the coordinates of the mid-span as $(L/2, f)$ into Eq. (2),

$$\frac{q}{H} = \frac{8fcos\alpha}{l^2} \tag{3}$$

The lineshape of the main cable can be calculated according to parabolic theory, as shown in Fig. 3.

3.2 Tractive Force Calculation

Before the catwalk system is completed, the traction system runs overhead between the three spans. To better set up the mid-span catwalk bearing cable and smoothly place the haulage cable into the catwalk portal guide wheel frame, the traction system maintains

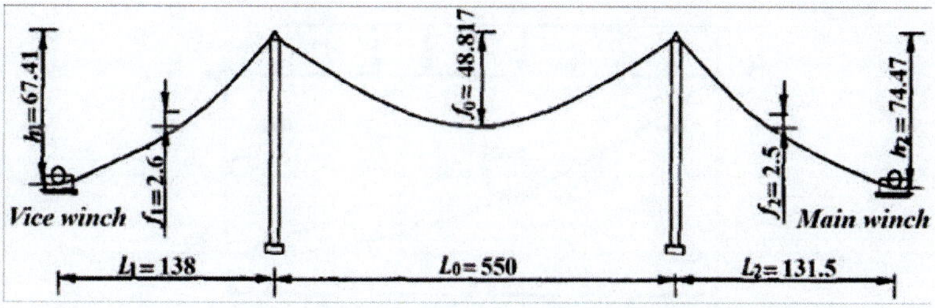

Fig. 3. The lineshape of the main cable

the linearity of the haulage cable during overhead operation. According to the parabolic theory, under uniform load, the single-span cable length S is calculated as shown in Eq. (4) to Eq. (8), and the simplified sketch is shown in Fig. 4.

$$ds = \sqrt{dx^2 + dy^2} = \sqrt{1 + (\frac{dy}{dx})^2}\,dx \tag{4}$$

The length of the whole suspension cable can be obtained by integrating Eq. (4),

$$S = \int_0^l [1 + \frac{1}{2}(\frac{dy}{dx})^2]dx \tag{5}$$

Substituting Eq. (2) into Eq. (5), and then integrating to obtain Eq. (6).

$$S = l + \frac{q^2 l^3}{24H^2 cos^2\alpha} + \frac{c^2}{2l} \tag{6}$$

Substituting Eq. (1) into Eq. (6), the original function of the length of the haulage cable can be obtained.

$$S = l + \frac{c^2}{2l} + \frac{8f^2}{3l} \tag{7}$$

The results of the cable length calculations are shown in Table 1.

During the erection of the catwalk, after the haulage cable has passed the tower with the anchor head of the support cable, the support cable is fixed on the side of the main tower to maintain the lineshape of the support cable. The pelican hook is used to erect the catwalk in the design, which can reduce the catwalk erection span from 550 m to 50 m and reduce the required tractive force of construction dramatically. When the track ropes are erected using the traction system, the system assumes the tension of the $\varphi 36$ wire rope of the traction system and the counterforce of the $\varphi 22$ wire rope of the track carrier cable. The maximum counterforce required by the winch is calculated as in Eq. (8).

$$T_0 = \frac{kqL_0}{2}\sqrt{1 + \frac{L_0^2}{16f_0^2}} \tag{8}$$

where the L_0 is the mid-span horizontal span, which is taken as 550 m. q is the unit weight of the wire rope of the track carrier cable, which is taken as 5.638 kg/m. f_0 is the

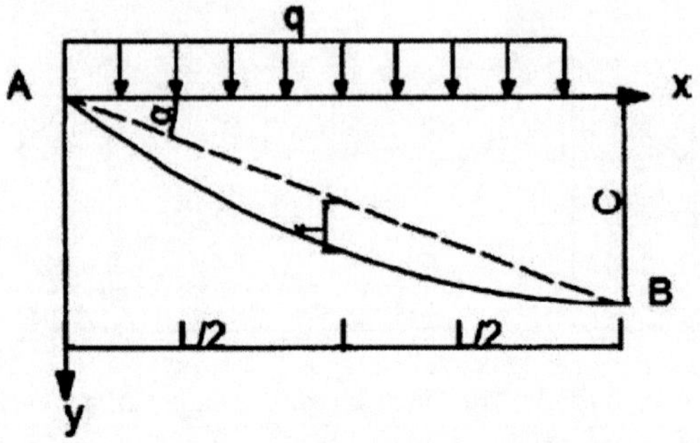

Fig. 4. The simplified sketch of the parabolic theory

Table 1. Cable length calculation results

	L (m)	c (m)	$\frac{c^2}{2L}$ (m)	f (m)	Cable length (m)	Total length (m)
East span	138	67.4	16.5	2.6	154.6	868.9
Mid-span	550	0.0	0.0	48.8	561.6	
West span	131.5	74.5	21.1	2.5	152.7	

Table 2. Tractive force calculation results

	Sag f (m)	Span L (m)	Counterforce T_0(t)	T_1(t)	T_2(t)	Maximum tractive force (kN)
East span	2.6	138.0	2.1	2.2	6.1	102.1
Mid-span	48.8	550.0	1.9	2.0	5.6	92.6
West span	2.5	131.5	2.0	2.1	5.9	98.1

allowable sag, taken as 48.817 m. The k is the dynamic amplification coefficient, taken as 1.2. The maximum tractive force of the winch is shown in Eq. (9).

$$T_{max} = k(T_1 + T_2)f^a \tag{9}$$

where,

$$T_1 = T_z f^b;\ T_2 = \frac{kqL_0}{2}\sqrt{1 + \frac{L_0^2}{16f_0^2}} \tag{10}$$

In this expression, f is the resistance coefficient of the fixed pulley, which is taken as 1.02. a is 2 and stands for the number of fixed pulleys for steering between the puller and the main winches at the splay saddle on the west bank. b is 2 and stands for the number of fixed pulleys for steering between the puller and the vice winches at the splay saddle on the east bank. q takes as 1.96 kg/m and represents the unit weight of the haulage cable. According to the calculations, the tractive force of the winches is shown in Table 2.

3.3 Calculation of Tractive Forces During the Main Cable Erection Stage

The main cable is erected by the two-wire reciprocation system. After the catwalk has been completed, the traction system is then converted. The two-wire reciprocation traction system consists of the portal and the guide-wheel group of the portal. The calculation diagram of the system is shown in Fig. 5.

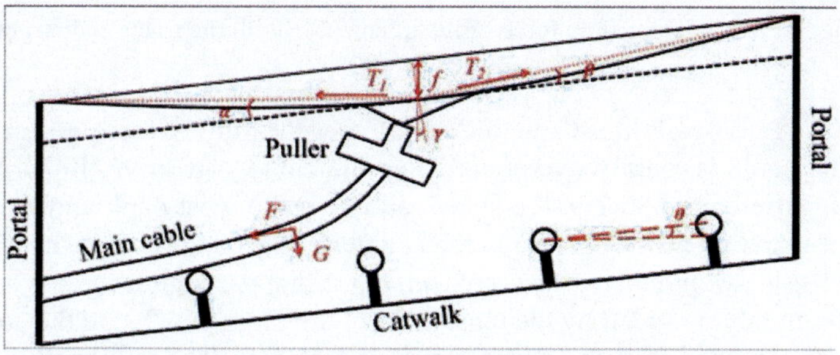

Fig. 5. The calculation diagram of the two-wire reciprocation traction system

The tension T of the winch can be shown by Eq. (11)–Eq. (13).

$$F(x) = \sum_{i=1}^{x/l} qgl(sin\theta_i + fcos\theta_i) \tag{11}$$

$$G = ql_{\text{free}} + g_{\text{puller}} \tag{12}$$

$$T = \frac{F(x)sin\alpha - Gcos(\alpha - \gamma)}{\sin(\alpha + \beta)} \tag{13}$$

where x is the distance during traction. q is the line density of the main cable. l is the distance between the two expansion cylinders of the catwalk. θ_i stands for the horizontal angle of the ligature between adjacent expansion cylinders, which depends on the line-shape of the catwalk. f is taken as 0.2 and represents the friction coefficient of expansion cylinders. l_{free} is the length of the free-end main cable. g_{puller} is 500 kg and the weight of the puller. α, β, γ are all functions about x, which depend on the height of the catwalk portal, the slag of the haulage cable, and the height difference between the adjacent portal. To analyze the effect of slag on tension, calculations were made using drapes of 0.3 m, 0.5 m, 0.8 m, 1 m, and 2 m, respectively. The variation of the counterforce force of the main cable with the traction distance is shown in Fig. 6. The change in the haulage force of the winch with the distance is shown in Fig. 7.

From Fig. 6 it can be obtained that the counterforce is about 1.3 t maximum when the cable is towed to the main tower. The changing trend during the traction process

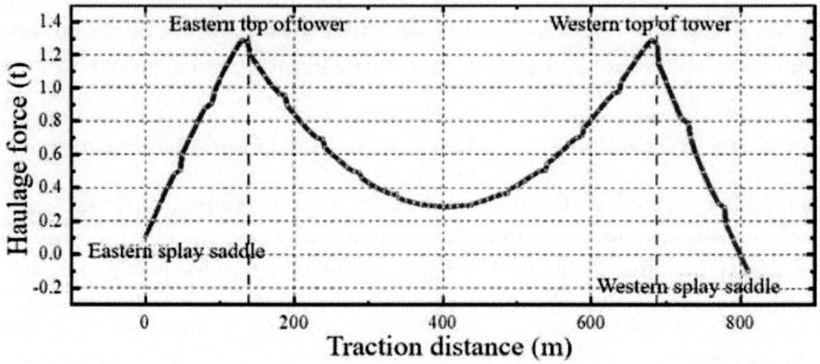

Fig. 6. The counterforce force of the main cable with the traction distance

is affected by the catwalk lineshape, indicating that the sliding force generated by the gravity of the cable is mainly overcome when the cable is tractive. In particular, due to the height difference between the splay saddle on the west bank and the tower top being greater than that on the east bank, the counterforce is slightly less than 0 under the influence of its self-weight. It can be seen from Fig. 5 that the counterforce F and the self-weight G are mainly provided by the tension T_1 of the main winch and the counterforce T_2 of the vice winch. When the counterforce is less than 0, T_2 will increase, so it is necessary to confirm that the counterforce T_2 of the vice winch does not exceed its rated tension.

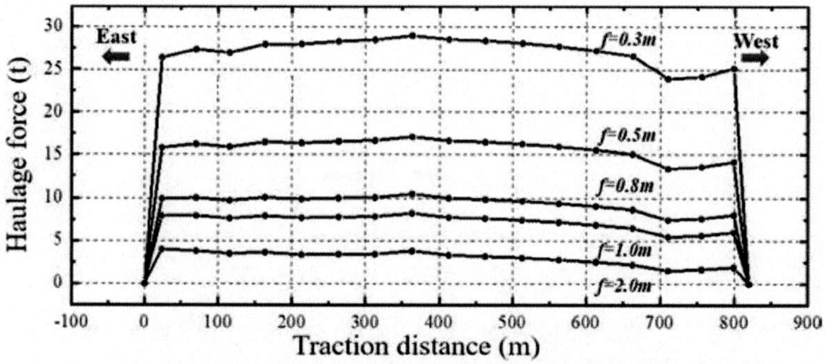

Fig. 7. The haulage force of the winch with the distance

Assuming a maximum traction force of 25 t for the winch, the maximum traction force of 29 t can be obtained from Fig. 7 when the slag is 0.3 m, which exceeds the rated tension. When the sag is 0.5 m, the maximum traction force is about 15 t. This shows that the mid-span sag between the portal will affect the traction force of the winch, but it has little effect on the variation trend of the traction process.

4 Force Analysis of the Catwalk Cable Hoisting System

4.1 Critical Parameters of the Cable Hoisting System

The cable hoisting system uses 2 × 10 φ60 mm wire ropes (6 × 36WS + IWR-1960) as carrier cables. The main cable anchorage point is set on the foundation of the splay saddle of the left and right anchorages. The anchorage form of an anchorage steel strip and pulley is adopted. Besides, adjusting the sag in the main cable installation process. The φ36 mm wire rope (6 × 37SW + IWR-1870) is used to take 12 lines as the hoisting cable. A hoisting system is set up on the left and right sides, including roadsters with 20 wheels. The φ36 wire rope (6 × 36SW + IWR-1870)is used, and four lines are taken on one side as the haulage cable. The specific wire rope parameters are shown in Table 3.

Table 3. Wire rope parameters

	Carrier cables	Hoisting cable	Haulage cable
Specification	φ60	φ36	φ36
	6 × 36WS + IWR	6 × 37WS + IWR	6 × 38WS + IWR
Wire rope routing	2 × 10	2 × 12	2 × 4
Unit weight (kg/m)	15	5.42	5.42
Nominal tensile strength(Mpa)	1960	1870	1870
Breaking force (kN)	2510	863	863
Length (m)	900	2500	4000
Modulus of elasticity (Mpa)	1.1 × 10^5	1.0 × 10^5	1.0 × 10^5
Cross-section area (mm^2)	1764	503	503

The main cable load of the cable crane includes two types, which are concentrated load and uniform load.

1. Concentrated load

The weight of the girder segment is mainly divided into five types. The first type is a mid-span unit, weighing 172.7 t; The second type is a near mid-span standard unit, weighing 206.6 t. The third type is the near tower unit, which weighs 215.2 t on the east bank and 195.8 t on the west coast. Because the erection of the main girder is a dynamic process. The girder type of representative girder erection stage is the midspan closure unit, which is taken as $Q_{1,Mid}$=172.7 t and the main tower closure unit, which is taken as $Q_{1,East}$=215.2 t, $Q_{1,West}$=195.8 t.

The load of the roadsters mechanism and the traction mechanism of the hoisting trolley is Q_2 = 9.487t = 93.0 kN. The load of upper pylons, lower pylons, hangers, and hoisting devices is Q_3 = 3.9 + 2.2 + 7.2 = 13.358 t = 130.9 kN. Each group of hoisting cables goes 12 lines, 2 single pieces. The formulated minimum working height is 18.4 m at mid-span, while the maximum is 62 m near the main tower. So, the maximum coiling length of the unilateral hoisting cable pulley group is 12 × 62 =

744.0 m, while the minimum length is 12 × 18.4 = 220.8 m. Then, the concentrated load of the carrier cable near the roadsters is,

$$Q_{4min} = \frac{5.42}{1000} \times 220.8 \times 2 \times 9.806 = 23.47\text{kN}$$

$$Q_{4max} = \frac{5.42}{1000} \times 744 \times 2 \times 9.806 = 79.08\text{kN}$$

According to the code, which names the Design Rules of Cranes [18], the hoisting dynamic load coefficient can be determined. Then, the concentrated load can be further obtained.

$$Q_{Z,Mid} = 1.05 \times (1727 + 130.9) + 93 + 23.47 = 2067.265\ \text{kN}$$
$$Q_{Z,East} = 1.05 \times (2152 + 130.9) + 93 + 79.08 = 2569.125\ \text{kN}$$
$$Q_{Z,West} = 1.05 \times (1958 + 130.9) + 93 + 79.08 = 2365.425\ \text{kN}$$

2. Uniform load
   The uniform load of the carrier cable is $q_1 = \frac{15}{1000} \times 2 \times 10 \times 9.806 = 2.942\text{kN/m}$. And the hoisting cable is $q_2 = \frac{5.42}{1000} \times 2 \times 12 \times 9.806 = 1.276\text{kN/m}$.

Table 4. Summary of load

Item		Parameters
Calculated length L		550 m
Maximum slag in the working conditions f		43.617 m
The tilt angle of the mid-span main cable		0
Concentrated load	The weight of the main segment girder Q_1	$Q_{1,Mid} = 2067.265\text{kN}$ $Q_{Z,East} = 2569.125\text{kN}$ $Q_{Z,West} = 2365.425\text{kN}$
	The roadsters mechanism and the traction mechanism of the hoisting trolley Q_2	93 kN
	upper pylons, lower pylons, hangers, and hoisting devices Q_3	130.9 kN
Uniform load	Carrier cable q_1	2.942 kN/m
	Hoisting cable q_2	1.276 kN/m
	Haulage cable q_3	0.425 kN/m
	Total	4.643 kN/m

Each group of tractive cable takes 4 lines, so the load can be obtained as $q_3 = \frac{5.42}{1000} \times 2 \times 4 \times 9.806 = 0.425\text{kN/m}$.
Finally, the total uniform load is calculated as $q = q_1 + q_2 + q_3 = 4.643\text{kN/m}$. Besides, the summary of the all load is listed in the Table 4.

4.2 Carrier Cable

The installation of the main girder of the cable hoisting system is a dynamic process. If the carrier cable is calculated for each position, it will only increase the workload and have little effect. So, two representative working conditions are selected for calculation, namely, the full-load condition and the roadsters in the mid-span condition.

The Mid-Span Full-Load Working Condition

1. Calculation of carrier cable self-breaking force

   Firstly, the horizontal component of the force of the carrier cable is,

$$H = \frac{qL^2}{8fcos\beta} + \frac{QL}{4f} = \frac{4.643 \times 550^2}{8 \times 43.617 \times 1} + \frac{2067.265 \times 550}{4 \times 43.617} = 10542.04\text{kN}$$

   And, the vertical component of the force of the carrier cable is,

$$V = \frac{qL}{2cos\beta} + \frac{Q}{2} + Htg\beta = \frac{4.643 \times 550}{2 \times 1} + \frac{2067.265}{2} + 0 = 2310.46\text{kN}$$

   The maximum tension can be calculated as follows,

$$T_{max} = \sqrt{H^2 + V^2} = 10792.26\text{kN}$$

   If the safety of the carrier cable is guaranteed, the safety factor corresponding to the breaking force needs to be guaranteed. The calculation of the breaking force of the carrier cable is as follows, and it can be seen that the safety factor can meet satisfaction. $K = \frac{[T]}{T_{max}} = \frac{20 \times 2510}{10792.26} = 4.651 > 3$, where the allowable force $[T]$ is the total breaking force of 20 carrier cables.

2. Calculation of carrier cable bending stress

   The maximum bending stress that the carrier cable material can bear when it breaks under the concentrated load or reaches the specified bending moment should be checked. The calculation formula is as Eq. (14).

$$\sigma_{\text{Bending}} = \frac{T_{max}}{A} + \frac{Q}{n}\sqrt{\frac{E_K}{T_{max}A}} \tag{14}$$

   where T_{max} is 10792.26 kN and represents the maximum tension of carrier cable. Q is the concentrated load of carrier cable, which is taken as 2067.27 kN according to Table 4. E_K is the modulus of elasticity of the carrier cable, with 1.1×10^5 Mpa. A stands for the cross-sectional area of the carrier cable, and it can be calculated as $A = 20 \times 1764 = 35280\ \text{mm}^2$. n is a total of 40 roadsters wheels of a single cable system. Then, the bending stress of the carrier cable can be obtained.

$$\sigma_{\text{Bending}} = \frac{10792.26 \times 10^3}{35280} + \frac{2067.27 \times 10^3}{40}$$
$$\sqrt{\frac{1.1 \times 10^5}{10792.26 \times 10^3 \times 35280}} = 333.682\ \text{Mpa}$$

Also, the bending stress needs to be verified through the safety factor, $K = \frac{[\sigma]}{\sigma_{max}} = \frac{1960}{33.682} = 5.874 > 2$, the safety factor meets satisfaction. Where the $[\sigma]$ stands for the nominal tensile strength of carrier cable in the above expression.

3. Calculation of carrier cable contact stress

The contact stress caused by the deformation of the pulley during contact is worthy of attention. The calculation equation of the contact stress is shown in Eq. (15).

$$\sigma_{\text{Contact}} = \frac{T_{max}}{A} + E_k \frac{D_W}{D} \tag{15}$$

In this expression, T_{max} is 10792.26 kN and stands for the maximum tension of the carrier cable. A stands for the cross-sectional area of the carrier cable, and it can be calculated as $A = 20 \times 1764 = 35280\ \text{mm}^2$. E_K is also taken as 1.1×10^5 Mpa, which stands for the modulus of elasticity of the carrier cable. Besides, D_W is the diameter of the carrier cable wire, which is taken as 2.4 mm. While D is 630 mm and is the diameter of the pulley. Take the above parameters into Eq. (15).

$$\sigma_{\text{Contact}} = \frac{10792.26 \times 10^3}{35280} + 1.1 \times 10^5 \times \frac{2.4}{630} = 724.95\text{Mpa}$$

In addition, the safety factor of contact stress should be verified, which can be calculated as $K = \frac{[\sigma]}{\sigma_{max}} = \frac{1960}{724.95} = 2.704 > 2$ and the safety factor meets satisfaction. Where the $[\sigma]$ is the nominal tensile strength of the carrier cable.

The Roadsters in the Mid-Span Working Condition

The maximum tension at the maximum hoisting weight in the span is substituted into the deformation coordination tension equation to solve the horizontal force and sag of the carrier cable when there is only a roadster. It should be noted that the carrier cable is under concentrated load at this time, the concentrated load is caused by the roadster with girders and part of the hoisting cable in the roadster.

1. The tension of the carrier cable

When the roadster acts in the mid-span, H_1 is the initial tension. The corresponding solving equation is shown in Eq. (16).

$$\begin{aligned} &H_1^3 + H_1^2 \left\{ \frac{E_K A cos^2\beta}{24H_m^2} \left[3Q(Q+G) + G^2 \right] - H_m \right\} \\ &-\frac{1}{8} Q_1 (Q_1 + G_0) E_K A cos^2\beta - \frac{G_0^2 E_K A cos^2\beta}{24} = 0 \end{aligned} \tag{16}$$

E_K is also taken as 1.1×10^5 Mpa and stands for the modulus of elasticity of the carrier cable. The meaning of A is similar to A in Eq. (15). Q is the total load when the full-load condition, which is taken as 2067.27 kN according to Table 4. Q_1 is the load when only the roadster and hoisting cable is located in the mid-span and the main girder segment has been completed. Q_1 can be calculated as $Q_1 = Q_2 + Q_{4min} = 116.47$kN. G is the self-weight of the carrier cable, hoisting cable, and haulage cable, this parameter is taken as 1618.10 kN. The maximum horizon component force of the carrier cable is 10542.04 kN, it notes as H_m. Because the height of the two towers between the mid-span is the same, $cos\beta = 1$.

Taking the above parameters into Eq. (16), it can calculate the initial tension of the carrier cable when the roadster is in the mid-span, $H_1 = 3098.90\text{kN}$.

2. The slag of the carrier cable

   The solving equation of the slag of the carrier cable is Eq. (17).

$$f_0 = \frac{q_1 L^2}{8H_1 cos\beta} + \frac{Q_1 L}{4H_1} \tag{17}$$

In this equation, q_1 is the uniform load and is taken as 2.942 kN/m. So the slag of the carrier cable can be obtained as $f_0 = \frac{2.942\times550^2}{8\times3098.9} = 35.90\text{m}$.

4.3 Hoisting Cable

1. Calculation of hoisting cable breaking force

   The rated hoisting weight of the dynamic load impact coefficient should be considered first in hoisting cable tension. Then the reduction effect of the pulley group on tension transmission is calculated. Finally, the hoisting cable tension required to complete the hoisting of the main girder segment is obtained. Among them, the rated hoisting weight includes the weight of the hoisting girder section, the weight of the hanger, and the weight of the hoisting cable below the hanger. Calculated according to the worst-case condition, the hoisting dynamic load impact coefficient should be considered when calculating the tension of the hoisting cable, which is taken as 1.05. Then, the unilateral hoisting load is

$$Q_{\text{Hoisting}} = \frac{1.05(Q_1 + Q_3 + Q_{4max})}{2} == 1459.03\text{kN}$$

The tension of hoisting cable around the winch is,

$$T_{max} = \frac{Q_{\text{Hoisting}}}{n\eta_1^{\mu}\eta_2^{\nu}} = \frac{1459.03}{12 \times 0.98^{15} \times 0.98^2} = 171.411\text{kN}$$

where n is 12 and stands for the number of lines of hoisting cable through the pulley group, η_1 is the efficiency of the pulley group, which takes 0.98. While η_2 is the efficiency of the steering pulley wheel, which takes 0.98. μ is 15 and stands for the number of pulley group wheels. v is 2 and stands for the number of the steering pulley wheel. Then the safety factor of hoisting cable breaking force can be verified as $K = \frac{[T]}{T_{max}} = \frac{863}{171.411} = 5.03 > 5$, which meets the satisfaction.

2. Calculation of hoisting cable contact stress

   The solving equation of the contact stress of the hoisting cable according to Eq. (15). Then the contact stress of the hoisting cable can be obtained as $\sigma_{max} = \frac{T_{max}}{A} + +E_k\frac{D_W}{D} = \frac{171.411\times10^3}{503} + 1.0 \times 10^5 \times \frac{1.8}{630} = 626.49\text{Mpa}$. Where the T_{max} is the tension of the hoisting cable around the winch, which has been calculated as 171.411 kN. A is the cross-sectional area of the hoisting cable, taking 503 mm^2. E_K is taken as 1.0×10^5 Mpa, which stands for the modulus of elasticity of the hoisting cable. D_W is the diameter of the hoisting cable wire, which is taken as 1.8 mm. While D is the diameter of the pulley, which is taken as 630 mm. Besides, the safety factor of the contact stress should be verified, it can be seen as $K = \frac{[\sigma]}{\sigma_{max}} = \frac{1870}{626.49} = 2.98 > 2$. The safety factor can meet satisfaction.

4.4 Haulage Cable

According to the position of the roadster, the inclination angle of the main cable is also different. The inclination angle of the main cable will affect the effectiveness of the haulage force. For example, near the tower, the larger the inclination angle, the greater the haulage force required to make the roadster rise. In the design of the bridge, 10.05 m in front of the tower is directly above the main girder segment assembly site, which is also the closure section of the main girder. Therefore, the hanging girder at 10.05 m in front of the east side tower is calculated as the worst-case position. The calculation force equation of haulage cable is shown in Eq. (18).

$$W = W_1 + W_2 - W_3 - W_4 \tag{18}$$

where W_1 is the weight component of the hoisting load in the tangential direction. W_2 is the natural tension caused by the self-weight section of the haulage cable opposite to the moving direction. W_3 is the needed force to overcome friction resistance when the hoisting roadster moves. W_4 is the motion resistance caused by the gravity of the hoisting cable.

1. Weight component W_1

   The weight component of the hoisting load in the tangential direction can be expressed in Eq. (19).

   $$W_1 = Qsin\gamma = 1284.56 \times \sin 29.36^\circ = 633.90\text{kN} \tag{19}$$

   where there are two haulage cables, so Q is half of the hoisting load, take 1284.56 kN. γ is the main cable inclination angle of 10.05 m in front of the tower, which is taken as 35.62°.
2. Natural tension W_2

   The natural tension caused by the self-weight section of the haulage cable opposite to the moving direction can be expressed in Eq. (20).

   $$W_2 = \frac{q_1 x_1^2}{8f} = \frac{0.2168 \times 50^2}{8 \times 5} = 13.55\text{kN} \tag{20}$$

   In this expression, q_1 is the self-weight uniform load of the haulage cable, the haulage cable has 4 lines, take $4 \times 0.0542 = 0.2168\text{kN/m}$. x_1 is the tension caused by the self-weight of the haulage cable in the opposite direction of movement. In the worst-case calculation, x_1 takes the empirical value of 50 m. f is the mid-span sag of the haulage cable in the opposite direction of movement under the action of self-weight, and the empirical value is 5 m.
3. Overcome friction resistance force W_3

   The needed force to overcome friction resistance when the hoisting roadster moves can be expressed in Eq. (21).

   $$W_3 = \mu Qcos\gamma = 0.01 \times 1284.56 \times \cos 29.36^\circ = 11.20\text{kN} \tag{21}$$

   where μ is 0.01 and stands for the coefficient of resistance between the main cable and wheels of the roadster.

4. Motion resistance W_4

The motion resistance caused by the gravity of the hoisting cable can be expressed in Eq. (22).

$$W_4 = \mathrm{T_{Hosting}}\left(1 - \eta^m\right) = 171.411 \times (1 - 0.98^{15}) = 45.60\mathrm{kN} \tag{22}$$

where $T_{Hosting}$ is the maximum tension generated during the hoisting process of the hoisting cable, it has been calculated as 171.411 kN in Sect. 4.3. η is taken as 0.98 and respects the penetration efficiency of hoisting cable in hoisting wheels. The number of pulleys that the hoisting cable passes through the roadster is 15, which is noted as m. Taking the calculated W_1, W_2, W_3, and W_4 into Eq. (18), it can be obtained that $W = W_1 + W_2 - W_3 - W_4 = 590.65\mathrm{kN}$.

Then the maximum tension of haulage can be obtained as follow, $\mathrm{T_{Haulage}} = (W + 2Lq) \times (2 - \eta^n) = (590.65 + 2 \times 550 \times 0.2168) \times \left(2 - 0.98^2\right) = 861.96\mathrm{kN}$. L is 550 m, which stands for the length. η is taken as 0.98 and respects the penetration efficiency of haulage cable in wheels. The number of pulleys that the haulage cable passes through the roadster is 2, which is noted as m. The haulage cable has 4 lines, so the maximum tension of a single haulage cable is $T_{max} = \frac{\mathrm{T}_{Haulage}}{4} = 215.49\mathrm{kN}$.

The breaking force of the single haulage cable is 863 kN. Using the safety factor to verify the reliability of haulage cable, $K = \frac{[T]}{T_{max}} = \frac{863}{215.49} = 4.005 > 4$, it can meet satisfaction.

The calculation equation for verifying the contact stress is shown in Eq. (15). Then the contact stress of the haulage cable can be calculated as $\sigma_{max} = \frac{T_{max}}{A} + E_k\frac{D_W}{D} = \frac{215.49\times10^3}{503} + 1.0 \times 10^5 \times \frac{1.8}{630} = 714.12\mathrm{Mpa}$. Besides, the safety factor can be verified as $K = \frac{[\sigma]}{\sigma_{max}} = \frac{1870}{714.12} = 2.62 > 2$, which meets satisfaction.

5 Conclusion

This research is based on the structural characteristics of large-span suspension bridges. Taking a double-tower single-span steel truss girder suspension bridge as the research object. The research on the main cable traction system and catwalk cable hoisting system is carried out. The following conclusions can be obtained by studying the corresponding key technologies.

1. The main cable has typical mechanical characteristics of flexible cable, and the parabolic theory can be used to calculate the lineshape of the main cable. In the design, the pelican hook is used to erect the catwalk, and the erection span of the catwalk is reduced from 550 m to 50 m, thus greatly reducing the tractive force required for construction. Similarly, according to the parabolic theory, the tractive force of the catwalk can be calculated.
2. In the process of hauling the main cable, the counterforce and self-weight are mainly provided by the tractive force of the winch. The variant trend of the counterforce is affected by the catwalk lineshape, indicating that the sliding force generated by the gravity of the cable is mainly overcome when the cable is hauled. The height difference between the cable saddle and the tower top will affect the magnitude of the

counterforce, which should be paid special attention to in similar projects. In addition, the mid-span sag between the portals will affect the tractive force of the hoist, but it has little effect on the changing trend of the traction process.
3. Through theoretical calculation, the stress of carrier cable, hoisting cable, and haulage cable in girder hoisting are verified. The safety of the cable hoisting system is guaranteed in the force calculation and analysis. The successful application of the cable hoisting system can provide methods and references for other similar types of cable hoisting design.

References

1. Huang, P., Li, C.: Review of the main cable shape control of the suspension bridge. Appl. Sci. **13**, 3106 (2023)
2. Wang, D., Ye, J., Wang, B., Wahab, M.A.: Review on the service safety assessment of main cable of long span multi-tower suspension bridge. Appl. Sci. **11**, 5920 (2021)
3. Kim, H.-K., Lee, M.-J., Chang, S.-P.: Determination of hanger installation procedure for a self-anchored suspension bridge. Eng. Struct. **28**, 959–976 (2006)
4. Yoo, H., Seo, J.-W., Lee, S.-H., Park, Y.-H.: High-strength prefabricated parallel wire strand for ulsan harbor bridge and its mass production system in Korea. Struct. Eng. Int. **24**, 293–297 (2014)
5. Lee, M., Kim, S., Seo, Y., Kim, J.: The Yi Sun-Sin bridge: innovative solutions for suspension bridges. Struct. Eng. Int. **22**, 32–35 (2012)
6. Ostenfeld, K.: Design of the great belt east bridge. Struct. Eng. Int. **5**, 218–220 (1995)
7. Pan, S., Cui, Y., Zhang, Z., Zhu, W.: Behaviour and design of three-tower, self-anchored suspension bridge with a concrete girder. Proc. Inst. Civil Eng. Bridge Eng. **172**, 190–203 (2019)
8. Son, Y., Lee, C., Yoo, D., Kim, J., Choi, J.: Cheon-sa bridge—the first sea crossing multi-span suspension bridge. Struct. Eng. Int. **31**, 431–434 (2021)
9. Yoo, H., Lee, S.-H., Lee, J.-K.: Mechanism of bonding between Zn–Al-coated wires and Zn–Cu alloy casting medium in hot casting socket terminations for large bridge cables. Constr. Build. Mater. **76**, 396–403 (2015)
10. Yoo, H., 이성형., Seo, J.: Effect of the tapered angle on the ultimate load factors of PPWS sockets in main cables of suspension bridges. KSCE J. Civil Environ. Eng. Res. **33**, 47–59 (2013)
11. Jiang, X., Yu, T.Q.: Construction technology and organization of Jiande Yangxi main bridge. AMR **671–674**, 1988–1992 (2013)
12. Zheng, S., Liao, H., Li, Y.: Stability of suspension bridge catwalks under a wind load. Wind Struct. **10**, 367–382 (2007)
13. Tarui, T., Konno, S., Takahashi, T.: High strength galvanized wire for bridge cables. MSF **426–432**, 829–834 (2003)
14. Li, Z.-G., Chen, F., Pei, C., Zhang, J.-M., Chen, X.: Comfort evaluation of double-sided catwalk for suspension bridge due to wind-induced vibration. Math. Probl. Eng. **2021**, 1–12 (2021)
15. Deng, N., Qin, Z., Liuyi, W.: Design and application of 9000 KN hydraulic lifting cable-borne crane for suspension bridge steel girder erection. IOP Conf. Ser.: Earth Environ. Sci. **510**, 052019 (2020)
16. Zhang, H., Zhang, Y.T., Zhu, H., Gao, J.B.: Research on integral lifting continuous steel box girder bridge construction technology. AMM **256–259**, 1674–1681 (2012)

17. Xia, X.N., Yang, Y.X., Yao, Y., Yue, Q.R., Zhang, P.Y., He, J.: FRP truss bridge construction technology research and application. AMM **275–277**, 966–971 (2013)
18. China Machinery Industry Federation. Design Rules for Cranes. GB/T 3811-2008 (2008)

Check for updates

Numerical Simulation Analysis of the Influence of Recharging Wells on the Settlement of Buildings Surrounding Deep Foundation Pits

Caihaiduojie[1,2](✉), Haifeng Tian[1], and Xugang Yin[2]

[1] Economic and Technological Research Institute of State Grid Qinghai Electric Power Company, Xining 810008, Qinghai, China
caihdj21@lzu.edu.cn

[2] School of Civil Engineering and Mechanics, Lanzhou University, Lanzhou 730000, Gansu, China

Abstract. In the process of excavating the deep pit, due to the change in groundwater level, will inevitably cause uneven settlement of the surrounding buildings. To prevent such uneven settlement and to protect the safety of surrounding buildings, recharging methods are often used in the project. In this paper, with the engineering example, the finite element software is used to simulate the deep excavation dewatering and recharge process, and the influence of different recharge volumes, different lengths of recharge wells, and recharge wells with different positions on the settlement of buildings is studied. The results show that the greater the amount of recharge, the better the control effect on building settlement, but the excessive amount of recharge will lead to the uplift of the surrounding soil of the building. With the same amount of recharge, the longer the length of the recharge well, the smaller the settlement of the building. The recharging well can restore the water level lost at the building due to dewatering. The larger the amount of recharge, the longer the length of the recharging well and the more the water level is restored. When the recharging well is between the building and the dewatering well, the settlement control effect of the recharging well is the best.

Keywords: Deep Foundation Pit · Dewatering · Uneven Settlement · Numerical Simulation Analysis

1 Introduction

With the rapid development of the city, a large number of pit-type buildings have appeared in the dense construction area of the original urban area. In the process of excavation and dewatering, it is inevitable that the buildings around the foundation pits will be unevenly settled due to the fall of the groundwater level. To prevent this uneven settlement and protect the safety of the surrounding buildings, the recharge method is often used in the project. The reason why the recharge method reduces the uneven settlement of the surrounding buildings is that the water level drop outside the pit due to dewatering is restored during the recharge, which also makes the ground settlement reduced [1]. The

W. Guo et al. (Eds.): GBCESC 2023, LNCE 328, pp. 184–195, 2024.
https://doi.org/10.1007/978-981-99-9947-7_19

main factors affecting the ground settlement around the pit are the soil modulus around the pit, the permeability coefficient and the depth of the dewatering wells [2].

In previous studies, the effects of dewatering and recharge on the surrounding surface settlement without surrounding building loads were mostly considered separately [3]. In this paper, take the actual project as an example and conduct finite element numerical simulation of the pre-dewatering process of a deep foundation pit project with buildings in the surrounding area, and analyze the effects of different recharge volumes, different recharge well lengths and different recharge well locations on the water level changes outside the pit and building settlement.

2 Project Overview and Hydrological Conditions

2.1 Project Overview

The project [4] is located in a busy area of the city, the pit plan area is about 27520 m^2, the shape is nearly rectangular, 183 m long and 171 m wide, the excavation depth of the pit is 6.3 m, there are more buildings around the pit (many buildings are sensitive to settlement), the construction of this project requires the groundwater level to be lowered to below 1.5 m of the pit floor.

2.2 Engineering Geological Conditions and Hydrogeological Conditions

The site geomorphic unit is a loess beam depression, and within 30 m depth, the site stratigraphy from top to bottom consists of Holocene artificial fill, Late Pleistocene loess, ancient soil, and Middle Pleistocene loess, with the following stratigraphic soil properties: ① fill, locally 6.9 m thick; ② loess, buried 6.3–11.9 m; ③ ancient soil, buried 14.2 m–16.8 m, ④ loess, buried 19.2–27.2 m. ⑤ powdered clay, for penetrating this layer.

The stable water level of groundwater at the site is 3.6 m–4.5 m, which is a diving type. Groundwater dynamic observation data, the annual variation of groundwater level is about 1.3 m, and the water level is close to the annual higher water level at the time of the survey. The thickness-weighted average of the permeability coefficient of ②–⑤ layers is 4.1 m/d. The construction requires the groundwater level to be lowered to 7.8 m below the current surface.

3 Dewatering Scheme Design

According to the engineering geology and hydrogeological conditions of the site, it is considered difficult to achieve the expected purpose of the pit dewatering by gravity type dewatering using drainage ditches and water collection wells, so, according to the construction experience, the pipe well point dewatering program is selected. Dewatering design parameters: pit dewatering area $A = 27520\ m^2$, pit water level drop depth $s = 3.7$ m, permeability coefficient $k = 4.1$ m/d, well radius $r_w = 0.4$ m, well tube radius $r_s = 0.24$ m, well length is 32 m. Backfilling well design parameters: well radius $r_w = 0.46$ m, well tube radius $r_s = 0.36$ m, well length is 18 m.

4 Calculation Model and Parameters

According to the engineering example, the calculation range of the model is set to 100 m × 40 m (length × width), and the specific model is shown in Fig. 1. 1 is the area of the foundation pit to be excavated, 2 is the dewatering well, 3 is the recharge well, and 4 is the precision machinery processing building.

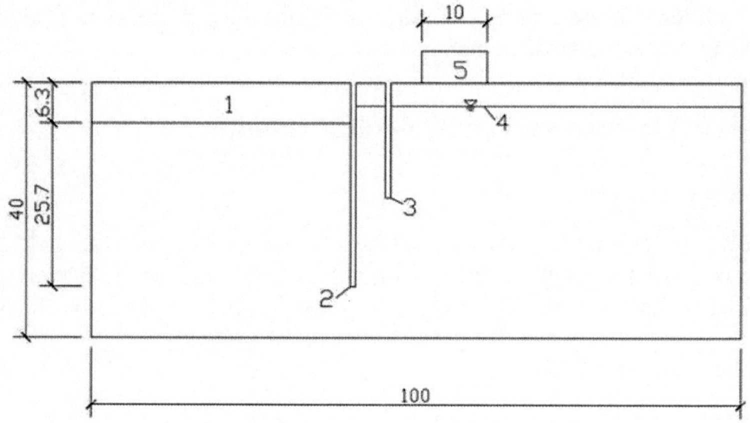

Fig. 1. Calculation model

It is assumed that the plastic yield criterion of the soil material is the Moore-Coulomb criterion, and the soil layer is homogeneous and isotropic, and the permeability of the soil layer is isotropic. The building is converted to a uniform load of 30 kPa and the distribution length is 10 m. The stratum in the area where the project is located is mainly a loess layer, and the soil parameters of the loess layer are used as the soil parameters of the whole model in order to simplify the calculation model. Soil layer calculation parameters are shown in Table 1.

Table 1. Soil calculation parameters

Modulus of deformation/MPa	Poisson's ratio	Internal friction angle/(°)	cohesion/kPa	Permeability coefficient/($m{\cdot}d^{-1}$)
18	0.32	30	17	4.1

In order to study the influence of different recharge volume, different recharge well length and different recharge well location on the building settlement, seven working conditions were designed, and the details are shown in Table 2. The building settlement under different recharge volume was studied by working condition 1, 2, 3 and 4. The effects of different recharge well lengths on the building settlement were studied in Case 4, Case 5 and Case 6. The effect of recharge well location on building settlement was investigated in Case 4 and Case 7.

Table 2. Condition design

Working condition	Dewatering	Recharge	Recharge well length/m	Recharge well location
1	q	–	–	–
2	q	$\frac{1}{3}$ q	18	M
3	q	$\frac{1}{2}$ q	18	M
4	q	q	18	M
5	q	q	25	M
6	q	q	32	M
7	q	q	18	F

Note: In the position of recharge well, M stands for the recharge well between the dewatering well and the building, and F stands for the recharge well on the side of the building away from the dewatering well

5 Numerical Simulation Analysis of Dewatering and Recharge

5.1 Effect of Different Recharge Volume Recharge Wells on Settlement

To be able to analyze the influence of recharge wells on settlement under different recharge volumes, numerical simulations of the pit dewatering recharge process under different recharge volumes were conducted, as shown in Figs. 2, 3, 4, 5, 6, 7, 8, 9, and 10.

The numerical simulation results of Case 1 are shown in Fig. 2 and Fig. 3. In this condition, the recharge well is not recharged, and only the dewatering well is used for dewatering work. Under the dewatering effect of the dewatering well, the pore pressure changes greatly, and a "landing funnel"-like dewatering curve is formed below the tube well. When only the dewatering well is precipitated, the soil on both sides of the dewatering well settles to a certain extent, and the soil on the side of the building is more settled than the soil on the side of the building. According to the engineering data, the settlement of the corner of the building near the pit is about 156 mm, and the numerical simulation results show that the settlement of the corner of the building near the pit is about 155 mm, which is basically similar to the two results and proves that the numerical simulation analysis process is consistent with the engineering reality.

The results of numerical simulation for Case 2 are shown in Fig. 4 and Fig. 5. In Case 2, a recharge well with a recharge volume of about 1/3q is added. The presence of the recharge well causes the settlement of the soil to the right of the dewatering well to be significantly reduced. Compared with Case 1, the settlement at the building location is reduced by nearly 56 mm, which indicates that the recharge well can well reduce the settlement at the building location. There is also a significant difference between the settlement of the soil on the left and right side of the dewatering well. The settlement of the soil on the left side without the recharge well is nearly 90 mm more than that on the right side.

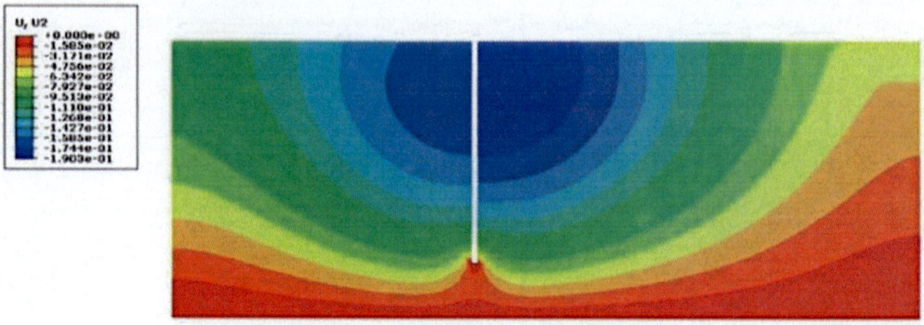

Fig. 2. Settlement change of condition 1

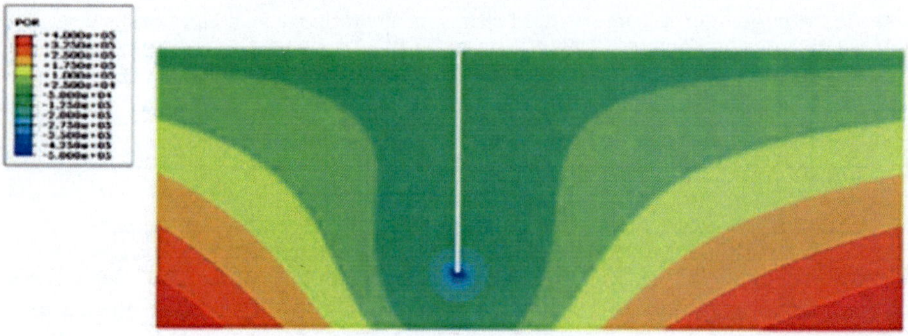

Fig. 3. Change in pore pressure of condition 1

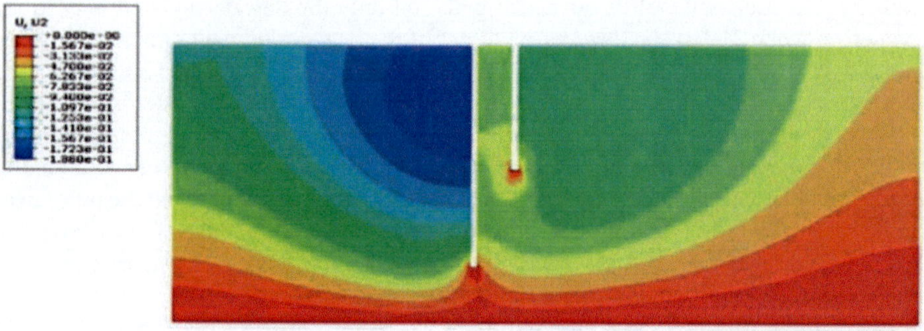

Fig. 4. Settlement change of condition 2

The numerical simulation results of Case 3 are shown in Fig. 6 and Fig. 7. In Case 3, the recharge volume of the recharge well is increased to 1/2q. The increase of the recharge volume also leads to further elevation of the groundwater level, which also leads to further reduction of the settlement. When the recharge volume was increased from 1/3q to 1/2q, the settlement at the building location was reduced by 15 mm.

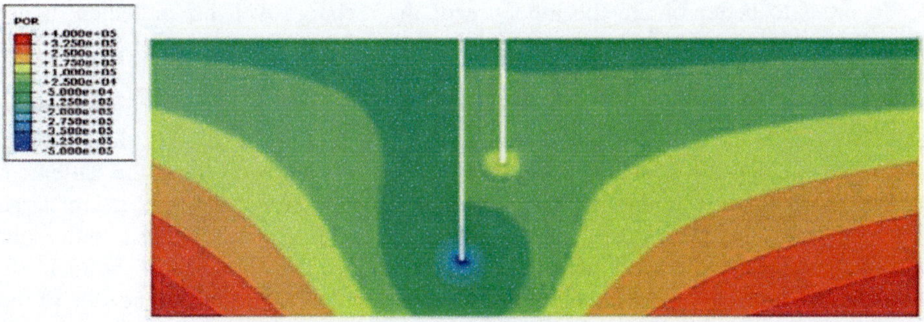

Fig. 5. Change in pore pressure of condition 2

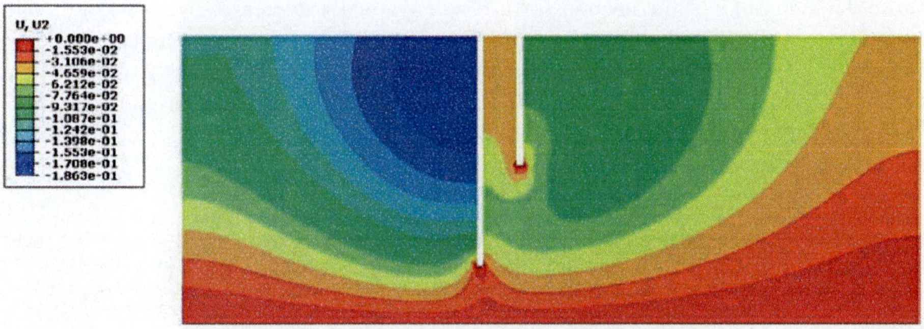

Fig. 6. Settlement change of condition 3

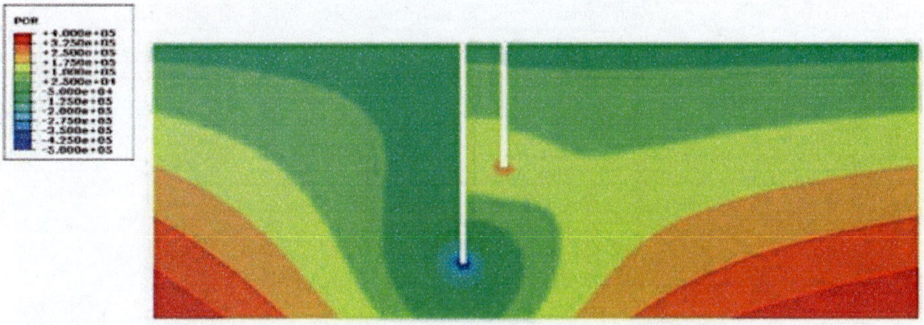

Fig. 7. Change in pore pressure of condition 3

The results of the numerical simulation for Case 4 are shown in Fig. 8 and Fig. 9. In this case, the recharge volume is set to q, i.e., the recharge volume is the same as the dewatering volume. Compared with Case 1, the settlement at the building location is reduced by 100 mm, and this shows that the recharge well has a great inhibiting effect on the settlement of the building around the foundation pit.

In order to better obtain the influence of the recharge well on settlement under different recharge volume, the surface settlement data from the right side of the recharge well to the 3 m range behind the building were selected for each working condition and analyzed as shown in Fig. 10. In the figure, the range of 0–5 m is the section from the recharge well to the building, and the range of 5–15 m is the location of the building. From the results, it can be seen that with the increase of recharge volume, the settlement of the soil around the foundation pit occurs a significant decrease, and when the recharge volume reaches the maximum, the settlement at the building is the smallest. In the four working conditions, the settlement values of the soil on the right side of the recharge well are 157 mm, 96 mm, 82 mm and 46 mm, the settlement values at the side of the building near the pit are 156 mm, 103 mm, 89 mm and 52 mm, and the settlement values at the side of the building far from the pit are 129 mm, 103 mm, 89 mm and 55 mm. Through the above results, it can be found that the further away from the recharge well, the settlement value of the ground surface will gradually decrease. In case 1, although the settlement is reduced in the whole process, the recovery of settlement at the building location is suppressed by the self-weight of the building. In conditions 2, 3 and 4, the settlement value at the building location increases abruptly, and this abrupt change will be recovered when moving away from the building.

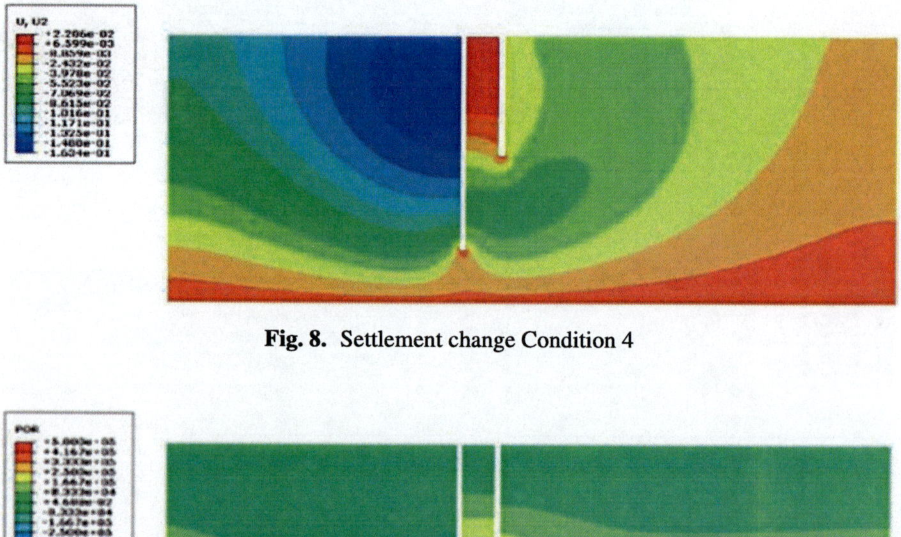

Fig. 8. Settlement change Condition 4

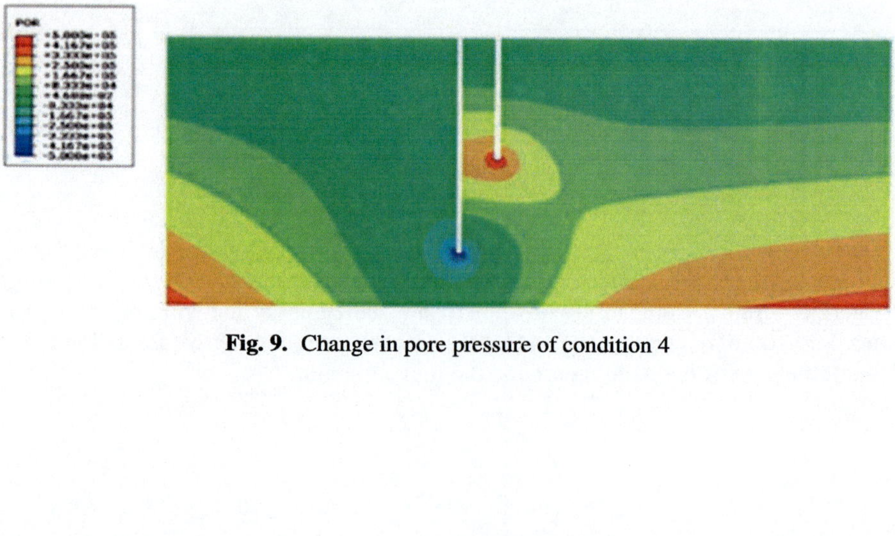

Fig. 9. Change in pore pressure of condition 4

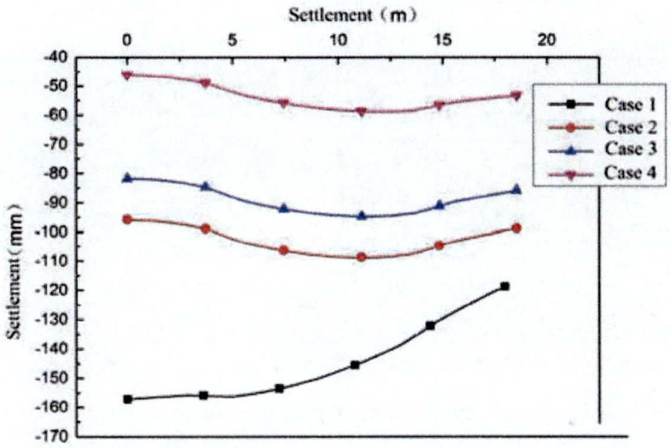

Fig. 10. Change in settlement of different amount of recharge

5.2 Effect of Different Lengths of Recharge Wells on Settlement

The effects of different lengths of recharge wells on building settlement were studied by comparing and analyzing Case 4, Case 5 and Case 6. The simulation results of Case 5 and Case 6 are shown in Figs. 11, 12, 13, 14, and 15.

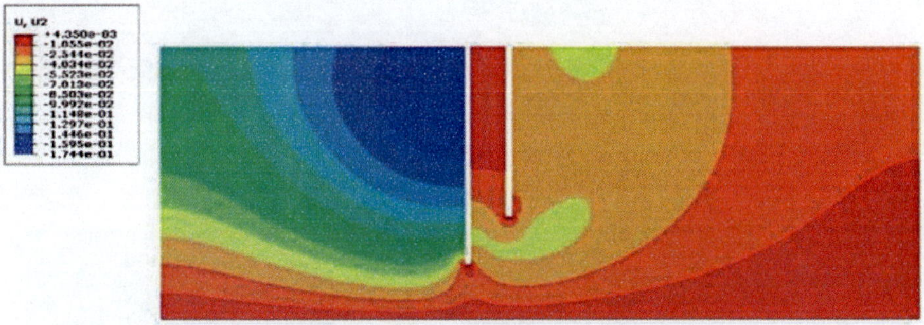

Fig. 11. Settlement change of condition 5

Comparing the above results, it can be found that the settlement at the building location has a significant rebound as the length of the recharge well increases. However, at the farthest end of the recharge well, Case 4 is significantly different from the other two cases. In Case 5 and Case 6, the settlement at the farthest end is smaller than that at the recharge well, by 2 mm and 3 mm respectively, while in Case 4, the settlement at the recharge well is 8 mm larger than that at the farthest end, although the increase in the length of the recharge well will reduce the surface settlement around the pit as well as the increase in the recharge volume, but the change in the length of the recharge well has a greater effect on the surface settlement far from the pit. In practice, attention should be paid to the selection of the length of recharge wells, as short recharge wells do not allow effective suppression of surface settlement, while long recharge wells are likely to cause surface uplift (Fig. 16).

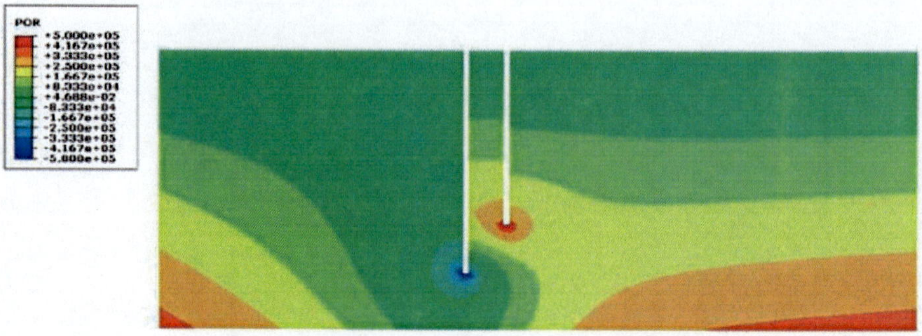

Fig. 12. Change in pore pressure of condition 5

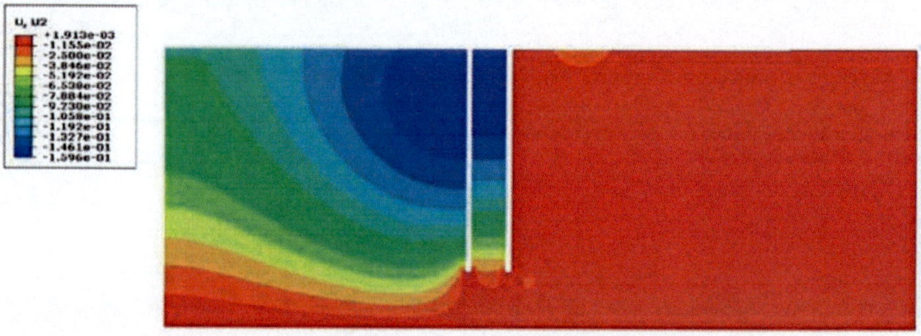

Fig. 13. Settlement change of condition 6

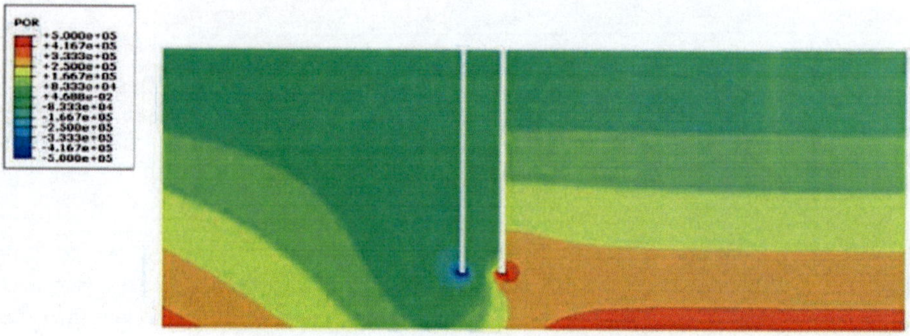

Fig. 14. Change in pore pressure of condition 6

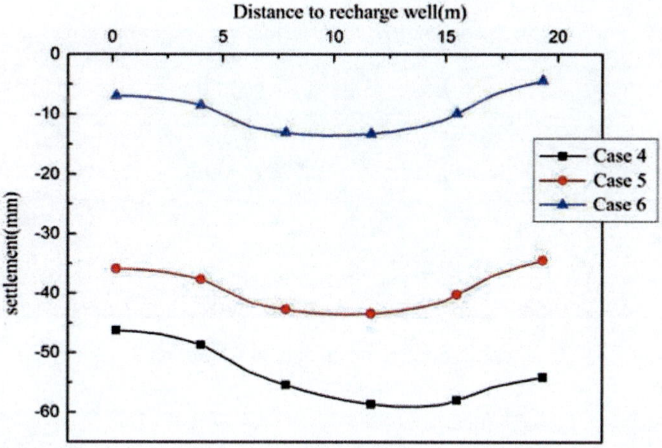

Fig. 15. Change in settlement of different recharge well length condition

5.3 The Influence of Reinjection Wells at Different Positions on Settlement

The Technical Specification for Building Foundation Pit Support [5] stipulates that the reinjection well should be arranged outside the dewatering well, and the distance between the reinjection well and the dewatering well should not be less than 6 m. However, in many projects, the reinjection well cannot be set between the foundation pit dewatering well and the building due to the close distance between the foundation pit and the building near the pit. In order to reduce the uneven settlement of buildings, the recharge wells are often set at other locations. In Condition 7, the reinjection well is set at the side of the building away from the drawdown well. Through comparative analysis of Condition 4 and Condition 7, the influence of different positions of reinjection wells on the building settlement is studied. See Fig. 17 and Fig. 18 for the numerical simulation results of Condition 7.

Compared with Condition 4 and Condition 7, it can be found that when the reinjection well is on the side of the building away from the drawdown well, the settlement value of the building decreases linearly along the direction away from Condition 4, and the difference between the settlement values on both sides of the building is 23 mm. When the reinjection well is between the dewatering well and the building, the settlement value of the building is basically stable between 55 and 58 mm, and the difference between the settlement values on both sides of the building is 4 mm. Through the above analysis, it can be found that the effect of restraining settlement is better when the reinjection well is between the building and the dewatering well. If the actual conditions do not allow, the reinjection well with smaller overall size can be set between the dewatering well and the building, and the main reinjection well can be set at a far place, or the reinjection well can be combined with the water stop curtain [6–8] to improve the reinjection effect.

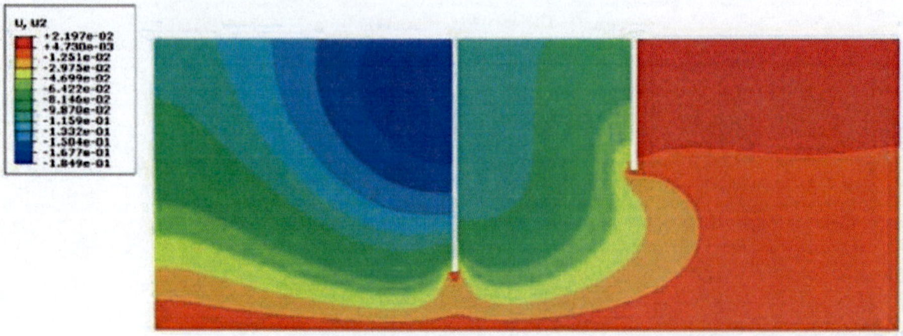

Fig. 16. Settlement change of Condition 7

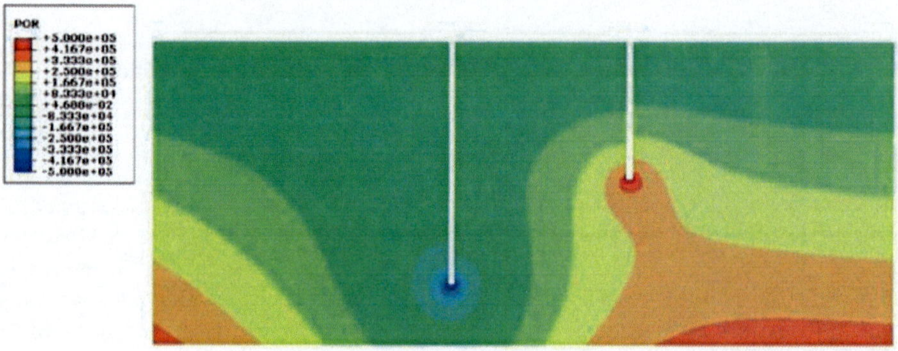

Fig. 17. Change in pore pressure of condition 7

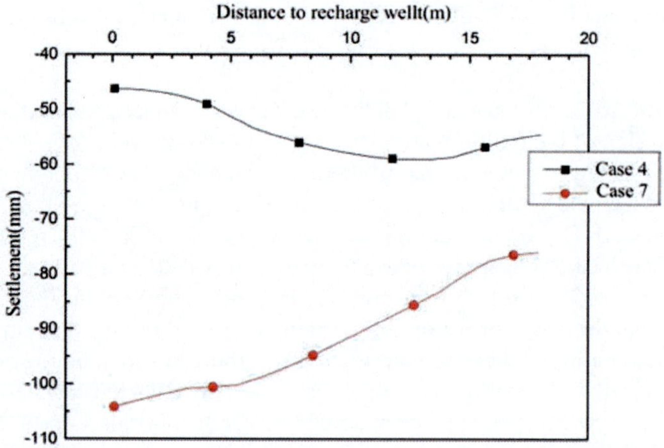

Fig. 18. Change in settlement of recharge wells at different locations

6 Conclusion

1) When only the dewatering well works, the dewatering work has a great impact on the settlement of the building outside the pit, and the dewatering will increase the settlement outside the pit, especially the location of the building, and it is easy to cause uneven settlement damage to the building.
2) The use of recharge method can significantly increase the groundwater level of buildings around the foundation pit. The existence of recharge wells can effectively control the settlement of the surface and buildings of the pit, and the amount of recharge is inversely proportional to the settlement amount, that is, the amount of reinjection increases and the settlement decreases.
3) The length of the recharge well has a great influence on the settlement of buildings outside the pit. As the length of the recharge well increases, the greater the amount of settlement that can be controlled, and the larger the radius of the surface that can be affected. However, if the length of the recharge well is too long, the soil will be uplifted far away from the recharge well.
4) When the recharge well is between the dewatering well and the building, the control effect on settlement is the greatest, and when it is necessary to set up recharge wells in other locations, other auxiliary measures need to be considered, so that the settlement effect of the recharge well can be optimized.

References

1. Liu, Y., Zhang, Q., Liu, R., et al.: Numerical simulation and field monitoring of deformation characteristics of TRD composite supporting structure for deep foundation pit in quaternary stratum: a case study in Qingdao. Geotech. Geol. Eng. **40**, 2691–2703 (2022)
2. Liu, F., Xing, H., Wu, B., et al.: Study on the effect of dewatering-recharge-excavation of deep foundation pits on soil-structure-groundwater under a deep confined water environment. Arab. J. Geosci. **15**, 710 (2022)
3. Watson, J., Thomas, S., Goodfellow, T.: Groundwater resource management during construction dewatering. Sustain. Water Resour. Manag. **8**, 91 (2022)
4. Kang, C.: Theory and Practice of Foundation Pit Engineering. Chemical Industry Press, Beijing (2009)
5. JGJ 120-2012. Technical Specification for Building Foundation Pit Support. Construction Industry Press, Beijing (2012)
6. Zeng, B., Zhen, Y., Zhang, D., et al.: A case study of vacuum tube-well dewatering technology for improving deep soft soil in Yangtze River floodplain. Environ. Earth Sci. **80**, 598 (2021)
7. Wang, J., Deng, Y., Wang, X., et al.: Numerical evaluation of a 70-m deep hydropower station foundation pit dewatering. Environ. Earth Sci. **81**, 364 (2022)
8. Liu, W., Zhu, J., Zhang, H., et al.: Geological conditions of saturated soft loess stratum and influence of tunnel excavation and dewatering system on its groundwater environment. Bull. Eng. Geol. Environ. **81**, 128 (2022)

Meso-Scale Study on Dynamic Shear Property and Size Effect of RC Beams Reinforced with CFRP

Dong Li, Bo Yang, Jiangxing Zhang, Liu Jin(✉), and Xiuli Du

Key Laboratory of Urban Security and Disaster Engineering of Ministry of Education, Beijing University of Technology, Beijing 100124, China
jinliu@bjut.edu.cn

Abstract. A three-dimensional mesoscale numerical analytical model for studying CFRP reinforced RC beams was developed using the ABAQUS finite element numerical calculation platform. The heterogeneity of concrete material and the interaction between rebar/CFRP and concrete was considered. Different strain rates were studied to explore the size effect behaviors and shear failure. The conclusions show that: 1) with the increase of the strain rate, the shear bearing capacity increases, and with the increase of the beam height, the nominal shear strength decreases; 2) increased strain rate can result in RC beams reinforced with CFRP having greater nominal shear strength, while it can also result in diminished size effect behaviors; 3) the nominal shear strength was predicted, a formula that considers both the size effect and the strain rate effect was proposed.

Keywords: CFRP · RC beam · Strain rate · Shear performance · Size effect

1 Introduction

Civil engineering plays a pivotal role in infrastructure construction. Reinforced concrete (RC) structures not only bear static loads, but also ocassionally suffer different types of dynamic loads during the service period. Normally, dynamic loads are classified by the strain rate, including the impact (10^{-4} /s ~ 10^{-1}/s), the explosion (10^{-2} /s ~ 10^{3} /s), the earthquake (10^{-4} /s ~ 10^{2}/s) and other types of dynamic load with different strain rates. Due to the strain rate effects, the mechanical property and failure mechanism of RC structure differ significantly under dynamic loads in comparison to static loads.

Due to the complexity of shear problems of RC beams reinforced with carbon fiber reinforced polymer (CFRP), the understanding of dynamic shear performance is still in its infancy. Specifically, there is a lack of quantitative analysis of the influencing mechanisms of loading rate, section size and other influence factors on dynamic shear performances of RC beam reinforced with CFRP. This study investigates the shear capabilities and size effects of RC beams reinforced with CFRP at varying loading rates. To accomplish this, the ABAQUS finite element numerical calculation platform was employed for thorough analysis. In addition, a quantitative prediction formula for RC beams reinforced with CFRP is raised for predicting the nominal shear strength by considering both the size effect and the strain rate effect.

W. Guo et al. (Eds.): GBCESC 2023, LNCE 328, pp. 196–203, 2024.
https://doi.org/10.1007/978-981-99-9947-7_20

2 Model Setup and Test Design

Concrete is considered to be a material composed of mortar, interface transition zones (ITZs) and aggregate. For the convenience of modeling and calculation, the aggregate was set to be spherical in the numerical model and its volume fraction is assumed to be fixed at 30%, and the "Monte-Carlo" method was adopted to randomly drop the aggregate particles into the mortar. ITZ was defined as a thin lamina area of 2mm around aggregate. Considering accuracy and efficiency of simulations, set the grid size to 2 mm.

2.1 Constitutive Relation

Concrete. Concrete is sensitive to strain rates [1, 2]. The factor of dynamic compressive strength (*CDIF*), the factor of dynamic tensile strength (*TDIF*) and modulus of elasticity dynamic increase factor (*MDIF*) proposed by reference [3] are adopted to obtain the compressive strength under dynamic load f_{cd}, the tensile strength under dynamic load f_{td} and the dynamic modulus of elasticity E_d. The specific expressions are listed as follows:

$$CDIF = \frac{f_{cd}}{f_c} = \begin{cases} (\dot{\varepsilon}_c/\dot{\varepsilon}_{c0})^{0.014} \dot{\varepsilon}_c \leq 30/s \\ 0.012(\dot{\varepsilon}_c/\dot{\varepsilon}_{c0})^{1/3} \dot{\varepsilon}_c > 30/s \end{cases} \tag{1}$$

$$TDIF = \frac{f_{td}}{f_t} = \begin{cases} (\dot{\varepsilon}_t/\dot{\varepsilon}_{t0})^{0.018} \dot{\varepsilon}_t \leq 10/s \\ 0.0062(\dot{\varepsilon}_t/\dot{\varepsilon}_{t0})^{1/3} \dot{\varepsilon}_t > 10/s \end{cases} \tag{2}$$

$$MDIF = \frac{E_d}{E_0} = (\frac{\dot{\varepsilon}}{\dot{\varepsilon}_0})^{0.026} \tag{3}$$

in which, f_c and f_t represent quasi-static compressive-trength and tensile-strength, respectively, $\dot{\varepsilon}_c$, $\dot{\varepsilon}_t$ and $\dot{\varepsilon}$ are the material's dynamic strain rate, $\dot{\varepsilon}_{c0}$, $\dot{\varepsilon}_{t0}$ and $\dot{\varepsilon}_0$ are the the material's quasi-static strain rate, $\dot{\varepsilon}_{c0} = 3.0 \times 10^{-5}/s$, $\dot{\varepsilon}_{t0} = 0.1 \times 10^{-5}/s$ and $\dot{\varepsilon}_0 = 3.0 \times 10^{-5}/s$. E_0 and E_d represent the static and dynamic elastic modulus, respectively.

The mechanical characteristic of concrete subjected to dynamic load is expressed by utilizing the plastic damage constitutive model which was initially suggested by reference [4] and later enhanced by Lee and Fenves [5].

CFRP Sheet. The behavior of CFRP sheet can be described by Lamina material in ABAQUS. To simplify calculations [6], the CFRP sheet can be set as the ideal linear elastic material.

For the mechanical properties of CFRP sheet under dynamic loading, according to Lu et al. [7], only if the strain rate is greater than 10.768 /s, the interface shear stress of CFRP sheet is going to be affected by strain rate. Herein this study, the largest strain rate is selected as 1 /s, thus the dynamic characteristics of the CFRP sheet are not considered in this study.

2.2 Interaction Behaviors

Interactions Between Concrete and Rebar. The interactions between concrete and rebar are described by the constitutive relationship model proposed by code (GB50010–2010) [8] of China.

Interactions Between Concrete and CFRP Sheet. CFRP sheets always present stripping failure in practice [9]. In order to meet the reality, Lu et al. [10]'s simplified bilinear model is adopted in the present study:

$$\tau = \begin{cases} \tau_{\max}(\frac{s}{s_0})\, 0 < s < s_0 \\ \tau_{\max}(\frac{s_u - s}{s_u - s_0})\, s_0 < s < s_u \\ 0\, s > s_u \end{cases} \tag{4}$$

in which, τ and $\tau_{\max}$ are stress and maximum stress, respectively, s, s_0 and s_u represent relative slip, initial slip and maximum slip, respectively.

2.3 Verification of Meso-Scale Numerical Models

Rationality of Shear Failure Mode. The BS-2.3–1 beam in ref. [11] was selected to verify RC beam's shear failure mode. The mechanical properties of longitudinal reinforcement and stirrup under quasi-static ($\dot{\varepsilon}_{\text{static}} = 10^{-5}$/s) conditions were selected [11].

Figure 1(a) displays a comparison of the failure modes between the test results and simulation results at three distinct loading rates. As can be observed, the test results [11] and the crack distribution at the three loading rates correspond rather well. In Fig. 1(b), there is a display of the comparison between the P-Δ curve obtained from the test results and simulation results. By comprehensively comparing the P-Δ curve and failure mode, the test results coincide with the simulation results [11].

Rationality of CFRP Sheet Refinforcing Method. In reference [12], 16 geometrically similar RC beams were designed for a four-point shear failure test. Herein, specimens of S-0 and S-0.0835% were used as benchmarks to confirm the rationality and applicability of RC beam reinforced with CFRP.

Figure 1(c) shows a comparison of failure modes. The approximate position, angle and shape of the oblique crack in the simulation results of specimen S-0.0835% agree well with the test [12]. The P-Δ curve is shown in Fig. 1(d), it also coincides with the test [12].

2.4 Test Design

To study the size effects and shear properties of RC beams at different strain rates, 12 numerical analysis models with proportional cross-sectional sizes and strain rates were established. The fiber ratio ρ_f of CFRP sheet is calculated by [12]:

$$\rho_f = \frac{2nw_f t_f}{bs_f} \tag{5}$$

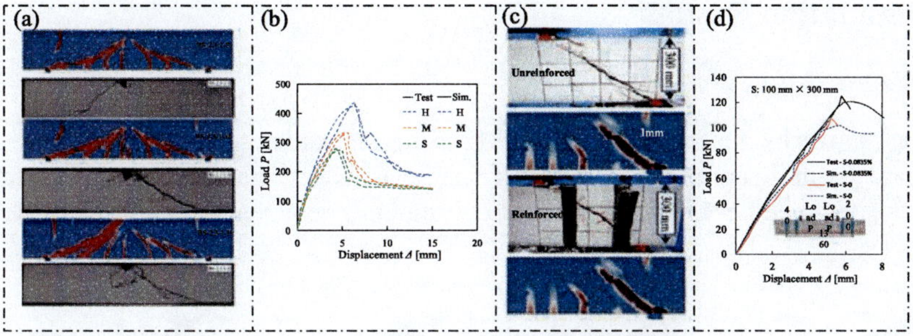

Fig. 1. Numerical simulation verification.

And the formula of strain rate is [13]:

$$\dot{\varepsilon} = \frac{6hv}{(L-a)(L+2a)} \tag{6}$$

where, t_f represents the thickness; s_f represents the distance between CFRP sheets; w_f represents the width of the CFRP sheet; b represents the width of the beam section; v represents the loading rate; h represents the beam height.

Physical parameters corresponding to the numerical analysis model are shown in Table 1. S, M and L represent the section size, A, B, C and D represent the strain rate (A: 1×10^0/s, B: 1×10^{-1}/s, C: 1×10^{-3}/s, D:1×10^{-5}/s).

Table 1. Physical parameters.

Specimen	Section size $b \times h$ (mm)	Loading rate v(m/s)	Strain rate $\dot{\varepsilon}$ (s^{-1})	Fiber ratioρ_f (%)
SA	100 mm × 300mm	1.40×10^0	10^0	0.1336
SB		1.40×10^{-1}	10^{-1}	0.1336
SC		1.40×10^{-3}	10^{-3}	0.1336
SD		1.40×10^{-5}	10^{-5}	0.1336
MA	200 mm × 600 mm	2.81×10^0	10^0	0.1336
MB		2.81×10^{-1}	10^{-1}	0.1336
MC		2.81×10^{-3}	10^{-3}	0.1336
MD		2.81×10^{-5}	10^{-5}	0.1336
LA	300 mm × 900 mm	4.21×10^0	10^0	0.1336
LB		4.21×10^{-1}	10^{-1}	0.1336
LC		4.21×10^{-3}	10^{-3}	0.1336
LD		4.21×10^{-5}	10^{-5}	0.1336

3 Simulation Results

3.1 Failure Mode and *P*-*Δ* Curve

Figure 2 shows the failure modes of the RC beams. We can observe that the failure mode is apparently in shear failure. With the strain rate increases, the crevices in the beam turn into spread out and evenly distributed, resulting in more extensive damage [13–15].

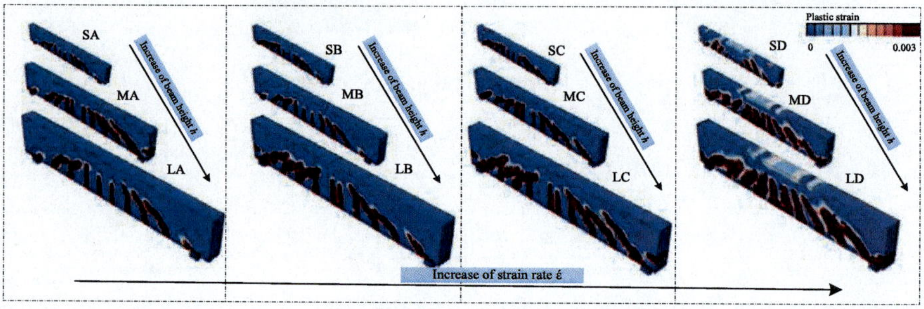

Fig. 2. Failure mode.

The *P*-*Δ* curves of beams with identical dimensions under four distinct strain rates are presented in Fig. 3. It can be observed that in the initial loading stage, all curves basically coincide. As the load continues, and the slope of the curve continues to decrease until it gets to the peak load. The shear strength with different dimensions increases with the increase of strain rate. [16].

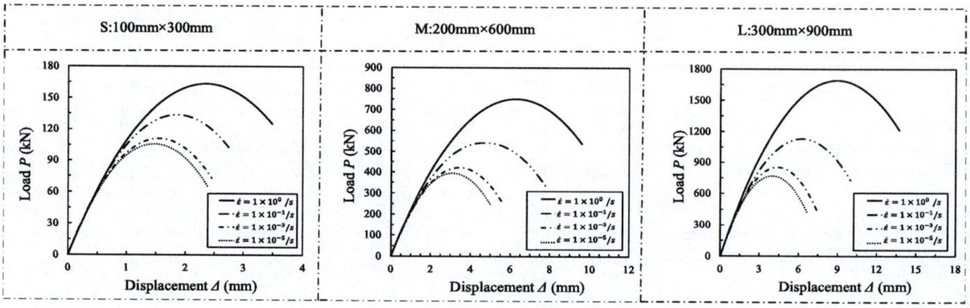

Fig. 3. *P*-*Δ* curves.

3.2 Size Effect Analysis

Nominal Shear Strength Trend. Use the nominal shear strength τ_u to analyze the shear properties and size effect. τ_u is defined as [17]:

$$\tau_u = \frac{P_u}{bh_0} \tag{7}$$

in which, P_u represents peak load, h_0 represents effective cross-sectional beam height, b represents width of beam.

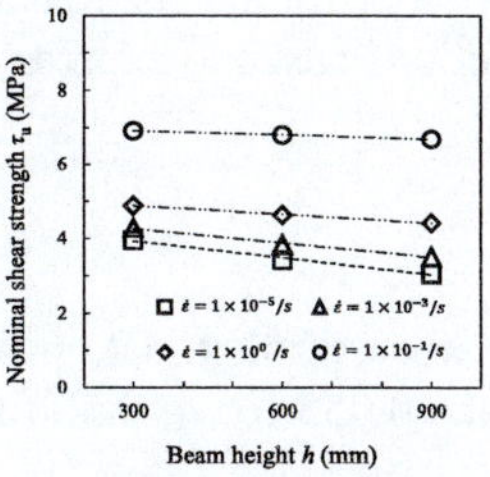

Fig. 4. Nominal shear strength trend.

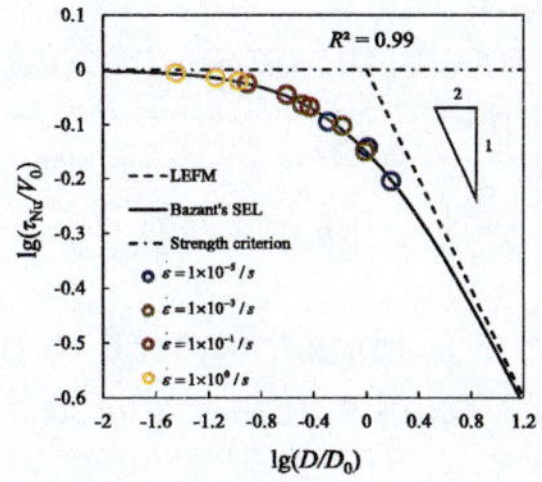

Fig. 5. Comparison with Bažant size effect law.

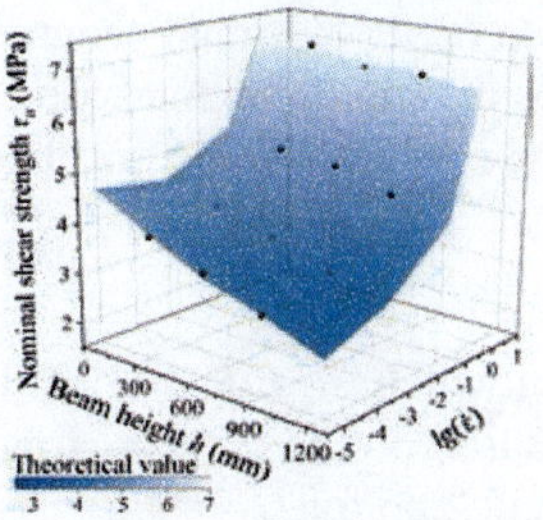

Fig. 6. Prediction method proposed.

In Fig. 4, the changes in nominal τ_u as the height of the beam h increases are illustrated. The nominal shear strength of RC beams reinforced with CFRP sheet has been found to decrease with increasing beam height, demonstrating a clear size effect behavior. Meanwhile, with an increase in strain rate, the nominal shear strength of RC beams increases to varied degrees, and the trend lines of the fitting slope become more level. Therefore, it can be inferred that the nominal shear strength of RC beams reinforced with CFRP sheets will increase with the increase of strain rate, whilst diminishing the size effect of the nominal shear strength. [13]. It is worth noting that even if the strain rate is 1/s, the slope of the trend line is not zero, indicating that within the range of strain rates given in this article, the size effect of nominal shear strength can only be diminished by raising the strain rate, it cannot be entirely eliminated.

Comparison with Bažant Size Effect Law. Bažant proposed the size effect law appropriate for quasi-brittle materials in accordance with the concept of fracture mechanics. [18]:

$$\tau_u = \frac{V_0}{\sqrt{1 + D/D_0}} \tag{8}$$

where, D represents the beam height, V_0 and D_0 represent empirical coefficients.

Regression analysis on the simulation data produced the logarithmic curves of nominal shear strength and cross-sectional size under various strain rates, as illustrated in Fig. 5. As it is observed, the data points rise with rising strain rate, suggesting that the size impact is lessened. This is coincident with the trend lines shown in Fig. 4.

3.3 Nominal Shear Strength Prediction

A formula has been established for predicting the nominal shear strength, considering both the strain rate and size effect. This formula is based on the classic Bažant size effect law and is semi-theoretical and semi-empirical in nature, which has the form as follows:

$$\tau_u = \frac{V_0}{\sqrt{1 + D/D_0}} \varphi_{\dot{\varepsilon}} \beta_{\dot{\varepsilon}} \tag{9}$$

where, $\varphi_{\dot{\varepsilon}}$ represents the strain rate enhancement coefficient, $\beta_{\dot{\varepsilon}}$ represents the weakening coefficient.

Determination of $\boldsymbol{\varphi_{\dot{\varepsilon}}}$. To sum up, under the same cross-sectional size, in a quadratic curve, the nominal shear strength of RC beams reinforced with CFRP sheets increases as the strain rate increases. Therefore, the quadratic form function is used to express this rule, which has the form as follows:

$$\varphi_{\dot{\varepsilon}} = A(\lg\dot{\varepsilon})^2 + B(\lg\dot{\varepsilon}) + \mathrm{C} \tag{10}$$

in which, *A*, *B* and *C* are the coefficients that need to be obtained through data fitting. Based on the current simulation results, it is suggested to take 0.055 for *A*, 0.50 for *B* and 2.86 for *C*.

Determination of $\boldsymbol{\beta_{\dot{\varepsilon}}}$. According to the current simulation results, the morphing trend of $\beta_{\dot{\varepsilon}}$ is expressed by the tangent curve of hyperbolic, which is as follows:

$$\beta_{\dot{\varepsilon}} = \begin{cases} 1, \ \dot{\varepsilon} \le \dot{\varepsilon}_{\text{static}} \\ (\beta_0 - 1) \cdot \tanh\left[\alpha(\lg\dot{\varepsilon} - \lg\dot{\varepsilon}_{\text{static}})\right] + 1, \ \dot{\varepsilon} > \dot{\varepsilon}_{\text{static}} \end{cases} \tag{11}$$

in which, $\beta_0 = \frac{\tau_0\sqrt{1+D/D_0}}{V_0}$; Under quasi-static stress, τ_0 denotes the nominal shear strength of the RC beam with the lowest size, herein, $\tau_0 = 3.93$; α represents the adjustment coefficient. It is discovered that the theoretical value and simulation results the coincide well when $\alpha = 0.2$, as shown in Fig. 6.

4 Conclusion

This article uses the ABAQUS finite element numerical calculation platform to develop a three-dimensional mesoscale numerical analysis model to investigate the size effect and shear performance of RC beams reinforced with CFRP sheet under different loading rates. Some closing remarks are as follows.

1) With the strain rate rises, the ability of the RC beam to withstand shear also increases; however, as the increase of beam height, the nominal shear strength decreases.
2) Increasing strain rate can improve the nominal shear strength, but it will weaken its size effect.
3) A formula for predicting the nominal shear strength of RC beams reinforced with CFRP sheets considering both strain rate effect and size effect is proposed.

References

1. Bischoff, P.H., Perry, S.H.: Compressive behaviour of concrete at high strain rates. Mater. Struct. **24**(6), 425–450 (1991)
2. Park, S.W., Xia, Z., Zhou, M.: Dynamic behavior of concrete at high strain rates and pressures: II. Numerical simulation. Inter. J. Impact Eng. **25**(9), 887–910 (2001)

3. International Federation for Structural Concrete. fib Model Code for Concrete Structures 2010. Lausanne: fib Bulletin (2010)
4. Lubliner, J., Oliver, J., Oller, S.: A plastic-damage model for concrete. Int. J. Solids Struct. **25**(3), 299–326 (1989)
5. Lee, J., Fenves, G.L.: Plastic-damage model for cyclic loading of concrete structures. Eng. Mech. **124**(8), 892–900 (1998)
6. Obaidat, Y.T., Heyden, S., Dahlblom, O.: The effect of CFFP and CFRP/concrete interface models when modelling retrofitted RC beams with FEM. Compos. Struct. **92**(6), 1391–1398 (2010)
7. Lu, S.Y., Chen, D., Wu, H.: Dynamic shear slip model for frp-concrete interface. Eng. Mech. **40**, 1–13 (2023)
8. GB50010–2010, Code for design of concrete structures. China Architecture & Building Press, Beijing (2010)
9. Jiang, X., Jin, L., Du, X.L.: Experimental study on the effect of fiber ratio on shear behavior of large size concrete beams wrapped with CFRP Cloths. J. Build. Struct. **27**, 1–12 (2022)
10. Lu, X.Z., Ye, L.P., Teng, J.G.: Bond-slip model for FRP-to-concrete interface. J. Build. Struct. **26**(4), 9 (2005)
11. Yuan, J., Yi, W.-J.: Tests for effects of loading rate on shear behaviors of RC beams. J. Vibrat. Shock **38**(07), 119–127 (2019)
12. Jin, L., Du, X., Li, D.: Seismic behavior of RC cantilever beams under low cyclic loading and size effect on shear strength: an experimental characterization. Eng. Struct. **122**, 93–107 (2016)
13. Jin, L.: Study on meso-scopic model and analysis method of concrete. Beijing University of Technology, Beijing (2014)
14. Adhikary, S.D., Li, B., Fujikake, K.: Dynamic behavior of reinforced concrete beams under varying rates of concentrated loading. Int. J. Impact Eng J. Impact Eng **47**(9), 24–38 (2012)
15. Xiao, S.-Y., Cao, W.-B., Pan, H.-H.: Experimental study on mechanical behavior of reinforced concrete beams at different loading rates. J. Build. Struct. **33**(12), 142–146 (2012)
16. Li, Z.B., Bao, X.C.: Effect of loading rate on the dynamic properties of reinforced concrete beams. J. Disaster Prev. Mitigation Eng. **31**(06), 627–631 (2011)
17. Godat, A., Labossiere, P., Neale, K.W.: Numerical investigation of the parameters influencing the behaviour of FRP shear-strengthened beams. Const. Build Mater **32**, 90–98 (2012)
18. Bažant, Z.P.: Size effect in blunt fracture: concrete, rock, metal. ASCE J. Eng. Mech. **110**(4), 518–535 (1984)

Experimental Investigation on the Interfacial Bond Failure Between FRP Bars and Sea Sand Concrete

Ben Yang, Chunheng Zhou, and Zihua Zhang(✉)

School of Civil and Environmental Engineering and Geography Science, Ningbo University, Ningbo 315211, China
zhangzihua@nbu.edu.cn

Abstract. The interface between the fiber-reinforced polymer (FRP) bars and sea sand concrete (SSC) is to connect the two materials, and the interfacial bonding performance (IBP) is crucial for the load-bearing capacity of structural members. To investigate the impacts of different factors, including bar type, bar diameter, and concrete strength, on the IBP between FRP bars and SSC, pull-out tests are performed on twenty-seven specimens.Based on the thick-walled cylinder theory, a semi-empirical model of interfacial ultimate bond strength at ambient temperature was proposed. The results show that the early interfacial bond strength of FRP bars-SSC specimens is slightly higher than that of steel rebar-SSC specimens. The profile of ribs has a greater impact on the IBP than the bar type and bar diameter. During the pull-out process, the surface of FRP bars is severely worn, and some fibers on the bar surface undergo brittle fracture, resulting in a reduction of interfacial bond strength. The interfacial bond strength estimated by the proposed model is in good agreement with the experimental results.

Keywords: Sea Sand Concrete · FRP Bar · Bond Performance · Bond Strength

1 Introduction

In recent decades, with the rapid development of urbanization, the dosage of concrete is increasing significantly, leading to the shortage of river sand. Taking sea sand instead of river sand is an effective way to solve this problem. However, the corrosion of steel bars will be accelerated by the chloride ions in sea sand, leading to an insufficient durability of reinforced sea sand concrete (SSC) structures. Fiber-reinforced polymers (FRP), owing to their lightweight, high-strength, and good corrosion resistance [1], can effectively tackle the problem of corrosion by replacing steel rebars in SSC structures. Therefore, FRP-reinforced SSC has broad application prospects in civil engineering, especially in offshore and coastal structures [2].

For the FRP-reinforced SSC structures, the interfacial bonding performance (IBP) between FRP bars and SSC is vital for their collaborative work. To date, there have been many works on the IBP between FRP bars and normal concrete [3–6], but concerns about the IBP between FRP bars and SSC are insufficient. Fahmy et al. [7] conducted pull-out

W. Guo et al. (Eds.): GBCESC 2023, LNCE 328, pp. 204–213, 2024.
https://doi.org/10.1007/978-981-99-9947-7_21

tests on the IBP between BFRP bars with four different surface characteristics (smooth, wrapped, ribbed, and frosted) and SSC. It is found that the surface characteristics of bars have a considerable impact on the bonding behavior and failure modes. Gao et al. [8] explored the bonding mechanism between CFRP bars and SSC through pull-out tests and SEM. It is found that the tensile properties of bars have no significant influence on the interaction between CFRP bars and concrete. During the pull-out process, some mortar particles replaced the bonding matrix on the bar surface, resulting in a macroscopic manifestation of significant friction between CFRP bars and concrete. Dan et al. [9] modified the formula in ACI440.1R-06 to evaluate the bond strength according to the pull-out tests of CFRP bar - SSC specimens. The above-mentioned works focus on the IBP between FRP bars and SSC at the macroscale level, while the relationship between bar ribs and the interfacial bond strength has not been established.

This work performed an experimental investigation on the IBP between FRP bars and SSC, with the bar diameter, bar type, and concrete strength grade as the experimental parameters. The focus was on analyzing the impact of different factors on the IBP between FRP bars and SSC, and a semi-empirical model considering the profile of bar ribs is developed to evaluate the interfacial bond strength.

2 Experimental Program

2.1 Materials

Three types of FRP bars, including BFRP bar, GFRP bar, and CFRP bar are used, as shown in Fig. 1. The bar surface is helical wrapping, and the bar diameters are 8 mm, 12 mm, and 16 mm, respectively. Crescent ribbed steel rebar is employed as a control group. Table 1 lists the details of different bars.

Table 1. Details of different bars.

Type	Diameter /mm	Surface	Rib spacing /mm	Rib height/mm	Tensile strength/MPa	Elastic modulus/MPa
BFRP	8	HW	9.06	0.08	1011.3	44.1
	12		7.06	0.17		
	16		6.48	0.22		
CFRP	8	HW	5.14	0.04	1801.6	105.5
	12		8.12	0.21		
GFRP	12	HW	8.61	0.11	854.3	21.3
Steel rebar	12	CR	3.48	1.35	540	200

Note: HW denotes helical wrapping, and CR denotes crescent ribbed

Three strength grades of concrete, including C30, C40, and C50, were prepared, using sea sand that meets the requirements of the construction sand specification as the fine aggregates, and continuously graded crushed stones sized from 5 mm to 20 mm as the coarse aggregates.

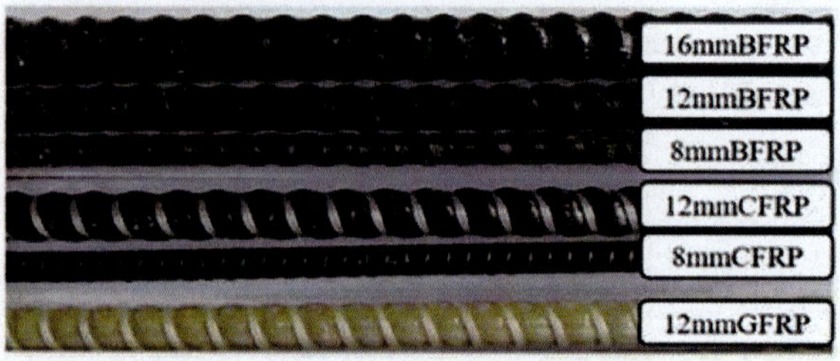

Fig. 1. Different types of FRP bars.

2.2 Experimental Scheme

The pull-out tests are performed on nine configurations of specimens, and three specimens with the same parameters for each configuration were tested. The specimen identification is defined as follows: C3, C4, and C5 denote the concrete strength grade of C30, C40, and C50, respectively; D denotes the bar diameter; CR denotes the crescent ribbed; B, C, and S denote the BFRP bar, CFRP bar, and steel rebar, respectively, as shown in Table 2.

Table 2. Configurations of pull-out tests.

Specimens	Concrete strength grade	Bar diameter/mm	Embedded length
C3D8B1–3	C30	8	5D
C3D16B1–3	C30	16	5D
C3D12B1–3	C30	12	5D
C4D12B1–3	C40	12	5D
C5D12B1–3	C50	12	5D
C3D8C1–3	C30	8	5D
C3D12C1–3	C30	12	5D
C3D12G1–3	C30	12	5D
C3D12S1–3	C30	12	5D

2.3 Test Setup

According to the Chinese code of "Test Method for Basic Mechanical Properties of Fiber Reinforced Polymer Bars" (GB/T30022–2013), the embedded length of bars is five times the bar diameter. The geometric dimensions of the specimen are shown in Fig. 2. The experiment was conducted using an electro-hydraulic servo universal testing machine with a measuring range of 60 T, as shown in Fig. 3. A displacement-control scheme is adopted with a loading rate of 0.5 mm/min, and the data acquisition frequency is 4 Hz.

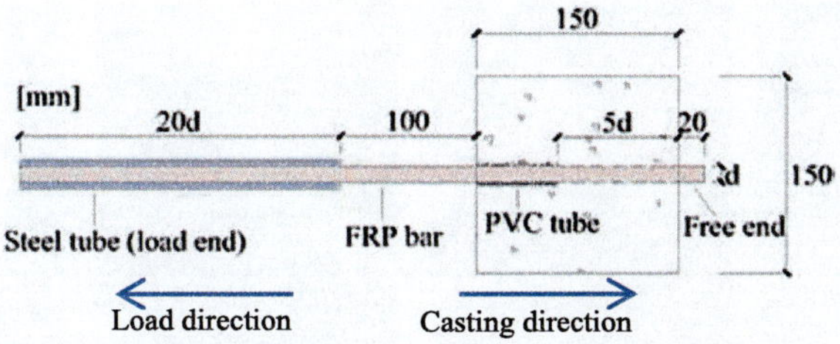

Fig. 2. Sketch of the pull-out specimen.

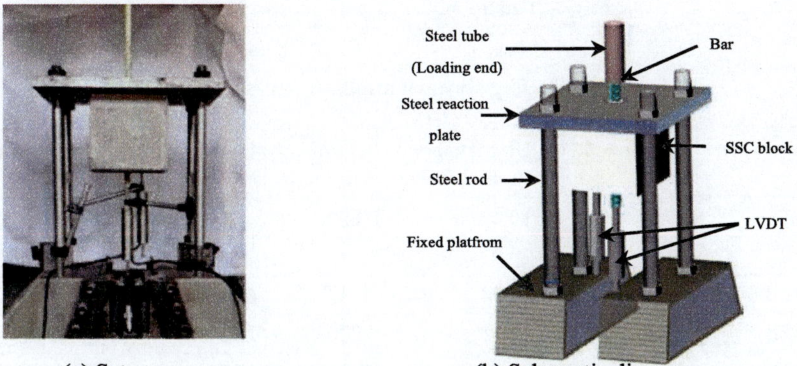

(a) Set up (b) Schematic diagram

Fig. 3. Experimental setup of the pull-out test.

3 Results and Discussion

3.1 Failure Modes

Four failure modes were observed after pull-out tests, including pull-out of the bar (P Mode), splitting of the concrete (CS Mode), and mixed mode (M Mode). Several BFRP bars split during the pull-out process (BS Mode). Table 3 lists the summary of experimental results, noting that the values are the mean of three specimens in one configuration (Fig. 4).

3.2 Bond-Slip Relationships

The global IBP between bars and concrete can be represented by the bond-slip curve, and some typical results are shown in Fig. 5. Generally speaking, the IBP between FRP bars and SSC is better than that between steel rebars and SSC. The bond-slip process of the interface between FRP bars and SSC can be roughly divided into four stages: **(i) Rising stage.** The bonding stress is mainly contributed by chemical adhesion, and the bond-slip relationship is approximately linear. As the load gradually increases, the chemical adhesion gradually disappears, and the slope of the curve decreases nonlinearly; **(ii) Slip stage.** As the load increases to near the ultimate load, the slip of the bars increases,

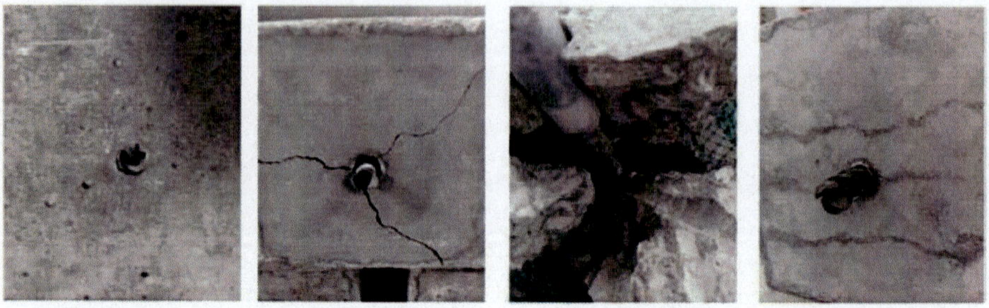

(a) P mode (b) M mode (c) CS mode (d) BS mode

Fig. 4. Failure modes of pull-out specimens.

Table 3. Experimental results of pull-out tests.

Configuration	Peak load/kN	Ultimate bond strength/MPa	Ultimate slip/mm	Ultimate bond stiffness	Failure modes
C3D8B	19.64	19.49	2.82	6.93	P
C3D12B	51.71	22.91	2.43	9.47	M
C3D16B	77.89	19.34	0.33	60.63	CS
C4D12B	57.48	25.41	2.20	11.57	M & BS
C5D12B	62.11	27.78	1.56	17.93	P
C3D8C	14.51	14.31	1.14	12.70	BS
C3D12C	84.34	37.65	2.68	14.13	M
C3D12G	23.03	9.81	2.03	4.90	P & M
C3D12S	50.44	22.13	1.69	13.17	P

the resin matrix on the bar surface is gradually worn out by the surrounding concrete, and the curve gradually trends to be horizontal; **(iii) Descending stage.** The mechanical interaction and friction gradually decrease with the wear of ribs, resulting in a rapid loss of bond strength, as shown in Fig. 5(a). It should be noted that some specimens split shortly after reaching the ultimate bond strength, thus no data was collected in this stage, as shown in Figs. 5 (b) and 5 (c). **(iv) Fluctuation stage.** As the load and slip increase continuously, the FRP ribs move in a channel with variable diameters, resulting in a change of interaction between the two materials, as shown in Fig. 5 (a).

3.3 Parameter Analysis

FRP Bar Diameter. Existed studies [10] have revealed that bleeding, Poisson's ratio effect, and shear-lag effect have a significant impact on the interfacial bond strength between concrete and FRP bars with large diameters, which means that the larger the bar diameter, the lower the interfacial bond strength. However, the phenomenon mentioned-above was not observed in this work. It may be caused that the rib height of bars with

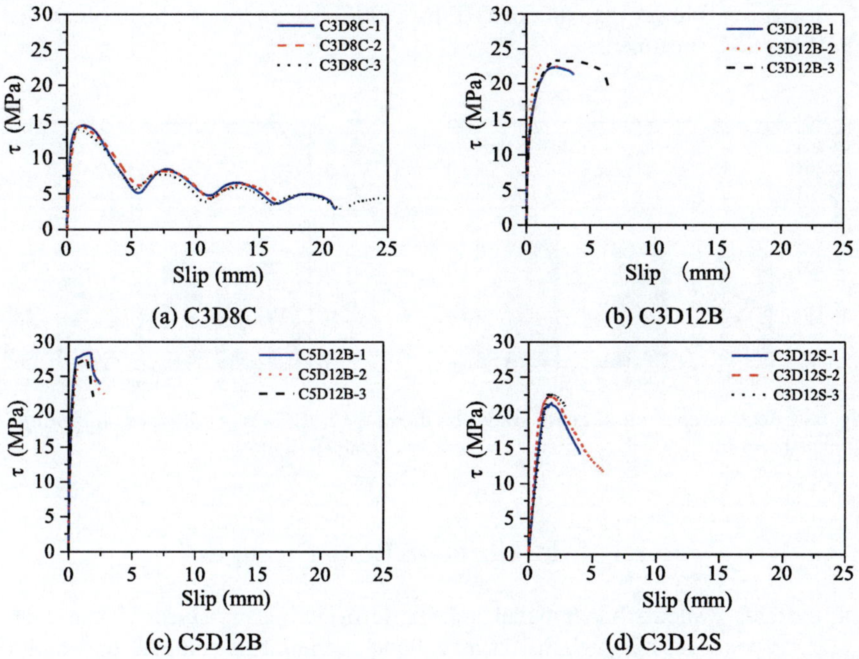

Fig. 5. Typical bond-slip curves of the pull-out specimens

a diameter of 8 mm is much smaller than that of bars with a diameter of 12 mm. The mechanical interaction between FRP bars and SSC increases significantly with the rib height.

Concrete Strength. From Fig. 6, we can see that the ultimate bond strength of the interface increases with the concrete strength grade. The improvement is related to the shear strength of SSC and bar ribs. As the shear strength of the concrete between ribs exceeds that of the ribs, the ribs are peeled off from the substrate. For the ribs along the embedded length of the specimens with CS Mode, the ribs of the C30 specimens were worn, partial ribs of the C40 specimens were worn and peeled off, and all ribs of the C50 specimens were peeled off. It is indicated that the higher the concrete strength grade, the more probably the specimen possesses CS Mode failure. The ultimate slip gradually decreases with the concrete strength grade. For smooth and sand-coating FRP bars, the impact of concrete strength on the interfacial bond strength is limited because the interfacial bond strength is mainly provided by chemical adhesion and friction [11].

FRP Type. As shown in Fig. 7, the ultimate bond strength of specimen C3D12C is significantly higher than that of specimen C3D12B, indicating that the bar type has a significant impact on the interfacial bond strength. The reason is that the fiber bundle of the CFRP bar is relatively dense, which can transfer the shear stress more effectively. Moreover, CFRP has better wear resistance than BFRP, thus the CFRP ribs can withstand more load from the concrete between the neighbouring ribs during the pull-out process. Compared to the configuration of steel rebar, the ultimate bond strength of steel rebar -

SSC specimens is lower than that of FRP bar - SSC specimens, but higher than that of GFRP bar - SSC specimens.

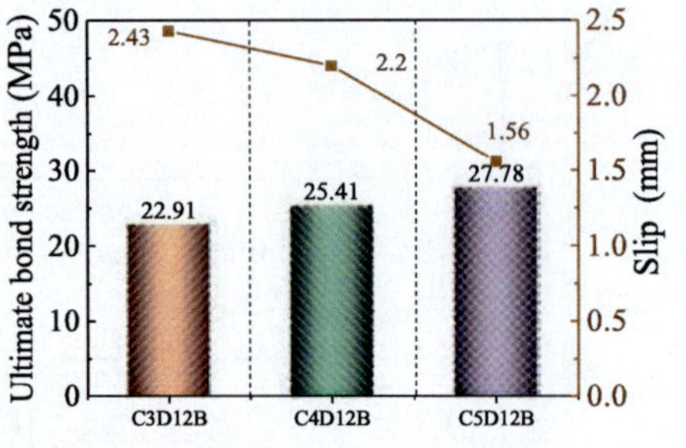

Fig. 6. Effects of concrete strength grades on the ultimate bond strength.

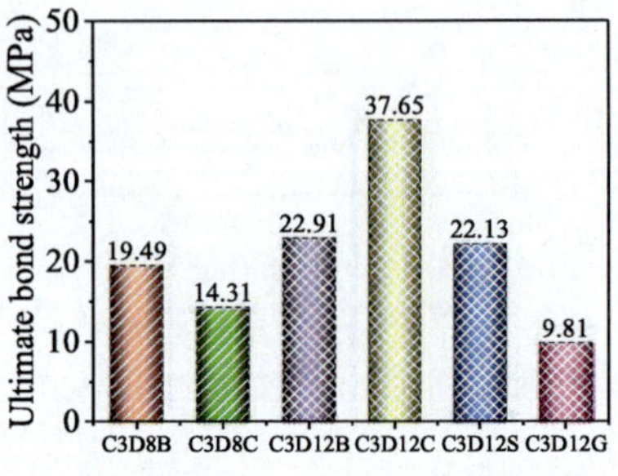

Fig. 7. Effects of bar types on the ultimate bond strength.

4 Analytical Model of the Ultimate Bond Strength

Using the thick-walled cylinder model under uniform internal pressure in elastic mechanics [12], as depicted in Fig. 8, the ultimate bond strength between FRP bars and SSC can be calculated analytically by

$$\tau_{\max} = \frac{(c + d/2)(\sin\theta + \mu\cos\theta)}{1.664d(\cos\theta - \mu\sin\theta)} f_t \tag{1}$$

where c is the thickness of concrete protective layer, d is the bar diameter, θ is the friction angle between FRP bars and concrete, μ is the coefficient of friction, and f_t is the tensile strength of concrete, respectively.

4.1 Model Parameter Correction

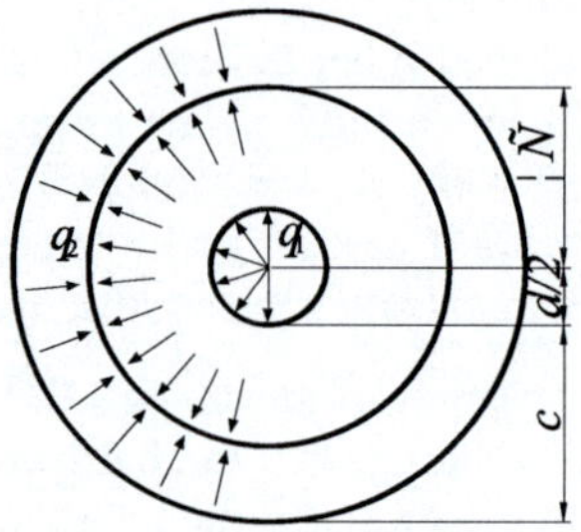

Fig. 8. Thick-walled cylinder model.

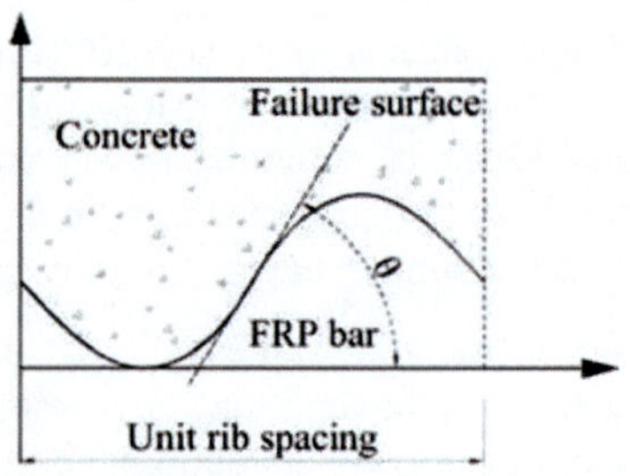

Fig. 9. Definition of the included angle.

From the bond-slip curves between FRP bars and concrete, we can see that before reaching the ultimate bond strength, the slip is relatively small. In this stage, the friction can be regarded as static friction with the static friction coefficient $\mu = 0.7$.

The projection of the failure surface in the plane is approximated to the tangent line at the inflection point of the hyperbola by measuring the shape of the BFRP ribs, and the included angle of the failure surface is defined as the included angle between the failure surface and the longitudinal line of the bar, as shown in Fig. 9. Typical included angles of the failure surface are shown in Fig. 10.

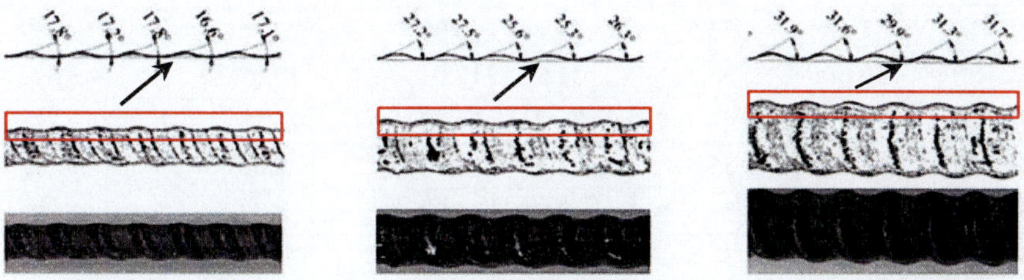

(a) 8mm BFRP (b) 12mm BFRP (c) 16mm BFRP

Fig. 10. Included angles of the failure surface.

Assuming $f_t = 0.395{f_{cu}}^{0.55}$, Eq. 1 becomes

$$\tau_{max} = \frac{0.237(c + d/2)\left(\sin\overline{\theta} + 0.7\cos\overline{\theta}\right)}{d\left(\cos\overline{\theta} - 0.7\sin\overline{\theta}\right)} f_{cu}^{0.55} \quad (2)$$

where $\overline{\theta}$ is the mean of included angles of the failure surface along the embedded length.

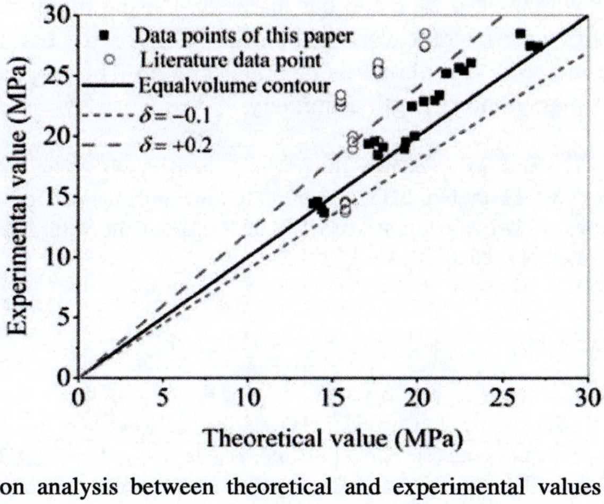

Fig. 11. Correlation analysis between theoretical and experimental values of ultimate bond strength.

The ultimate bond strength achieved by Eq. 2 is compared with the experimental values in Table 4 and Fig. 11, indicating that the model can estimate the ultimate bond strength with acceptable accuracy.

Table 4. Comparison between theoretical and experimental results.

Specimen	Bar diameter	f_{cu}/MPa	θ/°	Theoretical value/MPa	Experimental value/MPa	Relative error
C3D8B	8 mm	33.5	16.6	19.34	19.54	1.0%
C3D12B	12 mm	33.5	27.2	20.39	22.86	10.8%
C3D16B	16 mm	33.5	31.9	18.01	19.09	5.6%
C4D12B	12 mm	42.7	26.7	21.72	25.18	13.8%
C5D12B	12 mm	55.4	28.1	26.60	27.46	3.1%
C3D8C	8 mm	33.5	7.9	13.88	14.43	3.8%
C3D12C	12 mm	33.5	32.8	26.57	37.29	28.8%

5 Conclusions

The IBP between FRP bars and SSC is investigated through pull-out tests on twenty-seven specimens. The impacts of different parameters on the IBP are discussed. A semi-empirical model of the ultimate bond strength between FRP bars and SSC is developped in the frame of the thick-walled cylinder theory. The results show that the ultimate bond strength of the FRP bar - SSC interface increases with the concrete strength grade. The ultimate slip decreases with the concrete strength grade. The influence of FRP ribs on the IBP is greater than the bar type and bar diameter. During the pull-out process, the surface of FRP bars was severely worn and partial fibers on the bar surface fractured brittlely, resulting in a reduction of interfacial bond strength. The proposed model is able to estimate the ultimate bond strength accurately.

Acknowledgment. The authors appreciate the financial support of the Natural Science Foundation of Zhejiang Province (Grant No. LHY21E090002), the Fundamental Research Funds for the Provincial Universities of Zhejiang (Grant No. SJLZ2023003), and the Natural Science Foundation of Ningbo, China (Grant No. 202003N4139).

References

1. Nepomuceno, E., Sena-Cruz, J., Correia, L., D'Antino, T.: Review on the bond behavior and durability of FRP bars to concrete. Constr. Build. Mater. **287**, 123042 (2021)
2. Hua, Y.T., Yin, S.P., Peng, Z.T.: Crack development and calculation method for the flexural cracks in BFRP reinforced seawater sea-sand concrete (SWSSC) beams. J. Build. Struct. **42**(02), 166 (2021)

3. Rolland, A., Quiertant, M., Khadour, A., Chataigner, S., Benzarti, K.: Argoul, P: Experimental investigations on the bond behavior between concrete and FRP reinforcing bars. Constr. Build. Mater. **173**, 136–148 (2018)
4. Shen, D., Li, C.C., Feng, Z.Z., Wen, C.Y., Ojha, B.: Influence of strain rate on bond behavior of concrete members reinforced with basalt fiber-reinforced polymer rebars. Constr. Build. Mater. **227**, 116755 (2019)
5. Basaran, B., Kalkan, L.: Investigation on variables affecting bond strength between FRP reinforcing bar and concrete by modified hinged beam tests. Compos. Struct. **242**, 112185 (2020)
6. Wei, W., Liu, F., Xiong, Z., Lu, Z.Y., Li, L.J.: Bond performance between fibre-reinforced polymer bars and concrete under pull-out tests. Constr Build Mater **227**, 116803 (2019)
7. Fahmy, M., Ahmed, S., Wu, Z.S.: Bar surface treatment effect on the bond-slip behavior and mechanism of basalt FRP bars embedded in concrete. Constr. Build. Mater. **289**, 122844 (2020)
8. Gao, J., Fan, L.Y.: Bond ding test and mechanism analysis between CFRP reinforcement and sea sand concrete. J. Compos. **39**(03), 1194–1204 (2022)
9. Dan, B., Dong, G.Q., Liu, Q.Y.: CFRP reinforcement and sea sand concrete. Build. Sci. Eng. **37**(05), 11 (2020)
10. Achillides, Z., Pilakoutas, K.: Bond behavior of fiber reinforced polymer bars under direct pullout conditions. J. Compos. Constr. **8**(2), 173–181 (2004)
11. Baena, M., Torres, L., Turon, A., Barris, C.: Experimental study of bond behaviour between concrete and FRP bars using a pull-out test. Compos. B Eng. **40**(8), 784–797 (2009)
12. Esfahani, M.R., Rangan, B.V.: Local bond strength of reinforcing bars in normal strength and High-Strength Concrete (HSC). ACI Struct. J. **95**(2), 96–106 (1998)

On the Finite Element Modelling of Long-Term Behavior of Pre-cracked RC Beams Strengthened with FRP

Weilai Yao[1], Tao Sun[1](✉), Yuanxue Liu[1], Junru Ren[1], Rui Mu[1], Xinlei Cheng[1], Yixin Lei[1], and Binghong Li[2]

[1] Military Installations Department, Army Logistics Academy, University Town, Shapingba District, Chongqing 401311, China
suntao_tju@126.com

[2] School of Civil Engineering, Chongqing Jiaotong University, No. 66, Xuefu Avenue, Nan'an District, Chongqing 400074, China

Abstract. The principal objective of this paper is to numerically study the time-dependent behavior of reinforced concrete (RC) beam externally bonded with an FRP system considering concrete pre-cracking and FRP *stress*-lagging. Load-sustaining experiments were performed on seven specimens with different loading paths before strengthening. Based on the authors' previous study, an advanced finite element (FE) model was presented. The FE model included the time-dependent behaviors of materials and the bond-slip response between concrete and steel bars. From the results of the FE model, the time-dependent deflections and strains were reasonably predicted, as well as the gradual cracking of concrete along with time. The complicated changes of structural stress/strain with time were investigated. Also, the sensitiveness of FRP-concrete interfacial creep on long-term deformation was discussed.

Keywords: Fiber-reinforced polymer (FRP) · Creep · Finite element analysis (FEA)

1 Introduction

Structural repair and strengthening have received considerable emphasis over the past decades. Applying fiber-reinforced polymer (FRP) to structural members as reinforcing materials bonded externally is a widely used retrofitting technology.

Much literature has been published on FRP-strengthened structures. Many have rightfully dealt with the instantaneous or time-independent responses, e.g., ultimate capacity, stiffness, and ductility, whereas the actual service state for most constructions is under sustained load. To investigate the long-term response, one of the main problems involves the creep behaviors of materials. For FRP-strengthened.

reinforced concrete (RC)structures, the evident viscoelasticity of the epoxy adhesive layer and concrete are of particular concern.

W. Guo et al. (Eds.): GBCESC 2023, LNCE 328, pp. 214–225, 2024.
https://doi.org/10.1007/978-981-99-9947-7_22

In recent years, there has been an increasing amount of literature on the time-dependent responses of FRP-strengthened RC structures [1–3]. Moawad [4] undertook a long-term experimental program using 28 RC beams that were un-strengthened and NSM FRP-strengthened. The findings demonstrated that increasing concrete strength improved the ratio of time-dependent deflection to instantaneous deflection for both strengthened and un-strengthened beams. Considering the creep effect, Hadjazi [5] created a cohesive zone model to examine intermediate crack debonding failure in an FRP-plated concrete beam. It has been discovered that the creep load can cause an early interfacial debonding between the FRP plate and concrete beam after a given period of operation.

Finite element (FE) modelling is a powerful tool to investigate the time-dependent response of FRP-strengthened RC structure. However, in the previous studies, a rational FE model is still lacking. Moreover, the former research has not considered a stress/strain-lagging phenomenon for an FRP-strengthened structure. In most actual engineering, the concrete structure has been in service for some time before strengthening, and the previous deformation occurs. Until the completion of structural strengthening, the FRP externally is non-stressed. When the load is further increased, and additional deformation occurs, the FRP laminate becomes tensioned. The FRP deformation lags behind the RC structure, which can be named strain/stress-lagging [6]. Nevertheless, most prior research on long-term behaviors has not addressed this issue. The loading path of the RC structure before rehabilitation has not been considered, i.e., the RC structures were directly bonded with FRP without subjecting any preload before strengthening.

In this paper, a set of long-term tests including FRP stress-lagging are introduced. Then, based on the previous study of the author [7, 8], an updated FE model was recommended, which successfully simulated the stress-lagging phenomenon. The complicated changes of structural stress and strain along with time are also investigated.

2 Test Specimen

There were seven test beams in four groups, as presented in Table 1. The identification of the tested elements was GX-q, with G standing for the test group (A, B, C, D), X for whether unloading the preload before strengthening (U = unloading the preload, K = keeping the preload), q for the level of preload (0, $15\%F_u$, $45\%F_u$, $65\%F_u$).

Group A contains one beam, i.e., A-0. The A-0 was directly bonded with the CFRP plate without bearing preload (q = 0). Groups B, C, and D contain two beams each, which bore preloads before strengthening but with different loading paths. The BU-15, CU-45, and DU-65 were strengthened after entirely removing the preloads, while the BK-15, CK-45, and DK-65 were strengthened by keeping the preloads. The same values of forces preloaded the two specimens in each group. Typical FRP stress-lagging occurred in BK-15, CK-45 and DK-65.

After strengthening, each specimen was finally loaded to the same level of force ($75\%F_u$), which was then maintained for 300 days. The Fu (156.95kN) herein is the theoretical ultimate capacity of an un-strengthened RC beam with the same material properties, geometric information, and steel deployment, which is predicted based on cross-section analysis without considering the tension-stiffening of concrete.

Table 1. Strengthening schemes

Test Group	Specimen	Preload (The first stage load)	Whether unload the preload before strengthening?	The second stage load	The final sustained load
A	A-0	0	--	$75\%F_u$	$75\%F_u$
B	BU-15	$15\%F_u$	Yes	$75\%F_u$	
	BK-15	$15\%F_u$	No	$60\%F_u$	
C	CU-45	$45\%F_u$	Yes	$75\%F_u$	
	CK-45	$45\%F_u$	No	$30\%F_u$	
D	DU-65	$65\%F_u$	Yes	$75\%F_u$	
	DK-65	$65\%F_u$	No	$10\%F_u$	

The seven test beams were identical except for the different strengthening schemes and loading paths. The specimen cross-section was 250 × 400 mm. The span of the test element was 3300 mm, with a net span of 2900 mm. In order to reflect the structural service state, the creep of concrete was designed within the scope of linear behavior, i.e., the instantaneous compressive stress was limited to 40% of ultimate strength. Thus, the compression zone of the test beam was adequately reinforced: three D28 rebars with a diameter of 28 mm each were deployed. Aiming to highlight the effect of bonding FRP, relatively less tensile steels were applied: three D14 rebars with a diameter of 14 mm each were deployed. The width of the bonded CFRP plate was 200 mm. Additional CFRP laminates were used to wrap each end of the CFRP plate to prevent premature debonding. The material properties are given in Table 2.

Table 2. Material properties

Materials	Properties	Values	Materials	Properties	Values
Concrete	Slump (mm) [b]	220	CFRP plate	Thickness (mm) [b]	1.4
	Unit weight (kg/m^3) [b]	2301		Ultimate tensile strength (MPa) [b]	2482
	28-day cylinder strength (MPa) [a]	34.0		Modulus of elasticity (GPa) [b]	174
Tensile steel	Yield strength (MPa) [a]	461.2	Stirrup	Yield strength (MPa) [a]	435.0
	Ultimate tensile strength (MPa) [a]	625.0		Ultimate tensile strength (MPa) [a]	610.0
Compressive steel	Yield strength (MPa) [a]	420.2	Epoxy resin	Ultimate tensile strength (MPa) [b]	39.6
	Ultimate tensile strength (MPa) [a]	570.0		Modulus of elasticity (GPa) [b]	3.468

Notes: [a] values measured experimentally by the authors of this paper; [b] values provided by the materials suppliers

Custom-designed test systems mainly consisting of reaction frames, force sensors, and screw jacks were used for applying and sustaining load (Fig. 1). With the increasing deflection due to material creep, the sustained force may present slight decays. As the percentage of dropped load reached 1 ~ 2%, the screw jack was further lifted to replenish the load to the designed level (75%F_u). The load-time curve measured by the force sensors shown in Fig. 1 demonstrates that the external force was well maintained during the long term.

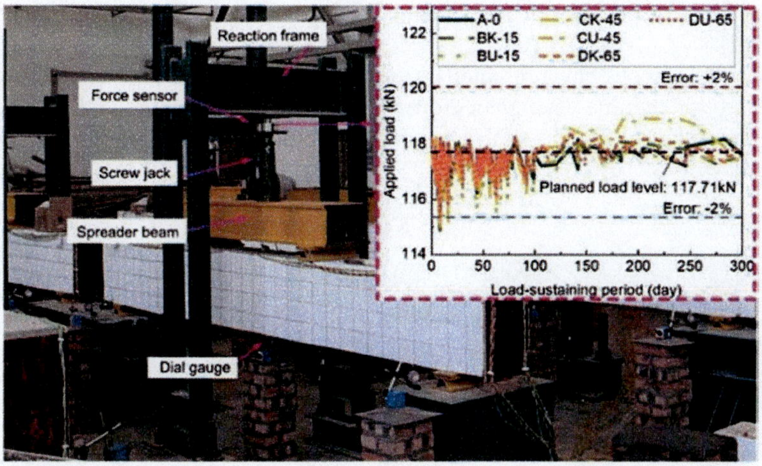

Fig. 1. Test setup

3 Finite Element Modeling

Time-dependent concrete cracking was observed during long-term test. Thus, one of the objectives of the FE model is to reasonably simulate the further propagation of cracks during the load-sustaining period. As noted by the previous studies, for accurate modeling of the localized cracking behavior, an FE model based on the smeared-crack approach must include an accurate bond-slip model to reflect the bond behavior between concrete and internal steels [9–11].

In the former research of the authors [7, 8], an FE model including the material creep and concrete shrinkage was established, which did not consider the concrete-steel bond-slip behavior. Thus, the previous model failed to reproduce the time-dependent cracking.

In this paper, an updated FE model based on ABAQUS was further developed, in which the concrete-steel bond-slip response was included as a significant improvement. Additionally, to further reduce the computational cost, the adhesive layer and the external FRP were no longer modeled as solid elements [7, 8]. The changing patterns of structural stress and strain with time were investigated based on the updated FE model.

3.1 Modeling of Materials

The detailed description of concrete mechanical properties can be found in the previous study of the author [7], which are not presented herein for brevity. User-defined subroutines (USDFLD and UEXPAN) were coded to model concrete's creep and shrinkage behaviors.

Both the steel and FRP were defined as linear elastic materials and were modeled by truss elements (T2D2). The bond behavior between concrete and internal steel, including the longitudinal bar and stirrup, was modeled by applying interfacial elements (C2DH4). The properties of the interfacial elements were defined using the CEB-FIP bond-slip model [12].

Due to the viscous flow presented by the epoxy adhesive, the FRP-concrete interface exhibits prominent time-dependent properties [13–15]. The bonding interface was modeled by interfacial element (C2DH4), and the viscoelastic constitutive model defined its creep behavior as [16]:

$$\tau = K(t, t_o) \cdot S \tag{1}$$

where τ and s are the interfacial shear stress and relative slip. The $K(t,t_0)$ is the effective stiffness, which can be expressed according to the effective modulus method as:

$$K(t, t_o) = \frac{k(t_o)}{1 + \varphi_k(t, t_o)} \tag{2}$$

where $K(t_0)$ is the stiffness at the time of loading t_0, and $\varphi_K(t, t_0)$ is the creep coefficient of the FRP-concrete interface, which can be described as [16]:

$$\varphi_k(t, t_o) = \varphi_u\left(1 - e^{\frac{-\Delta t}{\tau^*}}\right) \tag{3}$$

where $\Delta t = t\text{-}t_0$ refers to the period of load sustaining, φ_u is the ultimate value of the creep coefficient, and τ^* is a material parameter.

3.2 Modeling of FRP Stress-Lagging

In this study, except for specimen A-0, there are loading or unloading paths for the RC beams before strengthening. Thus, the external FRP does not work synchronously with the RC beams, which makes FE modeling difficult. This paper recommends an *air-tracking element* method to solve the problem.

Firstly, the adhesive layer and CFRP plate were typically modeled with real mechanical properties. Secondly, using the ABAQUS keyword **Elcopy*, the adhesive and CFRP elements were duplicated in situ. The elastic modulus of the duplicated elements was set to 10^{-5} times the modulus of the real elements. Since the duplicated elements had minimal stiffness, they were named *air elements*. The real elements were deactivated during preloading and unloading using the keyword **Model change, remove*. Thus, the RC beam was only bonded with the *air elements*, which hardly influenced the mechanical behavior of the RC beam due to their ignorable stiffness. In this way, the preloading and unloading stages were simulated. Thirdly, when simulating the bonding of the real

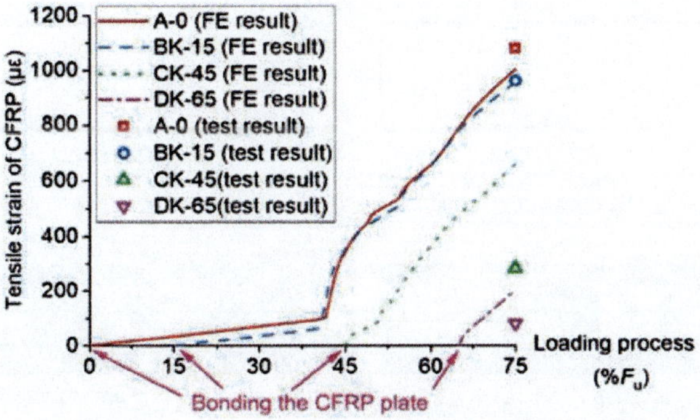

Fig. 2. The strains in the CFRP plates during the short-term loading (at the loading point section)

CFRP, the *air elements* were deactivated, while the real elements were activated using the keyword **Model change, add = strain free.* Since the *air* and real elements share common nodes, the activated real adhesive and FRP tracked the deformation of the RC beam but with zero initial stress. Figure 2 presents the strains in CFRP plates during the short-term loading, demonstrating that the phenomenon of FRP *stress/strain-lagging* is successfully reproduced.

4 Results of the FE Model

4.1 Cracking Patterns

The FE models predicted the concrete cracking patterns, as shown in Fig. 3. The cracking region was marked by black. For brevity, the FE results of A-0 and DU-65 are shown herein. As presented, after sustaining load for 300 days, black regions had extended (e.g., Fig. 3(c) and (d)), which reflects the further development of cracks.

The cracking patterns obtained from the test and the ones by the FE model for specimen A-0 are compared herein. For the instantaneous cracking state (by comparing Fig. 3(a) and (c)), the FE results basically reproduce the locations of cracks, as well as the length of cracks. The number of cracks also basically matches: 5 cracks in the experimental observation and 4 in the FE results. For the long-term cracking state (by comparing Fig. 3(b) and (d)), the FE results have successfully simulated the further propagations of the existing cracks, while the FE model did not ideally reproduce some newly appeared cracks in the test. Nevertheless, considering the applied smeared cracking model is still based on the continuum mechanics, the FE cracking patterns are actually acceptable.

It should be noted that the further development of longitudinal bonding cracks near the surface of tension steel took place in specimen DU-65 (Fig. 3(f)). This region presented the peak values of bond stresses, as shown in Fig. 3(h). However, they were less than $2\text{MPa} < 0.2\tau_u = 2.38\text{MPa}$, which reveals that the bond behavior was still in the linear stage nearly (Fig. 3(i)). The longitudinal cracks herein may be judged as the bonding cracks at the early loading phase (also may be called *Goto cracks* [17]) (Fig. 3(i)), which does not mean the loss of concrete-steel interfacial performance.

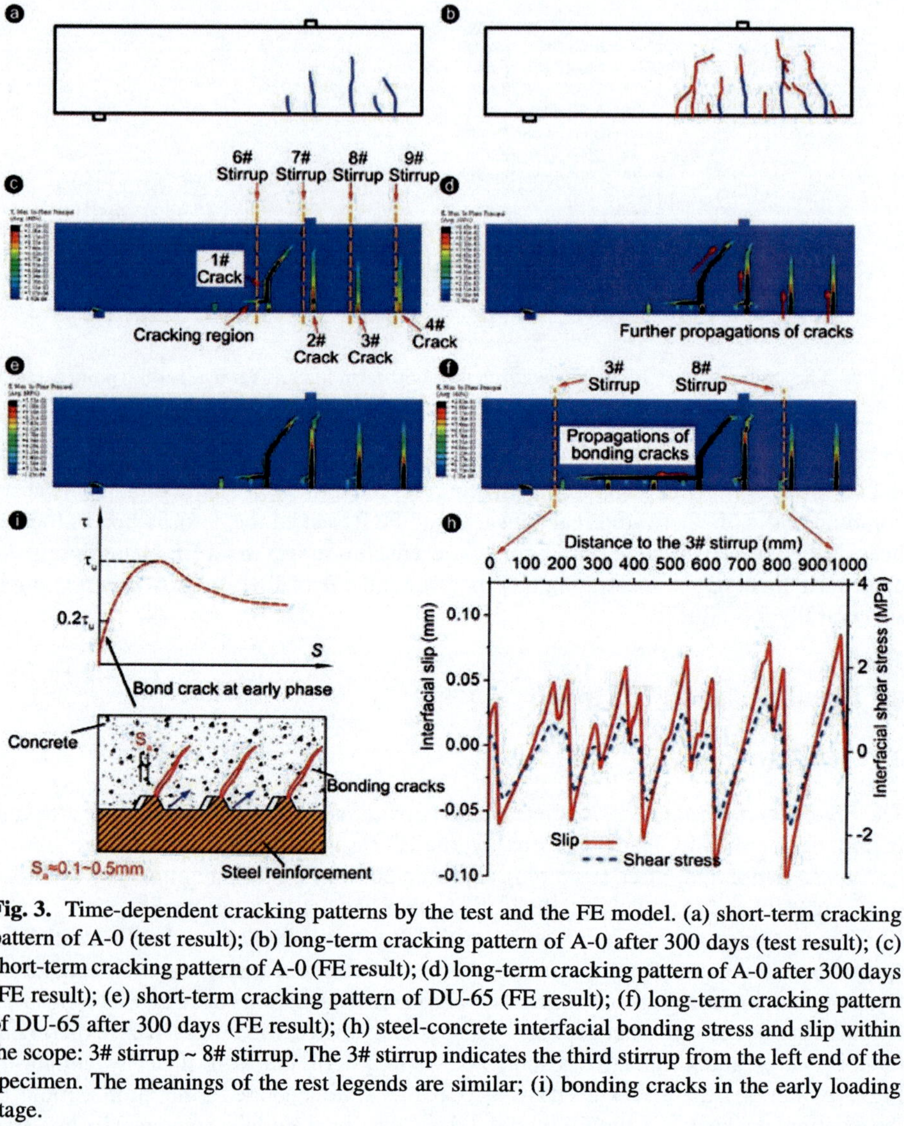

Fig. 3. Time-dependent cracking patterns by the test and the FE model. (a) short-term cracking pattern of A-0 (test result); (b) long-term cracking pattern of A-0 after 300 days (test result); (c) short-term cracking pattern of A-0 (FE result); (d) long-term cracking pattern of A-0 after 300 days (FE result); (e) short-term cracking pattern of DU-65 (FE result); (f) long-term cracking pattern of DU-65 after 300 days (FE result); (h) steel-concrete interfacial bonding stress and slip within the scope: 3# stirrup ~ 8# stirrup. The 3# stirrup indicates the third stirrup from the left end of the specimen. The meanings of the rest legends are similar; (i) bonding cracks in the early loading stage.

4.2 Deformations

Time-Dependent Deflections and Strains/Stress. Figure 4 shows the comparisons between the FE and test results of the time-dependent deflections. Generally, the FE results agree with the experimental results well. The predicting errors are within 0.5 mm or less for the time-dependent deflections. Figure 5 compares the FE and test results of the long-term strains of the beam top (measured by 1# vibrating wire gauge, VWG1) and the concrete near the compression steel (measured by 2# vibrating wire gauge, VWG2). As shown, the predicted results basically agree with the test data.

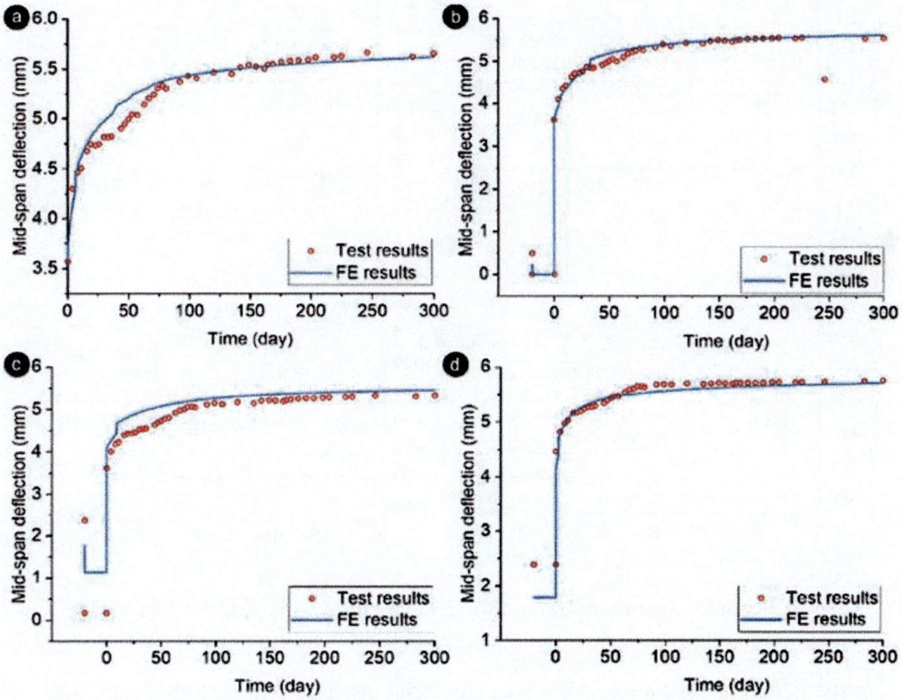

Fig. 4. Predicted long-term deflections of (a) A-0; (b) BU-15; (c) CU-45; (d) CK-45

This paper investigates the time-dependent strains of 5 locations at the tensile bar. As shown in Fig. 6, locations 1, 3, and 5 are at the cracks, while locations 2 and 4 are between the cracks. The strain-time relationships are curves for these locations as the deflection-time relation (Fig. 6(a)). Whereas it is found that the deflection-strain relationship for each site is nearly a line, and the slope (k) is close to each other (Fig. 6(b)). This finding may conduce to the analyses of time-dependent structural deformations.

The time-dependent stress in the stirrup, which is hardly addressed by previous research, has been numerically explored in this study. Based on the cracking pattern, the diagonal compression field model of the test beam can be developed, as shown in Fig. 7. The 8# and 9# stirrup are in the web members. The upper part of the 6# stirrup is located in the diagonal compressive zone, and the rest is in the tensile area. Under instantaneous loading, the 8# and 9# stirrup in the pure bending section nearly do not present stresses, the upper part of the 6# stirrup presents compressive stress, and the rest shows tensile stress. After 300 days of sustaining load, if both the material creep and concrete shrinkage are considered, all the 6#, 8#, and 9# stirrups show increased compressive stresses, as presented by the dashed lines in Fig. 7.

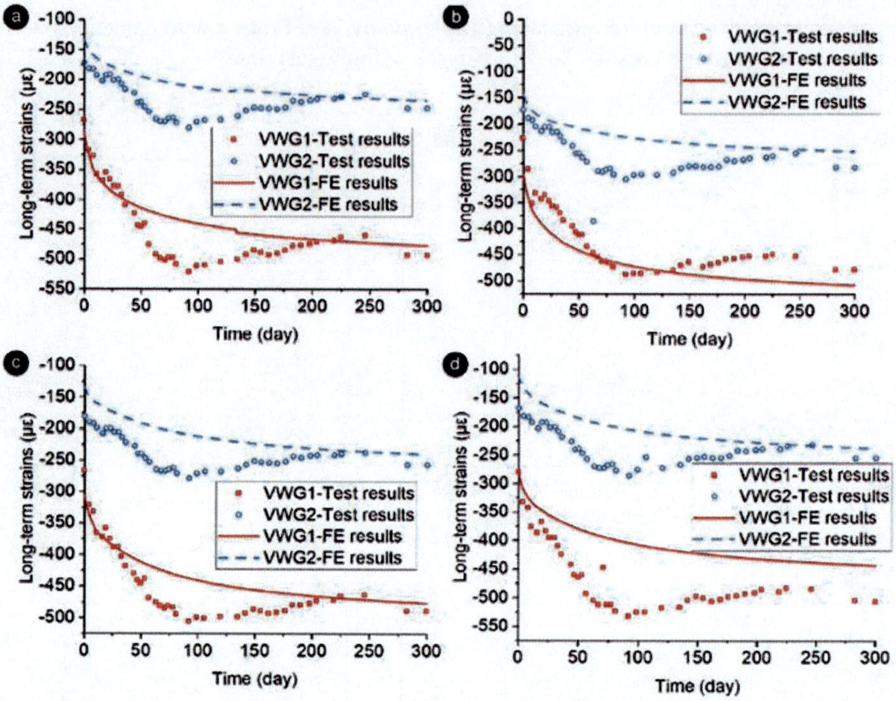

Fig. 5. Predicted long-term strains of (a) BK-15; (b) BU-15; (c) CU-45; (d) DU-65

If the concrete shrinkage is ignored (material creep is still included), the 300-day stresses almost remain unchanged compared to the instantaneous stresses, as shown by the dotted lines in Fig. 7. This result indicates that long-term stirrup stress is mainly influenced by the concrete shrinkage. To be more specific, it is believed that the concrete shrinkage in the vertical direction causes the further compression of stirrups.

Sensitiveness of FRP-Concrete Interfacial Creep. The effect that the interfacial creep leads to additional long-term deformation has been confirmed previously [18], whereas its sensitiveness remains unclear, which is discussed in this paper.

In this study, as shown in Fig. 8(a), the effect of interfacial creep is represented by the difference between the predicted 300-day deflections of the two cases neglecting adhesive creep or not, i.e., (f1-f2). Parameter analyses altering FRP amounts were carried out (Fig. 8(b)). As shown, when increasing the amount of FRP, the predicting difference (f1-f2) also increases, which indicates the increased effect of interfacial creep. Other parameter analyses altering tensile steel amounts were also conducted (Fig. 8(c)). As presented, when increasing the amount of tensile steel, the predicting difference (f1-f2) decreases, indicating that the effect of interfacial creep becomes less sensitive.

The above parameter investigations demonstrate that the relative amount of tensile steel and FRP influences the sensitivity of interfacial creep. Essentially, the FRP-concrete interface's creep that induces the gradual unloading of FRP undermines the strengthening

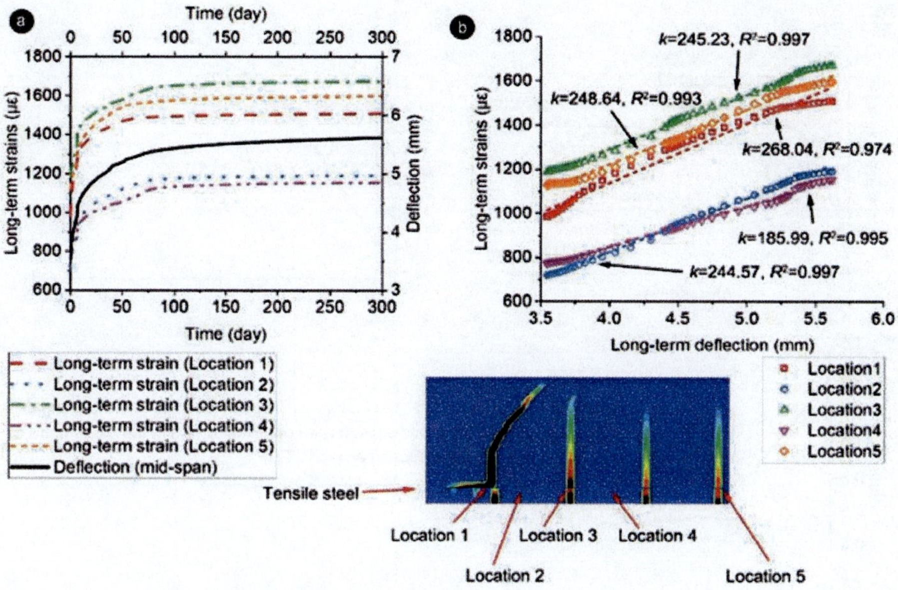

Fig. 6. (a) long-term strains in tensile steel (specimen A-0); (b) relationship between time-dependent strain and deflection (specimen A-0)

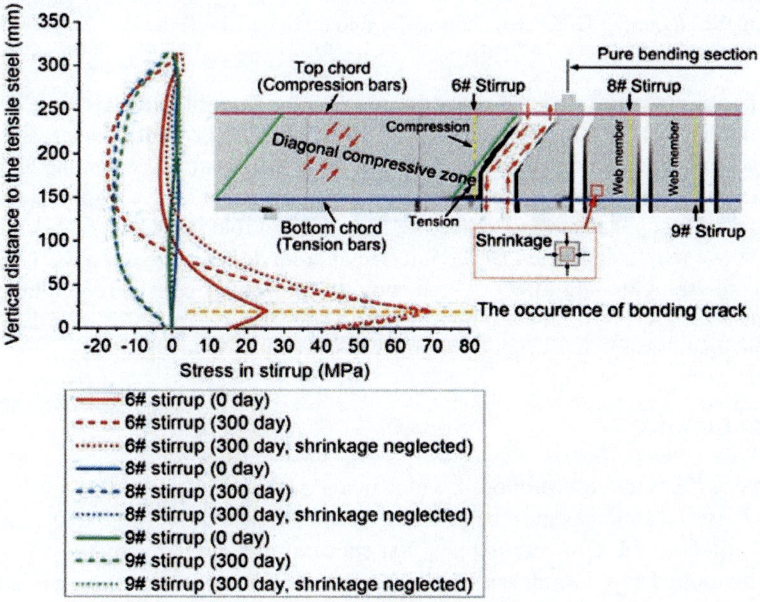

Fig. 7. Time-dependent stress in stirrups (specimen A-0)

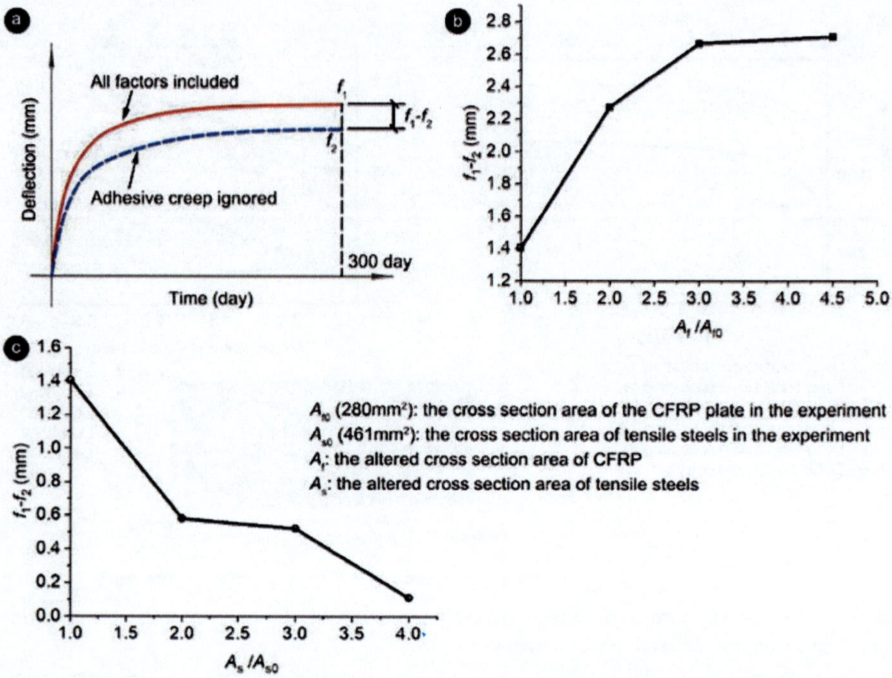

Fig. 8. Sensitiveness of FRP-concrete interfacial creep, (a) f_1-f_2, the effect of adhesive creep; (b) altering the amount of CFRP; (c) altering the amount of tensile steels

effect. The strengthening effect is pronounced when the amount of FRP is large or tensile steel is small. Thus, the deactivation of strengthening caused by adhesive creep leads to a clear response, i.e., a considerable value of (f_1-f_2). In contrast, when the amount of tensile steel is large or FRP is small, the strengthening effect is less significant. Hence, the interfacial creep leads to less response, i.e., an ignorable value of (f_1-f_2).

The above results indicate that the interfacial creep is sensitive to an RC beam with a significant strengthening effect, which should be carefully considered. Whereas, for a specimen with a less strengthening effect, the adhesive creep is not sensitive, which may be neglected when calculating the time-dependent deformations.

5 Conclusions

An advanced FE model was proposed, which included the shrinkage, creep, the bond-slip behavior between concrete and steel, and the viscous flow of the FRP-concrete interface. The phenomenon of FRP *stress-lagging* was successfully simulated by the *air-tracking element* method. The FE model rationally predicted the long-term responses, including the propagation of concrete cracks along with time. Numerical investigations were further performed. The time-dependent strain and deflection present a linear relationship for the location on the tensile steel. Although the analyzing location differs, the slope k is close to each other. The time-dependent stress in the stirrup was analyzed within the framework of the diagonal compression field model, whose variations with time were mainly induced by the concrete shrinkage. The effect of FRP-concrete interfacial creep is further confirmed, and its sensitiveness is found to be related to the relative amount of tensile steel and FRP externally, i.e., the level of strengthening effect.

Acknowledgement. This work was financially supported by the Natural Science Foundation of Chongqing (No. CSTB2022NSCQ-MSX0194 and No. CSTC2019JCYJ-MSXMX0585), the National Natural Science Foundation of China (Grant No. 41877219) and the Doctoral project of Chongqing Science and Technology Bureau (No. CSTB2022BSXM-JSX0023).

References

1. Marí, A.R., Oiler, E., Bairan, J.M., et al.: Simplified method for the calculation of long-term deflections in FRP-strengthened reinforced concrete beams. Compos. B Eng. **45**(1), 1368–1376 (2013)
2. Oller, E., Marí, A.R.: Long-term bond stresses and debonding failure of FRP-strengthened RC cracked members. Compos. B Eng. **52**(9), 30–39 (2013)
3. Jin, F., Lees, J.M.: Experimental behavior of CFRP strap-strengthened RC beams subjected to sustained loads. J. Compos. Constr. **23**(3), 04019012 (2019)
4. Moawad, M., Baena, M., Barris, C., et al.: Time-dependent behavior of NSM strengthened RC beams under sustained loading. Eng. Struct. **247**, 113210 (2021)
5. Hadjazi, K., Sereir, Z., Amziane, S.: Creep response of intermediate flexural cracking behavior of reinforced concrete beam strengthened with an externally bonded FRP plate. Inter. J Solids Struct. **94–95**, 196–205 (2016)
6. Saadatmanesh, H., Malek, A.M.: Design guidelines for flexural strengthening of RC beams with FRP plates. J. Compos. Constr. **3**(4), 158–164 (1998)
7. Jiang, S.Y., Yao, W.L., Chen, J., et al.: Time dependent behavior of FRP-strengthened RC beams subjected to preload: experimental study and finite element modeling. Compos. Struct. **200**, 599–613 (2018)
8. Jiang, S.Y., Yao, W.L., Chen, J., et al.: Finite element modeling of FRP-strengthened RC beam under sustained load. Advan. Mater. Sci. Eng. (2018)
9. Chen, G.M., Teng, J.G., Chen, J.F.: Finite element modeling of intermediate crack debonding in FRP-plated RC beams. J. Compos. Constr. **15**(3), 339–353 (2010)
10. Chen, G.M.: Behaviour and strength of RC beams shear-strengthened with externally bonded FRP reinforcement. Hong Kong Polytechnic University (2010)
11. Chen, G.M., Chen, J.F., Teng, J.G.: On the finite element modelling of RC beams shear-strengthened with FRP. Constr. Build. Mater. **32**(4), 13–26 (2012)
12. CEB-FIP. CEB-FIP Model Code 90. Thomas Telford, London (1993)
13. Diab, H., Wu, Z.: Nonlinear constitutive model for time-dependent behavior of FRP-concrete interface. Compos. Sci. Technol. **67**(11–12), 2323–2333 (2007)
14. Diab, H., Wu, Z.: A linear viscoelastic model for interfacial long-term behavior of FRP-concrete interface. Compos. B Eng. **39**(4), 722–730 (2008)
15. Wu, Z., Diab, H.: Constitutive model for time-dependent behavior of FRP-concrete interface. J. Compos. Constr. **11**(5), 477–486 (2007)
16. Mazzotti, C., Savoia, M.: Stress redistribution along the interface between concrete and FRP subject to long-term loading. Advances Struct Eng. **12**(5), 651–661 (2009)
17. Goto, Y.: Cracks formed in concrete around deformed tension bars. ACI J. Proc. **68**(4), 244–251 (1971)
18. Taha, M.M.R., Masia, M.J., Choi, K.K., et al.: Creep effects in plain and fiber-reinforced polymer-strengthened reinforced concrete beams. ACI Struct. J. **107**(6), 627–635 (2010)

Simulation Analysis of Reflection Crack Propagation Path of Asphalt Overlay Under Coupling Load

Qinshou Huang(✉)

Guangxi Communications Design Group Co. Ltd., Guangxi 530029, China
kz101216@163.com

Abstract. At present, the widely used scheme in the cement concrete pavement reconstruction and extension project is to add asphalt surface layer. It is urgent to elaborate the spreading principle of reflection crack. In this paper, the typical asphalt overlay structure of an urban road reconstruction and expansion project in Nanning, Guangxi is taken as an example. Combined with the theory of fracture mechanics, the plane strain model is established by ABAQUS software to analyze the changes of stress, strain field and expansion path of each reflecting crack propagation step under coupling load. The results show that the stress and strain field of the crack tip increase with the upward expansion of the reflected crack, and develop rapidly in the late expansion period. The crack propagation Angle decreases with the crack propagation. The expansion path tends to extend upward on the side without load, and the path gradually deviates from the side under load with the expansion of cracks, and the extension length also increases gradually.

Keywords: Asphalt overlay · Reflection crack · Expansion path

1 Introduction

With the increasing economic development and urban traffic volume, the contradiction between road standards and traffic volume demand is becoming increasingly prominent, and a large number of urban road reconstruction and expansion projects have emerged. Because asphalt overlay can effectively improve the performance of old cement concrete pavement, it is widely used in the reconstruction of old cement concrete pavement at home and abroad. How to prevent reflection cracks is one of the main problems to be solved. Lin Juan [1] used ABAQUS software to simulate and analyze the influence of the change of asphalt concrete parameters on stress value and crack propagation of asphalt concrete structure layer coated with old concrete pavement under load. Sun Gan [2] used a numerical simulation method to compare the propagation rules of reflection cracks under five inclination conditions at low temperature. Based on the extended finite element method, Huang Likui [3] conducted dynamic simulation on the fatigue propagation process of AC + CRC composite pavement temperature-shrinkage reflection cracks. At present, scholars' simulation and analysis of the propagation mechanism of

W. Guo et al. (Eds.): GBCESC 2023, LNCE 328, pp. 226–234, 2024.
https://doi.org/10.1007/978-981-99-9947-7_23

reflective cracks under coupled loads are still relatively few. This paper takes the typical asphalt overlay structure of an urban road reconstruction and expansion project in Nanning, Guangxi as the research background, and simulates the expansion path of asphalt overlay structure with reflective cracks under coupled loads. The stress-strain, Angle and coordinate values of each crack propagation step are obtained, and the variation law of stress-strain and propagation path is analyzed.

2 Basic Theory

2.1 Fracture Propagation Criteria

According to the maximum circumferential normal stress theory of fracture mechanics, crack propagation direction and critical length are determined, as shown in Eqs. (1) and (2).

The propagation direction of I-II composite fractures is determined by Formula (1) [4].

$$\theta_0 = \cos^{-1}\left(\frac{3K_{II}^2 + \sqrt{K_I^4 + 8K_I^2K_{II}^2}}{K_I^2 + 9K_{II}^2}\right) \tag{1}$$

(1) Where, θ_0 is the fracture propagation Angle.

The critical stress intensity factor of crack propagation is shown in Eq. (2) [4]:

$$K_\theta = \cos\frac{\theta_0}{2}\left(K_I\cos^2\frac{\theta_0}{2} - \frac{3}{2}K_{II}\sin\theta_0\right) = K_{IC} \tag{2}$$

(2) Where, K_θ is the effective stress intensity factor;, K_I and, K_{II} were type I and type II stress intensity factors, respectively. K_{IC} is the fracture toughness of the material.

2.2 Selection of Cracking Step Size

The crack develops in any direction with a certain crack step size, and generally does not extend along the original direction. The selection of expansion increment has a direct impact on the simulation results. If the selection of expansion increment is too large, it will produce a large error and even make the results deviate greatly from the actual situation. If the increment is too small, although it can improve the calculation accuracy, it will greatly reduce the calculation efficiency. At the same time, it can be seen from Formula 2 that the larger the crack is, the easier it is K_θ to expand. In this paper, Formula (3) below is adopted to define the cracking step size of each step [5].

$$\Delta b_n = \Delta b_0\frac{K_\theta^{(n)}}{K_\theta^{(0)}} \tag{3}$$

(3) Where, Δb_0 is the initial crack cracking step length; Δb_n is the crack step length of the NTH step; $K_\theta^{(0)}$ is the initial effective stress intensity factor of crack; $K_\theta^{(n)}$ is the effective stress intensity factor when the crack reaches the first step.

3 Calculation Parameters and Calculation Models

3.1 Typical Pavement Structure and Material Parameters

The current pavement structural of an urban road reconstruction and expansion project in Nanning, Guangxi is cement concrete pavement, which is expanded from the current urban branch road to the main road. The current pavement is less damaged, and the road evaluation grade is good. The design considers the use of asphalt overlay cladding treatment after the treatment of the old road diseases, and typical pavement structural parameters are shown in Table 1 [6].

Table 1. Main material parameters

Structural layer	Thickness h/cm	Modulus of elasticity E/MPa	Poisson's ratio μ	Thermal conductivity /w/m°C	Temperature shrinkage factor /1/°C
Asphalt paving	12	1200	0.25	1.2	2.1 × 10–5
Stress absorbing layer	2.5	800	0.25	1.2	2.1 × 10–5
Old cement concrete slab	24	30000	0.15	1.5	1.0 × 10–5
Base	—	100	0.3	1.0	0.5 × 10–5

3.2 Calculate the Load

The vehicle load adopts standard axle load (BZZ-100) [7]. The temperature load assumes that the initial temperature of the pavement structure is 10°C and the cooling range of the road surface is 15°C. The distribution of the pavement temperature field follows the mathematical model of exponential attenuation, as shown in Eq. (4) [8, 9].

$$T(t, z) = T_0 + \Delta T_m \times \frac{t}{t_m} \times \exp(-z/z_0) \tag{4}$$

(4) Where, z is the depth of the pavement, ΔT_m is the continuous cooling amplitude, t_m is the cooling experience time, z_0 is the initial thickness and takes the value of 0.2 to 0.3 m.

3.3 Calculation Model and Fracture Region Simulation

The calculation model consists of asphalt overlay, stress absorption layer, old cement concrete pavement and foundation. The plane strain finite element model is established by ABAQUS, a large commercial software. It is assumed that each structural layer is homogeneous and isotropic linear elastic material. The foundation and the old cement concrete plate are in friction contact, and the other layers are completely continuous contact; The joint width of the old cement concrete slab is 1cm, and there is no load transfer capacity. The bottom of the foundation is completely restrained, and both sides of the foundation and asphalt overlay are restrained in the normal direction. The initial length of the reflection crack is 5 mm [10].

The expanded size of the foundation was used for simulation. After taking different foundation sizes for error comparative analysis, the proposed expanded size of the foundation was 16.01 m × 8.5 m, the length of asphalt overlay was 10.01 m, the length of stress absorption layer was 10.01 m, and the length of old cement concrete slab was (5 + 5) m. Singular elements were used to simulate the stress and strain field at the crack tip [11], and the grid division near the crack tip was shown in Fig. 1. The asphalt overlay structure diagram is shown in Fig. 2.

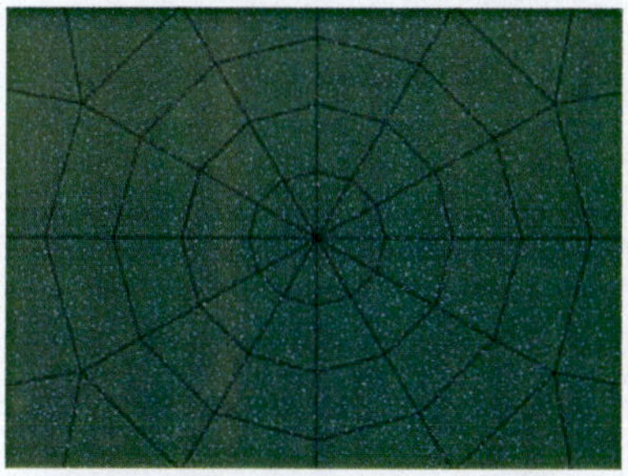

Fig.1. Crack tip meshing diagram

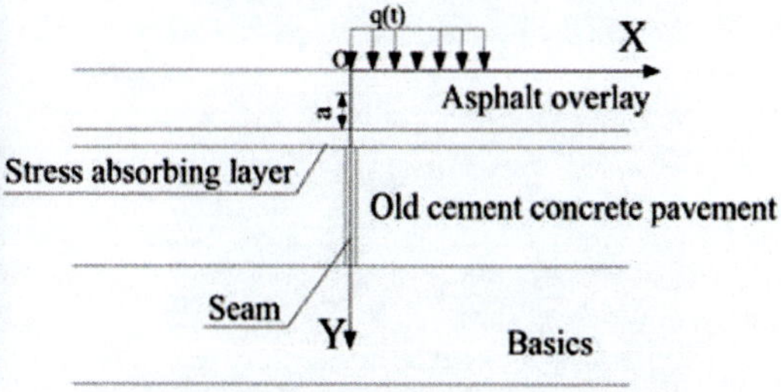

Fig.2. Asphalt overlay structure diagram

4 Reflection Crack Propagation Path Simulation Analysis

The calculation results are shown in Figs. 3, 4, 5, 6, 7, 8, 9, 10, 11, 12 and Table 2.

Figure 3 shows the finite element mesh calculated from step 1 to step 4 of crack propagation and the calculated stress nephogram. Figure 4 shows the Mises stress and strain nephogram at the fifth extended step. The stress changes mainly focus on the crack tip. As the crack spreads upward, the stress and strain gradually extend from the tip down. The propagation path of reflection crack expands upward from the side of load under coupling load.

Figure 5 shows the change of circumferential stress distribution near the crack tip r = 0.2 cm during each crack propagation step. The propagation Angle of each crack step can be calculated from the maximum circumferential tensile stress criterion, and the calculation results are shown in Table 2. The stress increases from 1.99 Mpa at the beginning to 4.61 Mpa at the end, with an increase of 132%.

Figures 6, 7, 8, 9 shows the changes of K*, J, equivalent stress and strain at the crack tip with the increase of crack propagation step. As can be seen from Fig. 6 and Fig. 7, the stress intensity factor and J-integral approximately go through three stages of change with the increase of crack expansion length, firstly increasing rapidly in the

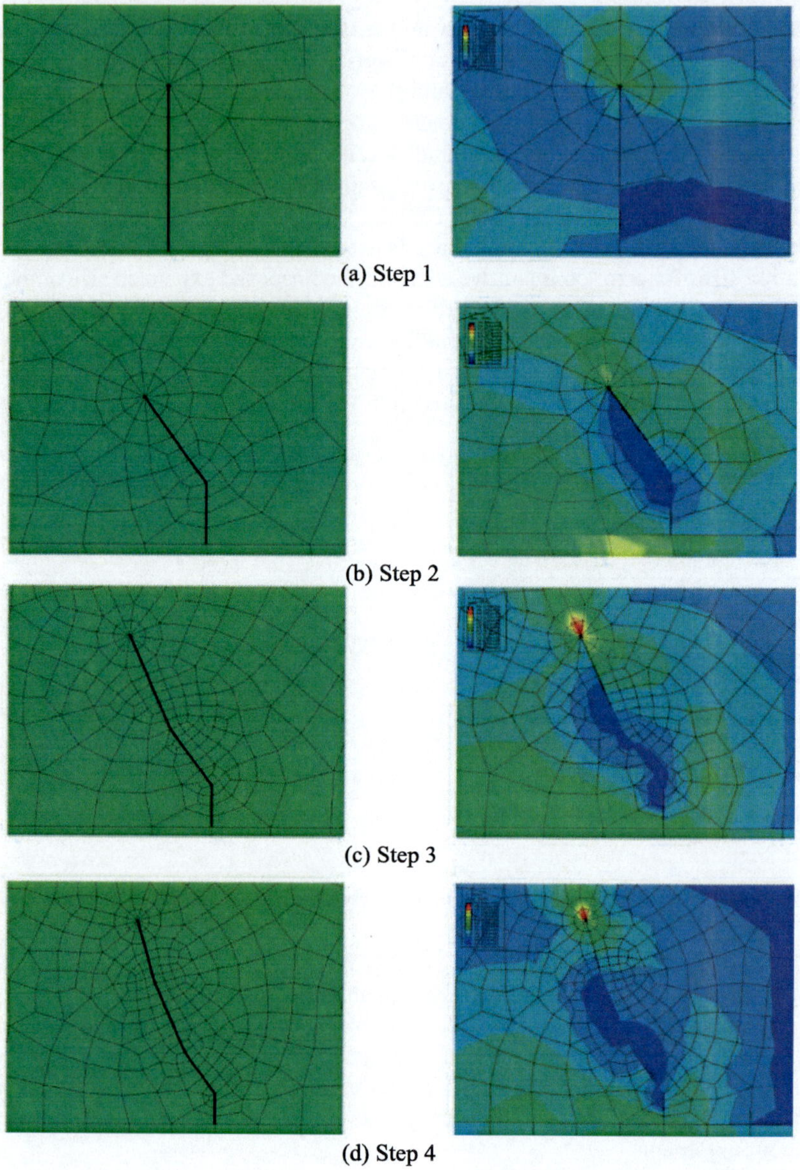

Fig. 3. The first step to the fourth finite element mesh and calculated stress nephogram

initial stage, then increasing slowly, and increasing the most in the later stage. As can be seen from Fig. 8 and Fig. 9, with the increase of crack propagation length, the equivalent stress and strain approximately undergo two stages of change, increasing slowly at the initial stage and increasing at the later stage. As can be seen from Fig. 6–9, the stress and strain field at the crack tip increase with the upward expansion of the reflective crack,

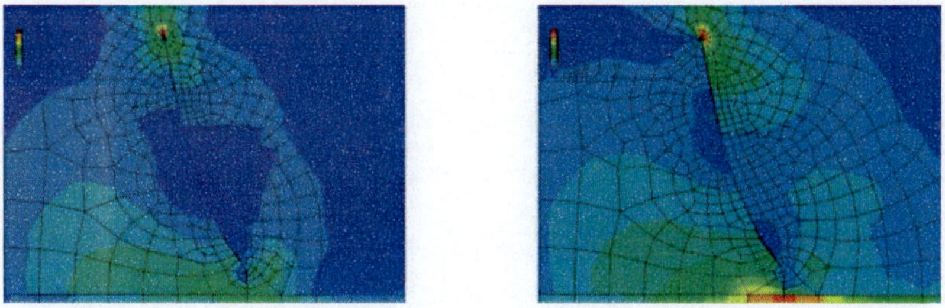

Fig. 4. The fifth step calculated Mises stress nephogram and strain nephogram

and develop rapidly in the late expansion period. When the crack extends to 6 cm, the K*, J, equivalent stress and strain at the crack tip are 2.77 times, 2.80 times, 1.58 times and 3.04 times of the initial crack growth (0.5 cm), respectively.

Figure 10 shows the change of fracture tip propagation Angle with the increase of fracture propagation step. Figures 11 and Fig 12 are the schematic diagram of the path of crack expansion to the top of the asphalt overlay layer and the coordinate diagram of the expansion path respectively. Table 2 shows the calculation results of crack expansion Angle and length at each step. It can be seen from Figs. 10, 11, 12 that with the expansion of the crack, the propagation Angle of the reflected crack gradually decreases, from 34.99° at the initial stage of the expansion to 1.88°, decreasing by 33.11°. With the expansion of the crack, the path gradually deviates from the side of the load, and the length also increases gradually, from 0.61 cm at the first expansion step to 1.90 cm at the seventh expansion step, with an increase of 211%. Because the tensile stress caused by temperature load is greater than the shear stress caused by vehicle load at the initial stage of expansion, the stress caused by vehicle load becomes larger and larger with the expansion of crack, so that the crack expansion Angle becomes smaller and smaller.

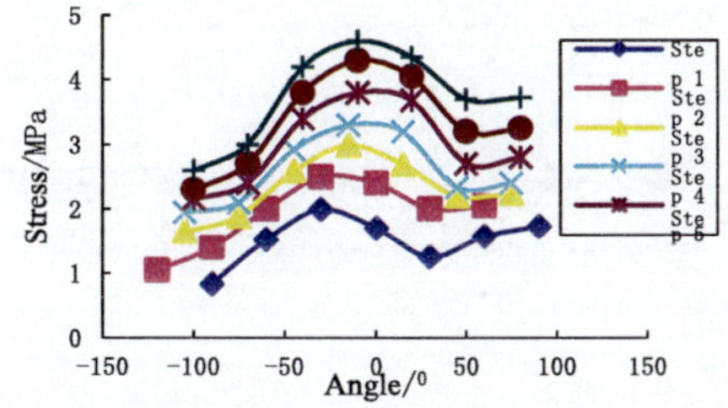

Fig. 5. Equivalent stress distribution near crack tip during crack propagation

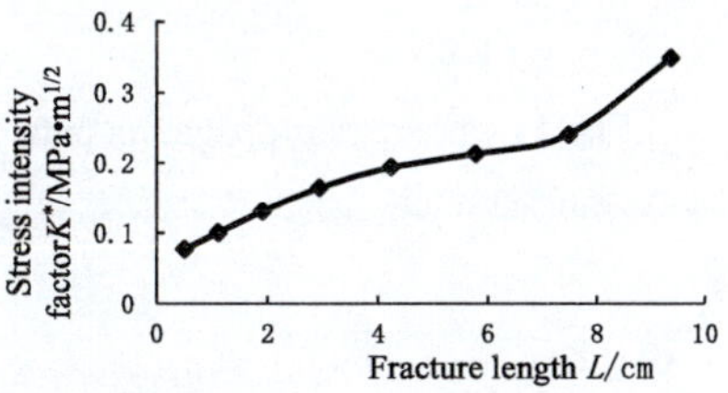

Fig. 6. Curve of relationship between crack propagation length and K*

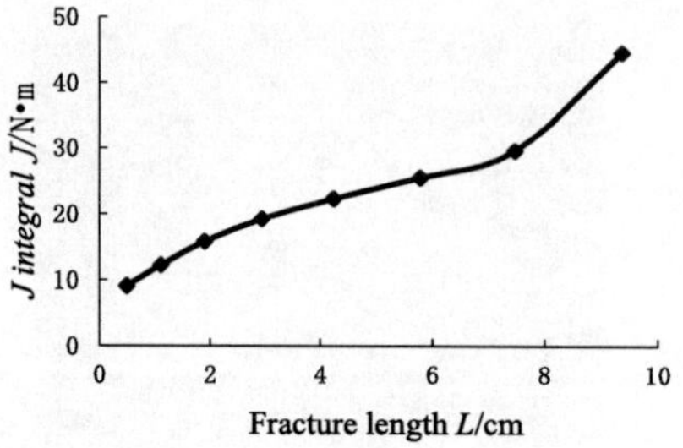

Fig. 7. Fracture extension length and J-integral relationship curve

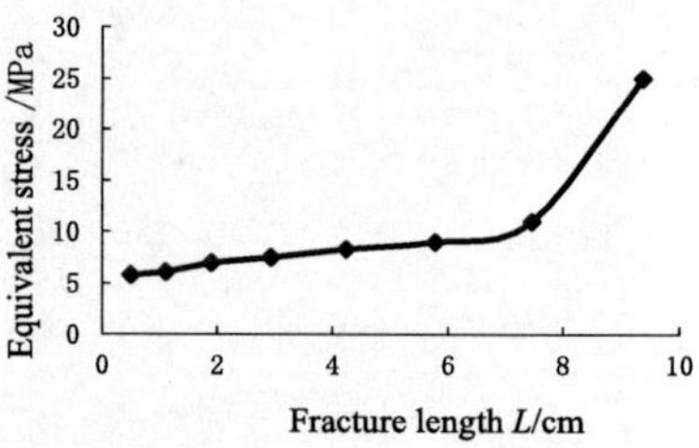

Fig. 8. Curve of the relationship between fracture extension length and equivalent stress at crack tip

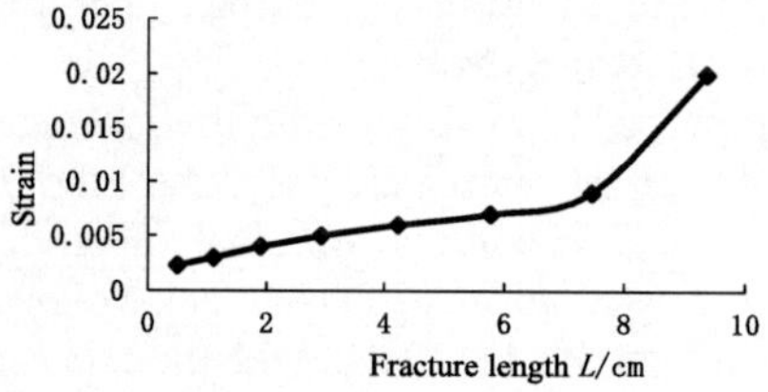

Fig. 9. Fracture extension length and strain curve of crack tip

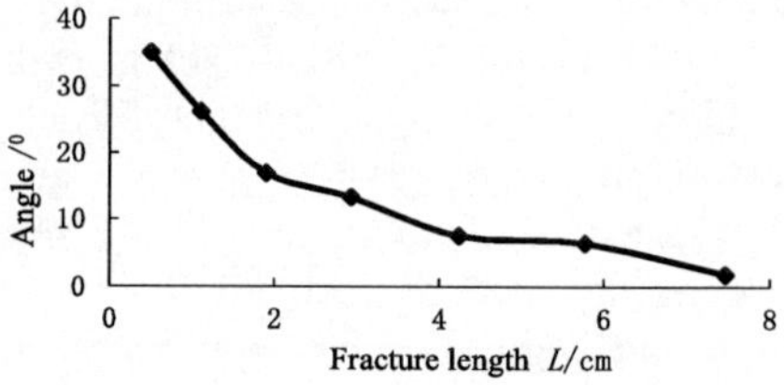

Fig.10. The relationship between the length and Angle of crack expansion

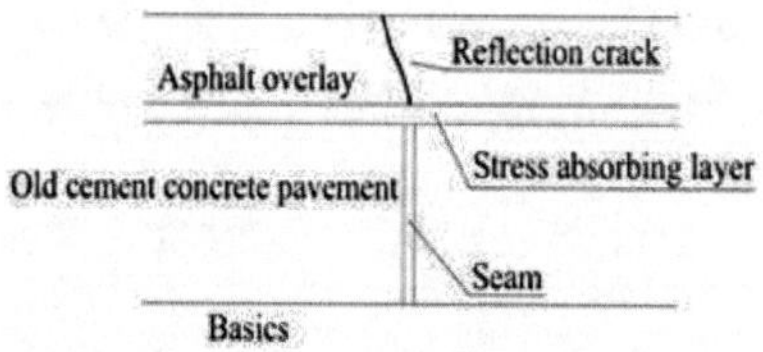

Fig. 11. Fracture propagation path

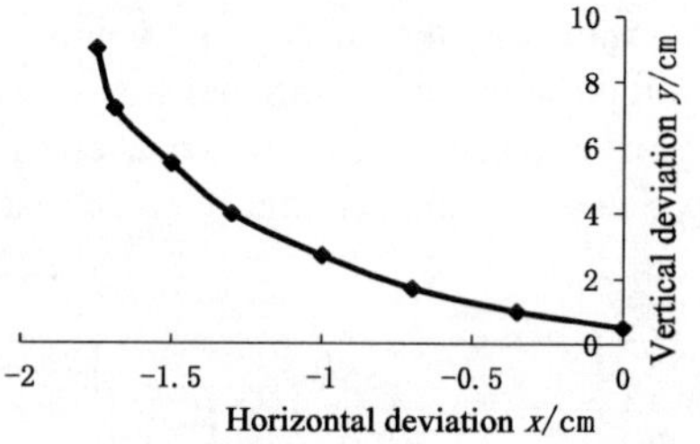

Fig. 12. Crack propagation path coordinates

Table 2. Crack propagation length and propagation Angle values at each step

Step	Fracture propagation length *L*/cm	Expansion Angle $\theta/^0$
initial	0.50	0.00
1	0.61	34.99
2	0.79	26.18
3	1.03	16.89
4	1.30	13.29
5	1.54	7.50
6	1.69	6.45
7	1.90	1.88
Total fracture length	9.36	——

5 Conclusion

With the increasing number of urban road reconstruction and extension projects, more and more asphalt concrete surfacing projects are added to the old cement concrete pavement. How to prevent reflection cracks is one of the main problems to be solved. In this paper, the asphalt overlay structure of a renovation and expansion project in Nanning City is taken as an example to simulate the propagation path of reflecting cracks under coupling load, and reveal the law of crack propagation. The analysis results show that:

(1) The stress change is mainly concentrated in the crack tip; The stress at the crack tip increases with the upward expansion of the reflected crack. The crack propagation Angle can be obtained from the maximum circumferential tensile stress criterion, and the propagation Angle decreases with the crack propagation.
(2) The stress intensity factor, J-integral, equivalent stress and strain of the crack tip increase with the upward expansion of the reflected crack, and develop rapidly in the late expansion period.
(3) The crack expansion path tends to extend upward on the side without load, and the path gradually deviates from the side under load with the crack expansion, and the length gradually increases.

Acknowledgement. The work was supported in part by NSFC project 51968006.

References

1. Juan, L.: Effect analysis of overlay stress absorbing layer on reflection crack based on extended finite element method. China Municipal Eng. **01**, 78–82 (2021)
2. Gan, S., Xinyu, T., Tao, Q.: Study on the propagation of reflective cracks on highway surface under low temperature. Smart City. **7**(23), 75–76 (2021)

3. Likui, H., Zijie, O., Qinghua, Z.: Fatigue propagation of temperature shrinkage reflection Crack in AC+CRC composite pavement. Highway Engineering. **48**(02), 62–67 (2023)
4. Suyuan, W.: Research on Fatigue Life Prediction Method of Asphalt Overlay on Old Cement Pavement [D], pp. 29–43. Changsha Jiaotong University, Changsha (2003)
5. Jianlong, Z., Zhigang, Z., Qisen, Z.: Theory and Method of Anti-crack Design of Asphalt Pavement [M], pp. 15–64. People's Communications Press, Beijing (2002)
6. Industry Standard of the People's Republic of China: Code for Design of Highway Asphalt Pavement (JTG D50–2006)[S]. People's Communications Press, Beijing (2006)
7. Jianlong, Z.; Qisen, Z.: The nolinear analysis of low temperature shrinkage cracking in ashpalt pavements. In: Proceedings of the Asian Pacific Conference on computational Mechanics, Hong Kong (1991)
8. Leslie Ann Myers.Development and propagation of surface-initiated longitudinal wheel path cracks in flexible highway pavements[D]. Dissertation of University of Florida.2000.6–9
9. Industry Standard of the People's Republic of China: Technical Specification for Highway Maintenance (JTJ H10–2009)[S]. People's Communications Press, Beijing (1996)
10. Dongwe, X.: Mechanical Analysis and Fatigue Life Prediction of Asphalt Pavement Overlay [D]. Master Thesis of Xi 'an: Chang 'an University (2005)
11. Weiyuan, Z., Xiaodong, K.: Element free method and its engineering application. Chinese J. Theor. Appl. Mech. **30**(2), 193–201 (1998)

Environmental Disturbance Analysis and Control in the Excavation of a Foundation Pit Near a Building Structure

Xitao Lin[1,2], Fan Mo[2], Yuebang Cui[3], Jinli Xie[3](✉), Gui Huang[1], Hailin Cheng[2], Zongli Gao[1], Shiying Lu[1], Qianwei Xu[3], and Hui Yan[3]

[1] China Construction Eighth Engineering Bureau Co., Ltd., Nanning 530029, China
[2] Guangxi Branch, China Construction Eighth Engineering Bureau Co., Ltd., Nanning 530029, China
[3] Key Laboratory of Road and TCic Engineering, Ministry of Education, Tongji University, Shanghai 201804, China

xiejinli@tongji.edu.cn

Abstract. This study focuses on the analysis and control of environmental disturbance resulting from the excavation of a foundation pit in proximity to an airport hotel. The research aims to investigate the impact of pit excavation on the hotel structure, surrounding ground, and the behavior of the pile support system. The settlement pattern indicates that the hotel structure experiences greater settlements on its southern side, near the excavation site. Moreover, the excavation of deeper layers leads to a noticeable rise in the ground level, with the bottom of the negative first floor exhibiting higher elevation compared to the negative ground floor. These observations highlight the need for careful consideration of excavation depth and unloading effects during construction. Furthermore, the study examines the displacement and deformation characteristics of enclosure piles. It demonstrates that the perimeter pile wall experiences convergence towards the excavation pit, particularly at the bottom and near the top of the piles.

Keywords: Foundation Pit Excavation · Numerical Simulation · Excavation Disturbance · building existing

1 Introduction

With the rapid progress of urbanization, deep foundation pit projects have garnered significant attention; however, their safety aspects are encountering novel challenges [1–3]. In recent years, instances of deep foundation pit instability have become increasingly prevalent, particularly in complex environments with challenging support conditions, resulting in substantial socio-economic repercussions for China [4–6].

Deep foundation pits exhibit notable spatial effects that profoundly impact the deformation and stability of such excavations [7–9]. Scholars have diligently conducted extensive research on the safety of deep foundation pits. For instance, Wang et al. [10] analyzed

W. Guo et al. (Eds.): GBCESC 2023, LNCE 328, pp. 235–243, 2024.
https://doi.org/10.1007/978-981-99-9947-7_24

deformation patterns, settlement causes, and control measures of subgrades and foundation pits based on on-site measurements. Liu et al. [11] constructed a three-dimensional model of a deep foundation pit to investigate the excavation-induced deformation and the accompanying ground settlement and retaining structure deflections. Other researchers have investigated the influence of deep foundation pit geometry, construction sequence, and other factors on deformation, focusing on the spatial effects [12–14]. Previous studies have demonstrated that the excavation of a foundation pit can disrupt the initial stress equilibrium of the surrounding soil [15, 16]. This excavation and unloading process redistributes the stress distribution in the adjacent rock, leading to various alterations in its mechanical behavior [17].

Despite the progress made in the practical analysis and numerical simulation of foundation pit engineering, further investigation is necessary to examine the effects of foundation pit excavation on adjacent buildings. The excavation of foundation pits for construction projects can have significant impacts on the surrounding environment and existing structures. Understanding and mitigating these impacts is crucial for ensuring the safety and stability of nearby buildings and infrastructure.

The objective of this research is to assess the horizontal and vertical deformation, settlement, and inclination of the airport hotel structure during the excavation process, as well as the force and deformation characteristics of the surrounding piles and anchor cables. By investigating the excavation process and its subsequent effects, valuable insights can be gained to inform engineering practices and improve the design and construction of foundation pits in similar geological conditions.

2 Three-Dimensional Finite Element Numerical Modelling

2.1 Model Calculation Parameters

The stratigraphy within the study area is primarily composed of three distinct soil layers: plain fill, clay, and dolomitic tuff, with intermittent occurrences of tuff. Table 1 provides detailed information on the materials and property parameters associated with the stratigraphy and structural units. To simulate the stratigraphic conditions, the Modified-MC Intrinsic Ontology model is employed. The foundation pit envelope is treated as a rigidly equivalent 2D slab unit for the 1000-diameter bored pile, while the pile wall is considered equivalent to a depth of 27 m. For the simulation of the anchor cable, a 1D implantable truss unit is employed, following the principles of linear elasticity. The airport hotel structure is designed as a frame structure. In the superstructure, the walls and upper beams are represented as 2D plate units, while the columns are depicted as 1D beam units. As for the substructure, the underground diaphragm wall and raft slab foundation are modeled using 2D plate units, with the reinforcement area of the CFG pile being treated as equivalent.

Table 1. Engineering components and stratigraphic materials and property parameters

Structures	Unit properties	Name of material	Severe kN/m^3	Poisson ratio	Modulus of elasticity MPa
Stratum I	3D solids	Vegetal fill	19.8	0.3	10.25
Stratum II	3D solids	Powdery clay and clay	19.5	0.3	9.75
Stratum III	3D solids	Dolomitic chert interbedded with tuff	24.8	0.3	1000
Surrounding pile wall	2D board	C30 concrete	24.0	0.2	30,000
Upper beam slab	2D board	C35 concrete	25.0	0.2	31500
Column	1D beam	C35 concrete	25.0	0.2	31500
Ground floor facade	2D board	C35 concrete	25.0	0.2	31500
Raft slab foundation	2D board	C35 concrete	25.0	0.2	31500
Slope protection concrete	2D board	C20 concrete	22.0	0.2	20,000
Anchor rods	1D implantable truss	Steel	76.0	0.27	250,000

2.2 Three-Dimensional Modeling

The model encompasses dimensions of 500 m in length, 350 m in width, and 60 m in height. Specifically, the pit exhibits an 11 m depth at the negative ground floor, situated to the east and south of the airport hotel structure. Furthermore, the negative first floor of the pit, located to the south of the airport hotel, reaches a depth of 24 m. The southern large slope release measures 11 m in height and spans 17 m in width, featuring a slope release ratio of 1:1.5. The northern portion consists entirely of pile-anchored support structures, while the southern section showcases a slope and pile-anchored support structures at the negative first floor.

The airport hotel model consists of a 9-storey frame structure, divided into three parts: the superstructure, the bottom part, and the substructure (Fig. 1). The storey height for the superstructure is 3.6 m, and it features a cut-out area in the middle. In the model, the above-ground portion of the building is considered a gravity-free structure, focusing solely on its stiffness. The self-weight of the building and the floor load is applied as an equivalent surface load on the F1 layer of the building structure. In accordance with China's Code of Structural Loads for Buildings (GB50009–2012), which applies to buildings such as shops, exhibition halls, stations, ports, airport halls, and passenger waiting rooms, the floor's live load is determined by selecting the standard value of 3.5 kN/m^2 for calculations. Considering the building's self-weight, floor load, and the weight of the decorations, and accounting for the structural characteristics of the airport hotel, a uniform load of 113.4 kPa is ultimately applied to the corresponding area of the F1 floor.

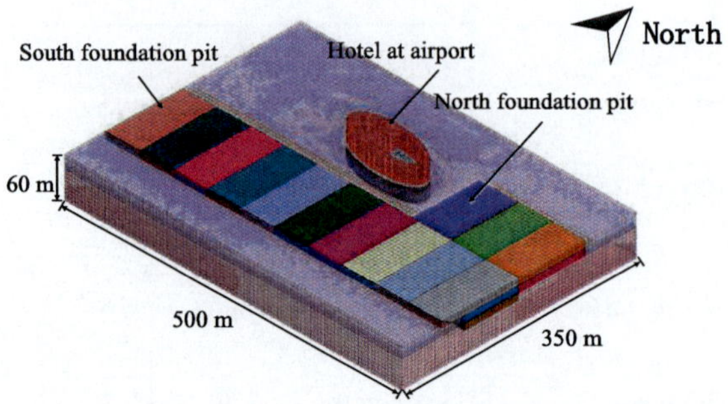

Fig.1. Schematic diagram of the 3D finite element calculation model

In the model, the boundary conditions are set as follows: the top of the model is free, while normal displacement constraints are imposed on the sides and bottom. To account for the bottom of the pit model being considered at infinity, a horizontal displacement constraint is applied to the bottom surface.

2.3 Construction Sequences

The excavation process utilizes a zoning method to divide the foundation pit, and the zoning of the foundation pit is illustrated in the provided figure. Both the southern and northern pits are partially excavated with a width of 50 m. The excavation proceeds by completing the negative layer excavation of the pit first, followed by the excavation of the negative second layer in the southern pit. Six distinct excavation conditions have been identified as effective for the study, as outlined below. The construction excavation sequence during the simulation is primarily illustrated in Table 2 and Table 3 and Fig. 2.

Table 2. Excavation Schedule of the Pit

Working conditions	South pit	Northern pit
Case 1	Minus one and minus two floors, excavation by subdivision ① to ⑩	Excavation ① to ③ in the south pit negative layer, corresponding to excavation ① to ③
Case 2		Excavation ① to ③ in the south pit negative layer, corresponding to excavation ③ to ①
Case 3		Excavation of the south pit, negative level 8 to ⑩, corresponding to excavation ① to ③

(*continued*)

Table 2. (*continued*)

Working conditions	South pit	Northern pit
Case 4	Minus one and minus two floors, excavated by subdivision ⑩ to ①	Excavation ③ to ① for the negative level of the southern foundation pit, corresponding to excavation ③ to ①
Case 5		Excavation ③ to ① in the south pit negative layer, corresponding to excavation ① to ③
Case 6		Excavation ⑩ to ⑧ for the south pit negative level, corresponding to excavation ③ to ①

Table 3. Simulation of the excavation step sequence for the construction phase

Construction steps	Contents	Description
S1 ~ S4	Construction around the foundation pit	Initial stress field analysis, hotel construction, displacement clearing and envelope construction
S5 ~ S14	Excavation of foundation pits in bins	Excavation of the southern pit and the negative level of the northern pit
S15 ~ S24	Excavation of foundation pits in bins	Excavation of the negative first floor of the southern foundation pit

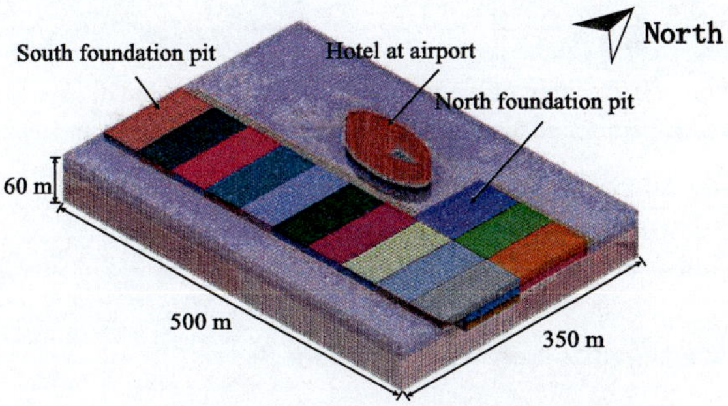

Fig. 2. Schematic diagram of the south pit and north pit zoning excavation area

3 Effects of Foundation Excavation Sequence

3.1 Impact of Pit Excavation on the Ground

The results of the foundation excavation ground displacement calculations are provided as an example for Case 1. Figure 3 displays the vertical displacement clouds representing the strata surrounding the foundation pit. From the figure, several observations can be made. Firstly, the most significant ground settlement occurs near the airport hotel, as well as on the north side of the southern pit and the west side of the northern pit. The maximum settlement increases from 17.9 mm to 19.4 mm after the completion of the negative layer excavation. Secondly, due to the absence of slope excavation on the east side of the hotel's enclosure structure, the ground surface in the vicinity primarily exhibits settlement. The settlement increases from 2.7 mm to 6.2 mm after the completion of the negative layer excavation. Thirdly, the excavation leads to an elevation of the pit bottom. The bottom of the negative first floor in the southern pit experiences a rise of 59.9 mm, exceeding the 39.3 mm rise observed in the bottom of the negative ground floor in the northern pit. This indicates that deeper excavations and unloading of large pit areas result in more pronounced elevation of the ground level.

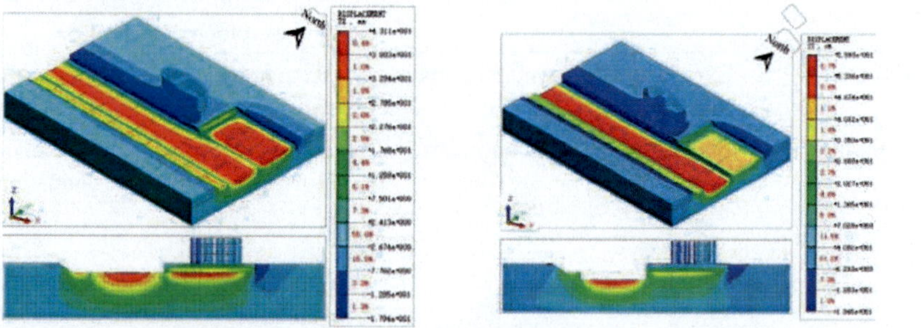

(a) After completion of excavation of the negative level of the southern foundation pit (b) After completion of excavation of the negative first floor of the southern foundation pit

Fig. 3. Vertical displacement clouds of the strata around the foundation pit

3.2 Impact of Pit Excavation on Adjacent Hotels

Figure 4 depicts the vertical settlement cloud of the Airport Hotel structure. In general, the Airport Hotel structure experiences settlement. After completing the overall excavation near the hotel, the maximum settlement of the structure occurs on the southern side. The settlement increases from 9.9 mm to 10.5 mm after the negative ground floor excavation. On the northern side, the settlement increases from 3.4 mm to 3.7 mm following the completion of the negative ground floor excavation. It is noteworthy that the excavation of the negative first floor in the southern pit has a relatively minor impact on the settlement of the hotel structure.

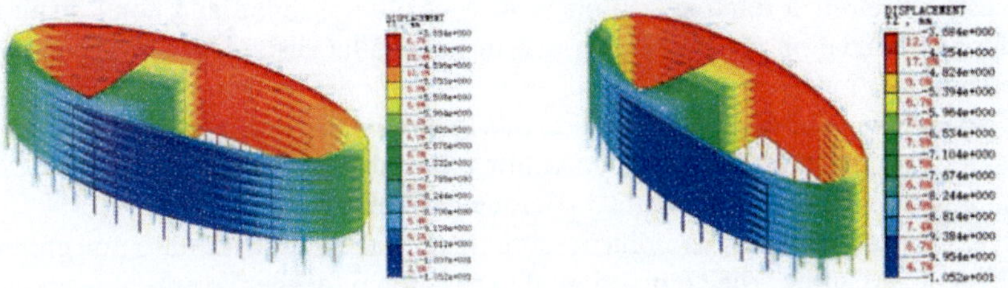

(a) After completion of excavation of the negative level of the southern foundation pit (b) After completion of excavation of the negative first floor of the southern foundation pit

Fig. 4. Vertical settlement clouds of the airport hotel structure

3.3 Horizontal Displacement of Enclosure Pile

Figure 5 displays a cloud of lateral horizontal displacement of the enclosure piles. After completing the excavation of the foundation pit soil, the enclosure piles primarily exhibit convergence towards the pit, indicating a movement inward. Considering the distribution characteristics of convergence during different construction stages, significant convergence areas are observed at the bottom and near the top of the fence pile after completing the excavation of the negative layer and negative layer two of the southern foundation pit, respectively. This suggests that the completion of the excavation of the negative layer triggers noticeable changes in the convergence areas of the fence pile. In terms of the magnitude of convergence during different construction stages, the maximum lateral horizontal displacement of the perimeter pile near the hotel measures 12.5 mm and 24.1 mm after completing the excavation of the first and second layers of soil in the southern foundation pit, respectively. These values indicate a substantial alteration in the deformation of the perimeter pile following the completion of the excavation of the negative second layer of soil.

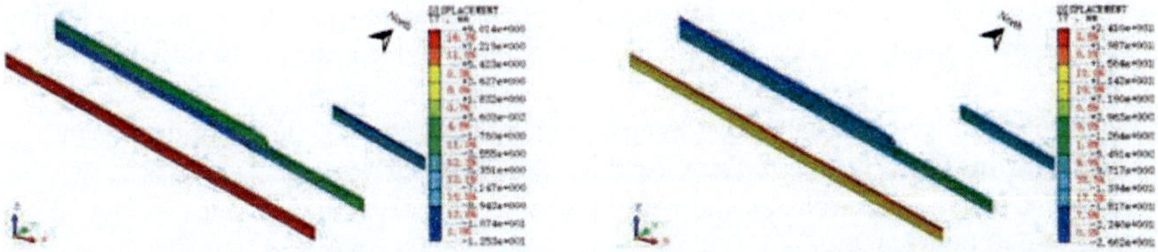

(a) After completion of excavation of the negative level of the southern foundation pit (b) After completion of excavation of the negative first floor of the southern foundation pit

Fig. 5. Lateral horizontal displacement cloud of the enclosure pile

4 Conclusion

This paper investigates the environmental disturbance caused by the excavation of a foundation pit near an airport hotel. The main conclusions are as follows.

(1) Case 1 exhibits a relative advantage in controlling ground and hotel disturbances during foundation pit excavation due to its smaller disturbance size compared to other conditions.
(2) The advantage of working condition one in perimeter pile disturbance control is not significant compared to other conditions. Further considerations are necessary to address the disturbance effects on perimeter piles.
(3) The excavation sequence influences the magnitude and distribution of ground settlement and uplift. The completion of excavation of deeper layers causes noticeable changes in settlement and uplift volumes.
(4) The maximum lateral horizontal displacement of the perimeter pile near the hotel measures 12.5 mm and 24.1 mm after completing the excavation of the first and second layers of soil in the southern foundation pit, respectively. The Airport Hotel structure experiences settlement, with the southern side showing greater settlement than the northern side. Excavation of deeper layers has a smaller impact on hotel settlement. The deformation conforms to the requirements specified in building foundation design codes.

Acknowledgement. This work is supported by Key R&D Program of Shandong Province (2021CXGC011203) and National Natural Science Foundation of China ((No.41672360).

References

1. Fei, Y., et al.: Research on deep foundation pit excavation based on data monitoring. In: IOP Conference Series: Earth and Environmental Science. 2020: IOP Publishing (2020)
2. Ye, S., Zhao, Z., Wang, D.: Deformation analysis and safety assessment of existing metro tunnels affected by excavation of a foundation pit. Underground Space **6**(4), 421–431 (2021)
3. Zhang, D.D., et al.: The finite element analysis of the excavation on adjacent buildings based on mohr coulomb model. Adv. Mater. Res. Trans Tech Publ **374** 2171–2175 (2012)
4. Dong, J., et al.: Test study on the influence of foundation pit excavation on the surface settlement of sandy soil natural foundation of adjacent buildings. Buildings **13**(5), 1293 (2023)
5. Wang, Z., Wang, C.: Analysis of deep foundation pit construction monitoring in a metro station in Jinan city. Geotech. Geol. Eng. **37**, 813–822 (2019)
6. Xie, J., Qin, Y.: Heat transfer and bearing characteristics of energy piles: review. Energies **14**(20), 6483 (2021)
7. Li, L.J., Liang, R.W.: Research on the spatial effect of double row piles structure system in deep foundation. Adv. Mater. Res. Trans Tech Publ **374**, 2367–2370 (2012)
8. Yang, X.H., et al.: Analysis of monitoring results of a deep foundation pit with pile-anchor retaining structure. Appl. Mech. Materials. Trans Tech Publ. **29**, 28–33 (2014)
9. Chen, A., et al.: Investigating pile anchor support system for deep foundation pit in a congested area of Changchun. Bull. Eng. Geol. Env. **80**, 1125–1136 (2021)
10. Pei-xin, W., et al.: Impacts of foundation pit excavation on adjacent railway subgrade and control. Rock Soil Mech. **37**, 469–476 (2016)
11. Liu, J., et al.: Deformation and numerical simulation analysis of deep foundation pit excavation of nanjing yangtze river floodplain metro station. In: IOP Conference Series: Earth and Environmental Science. 2021: IOP Publishing

12. Zhou, Y., Zhu, Y.: Interaction between pile-anchor supporting structure and soil in deep excavation (2018)
13. Cao, H.Y., Jia, D.B., Chen, T.J.: Study on deformation characteristics of deep foundation pit in unsaturated soil. Adv. Materials Res. Trans Tech Publ **374**, 1809–1812 (2012)
14. Tong, S.: Research on the application of deep foundation pit support construction technology in civil engineering. Int. J. Educ. Econ, **3**(3) (2020)
15. Goh, A., et al.: A simple estimation model for 3D braced excavation wall deflection. Comput. Geotech. **83**, 106–113 (2017)
16. Hou, Y.M., Wang, J.H., Zhang, L.L.: Finite-element modeling of a complex deep excavation in Shanghai. Acta Geotech. **4**, 7–16 (2009)
17. Liang, R., et al.: Simplified method for evaluating shield tunnel deformation due to adjacent excavation. Tunn. Undergr. Space Technol.Undergr. Space Technol. **71**, 94–105 (2018)

Topology Optimization Design of Liquid-Cooled Radiator Based on Variable Density Method

Kaixun Jia and Bin Zhang(✉)

Department of Engineering Mechanics, School of Mechanics, Civil Engineering and Architecture, Northwestern Polytechnical University, 127 West Youyi Road, Xi'an 710072, Shaanxi, People's Republic of China

binzhang@nwpu.edu.cn

Abstract. Liquid-cooled radiator has the advantages of high heat exchange efficiency and uniform temperature distribution, which makes it widely used in the thermal management system of electronic equipment. The structure of the cooling channels in the liquid-cooled radiator is a significant factor affecting the cooling performance of the radiator. In order to improve the comprehensive heat exchange efficiency of the liquid-cooled radiator, this paper uses the topology optimization method to design the layout of the flow channels. The variable density method is employed to study the problem of heat transfer maximization. Moreover, the effects of the pressure difference and heat generation coefficient of heat source on the optimization results are investigated. The results show that in the optimization design, with the increase of the pressure difference or the heat transfer coefficient of the heat source, the flow channel structure becomes more complex, and the amount of heat transfer of the radiator is increased.

Keywords: Topology optimization · heat sink · laminar flow

1 Introduction

With the improvement of equipment integration and operational efficiency, there is an urgent need for thermal control of electronic devices. The cooling plate is a commonly used radiator with advantages of good cooling performance, uniform temperature distribution, easy maintenance, etc. This makes it an efficient cooler for the electronic device. The heat produced by the heat source is removed from the cooling plate by coolant flowing via the cooling channel. The layout of cooling channels has a considerable impact on heat transfer performance. Numerous academics have conducted optimization design studies to establish the architecture of cooling channels in order to increase the comprehensive performance of heat sinks. These research have frequently been carried out using size or shape optimization techniques.

Topology optimization is an optimization method that, by enhancing the objective function, distributes materials in the design domain. The topology optimization method offers a higher degree of freedom, and does not rely on the experience of designers which makes it possible to obtain innovative optimized structures. This method has widely used

W. Guo et al. (Eds.): GBCESC 2023, LNCE 328, pp. 244–250, 2024.
https://doi.org/10.1007/978-981-99-9947-7_25

in multidisciplinary optimization problems. Topology optimization method was initially introduced in the field of solid mechanics in which the topology optimization method is well developed. However, it still needs more development for coupled problems involving fluid flow and heat transfer. Yoon [1] applied the topology optimization method to the design of forced convection heat exchanger. Yaji et al. [2] applied the level set method to the topology optimization design of 2D and 3D radiators. Alexandersen et al. [3] utilized the density-based topology optimization for the design of natural convection radiator. This study applies the variable density method to design the cooling plate. In addition, the influence of pressure difference and heat source coefficient on the optimization results was studied, which is the main contribution of this article.

2 Optimization Model

In the variable density method, a continuous material density γ between 0–1 is adopted as the design variable. $\gamma = 1$ represents a solid, and $\gamma = 0$ represents a fluid. The physical property parameters of the fictitious material can be represented by the assumed material density value γ. In the final optimization result, it only has clear physical meaning if the material density γ value is 0 or 1. In this study, the intermediate values of γ are reduced by using a projection method.

2.1 Fluid Field Modeling

The continuity equation and the momentum conservation equation for the fluid flow are expressed as

$$\begin{cases} \nabla \cdot \boldsymbol{u} = 0 \\ \rho \boldsymbol{u} \cdot \nabla = -\nabla p + \mu \nabla^2 \boldsymbol{u} + f \end{cases} \tag{1}$$

where $\boldsymbol{u}$ is the flow velocity, μ is the fluid dynamic viscosity, p is the pressure, and $\boldsymbol{f}$ is the fictitious body force. Within the design domain, the fictitious body force is utilized to punish the flow through solid domain. The fictitious body force is written as:

$$\boldsymbol{f} = -\alpha \boldsymbol{u} \tag{2}$$

where α is the inverse permeability of the porous media and the expression of the α is.

$$\begin{cases} \alpha(\gamma) = \alpha_s + \left(\alpha_f - \alpha_s\right) \frac{q(1-\gamma)}{q+\gamma} \\ \alpha_s = \frac{\mu}{D_a L^2} \\ \alpha_f = 0 \end{cases} \tag{3}$$

where D_a is the Darcy number and L is the characteristic length. The value of the Darcy number in this study is 1×10^{-5}, the value of L is the length of the inlet channel. α_f is the inverse permeability of solid phases, α_s is the inverse permeability of fluid phases, α_s is a large enough value.

2.2 Thermal Field Modeling

The governing equations for the heat transfer in solids and fluids are written as

$$\begin{cases} k_s \nabla^2 T + Q = 0 \\ \rho C_p (\boldsymbol{u} \cdot \nabla) T = k_f \nabla^2 T \end{cases} \tag{4}$$

where ρ is the density of the fluid, Q is the heat source term, T is the temperature of a point in the design domain, C_p is the heat capacity of the fluid, and k_s and k_f are the thermal conductivity of the solid and the fluid.

The unified equation after introducing the design variables as

$$\rho C_p (u \cdot \nabla) T = k_p \nabla^2 T + Q \tag{5}$$

where ρ, C_p and k_p are the density, constant pressure specific heat capacity and heat transfer coefficient of porous media, respectively, which are determined by the interpolation of the density, constant pressure specific heat capacity and heat transfer coefficient of the selected solid material and liquid material.

$$\begin{cases} k(\gamma) = k_s + \left(k_f - k_s\right) \frac{q(1-\gamma)}{q+\gamma} \\ C_p(\gamma) = C_{ps} + \left(C_{pf} - C_{ps}\right) \frac{q(1-\gamma)}{q+\gamma} \\ \rho(\gamma) = \rho_s + \left(\rho_f - \rho_s\right) \frac{q(1-\gamma)}{q+\gamma} \end{cases} \tag{6}$$

where subscript f represents the fluid material, subscript s represents the solid material, and q is the tuning parameter. The tuning parameter is set at 0.01 to make the interpolation function closely match the 0–1 distribution.

The ideal heat source constant temperature in the solid region is set to T_Q, and its heat output is relative to the temperature difference between T_Q and the local temperature T, the heat source term Q can be expressed as:

$$Q = \theta\left(T_Q - T\right)(1 - \gamma) \tag{7}$$

where θ is the heat generation coefficient.

The goal of the optimized design is to enhance heat transfer in the cold plate, and the objective function is the total heat production in the solid domain.

$$C = \int_{\Omega} \theta\left(T_Q - T\right)(1 - \gamma) \tag{8}$$

2.3 Geometry and Initial Condition Settings

The geometric model of the two-dimensional cooling plate is shown in Fig. 1, the size of the design domain is 3 mm × 3 mm, the width of the inlet and outlet is 0.5 mm. The initial temperature of the coolant is set to 293.15 K.The fluid inlet adopts fixed pressure boundary conditions, the outlet adopts static pressure conditions, i.e., $P_{out} = 0P_a$. The temperature T_Q of the heat source is set to 340 K, and the heat generation coefficient θ is set to $1 \times 10^3 \text{w/m}^3$. The boundary is adiabatic without slip. The physical properties of the solid material and liquid material are shown in Table 1.

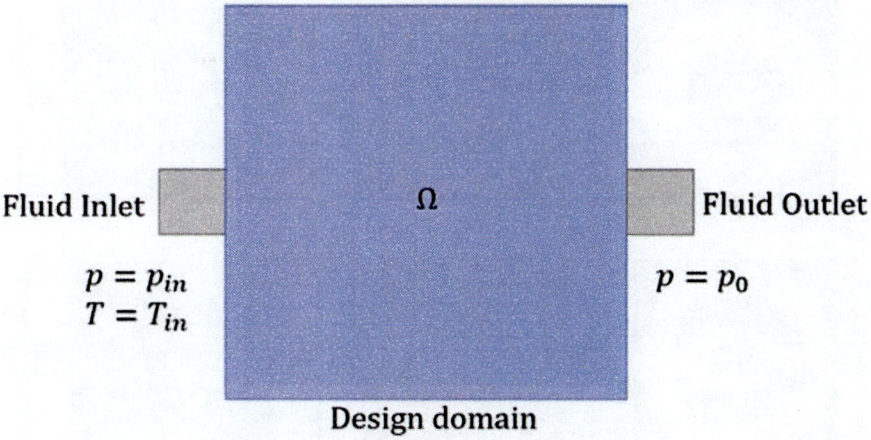

Fig. 1. 2D model of a cold plate

Table 1. Physical properties of solid and fluid materials

material	k[W/m · k]	c[J/kg · k]	ρ[kg/m^3]
solid	44	7800	7800
fluid	0.6	4200	1000

3 Results and Discussion

3.1 Influence of Differential Pressure on Optimal Results

The impact of different differential pressure on the optimization design results is studied. Figure 2 displays the optimized flow channels and associated temperature and flow velocity distributions for various differential pressure instances.

The results in Fig. 2 indicate that the number of flow channels steadily grows as the inlet pressure increases, and the structure of the flow channels becomes more complex. At the same time, the length of the heat transfer boundary significantly increases, and the fluid velocity increases. An increase in the objective function value and a decrease in the maximum temperature difference and average temperature within the design domain represent an enhancement in the cooling performance of the cooling plate. However, the increase of differential pressure will also lead to an increase in the number of iterations and calculation time. In practical engineering applications, a higher differential pressure also means higher energy consumption, which should be comprehensively considered based on performance requirements.

3.2 Influence of Heat Generation Coefficient on Optimal Result.

The impact of different heat generation coefficient on the optimization design results is studied. Here, the inlet pressure is set to $3P_a$. Figure 3 displays the optimized flow

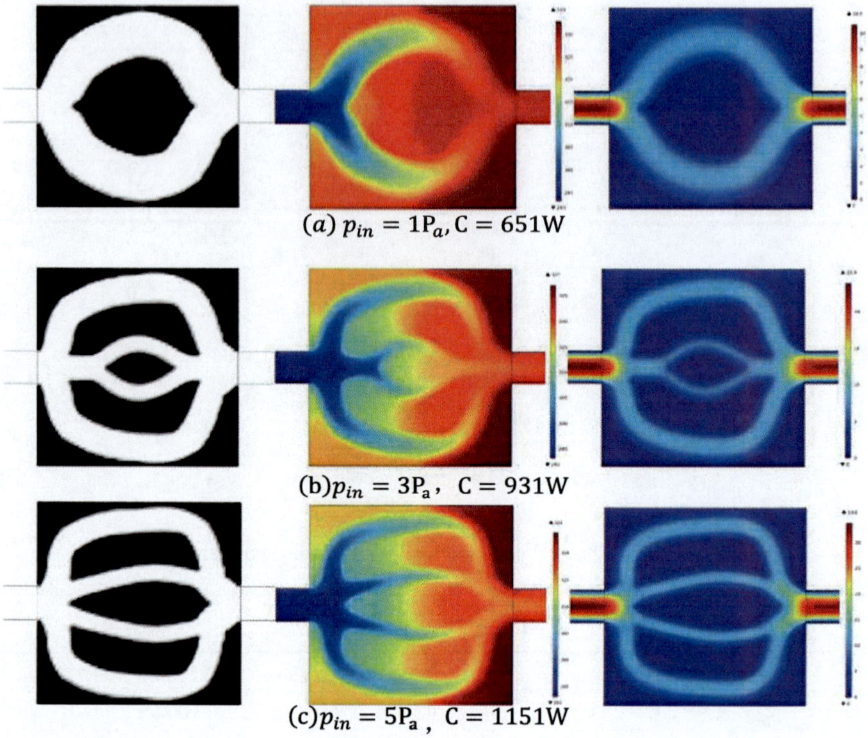

Fig. 2. The optimized flow channels and associated temperature and flow velocity distributions for various differential pressure instances.

channels and associated temperature and flow velocity distributions for various heat generation coefficient.

Figure 3 shows that with the heat generation coefficient increases, more thin channel branches appear in the optimization results, the length of the heat generation boundary increases, and convective heat transfer is enhanced. In addition, the volume fraction of the fluid gradually increased to 0.65, 0.71, and 0.75, respectively. Overall, the heat transfer performance of the cold plate has been improved.

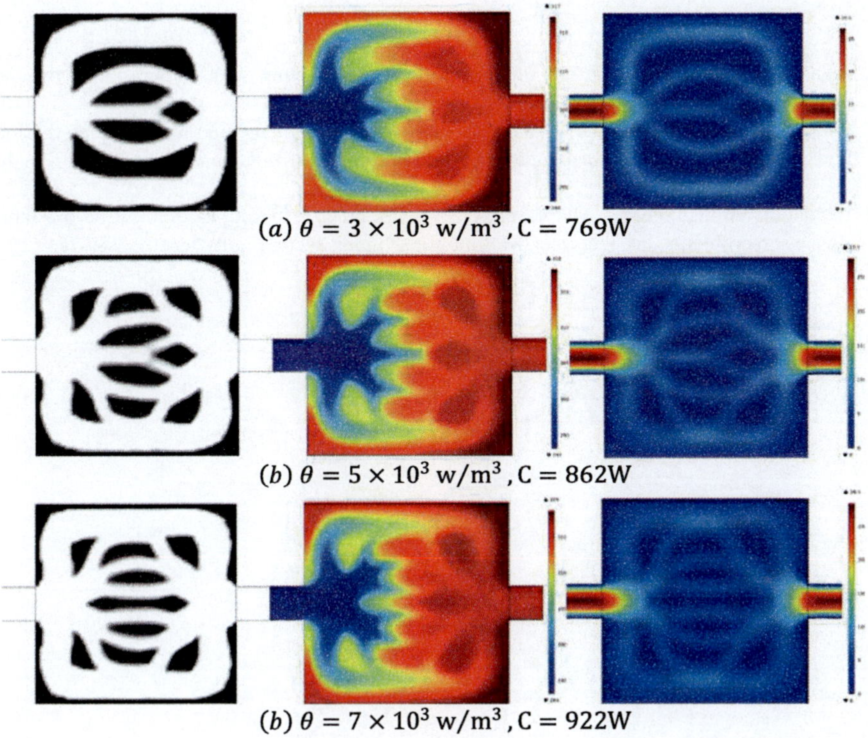

Fig. 3. The optimized flow channels and associated temperature and flow velocity distributions for various heat generation coefficient.

4 Conclusions

In this work, the topological optimization design of the channel structure of the cooling plate was accomplished by the variable density method, and the influence of differential pressure and heat generation coefficient on the optimization designs was studied. The numerical results indicate that as the pressure difference increases, more channel branches appear in the optimized designs resulting in an increase in flow velocity and heat transfer. In addition, with the increase of heat transfer coefficient, the optimized result became more complex, the fluid volume fraction increased, and the cooling performance of the heat sink was improved.

Acknowledgement. The work was supported by the National Natural Science Foundation of China (Grant No. 52275272 and No. 51905435), China Postdoctoral Science Foundation (Grant No. 2020M683550 and No. 2022T150533) and Fundamental Research Funds for the Central Universities (Grant No. G2018KY0306).

References

1. Yoon, G.H.: Topological design of heat dissipating structure with forced convective heat transfer. J. Mech. Sci. Technol. **24**(6), 1225–1233 (2010)
2. Yaji, K., Yamada, T., Kubo, S., et al.: A topology optimization method for a coupled thermal–fluid problem using level set boundary expressions. Int. J. Heat Mass Transf. **81**, 878–888 (2015)
3. Alexandersen, J., Aage, N., Andreasen, C.S., Sigmund, O.: Topology optimisation for natural convection problems. Int. J. Numer. Meth. FluidsNumer. Meth. Fluids **76**(10), 699–721 (2014)

Simulation Analysis of Long-Span Single-Tower Hybrid Beam Cable-Stayed Bridge

Tonghui Jiang[1](✉), Jiading Yang[1], Dequan Zhu[1], Yufeng Xu[2], and Mengyang Zhu[2]

[1] Guangdong Foying Huijian Engineering Management Co, Foshan, China
446998826@qq.com

[2] South China University of Technology, Guangzhou, China

Abstract. The bridge in Foshan is a single-tower hybrid beam cable-stayed bridge with a main span of 268m and a tower height of 151m. The arrangement of the main bridge spans is (65 + 75 + 268) meters. Using the finite element simulation analysis method, the deformation, stress, and cable forces of the main bridge structure under permanent, variable, and combined loads were calculated. The results indicate that the deformations, stresses, and cable forces of the structure in the permanent load condition are within a reasonable range, satisfying the requirements of construction control. Under combined loads, the compressive stress on the upper surface of the concrete main beam near the main tower is relatively high. It is recommended to reinforce the reinforcement or take structural measures based on the direction of compressive stress in that area. The deformations, stresses, and cable forces at other locations are within a reasonable range and generally meet the design requirements.

Keywords: Cable-Stayed Bridge · Hybrid Beam · Long-Span · Finite Element · Simulation Analysis

1 Introduction

The hybrid beam cable-stayed bridge has the advantage of large span capability due to the use of steel beams in the main span, while concrete beams are used in the side spans, providing effective anchorage and reducing bridge construction costs [1]. Since the 1970s, the hybrid beam cable-stayed bridge has become one of the competitive options for long-span bridge designs due to its good span capacity, balanced load distribution, and favorable economic performance [2–4].

The long-span hybrid beam cable-stayed bridge experiences complex load conditions, and numerous scholars have conducted research and analysis on it. Qin et al. [5], Wang et al. [6], introduced intelligent algorithms to optimize the calculation and testing process of the cable force, which improved the calculation efficiency and construction control accuracy. Lin et al. [7] introduced the importance rate analysis for durability failure risk assessment and proposed the importance analysis method for durability failure

W. Guo et al. (Eds.): GBCESC 2023, LNCE 328, pp. 251–258, 2024.
https://doi.org/10.1007/978-981-99-9947-7_26

risk, which provides a reference for safe operation and preventive maintenance of large-span hybrid beams. Some scholars have also conducted fine finite element simulation analysis of steel-hybrid combined section. Zhang et al. [8], Guo et al. [9], in order to study the stress performance of the steel-hybrid combined section of the hybrid girder cable-stayed bridge, a fine finite element simulation analysis was carried out locally on the steel-hybrid combined sections.

To further investigate the load characteristics of the hybrid beam cable-stayed bridge, this study focuses on a single-tower hybrid beam cable-stayed bridge in Foshan's Chancheng district with a main span of 268 m. Finite element simulation analysis is conducted to analyze the bridge's structural behavior.

2 Project Overview

A large-span single-tower cable-stayed bridge located in the main urban area of Foshan City. The main bridge is a single-tower double-cable-plane hybrid beam structure, with a tower height of 151 m. The arrangement of the main bridge spans is (65 + 75 + 268) meters. The main beam is a hybrid box girder, with a steel box girder for the central span and a prestressed concrete box girder for the side spans. The bridge layout diagram is shown in Figs. 1 and 2.

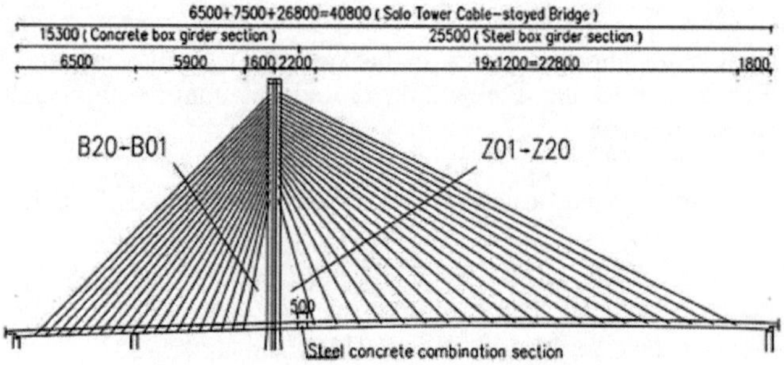

Fig. 1. Bridge elevation layout diagram (cm)

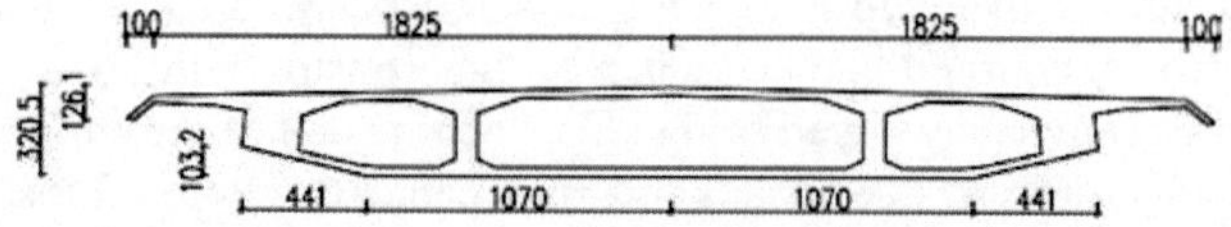

(a) Cross-section of the standard segment of the concrete beam

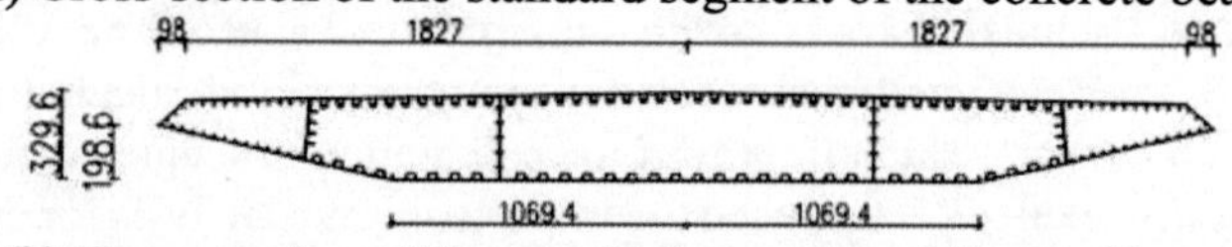

(b) Cross-section of the standard segment of the steel beam

Fig. 2. Cross-sectional layout diagram (cm)

3 Establishment of the Simulation Model

3.1 Structural Model

The entire bridge model was established using the Midas Civil finite element analysis software. The main tower, main beams, and substructure foundations were simulated using beam elements. The stay cables were simulated using truss elements. The elastic modulus was updated and calculated based on the cable forces from the previous stage using the Ernst formula [10]. The nodes of various components were not shared and were connected with rigid connections. The supports were simulated using only compression-only elastic supports. The entire bridge model consisted of 883 nodes and 788 elements. The computational model is shown in Fig. 3.

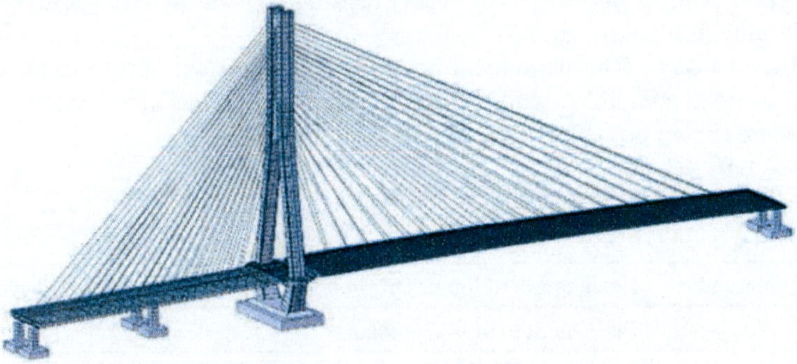

Fig. 3. Bridge finite element simulation analysis model

3.2 Construction Process

The load information for the simulation analysis of the main bridge includes permanent loads, variable loads, and combined loads.

Permanent Loads

First Phase Constant Load. The dead load consists of the self-weight of various structural components of the bridge. It is added based on the design drawings and construction plans, taking into account the actual weight of each component. The self-weight of concrete elements is adjusted by modifying the material density according to the actual statistical weight. The self-weight of steel main girders is replaced by a uniformly distributed load with a material density set to zero. The self-weight of cross beams is represented by nodal loads.

Second Phase Constant Load. The second phase includes the permanent loads added after the formation of structures such as pavement and railings. These loads are added according to the weight provided in the design drawings and are applied to the main girder elements as uniformly distributed loads.

Variable Loads

Car Live Load. According to the Class I highway standards, considering a dual carriageway with six lanes, lane loads should be added.

Braking Force Load. The vehicle loads should be calculated based on the loads generated by vehicles traveling in the same direction. The longitudinal reduction should be applied based on the most unfavorable loading length determined by the bridge piers.

Temperature Load. Adding according to the overall heating and cooling of the structure, the overall heating and cooling of the main girders, the overall heating and cooling of the diagonal cables, the temperature gradient of the main girders, and the temperature gradient of the bridge towers.

Wind Load. Adding in the form of evenly distributed loads by cross-bridge direction and longitudinal bridge direction.

Combined Loads. The permanent action is combined with the variable action to obtain the most unfavorable state of the structure, and the load combination working conditions are shown in Table 1 below.

Table 1. Load combination scenarios

Combination number	Load combination description	Type
1	Permanent load + live load	Summation
2	Permanent load + live load + braking force load	Summation
3	Permanent load + live load + temperature load	Summation
4	Permanent load + live load + wind load	Summation
5	Combination 1 + Combination 2 + Combination 3 + Combination 4	Envelope

4 Simulation Analysis Results

4.1 Results of Permanent Loads Calculation

The simulation analysis results of deformation, stress, and cable forces under permanent loads are shown in Figs. 4, 5, 6, 7, 8.

Based on the aforementioned calculations under permanent loads, the following conclusions can be drawn:

1. The structural deformations, stresses, and cable forces of the main bridge in its completed state under permanent loads are within a reasonable range, meeting the requirements of construction control.

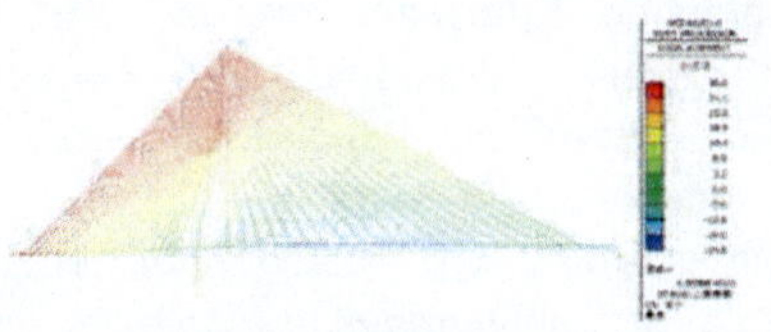

Fig. 4. Vertical deformation under permanent loads (mm)

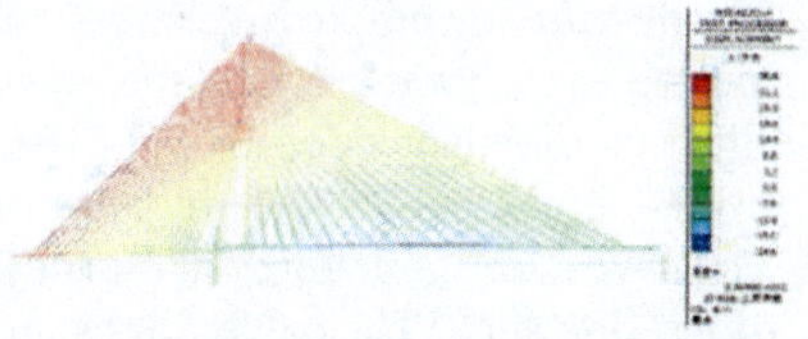

Fig. 5. Longitudinal deformation under permanent loads (mm)

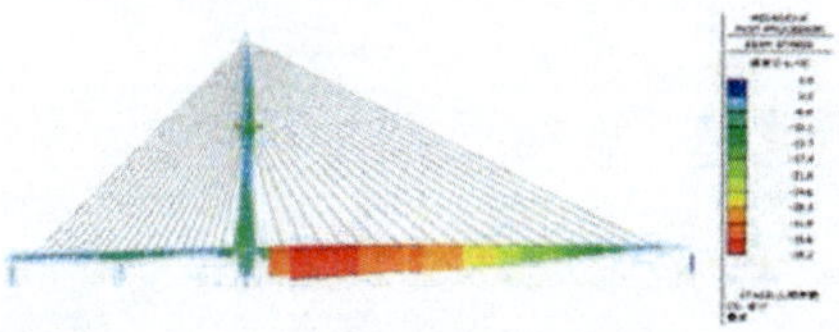

Fig. 6. Upper surface stress of main beam and outer side stress of main tower under permanent loads (MPa)

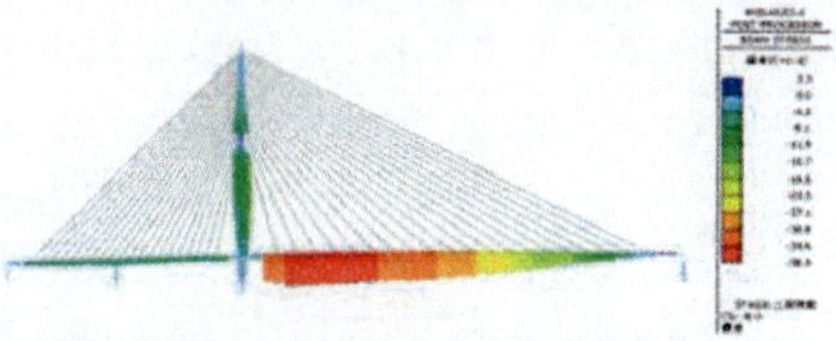

Fig. 7. Lower surface stress of main beam and inner side stress of main tower under permanent loads (MPa)

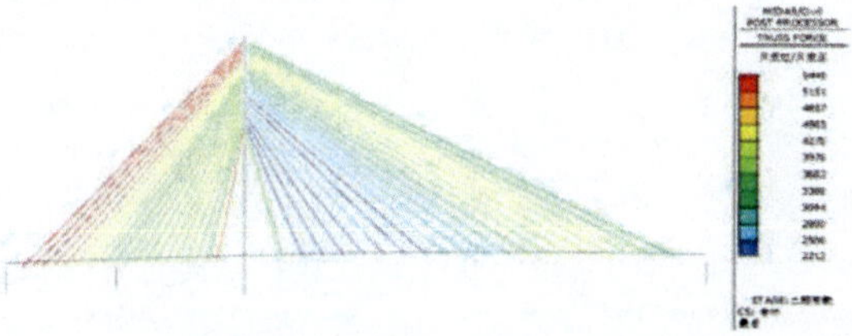

Fig. 8. Cable forces under permanent loads (kN)

2. The maximum vertical deformation of the steel main beam is 230 mm, located at the end of the 7th segment of the steel box beam at mid-span. The vertical deformations of the concrete beams are within 12 mm, and the vertical deformations of the main beams are within a reasonable range.
3. The maximum longitudinal deformation of the steel main beam is − 24 mm, located at the end of the 7th segment of the steel box beam at mid-span. The maximum longitudinal deformation of the concrete beams is 35 mm, located at the pier position, and the vertical deformation of the tower top is 37 mm. The longitudinal deformations of the main beam and the main tower are within a reasonable range.
4. The maximum compressive stress on the upper surface of the steel main beam is -39.2 MPa, and the maximum compressive stress on the lower surface is − 38.4 MPa, both located near the main tower. The overall stress level is generally within a reasonable range.
5. The maximum compressive stress on the upper surface of the concrete beam is − 14.3 MPa, located near the main tower, and the maximum compressive stress on the lower surface is − 7.6 MPa, located at the top of the auxiliary pier. The overall stress level is generally within a reasonable range.

6. The maximum compressive stress on the outer side of the main tower is – 14.5 MPa, located at the joint between the tower and the beam, and the maximum compressive stress on the inner side is – 9.3 MPa, located at the middle of the tower column. The overall stress level is generally within a reasonable range.
7. The maximum cable force in the inclined cables is 4836 kN, located in the 20th cable of the side span. The cable forces are within a reasonable range.

4.2 Results of Combined Loads Calculation

The simulation analysis results of deformation, stress, and cable forces under combined loads are shown in Figs. 9, 10, 11, 12 and 13.

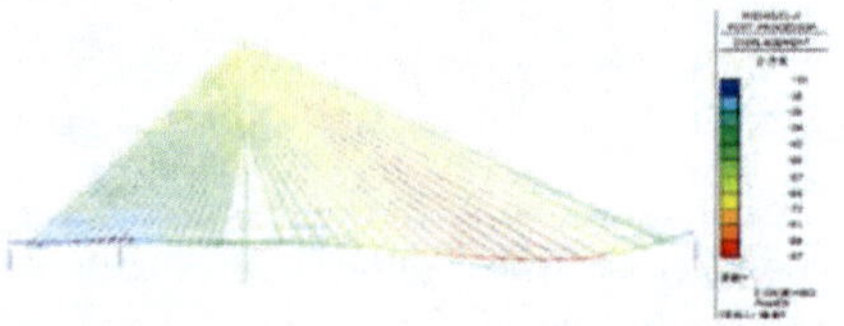

Fig. 9. Vertical deformation under combined loads (mm)

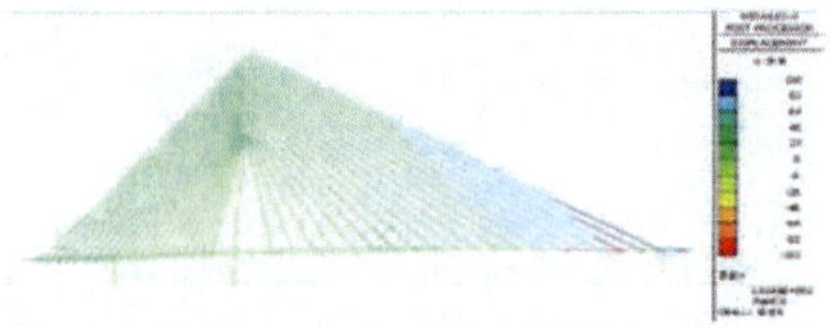

Fig. 10. Longitudinal deformation under combined loads (mm)

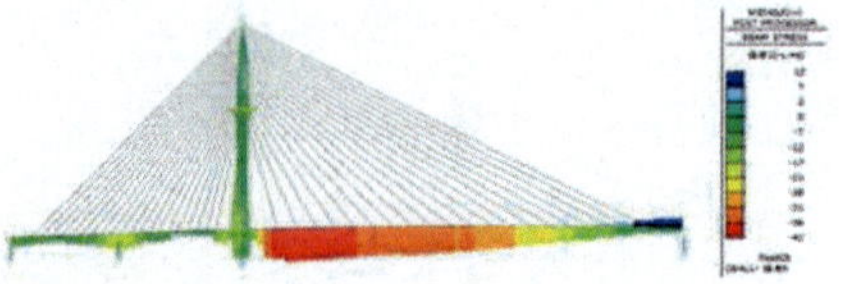

Fig. 11. Upper surface stress of main beam and outer side stress of main tower under combined loads (MPa)

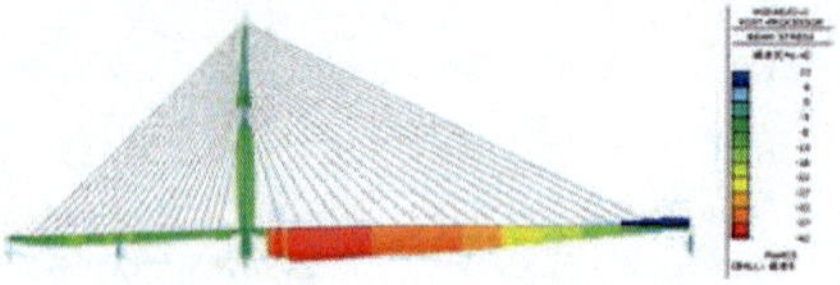

Fig. 12. Lower surface stress of main beam and inner side stress of main tower under combined loads (MPa)

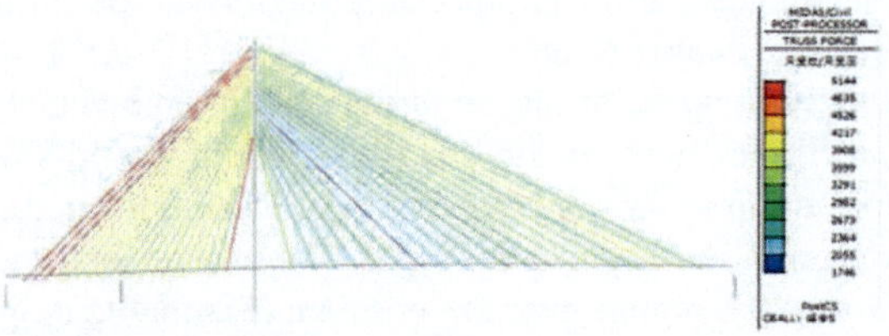

Fig. 13. Cable forces under combined loads (kN)

Based on the calculation results of the combination loads mentioned above, the following can be concluded:

1. Under the combination loads, the structural deformations, stresses, and cable forces of the main bridge are within a reasonable range, meeting the design requirements.

2. The maximum vertical deformation of the steel main beam is 97 mm, located at the mid-span of the main span. The maximum vertical deformation of the concrete beam is 39 mm, located at the mid-span between the auxiliary pier and the main tower. The vertical deformations of the main beam are within a reasonable range, indicating good vertical stiffness of the main beam.
3. The maximum longitudinal deformation of the steel main beam is 100 mm, located at the abutment. The maximum longitudinal deformation of the concrete beam is 43 mm, also located at the abutment. The vertical deformation at the top of the tower is 54 mm. The longitudinal deformations of the main beam and the main tower are within a reasonable range, meeting the design requirements for expansion joint allowances.
4. The maximum compressive stress on the upper flange of the steel main beam is − 40.5 MPa, and the maximum compressive stress on the lower flange is − 41.6 MPa, both located near the main tower. The overall stress levels are within a reasonable range.
5. The maximum compressive stress on the upper flange of the concrete beam is − 22.3 MPa, located near the main tower. It is recommended to reinforce the area with additional reinforcement based on the direction of compressive stress. The maximum compressive stress on the lower flange is − 14.7 MPa, located at the mid-span between the abutment and the auxiliary pier, with no tensile stress observed. The overall stress levels are within a reasonable range.
6. The maximum compressive stress on the outer side of the main tower is − 15.6 MPa, located at the junction of the tower and the beam. The maximum compressive stress on the inner side is − 11.9 MPa, located in the middle of the tower column. No tensile stress is present. The overall stress levels are within a reasonable range.
7. The maximum cable force in the stay cables is 5144 kN, located in the 18th cable span. The cable forces are within a reasonable range.

5 Conclusion

The study focuses on a large-span single-tower cable-stayed bridge in Foshan. Finite element simulation analysis was conducted to calculate the structural deformations and forces under permanent loads, variable loads, and combination loads. The following conclusions can be drawn:

1. Under permanent loads, the deformations, stresses, and cable forces of the structure in the bridge state are within a reasonable range, meeting the requirements of construction control.
2. Under combination loads, the deformations, stresses, and cable forces of the structure in the bridge state are within a reasonable range, generally satisfying the design requirements.
3. Under combination loads, the upper flange compressive stress of the concrete main beam near the main tower is relatively high. It is recommended to reinforce the area with additional reinforcement or consider structural measures based on the direction of compressive stress in that region.

References

1. Review on China's Bridge Engineering Research: 2014. China J. Highway Transp. **27**(5), 1–96 (2014)
2. Hu, N., et al.: Recent development of design and construction of medium and long span high-speed railway bridges in China. Eng. Struct. **74**, 233–241 (2014)
3. Hui, M.C., Yau, D.: Major bridge development in Hong Kong, China-past, present and future. Front. Archit. Civil Eng. China **5**(4), 405–414 (2011)
4. Qin, S.Q., Gao, Z.Y.: Developments and prospects of long-span high-speed railway bridge technologies in China. Engineering **3**(6), 787–794 (2017)
5. Qin, S.Q., et al.: Improved metaheuristic algorithm based finite element model updating of a hybrid girder cable-stayed railway bridge. Buildings **12**(7), 958 (2022)
6. Wang, D., et al.: Optimization and control of cable forces in a hybrid beam cable-stayed bridge based on a distributed algorithm. Eng. Optim. (2023)
7. Lin, J., Xiao, R.: Risk assessment of durability failure for large-span hybrid girder cable-stayed bridge. J. Tongji Univ. (Nat. Sci.) **43**(3), 364–370 (2015)
8. Zhang, D.L., et al.: Research on load transfer mechanism of steel-concrete joint section of hybrid beam cable-stayed bridge. In: 1st International Conference on Advances in Civil Infrastructure Engineering (ICACIE 2012), Changsha, PEOPLES R CHINA (2012)
9. Junfeng, G.: Analysis on steel-concrete joint section of hybrid girder cable-stayed bridge. In: IOP Conference Series: Materials Science and Engineering, vol. 490, p. 032022 (2019)
10. Wen, Y., Zhou, Z.W.: Qualification of the Ernst formula for modeling the sag effect of super-long stay cables in the long-span railway cable-stayed bridges. Structures **45**, 99–109 (2022)

The Influence of Multi-level Loading on Cracking Behavior of Sandstone with a Single Flaw

Yuxin Li[1,2], Pengzhi Pan[1](✉), Shuting Miao[1,2], and Yujie Feng[1,2]

[1] State Key Laboratory of Geomechanics and Geotechnical Engineering, Institute of Rock and Soil Mechanics, Chinese Academy of Sciences, Wuhan 430071, Hubei, China
pzpan@whrsm.ac.cn

[2] University of Chinese Academy of Sciences, Beijing 100049, China

Abstract. Underground engineering rocks are highly inclined to cyclic loading caused by various engineering activities such as earthquakes, excavation or blasting. Understanding the process of crack evolution in rocks under cyclic loading is crucial for assessing the stability and safety of engineering structures. Multi-level cyclic loading and unloading tests on flawed sandstone with varying flaw inclinations was conducted to investigate the crack propagation behavior under complex loading conditions. Digital Image Correlation (DIC) technique was utilized in tracking the process of cracking. According to results of the test, during the multiple cyclic loading levels, crack propagation was observed to go through two stages, i.e. stable propagation and unstable propagation. The behavior of cracks in these two stages was influenced by the number of cyclic loads and the upper limit stress of loading.

Keywords: Prefabricated fractured rock · Cyclic loading and unloading · Digital image correlation · Crack propagation

1 Introduction

Rock is the primary component of engineering rock mass in the most underground engineering projects. The mechanical behavior of rock under complex stress states presents a theoretical problem due to its inherent brittleness and the presence of flaws. For example, in tunnelling projects, the engineering rock experiences cyclic loads during the excavation; drilling and earthquakes. The deformation characteristics of rock, particularly during the operation phase of rock mass engineering, directly impact the project's usability. Engineering rocks produce several defects, fractures and even faults owing to the complexity of internal structures and tectonic history, resulting in variable macro-mechanical responses under various external stress levels Understanding the propagation of internal cracks within rocks is essential for assessing the stability and safety of rocks in engineering construction and subsequent operation, enabling a greater comprehension of rock mechanical properties. The cyclic loading and unloading test severs as a useful tool for analyzing the propagation behavior of internal crack in rocks under complex stress conditions.

W. Guo et al. (Eds.): GBCESC 2023, LNCE 328, pp. 259–266, 2024.
https://doi.org/10.1007/978-981-99-9947-7_27

The crack propagation of rocks under dynamic disturbance have been largely studied in the previous research. Eberhardt [1] investigated criteria for rock crack propagation under cyclic loading and unloading tests on rocks. Additionally, Wong and Einstein [2] conducted uniaxial compression tests on fractured rock type materials to determine their mechanical response and damage characteristics.

Researchers such as Sangong and Bobet [3] conducted on the compression tests on rock type specimens containing multiple flaws, and the effect of flaw geometry, such as dip, spacing, rock bridge length, and overlap ratio on the cracking behavior were symmetrically examined. In addition, Park and Bobet [4] conducted the uniaxial compression tests on rock type specimens, focusing on the coefficient of friction and its influence on crack coalescence stresses and damage patterns. Recently, Pan and Miao [5, 6] studied the effect of infilling on the crack evolution of flawed red sandstone and marble specimens with a novel DIC technique. Previous studies have clearly proved that DIC is a reliable method for rock damage and fracturing monitoring [7].

Although these earlier investigations have provided valuable insights, further research is still needed to understand the initiation, propagation, and coalescence of rocks under complex stress conditions. In this study, uniaxial cyclic loading tests were conducted on flawed rocks containing a single fracture with varying flaw inclination angles. DIC technique was employed to track the propagation of initiated cracks around the pre-existing flaw subjected to cyclic loads with varied stress amplitude.

2 Specimen Preparation and Testing

The sandstones specimens in this study were collected from Longchang, Neijiang, Sichuan. Uniaxial compression testing showed that the sandstone has a Young's modulus of 12.5–13.6 GPa, Poisson's ratio of 0.19–0.22, uniaxial compressive strength of 70–75 MPa, and the density of 2132–2287 kg/m^3. The specimen tested was collected from the same rock block with homogeneity. A linear flaw is prefabricated in the center of the tested specimens, which are 100 × 50 × 30 mm and cut using a high-pressure water-jet cutting machine. The length error of all specimens was less than 2 mm, and the unevenness of surfaces after polishing was less than 0.05 mm, and the deviation between the adjacent surfaces perpendicular to each other does not exceed 0.25°. The flaw is 20 mm in length and has an aperture of 2 mm. Varying flaw inclinations were taken into consideration, ranging from 0° to 90° with a 15° interval, shown in Fig. 1(a). Three rock specimens were tested for a given crack geometry and loading scheme. Black and white speckle patterns were randomly spray-painted on the surface of the specimens for DIC analysis. These speckles will deform along with the sample during deformation as a deformation information carrier. Petroleum jelly was applied to the top and bottom ends of the specimens in order to reduce friction with the loading plate.

Uniaxial cyclic loading tests were conducted on pre-cracked specimens using the RMT-150 C test system shown in Fig. 1(b). A high-speed camera with a resolution of 3376*2704 pixels was employed to capture images of specimens during the entire testing process, including crack initiation, propagation, and coalescence on the specimen's surface. Images were captured at a rate of 9 frames per second during the tests. Stable illumination was achieved by using two LED white light sources. The loading device and the image acquisition device were simultaneously started to ensure the consistency between the mechanical data and digital image. The specimens underwent following the stress path shown in Fig. 1(c) until fatigue damage occurred. The displacement control mode was employed in the testing process, ensuring a consistent loading and unloading rate of 0.02 mm per second. The loading peak stress increased in increments of 5 MPa until the specimen failed. During each loading level, 5 cycles were repeated with a constant loading peak stress. This multi-level loading and unloading approach allowed for the observation of crack behavior under different stress conditions and the study of fatigue damage accumulation.

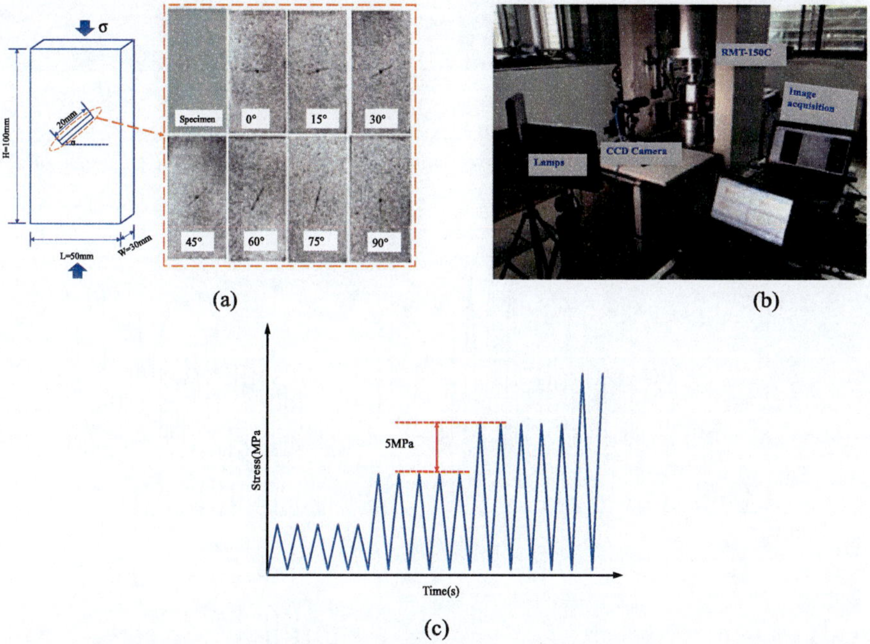

Fig. 1. Specimen preparation and testing: (a) geometry of specimen with different pre-existing flaw; (b) rock mechanical test system and digital image acquisition system; (c) loading path

3 Results

The deformation of full-field revealed by DIC technique is presented to quantitatively study the initiation and propagation of each crack under cyclic loading. To ensure the consistency and mitigate subjective factors, a strain value of 3% was established as the threshold for crack initiation. Consequently, the legend range for the major strain contours was set from 0% to 3% in order to identify cracks according to the full-field deformation. Using this method, the final crack pattern of flawed rock specimens under cyclic loading, with the serial numbers indicating the sequence in which they initiated (Fig. 2b).

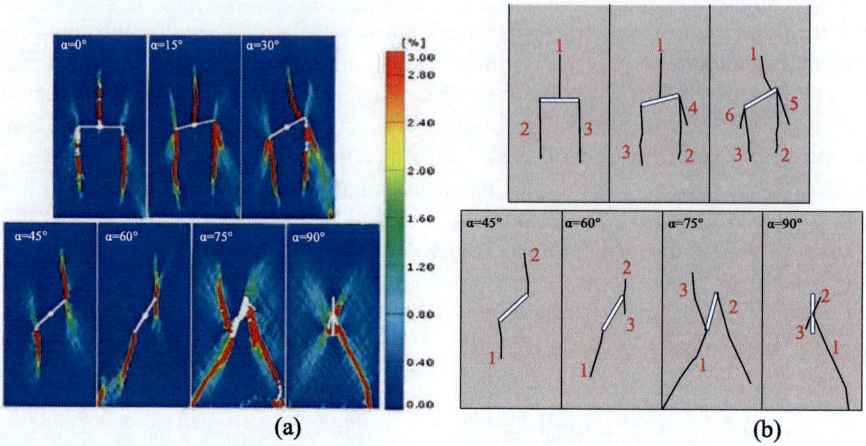

Fig. 2. (a)Strain contours for identifying cracks. (b)Crack patterns of flawed rock specimens with varying flaw inclination angles under cyclic loading (number denotes the sequence of crack initiation).

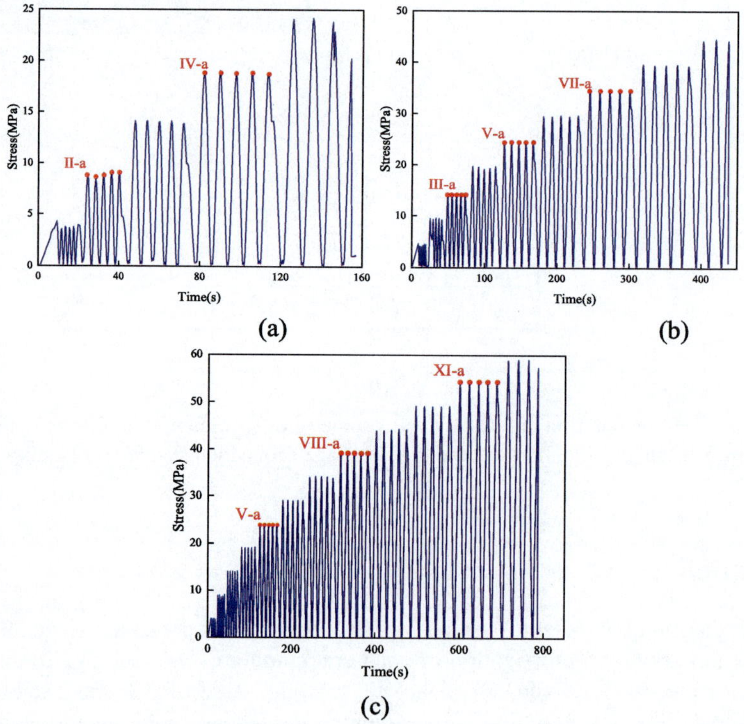

Fig. 3. Selection of key points of the specific specimen: (a) α = 0°; (b) α = 60°; (c) α = 90°

To comprehensively clarify the influence of varying inclination angle on cracking behavior during the cyclic loading and unloading process, specimens with inclinations of $\alpha = 0°$, $60°$, and $90°$ are analyzed in this study. In addition, the crack developments are further examined by observing full-field strains at several key points in Fig. 3. For example, III-a represents the first key point selected in the third loading level. In order to clarify the development of cracks throughout the full test process better, select key points as even distributed as possible.

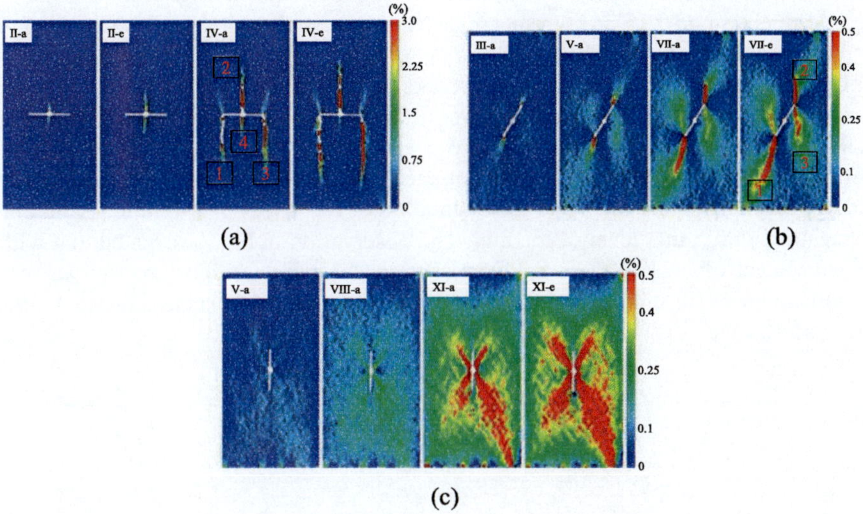

Fig. 4. Strain clouds for specimen at keypoints: (a) $\alpha = 0°$; (b) $\alpha = 60°$; (c) $\alpha = 90°$

Figure 4 showed the strain localization evolution of specimens with varying flaw inclination. In Fig. 4a, a relatively small strain localization zone began to initiate from the middle of pre-fabricated flaw at II-a of specimen with inclination of $\alpha = 0°$ due to the stress concentration. As the normalized cycle number increases, the microcracks in strain localization zone developed continuously. Two high-strain bands are distributed at the center of the flaw, resulting from the appearance of macroscopic cracks. It can be observed at IV-a and IV-e that 6 strain localization bands appeared along the axial at the middle and ends of flaw. The strain localization band located in the middle below the flaw gradually closed as the cycle proceeded, which may be the result of the transition of the stress state from tensile to compressive in this band. In Fig. 4b-c, regions with strain larger than average strain distribute more diffusely in the early loading phase for specimens with inclinations of $\alpha = 60°$ and $90°$. Those regions with ill-defined boundaries distribute around the flaws and shows the potential direction of upcoming crack propagation. In Fig. 4(b), the strain localization zones present in an elliptical shape at the tip of the flaw which indicates that more microcracks initiated. These microcracking zones are named as fracture process zone, which is the inherent properties of quasi-brittle materials. It can be seen that several fracture process zones with large size are observed near the flaw tips, indicating the presence of high compressive stress. Besides, the increment of the stress level and number of cycles promotes the extension of high-strain bands within the fracture process zone. In Fig. 4(c), regions with larger strain obviously distribute over the surface of specimen. Strain localization band appeared in X shape, which is similar to strain field of rock specimen under uniaxial compression.

By extracting the crack length at key points, the crack evolution of each specimen during the loading and unloading process was examined (Figs. 5–6). In addition, each strain localization bands are labeled based on their location in Fig. 4. It showed that the strain localization band is stable during the early loading stages (normalized cycle number around 0.25) and becomes unstable during the later loading stages (normalized cycle number around 0.75) in both specimens. It can be implied that the strain localization band correspond to the unstable development of microcracks. This phenomenon can be attributed to the accumulation of microcracks in the rocks in the later loading stages. Even with the same level of applied stress, the specimen experiences a larger strain localization due to the interaction of more microcracks.

Figure 6 displays the crack lengths at each of the five loading peaks of a specific loading level, providing a clearer understanding of the effect of cyclic loads on crack propagation within that level of loading. The observation in Fig. 6 revealed that within the same level of cycles, the crack propagation can be classified into two stages as the number of cycle increases, i.e. unstable crack propagation stage and stable crack propagation stage.

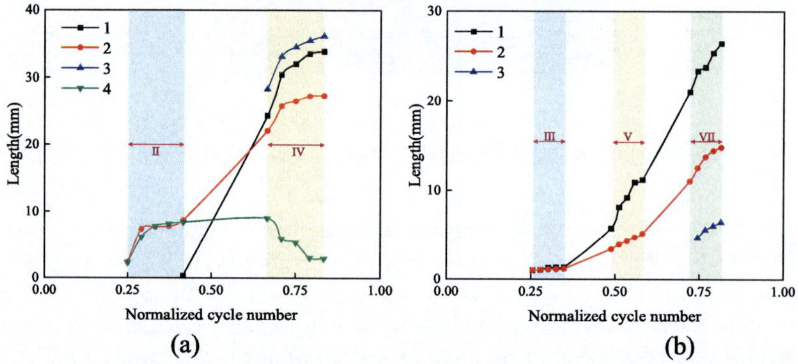

Fig. 5. Localized strain band analysis during the loading process: (a) $\alpha = 0°$; (b) $\alpha = 60°$

When $\alpha = 0°$, the crack rapidly initiates and propagates during the first cycle, indicating that upper limit stress of loading can influence the crack evolution. When $\alpha = 60°$, the crack also initiates and propagates in the first cycle; however, the propagation speed is slower, suggesting that cracking behavior such as initiation and propagation of cracks is affected not only by upper limit stress at loading but also by the number of cycles. In contrast to the previous two cases, When $\alpha = 90°$, the cracks do not initiate during the first cycle but rather in subsequent cycles, entering a phase of unstable propagation. This suggests that the crack evolution behaviors involving initiation and propagation are influenced by the number of cycles. It is hypothesized that there is a competitive relationship between the number of cyclic loads and the upper limit of loading stress on

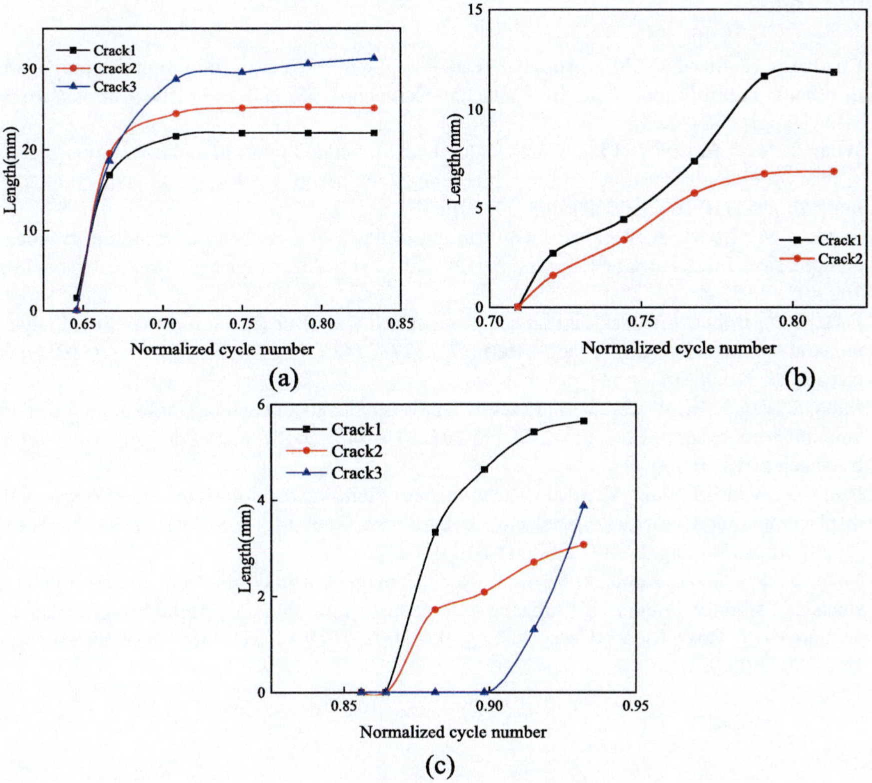

Fig. 6. Crack evolution in the same stress level cycle: (a) $\alpha = 0°$; (b) $\alpha = 60°$; (c) $\alpha = 90°$

crack evolution during the cyclic loading and unloading test. The upper limit of loading stress has a greater influence on crack evolution with smaller normalized cycle number.

4 Conclusion

During the entire process of cyclic loading and unloading, DIC was utilized to track the evolution of crack in the specimen. Based on the analysis, it is concluded that during multiple cycles of loading and unloading at the same stress level, crack propagation can be classified into two stages: stable propagation and unstable propagation. These stages are found to be strongly linked to the normalized cycle number of the cracked sandstone specimen. The findings suggest that the evolutions of crack are not solely dependent on the number of cyclic loads, but also on the upper limit stress of loading. The influence of both factors on the crack propagation differs at various stages.

Acknowledgments. This work was supported by National Natural Science Foundation of China (Grant No. 52125903).

References

1. Eberhardt, E., Stead, D., Stimpson, B., Read, R.S.: Identifying crack initiation and propagation thresholds in brittle rock. Can. Geotech. J.an Geotech. J. **35**, 222–233 (1998). https://doi.org/10.1139/cgj-35-2-222
2. Wong, L.N.Y., Einstein, H.H.: Crack coalescence in molded gypsum and carrara marble: part 1. macroscopic observations and interpretation. Rock Mech. Rock Eng. **42**, 475–511 (2009). https://doi.org/10.1007/s00603-008-0002-4
3. Sagong, M., Bobet, A.: Coalescence of multiple flaws in a rock-model material in uniaxial compression. Int. J. Rock Mech. Min. Sci. **39**, 229–241 (2002). https://doi.org/10.1016/S1365-1609(02)00027-8
4. Park, C.H., Bobet, A.: Crack initiation, propagation and coalescence from frictional flaws in uniaxial compression. Eng. Fract. Mech. **77**, 2727–2748 (2010). https://doi.org/10.1016/j.engfracmech.2010.06.027
5. Miao, S., Pan, P.-Z., Wu, Z., et al.: Fracture analysis of sandstone with a single filled flaw under uniaxial compression. Eng. Fract. Mech. **204**, 319–343 (2018). https://doi.org/10.1016/j.engfracmech.2018.10.009
6. Pan, P.-Z., Miao, S., Jiang, Q., et al.: The influence of infilling conditions on flaw surface relative displacement induced cracking behavior in hard rock. Rock Mech. Rock Eng. **53**, 4449–4470 (2020). https://doi.org/10.1007/s00603-019-02033-x
7. Miao, S., Pan, P.-Z., Konicek, P., et al.: Rock damage and fracturing induced by high static stress and slightly dynamic disturbance with acoustic emission and digital image correlation techniques. J. Rock Mech. Geotech. Eng. **13**, 1002–1019 (2021). https://doi.org/10.1016/j.jrmge.2021.05.001

Study on Seismic Damage Mode and Key Construction Damage Mechanism of Highway Pile-Plate Structure

Xiaoming Liu(✉) and Feng Xue

Anhui Transportation Holding Construction Management Co., Ltd., Hefei 241199, China
1398833475@qq.com

Abstract. The seismic design concept based on performance for bridge structures is an important direction for the continued improvement and development of bridge seismic design specifications. One of the key aspects is to choose reasonable indicators for defining and quantifying the seismic performance level of bridge structures. In this article, the Incremental Dynamic Analysis (IDA) method was used to simulate the entire earthquake damage and failure process of a rigid frame bridge with a high pier. The findings reveal that, in sleeve-type and bolted-type connection pile-plate structures under far-field and near-field seismic motions, different prefabricated pile positions and joints carry the same seismic force during the earthquake process. However, under far-field and near-field earthquakes, the state of the prefabricated pile pile bottom and connection structure differs under different Peak Ground Acceleration (PGA) levels.

Keywords: Seismic performance of bridge structures · high-pier rigid-frame bridge · panel pile structure · earthquake analysis

1 Introduction

In the context of current road construction, land resources continue to become increasingly scarce. The contradiction between the lack of available land and soil is extremely prominent. As a result, pile-plate roadbed has emerged as a new type of pile-plate beam structure, consisting of prefabricated plate beams and pipe piles as part of its framework system. Initially, pile-plate roads were mainly used in the field of railways. However, with technological advancements and innovative developments in construction techniques, pile-plate structures have since been widely applied in both the expansion and widening of highways, as well as in new construction projects [1–3]. Midas was used by Zhu Jun [4] to establish an entire model of the pile-plate roadbed. While selecting fatigue analysis points and analyzing the pile-top influence surface, preliminary calculations of the pile-plate connection fatigue performance were conducted. Zheng Wucong [5] explored various multi-scale modelling methods for pile-plate roadbeds in highways. By establishing multiple connection methods including CERIG connection, RBE3 connection, MPC method and common node coupling method, beam elements,

W. Guo et al. (Eds.): GBCESC 2023, LNCE 328, pp. 267–273, 2024.
https://doi.org/10.1007/978-981-99-9947-7_28

plate elements and solid elements could be connected through different ways to construct multi-scale models. Lei Jin [6] established a nonlinear finite element model of the pile-plate connection construction based on the engineering design example of this research project and using ANSYS finite element software.

The present study focuses on the pile-plate structure of a certain highway and aims to establish a nonlinear finite element model of the structure for seismic analysis.

2 Project Overview

The pile plate connection with the pre-stressed tube pile has a regular circular cross-section, with a diameter of 516 mm for the pile plate connection structure. The steel casing has a thickness of 8 mm. It is divided into two non-linear beam-column units along the height. The hoop reinforcement uses 10 mm HPB300 steel bars, and the longitudinal reinforcement uses 12 HRB400 steel bars with a diameter of 25 mm. The internal grouting material uses C50 compensating shrinkage concrete. The pre-stressed tube pile has a diameter of 500 mm and a height of 10 m. It is divided into 8 non-linear beam-column units along the height. The hoop reinforcement uses 10 mm HPB300 steel bars, and the longitudinal reinforcement uses 12 Φ25 mm bars for the filling section. The longitudinal reinforcement of the prefabricated pile uses 12 Φ12 mm and 12 pre-stressed steel bars with a diameter of 12.6 mm. The effective pre-pressure stress of concrete is 6 Mpa.

3 Establish a Finite Element Model of the Sheet Pile Structure

This paper presents the use of OpenSees software to create a nonlinear finite element model of the sheet pile structure and to conduct seismic analysis as shown in Fig. 1. The pile-plate connection structure has a steel casing thickness of 8 mm and its concrete grade is C50 to account for the constraining effects of the steel casing. The confinement concrete model of the steel pipe concrete is referenced. The material for the pre-stressed tube pile is C80 high-strength concrete, and its material model adopts the Kent-Park model due to its clear physical meaning, numerical stability, and ability to consider the stiffness degradation of concrete when subjected to repeated loads. The Steel02 model is employed in this study to simulate the steel casing, ordinary reinforcement, and pre-stressed reinforcement. The kinematic hardening rule with isotropic hardening in the tensile and compressive directions is adopted for all types of steel. Pre-stressing is applied by using the initial stress method. HRB400 steel bars and Q245 steel are used for the steel reinforcement and steel casing material, respectively.

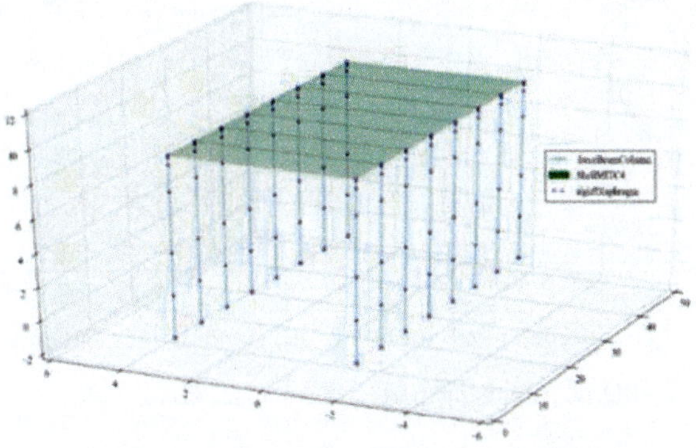

Fig. 1. OpenSees Finite Element Model.

4 Result Analysis

By selecting seismic waves from far and near fields, dynamic elastic-plastic time-history analysis of the sheet pile structure is conducted with increasing PGA to obtain a response of the entire failure process of the structure from the initial elastic state to gradual degradation and collapse. The bending moment-curvature changes of the pile bottom and joint positions, as well as the damage mechanisms of the protective layer concrete, core concrete, and longitudinal reinforcement strain, are analyzed. The analysis is carried out at three representative PGA-states of 0.3 g, 0.6 g and 0.9 g for the pile bottom of the prefabricated pile.

At PGA = 0.3 g, the bending moment curvature diagram, longitudinal bars, and strain-time relationship table (Fig. 2 and Table 1) of the prefabricated pile's base demonstrate a significant non-linear relationship between the bending moment curvature and the bottom of the pile, confirming its entry into the yield plastic stage. The longitudinal bars at the bottom have gradually shifted from compression to tension, with a maximum tensile strain of 0.00068, still within the elastic range. The protective and core concrete experience maximum compressive strains of −0.00061 and −0.00047, respectively. Small, barely visible cracks have appeared on the concrete's surface. Similarly, the joints exhibit a linear relationship in the bending moment curvature (Fig. 2 and Table 1), indicating that the external steel casing enhances lateral restraint. The joints remain in the linear elastic stage, with a maximum tensile strain of 0.00016 in the joints' longitudinal bar, while the steel bars primarily remain elastic. The core concrete shows a maximum compressive strain of −0.00027, indicating that the joint concrete remains in the elastic stage and has not reached the plastic stage.

Under PGA = 0.6 g, the bending moment curvature diagram, longitudinal bars, and strain-time relationship table (Fig. 3 and Table 1) at the bottom of the prefabricated pile show that the bottom of the pile has entered the plastic yield stage. The longitudinal bars at the bottom exhibit a maximum tensile strain of 0.00178, nearing the yield stage. The maximum compressive strains of the protective concrete and core concrete are −0.001 and −0.00072, respectively, with visible surface cracks indicating minor damage. Similarly, the joints' relationships (Fig. 3 and Table 1) suggest that the bending moment

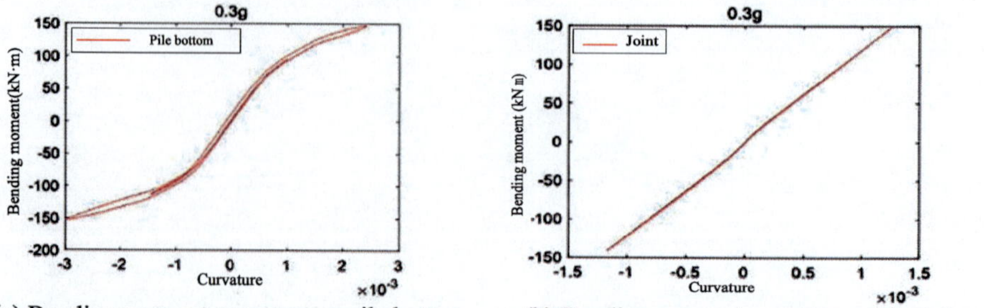

(a) Bending moment curvature at pile bottom (b) Bending moment curvature of the joint.

Fig. 2. Mechanical performance at PGA = 0.3 g

curvature remains linearly related, indicating a linearly elastic state under PGA = 0.6 g. The joints' longitudinal bar experiences a maximum tensile strain of 0.00027, while the steel bars primarily remain in the linear elastic stage. The core concrete experiences a maximum compressive strain of −0.00043, and the joint concrete remains fundamentally intact.

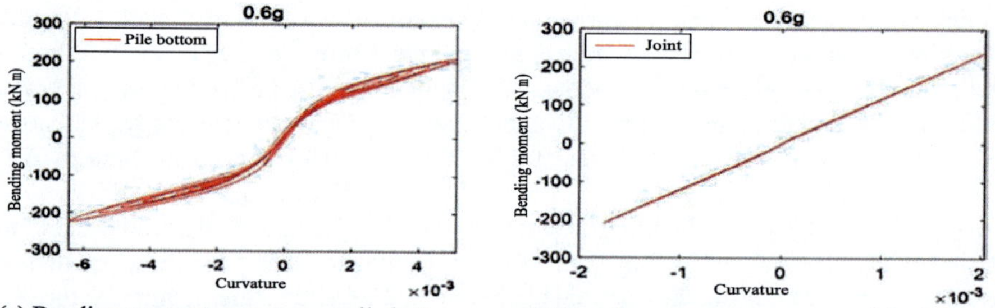

(a) Bending moment curvature at pile bottom (b) Bending moment curvature of the joint.

Fig. 3. Mechanical performance at PGA = 0.6g

Under PGA = 0.9 g, the bending moment curvature relationship diagram, longitudinal steel bars, and strain-time relationship diagram for the protective and core concrete of the prefabricated pile were investigated (Fig. 4 and Table 1). The results revealed that the pile bottom experienced plastic yielding stages, with a maximum tensile strain of 0.00295 in the longitudinal steel, indicating yielding. The maximum compressive strains of the protective layer and core concrete were −0.00138 and -0.00091, respectively. Surface cracks in the concrete continued to propagate, resulting in mild damage. Similarly, the analysis of joint behavior under similar conditions (Fig. 4 and Table 1) showed that the bending moment curvature of the joint remained linear under PGA = 0.9 g. The joint remained in the linear elastic stage, with a maximum tensile strain of 0.00036 in the longitudinal rebar, indicating the steel bars were still in the linear elastic stage. The maximum compressive strain of the core concrete was −0.00055, and the joint concrete exhibited minimal or no damage.

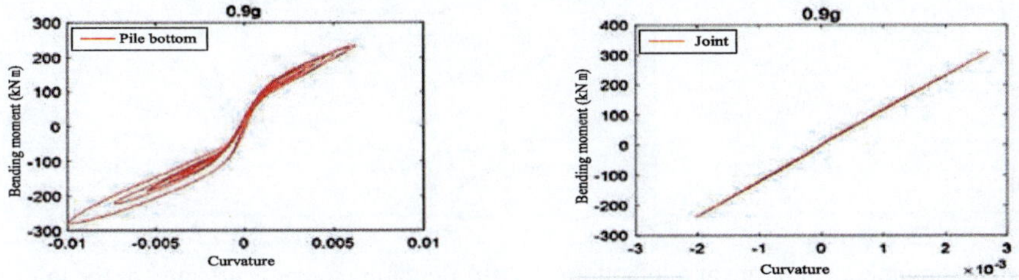

(a) Bending moment curvature at pile bottom (b) Bending moment curvature of the joint.

Fig. 4. Mechanical performance at PGA = 0.9g

Table 1. Strain-time relationship table under different PGAs in the far-field condition.

PGA(Unit: g)	The bottom of the prefabricated pile			The joint	
	Longitudinal bars	Protective layer concrete	Core concrete	Longitudinal bars	Core concrete
0.3	–0.00056	–0.00061	–0.00047	–0.00011	–0.00027
0.6	–0.00090	–0.00102	–0.00072	–0.000157	–0.00042
0.9	–0.00119	–0.00138	–0.00091	–0.000196	–0.00055

In the near-field region, the bending moment-curvature diagram, longitudinal reinforcement, and strain-time diagram of protective and core concrete at the bottom of prefabricated piles under PGA = 0.3 g indicate that the bottom of the piles has entered the plastic yielding stage. The maximum tensile strain in the longitudinal reinforcement is 0.0021, reaching the yield stage. The maximum compressive strains for the protective and core concrete are –0.00124 and –0.00087, respectively. Cracks gradually develop in the protective concrete, indicating mild damage. Similarly, the joint relationship reveals that the bending moment-curvature of the joint remains linear due to the reinforcing effect of the external steel sleeve. The joint remains in a linear elastic stage with a maximum tensile strain of 0.00030 in the longitudinal reinforcement. The steel bars remain in the linear elastic stage, while the maximum compressive strain in the core concrete is –0.00051. The joint concrete remains elastic and does not undergo plastic deformation (Fig. 5).

At PGA = 0.6 g, the bottom of the prefabricated piles entered the plastic stage. The longitudinal reinforcement reached a maximum tensile strain of 0.0061, indicating steel yielding. The protective and core concrete experienced maximum compressive strains of –0.00217 and –0.00129 respectively. The structure exhibited non-linear deformation with visible cracks and delamination in the protective concrete, resulting in moderate damage. However, the joint remained in a linear elastic state under the same conditions. The bending moment-curvature relationship of the joint remained linear, with a maximum tensile strain of 0.0005 in the longitudinal reinforcement. The steel in the joint maintained its linear elastic stage. The core concrete exhibited a maximum compressive strain of –0.00074, and there was no significant damage to the joint concrete (Fig. 6).

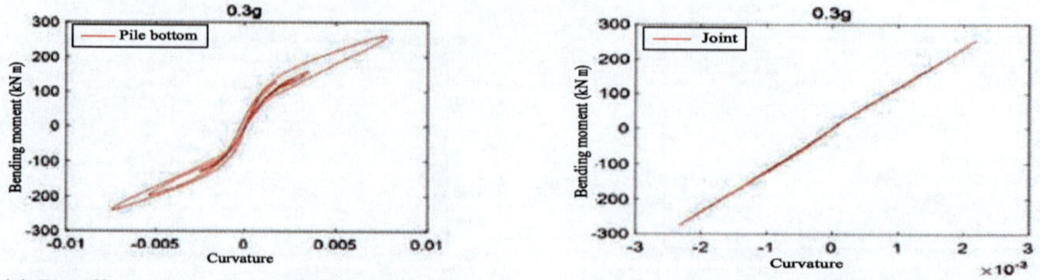

(a) Bending moment curvature at pile bottom (b) Bending moment curvature of the joint.

Fig. 5. Mechanical performance at PGA = 0.3 g

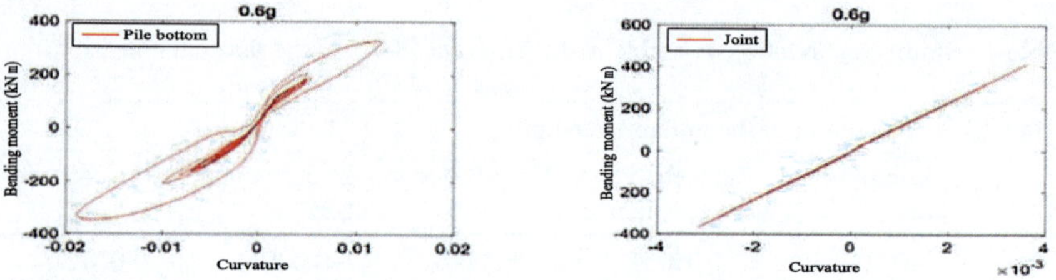

(a) Bending moment curvature at pile bottom (b) Bending moment curvature of the joint.

Fig. 6. Mechanical performance at PGA = 0.6g

At PGA = 0.9 g, the bottom of the prefabricated pipe pile entered the plastic yielding stage. The longitudinal reinforcement exhibited a maximum tensile strain of 0.0118, indicating yielding. The protective layer concrete and core concrete experienced maximum compressive strains of −0.0035 and −0.0019 respectively. This led to the formation of a structural plastic hinge with wider cracks (at least 2 mm in width). The concrete in the entire plastic hinge area peeled off, resulting in severe structural damage. Conversely, the joint remained in a linear elastic state under the same conditions. The bending moment-curvature relationship of the joint remained linear, with a maximum tensile strain of the longitudinal reinforcement at 0.00067. The steel bar in the joint remained in the linear elastic stage, and the core concrete experienced a maximum compressive strain of −0.00097. Overall, there was minimal damage to the joint concrete (Fig. 7 and Table 2).

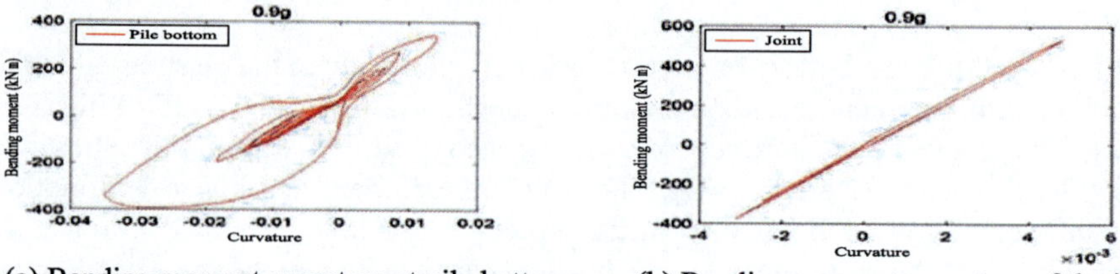

(a) Bending moment curvature at pile bottom (b) Bending moment curvature of the joint.

Fig. 7. Mechanical performance at 0.9 g

Table 2. Strain-time relationship at various PGAs.

PGA(Unit: g)	The bottom of the prefabricated pile			The joint	
	Longitudinal bars	Protective layer concrete	Core concrete	Longitudinal bars	Core concrete
0.3	−0.00110	−0.00124	−0.00087	−0.000198	−0.00051
0.6	−0.00181	−0.00217	−0.00129	−0.000258	−0.00074
0.9	−0.00289	−0.00353	−0.00197	−0.000329	−0.000968

5 Conclusion

This paper analyzes the effects of far-field and near-field earthquake ground motions on highway pile-slab structures through time-domain analysis. The following conclusions can be drawn:

(1) Far-field earthquakes gradually damage prefabricated pipe piles with increasing PGA. At 0.1 g, the pile is elastic. At 0.2 g, plastic deformation starts, and at 0.3 g, cracks appear. At 0.7 g, bars yield and cracks widen. Up to 1.0 g, slight damage occurs. Connections remain undamaged due to steel reinforcement.

(2) Near-field earthquakes cause progressive damage to prefabricated pipe piles. At 0.1 g, the pile remains elastic. At 0.3 g, plastic deformation starts, yielding reinforcement and developing cracks. At 0.6 g, non-linear deformation occurs with visible cracks and peeling, causing moderate damage. At 0.9 g, severe damage arises with plastic hinges and wide cracks. Connection structures exhibit minimal damage due to steel casing restraint, even at PGA = 1.0 g.

References

1. Wang, H., Liu, W., Huang, X.: Comparative Research and Analysis of Prefabricated Buildings and Cast-in-Place Buildings. Special Materials Compilation of Industrial Architecture Magazine (2017)
2. Fang, Z., Dou, W., Zhang, H.: Seismic performance analysis of pile-plate roadbed structure. Anhui Archit. **28**(12), 136–138 (2021)
3. Tao, C.: Roadbed detection technology and its application. Constr. Technol. (23), 101–104 (2016)
4. Jun, Z., Nan, D.: Fatigue performance analysis of pile-plate connection joints for pile-plate roadbed. Eng. Constr. **32**(05), 673–675 (2018)
5. Zheng, W.: Study on Characteristics of Pile-Plate Roadbed Structure Based on Multiscale Finite Element Model. Hefei University of Technology (2018)
6. Jin, L.: Nonlinear finite element analysis of highway pile-plate structure connection structure based on ANSYS. Eng. Constr. **33**(05), 780–782 (2019)

Free Vibration and Tension-Bending Coupling Behaviors of Sandwich Panels with Novel Tri-Chi Honeycomb

Minfang Chen[1,2], Yifeng Zhong[1,2](✉), Irakoze Alain Evrard[1,2], and Xiaoquan Liu[1,2]

[1] School of Civil Engineering, Chongqing University, Chongqing, People's Republic of China
zhongyf@cqu.edu.cn

[2] Key Laboratory of New Technology of Construction of Cities in Mountain Area (Chongqing University), Ministry Of Education, Chongqing 400045, People's Republic of China

Abstract. The triangular chiral honeycomb is inspired by the anti-tetra chiral design and consists of fully triangular cells, which is a novel structure that exhibits a negative Poisson's ratio. In this study, the effective plate properties of sandwich panels with triangular chiral honeycombs (SP-TCH) are evaluated using the variational asymptotic method. Based on this, a two-dimensional reduced-order plate model (2D-RPM) is developed. The accuracy and effectiveness of the 2D-RPM is verified by comparing its analysis of the tension-bending coupling and free vibration of SP-TCH against 3D finite element (FE) model results.

Keywords: Sandwich plate · Triangular chiral honeycomb · Effective plate properties

1 Introduction

Chiral metamaterials, a type of NPR structure that utilizes rotational mechanisms [1, 2], were first proposed by Kelvin and William and belong to the class of NPR metamaterials characterized by periodicity and internal spin overlap [3].Given that chirality is a common symmetry feature found in nature and the excellent performance of its features, chiral metamaterials are frequently applied for design optimization of structures.

Unlike involute metamaterials that exhibit internal concave deformation, the rotational deformation of chiral metamaterials occurs through rotating the ligaments around the central cylinder [4]. This rotational deformation of the unit cell induces a chain reaction in the entire structure, resulting in considerably enhanced energy absorption properties [5]. Given the unique structural properties of these materials, they have drawn significant interest. For instance, Abdeljaber et al. [6] developed an optimization method using an automatic genetic algorithm with nine independent geometrical parameters to control the structure's shape and gradient. Their numerical findings suggested that the chiral honeycomb structure's plastic deformation is a function of both its structural parameters and impact velocity.

W. Guo et al. (Eds.): GBCESC 2023, LNCE 328, pp. 274–281, 2024.
https://doi.org/10.1007/978-981-99-9947-7_29

Grima [7] also theoretically examined the "rotating triangles" mechanism and found it to be an incredibly potent way of introducing the NPR effect. Initially, the tetra-chiral cell's circular part underwent replacement with two triangles, resulting in a triangular chiral cell. The members were subsequently resized to form a fully triangulated chiral structure (Tri-Chi) by replicating cells, as demonstrated in Fig. 1. Sandwich panels with Tri-Chi honeycomb (SP-TCH) are a relatively new structure, and as a result, there has been limited research on their effective performance. Therefore, it is crucial to adopt a simplified analysis method that can predict their global and local behavior with both speed and accuracy.

2 VAM-Based Reduced-Order Model for SP-TCH

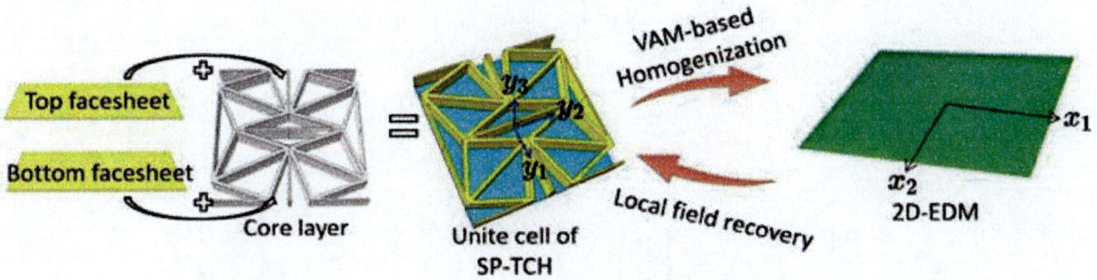

Fig. 1. Flow Chart of the 2D-RPM for SP-TCH.

To derive the 2D-RPM for SP-TCH, the 3D displacement field is represented using 2D plate variables, such as

$$\begin{aligned} u_1(x_\alpha, y_i, t) &= \overline{u}_1(x_\alpha, t) - \zeta y_3 \overline{u}_{3,1}(x_\alpha, t) + \zeta \omega_1(x_\alpha, y_i, t) \\ u(x_\alpha, y_i, t) &= \overline{u}_2(x_\alpha, t) - \zeta y_3 \overline{u}_{3,2}(x_\alpha, t) + \zeta \omega_2(x_\alpha, y_2, t) \\ u_3(x_\alpha, y_i, t) &= \overline{u}_3(x_\alpha, t) + \zeta \omega_3(x_\alpha, y_i, t) \end{aligned} \tag{1}$$

where u_i and $\overline{u}_i$ respectively represent the displacement of the original 3D plate and the 2D-RPM, w_i is the warping function, and the underlined terms represent the reference surface deformations of SP-TCH based on the classical plate theory, which subject to the conditions of

$$h\overline{u}_\alpha(x_\alpha) = \langle u_\alpha \rangle + \langle \eta y_3 \rangle \overline{u}_{3,2}, \quad h\overline{u}_3(x_\alpha) = \langle u_3 \rangle, \tag{2}$$

Equation (2) indicates that the warping function is constrained by

$$\langle \zeta \omega_i \rangle = 0 \tag{3}$$

For small local rotation deformation, the 2D generalized strain and warping function can be used to approximate 3D strain based on rotation tensor decomposition, such as

$$\varepsilon_{ij} = \frac{1}{2}\left(u_{i,j} + u_{j,i}\right) - \delta_{ij} \tag{4}$$

Substituting Eq. (1) into Eq. (4) and disregarding the high-order terms that have a negligible impact on the energy, we can derive 3D strain field as

$$\begin{aligned}&\varepsilon_{11} = \epsilon_{11} + \zeta y_3 \kappa_{11} + \omega_{1,1},\ 2\varepsilon_{13} = \omega_{1,3} + \omega_{3,1}\\ &2\varepsilon_{12} = 2\epsilon_{12} + 2\zeta y_3 \kappa_{12} + \omega_{1,2} + \omega_{2,1},\ 2\varepsilon_{23} = \omega_{2,3} + \omega_{3.2},\\ &\varepsilon_{22} = \epsilon_{22} + \zeta y_3 \kappa_{22} + \omega_{2,2},\ \varepsilon_{33} = \omega_{3,3}\end{aligned} \tag{5}$$

where the in-plane strain $\epsilon_{\alpha\beta}$ and bending curvature $\kappa_{\alpha\beta}$ of the 2D-RPM are

$$\epsilon_{\alpha\beta}(x_1, x_2) = \frac{1}{2}\left(\overline{u}_{\alpha,\beta} + \overline{u}_{\beta,\alpha}\right),\quad \kappa_{\alpha\beta}(x_1, x_2) = -\overline{u}_{3,\alpha\beta}. \tag{6}$$

The 3D strain field can be represented in the form of a matrix, which is given by

$$\begin{aligned}&\mathcal{E}_e = \begin{bmatrix}\varepsilon_{11}\ \varepsilon_{22}\ \varepsilon_{33}\end{bmatrix}^{\mathrm{T}} = \boldsymbol{\epsilon} + x_3 \boldsymbol{\kappa} + \boldsymbol{I}_\alpha \boldsymbol{\omega}_{\|,\alpha}\\ &2\mathcal{E}_s = \begin{bmatrix}2\varepsilon_{13}\ 2\varepsilon_{23}\end{bmatrix}^{\mathrm{T}} = \boldsymbol{\omega}_{\|,3} + \boldsymbol{e}_\alpha \omega_{3,\alpha}\quad ,\\ &\mathcal{E}_n = \varepsilon_{33} = \omega_{3,3}\end{aligned} \tag{7}$$

where $()_{\|} = [()_1 ()_2]^{\mathrm{T}}, = \begin{bmatrix}\epsilon_{11}\ 2\epsilon_{12}\ \epsilon_{22}\end{bmatrix}^{\mathrm{T}}$, $\kappa = \begin{bmatrix}\kappa_{11}\ \kappa_{12} + \kappa_{21}\ \kappa_{22}\end{bmatrix}^{\mathrm{T}}$, and

$$e_1 = \begin{Bmatrix}1\\0\end{Bmatrix},\quad e_2 = \begin{Bmatrix}0\\1\end{Bmatrix},\quad \boldsymbol{I}_1 = \begin{bmatrix}1\,0\\0\,1\\0\,0\end{bmatrix},\quad \boldsymbol{I}_2 = \begin{bmatrix}0\,0\\1\,0\\0\,1\end{bmatrix}. \tag{8}$$

The strain energy of the SP-TCH can be obtained as

$$\mathcal{U} = \frac{1}{2}\left\langle \mathcal{E}^{\mathrm{T}} \boldsymbol{D}\mathcal{E}\right\rangle = \frac{1}{2}\left\langle \begin{Bmatrix}\mathcal{E}_e\\ 2\mathcal{E}_s\\ \mathcal{E}_n\end{Bmatrix}^{\mathrm{T}} \begin{bmatrix}\boldsymbol{D}_e & \boldsymbol{D}_{es} & \boldsymbol{D}_{en}\\ \boldsymbol{D}_{es}^{\mathrm{T}} & \boldsymbol{D}_s & \boldsymbol{D}_{sn}\\ \boldsymbol{D}_{en}^{\mathrm{T}} & \boldsymbol{D}_{sn}^{\mathrm{T}} & \boldsymbol{D}_n\end{bmatrix} \begin{Bmatrix}\mathcal{E}_e\\ 2\mathcal{E}_s\\ \mathcal{E}_n\end{Bmatrix}\right\rangle, \tag{9}$$

The virtual work can be expressed as

$$\delta\overline{\mathcal{W}_{3D}} = \delta\overline{\mathcal{W}_{2D}} + \delta\overline{\mathcal{W}^*} \tag{10}$$

where $\delta\overline{\mathcal{W}_{2D}}$ and $\delta\overline{\mathcal{W}^*}$ represent 2D-RPM and the remaining virtual work, respectively.

The kinetic energy of the panel can be written as

$$\mathcal{K} = \frac{1}{2}\int_{\mathcal{V}} \rho^* v^{\mathrm{T}} v d\mathcal{V} = \mathcal{K}_{2D} + \mathcal{K}^*, \tag{11}$$

where $\rho*$ represents the mass density of the panel.

The elastodynamic behavior of SP-TCH can be determined through the extended Hamilton principle, such as

$$\int_{t_1}^{t_2} \left[\delta\left(\mathcal{K}_{2D} + \mathcal{K}^* - \mathcal{U}\right) + \delta\overline{\mathcal{W}_{2D}} + \delta\overline{\mathcal{W}^*}\right] dt = 0, \tag{12}$$

where only w_i are variable.

By neglecting asymptotically smaller terms, we can express the zeroth-order approximation of Eq. (12) as

$$\int_{t_1}^{t_2}\left[\delta\left(\mathcal{K}_{2D}-\int_{\Omega}\mathcal{U}_0 d\Omega\right)+\overline{\delta\mathcal{W}_{2D}}\right]dt=0, \tag{13}$$

where $\mathcal{U}_0$ can be obtained as

$$2\mathcal{U}_0=\left\langle\begin{array}{l}(\boldsymbol{\epsilon}+x_3\boldsymbol{\kappa})^{\mathrm{T}}\boldsymbol{D}_e(\boldsymbol{\epsilon}+x_3\boldsymbol{\kappa})+\boldsymbol{w}_{\parallel,3}^{\mathrm{T}}\boldsymbol{D}_s\boldsymbol{w}_{\parallel,3}+w_{3,3}^{\mathrm{T}}\boldsymbol{D}_n w_{3,3}\\+2(\boldsymbol{\epsilon}+x_3\boldsymbol{\kappa})^{\mathrm{T}}\left(\boldsymbol{D}_{es}\boldsymbol{w}_{\parallel,3}+\boldsymbol{D}_{en}w_{3,3}\right)+2\boldsymbol{w}_{\parallel,3}^{\mathrm{T}}\boldsymbol{D}_{sn}w_{3,3}\end{array}\right\rangle. \tag{14}$$

The corresponding Euler-Lagrange equations can be obtained by introducing Lagrange multipliers λ_i, such as

$$\begin{array}{l}\left[(\boldsymbol{\epsilon}+x_3\boldsymbol{\kappa})^{\mathrm{T}}\boldsymbol{D}_{es}+\boldsymbol{w}_{\parallel,3}^{\mathrm{T}}\boldsymbol{D}_s+w_{3,3}\boldsymbol{D}_{sn}\right]_{,3}=\lambda_{\parallel}\\\left[(\boldsymbol{\epsilon}+x_3\boldsymbol{\kappa})^{\mathrm{T}}\boldsymbol{D}_{en}+\boldsymbol{w}_{\parallel,3}^{\mathrm{T}}\boldsymbol{D}_{sn}+w_{3,3}\boldsymbol{D}_n\right]_{,3}=\lambda_3\end{array} \tag{15}$$

where $\lambda_{\parallel}=[\lambda_1\ \lambda_2]^{\mathrm{T}}$.

The square bracketed expression in Eq. (15) should equal zero at the bottom and top surfaces of the panel under the free surface conditions, such as

$$\begin{array}{l}\left[(\boldsymbol{\epsilon}+x_3\boldsymbol{\kappa})^{\mathrm{T}}\boldsymbol{D}_{es}+\boldsymbol{w}_{\parallel,3}^{\mathrm{T}}\boldsymbol{D}_s+w_{3,3}\boldsymbol{D}_{sn}^{\mathrm{T}}\right]^{+/-}=0\\\left[(\boldsymbol{\epsilon}+x_3\boldsymbol{\kappa})^{\mathrm{T}}\boldsymbol{D}_{en}+\boldsymbol{w}_{\parallel,3}^{\mathrm{T}}\boldsymbol{D}_{sn}+w_{3,3}\boldsymbol{D}_n\right]^{+/-}=0\end{array} \tag{16}$$

Using these conditions, we can obtain the solutions for $w_{\parallel}$ and w_3,

$$\boldsymbol{w}_{\parallel}=-\left\langle(\boldsymbol{\epsilon}+x_3\boldsymbol{\kappa})\overline{\boldsymbol{D}}_{es}\boldsymbol{D}_s^{-1}\right\rangle^{\mathrm{T}},\quad w_3=\left\langle-(\boldsymbol{\epsilon}+x_3\boldsymbol{\kappa})\overline{\boldsymbol{D}}_{en}\boldsymbol{D}_n^{-1}\right\rangle^{\mathrm{T}}, \tag{17}$$

where

$$\begin{array}{l}\overline{\boldsymbol{D}_{es}}=\boldsymbol{D}_{es}-\overline{\boldsymbol{D}_{en}}(\boldsymbol{D}_{sn})^{\mathrm{T}}\left(\overline{\boldsymbol{D}_n}\right)^{-1},\quad \overline{\boldsymbol{D}_{en}}=\boldsymbol{D}_{en}-\boldsymbol{D}_{es}(\boldsymbol{D}_s)^{-1}\boldsymbol{D}_{sn},\\\overline{\boldsymbol{D}_n}=\boldsymbol{D}_n-(\boldsymbol{D}_{sn})^{\mathrm{T}}(\boldsymbol{D}_s)^{-1}\boldsymbol{D}_{sn}\end{array} \tag{18}$$

The stain energy of 2D-RPM can be determined by substituting Eq. (17) in Eq. (14),

$$U_{2D}=\frac{1}{2}\left\langle(\boldsymbol{\epsilon}+x_3\boldsymbol{\kappa})^{\mathrm{T}}\overline{\boldsymbol{D}_e}(\boldsymbol{\epsilon}+x_3\boldsymbol{\kappa})\right\rangle=\frac{1}{2}\begin{Bmatrix}\boldsymbol{\epsilon}\\\boldsymbol{\kappa}\end{Bmatrix}^{\mathrm{T}}\begin{bmatrix}\boldsymbol{A}&\boldsymbol{B}\\\boldsymbol{B}^T&\boldsymbol{D}\end{bmatrix}\begin{Bmatrix}\boldsymbol{\epsilon}\\\boldsymbol{\kappa}\end{Bmatrix}, \tag{19}$$

where the 3 × 3 sub-matrices $\boldsymbol{A}$, $\boldsymbol{B}$, and $\boldsymbol{D}$ can be determined as

$$\begin{array}{l}\boldsymbol{A}=\left\langle\overline{\boldsymbol{D}_e}\right\rangle,\ \boldsymbol{B}=x_3\overline{\boldsymbol{D}_e},\ \boldsymbol{D}=\left\langle x_3^2\overline{\boldsymbol{D}_e}\right\rangle\\\overline{\boldsymbol{D}_e}=\boldsymbol{D}_e-\overline{\boldsymbol{D}_{es}}\boldsymbol{D}_s^{-1}\boldsymbol{D}_{es}^{\mathrm{T}}-\overline{\boldsymbol{D}_{en}}\boldsymbol{D}_{en}^{\mathrm{T}}/\overline{\boldsymbol{D}_n}\end{array} \tag{20}$$

As a result of the axis and plane symmetry inherent in SP-TCH, specific stiffnesses are not present in $\overline{\boldsymbol{D}_e}$, and the compliance matrix can be obtained by taking the inverse of stiffness matrix, such as

$$\bar{k}=\left[\begin{array}{ccc|ccc} A_{11} & A_{12} & 0 & B_{11} & B_{12} & 0 \\ A_{12} & A_{22} & 0 & B_{12} & B_{22} & 0 \\ 0 & 0 & A_{66} & 0 & 0 & B_{66} \\ \hline 0 & 0 & 0 & D_{11} & D_{12} & 0 \\ 0 & 0 & 0 & D_{12} & D_{22} & 0 \\ 0 & 0 & 0 & 0 & 0 & D_{66} \end{array}\right]^{-1}=\left[\begin{array}{ccc|ccc} a_{11}^{*} & a_{12}^{*} & 0 & b_{11}^{*} & b_{12}^{*} & 0 \\ a_{12}^{*} & a_{22}^{*} & 0 & b_{12}^{*} & b_{22}^{*} & 0 \\ 0 & 0 & a_{66}^{*} & 0 & 0 & b_{66}^{*} \\ \hline 0 & 0 & 0 & d_{11}^{*} & d_{12}^{*} & 0 \\ 0 & 0 & 0 & d_{12}^{*} & d_{22}^{*} & 0 \\ 0 & 0 & 0 & 0 & 0 & d_{66}^{*} \end{array}\right]. \tag{21}$$

and the tension-bending coupling coefficients can be defined as $\xi_{x,x}^{0}=b_{11}^{*}/a_{11}^{*}$, $\xi_{y,x}^{0}=b_{12}^{*}/a_{11}^{*}$, $\xi_{x,y}^{0}=b_{21}^{*}/a_{22}^{*}$, and $\xi_{y,y}^{0}=b_{22}^{*}/a_{22}^{*}$.

3 Model Verification

3.1 Model Parameters

The cell core of the structure is characterized by biaxial symmetry and is defined by structural parameters that are illustrated in Fig. 2. Specifically, $L_x=L_y=30$ mm, $t_1=0.4$ mm, $t_2=0.8$ mm, $h_c=4$ mm and $\alpha=20°$. The 3D-FEM has 10 and 15 unit cells in the x_1 and x_2 directions, respectively. The material properties of the aluminum core are $\rho=2.7$ g/cm^3, $E_{AL}=70$ GPa, and $\nu_{\mathrm{AL}}=0.3$. The facesheet, on the other hand, is constructed from a carbon fiber (T800) reinforced composite, with material properties that include $\rho=1.49$ g/cm^3, $E_1=105.5$ GPa, $E_2=E_3=11.3$ GPa, $G_{12}=3.23$ GPa, $G_{13}=G_{23}=3.18$ GPa, $\nu_{12}=\nu_{13}=0.28$, and $\nu_{23}=0.53$. The layup of the laminated facesheet is indicated as $[45/-45/0/90]_s$ in Fig. 2(c).

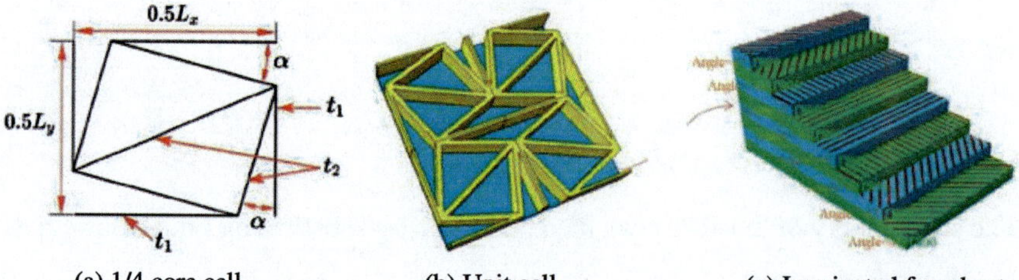

(a) 1/4 core cell (b) Unit cell (c) Laminated facesheet

Fig. 2. Geometry of SP-TCH and its unit cell.

3.2 Tensile-Bending Coupling Verification

Figure 3 compares the deformation clouds of SP-TCH under uniaxial tension ($q=100$ N), bending moment ($M=500$ N·mm), and tension-bending coupling ($q=100$ N and $M = 500$ N·mm). The results indicate that the deformation clouds of both models matched closely. Specifically, the deformation under uniaxial tension only occurred along the stretching direction and demonstrated almost zero lateral deformation. Additionally, the deformation clouds under bending moment showed a downward protruding tensile deformation. Similarly, the deformation clouds under tension-bending coupling exhibited remarkable coherence with the bending displacement cloud. This observation suggests that, for the bending load $M = 500$ N·mm and tensile load $q = 100$ N, the structural response is primarily governed by bending.

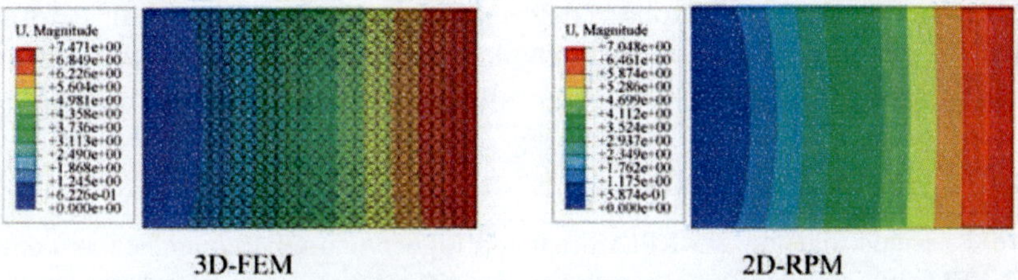

(a) Case 1: tension along the x_1 axis

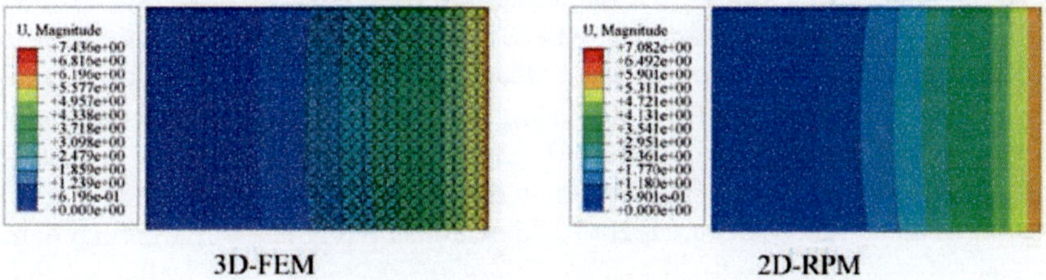

(b) Case 2: bending around the x_1 axis

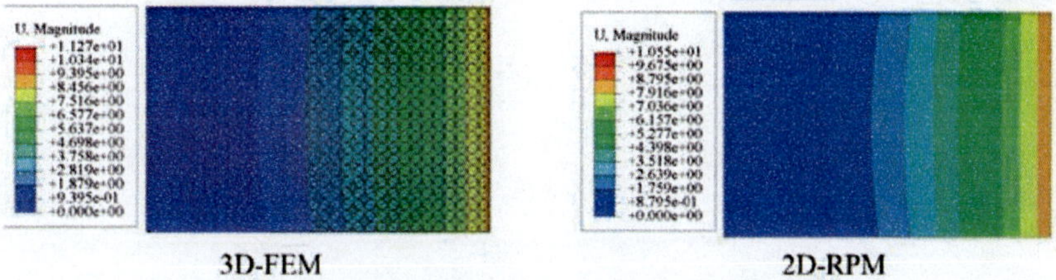

(c) Case 3: tension-bending coupling about the x axis

Fig. 3. Comparison of the uniaxial tension, bending, and tension-bending coupling behaviors of SP-TCH under CFFF BCs.

Table 1 reveals that the large disparity between the maximum displacement of the panel subjected to tension-bending coupling and that under separate loading conditions of tension and bending moment, was due to the significant tension-bending coupling effect. However, the displacement observed was not simply a linear summation of the displacements seen under the latter two loading conditions. The maximum displacement error under tension-bending coupling was found to be 6.83%, which fell within the engineering error range. These results indicate that the VAM-based 2D-RPM model not

Table 1. Comparison of maximum displacement predicted by different models

Cases	Case(FFCF)		
	3D-FEM	2D-RPM	Error/%
Case 1	7.471	7.048	6.00%
Case 2	7.436	7.082	4.99%
Case 3	11.27	10.55	6.83%

only provides accurate predictions of deformation clouds, but also specific deformation values, and may prove to be a suitable replacement for 3D-FEM in static analyses of SP-TCH, especially the complex tension-bending coupling behavior.

3.3 Free Vibration Verification

Figure 4 reveals that the 2D-RPM predicted higher natural frequencies as compared to 3D-FEM, which could be attributed to the differences in meshing between the two models. However, the errors were limited within the engineering error range (less than 3%). As the natural frequency order increased, the vibration mode became more intricate. It's worth noting that, except for the 3rd, 5th, and 8th vibration modes which were along the x_1 direction, all vibration modes were along the x_2 direction, given that the size of the panel was greater along the x_2 direction. This design should avoid strong resonance by applying the load on the short side. Based on the results of the free vibration analysis, it can be inferred that the VAM-based 2D-RPM exhibits favorable accuracy and reliability.

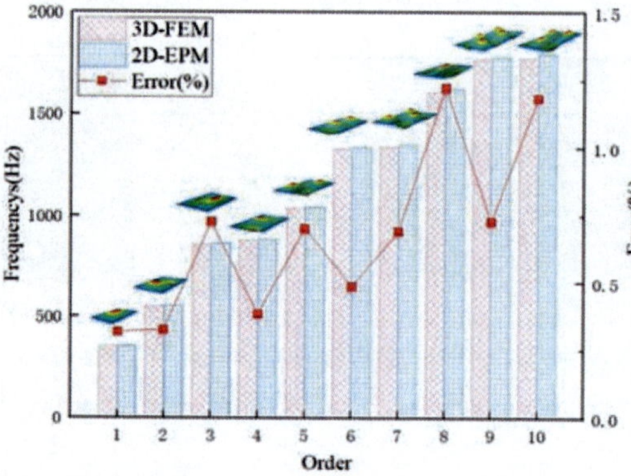

Fig. 4. Comparison of the first ten natural frequencies and corresponding vibration modes predicted by different models.

Acknowledgement. Thanks for the financial support from the National Natural Science Foundation of China (51778088, 52073036).

References

1. Alderson, A., Alderson, K.L., Attard, D., Evans, K.E., Gatt, R., Grima, J.N., et al.: Elastic constants of 3-, 4-and 6-connected chiral and anti-chiral honeycombs subject to uniaxial in-plane loading. Compos. Sci. Technol. **70**(7), 1042–1048 (2010)
2. Hu, C., et al.: 3D printing of chiral carbon fiber reinforced polylactic acid composites with negative poisson's ratios. Compos. B Eng. **201**, 108400 (2020)
3. Imbalzano, G., Linforth, S., Ngo, T.D., et al.: Blast resistance of auxetic and honeycomb sandwich panels: comparisons and parametric designs. Compos. Struct. **183**, 242–261 (2017)
4. Mozga, T., Prokop, Z., Chaloupková, R., et al.: Chiral aliphatic hydroxy compounds in nature: a review of biological functions and practical applications. J. Cheminformatics **41**(7), 1195–1278 (2010)
5. Li, S., Jiang, S., Wang, Y., et al.: Study on "Metamaterial" structural absorbing composite technology. J. Mater. Eng. **45**(11), 10–14 (2017)
6. Abdeljaber, O., Avci, O., Inman, D.J.: Optimization of chiral lattice based metastructures for broadband vibration suppression using genetic algorithms. J. Sound Vib. **369**, 50–62 (2016)
7. Grima, J.N.: Auxetic behavior from rotating triangles. J. Mater. Sci. **41**, 3193–3196 (2006)

Research on Construction Scheme for a Four-Span Continuous Slanting Heterotypic Stay Cable Arch Bridge

Yufeng Xu(✉), Zihui Li, and Zhantao Zhang

South China University of Technology, Guangzhou 510641, China
xuyf@scut.edu.cn

Abstract. The complex shape of heterotypic stay cable arch bridge have led to challenges in structural analysis and construction. This paper, based on the Echeng Bridge project in Huizhou City, introduces a bridge construction scheme. Finite element software is used to simulate the construction process. The analysis results indicate that the stress, deformation, and cable forces in the bridge during the construction process meet the requirements specified by materials and regulations. The proposed bridge construction scheme is considered feasible, and the simulation analysis results can serve as theoretical reference values for bridge construction monitoring. The findings provide valuable insights and guidance for future similar bridge construction projects.

Keywords: Heterotypic Stay Cable · Construction Scheme · Simulation Analysis · Asymmetric Support

1 Introduction

With the enhancement of China's economic strength and the advancement of urbanization, the position of bridges in urban landscapes has been gradually elevated, and people have higher aesthetic pursuits for urban bridges [1–3]. Heterotypic arch bridges possess both aesthetic and practical value, and their unique design has gained popularity among people. However, the structural behavior of heterotypic arch bridges differs from that of traditional bridges, posing new challenges in construction [4–6]. The construction of heterotypic arch bridges involves multiple stages and system transitions, leading to potential errors during the construction process and resulting in deviations from the ideal bridge state. Therefore, research on construction monitoring of heterotypic arch bridges is necessary [7–10]. This paper takes the Echeng Bridge, a four-span continuous skew heterotypic stay cable arch bridge in Huizhou City, as the background to study the construction monitoring techniques of heterotypic arch bridges and analyze the feasibility of the construction scheme through simulation and modeling.

W. Guo et al. (Eds.): GBCESC 2023, LNCE 328, pp. 282–291, 2024.
https://doi.org/10.1007/978-981-99-9947-7_30

2 Engineering Overview

The Echeng Bridge in Huizhou City spans the Dong River. The main bridge is symmetrically arranged with vertical curved lines within its scope. It has a maximum longitudinal slope of 1.113%. Positioned as a enhanced bridge in the overall urban planning of Huizhou City, it serves as a transportation and scenic urban arterial road. The mainline consists of dual-directional 6 lanes with a design speed of 50 km/h. Non-motorized vehicle lanes and pedestrian walkways are provided on both sides of the main bridge. The Echeng Bridge is a four-span continuous skew heterotypic stay cable arch bridge with a layout of 50m + 180m + 180m + 50m = 460m. The arch beams are made of steel structure, and the arch beams are fixed while the pier beams are separated. The steel tower in the middle of the main bridge is connected to the main beams. The overall elevation plan of the main bridge is shown in Fig. 1, and the overall layout plan is shown in Fig. 2.

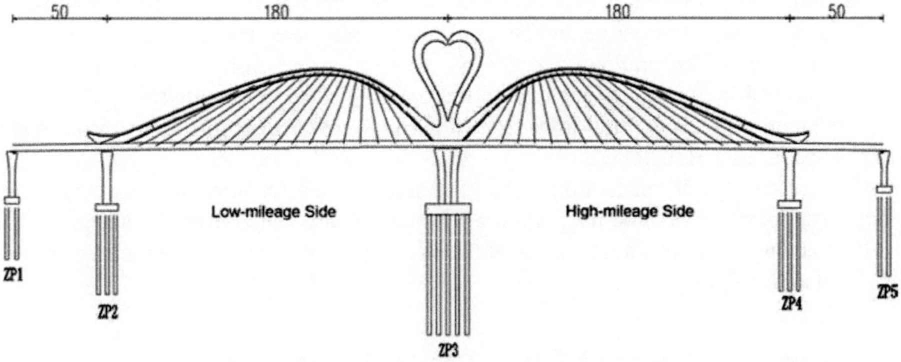

Fig. 1. Overall Elevation Plan of the Main Bridge

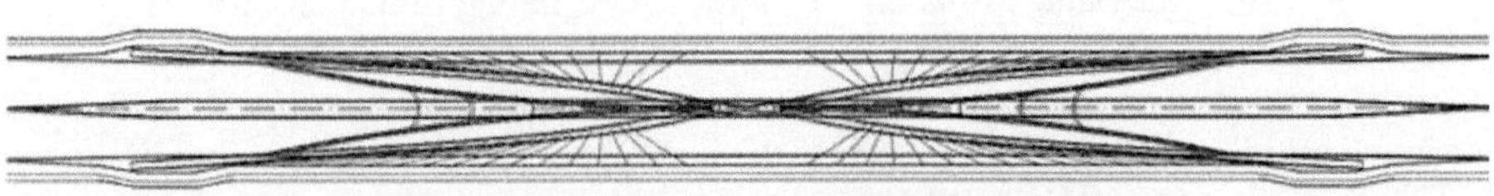

Fig. 2. Overall Plan View of the Main Bridge

3 Introduction of Construction Scheme

After fully considering the characteristics of the bridge structure, road transportation, construction site layout, surrounding environment of the bridge site, and traffic conditions, the final construction process involves completing the construction of the central piers, side piers, and temporary support of the main girders. Then, the main girders are installed and assembled, with welding performed between different segments of the steel main girders. Support brackets for the arch ribs are installed on the steel main girders, followed by the installation of the arch ribs and their welding and assembly between different segments. The central pier bridge towers are installed, and the support brackets for the arch ribs on the steel main girders are removed. Tensioning cables are applied between the bent shaft on the arch ribs, followed by tensioning of cable-stayed cables. The temporary support under the steel main girders is removed, and finally, the construction of the secondary permanent load is carried out to complete the bridge.

3.1 Installation of Steel Main Girders

The installation of the steel main girders involves the following steps: First Installing the waterborne support for the steel box beams and carrying out the construction of the lower structure of Piers ZP1-ZP5. Clearing the sedimentation in the upstream side of the riverbed between Piers ZP4-ZP5 and installing the steel beam supports for Piers ZP1-ZP5. Simultaneously, processing the steel girders, steel arches, and steel tower for the bridge in the steel structure fabrication plant.

Second, using a 660-ton floating crane, the ZP3 pier top steel beam is lifted and installed, along with the permanent bearings, as the reference installation section.

Third, using a 660-ton floating crane, all the steel girders from the ZP1 pier on low-mileage side of the bridge to the navigation opening one before (the three navigation openings named in sequential order from the bridge tower to high-mileage side of the bridge) are lifted and installed.

Forth, using a 660-ton floating crane, the steel girders for navigation opening one, navigation opening two, and navigation opening three are installed step by step.

Fifth, using a 660-ton floating crane, the steel box girders for the ZP4-ZP5 pier area are lifted and installed.

3.2 Installation of Steel Arch Ribs and Bridge Towers

The installation of steel arch ribs involves the following steps: First, remove the single-row column baseplates and shims on the right side of navigation span three to exclude them from subsequent load-bearing. Use a 660-ton floating crane to install the arch rib supports. Second, use the 660-ton floating crane to install the middle support and the arch ribs at the middle support location. Begin the installation from the side bridge abutment towards the closure point, installing the segments one by one. Finally, install the closure segment. Third, Use the 660-ton floating crane to lift and install the goose towers.

3.3 Tensioning of Cable-Stayed Cables

After the cables between the bent shaft on the arch ribs are tensioned, the following steps are taken: tensioning of the cable-stayed cables. The bridge has a total of 68 cable-stayed cables, symmetrically arranged with respect to both the bridge centerline and the tower centerline. A total of 17 groups are formed, with each group consisting of four cables: the left side cable of the high-mileage, the right side cable of the high-mileage, the left side cable of the low-mileage, and the right side cable of the low-mileage. The tensioning process starts from the towers and progresses towards both sides of the bridge. Each step is tensioned one group at a time, totaling 17 tensioning operations. Once the tensioning of cable-stayed is completed, the temporary supports for the main girders are removed, and the installation of the permanent load for the second phase begins.

4 Construction Process Simulation Analysis

4.1 Structural Model

The entire bridge model was created using the finite element analysis software Midas Civil. It consists of 3862 nodes and is discretized into 6736 elements. Beam elements are used to simulate the main girder, arch ribs, and bridge towers, while only tension elements are used to simulate cable-stayed cables. The connections between the main girder and arch ribs, diagonal braces and arch ribs, main and auxiliary arch ribs, cables and main girder, and cables and arch ribs are modeled using rigid-type connections within the elastic connections. Refer to Fig. 3.

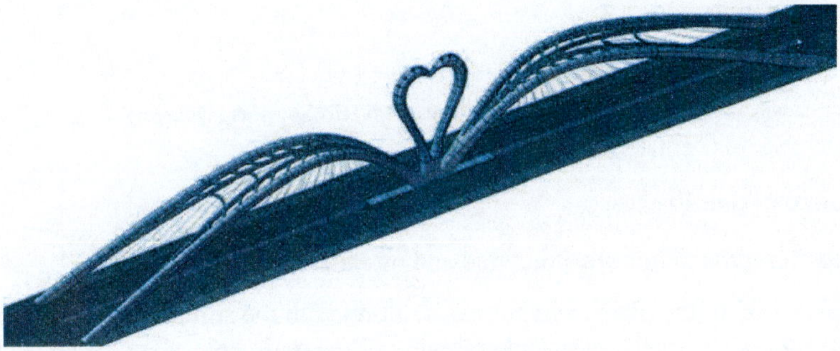

Fig. 3. Bridge Finite Element Model

Considering the actual construction process, it is unlikely for the steel beam supports to experience tensile forces. Therefore, the steel beam supports are simulated using compression-only elements within the node-based elastic supports. Refer to Fig. 4 and Fig. 5. The steel arch rib supports are simulated using beam elements and one set for each arch, for total of four groups. A quarter part refer to Fig. 6.

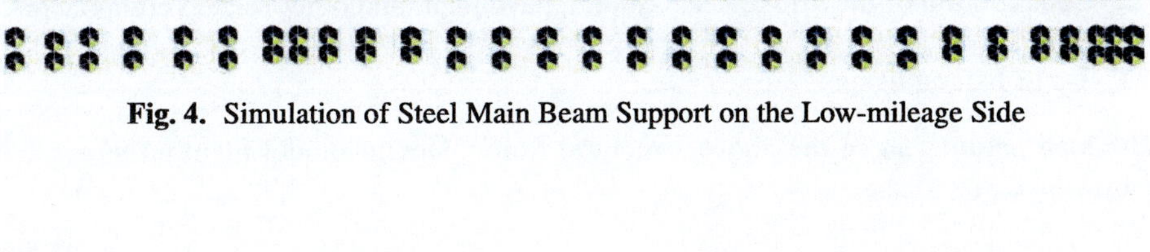

Fig. 4. Simulation of Steel Main Beam Support on the Low-mileage Side

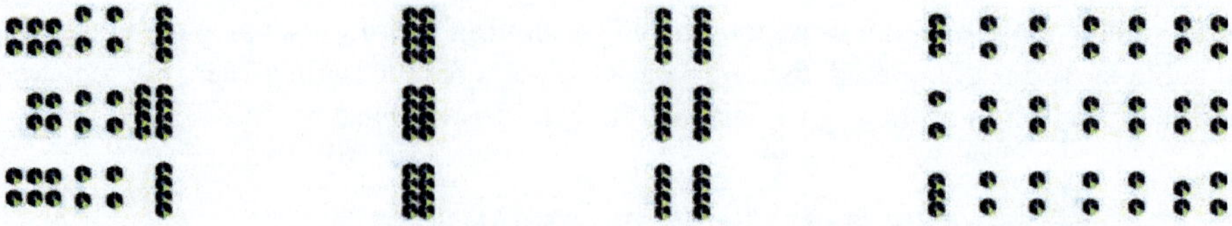

Fig. 5. Simulation of Steel Main Beam Support on the High-mileage Side

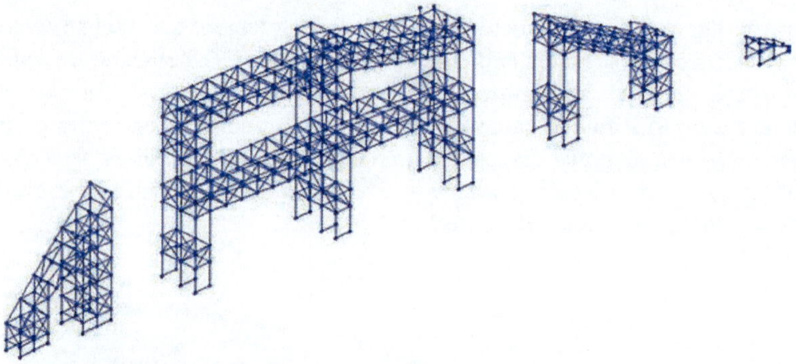

Fig. 6. Simulation of Steel Arch Rib Support (Quarter)

4.2 Construction Process

The following procedures are simulated and modeled for the bridge model:

(1) Construction of the piers and abutments, along with the simultaneous installation of temporary supports. Installation and assembly of the main beams are carried out, while preserving the closure sections.
(2) Welding and closure of the steel main beams.
(3)The installation of arch rib supports is carried out on the main beams.
(4) Installation and assembly of the main and auxiliary arch ribs.
(5) Welding and closure of the arch ribs.
(6) Installation of the bridge towers at the middle piers.
(7) Removal of the arch rib supports from the steel main beams.
(8) Tensioning of the cables between the bent shaft on the arch ribs.
(9) Sequential tensioning of the 17 sets of cable−stayed cables.
(10) Removal of the temporary supports under the steel main beams.
(11) If there is a significant deviation between the measured cable forces and the theoretical cable forces, secondary cable adjustment is performed.
(12) Construction of the bridge deck system, pavement, and other related components.

4.3 Load Information

The load parameters in the above structural finite element model mainly include the following categories:

Phase 1 Dead Load: This includes the weight of structural elements such as steel beams, steel arches, bridge towers, and cable-stayed cables. The structural weights of each segment are calculated to apply the self-weight load accordingly.

Phase 2 Dead Load: The corresponding loads specified in the design drawings are applied. Including asphalt concrete pavement, crash barriers, sidewalk bases and slabs, guardrails.The total loads are applied as element pressure loads.

Construction Load: Temporary loads during the construction process are considered and simulated based on actual conditions.

Temperature Load: Temperature effects are considered by applying an overall temperature rise of 30 °C, an overall temperature drop of 25 °C, and temperature gradients.

5 Simulation Analysis Results

After a comprehensive simulation analysis of the bridge, the final results include stress, deformation, and cable forces for each construction stage are below.

5.1 Stress Result

Simulating the construction process of the main bridge involves activating or deactivating elements, applying loads, and setting boundaries step by step, which allows for the calculation of the structural response at each construction stage. During erection, the structural stress result reveals that the maximum stress in the arch rib is 110 MPa, and the maximum stress in the main beam is 112 MPa during the construction process. These values comply with the requirements specified by the materials and design codes. The results are illustrated in Fig. 7.

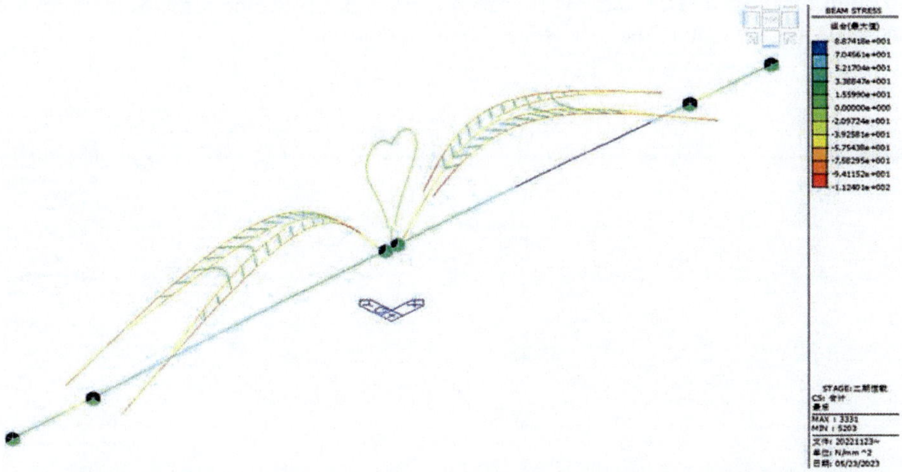

Fig. 7. Stress Results after Application of Phase 2 Dead Load on the Main Bridge

During the lifting process of the steel arch ribs, the maximum stress in the support structure is 138 MPa, which satisfies the material and code requirements. The results are shown in Fig. 8.

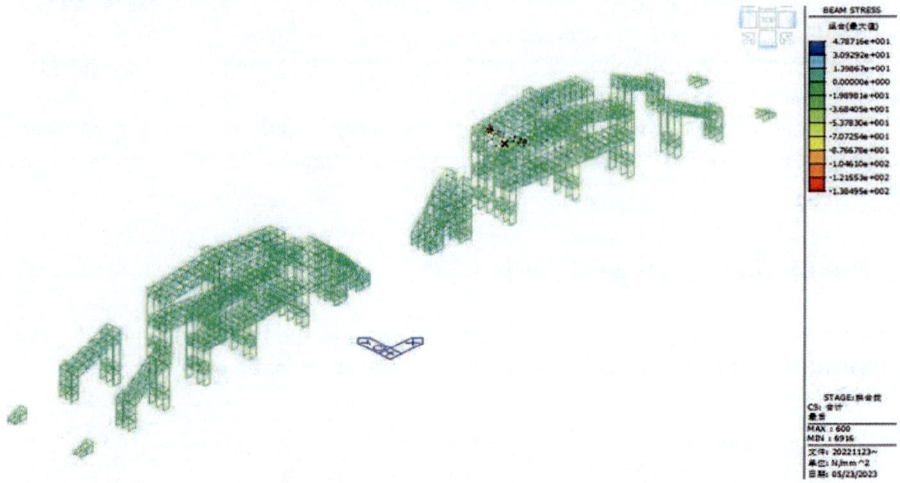

Fig. 8. Stress Results of Steel Arch Rib Supports after Completion of Erection

5.2 Deformation Results

By simulating the construction process of the main bridge, the structural deformations for each construction step can be calculated. The maximum deformation observed is 167 mm for the arch ribs and 134 mm for the main beam, which satisfies the material and code requirements. The results are shown in Fig. 9.

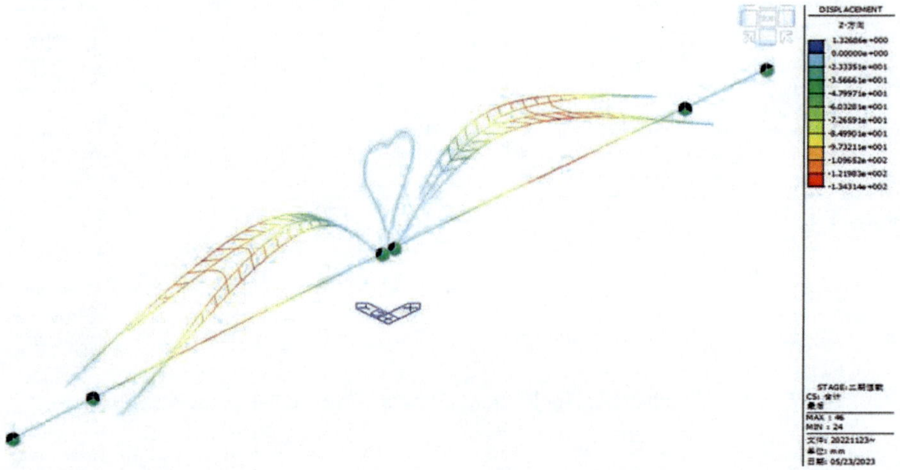

Fig. 9. Deformation Results after Application of Phase 2 Dead Load on the Main Bridge

In the process of lifting the steel arch ribs, the maximum deformation observed in the supports is 45 mm, which meets the material and code requirements. The results are shown in Fig. 10.

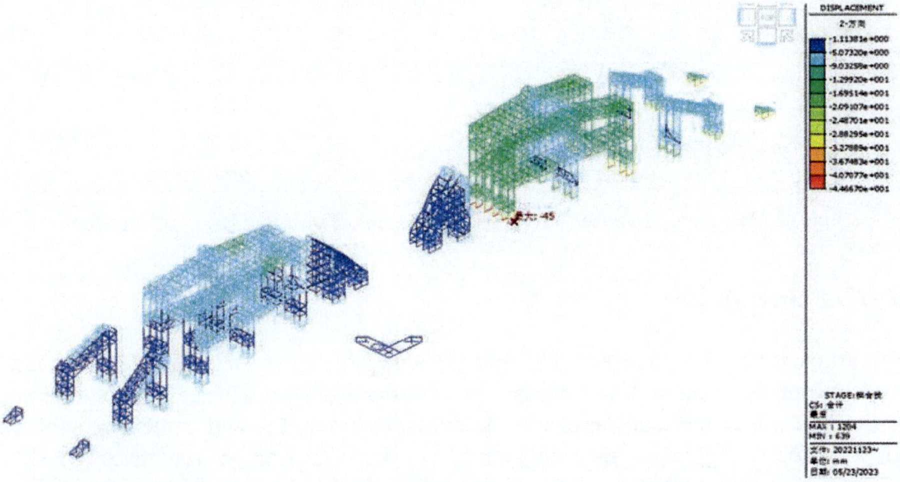

Fig. 10. Deformation Results of Steel Arch Rib Supports after Completion of Erection

5.3 Cable Force Results

Since the cable arrangement is axially symmetric about the bridge's centerline, the cable forces on both sides of the bridge are the same. Therefore, here only present the cable forces for a single side. The cables are numbered from the tower to the low-mileage side as 1 to 17#, and from the tower to the high-mileage side as 18 to 34#. The final cable forces in the completed bridge state are close to the target cable forces, satisfying the design requirements. The results and differences are shown in Figs. 11 to Fig. 12.

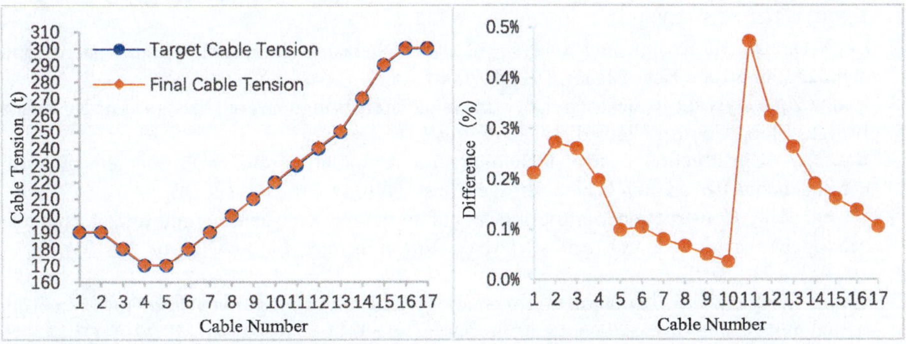

Fig. 11. Difference between Target and Final Cable Tension in Low-mileage Side

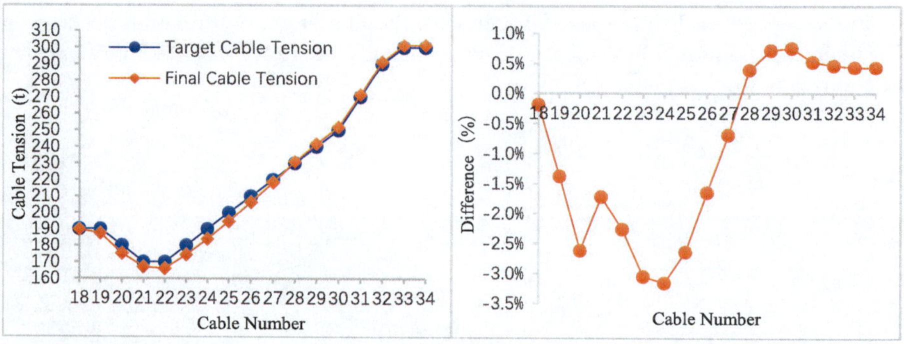

Fig. 12. Difference between Target and Final Cable Tension in High-mileage Side

6 Conclusion

This article is based on the Huizhou Echeng Bridge project, which involves the structural modeling of the steel arch rib support and the main bridge. Through simulation and analysis of the construction process, the stress, deformation, and cable forces of the support, steel main beams, and steel arch ribs during construction were investigated.

(1) During the steel arch rib lifting process, the stress and deformation of the arch rib support on the main beam were simulated, and the results met the requirements specified by the materials and standards.
(2) The simulation of the entire construction process of the main bridge showed that the stress, deformation, and cable forces in the completed bridge state met the design and code requirements. The overall structural behavior was reasonable.
(3) The proposed construction scheme is feasible and can be implemented. The simulated analysis results of the steel arch rib support and the steel arch bridge can serve as theoretical reference values for bridge construction monitoring.

References

1. Lou, T.H., Zhang, G.L.: Design of a landscape beam-arch composite bridge. Highw. Eng. **128**(01), 101–105 (2008)
2. Li, X.Q., Li, Y.: Design and analysis of two-arch beam composite bridge systems and comparison. Struct. Eng. **24**(06), 19–22 (2008)
3. Cao, J.Z.: Cityscape pedestrian space irregular arch bridge overall design. Urban Roads Bridges Flood Control **194**(06), 77–80+11 (2015)
4. Bai, J.C.: Construction control technology for asymmetric reinforced concrete arch rib heterogeneous tie rod arch bridge. Bridge Constr. **48**(03), 116–120 (2018)
5. Zhang, Z.J.: Construction monitoring technology for lower-bearing reinforced concrete heterogeneous tie rod arch bridge. Urban Roads Bridges Flood Control **229**(05), 193–195+222+21 (2018)
6. Wang, J., Qian, Y.W.: Construction technology and monitoring analysis of lower-bearing spatial multi-cable surface heterogeneous tie rod arch bridge. Anhui Archit. **29**(01), 142–146 (2022)

7. Zhu, W.G., et al.: Flexible suspension cable force testing of beam-arch composite bridge system. Highway Eng. (01), 21–23+36 (2004)
8. Guo, A.P., Jiang, A.J., Wang, L.G.: Analysis of prestressed tendon tensioning construction monitoring for irregular arch bridge main girder. Transp. Constr. Manage. **348**(04), 112–113 (2011)
9. Li, X.P., Luo, J.Q.: Research on construction and key load-bearing technology of inclined irregular arch bridge. Low Temp. Architect. Technol. **36**(05), 65–67 (2014)
10. Zhou, W.M.: Design of assembly support and key construction technology for beam-arch composite bridge. Railway Eng. **516**(02), 16–20 (2017)

Study on Seismic Reduction Effect of Friction Pendulum Isolation Bearing in Curved Beam Bridge with Variable Height Pier

Jiada Guan(✉), Xiyin Zhang, Xingchong Chen, and Yongliang Zhang

School of Civil Engineering, Lanzhou Jiao Tong University, Lanzhou 730070, Gansu, China
guanjiada@126.com

Abstract. This paper investigates the seismic response of curved beam bridges equipped with friction pendulum isolation bearings through nonlinear time history analysis. Various influencing factors are considered, including different bearing schemes, friction coefficients, and slide radii of the friction pendulum isolation bearings. The results demonstrate that the use of friction pendulum bearings significantly reduces the internal force and displacement responses of the curved beam bridge. The effectiveness of isolation is closely related to the location and number of bearings. For curved beam bridges with variable height piers, the friction pendulum bearings exhibit better isolation performance in the tangential direction compared to the radial direction. Both the friction coefficient and slide radius of the bearings influence the seismic response of the curved bridge. The influence of the friction coefficient on the seismic performance of the bridge is more pronounced compared to the slide radius of the bearing. The selection of an appropriate friction coefficient not only controls the internal force of the bridge but also mitigates the displacement response.

Keywords: curved beam bridge · friction pendulum isolation bearing · friction coefficient · slide radius · seismic response

1 Introduction

With the rapid development of expressways in western regions of China, numerous multi-span curved bridges suitable for mountainous terrains have been constructed. Curved bridges exhibit a phenomenon of bending-torsion coupling due to the curvature of the structures, which leads to complex dynamic characteristics and energy dissipation mechanisms compared to straight bridges under similar conditions. The internal forces and deformations of the structures are also affected by curve radius, center angle, and other factors.

In order to mitigate the seismic response of bridges, seismic reduction and isolation technology has been widely employed in bridge construction. Among the various isolation devices, friction pendulum isolation bearings have emerged as stable and reliable devices, effectively distributing seismic forces to each pier. These bearings, which integrate the concepts of sliding bearings and pendulums [1], can swing the superstructure

W. Guo et al. (Eds.): GBCESC 2023, LNCE 328, pp. 292–299, 2024.
https://doi.org/10.1007/978-981-99-9947-7_31

according to the designed sliding surface with a specific isolation period during strong earthquakes. Energy dissipation occurs through PTFE friction damping between spheres. Extensive studies have demonstrated that friction pendulum bearings can enhance the seismic performance of bridge structures by rapidly restoring stiffness and prolonging the natural vibration period. Considering the advantages of friction pendulum bearings, numerous studies have focused on the seismic response of isolated bridges. Zayas et al. [2] conducted research on a highway steel truss bridge with friction pendulum bearings, showing that no post-earthquake repairs were required after a rare earthquake, as the bearings automatically reset. Tongaonkar et al. [3] evaluated the seismic performance of a three-span continuous beam bridge with friction pendulum bearings using linear and nonlinear models, revealing significant reduction in the seismic response of the substructure compared to non-isolated bridges. Studies by Li [4] and Dong [5] demonstrated that friction pendulum bearings can amplify the displacement of the bridge's main beam during rare earthquakes.

The studies on the impact of friction pendulum bearings on the seismic response of bridge structures have primarily focused on conventional straight bridges. In this paper, we extend the research to include a practical project involving a curved bridge with variable height piers. A finite element model of the bridge is developed for nonlinear time history analysis. The main objective is to investigate the effects of bearing layout configurations and mechanical parameters on the seismic response of curved bridges. This research aims to provide a theoretical foundation for the future implementation of friction pendulum isolation bearings in curved bridge applications.

2 Influence of Bearing Layout Scheme on Seismic Performance of the Curved Bridge

2.1 Engineering Background and Finite Element Model

This study is based on the Shayipo overpass bridge located in Yunnan Province and focuses on the impact of the bearing layout scheme on the seismic performance of the curved bridge. The bridge has a span layout of $(3 \times 17) + (5 \times 16) + (3 \times 30)$ m, with the third span selected as the subject of analysis. The construction site of the Shayipo overpass bridge is classified as a class II site, with a design basic seismic acceleration peak of 0.15g and a seismic response spectrum period of 0.45s. The three-span curved bridge has a plane radius of 50m, and its superstructure consists of cast-in-place reinforced concrete main girders. The cross-section features a single-box with a double-chamber design, as illustrated in Fig. 1 (a typical section), constructed using C40 concrete. The substructure is supported by double-column piers made of C30 concrete. Due to the unique structural form, the displacement responses and internal force distribution of the beam and piers are complex under seismic actions. The original design scheme employs GPZ (II) basin-shaped rubber bearings, which fail to meet the seismic fortification requirements. Therefore, this paper proposes the introduction of friction pendulum isolation bearings for seismic reduction and isolation design. The bearing layout scheme for the bridge is discussed based on the utilization of friction pendulum isolation bearings.

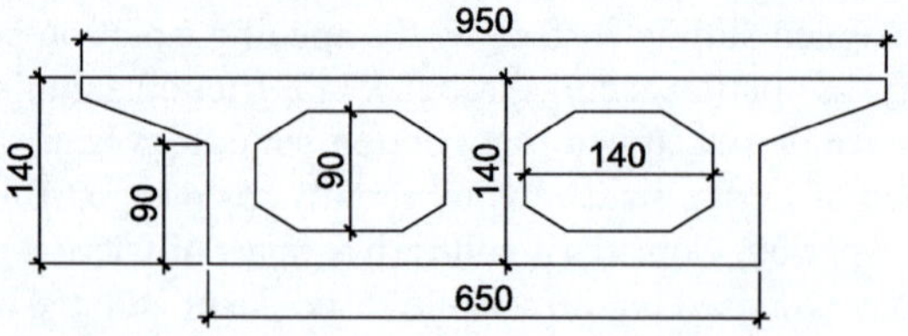

Fig. 1. Box beam section dimensions (unit: cm)

The finite element model of the bridge, depicted in Fig. 2, is established using MIDAS/civil software. In the model, the beams and piers are simulated using three-dimensional beam elements, while the main beam is modeled using the beam grid method. The fixed basin-shaped bearing is simulated through an elastic connection, while the movable basin-shaped bearings are represented by ideal elastic-plastic connection elements, incorporating hysteresis systems typically found in such connections. The friction pendulum isolation bearings are simulated using specialized elements specific to friction pendulum isolation bearings within the general connection framework. To account for pile-soil interaction, an equivalent soil spring is employed, with the dynamic calculation value (m) set at twice the static analysis value. According to the Seismic Code for Highway Engineering (JTG/T 2231–01-2020), the bridge is classified as a class B structure in terms of seismic fortification. In this study, the seismic performance of the bridge is analyzed using the E2 earthquake scenario. The nonlinear time history analysis is conducted using the El-Centro wave recorded during strong earthquakes, with the peak acceleration adjusted to 0.255g. According to the Specifications of Detailed Rules for Seismic Design of Highway Bridges (JTG/T B02–01-2008), when simulating curved bridges using curved beam elements, it is sufficient to calculate the seismic input along the secant direction at both ends and the vertical secant. Hence, this study considers only one direction of horizontal seismic ground motion along the secant (the direction connecting piers 1# and 4#), with the maximum calculated value regarded as the theoretical calculation value for the bridge's seismic response.

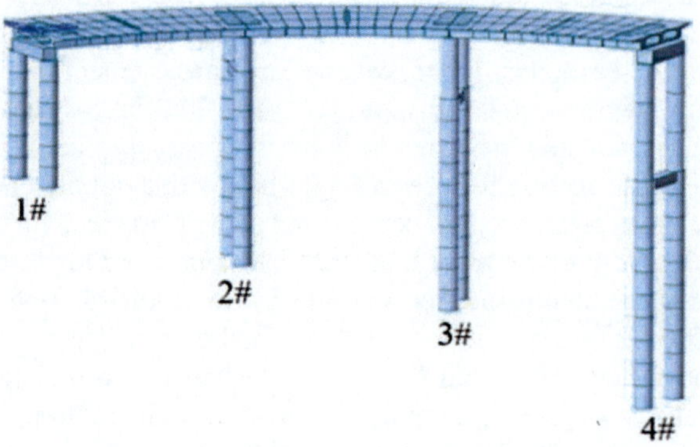

Fig. 2. Finite element model of Shayipo bridge

2.2 Bearing Layout Scheme

The bearing serves as a vital connecting element for the bridge, ensuring the transfer of forces from the superstructure and facilitating the free deformation of the beam body [6]. In the case of bridges, the fixed bearing primarily carries the seismic inertia forces of the superstructure along the bridge direction, making it crucial to reduce the seismic forces borne by the fixed bearing. The bearing arrangement encompasses fixed bearings, one-way movable bearings, and two-way movable bearings. The specific arrangement within the pier is illustrated in Fig. 3. By combining the layout principles of curved bridge bearings and the characteristics of friction pendulum seismic isolation bearings, six friction pendulum seismic isolation bearing schemes are formulated, as presented in Table 1. Scheme 0 serves as the non-isolated layout scheme control group, utilizing the original basin-shaped rubber bearing layout scheme. The remaining schemes are isolation schemes employing friction pendulum bearings.

The design parameters of the friction pendulum isolation bearing are selected based on the calculation results of the bearing reaction. The bearing's isolation period is approximately 3–5 s, which is taken as 4.5 s. The corresponding sliding radius is assumed to be 5 m, and the friction coefficient is set at 0.05.

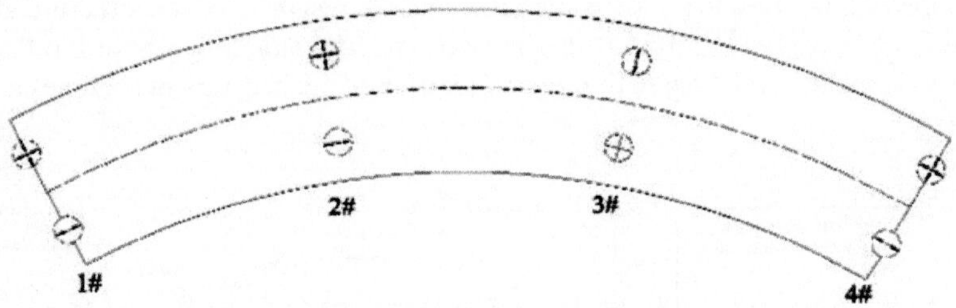

Fig. 3. Layout of basin type rubber bearing or friction pendulum bearing

Table 1. Bearing Layout Schemes

Layout scheme	Pier 1		Pier 2		Pier 3		Pier 4	
	Inside	Outside	Inside	Outside	Inside	Outside	Inside	Outside
0	GPZ(ZX)	GPZ(SX)	GPZ(ZX)	GPZ(SX)	GPZ(GD)	GPZ(HX)	GPZ(ZX)	GPZ(SX)
1	GPZ(ZX)	GPZ(SX)	GPZ(ZX)	GPZ(SX)	FPB(GD)	FPB(HX)	GPZ(ZX)	GPZ(SX)
2	FPB(ZX)	FPB(SX)	GPZ(ZX)	GPZ(SX)	FPB(GD)	FPB(HX)	GPZ(ZX)	GPZ(SX)
3	GPZ(ZX)	GPZ(SX)	FPB(ZX)	FPB(SX)	FPB(GD)	FPB(HX)	GPZ(ZX)	GPZ(SX)
4	FPB(ZX)	FPB(SX)	FPB(ZX)	FPB(SX)	FPB(GD)	FPB(HX)	GPZ(ZX)	GPZ(SX)
5	GPZ(ZX)	GPZ(SX)	FPB(ZX)	FPB(SX)	FPB(GD)	FPB(HX)	FPB(ZX)	FPB(SX)
6	FPB(ZX)	FPB(SX)	FPB(ZX)	FPB(SX)	FPB(GD)	FPB(HX)	FPB(ZX)	FPB(SX)

Notes: GPZ represents the basin-shaped rubber bearing, FPB represents friction pendulum bearing, GD represents fixed bearing, ZX represents a longitudinal movable bearing, HX represents a laterally movable bearing, SX represents a two-way movable bearing.

2.3 Analysis of Calculation Results of Each Layout Scheme

Nonlinear time history analysis of the structure is performed by inputting seismic waves, thereby obtaining the dynamic response of the bridge structure under uniform seismic excitation. This study considers the maximum value of the response calculation result under the seismic wave action as the theoretical calculation value for the bridge's seismic response. Extensive theoretical analysis and seismic damage investigations indicate that when bearing seismic loads, the majority of seismic actions concentrate on piers with fixed bearings. Hence, this section analyzes the seismic performance of the bridge under different bearing layout schemes based on the seismic response of the fixed pier.

To compare the seismic reduction and isolation effects of friction pendulum isolation bearings and basin-shaped rubber bearings with different layout schemes, the isolation rate is introduced to analyze the isolation effect of the six layout schemes employing friction pendulum isolation bearings. The expression for the seismic isolation rate is given by Eq. (1).

$$\lambda = \frac{K_0 - K_1}{K_0} \times 100\% \quad (1)$$

where, K_0 is the peak seismic response of the structure when all piers are equipped with basin-shaped rubber bearings; K1 is the peak seismic response of the structure after the application of friction pendulum isolation bearings. The isolation rate of the fixed pier (Inner pier 3) of the curved beam bridge under different bearing layout schemes is shown in Fig. 4.

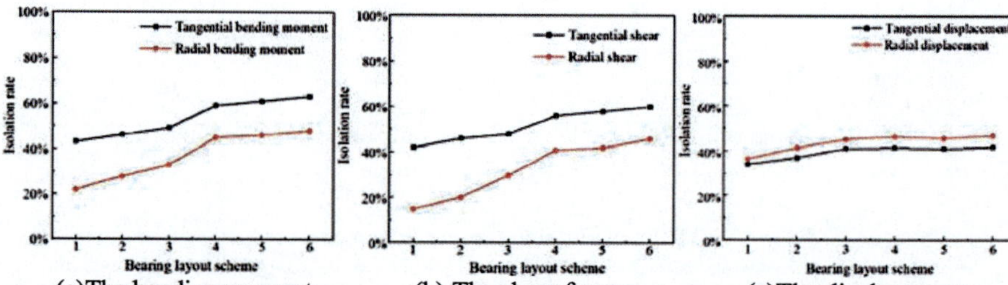

(a)The bending moment (b) The shear force (c)The displacement

Fig. 4. Seismic effects of each bearing layout scheme

Figure 4 illustrates that the friction pendulum isolation bearing exhibits noticeable isolation effects compared to the conventional basin-shaped rubber bearing. The isolation rate is closely related to the layout position and quantity of the friction pendulum bearings. The friction pendulum bearing demonstrates excellent seismic isolation performance in the tangential direction, achieving an isolation rate of 40% to 60% for the tangential internal forces of the fixed pier. In the radial direction, the isolation effect of the friction pendulum bearing is significantly influenced by the bearing layout scheme, resulting in a radial internal force isolation rate of approximately 20% to 50%. The displacement isolation rate of the fixed pier is minimally affected by different layout schemes of the friction pendulum bearing. Additionally, starting from scheme 4, as the number of friction pendulum bearings increases, the isolation rate exhibits a decreasing trend, with the isolation effect becoming stable.

In summary, for the curved beam bridge with variable height piers examined in this study, the arrangement of friction pendulum bearings can utilize the excess bearing capacity of the movable piers and prevent excessive internal forces on fixed piers during earthquakes.

3 Study on the Influence of Bearing Parameters on the Seismic Performance of Curved Bridges

The test results demonstrate that the friction coefficient of the contact surface and the slide radius of the arc surface are the most critical parameters of the friction pendulum bearing, playing a vital role in the bearing's seismic reduction and isolation effect [7]. Building upon scheme 6 of friction pendulum bearings from the previous section, this section explores the influences of the friction coefficient and slide radius on the seismic response of the curved beam bridge with friction pendulum bearings.

3.1 Influence Analysis of Friction Coefficient

In order to investigate the impact of the friction coefficient value on the seismic reduction and isolation effect of curved bridges, the friction coefficient is divided into five working conditions for nonlinear time history analysis: 0.01, 0.03, 0.05, 0.08, and 0.1. The aforementioned seismic wave is selected for time history analysis. The slide radius of 5 m remains constant from the previous section, and the influence of sliding speed on the friction coefficient is not considered. Figure 6 presents the seismic response calculation results of the fixed pier (Inner pier 3) as the friction coefficient varies.

Figure 5(a) and (b) display the extreme values of internal force in the two directions with changes in the friction coefficient. The bending moment and shear force at the bottom of the fixed pier exhibit noticeable variations, with similar trends observed in both directions. As the friction coefficient increases from 0.01 to 0.05, the tangential bending moment and shear force show a decreasing trend. At a friction coefficient of 0.05, the tangential internal force reaches its minimum value. Subsequently, as the friction coefficient increases from 0.05 to 0.1, the tangential bending moment and shear force increase. The radial bending moment and shear force at the pier bottom also increase with the friction coefficient but not significantly. For the tangential internal force of the fixed pier, the friction pendulum bearing's equivalent stiffness and energy dissipation capacity continue to increase with the friction coefficient, resulting in a weakening of the bearing's isolation capacity. Figure 5(c) presents the maximum relative displacements between the superstructure and the fixed pier as the friction coefficient varies. The figure illustrates that the relative displacement gradually decreases as the friction coefficient increases. This is attributed to the maximum friction displacement of the friction pendulum bearing continuously decreasing with the friction coefficient. In conclusion, the friction coefficient significantly influences the tangential internal force. A friction coefficient of 0.05 results in the minimum seismic response for the fixed pier.

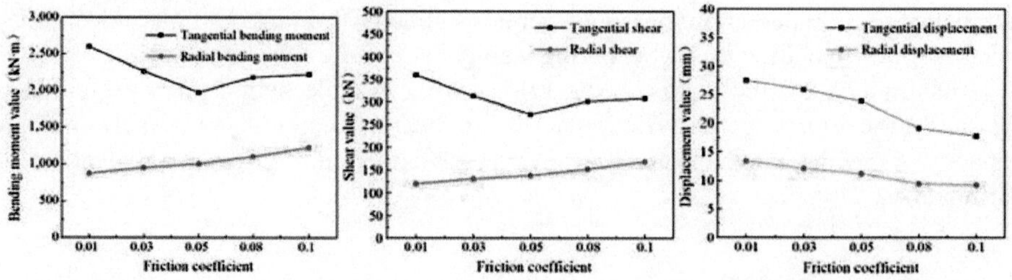

(a) Maximum bending moment (b) Maximum shear force (c) Relative displacement

Fig. 5. Influence of friction coefficient on internal force and deformation of the fixed pier

3.2 Influence Analysis of the Slide Radius

The slide radius directly correlates with the isolation period of the friction pendulum bearing and indirectly affects the seismic response of the structure. Considering the typical isolation period value of friction pendulum bearings, scheme 6 from the previous section is retained as the bearing layout scheme. The friction coefficient is set to 0.05, while the slide radius is varied as 1.0m, 2.0m, 3.0m, 5.0m, and 6.0m, respectively. Figure 6 presents the seismic response calculation results of the fixed pier (Inner Pier 3) as the slide radius changes.

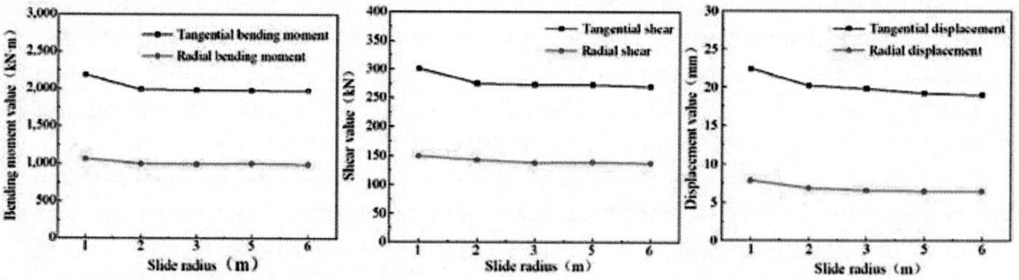

a) Maximum bending moment (b) Maximum shear force (c) Relative displacement

Fig. 6. Influence of slide radius on internal force and deformation of the fixed pier

Figure 6(a) and (b) depict the relationship between the extreme values of internal force and the slide radius of the friction pendulum bearing at the fixed pier. Figure 6(c) displays the maximum relative displacements between the superstructure and the fixed pier as the slide radius changes. It can be observed from Fig. 6 that the maximum internal force at the pier bottom and the maximum relative displacements between the superstructure and the fixed pier exhibit similar trends with the slide radius. As the slide radius increases from 1m to 2m, there is a significant reduction in the maximum bending moment, maximum shear force, and maximum relative displacement in both directions. Increasing the slide radius also leads to an extended natural vibration period of the structure, further decreasing the seismic response. Once the slide radius reaches 2m, the decline in the internal force at the bottom of the fixed pier and the relative displacement between the superstructure and the fixed pier becomes more gradual. For the isolation scheme employed in this study, a slide radius exceeding 2 m does not yield further improvement in the bearing's isolation performance.

4 Conclusions

This paper investigates the isolation effect of friction pendulum bearings in curved bridges through nonlinear time history analysis, leading to the following main conclusions:

(1) For the three-span curved beam bridge, the tangential direction demonstrates superior isolation effectiveness compared to the radial direction when employing friction pendulum bearings. The isolation effect in the radial direction is significantly influenced by the number and placement of the bearings. Selecting an appropriate bearing layout scheme effectively enhances the seismic isolation performance of the curved beam bridge.
(2) The seismic response of the bridge is greatly influenced by changes in the friction coefficient. The optimum isolation effect is achieved when the friction coefficient reaches 0.05. Selecting the friction coefficient based on the superstructure's weight not only controls the internal force response of the bridge but also mitigates the displacement response.
(3) The bearing's isolation effect is maximized when the slide radius reaches 2m. Once the slide radius of the friction pendulum bearing exceeds a certain range, its impact on the isolation effect of the curved beam bridge gradually diminishes. Compared to the friction coefficient, the slide radius exerts less influence on the seismic performance of the curved bridge.

Acknowledgement. This work is supported by NSFC (Grant Nos. 51808273, 52068045), General program of China Postdoctoral Science Foundation (Grant No. 2018M643767).

References

1. Ye, A.J., Guan, Z.G.: Seismic Design of Bridges. China Communications Press, Beijing (2002)
2. Zayas, V.A., et al.: Seismic Isolation of Benicia-Martinez Bridge. Bridges to the Future. Earthquake Protection Systems, Inc. Richmond, CA (2000)
3. Tongaonkar, N.P., Jangid, R.S.: Seismic response of isolated bridges with soil–structure interaction. Soil Dyn. Earthq. Eng. **23**(4), 287–302 (2003)
4. Li, Z.Y., Jiang, L.J., Li, Z.L.: Comparative analysis of seismic control schemes for continuous curved girder bridges. J. Vibr. Shock **35**(10), 157–161 (2016)
5. Dong, J., et al.: Study on vibration mitigation property of railway 32m prestressed concrete simply-supported gierder with friction pendulum bearings in highly seismic area. Railway Standard Des. **64**(S1), 63–68 (2020)
6. Wang, J.Q., et al.: Influence of friction coefficient of bearing on seismic response of base-isolated structure with friction pendulum system. World Earthq. Eng. **28**(2), 98–102 (2012)
7. Du, X.L., et al.: Seismic mitigation effect analysis on friction pendulum bearing applied in the underground subway station. Eng. Mech. **36**(09), 60–67+88 (2019)

Experimental Study on Direct Tensile Properties of UHPC

Huiqing Xue(✉)

Beijing Municipal Engineering Research Institute, Beijing 100037, China
1602651002@qq.com

Abstract. This article studied the direct tensile properties of UHPC with different ratios and found that after reaching the maximum tensile load, the load gradually began to decrease and all showed a decrease in load fluctuations, with a certain strain hardening effect; As the crack width of the specimen increases, the load begins to steadily decrease. With the same ratio, as the cement strength increases, the initial cracking strength of UHPC increases by 65% and the tensile strength increases by 14%; The initial cracking strength of UHPC using cement from different origins did not change, but the tensile strength slightly increased; With the increase of fiber content, the initial cracking strength of UHPC increased by 39%, and there was no significant change in tensile strength.

Keywords: Direct tension · Strain hardening · Crack width · Tensile strength

1 Preface

Ultra high performance concrete (UHPC), as a new generation of building materials born in the late 20th century, has excellent properties such as ultra-high strength, high toughness, and high durability. The key to preparing UHPC is to create a compact solid particle packing body that theoretically approaches the maximum density [1]; Compared with ordinary cement-based materials, UHPC exhibits better compressive, tensile, flexural, and impact and explosion resistance [2]. In addition, the addition of fibers has a significant impact on the overall strength improvement of UHPC, and due to its low water cement ratio, microcrack effect, and self-healing effect, UHPC also exhibits good durability [3]. The tensile strength of ordinary concrete is lower than compressive strength, and its role in the structure is relatively small. However, UHPC has a tensile strength of 10 MPa or even higher, which can play a certain structural role. The addition of steel fibers can significantly improve the tensile performance of UHPC [4]. UHPC has good crack control ability, and its ultimate tensile strain can reach over 3%; UHPC is not sensitive to notches, and even when the crack height ratio is 0.5, UHPC exhibits excellent crack harmless dispersion ability [5]. Research has shown that under the optimal steel fiber content, UHPC can meet the material performance requirements of large building main projects and exhibit good application prospects [6].

This article is based on the closest packing theory to prepare UHPC, and conducts a study on the direct tensile properties of UHPC. The tensile properties and strain hardening

W. Guo et al. (Eds.): GBCESC 2023, LNCE 328, pp. 300–306, 2024.
https://doi.org/10.1007/978-981-99-9947-7_32

behavior of UHPC under different ratios are studied, providing a basis for the preparation of UHPC.

2 Experimental Design

2.1 Test Materials

A total of 4 sets of ratios were designed, with the first ratio using locally produced P.O42.5 ordinary Portland cement; The ratio of No. 2 and No. 3 adopts P.O42.5 ordinary Portland cement produced in Yunnan; The No. 4 ratio uses locally produced P.O52.5 ordinary Portland cement. Natural medium sand is used as the sand; Fly ash is the first grade low calcium ash; The specific surface area of silica fume is 20000 m^2/kg; The water reducing agent adopts high-performance polycarboxylic acid water reducing agent; The steel fiber is made of 12mm copper plated steel fiber.

2.2 Test Mix Proportion

Based on DSP theory, optimize the design mix proportion, optimize the particle size distribution of powder particles (including cement, active and non active admixtures), reduce the porosity of particle stacking, and add modern high-performance water reducing agents. The experimental mix proportion of UHPC is shown in Table 1.

Table 1. Mix proportion of UHPC Unit: kg/m^3

Ratio	Cement	Sand	Fly ash	Silica fume	Water	Steel fiber	Retarder
1	800	1000	100	100	165	156	20
2	800	1000	100	100	170	156	20
3	800	1000	100	100	175	195	20
4	800	1000	100	100	165	156	20

2.3 Mixing Process

1) Preparation of cement matrix: Firstly, dry materials such as cement and sand are poured into a mixer and stirred at low speed for 2 min. Then, the water reducer and water are slowly poured into the mixer and stirred for 3 min to obtain a uniform substrate flow rate.
2) Adding steel fibers: While stirring at low speed, use a wide sieve to evenly sieve the steel fibers into the matrix, stirring for about 5 min to ensure that the steel fibers can be evenly dispersed during stirring.
3) Pouring: All specimens should be poured in two layers, with 2/3 of the mold poured first and shaken several times to ensure the material is dense. Then, pour the remaining 1/3 of the mold in the same way.

4) Maintenance: Use a trowel to smooth the surface and cover the mold surface with plastic film or waterproof cloth to prevent moisture from evaporating. After 24 h of sample curing, the template should be removed and placed in a standard curing room until the required curing time is reached.

2.4 Test Method

The uniaxial direct tensile test of ultra-high performance concrete shall be conducted in accordance with the tensile performance test method in T/CBMF37–2018 "Basic Properties and Test Methods of Ultra High Performance Concrete".

3 Test Results and Analysis

3.1 Test Phenomenon and Failure Mode

The failure modes of different proportions of direct tension specimens are shown in in Fig. 1.

As shown in Fig. 1, the loading rate of the UHPC specimen should not be too high. Before the concrete specimen is disconnected (i.e. the specimen reaches its maximum bearing capacity, cracks appear in the UHPC, but the steel fibers still work), there is no reaction on the specimen; When the load reaches its maximum, accompanied by a crisp sound, cracks appear in the test block, but due to the presence of steel fibers, the test block does not break and can still withstand the load; Continuing to stretch, the bearing capacity of the test block gradually decreases, accompanied by the sound of steel fibers pulling out of the concrete and breaking. As the displacement increases, the steel fibers continue to pull out and the cracks continue to increase. The tensile test blocks of each ratio showed only one crack, with the crack located in or near the middle section and not fixed.

3.2 Tensile Strength

The tensile strength of different ratios is shown in Fig. 2.

From Fig. 2, it can be seen that the analysis of UHPC tensile performance under different influencing factors is as follows:

(1) Cement grade

With the same ratio, as the cement strength increases, the initial cracking strength of UHPC increases by 65% and the tensile strength increases by 14%.

(2) Cement variety

The initial cracking strength of UHPC using cement from different origins did not change, but the tensile strength slightly increased.

(3) Fiber dosage

With the increase of fiber content, the initial cracking strength of UHPC increased by 39%, and there was no significant change in tensile strength.

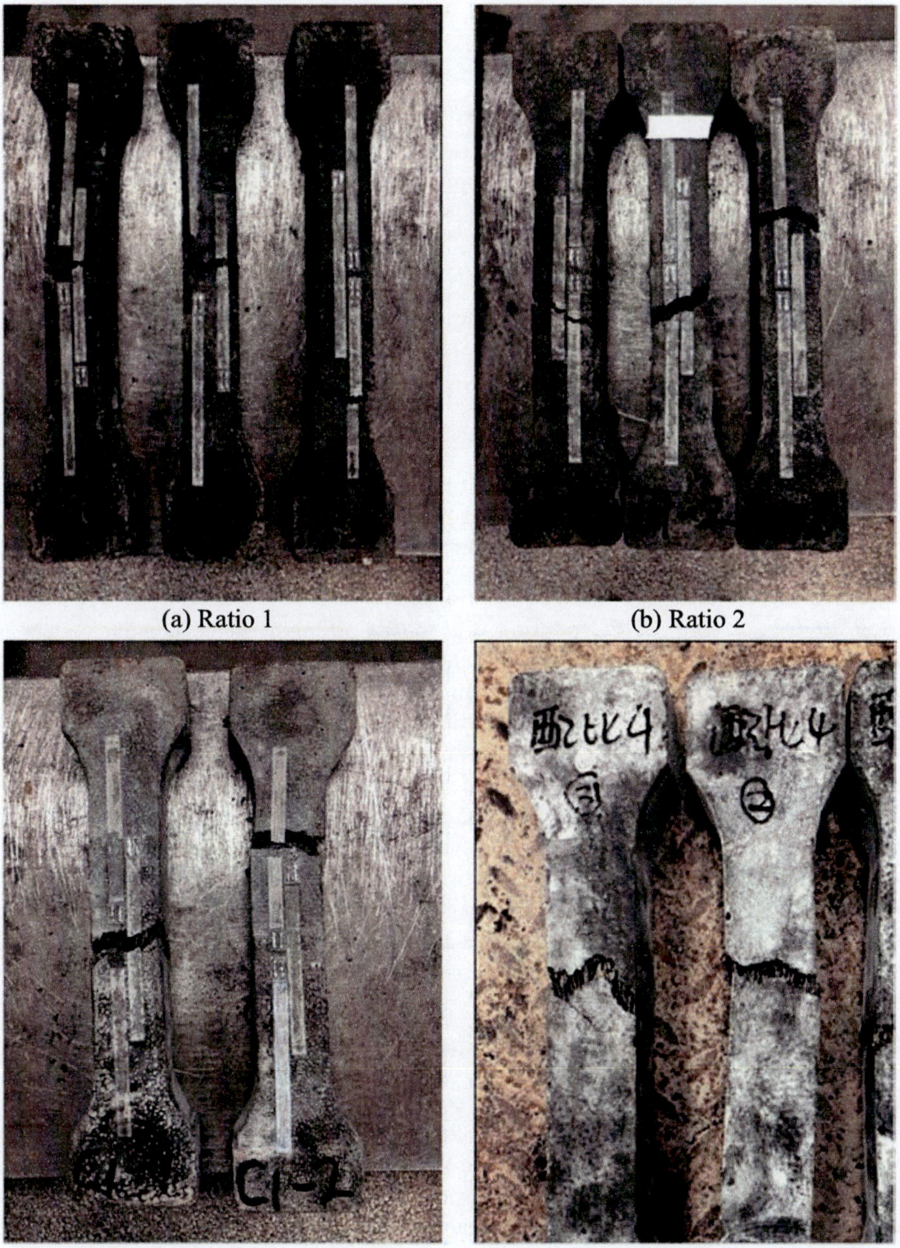

(a) Ratio 1 (b) Ratio 2

(c) Ratio 3 (d) Ratio 4

Fig. 1. Failure morphology diagram of straight tensile specimens

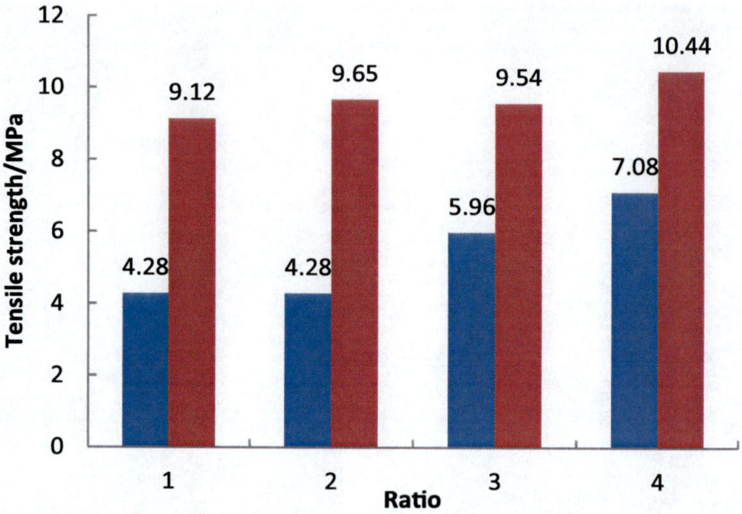

Fig. 2. Initial crack strength and tensile strength of different ratios

Within 24 h after UHPC pouring, the tensile strength of the cement matrix is extremely low, and a large number of cracks are easily generated when the moisture in the matrix evaporates rapidly. At this time, the steel fibers in the cement matrix can withstand the tensile stress caused by plastic shrinkage, thereby preventing or reducing the generation of cracks; After the cement matrix hardens, due to changes in humidity and environmental temperature around the specimen, it can cause dry shrinkage. When the tensile force caused by this exceeds the tensile strength of the matrix, it is also easy to generate a large number of cracks. The evenly distributed steel fibers can also prevent or reduce the generation of cracks at this time. Therefore, the steel fibers in UHPC can not only prevent the generation and propagation of microcracks in the cement matrix during the non hardening plastic stage, but also exert this crack resistance effect during the hardening stage of the cement matrix.

3.3 UHPC Tensile Load and Displacement Curve

The load displacement curve of UHPC direct Tensile testing is shown in Fig. 3.

From Fig. 3, it can be seen that after the load reaches its maximum, the load gradually begins to decrease and all show a decrease in load fluctuations, with a certain strain hardening effect; As the crack width of the specimen increases, the load begins to steadily decrease. This is because after the initial crack of the specimen, as the load increases, the crack width gradually increases. Due to the interfacial bonding performance between the cement matrix and the steel fiber, as well as the pulling force of the steel fiber itself, a short-term strain hardening effect appears, causing the bearing capacity of the specimen to slowly decrease and ultimately the specimen to fracture. The ultimate tensile strain of cement matrix is usually about 150–200 microstrain, while the ultimate tensile strain of steel fiber is about 3–4%; Therefore, when subjected to tension, the cement matrix

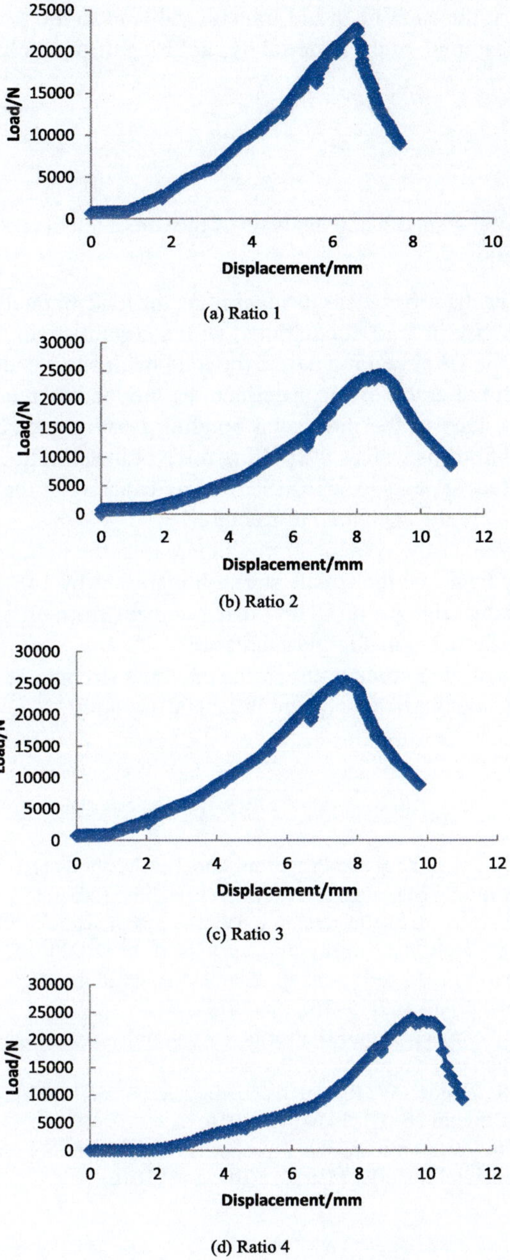

(a) Ratio 1

(b) Ratio 2

(c) Ratio 3

(d) Ratio 4

Fig. 3. UHPC tensile load and displacement curves

will first crack; When the crack extends to the surface of the steel fiber, due to the significant difference in elastic modulus between the two, the foundation will undergo shear deformation at the steel fiber and transfer the load to the steel fiber; Steel fibers improve the tensile strength of the material by capturing these cracks and hindering their deep propagation.

4 Conclusion

Through experimental research and analysis of the direct tensile properties of UHPC, it can be concluded that:

1) After the load of each ratio reaches its maximum, the load gradually begins to decrease and all show a decrease in load fluctuations, with a certain strain hardening effect; As the crack width of the specimen increases, the load begins to steadily decrease. This is because after the initial crack of the specimen, as the load increases, the crack width gradually increases. Due to the interfacial bonding performance between the cement matrix and the steel fiber, as well as the pulling force of the steel fiber itself, a short-term strain hardening effect appears, causing the bearing capacity of the specimen to slowly decrease and ultimately the specimen to fracture.
2) With the same ratio, as the cement strength increases, the initial cracking strength of UHPC increases by 65% and the tensile strength increases by 14%;
3) The initial cracking strength of UHPC using cement from different origins did not change, but the tensile strength slightly increased;
4) With the increase of fiber content, the initial cracking strength of UHPC increased by 39%, and there was no significant change in tensile strength.

References

1. Zhao, J., Shi, H.X., Lu, X.Y.: Basic Properties and Test Methods of Ultra High Performance Concrete. China Building Materials Industry Press, Beijing (2019)
2. Wu, Y.K., Yao, Y.M.: A review of dynamic damage mechanisms of ultra high performance concrete (UHPC) concrete and cement products **4**(1–5), 16 (2021)
3. Zhang, et al.: Research on the self shrinkage characteristics of ultra-high performance concrete concrete and cement products **7**(13–16), 22 (2019)
4. Zhong, Y.W., et al.: Overview of basic performance research on ultra high performance concrete (UHPC) concrete and cement products **09**(1–4) (2021)
5. Wang, T., Caijun, S., Linmei, W.: Research and application of ultra-high performance concrete in China Silicate. Bulletin **35**(01), 141–149 (2016)
6. Yang, J.J., et al.: Research on mechanical properties of high fluidity ultra high strength steel fiber reinforced concrete. J. Build. Mater. **13**(01), 1–6 (2010)

Attitude Adjustment Technology of Rectangular Pipe Jacking

Jiangsheng Xie[1] and Hongbin Guo[2,3](✉)

[1] China Railway 20Th Bureau Group Corporation Limited, Xi'an, China
[2] School of Civil Engineering, Xi'an University of Architecture and Technology, Xi'an, China
1720514047@qq.com
[3] Shaanxi Key Laboratory of Geotechnical and Underground Space Engineering, Xi'an, China

Abstract. The factors that affect the posture control of long distance and large section rectangular pipe jacking are very complex, including engineering geological factors, errors of pipe jacking supporting facilities, difficulties in controlling the posture of pipe jacking during the jacking process, etc., which greatly increases the construction risk of long distance and large section pipe jacking. Suzhou Cheng Bei Road Comprehensive Pipe Rack Project is a long distance and large cross-section rectangular pipe jacking construction project. In the process of pipe jacking and launching, aiming at the attitude control during jacking, this process has taken comprehensive control measures, including reinforcement and precipitation during the starting process, drainage during construction, axis control, deflection control, attitude control during jacking and launching, and "bump prevention".

Keywords: rectangular pipe jacking · posture adjustment · long distance and large section

1 Introduction

The rectangular pipe jacking construction technology is more and more applied to the construction of underground projects, which has played a positive role in promoting the development of urban economy and the development of underground space.However, due to the increasing cross section of rectangular pipe jacking, complex and diverse geological conditions, and the problems of pressure difference and uneven lateral force at each point of the excavation face during the construction process, it is difficult to control the construction posture of the super large section rectangular pipe jacking tunnel.Pipe jacking attitude control technology has become a hotspot in engineering practice research in recent years. Lin Xiugu [1] analyzed the engineering example of the second bid section of the Luxiang Qilianshan Road Connection Project in Baoshan District, Shanghai. In the process of pipe jacking, he took comprehensive control measures, including reinforcement and precipitation of the tunnel entrance and exit, hinge rectification, prevention of "knock", rotation deviation control, longitudinal bolt connection, and achieved good results in the construction process. Rong Liang and Yang Hong Jun [2, 3] introduced pipe

W. Guo et al. (Eds.): GBCESC 2023, LNCE 328, pp. 307–313, 2024.
https://doi.org/10.1007/978-981-99-9947-7_33

jacking attitude control and deviation correction measures in an all-round way in combination with the Zhongzhou Avenue Undercrossing Project of Weisi Road, Zhengzhou City, and elaborated on key technologies such as articulated deviation correction, double screw machine excavation, cutter head control, grouting deviation correction and small pitch propulsion control; Yang Hong Jun et al. [4, 5] studied pipe jacking drag reduction technologies such as the layout of grouting holes, optimal design of slurry pipelines, and thixotropic slurry configuration in combination with the tunnel project, and made targeted analysis on the drag reduction technology used in the tunnel.

2 Project Overview and Construction Equipment Supporting

2.1 Project Overview

The length of the main corridor of Suzhou Cheng Bei Road Comprehensive Pipe Gallery Project is about 1365.079m, and the section is distributed along the ground.This section of the pipe gallery of Youyuan and Tang River Road adopts pipe jacking construction, with a jacking length of 233.6m and a section size of 9.1 × 5.5m.The average thickness of the pipe jacking construction section is 9m, and the section size and jacking length are among the most domestic comprehensive pipe gallery projects. According to the geological survey data, the top pipe crosses the formation of silt and silt, and partially penetrates the river There is silt at the top of the top pipe and groundwater is abundant.

2.2 Construction Equipment and Supporting Facilities

This project adopts earth pressure balance type 9100 mm × 5500 mm rectangular pipe jacking machine for tunneling construction, as shown in Fig. 1. The pipe jacking equipment and supporting facilities are as follows:

(1) A total of seven cutter heads are arranged in the cutter head system, with 1 large, 4 medium and 2 small designs.
(2) Electrical system and rectifying oil pump are arranged in the nose shell;
(3) There is a sealing device between the spindle and the shaft sleeve to ensure that the soil and water do not invade the machine during the working process.
(4) The measuring system is composed of two parts, one is installed on the front shell of the measuring target, the other is installed in the front shell of the tipper, soil pressure gauge.
(5) There are 4 unloading oil cylinders at the connection between the rear shell of the pipe jacking machine and the rectangular concrete pipe for pipe connection and separation.

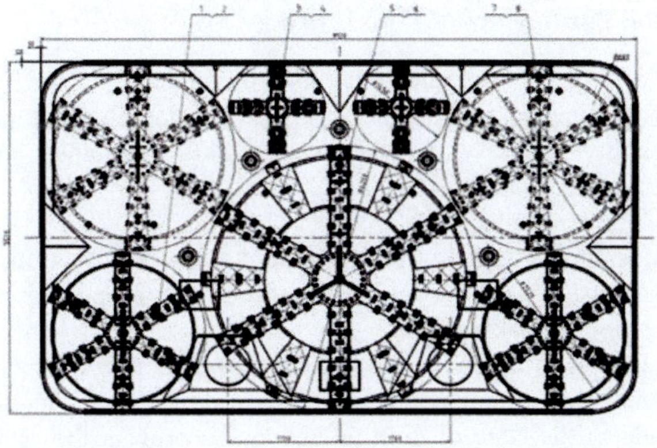

Fig. 1. 9100 mm × 550 mm rectangular pipe jacking machine

3 Analysis of Influencing Factors of Pipe Jacking Attitude Control

Combined with relevant engineering experience and comprehensive analysis of the characteristics of this project [6–8], it is found that the factors affecting the control of pipe jacking posture in the long distance jacking process of rectangular pipe jacking with large section mainly include engineering geological factors, supporting facilities for pipe jacking, construction errors and construction parameters of pipe jacking.

3.1 Engineering Geological Factors

Due to its own factors, the pipe jacking machine is easy to bend down in the soil layer, leading to the deviation of the jacking axis. This unfavorable condition is easy to produce the accumulation of attitude deviation in the process of long-distance jacking [9, 10].

3.2 Construction Errors of Supporting Facilities for Pipe Jacking

(1) the guide rail installation deviation is large, easy to lead to the top pipe kowtow, head up and other elevation deviation.
(2) The deviation of the back and the jacking force will lead to the eccentricity of the jacking pipe, and then lead to the axis deviation.
(3) The jack and oil route arrangement is unreasonable, and there is a jacking time difference between the jacks, which makes the jacking of the pipe offset, and then leads to the rotation of the pipe jacking machine.

3.3 Construction Parameters of Pipe Jacking

(1) The jacking footage does not match the unearthed quantity

In the process of pipe jacking and jacking, when the jacking footage does not match the unearthed quantity, it is easy to cause ground uplift or collapse, and then cause the deviation of pipe jacking axis.

(2) Attitude measurement in jacking

Measurement is the key to long distance pipe jacking construction. The accuracy of pipe jacking axis measurement is related to the accuracy of pipe jacking correction.

(3) Correction method of jacking in

The attitude control principle of pipe jacking construction is "rectify the elevation first, then the center line, small Angle continuous deviation correction", and always control the attitude in the state of high precision. To prevent the rotation of the pipe jacking machine due to the large offset size.

4 Tube Jacking Attitude Control Technology

4.1 Axis Control

During the construction process of pipe jacking, the force constantly changes, which will cause the pipe jacking machine to deviate from the designed axis and elevation during the jacking process. Therefore, in order to ensure the quality of construction, the pipe jacking machine needs to be continuously corrected. Specific preventive measures are:

(1) The guide rail should be repeatedly detected when installed, so that the center line, elevation and slope must meet the requirements.
(2) The geological conditions of the pipeline construction area should be investigated in detail before the pipe jacking construction.
(3) Monitor the actual pushing force on each joint of the pipeline and the eccentricity of the pipeline axis.
(4) Installation of automatic guide measurement system. The total station measuring table is installed at the entrance of the hole, and a special prism is installed at the pipe wall and the head of the pipe jacking machine, which is used to conduct precise orientation of the total station on and under the well. According to the measurement system display results and deviation, automatic correction.

4.2 Deflection Control

For the rectangular tunnel, the lateral turning of the rectangular pipe jacking machine will directly affect the segment assembly, and at the same time, it will cause the inclination of the bottom surface of the tunnel, which will affect the normal use of the tunnel. Therefore, it is necessary to correct the deviation in time to prevent lateral deflection. The main reasons for lateral deflection of rectangular pipe jacking machine are as follows:

(1) Geological reasons: uneven soft and hard soil layer, different soil properties of excavation surface and uneven mixing in the soil chamber, these external forces on the cutter head cause left-right deviation, resulting in the corner of the pipe jacking machine when it is pushed.

(2) Manufacturing error of the pipe jacking machine: there may be some asymmetry in the rectangular pipe jacking machine, such as the quality imbalance on both sides or the appearance deviation, which will cause the pipe jacking to deflect to one side.
(3) Unreasonable construction arrangement or improper operation: if the main jack and the pipe jacking axis are not parallel, a torque is formed between the two, which causes the pipe jacking machine to deflect; When axis deviation of pipe jacking machine occurs, the amount of correction is too large.

4.3 Jack-In-Start Attitude Control

During construction, the control of the initial and final posture of the pipe jacking machine is of great significance for the successful jacking of pipelines. Therefore, it is necessary to make full preparations before the arrival of the pipe jacking machine, including:

(1) Preparation before jacking, including checking the guide rail before arrival, adjusting the posture of the pipe jacking machine before arrival and recording the initial value of each instrument.
(2) Production and installation of extension guide rail. In order to avoid or slow down the phenomenon of "kowtow" during the arrival of the pipe jacking machine, the length of the pipe jacking machine to reach the support is extended, and the trend of the pipe jacking machine to reach the "kowtow" is delayed.

5 Attitude Control Effect After Completion of Pipe Jacking Construction

The single jacking length of Suzhou Cheng Bei Road comprehensive pipe corridor project is 233.6m, the section size is 9.1 × 5.5m, and the average soil cover thickness of the pipe jacking construction section is 9m. The section size and jacking length are the largest among domestic comprehensive pipe corridor projects. The notch plane, elevation deviation and nose deviation in the whole process of pipe jacking construction are shown in Fig. 2. According to the construction design axis, the deflection of the pipe jacking machine is negative on the left and positive on the right. Deviation to the left is negative, deviation to the right is positive; The elevation deviation is negative below and positive above. It can be seen from the figure that the horizontal and elevation deviation of nose incision is controlled within ± 10cm, and the rotation deviation is controlled within ± 0.1°.When jacking the pipe into the hole, the horizontal deviation of the pipe jacking machine head incision is −20 mm, the elevation deviation is + 30 mm, and the rotation deviation is −0.03°.

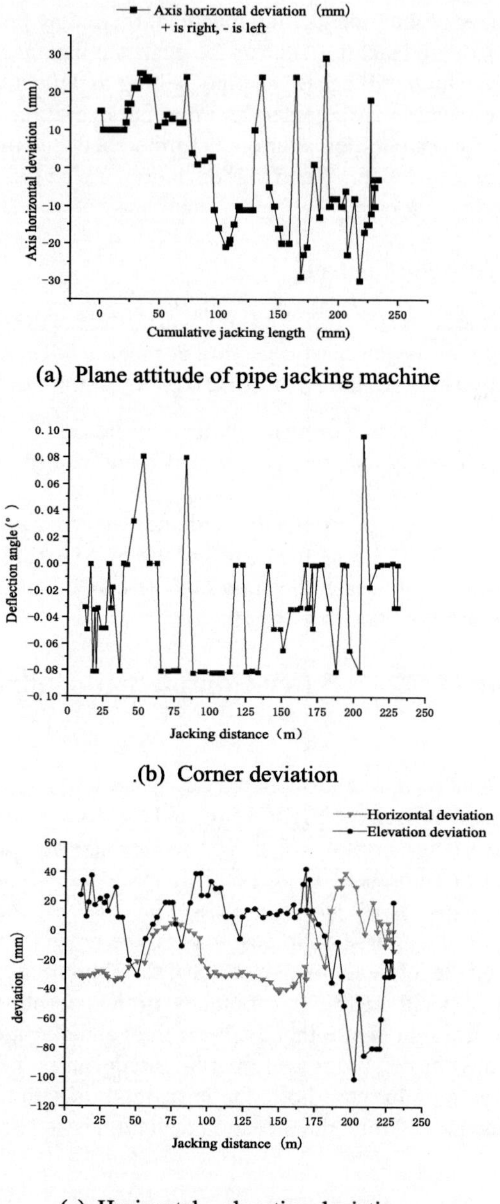

(a) Plane attitude of pipe jacking machine

.(b) Corner deviation

(c) Horizontal、 elevation deviation

Fig. 2. Attitude deviation of the pipe jacking machine head

6 Conclusion

(1) This paper first analyzes the construction difficulties of the pipe jacking project of Wu Biao Yuan and Tang section of the comprehensive pipe corridor project of Cheng Bei Road in Suzhou City, and concludes the factors affecting the construction progress, such as large section, long jacking distance and sudden address problem. According to the above factors, the solutions of axis control, deflection control and jacking and starting attitude control are put forward.
(2) According to the monitoring results of the pipe jacking machine in the actual construction process, in the deflection attitude curve diagram, the nose deflection is controlled within $\pm 0.1°$, and the pipe jacking machine is basically kept level. In the plane attitude control chart, there was a slight deviation at 75m jacking. Before crossing Yuanhetang, the nose position was very close to the designed axis. After crossing the river, the nose position was about $+ 5$ mm away from the designed axis, which was adjusted to be consistent with the designed axis.In the elevation attitude monitoring, after several adjustments, the pipe jacking machine finally arrived successfully.

References

1. Lin, X., Cao, Y., Xie, D.: Attitude control technology of long-distance large-section rectangular pipe jacking in soft ground. Modern Tunn. Technol. **57**(S1), 1007–1014 (2020)
2. Rong, L., Yang, H.: Attitude control technology for super-large cross-section rectangular pipe-jacking machine; case studly on tunneling crossing under Zhong Zhou avenue in Zheng Zhou. Tunnel Constr. **35**(10), 1097–1102 (2015)
3. Yang, H., Rong, L., Hucheng, X.: Application of drag reduction technology to extra-large cross-sectional rectangular pipe jacking: case study of tunnel crossing undermeath Zhongzhou road in Zhengzhou. Tunnel Constr. **36**(04), 458–464 (2016)
4. Bao, Y., Zheng, Q., Tang, J.: Analysis on construction measures used in jacking of rectangular-shapedstructure in soft soil. Railw. Eng. **08**, 75–78 (2009)
5. Bao, Y., et al.: Construction technique of jacking of rectangular-shape structure in shanghai soft soil. Railw. Eng. **09**, 68–70 (2009)
6. Liu, J., Liang, M.: Construction technology of rectangular jacking of multiple pipes with small interval and large section in soft soil strata. Subgrade Eng. **05**, 160–163 (2015)
7. Zheng, J.: Studly on large section rectangular pipe jacking construction method in soft-soil areas. Urban MassTransit. **14**(11), 93–96 (2011)
8. Zhu, J.: The underground passage incremental launching technology of the complex building of Zhongshan hospital in Shanghai. Zhejiang Constr. **29**(10), 38–42 (2012)
9. Guo, H.: Launching and receiving construction technology of large-section rectangular pipe jacking machine. Undergr. Eng. Tunnels **03**, 22–25 (2014)
10. Jia, L.: Key technologies for design of super-jarge rectangular pipe jacking machine. Tunnel Constr. **34**(11), 1098–1106 (2014)

Numerical Study on Performance of Single-Keyed Epoxy Joint of Ultra-high Performance Concrete (UHPC) Under Combined Shear and Torsion Load

Zhe Li, Yun Shen, and Lei Sun(✉)

Anhui Transportation Holding Group Co., Ltd., Hefei 230000, China
1968457282@qq.com

Abstract. Under the vehicle load, segment joints are subjected to coupling effects of bending, shear, and torsion, while epoxy joints are mainly subjected to a combination of shear and torsion, making them more prone to failure. In this study, to investigate the torsion-shear performance of single-keyed epoxy joints of ultra-high performance concrete (UHPC), a refined finite element model (FEM) was carried out considering the effect of confining pressure in this study. The results of FEM indicated that the increase of confining pressure can effectively improve the shear-torsional load capacity of the epoxy joints. Whereas the increase of confining pressure has little effect on the improvement of stiffness. However, in view of the high confining pressure (when the confining pressure is greater than 18 MPa), the damaged surface of the specimen changes from the root of the male key to the feminine key.

Keywords: Precast bridge · Ultra-high performance concrete (UHPC) · Cpoxy joint · Finite element model (FEM) · Combined shear and torsion load

1 Introduction

With the advancement of bridge industrialization, assembled precast bridges have been promoted and applied widely [2]. As a new type of bridge, Ultra high performance concrete (UHPC) segmental girder bridge is an important study direction to promote the industrialization of bridges. However, the structural integrity of segmental bridges is the main issue affecting its flexural and shear performance [2]. In actual bridge operation, segment joints are subjected to coupling effects of bending, shear, and torsion, while for epoxy joints, they are mainly subjected to a combination of shear and torsion, making them more prone to failure. Therefore, for segmental girder bridges, joints are fragile structural components that require special treatment, especially in terms of shear-torsion load capacity [2, 3].

W. Guo et al. (Eds.): GBCESC 2023, LNCE 328, pp. 314–321, 2024.
https://doi.org/10.1007/978-981-99-9947-7_34

Compared with UHPC segmental dry joints, adhesive epoxy joints have more excellent bending shear properties [4], and the research on the shear capacity of UHPC epoxy joints has been focused by major well-known UHPC research teams. To research the shear damage mode of epoxy joints, Yan Zeyu [5] concluded that the damage form of UHPC glue joints is brittle damage by two groups of UHPC epoxy joint direct shear tests. Li Xing heat [6] found that the damage mode of keyed epoxy joints can be divided into slip damage and shear damage by direct shear test. Kim et al. [7] found that the ultimate load of epoxy joints increased with the increase of the number of keys, confining pressure and key depth by shear test of UHPC epoxy joints. However, the study of joints under shear-torsion loads is limited.

Based on the above discussion, to investigate the performance of single-keyed epoxy joint of ultra-high performance concrete (UHPC) under combined shear and torsion load, a numerical study was carried out considering the effect of confining pressure.

2 Specimen Design

Based on the literature [1], considering the load moving space of shear-torsion compound action, the upper part of the specimen is set 200 mm thick area, and a 200 mm × 200 mm × 200 mm trigonal concrete axil is designed to strengthen the projection area; the size of the lower part of the specimen is 500 mm × 1250 mm × 200 mm, and the size of this part is increased to prevent the whole load process. In order to ensure that the joint position is destroyed before the non-key area, the specimen is equipped with HRB400 reinforcement, except for two φ18 mm bars in the lower part of the specimen, the rest are φ16 mm bars in each group, the specimen size are shown in Figs. 1 and 2. Of note, load eccentricity is 200 mm in specimens (see Figs. 1 and 2).

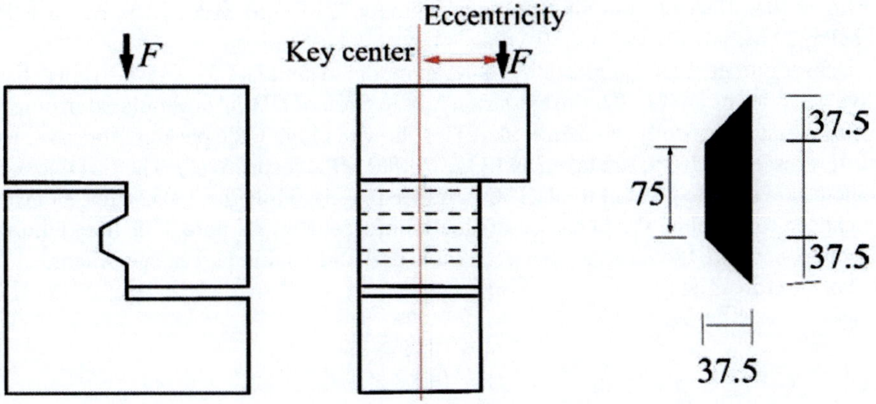

Fig. 1. Structural drawing of specimens [1]

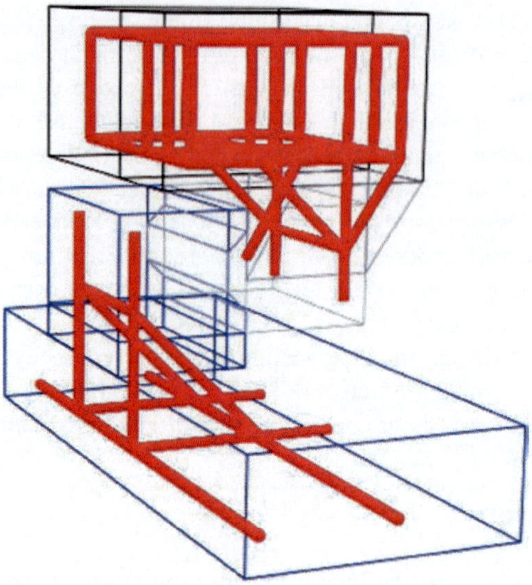

Fig. 2. 3D rendering of specimens

3 Finite Element Analysis

3.1 Finite Element Model (FEM)

To understand the shear-torsion behavior of UHPC epoxy joint interface, finite element modeling and analysis were conducted based on the Abaqus platform (2020). 3D finite element models of the tested specimens were established in Abaqus (see Fig. 4), which considers the effect of confining pressure, namely, 2 MPa, 6 MPa, 10 MPa, 14 MPa, 18 MPa, 20 MPa, 25 MPa, and 30 MPa.

Using concrete damage plasticity material model to simulate the stress-strain constitutive behavior of UHPC. The stress-strain relationship of UHPC is simulated in consistent with that in the study of Chen et al. [2], as shown in Fig. 3, the peak compressive and tensile stresses of UHPC are taken as 133 MPa, 8.0 MPa, respectively. The C3D8R solid elements are used to present the UHPC part (see Fig. 5), while the T3D2 truss elements are chosen to simulate the behavior of steel reinforcement. Of note, TIE interaction is chosen to simulate the interface between male part and female part in specimens.

For reinforcement,

$$\sigma_s = \begin{cases} E_s \varepsilon_s \ (0 \le \varepsilon_s \le \varepsilon_y) \\ f_y \quad (\varepsilon_y \le \varepsilon \le \varepsilon_u) \end{cases} \tag{1}$$

For the compression model of UHPC,

$$y = \begin{cases} ax + (6-5a)x^5 + (4a-5)x^6 & 0 \le x \le 1 \\ \frac{x}{b(x-1)^2+x} & x \ge 1 \end{cases}$$

$$y = \sigma/f_c, x = \varepsilon/\varepsilon_0, \varepsilon_0 = 3500\mu\varepsilon,$$

$$a = \frac{E_0}{E_c} \tag{2}$$

For tension model of UHPC,

$$\begin{cases} E_c\varepsilon_t & 0 \le \varepsilon_t \le \varepsilon_{t0} \\ f_1 & \varepsilon_{t0} < \varepsilon_t \le \varepsilon_{tp} \\ \frac{f_1}{1+(\varepsilon_t-\varepsilon_{tp})l_c/\omega_p} & \varepsilon_{tp} < \varepsilon_t \end{cases} \tag{3}$$

In which, $f_t = 8$ MPa, $\varepsilon_{tp} = 0.002$, $\omega_p = 1.0$ mm, $l_c = 400$ mm, $p = 0.95$.

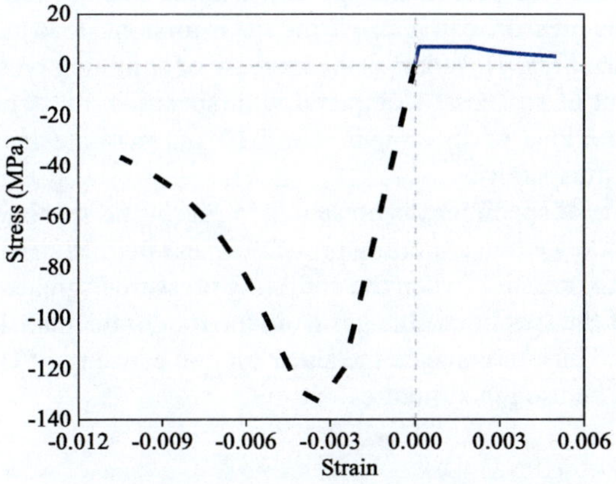

Fig. 3. The stress-strain relationship of UHPC

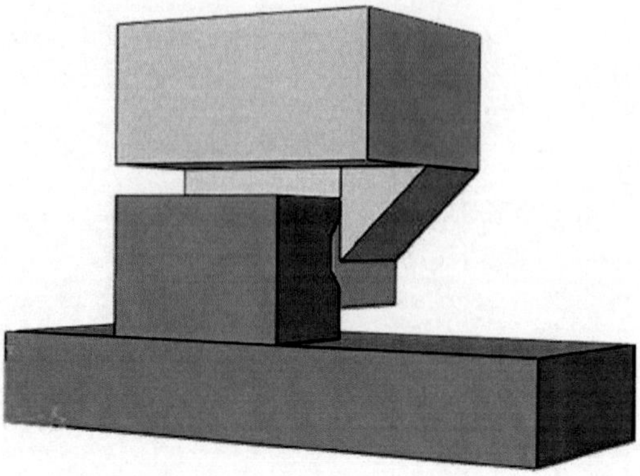

Fig. 4. Model schematic

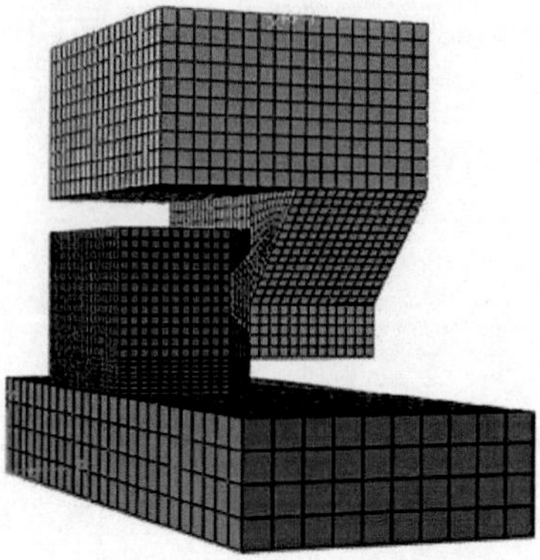

Fig. 5. Mesh schematic of FEM

3.2 FEM Results

Figure 6 plots the load-displacement curves of finite element model, obviously, the increase of confining pressure can improve the shear-torsional load capacity of the epoxy joints effectively (see Figs. 7). Whereas the increase of confining pressure has low effect on the improvement of stiffness. Compared with specimen m2 (confining pressure is 2 MPa), the ultimate load of specimens m6, m10 and m14 are increased by 108 kN, 177 kN and 216 kN respectively.

The failure mode of specimens is presented in Fig. 7, the specimens are damaged in the root of the shear key when confining pressure is less than 18 MPa. However, in view of the high confining pressure (when the confining pressure is greater than 18 MPa), the damaged surface of the specimen changes from the root of the male key to the feminine key (see Fig. 8f-g). Thus, to enhance the shear-torsion capacity of UHPC epoxy joints, reinforcements can be used in structures.

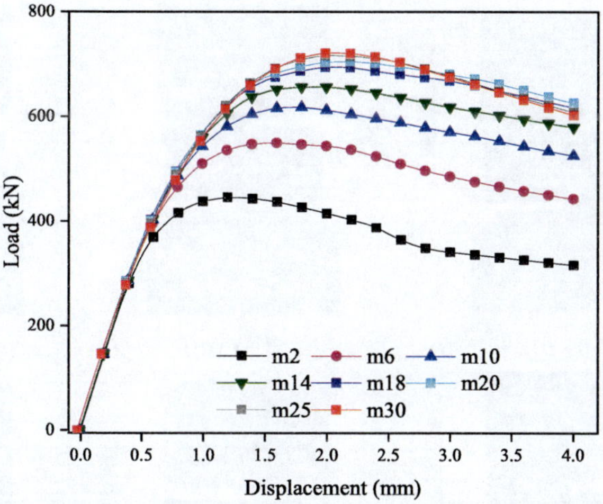

Fig. 6. Load-displacement curves of UHPC single-keyed epoxy joint

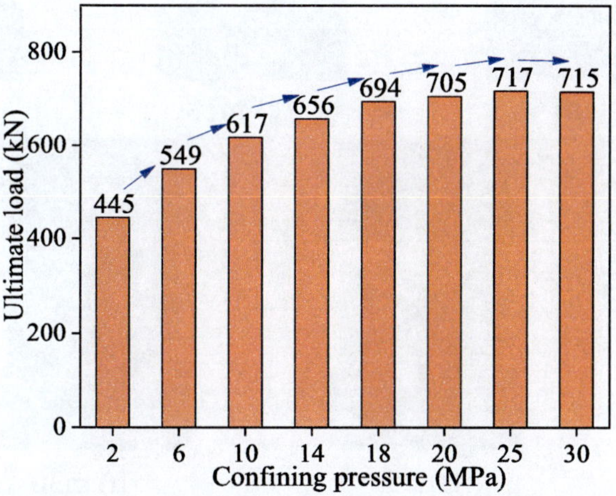

Fig. 7. Ultimate load of UHPC single-keyed epoxy joint

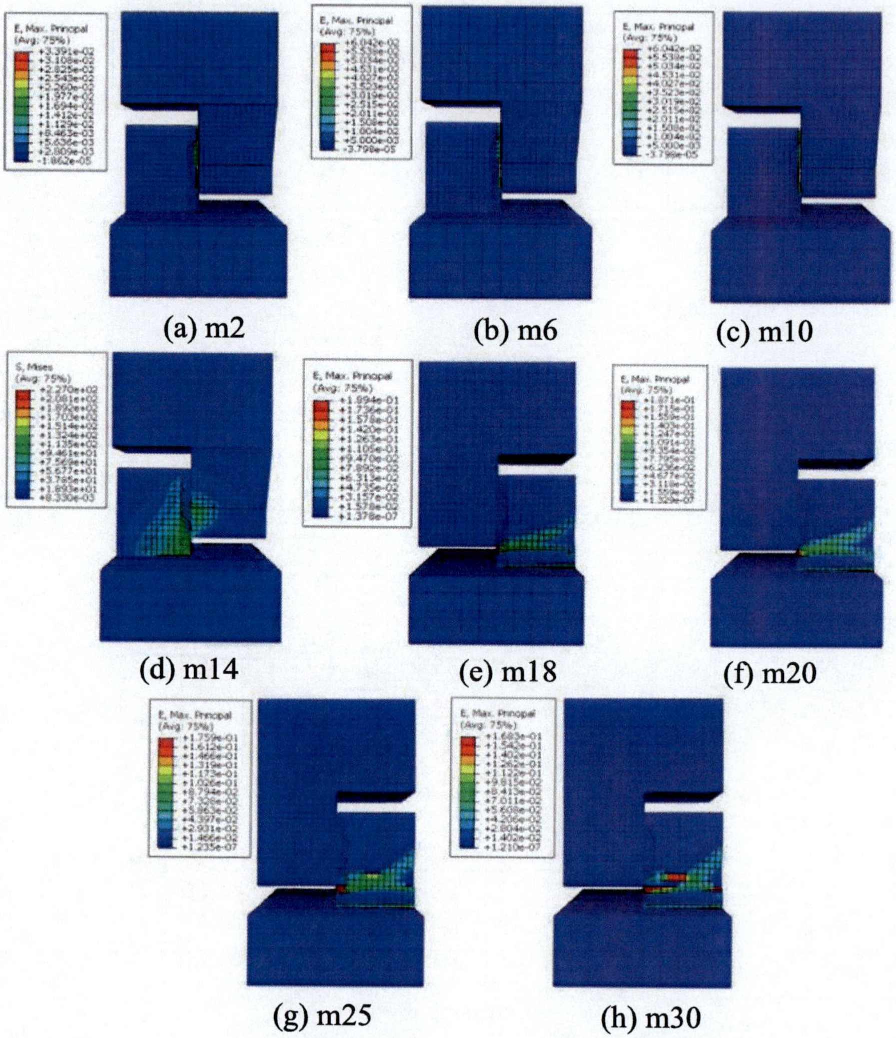

Fig. 8. Failure mode of epoxy joint specimens.

4 Conclusion

Based on the FE results of UHPC single-keyed epoxy joints under shear-torsion load, the following conclusions are obtained.

(1) The increase of confining pressure can effectively improve the shear-torsional load capacity of epoxy joints. Whereas the increase of confining pressure has little effect on the improvement of stiffness.

(2) From the failure mode of specimens, the specimens are damaged in the root of the shear key when confining pressure is less than 18 MPa. However, in view of the high confining pressure (when the confining pressure is greater than 18 MPa), the damaged surface of the specimen changes from the root of the male key to the feminine key.

References

1. Wang, H.L., Li, B.H., Guo, X., et al.: Experimental study on shear behavior of single-keyed dry joint of ultra-high performance concrete (UHPC) under combined shear and torsion. Bridge Constr. **52**(02), 31–38 (2022). (in Chinese)
2. Chen, L., et al.: Shear performance of ultra-high performance concrete multi-keyed epoxy joints in precast segmental bridges. Structures **46**, 1696–1708 (2022)
3. Jiang, H., et al.: Shear strength of steel fiber-reinforced concrete dry joints in precast segmental bridges. J. Bridge Eng. **21**(11), 04016085 (2016)
4. Buyukozturk, O., Bakhoum, M.M., Michael Beattie, S.: Shear behavior of joints in precast concrete segmental bridges. J. Struct. Eng. **116**(12), 3380–3401 (1990)
5. Yan, Z.: Experimental and numerical analysis on shear behavior of epoxied joints in precast UHPC segmental bridges. Highway Eng. **44**(6), 228–233 (2019)
6. Li, X.: Research on Structural Types and Mechanical Properties of Joints in Precast UHPC Segmental Box Girder. Hunan University, Changsha (2021)
7. Kim, Y.-J., Chin, W.-J., et al.: Interface shear strength at joints of ultra-high performance concrete structures. Int. J. Concr. Struct. Mater. **12**(1), 1–14 (2018)

Simulation Analysis of the Construction Process of a Hybrid Girder Cable-Stayed Bridge with Profiled Towers

Tonghui Jiang[1](✉), Jiading Yang[1], Dequan Zhu[1], and Yufeng Xu[2]

[1] Guangdong Foying Huijian Engineering Management Co., Foshan, China
446998826@qq.com

[2] South China University of Technology, Guangzhou, China

Abstract. In order to study the mechanical state of the construction process of the hybrid girder cable-stayed bridge with profiled towers, this paper takes a bridge in Foshan as the research background, which is a 200 m main span hybrid girder cable-stayed bridge with 125 m high main towers and a span arrangement of (200 + 68 + 46) m. The research method of finite element simulation analysis is used to calculate the mechanical state of the main bridge structure under load. The results show that: the main tower under permanent action and combined action are maintained in a reasonable range, the structure under permanent action and combined action into the bridge state of the main beam stress, main beam deformation and cable force are located in a reasonable range, for the development of construction plans to provide effective reference value.

Keywords: cable-stayed bridge · profiled towers · hybrid beam · finite element · simulation analysis

1 Introduction

The construction of a hybrid girder cable-stayed bridge with shaped towers is a complex process with many influencing factors [1]. During the construction of a hybrid girder cable-stayed bridge with shaped towers, detailed analysis and calculations are required for each construction stage to obtain theoretical calculations of the tension tonnage, cable force, main girder deflection, tower top displacement and structural internal forces of the cable-stayed bridge, and to control them effectively during construction.

The mechanical behaviour of hybrid girder cable-stayed bridges with shaped towers during construction is complex and has been the subject of much research. Luo Lei [2] and others based on the finite element analysis of the design phase of the shaped tower cable-stayed bridge and the analysis of the monitoring data during the construction process, so that the finished state of the bridge meets the expected requirements. Yang Jiading [3] and others used ANSYS finite element model to analyse the mechanical state of the towers during the construction of an asymmetric single-tower hybrid girder cable-stayed bridge. Guo Fei [4] and others have effectively controlled the steel tower alignment of cable-stayed bridges based on steel tower segmentation and steel tower

W. Guo et al. (Eds.): GBCESC 2023, LNCE 328, pp. 322–333, 2024.
https://doi.org/10.1007/978-981-99-9947-7_35

section fabrication and pre-assembly. At the same time, some scholars have also analysed the mechanical behaviour of the main girder structure of cable-stayed bridges. Zhang Shoufeng [5] and others established a refined three-dimensional model of the steel-hybrid combination based on large finite element software to study the force transfer mechanism and deformation characteristics of hybrid girder cable-stayed bridges. Xu Yufeng [6] et al. carried out a third-level simulation analysis of the construction process of cable-stayed bridges, with high accuracy of the analysis results, and the simulation values can effectively guide the design work.

In order to further grasp the mechanical state of the construction process of a hybrid girder cable-stayed bridge with shaped towers, finite element simulation analysis is carried out in this paper with a 200 m main span hybrid girder cable-stayed bridge with shaped towers in Foshan City as the background.

2 Project Overview

Foshan City, a shaped tower hybrid girder cable-stayed bridge, the main tower height of 125 m, the main bridge span arrangement for (200 + 68 + 46) m, the main tower using C50 concrete, the main beam for the use of "PK" section box girder, the main beam side span using concrete box girder, the main beam in the span using steel box girder. The bridge layout is shown in Fig. 1 below.

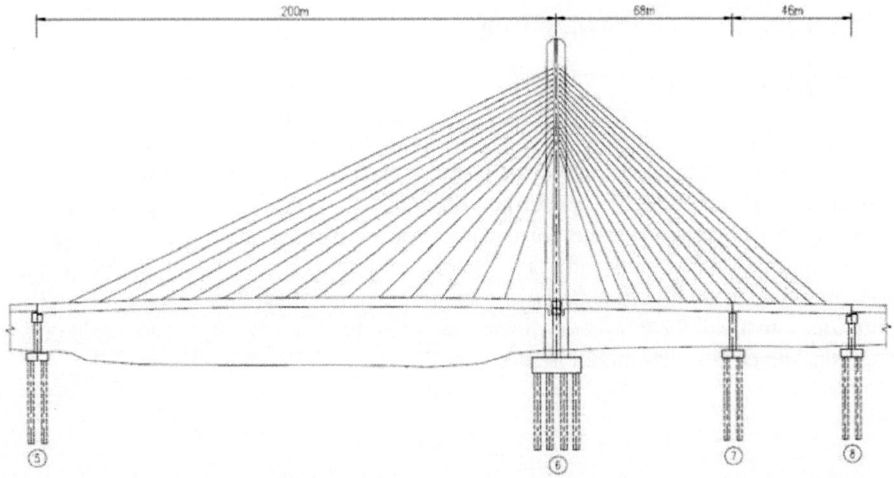

Fig. 1. Bridge type elevation layout (m)

3 Construction Programme

The construction sequence of the main bridge is as follows:

(1) The main tower, side piers, and auxiliary pier foundations are all located on shore, and the bored and grouted piles and bearing platform foundations are constructed using land-based construction methods.
(2) The main tower was poured in sections, the side span concrete cast-in-place sections were poured simultaneously, the steel box girder sections 1 to 2 and sections 14 to 17 were erected on supports.
(3) The main tower was topped, the side span concrete cast-in-place sections were closed and tensioned, the steel box girders were erected and welded, and the bridge deck crane was installed to the 2nd section of steel box girders.
(4) Start tensioning No. 1 diagonal cable, follow the sequence of hoisting steel girders, steel girder installation, tensioning diagonal cable of corresponding girder section, crane forward for the construction of section 3 to 11 steel box girders.
(5) Section 12 is used as the main span closure section and the main span closure is completed by the temperature matching cut method.
(6) Tensioning of subsequent diagonal cables in sequence.
(7) Removal of the deck crane and main girder supports.
(8) Complete the construction of the bridge deck system and ancillary works, and carry out the cable adjustment work of the diagonal cables.

4 Simulation Model Modelling

4.1 Structural Models

Midas Civil finite element analysis software was used to establish the full bridge model. The calculation model divided the main tower, bearing platform, pile foundation and concrete box girder, steel box girder and diagonal cable into 868 nodes and 761 units according to the division of the construction sections of the main bridge, the support points, section change points and other control sections. The main tower, main beam and lower foundation are all simulated using beam units, the diagonal cables are simulated using truss units and the temporary braces are simulated using compression-only elastic supports. The calculation model is shown in Fig. 2.

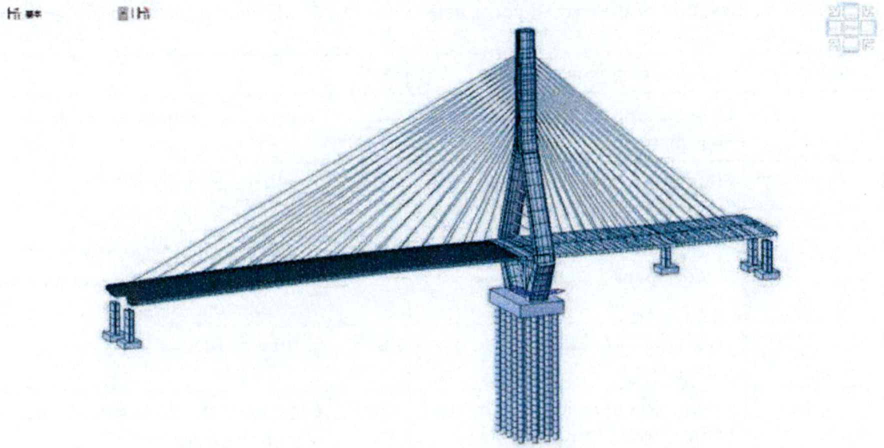

Fig. 2. Bridge finite element simulation analysis model

4.2 Construction Process

The main bridge will go through the process of construction of the main tower, construction of the side piers and auxiliary piers, construction of the concrete box girders for the side spans and lifting of the steel box girders for the main spans. The individual calculation formulae are based on the actual structural system and load conditions provided by the relevant units.Based on the design drawings and the actual construction organisation scheme, the main bridge construction process is simulated and analysed using the positive assembly analysis method. The calculation conditions of the finite element model are divided into the following Table 1.

4.3 Load Information

The load information for the main bridge simulation analysis includes permanent, variable and combined actions.

(1) Permanent action

Phase I constant load: the self-weight of each part of the bridge structure. Second phase constant load: constant load added after the second phase paving, railings and other structures have taken shape.

(2) Variable action

Variable action includes vehicle live load, temperature load and wind load.

(3) Combined action

The permanent and variable actions are combined to obtain the most unfavourable state of the structure, and the load combination conditions are shown in Table 2 below.

Table 1. Division of the overall construction process conditions of the main bridge

Condition No.	Construction work content		
1	Construction of piers and main pier abutments	11	Tensioning of diagonal cable No. 2
2	Pouring of main tower in sections with cross bracing	12	Deck crane forward
3	Pouring of cast-in-place sections of side spans	13–45	Construction of steel box girder sections 4 to 11 in the sequence of working conditions 9 to 12
4	Tensioning of side span prestressing	46	Lifting of section 12
5	Installation of sections 1 to 2 and 14 to 17 steel box girders on supports	47	Closing of the main span of the steel box girder
6	Casting of steel and concrete sections	48–50	Tensioning of diagonal cables No. 13 to No. 15
7	Installation of bridge deck cranes	51	Removal of bridge deck cranes
8	Tensioning of diagonal cable No. 1	52	Removal of main girder supports
9	Lifting of section 3 steel box girder	53	Phase II constant load construction
10	Installation of section 3 steel box girder		

Table 2. Load combination conditions

Combination number	Description of load combinations	Type
1	Constant load + live load	summation
2	Constant load + live load + braking force load	summation
3	constant load + live load + temperature load	summation
4	Constant load + live load + wind load	summation
5	Combination 1 + Combination 2 + Combination 3 + Combination 4	envelope

5 Simulation Results

5.1 Calculation of the Permanent Effect

The results of the simulation analysis of deformation, stress and cable force under permanent action are shown in Figs. 3, 4, 5, 6, 7, 8 and 9.

Combining the results of the permanent action calculations above, it follows that.

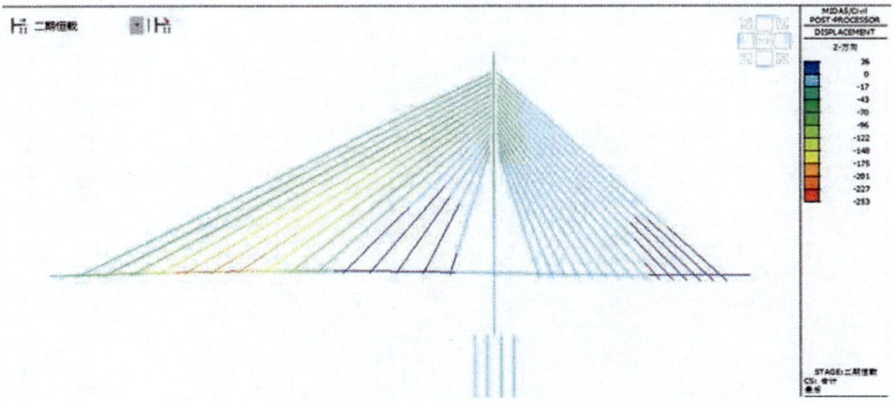

Fig. 3. Vertical deformation under permanent action (mm)

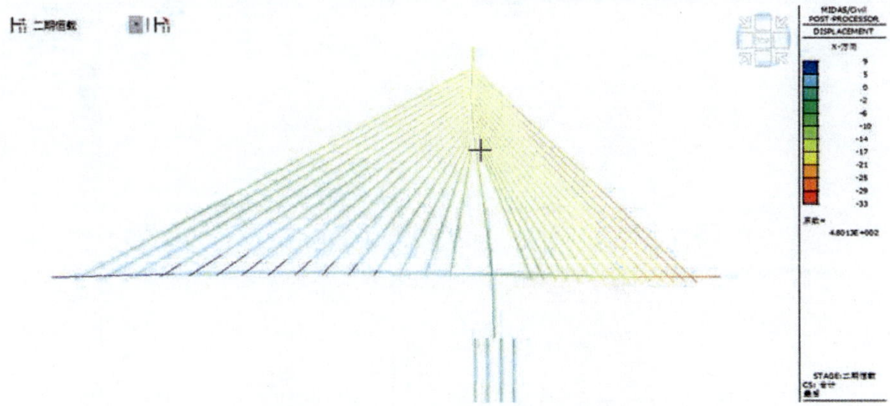

Fig. 4. Longitudinal bridge deformation under permanent action (mm)

(1) The structural deformation, stresses and cable forces of the bridge in its permanent state are all within a reasonable range and meet the requirements of construction control.
(2) The maximum vertical deformation of the steel main girder is −253 mm, and the vertical deformation of the concrete girder is within −11 mm.
(3) The maximum longitudinal deformation of the steel main beam is −9 mm, the maximum longitudinal deformation of the concrete beam is −33 mm, the longitudinal deformation of the top of the tower is −19 mm, the longitudinal deformation of the main beam and the main tower are within reasonable range.
(4) The maximum compressive stress at the upper edge of the steel main beam is −45.2 MPa, and the maximum compressive stress at the lower edge is −125.5 MPa, the overall stress level is basically within a reasonable range.

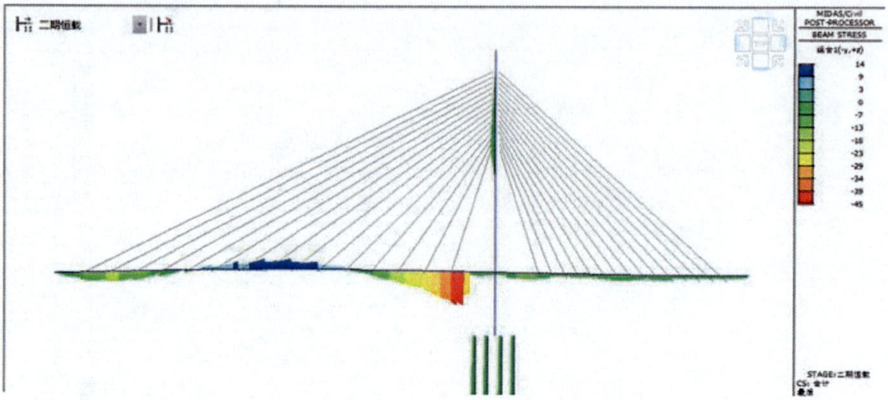

Fig. 5. Stress at the upper edge of the main beam under permanent action (MPa)

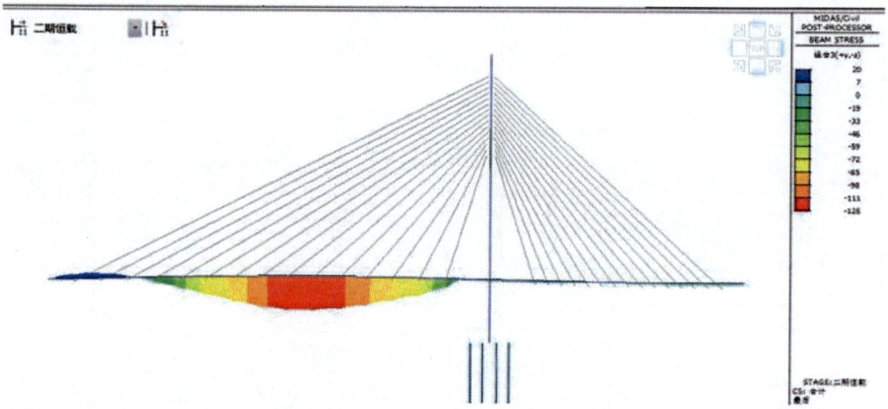

Fig. 6. Stress at lower edge of main beam under permanent action (MPa)

(5) The maximum compressive stress at the upper edge of the concrete is −8.6 MPa, and the maximum compressive stress at the lower edge is −8.6 MPa, the overall stress level is basically within a reasonable range.
(6) The maximum compressive stress on the outer side of the main tower is −8.1 MPa, and the maximum compressive stress on the inner side is −7.8 MPa, the overall stress level is basically within a reasonable range.
(7) The maximum cable force of the diagonal cable is 4836kN, the cable force result is basically within the reasonable range.

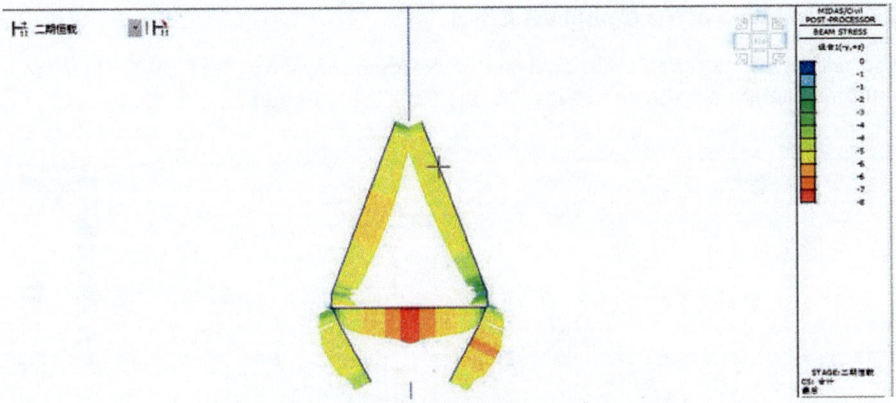

Fig. 7. Stress on outer side of main tower (MPa)

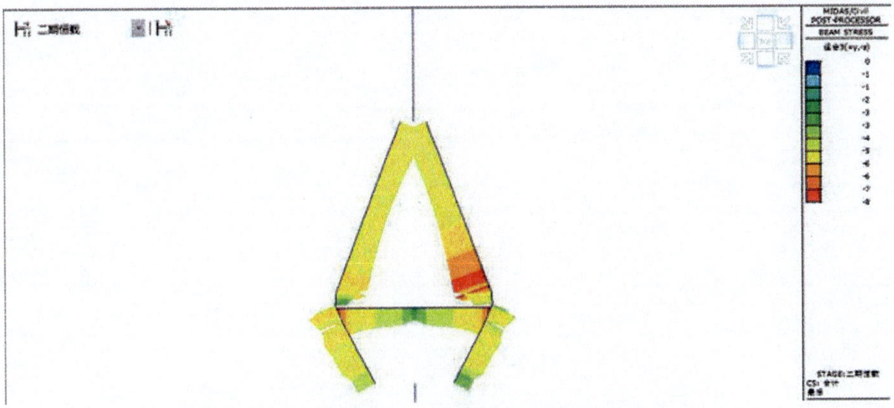

Fig. 8. Stress in the inner side of the main tower (MPa)

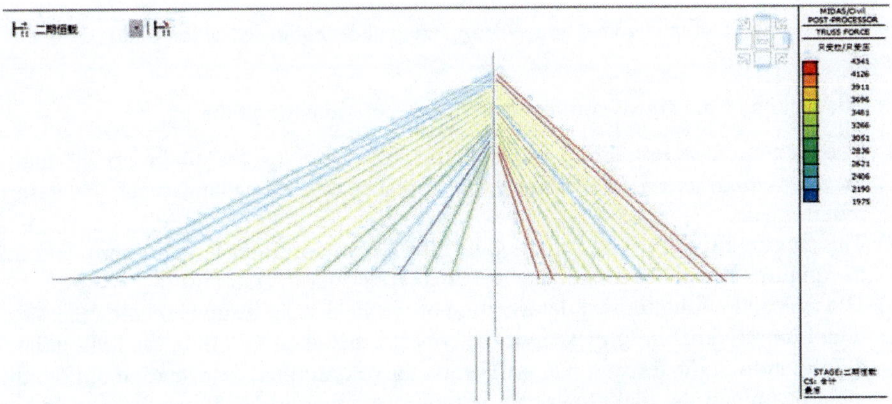

Fig. 9. Permanently acting cable force (kN)

5.2 Calculation of the Combined Effect

The results of the simulation analysis of deformation, stress and cable force under combined action are shown in Figs. 10, 11, 12, 13, 14, 15 and 16.

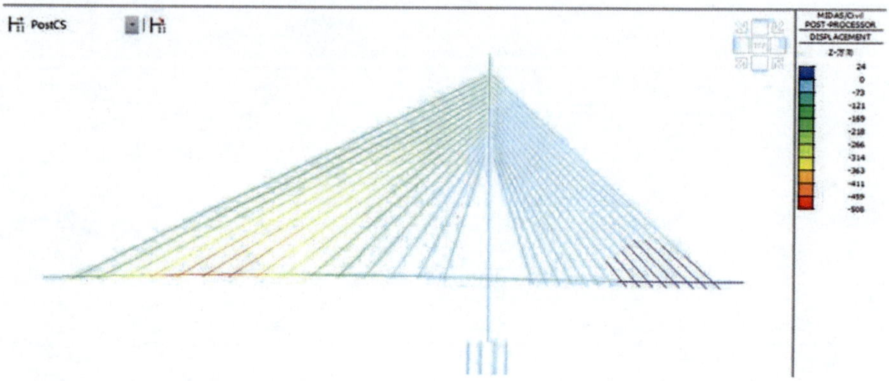

Fig. 10. Vertical deformation under combined action (mm)

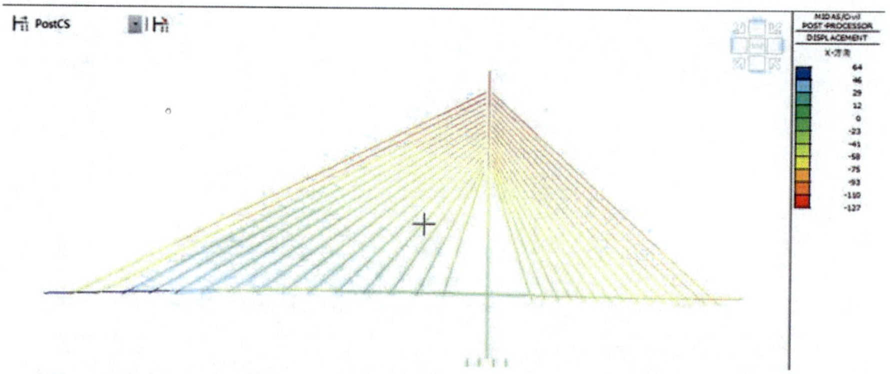

Fig. 11. Longitudinal bridge deformation under combined action (mm)

The results of the above combined action calculations show that:

(1) The structural deformation, stress and cable force of the main bridge under the combined action are within a more reasonable range and meet the design requirements.
(2) The maximum vertical deformation of the steel main girder is −508 mm, and the maximum vertical deformation of the concrete girder is −35 mm.
(3) The maximum longitudinal deformation of the steel main beam is 63 mm, the maximum longitudinal deformation of the concrete beam is −88 mm, the longitudinal deformation of the tower top is −127 mm, the longitudinal deformation of both the main beam and the main tower are within a reasonable range, meeting the design expansion joint reserve length.

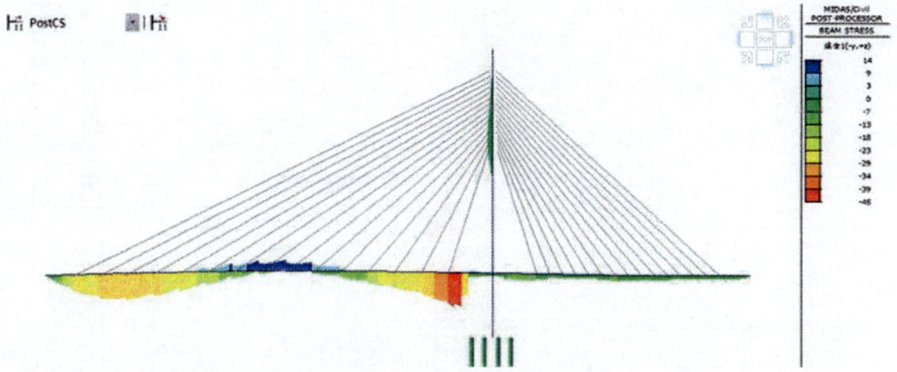

Fig. 12. Stress at the upper edge of the main beam under combined action (MPa)

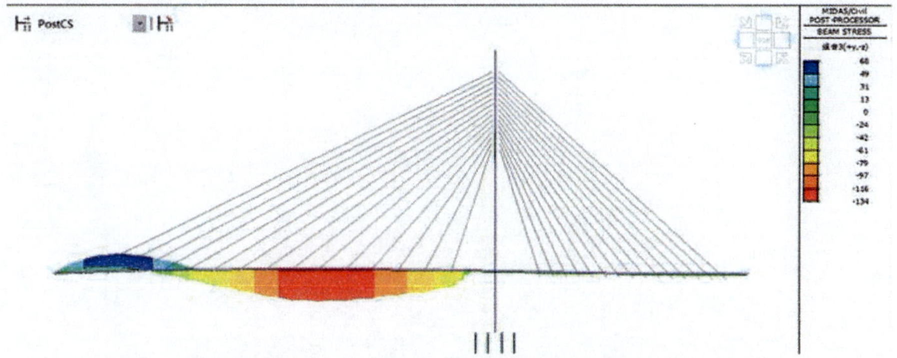

Fig. 13. Stress at lower edge of main beam under combined action (MPa)

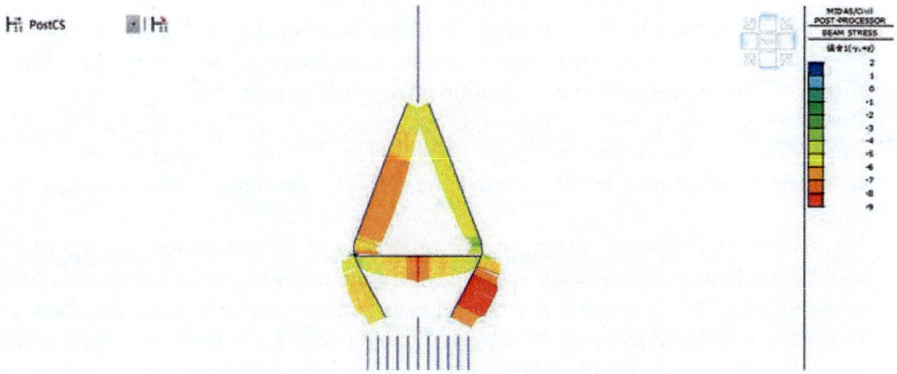

Fig. 14. Stress on outer side of main tower (MPa)

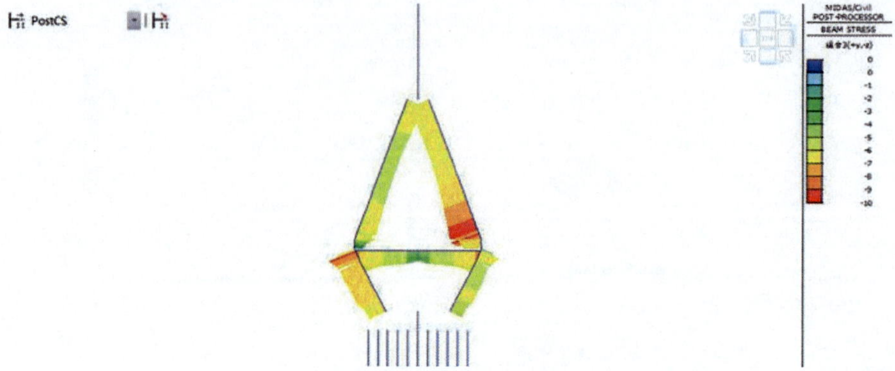

Fig. 15. Stress in the inner side of the main tower (MPa)

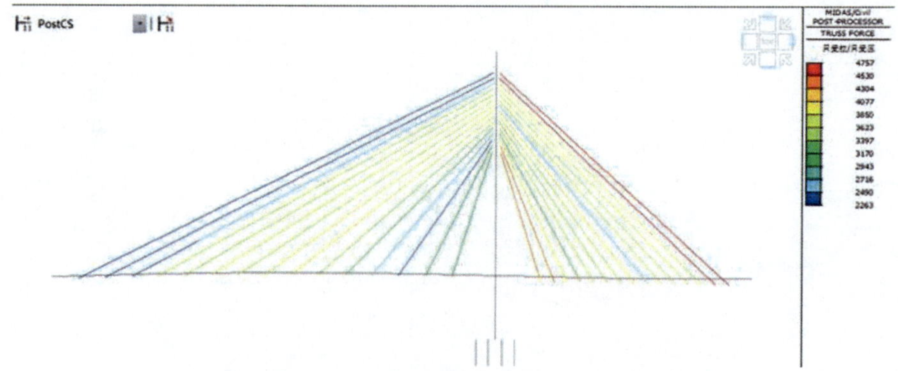

Fig. 16. Cable force under combined action (kN)

(4) The maximum compressive stress at the upper edge of the steel main beam is −44.6 MPa and the maximum compressive stress at the lower edge is −134.2 MPa, the overall stress level is basically within a reasonable range.
(5) The maximum compressive stress at the upper edge of the concrete main beam is −8.9 MPa, and the maximum compressive stress at the lower edge is −8.7 MPa, no tensile stress occurs, and the overall stress level is basically within a reasonable range.
(6) The maximum compressive stress on the outer side of the main tower is −9.8 MPa, and the maximum compressive stress on the inner side is −10.3 MPa, no tensile stress appears, and the overall stress level is basically within a reasonable range.
(7) The maximum cable force of the diagonal cable is 4757 kN, the cable force result is basically within the reasonable range.

6 Conclusion

This paper takes a shaped tower hybrid girder cable-stayed bridge in Foshan as the research object, and calculates the main tower stress, main girder stress, main girder deformation and cable force of the main bridge structure under load through fine finite element simulation analysis, and concludes as follows:

(1) The main tower of the cable-stayed bridge does not show any tensile stress under permanent action and combined action, and the stress state of the main tower is kept within a reasonable range.
(2) The main girder stresses, deformation of the main girder and the cable force of the structure under permanent action are all within a reasonable range, providing effective reference values for the formulation of construction plans.
(3) The main girder stresses, deformation of the main girder and the cable force of the structure under combined action are all within a reasonable range, which is basically in line with the design intention.

References

1. Wu, K.: Construction control and monitoring analysis of shaped single tower cable-stayed bridges. Urban Roads Bridges Flood Control (07), 212–215+22 (2016)
2. Lei, L., Kai, L.: Simulation and monitoring of construction process of shaped tower cable-stayed bridge. Build. Mater. World **39**(05), 70–74 (2018)
3. Yang, J., Li, J., Mei, J., Zhou, R., Zhu, D.: Analysis of mechanical properties of shaped towers of asymmetric single-tower hybrid girder cable-stayed bridges in different states. Guangdong Civil Eng. Constr. **29**(01), 50–52+60 (2022)
4. Fei, G., Li, P., Zhu, Y.: Study on the control technology of shaped steel tower alignment of cable-stayed bridges. Spec. Struct. **37**(04), 102–106 (2020)
5. Zhang, S., Fang, S., Hao, D.: Finite element analysis of the mechanical behavior of steel-hybrid combination in hybrid girder cable-stayed bridges. Eng. Constr. **36**(05), 1398–1403 (2022)
6. Xu, Y.F., Han, D.J., Liang, L.N.: Finite element simulation analysis of the Gaozan Bridge. China Port Harbor Constr. **03**, 9–14 (2008)
7. JTG D60-2015: General specification for the design of highway bridges and culverts

Calculation and Analysis of Embodied Carbon Emissions in Open Cut Foundation Pits

Lianjin Tao, Kaiyue Sun(✉), and Xu Zhao

Civil and Transportation Engineering, Beijing University of Technology, Beijing 100124, China
kaiyuesun0119@163.com

Abstract. In order to achieve the goal of "double carbon" in China, to provide data support for energy saving and emission reduction of foundation pit projects, to solve the problem of complicated and tedious process of quantifying greenhouse gases in foundation pit projects, to improve the engineering quality of foundation pit projects, and to increase carbon emission as one of the evaluation indexes of foundation pit projects, this paper establishes a modular calculation method of embodied carbon emission for foundation pit projects based on the whole life cycle theory. Based on the whole life cycle theory, this paper establishes a modular calculation method for the embodied carbon emission of foundation pit projects, considers the characteristics of "temporary" foundation pit projects, defines that the embodied carbon emission of foundation pit projects includes carbon dioxide generated in the building materials production stage, construction stage and transportation stage, and calculates the carbon dioxide emissions of two open-cut underground stations and one open-cut interval of the western section of Beijing Railway Line 11 (Winter Olympic Branch Line). Carbon dioxide emissions were calculated for three foundation pit projects in the western section of Beijing Railway Line 11 (Winter Olympic Branch), and the embodied carbon emissions of foundation pits were analyzed from three levels: element level, module level and phase level. Considering the influencing factors of excavated earth volume, backfilled earth volume and number of bored piles, the main influencing factors of embodied carbon emissions of foundation pit projects were analyzed.

Keywords: Carbon emission calculation · Whole life cycle · Embodied carbon emissions · Foundation pit engineering · Influencing factors

1 Introduction

Building and construction accounts for 38% of global carbon emissions [1]. As one of the main types of consumption of carbon emissions, buildings need to be green to address the threat of energy shortages and environmental degradation.

As a large developing country, China accounts for half of the growth in global carbon emissions and is the world's top carbon emitter, emitting around 10 billion tones of carbon annually [2]. China is urbanizing rapidly and the total energy consumption of buildings continues to grow. At the 75th session of the United Nations General Assembly, President Xi Jinping pledged to achieve "peak carbon" by 2030 and "carbon neutrality"

W. Guo et al. (Eds.): GBCESC 2023, LNCE 328, pp. 334–348, 2024.
https://doi.org/10.1007/978-981-99-9947-7_36

by 2060. "Achieving the 'double carbon' target will require the concerted efforts of all sectors. The construction sector is one of the three main areas of energy consumption (industry, transport and construction) and is also one of the main areas responsible for direct and indirect carbon emissions, accounting for 35% to 50% of China's total carbon emissions. [3] It has a great potential to reduce emissions.

The carbon emissions from the construction of pits and earthworks account for a large proportion of the total construction period. Pit construction is an important part of the construction process, and the demands of engineering practice have led to advances in pit design and construction technology, while at the same time the carbon emissions from pit construction are considerable. With the implementation of the "double carbon" objective, ecological balance and the concept of green engineering and green development in China, the indicators for evaluating pit construction should not be limited to technical and economic aspects, but the carbon emission should also become one of the indicators for pit evaluation.

2 Literature Review

Currently, the actual measurement method, the material balance method, and the emission factor method (IPCC inventory method) are the three major approaches utilized to account for Carbon Emission (CE) [4]. 2006 IPCC National Greenhouse Gas Guidelines were published by the IPCC [5]. The primary concept is to gather activity data based on the carbon inventory; activity data and carbon emission factor are then combined to produce carbon emission, which is currently the most popular technique.

The Life Cycle Assessment (LCA) method quantifies the entire process of a product, process, or activity, including raw material extraction, production, transportation, use, and disposal of waste, resources consumed, and the potential environmental impact of pollutant emissions. [6] Life Cycle Theory (LCT) is now used in a variety of industries to assess a product's or facility's total environmental impact from "cradle to grave."

The whole life cycle of a building can be divided into the embodied, operational and demolition phases, as shown in Fig. 1. There is no clear definition of Embodied Carbon Emission (ECE), Li Xiaodong [7] et al. define ECE as the life-cycle greenhouse gas emissions that occur during the production and transportation of building materials and components and during the construction process. Teng, Y [8] et al. explain ECE as the CE from the manufacture, transportation and construction of materials. In Rahman Azari's [9] study, ECE are divided into two categories: initial embodied carbon emissions, which are the CE used to extract raw materials, manufacture and transportation, and construct the building, and are all the energy used before the building is occupied, i.e. during the pre-use phase of the building's life cycle. Recurrent ECE, which are the carbon emissions generated by maintaining the building while it is in use, which is embodied in the phase when damaged materials and components are repaired or replaced. It is also suggested that CE from the demolition phase are uncertain and it is not clear whether they should be included in the ECE.

Existing examples of carbon emission calculations are mostly focused on the whole building [10, 11], rail engineering [12–14] and bridge engineering [15], with little research has been done on CE from foundation pit projects. Cai Min [16] et al. compared and analyzed the CE and energy consumption of two forms of support, namely

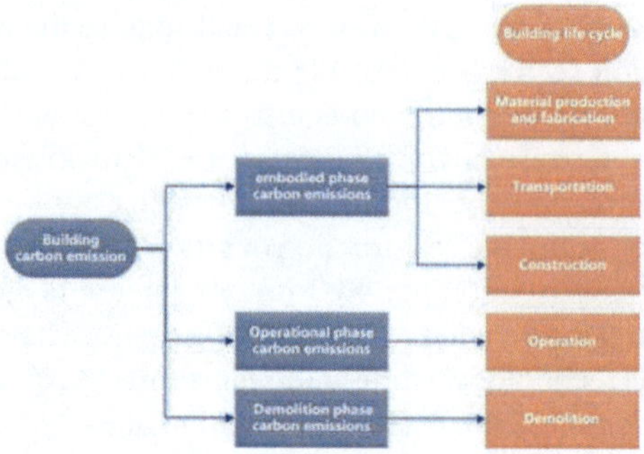

Fig. 1. The whole life cycle of the building

reinforced concrete support and steel structure support, and concluded that the CE of steel structure support is about 14.3% of that of reinforced concrete support, and the energy consumption is about 16.2%. From Yuan Qingyun's [17] research, the results show that the production and transportation of raw materials are the main sources of CE of concrete and steel supports respectively, and the total CE of steel supports are about 20% of those of concrete supports. Dong Yanhui [18] et al. proposed that the integration of permanent structures in underground engineering should be strongly advocated in the context of dual carbon to reduce material waste and the impact on the surrounding environment. Bie Xiaoyong [19] et al. introduced the practice of some environmental protection and energy saving design schemes in the design of foundation pit engineering support in the Wuxi area, and proposed to try to establish an evaluation system with CE as the target function in the design of foundation pit engineering support as one of the criteria for evaluating the scheme's energy saving and environmental protection indicators, which is of great significance.

It can be seen that the existing literature on the environmental impact of pit construction can be divided into two categories. One category is about reducing the environmental impact of pit construction and putting forward suggestions for energy saving and environmental protection, while the other category calculates and analyses the carbon emission and energy consumption of pit construction, but the research object is limited to the supporting structure of pit construction and does not calculate and analyse the whole process of pit construction.

3 Carbon Emission Calculation Model

Based on the LCT, the carbon emission factor method for unit work volume is used to calculate the embodied carbon emissions of open cut pits. Figure 2 illustrates the calculation process.

3.1 Carbon Emission Calculation Boundary

Gas Boundaries. According to the definition of greenhouse gases in the Kyoto Protocol, they consist mainly of CO_2, CH_4, N_2O, HFC_S, PFC_S, SF_6. It contributes up to 60% to

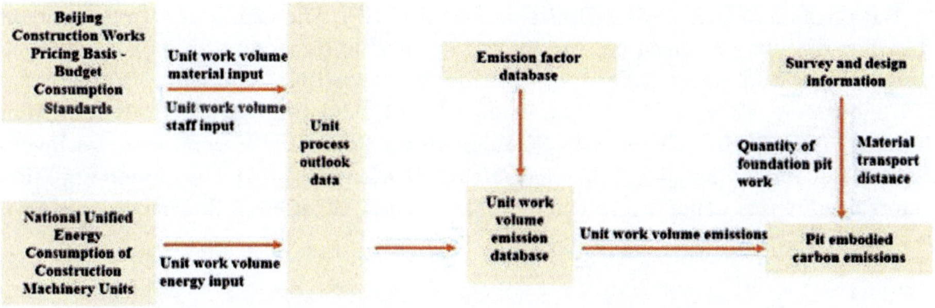

Fig. 2. Modular calculation process of foundation pit construction

global warming. As a result, **carbon** dioxide is considered to be the main greenhouse gas, and emissions of **greenhouse** gases are also referred to as carbon emissions. This calculation of carbon emissions considers six greenhouse gases and is expressed in terms of carbon dioxide equivalents (tCO_2e).

Spatial Boundaries. The full life cycle of a typical construction project consists of the production phase of building materials, the construction phase, the operational use phase and the structural dismantling phase. However, pit construction has a temporary character and does not include the subsequent operation and demolition stages, so the lifecycle of pit construction includes the production, construction and transportation stages, i.e. the embodied stage. Therefore, the whole life cycle carbon emission of foundation pit project is the embodied carbon emission of foundation pit project. Therefore, the spatial boundary of foundation pit construction contains three phases, namely upstream material production, foundation pit site construction and material transportation.

Time Boundaries. The time boundary for the calculation of embodied carbon emissions from the pit is from the start of the project to its completion.

3.2 Inventory Data

Inventory data includes foreground data and background data.

The prospective data includes the amount of materials, energy and labor. The amount of materials and labor per unit of work can be found in the "Beijing Construction Project Pricing Basis - Budget Consumption Standards [20] available in The amount of energy refers to the amount of energy consumed by machinery during construction, which can be obtained from the "Beijing Construction Project Pricing Basis - Budget Consumption Standard". [20] The number of shifts of various types of machinery used in the unit construction volume can be obtained from the Beijing Construction Project Pricing Basis - Budget Consumption Standard, and then through the Standard for Calculating Carbon Emissions in Construction (GB/T51366-2019) [21] and the National Uniform Machinery Shift Cost Quotas [22] to obtain the energy consumption of construction machinery shifts. Energy consumption of the unit project = number of shifts × energy consumption of shifts.

Background data is Carbon Emission Factor (CEF). The CEC refers to the amount of carbon dioxide produced per unit of energy used in the combustion or use of each energy source and material. It is derived from the 2006 IPCC Guidelines for National Greenhouse Gas Inventories [5], the Standard for Calculating Carbon Emissions from Buildings (GB/T51366-2019) [21], "Guidelines for Provincial Greenhouse Gas Inventories (Trial Implementation) [23] and related literature [24–26] The carbon emission factors used in this paper are shown in Table 1. The CEC used in this paper are shown in Table 1.

Table 1. Carbon emission factors

Materials/Energy	Emission factors	Unit
Diesel	3.121	tCO_2e/t
Electricity	0.610	t/MWh
Steel	2.309	tCO_2e/t
Alloy	9.530	tCO_2e/t
Concrete	0.295	$tCO2e/m^3$
Soil	0.00269	tCO_2e/t
Cement	0.702	tCO_2e/t
Wood	0.146	tCO_2e/t
Water	0.000168	tCO_2e/t
Steel swing material	0.262108	tCO_2e/t
Labour	0.00046	tCO_2e/working day

3.3 Calculation Formula

The carbon emissions formula from IPCC [5] proposed is: carbon emissions = activity data × activity factor, see Eq. (1). It is a widely used formula for calculating carbon emissions in the some guidelines [23], standards [23] and extant literatures. Therefore, it is chosen as the basic formula for ECE calculations in this paper.

$$GHG = AD \times EF \tag{1}$$

where:

GHG–Greenhouse Gas Emissions.

AD–The amount of activity that leads to the production or consumption of greenhouse gas emissions.

EF–Greenhouse gas emission factor corresponding to activity level data.

Embodied Carbon Emissions from the Whole Pit. According to the construction process of the foundation pit project, it is divided into different modules, such as earthwork, concrete and reinforced concrete works, **foundation** treatment and slope support works, demolition works and measure projects, etc. The carbon emission of the foundation pit project as a whole is the sum of carbon emission of each module. Calculated according to Eq. (2).

$$C = \sum\nolimits_{k}^{n} C_m \tag{2}$$

where:

C—ECE from the whole pit
C_m—ECE from a module

Embodied Carbon Emissions at the Module Level. According to the construction steps of each module, each module will continue to be subdivided into different elements, the element is the volume of each unit, and the number of each element will be counted, and the carbon emission of each module will be the sum of the product of its element and the number of elements. For example, the earthwork can be divided by construction steps into element E1 1 m^3 open cut earth excavation, element E2 1 m^3 fill. Calculated according to formula (3).

$$C_m = \sum\nolimits_{k}^{n} C_e \times a_k \tag{3}$$

where

C_m–ECE from a module
C_e–ECE of the kth element
a_k–Number of kth elements.

Embodied Carbon Emissions at the Element Level. The scope of calculation of carbon emissions of the element shall include carbon emissions from the production phase of building materials, the construction phase and the transportation phase of building materials, calculated according to Eq. (4).

$$C_e = C_{ep} + C_{ec} + C_{et} \tag{4}$$

where

C_e–ECE of elements.
C_{ep}–ECE from the production phase of the element building material.
C_{ec}–ECE during the construction phase of the element.
C_{et}–ECE from the transportation phase of the element building materials.

Building Materials Production Phase. Carbon emissions from the production phase of building materials are the carbon emissions generated during the production of building materials at the site of production. The carbon emissions from the production phase of building materials are calculated according to (5).

$$C_{ep} = \sum\nolimits_{i} m_i \times EF_i \tag{5}$$

where

C_{ep}–ECE from the production phase of the element building material.
m_i–the amount of input of material of category i.
EF_i–CEF for material category i.

Construction Phase. Carbon emissions during the construction phase include carbon emissions generated by machinery and labor during the construction process. The carbon emissions from machinery come from the energy consumed by machinery in each process during the construction of the pit, mainly diesel, petrol and electricity, and are expressed as the product of the number of construction machinery shifts and energy consumption per unit shift in each base element. Because of the 'temporary' character of pit construction, i.e. the life cycle of a pit project begins with the production of building materials and ends with the completion of the pit project, and does not involve an operational phase, the demolition of the pit project is also included in the construction phase, such as the removal of steel supports and concrete supports. The carbon emissions generated by labor come from the carbon emissions generated by workers during production work, regardless of the type of work and skill level, and are expressed in terms of combined man-days, each calculated on the basis of an 8-h working day. Carbon emissions from the construction phase should be calculated according to Eqs. (6) and (7).

$$E_{ec} = \sum_{j} T_j \times R_j + \sum L \tag{6}$$

$$C_{ec} = \sum_{j} T_{uj} \times R_j \times EF_j + \sum L \times EF_l \tag{7}$$

where

E_{ec}—Energy and labour inputs during the construction phase of the construction of the element.

C_{ec}—ECE from labour and energy during the construction phase of the element.

T_j—The number of shifts of construction machinery in category j.

R_j—Energy consumption per unit shift of construction machinery in category j.

EF_j—CEF for construction energy use in category j.

L—The input of personnel of various trades during the construction phase of the construction of the element.

EF_l—CEF for a combined manual workday.

Building Materials Transportation Phase. Carbon emissions from the transport phase of building materials come from the direct carbon emissions of the transport process of building materials from the place of production to the construction site and from the carbon emissions of the production process of the energy consumed during transport. It is appropriate to use the actual transport distance of building materials. When the actual transport distance of the building material is unknown or not available, the default transport distance for concrete is 40 km and for other building materials is 500 km. [21] The default transport distance for other building materials is 500 km. Carbon emissions from the transportation of building materials are calculated according to Eq. (8).

$$C_{et} = \sum_{i} m_i \times D_i \times T_i \tag{8}$$

where

C_{et}–ECE from the transportation phase of the element building materials.

m_i–The amount of input of material of category i.

D_i–The average transport distance of the ith building material.

T_i–CEF per unit weight of transport distance for the ith mode of transport of building materials.

4 Case Studies

4.1 Project Overview

The civil construction section 02 of Beijing Railway Line 11 West (Winter Olympic Branch) Project consists of two stations and two intervals, including Jinanqiao Station, Jinanqiao Station ~ Beixinan Road Station interval, Beixinan Road Station, Beixinan Road Station ~ Shougang Station interval, Fig. 3 is the schematic diagram of Beijing Railway Line 11 Section 02 Contract.

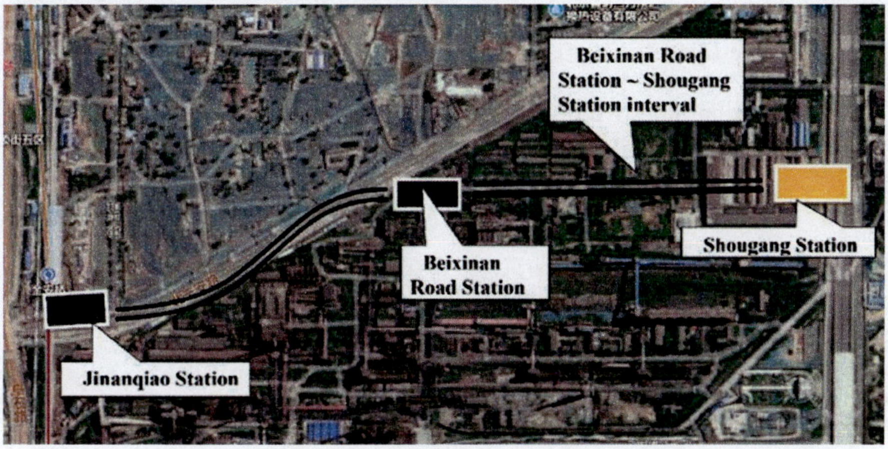

Fig. 3. Schematic diagram of the 02 contract section of Beijing Rail Transit Line 11

Jinanqiao Station. This station is an underground three-storey island-type double column three-span station with an effective platform width of 15 m. The total length of the station is 192.000 m, the depth of the bottom slab is 27.62 m–29.14 m, the thickness of the top slab overburden is 3.91 m–5.76 m, the width of the standard section is 24.30 m, the width of the shield **shaft** at the north end is 28.2 m, the width of the shield shaft at the south end is 30.6 m. Except for the B entrance which contains a concealed excavation section, the station and The foundation pit is constructed by $\varphi 1000@1300$ bored piles + 5 steel support system.

Jinanqiao Station ~ Beixinan Road Station Interval. Jinanqiao Station ~ Beixinan Road Station, 639.216 m long, is a shield method interval.

Beixinan Road Station. This station is an underground two-storey island type double column three span station, the effective platform width is 16 m, the total length of the station is 180.8 m, the width of the standard section is 25.3 m, the overburden thickness is 3 m, the depth of the bottom slab is 17.89 m, the width of the end of the shield shaft is 29.74 m, the width of the end of the open excavation section is 29.2 m, except for the concealed excavation method for some entrances and exits through Beixinan Road, the station and the rest of the accessories are constructed by the open excavation method.

The main pit enclosure structure adopts the slope release above the negative first floor roof structure + φ800@1400 + 2 steel support support system.

Beixinan Road Station ~ Shougang Station Interval. Beixinan Road Station ~ Shougang Station interval, the total length of 393.346 m, the interval tunnel open cut three-storey rectangular frame structure (100 m at the mileage end of Beixinan Road is a two-storey rectangular frame structure), surrounded by piles ~ internal support form. The width of the foundation pit is 29.35 m depth 15.8 m–25.8 m. After this interval is connected from the mileage end of Beixinan Road, the vertical slope is 18‰ slope down, and the plane is laid from south to north along the current West Road of the repair factory, the line spacing is 18.2 m, the overburden thickness of the interval is 12.26 m–25.58 m, the foundation pit adopts φ1000@1400 bored pile + 5 steel support system (the first 100 m is double-layer structure). The first 100 m is a double-layered structure with 3 steel supports).

The calculation objective of this paper is the open cut foundation pit, Jinanqiao Station to Beixinan Road Station is a shield interval, which is not included in the carbon emission calculation. Therefore, the calculation objects of this paper are the open cut foundation pit works of Jinanqiao Station, Beixinan Road Station and Beixinan Road.

4.2 Carbon Emission Calculation Results

According to the CE calculation method for unit work volume in the CE calculation model in Sect. 2, the CE calculation was carried out for three open cut pits at Jinanqiao Station, Beixinan Road Station and Beixinan Road Station ~ Shougang Station interval, and the calculation results are shown in Tables 6–12.

Element Level. The element ECE for this accounting are shown in Table 2.

Table 2. Embodied carbon emissions of elements

Element number	Name of primitive	Carbon emissions (tCO_2 e)
E1	1 m^3 Open cut earthwork excavation	0.006
E2	1 m^3 Filling	0.008
E3	1 m^3 Rotary Drilling Rig Drilling (1000 mm)	0.032
E4	1 m^3 Pile concrete (rotary drilling)	0.353
E5	1 m^3 Rotary Drilling Rig Drilling (800 mm)	0.090
E6	1t steel support pipe fabrication	0.484
E7	1t steel support removal (open cut)	0.279

ECE from the production phase of concrete and steel are calculated in the E4 and E7 elements, hence their larger values.

Modular Level. The modular ECE for this accounting are shown in Table 3.

Whole Pit. The whole pit ECE for this accounting are shown in Table 4.

Table 3. Embodied carbon emissions of modulars

Modular number	Name of modular	Carbon emissions (tCO_2e)
1	Jinanqiao Station Earthworks	1628.16
2	Jinanqiao Station Containment Works	5931.77
3	Earthworks at Beixinan Road Station	2198.93
4	Beixinan Road Station Containment Works	2312.3
5	Beixinan Road Station ~ Shougang Station Earthworks	1954.94
6	Beixinan Road Station ~ Shougang Station Interval Containment Project	7245.38

Table 4. Whole pit embodied carbon emissions

Projects	Carbon emissions from earthworks	Carbon emissions from enclosure works	Total carbon emissions
Jinanqiao Station	1628.16	5931.77	7433.01
Beixinan Road Station	2198.93	2312.29	4511.22
Beixinan Road Station ~ Shougang Station	1954.94	7245.38	9200.32

4.3 Analysis of Calculation Results

Figure 4 shows the energy flow diagram of the emission sources and tools at Jinanqiao Station. It can be clearly seen that the main source of ECE at Jinanqiao Station is concrete, with all the greenhouse gases produced by concrete going to the slurry wall forming and piling process. The second source of emissions comes from diesel fuel, which is consumed by a large number of machines on the construction site.

The three open cut pits are analyzed using a calculating method focused on unit quantities on three levels: elements, modular, and phase level. Seven elements, numbered E1–E7, make up the element level. The earthwork and enclosure work are included in the modulars.The manufacturing of building supplies, the building stage, and the transportation stage are all included in the phase level. Figure 5 displays the analysis findings.

The element E4 1 m^3 pile concrete (rotary drilling) at Jinanqiao Station produced the most embodied carbon emissions, emitting 4820,5 tCO2e, or 63.8% of the station's overall carbon emissions. The enclosure works produced 5931.8 tCO_2e at the modular level, or 78.5% of the total emissions. The manufacturing of building materials was the phase with the highest emissions, 5436.3 tCO_2e, accounting for 71.9%.

Beixinan Road Station's element E4 1 m^3 pile concrete (rotary drilling) had the highest (ECE), emitting 1705.8 tCO_2e, or 37.8% of all emissions, while element E1 1 m^3 open cut earth excavation had the second-highest emissions, emitting 1384.9 tCO2e, or

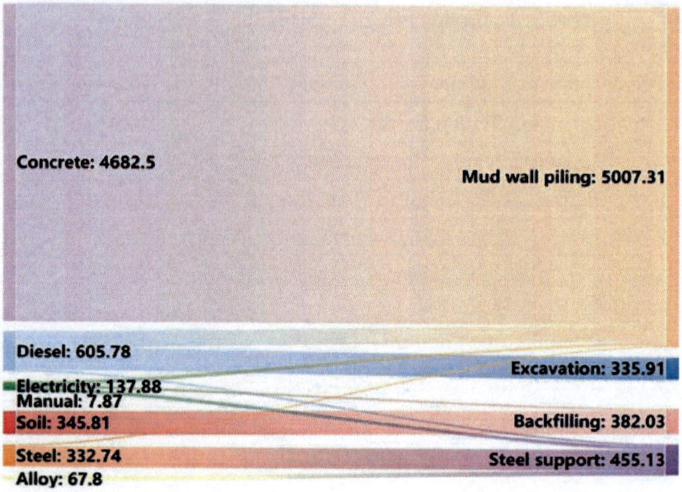

Fig. 4. Energy flow between emission sources and processes at Jinanqiao Station (unit: tCO_2e)

30.7% of all emissions. At the modular level, the containment works and the earthwork both produce similar amounts of carbon dioxide, with the confinement works emitting slightly more (2312.3 tCO2e, or 51.3%), and the earthwork emitting 2198.9 tCO2e, accounting for 48.7%. Building materials manufacturing emits 2547.7 tCO2e at the phase level, or 56.5% of the total ECE.

The element E4 1 m^3 pile concrete (rotary drilling) has the greatest (ECE) at the element level, emitting 5947.7 tCO2e, or 63.5% of the total carbon emission. At the modular level, enclosure engineering emits 7416.8 tCO2e, or 79.1% of all emissions, whereas the building materials production stage emits 6748.5 tCO2e, or 72.0% of all emissions, at the phase level.

The three open cut foundation pit projects that emit the most at the level of element is E4, according to the calculation results for those three projects. The concrete used in the slurry retaining wall piles is the main source of embodied carbon emissions from the foundation pit. The ECE from the earthwork and perimeter protection works at Beixinan Road Station are comparable, but the modular level carbon emissions from the perimeter protection works at Jinanqiao Station and Beixinan Road Station to Shougang Station are significantly higher than those from the earthwork.

From the calculation results of the three open cut foundation pits, it can be seen that the three open cut foundation pit projects emit the most at the level of the foundation element is the E4 foundation element, and the concrete used in the slurry retaining wall piles is the main source of embodied carbon emissions from the foundation pit. At the modular level, the carbon emissions from the perimeter protection works at Jinanqiao Station and Beixinan Road Station to Shougang Station are much greater than those from the earthwork, and the ECE from the perimeter protection works and earthwork at Beixinan Road Station are comparable. The manufacture of building materials accounts for the majority of the ECE of the three pits at the phase calculation stage; the ECE of the construction and transportation stages are comparable.

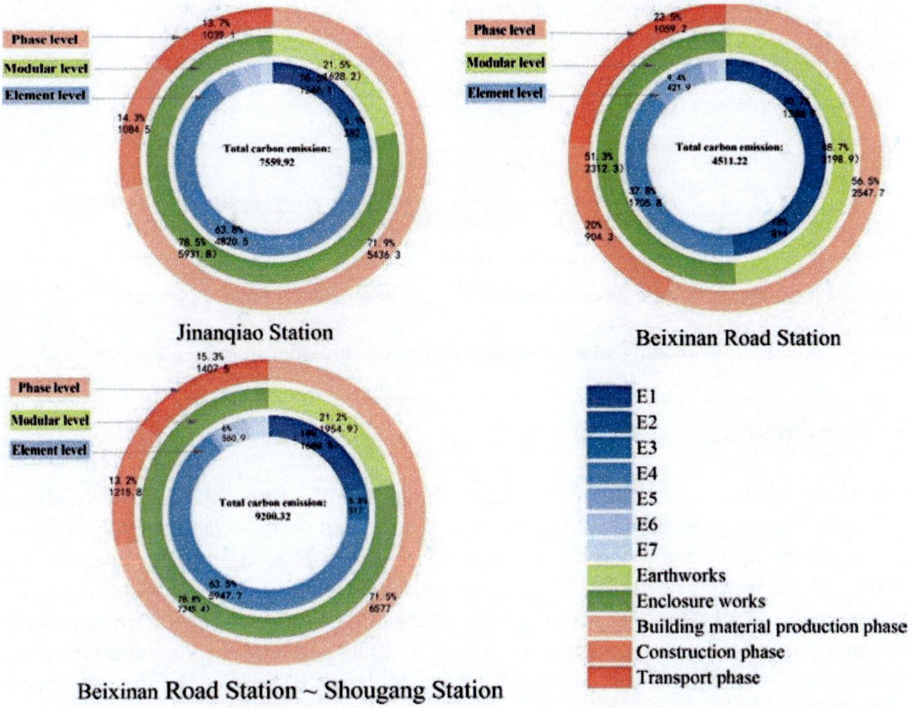

Fig. 5. Three-level embodied carbon emission percentage of three pits (unit: tCO_2e)

The relationship between the amount of excavated earth and backfill and the number of bored piles and the embodied carbon emissions from the three open cut pits was analyzed, and the influence of the three parameters of the amount of excavated earth, backfill and bored piles on the embodied carbon emissions from the pits was analyzed. As shown in Fig. 6, the excavated earth and backfilled earth indicate that the main influencing factor for the embodied carbon emissions from the pit is the form of support for the pit.

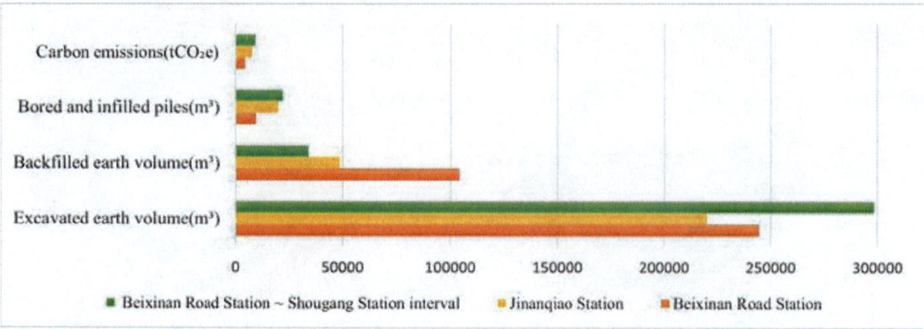

Fig. 6. Influencing factors of embodied carbon emissions in foundation pits

5 Conclusion

This paper establishes a modular calculation method for the embodied carbon emissions of foundation pit projects based on the LCA theory, and defines that the ECE of foundation pit projects include the carbon emissions of the production phase, construction phase and transportation phase. The carbon emissions from Jinanqiao Station were calculated to be 7559.92 tCO_2e, from Beixinan Road Station 4511.22 tCO_2e, and from Beixinan Road Station to Shougang Station 9200.32 tCO_2e. The carbon emissions from the elements in this paper can be used for the calculation of the ECE of the foundation pit projects of similar project profiles in Beijing.

The three open cut foundation pits' ECE composition was examined at three different levels, including the element, engineering module, and calculation phase stage, according to the results. In other words, pouring concrete for the slurry retaining wall bored piles emits the majority of the carbon dioxide, and the concrete used in the concrete slurry retaining wall piles is the main source of ECE from the foundation pit. E4 at Beixinan Road Station to Shougang Station emitted 5947.7 tCO_2e, accounting for 63.5% of the total carbon emissions. When compared to the earthwork at the modular level, the carbon emissions from the perimeter work at Jinanqiao Station and Beixinan Road Station to Shougang Station are significantly higher. At the modular level, the perimeter works at Jinanqiao Station and Beixinan Road Station to Shougang Station emit 5931.8 tCO2e and 7416.8 tCO_2e, respectively, accounting for 78.5% and 79.1% of the emissions. These perimeter works also emit 7416.8 tCO_2e, which is significantly more than the emissions from the earthwork. At Beixinan Road Station, the ECE from the earthwork and enclosure work were equivalent, with the enclosure work producing more carbon emissions. 51.3% of the total carbon emissions, or 2312.3 tCO_2e, were produced by the enclosure works. More than 51.3% of the ECE from the foundation pits came from the enclosing works. At the phase level, the three pits' embodied carbon emissions primarily came from the production of building materials, with the Jinanqiao Station building materials production stage emitting 5436.3 tCO_2e, accounting for 71.9%; the Beixinan Road station building materials production stage emitting 2547.7 tCO_2e, accounting for 56.5%; and the Beixinan Road station to Shougang station building materials production stage emitting 6748.5 tCO_2e More than 56.5% of the carbon emissions from the entire foundation pit operation came from the manufacturing of building materials. Construction and transportation stages' ECEs are similar.

The trend of ECE from the pit and the number of bored piles are similar when the influencing factors of excavation volume, backfill volume, and the number of bored piles are taken into account, indicating that the main driver of ECE from the pit is the form of support of the pit.

References

1. United Nations Environment Programme: Global Status Report for Buildings and Construction: Towards a Zero-emission, Efficient and Resilient Buildings and Construction Sector. Global Alliance of Buildings and Construction, Nairob (2021)
2. China CO2 emission accounts 2016–2017 | Scientific Data. https://www.nature.com/articles/s41597-020-0393-y. Accessed 18 Oct 2022
3. Tsinghua University Building Energy Efficiency Research Center: China Building Energy Efficiency Annual Development Research Report 2021. China Construction Industry Press (2021)
4. Li, X.D., Zhu, C.: Summary of research on account of carbon emission in building industry and analysis of its influential factors. J. Saf. Environ. **20**, 317–327 (2022)
5. Intergovernmental Panel on Climate Change (IPCC): IPCC Guidelines for National Greenhouse Gas Inventories. IGES, Japan (2006)
6. International Organization for Standardization (ISO): ISO 14040 Environmental Management Life Cycle Assessment General Principles and Framework. ISO, Geneva (2006)
7. Li, X., Yang, F., Zhu, Y., Gao, Y.: An assessment framework for analyzing the embodied carbon impacts of residential buildings in China. Energy Build. **85**, 400–409 (2014)
8. Teng, Y., Pan, W.: Estimating and minimizing embodied carbon of prefabricated high-rise residential buildings considering parameter, scenario and model uncertainties. Build. Environ. **180**, 106951 (2020)
9. Azari, R., Abbasabadi, N.: Embodied energy of buildings: a review of data, methods, challenges, and research trends. Energy Build. **168**, 225–235 (2018)
10. Liu, Y., Liu, N., Xu, P.: Carbon emission prediction model during the material production stage for cold zone residential buildings. J. Tsinghua Univ. (Sci. Technol.) **63**(1) (2022)
11. Wang, J.: Calculation and analysis of life-cycle CO2 emissions of Chinese urban residential communities. Tsinghua University (2009)
12. Environmental impact analysis for the construction of subway stations: Comparison between open-excavation and underground-excavation scheme. ScienceDirect. https://www.sciencedirect.com/science/article/pii/S0195925521000949. Accessed 07 Nov 2022
13. Evaluation of mitigation potential of GHG emissions from the construction of prefabricated subway station – ScienceDirect. https://www.sciencedirect.com/science/artcle/pii/S0959652619325508?via%3Dihub. Accessed 07 Nov 2022
14. Chen, K.Y., Zhou, D., Su, Y.H., et al.: Research on carbon emission intensity and reduction potential of urban rail transit life circle. Railway Standard Design (005), 066 (2022)
15. Li, X.J.: Research on life-cycle carbon emissions calculation of long-span suspension bridges in mountainous areas. Southwest Jiaotong University (2019)
16. Cai, M., Zhu, X.J., Shan, C.C., Sun, C.X., Zheng, Y.L.: Calculation of carbon emission and energy consumption of assembled recyclable pit support structures. Building Energy Efficiency Anhui Building (2022)
17. Yuan, Q.Y., Sun, W.: Calculation and analysis of carbon emissions in whole life cycle of concrete bracing support and the H-shaped steel bracing system for foundation pit. Chongqing Archit., 33–37 (2022)

18. Dong, Y.H., You, X.H.: Technical practice of permanent and temporary structural integration in underground engineering under the carbon peaking and carbon neutrality background. J. Eng. Geol., 65–269 (2022)
19. Bie, X.Y., Wu, L., Chen, J.D.: Practice of energy-saving and surrounding protection technology during design of foundation excavation support in Wuxi region. Constr. Technol. **0**(4) (2016)
20. Beijing Municipal Housing and Urban Construction Commission: Beijing construction project pricing basis - budget consumption standard, Beijing (2021)
21. Building carbon emission calculation standard: GB/T51366-2019. China Industry Press, Beijing (2019)
22. National unified machinery shift cost quota: JTG/T 3822-2018. People's Communications Press, Beijing (2018)
23. Department of Climate Change, National Development and Reform Commission. Guidelines for the preparation of provincial greenhouse gas inventories (for trial implementation) (2011)
24. Wei, J.X., Geng, Y.B., Wang, S.: Analysis of influencing factors and uncertainty calculation of cement carbon emission measurement in China. J. Environ. Sci. **36**(11), 4234 (2021)
25. Zhang, Z.H., Liu, M.H., Chen, W.: Analysis on the uncertainty of carbon emissions during the bridge construction. Transp. Sci. Eng. **35**(4), 57 (2019)
26. Huang, N., Wang, H.T., Fan, C.D., et al.: LCA data quality assessment and control method based on uncertainty and sensitivity analysis. J. Environ. Sci. **32**(6), 1529 (2012)

A Design Method Based on 3D Printing for the Integration of Human Computer Dynamic Interaction and Digital Sculpture

Zhen Zheng(✉)

Minjiang University, Fuzhou 350108, China
Shuangyue628@163.com

Abstract. The continuous popularization and application of computer 3D modeling and 3D printing technology in the field of art design has improved and changed the traditional design and production mode to a certain extent. In order to solve the shortcomings of the existing research on human-machine dynamic interaction of 3D printing and digital sculpture, this paper discusses A3D digitalization, the mainstream technology of 3D printing, human-machine dynamic interaction and traditional sculpture making, and briefly discusses the image preprocessing and human-machine interaction and digital design software of 3D printing based on human-machine interaction. In addition, it designs the technology of human-computer interaction information extraction, visual interaction key frame selection, equipment operation and 3D printing body structure digital molding technology. Finally through the concrete experiment analysis. The results show that the average inter-frame distance error of the key frame sequence extracted based on the visual interaction method is the smallest. The calculated frame spacing error is as low as 0.029. The minimum error of manual extraction was 0.557. The minimum X-means clustering error is 0.047. Therefore, it is verified that the integration method of human-machine dynamic interaction and digital sculpture based on 3D printing designed in this paper has high practical value.

Keywords: 3D Printing · Human-computer Interaction · Digital Sculpture · Integrated Design

1 Introduction

3D printing technology, as a new thing that all industries are eager to care about, is also developing at an unprecedented speed. With the advent of 3D printing technology, 3D modeling in virtual space has entered the real world and is displayed in the form of physical materials.

Nowadays, more and more scholars have done a lot of research on human-computer dynamic interaction and digital sculpture of 3D printing through various technologies and system tools, and have also achieved certain research results through practical research. Vavrik D took the tomb murals as the research object, and realized the development of the costume pattern of the Tang Dynasty tomb murals and the three-dimensional simulation

W. Guo et al. (Eds.): GBCESC 2023, LNCE 328, pp. 349–356, 2024.
https://doi.org/10.1007/978-981-99-9947-7_37

repair of the costume through the 3D interactive costume pattern making technology and virtual simulation technology. First, build a 3D clothing model in the virtual environment. Draw the structural curve of the garment on the 3D garment, expand the 3D surface and enclose it in the 2D garment plane by these curves, and adjust the expanded surface to obtain the 2D garment plane pattern. By using 3D virtual simulation technology, the patterns of Tang Dynasty tomb mural costumes were stitched, and the virtual simulation repair of Tang Dynasty tomb mural costumes was realized [1]. In order to improve the quality of sculpture and overcome the shortcomings of existing schemes, Belikovetsky proposed a data optimization method for portrait sculpture by combining mobile edge computing and 3D images. This paper analyzes the portrait data acquisition technology of 3D scanning and image reconstruction, and points out the blind spots in the application of this technology. A data acquisition method of human sculpture based on feature description is proposed. After determining the data optimization method, the data optimization architecture of portrait sculpture is constructed through mobile edge computing technology [2]. Srinivasa P provides a user-friendly interface for people pursuing new artistic experience, realizes the parametric design method often used in architecture, and uses it as a tool to explore the attributes and representation of 3D modeling in costume sculpture design. The data of biomorphic costume sculpture based on knowledge is proposed, which can predict the results of generative design. Through case studies, the intermediary variables and attributes of the parametric process related to the modeling of biomorphic costume sculpture are determined. The knowledge-based biological costume sculpture data and biological costume sculpture modeling intermediary variables discovered during the research are applied to the 3D modeling process as visual data [3]. Although the existing research on human-computer dynamic interaction and digital sculpture of 3D printing is very rich, there are still many problems in its real practical application.

This paper mainly defines and explains the concepts related to 3D digitization, 3D printing mainstream technology, human-computer dynamic interaction and traditional sculpture making, and expounds the image preprocessing and 3D printing digital design software based on human-computer interaction. On the basis of clarifying that the integration modeling of human-computer dynamic interaction and digital sculpture includes 3D digital design and human-computer dynamic interaction design, the paper analyzes the 3D structure information extraction technology of human-computer dynamic interaction sculpture and the selected key frames of visual interaction, and analyzes in detail the equipment operation and 3D printer body structure digital molding technology included in the 3D printing digital sculpture molding technology.

2 Integrated Design Method of Human-Computer Dynamic Interaction and Digital Sculpture Based on 3D Printing

2.1 3D Digitization

The processing and forming methods of materials in the real world can be roughly summarized as: additive forming, subtractive forming, compression forming, growth forming. The corresponding specific technical means are divided into: 3D printing, numerical control sculpture and digital progressive molding. These three technical means all belong to the concept of three-dimensional digital forming, and their essence and purpose are to use computer numerical control technology to show virtual three-dimensional modeling information through real object materials [4].

3D Printing. The main technologies commonly used in 3D printing are as follows.

(1) Laser sintering technology. First, the 3D data of the object is sliced into a whole set of slices, and each slice determines the height of the cross section of the part. Then the laser sintering equipment is used to accumulate the slices layer by layer to obtain the desired object [5].
(2) UV curing molding technology. The technological process is as follows: First, the 3D model is designed by the graphic design software, and then the specific data is obtained by slicing. Secondly, the laser beam emitted by the machine is used to solidify the resin on the path according to the planned path in advance, followed by the second layer of curing, and so on. Finally, take the mold out of the resin for final curing and finished product treatment [6].
(3) Melt deposition molding technology. The basic principle is that during the forming process, the filament consumables are melted by heating the nozzle. The extruder at the top of the nozzle and other axial stepping motors move together to form the melted filament layer by layer [7].

2.2 Man Machine Dynamic Interaction

Human computer interaction, also known as the interaction between people and computers, is a behavior driven by the realization of a specific task and using a specific method to achieve data interaction with computers [8]. The process of human-computer interaction is usually realized with the help of corresponding programs or external IO systems. People are the end users of computer programs, while computers are the main body of software. The interaction between the two requires people to tell the computer what to accomplish and feed back the results of tasks to the computer.

2.3 Traditional Sculpture Production

(1) Scheme making: designers need to express their creative intention to partners through scheme design, and determine the style, shape, texture, materials, layout, etc. of works according to the scheme information [9].
(2) The internal skeleton can determine the dynamics of the sculpture and support the internal structure and surface mud of the sculpture.
(3) The enlargement of clay sculpture is to make the work according to the actual size after the original small draft is finalized.
(4) In order to further shape, we need to conduct in-depth portrayal, starting from the portrayal of local details, and then to the whole, observe and modify, adjust the relationship between various parts, until the final shape.
(5) Sculpture turnover. After the clay sculpture is completed, it needs to be preserved for a long time, which requires turning the work into harder materials.
(6) After the reproduction is completed, the surface shall be manually treated: polished, repaired, polished, etc. Finally, the art creators shall color the works according to the design requirements [10].

3 Investigation and Research on the Integration Design Method of Human-Computer Dynamic Interaction and Digital Sculpture Based on 3D Printing

3.1 Image Preprocessing Based on Human-Computer Interaction

(1) Firstly, several consecutive frames are extracted and recorded as $k_{tsa}(u, v)$. For each frame $k_{tsa}(u, v)$, we can convert RGB image to grayscale image $k_{tsa}(u, v)$, which can be directly processed through Opencv.
(2) Add the gray value $k_s(u, v)$ of the gray image of the picture, and take the average value of its gray image. The gray image $k_a(u, v)$ of the image background image is:

$$k_a(u, v) = \frac{1}{G}\sum_{x=1}^{G} k_s(u, v) \tag{1}$$

(3) By making a difference between each gray image $k_s(u, v)$ and the background gray image $k_a(u, v)$ of the scene, we can find the area $k_c(u, v)$ in the current frame that is different from the background, namely:

$$k_c(u, v) = |k_s(u, v) - k_a(u, v)| \tag{2}$$

(4) Compare $k_c(u, v)$ with the predetermined threshold value, and then convert it into a binary chart, as shown in Formula 3.

$$k_e(u, v) = \begin{cases} 1.\mathit{if} k_c(u, v) > \eta \\ 0.\mathit{if} k_c(u, v) < \eta \end{cases} \tag{3}$$

wherein, $k_c(u, v)$ represents the binary graph of the extracted figure contour, and η represents the threshold.

3.2 3D Printing Human-Computer Interaction and Digital Design Software

(1) 3dsMax: It is a digital solution for advanced 3D modeling, rendering and animation developed by Discreet. The latest version is 3dsMax2017. This software has a high cost performance ratio, is very simple and easy to learn, is very convenient, and has powerful rendering functions [11].
(2) OpenGL: It is an open 3D graphics software package, which can provide a programming interface to access the characteristics of graphics hardware devices, and support the call of multiple programming languages. With the powerful basic class function library of OpenGL, we can draw and generate realistic 3D scenes. In addition, OpenGL provides a 3D graphics accelerator card. Therefore, we can develop 3D graphics image software on Windows system with OpenGL.

4 Research on the Application of the Integration Design Method of Human-Computer Dynamic Interaction and Digital Sculpture Based on 3D Printing

4.1 Integrated Design Method of Human-Computer Dynamic Interaction and Digital Sculpture Based on 3D Printing

Design of Human-Computer Interaction Method Based on Monocular Camera
This paper presents a method of human-computer interaction based on monocular camera. Firstly, the 3D structure information of sculpture is extracted from the video stream captured by the monocular camera. Through the analysis of the sculpture structure, the interaction with the virtual environment is completed. The specific process is shown in Fig. 1.

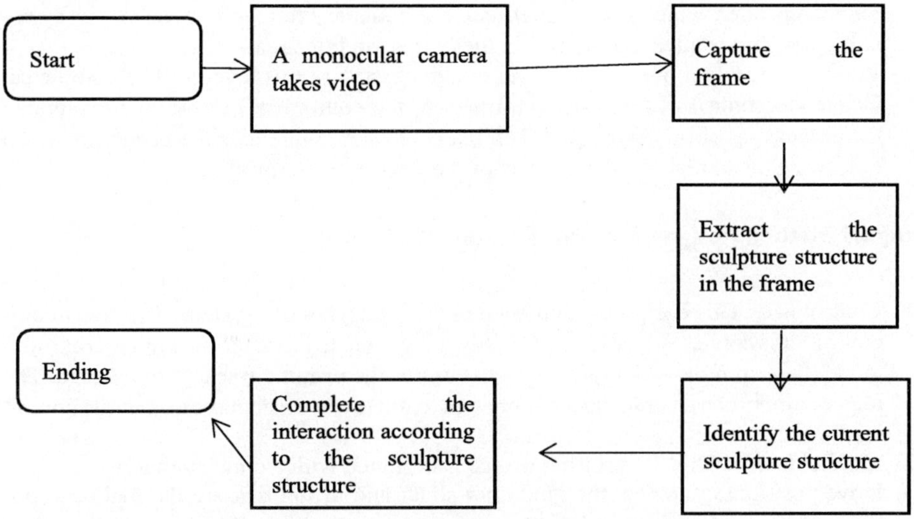

Fig. 1. General process of human-computer interaction

Visual Interactive Selected Keyframes
In this paper, after obtaining candidate key pose frames through simple error threshold adjustment, the candidate key pose frame sequence is intuitively displayed to users in the form of graphic images using visual display technology.

1) Humanized interaction design

In this paper, we first design the window sliding button, and design an interactive mode for a single structural sculpture figure to selectively delete the redundant frames. After the redundant frames are deleted, the subsequent structural frames can automatically fill the deleted vacancy and form a new key frame sequence.

2) Generation of attitude standard library
   The structure semantics are named by users in the form of strings according to the structure content and application scenarios. After entering the self named structure name through the 3D-HMVICS system interface, the system background will automatically generate the structure semantic sequence and store it in the standard action database.

This paper focuses on the operation of 3D printing equipment and the digitization process of 3D printer body structure.

Digital Design Positioning of 3D Equipment Operation is as Follows:

1) Establish an X/Y plane, and establish a coordinate in the X axis and Y axis area, so that each point on the print plane has its uniqueness and corresponds to the coordinate value.
2) The model is layered, and then the STL file generated by the file is transferred to the microcontroller, which is converted into the boundary function of each layer by the computer, and printed one by one in the corresponding area.
3) When the nozzle moves to the edge, the processing of this layer will be completed. Under the control of feedback information, the equipment can automatically stop the processing of this layer, and then the controller commands the printer to shift a indexing unit so that the Z axis reaches the next layer for printing.

Digital Method Design of 3D Printer Body Structure

1) Control part: This design uses embedded PC digital control system. The AVI microcontroller is connected with the PC hardware system to make the AVI microcontroller guide the step motor to rotate, so as to control the printing track of the nozzle. The upper computer transmits data parameters to the lower computer through the control window.
2) Mobile device: This 3D printing device is designed with a crawler chassis.
3) Drive positioning device: the guide rail slider and lifting link are the main components of the positioning device. The straightness of guide rail is generally less than 0.10 mm/m. MP series vertical hydrostatic guideway is adopted in this design.
4) On board equipment: equipped with a 320 W mobile power supply equipment, the transformer and high and low voltage electrical equipment are combined into a set of mobile power distribution device.
5) Extrusion device: the extrusion device consists of a feeding device and a basic nozzle. It is necessary to use the elasticity of the strip itself to ensure the grasping force of the feeding gear. The accuracy of feeding can be ensured by controlling the step angle.

4.2 Verification of the Integration Design Method of Human-Computer Dynamic Interaction and Digital Sculpture Based on 3D Printing

In order to verify the effectiveness of the key frame extraction algorithm in this paper, we calculated the average distance between the key frame sequence extracted by the visual interaction method in this paper and the key frame sequence obtained by manual extraction and X-means clustering. In this paper, the visual interaction method and manual extraction were adopted, as well as X-means clustering to extract key frames for the three sculptural gestures of D, E and F successively, and the frame spacing error Fd of the three methods was calculated. The calculation and comparison results of the three methods are shown in Table 1 below.

Table 1. Key frame extraction algorithm error Fd result

Algorithm	Extraction by hand	X-means clustering	Visual interaction
Posture D	0.0557	0.047	0.031
Posture E	0.075	0.068	0.029
Posture F	0.088	0.058	0.030
Error of mean	0.075	0.059	0.031

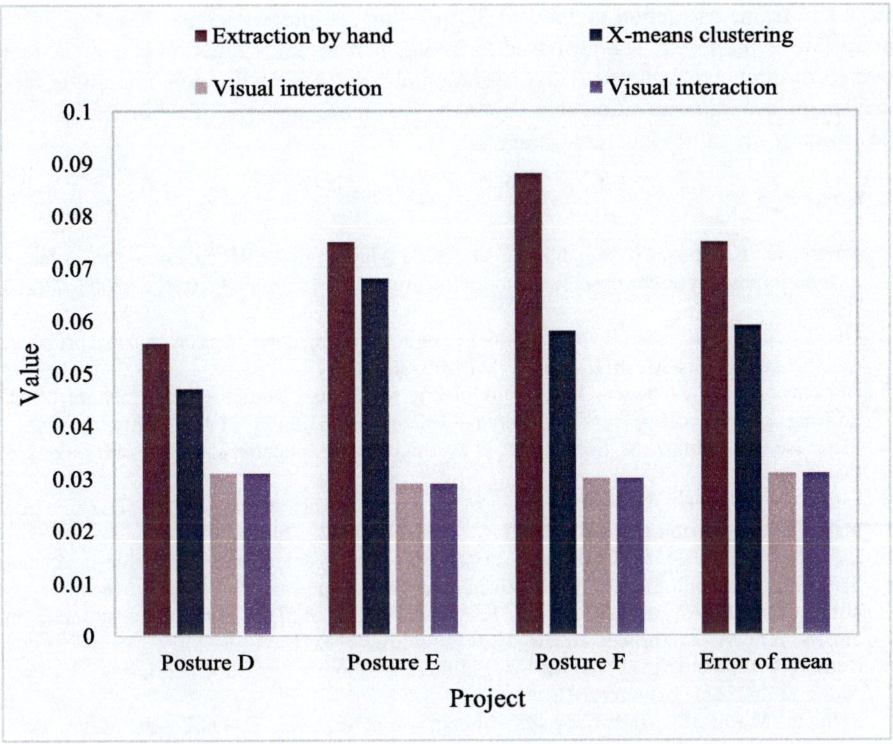

Fig. 2. Key frame extraction algorithm error Fd comparison

As can be seen from the comparison data in Fig. 2, the average inter-frame distance error between the key frame sequence extracted based on the visual interaction method and the key frame pose frame sequence of X-means clustering is the smallest, indicating that the method in this paper can more effectively summarize the original sculpture structure sequence. In addition, for the three different sculpture pose types, the inter-frame distance error is close to 0.3, which is 0.031, 0.029 and 0.030 respectively. There is little change, indicating that the accuracy of the results is relatively stable when using the proposed method for key frame extraction. Based on the above analysis, the method of visual interactive key frame extraction in this paper can better retain the transition attitude and boundary frame of each structure, and the extracted key frame set can fully reflect the original structure content.

5 Conclusion

Starting from the human-computer dynamic interaction of 3D printing, this paper expounds the key frame and feature extraction method of 3D sculpture structure data, gives the extraction method of pose frame of sculpture structure key frame, and focuses on the operation of 3D printing equipment and the digitization process of 3D printer body structure. Based on the human-computer interaction method of monocular camera, a key frame extraction method of 3D printing sculpture structure based on visual interaction is proposed. The proposed technology of visual interaction is analyzed and verified through experiments. The experimental results show that this method is more convenient, accurate and efficient to obtain the key frame sequence of action and generate the standard sculpture structure library.

References

1. Vavrik, D., Kumpova, I., Vopalensky, M., et al.: Mapping of XRF data onto the surface of a tomographically reconstructed historical sculpture. J. Instrum. **14**(02), C02003–C02003 (2019)
2. Ota, T., Yoshida, K., Tase, T., et al.: Influence of 3D-printing conditions on physical properties of hydrogel objects. Mech. Eng. J. **5**(1), 17-00538 (2018)
3. Belikovetsky, S., Solewicz, Y.A., Yampolskiy, M., et al.: Digital audio signature for 3D printing integrity. IEEE Trans. Inf. Forensics Secur. **PP**(5), 1127–1141 (2018)
4. Srinivasa, P., Abdul, K.N., Sujatha, G., et al.: 3D printing in dentistry. J. 3D Print. Med. **2**(3), 89–91 (2018)
5. Lille, M., Nurmela, A., Nordlund, E., et al.: Applicability of protein and fiber-rich food materials in extrusion-based 3D printing. J. Food Eng. **220**, 20–27 (2018)
6. Aquino, R.P., Barile, S., Grasso, A., et al.: Envisioning smart and sustainable healthcare: 3D-printing technologies for personalized medication. Futures **103**, 35–50 (2018)
7. Tohic, C.L., O'Sullivan, J.J., Drapala, K.P., et al.: Effect of 3D printing on the structure and textural properties of processed cheese. J. Food Eng. **220**, 56–64 (2018)
8. Benson, C.L., Triulzi, G., Magee, C.L.: Is there a Moore's Law for 3D printing? 3D Print. Addit. Manuf. **5**(1), 53–62 (2018)
9. Rehman, M.M.: 3D printing for soft robotics - a review. Sci. Technol. Adv. Mater. **19**(1), 243–262 (2018)
10. Nagarajan, N., Dupret-Bories, A., Karabulut, E., et al.: Enabling personalized implant and controllable biosystem development through 3D printing. Biotechnol. Adv. **36**(2), 521–533 (2018)
11. Kunchala, P., Kappagantula, K.: 3D printing high density ceramics using binder jetting with nanoparticle densifiers. Mater. Des. **155**, 443–450 (2018)

Study on the Mix Proportion of Waste Marble Powder-Ground Granulated Furnace Slag-Based Alkali-Activated Ultra-high Ductility Concrete

Yi Zhang[1], Ruihao Ren[1,2], Binyu Mo[1,2], Rongcun Mu[1,2], Ting Huang[1,2], and Bing Liu[1,2(✉)]

[1] Guangxi Key Laboratory of Green Building Materials and Construction Industrialization, Guilin University of Technology, Guilin 541004, China
cnulb@glut.edu.cn

[2] College of Civil Engineering and Architecture, Guilin University of Technology, Guilin 541004, China

Abstract. The engineered cementitious composites (ECC) has an ultimate tensile strain more than 300 times that of regular concrete, making it highly promising for improving the seismic energy dissipation capacity of buildings, repairing damaged structures, and bridge expansion joints. However, the high CO_2 emissions associated with the use of high cement content in ECC contradict the low-carbon and sustainable development principles of the building materials industry. In light of this, this paper investigates the use of solid waste-based alkali-activated cementitious materials to replace cement to prepare ground granulated blast furnace slag (GGBFS)-waste marble powder (WMP) -based alkali-activated ultra-high ductility concrete (AUHDC), achieving high-value utilization of solid waste and low-carbon preparation of ultra-high ductility engineering materials. This study assesses the key parameters of the WMP percentage in the precursor mixture, sand ratio, thickener content, and fiber composition on the tensile properties of AUHDC through uniaxial tensile tests. The results indicate that the ultimate tensile strain initially increases and then decreases with increasing WMP percentage in the precursor mixture, and it increases with increasing sand ratio. The thickener content does not significantly affect the ultimate tensile stress and strain, but the flowability decreases with increasing thickener content. The ultimate tensile strain of AUHDC with 2% polyvinyl alcohol (PVA) fiber and 2% polypropylene (PP) fiber is much lower than that of 1% ultra-high molecular weight polyethylene (PE) + 1% PP fiber and 2% PE fiber AUHDC. The ultimate tensile strain of AUHDC prepared with a WMP percentage of 50%, sand ratio of 55%, thickener content of 0.1%, and 1% PE + 1% PP fiber is 7.4%, which is 700 times that of regular concrete, and the ultimate tensile stress can also reach 6.2 MPa.

Keywords: Solid waste recycling · Waste marble powder · Ultra-high ductility · Alkali-activated concrete · Low carbon

W. Guo et al. (Eds.): GBCESC 2023, LNCE 328, pp. 357–368, 2024.
https://doi.org/10.1007/978-981-99-9947-7_38

1 Introduction

The ductility of engineered cementitious composites (ECC) is over 300 times that of regular concrete, as the ultimate tensile strain in uniaxial tension can reach over 3% [1]. Moreover, ECC have superior fatigue resistance and a longer service life [2, 3]. Due to these properties, ECC have broad application prospects, such as improving the seismic energy dissipation capacity of buildings and repairing damaged structures. However, ECC requires a large amount of cement, accounting for 50% or more, which raises concerns about the high CO_2 emissions and energy consumption related with cement manufacture. These concerns contradict the low-carbon sustainable development concept advocated by today's building materials industry [4–6].

To reduce the cement content in ECC, scholars have attempted to use other materials as cement substitutes in the ECC matrix [7–11]. Victor C. Li used fly ash (FA) to displace cement in the development of ECC and evaluated the effect of aging on the mechanical properties of ECC with a large amount of FA [7]. Avanaki used metakaolin to replace cement and produce fiber-reinforced geopolymer concrete (GPC) and found that fibers improved the physical and mechanical properties of GPC [9]. Chen used silica fume and fly ash to achieve green environmental protection and replaced oiled PVA fiber and silica sand with mixed fibers (unoiled PVA fiber and basalt fiber) and river sand to reduce ECC costs [11].

One of the most common measures to decrease the environmental effect of cement is the use of alkali-activated cementitious materials [8, 12–16]. Rishabh Bajpa produced alkali-activated concrete (AAC) using fly ash and silica fume and compared its environmental impact with ordinary cement concrete. AAC was found to have a lower impact on global greenhouse effects than cement concrete, with a 10.87%–17.77% decrease in cost per unit volume [12]. Togay Ozbakkaloglu used marble waste and zeolitic tuff to produce AAC and added viscose fibers and cotton fibers to compare their properties. It was found that viscose fiber AAC had higher thermal conductivity, elastic modulus, and compressive strength but under dry density than cotton fiber AAC [13]. Ma partially replaced metakaolin with crushed concrete waste to produce AAC, and up to 75% metakaolin replacement resulted in a 4.75% increase in compressive strength [14]. Numerous studies have shown that alkali-activated cementitious materials have lower carbon emissions and energy consumption than cement, making them a green cementitious material with great potential.

China is among the world's leading countries in marble production, consumption, and export, with an output of over 250 million m^3 of marble slabs in recent years. However, mining and processing marble generates a considerable amount of waste marble powder (WMP), which accounts for approximately 40% of the mining volume and poses a challenge for waste management [17, 18]. To address this issue, our research group has developed alkali-activated cementitious materials using ground granulated blast furnace slag (GGBFS) and WMP as precursors, which exhibit excellent properties [19]. By using WMP-GGBFS-based alkali-activated cementitious materials to replace cement in the manufacture of concrete with ultra-high ductility, we can not only achieve the efficient utilization of WMP but also provide a low-carbon and sustainable alternative material for ECC.

Based on the background described above, this study investigates the complete replace of cement with alkali-activated cementitious materials based on WMP and GGBFS, as well as the partial replacement of sand with WMP to develop alkali-activated ultra-high ductility concrete (AUHDC). To reinforce the AUHDC, several fiber compositions, ultra-high molecular weight polypropylene (PP) fibers, polyvinyl alcohol (PVA) fibers, polyethylene (PE) fibers, PE + PVA hybrid fibers, and PE + PP hybrid fibers, were selected. The research parameters include the percentage of WMP in the precursor, sand ratio, thickener content, and fiber composition. This study successfully produces AUHDC with an ultimate tensile strain of up to 7.4% and investigates the influence of key parameters on the ultimate tensile strain and stress of AUHDC.

2 Experimental

2.1 Raw Materials

The study utilized a range of raw materials, including WMP, GGBFS, sodium silicate solution, flake sodium hydroxide, hydroxypropyl methylcellulose (HPMC), PE fiber, PP fiber, PVA fiber, and aggregates. HPMC with a viscosity of 200,000 was utilized as a thickener, while the flake sodium hydroxide used was of 99% purity. PE fibers were sourced from Solvay, PP fibers from Shandong Taian Composite Material Co., Ltd., and PVA fibers from Japan. The aggregate is composed of 50% quartz sand and 50% WMP, of which the particle size range of quartz sand is 0.18–0.45 mm and 0.125–0.18 mm, accounting for 20% and 30%. Detailed parameters of WMP and GGBFS are presented in Table 1, while the detailed parameters of fibers are presented in Table 2.

Table 1. Chemical compositions of WMP and GGBFS

Raw material	Oxide (% by weight)						
	CaO	SiO_2	Al_2O_3	TiO_2	Fe_2O_3	P_2O_5	Others
WMP	54.09	1.07	1.14	0.04	0.39	0.35	0.39
GGBFS	56.90	25.55	12.07	2.68	0.73	0.49	1.96

Table 2. Performance parameters of fiber materials

Fiber	Length/mm	Diameter/μm	Density/(g·cm^3)	Modulus of elasticity/GPa	Tensile strength/MPa
PE	12	20	0.97	84	2800
PVA	12	39	1.3	36	1260
PP	20	48	0.91	3.8	460

2.2 Designing Mix Proportion

Based on previous literature [7, 10, 12, 14, 19] and extensive preliminary experimentation, the research group determined the alkali activator modulus to be 1.2, with a 7% mass fraction of Na_2O and a 0.34 water-to-binder ratio. The reinforcing fibers were added at a volume fraction of 2%. The study focused on the percentage of WMP in the precursor (WMP/P), sand ratio, thickener (HPMC) content, and fiber composition. The specific mix proportions used in the experiments are listed in Table 3.

Table 3. Mix proportions of AUHDC

Number	WMP/P %	Sand ratio %	HPMC %	Fiber composition	Ultimate tensile Strain %	Ultimate tensile Stress MPa
AUHDC1	40	45	0.2	2%PE	6.7	5.3
AUHDC2	30	45	0.2	2%PE	4.4	2.8
AUHDC3	50	45	0.2	2%PE	7.4	4.4
AUHDC4	60	45	0.2	2%PE	6.1	4.6
AUHDC5	40	40	0.2	2%PE	1.7	3.2
AUHDC6	40	50	0.2	2%PE	6.2	5.1
AUHDC7	40	55	0.2	2%PE	7.1	7.3
AUHDC8	40	45	0.1	2%PE	7.2	5.9
AUHDC9	40	45	0.15	2%PE	6.1	5.1
AUHDC10	40	45	0.25	2%PE	6.4	4.8
AUHDC11	40	45	0.3	2%PE	7.1	5.2
AUHDC12	40	45	0.2	2%PVA	3.3	4.8
AUHDC13	40	45	0.2	2%PP	5.3	3.7
AUHDC14	40	45	0.2	1%PE + 1% PVA	2.7	3.5
AUHDC15	40	45	0.2	1%PE + 1% PP	6.8	5.1

The experimental groups were compared to the reference group (AUHDC1), with each group having a single variable change relative to the reference group. The effect of each variable on the ultimate tensile stress and strain of the AUHDC was analyzed and compared.

2.3 Testing Method

This study followed the specimen preparation process shown in Fig. 1 to prepare the specimens presented in Fig. 2 for uniaxial tensile tests. The testing was conducted according to the standard "Mechanical properties test methods for high-ductility fiber-reinforced cement-based composites" (JC/T 2461—2018) using an AGX-100KN electronic universal testing machine with displacement-controlled loading at a rate of 0.05%/min. During the tests, the crack width and ultimate tensile strain were measured using a digital image correlation (DIC) system. The loading schematic diagram is illustrated in Fig. 3.

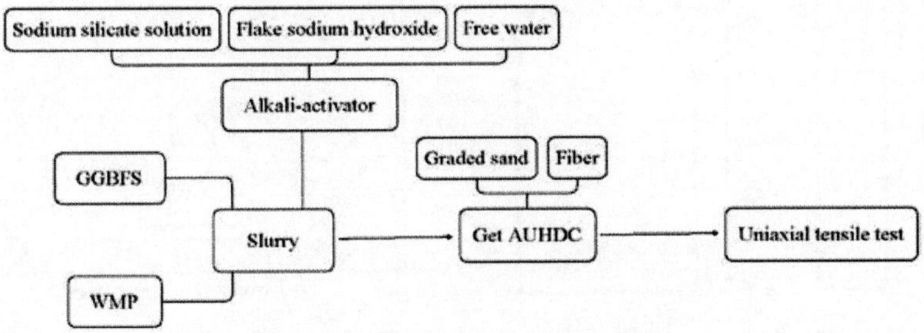

Fig. 1. Preparing process of AUHDC

3 Results and discussion

3.1 Failure Phenomenon

Partial specimen failure was observed in Fig. 4, revealing numerous surface cracks. The crack width and quantity were further analyzed using DIC cloud images, as illustrated in Fig. 5. The cloud image analysis showed that the crack width ranged from 0.04 mm to 0.07 mm, with the number of cracks typically between 80 and 115. Comparing the DIC cloud image with the actual image, it can be noted that analyzing the failure phenomenon of AUHDC using DIC provided a clearer observation of the saturated multiple cracking of the specimen.

Based on the analysis of DIC cloud picture and photos, it was observed that the increase in WMP percentage in the precursor resulted in an increase in the quantity of cracks in the AUHDC matrix, with smaller crack widths. Increasing the sand ratio resulted in an increase in the quantity of cracks in the AUHDC matrix, but the crack width did not change significantly. However, an increase in HPMC content did not significantly affect the quantity and width of cracks in the AUHDC matrix. For the fiber components,

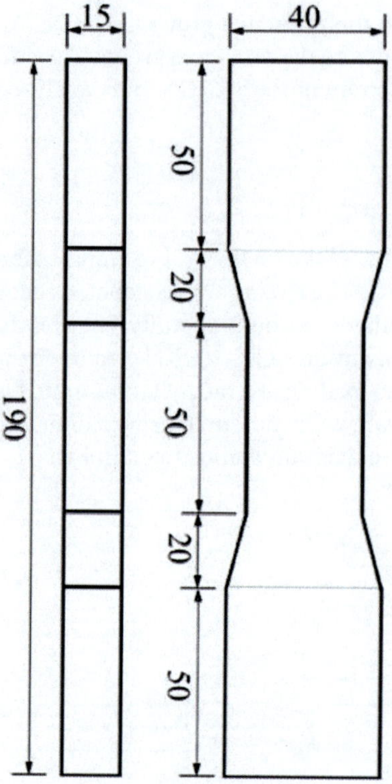

Fig. 2. Specimen design

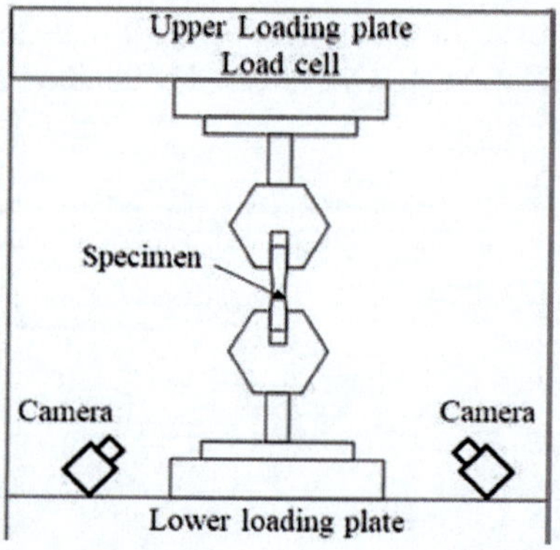

Fig. 3. Schematic diagram of uniaxial tensile loading

(a) AUHDC 1 (b) AUHDC 3

Fig. 4. Failure phenomenon of specimen

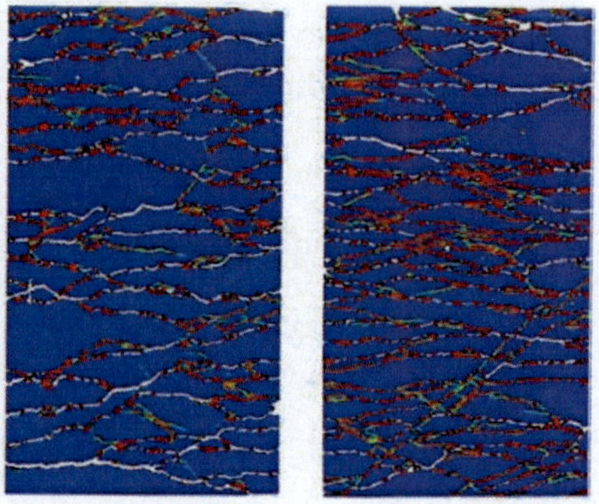

(a) AUHDC 1 (b) AUHDC 3

Fig. 5. DIC cloud diagram of specimen

the 2% PE fiber group exhibited good multiple cracking behavior with a smaller crack width, while the 2% PVA fiber group had fewer cracks with a wider crack width. The 2% PP fiber group had few cracks but with smaller widths, while the 1% PE + 1% PP fiber group had more cracks with smaller widths. This indicates that most of the AUHDC exhibits good saturation and multiple cracking behavior, with fine and closely spaced cracks in the matrix. When a tensile force acts on the matrix, the fibers play an significant role in blocking and absorbing some of the energy. The saturation and multiple cracking of the AUHDC matrix are related to the matrix defect effect [20].

3.2 Tensile Stress–Strain Curves

Figure 6 exhibition the stress–strain curves of all specimens subjected to tensile loads. It can be observed that for the majority of curves, the stress increases with increasing strain. In Fig. 6(b), the initial stress increases with increasing sand ratio, and the maximum stress occurs at a sand ratio of 55%. In Fig. 6(c), the stress–strain curves of the thickener group show no significant changes. In Fig. 6(d), the stress–strain curve of the 2% PP fiber group is almost linear, as the material strength and tensile modulus of the PP fiber are relatively low. Table 3 presents the ultimate tensile stress and strain extracted from the curves.

3.3 Factor Analysis

Effect of WMP Percentage in the Precursors (WMP/P). The impact of the WMP/P on the ultimate tensile stress and strain is shown in Fig. 7. When other factors are kept constant, the ultimate tensile strain increases first and then decreases with increasing WMP content in the precursor. The ultimate tensile strain reaches a maximum of 7.4% when the WMP percentage in the precursor reaches 50%. The ultimate tensile stress reaches a maximum when the WMP percentage in the precursor is 40%. Considering all factors, a WMP percentage of 50% in the precursor is chosen, as it provides the best ultimate tensile strain of AUHDC and ultimate tensile stress of 4.4 MPa, and it also has a relatively high utilization rate of WMP.

Effect of Sand Ratio. The impact of the sand ratio on the ultimate tensile stress and strain is presented in Fig. 8. When other factors remain constant, the ultimate tensile strain increases with increasing sand ratio and reaches its maximum of 7.1% at a sand ratio of 55%. Similarly, the ultimate tensile stress also increases with increasing sand ratio, indicating that higher sand ratios lead to higher ultimate tensile stress and strain. Thus, a sand ratio of 55% is considered the optimum sand ratio for AUHDC under laboratory environment.

Effect of Thickener Contents. The impact of the thickener content on the ultimate tensile stress and strain relationship is shown in Fig. 9(a). From the figure, it can be seen that the change in thickener content has no significant effect on the ultimate tensile stress and strain relationship. At 0.1% and 0.3%, the ultimate tensile strain is greater than 7%, and the ultimate tensile stress is greater than 5 MPa. Previous literature has shown that thickeners have a significant impact on the flowability of ECC [21]. In this research, the flowability of different thickener groups was also tested, and the effect

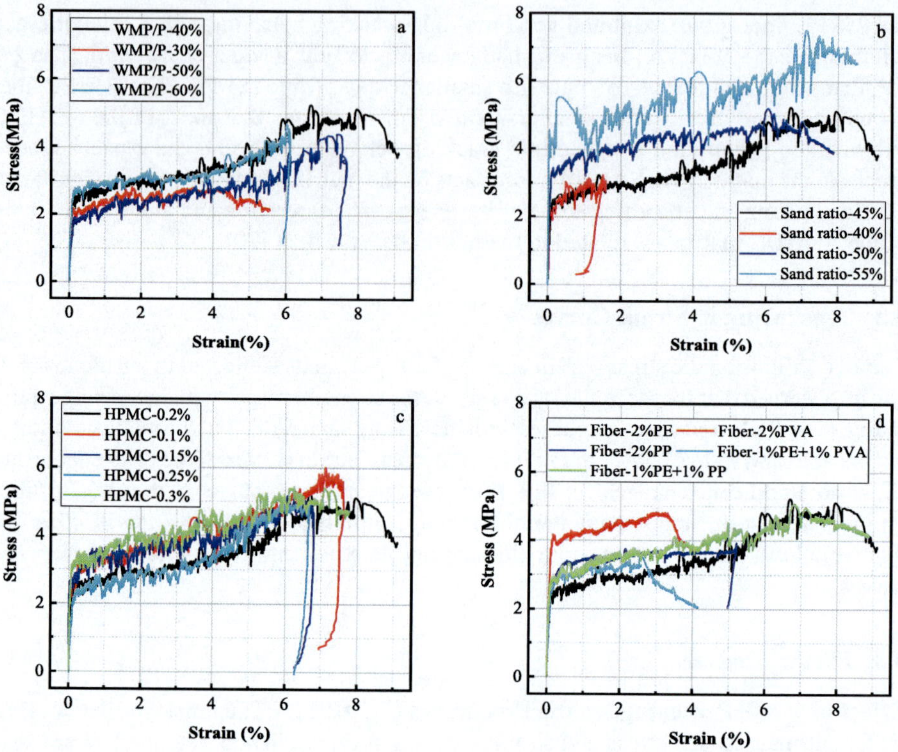

Fig. 6. Effect of various factors on the stress–strain curves: (a) WMP/P; (b) sand ratio; (c) thickener contents; (d) fiber composition.

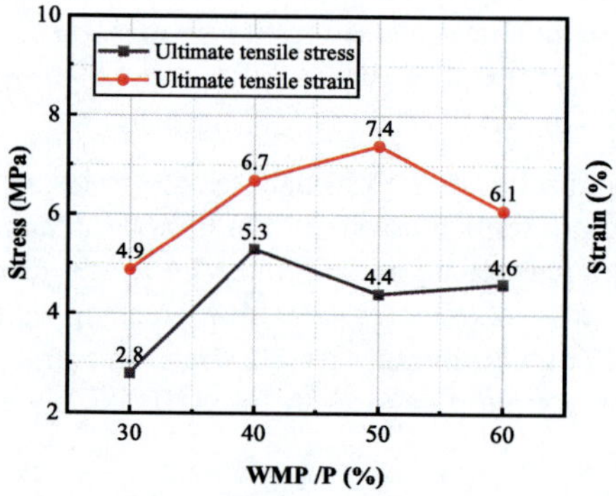

Fig. 7. Effect of WMP/P on the ultimate tensile stress and strain

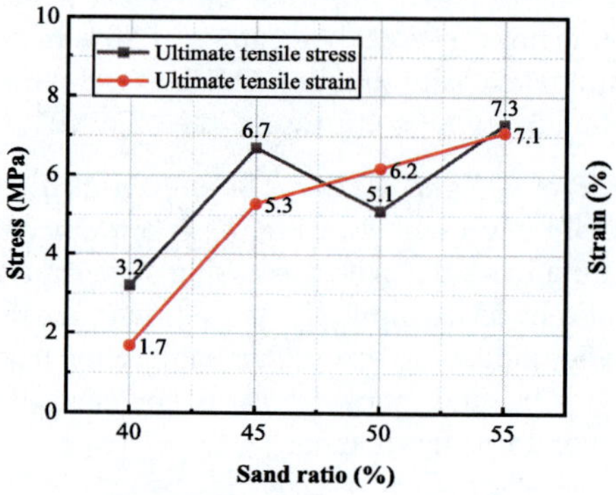

Fig. 8. Effect of sand ratio on the ultimate tensile stress and strain

of thickener on flowability was obtained as shown in Fig. 9(b). It was found that the flowability decreased with increasing thickener content. Considering the application in actual engineering, a thickener content of 0.1% is considered the optimum content.

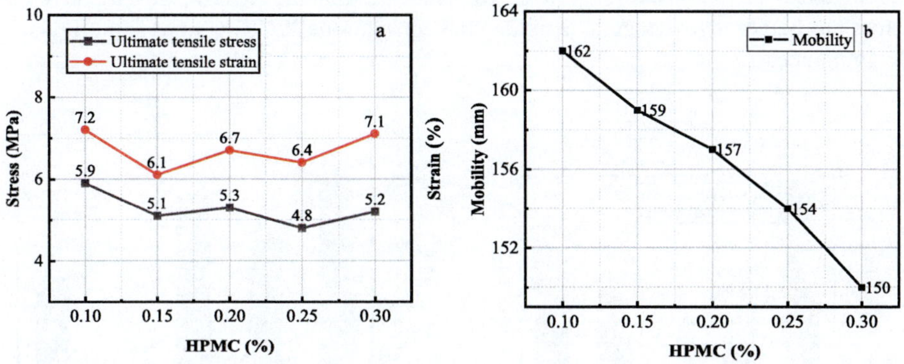

Fig. 9. Effect of thickener (HPMC) content on: (a) the ultimate tensile stress and strain; (b) the fluidity

Effects of Fiber Composition. The impact of the fiber composition on the ultimate tensile stress and strain curve is illustrated in Fig. 10. The figure demonstrates that PVA has a negligible effect on the enhancement of the ultimate tensile strain of AUHDC compared to PE and PP, and its benefit is insignificant. Conversely, 2% PP exhibits good ultimate tensile strain. However, due to the low material strength and tensile modulus of PP, its tensile stress–strain curve does not demonstrate an upward trend and is essentially a straight line, as presented in Fig. 6(d), leading to a lower ultimate tensile stress. The 1% PE + 1% PP hybrid fiber fully exploits the synergistic blockage effect between fibers, with an ultimate tensile strain of 6.8% and an ultimate tensile stress exceeding 5 MPa. Furthermore, the 1% PE + 1% PP hybrid fiber can lower fiber costs by 46.8% compared to 2% PE fiber while sustaining excellent ultimate tensile stress and strain characteristics. Figure 11 depicts the fiber cost of 1 m3 of AUHDC. Based on the comprehensive analysis, the 1% PE + 1% PP hybrid fiber is considered the optimum fiber composition.

3.4 The Optimum Mix Proportion

Based on the experimental results discussed earlier, the optimal mix proportion was determined considering the cost and WMP utilization ratio, using the ultimate tensile strength and strain of AUHDC as the standard. The optimal mix proportion comprises 50% WMP in the precursor, 55% sand, 0.1% thickener content, and a hybrid fiber component of 1% PE + 1% PP hybrid fiber. In this mix proportion, WMP accounts for 39.3% of the mass of the AUHDC matrix (including water). The uniaxial tensile stress–strain curve of this mix proportion is presented in Fig. 12, with an ultimate tensile strain of up to 7.4% and a corresponding ultimate tensile strength of 6.2 MPa. Figure 13 shows the specimen failure and DIC strain cloud map. It can be clearly observed that there are many thin cracks, and the average crack width is not more than 0.05 mm. It is noteworthy that one ton of conventional ECC (with cement accounting for 54.6% [22]) emits 340 kg of CO_2 [6], while one ton of AUHDC (with alkali-activated cementitious materials accounting for 45%) emits approximately 80 kg of CO_2, which is only 24% of the carbon emission of conventional ECC. This indicates that the AUHDC developed in this paper has better environmental benefits than conventional ECC without compromising performance.

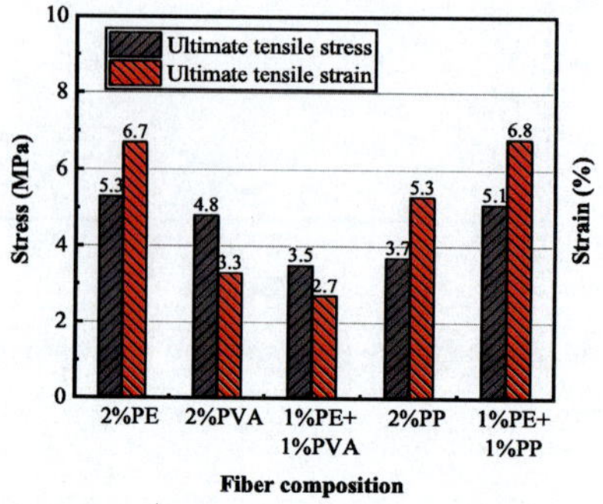

Fig. 10. Effect of fiber composition on ultimate tensile stress and strain

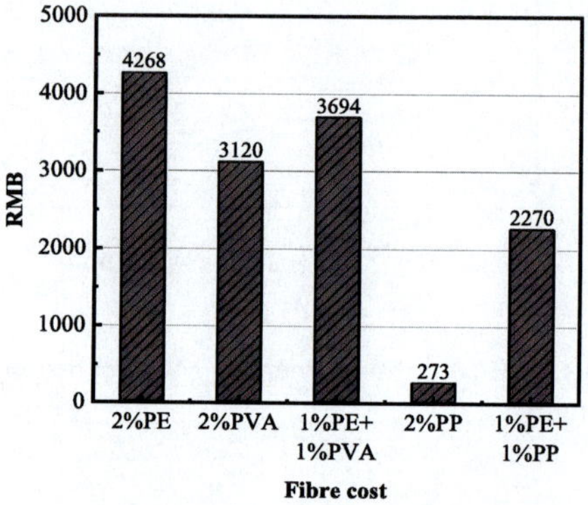

Fig. 11. Fiber cost of 1m^3 AUHDC

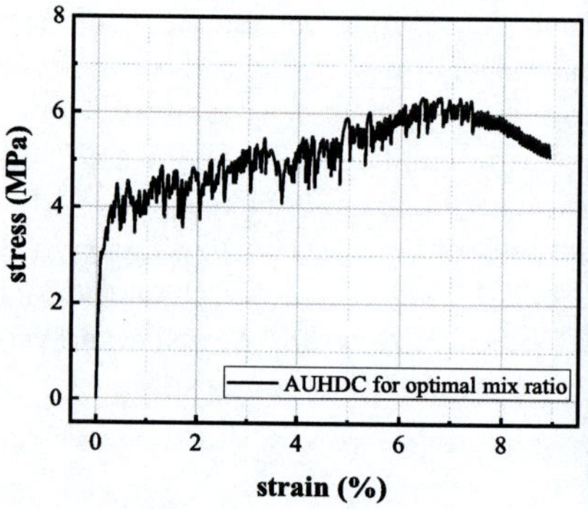

Fig. 12. Tensile stress–strain curve of the optimum mix proportion

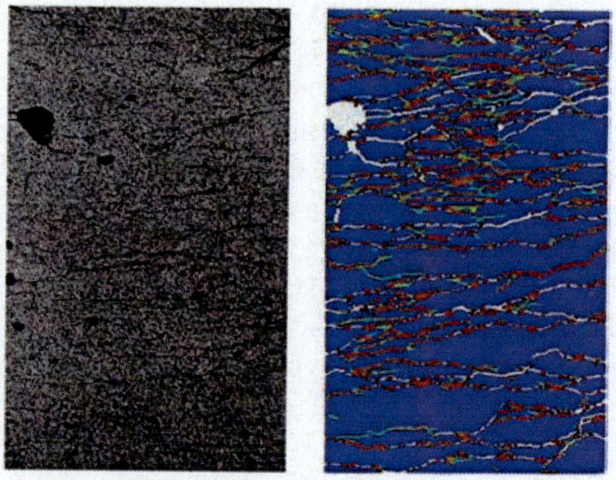

Fig. 13. Failure photo and strain cloud map of the optimum mix proportion

4 Conclusion

This study aimed to develop a low-carbon green ultra-ductility concrete alternative to traditional ECC by using a WMP-GGBFS-based alkali-activated cementitious material to completely replace cement. The effects of various factors, including the WMP percentage in the precursor, sand ratio, thickener content, and fiber composition, on the ultimate tensile stress and strain of AUHDC were investigated through the uniaxial tensile test. The key findings are as follows:

1) The ultimate tensile strain of AUHDC increased first and then decreased with increasing WMP content, reaching a maximum at a percentage of 50%, with a corresponding ultimate tensile strain of 7.4% and ultimate tensile stress of 4.4 MPa.
2) The ultimate tensile stress and strain of AUHDC increased with increasing sand ratio, reaching a maximum at a ratio of 55%, with a corresponding ultimate tensile strain of 7.1% and ultimate tensile stress of 7.3 MPa.
3) The thickener content did not have a significant effect on the ultimate tensile strain stress, but it was observed that flowability decreased with increasing thickener content, and 0.1% thickener content was determined to be optimal.
4) Among the tested fiber compositions, 1% PE+1% PP fiber and 2% PE fiber showed the most effective reinforcement effect, while 2% PVA fiber and 2% PP fiber exhibited weaker reinforcing effects. The 1% PE+1% PP fiber AUHDC was found to be the most cost-effective option.
5) With a high percentage of WMP accounting for 39.3%, the developed AUHDC achieved an ultimate tensile strain of 7.4% and an ultimate tensile stress of 6.2 MPa, which is not weaker than the performance of traditional ECC but with a significantly lower carbon emission of only 24% of that of traditional ECC. These findings demonstrate that the developed AUHDC has great potential for widespread application as a low-carbon and environmentally friendly alternative to traditional ECC.

Acknowledgment. The authors gratefully acknowledge the financial support by the Guangxi Science and Technology Base and Special Fund for Talents Program (Grant No. GuikeAD22035999), the Natural Science Foundation of Guangxi (Grant No. 2021GXNSFBA220049), the National Natural Science Foundation of China (Grant Nos. 52108201 and U22A20244), the Guangxi Key Laboratory of Green Building Materials and Construction Industrialization (grant No. 22-J-21–9), the Guangxi Science and Technology Major Project (grant No. GuikeAA22068073–3) ,and the Innovation Project Guangxi Graduate Education (grant No. YCSW2023339).

References

1. Meng, D., Huang, T., Zhang, Y.X., Lee, C.K.: Mechanical behaviour of a polyvinyl alcohol fiber reinforced engineered cementitious composite (PVA-ECC) using local ingredients. Constr. Build. Mater. **141**, 259–270 (2017)
2. Alam, B., Yaman, I.O.: Stress-based fatigue performance and fatigue life prediction of engineered cementitious composites. J. Mater. Civ. Eng. **33**, 2 (2021)
3. Meng, D., Lee, C., Zhang, Y.: Flexural fatigue properties of a polyvinyl alcohol-engineered cementitious composite. Mag. Concr. Res. **71**, 1130–1141 (2019)
4. Favier, A., De Wolf, C., Scrivener, K., Habert, G.: A sustainable future for the European cement and concrete industry, ETH Zurich (2018)
5. Uratani, J.M., Griffiths, S.: A forward looking perspective on the cement and concrete industry: implications of growth and development in the Global South. Energy Res. Soc. Sci. **97** (2023)
6. Wang, B., Wu, Y., Zhang, W., Zhang, J.: Green technology innovation path and policy paradigm transformation under the background of "dual carbon" goals. Sci. Manage. Res. **40**, 2–6 (2022)

7. Huang, X., Ranade, R., Zhang, Q., Ni, W., Li, V.C.: Mechanical and thermal properties of green lightweight engineered cementitious composites. Constr. Build. Mater. **48**, 954–960 (2013)
8. Ma, K., Deng, H., Guo, X., Yin, J., Zhao, Y.: The investigating on mechanical properties of engineered cementitious composites with high ductility and low cost. J. Build. Eng. **57**, 104873 (2022)
9. Moradikhou, A.B., Esparham, A., Jamshidi Avanaki, M.: Physical & mechanical properties of fiber reinforced metakaolin-based geopolymer concrete. Constr. Build. Mater. **251**, 118965 (2020)
10. Hmaran, M., Li, V.C.: Durability properties of microcracked ECC containing high volumes fly ash. Cem. Concr. Res. **39**, 1033–1043 (2009)
11. Zhu, M., Chen, B., Wu, M., Han, J.: Effects of different mixing ratio parameters on mechanical properties of cost-effective green engineered cementitious composites (ECC). Constr. Build. Mater. **328**, 127093 (2022)
12. Bajpai, R., Choudhary, K., Srivastava, A., Sangwan, K.S., Singh, M.: Environmental impact assessment of fly ash and silica fume based geopolymer concrete. J. Clean. Prod. **254**, 120147 (2020)
13. Tekin, I., Gencel, O., Gholampour, A., Oren, O.H., Koksal, F., Ozbakkaloglu, T.: Recycling zeolitic tuff and marble waste in the production of eco-friendly geopolymer concretes. J. Clean. Prod. **268**, 122298 (2020)
14. Liu, M., Hu, R., Zhang, Y., Wang, C., Ma, Z.: Effect of ground concrete waste as green binder on the micro-macro properties of eco-friendly metakaolin-based geopolymer mortar. J. Build. Eng. **68**, 106191 (2023)
15. Rattanasak, U., Chindaprasirt, P.: Influence of NaOH solution on the synthesis of fly ash geopolymer. Miner. Eng. **22**, 1073–1078 (2009)
16. Wu Chen, Z.Z. in Proceedings of the 2nd International Symposium on Asia Urban GeoEngineering (2018)
17. Aguiar, L.L., Tonon, C.B., Nunes, E.T., Braga, A.C.A., Neves, M.A., de Oliveira David, J.A.: Mutagenic potential of fine wastes from dimension stone industry. Ecotox. Environ. Safe. 125, 116–120 (2016)
18. Tong, J.: Analysis of the development environment of stone industry in the 14th Five-Year Plan. Stone, 2–8 (2022)
19. Liu, B., et al.: A preliminary study on waste marble powder-based alkali-activated binders. Constr. Build. Mater. **378**, 131094 (2023)
20. Li, V.C., Wang, S.: Microstructure variability and macroscopic composite properties of high performance fiber reinforced cementitious composites. Probab. Eng. Eng. Mech. **21**, 201–206 (2006)
21. Cao, M., Xu, L., Zhang, C.: Rheological and mechanical properties of hybrid fiber reinforced cement mortar. Constr. Build. Mater. **171**, 736–742 (2018)
22. Qian, H., Ye, Y., Yan, C., Jin, G., Li, C., Shi, Y.: Experimental study on the seismic performance of self-centering bridge piers incorporating ECC and superelastic SMA bars in the plastic hinge regions. Structures **46**, 1955–1967 (2022)

Effect of Waterborne Epoxy Resin on the Shrinkage and Mechanical Properties of Geopolymer Material

Huachong Cai[1], Hanqing Liu[2], Xiongfei Liu[1(✉)], and Yaoyao Wu[1]

[1] School of Civil Engineering and Transportation, Hebei University of Technology, Tianjin 300401, China
liuxfking@foxmail.com

[2] SASAC Metallurgical Authority Service Center, Beijing 100010, China

Abstract. Due to the drawbacks of high shrinkage and lots of cracks, geopolymer material shows poor brittle performance. The waterborne epoxy resin is added to address these issues for geopolymers. The shrinkage and mechanical properties of modified geopolymers with aqueous epoxy resin at 10 wt %, 20 wt %, and 30 wt % are investigated in this paper. Furthermore, SEM is used to evaluate the effect of resin on the hydration process of geopolymers. The results show that the shrinkage and toughness of modified geopolymers with waterborne epoxy resin are reduced and improved, respectively. The shrinkage of the geopolymer with 10 wt% epoxy resin is reduced by 25% in 28 days compared to that of the control, with the flexural strength increasing by 17.5%. Simultaneously, the bonding strength at the concrete interface is higher than 2.0 MPa. This paper describes an efficient method to optimize the brittleness of geopolymers.

Keywords: Geopolymer · Shrinkage · Mechanical properties · Waterborne epoxy resin

1 Introduction

Geopolymer is a type of green building material with an amorphous aluminosilicate network structure created by the reaction of various solid waste materials, which has the advantages of high bonding strength with the substrate interface, good chemical corrosion resistance [1], fast hardening and early strength [2], low energy consumption, etc. However, the engineering application of geopolymers is severely limited due to their brittleness and cracking [3]. As a result, the focus of research has shifted to improving the toughness of geopolymers.

At the moment, there are two common methods for increasing the toughness of geopolymers: mixing with fibers or epoxy resin. The fibers include glass fiber, carbon fiber, and plant fiber. However, there are some evident drawbacks to this type of fiber incorporation, including its high cost and uneven dispersion in the paste. Contrarily, organic resin can speed up the polymerization reaction rate by delaying the evaporation

W. Guo et al. (Eds.): GBCESC 2023, LNCE 328, pp. 369–375, 2024.
https://doi.org/10.1007/978-981-99-9947-7_39

of water molecules, effectively inhibiting the initiation and propagation of microcracks, and thus enhancing the toughness and durability of geopolymers. Saludung et al. [4] created geopolymer composites of ground blast furnace slag and fly ash with varying amounts of epoxy resin. The outcomes demonstrated that epoxy resin can reduce the brittleness of geopolymer. Du et al. [5] modified a metakaolin-slag based geopolymer with aqueous epoxy resin, and discovered that epoxy resin could shorten the setting time. Chen et al. [6] investigated the impact of polyacrylic acid resin on the mechanical properties of slag geopolymer. The bending toughness of samples containing 1wt % polyacrylic acid resin at 28 days was 2.05 times that of pure geopolymer, indicating that the toughness was significantly improved.

In this study, the geopolymer is modified with different contents of aqueous epoxy resin (10 wt%–30 wt%) to optimize shrinkage and mechanical properties. Then, the mechanism of adding aqueous epoxy resin to the geopolymers to increase toughness and prevent shrinkage cracking is investigated.

2 Experiments

2.1 Raw Materials

Resin-modified geopolymer is made up of aqueous epoxy resin, sodium silicate, sodium hydroxide solution, metakaolin, and quartz sand. Aqueous epoxy resin consists of bisphenol-A epoxy resin (E51), benzyl glycidyl ether (BGE), and a water-borne epoxy curing agent (CA). The alkali activator of the geopolymer is composed of sodium silicate (solid content of 42%, $Na_2O/SiO_2 = 2.0$) and a 50% concentration of sodium hydroxide solution. Quartz sand details are particle size = 40–80 mesh, fineness modulus = 1.8, and bulk density = 1620 kg/m^3.

2.2 Sample Preparation and Tests

Sample Preparation. Different mass amounts of aqueous epoxy resin (10 wt%, 20 wt%, 30 wt%) are used to explore the impact of resin on the shrinkage and mechanical properties of geopolymers. Table 1 shows the material mix design. CA is poured into diluted E51 and thoroughly mixed to make an aqueous epoxy resin. BGE has a mass ratio of 1:100 to E51. CA/E51 has a mass fraction of 0.6. Geopolymer is combined with metakaolin, alkali activator, and quartz sand. The mass ratio of metakaolin to alkali activator is 0.32: 0.68. At the same time, the alkali activator contains sodium silicate and sodium hydroxide solutions in a 5:1 mass proportion.

Initially, a uniform mixture of metakaolin, quartz sand, and an alkaline activating solution is made. The combined slurry is then blended with the aqueous epoxy resin, which is thoroughly mixed to create the resin geopolymer composite. To get a homogenous sample, the cast sample is vibrated for 2 min to make it evenly distributed. After one day of laboratory curing, the samples are placed in the standard curing room to reach the required age of 28 d.

Test Methods. The specimen size is 40 mm × 40 mm × 160 mm. Constructed to test the flexural strength in accordance with the specification GB/T17671–1999 'Test

Table 1. Mix design proportion of the materials.

Samples	Metakaolin (wt %)	ER (wt %)	CA (wt %)	Sand (wt %)
E0	66.7	0	0	33.3
E10	60	4.2	2.5	33.3
E20	53.4	8.3	5.0	33.3
E30	46.7	12.5	7.5	33.3

method of strength of cement mortar (ISO method)'. Samples measuring 25 mm × 25 mm × 280 mm are tested in accordance with the specification GB/T 29417–2012 'Test method for dry shrinkage cracking performance of cement mortar and concrete'. The multi-function adhesion strength tester (HICHANCE, HC-40) is used to carry out the tensile test to assess the interface bonding strength, as shown in Fig. 1, according to the specification GB/T 23445–2009 'Polymer cement waterproof coating'. Scanning electron microscopy (SEM, Hitachi, SEU8010) was used to characterize the microscopic structure of samples.

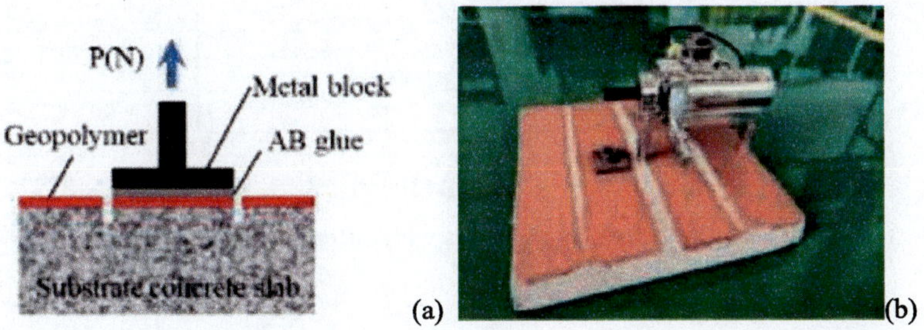

Fig. 1. (a) Schematic view of the interface bonding strength test; (b) Pull-off test.

3 Results and Discussion

3.1 Shrinkage Property of Resin Geopolymer

The shrinkage rate of geopolymer gradually increases with age, presenting roughly three stages of trend, as illustrated in Fig. 2. The sample (E0) is given a high shrinkage rate at 7 days for the initial stage. The drying shrinkage of E0 at 7 d is 70.18% of that at 28 d. The second stage lasts from 7 to 14 days and is known as the gradual shrinkage period. The final stage is a relatively stable shrinkage period from 14 d to 28 d. Aqueous epoxy resin, as seen in Fig. 2, can greatly lessen the drying shrinkage of the geopolymer. The development of shrinkage is consistent with that of E0 at the same time. The shrinkage of geopolymer modified with 10 wt% epoxy resin (E10) yields an ideal response. The shrinkage of E10 is reduced by 25%, compared to that of E0 in 28 d.

The shrinkage of geopolymer is caused by the formation of a meniscus between water and air as a result of pore water evaporation from capillary pores, which generates capillary pressure and causes shrinkage. The gel structure of geopolymers containing alkali ions is also a key factor in the shrinkage of geopolymers. Water will not be directly involved in the formation of aluminosilicate gel. The presence of large amounts of water increases the probability of shrinkage due to water evaporation. Because of the polar active groups in epoxy resin, such as hydroxyl, organic particles can be adsorbed on the surface of the geopolymer gel particles, forming an adsorption hydration film to bind more free water. Epoxy resin has the ability to fix water molecules and effectively delay water evaporation [7], reducing sample shrinkage. As shown in Fig. 6, pore water has a higher tendency to evaporate and cause shrinkage. When compared to E20 and E30, E10 has a smaller pore structure, which limits water evaporation during drying. E10 has a better shrinkage performance than E20 and E30.

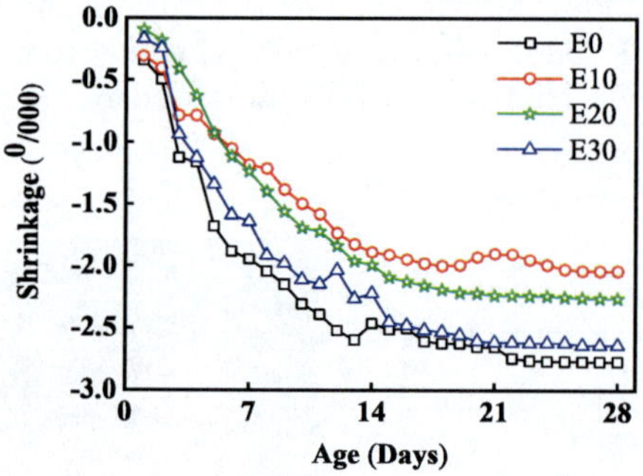

Fig. 2. Shrinkage of geopolymers with different contents of resin.

3.2 Flexural Strength of Resin Geopolymer

Epoxy resin can greatly enhance the 7d flexural strength of the geopolymer, but it has no discernible impact on the strength at 28 d, as demonstrated in Fig. 3. The flexural strength of geopolymer increased steadily at 7 d with the increase in epoxy resin dosage. The flexural strength reaches its peak value at 28 d when the epoxy resin concentration is 10 wt%. The flexural strength of E10 improved by 41.7% and 17.5% at 7 d and 28 d, respectively, compared to that of E0.

Griffith crack propagation energy criterion is used to assess the flexural strength of resin geopolymer. With the addition of resin to the geopolymer, more interfaces are formed. The hydrophilic functional groups in the epoxy resin, for example, the carboxyl and hydroxyl groups, interact with the hydrophilic functional groups in the geopolymer gel. The surface energy of the composite material is then increased [7]. The retardation effect is generated as crack growth resistance increases, preventing or delaying the development of cracks and thus improving flexural strength. However, if the content of organic resin exceeds 10 wt %, the crack propagation resistance is reduced due to the uneven structure, which decreases the flexural strength.

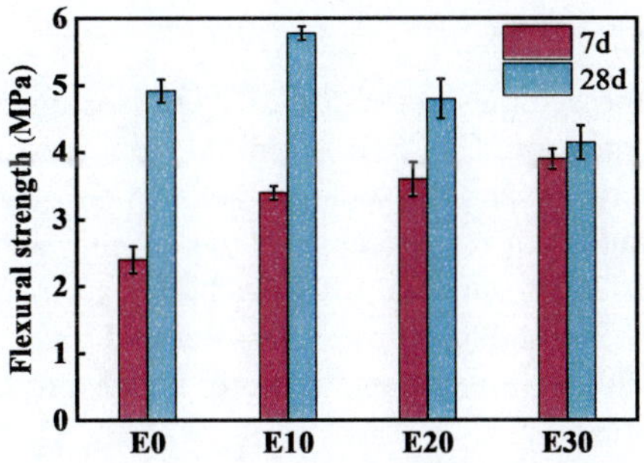

Fig. 3. Flexural strength results of geopolymers with different resin contents at 7d and 28 d.

3.3 Interface Bonding Strength of Resin Geopolymer

The addition of resin improves the interfacial bonding strength between the sample and the concrete base, as shown in Fig. 4. At the same time, as the resin content increases, so does the interface bonding strength. Except for the E0 sample, the interface bonding strength of resin geopolymer at 28 d is all greater than 2.0 MPa. The sample interface bonding failure mode [8] can also be used to determine bonding quality. Figure 5 shows that the failure mode of the pull-off test is concrete substrate damage. This failure mode demonstrates that the resin geopolymer has a high adhesive property and a strong affinity for concrete.

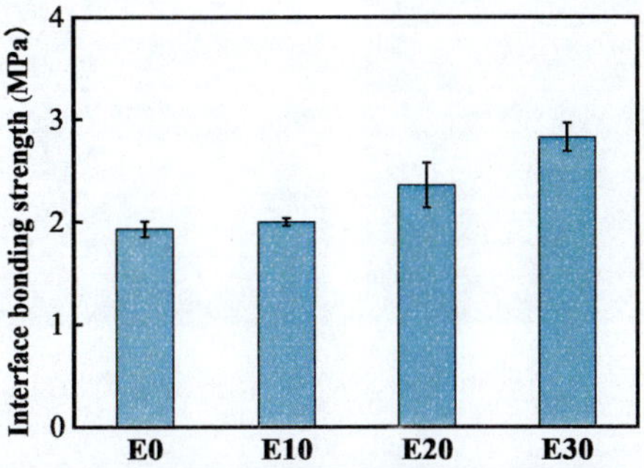

Fig. 4. Interface bonding strength between resin modified geopolymer and concrete.

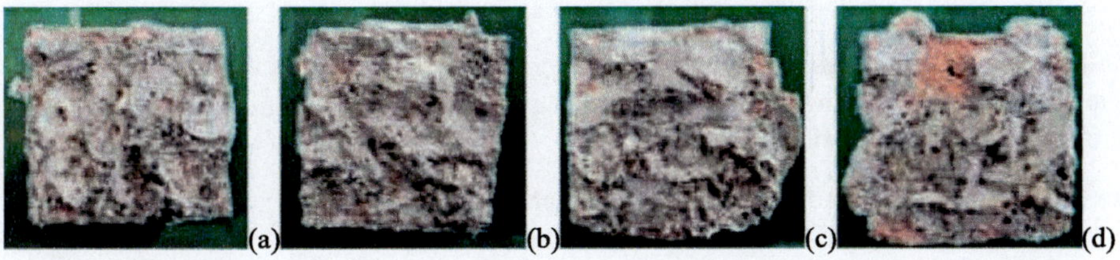

Fig. 5. Morphology of fracture surface after pull-off test: (a) E0; (b) E10; (c) E20; (d) E30.

3.4 Microscopic Properties

The microstructure of the samples cured for 28 d is investigated by scanning electron microscopy (SEM), as shown in Fig. 6. E0 exhibits the lowest shrinkage performance, the greatest number of cracks, and the widest cracks among the examined samples. As shown in Fig. 6, the addition of resin decreases the number and width of microcracks, compared to those of E0. The resin is critical to reducing shrinkage. There is no obvious crack, especially in the E10 sample. The geopolymer and resin are more tightly bonded to create a dense and homogenous microstructure, which produces the best shrinkage performance of the samples.

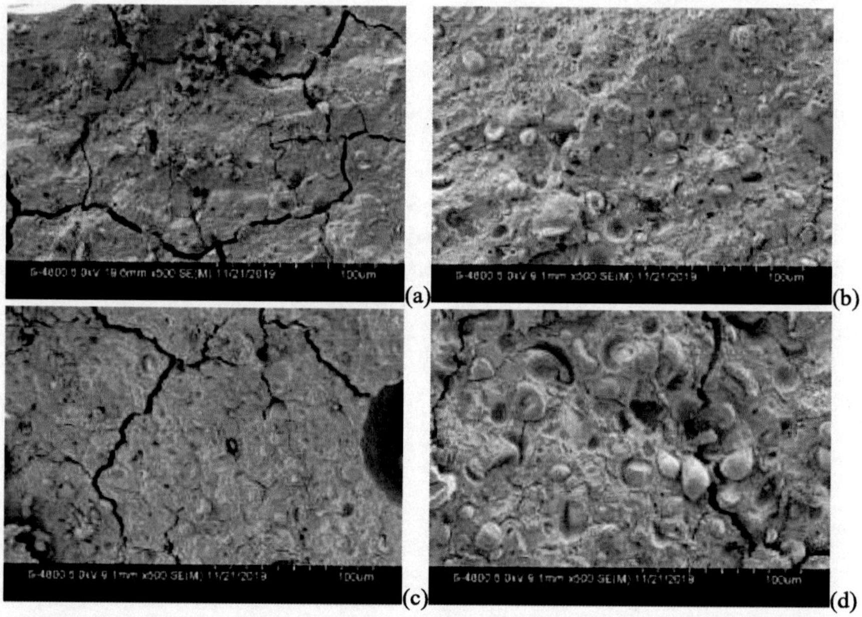

Fig. 6. SEM micrographs (amplification is 5×10^2 in a, b, c, d): a) E0; b) E10; c) E20; d) E30.

4 Conclusion

In this study, epoxy resin is used to modify geopolymers. Also, a thorough study of the mechanical and shrinkage characteristics of geopolymers is conducted. The following are the primary conclusions:

1) Aqueous epoxy resin improves the mechanical and shrinkage characteristics of geopolymers. The flexural strength of E10 at 7 d and 28 d was 41.7% and 17.5% higher than that of the pure geopolymer E0, respectively. At the same time, the interface bonding strength between the sample and the concrete base is greater than 2 MPa.

2) The shrinkage of E10 can be reduced by 25% in 28 d compared to E0. The geopolymer achieves a balance between shrinkage and mechanical strength.

Acknowledgements. The authors gratefully thank the support by the National Natural Science Foundation of China (Nos. 52278252 and 520781781), Natural Science Foundation of Hebei Province (Nos. E2020202043, and E2022202041).

References

1. Huang, T., Zhang, S., Liu, L., et al.: Green rust functionalized geopolymer of composite cementitious materials and its application on treating chromate in a holistic system. Chemosphere **263**, 128319 (2021)
2. Chen, K., Wu, D., Chen, H., et al.: Development of low-calcium fly ash-based geopolymer mortar using nanosilica and hybrid fibers. Ceram. Int. **47**(15), 21791–21806 (2021)
3. Archez, J., Farges, R., Gharzouni, A., et al.: Influence of the geopolymer formulation on the endogeneous shrinkage. Constr. Build. Mater. **298**, 123813 (2021)
4. Saludung, A., Azeyanagi, T., Ogawa, Y., et al.: Alkali leaching and mechanical performance of epoxy resin-reinforced geopolymer composite. Mater. Lett. **304**, 130663 (2021)
5. Du, J., Bu, Y., Shen, Z., et al.: Effects of epoxy resin on the mechanical performance and thickening properties of geopolymer cured at low temperature. Mater. Des. **109**, 133–145 (2016)
6. Chen, X., Zhu, G., Wang, J., et al.: Effect of polyacrylic resin on mechanical properties of granulated blast furnace slag based geopolymer. J. Non-Cryst. Solids **481**, 4–9 (2018)
7. Zhang, Y., Wang, Y., Xu, D., et al.: Mechanical performance and hydration mechanism of geopolymer composite reinforced by resin. Mater. Sci. Eng. A **527**(24), 6574–6580 (2010)
8. Cui, E., Jiang, S., Wang, J., et al.: Bond behavior of CFRP-concrete bonding interface considering degradation of epoxy primer under wet-dry cycles. Constr. Build. Mater. **292**, 123286 (2021)

Preparation and Performance Study of Slag-Waste Marble Powder Based Alkali-Activated High Performance Concrete

Xiaofang Deng[1,2], Weixin Lin[1,2], Hongtao Li[1,2], Yuanju Li[1,2], Yunhao Weng[1,2], and Bing Liu[1,2](✉)

[1] Guangxi Key Laboratory of Green Building Materials and Construction Industrialization, Guilin University of Technology, Guilin 541004, China
cnulb@glut.edu.cn

[2] College of Civil Engineering and Architecture, Guilin University of Technology, Guilin 541004, China

Abstract. In this study, ground granulated blast-furnace slag (GGBS) and waste marble powder (WMP) were used as precursors to prepare slag-waste marble powder-based alkali-activated high-performance concrete (SWAHPC), and the effects and rules of different alkali activator modulus (1.0, 1.2, 1.4, 1.6, 1.8), Na_2O dosages (4%, 7%, 10%), and WMP content (replacing 20%, 25%, 30%, 35%, 40% of GGBS) on the performance of SWAHPC were investigated. The results showed that excellent working performance and strength grade not lower than C85 high-strength concrete can be easily prepared when the WMP content does not exceed 40%. With increasing alkali activator modulus, the compressive strength of SWAHPC showed an increase followed by a decrease, while the flexural strength showed a decreasing trend, and the setting times and fluidity all exhibited an increasing trend. With increasing Na_2O dosage, the compressive strength of SWAHPC showed an increasing and then decreasing trend, while the flexural strength showed a decreasing trend, the setting time showed an increasing trend, and the fluidity showed an increasing and then decreasing trend. With increasing WMP content, both the compressive and flexural strengths of SWAHPC showed an increase followed by a decrease, the setting times showed an increasing trend, and the fluidity showed a decreasing trend. The optimal values for the alkali activator modulus, Na_2O dosage, and WMP content were 1.6, 7% and 25%, respectively for the compressive strength.

Keywords: Waste marble powder · Alkali-activated concrete · Working performance · Mechanical performance

1 Introduction

The promotion of low-carbon and green development concepts has brought significant attention to the sustainable development of the field of construction. One of the largest consumables in the world, concrete accounts for 8% of all CO_2 emissions globally [1].

W. Guo et al. (Eds.): GBCESC 2023, LNCE 328, pp. 376–386, 2024.
https://doi.org/10.1007/978-981-99-9947-7_40

However, alkali-activated concrete (AAC) has lower production energy consumption and carbon emissions rather than ordinary Portland cement (OPC), with energy consumption during the preparation process being only approximately 30% of OPC and CO_2 emissions being only approximately 20% of OPC [2]. In addition, most of the raw materials used in AAC are common industrial waste or byproducts such as slag, red mud, and fly ash, making the prepared concrete environmentally friendly, cost-effective, and sustainable with excellent mechanical and durability properties. Therefore, the research significance and development potential of AAC are significant.

In recent years, numerous scholars have conducted research on the performance and influencing factors of AAC, leading to significant progress in the field. Wang et al. [3] prepared SG-based AAC using slag (SG) and found that the concrete strength was relatively high when the sodium oxide content in the activator was between 3.00% and 5.55% and the modulus of water glass was 1.0. Barbosa [4] utilized metakaolin (MK) as a raw material to prepare an MK-based AAC and found that it possessed remarkable early strength characteristics after high-temperature curing. Zhang [5] discovered that for MK-SG-based AAC, the alkaline activator modulus (SiO_2 /Na_2O) had a significant influence on its mechanical and durability properties, and an optimal modulus of 1.2 could result in superior strength, crack resistance, and water permeability. Polomo [6] prepared an FA-based AAC using NaOH and water glass and reported that its compressive strength could reach 60 MPa after curing at 85 °C for 5 h. The SG content was found to have a significant effect on the mechanical performance of FA-SG-based AAC [7]. Pusty found that increasing the SG content would lead to the formation of more N-(C)-A-S-H and improve the strength of AAC, but it would reduce the setting time [8]. Factors that influence the polymerization process of AAC include the type of alkaline activator, alkaline activator modulus, Na_2O dosage, chemical composition of raw materials, and curing temperature, and these factors are closely related to the performance of AAC [2, 9].

While research on AAC has primarily focused on active materials rich in silicon and aluminum like fly ash, metakaolin and slag, the research value of less active inert materials has often been overlooked. Waste marble powder (WMP) is a significant issue, as approximately 70% of resources generated from mining, cutting, and polishing are converted into waste, leading to severe resource waste and environmental pollution [10]. Therefore, it is crucial to effectively handle and utilize WMP. Some researchers have used WMP to replace a portion of the cement to prepare concrete and found that an appropriate amount of WMP could enhance the mechanical and durability performance of concrete [11]. Kumar [12] used waste marble as an aggregate to prepare FA-based AAC and reported that the best mechanical and durability performance was obtained when waste marble replaced 50% of the natural aggregate. As a major building decoration material, WMP has considerable potential as a sustainable alternative material for AAC production.

Research on the application of WMP as a precursor in AAC is relatively scarce. Coppola et al. [13] used WMP to prepare AAC and found that the curing effect was best in air, and the compressive strength reached 45 MPa. Thakur et al. [14] prepared composite bricks using WMP and FA and reported that the addition of WMP resulted in higher compressive strength and lower water absorption. The maximum compressive strength was achieved at a WMP content of 60%. Liu et al. [15] fully verified the possibility of using WMP in AAC by preparing 100% WMP-based AAC, but the strength of the prepared WMP-based AAC was low. In this paper, GGBS and WMP were used as precursors to prepare SA-WMP-based alkali-activated high-performance concrete (SWAHPC), and the study focuses on the effects of different alkaline-activator modulus, Na_2O dosages, and WMP content on its workability and mechanical properties.

2 Experimental Investigations

2.1 Raw Material

The GGBS used in this study was a commercially available S95 grade slag powder with a fineness of over 90% passing through a 200 mesh sieve. The WMP was provided by a local marble slab processing plant in Guilin city, which was the waste product from processing. The waste marble was dried and sieved to obtain the required WMP with a diameter of 0.9 mm. The sodium hydroxide (NaOH) used was a commercially available flake sodium hydroxide with a purity of 96%. The water glass used is a solution produced by Guilin City Pohua Alkali Factory, with a SiO_2 dosage of 27.6%, a Na_2O dosage of 8.75%, and a modulus (molar ratio of SiO_2/Na_2O) of 3.26. The water used was ordinary tap water from Guilin city. The sand used was local quartz sand in Guilin city, with a fineness of 20% for 10–20 mesh, 30% for 20–40 mesh, and 50% for 80–120 mesh. Table 1 shows the chemical compositions and specific surface areas of GGBS and WMP detected by X-ray fluorescence spectroscopy.

Table 1. Chemical composition and specific surface area of GGBS and WMP.

Material	Composition									Specific Surface area (m^2/g)
	CaO	SiO_2	Al_2O_3	TiO_2	Fe_2O_3	P_2O_5	Others	LOI	Total	
GGBS	56.90	25.55	12.07	2.68	0.73	0.49	1.96	0.72	99.14	0.43
WMP	54.09	1.07	1.14	0.04	0.39	0.35	0.39	42.48	99.95	1.10

2.2 Designing Mix Proportions

Based on the research of other scholars and our research group [5, 14, 15], this paper investigates the fundamental law of the effects of different alkaline-activator modulus (1.0, 1.2, 1.4, 1.6, 1.8), Na_2O dosages (4%, 7%, 10%), and WMP content (replacing 20%, 25%, 30%, 35%, 40% of GGBS) on SWAHPC. Table 2 shows the mix design for each group of experiments. The water-binder ratio for each group of experiments is 0.3, and the total amount of binder material is 1499.8 kg, with a binder-to-aggregate ratio of 2:3.

Table 2. Mix design of WSAHPC.

Group	$\%SiO_2/Na_2O$	% Na_2O	%WMP	Raw material/(kg/m^3)					
				Slag	WMP	NaOH	Na_2SiO_3	Water	Sand
M100730	1.0	7	30	905.3	387.9	97.9	368.1	216.8	2249.6
M120730	1.2	7	30	891.1	381.8	89.2	441.8	176.2	2249.6
M140730	1.4	7	30	876.7	375.8	80.4	515.3	135.8	2249.6
M160730	1.6	7	30	862.6	369.7	71.9	588.9	95.2	2249.6
M180730	1.8	7	30	848.4	363.5	63.1	662.6	54.8	2249.6
M120430	1.2	4	30	959.0	411.0	51.0	252.5	293.5	2249.6
M121030	1.2	10	30	823.0	352.7	127.3	630.9	58.9	2249.6
M120720	1.2	7	20	1018.4	254.5	89.2	441.8	176.2	2249.6
M120725	1.2	7	25	954.6	318.3	89.2	441.8	176.2	2249.6
M120735	1.2	7	35	827.4	445.6	89.2	441.8	176.2	2249.6
M120740	1.2	7	40	763.6	509.1	89.2	441.8	176.2	2249.6

2.3 Testing Method

The experimental procedure involved mixing NaOH, Na_2SiO_3, and water based on the mix design determined for the experiment. The resulting mixture was then cooled at room temperature for six hours before use. The alkaline activator, GGBS, WMP, and sand were mixed uniformly to form mortar and make 40 mm × 40 mm × 100 mm specimens. The specimens were placed in a standard curing room. Figure 1 shows the preparing process of SWAHPC specimens. The specimens were tested for compressive and flexural strengths after 3, 7 and 28 days of curing in accordance with GB/T 17671–1999. Three randomly selected specimens were tested for three-point flexural strength and the average value was used as the experimental result. After that, the six fractured specimens were tested for compressive strength and the average value was used as the experimental result. The setting times of the paste were measured using a standard Vicat apparatus according to GB/T 1346–2001. The fluidity of the cement mortar was assessed using a cement mortar flow table according to GB/T 2419–2005.

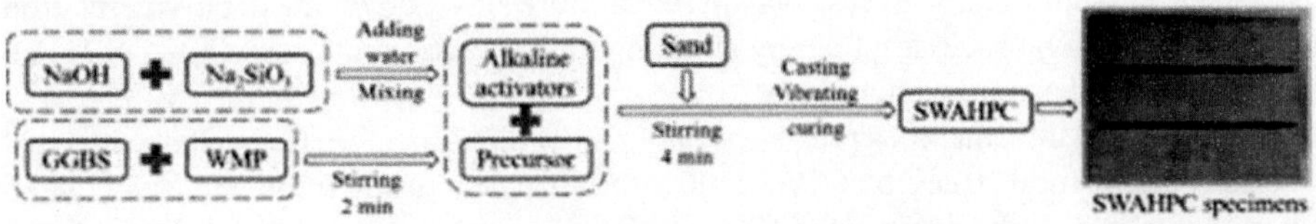

Fig. 1. Preparing process of SWAHPC

3 Results and Discussion

The mechanical and workability performance of SWAHPC are dependent on several factors, including the alkaline-activator modulus, Na_2O dosage, and WMP content. The mechanical and workability performance of SWAHPC exhibit distinct trends under varying alkaline-activator modulus (ranging from 1.0 to 1.8), Na_2O dosages (4%, 7%, and 10%), and WMP content (ranging from 20% to 40%).

3.1 Effect of the Alkaline-Activator Modulus

Effect of Alkaline-Activator Modulus on the Workability Performance. With the increase in the modulus, the setting times of SWAHPC show an increasing trend. As shown in Fig. 2(a), when the modulus increases from 1.0 to 1.2, 1.4, 1.6, and 1.8, the increase rates of the initial setting time are 26.4%, 35.8%, 81.1%, and 113.7%, respectively. The increase rates of the final setting time are 11.3%, 19.7%, 57.7%, and 93.0%, respectively. As shown in Fig. 2(b), when the modulus increases from 1.0 to 1.2, 1.4, 1.6, and 1.8, the increase rates of fluidity are 11.0%, 11.5%, 18.7%, and 21.4%, respectively. This is because a higher alkaline-activator modulus represents a relatively lower alkali content, which leads to a slower polymerization of SWAHPC. At the same time, a higher modulus results in a lower viscosity of SWAHPC, leading to an increase in the setting times and fluidity.

Effect of the Alkaline-Activator Modulus on the Mechanical Performance. According to the data presented in Fig. 3(a), the compressive strength of SWAHPC at different ages exhibits an increasing and then decreasing trend as the alkaline-activator modulus increases. At a modulus of 1.6, the compressive strength of SWAHPC is 81.6 MPa, 97.8 MPa, and 106.9 MPa at 3 d, 7 d, and 28 d, respectively. The compressive strength increases by 5.3%, 21.3%, and 8.5% at three curing periods, respectively, when the modulus rises from 1.0 to 1.6. Conversely, when the modulus increases from 1.6 to 1.8, the compressive strength at three curing periods decreases by 14.1%, 8.8%, and 7.9%, respectively. For modulus less than 1.6, the increase in the alkaline-activator modulus leads to an increase in the $[SiO_4]^{4-}$ concentration, promoting the polymerization and generation of more C-(N)-A-S-H, increasing compressive strength. However, as the modulus increases, the OH^- concentration decreases, inhibiting the polymerization reaction. When the modulus exceeds 1.6, the relatively insufficient alkaline content of the concrete further inhibits the polymerization reaction, which reduces SWAHPC's compressive strength. Additionally, excess Na^+ in the mixture slows down the depolymerization and polymerization processes, thereby affecting the generation rate of C-(N)-A-S-H and leading to a decrease in strength [16].

According to the outcomes presented in Fig. 3(b), With an increase in alkali activator modulus, the flexural strength of SWAHPC exhibits a decreasing trend. However, when the modulus is 1.4, the early-stage (3 d and 7d) flexural strength of SWAHPC shows an increase. The flexural strength of SWAHPC is found to be the highest when the modulus is 1.0, with strengths of 8.8 MPa, 7.5 MPa, and 6.9 MPa at three curing periods, respectively. With an increase in the modulus from 1.0 to 1.8, the growth rates of the flexural strength are observed to be −19.3%, −14.7%, and −11.6% at three curing

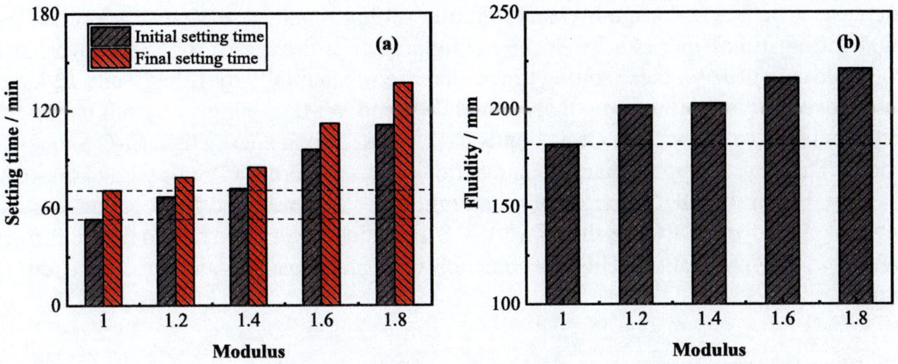

Fig. 2. Working performance of WSAHPC in different modulus

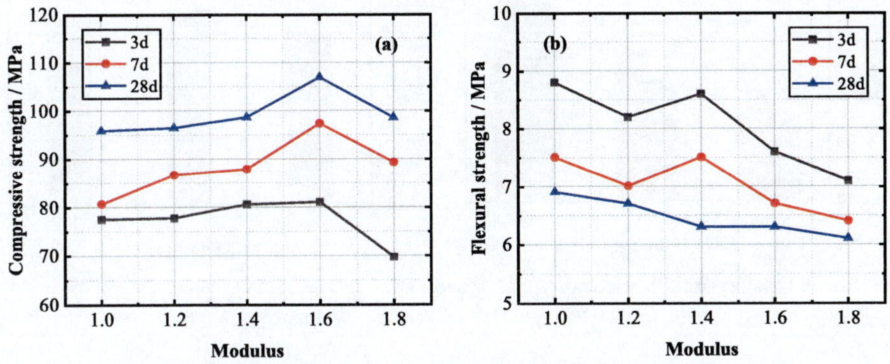

Fig. 3. Mechanical performance of WSAHPC in different modulus

periods, respectively. However, when the modulus increases from 1.2 to 1.4, the flexural strength at early stage is observed to increase, with growth rates of 4.9% and 4.3%, respectively. This could be attributed to the fact that a modulus of 1.4 is more conducive to the development of the early-stage adhesion strength of SWAHPC, increasing its early-stage flexural strength in the process. Moreover, with an increase in the curing age, the effect of modulus on the flexural strength of SWAHPC gradually decreases, and the flexural strength curve at 28 d is the gentlest. However, it is noteworthy that the flexural strength of SWAHPC decreases to varying degrees with an increase in curing age, which is significantly different from the development trend of OPC flexural strength. The internal adhesion strength of SWAHPC decreases with increasing curing time, resulting in a decrease in flexural strength. However, further research is required to elucidate the specific reasons behind this observation.

3.2 Effect of Na_2O Dosages

Effect of Na_2O Dosages on the Workability Performance. According to the data presented in Fig. 4(a), With an increase in Na_2O dosage, the setting times of SWAHP increase. At a Na_2O dosage of 7%, the initial setting times increase by 53.3% and the final setting times increase by 36.2%, compared to a dosage of 4%. Moreover, at a Na_2O dosage of 10%, these setting times increase substantially by 128.9% and 124.1%, respectively. It is worth noting that both GGBS and WMP are high-calcium materials, and studies have shown that Ca-O bonds have weaker bond energy than Si-O and Al-O bonds. Thereby, during the hardening reaction process, more Ca^{2+} ions tend to dissolve quickly. When the OH^- environment is high, Ca^{2+} ions released from precursors are more likely to form $Ca(OH)_2$ than C-(N)-A-S-H, which generates a film to inhibit further hydration polymerization. This phenomenon ultimately leads to an increase in setting times.

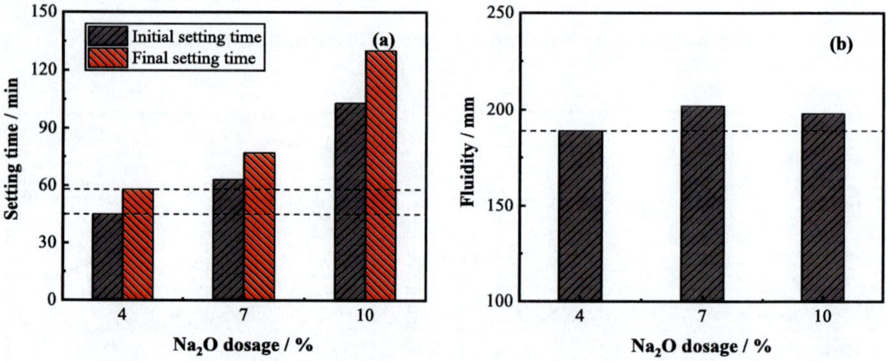

Fig. 4. Working performance of WSAHPC in different Na_2O dosages

According to the data presented in Fig. 4(b), the fluidity of SWAHP exhibits a nonlinear relationship with Na_2O dosage. Specifically, the fluidity increases firstly and then decreases, achieving a its peak at 7%. In comparison to a Na_2O dosage of 4%, the fluidity of SWAHP increases by 6.9% and 4.8% at Na_2O dosages of 7% and 10%, respectively. This trend can be attributed to the decrease in polymerization hydration efficiency as the Na_2O dosage increases, resulting in increased fluidity. However, when the Na_2O dosage reaches 10%, the proportion of water glass and sodium hydroxide increases, while the amount of water decreases, leading to a subsequent increase in SWAHPC viscosity and decrease in fluidity. It is important to note that these findings hold significant implications for the optimal design and formulation of SWAHPC, particularly in relation to its fluidity and Na_2O dosage.

Effect of Na_2O Dosage on the Mechanical Performance. According to Fig. 5(a), as the Na_2O dosage increases, the compressive strength of SWAHP show an increase followed by a decrease with age. When the Na_2O dosage is 7%, the compressive strength of all ages is the best, with compressive strengths of 77.8 MPa, 86.6 MPa, and 96.4 MPa at three curing periods, respectively. Compared to Na_2O dosages of 4% and 10%, the compressive strength growth rates at 3 d are 8.5% and 13.2%, at 7 d are 2.3% and 14.1%, and at 28 d are 5.5% and 3.1%, respectively, when the Na_2O dosage is 7%.

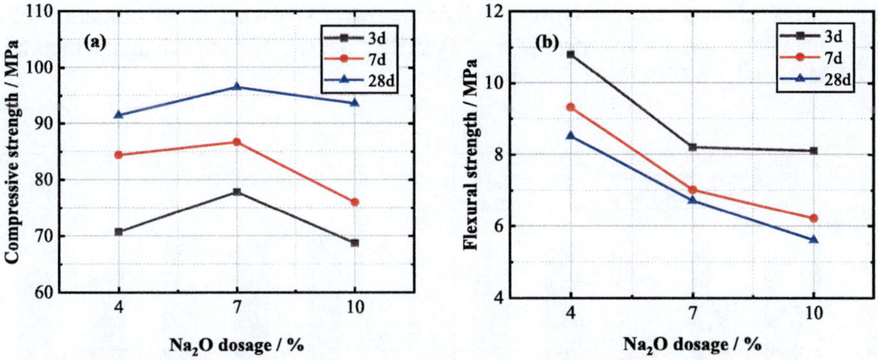

Fig. 5. Mechanical performance of WSAHPC in different Na_2O dosages

According to Fig. 5(b), as the Na_2O dosage increases, the flexural strength of SWAHPC shows a decreasing trend at all ages. When the Na_2O dosage is 4%, the flexural strength at all ages is the best, with strengths of 10.8 MPa, 9.3 MPa, and 8.5 MPa at three curing periods, respectively. Compared to Na_2O dosages of 7% and 10%, the compressive strength growth rates at 3 d are 8.5% and 13.2%, at 7 d are 2.3% and 14.1%, and at 28 d are 5.5% and 3.1%, respectively, when the Na_2O dosage is 4%. It is important to note that the early precursor polymerization rate is rapid when the Na2O dosage is too high, which leads to the production of a large amount of gel-coated particles that form a film between particles, which inhibits the subsequent polymerization reaction and causes a decrease in the strength of SWAHPC. Excessive Na_2O dosage is not conducive to the mechanical performance of SWAHPC. The optimal compressive strength is achieved at a Na_2O dosage of 7%, and the optimal flexural strength is achieved at a Na_2O dosage of 4%.

3.3 Effect of WMP Content

Effect of WMP Content on the Workability Performance. The results presented in Fig. 6(a) show that with an increase in WMP content from 20% to 25%, 30%, 35%, and 40%, the initial setting time of the mixture increases by 3.3%, 9.8%, 13.1%, and 11.5%, respectively. Similarly, the final setting time of the mixture also increases by 1.3%, 3.9%, 5.2%, and 6.6%, respectively, under the same conditions. The lower activity of WMP compared to GGBS leads to a delay in the hydration reaction and consequently prolongs the setting time. Thus, as the WMP content increases, the setting time of the mixture also increases. Moreover, Fig. 6(b) illustrates that as the WMP content increases from 20% to 25%, 30%, 35%, and 40%, the fluidity of the mixture decreases by −1.4%, −0.4%, −5.3%, and −9.1%, respectively. The larger specific surface area of WMP compared to GGBS necessitates more water for the formation of a water film, resulting in less free water. Consequently, as the WMP content increases, the fluidity of the mixture decreases.

Effect of WMP Content on the Mechanical Performance. According to Fig. 7(a) and 7(b), with increasing WMP content (20% – 40%), the compressive and flexural strengths of SWAHPC show an increase followed by a decrease. When the WMP content is 25%, the compressive and flexural strengths of SWAHPC are the best at each age, which were 101.1 MPa and 7.2 MPa at 28 d, respectively.

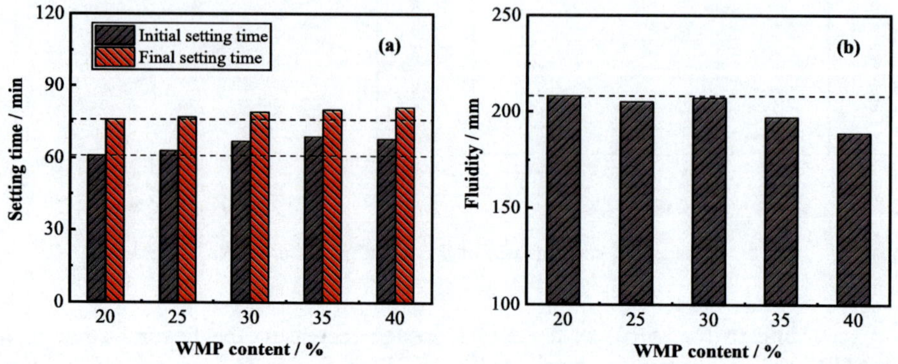

Fig. 6. Working performance of WSAHPC in different WMP content

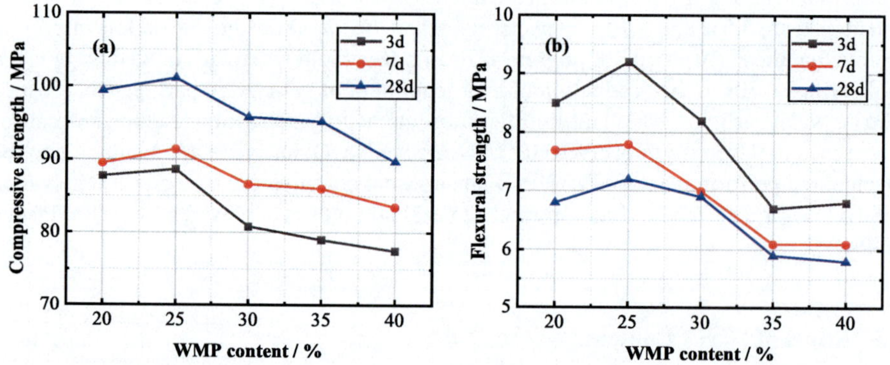

Fig. 7. Mechanical performance of WSAHPC in different WMP content

Compared with the WMP content of 20%, when the WMP content is 25%, the growth rates of compressive strength and flexural strength are 1.0% and 8.2% at 3 d, respectively, 5.7% and 1.3% at 7 d and 6.9% and 5.9% at 28 d, respectively. When the WMP content is larger than 25%, the mechanical performance of SWAHPC decrease. Compared with the WMP content of 25%, when the WMP content is 40%, the growth rates of strength are −12.5% and −26.1% at 3 d, respectively, −8.6% and −21.8% at 7 d, and −11.3% and −19.4% at 28 d, respectively. WMP is an inert material with insufficient activity due to its low Si and Al content. An appropriate amount of WMP can promote hydration and improve the internal structure of concrete, playing a positive role in the mechanical properties of SWAHPC. However, when the WMP content is too high, the low activity of WMP reduces the mechanical performance of SWAHPC.

4 Conclusions

This study investigates the influence and rules of the alkali activator modulus, Na_2O dosage, and WMP content on the flexural and compressive strength, setting time, and fluidity of SWAHPC prepared from GGBS and WMP as precursors through workability tests and mechanical performance. The conclusions are as follows:

(1) When the alkali activator modulus is 1.0–1.8, the Na_2O dosage is 4%–10%, and the WMP content is 20%–40%, SWAHPC with a compressive strength of 89.7 MPa–106.9 MPa, a flexural strength of 5.6 MPa–8.5 MPa, an initial setting time of 45 min–111 min, a final setting time of 58 min–130 min, and a fluidity of 182 mm–221 mm can be easily prepared. With a WMP content not exceeding 40%, high-strength concrete with excellent workability and strength grade not lower than C85 can be readily produced.
(2) With an increase in in alkali activator modulus, the compressive strength of SWAHPC shows an increase followed by a decrease, with the optimum modulus being 1.6; the flexural strength generally shows a decreasing trend, with the optimum modulus being 1.0; the setting time and fluidity are consistent, showing an increasing trend.
(3) With an increase in Na_2O dosage, the compressive strength of SWAHPC shows an increase followed by a decrease, with the optimum dosage being 7%; the flexural strength shows a decreasing trend, with the optimum dosage being 4%; the final setting time shows an increasing trend, and the fluidity is consistent with compressive strength.
(4) With an increase in WMP content, the compressive and flexural strengths of SWAHPC show a trend of increasing and then decreasing, with the optimum WMP content being 25%; the final setting time shows an increasing trend, and the fluidity shows a decreasing trend.
(5) With an increase in curing age, the flexural strength of SWAHPC decreases to varying degrees, which is significantly different from the development law of the flexural strength of OPC.

Acknowledgments. The authors gratefully acknowledge the financial support by the Natural Science Foundation of Guangxi (Grant No. 2021GXNSFBA220049), the Guangxi Science and Technology Base and Special Fund for Talents Program (Grant No. GuikeAD22035999), the National Natural Science Foundation of China (Grant Nos. 52108201 and U22A20244), the Guangxi Key Laboratory of New Energy and Building Energy Saving (grant No. 22-J-21-9), the Guangxi Science and Technology Major Project (grant No. GuikeAA22068073-3), and the Innovation Project Guangxi Graduate Education (grant No. YCSW2023339).

References

1. Amiri, M., Hatami, F., Golafshani, E.M.: Evaluating the synergic effect of waste rubber powder and recycled concrete aggregate on mechanical properties and durability of concrete. Case Stud. Constr. Mater. **15**, e00639 (2021)
2. Deventer, J., Provis, J.L., Duxson, P.: Technical and commercial progress in the adoption of geopolymer cement. Miner. Eng. **29**, 89–104 (2012)

3. Wang, S.D., Scrivener, K.L., Pratt, P.L.: Factors affecting the strength of alkali-activated slag. Cem. Concr. Res. **24**, 1033–1043 (1994)
4. Barbosa, V., Mackenzie, K., Thaumaturgo, C.: Synthesis and characterisation of materials based on inorganic polymers of alumina and silica: sodium polysialate polymers. Int. J. Inorg. Mater. **2**, 309–317 (2000)
5. Zhang, P., Kang, L., Zheng, Y., Zhang, T., Zhang, B.: Influence of SiO_2/Na_2O molar ratio on mechanical properties and durability of metakaolin-fly ash blend alkali-activated sustainable mortar incorporating manufactured sand. J. Mater. Res. Technol. **18**, 3553–3563 (2022). https://doi.org/10.1016/j.jmrt.2022.04.041
6. Palomo, A., Grutzeck, M.W., Blanco, M.T.C., Blanco, M.T.C.: Alkali-activated fly ashes: a cement for the future. Cem. Concr. Res. **29**, 1323–1329 (1999)
7. Puertas, F., Martnez-Ramrez, S., Alonso, S., Vázquez, T.: Alkali-activated fly ash/slag cements: strength behaviour and hydration products. Cem. Concr. Res. **30**, 1625–1632 (2000)
8. Prusty, J.K., Pradhan, B.: Multi-response optimization using Taguchi-Grey relational analysis for composition of fly ash-ground granulated blast furnace slag based geopolymer concrete. Constr. Build. Mater. **241**, 118049 (2020)
9. Podolsky, Z., Liu, J., Dinh, H., Doh, J.H., Fragomeni, S.: State of the art on the application of waste materials in geopolymer concrete. Case Stud. Constr. Mater. **15**, e00637 (2021)
10. Ince, C., Hamza, A., Derogar, S., Ball, R.J.: Utilisation of waste marble dust for improved durability and cost efficiency of pozzolanic concrete. J. Clean. Prod. **270**, 122213 (2020)
11. Erguen, A.: Effects of the usage of diatomite and waste marble powder as partial replacement of cement on the mechanical properties of concrete. Constr. Build. Mater. **25**, 806–812 (2011)
12. Kumar, J., Jatin, Jangra, P., Pham, T.M., Lim, Y.Y.: Sustainable alkali activated concrete with fly ash and waste marble aggregates: strength and durability studies. Constr. Build. Mater. **283**, 122795 (2021)
13. Coppola, B., Palmero, P., Montanaro, L., Tulliani, J.M.: Alkali-activation of marble sludge: Influence of curing conditions and waste glass addition. J. Eur. Ceram. Soc. **40**, 3776–3787 (2020)
14. Thakur, A.K., Pappu, A., Thakur, V.K.: Synthesis and characterization of new class of geopolymer hybrid composite materials from industrial wastes. J. Clean. Prod. **230**, 11–20 (2019)
15. Liu, B., Geng, S., Ye, J., Liu, X., Lin, W., Wu, S., Qian, K.: A preliminary study on waste marble powder-based alkali-activated binders. Constr. Build. Mater. **378**, 131094 (2023)
16. Murayama, N., Yamamoto, H., Shibata, J.: Mechanism of zeolite synthesis from coal fly ash by alkali hydrothermal reaction. Int. J. Miner. Process. **64**, 1–17 (2002)

Effect of Sintering Temperature on Properties of Regenerated Sintered Sheet Brick

Bing-zhang Huang[1] (✉), Guang-feng Li[2], Li-hua Pan[2], Yu Zhang[2], and Bang-biao Huang[2]

[1] College of Civil Engineering and Architecture, Liuzhou Institute of Technology, Liuzhou 545616, China
71889671@qq.com

[2] College of Civil Engineering and Architecture, Guangxi University of Science and Technology, Liuzhou 545006, China

Abstract. In order to study the effect of sintering temperature on the performance of recycled shale bricks, in this paper, a total of 8 groups of mix proportions were designed by changing the proportion of shale, waste shale brick powder, and waste concrete powder to prepare recycled sintered shale bricks. The sintering temperature was increased at a rate of 1 °C/min to 900 °C–1000 °C, and performance tests were conducted with the requirements of the "Test Methods for Wall Building Bricks" (GB/T 2542–2012). Results of test indicate that when the sintering temperature of the brick is 1050 °C, the minimum amount of waste mixture can be used to produce regenerated sintered shale bricks with excellent strength and density. By reducing the sintering temperature and increasing the content of waste mixture, the thermal conductivity of recycled sintered bricks can be improved. In practical applications, waste mixtures can be added and the sintering temperature can be adjusted according to actual engineering requirements to further encourage the recycling and utilization of construction waste.

Keywords: Recycled sintered shale brick · Waste mix · Sintering temperature · Mechanical properties · Physical properties

1 Introduction

Older buildings will be demolished as part of plans for increased urbanization and the promotion of urban regeneration. This will result in a considerable volume of solid building debris, and environment will be harm with improper handling [1]. The conventional method of solid waste accumulation and treatment in construction is no longer consistent with China's green growth policy. The report of the 19th National Congress of the CPC makes recommendations to "improve solid waste and waste disposal", "promote comprehensive conservation and recycling of resources, reduce energy consumption and material consumption, and realize the recycling link between the production system and living system" [2], with the recycling and utilization of solid waste in construction being the future development trend.

W. Guo et al. (Eds.): GBCESC 2023, LNCE 328, pp. 387–396, 2024.
https://doi.org/10.1007/978-981-99-9947-7_41

The recycling of waste brick has been one of the major areas of academics in recent years because it is a significant component of solid waste in structures. Waste clay brick was used as the fine aggregate in the mortar created by Bektas F et al. [3], who reported that doing so improved the mortar's resilience to freeze-thaw. In a comparison study of the mechanical qualities of asphalt mixtures made from waste brick powder and limestone powder, Chen M et al. [4] reported that the asphalt mixture made from waste brick powder had better mechanical properties. Ge Zhi et al.'s [5] experimental study on concrete made by partially substituting cement with leftover clay brick powder revealed that doing so would not harm the concrete's performance. The composite of construction waste brick powder and other building powder materials has good activity, and the compressive strength of 28d mortar with a content of no more than 40% can reach 50 MPa, according to Xue Cuizhen et al.'s [6] study of the relationship between construction waste brick powder, alkali excitation, and compound excitation. Che Fa et al. [7] prepared recycled concrete using waste clay brick powder rather than cement and discovered that the early compressive performance and elastic modulus of recycled concrete were primarily impacted by the waste clay brick powder. Letelier V et al.'s research [8] on concrete made using recycled aggregate and waste brick powder demonstrated that the percentages of 15% waste brick powder and 30% recycled aggregate have no discernible impact on the strength of the concrete. Chen Wen et al. [9] used waste clay brick powder to prepare foamed concrete. Through study, they reported that substituting leftover clay brick powder for cement and sand might improve the compressive strength and thermal insulation properties of foamed concrete, as well as lower the concrete's dead weight. The brick powder was employed to create new brick powder blocks, and Chen Bo et al. [10] discovered that the block with a 50% brick powder content had the best compressive strength. Instead of using fly ash, Zhao Yahui et al. [11] created mortar using superfine brick powder. They discovered that while the inclusion of brick powder improved the mortar's ability to retain water and resist freezing, using too much brick powder reduced the mortar's compressive strength. Different techniques were employed by Yuan C et al. [12] to increase the activity of brick powder, and they discovered that mechanical excitation during a 45-min ball milling process and high temperature at 800 °C had the best results, with the greatest activity index reaching 71%. The recycled brick powder was utilized to make phosphorus building gypsum by Ma Qian et al. [13], who discovered that 10% was the ideal amount to use. The mechanical, shrinkage, and freeze-thaw properties of the new type of subgrade stabilizing material were found to be best when the brick powder content was 30% and the sulfur fixing ash content was 70% by Zhang Rui et al. [14] using waste brick powder. Numerous researchers [15–17] have conducted a study on recycled brick aggregate in addition to that on discarded brick powder, with different findings.

In this work, we propose that regenerated sintered shale rock brick be prepared using a waste mixture made of leftover waste sintered shale rock brick powder and leftover waste mortar powder in building waste. We also investigate the impact of the amount of leftover waste sintered shale rock brick powder and leftover waste mortar powder on the functionality of regenerated sintered shale rock brick.

2 Experiment

2.1 Raw Materials

Shale. The shale selected for the test is from brick factories around Liuzhou. Table 1 displays the compositional content of the chosen shale. The main component is SiO_2, and the shale firing loss is 9.51%.

Table 1. Chemical composition of shale

composition	SiO_2	MgO	CaO	Fe_2O_3	Al_2O_3	K_2O	Na_2O	Firing loss
Content (%)	63.86	1.47	5.9	6.71	14.17	1.73	0.87	9.51

Waste Sintered Shale Brick. The average mass of the ten abandoned sintered shale rock bricks that were chosen for weighing was 3337 g. The abandoned sintered shale rock bricks' primary component, which can be seen from the XRD pattern of their composition in Fig. 1, is SiO_2.

Waste Mortar. The test cement mortar is M5 cement mortar constructed with medium sand and 32.5 grade cement. The ratio of cement, sand, and water is 1:5.23:0.53, the density of cement mortar is up to 2100 kg/m3, and this mixture complies with JGJT 98–2010's standards for the design of masonry mortar mix ratio [18]. Before usage, the leftover mortar is broken up and put through a vibrating screen machine after drying. Figure 2 depicts the waste mortar's XRD analysis pattern, which reveals that SiO_2 and $CaCO_3$ make up the majority of the powder.

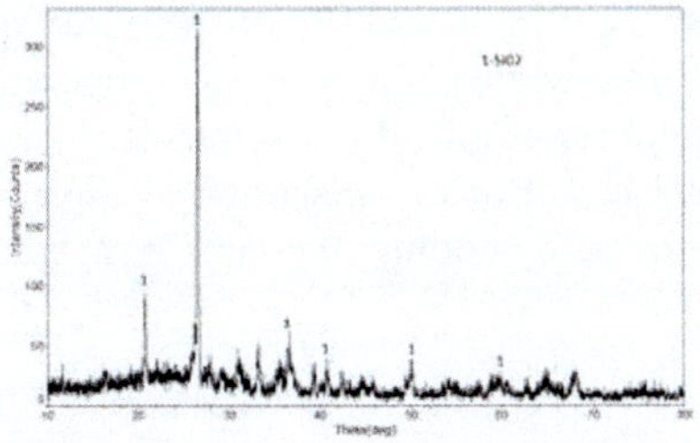

Fig. 1. XRD pattern of abandoned sintered brick

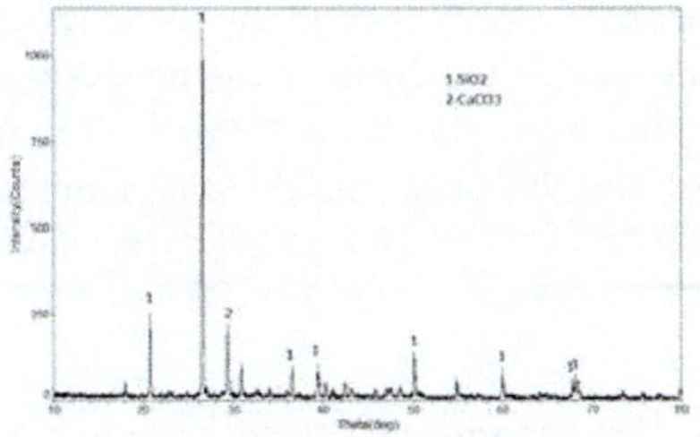

Fig. 2. XRD pattern of Mortar powder

It is possible to make recycled sintered shale brick using a waste mixture of waste sintered shale brick and waste mortar rather than shale because the principal ingredient in all three is SiO_2.

2.2 Experimental Mix Design

The test block is made using 8 combinations of shale, waste sintered pagination brick powder, and waste mortar powder following the general proportioning theory of sintered pagination brick. The mixing amount of waste mixture is 0%, 10%, 20%, 30%, 40%, 50%, 60%, and 70%. See Table 2 for specific proportioning.

Table 2. Mix proportion data

Specimen number	shale/g	Waste sintered brick powder/g	Waste mortar powder/g	Sintering temperature/°C
A0	10000	0	0	900–1150
A10	9000	776	224	900–1150
A20	8000	1552	448	900–1150
A30	7000	2327	673	900–1150
A40	6000	3103	897	900–1150
A50	5000	3879	1121	900–1150
A60	4000	4655	1345	900–1150
A70	3000	5430	1570	900–1150

2.3 Specimen Making

According to Table 2, the regenerated sintered shale rock block test block was prepared using a comparison of the raw materials through aging, forming, drying, roasting, and other procedures. According to the research team's earlier work [19], the prepared specimen was heated during roasting at a rate of 1 °C/min while being raised to temperatures ranging from 900 °C to 1100 °C. The temperature range was divided into 50 °C intervals, and the firing period was maintained for 8 h. All of the specimen's preparation procedures were finished after the calcination process by cooling the sample repeatedly for 12 h.

2.4 Test Method

The physical and mechanical characteristics of recycled sintered shale brick, including compressive strength, volume density, volume sintering shrinkage, and thermal conductivity, were tested following the specifications of the Test Method for Wall Building Bricks (GB/T 2542–2012) [20]. (1) Compressive strength measurement.

The Test Method for Wall Bricks (GB/T 2542–2012) for the compressive strength test was followed. The following actions are detailed:

Each group of 10 pieces of the appearance of the complete sample was sawn into two half bricks, with the superimposed part being no less than 100 mm cutting port opposite stacked, and the length and width of each group of 2 samples of the connection surface or compression surface were measured with an accuracy of 1 mm. The sample was then placed in the center of the pressing plate, parallel to the compression surface, and compressed at a speed of 2–6 kn/s. We recorded the maximum failure load P using Eq. (1) to get the compressive strength:

$$R_p = \frac{P}{L \times B} \tag{1}$$

where:

Rp—Compressive strength, unit: MPa;

P—maximum failure load, the unit is Newton (N);

L—length of compression surface (connecting surface), the unit is (mm);

B—Width of the compression surface (connecting surface), expressed in millimeters (mm).

(2) Volume density measurement.

According to Test Method for Wall Bricks (GB/T 2542–2012), the volume density test was carried out as follows:

From each group, 5 complete samples were selected, and they were dried to a consistent quality in a blast dryer at 105 °C to 5 °C (the difference between the two weights was less than 0.2%, and the time between the two weights was 2 h). Their mass was described as m, and their appearance was checked. No missing edges or angles were allowed. After drying, the sample size was measured twice, with the average value used to calculate the volume V, with the volume density was determined using Eq. (2):

$$\rho = \frac{m}{V} \times 10^9 \tag{2}$$

where:

ρ—volume density, the unit is kg per cubic meter (kg/m^3);

m—the dry mass of the sample, expressed in kg (kg);

V—sample volume, in cubic millimeters (mm^3).

(3) volume sintering shrinkage rate.

After drying and shaping, using a brick caliper, we measured the brick billet's length, width, and height (designated as L0, B0, and H0, respectively). We measured the same location after sintering, designated as L1, B1, and H1 accordingly, where the volume sintering shrinkage rate is computed using Eq. (3):

$$V_S = \frac{L_0 \times B_0 \times H_0 - L_1 \times B_1 \times H_1}{L_0 \times B_0 \times H_0} \tag{3}$$

(4) Thermal conductivity test.

In the unsteady state method of thermal conductivity, the hot disk with a transient plate heat source approach was used to measure the thermal conductivity. The test probe is then positioned between two building waste recycled sintered shale bricks, the thermal resistance is first attached to the data acquisition device, and the acquired temperature curve is screened to determine the proper thermal conductivity.

3 Test Results and Analysis

When compared to pure shale sintered brick, the performance of recycled sintered brick made from the waste mixture is different. The degree of partial performance of recycled sintered brick will vary depending on the amount of waste mixture and sintering temperature.

3.1 Influence of Sintering Temperature on Compressive Strength of Regenerated Sintered Brick

One crucial metric to assess the fundamental functionality of sintered shale brick products is compressive strength. Figure 3 depicts the variations in compressive strength of recycled sintered shale brick specimens at various sintering temperatures.

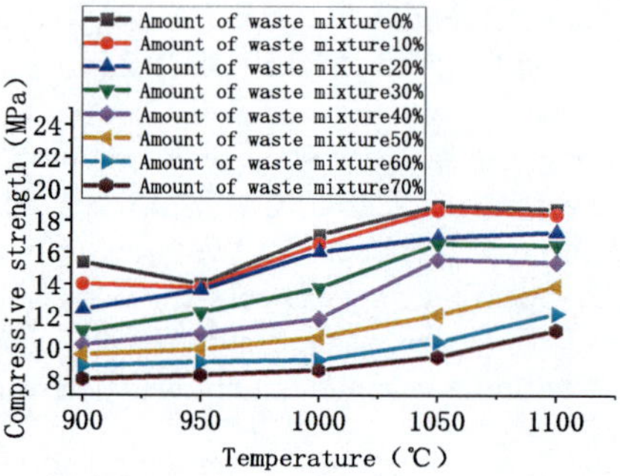

Fig. 3. Effect of sintering temperature on compressive strength of recycled sintered brick

As seen in Fig. 3, when the waste mixture is fixed, the compressive strength of regenerated sintered shale brick generally increases with temperature increases between 900 °C and ~1100 °C. This finding suggests that increasing the sintering temperature is beneficial for improving the compressive performance of regenerated sintered shale brick. The compressive strength of the brick body is more susceptible to the sintering temperature range of 900 °C–1050 °C when the waste mixture content of resintered shale brick is less than 40%. The compressive strength of the brick body is more sensitive to the sintering temperature of 1000 °C–1100 °C when the waste mixture content of the regenerated sintered shale brick is greater than 50%, indicating that the sensitivity of the shale and waste mixture to the sintering temperature is different and the regenerated sintered shale with more waste mixture requires a higher sintering temperature. It is not possible to improve the compressive strength of the regenerated sintered shale revolution because, at the same temperature, the amount of waste mixture causes the compressive strength of the regenerated sintered shale revolution to fall. The compressive strength of the brick body is 9.47 MPa when the waste mixture content is 70% and the sintering temperature is 1050 °C, which is insufficient to meet the criteria of the strength standard MU10 sintered pagination brick. The compressive strength of the regenerated sintered pagined brick is 11.13 MPa when the waste mixture content is 70% and the sintering temperature is increased to 1100 °C. However, the brick itself exhibits over burning phenomenon and the quality is subpar, as shown in Fig. 4, indicating that the sintering temperature will limit the improvement of the compressive strength of the regenerated sintered pagined brick. Regenerated sintered shale bricks are better suited for sintering at a temperature of 1050 °C.

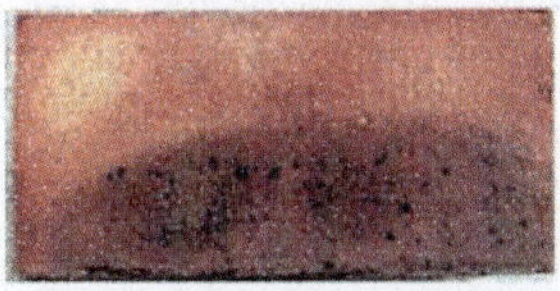

Fig. 4. Over burned recycled sintered shale brick

3.2 Influence of Sintering Temperature on Volume Density of Regenerated Sintered Shale Brick

Figure 5 shows the volume density changes of regenerated sintered shale brick specimens at different sintering temperatures.

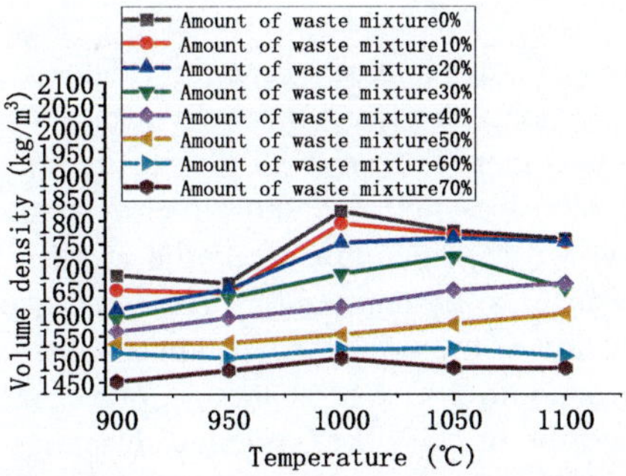

Fig. 5. Effect of sintering temperature on volume density of a regenerated sintered brick

As observed in Fig. 5, the volume density of regenerated sintered shale brick typically increases with the increase in temperature between 900 °C and 1100 °C when the dosage of the waste mixture is constant. When the sintering temperature is constant, adding more waste mixture causes the volume density of recycled sintered paginite to drop. Volume density change trends often resemble those of compressive strength. The compressive strength increases with increasing volume density. As the temperature rises, the shale and waste combination will melt, improving the brick's ability to be compacted. This is the reason why the volume density rises as the temperature rises. But if the temperature is too high, too much molten material will be produced, and this will cause tiny holes to appear inside the brick. As a result, when the temperature is too high, the volume density of regenerated sintered shale rock brick will drop. To guarantee the bulk density and compressive strength of the regenerated sintered shale brick, the sintering temperature of 1050 °C should be chosen.

3.3 Influence of Sintering Temperature on Volume Sintering Shrinkage of Regenerated Sintered Sheet Brick

Figure 6 shows the volume sintering shrinkage of regenerated sintered shale brick specimens at different sintering temperatures.

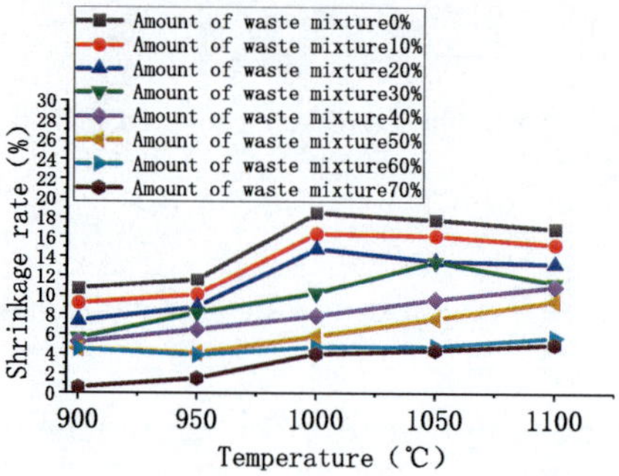

Fig. 6. Effect of sintering temperature on volume shrinkage of regenerated sintered sheet brick

The volume sintering shrinkage rate of regenerated sintered bricks generally exhibits an increasing trend with temperature rise, as shown in Fig. 6. When the waste mixture percentage is less than 40%, there is an overall increasing tendency in the volume sintering shrinkage, which increases first and subsequently reduces. The rate of volume sintering shrinkage grows with temperature when the waste mixture content is greater than 50%, primarily because as the temperature rises, the gap between the shrinkage occurring within and outside of the brick widens and intensifies. In the meantime, the high temperature-induced liquid phase of shale and other materials will produce gas and other byproducts, and the brick will experience shrinkage and breaking under the effect of gas and temperature. When more waste mixture is added, the recycled sintered shale brick shrinks in volume at a slower pace while the sintering temperature stays the same. The main reason is that the chemical stability of the waste mixture is better than that of shale. Therefore, more waste mixtures can be used to maintain a low sintering temperature if it becomes required to minimize the volume sintering shrinkage rate of brick.

3.4 Influence of Sintering Temperature on Thermal Conductivity of Regenerated Sintered Shale Brick

Figure 7 shows the thermal conductivity of regenerated sintered shale brick specimens at different sintering temperatures.

Figure 7 illustrates how brick's thermal conductivity normally rises as temperature rises and falls as waste mixture content rises. The pore structure within the brick will

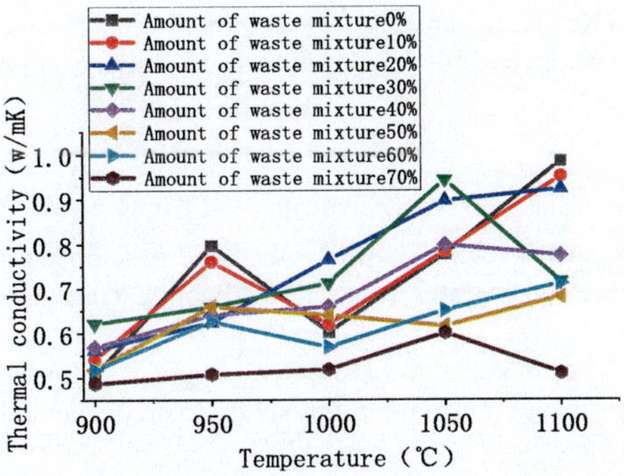

Fig. 7. Effect of sintering temperature on thermal conductivity of regenerated sintered brick

be reduced as a result of the movement of the liquid phase brought on by the rise in temperature. It is typically difficult to create dense brick out of waste mixture since it shares a boundary with other materials, and the pore structure is typically greater. It demonstrates how brick's thermal conductivity is somewhat decreased by the porous material. The thermal conductivity decreases significantly when the waste mixture percentage is between 30% and 70% and the sintering temperature is 1100 °C. The internal pore structure of the brick limits its thermal conductivity at this point because the waste mixture and shale cannot be well integrated.

4 Conclusion

(1) The compressive strength of regenerated sintered shale has a similar pattern concerning compactness. As the sintering temperature rises, the brick's compressive strength and bulk density both increase, but a high sintering temperature will cause the brick's bulk density to drop and cause the over-burning phenomenon. Compressive strength and volume density of sintered shale that has been rejuvenated lose strength when waste mixture content rises. Therefore, a sintering temperature of 1050 °C and a low waste mixture content can be chosen if a regenerated sintered pagination brick with greater strength and density is required. Thus, one might decide to use more waste mixture to get lighter regenerated sintered shale brick.

(2) With an increase in temperature and a decrease in the waste mixture, the volume sintering shrinkage rate of recycled sintered shale brick increases. The brick body's volume sintering shrinkage rate can be increased by adding a waste mixture. Therefore, by selecting a lower sintering temperature or increasing the amount of waste mixture, the volume sintering shrinkage rate of recycled sintered shale brick can be enhanced.

(3) Regenerated sintered shale brick's thermal conductivity rises with temperature and falls with waste mixture content as the temperature rises. To increase the thermal insulation capabilities of brick, the thermal conductivity of recycled sintered shale brick can be decreased by selecting a lower sintering temperature and increasing the waste mixture content.

Fund Project. Guangxi Key Research and Development Project (Guike AB22080083); Liuzhou Science and Technology Research and New Product Trial Production Project (2022CCC0101).

References

1. Mahmood, W., Khan, A.R., Ayub, T.: Mechanical and durability properties of concrete containing recycled concrete aggregates. Iran. J. Sci. Technol. Trans. Civ. Eng. **46**(3), 2111–2130 (2022)
2. Yang Ming, G., Yufeng, Y.W., et al.: Research progress of energy-environment-economy comprehensive performance evaluation for solid waste reclamation. Mater. Rev. **35**(17), 17103–17110 (2021)
3. Bektas, F., Wang, K., Ceylan, H.: Effects of crushed clay brick aggregate on mortar durability. Constr. Build. Mater. **23**(5), 1909–1914 (2009)
4. Chen, M., Lin, J., Wu, S., et al.: Utilization of recycled brick powder as alternative filler in asphalt mixture. Constr. Build. Mater. **25**(4), 1532–1536 (2011)
5. Zhi, G., Hao, W., Li, Z., et al.: Study on properties of waste clay brick powder concrete. J. Shandong Univ. Eng. Sci. **42**(1), 104–105 (2012)
6. Cuizhen, X., Aiqin, S., Yin-chuan, G., et al.: Study on powder activity of construction waste brick under alkali excitation and compound excitation. Mater. Rev. **30**(10), 130–134 (2016)
7. Che Fa, X., Shiwei, Z.H.: Experimental study on working and mechanical properties of waste clay brick powder concrete. Railway Constr. **9**, 139–142 (2016)
8. Letelier, V., Ortega, J.M., Muñoz, P., et al.: Influence of waste brick powder in the mechanical properties of recycled aggregate concrete. Sustainability **10**(4), 1037 (2018)
9. Wen, C., Chunbao, X., Guiyin, C.A.I., et al.: Experimental study on the effect of waste clay brick powder on the performance of foamed concrete. Concrete **9**, 79–82 (2019)
10. Bo, C., Ailiang, Z., Jinlong, W.: Preparation and strength of new brick powder block materials. Journal of Shandong Agricultural University (Natural Science Edition) (2020)
11. Yahui, Z., Yue-dong, S., Guang-fu, C.: Effect of superfine brick powder on properties of cement fly ash mortar. Chinese Science and Technology Paper (2020)
12. Yuan, C., Wang, D., Setiawan, H., et al.: Effect and mechanism of different excitation modes on the activities of the recycled brick micropowder. Sci. Eng. Compos. Mater. **28**(1), 676–688 (2021)
13. Ma Qian, Li Yuxiang, Tan Hongbin, et al. Effect of recycled brick powder on properties of phosphorus building gypsum. Non-metallic Mines, (2022)
14. Rui, Z., Wenhuan, L., Hao, Z., et al.: Research on the performance of new subgrade stabilization material prepared by waste brick powder and sulfur fixing ash. Mater. Rev. **36**(12), 21020154–21020155 (2022)
15. Mohammed, T.U., Das, H.K., Mahmood, A.H., et al.: Flexural performance of RC beams made with recycled brick aggregate. Constr. Build. Mater. **134**, 67–74 (2017)
16. Qingdong, L., Xinlong, Z., Wenping, Q., et al.: Research on strengthening and application of recycled aggregate of waste brick. Concrete **2**, 42–45 (2018)
17. Chen, F., Wu, K., Ren, L., et al.: Internal curing effect and compressive strength calculation of recycled clay brick aggregate concrete. Materials **12**(11), 1815 (2019)
18. Design specification for mixing ratio of masonry mortar: JGJT 98-2010. China Architecture and Building Press, Beijing (2010).
19. Hao, L., Jizhen, Z.H.U., Bingzhang, H., et al.: Research on firing technology of recycled sintered shale brick. New Build. Mater. **46**(11), 60–62 (2019)
20. Test method for wall bricks: GB/T 2542–2012. China Building Materials Federation, Beijing (2011)

Research on Crack Control Method of Girder End Anchorage Zone Based on Nonlinear Finite Element Analysis

Tongyi Wang[1,1](✉), Jinjian Gu[2], and Jianrong Xu[2]

[1] Hohai University, No. 1 Xikang Road, 210098 Nanjing, People's Republic of China
tongtizhetong@163.com

[2] Huadong Engineering Corporation Limited, 201 Gaojiao Road, Zhejiang 311122, People's Republic of China

Abstract. A new, contemporary post-tensioned prestressed concrete I girder is created for a hydropower facility to accommodate larger loads in services. The increased transverse tensile loads caused by the enhanced I-girders with more tendons, however, have raised the possibility of girder end cracking. In this study, solutions for controlling cracks on the targeted girders were investigated. A modified finite element analysis model was used to simulate girder end behaviors and investigate effective crack management approaches. Two solutions were investigated: altering the girder's anchoring end section and enhancing the reinforcement. The findings indicated that the girder end cracks, which included horizontal cracks under the anchorage plate and inclined cracks at the upper corner of the anchoring plate, followed predictable patterns. Excessive tensile stresses generated by the various tendon tensioning sequences, as well as excessive initial tensioning control stresses, all contributed to the cracks that formed during the prestress tensioning procedure. The first method could not fundamentally eliminate the appearance of cracks, while the second method modifying the reinforcement could control the appearance of cracks.

Keywords: Girder end cracks · Anchorage zone · Finite element analysis · Crack control methods

1 Literature Reviews

Post-tensioned or pre-tensioned prestressed concrete girders are widely employed in bridges both at home and abroad because they are simple to build, easy to erect, and have a high load capacity [1–3]. The height of the girders and the number of prestressing tendons rise as the span of the bridge and the weights it carries increase, which frequently leads to cracking at the girder ends of such girders [4]. These cracks are mostly caused by the concentration of prestressing force during tension [5–8]. They have distinctive patterns that include horizontal and inclined cracks in the web, Y cracks in the bottom flange [9, 10], and vertical cracks at the girder end [11].

W. Guo et al. (Eds.): GBCESC 2023, LNCE 328, pp. 397–404, 2024.
https://doi.org/10.1007/978-981-99-9947-7_42

Several end zone crack control techniques have been researched. In the past, the most common crack management approach employed and researched by researchers was adjusting the configuration of the end reinforcement. Other strategies for controlling the formation of girder end cracks in pretensioned concrete girders were also examined. Many researchers [6, 12–15] have researched the effects of different strand debonding ratios and debonding patterns on the vertical stress distribution at the girder ends, and as a result, reinforcement methods to control girder end cracking have been implemented. Changing the cutting sequence of pretensioned strands can substantially affect the stress field in the girder end, and such a strategy may be possible to lessen or eliminate girder cracking, according to some research results [6, 16].

According to the findings of the preceding literature reviews, with the adoption of new section forms and the growing quantity of prestressing reinforcement in prestressed concrete I-shaped girders, the technique of girder end crack control techniques for prestressed concrete I-shaped girders remains one of the hot topics for research in this field. The numerical simulation results are contrasted and evaluated with the field-measured values to verify the validity of the model, which revealed the genuine causes of girder end cracking in the originally proposed scheme. On this basis, the impacts of two ways on crack prevention were investigated: modifying the girder's anchoring end section and improving the anchorage zone reinforcement. Some useful suggestions are made.

2 Background

The I-girder under consideration in this study comes from an actual bridge that acts as a pathway for a portal crane at a hydropower station. During the process of fabrication, the ends of the designated girders showed signs of cracking. Figure 1 shows the location and length of three typical cracks on a 2200 mm depth I-girder. The cracks are classified into three categories: horizontal web cracks under the anchorage plate with a maximum width and length of 0.20 mm and 470 mm, respectively (behavior 1); inclined cracks located near the upper corner of the N2 anchorage plate with a maximum width and length of 0.28 mm and 650 mm, respectively (behavior 2); and horizontal cracks located between the N1 and N2 anchorage plates with a maximum width and length of 0.20 mm and 380 mm, respectively (behavior 3).

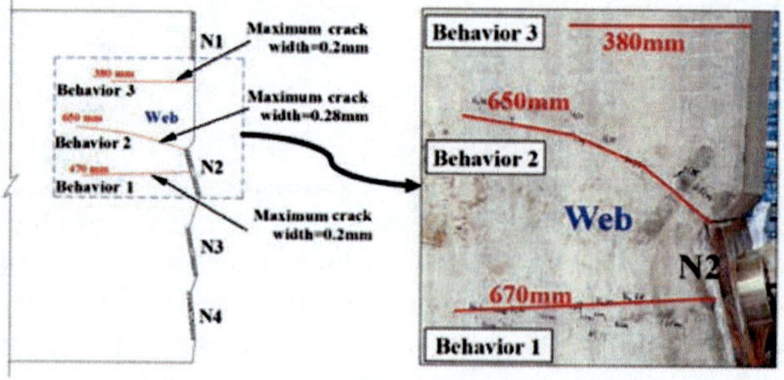

Fig. 1. Typical inclined and horizontal cracks.

3 Non-linear Finite Element Modeling

3.1 Material Characteristic

A nonlinear three-dimensional (3-D) finite element model was built using ABAQUS software to analyze the behaviors of the targeted I-girder ends. To improve computational efficiency, just one-fourth of the girder was modeled with symmetry along its length and width, as illustrated in Fig. 2(a).

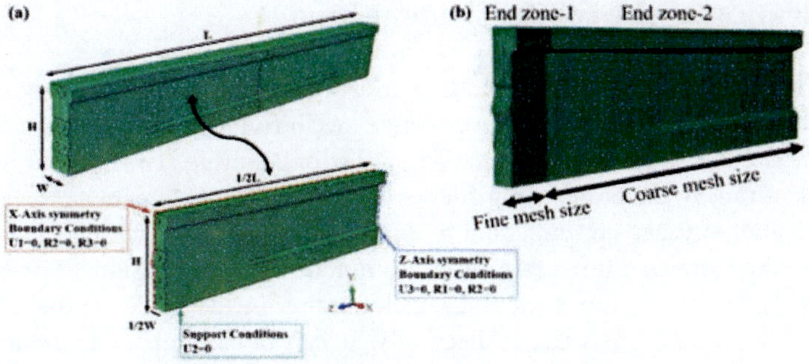

Fig. 2. Analysis model of targeted I-girder.

The constitutive relationship and damage of concrete were modeled using the concrete damaged plasticity (CDP) model, which is frequently employed in existing studies [19]. The stress-strain relationship was based on the China Code for Design of Concrete Structures [20]. The compression properties of concrete consist of three stages: the linear elastic stage, the nonlinear elastic-plastic ascending stage, and the nonlinear plastic-damage descending stage. The tensile properties of concrete are composed of two stages: the linear elastic phase and the stress-crack opening stage.

Because previous research showed the reinforcement bar stresses far below the yielding strength, therefore the reinforcement bars were modeled as linear elastic elements, and the elastic modulus of reinforcement rebars was taken 200 GPa.

3.2 Mesh Size and Element Types

The modeled concrete girder was made of two regions as shown in Fig. 2 (b). In End Zone-1, the size of the elements used was 22 mm while in End Zone-2, the dimension of the elements was 50 mm. Between End Zone-1 and End Zone-2, the size of a layer of elements was gradually increased from 22 mm to 50 mm. Reinforcements and tendons were also seeded to 22 mm to allow for the exact modeling of embedded parts in the concrete girder end.

The concrete components in the linear and nonlinear regions were both modeled using the eight-node linear reduction integration C3D8R elements. The reinforcement and tendons were built using two-node linear three-dimensional T3D2 truss elements.

3.3 Boundary Condition and Interaction

Figure 2 (a) also shows the midspan symmetry constraints applied to the X-Y plane and the Y-Z plane in the middle of the cross-section. The boundaries of the sections on the symmetry planes are indicated by colored lines.

Since the reinforcements were poured inside the framework, all reinforcements were assumed fully bonded in the concrete. The anchorage heads and tendons in this study were similarly totally bonded in concrete.

4 Verification of the Finite Element Model

To show the location of concrete cracks appearing during the prestressing tensioning process, the principal tensile strains in concrete are introduced in the concrete damaged plasticity model to characterize the cracking index of concrete. The contour of the principal tensile strains in the targeted I-girder end from the finite element analysis is shown in Fig. 3. Higher stresses are indicated by grey and red colors, and places with tensile strains exceeding the fracture strain limit are noted in chosen regions. For behavior 1 to behavior 3, the maximum principal tensile strains in the girder end surface reach 2490 με, 6273 με and 5516 με, respectively. It can be seen that the position where high-tension strains were predicted using FEA coincides well with the location where the cracks occurred on the girder as shown in Fig. 1. The FEA results of this targeted I-girder will be utilized as a standard girder for further comparative analysis to evaluate the effectiveness of additional crack control techniques.

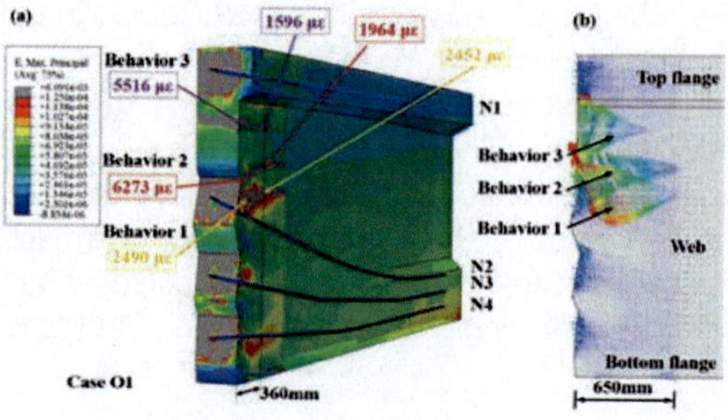

Fig. 3. Crack patterns and Principal tension strain contours of the targeted I-girder.

5 Evaluation of Crack Control Methods

5.1 Effect of Girder End Form

As the inclined cracks are not presented in many other I-girder with different girder end forms, the girder end forms are possibly the major factors that lead to concentrated tensile stresses and cause the diagonal crack that was observed at the upper corner of the

N2 anchorage plate in the actual girder. So the designer proposed an improved girder end form without a large turning angle at the upper corner of the N2 anchorage plate as shown in Fig. 4. The new FEA model was identical to the previous FEA model used for the original I-girder except for the details of girder end form.

As shown in Fig. 4, it can be observed that the maximum principal tensile strain decreases compared with the results obtained from the original I-girder model (see Fig. 3). The maximum principal tensile strain in behavior 1 to behavior 3 decreases from 2490 με, 6273 με, and 5516 με to 1958 με, 2840 με, and 2202 με, respectively. All values are still higher than the cracking strain (125 με). The maximum principal tensile strain was located at the upper corner of the N2 anchorage plate in the original model, where an inclined crack was observed in the actual I-girder. But in the girder end with an improved girder end form, the inclined crack disappeared. Between the strands N1 and N2 anchorage plates, there are still two regions with high principal tensile strain, which implies these areas are most likely to form two horizontal cracks. The analysis's findings show that while the improved girder end form cannot prevent girder ends from cracking, it can alter the patterns of those cracks. It is still one of the most important factors contributing to the crack patterns in the I-girder with the improved girder end form.

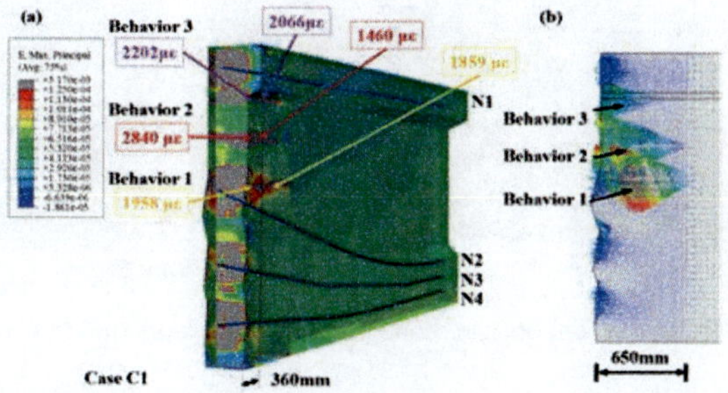

Fig. 4. Principal tension strain and principal tension strain directions contours of I-girder with improved girder end form.

5.2 Effect of Reinforcement

The original anchorage zone at the girder's end had insufficient reinforcement for stripping steel bars just below the anchor, making it incapable of successfully resisting cracks induced by high stripping force. Furthermore, the inappropriate placement of cross-section forms allows oblique cracks to form at the corner and progress through the girder end along the oblique direction. At the same time, the original reinforcement design's split steel bars are loosely distributed, and there is no steel bar near the beam end section to prevent cracking, consequently, horizontal cracks in the concrete between N1 and N2 are easily caused and develop horizontally. As a result, the reinforcement in the anchorage zone must be modified. Figure 5 depicts the configuration of bars with modified reinforcing.

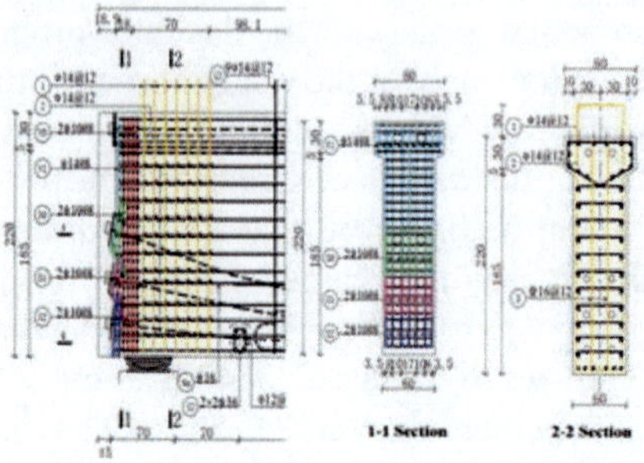

Fig. 5. Schematic diagram of steel bar layout section in the reinforced area under anchor.

Two models were created to study the effect of reinforcement: Case E1 with modified reinforcement and Case F1 with modified reinforcement and better anchorage end section. Figure 6 shows the strain situation in the anchorage zone of the girder end under the two conditions.

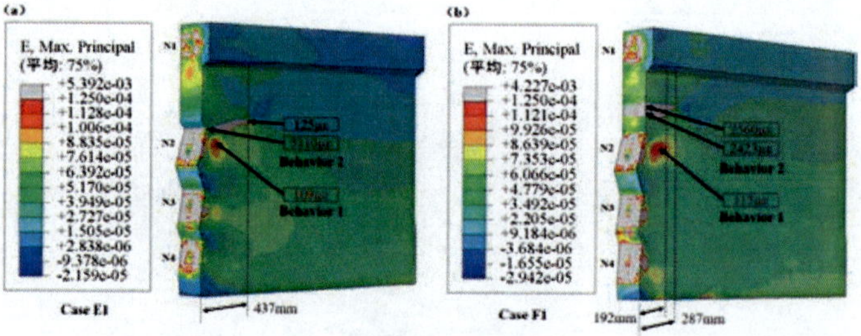

Fig. 6. Strain situation in the anchorage zone of the beam end under Case E1 and Case F1 conditions.

The comparison of Case E1 and Case O1 demonstrates that the maximum principal strain at the girder end of a prestressed concrete girder decreases with the improvement of reinforcing form, from 5366 $\mu\varepsilon$ to 5110 $\mu\varepsilon$. The strain at the original position of crack 1 is less than the cracking strain after changing the reinforcement form, so it can be said that the occurrence of crack 1 is successfully prevented, and the length of crack 2 is reduced from the original 650 mm to 437 mm, indicating that after improving the reinforcement form, the length of crack 2 is mainly controlled.

The comparison between Case F1 and Case O1 shows the maximum principal strain is still above the cracking strain after improving the section and reinforcement. The maximum principal strain of the beam end section has decreased from the original 6273 $\mu\varepsilon$ to 2560 $\mu\varepsilon$. In addition, another new horizontal crack was generated at this location, with a maximum principal strain of 2423 $\mu\varepsilon$. After changing the reinforcement form, the crack at the location of crack 1 was successfully eliminated. Two cracks appeared in the N1 and N2 anchor plates, but the length of the cracks was shortened compared to Case O1, from 650 mm to 287 mm.

To further investigate the effect of this control method, the stresses of reinforcement bars were presented in Fig. 7. The principal tension stresses in the rebars also predict the position of different crack patterns. The web rebars were laid almost parallel to principal tensile strains in the web cracking regions. Therefore, they will be the most effective in restraining these crack widths. The stresses of vertical rebars exhibit the maximum principal tension stress (138 MPa) means cracking.

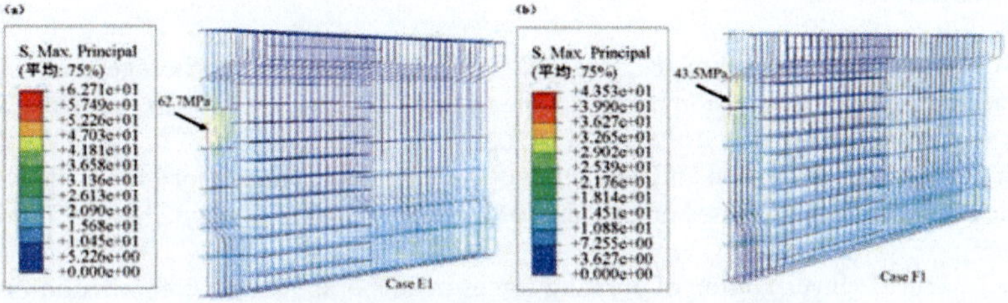

Fig. 7. Stresses in the stirrup under Case E1 and Case F1.

As seen from Fig. 7 (a), it can be found that the maximum stress of stirrup reinforcement at the location of behavior 2 is 62.7 MPa and does not exceed 138 MPa, which means that the crack width of behavior 2 is acceptable according to the AASHTO LRFD Bridge Design Specification. And it can be found that the maximum stress of stirrup reinforcement at the location of behavior 2 is 43.5 MPa and also does not exceed 138 MPa, as seen in Fig. 7 (b). Although the surface of the girder end has large strains beyond the concrete tensile strain limit (125 $\mu\varepsilon$) in Case E1 and Case F1, they do not propagate to the girder end because the anchorage plate and their effects on the cracking can be ignored. The method of modifying reinforcement can prevent the appearance of cracks at the girder end and control the length and width of the cracks effectively.

According to the analysis presented above, directly densifying the anti-splitting and anti-peeling steel bars in the anchoring zone can help the concrete itself resist cracking to a certain extent, effectively reducing the length and width of cracks, especially for controlling those under the anchor.

6 Conclusions

This paper presents an analytical study to develop practical solutions to minimize typical end zone cracking in a practical I-girder. The following conclusion and recommendations are drawn based on the FEA results:

1. An improved girder end form without a large turning angle at the upper corner of the N2 anchorage plate could not overcome the cracking of the girder end, but it could change the crack patterns. Though cracks still occurred, the values of the maximum principal tensile strains were lower than the initial I-girder. Since the girder end form plays an important role in the mechanism of girder end cracking, it is recommended that the girder end shape be designed carefully.

2. According to the AASHTO LRFD Bridge Design Specification, the tensile stress in the stirrup close to the girder end is within the rebar limit even though the principal tensile strains of behavior 2 remain above the concrete cracking limit. The method of adjusting reinforcement could effectively regulate the length and width of the cracks as well as prevent the appearance of cracks at the girder end.

References

1. De Corte, W., Van Meirvenne, K., Boel, V., Taerwe, L.: Design of anchorage zones of pretensioned concrete girders: a comparison of nonlinear 3D FEM results with measurements on a full scale beam. Appl. Sci. **10**(22), 8221 (2020)
2. Yuan, A.: Model derivation and validation of spalling-force calculations for prestressed concrete bridge girder ends based on a modified G-S model. 2021. J. Bridge Eng. **24**(3), 04018122 (2019)
3. He, Z., Liu, Z.: Investigation of bursting forces in anchorage zones: compression-dispersion models and unified design equation. J Bridge Eng. **16**, 820–827 (2011)
4. Weizhong, M., Hongye, G., Yannian, H., Qianhui, P.: Local stress behavior of post-tensioned prestressed anchorage zones in continuous rigid frame arch railway bridge. Appl. Sci. **8**(10), 1833 (2018)
5. Jirawattanasomkul, T., Kongwang, N., Likitlersuang, S., Yodsudjai, W., Charuvist, S., Sato, Y.: Failure analysis of dapped-end cracking in post-tensioned bridge girder. J Bridge Eng. **26**(11), 04021082 (2021)
6. Kannel, J., French, C., Stolarski, H.: Release methodology of strands to reduce end cracking in pretensioned concrete girders. PCI J. **42**(1), 42–54 (1997)
7. Marshall, W.T., Mattock, A.H.: Control of horizontal cracking in the ends of pretensioned prestressed concrete girders. PCI J. **7**, 56–74 (1962)
8. Tuan, C.Y., Yehia, S.A., Jongpitaksseel, N., Tadros, M.K.: End zone reinforcement for pretensioned concrete girders. PCI J. **49**(3), 68–82 (2004)
9. Okumus, P., Oliva, M.G., Becker, S.: Nonlinear finite element modeling of cracking at ends of pretensioned bridge girders. Eng. Struct. **40**, 267–275 (2012)
10. Hamilton, H.R., Consolazio, G.R., Ross, B.E.: End region detailing of pretensioned concrete bridge girders. University of Florida Civil and Coastal Engineering, Gainesville, Florida (2013)
11. Jia, J., Zhang, K., Wu, S., Xiong, T., Bai, Y., Li, W.: Vertical cracking at girder ends during post-tensioning of prefabricated prestress concrete bridge T-girders. Structur Concr. **22**(5), 3094–3108 (2021)
12. Breen, J.E., Burdet, O., Roberts, C., Wollmannn, G.: Anchorage zone reinforcement for post-tensioned concrete girders. Rep. For NCHRP 10–29. Austin (TX): The University of Texas at Austin (1991)
13. Okumus, P., Oliva, M.G.: Strand debonding for pretensioned bridge girders to control end cracks. ACI Struct. J. **111**(1), 201–210 (2014)
14. Yuan, A., Yang, D., Miao, X.: Strand debonding for prestressed concrete girders to control end horizontal web cracks based on a modified G-S model. J. Bridge Eng. **27**(2), 04021109 (2022)
15. Ronanki, V.S., Burkhalter, D.I., Aaleti, S., Song, W., Richardson, J.A.: Experimental and analytical investigation of end zone cracking in BT-78 girders. Eng. Struct. **151**, 503–517 (2017)
16. Okumus, P., Oliva, M.G.: Evaluation of crack control methods for end zone cracking in prestressed concrete bridge girders. PCI J. **58**(2), 91–105 (2013)

Optimization of Optimal Pre Maintenance Timing Decision for Asphalt Pavement Based on Matter Element Analysis and Combination Weighting

Ying Li(✉) and Qiangnian Li

School of Civil Engineering, Lanzhou University of Technology, Lanzhou, China
wsd1998202 2@163.com

Abstract. With the increasing number of highways in China, the demand for highway maintenance is rising. Selecting the right timing for maintenance is crucial. This study uses matter element analysis to determine the best preventive maintenance opportunity for asphalt pavement. Factors such as maintenance costs, benefits, energy consumption, and carbon emissions are considered. The entropy weight method and Analytic Hierarchy Process are used to assign weights to these indicators, improving the rationality and reliability of the decision-making process. An actual maintenance case in Gansu Province is analyzed and verified. The research results indicate that the traditional method with the lowest cost suggests 2021 as the best maintenance time. However, the matter element Decision model shows that pavement maintenance starting from 2020 achieves the maximum comprehensive benefit across all indicators. This multi-indicator model provides a comprehensive and reasonable approach for decision-making in asphalt pavement maintenance timing, optimizing the management level of pavement maintenance.

Keywords: Maintenance timing · Matter element Decision model · Cost benefit analysis · Analytic Hierarchy Process (AHP) · Entropy weight method First Section

1 Introduction

In recent years, the continuous acceleration of highway construction in our country has made highway maintenance in the later stage a hugely expensive project, with increasingly serious environmental impacts. Therefore, the country has put forward the concept of low-carbon environmental protection and issued a series of policies [1]. Currently, preventive maintenance is the main method for asphalt pavement maintenance in China. The key issue in the preventive maintenance process is when to carry out maintenance. However, the decision on maintenance timing mostly relies on experience and subjective awareness, lacking universality and scientific basis. Performing maintenance too early or too late will result in waste of maintenance costs and resources, and will also have a significant impact on the environment [2]. Domestic and foreign scholars have conducted

W. Guo et al. (Eds.): GBCESC 2023, LNCE 328, pp. 405–413, 2024.
https://doi.org/10.1007/978-981-99-9947-7_43

extensive research on the rational selection of preventive maintenance timing. Literature [3] established a decision model based on the cost-effectiveness ratio, using the cost-effectiveness ratio to determine the optimal timing for preventive maintenance. Literature [4] based on cost-benefit analysis, analyzed the comprehensive benefits and costs, and formed four different decision indicators: comprehensive benefit value, equivalent annual cost value, benefit index, and pavement service life change, to determine the optimal timing for preventive maintenance. Literature [5] developed a method for selecting maintenance timing using decision trees and decision matrices, which only established corresponding rules based on pavement performance indicators for maintenance timing decision-making. In summary, existing decision-making methods only consider a single perspective and lack consideration of various aspects such as the environment and resources [6]. They rarely consider the impact on the environment, and there are certain conflicts and contradictions among the indicators. This paper, based on the theory of elemental analysis, conducts research and analysis using qualitative and quantitative methods. It establishes four research indicators from different perspectives: cost, benefit, energy consumption, and carbon emissions. The analytic hierarchy process and entropy weight method are used to combine and assign weights to the indicators, improving the rationality and credibility of the indicator weights [7]. A elemental decision-making model is established to determine the optimal timing for preventive maintenance of asphalt pavement using the comprehensive relevance value in the model.

The Matter-Element Analysis Method is a fuzzy mathematics-based evaluation method that handles uncertainty and vagueness problems. It converts evaluation indicators into fuzzy numbers, establishes a fuzzy evaluation matrix, and uses operational rules of fuzzy numbers for comprehensive evaluation. Compared to traditional methods, it considers interrelationships and uncertainties among indicators, improving accuracy and reliability of evaluation results. The Entropy Weight Method is an information entropy theory-based weight determination method. It determines weights of evaluation indicators and allocates their importance appropriately. It calculates information entropy value of each indicator to objectively determine weights. Compared to subjective methods, it objectively reflects indicator importance, enhancing scientificity and credibility of evaluation results. The Analytic Hierarchy Process is a multi-criteria decision-making method based on expert judgments. It determines importance and priority of factors in asphalt pavement maintenance. It constructs a hierarchical structure, uses expert judgments and quantitative analysis to calculate factor weights, and conducts consistency checks. It decomposes complex problems into hierarchical structures, clarifying relationships and improving rationality and feasibility of evaluation results.

2 Construction of Multi-attribute Decision Model Based on Matter Element Analysis

2.1 Determine the Indicator Evaluation System

The evaluation system for optimal maintenance timing should be comprehensive, considering factors like the environment and transportation. Including carbon emissions and energy consumption as research indicators has led to a decision-making system that considers benefits, costs, carbon emissions, and energy consumption [8]. The cost includes management expenses and daily maintenance expenses.

2.2 Quantification and Standardization of Indicators

Quantification of Indicators. Quantifying research indicators is necessary to highlight differences among alternative solutions [9]. Benefit refers to the quantitative response of the improvement in pavement performance caused by the implementation of preventive maintenance throughout the entire life cycle, and the pavement condition index (PCI) is selected for benefit evaluation. Maintenance costs are quantified using the present value of costs to reflect dynamic economic effects. Carbon emissions and energy consumption are quantified using life cycle assessment theory and relevant construction quotas [10]. Using the maintenance construction quota of emulsified asphalt slurry seal layer to estimate the energy consumption and carbon emissions required for the 1-cm thick micro surface maintenance project.

Standardization of Indicators. When conducting matter element analysis on the evaluated project case, it is necessary to normalize the quantified indicators and transform the normative values to the [0,1] interval. The calculation formula is as follows:

$$a_{ij} = \frac{x_{ij} - min\{x_{ik}\}}{max\{x_{ik}\} - min\{x_{ik}\}} \tag{1}$$

$$a_{ij} = \frac{max\{x_{ik}\} - x_{ij}}{max\{x_{ik}\} - min\{x_{ik}\}} \tag{2}$$

$((i = 1, 2 \cdots \leq, m, \ \ j = 1, 2 \ldots, n, \ \ 1 \leq k \leq n)$

The benefit indicators are normalized using Eq. (1), while the cost indicators are normalized using Eq. (2). X_{ij} represents the quantitative values of each indicator, and a_{ij} is the standardized result, which is within the range of [0,1].

2.3 AHP Entropy Weight Method for Determining the Weight of Indicator System

This article utilizes the AHP-entropy weight method to assign weights to indicators objectively and subjectively. This method can to some extent avoid the strong subjectivity of the Analytic Hierarchy Process, and combine the objective weighting advantage of entropy weighting method, so that each indicator weight has credibility. The analysis steps of the combination weight method are as follows:

(1) Building a hierarchical model.
(2) Through expert consultation and combining the importance scale (as shown in Table 1), score the same level of indicators by comparing them in pairs, and establish a judgment matrix B.
(3) Calculate the maximum eigenvalue of the judgment matrix λ Max and the feature vector, then normalize the feature vector to obtain the weight vector, and perform consistency testing. When the consistency ratio *C.R* is less than 0.1, it indicates that the weight is valid. Otherwise, it is necessary to rebuild the judgment matrix, as shown in the following formula.

$$\lambda max = \frac{1}{n}\sum_{i=1}^{n}\frac{\sum_{j=1}^{n} a_{ij}w_j}{w_i} \tag{3}$$

$$C.I = \frac{\lambda_{\max} - n}{n-1} \tag{4}$$

$$C.R = \frac{C.I}{RI} \tag{5}$$

Normalize the judgment matrix obtained from the Analytic Hierarchy Process, and use the entropy weight method to modify the indicator weight T obtained from the Analytic Hierarchy Process. The formula is as follows:

$$H_j = -\frac{1}{\ln n}\sum\nolimits_{i=1}^{n} p_{ij}\ln p_{ij} \tag{6}$$

$$E_j = (1 - H_j)/\sum\nolimits_{j=1}^{n}(1 - H_j) \tag{7}$$

$$W_j = T_jE_j/\sum\nolimits_{i=1}^{n} T_jE_j \tag{8}$$

Subjectively speaking, due to the significant proportion of benefits and costs in the maintenance process of road surfaces, with a greater proportion of energy consumption and carbon emissions, the weight of benefits and costs is greater.

2.4 Decision Evaluation Model Based on Matter Element Method

Matter elements are the basic unit in the theory of matter element analysis to analyze and solve contradictions. Things have multiple different features, and each feature can be quantified numerically, with N representing the name of the thing, C represents the characteristics of a thing, V is the quantitative value of a thing, $R = \{N, C, V\}$ is the matter element composed of the name, feature, and feature quantity of a thing, serving as the basic unit for describing things. The indicator system constitutes a feature set $C = \{C_1, C_2, \ldots, C_m\}$, and N comparison schemes D1, D2,..., D_n are formulated. The evaluation indicators of N schemes are quantified and normalized, and the normalized values a_{ij} are finally obtained, which form the material element R, as shown in formula (9).

$$R = \left\{ \begin{matrix} & D \\ C_1 & a_1 \\ C_2 & a_2 \\ L & L \\ C_m & a_m \end{matrix} \right\} \tag{9}$$

N elements R form a composite element R_n. As shown in Eq. (10), R_0 is the optimal solution material element composed of selecting the maximum value of a_{ij} from each indicator of R_n, as shown in Eq. (11).

$$R_n = \left\{ \begin{matrix} & D_1 & D_2 & L & D_n \\ C_1 & a_{11} & a_{12} & L & a_{1n} \\ C_2 & a_{21} & a_{22} & L & a_{2n} \\ L & L & L & L & L \\ C_m & a_{m1} & a_{m2} & L & a_{mn} \end{matrix} \right\} \tag{10}$$

$$R_0 = \left\{ \begin{matrix} & D \\ C_1 & a_{10} \\ C_2 & a_{20} \\ L & L \\ C_m & a_{m0} \end{matrix} \right\} \tag{11}$$

Based on the data in the material element R_{n} and Eq. (13), the correlation coefficient L_{ij} is calculated, which represents the degree of correlation between each scheme and the optimal scheme on the same indicator. All L_{ij} values form the composite material element R_{ln} of the correlation coefficient, as shown in Eq. (12). It is necessary to weight and average the L_{ij} corresponding to all indicators of each scheme according to their respective weights, and calculate the comprehensive correlation coefficient L_{0j} for each scheme. The comprehensive correlation coefficient of each selected scheme constitutes the composite RD_{ln} of the comprehensive correlation degree, as shown in Eq. (14).

$$R_{\text{ln}} = \left\{ \begin{matrix} & D_1 & D_2 & L & D_n \\ C_1 & L_{11} & L_{12} & L & L_{1n} \\ C_2 & L_{21} & L_{22} & L & L_{2n} \\ L & L & L & L & L \\ C_m & L_{m1} & L_{m2} & L & L_{mn} \end{matrix} \right\} \tag{12}$$

$$L_{ij} = \frac{\Delta \min + \Delta \max \rho}{\Delta_{ij} + \Delta \max \rho} \tag{13}$$

$$RD_{\text{ln}} = W * R_{\text{ln}} = \left\{ \begin{matrix} & D_1 & L & D_n \\ L_{0j} & L_{01} = \sum_{i=1}^{m} W_i * L_{i1} & L & L_{0n} = \sum_{i=1}^{m} W_i * L_{in} \end{matrix} \right\} \tag{14}$$

3 Case Analysis

3.1 Project Overview

The data for case analysis and model validation comes from a highway in Gansu Province. The basic information of the highway is as follows:

It was opened to traffic at the end of 2015, with a total length of 47.92km and a design speed of 100km/h. The terminal point is K11 + 474~K59 + 400. Lane width 3.75 m; Since the completion of the highway, there has been no large-scale maintenance. Using PCIe = 90, 87, 83, and 80 as thresholds, as the starting time points for the first pre maintenance plan D_1, D_2, D_3, and D_4, the maintenance start times for D_1, D_2, D_3, and D_4 are 2020, 2021, 2022, and 2023, respectively. The full life cycle analysis period of the road surface is 15 years. During the analysis period, after the first maintenance of each plan, the road surface indicators will decrease to the threshold again with the use and damage of the road surface, and then maintenance will be carried out until the analysis period. The quantification value of each indicator corresponding to each plan depends on this process.

3.2 Quantification and Standardization of Indicators

Quantification of Indicators. Quantify the benefits, costs, energy consumption, and carbon emissions, as shown in Table 1. Cost/(10000 yuan/km); Energy consumption/(functional unit/MJ); CO_2/(functional unit/kg).

Table 1. Quantification of indicators

Index	2020	2021	2022	2023
benefit	141. 126	122.474	108.797	138
cost	4070.07	4062.87	4068.87	4087.21
energy consumption	189541.69	964073. 44	642715. 63	480677.85
carbon emission	1697. 19	32248. 09	21498. 73	31741.88

Standardization of Indicators. Standardize the results of quantifying benefits, costs, carbon emissions, and energy consumption according to Eqs. (1) and (2), as shown in Fig. 1.

3.3 AHP Entropy Weight Method for Determining Indicator Weights

Remove the highest and lowest scores, and take the average value of the remaining data to the judgment matrix *B*, as shown in Eq. (15).

$$B = \begin{bmatrix} 1 & 3 & 6 & 6 \\ 1/3 & 1 & 5 & 5 \\ 1/6 & 1/5 & 1 & 1/3 \\ 1/6 & 1/5 & 3 & 1 \end{bmatrix} \tag{15}$$

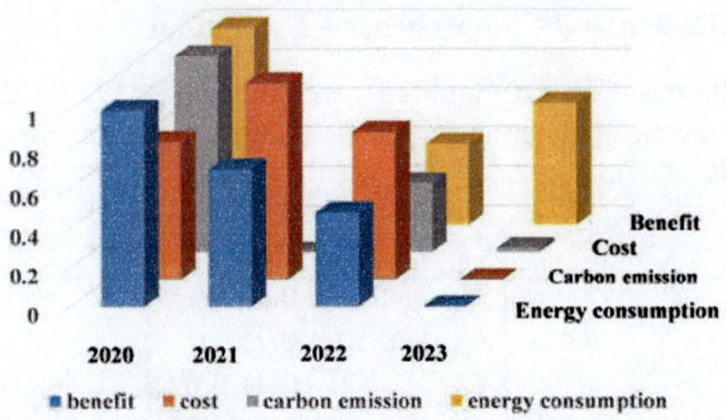

Fig. 1. Three dimensional bar chart of normalized values of indicators

Use the root method to calculate the matrix eigenvectors, which are the corresponding weights of each indicator, as shown in Table 2.

Table 2. Indicator weight

index	benefit	cost	energy consumption	carbon emission
weight	0.5549	0.2924	0.0559	0.0968

Using Eq. (8), calculate the correction weights as $W_1 = 0.460$, $W_2 = 0.416$, $W_3 = 0.024$, and $W_4 = 0.100$, respectively.

$$B_1 = \begin{bmatrix} 0.600 & 0.683 & 0.400 & 0.486 \\ 0.200 & 0.227 & 0.333 & 0.405 \\ 0.100 & 0.045 & 0.067 & 0.027 \\ 0.100 & 0.045 & 0.200 & 0.081 \end{bmatrix} \tag{16}$$

Calculate the maximum eigenvalue using formula (3) $\lambda_{max} = 4.261$, and finally calculated from Eqs. (4) and (5), $C.R = 0.098 < 0.1$, indicating that the weight is valid through consistency testing.Normalize the judgment matrix B to obtain matrix B_1, as shown in Eq. (16). Use Eq. (6) to obtain the information entropy of each indicator as $H_1 = 0.785$, $H_2 = 0.632$, $H_3 = 0.891$, and $H_4 = 0.734$, respectively; Using Eq. (7), obtain the entropy weights of each indicator as $E_1 = 0.224$, $E_2 = 0.384$, $E_3 = 0.114$, $E_4 = 0.278$;

3.4 Model Application and Comprehensive Evaluation

Substituting results into Eqs. (10) and (13), we obtain Eq. (17) for correlation degree Lij. From Eq. (18), we calculate the composite matter element using corrected weights and data. This is Eq. (19).

$$R_4 = \begin{Bmatrix} & D_1 & D_2 & D_4 & D_n \\ C_1 & 1 & 0.70 & 0.48 & 0 \\ C_2 & 0.70 & 1 & 0.75 & 0 \\ C_3 & 1 & 0 & 0.41 & 0.62 \\ C_4 & 1 & 0 & 0.35 & 0.02 \end{Bmatrix} \quad (17)$$

$$R_{l4} = \begin{Bmatrix} & D_1 & D_2 & D_4 & D_n \\ C_1 & 1 & 0.63 & 0.49 & 0.33 \\ C_2 & 0.63 & 1 & 0.67 & 0.33 \\ C_3 & 1 & 0.33 & 0.46 & 0.57 \\ C_4 & 1 & 0.33 & 0.43 & 0.34 \end{Bmatrix} \quad (18)$$

$$RD_{l4} = \begin{Bmatrix} 0 & D_1 & D_2 & D_3 & D_4 \\ L_{0j} & 0.89 & 0.69 & 0.54 & 0.34 \end{Bmatrix} \quad (19)$$

The analysis using the matter-element model shows that delaying pavement maintenance leads to lower overall benefits. Starting maintenance in 2020 achieves maximum economic, social, and environmental effects. The multi-index matter-element decision model provides more accurate and comprehensive results compared to the traditional single index principle.

The Matter-Element Analysis Method has several advantages over other methods, but also has some limitations. It effectively handles uncertainty and vagueness problems, making it suitable for evaluating complex systems with incomplete or imprecise information.It considers interrelationships and uncertainties among indicators, leading to more accurate and reliable evaluation results.It allows for a comprehensive assessment by converting evaluation indicators into fuzzy numbers and utilizing operational rules of fuzzy numbers.It improves the scientificity and credibility of evaluation results by objectively reflecting the importance of indicators.It relies on subjective judgments in determining the membership functions and weights of fuzzy numbers, which may introduce bias into the evaluation process.In summary, the Matter-Element Analysis Method is advantageous in handling uncertainty, considering interrelationships, and providing comprehensive evaluations. However, it has limitations in terms of complexity, subjectivity, and potential oversimplification..

4 Conclusion

The Research Conclusions of This Article Are as Follows:

(1) From an integrated perspective of economics, society, and environmental protection, the optimal maintenance timing for roads is determined by maximizing benefits while

minimizing costs, energy consumption, and carbon emissions. Traditional cost-based decision-making methods suggest that 2021 is the best time for road maintenance. However, according to the material-element model, conducting road maintenance in 2020 can achieve the optimal state for all indicators. Therefore, 2020 is considered the best time for maintaining this road.

(2) Performing preventive maintenance on roads with high service levels and good pavement conditions can achieve comprehensive optimization in terms of funds, environment, and society. When the pavement condition is at a lower level, delaying the start of maintenance will result in lower comprehensive maintenance benefits for the pavement.

References

1. China Building Energy Conservation Association China Building Energy Consumption Research Report (2020)
2. Jiwei, Z., Haiting, P., Jichao, Z.: Development trends and impacts of green, low carbon, and energy efficient buildings. Const. Technol. (Chin. English) **50**(16), 92–94 (2021)
3. Li, F., Feng, J., Li, Y., Zhou, S.: Preventive maintenance technology for asphalt pavement. Springer Singapore, Singapore. (2021). https://doi.org/10.1007/978-981-15-6206-8
4. Xiao, F., Chen, X., Yang, S., et al.: Bi-objective pavement maintenance and rehabilitation optimization decision-making model incorporating the construction length of preventive maintenance projects. Struct. Infrast. Eng., 1–15 (2023)
5. Michigan DOT. Capital Preventive Maintenance Program Guidelines, Michigan (March 1999)
6. Wang, Z., Guo, N., Wang, S., et al.: Prediction of highway asphalt pavement performance based on Markov chain and artificial neural network approach. J. Supercomput. **77**, 1354–1376 (2021)
7. Xu, X., Wei, J., Yan, X., Zhang, Z.: Current status and prospects of long life asphalt pavement research. Sino Foreign Highway **43**(01), 36–43 (2023)
8. Zhu, M., Sun, S.: Analysis of China's green and low-carbon building research knowledge graph. Project Manag. Technol. **21**(02), 93–100 (2023)
9. Dong, M., Zhou, X., Hou, Y., Zhang, Y.: Optimization of pre maintenance timing decision for asphalt pavement based on life cycle. Highway **65**(4), 325–331 (2020)
10. Wang, Y., Feng, Q.: Research on the whole life cycle energy consumption of buildings based on LCA. Project Manag, Technol. **21**(02), 70–74 (2023)

Research on Reinforcement Cage Connection Techniques for Cast-in-Place Concrete Piles

Haijun Wang[1], Weiqiang Chen[2], Hongjun Lv[3], Wenxian Yang[3], and Minting Zhong[4](✉)

[1] China Railway Construction South China Construction Co. LTD., Guangzhou 511458, China
[2] Guangzhou Metro Group Co. LTD., Guangzhou 510330, China
[3] Guangzhou Engineering Co., Ltd. of China Railway 19th Bureau Group, Guangzhou 511455, China
[4] School of Civil Engineering, Guangzhou University, Guangzhou 510006, China
MintingZhong@e.gzhu.edu.cn

Abstract. In this study, various techniques for connecting reinforcement cages in cast-in-place concrete piles are being investigated with the aim of enhancing their overall structural integrity and performance. The traditional method used for connecting steel bars in construction is through overlap. However, there are several issues associated with the overlap or welding shear splicing of steel bars, including insufficient overlap length, poor welding quality, increased labor costs, and joint failures. To address these challenges, a new technique for splicing steel bars is being researched. The focus of the study is on two main aspects: the tendency of loosening at the coupler connection area and the difficulty in aligning the two connected steel bars. To tackle these problems, the use of a shape memory alloy self-repairing coupler for steel bars is proposed. This innovative coupler has the capability to strengthen the connection between the steel bars and the coupler, improve the quality and speed of the connection, and also offer real-time monitoring and recoverable advantages.

Keywords: Reinforcing cage · Connection method · SMA coupler

1 Introduction

With the progress of urbanization in our country, an increasing number of reinforced concrete structures have emerged. During the construction process of reinforced concrete, due to practical limitations, the reinforcement is often transported in sections and then extended by splicing on-site. Currently, common methods of steel bar connections include welding and coupler connections, among others. However, there are several issues with the processing of steel cages, including:

(1) The main reinforcing bars of the steel cage are not on the same cross-section, resulting in large gaps between the main bars.
(2) The thread of the steel bar joint does not meet the requirements, making it impossible to fit the coupler.

W. Guo et al. (Eds.): GBCESC 2023, LNCE 328, pp. 414–420, 2024.
https://doi.org/10.1007/978-981-99-9947-7_44

(3) Conflicts between the stirrups and core extraction pipes and acoustic monitoring pipes prevent installation.

To address the issues in the processing of steel cages, scholars at home and abroad have conducted research on the strategies of connecting steel bars. George I. Kalogeropoulos et al. [1] assessed the effectiveness of welded reinforcement and mechanical splice joints in ensuring load transfer between steel rebars through experimental evaluations. The study found that mechanical splice joints were more prone to rebar slippage. Bilić Željko et al. [2] investigated the influence of basic welding parameters and calculated their direct or indirect effects to provide sufficient accuracy in achieving high-quality welded joints. They also assessed the mechanical properties of the reinforced steel after welding. Fawzy Tarek et al. [3] proposed and studied a newly developed tension lap splicing technique for reinforcing bars to validate its effectiveness. Experimental tests revealed that by applying the developed technique, higher tensile stress was obtained not only at failure but also with a transition from bond-induced failure to bending failure. Stavroula J. Pantazopoulou et al. [4] examined the failure of the lap splicing zone in reinforcing bars and evaluated the residual anchorage capacity of the spliced region. The bonding behavior and failure mechanisms of lap splices with multiple deformed bars were assessed. Coogler, Keith L et al. [5] investigated the behavior of offset mechanical splice systems. Two commercially available systems were tested under direct tension, with the splice restrained or unrestrained from rotation. Batista RJS et al. [6] studied the deformation modes, forming limits, forming force, and resistance before failure of a novel dual-material cylindrical insert splice joint using a combination of experimental and finite element methods. This research enabled the design of a simple and easily processable dual-material cylindrical insert to overcome previous issues with conflicting rebar, core tube, and piezometer installation.

This article focuses on the production of actual cast-in-place concrete piles as the research object. It provides a detailed description of the production process for pile foundation reinforcement cages and quality control methods for reinforcement cages. Furthermore, it proposes a self-healing steel coupler based on shape memory alloy to address processing issues with the reinforcement cage, and optimizes the manufacturing process of the reinforcement cage.

2 Construction Process of Reinforcement Cages

2.1 Construction Preparation

The production site should be chosen in a convenient location for transportation and installation. The site should be close to the construction area and have sufficient space for stacking and processing the reinforcement. Good drainage conditions, proper hardening of the ground, and the availability of simple steel processing sheds and storage sheds are necessary. Technical instructions for the reinforcement cage or a reinforcement cutting list should be prepared according to the design drawings and specifications.

2.2 Production of Reinforcement Cages

The reinforcement cage must be securely fixed to prevent floating and should be fixed at the designated elevation position. Additionally, attention should be paid to fixing them to the sound detection pipes of the reinforcement cage as required and protecting their ports, ensuring smooth future testing of the pile foundation. The processing sequence can be from the top cage to the bottom cage or from the bottom cage to the top cage. The reinforcement cage should be processed using a continuous, multi-section water flow method, and adjacent reinforcement cage joints should be marked as specified. After completion, they should be stacked in order. The construction process flow of the reinforcement cage is shown in Fig. 1.

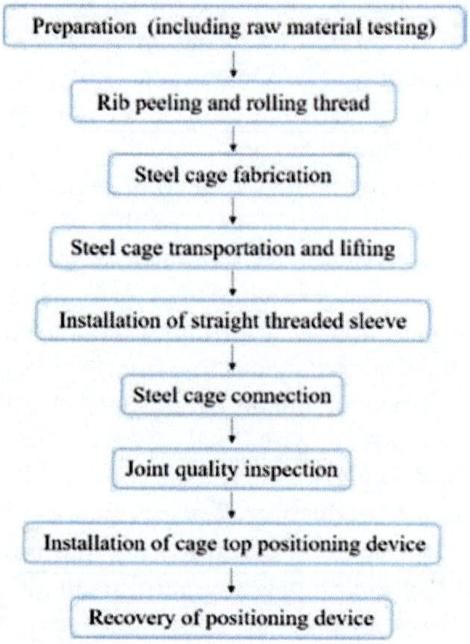

Fig. 1. Steel cage construction process

3 Reinforcement Cage Quality Control

In the connection of reinforcement cages, the efficiency and quality of the connections are the main factors that affect the successful implementation of this process.

3.1 Connection Efficiency and Connection Quality

There are two traditional methods for connecting reinforcement cages: welding (Fig. 2(a)) and mechanical connection (Fig. 2(b)). Welding involves connecting with the help of bars, while mechanical connection is achieved through steel couplers.

When connecting the main bars of a steel cage, the welding time for 16 main reinforcing bars with a diameter of 22 mm is 66 min. However, when using steel couplers

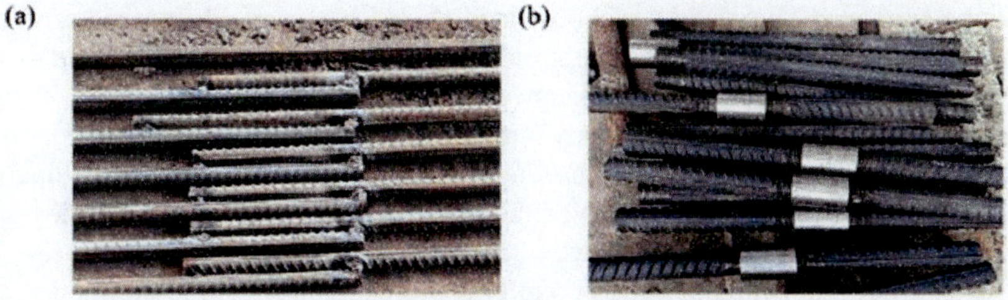

Fig. 2. Connection of rebars (a) Welding (b) Mechanical connection

for connections, the connection time is only 5–8 min. Compared to welding, mechanical connections are faster and more efficient. During the construction process of steel cage joints, it is important to avoid additional factors that can interfere with the progress. For example, insufficient reserved reinforcing bars, unstable core tube installation, and lack of pre-positioning and fixing of sound measurement tubes and core tubes can all reduce the efficiency of steel cage joints. Construction procedures should be classified and sorted, the main line of construction should be identified, secondary tasks should be completed in advance without affecting the construction nodes. Strictly follow the construction sequence, ensure good coordination between construction processes, and improve construction efficiency.

In the mechanical connection process, it is common to have an unacceptable number of exposed steel bar ends. When using threaded sleeve connections during construction, the short end of the thread should be fully utilized, leaving half of the thread on the long end. However, in actual mechanical connections, this requirement is seldom met. The empty length of joint socket exceeds 10mm, accounting for 37.5% of the total joint sockets used in the project. It becomes difficult to completely tighten the joint, and the positioning of the steel bars within the coupler cannot be guaranteed.

To address this issue, it is recommended to use a torque wrench to measure the torque value of steel bar connections during construction, and take samples of mechanical connections for quality inspection to verify the quality of steel bar connections. In large-scale construction, the efficiency and quality of steel cage connections need to be fully ensured. Strict control should be implemented throughout the entire construction process, including materials, steel cage processing and acceptance, and quality control during construction. This will guarantee qualified steel cage joints, improve connection efficiency and quality, enforce strict on-site quality control, prevent shortcuts during construction, and ensure construction quality.

3.2 Self-repairing Steel Coupler Technology Based on Shape Memory Alloy

As mentioned in the previous section, commonly used methods for steel bar connection include welding and coupler connections. Coupler connections are widely used in field construction due to their convenience and efficiency. However, coupler connections can easily result in loose joint areas and misalignment of the connected steel bars, making it difficult to accurately control the quality of steel cage connections.

Shape Memory Alloy (SMA), abbreviated as SMA, can generate a restoring force through electrical activation under pre-stretching. Additionally, when subjected to deformation under external forces, its resistance changes. After the deformation, it can be restored by electrical heating, with only a minimal residual deformation [7, 8].

By incorporating SMA into the currently used coupler connections, the quality of the coupler connection can be improved, allowing it to possess the capabilities of pre-tensioning, monitoring, and recovery.

The external shape memory alloy coupler is pre-expanded and pre-strained to correspond to the desired dimensions for practical engineering. Four internal steel couplers are installed inside the external shape memory alloy coupler, matching the raised edges of the inner coupler with the grooves on the outer shape memory alloy coupler. Circular plastic washers are inserted between the two ends of the internal steel couplers, and rectangular plastic washers are placed in the gaps between the internal steel couplers after installation. The self-repairing steel coupler assembly is then completed.

During use, the steel bars to be connected are processed into threaded shapes, corresponding to the threads on the internal steel couplers, and screwed into the couplers. After screwing the steel bars into the couplers, a controllable DC power supply is connected to the external shape memory alloy coupler using wires, forming a closed circuit. Electrifying the external shape memory alloy coupler heats it up. When the temperature of the external shape memory alloy coupler reaches the austenite phase transition temperature, it begins to contract both longitudinally and in diameter, exerting a restoring force on the steel bars and reinforcing the connection between them. Simultaneously, it forcibly corrects any misaligned steel bars, completing the forced alignment of the steel bars. Refer to Fig. 3 for the specific composition of the shape memory alloy coupler.

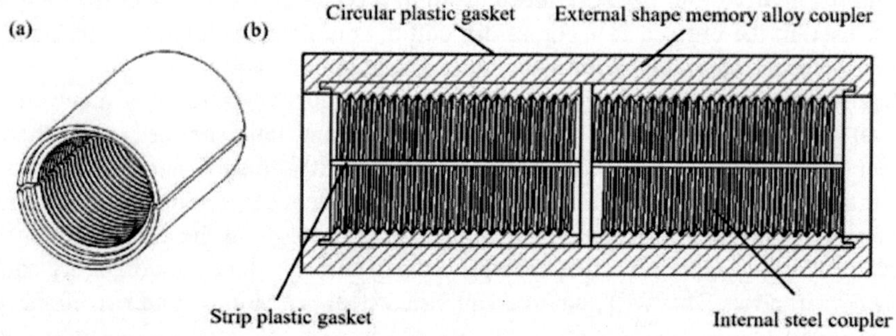

Fig. 3. Schematic diagram of self-healing reinforced steel coupler (a) Overall schematic diagram of internal steel coupler (b) Overall sectional view.

After the reinforcement is threaded into the coupler for use, controlled direct current is applied to the external shape memory alloy coupler by an electric power supply. The external shape memory alloy coupler restores its original shape, creating a pre-stressing force between the coupler and the reinforcement, reinforcing the connection between them. Simultaneously, under the restoring force of the shape memory alloy coupler, the spacer in the middle of the coupler deforms under the pressure from the reinforcement, eliminating the influence of unevenness of the reinforcement ends on the alignment of the reinforcement. This technique has advantages such as easy alignment of the reinforcement and strengthening the connection between the reinforcement and the coupler.

After the coupler is put into formal use, the resistance of the coupler can be monitored using an alloy resistance meter (Fig. 4). During normal usage, if there are no major deformations happening to the coupler, its resistance will not fluctuate significantly. When the coupler deforms due to applied forces, the shape memory alloy coupler also deforms, causing a change in resistance. By using the alloy resistance meter, the deformation of the coupler can be monitored. Once the deformation is detected, the control computer connected to the alloy resistance meter will regulate the controlled direct current power supply to heat the external shape memory alloy coupler, enabling it to restore its original shape and thus restoring the connection between the reinforcement and the coupler. This method provides the advantages of detectability and recoverability.

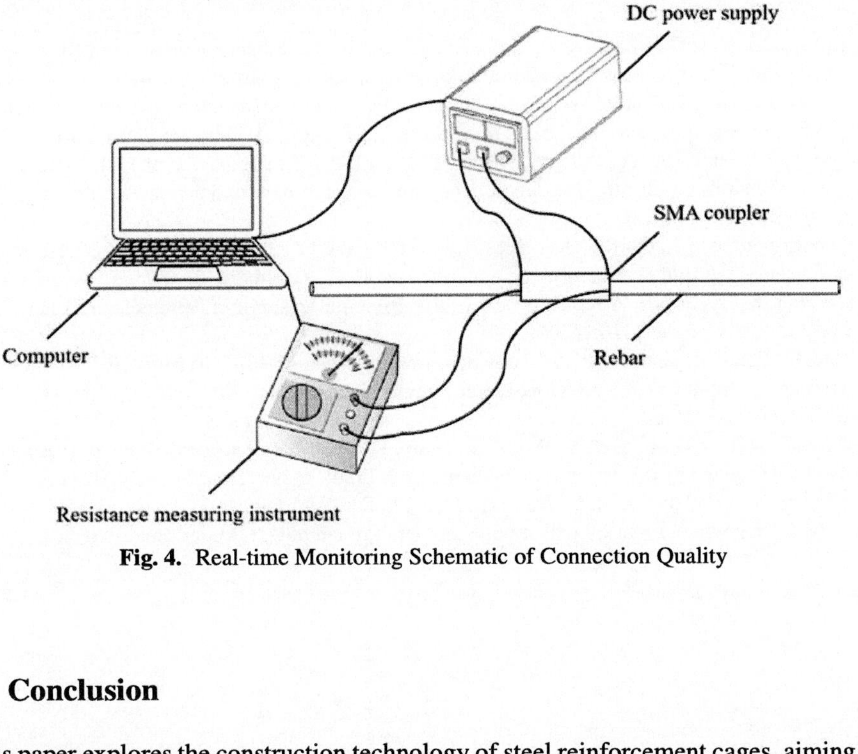

Fig. 4. Real-time Monitoring Schematic of Connection Quality

4 Conclusion

This paper explores the construction technology of steel reinforcement cages, aiming to control the quality and efficiency of rebar connections and ensure the smooth implementation of rebar cage alignment. Furthermore, it introduces a self-repairing steel reinforcement coupler based on shape memory alloy. The specific conclusions are as follows:

By pre-processing the rebar cage before construction and using steel reinforcement couplers to accelerate the docking speed of the rebar cage, the connection efficiency, accuracy, and quality of the rebar cage can be ensured by controlling the torque applied to the steel reinforcement couplers.

Introducing shape memory alloy into the steel reinforcement couplers and applying heat to generate pre-stress force during coupler deployment enhances the connection between the coupler and rebar, achieving forced alignment of the rebar.

By monitoring the resistance changes of the shape memory alloy coupler in real-time using an alloy resistance measuring device, any deformation or fluctuation in resistance can be detected. In such cases, direct current can be applied to heat the connection area, restoring it to its original state.

Compared to ordinary steel couplers, this method offers the advantages of convenient rebar alignment, strengthened connection between rebar and couplers, improved connection quality and speed, as well as real-time monitoring and recoverability.

References

1. Kalogeropoulos, G.I., Tsonos, A.-D.G., Konstantinidis, D.: Seismic behaviour of RC columns with welded rebars or mechanical splices of reinforcement. Earthquakes Struct. **17** (2019)
2. Željko, B., Ivan, S., Nedjeljko, M., Katarina, S.: Preliminary research on influence of welding parameters strength at welded joints in ribbed reinforcing steel. Tehnički glasnik **14** (2020)
3. Fawzy, T., Khalil, A.-H., Atta, A., Afefy, H.M., Ellithy, M.: Behavior of high-performance concrete beams having tension lap spliced anchor-ended bars under repeated loading. Eng. Struct. **274**, 115140 (2023)
4. Pantazopoulou, S.J., Petrou, M.F., Spastri, V., Archontas, N., Christofides, C.: The performance of corroded lap splices in reinforced concrete beams **37**, 31–44 (2019)
5. Coogler, K.L., Harries, A.K., Gallick, M.: Experimental study of offset mechanical lap splice behavior. ACI Struct. J. **105** (2008)
6. Batista, R.J.S., Pragana, J.P.M., Bragança, I.M.F., Silva, C.M.A., Martins, P.A.F.: Invisible mechanical lap joints for metal–polymer laminates. Proc. Inst. Mech. Eng. Part L J. Mater. Des. Appl. **235** (2021)
7. Yang, C., Liu, T., Zhong, M.T., Wu, Z.G., Deng, J., Du, Y.L.: Enhancement of recovery stress by annealing and prestraining of NiTiNb wires. J. Intell. Mater. Syst. Struct. (2023)
8. Yang, C., Liu, T., Zhong, M.T., Wu, Z.G., Deng, J., Du, Y.L.: High recovery stress performance of NiTi shape memory alloy with a wide temperature window. J. Mater. Eng. Perform. (2023)

Study on Hole Cleaning Construction Technology of Bored Cast-in-Place Pile

Xuefeng Shi[1], Weiqiang Chen[2], Hongyan Sun[3], Shuqiang Cao[3], Peng Sun[3], and Xingpei Wu[4(✉)]

[1] China Railway Construction South China Construction Co. LTD., Guangzhou 511458, China
[2] Guangzhou Metro Group Co. LTD., Guangzhou 510330, China
[3] Guangzhou Engineering Co. LTD. of China Railway 19th Bureau Group, Guangzhou 511455, China
[4] School of Civil Engineering, Guangzhou University, Guangzhou 510006, China
xingpeiwu@qq.com

Abstract. Bored pile is a fundamental construction technique widely employed in modern engineering projects, including high-rise buildings, highway bridges, and other infrastructure constructions. Within the bored pile construction process, hole cleaning plays a critical role. By optimizing the hole cleaning procedure, it is possible to ensure hole quality, improve construction efficiency, reduce environmental impact, and enhance construction safety. This paper focuses on large pile diameters or super long pile foundations as the research object and presents a detailed introduction to the air lift reverse circulation secondary hole cleaning method. The construction parameters for air lift reverse circulation hole cleaning of large pile diameters and super long pile foundations are optimized through pile tests, offering a new perspective on hole cleaning construction for super long and large diameter bored piles.

Keywords: Construction Technology · Air-lift Reverse Circulation · Hole Cleaning · Bored Cast-in-place Pile

1 Introduction

The bored pile is a commonly used method for foundation treatment. During the construction process, there is a possibility of residual soil and other impurities in the pile holes. The presence of these substances can have negative impacts on the quality and performance of the pile foundation. Therefore, it is crucial to ensure high-quality construction by implementing effective hole cleaning techniques [1]. Improving the quality and performance of bored piles requires focused research on hole cleaning construction methods. Wu et al. [2] proposed a hole cleaning construction technology specifically designed for large-diameter bored piles. Their approach aimed to enhance the interaction between the pile body and the surrounding soil. In a similar vein, Guo et al. [3] emphasized the significant impact of hole cleanliness on drilling extension (ERD). They employed computational fluid dynamics (CFD) to simulate the two-phase flow of

W. Guo et al. (Eds.): GBCESC 2023, LNCE 328, pp. 421–429, 2024.
https://doi.org/10.1007/978-981-99-9947-7_45

solid-liquid in the annulus of the wellbore. Their study demonstrated the crucial role of wellbore cleaning in achieving optimal drilling extension. Al Rubaii et al. [4] proposed a model for assessing hole cleaning called the hole cleaning index (HCI). This model offers the potential for automated and real-time evaluation of the degree of hole cleanliness. In a similar vein, Al Malki et al. [5] developed and assessed a hole cleaning factor (HCF) model, utilizing it to appraise hole cleaning efficiency in vertical and deviated wells.

The traditional hole cleaning technologies for bored piles include positive circulation, reverse circulation, and impact reverse circulation. Reverse circulation hole cleaning construction technology is known for its high efficiency, cost-effectiveness, and effective removal of debris [6, 7]. Ma et al. [8] successfully applied air-lift reverse circulation drilling technology in a geothermal well construction project in Dalian, resolving the issue of wellbore stability that remained unresolved with conventional positive circulation methods. Boyou et al. [9] developed a hole cleaning fluid with enhanced rheological properties to improve the efficiency of hole cleaning with air-lift reverse circulation. Wang et al. [10] employed air-lift reverse circulation for dual hole cleaning in deep foundation pit bored piles, and assessed the benefits and applicability of this technique.

The pile foundations in the Wanqingsha depot of Guangzhou Metro Line 18 are constructed using concrete bored piles. These bored piles consist of frictional end-bearing piles and end-bearing friction piles, both of which are considered as ultra-long bored piles. The positive circulation borehole cleaning method is generally reliable and stable. However, when constructing with larger diameters or ultra-long piles, the cleaning effectiveness may decrease unless a larger pump is utilized to enhance pumping capacity. Additionally, in load-bearing layers composed of gravel or pebbles, completely removing larger particles like pebbles or crushed stones becomes more difficult with the positive circulation method. On the other hand, the air-lift reverse circulation method provides several advantages, such as a rapid mud removal flow rate and high cleaning efficiency. It is especially suitable for larger diameter piles and deeper bores. Moreover, compared to the positive circulation method, it demonstrates superior cleaning results when dealing with thicker sediments or sediments containing larger particles.

Currently, the commonly used method for cleaning holes during bored pile construction is positive circulation. However, in large-scale construction projects, the air-lift reverse circulation method is not frequently used for hole cleaning. There is no previous record of applying the air-lift reverse circulation method for cleaning holes in large diameter or ultra-long piles. This paper focuses on the pile foundation in the cover area of Wanqingsha Depot of Guangzhou Metro Line 18 as a case study. It introduces the construction technology and key operational points of air-lift reverse circulation hole cleaning in this project. Additionally, it enhances and refines the construction technology of air-lift reverse circulation hole cleaning through pile testing. The findings of this study provide a new approach and method for hole cleaning in the construction of large diameter or ultra-long piles.

2 Principle and Process of Air Lift Reverse Circulation Borehole Cleaning

2.1 The Principle of Air-Lifting Reverse Circulation Hole Cleaning

The air-lifting reverse circulation hole cleaning method, depicted in Fig. 1, utilizes compressed air derived from an air compressor. This compressed air is directed through a windpipe situated within the conduit and subsequently delivered into the pile hole. As the high-pressure air mixes with the mud, a slurry-air mixture materializes within the conduit. Given its lower density relative to the mud, this slurry-air mixture ascends due to its decreased weight. At the lower end of the mixer, a negative pressure is created within the conduit, prompting the mud below to rise. Through the combined action of this negative pressure and the momentum generated from the air pressure, a continuous replenishment of mud is facilitated, thereby causing its ascent to the mixer where it combines with the air to produce a slurry-air mixture. It should be noted that this mixture continues to rise, thus generating a flow. Owing to the larger cross-sectional area present within the conduit as compared to the annular cross-sectional area existing between the conduit and the pile wall, a reverse circulation with a high flow speed and flow rate is achieved. Consequently, the sediments are effectively transported out of the conduit, successfully eliminating them from the bore-hole.

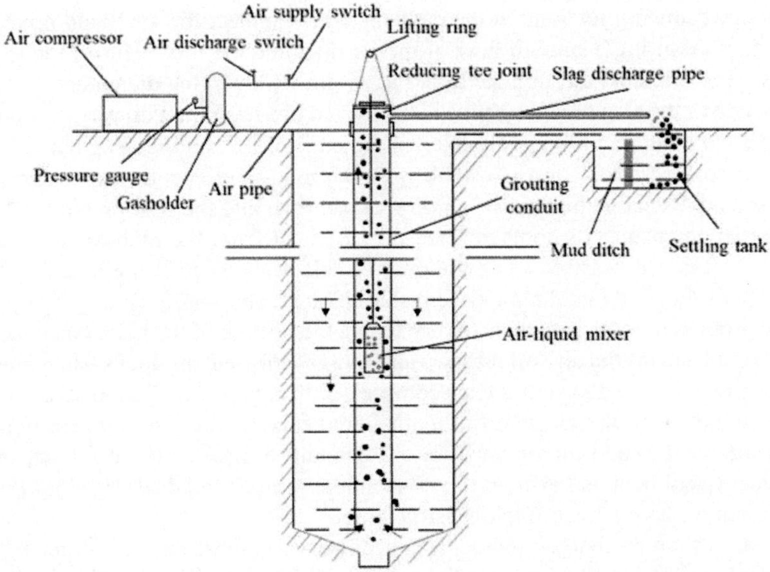

Fig. 1. Schematic diagram of air-lifting reverse circulation hole cleaning principle

2.2 Air-Lifting Reverse Circulation Hole Cleaning Equipment

In addition to the common equipment such as the drilling rig, mud pump, and grouting conduit, the air lift reverse circulation hole cleaning process requires the inclusion of the following equipment:

- VF-6/7 Air Compressor: Provides compressed air.
- Ducting: Transports compressed air.
- Air-Liquid Mixer: Mixes compressed air with mud inside the conduit to produce an air-liquid mixture slurry.
- Reducing Tee Joint: Modified from a φ258 mm grouting conduit.
- Slag Discharge Pipe: Made of φ6" high-pressure rubber hose.

2.3 Air-Lifting Reverse Circulation Hole Cleaning Process and Operation Method

The air-lifting reverse circulation hole cleaning process is illustrated in Fig. 2.

The operational steps for the air-lift reverse circulation hole cleaning process are outlined below:

(1) After reaching the designated depth of the pile hole, begin the initial hole cleaning using drilling equipment and a mud pump. Once the initial hole cleaning is finished, promptly lift the drilling tool and continue with the installation of the steel cage and insertion of the grout conduit.
(2) Install a reducing tee joint at the catheter port. Connect the air-liquid mixer to the air duct, ensuring a smooth flow of the air duct and the mixer. Insert the air-liquid mixer into the duct through the three-way joint with varying diameters.
(3) One end of the slag discharge pipe is connected to a tee joint with varying diameters, while the other end is securely fixed at the edge of the mud settling tank. Once the air-lift reverse circulation is established, the mud is efficiently expelled, with the fixed slag discharge pipe ensuring direct discharge into the settling tank.
(4) In order to ensure a smooth reverse flow of mud from the mud ditch to the hole mouth, it is necessary to clean the mud circulation system at the hole mouth. The length of the mud circulation system should be greater than 10 m.
(5) The grout conduit should be lifted 1m above the bottom of the hole. Once the safety and reliability of the air compressor, air storage tank, and air duct system have been confirmed, proceed to start the air compressor to supply air. Compressed air is then directed through the air duct to the air-liquid mixer, where it is mixed with mud, forming an air-mud mixture within the conduit. Subsequently, the mud, together with any sediment, is swiftly returned from the conduit and discharged into the mud purification pool via the slag discharge pipe.
(6) To control the discharge volume, it is necessary to adjust the air volume of the air compressor. During construction operations, the air pressure is typically maintained between 0.6–0.7 MPa.
(7) To ensure effective sediment flushing at the bottom of the hole, it is recommended to frequently adjust the distance between the bottom end of the conduit and the hole bottom. Additionally, continuously move the position of the conduit during the hole cleaning process.

(8) Once the mud performance indicators meet the specified requirements, the air compressor should be closed, and the sediment thickness at the bottom of the hole should be measured. If the sediment thickness is deemed satisfactory, the air lift reverse circulation hole cleaning equipment can be removed, and the grouting funnel can be connected to proceed with the underwater concrete pouring process. However, if the sediment thickness does not meet the design requirements, it is necessary to continue the hole cleaning until the sediment thickness reaches the specified standards.

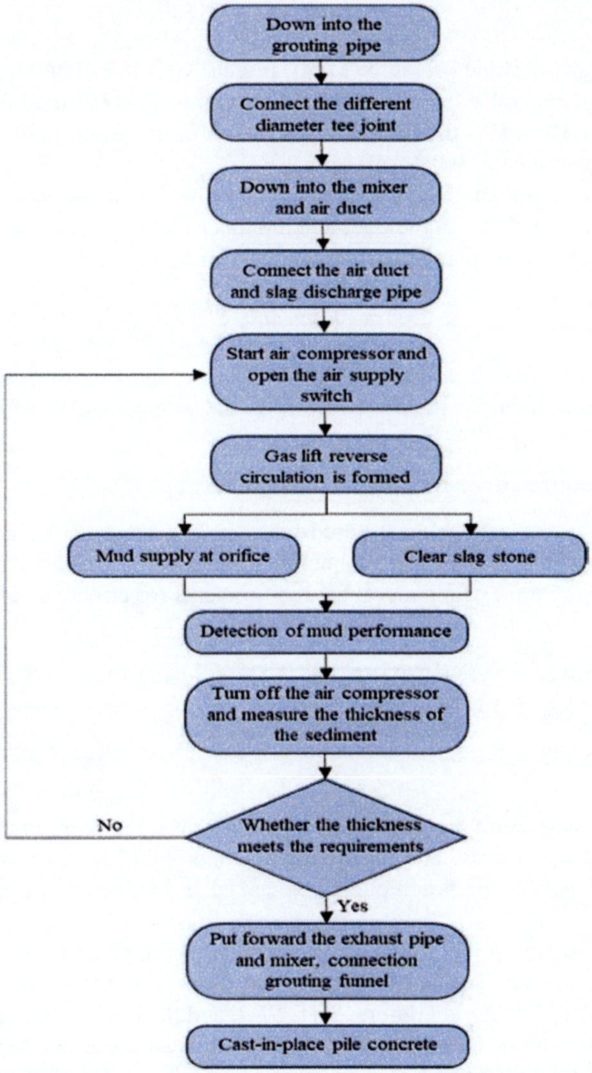

Fig. 2. Air-lifting reverse circulation hole cleaning process flow chart

2.4 Selection of Air-Lifting Reverse Circulation Hole Cleaning Process Parameters

(1) Air pressure (P), air flow rate (Q)

When selecting the air compressor for gas-lift anti-cyclic cleaning, the required air pressure (P) and air flow rate (Q) during the hole cleaning process should be taken into consideration. The pressure (P) of the air compressor can be expressed using the following equation:

$$P = 10Hrh - 2 + \Delta P \tag{1}$$

where, P is the air machine pressure (MPa); h is the best installation depth of the air-liquid mixer (m), generally 0.6 times of the hole depth; rh is the mud density (g/cm^3), generally 1.15–1.18; ΔP is the pressure loss (MPa) in the ground air supply pipeline, which is generally taken as 0.04–0.10 MPa.

The air volume (Q) of the air compressor can be calculated based on the return velocity of the mixed slurry in the duct and the inner diameter of the duct. This can be mathematically expressed using the following empirical formula:

$$Q = (0.5 \sim 0.6)\, d^2 \nu \tag{2}$$

where Q is the required air volume (m^3/min); d is the diameter of the catheter (m); ν is the return flow velocity of the mixed slurry in the catheter (m/s), which is generally 1.5–2.0.

(2) Air-lifting reverse circulation hole cleaning time (T)

In construction, typically when the mud density is below 1.2 g/cm^3, the mud viscosity ranges from 22 to 28 s, and the sand content is less than 5%, the hole cleaning time (T) is approximately 45 min, which meets the requirements for effective hole cleaning.

3 Experimental Investigation of Hole Cleaning for Bored Cast-in-Place Pile

3.1 Pile Test Conditions

Improved technology is employed to clean the holes of three representative pile foundations in geological aspects. The construction technology of air-lift reverse circulation hole cleaning for super-long bored piles is enhanced and refined through the utilization of the test pile.

The pile foundation project for the Wanqingsha Depot of Guangzhou Rail Transit Line 18 is situated in the Nansha District of Guangzhou City. The project involves three test piles with the following identifiers: SJ-1, SJ-2, and SJ-3. Bored piles are employed, utilizing concrete with a strength grade of C35. The designed pile length ranges from 47 to 60 m, with a diameter of 1 m. The following Table 1 gives all the information of bored cast-in-place piles:

Table 1. Test Pile Parameters for bored Cast-in-place Pile.

Heading level	Design pile length (m)	Pile diameter (m)	Bearing stratum
SJ-1	43.7	1.0	Micro-weathering
SJ-2	59.9	1.0	Strong weathering
SJ-3	41.7	1.0	Mid-weathering

SJ-1, SJ-2, and SJ-3 serve as representatives in the geological selection, with each bearing different geological conditions: slightly weathered, strongly weathered, and moderately weathered, respectively. Upon conducting geological sampling of the three test piles, it was observed that the geological conditions were relatively similar. The sequence of soil layers is as follows: a backfill soil layer of less than 2 m, a silt layer ranging from 2 to 20 m, a medium sand layer from 20 to 24 m, a silt soil layer spanning from 24 to 29 m (24 to 38 m for SJ-3), a silty clay layer between 29 and 33 m, a coarse round gravel soil layer extending from 33 to 50 m, and finally, the weathering bearing layer.

3.2 Test Pile Effect

Prior to concrete pouring, it is necessary to address the buildup of sediment resulting from the disassembly of the drill pipe, drill bit, lower steel skeleton, and conduit. Adequate hole cleaning is required in such cases. For the test piles, air lift reverse circulation hole cleaning is employed. The construction technology for air lift reverse circulation has been continuously enhanced and perfected within the pile testing process. Table 2 provides detailed information on the air lift reverse circulation construction and the results of the pile tests.

Table 2. Construction status of air-lift reverse circulation process.

Pile number	Hole depth (m)	Hose length (m)	Pipe diameter	Number of floral tube holes	Air pressure (P)	air flow rate (Q)	Pile test results
SJ-1	45.26	36	DN20	None	0.8 MPa	6 m^3/min	failed
SJ-2	61.28	49	DN20	12	0.8 MPa	6 m^3/min	failed
SJ-3	44.10	27	DN20	48	1.45 MPa	14 m^3/min	succeeded

Based on the results presented in Table 2, it is evident that both pile SJ-1 and pile SJ-2 tests resulted in failure, whereas the pile SJ-3 test was successful. The failure of the air lift reverse circulation in pile SJ-1 and pile SJ-2 can be attributed to the improper adherence to the specified length of the air pipe and flower pipe during hole cleaning, as per the requirements of the air lift reverse circulation construction process. Conversely, the successful air lift reverse circulation hole cleaning in pile SJ-3 was achieved by strictly adjusting the pipe length and the number of holes in the flower tube to align with the specifications of the air lift reverse circulation process. Additionally, the power of the air compressor was enhanced, wherein the rated pressure was increased from 0.8 MPa to 1.45 MPa, and the air flow rate was raised from 6 m^3/min to 14 m^3/min. Consequently, the successful implementation of the air-lift reverse circulation hole cleaning technique was ensured.

4 Conclusions

This paper focuses on the pile foundation in the cover area of Wanqingsha Depot of Guangzhou Metro Line 18. The research objective is to investigate the construction technology of super-long pile hole cleaning and optimize it through the analysis of three test piles. The specific conclusions drawn from this study are summarized as follows:

(1) The installation of air-lifting reverse circulation must strictly adhere to the construction process requirements, with a length ranging from 0.55 to 0.65 of the depth from the mud level to the bottom of the hole. For the air-liquid mixer (floral tube), it is recommended to use DN20 galvanized pipe, measuring 1.2 m in length. The upper end should be connected to the air duct, while the lower end orifice must be sealed. The hole should be evenly drilled around the pipe, with a pore diameter of 10 mm and a hole spacing of 30 to 40 mm.
(2) To ensure optimal performance, it is recommended to select an air compressor with a pressure range of 0.6 to 1 MPa and a flow rate between 8 and 12 m^3/min. The specific power of the air compressor can be determined through pile testing, taking into account the actual site conditions. This will allow for the selection of the most suitable and efficient air compressor for the application.
(3) Prior to conducting the air-lift reverse circulation, it is necessary to replace the viscous and agglomerated mud in the hole. This step aims to enhance the fluidity of the mud, resulting in easier entry of the mud at the bottom of the hole into the pipe under the negative pressure created by the air lift.
(4) In order to minimize air pressure loss from the air compressor, it is crucial to ensure proper sealing during the installation of the tubing and the trachea. During the air-lift reverse circulation process, it is recommended to elevate the catheter and position it approximately 0.2 to 0.4 m above the sediment surface. This arrangement promotes more effective hole cleaning and improves overall cleaning efficiency.

References

1. Dzagov, A.M., Kitaikin, V.A., Chernov, R.I.: On the impact of cleaning widened sections of boreholes on the bearing capacity of bored situ-cast piles. Soil Mech. Found. Eng. **53**(4), 274–281 (2016)
2. Wu, X., Lu, J., Li, X.: Construction process of large-diameter drilled displacement piles based on grouting clearing and vibration device. J. Geomech. **40**(6), 2319–2326 (2019)
3. Guo, Z.M., Xiao-Le, L.I., Hong, M., et al.: Effect of drill pipe rotation on borehole cleaning for extended reach well. J. Hydrodyn. **21**(3), 7 (2009)

4. Al Rubaii, M., Al Shargabi, M., Al, S.D.: A novel model for the real-time evaluation of hole-cleaning conditions with case studies. Energies **16**, 4934 (2023)
5. Al Malki, M., Mahmoud, A.A., Elkatatny, S.: Performance of the hole cleaning factor in predicting the hole cleaning conditions in vertical and deviated wells: real case studies. Arab J. Sci. Eng. (2023)
6. Wang, Q., Zhao, C.F., Zhao, C., Xue, J.X., Lou, Y.: Study on the influence of positive and negative circulation mud retaining wall of bored pile on pore diameter. J. Geotech. Eng. **33**(S2), 205–208 (2011)
7. Wang, S.H., Sun, K.Q., Liu, G.Y., et al.: Construction technology and intelligent equipment of bridge squeezed branch pile. Guangdong Highway Transp. **45**(05), 55–59 (2019)
8. Ma, X.M., Chen, Y., Qi, L.H.: Research and application of gas-lift reverse circulation drilling technology to geothermal well construction in Dalian Jiaoliu Island. Procedia Eng. **73**, 252–257 (2014)
9. Boyou, N.V., Ismail, I., Wan Sulaiman, W.R., et al.: Experimental investigation of hole cleaning in directional drilling by using nano-enhanced water-based drilling fluids. J. Petroleum Sci. Eng. (2019)
10. Wang, Y., Yang, J., Shen, Y., Liu, G.: Research on construction technology for air lift and reverse circulation double clear hole in bored pile of deep foundation pit. J. Eng. Sci. Technol. **13**(5), 1236–1247 (2018)

Design and Experimental Study on a Novel Direct Measuring Force Device for Rod-Cable Structure Bridge Cables

Jinggan Shao[1], Tingdong Shang[2], Genshang Wu[1], Wei Liu[3], Le Bo[3], and Xuling Liu[3](✉)

[1] School of Highway, Henan College of Transportation, Zhengzhou 451450, China
[2] Zhengzhou Dongchen Science and Technology Co., LTD, Zhengzhou 450000, China
[3] School of Mechatronics Engineering, Zhengzhou University of Light Industry, Zhengzhou 450002, Henan, China
liuxuling@zzuli.edu.cn

Abstract. Able force testing plays an important role in the daily maintenance of cable-stayed Bridges, which is conducive to ensuring the safe and stable operation of cable-stayed Bridges. A novel direct measuring force device for cable forces based on Newton's third law is presented in this paper. The installation method and experimental platform construction method of the measuring device are described. The experimental results show that the designed measuring device has the advantages of compact structure, lightweight, low operating strength, no electricity requirement, and is suitable for on-site operation. Moreover, the detection method does not need to change the installation and connection relationship of the cables, and can detect the cable force without affecting the normal use of the bridge.

Keywords: Cable force measurement · direct measuring force device · structural design · perforated measuring force sensor · adjusting screw nut

1 Introduction

With the rapid development of society, the development of traffic has been driven up, and Bridges are an indispensable part of the traffic system, and cable-stayed Bridges are extremely important among them [1, 2]. Therefore, it is necessary to test the cable force many times during the construction of the bridge, and the measurement results determine whether the construction of the bridge can continue smoothly, and whether it is close to the design state after the completion of the bridge construction [3, 4]. Furthermore, due to the influence of various external factors leading to changes in the cables during operation, it is also necessary to measure the cable force again to improve the stress of the cable, ensure the health of the bridge, and protect the safety of people's lives, so the cable force measurement is extremely important.

W. Guo et al. (Eds.): GBCESC 2023, LNCE 328, pp. 430–438, 2024.
https://doi.org/10.1007/978-981-99-9947-7_46

Currently, the detection of cable force predominantly employs methods such as jack pressure gauge detection, grating pressure sensor detection, resistance strain gauge detection, magnetic flux detection, and vibration frequency detection [5]. The jack pressure gauge detection method to test the cable force will become very difficult and inconvenient if the cable installation is completed [6]. For the detection method, the grating pressure sensor its performance and measurement accuracy are also constantly changing, so it needs to be calibrated regularly [7]. The disadvantages of the resistance strain gauge detection method are that the strain gauge is not easy to protect, the survival rate of the patch is low, and because of the large elongation of the cable after the tension, the zero point should be selected very carefully. Otherwise, it is easy to exceed its range [8]. The electromagnetic sensor needs to be set in the cable production, the operation procedure is a troublesome, complicated structure [8]. In addition, the resistance strain gauge measurement and fiber grating strain gauge measurement, although mature in theory, are less frequently used in practical engineering applications because they are easy to be affected by environmental factors and other factors [9, 10].

At present, the vibration frequency method is the most practical and economical detection method. Furthermore, the test process required equipment is simple, easy to achieve, the result is reliable and accurate, and can meet the requirements of repeated testing during the construction process. Shao S et al. used the holographic vision sensor system to obtain cable digital image data, calculate and analyze the spatial geometry, holographic mode, dynamic displacement vector, and other holographic characteristic parameters, and proposed a cable force calculation method based on dynamic/static morphology, with the maximum relative error of 7.31% and root mean square error of 1.79% [11]. Chen Y et al. proposed a vibration fundamental frequency identification method based on multiple autocorrelation, and verified through experiments that the method could accurately identify the fundamental frequency of cable vibration in the presence of interference signals [12]. The above research results improve the theory of the cable force testing system, and provide a reference for the actual cable force testing methods and devices. However, the vibration frequency method is an indirect force measurement method. Furthermore, because of the different angle, length, and anchoring modes of each rod and cable, different calculation models, the measurement accuracy and accuracy are unstable.

According to the demand for the cable force detection in the operation stage of cable-stayed bridges and the current situation of cable force detection of cable-stayed bridges, a direct measuring force device is proposed with a compact structure, light weight, low operation strength, no electricity requirement, and suitable for on-site operation, evolution, and modeling process in this paper. In addition, it does not need to change the installation and connection relationship of the cable members, and can detect the cable force without affecting the normal use of the bridge of this direct detection device. The performance simulation and experimental research on the direct detection device of cable force of cable-stayed bridges in the operation stage are mainly studied, aiming at strengthening the monitoring of cable force and further ensuring the safe operation of cable-stayed bridges.

2 Structure and Working Principle of the Cable Force Mearing Device

2.1 Bridge Cable Anchorage Structure

The block diagram of the direct measuring force device is shown in Fig. 1.

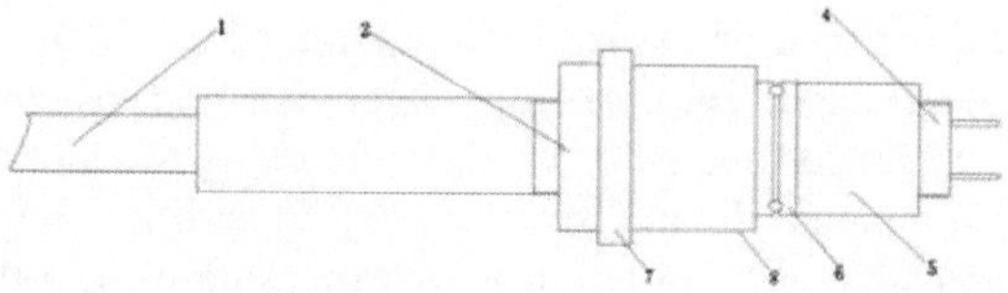

1. Bridge cable; 2. Bridge anchor gasket; 3. Initial adjustment nut (not marked in the figure, covered by the drive sleeve 8); 4. Force transfer cylinder; 5. Adjusting nut; 6. Rotating bearing; 7. Perforated measuring force sensor; 8. Transmission sleeve.

Fig.1. Block diagram of the direct measuring force device

The initial adjustment nut 3 is not marked in the figure and is covered by the drive sleeve 8. The perforated measuring force sensor 7 is not directly pushed on the adjusting nut, the measuring force sensor 7 is set outside the initial adjusting nut 3 and pushed on the anchor gasket 2, and the anchor gasket plate forms the reaction frame of the entire cable measuring force device. At this time, due to the increase of axial transmission stroke, a transmission sleeve 8 needs to be arranged between the rotating bearing 6 and the measuring sensor. In this embodiment, the transmission sleeve is a transmission sleeve. When the adjusting nut 5 is twisted, the mechanism seat gives the adjusting nut a force in the direction away from the anchor pad plate 2 through the supporting cylinder. When the adjusting nut is detached from the anchor pad plate 2, the whole cable direct measuring force device is completely strung on the tension path of the cable. At this time, the indicating value of the measuring force sensor is the tension value of the rope.

3 Simulation Model

In this model, 8 components need to be built, which are the cable, anchor gasket, initial adjustment nut, transmission cylinder, adjusting nut, rotating bearing, perforated force sensor, and the drive sleeve, as shown in Fig. 2 a). It is noted that the beads and the bearing housing together form the rotating bearing. The complete configuration after assembly is shown in Fig. 2 b).

After the modeling was completed, the model is imported into the Workbench of ANSYS finite element software for grid division. The grid division is shown in 2 c). The grid format adopted in this study is the span angle center of large scale, the border diagonal reaches 0.56796 m, the average surface area reaches 2.2572e-003 m^2, and the minimum edge length is 5.e-003m. For the mesh quality, the error limit is the strong machine; the target quality is the default value (0.050000), and then the medium smooth.

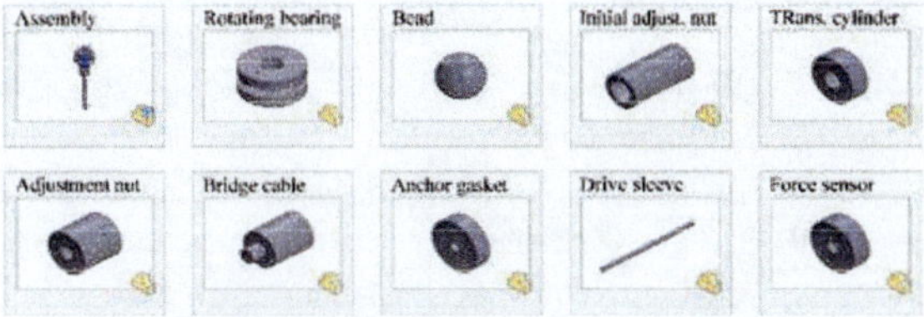

a) Model parts

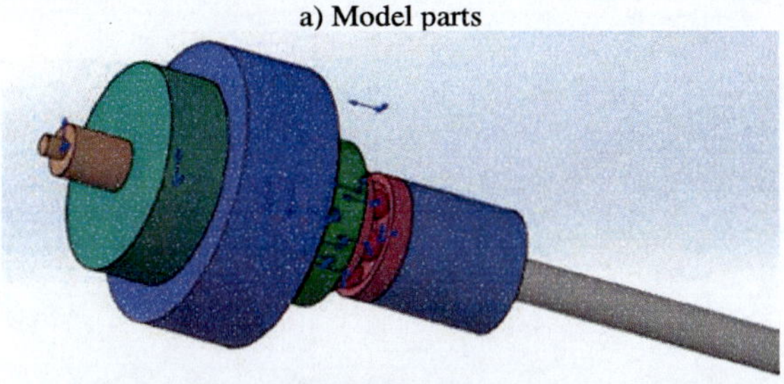

b) Assembly

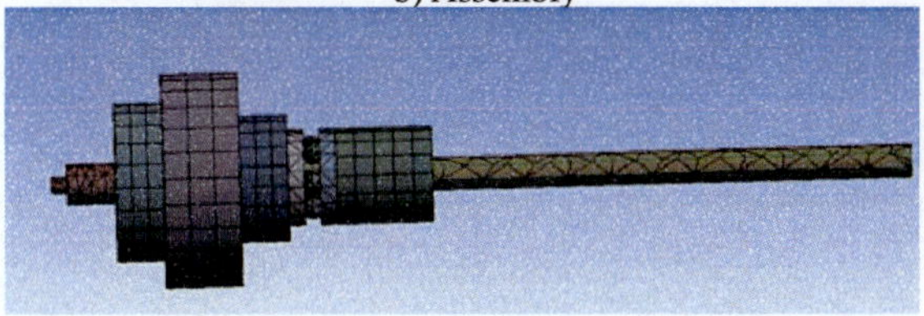

c) Meshing

Fig. 2. Parts, assembly, and meshing of the direct measuring force device

In the expansion set, the expansion option selects smooth transition, and the transition ratio is 0.272. In the number of expansion layers, the maximum number of expansion layers is 5, the growth rate is 1.2, and the expansion algorithm used is Pre.

4 Characteristics Simulation Results Analysis

The simulation results include three parts: total deformation, equivalent elastic strain, and equivalent stress, as shown in Fig. 3.

In the simulation process, the cable is regarded as a cylinder with a diameter of 1cm, the cable is a cylinder with a diameter of 2.5cm, and the material is selected as structural steel. The role of the anchor gasket is to fix the entire anchoring area, because the anchor gasket is connected to the bridge, so the anchor gasket should be set as a fixed object in the simulation.

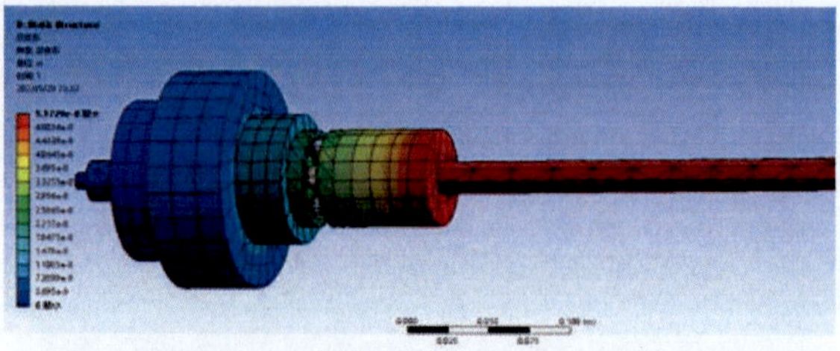

a) Total deformation

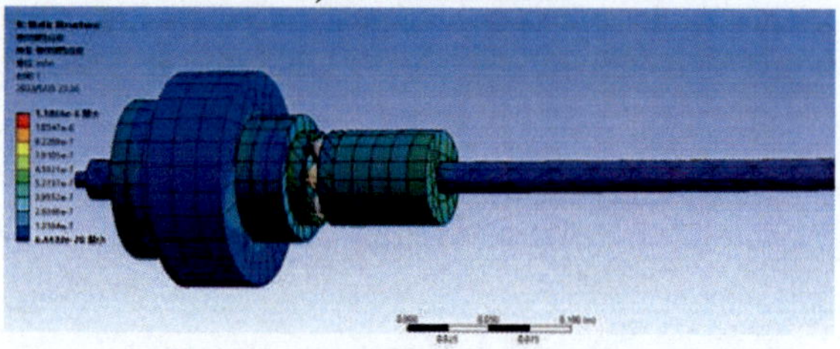

a) Equivalent elastic strain

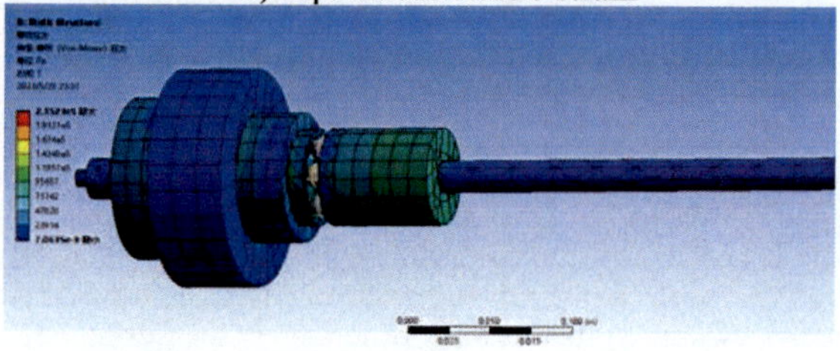

a) equivalent stress

Fig. 3. Performance simulation results of the direct measuring force device

5 Performance Test

5.1 Force Sensor

The pictures of the perforated measuring force sensor are shown in Fig. 4. Its model is MKY-JMZX-3102AT, this sensor belongs to the three-string intelligent type, the measuring range is 200 kN, and the sensitivity is 0.1 kN. The function of the perforated measuring force sensor is to collect the force of the adjusting nut. In the direct tension measuring system, when the adjusting nut is separated from the anchor gasket, the whole cable measuring force device is completely strung on the tension path of the cable, and the indicating value of the measuring force sensor is the tension value of the cable.

The experimental dates are acquired by the perforated measuring force sensor, are collected by the NIELVIS teaching platform. Finally, calculated, and analyzed by the Labview software.

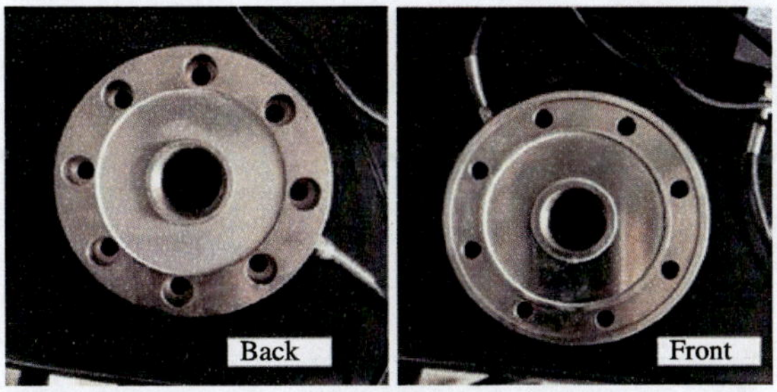

Fig. 4. Pictures of the perforated measuring force sensor

5.2 Force Sensor

The contours of outlet pressure and contours of outlet flow rate are shown in Fig. 5. The anchor gasket is a structural steel connector with an outer diameter of 10 cm and an inner diameter of 2 cm and a thickness of 3 cm. The initial adjustment nut is pushed and the anchor washer is used to support the force cylinder and sensor. The part size of the initial adjustment nut is 3 cm outer diameter, 2 cm inner diameter and 6 cm length. The role of the support cylinder is to support all parts to achieve a fixed support and connect all parts, The outer diameter of the support cylinder is 2 cm, the inner diameter is 1 cm and the length is 50 cm. The role of rotating bearings is to reduce friction and resistance, so that the machine or equipment can run smoothly. It transfers the rotating force of the shaft to the bearing by forming a contact surface with low friction and free rotation between the shaft and the bearing, thereby reducing friction and energy loss. Because of the increase in the axial transmission stroke, it is necessary to set a transmission component, that is, the force transfer cylinder, between the rotating bearing and the force measurement sensor.

This experiment is a full-scale model. Therefore, the steel wire is used instead of the cable, the jack is used to tighten the steel wire to provide the cable force. The tensile force value of the jack in this experiment is 180N. Furthermore, a piece of paper is placed under the jack and the measuring force device to keep the whole system at the same height to ensure the reliability of the experiment.

At the beginning of screwing the mechanism nut, the force value of the adjusting nut acting on the anchor plate is the largest, and no force is transmitted to the mechanism nut, that is, the origin of Fig. 6. In the process of screwing the nut of the mechanism, the force value of the adjusting nut acting on the anchor plate gradually decreases, and the force value transferred to the nut of the mechanism gradually increases, that is, the first half of the figure. When the adjusting nut and the anchor plate are just separated, the force value of the adjusting nut acting on the anchor plate is zero, that is, the watershed point in the figure. At this time, the force value of the adjusting nut is transferred to two parts, the mechanism nut bears a part of the force value, and the thread and the nut bear a part of the force value, so the nut force value of the screwing mechanism continues to increase, that is, the second half of the figure. Figure 6 shows the change process of the force values in the whole operation. The turning point force value is 187.53N, slightly higher than the cable force value provided by the jack.

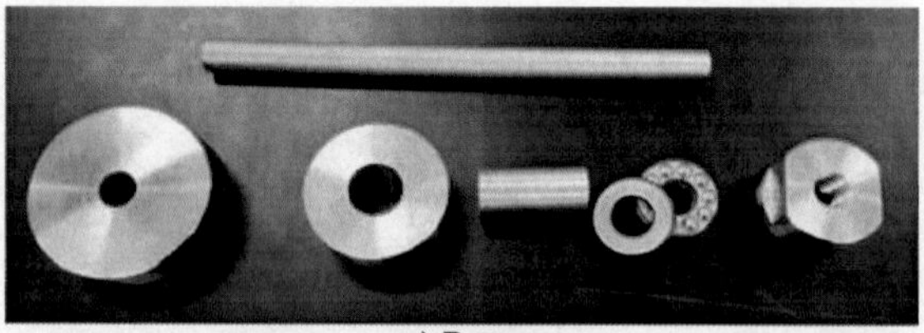

a) Parts

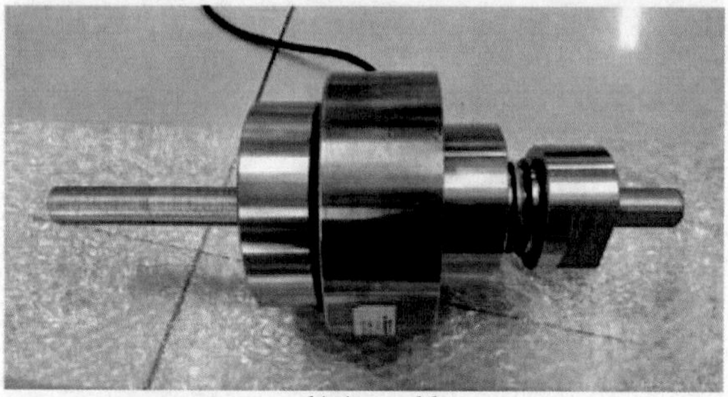

b) Assembly

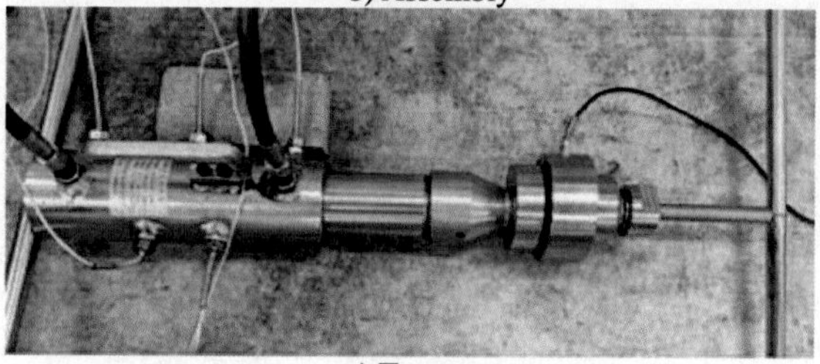

c) Test setup

Fig. 5. Experimental setup for testing the simulated bridge cable forces

There are three main reasons for the deviation in the analysis experiment. Firstly, due to the limited experimental site, the experiment adopts a full-scale model for force measurement, and can only exert a force of several hundred N, which has a great deviation from the actual bridge of tens of thousands of tons, so the data obtained from the experiment deviates from the theoretical value. Secondly, the tension jack used in the experiment is not stable, which also caused some experimental deviations. Last but not least, the direct measuring force device designed in this experiment is also slightly different from the ideal model used in numerical simulation.

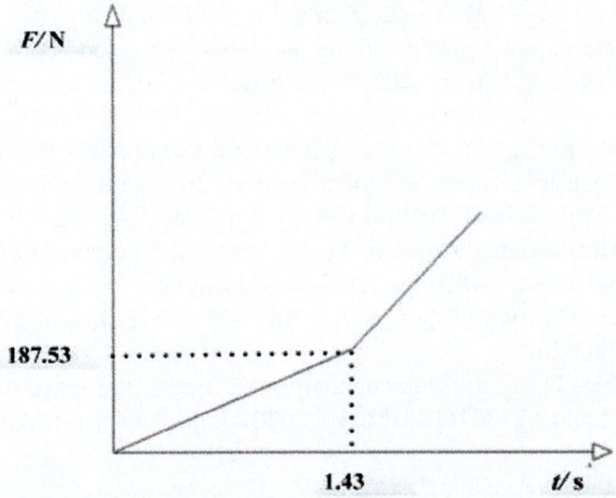

Fig. 6. Fitting curve of the experimental result

6 Conclusion

In this paper, the novel direct measuring force device for bridge cables is designed first. The dynamic measuring characteristics of the direct measuring force device are simulated, and its total deformation, equivalent elastic strain, and equivalent stress are got by ANSYS software. Secondly, according to the simulated results in this work, with the same conditions, the fabrication, and experimental results of the direct measuring force device are presented. Finally, the influencing factors of the measuring results are discussed and analyzed. The proposed direct measuring force device can be directly applied to the cable bridge with the cable installed, and there is no need to install the force sensor on each cable, thus reducing the construction cost of the cable-stayed bridge.

Our method provides a simple, cheap, and small instrument that avoids the bulky and expensive traditional measuring force devices for the cables of cable-stayed bridges. Based on the research and development of the high precision and reliability of the direct measuring force device, it satisfies the market the huge demand for new bridge monitoring equipment, and has great realistic promoting significance for global monitoring of key parts of bridge technology products.

Acknowledgement. This article is funded by the National Natural Science Foundation of China (Project No. 52005453), the Key Science and Technology Research Project of Henan Province (NO: 232102320056), and the Science and Technology Project of the Henan Provincial Department of Transportation (NO: 2020J1 and NO: 2020J2).

References

1. Li, J.-X., Yi, T.-H., Qu, C.-X., Li, H.-N., Liu, H.: Early warning for abnormal cable forces of cable-stayed bridges considering structural temperature changes. J. Bridge Eng. **28**(2), 04022137 (2023). https://doi.org/10.1061/JBENF2.BEENG-5797

2. Song, C., Xiao, R., Sun, B., Wang, Z., Zhang, C.: Cable force optimization of cable-stayed bridges: a surrogate model-assisted differential evolution method combined with B-Spline interpolation curves. Eng. Struct. **283**, 115856 (2023). https://doi.org/10.1016/j.engstruct.2023.115856
3. Wang, L., Xiao, Z., Li, M., Fu, N.: Cable force optimization of cable-stayed bridge based on multiobjective particle swarm optimization algorithm with mutation operation and the influence matrix. Appl. Sci. **13**(4), 2611 (2023). https://doi.org/10.3390/app13042611
4. Khalili, S.M.R., BotshekananDehkordi, M., Carrera, E., Shariyat, M.: Non-linear dynamic analysis of a sandwich beam with pseudoelastic SMA hybrid composite faces based on higher order finite element theory. Compos. Struct. **96**, 243–255 (2013). https://doi.org/10.1016/j.compstruct.2012.08.020
5. Singh, A.P., Pradhan, D.: Finite element analysis of square ring shape force transducer. IOP Conf. Ser. Mater. Sci. Eng. **872**(1), 012039 (2020). https://doi.org/10.1088/1757-899X/872/1/012039
6. Du, W., Dong, L., Bai, P., Zhu, F., Huang, Z.: Dynamic measurement of stay-cable force using digital image techniques. Measurement **151**, 107211 (2020). https://doi.org/10.1016/j.measurement.2019.107211
7. Wang, Y., Li, K., Chen, Y., Xu, S., Shou, W.: Research on non-contact and non-fixed cable force measurement based on smartphone. Appl. Sci. **11**(19), 8902 (2021). https://doi.org/10.3390/app11198902
8. Tan, H., Lin, Y., Jiang, Y., Chen, F.: Application of Fiber Bragg grating sensor in cable force monitoring of cable-stayed bridges. Railw. Constr. **60**(10), 51–55 (2020). (in Chinese)
9. Zhang, H., Wei, L., Wang, S.: Review of cable force testing methods and principles of cable-stayed bridges. J. Highw. Transport. Res. Dev. App. Techn. Edn. **10**, 268–269 (2011)
10. Jiang, J., Tang, D., Zhou, J.: Measurement methods and development trend of bridge cable force. J. Chongqing Jiaotong Univ. (Nat. Sci. Edn.). **27**(3), 379–382, 466 (2008)
11. Shao, S., Wu, T., Guo, J., et al.: Experimental research on cable force test of cable-stayed bridge with computer vision aid. In: Proceedings of the World Transport Congress 2022 (WTC2022), pp. 339–349 (2022)
12. Chen, Y., Yu, X.: Fast data acquisition and recognition method for cable force detection of expressway cable-stayed Bridges. J. State Norm. Univ. (Nat. Sci. Edn.) **31**(04), 92–95 (2013)

Effect of Embedded Filament Fibers on Mechanical Properties of 3D Printing Cement-Based Materials

Weihong Li[1], Xuhao Chen[1], Yaoyu Wang[1], Detian Wan[2], Nan Li[1], and Fenghai Ma[1](✉)

[1] College of Architectural Engineering, Dalian University, Dalian 116000, China
liweihong714@163.com

[2] State Key Laboratory of Green Building Materials, China Building Materials Academy, Beijing 100024, China

Abstract. 3D printing technology has special printing process, could not print both cement-based materials and steel bars in practical engineering, affects the mechanical properties of concrete. In order to solve the problem, three kinds of embedded filament fibers (Basalt filament fiber, Glass filament fiber, and Polyethylene filament fiber) are selected to add into the cement mortar to improve the bond and hardness strength of ITZ, which could be characterized by compressive test, flexural text, and bond strength test, and study the influence rule of different embedded filament fiber on the mechanical properties of 3D printing mortar by scanning electron microscope (SEM). The results show that all of the three kinds of embedded filament fibers can significantly improve the compressive strength, flexural strength, and interlayer bond strength of the 3D printed mortar, and the three kinds of fibers can have good bond with cement mortar. The Polyethylene fiber is the best, improving the flexural strength, compressive strength, and interlayer bond strength of the printed mortar by 36.4%, 29.9%, and 36.1%, respectively.

Keywords: 3D printing cement mortar · Embedded filament fibers · Mechanical properties · Bonding strength

1 Introduction

3D printing of building is a kind of "additive manufacturing" technology, which has the advantages of high intelligence, rapid construction, reduced labor demand, improved solid waste utilization, low carbon, and environmental protection [1–4]. However, some technical difficulties in 3D printing construction still need to be resolved. For example, it is not possible to realize synchronous printing of steel rods, which leads to low flexural strength and brittle failure of 3D printed components, thus hindering the application and generalization of this technique in construction engineering.

At present, the mechanical properties of 3D printing concrete are mainly enhanced by short fibers with high strength and high modulus, continuous microtubules, wires, and ropes [5–7]. The above different reinforcement methods have achieved good results

W. Guo et al. (Eds.): GBCESC 2023, LNCE 328, pp. 439–446, 2024.
https://doi.org/10.1007/978-981-99-9947-7_47

in improving the mechanical properties of 3DPC structures. However, the impact of discontinuous fibers on the mechanical properties of 3D printed components is still limited to the control of micro-cracks, and it is difficult to play a toughening effect similar to that of steel reinforcement, while the simultaneous arrangement of reinforcing materials such as tendons, wires and cables during the printing process has a good effect on the mechanical properties of the printed components, but if arranged in the weak area between layers. It will lead to a slight lack of load-bearing capacity of the structure, and there are also problems such as slippage with the concrete matrix and poor bonding.

In this study, a concrete 3D printer with the integration of embedded filamentary fiber and material mixing and extrusion function, which was developed by the group, was used to successfully realize the 3D printing of concrete with embedded filamentary fiber. On this basis, the impact pattern of embedded filamentary fibers on the mechanical properties of 3D printed mortar components is studied. This simultaneous printing technology of embedded filamentary fibers and cement mortar can increase the bonding area between filamentary fibers and cement mortar, and effectively avoid the bonding problem of filamentary fibers and matrix, as well as is not restricted by bar meshes and fiber fabric so that each printed mortar is wrapped with filamentary fibers, which further enhance the mechanical properties of the printed components and reduce the dependence of the printed components on steel reinforcement. It can provide a new method and new idea for the application of concrete 3D printing technology in engineering.

2 Materials and Experiments

2.1 Raw Material

In this experiment, sulfate aluminate composite cement (SAC), class I fly ash (FA), and silica fume (SF) are used as cementitious materials. The main chemical composition is shown in Table 1. Aggregate: silica sand, with particle sizes of 20 ~ 40 mesh, 40 ~ 70 mesh, and 70 ~ 120 mesh, respectively, the main indices are shown in Table 2. Admixtures use water reducing agents (SP) with a water reduction rate greater > 30%, thickeners (TA), thixotropic agents (TK) and setting accelerator (SA). Three kinds of filament fiber: Basalt fiber (BF) tensile fracture strength of 70 MPa, Polyethylene fiber (PE) tensile fracture strength of 343.58 MPa, and Glass fiber (GF) tensile fracture strength of 150 MPa were selected.

Table 1. Cementitious materials chemical composition

Sample	SiO_2	MgO	Cao	Fe_2O_3	SO_3	Al_2O_3	TiO_2	Loss
SAC	6.13	1.24	42.54	1.53	10.79	35.17		2.10
FA	52.97	1.71	6.27	5.27	1.46	28.77	1.66	1.23
SF	92.46							2.2

2.2 Mix Proportion

In this study, four groups of test schemes were designed to study the influence of different embedded filament fiber types on the mechanical properties of 3D printing mortar. The specific schemes are shown in Table 2. Among them, the first group was the blank group, without filament fiber, and the other three groups were mixed with BF, PE, and GF.

Table 2. Mix proportions of 3D printing mortar test scheme (g)

Group	SAC	Sand	FA	SF	SP	TK	TA	SA	Water	Fiber
C	1000	1500	250	100	5	1	10	0.6	160	\
BF	1000	1500	250	100	5	1	10	0.6	160	Basalt
PE	1000	1500	250	100	5	1	10	0.6	160	Polyethylene
GF	1000	1500	250	100	5	1	10	0.6	160	Glass

2.3 3D Concrete Printer

The embedded filament fiber and material mixed extrusion function integrated concrete 3D printer used in this study is shown in Fig. 1. The specific parameters of the printer are detailed in the literature [15].

To realize the 3D printing of embedded filament fiber-reinforced cement-based materials, the printing system in the printing equipment adopts a rotating 3D printer head with mixing and extruding functions and continuous wire feeding. The mixing and extruding of the printing material and water are successfully realized in the printer head at the same time, and the filament fiber is transported to the nozzle through the central pipe of the printer head. Due to the rotating function of the printer head, the filament fiber forms a spiral shape at the nozzle, which greatly increases the bonding area of the filament fiber and the cement mortar, and is accompanied by the simultaneous extrusion of the material.

2.4 Experimental Methodology

The 3D printed mortar specimens after standard curing to a predetermined age (flexural and compressive strength ages are 7 d and 28 d) were used to determine the flexural strength and compressive strength according to the "Test method of cement mortar strength (ISO method)" (GB / T17671-2021). The test direction is perpendicular to the printing direction (Y direction). The average value of the interlayer strength test results of each group of 6 specimens is used as the test result. The interlayer bond strength test method is based on the "Standard for test method of mechanical properties on ordinary concrete" (GB/T 50081-2016), and the standard mortar test block with a size of 40 mm × 40 mm × 40 mm is tested for unidirectional splitting tensile strength.

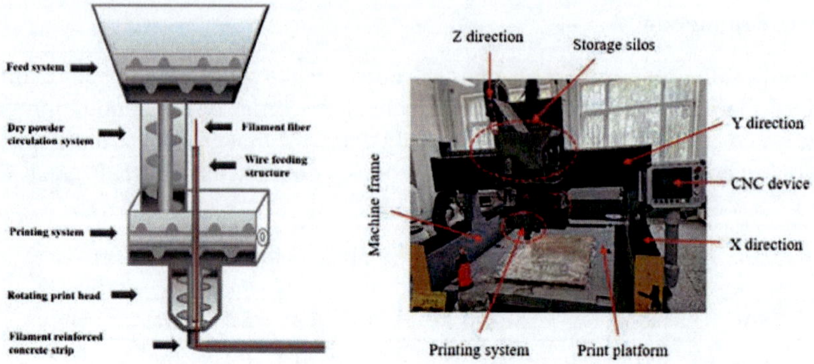

Fig. 1. 3D concrete printer

3 Results and Discuss

3.1 Effect of Filament Fiber Types on the Mechanical Properties of 3DPC

Figure 2(a) shows the test results of 7d and 28d flexural strength of 3D printed mortar specimens with different filament fiber types. As can be seen from Fig. 2(a), the flexural strength of embedded BF, GF, and PE reinforced 3D printing specimens increased significantly compared with the control group (Group C) regardless of the age of 7 days or 28 days, which increased by 19.2%, 21.7%, and 36.4% respectively at 7 days and 15.5%, 19.3% and 24.8% at 28 days. It indicates that the type of embedded filament fiber selected in this study can significantly improve the flexural strength of 3D printed cement mortar test block. This is because the long fibers are continuous and uniform materials, which are oriented in the cement-based material parallel to the printing direction and play a pulling role in the matrix, slowing down the cracking and failure of the specimen, and thus enhancing the bending strength of the printing component in the Y direction.

At the same time, it can be seen from the above results that compared with the BF and GF-reinforced 3D printing mortar specimens, the flexural strength lifting rates of PE-reinforced 3D printing mortar components are 17.2% and 14.7% higher, respectively. This indicates that among the three kinds of embedded filament fibers selected, PE has the best effect on the bending strength of the 3D printed component, while BF and GF have the same effect on the bending strength of the 3D printed mortar component. This is because PE has the highest tensile fracture strength and can withstand higher energy. At the same time, as can be seen from the damage pattern of the flexural test of specimen block in Fig. 3, only the long fibers in PE reinforced test block were not broken, while the filament fibers in BF and GF test blocks were broken, indicating that PE could improve the integrity of 3DPC specimen more effectively than BF and GF.

Figure 2(b) shows the test results of 7 d and 28 d compressive strength of 3D printed mortar specimens with different filament fiber types. As can be seen from Fig. 2(b), compared with the control group, the improvement of compressive strength of filament fiber-reinforced print test blocks is similar to the bending strength, with varying degrees of improvement. Compared with the control group, the 7d compressive strength enhancement rates of BF, GF, and PE fiber were 12.1%, 9.4%, and 29.9%, respectively. The 28 d

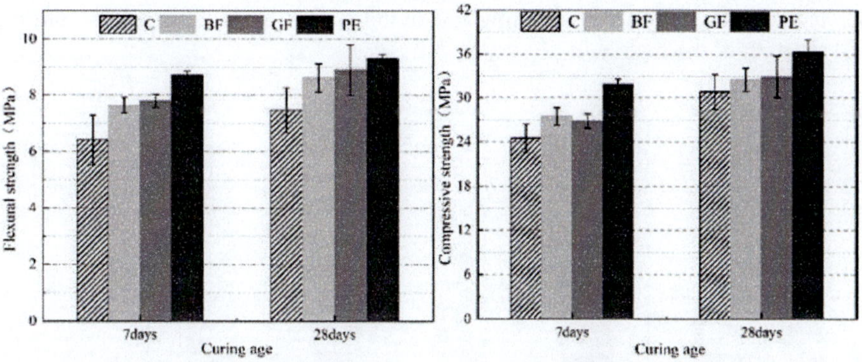

(a) Flexural strength (b) Compressive strength

Fig. 2. Effect of different filament fibers on mechanical property of 3D printed specimen

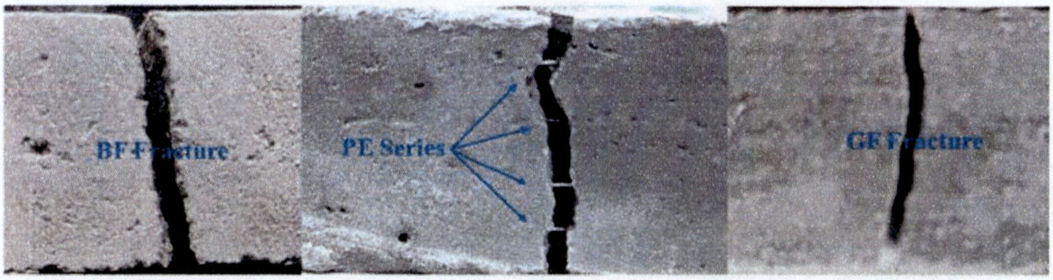

Fig. 3. Failure morphology of three filament fiber reinforced 3DPC specimens

values were 5.5%, 6.8%, and 16.8%, respectively, and the improvement effect of PE fiber on the compressive strength of 3D printed components was 2.47 times and 3.18 times that of BF fiber and GF fiber, respectively. The above results once again show that the three different kinds of filament fibers selected have improved the compressive strength of the 3D printed mortar test block, among which PE fiber has the best enhancement effect, BF fiber and GF fiber have the same enhancement effect. The results show again that the directional arrangement of filament fibers with low tensile fracture strength in the matrix has no obvious improvement effect on the compressive strength of print mortar specimen, and the main enhancement effect is mainly reflected in the improvement of the flexural strength of print mortar specimen.

The 28 d interlayer bond strength test results of embedded filament fiber reinforced 3D printing mortar specimens are shown in Fig. 4. Compared with the control group (group C), the addition of three filament fibers significantly enhanced the interlayer bond strength of the printing specimens. Among them, the interlayer bond strength of PE-reinforced 3D printing mortar specimens reached 2.56 MPa, with the best lifting effect, and the lifting rate reached 36.1%. BF and GF were followed by 24.8% and 19.1% respectively. The results show that the interlayer strength can be improved significantly by adding embedded filament fibers into the matrix of printed mortar. This is because when inserted into the 3D printed cement-based matrix, the embedded filament fiber can firmly adsorb the slurry around the fiber, increase the bond between the slurry around the fiber and the matrix, and play a pulling role in the matrix, thus increasing the interlayer bond strength of the matrix.

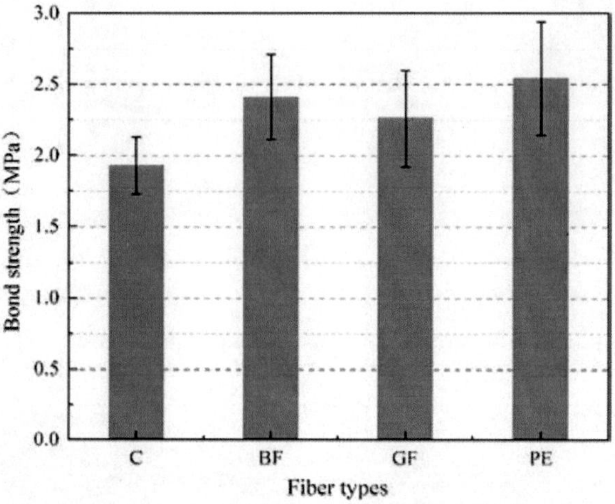

Fig. 4. Effect of different filament fibers on bond strength of 3D printed mortar specimen

Based on the above test results of mechanical properties, it can be seen that the flexural strength, compressive strength, and interlayer bond strength of printing specimens are significantly improved after the addition of different types of filament fibers, and the improvement law is consistent, that is, the three kinds of embedded filament fibers improve the flexural strength, compressive strength and interlayer bond strength of 3D printing components as PE > BF > GF. This shows that the embedded filament fiber and cement mortar synchronous printing technology can significantly improve the overall mechanical properties of the printing component.

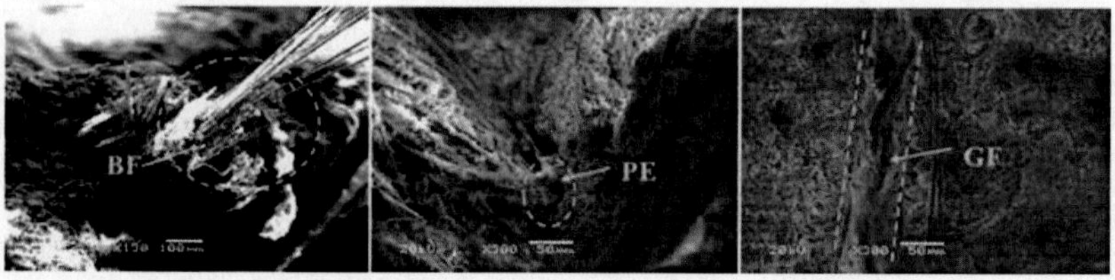

Fig. 5. Micromorphology of different filament fibers in a cement mortar matrix

Figure 5 shows the microscopic morphology of different embedded filament fibers in the cement mortar matrix, from which it can be seen that the three fibers are tightly bonded to the matrix, which indicates that the above three embedded filament fibers have good bonding with the cement mortar inside the matrix. This result proves that the synchronous printing process of filamentary fibers and printed mortar by this 3D concrete printer can increase the bonding area of filamentary fibers and printed mortar.

The rotating print head has an extrusion effect on the extruded cement mortar during the printing process, which enhances the bonding of the filamentary fibers with surrounding mortar, and plays a pulling role inside the matrix to improve the integrity of the specimen, thus enhancing the overall mechanical properties. This also proves the superiority of the printing process of simultaneous printing of embedded filamentary fiber reinforced concrete, with embedded filamentary fibers instead of steel reinforcement from another perspective.

Simultaneously, the consistency of the flexural strength, compressive strength, and interlayer bond strength of the 3D printed component on the above embedded long fiber fibers also proves that the interlayer bond performance will directly affect the overall strength of the specimen. This is because the particularity of 3D printing technology will cause the specimen to form a weak link between layers in the printing process, and the failure of the specimen usually starts at the weak point between layers. Therefore, the level of the bond strength between layers will directly affect the overall mechanical properties of the specimen.

4 Conclusion

(1) After mixing three different kinds of filament fibers (BF, GF, and PE) into 3D-printed cement-based materials, filament fibers were printed synchronously with cement mortars. The research showed that filament fibers improved the flexural strength and compressive strength of 3D printed cement-based materials. Under the same diameter, embedded filament fiber's flexural strength is better than printed mortar's compressive strength. This is because the long fiber is a continuous uniform material, and the cement-based material has a directional arrangement parallel to the printing direction, which exerts a pulling effect on the cement mortar to delay the cracking and damage and improve the flexural strength of specimens.

(2) BF, GF, and PE all enhance the interlayer bond strength of 3D printing components. They could significantly increase the bond strength by 36.1%, 24.8%, and 19.1%, respectively. PE has the best reinforcement effect.

(3) PE has the best enhancement effect on the mechanical properties of 3D printing cement-based materials. The compressive strength enhancement rates at 7 days and 28 days increased by 29.9% and 16.8%, respectively. The flexural strength enhancement rates at 7 days and 28 days increased by 36.4% and 24.8%. This is because PE fiber itself has the highest tensile fracture strength and can withstand higher energy; At the same time, only PE fibers in the PE fiber reinforced specimens were not broken, while the filament fibers in the BF and GF specimens were all broken, indicating that PE improved the integrity of specimens and the overall strength was further improved.

Acknowledgment. The authors would like to acknowledge the support of the National Nature Foundation Major Fund of China (52032011).

References

1. Panda, B., Paul, S.C., Ming, J.T.: Anisotropic mechanical performance of 3D printed fiber reinforced sustainable construction material. Mater. Lett. **209**, 146–149 (2017)
2. Nematollahi, B., Vijay, P., Sanjayan, J., Nazari, A., Xia, M., Nerella, V.N., Mechtcherine, V.: Effect of polypropylene fibre addition on properties of geopolymers made by 3D printing for digital construction. Materials **11**(12), 2352 (2018). https://doi.org/10.3390/ma11122352
3. Ara, A., Bn, A., Rr, B., et al.: Development of 3D-printable ultra-high performance fiber-reinforced concrete for digital construction - ScienceDirect. Constr. Build. Mater. **257**, 119546 (2020)
4. Salet, T.A.M., Ahmed, Z.Y., Bos, F.P., et al.: Design of a 3D printed concrete bridge by testing. Virtual Phys. Prototyping **13**(3), 1–15 (2018)
5. Vantyghem, G., Corte, W.D., Shakour, E., et al.: 3D printing of a post-tensioned concrete girder designed by topology optimization. Autom. Constr. **112**, 103–084 (2020)
6. Manuel, H., Dirk, V.: Properties of 3D-printed fiber-reinforced Portland cement paste Manuel Hambach, Dirk Volkmer. Cement Concr. Compos. **79**, 62–70 (2017)
7. Li, W., Chen, X., Li, N.: 3D printing of fast setting basalt fiber reinforced cement-based materials, lecture notes in civil. Engineering **211**, 57–64 (2022)

Frictional Rotation Performance Study Based on Non-Standard Bolt Hole Nodes

Qikai Liu[1](✉), Yueguo Zhang[1], and Xuyu Cheng[1,2]

[1] Shenyang Jianzhu University, Shenyang 110168, China
ceygzhang@sjzu.edu.cn

[2] Liaoning Vocational College of Light Industry, Dalian 116100, China

Abstract. In order to study the frictional energy dissipation of steel structure nodes under the condition of small rotation, a frictionally rotatable node is proposed based on non-standard bolt holes, which can effectively increase the energy dissipation capacity of the node and reduce the damage of the node, so that the node can better meet the current requirements. The seismic performance of the node is studied by means of quasi-static test and finite element analysis. The displacement-load hysteresis curves and skeleton curves of the node were obtained by varying the size of the bar holes, the number of bolts and the position of the curved holes under different parameters, and the hysteretic performance, failure mode and energy dissipation mechanism of the joint are analyzed, which provides a basis for further research on non-standard hole joints. The results show that the length of the energy dissipating slip section of the node depends on the size of the flange bar bolt hole, the magnitude of the slip load is directly influenced by the friction between the plates, and the node can effectively achieve energy dissipation during frictional rotation, which can effectively prevent the premature failure of the connecting welds of the joint plate.

Keywords: frictional energy dissipation · nodal rotation · non-standard bolt holes · seismic performance

1 Introduction

At present, with the increase of civil buildings of steel structure, the research on the seismic performance of steel structure is more and more in-depth, and the seismic performance of steel constructions mainly depend on the seismic performance nodes. In the process of continuous development and research of steel structure connection technology, the concept of node energy consumption is one of the main research contents, while the concept of rotational friction energy consumption is the priority. Ma et al. conducted low cyclic loading tests on three high-strength bolted nodes with long round holes an into a contrast node specimen, and compared them with traditional bolted welded nodes. The results show that the ductility of this node is better than that of the traditional bolt-welded node, and it can be obviously improved by the slip of the bolt in the long hole[1].

Zhang et al. proposed a new type of reaming bolt connection energy dissipation beam section. The results show that increasing the anti-slip bearing capacity of the

W. Guo et al. (Eds.): GBCESC 2023, LNCE 328, pp. 447–457, 2024.
https://doi.org/10.1007/978-981-99-9947-7_48

reaming bolt connection and the size of the long round hole can improve the dissipation capacity of the reaming bolt connection energy dissipation beam section [2]. Liu et al. conducted cyclic loading tests on six truss-column node specimens with super-large bolt holes or slotted bolt holes. Studies show that increasing the bolt holes will slightly reduce the anti-slip ability of the nodes, but will significantly improve the ductility and energy dissipation capacity of the nodes [3]. Rojas P et al. designed a friction energy dissipation joint that is prestressed by steel strands and dissipates energy through friction between the upper and lower flanges of the beam. The results show that the steel structure frame equipped with this node has good energy dissipation capacity [4]. Law S S et al. found that the size of the strip hole is an important parameter affecting the ductility and hysteretic performance of the reaming bolt connection [5]. Han et al. designed a friction damper composed of steel plate and four kinds of friction plate materials for sliding slotted hole high-strength bolt friction damper, and carried out low cyclic loading tests under different bolt torsion forces. The hysteretic stability and energy dissipation capacity of dampers with different friction plate materials are well [6]. In order to study the hysteretic performance of rotational friction damper under static repeated load effect, Yang et al. conducted quasi-static test on six rotational friction damper specimens, and used ABAQUS finite element software for numerical simulation. The results show that the hysteretic curve of rotational friction damper is full and has good energy dissipation capacity [7].

It can be seen from the existing research results that the ductility of friction sliding nodes is better than that of traditional bolt-welded nodes, and it has good seismic performance. Therefore, on the basis of relevant research, a steel structure node connected by high-strength bolt friction connection is proposed. Using strip holes and arc bolt holes, the node can rotate under controllable small angle conditions to achieve energy dissipation. Through quasi-static test and numerical simulation, the hysteretic performance, energy dissipation capacity and failure mode of the nodes are analyzed and studied, which provides a theoretical basis for further research and engineering application of such nodes. This work was financially supported by Basic scientific research Project of Education Department of Liaoning Province (LJKZ0567).

2 Node Test Model

In order to fully study the bearing performance of the whole process of friction rotation of nodes, the design of node test model follows the principle of ‘weak joint and strong componen’, so as to prevent premature failure of components during the test. The node is connected by 8.8 grade M20 high strength bolt friction type. The steel is Q235 B, the column length is 2.0 m, and the beam length is 1.2 m. The specific size are shown in Table 1.

Considering the change of parameters, four nodes are designed, numbered BGJD-1 ~ BGJD-4, in which the number of bolts, spacing, bar bolt hole and arc bolt hole size are shown in Fig. 1 and Table 2.

Table 1. Node model part sizes

Name		Member size(mm)
Column		H400 × 400 × 10 × 14
beam		H400 × 200 × 10 × 12
Flange connection plate	BGJD-1、3、4	200 × 255 × 14
	BGJD-2	200 × 255 × 10、90 × 255 × 6
web connecting plate		300 × 255 × 10
column web stiffener		190 × 355 × 10
Stiffening rib of beam web		95 × 375 × 10

Table 2. Node model parameters mm

Node number	The value of L(mm)	The value of A(mm)	The value of B(mm)	The value of C(mm)
BGJD-1	9	100	5	2
BGJD-2	9	100	5	2
BGJD-3	6	100	3	1.5
BGJD-4	9	70	4	1

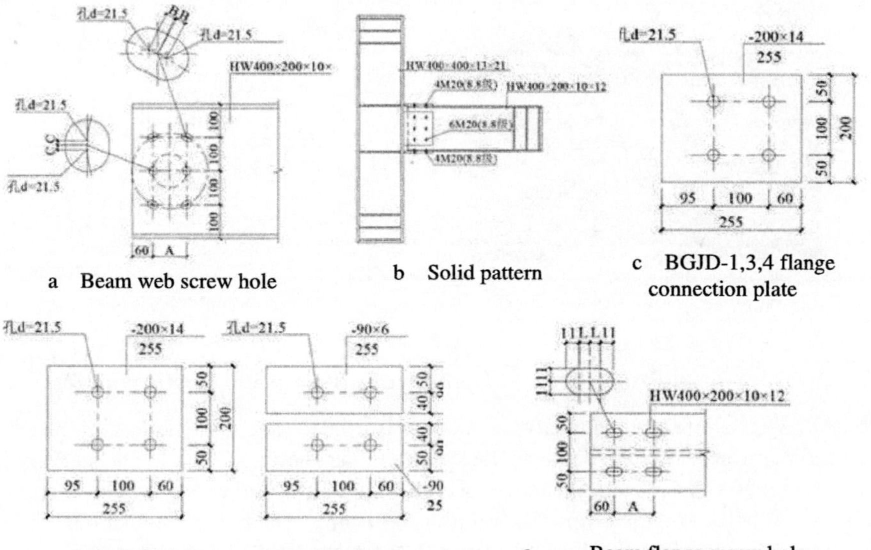

a Beam web screw hole
b Solid pattern
c BGJD-1,3,4 flange connection plate
d BGJD-2 flange connection plate
e Beam flange screw hole

Fig. 1. Node model

3 Connection Test and Finite Element Model

3.1 Steel Material Properties and High Strength Anti-Slip Coefficient Test

According to the relevant provisions of the 'Metallic Materials Tensile Test Part 1: Room Temperature Test Method' (GB/T 228.1-2010), the steel tensile test was carried out. For the steel plates with thicknesses of 6 mm, 10 mm, and 14 mm, it takes three groups of each for tensile test. The results were averaged. The experimentally measured yield strength of Q235B steel was 280.7 MPa, the ultimate strength was 464.2MPa, and the elastic modulus E was 1.87×105 MPa. The steel plate contact surface is only treated with floating rust. According to the relevant provisions of the 'Code for Acceptance of Construction Quality of Steel Structure Engineering' (GB50205-2001), three sets of double shear anti-slip tests are carried out to measure the average value of anti-slip coefficient, and the figure is 0.203.

3.2 Steel Material Properties and High Strength Anti-Slip Coefficient Test

The hysteretic loading device is composed of reaction wall, MTS loading system and data acquisition device. During the test, the column is placed horizontally on the loading device, and the reaction beam is placed at both ends of the column and fixed, as shown Fig. 2. The test adopted the displacement loading model. Before the formal loading, the pre-loading 10 kN cycle is performed twice to detect the correctness of the loading device and the data acquisition system. The loading system is shown in Fig. 3, and the Δ is taken as 5mm.

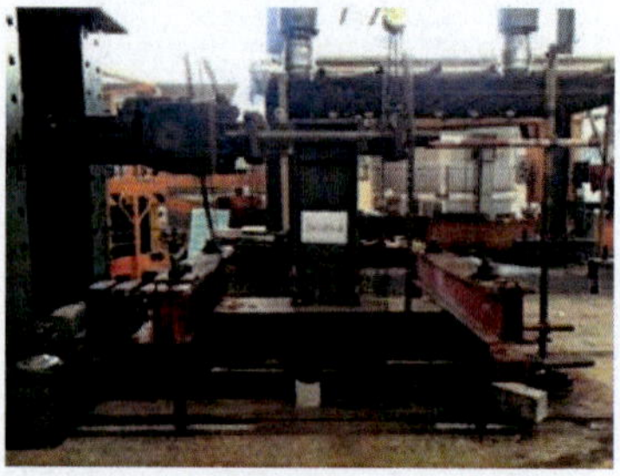

Fig. 2. Loading device diagram

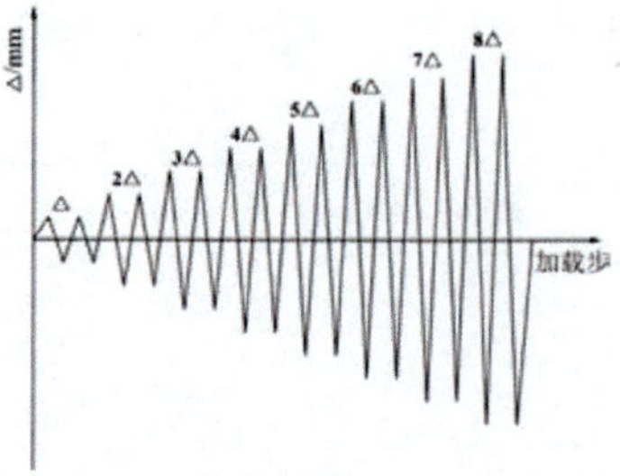

Fig. 3. Loading system

3.3 Steel Material Properties and High Strength Anti-Slip Coefficient Test

The ABAQUS finite element analysis software is used to model the node, and the model size and parameters are the same as the specimen, as shown in Fig. 4. The steel performance is measured according to the measured value. The yield strength of high-strength bolts is 640 MPa, and the ultimate strength is 840 MPa. The bolt pre-tension is 125 kN, the anti-slip coefficient is taken according to the test value, and the contact surface of the bolt connection plate is tangential friction and normal hard contact. The interaction master-slave surface and mesh size are determined according to the stiffness. The column end is hinged to restrain the lateral displacement and torsion of the beam. The beam end face is coupled to the reference point and the vertical displacement is applied. The loading method is the same as the test.

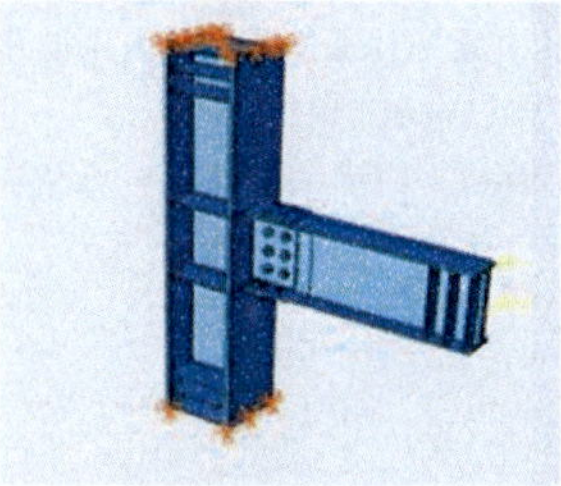

Fig. 4. Finite element model

4 Seismic Performance Analysis of Joints

4.1 Failure Form of Specimen

At the initial stage of loading, the node action of shear force and bending moment at the beam end under load is not enough to overcome the friction between the flange and web and the node connecting plate. The beam end does not rotate, and the bearing performance of the node is the same as that of the beam end fixed node, maintaining the elastic working performance. As the load increases, when the internal force of the beam end overcomes the friction and enters the friction rotation stage, the energy is dissipated during the friction sliding process of the flange and web and the node connection plate. Since the pressure and anti-slip coefficient of the high-strength bolt reaction to the plate remain basically unchanged during the friction rotation stage, the node resistance does not change much during the friction rotation stage, and the load-displacement curve shows a horizontal line. When the node rotates to the contact between the bolt rod and the hole wall, the node force enters the shear force transmission stage of the bolt rod. The initial stage of the stress stage belongs to the elastic working state, and the later stage increases with the load, the internal force after overcoming the friction will make the bolt connection enter the elastic-plastic stress process, and the tangent slope of the load-displacement curve decreases continuously, that is, the load increase rate is less than the displacement increase rate. When the load increases to the shear bearing capacity of the bolt rod is not enough to bear, the bolt rod shear failure occurs, and the flange connection plate has a large bending deformation when the failure occurs. Since the bolt rod is located at the center of the bar or arc bolt hole when the node is installed, as a result, the bearing performance of the node is basically consistent when it rotates in both directions.

After observing the failure state of the node, it can be seen that the end of the row bolt hole on the beam flange and the connecting plate has obvious hole wall extrusion deformation, the shear deformation of the flange connecting bolt is obvious, and part screws have shear failure. During the reciprocating loading process, the plate near the arc hole of the beam web is seriously worn, which is caused by the friction movement between the nut and the plate during the two-way rotation of the node. During the test, no shear failure occurred in the web connection bolts, but the corner bolts also experienced obvious deformation. In the later stage of reciprocating load, with the increase of wear between nut and plate, the nut has a tendency to loosen, which is manifested as the decrease of load in the friction rotation stage.

The numerical simulation analysis of four node specimens is carried out. The stress process and destruction form are consistent with the test. The final failure of the joint is also due to the fact that the flange bolt first reaches its shear bearing capacity limit. The failure mode of BGJD-3 is shown in Fig. 5.

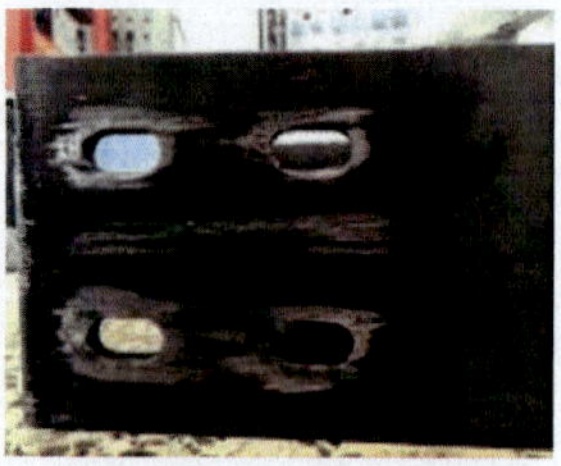

a BGJD-3 node beam flange

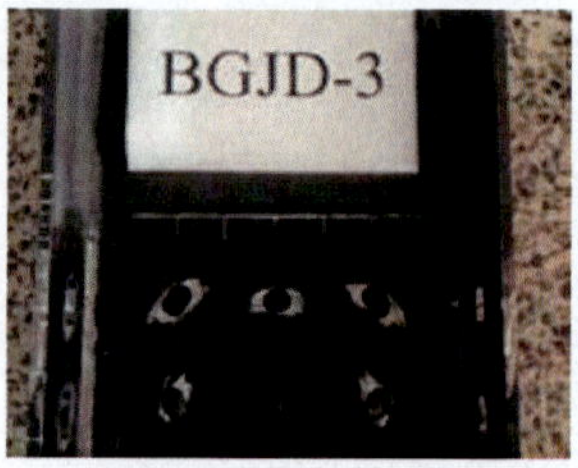

b BGJD-3 joint beam web

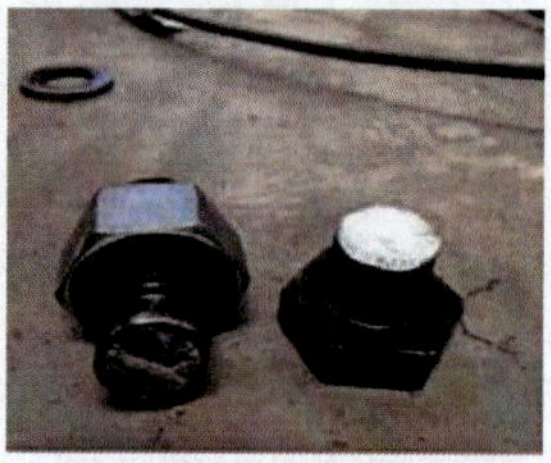

c Bolt failure diagram

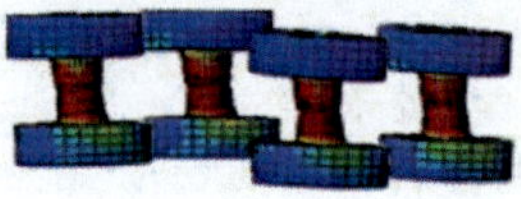

d Simulation of bolt deformation diagram

Fig. 5. BGJD-3 Nodal damage patterns

4.2 Hysteretic Performance

The hysteretic curves of node test and numerical simulation under cyclic loading are shown in Fig. 6. It can be seen from the hysteresis curves of the four specimens that the stress process of the nodes can be divided into four stages: elastic force, friction rotation, bolt rod shear and bolt rod failure under each load. In the loading stage of both positive and negative directions, the load increases slowly with the increase of displacement. This is due to the deformation of the flange connecting plate and the beam flange, which causes the pressure of the contact surface of the plate to increase and the friction force to increase. Under different load cycles, the load in the friction rotation stage is not the same, resulting in a small fluctuation phenomenon. This is due to the change of friction resistance of the contact surface caused by the deformation and stress recovery process of the flange bolt connection plate.

The hysteresis loop envelope of the node is shown in Fig. 7. The abcd region in the diagram is due to friction rotation. Obviously, the larger the allowable rotation angle of the node, the greater the shear capacity of the node bolt group, the larger the area and the stronger the energy dissipation capacity. The b and d points in the positive and negative directions are the termination points of the friction rotation, that is, the starting point of the shear of the bolt rod, and the subsequent stress characteristics will be consistent with the standard hole connection node.

It should be noted that in order to sufficiently understand the bearing properties of such connections, the design of the node model follows the principle of ‘weak connection and strong componen’. The final failure of the model is bolt cutting, and the connecting plates and components do not yield, so the hysteresis curve does not appear descending section. However, the practical engineering applications should follow the principle of ‘strong connection and weak componen’ to prevent premature failure of bolt connections.

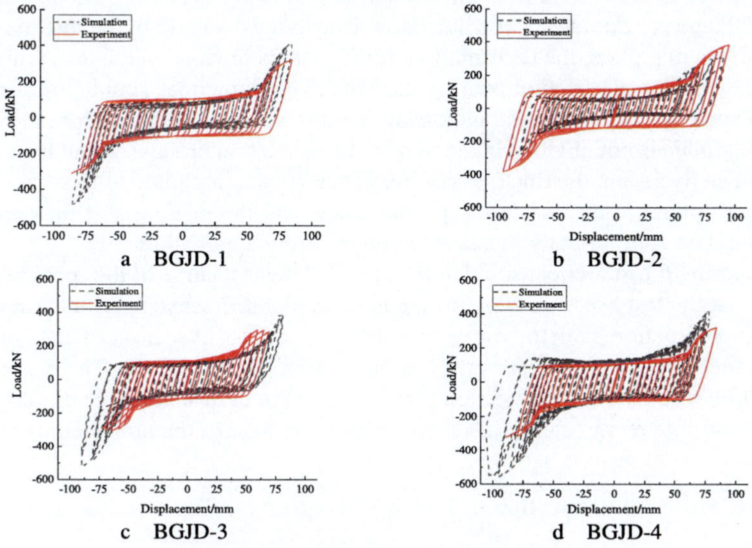

Fig. 6. New hysteretic curve of joints

4.3 Steel Material Properties and High Strength Anti-Slip Coefficient Test

The test and numerical simulation skeleton curves of the specimens are shown in Fig. 8. The displacement and load of the feature points of the node model are shown in Table 3. It can be seen from Fig. 8 and Table 3 that the maximum load of the elastic section at the initial stage of each stage of loading, that is, the anti-slip load, mainly depends on the frictional shear capacity of the bolt group at the beam end.

Compared with BGJD-3, the length of the flange strip hole and the web arc hole of BGJD-1 is longer than that of BGJD-3, the slip section of BGJD-1 is significantly larger than that of BGJD-3, and the rotational energy dissipation process is longer. The slip load

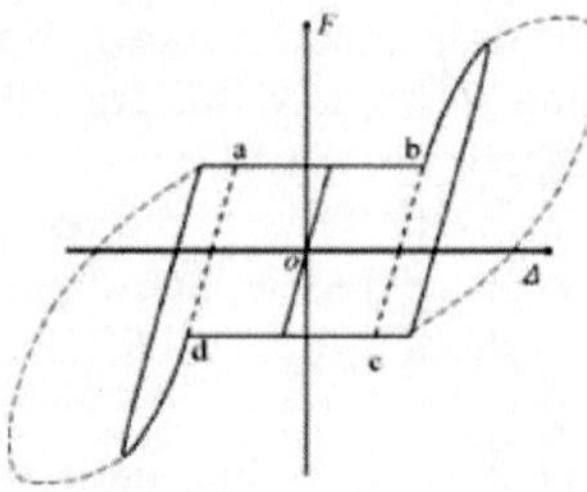

Fig. 7. Finite element model

test and numerical simulation of the two nodes are basically the same.The bolt spacing of BGJD-1 and BGJD-4 is different. BGJD-4 reaches the shear stress stage of the bolt rod earlier than BGJD-1, and the curve enters the rising section earlier. The horizontal slip section of BGJD-4 is shorter, but the anti-slip load of the two is basically the same. BGJD-2 flange is a double shear connection. Due to the decrease of the thickness of the flange connecting plate, the deformation of the plate is obvious in the test, which leads to the decrease of the friction bearing capacity. The numerical simulation results are larger than the measured values, indicating that the influence of the deformation of the connecting plate is not obvious in the numerical simulation process. It can be seen from the above analysis that the thickness of the flange connecting plate of this kind of node has a great influence on its bearing performance, and the thickness of the connecting plate should be appropriately increased in engineering application.

Comparing the numerical simulation curve and the test curve of the specimen, it can be seen that the test skeleton curve of the node model is consistent with the law of the numerical simulation skeleton curve, but the horizontal slip section of the simulation curve is smoother, and the test curve has fluctuation characteristics. This is due to the deformation of the bolted connection plate under large deformation conditions and the effect of multiple reciprocating loads on the anti-slip load of the bolted connection.

4.4 Steel Material Properties and High Strength Anti-Slip Coefficient Test

The calculation method of energy dissipation capacity is given according to the Code for Seismic Test Methods of Buildings (JGJ / T101–2015). The energy dissipation coefficient E of the node model under different loading displacements is calculated as shown in Table 4. The energy dissipation coefficient of the four nodes gradually increases with the increase of displacement in the friction rotation stage, and decreases with the increase of displacement in the bolt shear section. The energy dissipation capacity of BGJD-1 is basically the same as that of BGJD-2 before the flange slip distance is 50 mm. After 50 mm, the bending deformation of the inner flange connecting plate of BGJD-2 leads to a significant decrease in energy dissipation capacity. The length of BGJD-3 flange bolt hole is smaller than that of BGJD-1, and the flange bolt rod contacts with the bolt hole earlier, and the energy dissipation coefficient also decreases before BGJD-1 node.

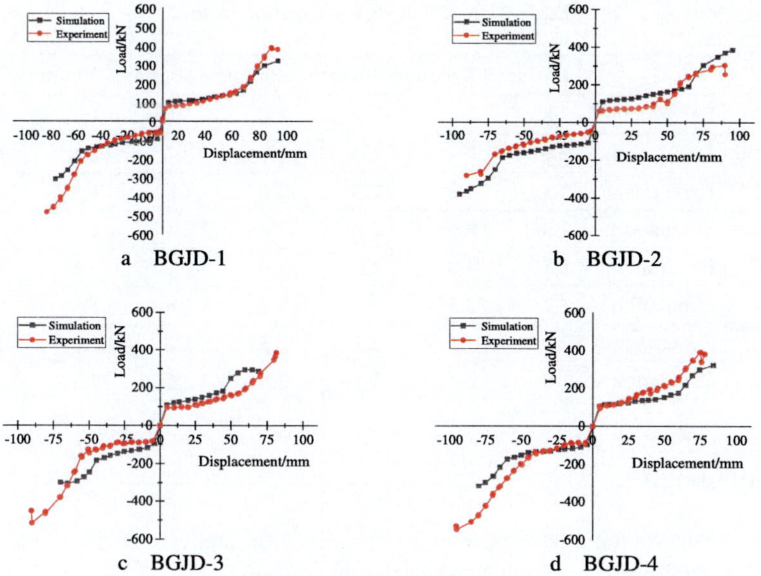

Fig. 8. Specimen skeleton curve

Table 3. New node load and displacement parameter table

Name		Slip load (kN)	Sliding displacement (mm)
BGJD-1	Experiment	78.55	10.23
	Simulation	105.1	9.11
BGJD-2	Experiment	54.96	5
	Simulation	115.68	9.23
BGJD-3	Experiment	93.68	9.79
	Simulation	109.5	9.63
BGJD-4	Experiment	109.13	9.71
	Simulation	110.98	8.35

Table 4. Nodal energy dissipation factor

Name		10mm	20mm	30mm	40mm	50mm	60mm	70mm
BGJD-1	Experiment	2.8	2.6	2.3	2.0	1.6	1.3	1.1
	Simulation	2.0	2.9	3.0	2.8	2.7	2.4	1.5
BGJD-2	Experiment	2.4	2.5	2.3	2.0	1.8	1.3	0.9
	Simulation	1.8	2.9	2.8	2.6	2.1	1.2	0.9
BGJD-3	Experiment	2.7	3.0	2.9	2.4	2.3	1.6	1.1
	Simulation	1.7	2.5	2.5	2.2	1.6	1.4	1.5
BGJD-4	Experiment	3.1	3.1	2.7	2.5	2.2	1.7	1.3
	Simulation	1.8	2.7	2.7	2.8	2.6	2.3	1.5

5 Conclusion

Through the quasi-static test and numerical simulation analysis of non-standard bolt hole nodes, the following conclusions can be drawn:

(1) For the rotatable nodes with strip holes in the beam flange connection and arc holes in the beam web connection, when the high-strength bolt friction connection is used, the rotation process has energy dissipation capacity, and its energy dissipation capacity is directly related to the allowable rotation angle.
(2) The friction rotation process of the node effectively prolongs the capacity dissipation process of the node under reciprocating load, and the friction rotation angle of the node can be realized by adjusting the length of the bolt hole.
(3) When the internal force of the joint is not enough to overcome the friction bearing capacity of the bolt group, the bearing performance of the node can be regarded as a general rigid nodes.
(4) When the internal force of the node overcomes the anti-friction bearing capacity of the node bolt group, the node rotates with friction. Under a certain number of reciprocating cyclic loads, the bearing capacity of the node does not change much, and the bearing performance is similar to 'plastic hinge ', which can effectively delay the yield of the connecting plate and component.

References

1. Ma, R.L., Yang, Y., Chen, Q.S., Lu, L.C.: Seismic performance testing study on high strength bolt connections with slotted holes. J. Build. Struct. **30**(01), 101–106 (2009)
2. Zhang, B., Hu, S.J., Wang, J.G.: Study of mechanical model and influential parameters for a new type of SSL-SSBC. World Earthq. Eng. **35**(03), 73–82 (2009)
3. Liu, X.C., Chen, M.L., Chen, X.: Seismic behavior of bolted truss-to-column joint with oversized or slotted bolt hole. Eng. Struct. **247**, 110–113 (2021)
4. Rojas, P., Ricles, J.M., Sause, R.: Seismic performance of post-tensioned steel moment resisting frames with friction devices. J. Struct. Eng. **131**(4), 529–540 (2005)

5. Law, S.S., Wu, Z.M., Chan, S.L.: Analytical model of a slotted bolted connection element and its behaviour under dynamic load. J. Sound Vib. **292**(3–5), 777–787 (2006)
6. Han, J.Q., Zhang, H.F., Qiao, Y.: Experimental study on hysteretic performance of friction dampers with sliding long hole and high strength bolt. J. Build. Struct. **39**(S2), 315–320 (2018)
7. Yang, X.Q., Wang, Y., Zhang, H.B., Ma, L.W.: Research on hysteretic behavior of rotational friction damper. Ind. Construct. **50**(05), 151–157+108 (2020)
8. Schneider, S.P., Teeraparbwong, I.: Inelastic behavior of bolted flange plate connections. J. Struct. Eng. **128**(4), 492–500 (2001)

Fragility Analysis of Pier-Tower-Girder Fixed Cable-Stayed Bridge Subjected to Near-Fault and Far-Fault Ground Motions

Wei Xia[1(✉)], Qiliang Si[2], and Nailiang Xiang[2]

[1] Anhui Provincial Transportation Survey and Design Institute Co. Ltd., Hefei 230011, Anhui, China
19253622@qq.com

[2] Hefei University of Technology, Hefei 230009, People's Republic of China

Abstract. The cable-stayed bridge (CSB) with a tower-pier-beam fixed system has gained wide recognition due to its aesthetically pleasing design, excellent wind resistance, convenient construction, and relatively lower cost. A pier-tower-girder fixed cable-stayed bridge (CSB) is constructed using the OpenSees nonlinear finite element model, and the bridge tower response under near-field (NF-GMs) and far-field ground motions (FF-GMs) is analyzed. A probabilistic seismic demand model (PSDM) is developed for cable-stayed bridge tower subjected to both NF-GMs and FF-GMs, using the peak ground velocity (PGV) and peak ground acceleration (PGA). The CSB tower's fragility curves for each section are obtained, and the fragility median points of each pylon section under various seismic excitations with slight damage are further contrasted. The study indicates that utilizing PGA as the measure of ground motion intensity, the lower portion of the bridge tower experiences the initial damage when subjected to GMs. Furthermore, in comparison to FF-GMs, CSB tower demonstrate higher vulnerability to damage when subjected to NF-GMs.

Keywords: cable-stayed bridge · fragility curves · near-fault ground motion · far-fault ground motion · PSDM · pier-tower-girder fixed

1 Introduction

The region within a 20 km radius of the fault is considered as the NF area. The nature and features NF-GMs are distinct from those of FF-GMs: forward-directivity effect, fling-step effect, and hanging wall effect [1, 2]. The features often cause structural deformation, which in turn causes serious damage to the building structure located in the NF area [3].

Three primary structural systems are used for CSB, classified according to the configuration of tower, pier, and beam: the full-floating system, semi-floating system, and pier-tower-girder fixed system. Relative to other systems, the pier-tower-girder fixed system is characterized by significant structural stiffness and a shorter natural vibration

W. Guo et al. (Eds.): GBCESC 2023, LNCE 328, pp. 458–469, 2024.
https://doi.org/10.1007/978-981-99-9947-7_49

period [4]. At present, there exists a scarcity of analyses and comparative studies investigating the dynamic response and fragility of CSB with a pier-tower-girder fixed system under FF-GMs and NF-GMs. Traditional seismic model underestimates the seismic risk of bridges located in NF regions during probabilistic seismic risk analysis due to its failure to consider the pulse GMs. Xu et al. [5] conducted a study on the susceptibility of CSB towers to seismic excitations. The results indicated that, at equivalent PGA values, CSB tower WAS more prone to failure under FF-GMs than NF-GMs. Zhong et al. [6] investigated how the spatial variability of GMs affects the seismic behavior in a CSB. The results revealed that considering spatial variability makes the components of cable-stayed bridges more susceptible to damage. Dong et al. [7] conducted a study using a rigid frame bridge to scrutinize the impact of NF-GMs and FF-GMs on the bridge structure's response. The findings indicate a conspicuously higher probability of damage for each component under NF-GMs compared to FF-GMs. Liang et al. [8] extensively examined the impact of velocity pulse effects in NF-GMs on the behavior of bridge structures. The findings indicated that accounting for velocity pulse effects increased the vulnerability of bridge components to damage when compared to FF-GMs. Jia [9] studied the sequence of damage of the cable-stayed members of the semi-floating system hybrid beam subjected to longitudinal NF-GMs.

Many researchers explored the seismic impact on structures in NF regions. However, the fragility research of tower-pier-beam fixed CSB tower under the consequence of NF-GMs and FF-GMs is not sufficient. Consequently, the objective of this study is to utilize PGA and PGV as ground motion intensity measures (IMs) to construct the fragility curves for bridge tower exposed to NF-GMs and FF-GMs [10]. The fragility of the curvature response regularity (the longitudinal bridge direction) of the CSB tower under the effect of NF-GMs and FF-GMs is analyzed and compared. These results are expected to serve as valuable guidelines for estimating the seismic demands of CSB under both NF and FF seismic conditions.

2 Cable-Stayed Bridge Model

Taking a CSB as an example, The span arrangement of the bridge is: 150 m + 96 m and the side span is built with auxiliary piers. The main girder material of the main bridge is steel within the length range of 138 m, and the material of the bridge within the length range of 108 m of the remaining 96 m side span and 12 m main span is reinforced concrete. For the purpose of connecting with the box girder of the approach bridge, the side pier (handover pier) adopts two frame piers on the left and right, and the beam is L-shaped. Friction pendulum bearings are used as the bridge's bearings. The general layout is shown in Fig. 1.

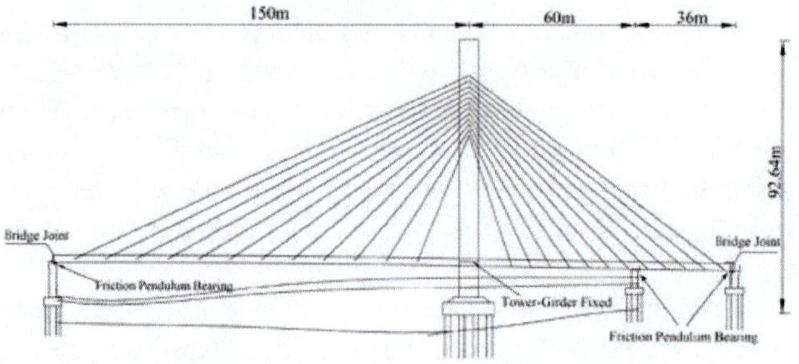

Fig. 1. Bridge general layout diagram.

2.1 Modeling of the Bridge

OpenSees software is used to analyze the response of the CSB's dynamic response. The bridge tower is 92.64 m high and the material is reinforced concrete. The section size and reinforcement layout are shown in Fig. 2. In general, the superstructure of the CSB will not be damaged under the action of earthquakes, and the elastic Beam Column elastic element is utilized to mimic the superstructure. Under the effect of GMs, the pier and bridge tower may enter the plastic deformation stage. The pier and bridge tower are simulated using OpenSees elastic-plastic fiber element. The segment is discretized into fibers. Steel02 is a material in OpenSees that simulates the steel bars of bridge towers and piers. The steel bars of bridge towers and piers are HRB400 steel bars. The strength standard value is Fy = 400 × 103 kPa, and the initial stiffness E0 = 2.0 × 108 kN/m. The bridge tower adopts C50 concrete, and the concrete material does not consider tensile properties. The selected concrete constitutive model is the Mander model. The parameters of the protective layer concrete and the core concrete of the bridge tower are shown in Table 1. For the stay cable of a CSB, always maintains a tensile state under the action of earthquakes, without considering its tensile state. Considering the prestress of the cable, the InitStrainMaterial is used to simulate it. The constitutive relationships for the protective layer concrete, core concrete, and steel bars are illustrated in Fig. 3(a) and (b), respectively. Friction pendulum bearings with longitudinal sliding are installed at the side pier and auxiliary pier, providing horizontal limitation. The constitutive relationship of the bearing is depicted in Fig. 3(c). The structure is anchored to the bottom of the pier. A large stiffening element connects the bridge tower to the main girder.

2.2 Ground Motions

For the NF-GMs, this paper selects 121 seismic waves of Dabaghi as the longitudinal seismic excitation input of the CSB [11]. The magnitude of the seismic waves in this group of NF areas ranges from 5.4 to 7.9. For the FF-GMs, this paper selects the FF-GMs records used by Jiang L and others [12]. Records 1 to 80 were selected from the PEER, and 80 to 100 documents were identified from the SAC database.

The magnitude is between 5.8 and 6.9, and the epicenter distance is between 10 and 60 km. The relevant NF-GMs and FF-GMs acceleration response spectra are presented in Fig. 4.

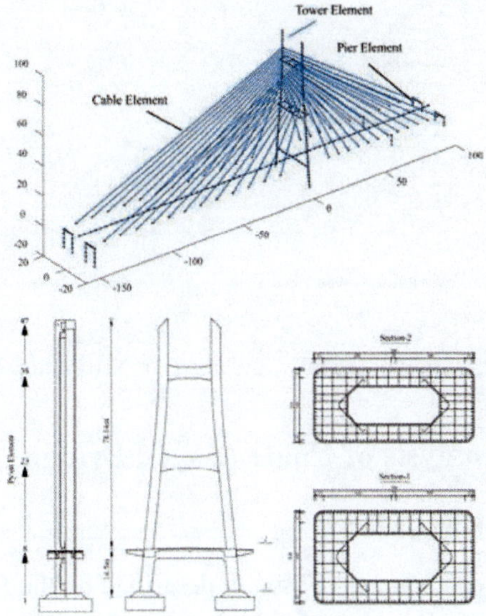

Fig. 2. Configuration and sections of the pylon.

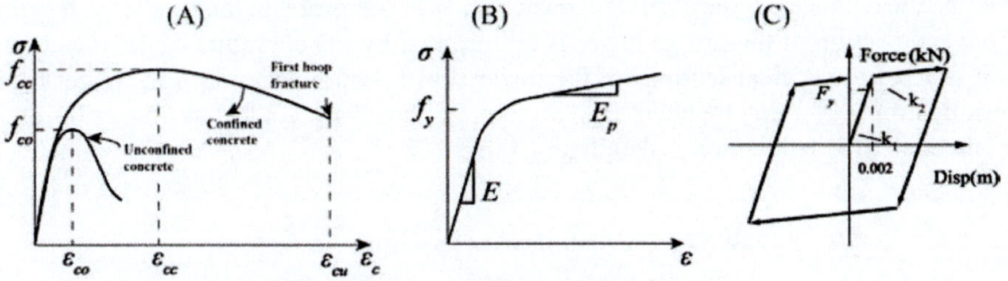

Fig. 3. Constitutive model of the (A) concrete force and deformation. (B) longitudinal rebar force and deformation. (C) friction bearing force and deformation.

Table 1. The parameters of concrete.

Concrete	f_{pc}(MPa)	ε_{cc}	f_{pcu}(MPa)	εcu
Cover	32.4	0.002	6.48	0.0035
Core	45.5	0.0052	36.3	0.017

The 100 FF-GMs and 121 NF-GMs were input along the longitudinal direction of the bridge, and dynamic time-history analyses were conducted for the tower to obtain the peak response values of the CSB.

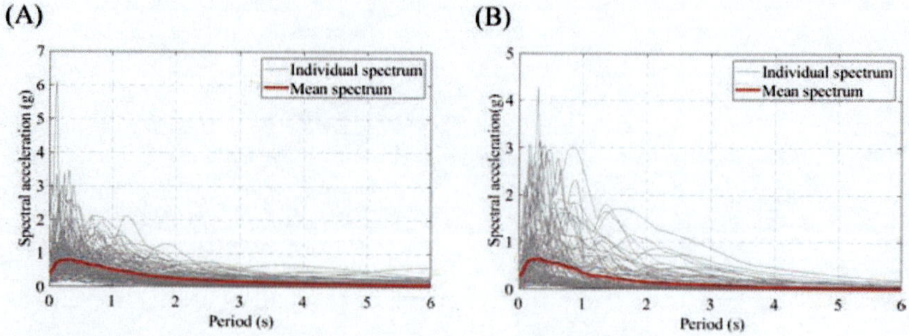

Fig. 4. (A) Distribution of the NF-GMs and (B) Distribution of the FF-GMs.

3 Seismic Risk Analysis of Cable-Stayed Bridge

3.1 Selection of the Damage Index

The damage of bridge components is usually described by the following four damage states: slight damage state (DS1), moderate damage state (DS2), severe damage state (DS3), and complete damage state (DS4), corresponding to the four performance levels of cable-stayed bridge. For the CSB, curvature ductility (μ_φ) is often used as EDPs to describe the damage degree of bridge tower [13, 14]. Therefore, in this study, the fragility of the key section of the bridge tower is represented by the curvature of the tower. The limit state of the critical sections of the tower (tower bottom section $\mu_{\varphi\text{-}1}$, tower beam consolidation $\mu_{\varphi\text{-}8}$, middle beam $\mu_{\varphi\text{-}33}$ and upper beam $\mu_{\varphi\text{-}38}$) follows the criterion of Zhong et al. [15, 16], which is detailed in Table 2.

Table 2. The limit states of the critical sections.

EDPs	yield curvature (10^{-4})	DS1	DS2	DS3	DS4
$\mu_{\varphi\text{-}1}$	4.43	1.0	1.12	8.79	33.4
$\mu_{\varphi\text{-}8}$	4.60	1.0	1.16	9.67	33.5
$\mu_{\varphi\text{-}33}$	4.74	1.0	1.21	10.72	35.2
$\mu_{\varphi\text{-}38}$	3.91	1.0	1.19	12.32	38.2

3.2 Probabilistic Seismic Demand Model (PSDM)

Earthquakes risk probabilistic evaluation methods have become essential in making risk reduction decisions for structural and infrastructure systems [17, 18]. By drawing fragility curves for components allows for an intuitive observation of the probability of component damage under the influence of ground motions. Assuming that the seismic demand (S_D) of the component and the capacity of the structural (S_C) component conform to the lognormal distribution, the fragility function P_f of the cable-stayed bridge component can be expressed as:

$$P_f = P[D \geq C|IM] = \Phi\left(\frac{\ln(S_D) - \ln(S_C)}{\sqrt{\beta_{D/IM}^2 + \beta_C^2}}\right) \tag{1}$$

In the formula, the cumulative normal distribution function is expressed as Φ [•]; C is the corresponding structural performance level of the component under different damage states. β_C means the uncertainty of the seismic capacity of the structural component, and the value in this paper is 0.35.

The PSDM usually assumes that there is a logarithmic linear correlation between the seismic response demand (SD) of structural members and the GMs intensity measure (IM) as shown in the following formula:

$$\ln(S_{\mathrm{D}}) = \ln a + b \ln(IM) \tag{2}$$

where a and b are logarithmic linear fitting parameters, which can be obtained by logarithmic linear fitting between the response peak value obtained by nonlinear dynamic analysis of structural members and a strength index IM of GMs. The deviation $\beta_{D|IM}$ of the component PSDM model can be calculated by the following formula:

$$\beta_{D/IM} = \sqrt{\frac{\sum_{i=1}^{N} [\ln D_i - \ln D]}{(N-2)}} \tag{3}$$

D_i refers to the structural response peak obtained by inputting the nonlinear dynamic response analysis of the GMs, and N represents the number of input seismic waves.

The PSDM of curvature ductility of the tower under NF-GMs and FF-GMs is established by using the PGV and PGA respectively. Limited to space constraints, this paper selects the critical sections of the bridge tower ($\mu_{\varphi\text{-}1}$、$\mu_{\varphi\text{-}8}$、$\mu_{\varphi\text{-}33}$ and $\mu_{\varphi\text{-}38}$) to establish PSDMs, as detailed in Fig. 5 and Fig. 6. For the PSDM model, the smaller the coefficient of variation $\beta_{D|IM}$ value is, the more efficient it is, and the larger the value of the fitting coefficient b is, the more practical it is. In addition, Zhong et al. [1] show that PGA is suitable for short-period structures. Due to the small natural vibration period of the CSB with a pier-tower-girder fixed system, PGA is used as the IM of the bridge tower, meeting the requirements of high efficiency and practicality.

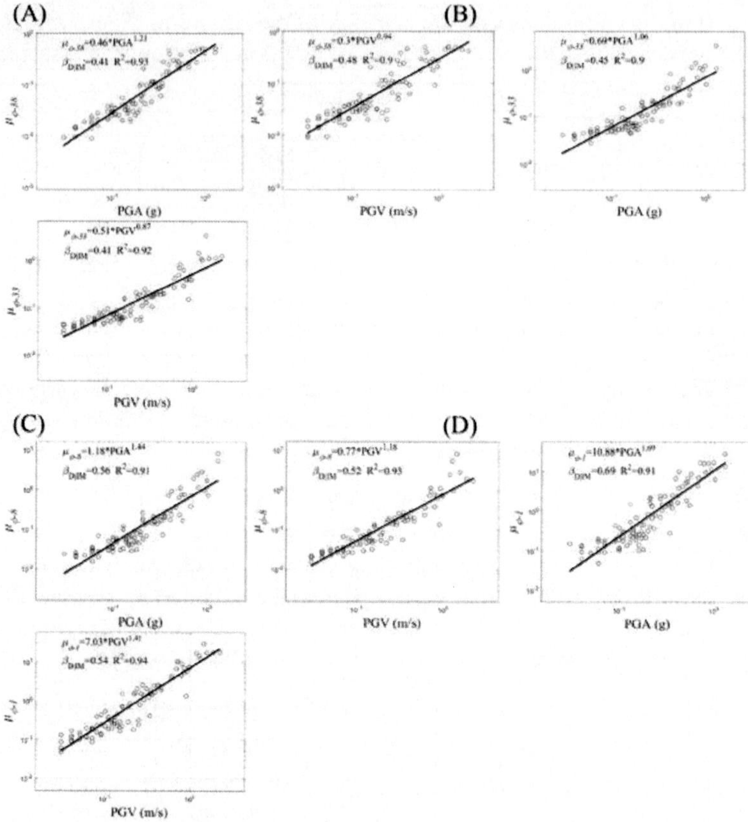

Fig. 5. (A) PSDMs of the upper beam section under FF-GMs. (B) PSDMs of the middle beam section under FF-GMs. (C) PSDMs of the tower beam consolidation section under FF-GMs. (D) PSDMs of the tower bottom section under FF-GMs.

3.3 Research on Fragility Analysis of Pylon

By convolving the PSDMs of each component with the limit state, fragility curves for the CSB tower can be constructed. The assumption of lognormal distribution for the engineering demand parameters (EDPs) and the limit state leads to damage probabilities following a lognormal distribution function. Utilizing the fragility function formula (1), fragility curves for the critical sections of the CSB tower are calculated under the four damage states, as depicted in Fig. 7 and Fig. 8. It can be concluded that under the same PGA value, the sections of the tower ($\mu_{\varphi\text{-}1}$、$\mu_{\varphi\text{-}8}$、$\mu_{\varphi\text{-}33}$) are more likely to be damaged under the NF-GMs, while the section of the upper beam ($\mu_{\varphi\text{-}38}$) is more likely to be damaged under the FF-GMs.

When PGA is used as the IM, the median fragility of the pylon under slight damage is compared. As shown in Fig. 8, the smaller the median value of the fragility of the sections, the higher the probability for damaged. It can be concluded that under the effect of NF-GMs and FF-GMs, the bottom of the bridge tower is the first to be damaged compared

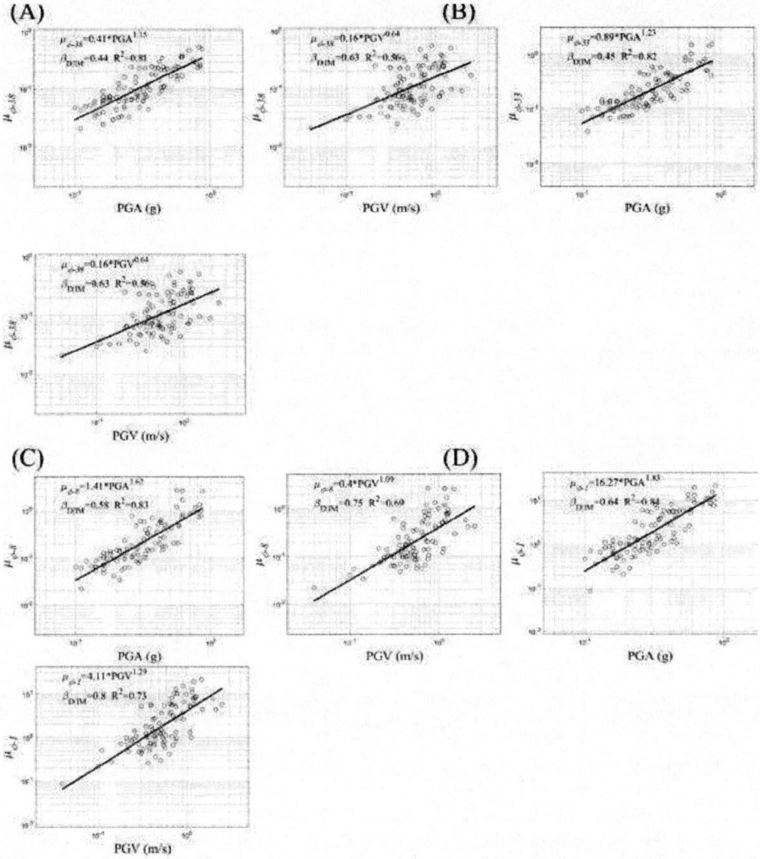

Fig. 6. (A) PSDMs of the upper beam section under NF-GMs. (B) PSDMs of the middle beam section under NF-GMs. (C) PSDMs of the tower beam consolidation section under NF-GMs. (D) PSDMs of the tower bottom section o under NF-GMs.

with other sections of the bridge tower, and some sections of the bridge tower from the middle beam to the tower beam are almost slightly damaged at the same time. When the value of PGA is the same, the sections from the bottom section to the middle beam section are more likely to be damaged under NF-GMs than under FF-GMs. However, for the sections around the upper beam, the results show that the sections are more likely to be damaged under the effect of the FF-GMs.

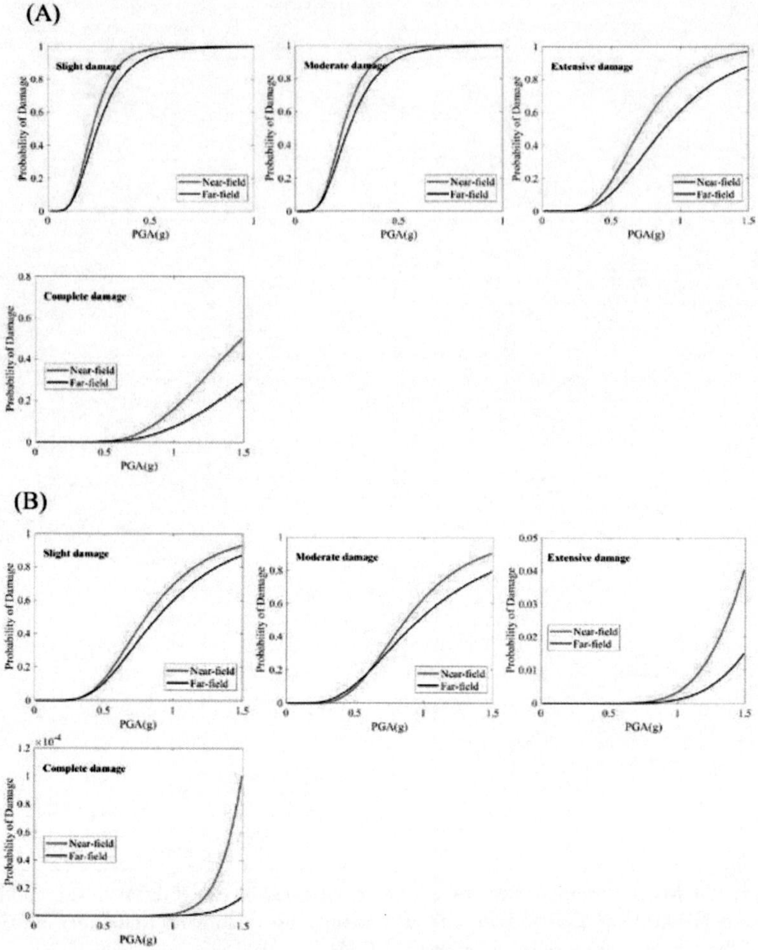

Fig. 7. Fragility curves with NF-GMs and FF-GMs:(A) bottom section. (B) tower beam consolidation Section. (C) middle beam section. (D) upper beam section.

(C)

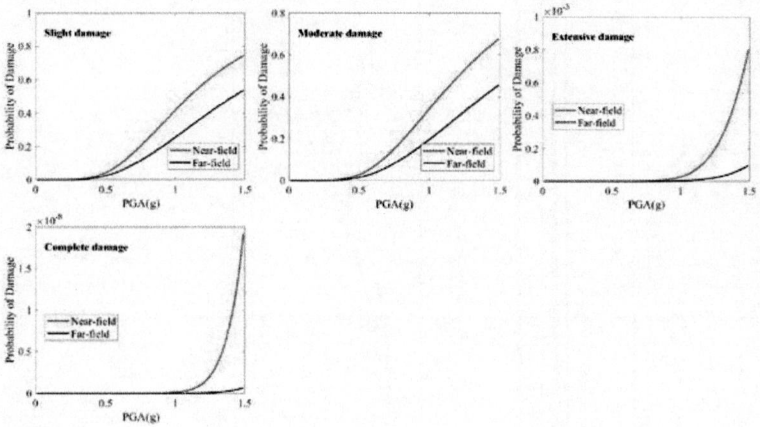

(D)

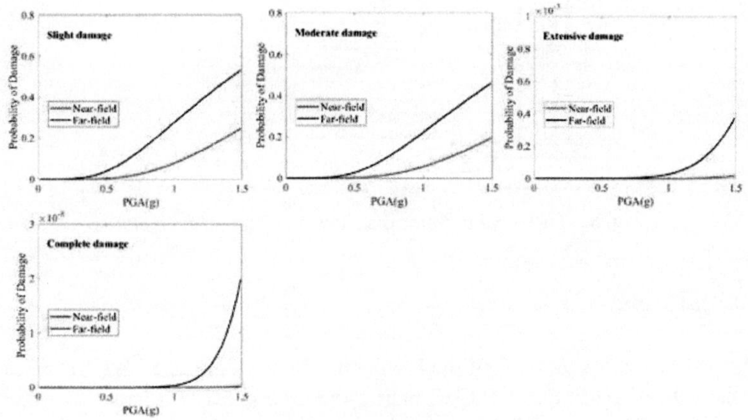

Fig. 7. (*continued*)

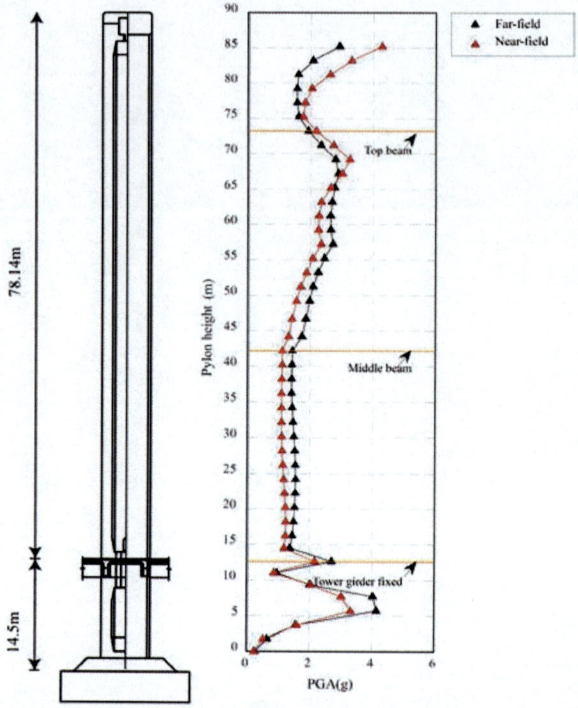

Fig. 8. The median point distribution of tower fragility.

4 Conclusions

This study conducted an analysis of the tower-pier-beam fixed system CSB model under the influence of NF-GMs and FF-GMs using the OpenSees software. To establish the PSDM for the tower, a suitable IM was determined. Furthermore, fragility curves for the structure were compared under NF-GMs and FF-GMs. The following conclusions were drawn:

(1) The CSB is established by using OpenSees analysis software, and the model is analyzed by inputting 121 NF seismic waves and 100 FF seismic waves in the longitudinal direction of the bridge. The results show that when the value of PGA is the same, the sections from the bottom section to the middle beam section are more likely to be damaged under NF-GMs than under FF-GMs. However, for the sections around the upper beam, the results show that the sections are more likely to be damaged under the effect of the FF-GMs.
(2) The PSDM and fragility curves for each section of the bridge tower are established. The bottom section of the pylon is identified as the most fragile section in the tower, and the section near the consolidation of the tower beam is also prone to damage. The sections of the tower from the middle beam to the tower beam show a tendency of minimal damage occurrence simultaneously.

References

1. Zhong, J., Jeon, J.S., Shao, Y.H., Chen, L.: Optimal intensity measures in probabilistic seismic demand models of cable-stayed bridges subjected to pulse-like ground motions. J. Bridge Eng. **24**(2), 04018118 (2019)
2. Shao, Y., Wei, Y., Yang, T., Ni, M., Zhong, J.: Empirical models of bridge seismic fragility surface considering the vertical effect of near-fault ground motions. Structures **34**, 2962–2973 (2021)
3. Zhao, X.F., Wen, Z.P.: Review on issues of near-fault velocity pulse-like ground motions. Acta Seismologica Sin. **44**(05), 765–782 (2022)
4. Guo, W.L.: Study on reasonable design states and dynamic characteristics of extradosed cable stayed bridges. Chang'an University (2011)
5. Xu, W., Wang, Y., Zhong, J.: Probabilistic seismic performance of pylons of a cable-stayed bridge under near-fault and far-fault ground motions. Earthq. Eng. Resil. **1**, 225–240 (2022)
6. Jian, Z., Jong-Su, J., Wei-Xin, R.: Risk assessment for a long-span cable-stayed bridge subjected to multiple support excitations. Eng. Struct. **176**, 220–230 (2018)
7. Dong, J., Shan, D.S., Zhang, E.H.: Near and far-field seismic fragility comparative analysis of irregular bridge. J. Harbin Inst. Technol. **48**(3), 159–165 (2016)
8. Liang, Y., Yan, S.C., Zhao, B.Y.: Seismic fragility analysis of rigid frame bridge near-fault high speed railway. J. Zhengzhou Univ. (Engineering Science) **2022**(004), 043 (2022)
9. Jia, W.T.: Seismic vulnerability analysis of the hybrid girder cable-stayed bridge under near-fault earthquake motion with consideration of uncertainties. Southwest Jiaotong University (2021)
10. Zhong, J., Pang, Y., Shen, G., Yuan, W.: Seismic fragility analysis of cable-stayed bridges retrofitted with cable-sliding friction aseismic bearing. J. Tongji Univ. (Natural Science) **42**(03), 351–357 (2014)
11. Dabaghi, M., Der Kiureghian, A.: Stochastic model for simulation of near-fault ground motions. Earthq. Eng. Struct. Dyn. **46**(6), 963–984 (2017)
12. Jiang, L., Zhong, J., Yuan, W.: The pulse effect on the isolation device optimization of simply supported bridges in near-fault regions. Structures **27**, 853–867 (2020)
13. Shahi, S.K.: A probabilistic framework to include the effects of near-fault directivity in seismic hazard assessment. Stanford University (2011)
14. Tothong, P., Cornell, C.A.: Structural performance assessment under near-source pulse-like ground motions using advanced ground motion intensity measures. Earthq. Eng. Struct. D. **37**(7), 1013–1037 (2008)
15. Zhong, J., Ni, M., Hu, H., Yuan, W., Yuan, H., Pang, Y.: Uncoupled multivariate power models for estimation performance-based seismic damage states of column curvature ductility. Structures **36**, 752–764 (2022)
16. Zhong, J., Yang, T., Pang, Y., Yuan, W.: A novel structure-pulse coupled model for quantifying the column ductility demand under pulse-like GMs. J. Earthq. Eng. **26**(15), 1–19 (2021)
17. Baker, J.W.: Quantitative classification of near-fault ground motions using wavelet analysis. B Seismol. Soc. Am. **97**(5), 1486–1501 (2007)
18. Zhong, J., Jiang, L., Pang, Y., Yuan, W.: Near-fault seismic risk assessment of simply supported bridges. Earthq. Spectra **36**(4), 1645–1669 (2020)

A Brief Review on Compression Strength Prediction Models of Alkaline-Activated Slag Concrete

Yeong Huei Lee[1(✉)], Yee Yong Lee[2], Siaw Fui Kiew[3], Yie Hua Tan[4], and Cher Siang Tan[5]

[1] Department of Civil and Construction Engineering, Faculty of Engineering and Science, Curtin University Malaysia, CDT 250, 98009 Miri, Sarawak, Malaysia
yeong.huei@curtin.edu.my

[2] Department of Civil Engineering, Faculty of Engineering, Universiti Malaysia Sarawak, 94300 Kota Samarahan, Sarawak, Malaysia

[3] Curtin Biovalley, Sarawak Biovalley Pilot Plant, Curtin University Malaysia, CDT 250, 98009 Miri, Sarawak, Malaysia

[4] Petroleum and Chemical Engineering, Faculty of Engineering, Universiti Teknologi Brunei, Bandar Seri Begawan BE1410, Brunei Darussalam

[5] Faculty of Civil Engineering, Universiti Teknologi Malaysia, UTM, 81310 Johor Bahru, Johor, Malaysia

Abstract. In the present construction sector, alkali-activated slag (AAS) concrete is classified as green concrete. However, it is still infrequently used because site engineers and workers have little expertise with the mix proportion of this novel material for achieving the appropriate concrete strength compared to ordinary Portland cement (OPC) concrete. Experiments and tests are thus required to determine an optimised AASC design mix for a specific design strength. A strength prediction model would be a valid way at this early stage to estimate the characteristic strength of concrete at different mixing ratios and to produce an optimised AASC mixing ratio with the desired concrete strength. As a result, developing a reliable approach to forecast concrete strength is critical. This paper summarises the prediction of concrete compression strength of AASC. The prediction model using the artificial neural network method is thoroughly explored. The determinants alternating AASC strength have also been highlighted for further study. Overall, a trustworthy concrete strength prediction model should be developed for use in the construction industry.

Keywords: alkali activated slag · concrete · artificial neural network · compression strength

1 Introduction

In the context of mitigating anthropogenic carbon dioxide's emission, one of the prospects that is currently practised is partially or fully replacement of Ordinary Portland Cement (OPC) with supplementary cementitious materials (SCMs), due to high carbon

W. Guo et al. (Eds.): GBCESC 2023, LNCE 328, pp. 470–485, 2024.
https://doi.org/10.1007/978-981-99-9947-7_50

footprint of cement production. Examples of SCMs are natural pozzolans, ground granulated blast furnace slag (GGBFS) [1], coal fly ash (FA) [2], rice husk ash (RHA), silica fume (SF) and etc. The OPC that are partially replaced with slag-based SCMs (e.g., GGBFS) is called slag concrete, while a complete substitution of OPC with GGBFS forms alkali-activated slag (AAS) concrete. Alkaline, as the name implies, is employed as an activator of the SCMs to increase their strength. The usage of AAS concrete as an alternative to OPC concrete has recently received a lot of interest [3]. This is due to the fact that the manufacturing of clinker-free cement - alkali-activated slag (AAS) [1] uses less energy and produces less CO_2. According to [4], producing one tonne of OPC requires 5000 MJ of energy and emits one tonne of CO_2, whereas producing a comparable quantity of slag requires just 1300 MJ of energy and emits 0.07 tonnes of CO_2. The use of AAS in concrete matrix involves transformation of the industrial by-products into value-added product and at the same time, alleviating the environmental burden for landfill.

AAS concrete is still infrequently employed in the building industry in comparison to OPC concrete since site engineers and site workers have limited experience in the mix proportion of this new material for a desired concrete strength. Experiments and tests are thus required to determine an optimised design mix of AAS concrete for a certain design strength. A strength prediction model would be a reliable way at this early stage to determine the characteristic strength of concrete at varied mixing ratios. The goal will be to find the best mixing ratio for making AASC with the appropriate concrete strength.

Compression strength prediction using a machine learning algorithm is a non-destructive technique that requires just minimal sample preparation [5]. In addition to reducing waste of raw materials and building failures, the prediction model saves both resources and time. There are extensively used prediction models for forecasting the concrete compression strength of OPC concrete, however these models may not be adequate for AAS concrete. The strength prediction of AAS concrete is summarised in this paper.

2 Binder in Concrete

Concrete is made up of a binder, aggregate, and water. OPC is typically employed as a binder. Further research and development on concrete materials stimulates the use of SCMs (e.g., slag and fly ash) as the concrete binder, which are made of a by-product or waste from industrial processes.

2.1 Ordinary Portland Cement (OPC)

In the construction industry, OPC is the most commonly used cement. According to [6], OPC is hydraulic in the sense that it acquires setting and hardening qualities when exposed to water. OPC is made by grinding, combining, and heating raw materials such as limestone, clay, and shale. The limestone contains a lot of lime, but the clay and shale include a lot of silica and alumina. Cement is made by calcining a mixture of calcareous components found mostly in limestone and argillaceous substances (e.g., clays, shale, and slags) at high temperatures (1450 °C) [7]. The clinker produced during the process will then be mixed with gypsum to obtained OPC as the final product [8]. The chemical composition of OPC includes a higher degree of major oxides (e.g., CaO, SiO_2, Al_2O_3 and Fe_2O_3), and a lower degree of minor oxides (e.g., MgO, SO_3), and some alkali oxides that might not occur in every slag type, such as K_2O and Na_2O [9]. Table 1 provides the compositional range for OPC.

Table 1. Summary of Portland cement compositional range

Component	[% mass]	
	[10]	[8]
CaO	60.00–67.00	58.10–68.00
SiO_2	17.00–25.00	18.40–24.50
Al_2O_3	3.00–8.00	3.10–7.56
Fe_2O_3	0.50–6.00	0.16–5.78
MgO	0.50–4.00	0.02–7.10
SO_3	2.00–3.50	0.00–5.35
K_2O	–	0.04–1.66
Na_2O	0.30–1.20	0.00–0.78

2.2 Alkali Activated Materials (AAM)

Alkali activated materials (AAM) are a type of green building material that has the potential to replace the usage of OPC in concrete. AAM is defined as the binder system or products that come from the reaction of a solid silicate powder with an alkali metal source, either solid or dissolution [11]. GGBFS, FA, metakaolin, and other aluminosilicate-rich precursors are the sources. The precursors or SCMs are industrial byproducts or wastes. Because of their lower energy use and carbon emissions, AAMs are more sustainable than OPC [12]. Any solution that may offer alkali metal cations to raise the alkalinity of the mixture and the rate of dissolution of the solid precursors can be used as an alkali source. Alkali hydroxide, silicates, carbonates, sulphates, aluminates, and oxides are examples of alkali solutions [13].

AAMs are often classified into two types of gel systems: high-calcium systems (e.g., alkali-activated slag) and low-calcium systems (e.g., alkali-activated fly ash) [14]. The principal product for a high-calcium gel system is calcium alumina silicate hydrate (CASH) gels with a tobermorite-like structure. While for a low-calcium structure, the resulting product is alkali aluminosilicate (NASH) gels with a pseudo-zeolitic structure [15]. The low-calcium system requires a high temperature to activate, whereas the high-calcium system can be activated at room temperature [4]. AAMs, such as alkali-activated slag, are manufactured at a cheaper cost and with less CO_2 emissions [16].

Ground Granulated Blast Furnace Slag (GGBFS)

GGBFS stands for industrial solid wastes from iron manufacture. It is formed in a blast furnace while iron is molten and is constituted of silicates and oxides of undesirable elements throughout the refining process [17]. In the creation of green high-performance concrete, slag is known as the "sixth ingredient" of concrete [18].

Blaine air permeability tests are used to determine the specific surface area of slag or the Blaine fineness [16]. GGBFS fineness must be greater than 350 m^2/kg [1]. Previous experimental research found that the Blaine fineness of slag ranged from 370 m^2/kg to 460 m^2/kg, and that the specific gravity of slag was between 2.88 and 2.9 (Table 2).

Table 2. Physical properties of GGBFS

GGBFS	[% mass]					
	[19]	[20]	[21]	[14]	[22]	[23]
Specific gravity	–	2.88	2.9	2.9	2.9	2.9
Blaine Fineness (m^2/kg)	460	435	460	383	370	370
Particle size	1–10 μm	–	1–10 μm	–	–	–

When finely crushed slag is activated by alkaline, it can completely replace OPC in concrete production since it has a comparable mineral composition to OPC and thus potential cementitious properties [24–26]. In comparison to OPC concrete, slag cement concrete demonstrated higher durability and reduced hydration heat. Table 3 summarises the chemical composition of slag from several researchers.

Table 3. Chemical composition of GGBFS

Chemical	GGBFS					
	[19]	[20]	[21]	[14]	[22]	[23]
CaO	39.4	40.7	41.7	33.7	33.7	37.3
SiO_2	35	34.6	33.5	33.9	32.4	37.7
Al_2O_3	13.9	13.7	13.5	20.4	16.7	14.4
Fe_2O_3	0.3	0.4	0.3	0.8	1.2	1.1
MgO	6.1	7.1	6	7.9	9.7	8.7
SO_2	2.4	0.6	2.7	0.9	0.9	–
K_2O	0.4	0.6	0.3	–	<0.1	–
Na_2O	0.3	0.5	0.2	0.1	0.2	–
TiO_2	0.4	–	0.8	–	–	–
P_2O_5	<0.1	–	0.1	–	–	–
MnO	0.4	–	0.4	–	–	–
Loss on ignition	1.45	1.13	–	3.6	<0.1	1.41

2.3 Chemical Reaction of AAS

Slag is a substance with latent hydraulic capabilities. When slag is dissolved in water alone, it forms a protective coating that prevents further reaction [27]. As a result, alkaline solution is utilised to start the slag's hydration process. When alkali-activated ground granulated slag is used, it has good cementitious properties [28].

GGBFS contains chemical compounds such as CaO, SiO_2, MgO and Al_2O_3, that contribute to the two types of network-forming ions - anions (e.g., $(SiO_4)^{4-}$, $(AlO_4)^{5-}$ $(MgO_4)^{6-}$) and cations (e.g., Ca^{2+}, Al^{3+} and Mg^{2+}) [29]. In the process of activating slag, the alkali will attack the slag particles through a polycondensation reaction. The alkaline cations act as a catalyst to speed up the reaction between the slag and alkali in the initial stage of hydration through the exchanging process with the cations (e.g., Ca^{2+}) from slag [30]. The following equations explain the alkaline activation of slag that is rich with SiO_2 and CaO, where R^+ represents the cation from alkali.

$$= \text{Si-O-} + R^+ \text{®} \text{Si-O-R}$$
$$= \text{Si-O-R} + OH^- \text{®} \text{Si-O-R-OH}^-$$
$$= \text{Si-O-R-OH}^- + Ca^{2+} \text{®} \text{Si-O-Ca-OH} + R^+$$

When GGBFS is activated by alkaline activators, a reaction product called calcium alumina silicate hydrate (C-A-S-H) gel is formed, which is identical to the calcium silicate hydrate (C-S-H) gel formed during the hydration of OPC [31]. The structures of C-A-S-H gel, as explored by [32], are produced from linear finite chains of silicate tetrahedra linked to core Ca-O layers, with aluminium assuming the bridging positions in the chains. The chemical displacements that occur throughout the process are generally determined by the silicon Qn environments, where n indicates the number of bridging oxygens that are coupled to other Si atoms for each Q (SiO_4) tetrahedral unit [33]. According to Puertas et al. [34], CSH gel from the OPC system had a mean chain length (MCL) of three to five tetrahedra and a low Al concentration. The gel structure of AAS has been found to alter depending on the nature of the alkali activator used. The C(A)SH gel is composed of MCL of nine tetrahedra and Al that blends into the structure bridging tetrahedra and was created with sodium hydroxide (NaOH) as activators. When sodium silicate (Na_2SiO_3) solution is used as the activator, the CASH gel is composed of MCL of 11 tetrahedra and has more Al in tetrahedral locations than the former. Figure 1 depicts the structural and Qn composition of the CASH gel system.

2.4 Hydration Process of Alkali Activated Slag (AAS)

The rate of reaction of a product is defined as hydration kinetics and is monitored by isothermal calorimetry by measuring the heat evolved during the hydration reaction [36]. [35] reported that the hydration kinetics of the AAS binder is different from OPC. The kinetics of the AAS binder are significantly depending on the activators used and the curing temperature. The AAS hydration reaction is divided into three stages, i. wetting and dissolution of particles, ii. gel formation and iii. the bulk reaction of slag product. [37] investigated the hydration kinetics of AAS with different activators, AAS-OPC and OPC. The results indicated that slag hydrates very slowly compared to OPC as shown

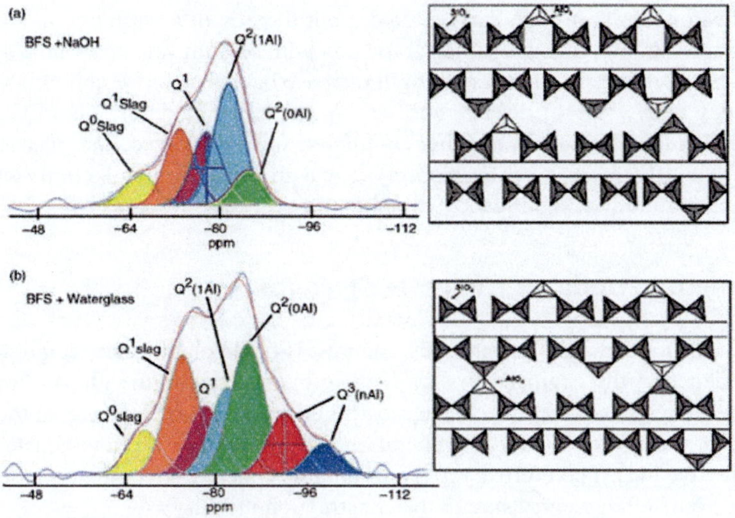

Fig. 1. Structural model for an Al-containing C-S-H gel: (a) linear chains C-(A)-S-H gel; (b) linear chains with occasional cross-linking, forming planes C-A-S-H gel [35]

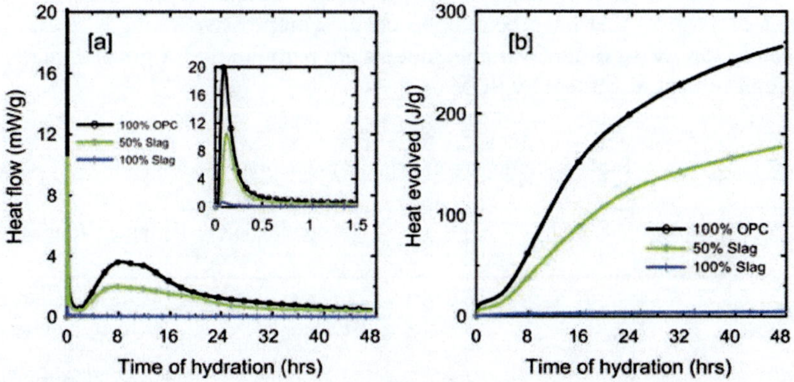

Fig. 2. (a) Hydration curve of and (b) Calorimetric curve of 100% OPC paste, 50/50 OPC/slag paste and 100% slag (sole slag) paste, each at 25 °C. [37]

in the calorimetric curve which depicts that there is no heat evolved during hydration of slag with water (Fig. 2).

Figure 2 shows that the hydration heat of the activated slag binder is approximately half that of the OPC. The duration of each stage in terms of wetting, acceleration, and deceleration in AAS hydration is shorter than in OPC hydration. The hydration curve revealed that heat generation in both alkali-activated slag and OPC occurred at relatively early stages. [38] discovered that sodium hydroxide (NaOH) activated slag reacted spontaneously after a short induction period, whereas sodium silicate activated slag required a longer induction period. This clearly illustrates that using NaOH solution as an activator results in high early strength, but there is little variation in strength as the hydration duration increases. Slag activation with sodium silicate as an activator, on the other hand, results in a slower strength increase but can achieve quite high strength after 28 days. [37, 39, 40] discovered that a larger silica modulus, Ms (SiO2/Na2O), encouraged greater hydration, a higher cumulative of heat evolved, and intense product production. Furthermore, it has been observed that greater activator alkalinity and higher curing temperature will significantly expedite the hydration process [26].

3 Prediction Model of Concrete Strength

Generally, the mix design of a particular concrete is based on the code recommendation and experience of the engineer on site. If there is any error in the preparation of mix design, the concrete would not achieve the desired strength which results in the need of repetition of the entire process of preparation and testing that are time-consuming and cost ineffective [41]. Therefore, a rapid and suitable concrete strength prediction model would be a cost-effective measure in the construction industry.

Recently, researchers have been contributing to anticipate, evaluate, and predict the properties of different types of concrete by utilising various methods. However, study involving the prediction of AASC has rarely been done. Table 4 shows the structural review of previous studies working on the prediction of concrete compressive strength. It reveals that works that have been done on the compressive strength of AASC are significantly low. Most of the available models are particularly for green concrete with partial replacement of cement by SCMs.

Table 4. Summary of existing prediction models

Previous work	Sample size	Method	R^2	Concrete type	Input variables
[42]	135	ANN	0.993	Concrete containing BFS and FA	Ultrasonic; pulse velocity; rebound number; curing age; cement; BFS; FA
		MNLR	0.899		
[43]	37	MNLR	–	AASC	water content; w/b ratio; fine aggregate to total aggregate ratio
[44]	32	ANN	0.868	Concrete containing Nano-silica	Superplasticizer; fine and coarse aggregates; w/b; cement
[45]	114	ANN	0.950	Self-compacting fly ash concrete	w/b; fly ash; fine and coarse aggregate; superplasticizer

(continued)

Table 4. (*continued*)

Previous work	Sample size	Method	R^2	Concrete type	Input variables
[46]	180	ANN	0.999	HPC containing nano-silica and copper slag	Cement; nano-silica; fine aggregate; copper slag; specimen age; superplasticizer
		MNLR	0.669		
[47]	125	ANN	0.860	HPC containing BFS and fly ash	BFS, FA, water, superplasticizer, fine and coarse aggregate
[48]	139	ANN	0.892	Concrete containing recycled aggregate	Fine aggregate; recycled and natural coarse aggregate; w/b ratio; water/total material ratio
[49]	49	ANN	0.898	Concrete without FA	Cement; fine and coarse aggregates; w/b
	27		0.996	Concrete with FA	Cement; FA; fine and coarse aggregates; w/b
	49	Genetic Programming	0.774	Concrete without FA	Cement; fine and coarse aggregates; w/b
	27		0.938	Concrete with FA	Cement; FA; fine and coarse aggregates; w/b
[50]	149	ANN	0.930	LWAC	Concrete density; cement; FA; SF; water; fine and coarse aggregates; w/b
		MNLR	0.841		
		MNLR	0.880		
[51]	335	DNN	0.970	Alkaline-activated Fly Ash Concrete (AAFC)	FA; water glass solution; sodium hydroxide solution; coarse aggregate; fine aggregate; water; sodium hydroxide concentration; curing time; curing temperature
[52]	180	RF	0.944	AAFC	FA specific surface area; chemical composition of FA; mixing procedure; curing regime; curing temperature; curing time

3.1 Codes of Practice

Mechanical properties of concrete can be predicted based on international codes such American Concrete Institute (ACI) code, Eurocode (EC) and other related codes. However, neither American nor European concrete or standardization institutes consist of integrated provisions for the design of Alkali Activated Slag concrete in their respective codes. The design procedures for AASC have been lacking.

According to [53], only high-strength and lightweight aggregate of conventional aggregate concrete has been included in the international codes. Intensive literature searches show that the code prediction is extensively used for anticipating creep and shrinkage of concrete but its use for compressive strength of concrete is rare. [54] have conducted code prediction of the concrete creep and shrinkage using different codes such as ACI209R-92 (ACI), CEN-EC2 (Eurocode 2) and CEB-FIP Model Code. The results obtained demonstrated that all codes resulted in overestimation except the ACI code seemed to provide a more accurate prediction. [55] investigated on the variations between experimental and analytical results obtained from code prediction. ACI 209R-92, BS EN-92, B3, and GL2000 codes were used to predict the shrinkage of self-compacting concrete and apparently, most models resulted in overestimation. In summary, the code prediction is usually utilised for shrinkage and creep prediction but its accuracy should be considered carefully.

3.2 Regression

The statistical model is designed for inferencing the relationships among variables. Before the advent of the new computational technology, ANN, other AI models, and statistical methods such as linear, non-linear, or multi-linear regression are utilised in function approximation and forecasting. Linear regression models are defined as empirical models which solely based on the observed data and the variables involved do not have relationship to any mechanism for predicting the data [56]. Whereas non-linear regression models involve some specific physical interpretations of the variables. Hence, prior knowledge of the functional relationships between the variables is a essential for making assumptions during the data-producing process [42]. Several researchers have carried out compressive strength prediction on concretes through the statistical method. [42] applied multiple non-linear regression (MNLR) on strength prediction of green concrete with partial replacement of cement by SCMs. 135 sets of the sample were used in the model. [46] adopted MNLR for high-performance concrete (HPC) that consists of nano-silica and slag. [50] utilised both linear- and non-linear regression for lightweight aggregate concrete. A compressive strength prediction on AASC with calcium hydroxide as activator based on a MNLR was carried out by [43]. The empirical equation for strength was developed depending on the MNLR analysis of 34 test data.

$$f'_c = 0.43\left[\alpha\beta\left(\frac{S}{A}\right)^{0.25}\left(\frac{w}{b}\right)^{-1.25}\right]^{1.4} f_{co} \tag{1}$$

where $\alpha = 1.1$ for $Ca(OH)_2$ and Na_2SiO_3; 1.0 for $Ca(OH)_2$ and Na_2CO_3; $\beta = 1.0$ for air-dried curing; $(1.4\ w/b/\alpha)^{0.23}$ for water curing; S/A = fine aggregate to total weight of aggregate ratio; $f_{co} = 10$ MPa (reference strength).

In summary, only a few regression methods that work on the compression strength prediction of AASC has been done.

3.3 Algorithm Prediction

Supervised machine learning is a promising strategy that employs algorithm prediction to determine the complex relationship between variables. In the realm of construction materials, algorithm prediction has been widely used to anticipate various compliance-relevant features of various types of concrete [50].

Some applications of machine learning methods in the context of compressive strength prediction are highlighted here. [5, 42, 44, 46, 48–50] used an artificial neural network (ANN) to forecast the compressive strength of green concretes based on compositional characteristics. The sample sizes for the models range from 27 to 180. [57] used random forest (RF), Nave Bayes, and k-nearest neighbour (KNN) to forecast the strength of metakaolin-based concrete. There were 71 distinct data sets used. [51] used a deep neural network (DNN) and a deep residual network (ResNet) with 9 input features to estimate the compressive strength of fly ash concrete. In recent investigations, [52] used random forest (RF) to forecast the strength of alkali-activated fly ash concrete. Previous research (Table 4) demonstrates that the algorithm-based ANN machine learning model is widely used for strength prediction. R2, or the correlation coefficient, is a crucial factor in determining a model's performance and accuracy [44]. Table 4 shows that prediction models utilising the ANN technique have a higher R2 value, indicating that the forecast is accurate.

3.4 Artificial Neural Network

An Artificial Neural Network (ANN) is a computational architecture that learns the non-linearity of variables, predicts the input–output relationship, and adapts to changes in the variables [46]. According to [51], an ANN is a collection of connected neurons (perceptron) that is used to learn, model, and predict the relationship between inputs and outputs. A neural network is made up of three primary layers of neurons: input layers, hidden layers, and output layers (Fig. 3). Any information accessible in the input layers will be sent to the output layer via hidden layers, with the number of hidden layers determining whether the network is a single- or multi-layer perceptron [45].

The back-propagation technique is employed in ANN to estimate the weight of each input and the bias values in order to minimise the prediction error, according to [50]. As shown in Fig. 3, back-propagation learning will be repeated from the output layer to the input layer in each running step until the prediction achieves a consistent mean square error (MSE). Any network mistake that happens throughout the learning process is commonly described as MSE, which can be calculated using Eq. 2 [45]. After obtaining the updated weight and bias from the training, new variables will be used as inputs to generate the corresponding outputs. The determination coefficient (R^2), root mean square error (RMSE), mean absolute error (MAE), mean absolute percentage error (MAPE) can be used to evaluate the accuracy of prediction as well. The equations for

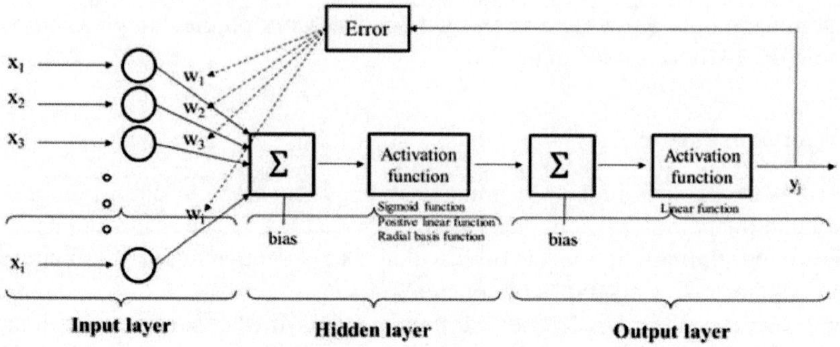

Fig. 3. Basic structure of Artificial Neural Network [50]

R^2, RMSE, MAE, and MAPE, Eqs. 3 to 6 corresponding are defined as below [47].

$$MSE = \frac{1}{P}\sum_{j=1}^{p}(t_j - O_j)^2 \tag{2}$$

$$R^2 = 1 - \left(\frac{\sum_{j=1}^{p}(t_j - O_j)^2}{\sum_{j=1}^{p}(t_j - \overline{t_j})^2}\right) \tag{3}$$

$$RMSE = \sqrt{\frac{\sum_{j=1}^{p}(t_j - O_j)^2}{P}} \tag{4}$$

$$MAE = \frac{1}{P}\sum_{j=1}^{p}|t_j - O_j| \tag{5}$$

$$MAPE = \left(\frac{1}{P}\sum_{j=1}^{p}\left|\frac{t_j - O_j}{t_j}\right|\right) \times 100\% \tag{6}$$

where the term t_j represents the target value of the jth pattern; O_j represents the network output value (predicted value) of the jth pattern; P represents the number of the pattern.

4 Discussion

Many experimental experiments have recently been conducted to investigate the characteristics of AASC. AASC has stronger early strength, superior long-term compressive strength behaviour, higher tensile strength, flexural strength, and better durability in an aggressive environment, according to past research. The compressive strength of the concrete is the most important mechanical feature to consider while using it. The compressive strength of AASC, on the other hand, is unpredictable because to a variety of elements that may impact the concrete strength. Despite the design mix of the concrete such as the water/binder ratio, fine and coarse aggregate ratio, major factors that influence the concrete strength are the fineness of slag, molarity, or concentration of NaOH and NaOH: Na_2SiO_3 ratio. Blaine fineness of slag or known as the specific surface area will affect the hydration of the slag and consequently, influencing the CASH gel formation during activation of AAS. According to prior research, the combination of sodium hydroxide and sodium silicate activated slag concrete is commonly utilised as the activators due to the extra reaction products created with these alkalis. The effect of activator molarity on AAS compressive strength was noticed, as the oxide ion provided affects gel formation similarly to slag fineness. The soluble silica in the activators aids in the dissolving of slag particles during the activation and hydration processes. As a result, one of the most important parameters to consider when calculating the compressive strength of concrete is the ratio of sodium hydroxide to sodium silicate.

The assessment of the compression strength of concrete is time-consuming and expensive since concrete cubes of various mix patterns must be manufactured, fabricated, and tested three days, seven days, and 28 days after the concrete is cast. If the necessary concrete strength is not obtained, the entire procedure must be repeated until it is. As a result, one of the scientific focus points has been a non-destructive method, which is a prediction model using international concrete code or a prediction model by statistical analysis and algorithm to identify the compressive strength of concrete. Although code prediction is commonly used to predict shrinkage and creep in concrete, the strength prediction using international codes is quite limited. In comparison to statistical analysis, the correlation coefficient that reflected the accuracy of the prediction model from the algorithm technique was shown to be better. Because it can forecast the complicated interaction between factors and may be trained to achieve a more accurate prediction output, Artificial Neural Network (ANN) is widely used to develop the concrete strength prediction model. Nonetheless, the AASC strength prediction model is rarely created. Except for alkali-activated slag concrete, the majority of the work has been done for conventional concrete, green concrete with partial substitution of OPC by fly ash or slag, and high-performance concrete. This could be because the AASC is rarely used in the building business.Therefore, the development of a prediction model for AASC is essential to promote the use of sustainable AASC in building and construction.

5 Conclusions

Moving towards new era, prediction through machine learning is dominant in current technology development trend. With the big data from various sources, concrete strengths prediction can be done with high accuracy using supervised and unsupervised machine learning algorithms. This paper specifically summarizes the compressive strength prediction models of AASC. Several conclusions can be drawn from the discussion,

i. Compressive strength of AASC is affected by the fineness of slag, molarity, or concentration of NaOH and NaOH: Na_2SiO_3 ratio
ii. Artificial neural network prediction is suitable to be used in predicting the compressive strength for AASC due to inter-related factors that contributing to the strength development

iii. AASC compressive strength prediction model with ANN can be applied in construction industry for more accurate non-destructive prediction.

Acknowledgements. The authors would like to acknowledge the technical supports provided by Curtin University Malaysia.

Conflicts of Interest. The authors have no conflicts of interest to disclose.

References

1. Amran, M., et al.: Slag uses in making an ecofriendly and sustainable concrete: a review. Constr. Build. Mater. **272**, 121942 (2021). https://doi.org/10.1016/j.conbuildmat.2020.121942
2. Hossain, M.B., Roknuzzaman, M., Biswas, M.A., Islam, M.: Evaluation of engineering properties of thermal power plant waste for subgrade treatment. J. Civ. Eng. Sci. Technol. **12**, 112–123 (2021). https://doi.org/10.33736/jcest.3975.2021
3. Mithun, B.M., Narasimhan, M.C.: Performance of alkali activated slag concrete mixes incorporating copper slag as fine aggregate. J. Clean. Prod. **112**, 837–844 (2016). https://doi.org/10.1016/j.jclepro.2015.06.026
4. Awoyera, P., Adesina, A.: A critical review on application of alkali activated slag as a sustainable composite binder. Case Stud. Constr. Mater. **11**, e00268 (2019). https://doi.org/10.1016/j.cscm.2019.e00268
5. Liu, G., Zheng, J.: Prediction model of compressive strength development in concrete containing four kinds of gelled materials with the artificial intelligence method. Appl. Sci. **9**, 1039 (2019). https://doi.org/10.3390/app9061039
6. Alaa, A.: Investigating mechanical properties of ordinary portland cement. Investigating improvements to the mechanical properties of Ordinary Portland Cement (OPC) bodies by utilizing the phase transformation properties of a ceramic (Zirconia) (2021)
7. Del Strother, P.: Manufacture of portland cement. In: Lea's Chemistry of Cement and Concrete, pp. 31–56 (2019). https://doi.org/10.1016/B978-0-08-100773-0.00002-2
8. Bediako, M., Amankwah, E.O.: Analysis of chemical composition of Portland cement in Ghana: a key to understand the behavior of cement. Adv. Mater. Sci. Eng. **2015** (2015). https://doi.org/10.1155/2015/349401
9. Punmathari, T., Rachakornk, M., Imyim, A., Wecharatan, M.: Co-processing of grinding sludge as alternative raw material in portland cement clinker production. J. Appl. Sci. **10**(15), 1525–1535 (2010)
10. Neville, A.M.: Chemical composition of portland cement. Prop. Concr. 8–12 (2011). https://doi.org/10.1016/s0016-0032(36)91056-6
11. Provis, J.L.: Alkali-activated materials. Cem. Concr. Res. **114**, 40–48 (2018). https://doi.org/10.1016/j.cemconres.2017.02.009
12. Ma, H., et al.: Preparation and reaction mechanism characterization of alkali-activated coal gangue-slag materials. Materials **12**, 2250 (2019). https://doi.org/10.3390/ma12142250
13. Provis, J.L., van Deventer Jannie, S.J.: RILEM report on alkali activated material (2014)
14. Thunuguntla, C.S., Gunneswara Rao, T.D.: Effect of mix design parameters on mechanical and durability properties of alkali activated slag concrete. Constr. Build. Mater. **193**, 173–188 (2018). https://doi.org/10.1016/j.conbuildmat.2018.10.189

15. Zhang, J., Shi, C., Zhang, Z., Ou, Z.: Durability of alkali-activated materials in aggressive environments: a review on recent studies. Constr. Build. Mater. **152**, 598–613 (2017). https://doi.org/10.1016/j.conbuildmat.2017.07.027
16. Dyer, T.D., McCarthy, M.J., Csetenyi, L.J.: Potential of weathered blast furnace slag for use as an addition in concrete. Mag. Concr. Res. **73**, 240–251 (2021). https://doi.org/10.1680/jmacr.19.00142
17. Akin Altun, I., Yilmaz, I.: Study on steel furnace slags with high MgO as additive in Portland cement. Cem. Concr. Res. **32**, 1247–1249 (2002). https://doi.org/10.1016/S0008-8846(02)00763-9
18. Dai, J., Wang, Q., Xie, C., Xue, Y., Duan, Y., Cui, X.: The effect of fineness on the hydration activity index of ground granulated blast furnace slag. Materials **12**, 2984 (2019). https://doi.org/10.3390/ma12182984
19. Bakharev, T., Sanjayan, J.G., Cheng, Y.B.: Sulfate attack on alkali-activated slag concrete. Cem. Concr. Res. **32**, 211–216 (2002). https://doi.org/10.1016/S0008-8846(01)00659-7
20. Chi, M.: Effects of dosage of alkali-activated solution and curing conditions on the properties and durability of alkali-activated slag concrete. Constr. Build. Mater. **35**, 240–245 (2012). https://doi.org/10.1016/j.conbuildmat.2012.04.005
21. Law, D.W., Adam, A.A., Molyneaux, T.K., Patnaikuni, I.: Durability assessment of alkali activated slag (AAS) concrete. Mater. Struct. Constr. **45**, 1425–1473 (2012). https://doi.org/10.1617/s11527-012-9842-1
22. Manjunath, R., Narasimhan, M.C.: An experimental investigation on self-compacting alkali activated slag concrete mixes. J. Build. Eng. **17**, 1–12 (2018). https://doi.org/10.1016/j.jobe.2018.01.009
23. Talkeri, A.H., Shankar, A.U.R.: A study on initial setting time and the mechanical properties of AASC using the PS ball as fine aggregate. Int. J. Pavement Res. Technol. **12**, 659–663 (2019). https://doi.org/10.1007/s42947-019-0078-0
24. Wang, W., Noguchi, T.: Alkali-silica reaction (ASR) in the alkali-activated cement (AAC) system: a state-of-the-art review. Constr. Build. Mater. **252**, 119105 (2020). https://doi.org/10.1016/j.conbuildmat.2020.119105
25. Jing, W., Jiang, J., Ding, S., Duan, P.: Hydration and microstructure of steel slag as cementitious material and fine aggregate in mortar. Molecules **25**, 4456 (2020). https://doi.org/10.3390/molecules25194456
26. Thomas, R., Ye, H., Radlinska, A., Peethamparan, S.: Alkali-activated slag cement concrete. Concr. Int. **38**, 33–38 (2016)
27. Narender Reddy, A., Rajesh, D.V.S.P., Reddy, A.N., Tilak, U.V., Raghavendra, M.: Performance of alkali activated slag with various alkali activators. Int. J. Innov. Res. Sci. Eng. Technol. **2**, 378–386 (2013)
28. Özbay, E., Erdemir, M., Durmuş, H.I.: Utilization and efficiency of ground granulated blast furnace slag on concrete properties - a review. Constr. Build. Mater. **105**, 423–434 (2016). https://doi.org/10.1016/j.conbuildmat.2015.12.153
29. Garcia-Lodeiro, I., Palomo, A., Fernández-Jiménez, A.: An overview of the chemistry of alkali-activated cement-based binders. In: Handbook of Alkali-Activated Cements, Mortars and Concretes, pp. 19–47 (2015). https://doi.org/10.1533/9781782422884.1.19
30. Chithiraputhiran, S.R.: Kinetics of alkaline activation of slag and fly ash-slag systems, p. 32 (2012)
31. Mavroulidou, M., Martynková, R.: A study of alkali-activated concrete mixes with ground granulated blast furnace slag. Glob. Nest J. **20**, 208–215 (2018). https://doi.org/10.30955/gnj.002370
32. Fernández-Jiménez, A., Puertas, F., Sobrados, I., Sanz, J.: Structure of calcium silicate hydrates formed in alkaline-activated slag: influence of the type of alkaline activator. J. Am. Ceram. Soc. **86**, 1389–1394 (2003). https://doi.org/10.1111/j.1151-2916.2003.tb03481.x

33. Ye, H., Radlińska, A.: Shrinkage mechanisms of alkali-activated slag. Cem. Concr. Res. **88**, 126–135 (2016). https://doi.org/10.1016/j.cemconres.2016.07.001
34. Puertas, F., Palacios, M., Manzano, H., Dolado, J.S., Rico, A., Rodríguez, J.: A model for the C-A-S-H gel formed in alkali-activated slag cements. J. Eur. Ceram. Soc. **31**, 2043–2056 (2011). https://doi.org/10.1016/j.jeurceramsoc.2011.04.036
35. Palomo, A., Krivenko, P., Garcia-Lodeiro, I., Kavalerova, E., Maltseva, O., Fernández-Jiménez, A.: A review on alkaline activation: new analytical perspectives. Mater. Constr. **64**, e022 (2014). https://doi.org/10.3989/mc.2014.00314
36. Deir, E., Gebregziabiher, B.S., Peethamparan, S.: Influence of starting material on the early age hydration kinetics, microstructure and composition of binding gel in alkali activated binder systems. Cem. Concr. Compos. **48**, 108–117 (2014). https://doi.org/10.1016/j.cemconcomp.2013.11.010
37. Gebregziabiher, B.S., Thomas, R., Peethamparan, S.: Very early-age reaction kinetics and microstructural development in alkali-activated slag. Cem. Concr. Compos. **55**, 91–102 (2015). https://doi.org/10.1016/j.cemconcomp.2014.09.001
38. Ben Haha, M., Le Saout, G., Winnefeld, F., Lothenbach, B.: Influence of activator type on hydration kinetics, hydrate assemblage and microstructural development of alkali activated blast-furnace slags. Cem. Concr. Res. **41**, 301–310 (2011). https://doi.org/10.1016/j.cemconres.2010.11.016
39. Ravikumar, D., Neithalath, N.: Effects of activator characteristics on the reaction product formation in slag binders activated using alkali silicate powder and NaOH. Cem. Concr. Compos. **34**, 809–818 (2012). https://doi.org/10.1016/j.cemconcomp.2012.03.006
40. Ravikumar, D., Neithalath, N.: Reaction kinetics in sodium silicate powder and liquid activated slag binders evaluated using isothermal calorimetry. Thermochim. Acta **546**, 32–43 (2012). https://doi.org/10.1016/j.tca.2012.07.010
41. Kabir, A., Hasan, M., Miah, K.: Predicting 28 days compressive strength of concrete from 7 days test result. In: Proceedings of the International Conference on Advances in Design and Construction of Structures, pp. 18–22 (2012)
42. Atici, U.: Prediction of the strength of mineral admixture concrete using multivariable regression analysis and an artificial neural network. Expert Syst. Appl. **38**, 9609–9618 (2011). https://doi.org/10.1016/j.eswa.2011.01.156
43. Yang, K.H., Song, J.K.: Empirical equations for mechanical properties of Ca(OH)2-based alkali-activated slag concrete. ACI Mater. J. **109**, 431–440 (2012). https://doi.org/10.14359/51683918
44. Gupta, S.: Using artificial neural network to predict the compressive strength of concrete containing nano-silica. Civ. Eng. Archit. **1**, 96–102 (2013). https://doi.org/10.13189/cea.2013.010306
45. Belalia Douma, O., Boukhatem, B., Ghrici, M., Tagnit-Hamou, A.: Prediction of properties of self-compacting concrete containing fly ash using artificial neural network. Neural Comput. Appl. **28**, 707–718 (2016). https://doi.org/10.1007/s00521-016-2368-7
46. Chithra, S., Kumar, S.R.R.S., Chinnaraju, K., Alfin Ashmita, F.: A comparative study on the compressive strength prediction models for High Performance Concrete containing nano silica and copper slag using regression analysis and Artificial Neural Networks. Constr. Build. Mater. **114**, 528–535 (2016). https://doi.org/10.1016/j.conbuildmat.2016.03.214
47. Siraj, N.B., Fayek, A.R., Tsehayee, A.A.: Development and optimization of artificial intelligence-based concrete compressive strength predictive models. Int. J. Struct. Civ. Eng. Res. **5**(3), 156–167 (2016). https://doi.org/10.18178/ijscer.5.3.156-167
48. Naderpour, H., Rafiean, A.H., Fakharian, P.: Compressive strength prediction of environmentally friendly concrete using artificial neural networks. J. Build. Eng. **16**, 213–219 (2018). https://doi.org/10.1016/j.jobe.2018.01.007

49. Rajeshwari, R., Mandal, S.: Prediction of compressive strength of high-volume fly ash concrete using artificial neural network. In: Das, B., Neithalath, N. (eds.) Sustainable Construction and Building Materials, vol. 25, pp. 471–483. Springer, Singapore (2019). https://doi.org/10.1007/978-981-13-3317-0_42
50. Yoon, J.Y., Kim, H., Lee, Y.J., Sim, S.H.: Prediction model for mechanical properties of lightweight aggregate concrete using artificial neural network. Materials **12**, 2678 (2019). https://doi.org/10.3390/ma12172678
51. Nguyen, K.T., Nguyen, Q.D., Le, T.A., Shin, J., Lee, K.: Analyzing the compressive strength of green fly ash based geopolymer concrete using experiment and machine learning approaches. Constr. Build. Mater. **247**, 118581 (2020). https://doi.org/10.1016/j.conbuildmat.2020.118581
52. Gomaa, E., Han, T., ElGawady, M., Huang, J., Kumar, A.: Machine learning to predict properties of fresh and hardened alkali-activated concrete. Cem. Concr. Compos. **115**, 103863 (2021). https://doi.org/10.1016/j.cemconcomp.2020.103863
53. Tošić, N., Marinković, S., Ignjatović, I.: A database on flexural and shear strength of reinforced recycled aggregate concrete beams and comparison to Eurocode 2 predictions. Constr. Build. Mater. **127**, 932–944 (2016). https://doi.org/10.1016/j.conbuildmat.2016.10.058
54. Bocciarelli, M., Cattaneo, S., Ferrari, R., Ostinelli, A., Terminio, A.: Long-term behavior of self-compacting and normal vibrated concrete: experiments and code predictions. Constr. Build. Mater. **168**, 650–659 (2018). https://doi.org/10.1016/j.conbuildmat.2018.02.139
55. Abdalhmid, J.M., Ashour, A.F., Sheehan, T.: Long-term drying shrinkage of self-compacting concrete: experimental and analytical investigations. Constr. Build. Mater. **202**, 825–837 (2019). https://doi.org/10.1016/j.conbuildmat.2018.12.152
56. Srivastava, S., Tripathi, K.: Artificial neural network and non-linear regression: a comparative study. Int. J. Sci. Res. Publ. **2**, 1–5 (2011)
57. Lahoti, M., Narang, P., Tan, K.H., Yang, E.H.: Mix design factors and strength prediction of metakaolin-based geopolymer. Ceram. Int. **43**, 11433–11441 (2017). https://doi.org/10.1016/j.ceramint.2017.06.006

Green Building

Energy Performance Optimisation of Low-Rise Lightweight Steel-Frame Houses by Evolutionary Approach

Yang Yang, Marco Cimillo(✉), and Xi Chen

Xi'an Jiaotong-Liverpool University, Suzhou 215123, Jiangsu, China
MARCO.CIMILLO@xjtlu.edu.cn

Abstract. China is in the process of expanding its stock of low-rise LSF buildings in suburban and rural areas. This construction system is underused compared with most developed countries, but the number of low-rise steel-framed residential buildings is projected to increase due to the current national policies promoting rural green steel-framed houses. The building envelope is the most influential design factor in energy performance, and to better understand the effects of design decisions, this article investigates its optimisation by jEPlus + EA (EnergyPlus simulation manager for parametric + Evolutionary Algorithms). The objective functions are to minimise heating and cooling needs. The variables describe the main envelope components of a low-rise lightweight steel-frame house. The results confirm a significant potential for energy saving in comparison to different benchmark specifications and demonstrate the effectiveness of the method tested in the study.

Keywords: Building Energy Performance · Low-rise Housing · Lightweight Steel-Frame Houses · Design Optimisation · Evolutionary Algorithms

1 Introduction

The lightweight steel frame (LSF) is one of the most commonly applied construction methods for low-rise residential buildings. China is in the process of developing low-rise LSF buildings in suburban and rural areas, although the technology is still underused compared with most developed countries. However, the number of low-rise steel-framed residential buildings in China is projected to increase due to the policies promoting the construction of a substantial number of green rural houses by 2030 [1] and the incentivisation of steel-framed houses by the government. At the same time, in February 2022, the national 'Guidance on Promoting High-quality Development of Iron and Steel Industry' postulated that China's steel industry should peak carbon emissions in 2030. The optimisation of LSF houses is crucial to reduce both energy consumption and carbon emissions.

Building performance simulation (BPS) has been applied extensively in evaluating low-rise residential building performance at the design stage. In the hot-summer-cold-winter (HSCW) zone in China, including some of the most developed regions in the

W. Guo et al. (Eds.): GBCESC 2023, LNCE 328, pp. 489–497, 2024.
https://doi.org/10.1007/978-981-99-9947-7_51

country, previous studies found that the most impacting parameters on the annual load of LSF houses include (in order of relevance): window-wall ratio (WWR), U-value of exterior wall, U-value of exterior window, U-value of roof, shading, solar absorptivity of roof, solar absorptivity of the exterior wall [2]. Researchers also simulated the optimal configuration of the envelope of the LSF house and established that with a WWR of 0.2, the U-value of the exterior wall, exterior windows, and roof should be respectively 0.4 W/(m^2·K), 3.0 W/(m^2·K), and 0.5 W/(m^2·K) to achieve an energy saving of 56.93% compared with the baseline standard JGJ134-2001 [2]. However, previous studies are model-based simulations based on single-variable case-by-case scenarios. Although the model-based simulation is relatively accurate with comprehensive actual data in BPS, it is hard to apply for general optimal design guidelines and optimisation problems, including a larger number of input variables and discrete and nonlinear objectives. Systematic optimisations with enough variable combinations for energy performance offer a better alternative but have received limited attention so far.

BPS using the evolutionary algorithm is one of the most popular optimisation methods to deal with a large number of parameters in the building sector [3]. An evolutionary algorithm, mimicking biological evolution, evaluates a population of potential solutions by using a fitness function to evolve over time and find better solutions [4]. Non-dominated sorting genetic algorithms (NSGA-II), part of the evolutionary algorithm family, are widely used for multi-objective optimisation. NSGA-II efficiently identifies trade-off relationships among different objective functions and has been successfully applied in building optimisation research [5]. Figure 1 depicts the standard process of a Genetic Algorithm (GA): 1) Creation of an initial population through random generation; 2) Computation and assessment of the fitness of each individual within the population; 3) Identification and selection of the best ranking chromosomes for replication predicated upon their fitness scores; 4) Generation of progeny through the application of crossover and/or mutation procedures; 5) Iterative repetition, commencing from step two, until a stipulated termination criterion is met, either in terms of temporal constraints or the attainment of an adequately optimal fitness level.

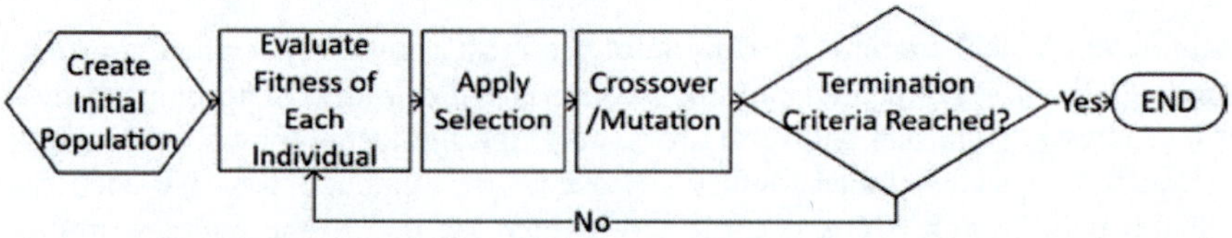

Fig. 1. Flow chart of typical genetic algorithm. Adapted from [6].

Notably, NSGA-II enables to manage an extensive array of input variables within the design process, accommodating both discrete and nonlinear objectives. Hence, incorporating the NSGA-II can be advantageous in optimising the design of LSF houses.

To better understand the effects of different parameters on energy performance and provide guidance for design decisions of LSF houses in HSCW, this paper studies the optimisations on the energy performance of a notional low-rise lightweight residential building methods by NSGA-II. The selected objective functions are to minimise heating loads and cooling loads.

The study will also test the optimisation method on this particular application and estimate the achievable energy use reduction.

2 Methodology

2.1 Tools and Workflow

The simulations are conducted using OpenStudio [6] and EnergyPlus [7]. OpenStudio is an open-source software to support thermal and energy balance simulation on buildings. OpenStudio covers the most critical factors of building with the geographical interface. EnergyPlus, developed mainly by the US Department of Energy and the National Renewable Energy Laboratory, constitutes the simulation engine of OpenStudio and has been extensively validated against a number of international standards [7].

JEPlus and jEplus+EA [8] are used for optimisation in this study. JEPlus is a parametric analysis and optimisation tool using energy simulation tools like EnergyPlus. jEPlus + EA, an extension of jEPlus using the NSGA-II algorithm, allow users to perform full-factorial, Monte Carlo, global sensitivity, and constrained multi-objective optimisation on existing jEPlus projects. Due to their user-friendly graphical interface and reliability of results, OpenStudio, EnergyPlus and jEPlus + EA are selected as the simulation and optimisation tools for this study. The workflow with the aforementioned tools is shown in Fig. 2.

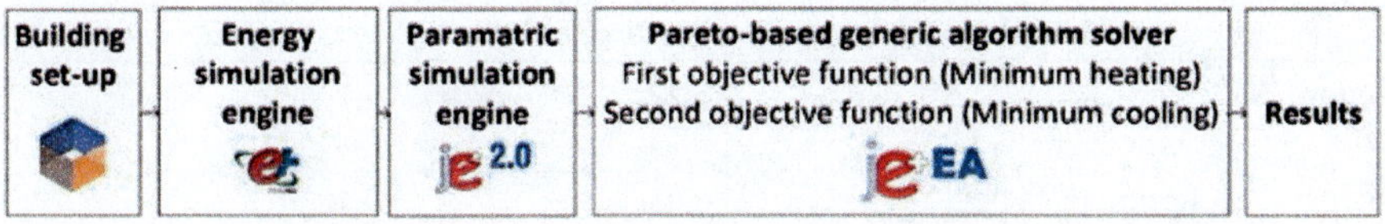

Fig. 2. Workflow of simulation-based energy performance optimisation.

To take into account the influence of the steel studs on the thermal resistance and the thermal mass of the wall, the composite structures have been modelled by combining the thermal properties of the different materials into a homogeneous layer with equivalent performance [9]. Such performance was computed with the software BuildDesk.

The climate selected for the study is Shanghai. It belongs to the HSCW areas of China, and its suburbs are among the most likely location for the development of a large number of low-rise residential buildings in the near future. Due to the compact spacing of residences in the suburban areas of Shanghai, surroundings are modelled as vertical obstructions of 6m of height, at a distance of 6m, and with a width corresponding to the simulated building's opposing façades. The climate data is provided by the Chinese Standard Weather Database and retrieved from the EnergyPlus website [10].

2.2 Benchmarks

The characteristics of the building envelope are fixed within a range including a baseline standard (BS) and the best international best practice (BI). The range also includes options complying with the best local practice (BL).

The BS was defined according to DGJ 08-205-2015 [11]. The standard stipulates the baseline requirements for all new residential buildings in Shanghai. According to DGJ 08-205-2015, the simulation's main characteristics are required as follows: The heating set-point temperature from December 1st to February 28th is 18°C. The cooling set-point temperature from June 15th to August 31st is 26°C. Performance time is 8760 h. Outdoor airflow is set at one change per hour. Internal heating gain is 4.3 W/m^2.

The BL complies with the requirements for the Three Star-level of GB/T 50378-2019 Assessment standard for green buildings [12]. As a three-star building, this is representative of the best local practice.

The high-performance benchmark was created to represent the BI based on BREEAM In-Use International Technical Manual: Residential SD243-V6.0.0. The Building Research Establishment's Environmental Assessment Method, or BREEAM [13], is one of the most comprehensive and widely used environmental assessment tools worldwide. The parameters of benchmarks are summarised in Table 1. The windows distribution in BS and BI has been modified according to relevant standards, expressed as WWRs by DGJ 08-205-2015, and as WFRs by BREAAM.

Table 1. Scenarios of baseline, best international practice and best local practice.

Design feature	BS (Baseline Standard)	BL (Best Local Practice)	BI (Best International Practice)
Window-Wall Ratio: North	0.35	0.27	0.26
Window-Wall Ratio: East and West	0.25	0.08	0.18
Window-Wall Ratio: South	0.50	0.25	0.36
Window-Floor Ratio	0.44	0.185	0.321
Roof U-value [W/(m^2 ·K)]	0.50	0.27	0.2
Exterior Wall U-value [W/(m^2 ·K)]	0.6	0.37	0.2
Window U-value [W/(m^2 ·K)]	2.20	1.772	0.7
SHGC	0.52	0.579	0.5

2.3 Input Parameters

Geometries and Orientations

Two geometries, square and rectangle, are studied in this paper due to their most common application in China. Both square and rectangle are defined as the same area of 144 m^2. The dimensions of the square and rectangles are, respectively, 12m by 12 m, 16 m by 9 m, and 9 m by 16 m. According to the recommendation of BS, three orientations, −30°, 0° and 30°, are studied in this paper.

Exterior Wall

The exterior wall with the insulation both on the external side and in between lightweight steel frames - hybrid frame - is selected to be optimised. Four kinds of assemblies are studied according to the position of the thermal mass (concrete). They are Assembly 1 with thermal mass on both sides, Assembly 2 with thermal mass on the inner side, Assembly 3 with thermal mass on the outer side and Assembly 4 without thermal mass. The inhomogeneous layer consists of 99.70% mineral wool and 0.3% lightweight steel frames. Its minimum thickness, 80 mm, is obtained when the U-value of the exterior wall is 0.29 W/(m^2·K), which is the thinnest section for the lightweight steel frame. Its maximum thickness, 200 mm, is obtained when the U-value of the exterior wall is 0.2 W/(m^2·K), which is aligned with BI. According to the literature, the thickness of the concrete wall is typically 150 mm in Shanghai, China [14]. The thermal mass is set as 50–100 mm, as efficiency diminishes beyond such range [15]. The thickness for other materials (with a lower impact on thermal performance) is based on products typically available on the market. The detailed order of the exterior wall layers and thickness are listed in Table 2.

Table 2. Thickness of LSF exterior wall materials. Unit: mm.

Assembly 1		Assembly 2		Assembly 3		Assembly 4	
Concrete finish	25	Concrete finish	25	Concrete finish	25	Concrete finish	25
Polyethylene	0.2	Polyethylene	0.2	Polyethylene	0.2	Polyethylene	0.2
OSB	12	OSB	12	OSB	12	OSB	12
Concrete	25–50	Polyurethane	50	Concrete	50–100	Polyurethane	50
Polyethene	0.2	Inhomogeneous layer	80–200	Polyethylene	0.2	Inhomogeneous layer	80–200
Polyurethane	50	Polyethylene	0.2	Polyurethane	50	Polyethylene	0.2
Inhomogeneous layer	80–200	Concrete	50–100	Inhomogeneous layer	80–200	Gypsum plasterboard	25
Polyethylene	0.2	Gypsum plasterboard	25	Polyethylene	0.2		
Concrete	25–50			Gypsum plasterboard	25		
Gypsum plasterboard	25						

Windows, Shading and Roof

The original window dimensions are established as 1.2 m in length and 0.3 m in height, representing the minimum WWR of 0.1. The maximum north WWR is 0.47, which is obtained from BL. The maximum east and west WWR and south WWR are respectively 0.2 and 0.5, as derived from BS. The maximum WWF is 0.44, which is obtained from BL. In terms of properties of the windows, three types of windows with U-value and G-value aligned with benchmarks are selected as parameters.

The shadings of windows are set as proportions, 0.2 to 1, of the window height.

The maximum U-value of the roof is 0.5, obtained from BS.

2.4 Optimisation Settings

As aforementioned, the optimisation was carried out using the evolutionary algorithm, NSGA-II algorithm. Based on previous studies [16], the population size of 100 with parallel simulations is deemed adequate for this research. The simulated binary crossover (Crossover is to exchange genetic information of two parents and recombine into a new generation) with a rate of 0.9 was selected. The mutation (Mutation is to randomly change the genetic information of a parent and maintain the diversity of the population) rate is determined to be 20%. The maximum generation number is set to 200.

3 Result and Conclusion

The results of the optimisation process in Fig. 3 show a significant potential for energy saving by design optimisation, particularly for cooling. In comparison with the BI (the highest performing), the energy use decreases by up to around 24% in total, 26% for heating and 56% for cooling, indicating that the current standards are best suited for heating-dominated climates, while there is a substantial margin of improvement in controlling overheating. The prevalence of cooling in determining the total energy need, even in the current climate conditions, is also evidenced in Table 3.

The series of optimised solutions (in blue) have a fairly constant total energy need but with different proportions for heating and cooling; within the series, heating-optimised options reduce winter consumption by up to 19% circa compared to the cooling-optimised solutions, and the latter save up to around 24% of cooling energy in summer compared with the former. This demonstrates good flexibility for decision-making within the process, allowing designers to select different strategies and priorities but also to adapt to slightly different climate conditions within the HSCW zone. Prioritising cooling demand reduction may also prove valuable in future, proving new buildings against rapidly rising temperatures and increasingly frequent heat waves.

A sensitivity analysis also allows to identify the most influential design variables in each case, namely geometry, south-west-east WWR, wall insulation and roof insulation for cooling and total energy needs; additionally, the position of the thermal mass within the wall, south shading, window U-value and g-value play an important role in determining heating needs.

The impact of the main parameters is also illustrated by the correlation coefficients listed in Table 3. The correlation has been calculated for heating, cooling and total

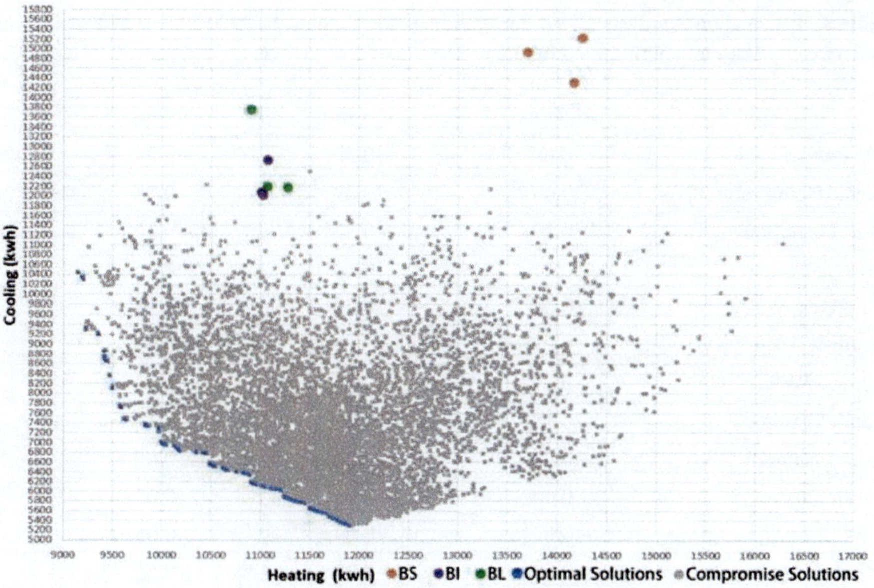

Fig. 3. Optimal solutions and benchmarks.

energy needs, assigning numerical values to all parameters. The building geometry and orientation have been expressed as a surface-to-volume (S/V) ratio and a ratio of south-facing façade surface. The position and thickness of the thermal mass were combined in one single parameter, multiplying the thickness by a factor expressing the influence on the internal space (lower for external mass and increasingly higher for layers placed on the interior and on both sides). The window types (which vary by U-value and SHGC) were expressed by a performance indicator, lower for the BS and increasingly higher for BL and BI.

Additionally, the analysis of the interplay among different parameters included in the simulation reveals the optimal combinations and the range of options available to reach the required performance targets. Figure 4 shows the performance outcomes of a sample of about 2,000 solutions. The best-performing solutions for both heating and cooling are highlighted, showing how the possible combinations are restricted. Solutions with low energy needs in both categories require windows compliant with the "Best International Practice", high WWRs on the South façades and low WWRs on all other orientations. Most solutions feature higher insulation on both walls and roofs; however, good performance can be achieved with a range of thicknesses, excluding only the worst options. No restrictions are identified regarding building geometry and the position of the thermal mass within the wall assembly; all options are included in some of the solutions reaching the highest performance.

Considering the above, the optimisation method tested in this study seems effective in identifying a set of design configurations able to yield considerable energy saving, and in producing data clarifying the trade-offs among alternative design options. The process also proved reasonably efficient, testing a total of 18,792 models in slightly more than 10

Table 3. Correlation Analysis: High correlation is highlighted by bold-underlined values, and moderate correlation is highlighted by underlined values.

	Heating	Cooling	Total H+C
SV Ratio	0.08	−0.33	−0.24
South-facing ratio	0.00	0.09	0.08
Wall Insulation	−0.29	−0.38	−0.50
Mass Influence	−0.11	−0.11	−0.16
East-West WWR	0.04	**0.74**	**0.65**
North WWR	0.27	0.56	**0.63**
South WWR	**−0.64**	0.48	0.02
South Shading	0.18	0.00	0.11
Roof Insulation	−0.44	−0.26	−0.48
North Shading	0.00	0.03	0.02
E-W Shading	−0.02	−0.04	−0.05
Window Type	−0.55	−0.25	−0.55
Heating	1.00	−0.09	0.53
Cooling		1.00	**0.79**

h of simulation time on a standard workstation, which is consider-ably more accurate than a more traditional scenario approach with a limited number of combinations modelled manually.

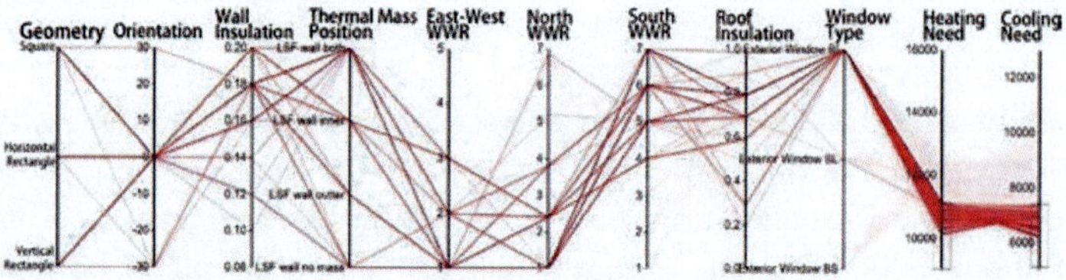

Fig. 4. Parallel Coordinates Chart Highlighting Selected Parameter Values of Solutions Optimized by both Heating and Cooling Needs.

While the preliminary results of optimisation in this study are encouraging, further research is needed for the validation of the results and the inclusion in the study of more representative typologies. A better understanding of the design parameters with nonlinear objectives may be beneficial to the development of design guidelines of LSF houses and provide valuable information to support China's plans for low-rise green houses.

References

1. Ministry of Housing and Urban-Rural Development of the People's Republic of China, National Development and Reform Commission: Implementation plan for carbon peaking in urban and rural construction. http://www.gov.cn/zhengce/zhengceku/2022-07/13/5700752/files/679a410848f04ea6bd2034faea9d4839.pdf. Accessed 30 June 2022
2. Li, F., Zhang, W., Wang, Y., Wu, J.: Study on rural prefabricated steel structure energy-saving buildings in hot summer and cold winter area. New Build. Mater. 150–156 (2021)
3. Hamdy, M., Nguyen, A.-T., Hensen, J.L.M.: A performance comparison of multi-objective optimization algorithms for solving nearly-zero-energy-building design problems. Energy Build. **121**, 57–71 (2016)
4. Leach, A.R.: 4.05 - ligand-based approaches: core molecular modeling. In: Taylor, J.B., Triggle, D.J. (eds.) Comprehensive Medicinal Chemistry II, pp. 87–118. Elsevier, Oxford (2007)
5. Shady, A., Mohamed, H., William, O., Salvatore, C.: Assessing gaps and needs for integrating building performance optimization tools in net zero energy buildings design. Energy Build. **60**, 110–124 (2013)
6. Pereira, G.C., de Oliveira, M.M.F., Ebecken, N.F.F.: Genetic optimization of artificial neural networks to forecast virioplankton abundance from cytometric data. J. Intell. Learn. Syst. Appl. **5**(1), 57–66 (2013)
7. OpenStudio (2022). https://openstudio.net/
8. EnergyPlus (2022). https://energyplus.net/testing
9. Zhang, Y.: jEPlus + EA (2020). http://www.jeplus.org/wiki/doku.php?id=docs:jeplus_ea_v2
10. Mahattanatawe, P., Puvanant, C., Mongkolsawat, D.: The energy performance of the cold-formed steel-frame and wood-frame houses developed for Thailand (2006)
11. EnergyPlus weather data. https://energyplus.net/weather. Accessed 21 Mar 2023
12. SRIBS: DGJ 08-205-2015 Design Standard for Energy Efficiency of Residential Buildings (2015)
13. CCSSI, SRIBS: GB/T 50378-2019 Assessment standard for green building (2019)
14. BREEAM - BRE Group. https://bregroup.com/products/breeam/. Accessed 13 July 2023
15. Li, X.: The climate-adaptability research on the exterior wall of prefabricated concrete residence in hot summer and cold winter area. Thesis, Hunan University, China (2019)
16. GreenSpec: Thermal Performance: Thermal Mass in Buildings. http://www.greenspec.co.uk/building-design/thermal-mass/. Accessed 19 Aug 2023
17. Li, T., Shao, G., Zuo, W., Huang, S.: Genetic algorithm for building optimization: state-of-the-art survey. In: Proceedings of the 9th International Conference on Machine Learning and Computing, pp. 205–210. ACM, Singapore (2017)

Check for updates

CFD Analysis of Thermal Comfort Condition Inside Malaysian Traditional House

Joristine Wong Yun Tong[1], Nur Hasyimah Binti Hashim[1], Yeong Huei Lee[1](✉), and Yee Yong Lee[2]

[1] Department of Civil and Construction Engineering, Faculty of Engineering and Science, Curtin University Malaysia, CDT 250, 98009 Miri, Sarawak, Malaysia
yeong.huei@curtin.edu.my

[2] Department of Civil Engineering, Faculty of Engineering, Universiti Malaysia Sarawak, 94300 Kota Samarahan, Sarawak, Malaysia

Abstract. It is crucial to ensure safe, healthy and comfortable indoor air conditions in buildings. Indoor thermal comfort is an essential aspect of sustainable architecture and is also the key to maintaining a safe indoor environment. The use of natural ventilation has been increasingly recognised as an energy-efficient method to establish thermal comfort. This study is to analyse the thermal comfort condition inside the traditional house using Computational Fluid Dynamics (CFD) software, ANSYS FLUENT and to compare the results with the thermal comfort recommended by ASHRAE standards. Previous studies of thermal comfort in hot and humid climate countries like Malaysia. The developed CFD model was validated by comparing experimental and simulated data. The experimental data are compared with the simulation results, which are in good agreement. From the results obtained, the confidence level is up to 99%. A series of simulations are also conducted to determine the air temperature and air velocity distribution inside the traditional house. The results show that the indoor air temperature of a traditional house is between 27.35°C and 27.60°C, and the air velocity is between 0.20 m/s and 0.68 m/s which are all within the thermal comfort range of ASHRAE standard. The obtained results indicate that the traditional Malay house's design effectively provides natural ventilation for thermal comfort. In addition, the research result will provide a reference for the modern architect to design houses with local architectural design characteristics.

Keywords: thermal environment · thermal comfort · traditional Malay house · natural ventilation

1 Introduction

Malaysia is a country with a humid tropical climate characterized by high temperatures and humidity. In order to solve the problem of high humidity, dehumidification is one of the methods with enough airflow through the building to restore comfort. Thus the most critical considerations in traditional house design are the provision of airflow (Nugroho, Ahmad, and Ossen 2007).

W. Guo et al. (Eds.): GBCESC 2023, LNCE 328, pp. 498–510, 2024.
https://doi.org/10.1007/978-981-99-9947-7_52

For most living spaces, thermal comfort is a essential requirement because it affects the occupants' health and thermal satisfaction. Thermal comfort is defined as the mental state of being satisfied with the thermal environment (Fanger, 1973). The proper thermal environment in the house will affect thermal comfort and building energy consumption. Research shows the importance of thermal comfort to students' performance (Jiang et al. 2019). Comfort is based on the individual's perception of the thermal environment, which is the neutral feeling experienced by people relative to the given thermal environment rather than sweating (Balbis-Morejón et al. 2020). Jiang et al. (2019) believe that participants' happiness will be affected by the discomfort related to feeling hot or cold.

Meanwhile, ventilation is the most traditional cooling method for providing thermal comfort. The ventilation system directly influences the comfort of occupants through temperature, airflow and indoor air quality (Andrzej Raczkowski, Suchorab, and Brzyski 2019). Ventilation can generally be categorized into three methods such as natural, mechanical and hybrid (mixed-mode) ventilation. In humid and hot climate areas like Malaysia, the recommended typical system is natural ventilation (Khalid and Yusup 2011).

Natural ventilation is a system that uses natural forces to provide fresh air for buildings. The air exchange is carried out through the air inlet and outlet designed in the building. Natural ventilation is also a very efficient management technology because it saves fans' costs and maintenance. Many factors influence natural ventilation, such as the roof design, occupants, main openings design, and building orientation. However, any mistake in designing a natural ventilation system may cause poor indoor air quality in a building or house. Moreover, the wrong location while installing the system may also impact indoor air quality (Andrzej Raczkowski, Suchorab, and Brzyski 2019).

In order to assess indoor thermal comfort due to the ventilation system, there are various computational tools and methods can be applied. The standard computations tools are ANSYS, COMSOL, TRNSYS, Openfoam and others. However, Computational Fluid Dynamics (CFD) program Ansys Fluent are often used (Andrzej Raczkowski, Suchorab, and Brzyski 2019). CFD method has been used in many studies to assess indoor thermal comfort.

2 Model Description

2.1 SM1 for Validation

SM1 was a traditional house selected as the model for validation purposes in the simulation setting to be used later in this study. SM1 was based on the research papers published by Tinker et al. (2004). This part is crucial because the modelling and simulation technique should be validated with published data to ensure the high reliability of the data collection in the parametric study. The views of SM1 is presented in Fig. 1.

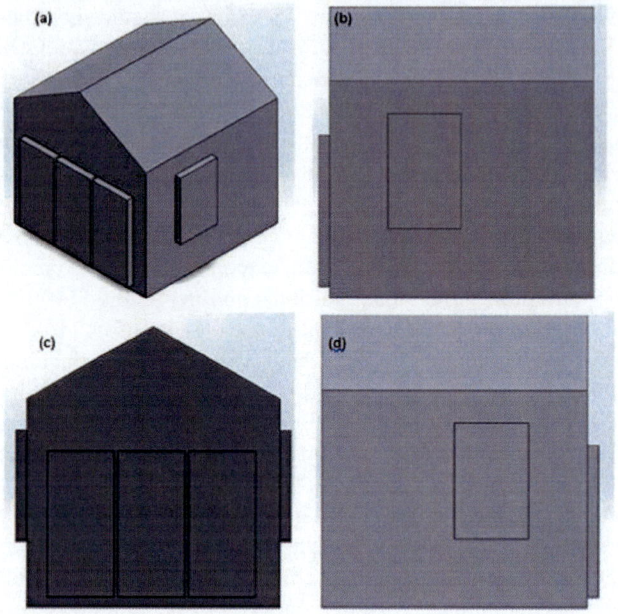

Fig. 1. Orientation of the traditional house model 1(a) Isometric view, (b) South elevation, (c) West elevation, (d) North elevation for SM1

2.2 SM2 for Current Study

SM2 is a basic Malay house. The house was raised on stilts and built with timber floors, timber-covered walls, and roofs. For SM2, the views were presented in Fig. 2. The dimensions of windows and doors were considered standard in size for all the traditional houses in Malaysia.

2.3 CFD Simulation

The simulation was carried out using the CFD software ANSYS Fluent. The house model was meshed and analyzed to determine the air temperature and velocity inside the house. The CFD program needs to express the inputs of the problem type, flow domain (type of flow, material, etc.), boundary conditions (inlet, outlet, walls), and the calculation method. The initial conditions have been set using the standard values from ASHRAE. Figures 3 and 4 show the meshed model for SM1 and 2. This study had three boundary conditions: wind velocity, pressure and other surfaces. Wind velocity was used as an inlet boundary, the pressure was used as an outlet condition, and the other surface was a condition that caused the fluid flow along the wall.

The orientation of SM2 from 0° to 90° and 180° is the main parameter for this parametric study with the validated modelling technique. The other parameters are the temperature and the air velocity distribution inside the traditional Malay house. These parameters are used to analyze the thermal comfort inside the traditional Malay house.

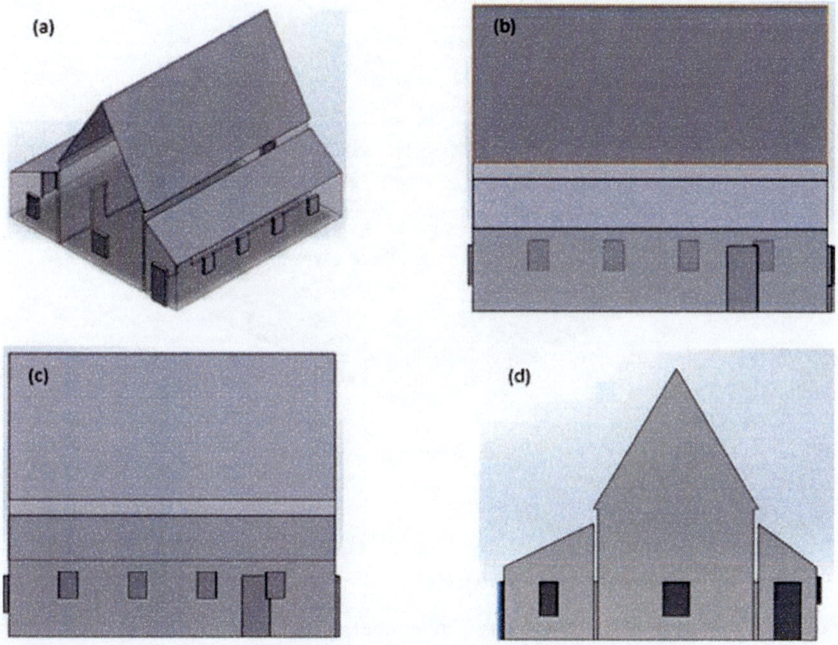

Fig. 2. Orientation of the traditional house model (a) isometric view, (b) from 0°, (c) from 90°, (d) from 180° for SM2

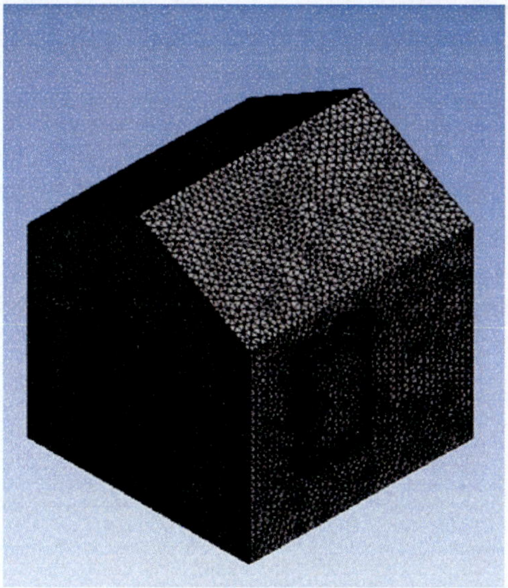

Fig. 3. Meshing o SM1

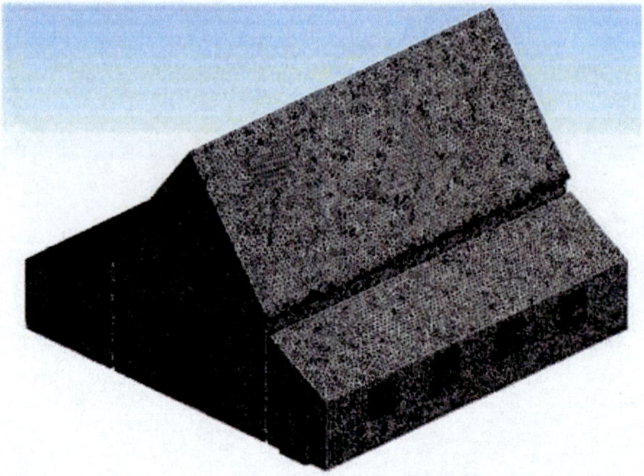

Fig. 4. Meshing of SM2

3 Results and Discussion

3.1 Validation of Developed CFD Model

Based on the experiment conducted by Tinker et al. (2004), the following window opening configurations were used in the experimental work for validation in Model 1. During the measurement, the window is fully opened from early morning and kept open throughout.

- Experiment A: Window (W1) open only
- Experiment B: Windows (W1) and (W5) open only
- Experiment C: Windows (W1) and (W3) open only

The measured and simulated values of the node points and their percentage differences are also shown in Tables 1 and 2. The measured air temperature and velocity values given in the table are the average values acquired from anemometers, which are integrated over a three-minute measurement period.

Table 1. Comparison of the air temperature and velocity of SM1

Exp	Node Point	Air Temperature (°C)			Air Velocity (m/s)		
		Measured	CFD	% diff	Measured	CFD	% diff
2	North	28.8	29.1	+1.0	0.40	0.37	−7.50
	South	28.7	29.1	+1.3	0.51	0.49	−3.92
	East	28.2	29.1	+3.0	0.27	0.25	−7.41
	West	28.8	29.1	+1.0	0.45	0.42	−6.67
	Middle	28.3	29.1	+2.0	0.48	0.44	−8.33

(*continued*)

Table 1. (*continued*)

Exp	Node Point	Air Temperature (°C)			Air Velocity (m/s)		
		Measured	CFD	% diff	Measured	CFD	% diff
3	North	28.5	29.1	+2.0	0.47	0.41	−12.77
	South	28.3	29.1	+2.8	0.54	0.51	−5.56
	East	28.1	29.1	+3.4	0.30	0.26	−13.33
	West	28.3	29.1	+2.8	0.50	0.54	+8.00
	Middle	28.6	29.1	+1.6	0.47	0.44	−6.38
4	North	28.7	29.1	+1.3	0.53	0.53	0.00
	South	29.0	29.1	+0.2	0.63	0.62	−1.59
	East	28.6	29.1	+1.6	0.41	0.39	−4.88
	West	29.0	29.1	+0.2	0.65	0.61	−6.15
	Middle	28.8	29.1	+0.9	0.60	0.59	−1.67

3.2 Evaluation of Thermal Comfort in SM2

As the accuracy of the CFD technique had been established, the technique was then used to analyse in order to determine the thermal comfort of SM2 under different angles:

Table 2. Comparison of CFD models between Tinker et al., 2004 and current study

Exp	Node Point	Air Temperature (°C)			Air Velocity (m/s)		
		CFD by Tinker et al. (2004)	CFD by current study	% diff	CFD by Tinker et al. (2004)	CFD by current study	% diff
2	North	29.1	28.85	−0.86	0.37	0.35	−5.41
	South	29.1	28.85	−0.86	0.49	0.47	−4.08
	East	29.1	28.85	−0.86	0.25	0.26	+4.00
	West	29.1	28.85	−0.86	0.42	0.40	−4.76
	Middle	29.1	28.85	−0.86	0.44	0.42	−4.55
3	North	29.1	28.85	−0.86	0.41	0.44	+7.32
	South	29.1	28.85	−0.86	0.51	0.51	0.00
	East	29.1	28.85	−0.86	0.26	0.26	0.00
	West	29.1	28.85	−0.86	0.54	0.45	−16.67
	Middle	29.1	28.85	−0.86	0.44	0.41	−6.82
4	North	29.1	28.85	−0.86	0.53	0.52	−1.89
	South	29.1	28.85	−0.86	0.62	0.62	0.00
	East	29.1	28.85	−0.86	0.39	0.39	0.00
	West	29.1	28.85	−0.86	0.61	0.62	+1.64
	Middle	29.1	28.85	−0.86	0.59	0.56	−5.08

- Experiment 1: Orientation from 0°
- Experiment 2: Orientation from 90°
- Experiment 3: Orientation from 180°

The results showed that Experiment 1 has the highest temperature, 27.6 °C. Experiment 3 has the lowest temperature, 27.35 °C, while Experiment 3 has 27.55 °C. Knowing that the outside temperature is set as 27 °C, it shows that the air temperatures remain similar to the outside air temperature. Some research also determined the same result: the air temperatures remain similar to outside air temperature (Ibrahim et al. 2013). Table 3 shows the air temperatures and velocity inside the Traditional Malay House of SM2.

Table 3. Air temperature and velocity of developed CFD model of SM2

Experiment	Node Point	Air Temperature (°C)	Air Velocity (m/s)
1	North	27.60	0.22
	South	27.60	0.48
	East	27.60	0.21
	West	27.60	0.36
	Middle	27.60	0.39
2	North	27.55	0.20
	South	27.55	0.58
	East	27.55	0.22
	West	27.55	0.44
	Middle	27.55	0.36
3	North	27.35	0.31
	South	27.35	0.54
	East	27.35	0.68
	West	27.35	0.64
	Middle	27.35	0.62

Comparison of Air Temperature

The data are shown in Figs. 5, 6 and 7, where it can be seen that by opening particular doors and windows in the house from different orientations, comfort conditions were provided. The air temperature inside the traditional house was in the comfort temperature range. In Experiments 1, 2, and 3, the average temperature was 27.60 °C, 27.55 °C and 27.35 °C, respectively. Jones et al. (1993) and Brager et al. (1998) agree that in tropical climates like Malaysia, the effective temperature range for thermal comfort rarely exceeds 28.5 °C, and the lowest temperature is not lower than 22 °C. According to ASHRAE standard, the thermal comfort range from 20 °C to 28 °C.

Comparison with Air Velocity

The air velocity is in the comfort condition. It is well-known and crucial that room air velocity plays a role in promoting thermal comfort. The averaged air velocity for the Experiment 1, 2, and 3 were 0.33 m/s, 0.36 m/s and 0.56 m/s respectively. Most researchers concur that air velocity should be between 0.15 m/s and 1.5 m/s based on the published data, which are highly consistent (Tinker et al. 2004). According to ASHRAE standards, the air velocity must be 0.15 m/s and not exceed 0.8 m/s to ensure the environment is comfortable.

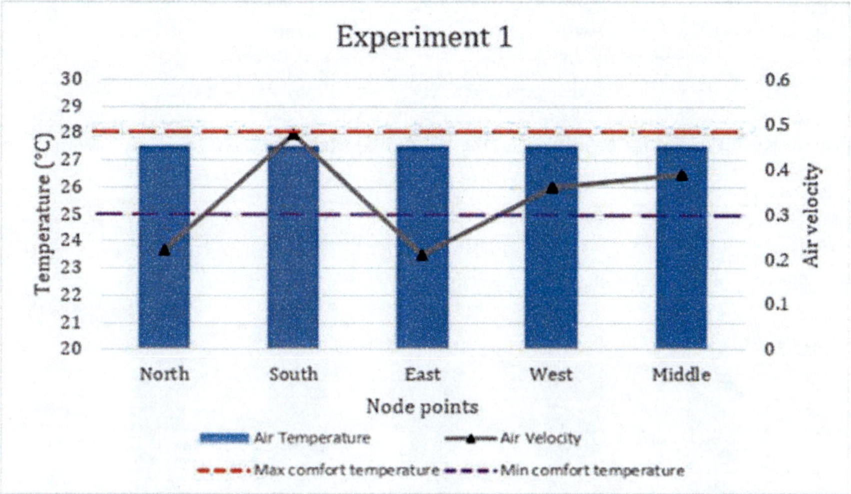

Fig. 5. Result of indoor conditions in the traditional Malay house (Orientation from 0°)

All the highest air velocity occurs at the opening, and the airflow is mainly focused on the middle of the house. By comparing, all the experiment results in SM2, Experiment 3, which at the orientation from 180° has the highest range of the velocity and the lowest air temperature. Experiment 3 has the highest velocity range because the airflow has not been disturbed by the barriers such as walls. This indicated that the design of the opening at the orientation from 180° is the best for achieving better thermal comfort.

When the air forming the boundary layer is changed, the faster the air flow rate, the easier it is for the surface to cool, and the temperature measured in this area will decrease (Chun 2018). Hence, the heat will be lost to achieve better thermal comfort due to the faster air flow rate. Nugroho et al. (2007) also stated that higher air velocity is required in order to obtain better thermal comfort. Therefore, indoor thermal comfort can be improved when the opening or the ventilation is modified by changing the size of the opening or increasing the number of openings.

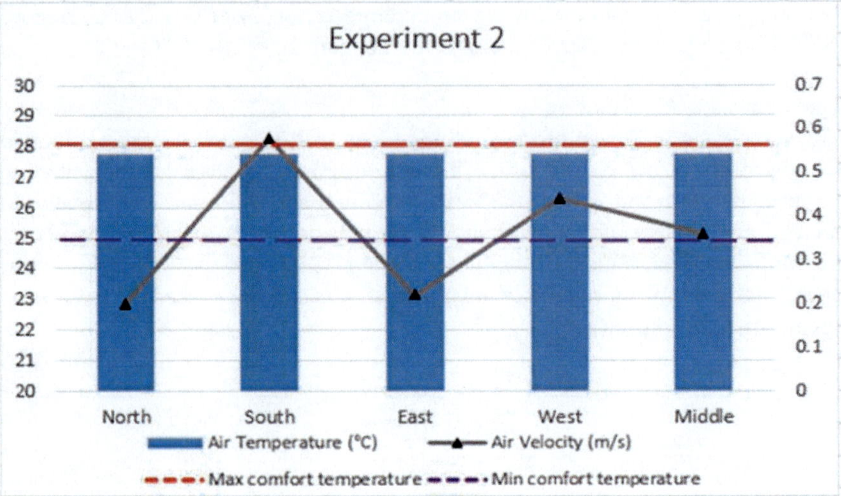

Fig. 6. Result of indoor conditions in the traditional Malay house (Orientation from 90°)

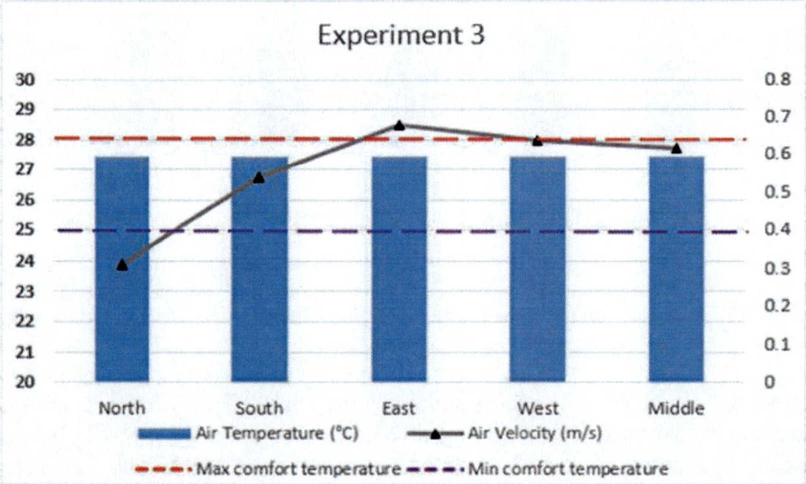

Fig. 7. Result of indoor conditions in the traditional Malay house (Orientation from 180°)

4 Conclusions

This research explains the validation of the developed CFD model and the results of the thermal comfort conditions inside the traditional house. The CFD model was verified with the experimental and simulation data carried out by other researchers. The CFD simulation model was used to evaluate thermal comfort. The mean percentage difference for the air temperature is 0.86% which indicates that the confidence level of temperature is about 99%. In comparison, the mean percentage difference for air velocity is 2.42%, and the confidence level of air velocity is about 98%. Therefore, it is confirmed that the simulation setting is successfully verified and can be used for further parametric study.

The research confirmed that the traditional house was thermally comfortable. The findings showed that high air temperatures occurred when there was only one opening for the air flow inside the traditional house in Experiment 1. While the opening of the traditional house increased, the air movement increased, and then the air temperature decreased. Hence, the results show that the air temperature at Experiments 2 and 3 is lower than in Experiment 1 since both experiments 2 and 3 have more openings than in Experiment 1. Moreover, the same theory goes for air velocity. The more the opening of the traditional house, the higher air velocity occurred inside the traditional house. The research also proved that a higher air velocity is required to achieve acceptable thermal comfort.

In addition, the results from the simulation showed that the indoor air temperature was between 25.5 °C and 28.5 °C, and the air velocity was between 0.2 m/s to 0.68 m/s. Both of the simulated results were within the recommended ranges. In Experiments 1, 2, and 3, the average temperature was 27.60 °C, 27.55 °C and 27.35 °C. Besides, the averaged air velocity for the Experiment 1, 2, and 3 were 0.33 m/s, 0.36 m/s and 0.56 m/s respectively. Hence, thermal comfort was achieved in the traditional Malay house and satisfied ASHRAE's standards. Due to the design of the traditional house, it can promote good air movement throughout the house regardless of any orientation. These findings were consistent with other researchers.

Acknowledgment. The authors would like to acknowledge the technical supports provided by Curtin University Malaysia.

Conflicts of Interest. The authors have no conflicts of interest to disclose.

References

Abas, N.F., Al-Absi, Z.A.: Subjective assessment of thermal comfort for residents in naturally ventilated residential building in Malaysia, vol. 401, p. 012009 (2018). https://doi.org/10.1088/1757-899X/401/1/012009

Razak, B.A., Mohamed, M.F., Mohammad Yusoff, W.F., Sulaiman, M.K.A.M., Ali, M.: Thermal comfort performance of naturally ventilated Royal Malaysian Police (RMP) lockup in hot and humid climate of Malaysia. Pertanika J. Sci. Technol. **29**(3), 1627–1654 (2021). https://doi.org/10.47836/pjst.29.3.35

Almhafdy, A., Ibrahim, N., Ahmad, S.S., Yahya, J.: Thermal performance analysis of courtyards in a hot humid climate using Computational Fluid Dynamics CFD method. Procedia Soc. Behav. Sci. **170**, 474–483 (2015). https://doi.org/10.1016/j.sbspro.2015.01.012

Alsmo, T., Alsmo, C.: Ventilation and relative humidity in Swedish buildings. J. Environ. Prot. **2014**(11), 1022–1036 (2014). https://doi.org/10.4236/JEP.2014.511102

Ariffin, S.I.: Order in Traditional Malay House Form. Oxford Brookes University, Doctor of Philosophy (2001)

ASHRAE. ANSI/ASHRAE Standard 55-2017: Thermal Environmental Conditions for Human Occupancy. ASHRAE Inc. (2017)

Balbis-Morejón, M., Rey-Hernández, J.M., Amaris-Castilla, C., Velasco-Gómez, E., San José-Alonso, J.F., Rey-Martínez, F.J.: Experimental study and analysis of thermal comfort in a university campus building in tropical climate. Sustainability **12**(21), 1–18 (2020). https://doi.org/10.3390/su12218886

Bouskill, L.M., Havenith, G., Kuklane, K., Parsons, K.C., Withey, W.R.: Relationship between clothing ventilation and thermal insulation. AIHA J. **63**(3), 262–268 (2017). https://doi.org/10.1080/15428110208984712

Brager, G.S., De Dear, R.J.: Thermal adaptation in the built environment: a literature review. Energy Build. **27**(1), 83–96 (1998). https://doi.org/10.1016/s0378-7788(97)00053-4

Callahan, C.W., Elansari, A.M., Fenton, D.L.: Psychrometrics. In: Postharvest Technology of Perishable Horticultural Commodities, pp. 271–310 (2019). https://doi.org/10.1016/B978-0-12-813276-0.00008-0

Chun, F.P.K.: Effects of Air Flow and Heating on Temperature: Experiment. UKEssays (2018). https://www.ukessays.com/essays/physics/effects-air-flow-heating-temperature-5612.php

Climate and Temperature Development in Malaysia. n.d. https://www.worlddata.info/asia/malaysia/climate.php. Accessed 31 Mar 2022

Eraña, J.: Satirical Regionalism. School of Architecture, McGill University, Montreal, Canada (2008)

Fitriaty, P., Antaryama Antaryama, I.G.N., Nastiti N.E., S.N.N.E.: Thermal performance of traditional house in the upland central celebes of Indonesia. IPTEK J. Technol. Sci. **22**(4) (2011). https://doi.org/10.12962/j20882033.v22i4.77

Goldman, R.F.: Assessment of thermal comfort. ASHRAE Trans. **84**(1), 1–6 (1978)

Harlina. n.d. Universiti Putra Malaysia the Interdependency of House Layout, Privacy and Culture from the User-Behavioural Perspective of Urban Malays Harlina Mohamad Ali Fem (2007)

Hensen, J.L.M., Lamberts, R.: Building Performance Simulation for Design and Operation (2012). https://doi.org/10.4324/9780203891612

Hosseini, E., Mursib, G., Shahminan, R.N.R., Shahedi, B.: Values in traditional architecture: Malay house. In: 6th International Seminar on Vernacular Settlements (2012)

Ibrahim, S.H., Baharun, A., Abdul Manan, M.D., Abang Adenan, D.A.: Importance of thermal comfort for library building in Kuching, Sarawak. Int. J. Energy Environ. **4**(6), 1003–1012 (2013)

Jadhav, R.P., Kengar, M.D., Nikam, N.R., Kumbhar, S.B., Devkar, S.B., Bhutkar, M.A.: Review on computational fluid dynamics and application. Asian J. Pharm. Res. **9**(4), 263 (2019). https://doi.org/10.5958/2231-5691.2019.00043.1

Jiang, H., Iandoli, M., Van Dessel, S., Liu, S., Whitehill, J.: Measuring students' thermal comfort and its impact on learning. In: EDM 2019 - Proceedings of the 12th International Conference on Educational Data Mining, pp. 89–98 (2019)

Jones, P.J., Alexander, D.K., Rahman, A.M.: Evaluation of the thermal performance of low-cost tropical housing (1993)

Kamarul, S.K., Lilawati, A.W., Asmalia, C.A.: Adaptation design of traditional Malay house to meet the requirements of comfort living in modern house. J. Malays. Surveyor **40**(1), 43–48 (2006)

Khairudn, K.F., Salleh, N.H., Aripin, S.B.: Construction techniques of traditional Malay houses (2018). https://doi.org/10.21834/e-bpj.v3i7.1131

Khalid, H.L., Yusup, Y.: Optimizing wind power for energy efficient building design in tropical hot-humid climate of Malaysia. VAWT turbulence measurements view project emission inventory development for the energy industry of Malaysia view project. J. Sustain. Dev. (2011) https://doi.org/10.5539/jsd.v4n2p217

Kwong, Q.J., Yang, J.Y., Ling, O.H.L., Edwards, R., Abdullah, J.: Thermal comfort prediction of air-conditioned and passively cooled engineering testing centres in a higher educational institution using CFD. Smart Sustain. Built Environ. **10**(1), 18–36 (2021). https://doi.org/10.1108/SASBE-08-2019-0115

Lim, J.Y.: The Malay house: rediscovering Malaysia's indigenous shelter system, pp. 74–79. Institut Masyarakat, Pulau Pinang (2009)

Lai, A., Maing, M., Ng, E.: Observational studies of mean radiant temperature across different outdoor spaces under shaded conditions in densely built environment. Build. Environ. **114**, 397–409 (2017). https://doi.org/10.1016/j.buildenv.2016.12.034

Nugroho, A.M., Ahmad, M.H., Ossen, D.R.: A preliminary study of thermal comfort in Malaysia's single storey terraced houses. J. Asian Archit. Build. Eng. **6**(1), 175–182 (2007). https://doi.org/10.3130/jaabe.6.175

Raczkowski, A., Wywiórka, A.: Assessment of perceived air quality for selected flat in the residential building, pp. 149–152 (2010)

Raczkowski, A., Suchorab, Z., Brzyski, P.: Computational fluid dynamics simulation of thermal comfort in naturally ventilated room. In: MATEC Web of Conferences, vol. 252, p. 04007 (2019). https://doi.org/10.1051/matecconf/201925204007

Ramli, N.H., Kassim, N., Taib, M.Z.M., Masri, M.H.: Re-adaptation of malay vernacular architecture thermal comfort elements: towards sustainable design in Malaysia (a literature review). In: International Conference on Innovation and Technology for Sustainable Built Environment (ICITSBE) (2012)

Salthammer, T.: Critical evaluation of approaches in setting indoor air quality guidelines and reference values. Chemosphere **82**(11), 1507–1517 (2011). https://doi.org/10.1016/j.chemosphere.2010.11.023

Sani, N.M.: Modern houses design for Malaysian **3**(6) (2012)

Schiavon, S., Lee, K.H.: Dynamic predictive clothing insulation models based on outdoor air and indoor operative temperatures. Build. Environ. **59**, 250–260 (2013). https://doi.org/10.1016/J.BUILDENV.2012.08.024

Seitablaiev, M.Ö., Umaroğulları, F.: Thermal comfort and indoor air quality. Reinforced Concrete Prefabrication, Sustainability, Reinforced Concrete Construction System View Project The Fire Regulation View Project (2018). https://www.researchgate.net/publication/326324068

Ju, S.R., Omar, S.: A typology of modern housing in Malaysia. Int. J. Hum. Ecol. **11**, 109–119 (2010)

Simion, M., Socaciu, L., Unguresan, P.: Factors which influence the thermal comfort inside of vehicles. Energy Procedia **85**, 472–480 (2016). https://doi.org/10.1016/J.EGYPRO.2015.12.229

Strauss, R.H., McFadden, E.R., Ingram, R.H., Deal, E.C., Jaeger, J.J.: Influence of heat and humidity on the airway obstruction induced by exercise in asthma. J. Clin. Investig. **61**(2), 433–440 (1978). https://doi.org/10.1172/JCI108954

Suhaila, H., Suhaimi, I., Mohamed, M.A.S., Mustaffa, M.T., Kartini, A., Atikah, A.: Thermal comfort study of a classroom in Northern Malaysia: a CFD approach. In: IOP Conference Series: Materials Science and Engineering, vol. 670, no. 1 (2019). https://doi.org/10.1088/1757-899X/670/1/012011

Kamal, K.S., Wahab, L.A., Ahmad, A.C.: Climatic design of the traditional Malay house to meet the requirements of modern living, pp. 10–12 (2004)

Szokolay, S.V., Auliciems, A.: Passive and low energy architecture international design tools and techniques (2007)

Tahir, M.M., Sopian, K., Surat, M., Tawil, N.: The development of a sustainably responsive ultra low energy terrace housing for the tropics incorporating the raised floor innovation. In: Recent Advances in Energy & Environment, pp. 36–45 (2010)

Tharim, A.H.A., Munir, F.F.A., Samad, M.H.A., Mohd, T.: A field investigation of thermal comfort parameters in green building index (GBI)-rated office buildings in Malaysia. Int. J. Technol. **9**(8), 1588–1596 (2018). https://doi.org/10.14716/ijtech.v9i8.2763

Tinker, J.A., Ibrahim, S.H., Ghisi, E.J.A.B.: Thermal comfort in modern low-income housing in Malaysia/Energy consumption assessment on industrialised building system (IBS) view project sustainable zero energy houses view project. An evaluation of thermal comfort in typical modern low-income housing in Malaysia (2004)

Toe, D.H.C., Kubota, T.: Potential application of vernacular passive cooling techniques to naturally ventilated modern terraced houses in hot-humid climate of Malaysia. Thermal functions of courtyard houses in Kabul view project potential application of vernacular passive cooling (2014)

Walker, A.R.: By Lim Jee Yuan, and Shah Badan Warisan. 1991. Malaysia - The Malay House: Rediscovering Malaysia's Indigenous Shelter System. By Lim Jee Yuan. Pulau Pinang: Institut Masyaraket, 1987. p. 152. Illustrations, Glossary, Bibliography. - Building a Malay House. By Phillip Gibbs. Singapore: Oxford University Press, 1987. p. 99. Illustrations, Glossary, Bibliography. - The Terengganu Timber Malay House. By Raja Bahrin Shah Bin Raja Ahmad Shah. Badan Warisan Malaysia, 1988. p. 83. Illustrations, Bibliography. - Shelter for the Homeless: Asian-Pa. Journal of Southeast Asian Studies **22**(1), 206–208. https://doi.org/10.1017/S0022463400005907

Yang, T., Clements-Croome, D.J.: Natural Ventilation in Built Environment Intelligent Solutions for Dependable Built Environment View Project Encyclopedia of Sustainable Technologies Edited by Martin Abraham (2018). https://www.elsevier.com/books/encyclopedia-of-sustainable-technologies/abraham/. https://doi.org/10.1007/978-1-0716-0684-1_488

Zhang, L., et al.: Outdoor thermal comfort of urban park—a case study. Sustainability **12**(5), 1961 (2020). https://doi.org/10.3390/SU12051961

Numerical Analysis of Improvement Effects on Summer Outdoor Thermal Environment Around Enclosed Teaching Buildings in the Hot-Humid and Less-Windy Climate

Xuexiu Zhao[1] and Yigang Li[2(✉)]

[1] Tokyo Institute of Technology, Kanagawa 226-8502, Japan
[2] Guangxi University, Guangxi 530004, People's Republic of China
lyg1023@foxmail.com

Abstract. Outdoor thermal environment around teaching buildings affects students' comfort and building energy consumption, especially in summer. This paper focuses on a teaching building with enclosed arrangement in a hot-humid and less-windy climate and analyzes the thermal environment in outdoor spaces around the building. The measurement results showed that average air temperatures and MRT (mean radiant temperature) in six outdoor spaces are higher than the ambient temperature by around 1 °C and 6 °C, respectively. Thermal improvement strategies for reducing surface temperatures in the analyzed spaces were proposed based on measurement results. A thermal simulation tool (ThermoRender) was used to quantify the improvement effects of the proposed strategies on the thermal environment. Simulation results revealed that the average values of MRT were reduced by 1.0 °C to 5.1 °C. It is noted that the reduction of received solar heat through sunshade devices and tree shading is an effective way to improve outdoor thermal environment, which can be considered as design guidance for creating a comfortable outdoor thermal environment around enclosed teaching buildings in the hot-humid and less-windy climate.

Keywords: Outdoor thermal environment · Enclosed building · Sun shading · Hot-humid and less-windy climate · Numerical analysis

1 Introduction

Teaching buildings are the main places for teachers and students to study, work and communicate in campuses. A comfortable thermal environment is conducive to improve the efficiency of work and study [1–3]. At present, the thermal environment in teaching buildings has been concerned by many researchers. However, most of the related researches focused on the thermal environment in indoor spaces [4–9]. Teaching buildings include not only the indoor space for work and study, but also the outdoor space for communication, rest and exercise. Appropriate rest and activities can relieve work stress and ensure the health of teachers and students. More importantly, the thermal

W. Guo et al. (Eds.): GBCESC 2023, LNCE 328, pp. 511–519, 2024.
https://doi.org/10.1007/978-981-99-9947-7_53

environment in outdoor spaces is an important factor for improving the willingness of students to conduct outdoor activities and reducing the building energy consumption, especially on summer days. Therefore, it is necessary to improve the outdoor thermal environment around teaching buildings in summer by taking measures. At present, thermal improvement strategies are universally adopted to different buildings [10]. It is an effective way to determine the improvement strategies according to the characteristic of analyzed spaces. Additionally, the hot-humid and less-windy climate is characterized by abundant solar radiation, low wind speed and prolonged discomfort period in summer. The improvement of outdoor thermal environment around buildings in the hot-humid and less-windy region is significantly urgent.

In terms of the outdoor space formed by enclosed buildings, it has more potential to enrich outdoor environments by adding pool, greening, exercise field, etc. However, this enclosed layout has some negative impacts on the thermal environment in outdoor spaces, such as the excessive solar heat and poor natural ventilation. It is necessary to improve the thermal environment in outdoor spaces around enclosed teaching buildings through appropriate improvement strategies. At present, there is still a lack of quantitative analysis on the outdoor thermal environment around enclosed teaching buildings in the hot-humid and less-windy climate.

Therefore, this study aims to analyze the characteristics of outdoor thermal environment around enclosed teaching buildings and quantify the improvement effects of the proposed strategies on the thermal environment. A teaching building in Nanning was selected as the research object. Thermal improvement strategies were proposed according to analysis results from field measurement, and their improvement effects were quantified using a thermal simulation tool (ThermoRender).

2 Methodology

2.1 Description of Study Building

The study building is the No.4 teaching building at a university which is located in Nanning (N22′ 13′ ~ 23′ 32′, L107′ 45′ ~ 08′ 51′), China. An aerial view of the building and the layout of the first floor are shown in Fig. 1(a). It is a typical enclosed building composed of two parts i.e. the north building with 7 floors and the south building with 2 floors. The two parts are connected by corridors. The study building covers an area of 4400 m^2 with a construction area of around 33600 m^2, and is mainly included classrooms, activity rooms, etc. The ground around the building is concrete-paved.

Two courtyards are formed by the north and south buildings, covering an area of about 1200 m^2 and 500 m^2, respectively. The large courtyard has two badminton courts with plastic-paved ground and a L-shaped pool, and most of its ground is paved by hard tiles. There is few vegetation (lawn and trees) in the large courtyard. On the east side of the small courtyard, there is a pool which is surrounded by lawn and several trees. Since the teaching building is frequently used and has various outdoor spaces (including exercise field, resting space, corridor, etc.), it was selected as the study object.

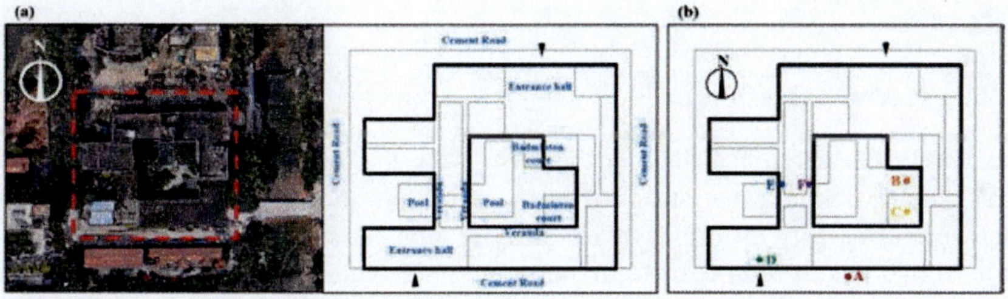

Fig. 1. (a) Aerial view of the building and layout of the first floor; (b) Locations of six measurement points.

2.2 Field Measurement

In order to analyze the thermal environment in outdoor spaces, six measurement points were selected according to their locations and surrounding conditions, as described in Sect. 2.1. The locations and detailed information of the six measurement points are indicated in Fig. 1(b) and Table 1 respectively.

Field measurement was conducted from 9:00 to 17:00 on a typical sunny day in the summer. Air temperature, mean radiant temperature (MRT) and wind speed at the six measurement points were monitored, and thermal images at the measurement points were captured by an infrared camera.

Table 1. Detailed information of the six measurement points.

Type	Point	Location	Surrounding condition
Outdoor	A	South sidewalk	No shading, concrete pavement
	B	Courtyard walking area	Trees shading, hard-tile pavement
	C	Badminton court	No shading, plastic-paved ground
Semi-outdoor (Transition space)	D	Entrance hall	Near concrete pavement, no trees
	E	West corridor	Near a pool
	F	East corridor	Near a pool

2.3 Numerical Simulation

Thermal improvement strategies for six outdoor spaces are proposed based on measurement results. Numerical simulation is used to quantify their improvement effects, which has advantages of accurate simulation results, low cost and without weather limitations. At present, there are some simulation tools for analyzing the thermal environment. A thermal simulation tool called ThermoRender is used for conducting this numerical analysis with consideration of outputting the surface temperature and MRT in the analyzed area. Its development principles and validation have been discussed in the previous study [11].

3 Results

3.1 Measurement Results

There was little difference of wind speeds at six measurement points, being around 0.8 m/s. On the contrary, variations of air temperatures and MRT at the six measurement points were obviously different, their hourly results are shown in Fig. 2 and Fig. 3. As seen in Fig. 2, the air temperatures in outdoor spaces (measurement points A, B and C) were higher than those in semi-open spaces (measurement points D, E and F). The same trend also appeared in the variations of MRT values, as seen in Fig. 3. In other words, the thermal environment in the outdoor spaces was hotter than in the semi-outdoor space. The main reason for this can be explained from the surrounding conditions around the six measurement points.

In the outdoor spaces (measurement points A, B and C), A and C are completely exposed to the sun, and their pavements absorb more solar heat. Therefore, uncomfortable thermal environments were formed at locations A and C. Due to shading from surrounding trees, air temperatures and MRT values at location B (outdoor space) were close to those in the semi-outdoor spaces.

In semi-outdoor space D, the shading area generated by upper sunshade was small, and both its ground and south road were paved with materials which are easy to absorb solar heat. Thus, the variation trends of air temperatures and MRT at point D were similar to that at point A. In addition, points E and F were in the space with long shading and near a pool, being relatively comfortable.

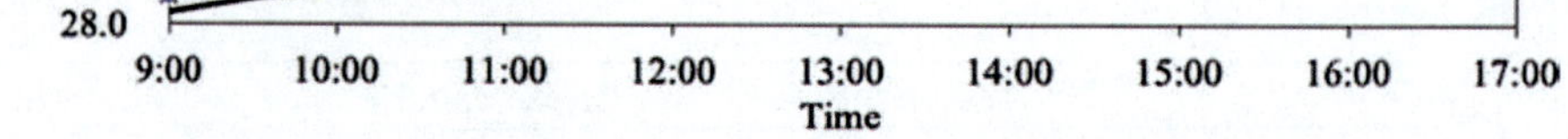

Fig. 2. Hourly variations of air temperatures at six measurement points.

Figure 4 presents thermography which was captured at 15:00. It can be seen that the surface temperatures of concrete pavement, building wall, plastic-paved ground and corridor ground were 50.2 °C, 39.4 °C, 50.8 °C and 38.5 °C, respectively. As shown in Fig. 4(b), the surface temperature of the hard tile under tree shading was 35.6 °C, more than 10 °C lower than that under sunlight. As a result, the surface temperature higher than the ambient temperature is the major cause for the formation of an uncomfortable thermal environment in the outdoor space.

The average air temperatures and MRT at the six measurement points were all higher than the ambient temperature by around 1 °C and 6 °C, respectively. It also can be found

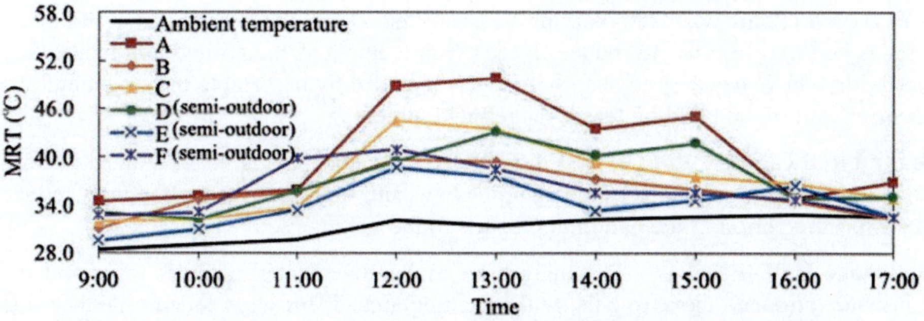

Fig. 3. Hourly variations of mean radiant temperatures at six measurement points.

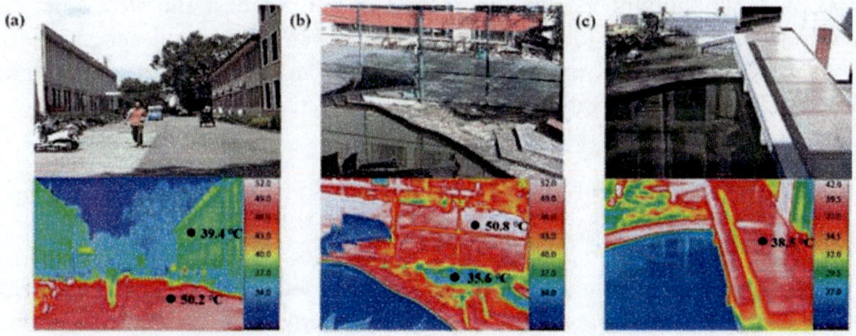

Fig. 4. Thermography captured at 15:00 in (a) location A (concrete pavement), (b) location C (badminton court), (c) location E (corridor).

that the surface temperature of the ground and building facades is the main factor affecting the thermal environment in outdoor and semi-outdoor spaces. Based on the above results, thermal improvement strategies are proposed for reducing the surface temperature in the analyzed area.

3.2 Thermal Improvement Strategies

Due to strong solar radiation and long summer in the hot-humid and less-windy climate, it is easy to increase surface temperatures in the analyzed outdoor space and form an uncomfortable thermal environment. There are some thermal improvement strategies for reducing surface temperature such as sunshade devices, planting trees and adding evaporative cooling equipment. The six measurement points covers different outdoor spaces including sidewalk, courtyard, badminton court, entrance hall and open corridor. Corresponding thermal improvement strategies were proposed based on the characteristics of these spaces.

Sidewalk (Point A). Trees are added along the sidewalks, which can provide shadow areas for pedestrians. And the original concrete pavement was replaced by permeable pavement to reduce the heat absorption, thereby decreased its surface temperature. Vertical greening was added on the building exterior walls near the sidewalks.

Courtyard (Point B). Tree-planting was selected to provide both shading and evaporative cooling. In order to reduce surface temperatures of the ground and facade, the hard title which was easy to absorb heat was replaced by permeable pavement and the exterior wall of the building was arranged with greening.

Badminton Court (Point C). It is difficult to change the paving material due to its use function. Shadow generated by surrounding trees and sun canopy can effectively reduce the exposure period of the badminton courts to the sun.

Entrance Hall (Point D). The main thermal improvement strategy is to extend the sunshade (cornice) above to enlarge the shading area. Trees were added on either side of the entrance hall to provide more shading and evaporative cooling.

Open Corridor (Points E and F). Horizontal sunshade devices were installed to reduce the effect of direct sunlight. For the open corridors on the east and west side, vertical sunshade devices were set up to reduce the influence of eastern and western sun exposure on its internal thermal environment. In addition, these sunshade devices were combined with water spraying, which could decrease the surface temperature of sunshade devices to a large extent.

3.3 Simulation Parameter Setting and Results

A thermal simulation tool (ThermoRender) [11] was used to calculate MRT distribution in the study area. Figure 5(a) is the 3D model of the study area. The same locations with the six measurement points were selected for simulation analysis. Meteorological data on a typical summer day in Nanning was used in the simulation, as shown in Fig. 5(b). The peak of outdoor air temperatures is above 33 °C. And the maximum intensity of solar radiation exceeds 800 W/m^2. Moreover, the average of wind speed is less than 1 m/s. Material parameters used in the simulation were set according to the actual condition. The evaporation of permeable pavement was set as 1.3 kg/(m$^2 \cdot$ d). In the simulation, the sunshade devices were maintained at the wet state by water spraying. This improvement strategy has obvious effect on humidity which is one of the factors affecting thermal comfort. However, this study neglected the humidity variation by water spraying with low intensity as much as possible.

Additionally, the width of the horizontal sunshade was set as 1.5 m, at a height of 3 m above the ground. The vertical sunshade was a shading grid with a size of 1×3 m (width $\times$ height) and its interval was 1 m. The sun canopy was 6 m above the ground. The height and crown diameter of the tree were all 6 m, and the tree-planting interval was 5 m.

MRT values at 1.5 m above the ground were used to analyze the improvement effect of the proposed strategies for the measurement locations. The comparison analysis of MRT variation between measurement results and simulation results was conducted before discussing the improvement effect. The results showed that there were different MRT variation trends among them. Nevertheless, the numerical simulation in this study aims to analyze the improvement effect of the proposed strategies by comparing simulation results with and without improvement strategies, the accuracy of simulation results is not discussed in this study.

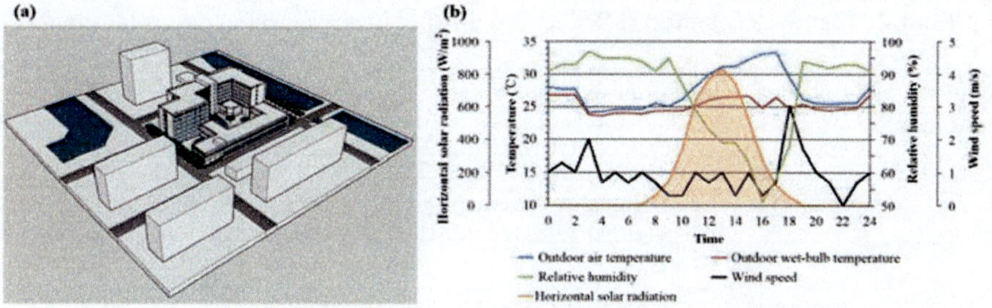

Fig. 5. (a) 3D model of the study area, (b) Meteorological data used in the simulation.

Hourly variations of MRT with and without improvement strategies are shown in Fig. 6. The average values of MRT from 9:00 to 15:00 for the six locations were used for comparison analysis, as shown in Table 2.

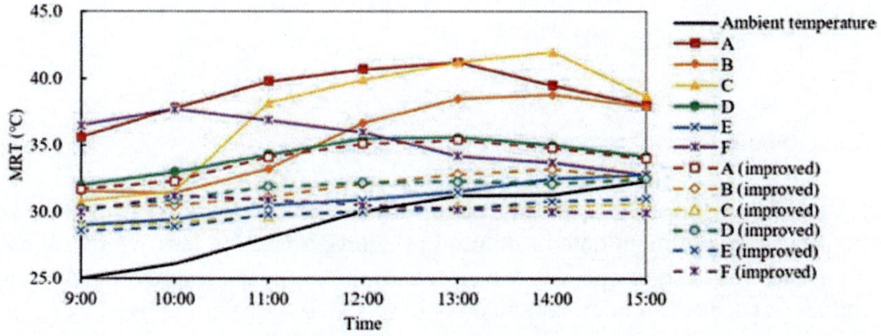

Fig. 6. Hourly variations of MRT in the six locations with and without improvement strategies.

In general, the thermal environment in the six locations has been effectively improved by the proposed strategies. The average MRT values in the sidewalk, courtyard, badminton court, entrance hall, west open corridor and east open corridor decreased by 5.0 °C, 3.6 °C, 7.6 °C, 2.5 °C, 1.0 °C and 5.1 °C, respectively. The maximum reduction of the MRT was found in the badminton court. The reason for this result was that the thermal environment was seriously hot before improvement, while the improvement effect of the sun canopy was obvious. The same improvement strategies were applied to the courtyard and sidewalk. The improvement effect was more obvious in the sidewalk due to more spaces for tree-planting. In the entrance hall, it is difficult to adopt more strategies due to the limitations of its location and function, thus the improvement effect was the least obvious. By comparing MRT reduction in the west open corridor and east open corridor, the sunshade was an effective measure to improve the thermal environment in the spaces exposed to solar radiation for a long time.

Various shading strategies were adopted in the targeted spaces, and the MRT reduction indicated their obvious improvement effects. Thus, reducing solar heat in these spaces is the most effective approach to improve outdoor thermal environment in the hot-humid and less-windy climate.

Table 2. Improvement effect (MRT reduction) after improvement in the six locations.

Analysis point	Location	Improvement strategy	Average MRT without improvement	MRT reduction
A	Sidewalk	Permeable pavement, trees, vertical greening of external wall	38.9 °C	5.0 °C
B	Courtyard	Permeable pavement, trees, vertical greening of external wall	35.4 °C	3.6 °C
C	Badminton court	Trees, sun canopy	37.5 °C	7.6 °C
D	Entrance hall	Sunshade extension, trees	34.2 °C	2.5 °C
E (West)	Open corridor	Horizontal and vertical sunshade, water spraying equipment	30.9 °C	1.0 °C
F (East)			35.4 °C	5.1 °C

4 Conclusion

This paper analyzed the outdoor thermal environment around an enclosed teaching building using field measurement data monitored in a hot-humid and less-windy climate. A simulation tool was used to quantify the effects of the proposed improvement strategies in the outdoor spaces. The main conclusions in this study are summarized as follows:

(a) The summertime thermal environment in the measured outdoor spaces was severe, and their average air temperatures and MRT were higher than the ambient temperature by about 1 °C and 6 °C, respectively.
(b) In the study area with hot-humid and less-windy climate, the surface temperature is the main factor affecting the formation of the thermal environment in outdoor and semi-outdoor spaces. The improvement strategies for reducing the surface temperatures in the measured spaces were proposed, including sunshade devices, greening and water spray cooling.
(c) The proposed improvement strategies effectively improved the thermal environment in the outdoor spaces, and the MRT value can be reduced by 1.0 °C to 5.1 °C on average.
(d) Solar shading is the most effective approach to improve summertime outdoor thermal environment in the hot-humid and less-windy climate.

References

1. Marchand, G.C., Nardi, N.M., Reynolds, D., et al.: The impact of the classroom built environment on student perceptions and learning. J. Environ. Psychol. **40**, 187–197 (2014)

2. Wang, D., Xu, Y., Liu, Y., et al.: Experimental investigation of the effect of indoor air temperature on students' learning performance under the summer conditions in China. Build. Environ. **140**, 140–152 (2018)
3. Akanmu, W.P., Nunayon, S.S., Eboson, U.C.: Indoor environmental quality (IEQ) assessment of Nigerian university libraries: a pilot study. Energy Built Environ. **2**, 302–314 (2020)
4. Zomorodian, Z.S., Tahsildoost, M., Hafezi, M.: Thermal comfort in educational buildings: a review article. Renew. Sustain. Energy Rev. **59**, 895–906 (2016)
5. Teli, D., Jentsch, M.F., James, P.A.B.: Naturally ventilated classrooms: an assessment of existing comfort models for predicting the thermal sensation and preference of primary school children. Energy Build. **53**, 166–182 (2012)
6. Heracleous, C., Michael, A.: Thermal comfort models and perception of users in free-running school buildings of East-Mediterranean region. Energy Build. **215**, 109912 (2020)
7. Huang, K., Huang, W., Lin, T., et al.: Implementation of green building specification credits for better thermal conditions in naturally ventilated school buildings. Build. Environ. **86**, 141–150 (2015)
8. Park, J.H., Yun, B.Y., Chang, S.J., et al.: Impact of a passive retrofit shading system on educational building to improve thermal comfort and energy consumption. Energy Build. **216**, 109930 (2020)
9. Synnefa, A., Saliari, M., Santamouris, M.: Experimental and numerical assessment of the impact of increased roof reflectance on a school building in Athens. Energy Build. **55**, 7–15 (2012)
10. Lin, Y., Zhao, L., Yang, W., et al.: A review on research and development of passive building in China. J. Build. Eng. **42**, 102509 (2021)
11. Asawa, T., Hoyano, A., Nakaohkubo, K.: Thermal design tool for outdoor spaces based on heat balance simulation using a 3D-CAD system. Build. Environ. **43**, 2112–2123 (2008)

Check for updates

Accounting for Carbon Emissions During the Building Phase of Academic Buildings

Jie Gao[1] and Shenqi Gan[2(✉)]

[1] School of Architecture and Planning, Anhui University of Architecture, Hefei 230601, China
[2] Key Laboratory of Built Environment and Health, Anhui University of Architecture, Hefei 230601, China
46726519@qq.com

Abstract. This work adopts the carbon emission accounting model developed for the teaching building in the buildingization stage and accounts for the carbon emission of the building in accordance with the construction list, using the construction of a concrete structure teaching building as an example. The calculated results indicate that the manufacture of building materials is where the majority of the carbon emission in the buildingization stage occurs, with less carbon emission occurring during the building's construction and transportation phases. The research also suggests three measures for reducing carbon emissions during the building phase: maximizing the use of building materials, minimizing travel distances, and improving construction techniques.

Keywords: Physicochemical stage · Carbon emissions · Academic building buildings · Optimize the strategy

1 Quotes

Due to the increasing climate change awareness and the need for energy-saving measures, various fields in China are now facing the challenges of implementing sustainable development. In 2020, a report released by committee of the buildings for saving energy in Chinese society revealed that the country's construction industry contributed over 50% to the country's total carbon emissions [1]. In China, several types of school buildings totaling more than 500,000,000 were built in 2021. Over 8 million square meters of floor space was available in the nation's elementary schools, an increase of 25 million square meters from the previous year. There will inevitably be an increase in carbon emissions related to building construction and operation as school floor space expands [2]. The life cycle of a building encompasses a number of stages, including the production of building materials, the transportation of building parts, and the upkeep and renovation of the structure. The building's demolition and scrapping also contribute significantly to the greenhouse gas emissions.

In this essay, carbon dioxide outputs from the building process of a primary school in Hefei, Anhui Province, are combined with a demonstration study. This study presents a more practical technical guideline for the measurement of the building's (structure's)

W. Guo et al. (Eds.): GBCESC 2023, LNCE 328, pp. 520–528, 2024.
https://doi.org/10.1007/978-981-99-9947-7_54

emissions of carbon and examines the method of measuring carbon emissions throughout the concrete building's construction phase of the teaching building.

2 Research Methodology

2.1 Carbon Emission Measurement Methods and Standards

The following techniques are most frequently used to study carbon emissions in the construction sector: measurement, mass balance, emission factor, life cycle assessment, etc. See Table 1. According to this study's emission factor technique, the emission factor methodology offers the benefit of ensuring the most precise measurement of carbon emissions [3, 4].

In this paper, we will explore the analysis of carbon emission sources in the construction phase of primary school buildings and calculate the carbon emission of primary school buildings in the materialization phase based on the emission factor method. The specific building materials used for construction and the use of machinery and equipment are mainly sourced from the itemization bill of quantities tariff. The main accounting standard selected for this study is the Standard for Calculation of Carbon Emissions from Buildings of the People's Republic of China (GB/T 51366-2019).

2.2 Analysis of Carbon Emission Sources

Carbon emissions from the manufacture of building materials, carbon emissions from their transportation, and carbon emissions from building operations are all included in the calculation of carbon emissions during the physical phase of construction [5]. Carbon emissions from the production, manufacture, and processing of building materials as well as carbon emissions from energy consumption during the transportation of building materials from the manufacturer to the construction site are sources of carbon emissions.

2.3 Inventory Analysis

The main function of inventory analysis is to quantify the inputs and outputs (including material and energy flows) of all unit processes within the system boundary, which in turn supports the measurement of carbon emissions. For the purpose of inventory analysis, this paper collects on-site data and emission factors for the physical phase of the building. The on-site data includes the usage of foundation materials (masonry, commercial concrete, steel reinforcement, etc.), the usage of construction machinery and equipment, and the working hours of site personnel. The validity of on-site data is closely related to the quality of engineering data acquisition, so attention should be paid to the reliability of each engineering data record, and on-site data should be obtained from first-hand engineering records (such as labour ledger, material ledger, energy ledger, etc.) as far as possible.

3 Case studies of School Building Architecture

The subject of this paper is the main teaching building of Lieshan Road Primary School, the overall structure of which is a concrete frame and frame structure. The total site area of the building is 44,877.7 square meters (approximately 67.32mu) and the total construction area is 29,782.1 square meters, of which 26455.7square meters is above ground and 3326.4 square meters is underground.

Table 1. Comparative analysis of carbon emission accounting methods

Calculation method	Advantages	Disadvantages	Application
Actual measurement method	Measured carbon emissions The most realistic and accurate	High demands on experimental conditions and data collection and processing High methodological requirements and high sample influence	Reliable data available and high level technology used countries or regions that have reliable data and use advanced technology
Mass balance method	A systematic and comprehensive study of the production of carbon emissions in the production process	Difficult activity data when chemical composition is complex High sorting costs for sorting and testing	Generally used in production processes with very small amounts of non-carbon dioxide containing gas emissions and accurate and reliable analysis of the chemical composition of this energy material
Emission factor method	Convenient, easy-to-access data	The carbon emission factor varies considerably with different levels of technology, production conditions, energy use, processes, etc. and there is uncertainty	Countries or regions and companies with less detailed statistics are now widely used to calculate and evaluate carbon emissions from different perspectives in different areas

(*continued*)

Table 1. (*continued*)

Calculation method	Advantages	Disadvantages	Application
Life Cycle Assessment Method	Measurement results Widely used by scholars	Relatively high data requirements, study subjects Data availability should be available over the life cycle	Apply to emissions of carbon across the whole product life cycle, from manufacturing to use to disposal, accounting for all resource and energy use as well as any environmental effects

4 Modelling and Measurement of Carbon Dioxide Emissions Produced While Constructing an Academic Building

The carbon emissions from the physical process of building construction are added up in three steps, in chronological order: the carbon emissions from the production phase of building materials, the carbon emissions from the transport phase of building materials an the carbon emissions from the construction phase of the building.

$$C = C1 + C2 + C3 \tag{1}$$

Where: C stands for carbon emissions during the building phase; C_1: for carbon emissions from the manufacture and transport of building materials; C_2: Carbon emissions from building material transportation; C_3: Carbon emissions from building construction.

The next step in the paper is a historical, step-by-step calculation of the carbon emissions during this building phase. The paper then proceeds to calculate the carbon emissions for this building phase in a chronological step-by-step manner.

4.1 Modelling and Measurement of Carbon Emissions Produced While Building Materials are Being Produced

In this paper's calculation of carbon emissions from the production phase of building construction materials, two issues need to be clarified in advance:

1. The building materials items given in the construction bill of quantities obtained in this paper are partly "ramp surface layer", "step connection platform layer", "fine stone concrete leveling layer" and so on. Although the specific amount of these building materials is given, such as the weight of the item or the area of the item, the exact proportion of the components is not known, so it is not possible to measure the carbon emissions of these building materials.
2. The concrete form work and bracing items in the building materials inventory do not give the specific composition of the components, and it is also not possible to determine their mass or volume, resulting in the carbon emissions of such building materials not being included in the accounting of carbon emissions.

Despite the incompleteness of the calculation results due to the above points, all building materials that could be calculated or approximated in the calculation process were included in the carbon emissions calculation, where the building materials that were not accounted for were used in very small quantities, not in major quantities, and their carbon emissions were almost negligible compared to the other building materials of the whole building. As a result, the carbon emission accounting results of the building material production phase in this study are more thorough. Based on the itemized bill of quantities tariff, the carbon emissions from the manufacturing process of the building materials for the school building were determined and are displayed in Table 2.

4.2 Modelling and the Shipping of Building Materials Results in Carbon Emissions

The following formula is used to determine the amount of carbon emissions generated during the transport of building materials;

$$C_{ys} = \sum\nolimits_{i=1}^{n} M_i D_i T_i \tag{2}$$

where C_{ys} -Carbon emissions from the transportation of building materials(kgCO2e). Mi - consumption of the ith major building material (t). Di -Average transport distance of the ith building material (km). Ti-Carbon emission factor per unit weight of transport distance for transport mode of the ith building material [kgCO2e/(t · km)].

The default transport length is given in the Standard for Calculating Carbon Emissions from Buildings (GB/T 51366-2019), where concrete is 40 km and the default transport length for other building materials is 500 km [5]. The carbon emission factors for each type of transport are obtained according to the carbon emission factors for each type of transport given in the Standard for Calculating Carbon Emissions from Buildings (GB/T 51366-2019) and calculated Table 3 is obtained.

Table 2. Carbon emissions produced during the building materials production process for academic buildings

Category	Name of material	Carbon emission factor	Number	Carbon emissions (kgCO2e)
Masonry	Solid bricks with coal gangue	22.8 kgCO2 e/m3	139.96m3	3,191.09
	Coal gangue hollow bricks	16kgCO2 e/m3	1,656m3	26,496
Commercial concrete	Commercial concrete C15	178kgCO2 e/m3	433.51m3	77164.78
	Commercial concrete C20	265kgCO2 e/m3	961.35m3	254,757.75

(continued)

Table 2. (*continued*)

Category	Name of material	Carbon emission factor	Number	Carbon emissions (kgCO2e)
	Commercial concrete C25	293kgCO2 e/m3	1529.82m3	448237.26
	Commercial concrete C30	316kgCO2 e/m3	10,009.02m3	3,162,850.32
Sand and gravel	Sand	3.49 kgCO2 e/m3	1652.6m3	5767.57
	Gravel	4.4kgCO2 e/t	310.28t	1365.23
Lime	Quicklime	1190kgCO2 e/t	42t	49980
Steel reinforcement	Φ10 Internal and external reinforcement	2137kgCO2 e/t	735.21t	1571143.77
	Steel bars over Φ10	2140kgCO2 e/t	1072.2t	2294508
Timber	Plywood	1160kgCO2 e/m3	77.7m3	90,132
Doors and windows	Grade A Wooden Fire Doors	48.28 kgCO2 e/m2	9.66m2	466.47
	Class C Wooden Fire Doors	35.12 kgCO2 e/m2	74.76m2	2625.57
	Aluminium fixed windows	46.3kgCO2 e/m2	2,388.39m2	110,582.45
Total				8099268.26

4.3 Modelling and Carbon Emissions During the Construction Phase of a Building

Carbon emissions from the construction phase of the building are modeled as follows:

$$C_{\text{build}} = \sum\nolimits_{i=1}^{n} E_{jii} EF_i + E_{artificial} \tag{3}$$

where: C_{build} – carbon emissions during the construction phase ($kgCO_2$);

$\sum_{i=1}^{n} E_{jii}$ – Total use of the ith energy source, (kW · h or kg);

EF_i – Carbon emission factor for the ith energy source, determined as per GB/T 51366-2019 Appendix A, ($kgCO_2$ /kW·h);

$E_{\text{artificial}}$ – Artificial carbon emissions, ($kgCO_2$ e).

Table 4 is based on the use of machinery and the work of personnel on the building construction site.

Table 3. Carbon emissions from the transport phase of construction materials

Construction materials	Actual consumption (t)	Distance (km)	Mode of transport	Carbon emission factor [kgCO2e/(t-km)	Carbon emissions (kgCO2 e)
Solid bricks with coal gangue	181.948t	500	Heavy duty (10t load)	0.104	9461.29
Coal gangue hollow bricks	1821.6t	500	Heavy duty (10t load)	0.104	94723.2
Commercial concrete	2503.76t	40	Heavy duty (18t load)	0.104	10415.64
Sand	2478.90t	500	Medium (8t load)	0.115	142536.75
Gravel	310.28t	500	Heavy duty (10t load)	0.104	16134.56
Quicklime	42t	500	Heavy duty (10t load)	1.104	552
Φ10 internal and external reinforcement	735.21t	500	Heavy duty (30t load)	0.078	28673.19
Steel bars over Φ10	1072.2t	500	Heavy duty (30t load)	0.078	41815.8
Plywood	54.390t	500	Heavy duty (30t load)	0.078	2121.21
Grade A Wooden Fire Doors	0.3381t	500	Heavy duty (10t load)	0.104	17.58
Class C Wooden Fire Doors	1.4952t	500	Heavy duty (10t load)	0.104	77.75
Aluminium fixed windows	6448.653t	500	Heavy duty (10t load)	0.104	335329.95
Total					681858.92

5 Conclusion

The production phase of building materials emits the most carbon (92.0%), the transportation phase of building materials emits 7.9% of the physical phase, and the construction phase of the building emits the least carbon (0.1% of the total carbon emissions of the physical phase), according to the carbon emissions of the physical phase of concrete teaching buildings measured in this paper.

The physical phase of the concrete teaching building's findings demonstrate that the building's primary method for reducing carbon dioxide emissions is the manufacture of building materials, which is closely related to the shape of the building's structure. The

Table 4. Carbon emissions during the construction phase of a building

Category	Dosage	Carbon emission factors	Carbon emissions (kgCO2 e)
Concrete smoothing machines	0.13 units	23.14 kW-h	3
Concrete transfer pumps	0.7 shifts	243.46kW-h	170.42
Steel processing equipment	0.82 units	32.1kW-h	26.32
Water	0.923m3	0.168kgCO2e/t	0.15
Artificial	10.425 man-days/m3	19.18 kg CO2 (d/person)	199.9515
Total			399.8415

book suggests three methods for the building to reduce its carbon footprint based on its materialization phase:

5.1 Maximizing the Use of Building Materials

The nationalization of building construction is an effective way of increasing the mutilation of building materials and reducing the carbon footprint in the building design process [6]. The rational and economic use of building materials reduces the carbon footprint of buildings through the use of material-saving features in the production of components. At the same time, the use of green building materials is increased and as many recycled building materials as possible are used.

5.2 Minimizing Travel Distances

In the course of delivery, including the second vertical delivery and a horizontal delivery. Horizontal transportation should take into account various factors such as transportation means, transportation paths and transportation timing, and design a scientific and reasonable plan to reduce unnecessary secondary transportation. Vertical transportation requires reasonable configuration of transport, reduction of lifting operations and optimization of spreader configuration.

5.3 Improving Construction Techniques

There are many types of risk factors for low-carbon pollution during construction and it is difficult to deal with them, so low-carbon construction has been considered from the following aspects: 1. Adopting a fully enclosed construction method can effectively reduce the ecological environment problems caused by the dust raised during construction, thus diminishing the environmental hazards to the surrounding residents [7]; 2.

Time, thus reducing the construction wastage rate; 3. Strengthen the supervision of the construction site, improve the rules and regulations, and enhance the ability of managers and construction personnel to save and protect the environment; 4. By introducing BIM technology and integrating it into the site management of infrastructure construction. It enables simulation control and fault point monitoring of construction projects and effectively achieves prior control, thus effectively reducing the risks in terms of raw materials, personnel and time due to rework issues. In addition, through BIM technology by optimizing the construction schedule and providing optimized material entry solutions, by improving the selection of construction equipment and reducing. Construction material accumulation, which in turn reduces construction material consumption as well as CO_2 emissions.

Acknowledgement. The work was supported in part by Anhui Province Housing and Urban-Rural Construction Science and Technology Program Project "Research and Application of Prefabricated Building Information Management Platform Based on BIM Technology" (2022-YF163); Anhui Provincial Department of Education Natural Science Research Key Project "Research and Application of Prefabricated Building Maintenance and Update Information Technology Based on BIM Technology" (KJ2021A0613).

References

1. China Building Energy Consumption Research Report 2020. Building Energy Efficiency (in English and Chinese) 49(2), 1–6 (2021)
2. Yueyan, L., Kai, Z., Jinlu, L.: Comparative analysis and carbon reduction strategies for the whole life cycle of residential buildings. J. Xi'an Univ. Archit. Technol. (Nat. Sci. Edit.) (005), 053 (2021)
3. Caiyun, M., Lin, X., Biangyu, L., et al.: Prediction of carbon emissions from construction industry in Anhui Province based on gray prediction model. J. Nantong Univ. Nat. Sci. Edit. **18**(1), 6 (2019)
4. Zhao, Y., Liu, L., Yu, M.: Comparison and analysis of carbon emissions of traditional, prefabricated, and green material buildings in materialization stage. J. Clean. Prod. **406**, 137152 (2023)
5. Zhenggen, F., Jun, H.: Research on carbon emission calculation methods in the construction phase of buildings. Constr. Econ. (11), 4 (2013)
6. Junxiao, W., Yuanbo, G., Song, W.: Analysis of influencing factors and uncertainty calculation of carbon emission measurement of cement in China. J. Environ. Sci. **36**(11), 11 (2016)
7. En, L.K., Rahiman, N.A., Othman, N., et al.: Quantification process of carbon emissions in the construction industry. Energy Build. 113025 (2023)

Research on Measurement and Optimization of an Old Building in Wuxi Based on Ultra-low-energy Consumption and Energy Saving Transformation

Wei Zhang[1,2], Jie Wu[1,2](✉), Jinghua Shen[1,2], and Zhijun Xue[1,2]

[1] School of Architecture, Soochow University, Suzhou, China
jwu@suda.edu.cn
[2] International Research Center for Green Building of Soochow University, Suzhou, China

Abstract. In order to improve the indoor heat and humidity environment of existing residences and reduce the total energy consumption of buildings, this paper takes an ultra-low-energy consumption reconstructed building in Wuxi as an example, and studies the influence of ultra-low-energy consumption reconstruction on the indoor heat and humidity environment in different spaces and the total energy consumption of heating and cooling through quantitative measurement and analysis and simulation verification with software. The analysis shows that the change range and fluctuation frequency of indoor heat and humidity environment can be reduced by increasing the thermal insulation performance of the external envelope, improving the air tightness of the building and reducing the thermal bridges, especially for the south room, and at the same time, the local heat accumulation in the room can be improved and the cooling and heating load of the building can be reduced.

Keywords: Existing buildings · Energy-saving renovation · Ultra-low-energy consumption · Indoor heat and humidity environment · Energy consumption

1 Introduction

On December 28th, 2022, the report "Series Research Report on Carbon Emission in Urban and Rural Construction in China in 2022" issued by China Building Energy Conservation Association pointed out that the total energy consumption of buildings and construction in China will reach 2.27 billion tce in 2020, accounting for 45.5% of the total energy consumption in China [1]. Among the total energy consumption in the operation stage, according to statistics, the energy consumption of building heating, air-conditioning and lighting accounts for more than 70%, so improving the indoor heat and humidity environment and improving the comfort of indoor personnel is the main strategy to reduce the operating energy consumption. Li Nianping [2] did an experimental research on the influencing factors of indoor heat environment of residence, and the results showed that the heat environment of residence was mainly influenced by the

W. Guo et al. (Eds.): GBCESC 2023, LNCE 328, pp. 529–539, 2024.
https://doi.org/10.1007/978-981-99-9947-7_55

building itself and building groups. Li Jianghe [3] researched the indoor heat environment of ultra-low-energy consumption residence under different ventilation modes, and analyzed the cross simulation experiments under nine air supply temperatures and wind speeds. Generally speaking, the research and thinking about indoor thermal comfort can be summarized as: measured investigation-analysis of influencing factors-evaluation of heat and humidity environment-optimization strategy. There is a lack of complete measurement and verification of the real improvement benefits before and after optimization. Therefore, this paper selects the ultra-low- energy consumption and energy-saving reconstructed residence (No.53 residence) built around 2000 and another residence (No.51 residence) which is basically the same as that before the reconstruction of No.53 in Dongquan Garden Community of Wuxi as the measured objects. Under the same test conditions and corresponding test spaces, we continuously monitor their indoor environmental data, then use DesignBuilder software to simulate annual energy consumption indicators, comprehensively analyze the improvement benefits of ultra-low-energy consumption renovation on different types of indoor spatial environments, and hope to provide reference for energy-saving renovation of residential buildings in southern Jiangsu.

2 Case Background and Test plan

Dongquan Huayuan Community is located on the west side of Xinan Road in Binhu District, Wuxi City. Both buildings face south, with similar locations. Before the renovation, No. 53 residence and No. 51 residence were built in the same batch of buildings in the community, and their structure, materials, and layout, etc. are basically the same. The two houses have a total of three floors, with a construction area of about 480 m^2 and a rectangular plane.

Up to now, No.51 residence (Fig. 1) have not been rebuilt, and their shapes and structures are well preserved with brick-concrete structure. There is no thermal insulation and steam insulation layer in the exterior envelopes. The window adopts single-layer aluminum alloy window and the door is ordinary wooden door. Windows and doors have certain gaps when they are closed. No.53 residence (Fig. 2) were renovated according to the German passive housing PHI certification standards, and specific renovation measures are shown in Table 1. The building has been converted to a reinforced concrete frame structure. The plane retains the original layout.

The testing time is from 9:00 on September 19th to 9: 00 on September 23rd, 2022, and each floor is monitored for 2 days, with two floors in total. The measuring point is shown in Fig. 3. According to the Evaluation Standard for Indoor heat and humidityity Environment in Civil Buildings, an automatic temperature and humidity recorder is arranged at the diagonal midpoint of the room, with a height of 1.2 m. Using weather station to test outdoor environmental data synchronously.

Table 1. Energy-saving measures for ultra-low-energy consumption of No.51 residence

Renovation category	Specific practices	
residence	No.53	No.51
Roofing	The roof adopts the integrated design of 4mm thick Vidal waterproof material and 250 mm thick graphite polystyrene board thermal insulation material. The structural layer retains 120 mm reinforced concrete. The overall heat transfer coefficient after renovation is 0.12 $W/(m^2 \cdot k)$	Reinforced concrete roof without insulation layer
Exterior wall	The exterior wall insulation material is 170 mm thick graphite polystyrene board and 170 mm thick XPS board. Heat transfer coefficient of exterior walls after reconstruction is 0.16 $W/(m^2 \cdot k)$	Non-thermal insulation concrete external wall
Exterior window	Windows are PHI certified plastic steel windows Glass is three-glass and two-cavity coated LOW-E glass The joint of window frames is sealed with sealing material. Overall U value of window 0.8 $W/(m^2 \cdot k)$	Aluminum alloy window: single glass
Thermal insulation design of heat bridge	Rock wool board, airtight film, waterproof material, etc. are used to focus on the connection part of insulation layer, the connection part between external window and wall, the wall-penetrating part such as pipeline and sunshade device	None
Air tightness design	The building adopts German permanent airtight system, such as steam barrier film and sealant tape	None
HVAC system design	The building adopts fresh air heat pump multifunctional unit, and its total heat recovery rate reaches 81%	None

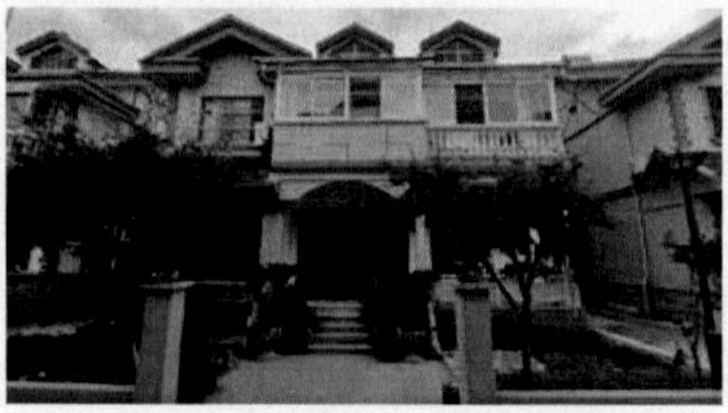

Fig. 1. No. 51 residence

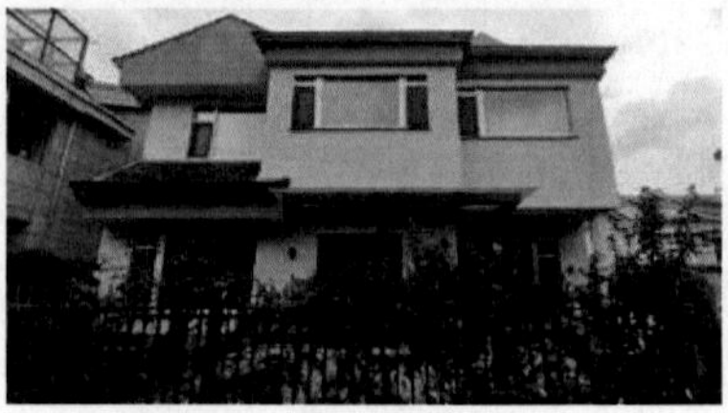

Fig. 2. No. 53 residence

3 Test Results and Analysis

3.1 Analysis of Outdoor Meteorological Environments

From Fig. 4, it can be seen that during the testing period in Binhu District, Wuxi City in September 2022, the temperature difference was about 11 °C, and the humidity difference was about 60%. Outdoor air temperature and relative humidity fluctuate greatly, and there is an obvious reverse fluctuation trend in numerical value. During the test, the average temperature was about 24 °C, and the highest temperature was 29.7 °C, which appeared at about 15:00 pm, which was also the time when the relative humidity was the lowest. The lowest temperature is 18.4 °C, which appears at around 6:00 am, which is also the time when the relative humidity is the highest. In the first three days of the measurement, the daily maximum temperature showed a downward trend, and the highest value was restored on the fourth day. On the sixth day, due to the influence of climate, the peak and valley values of temperature decreased as a whole. We can see from the figure that the outdoor temperature rises period in this area was slightly shorter than the temperature drop period, with a difference of 6h. Correspondingly, the relative humidity of the air showed a slower rise and faster decline trend. This is because after the summer solstice, the sunrise time becomes later and the sunset is earlier, which leads to a shorter solar radiation time throughout the day. Therefore, the temperature rises rapidly with the sunrise, and the solar altitude angle decreases in the afternoon, so that the temperature peaks at 15:00 and then drops until the early morning. It can be seen that during the test, the temperature in this area is high during the day, besides, it rises quickly and cools slowly. Secondly, when the temperature is low at night, the relative humidity is high and the humidity feeling is obvious.

3.2 Comparative Analysis of Heat and Humidity Environment of Energy-saving Renovation in Open Space with Ventilation

The test condition of the first floor of the building is the normal use of the two houses before and after the renovation. As shown in the Fig. 3, the tea rooms of the two buildings are located in the southwest corner of the building. The lobby is located in the middle of the two rooms facing north. During the test, 2–3 people were active from 9: 00 to 20: 00, and no one was there the rest of the time. 51 buildings open windows at 9–20 o'clock for natural ventilation and close windows at night. 51 buildings open the fresh air system during the day and close the windows all day.

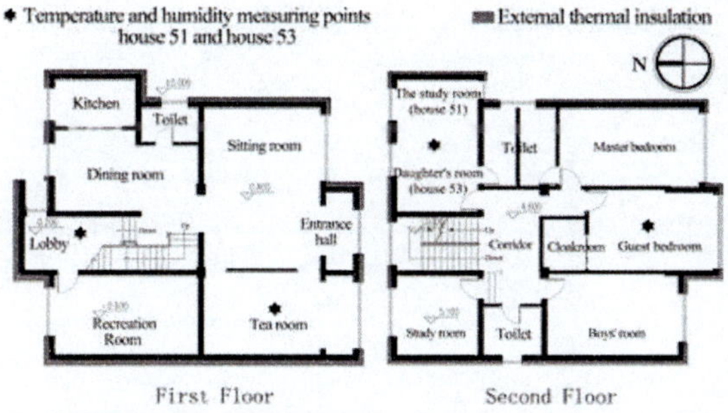

Fig. 3. Measuring points on each floor

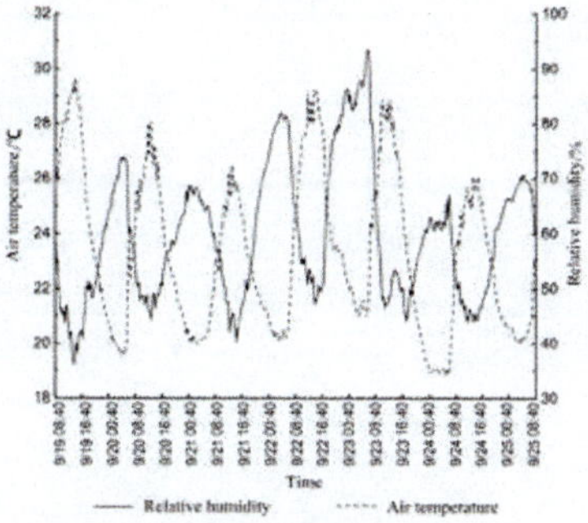

Fig. 4. Outdoor air temperature and relative humidity

Comparative Analysis of Heat Environment. As can be seen from Fig. 5, the average temperature of No.51 tearoom is 27.2 °C within 24 h during the test period, and that of the lobby is 27 °C. The average temperature of No.53 tea room is 24 °C, and that of the lobby is 24 °C. The temperature difference in the interior space of the two house is small. During the test of No.53 residence, the fresh air system on the first floor runs normally, and the indoor air supply outlet is designed evenly, so the temperature difference between the tea room and the lobby is basically zero. At the same time, during the test of No.51 residence, the inner door, south window and north window of the tea room are all open, so the temperature difference is not big. Because the tea room is located in the south, but the window is shaded by trees, its indoor temperature is only slightly higher than that of the north lobby. Compared with the first floor test room, in the open indoor space, the fresh air system or natural ventilation can effectively eliminate the interior space temperature difference and reduce the temperature variation range and fluctuation frequency. However, the indoor temperature of No.51 residence is still higher than the upper limit of summer comfort temperature in Code for Design of Heating, Ventilation and Air Conditioning [4], while the indoor temperature of No.53 residence is about 2 °C lower than the limit, which is more stable than that of No.51 residence.

Comparative Analysis of Humidity Environment. Air relative humidity is an important parameter that affects human comfort. GB/T 18049-2017 (heat environment analysis and determination) stipulates that indoor relative humidity should be set at 30% ~ 70% [5]. Taking RH70% as the upper limit of indoor humidity, the indoor humidity level is divided into four grades: no humidity ($RH \le 70\%$), slightly humid ($70\% < RH < 80\%$), humid ($80\% < RH < 90\%$), very humid ($RH > 90\%$) [6].

Figure 6 shows the relative humidity change curve of the first floor test space of two buildings. From the temperature and humidity curve of the first floor, it can be seen that the changing tendency of outdoor relative humidity is negatively correlated with the outdoor air temperature during overseason. When the outdoor air temperature reaches the maximum value, the relative humidity then reaches the minimum value, and vice versa. Because the first floor of No.51 residence is ventilated by windows in the daytime, the change trend of its relative humidity before 17:30 is influenced by

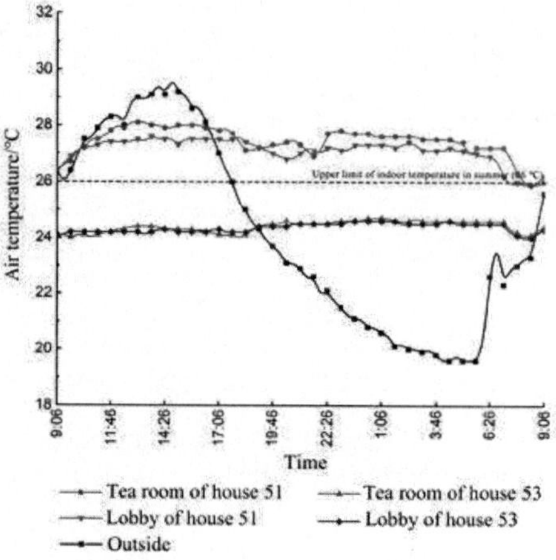

Fig. 5. Temperature changes on the first floor

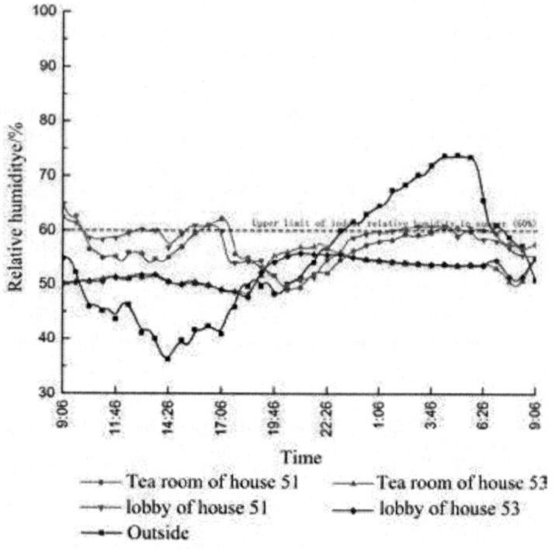

Fig. 6. Humidity Changes on the first floor

outdoor relative humidity, but the amplitude is smaller than that of outdoor relative humidity. After 17:30, the outdoor relative humidity continues to rise. Because the householder closes the doors and windows, the indoor and outdoor convection decreases, at the same time, the indoor walls and clothes absorb the moisture in the air, so the indoor relative humidity shows a short-term downward trend. But due to the continuous increase of outdoor relative humidity, No.51 residence have not been retrofitted with exterior envelopes and air tightness. However, the fresh air dehumidification system of No.53 residence was turned on before 17:30, so its relative humidity change curve was relatively stable. After 17:30, the householder turned off the fresh air dehumidification. Because the wall has certain humidity regulation and permeability, when the moisture in the air decreases, indoor moisture sources such as walls will release moisture into the air. After the dehumidification system is turned off, the water vapor content in the air will increase and gradually reach a new balance, and it will not be affected by the continuous increase of outdoor relative humidity. Therefore, the renovation of building insulation, air tightness and no thermal bridge can effectively maintain the stability of indoor humidity, secondly, the fresh air system can control the indoor humidity level to reach the ideal value. The buildings without thermal insulation and low air tightness can keep the temperature stable through natural ventilation, but at the same time, it will also lead to frequent fluctuations in humidity, and the change range is large.

3.3 Comparative Analysis of Heat and Humidity Environment of Energy-saving Renovation in Closed Space without Ventilation

Comparative Analysis of Heat Environment. The renovation of external walls mainly involves the renovation of external wall material, window, air tightness design and heat-less bridge design. In order to explore the thermal performance of buildings, the test condition of the second floor of the building is closed and unmanned. The guest bedrooms of the two buildings are located in the middle of the south direction of the building, and No.51 study room and No.53 daughter room are located in the northeast corner of the building.

According to the indoor temperature curves of the second floor's rooms of two houses shown in Fig. 7, the average indoor air temperature of the second floor of No.51 residence is 27 °C, that of the guest bedroom is 27.8 °C and that of the study room is 26.2 °C within 24 h of the test period. The average indoor air temperature of the second floor of No.53 residence is 23.8 °C, that of the guest bedroom is 23.7 °C, and that of the daughter room is 23.8 °C. Compared with the temperature curve of the first floor when the fresh air system is turned on, the temperature curve of No.53 rooms is more prone to fluctuation when the second floor only relies on the external wall, but the temperature difference is smaller and the indoor temperature distribution is more uniform. However, the average temperature of No.51 south guest bedroom is 1.6 °C higher than that of the north study. This is because the walls of No.51 and No.53 guest bedrooms are only the south wall, and the ratio of their windows to the external wall area is 0.65. The windows and their installation methods are the weak points of heat conduction, so the temperature of No.51 south guest bedroom with ordinary single glass aluminum alloy windows and five insulation walls is higher than that of the north study. The windows of 53 guest bedroom is external plastic-steel windows, and low-e coated glass with three glasses and two cavities is used to reduce the influence of direct sunlight, in addition, thermal insulation materials are used to reduce the generation of thermal bridges during installation. Meanwhile, the external wall is made of 170 mm graphite polystyrene board, so its indoor air temperature is 4.1 °C lower than that of 51 guest bedroom. The ratio of window to wall in the north test room of the two buildings is 0.4, and there is no window in the east, so the exterior envelopes is mainly the wall. In the case of the same room area, window to external wall area ratio and other factors, the temperature of No.53 daughter room is 2.4 °C lower than that of No.51 study room, and the temperature difference between the north test rooms of the two buildings is 1.7 °C lower than that of the south test rooms. It can be seen that the effect of south windows on indoor temperature is more considerable. From the above, it can be seen that in the closed indoor room, the low air tightness and non-thermal insulation residence will lead to a large temperature difference between the north and south rooms, although the frequency of temperature fluctuation decreases, the daytime variation amplitude increases. In contrast, the temperature difference between the north and south rooms will not occur in a short time after the renovation of the house in a closed environment, and the temperature change range is still small, but the fluctuation frequency is relatively increased.

Impact Analysis of Humidity Environment. Figure 8 shows the indoor relative humidity variation curves of the test space on the second floor of two buildings. It can be seen from the figure that Building 51 exceeded the standard limit for most of the day. During the period of high outdoor humidity at night, the indoor relative humidity was close to 70%, which was in the slightly humid range. The overall trend of the curve was the same as that of the outdoor, but the amplitude was different. This indicates that closing doors and windows can block most of the outdoor moisture, but compared to the curve of Building 53, it can be seen that, when the outdoor humidity rises slowly, the indoor temperature of two houses changes little. At this time, the humidity curve of Building 53 is relatively stable, while the humidity of Building 51 rises slowly, indicating that

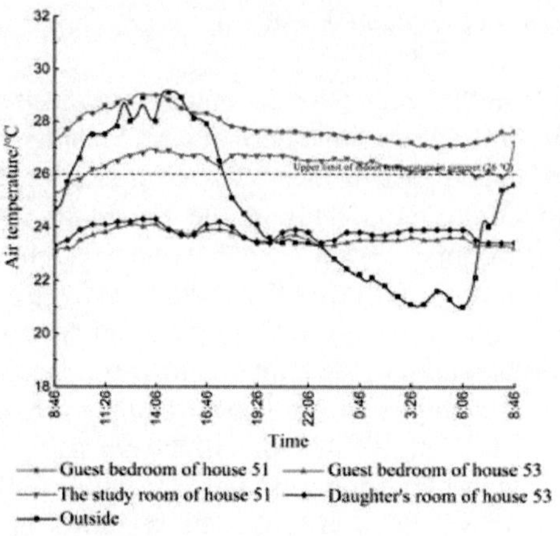

Fig. 7. Temperature changes on the second floor

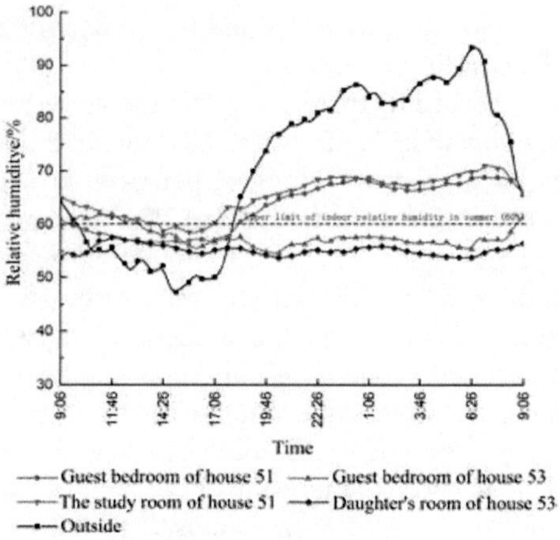

Fig. 8. Humidity Changes on the second floor

some water still seeps into the room. The indoor relative humidity of No.53 residence is within 52%~58% most of the time, with a small change range, while the humidity distribution of 51 buildings is relatively more dispersed, and more than half of the values are above the limit value of 60%. Therefore, in the indoor closed and unventilated environment, there is no significant difference in humidity between the north and south rooms. However, the humidity variation in unmodified residential buildings is greater. The renovation of external insulation and air tightness can effectively reduce the impact of outdoor humidity on the indoor environment.

4 Numerical Simulation and Analysis

4.1 Simulation Scheme

In order to further explore the improvement benefits of the renovation to the building, the following two houses before and after the renovation are numerically simulated in DesignBuilder. According to the original building information of the two houses, the annual energy consumption of two buildings under normal operation is simulated, and the simulation parameters are set as shown in Table 2, so as to analyze the influence of energy-saving renovation on energy consumption on the basis of maintaining indoor comfort. Finally, the indoor heat and humidity environment changes of No.53 residence with natural ventilation during the day and closed doors and windows at night were simulated during the test period (transition period from summer to autumn), among which 4 residents were designed, and the windows were opened from 9: 00 to 20: 00, while the rest of the time was closed, so as to explore whether the fresh air system of ultra-low energy-consumption buildings is suitable for intermittent operation and management in this transitional season, in this manner to further reduce energy consumption. This time, the indoor temperature changes of 51 buildings under natural ventilation are simulated (Fig. 10), and the results show that the error between the curves and the measured curves is small, which indicates that the simulation results are accurate to some extent.

Table 2. Annual energy consumption simulation parameters for residential buildings

Simulated object	No.51 residence	No.53 residence
Device type	Unit air conditioning system	Heat recovery fresh air dehumidification integrated machine
Air distribution mode	Upper air supply and lower air return	Side air supply top return air
Equipment parameters	Cooling power is 2.5 kw Heating power is 3 kw	Cooling power is 2.7kw Heating power is 2.9kw The total heat recovery rate is ≥78%
Indoor design temperature	26 °C in summer 20 °C in winter	26 °C in summer 20 °C in winter
Number of users	4 people	4 people

4.2 Analysis of Simulation Results

Comparative Analysis of Annual Energy Consumption. In addition, the annual energy consumption data of two buildings are calculated in DesignBuilder. As shown in Fig. 9, the annual energy consumption per unit area of No.51 residence is 45.79 kW · h/m^3, the annual cold consumption per unit area is 73.38 kW · h/m^2, and the annual heat consumption per unit area is 31.92 kW · h/m^2, which is far from meeting the "Design Standard for Heat Environment and Energy Conservation of Residential Buildings in Jiangsu Province". The annual power consumption per unit area of No.53 residence is 16.46 kW · h/m^2, which is 29.33 kW · h/m^2 less than that of No.51. The annual cold consumption per unit area is 29.78 kW h/m^2, which is 43.6 kW h/m^2 lower than that of No.51 residence, and the heat consumption per unit area is 7.83 kW · h/m^2, which is 24.09 KW · h/m^2 lower than that of No.51 residence. It can be seen that the total annual power consumption of No.53 residence is reduced by about 2/3 after the energy-saving renovation, and the energy-saving effect is more significant in winter. The heating demand is reduced by 3/4 and the cooling demand is reduced by about 3/5 compared with that before the renovation.

Simulation Analysis of Natural Ventilation in Ultra-low-energy Consumption Residence. No.53 windows are set to open during the day and close at night, and there is no other ventilation mode. As can be seen from Fig. 10, among the three modes, the indoor temperature of No.53 residence with fresh air system is lower than the summer limit all day, and that of No.51 residence with natural ventilation is higher than the limit all day, while that of No.53 residence with natural ventilation only exceeds the limit by about 0.6 °C from 13: 00 to 16: 30 and the rest of the time meets the comfort requirements. Different from the fresh air system, the indoor temperature rises slowly with the increase of outdoor temperature during the day, and the indoor temperature fluctuates slightly after the doors and windows are closed at night. From the foregoing, we can see that No.51 residence with natural ventilation have improved the indoor heat environment to

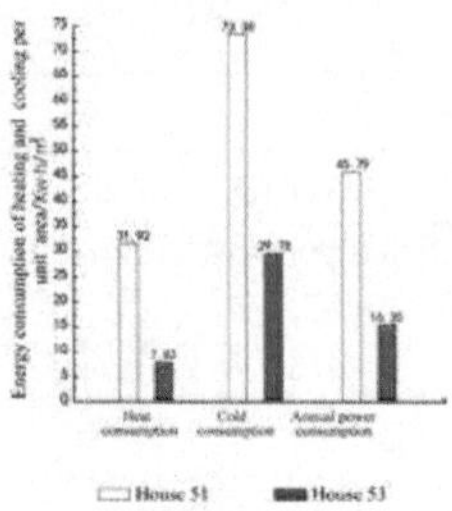

Fig. 9. Annual energy consumption

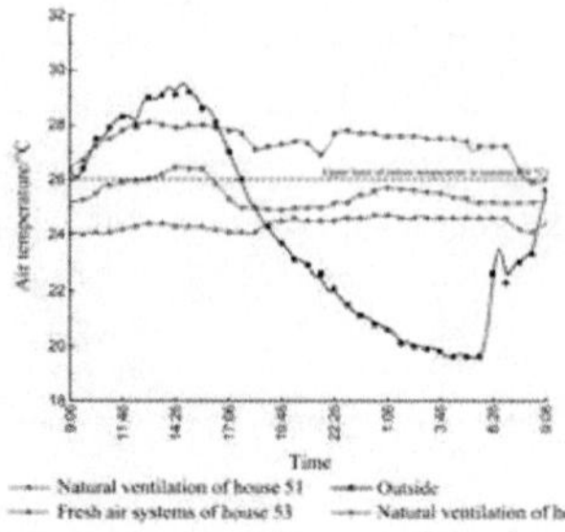

Fig. 10. Temperature changes

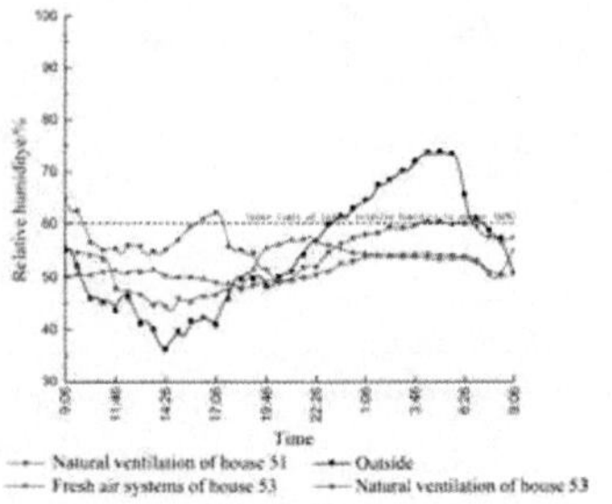

Fig. 11. Humidity changes

a certain extent, but at the same time, they have also increased the fluctuation of indoor humidity environment. From Fig. 11, we can see that the variation range and fluctuation frequency of indoor relative humidity in No.53 residence with natural ventilation are less than No.51 residence. Compared with the fresh air system, although the indoor relative humidity fluctuates more during the day, it is still below the limit, and it slowly returns to a steady state after closing the doors and windows. Therefore, under the condition of natural ventilation from summer to autumn, compared with No.51 residence, the indoor temperature of the reconstructed No.53 residence is lower, and the relative humidity is kept within the comfortable range. Only when the outdoor temperature is high in the afternoon, the indoor temperature exceeds the comfort limit. Therefore, it can be inferred that the ultra-low-energy consumption building can open the doors and windows when the sun goes down, store the cold energy through natural ventilation, and close the doors and windows to open the air-conditioning system when the outdoor temperature is high from 12: 00 to 17: 00, so as to maintain the comfortable indoor temperature and reduce the energy consumption of mechanical operation.

5 Conclusion

Through the measurement and simulation analysis of indoor heat and humidity environment of No.51 and No.53 residence, the following conclusions can be drawn:

1. Natural ventilation in No.51 residence can reduce the problem of excessive local indoor temperature, but the temperature fluctuates easily and the relative humidity changes unsteadily. Compared with No.53 residence, its temperature and humidity fluctuate more frequently and the indoor thermal sensation is stronger.
2. When the test space is a closed room without ventilation, the indoor humidity in the north-south direction of the 51 houses is not much different, but the temperature difference is large, and the thermal sensation is strong all day. No.53 residence reduces the indoor temperature distribution difference and the temperature difference between day and night, meanwhile, it reduces the variation range of indoor relative humidity compared with the 51 houses, but the temperature and humidity curve is more prone to fluctuation compared with the fresh air opening state. The upgrading effect of the south-facing room is higher than that of the north room, and the transparent envelope is the focus of the south room renovation.

3. The annual power consumption of No.53 residence is 2/3 lower than that of No.51 residence, and the reduction of heating demand is higher than that of cooling demand. In the summer to autumn transition season, the ultra-low-energy consumption residence can adopt intermittent operation management to reduce the energy consumption of building operation under the premise of maintaining the indoor heat and humidity environment in a comfortable range.
4. Comparing the data before and after the renovation, it shows that the renovation of the thermal insulation and air tightness of the envelopes can effectively improve the indoor heat and humidity environment and reduce the energy consumption.

Acknowledgments. This writing and research of this paper are obtained from the National Key R&D Program of China (2021YFE0200100) and the 2021 Jiangsu Construction System Science and Technology Project (2021ZD03).

References

1. Report of China Building Energy Conservation Association: 2022 Series Research Report on Carbon Emissions in Urban and Rural Construction in China, Database of Energy Consumption and Carbon Emissions in Urban and Rural Construction, 30 December 2022. https://www.cabee.org/site/content/24420.html. Accessed 30 Dec 2022
2. Li, N.P., Liu, P.L., Wu, Z.B.: Field research on thermal comfort and thermal adaptation of Changsha university dormitory in summer. J. Harbin Inst. Technol. **51**(04), 194–200 (2019)
3. Li, J.H.: Monitoring and Simulation Research on Indoor Thermal Comfort of Passive Ultra-low-Energy Consumption Buildings in Zhengzhou, Zhengzhou University (2021)
4. GB50019-2003. Code for Design of Heating, Ventilation and Air Conditioning. China Standards Press, Beijing (2003)
5. GB/T 18049-2017. Ergonomics of the heat environment analyzes, measures, and explains heat comfort by calculating PMV and PPD indices and local heat comfort criteria. China Standards Press, Beijing (2017)
6. Huang, Z.J., Liu, J.J., Zhang, H., et al.: Indoor humid environments of Huizhou dwellings in winter. Civ. Archit. Environ. Eng. **40**(02), 109–115 (2018)

Development Research on Openness Evaluation Factors for Pocket Parks

Yunjie Sun(✉)

Qingdao Technical College, Qingdao 266555, Shandong, China
775821992@qq.com

Abstract. This research aims to conduct an in-depth study on the open evaluation factors of urban pocket parks with the objective of improving their service to users. The research methodology includes literature review, interviews, and questionnaire surveys. The study begins by examining the concepts of open parks, pocket parks, and openness, starting from open parks. It then explores the open evaluation factors of pocket parks based on previous research. A comparative analysis is conducted, leading to proposed improvement strategies. The research findings present an evaluation scheme for the open evaluation factors of pocket parks. In conclusion, while previous studies have acknowledged the need for in-depth research on the openness of pocket parks, specific investigations on this aspect are lacking. This study fills that gap by developing open evaluation factors for pocket park users and proposing tailored strategies to enhance the openness of pocket parks based on user types.

Keywords: Pocket Park · Openness · Evaluating elements

1 Introduction

City pocket parks are vital green infrastructure in urban areas, serving as important public open spaces (Sun Yunjie, 2021) [1]. They provide functions for citizens and are not only sports and leisure spaces but also important social spaces. Many small-scale pocket parks have been built and maintained by local governments, reflecting the essence of opening public green spaces to the community. Pocket parks have the potential to improve the ecological environment, provide recreational spaces, and enhance the sense of belonging and community cohesion (Alexander, 1997; Warrick, B., 1997) [2, 3]. As the elderly increasingly utilize pocket parks, the demand for higher quality pocket parks is rising. Openness is an important evaluation factor, with attention paid to the preferences of different user groups, including the elderly (Sun Yunjie, 2021; Li Deming, 2006) [1, 4]. Closure of pocket parks can positively impact psychological recovery, and elderly individuals often seek green rest spaces with solitude and sunshine (Nordh, H, 2013; Rodiek, S. 2009) [5, 6]. While factors influencing the space design of pocket parks for the elderly are multifaceted, the discussion on openness and closure is significant (Liu Boxin, 2016; Kellert, 2011) [7, 8]. Despite extensive research on pocket parks in areas such as space and landscape design, accessibility, and satisfaction, the topic of openness remains understudied.

W. Guo et al. (Eds.): GBCESC 2023, LNCE 328, pp. 540–550, 2024.
https://doi.org/10.1007/978-981-99-9947-7_56

2 Conceptual Exposition

2.1 The Concept of Pocket Parks

Pocket Park, also known as Mini Park, refers to a small urban green open space. City Pocket Park is distributed in spots, which directly provides services for local residents and is an important green open space in the city. The concept of pocket park was first put forward by Robert Zion in 1963 (Zhang Luyan, 2011) [9]. The Pocket Park defined in this study is modeled on a small park with a speckle distribution in a high-density urban center.

2.2 Pocket Park Related Research

Pocket Park is a significant green open space in modern cities, originating from the United States (Houstoun, L., 2009) [10]. These parks are developed due to the difficulty of constructing large parks in cities, leading to the creation of scattered small parks throughout the urban landscape. Building pocket parks in inconvenient spaces increases the city's open space (LeFlore, A.J., 2012) [11] and contributes to creating a psychologically secure environment during tropical nights (Iwashita Zhao and Fujimoto Nobuyuki, 1991) [12]. These parks not only enhance the city's aesthetics but also provide comfortable leisure areas for residents.

In the United States, residential developers are required to incorporate a certain number of pocket parks into residential areas (National Association of Homebuilders, 2012), while European countries emphasize the concept of "Country in the Doorstep" to enhance green spaces and protect wildlife and cultural heritage through flexible site selection. London focuses on accessibility and participation, connecting gardens to parks via pocket parks and establishing a green network (Zhang Wenying, 2007) [13]. Spain introduced the "Urban Acupuncture" approach in the 1970s, transforming abandoned sites into pocket parks to activate open spaces (Wu Qiao, 2015) [14]. Italy combines pocket parks with existing urban open spaces, such as city squares, to provide green spaces with a scattered layout (Huang Xumin, 2010) [15]. Paris has successfully built numerous pocket parks in its city center (Li Jingjing, 2015) [16], and examples of well-designed pocket parks include the Royal Square in Spain, the Sensory Pocket Park in Italy, and the Vantbury Water Plaza in the Netherlands.

Japan introduced the "Urban Planning Law" in 1919, which laid the foundation for a modern park system with small parks as the main component (Ma Jie, 2012) [17]. Japanese parks, especially Pocket Parks, emphasize structure and efficiency, combining natural landscapes with leisure facilities. They are highly valued, known for their unique humanized design and creative appeal (Iwashita Zhao, 1991) [12].

China first proposed the concept of pocket parks in 1989. From 1990 to 1999, the development of green spaces in small cities was slow. However, from 2000 to 2009, due to global informationization, the adoption and implementation of pocket parks accelerated, leading to rapid growth (Zhang Luyan, 2011) [9]. From 2010 onwards, the localization process of Pocket Parks suitable for all regions has been ongoing. It is worth noting that while various aspects of pocket parks have been researched, such as space design, landscape design, accessibility, and satisfaction, the aspect of openness remains unexplored.

3 Empirical Research

3.1 Investigation on the Evaluation Elements of Pocket Park Openness

There are too few references about Pocket Park, and there are no papers about the openness of Pocket Park. Therefore, there is no way to use other people's existing evaluation elements for investigation. Therefore, face-to-face interviews with users about the openness of pocket parks are conducted in actual pocket parks. To determine the influencing factors of openness.In August, 2022, 30 elderly people and 30 ordinary people were interviewed face to face in three pocket parks in Qingdao, and 30 students majoring in design were discussed. The question is, "What do you think are the factors that affect the openness of Pocket Park?" In the process of collection, no filtering process is carried out until no new keywords appear, and the collection is stopped. Then the results are sorted out, open positive (+) open negative (−) (Table 1).

3.2 Arrangement of Factors Affecting the Opening Evaluation of Pocket Park

After that, by sorting out the results of direct investigation, graduate students and professors majoring in design are invited to sort out the oral language, integrate the repeated and similar interview records, and remove the factors unrelated to design. There are 25 influencing factors on the openness of Pocket Park, as shown in Table 3 below (Table 2).

3.3 Factor Analysis

Through the questionnaire survey to obtain the evaluation information of pocket park users, and then, using SPSS software, using the statistical method of factor analysis, quantitative analysis of the evaluation elements of pocket park openness, open evaluation elements in the importance of pocket park openness evaluation. The questionnaire involves 25 evaluation elements mentioned above, and uses 5 scale scores to evaluate: very unimportant, unimportant, average, important and very important. The scores are-2,-1, 0, 1 and 2 in turn. Respondents were asked to evaluate in Pocket Park, and 150 valid questionnaires were collected. Among them, there are 75 elderly people and 75 ordinary people (Table 4).

KMO and Bartlett's Test

KMO and Bartlett spherical test were used in the experiment, and the KMO value was 0.901, which was obviously greater than 0.5, which indicated that the data was suitable for factor analysis. At the same time, the Sig value of Bartlett spherical test was 0.000, which indicated that there was a correlation between variables, which also indicated that the data was suitable for factor analysis (Table 5).

Determination of Factor Number

In this study, principal component analysis was used in factor analysis. According to the results, when the number of extraction factors is 5, the cumulative contribution rate of

Table 1. Examples of positive (+) and negative (–) interviews with respondents

Interview record	
+ There are many entrances and exits in Pocket Park, which is more convenient to enter and exit	– I don't like iron fences on the side of the road. It's deserted
+ Pocket Park does not have a high fence	– I don't like cars parked on the side of the road and blocking the exits
– The enclosure of plants around Pocket Park is too high	+ The leisure space is large enough to communicate while resting
+ No tickets	+ There is space to chat and walk with friends
– The park is separated from the road outside, and there is a protective barrier	+ The seats are relatively long, which can provide many people to rest, or there is a pavilion
– There are too many facilities in the park and the venue is crowded	+ There is a place to provide communication
+ Plants are relatively low	+ Be able to see other people's activities
+ Pocket Park has a large area	+ Exercise space and rest space can be converted to each other
+ Park site leveling	+ The permeability of landscape sketches is high
+ The size of the entrance and exit is relatively wide	+ Can chat with friends
+ I am in a good mood and feel open now	+ You can exercise, rest and watch children
+ I like the low boundary fence, which can see the inside of the park from the outside	+ Plant leaves are relatively small and have high permeability
+ The internal facilities of the park are relatively low, so you can see the panoramic view of the park	– It is better to have a clear distinction between different functional intervals
+ Can dance with friends, keep fit and enhance feelings	+ There is a place to walk the dog, which feels very open
+ The width of the walkway is wider	– The density of landscape sketches is high
– There are tall trees in the park and there is a sense of oppression	+ Suitable for different groups of people to carry out activities
– The park boundary fence is too high and oppressive	+ There are more places to rest
– The lights are too dim at night	+ Able to perform or exhibit activities
– too many people	+ Able to play ball, dance and other activities
– Too many plants	+ The area of transition space is relatively large

Table 2. Expert consultation and survey objects

Crowd	Selection criteria	Number of people
Undergraduate teachers majoring in design	Current university lecturer and professor	12
Pocket Park researcher	Researchers with experience in pocket park design	12

Table 3. Organize the evaluation items affecting openness (25 items)

Arrangement of factors affecting openness		
The area of the park	Size of entrance and exit	Width of vision
Number of entrances and exits	Enclosure degree of plants	Enclosure degree of fence
The height of a plant	Park ground flatness	Is the periphery of the park empty
Number of facilities	Height of facility	Permeability of plants
Permeability of fence	The brightness of night lights	Height of fence
The number of plants	Number of rest spaces	Size of rest area
Barrier-free design degree	Types of functional spaces s	Width of walkway
Possibility of functional space transformation	Can you see other people's activitie	Differentiation degree of functional space
	Area of transition space	

Table 4. Investigation on Evaluation Factors of Pocket Park Openness

Place	Time	Survey object (Number)
Three Pocket Parks in Qingdao	November 2022	Older people (75) Ordinary people (75)

Table 5. KMO and Bartlett's Test

KMO and Bartlett's Test		
Kaiser-Meyer-Olkin Measure of Sampling Adequacy		0.901
Bartlett's Test of Sphericity	Approx. Chi-Square	1925.555
	df	300
	Sig.	0

Total Variance Explained

Component	Initial Eigenvalues			Rotation Sums of Squared Loadings		
	Total	% of Variance	Cumulative %	Total	% of Variance	Cumulative %
1	9.769	39.074	39.074	4.538	18.151	18.151
2	2.212	8.849	47.923	3.850	15.401	33.553
3	1.293	5.174	53.097	2.666	10.665	44.217
4	1.189	4.755	57.852	2.606	10.424	54.641
5	1.123	4.491	62.342	1.925	7.701	62.342
6	.963	3.851	66.193			
7	.856	3.425	69.618			
8	.749	2.998	72.615			
9	.678	2.713	75.329			
10	.651	2.605	77.933			
11	.597	2.390	80.323			
12	.587	2.347	82.669			
13	.553	2.213	84.882			
14	.467	1.867	86.749			
15	.441	1.764	88.513			
16	.416	1.666	90.179			
17	.400	1.602	91.781			
18	.340	1.362	93.142			
19	.310	1.239	94.381			
20	.287	1.149	95.530			
21	.276	1.103	96.634			
22	.254	1.018	97.651			
23	.230	.921	98.572			
24	.196	.782	99.354			
25	.162	.646	100.000			

Extraction Method: Principal Component Analysis.

Fig. 1. Total Variance Explained

principal components reaches 62.342%, and the effect is obvious. It can explain most of the differences in variables and show that these five factors are meaningful (Fig. 1).

According to the questionnaire results, 25 open related evaluation items are divided into 5 factors, and the results are shown in Table 6.

Reliability Analysis

The results of reliability analysis of the five open elements showed that Cronbach's α values were 0.88 for morphological openness, 0.862 for functional openness, 0.711 for visual openness, 0.781 for social interaction and 0.627 for entrance and exit openness. The results showed that Cronbach's α values were all above 0.6, so the reliability of determination was significant (Table 7).

Extraction and Arrangement of Open Evaluation Elements

"Rotated Component Matrixa" is obtained through analysis. Looking at the size of coefficients in each factor of each variable, it indicates the load size of variable factors. Generally, if the universality is above 0.4, it can be called good fitting degree of factor analysis. The evaluation items of the first factor include "plant enclosure", "fence permeability", "facility height", "plant height", "fence height", "plant number" and "whether

Table 6. Rotated Component Matrixa

Evaluation project	Component				
	1	2	3	4	5
Enclosure degree of plants	**.760**	.071	.119	.211	.089
Permeability of fence	**.748**	.140	.119	.158	.152
Height of facility	**.747**	.153	.205	.023	.036
The height of a plant	**.664**	.239	.089	.390	−.011
Height of fence	**.660**	.192	.149	−.014	.390
The number of plants	**.622**	.179	.159	.349	.071
Is the periphery of the park empty	**.516**	.227	.279	.330	.120
Differentiation degree of functional space	.108	**.822**	.092	.189	−.039
Possibility of functional space transformation	.327	**.734**	.029	.286	.058
Types of functional spaces	.060	**.676**	.475	−.007	.087
The area of the park	.027	**.600**	.345	.277	.225
Barrier-free design degree	.351	**.600**	−.069	−.019	.207
Area of transition space	.306	**.509**	.358	.209	.127
Number of facilities	.310	**.478**	.445	.117	.257
Ground flatness	.304	.047	**.712**	.325	−.099
Width of walkway	.426	.130	**.669**	−.080	.147
Number of rest spaces	.059	.484	**.582**	.182	.256
Size of rest area	.248	.409	**.417**	.375	.077
Can you see other people's activities	.055	.328	**.406**	.352	.189
The brightness of night lights	.179	.216	.087	**.713**	.228
Width of vision	.288	.240	.122	**.672**	−.008
Permeability of plants	.531	.036	.135	**.558**	.154
Enclosure degree of fence	.304	.038	-.037	.005	**.812**
Number of entrances and exits	−.030	.324	.190	.214	**.605**
Size of entrance and exit	.211	.069	.278	.418	**.543**

Extraction Method: Principal Component Analysis.
Rotation Method: Varimax with Kaiser Normalization.
a. Rotation converged in 9 iterations.

the park periphery is empty", among which "plant enclosure" has the highest value of 0.760. These seven evaluation items reflect the specific morphological design, so the factors are named as morphological openness. The evaluation items of the second factor include "functional space discrimination", "functional space transformation possibility", "functional space type", "park area", "barrier-free design degree", "transitional space area" and "facilities quantity", among which "functional space discrimination" is

Table 7. Reliability analysis resultsa

Elements	Evaluation project	Cronbach's Alpha
Morphological Openness	Enclosure degree of plants	0.880
	Permeability of fence	
	Height of facility	
	The height of a plant	
	Height of fence	
	The number of plants	
	Is the periphery of the park empty	
Functional Openness	Differentiation degree of functional space	0.862
	Possibility of functional space transformation	
	Types of functional spaces	
	The area of the park	
	Barrier-free design degree	
	Area of transition space	
	Number of facilities	
Social interaction Openness	Ground flatness	0.781
	Width of walkway	
	Number of rest spaces	
	Size of rest area	
	Can you see other people's activities	
Visual Openness	The brightness of night lights	0.711
	Width of vision	
	Permeability of plants	
Entrance and exit Openness	Enclosure degree of fence	0.627
	Number of entrances and exits	
	Size of entrance and exit	

the highest, reaching 0.822. The seven evaluation items in this way reflect the division and mutual transformation of functional space, the functionality of pocket park, etc., so they are named functional openness. The evaluation items of the third factor include "ground smoothness", "width of walkway", "number of rest spaces", "size of rest area", "whether other people's activities can be seen", etc. Among them, "ground smoothness" has the highest value of 0.712. The five evaluation items are named as openness of social communication, which is equivalent to the communication between users and the significance of social communication. The evaluation items of the fourth factor include "brightness of night light", "visual breadth" and "permeability of plants", among which "brightness of night light" has the highest value of 0.713. These three evaluation items contain visual meanings, so they are named visual openness.

The evaluation items of the last factor include "the enclosure degree of fence", "the number of entrances and exits" and "the size of entrances and exits", and the value of "the enclosure degree of fence" is the highest, which is 0.812. These three evaluation items contain contents related to specific access, so the elements are named as access openness.

The five openness named in this way are morphological openness, functional openness, social interaction openness, visual openness and entrance and exit openness. The explanation of each factor is compiled on the basis of factor analysis. After expert consultation, the final arrangement is made, and the results are shown in Table 8.

Table 8. Openness evaluation elements

Type	Definition of evaluation elements	Evaluation elements
Morphological Openness	Through the opening of form design, it is a kind of space opening	Enclosure degree of plants
		Permeability of fence
		Height of facility
		The height of a plant
		Height of fence
		The number of plants
		Is the periphery of the park empty
Functional Openness	The openness of functional areas, pocket park sports space, rest space, exchange space of the mutual conversion of openness	Differentiation degree of functional space
		Possibility of functional space transformation
		Types of functional spaces
		The area of the park
		Barrier-free design degree
		Area of transition space
		Number of facilities
Social interaction Openness	The openness of social contact space is used to explain the openness of communication and contact between people	Ground flatness
		Width of walkway
		Number of rest spaces
		Size of rest area
		Can you see other people's activities

(*continued*)

Table 8. *(continued)*

Type	Definition of evaluation elements	Evaluation elements
Visual Openness	The user's visual openness	The brightness of night lights
		Width of vision
		Permeability of plants
Import and export Openness	Opening degree of access and entry to Pocket Park	Enclosure degree of fence
		Number of entrances and exits
		Size of entrance and exit

4 Conclusion

In this study, in the various influencing factors of Pocket Park, the openness is the center for in-depth study. Firstly, in order to extract the evaluation items that affect the openness of Pocket Park, a survey on the opening factors of Pocket Park was carried out, and 25 evaluation items were selected through expert consultation. Secondly, through factor analysis, the classification and reliability are verified from five factors: "morphological openness", "functional openness", "social interaction openness", "visual openness" and "entrance and exit openness". Future research shows that there are differences in the importance of open evaluation items in this study, and through the "research on the application of weight of evaluation items centered on the openness of Pocket Park", a more objective evaluation index is formulated.

References

1. Sun, Y., Zhao, G., Jin, R.: Development of evaluation elements for the elderly in urban pocket parks and research on the cognition of pocket parks to the elderly. Industr. Design Res. **15**(4), 1–11 (2021)
2. Alexander, G., Gayle, B.: Urban parks and OpenSpace. Urban Land Institute, Washington DC (1997)
3. Warrick, B., Alexander, T.: Looking for hometown America. Urban Land **56**, 27–31 (1997)
4. Li, D., Chen, T.: Cognitive aging and mental health of the elderly (2006)
5. Nordh, H., Østby, K.: Pocket parks for people–a study of park design and use. Urban For. Urban Green. **12**(1), 12–17 (2013)
6. Rodiek, S.: Access to nature for older adults [Educational video series]. Available from Center for Health Systems & Design, College of Architecture, Texas A & M University, College Station, TX, 77843 (2009)
7. Liu, B., Xu, Y.: Study on the influence of different landscape types on the physical and mental health of the elderly. Landsc. Archit. (7), 113–120 (2016)
8. Kellert, S.R., Heerwagen, J., Mador, M.: Biophilic Design: The Theory, Science and Practice of Bringing Buildings to Life. Wiley, Hoboken (2011)
9. Zhang, L.: Analysis of landscape planning and design of Shanghai Pocket Park. Doctoral Dissertation, 11–12, 41–57 (2011)
10. Houstoun, L.: Unexpected parks in public spaces. Urban Land 100–103 (2009)

11. LeFlore, A.J.: Increasing urban open space through pocket parks. Tufts University (2012)
12. Zhao, I., Yang, L.: Pocket park-a place for "rest" and "use". New Archit. (4), 54–57 (1991)
13. Zhang, W.: Pocket Park-an oasis to avoid the hustle and bustle of the city. Chin. Garden **23**(4), 47–53 (2007)
14. Wu, Q.: Pocket park-green antidote to high density city. Garden (2), 45–49 (2015)
15. Huang, X., Pan, S., Wang, H.: On the present situation and development of pocket parks. Shandong For. Sci. Technol. **40**(5), 96–97 (2010)
16. Li, J.: Flexible space of city-pocket park. Art Life (5), 52–53 (2015)
17. Ma, J.: Design and research of hangzhou pocket park. Zhejiang University, Hangzhou (2012)

Study on Landscape Characteristics and Formation Mechanism of Chinese Traditional Settlements Based on Niche Theory

Jianfu Chen[1,2](✉), An Yan[3], and Hailing Sun[2]

[1] Graduate School of Environmental Engineering, The University of Kitakyushu, 1-1 Hibikino, Kitakyushu City, Fukuoka, Japan
acjeff77500@163.com
[2] West AnHui University, Lu'an 237012, Anhui, China
[3] Qufu Normal University, Qufu 273165, Shandong, China

Abstract. The Anhui settlement is one of the best preserved in China. Its conservation research is a typical case of sustainable research and an essential component of ecosystem research. In this paper, six typical Anhui communities were considered. Combining traditional Fengshui environmental studies with modern niche-related ideas and using techniques of entity study, including qualitative analysis, quantitative research, and data retrieval, this paper analyzes the landscape characteristics at the macro, medium, and micro levels. The research results show that the conventional settlements in Huizhou are built on the mountains, forming a geographical space model with mountains on four sides or three sides in the east, south, west, and north and open halls, which form an ideal settlement model. At the same time, the rational use of land, resources, and space should be promoted according to the development conditions of settlements to steadily realize the spatial succession of traditional settlements and walk off the road of characteristic ecological development.

Keywords: Traditional Settlement · Ecology · Landscape feature · Formation mechanism

1 Introduction

1.1 A Subsection Sample

Chinese villages have existed for a long time, and most buildings were built in the Song, Ming, and Qing dynasties. Since 1980, domestic and international research on traditional Anhui settlements has led to useful findings. Most research has focused on the shapes of settlements and individual buildings [1]. Based on a study of how people live in formal settlements, the laws of interaction between human society, nature, and the ecosystem help people live in harmony with the land. It gives a scientific basis for the fair distribution of regional populations, the overall development of urban and rural areas, the fair distribution of resources, and the effective use of territorial space [2]. In the

W. Guo et al. (Eds.): GBCESC 2023, LNCE 328, pp. 551–559, 2024.
https://doi.org/10.1007/978-981-99-9947-7_57

analysis of Chinese settlement-related studies, it is known that there is a lack of research on landscape characteristics and discussion on their formation mechanism. Therefore, based on previous studies, the author analyzes multiple settlements' forms and landscape patterns from a multi-dimensional perspective and examines the formation mechanism of the settlement landscape.

2 Research Area and Research Method

2.1 Study Area

According to the typicality and uniform distribution principle, we selected six traditional villages in Southern Anhui Provience in China: Xixinan Village, Hongcun Village, Xidi Village, Yuliang Village, Chengkan Village, and Village of Zhanqi Xiaonu (see Fig. 1). The settlements are distributed in the buffer area of the river water network, with an elevation of 200–300 m and the number of households ranging from 200–600 m. With the development of the integration of urban and rural areas in China, the disharmony between the landscape of counties and cities and the landscape of ancient towns is more frequent (Fig. 2).

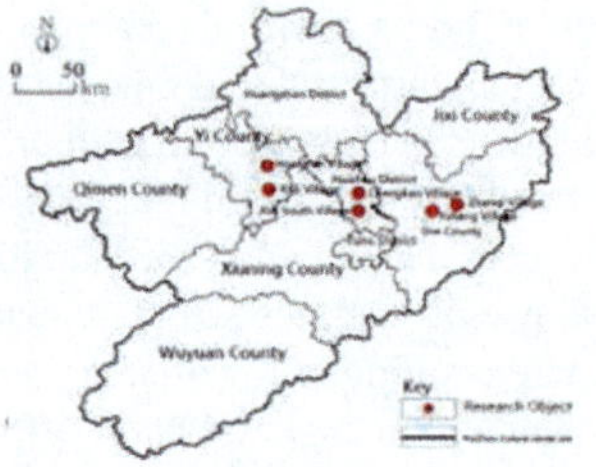

Fig. 1. Location of the research object. Picture source: modified from [3].

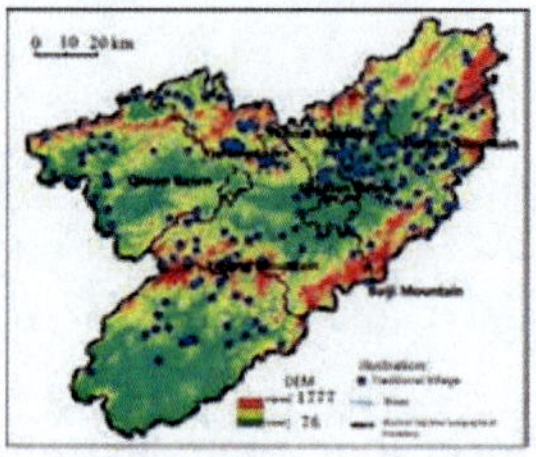

Fig. 2. Distribution of villages in the Huizhou region. Picture source modified from [4].

2.2 Theoretical Principles and Methods

This study adopts the method of field investigation. ArcGIS software is used to analyze the elevation and village distribution in this area to show that the surveyed villages can reflect the regional landscape characteristics in a concentrated way (see Fig. 1). Relevant data are based on survey analysis and literature research [5].

Niche Theory mainly describes the interaction between different ecosystems. Based on this theory, the ideas of niche fitness and niche expansion were derived, explaining the species unit mechanism. Its theoretical significance is constantly being expanded. For traditional settlements, niche-related theories have an apparent coupling relationship between economy, society, environment, and production. From a macro perspective, traditional villages can be regarded as point-like distribution. The nearest distance is a geographical index indicating the degree of mutual proximity of point-like things in geographical space. The spatial distribution of point elements is uniform, random, and clustered, so the nearest distance formula and nearest point index can be used to measure the spatial distribution of each traditional village. The calculation formula is as follows:

$$R = \frac{\bar{r}_1}{\bar{r}_E}\sqrt{\rho} \times \bar{r}_1 \quad (1)$$

$$\bar{r}_E = \frac{1}{2\sqrt{n/s}}x = \frac{1}{2\sqrt{\rho}} \quad (2)$$

where R is the nearest point index, R is the theoretical nearest distance, and p

denotes the density of traditional villages. S represents the region's area, n is the number of research objects and is the average distance value of the nearest distance r1. When R = 1, it indicates that the dotted villages are distributed along with the model. When R< 1 indicates that villages are distributed in an agglomeration type When R> 1 means that villages are evenly distributed. According to Formula (1 and 2), the spatial statistical tool in ArcGIS9.3 was used to calculate the index of the nearest point of These traditional villages. The average observation distance was 3678.03 m, the expected average distance was 4479.13 m, the Z score was −4.6, and the significance level was P&lt. 0.01, the index of the nearest point R = 0.82, indicating that villages in Huizhou present an agglomeration distribution on the whole.

The overall distribution of traditional villages in Southern Anhui is concentrated. Still, their distribution has significant spatial differences, generally showing a spatial distribution pattern of "less core and more edges". To further explain the spatial agglomeration characteristics of Anhui traditional villages, kernel density was used for the analysis. The kernel density estimation method is used to calculate the density of elements in the surrounding domain, which greatly analyses the spatial distribution of point data. The calculation formula is:

$$f(x) = \frac{1}{nh}\sum_{i=1}^{n} k\left(\frac{x - x_i}{h}\right) \quad (3)$$

where f (x) represents the kernel density function; k() denotes the kernel density equation; h is the search radius (h> 0), indicating the distance from the core point; n is the number of traditional villages, and I represent the value between 1 and n. The kernel density of traditional villages was calculated using Eq. (3). The central portion of Yi County has the highest kernel density value, forming the main agglomeration area of traditional villages in Southern Anhui province, which also depicts two agglomeration areas in space. Traditional villages at the edge of Anhui Provience have a relatively high kernel density value. The core density values of Huizhou District's central part and She County's central part are above 0.05, forming the secondary core area of Anhui traditional village gathering. Xiuning County and Tunxi District, located in Anhu's core area, have the smallest kernel density, and few traditional villages remain. Therefore, traditional villages show a spatial distribution pattern of "less core and more edges" overall, and the trend of marginalization distribution of traditional villages is apparent. According to the Lorentz curve of spatial distribution of traditional villages in various counties, nearly 90% of traditional villages were distributed in Yixian County, Shexian County, Wuyuan County, Xiuning County, Jixi County and Qimen County, but the number of traditional villages in Chinese cities was relatively small.

3 Multi-scale Landscape Features of Chinese Settlements

3.1 Macro-scale Landscape Environment Analysis of Settlements

Traditional settlements in China adhere to the macro-regional idea of "living with nature, living with mountains and rivers", which is the ecological idea of respecting nature and wisely using the environment. This is precisely the outstanding idea lacking in constructing Chinese villages and towns against the backdrop of urban development in recent years. Chinese traditional villages pay attention to "harmony between nature and man". Residents hope to get close to this ideal living state through the construction of architecture and landscape, settlement, and ecological environment.

The Influence of Site Selection and Feng Shui. The unique natural environment results in its unique living pattern. The traditional landscape pattern of settlement is an active adaptation to the environment. Residents not only protect many forest resources and regard them as the blood of life but also improve the living environment by planting forests to form "Feng Shui forest", to achieve sustainable development and maintain ecological balance [6]. The construction of the macro settlement space technique is mainly based on the pattern of the mountain and the "Feng Shui" layout.

The Impact of Landform. Most of the traditional villages in the region are distributed in the gradient level of 1–4, and 90% of the traditional villages with an elevation of 60 to 500 m are in She County, Yixian County, and Xiuning in Huizhou. She county has the most abundant landscape resources, among which She county's mountain and hill area accounts for 90% of the country's total area. According to relevant data on traditional national villages, the number of traditional villages in Shexian County is the largest, followed by Yixian and Xiuning. Based on the consideration of regional scale, topography, geomorphology, and water system are selected as factors affecting the distribution of villages. It is found that most villages are distributed around the water system and in hilly areas, and few villages with very low or very high elevations are distributed, showing a marginal distribution from high elevation to low elevation [7].

In the past 300 years, the altitude of newly added traditional villages has generally shown an increasing trend. And 78% of them are distributed at an altitude of 100–300 m. In the research sample, only 20 villages are distributed at an altitude of more than 500 m, and only 18 villages are distributed at an altitude of less than 100 m, accounting for less than 4% (Table 1). The selection of ecological adaptability of settlements is directly related to geomorphic features.

Table 1. Distribution of villages at different elevations in urban areas of Anhui

City/county	village distribution			
	<100 m	100–300 m	300–500 m	>500 m
Xiuning County	0	47	3	3
Jixi County	0	32	8	1
She County	0	189	38	14
Qimen County	11	35	0	1
Wuyuan County	7	18	2	1
Yi Xian	0	47	13	0
Huangshan City	0	40	4	0
Accounting for	3.5%	48%	44.7%	3.8%

In contrast, the basin is suitable for grain planting, urban construction, economic development, and commercial transportation. Settlements, farmland, tea gardens, rivers, and other ecological landscapes are organically distributed here, giving birth to many famous towns and ecological and cultural villages. Since the Ming and Qing Dynasties, most cities in Anhui have been arranged in relatively flat basin areas, and Anhui continues this traditional ecological construction pattern today.

3.2 Meso-Scale Landscape Environment Analysis of Settlement

Land resource utilization types and natural landscapes such as agriculture, water network, rivers, and woodland identify the mesoscale environmental characteristics of Huizhou settlements. The Xidi Ancient Village, for instance, has an average elevation of 260 m and is situated in a basin at the southern foot of Huangshan Mountain. The north and south mountains sandwich the village, which Luofeng Mountain and Shishi Mountain back. Three streams run through the village and into the outlet of Shuikou, making it an ideal place for residents to live. Among the six traditional villages, Xidi Village, Hongcun Village, and Chengkan Village are built at the foot of the mountain, Xixinan Village is built on the plain, Lianjiang River builds Yuliang Village, and Zhanqi Xiaonu Village is built on the mountain. Its natural slope is quite steep, nearly 15°, while the slope of the other villages is not more than 10°. As early as in the primitive period, Huizhou people chose the site at the foothills with gentle terrain and diversified base locations, which differed from the site selection of other ethnic villages [8].

As mentioned above, the scale of villages is restricted by a river system, terrain, land type, and other factors. Among these constraints, rivers have a significant impact on the form of nearby traditional villages. According to the GIS analysis, most villages are near the river system. Due to different river trends, rural settlement patterns of various geometric forms can be generated, mainly summarized as flat forms such as clumps, bands, scattered patterns, and accessible villages. Through such flow trend and design, Chengkan village integrates ecological irrigation and breeding, and the settlement form shows adaptability to the water environment. The three can be found everywhere in the landscape composition of settlements, farmland, and rivers. They cooperate to build the water landscape pattern of settlements, representing the typical water cultural landscape.

3.3 Landscape Environment Analysis at the Micro-Scale of Settlement

In environmental science, "back the mountain and face the water" can play the role of gathering "qi" and generating "qi". It should be the environmental science that "mountain pipe people Ding water pipe wealth, mountains and rivers love wealth" words. In the eyes of local housebuilders and craftsmen, the flood ensured that the family's wealth would not leave the country. From the perspective of human settlements, residential dwellings are distributed in mountainous areas, and the sunshine time is not as sufficient as that in plain areas. The distribution of the houses on the sunny slope is conducive to the radiant surface of the sun to accelerate the flow of gases, allowing miasma and fog in the mountains to disperse out in time. Chinese folk houses mainly conform to the south wind direction in summer. According to the climate conditions, folk houses usually improve the natural ventilation and microclimate in residence through reasonable functional layout, structural material type, indoor and outdoor landscape, and other ways. At the same time, doors, Windows, corridors, and other openings are used to introduce drafts. The layout of residential houses can create a more comfortable indoor environment to adapt to hot summer environment.

4 Analysis of the Formation Mechanism of Traditional Settlement View from a Modern Ecological Perspective

A settlement system is a composite system composed of closely related economic, social, and historical elements. It has already been discussed that natural elements are jointly affected by site selection, landform, and altitude in traditional geomantic theory, so the mechanism under the influence of the natural environment is not stated here [9] (see Fig. 1) (Fig. 3).

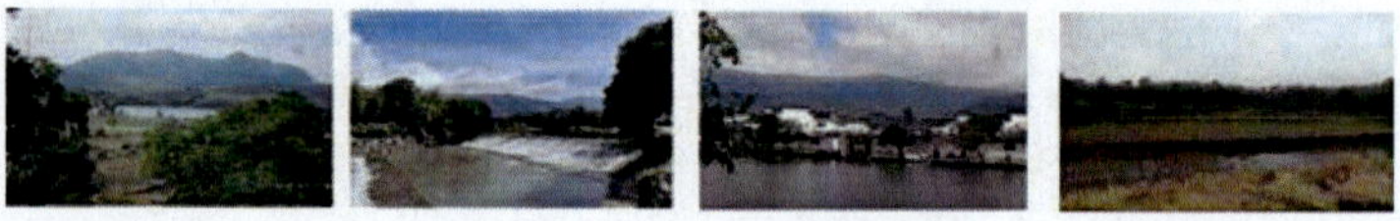

Fig. 3. Environment of Hongcun village in Anhui: "Mountain-Water-Field-Forest" diagram

4.1 Social Clan Influence

The ancient people moved down from the Central Plains. The spatial style of traditional ecological landscape is very typical in China. Compared with other traditional villages in China, traditional villages have strong identification and significant differences in external characteristics and internal construction ideas. The reason behind this phenomenon is closely related to its clan culture. When Chinese ancestors immigrated, they also brought Confucian thought, including the ethical concepts of the "Three Principles" and "Six Disciplines". These ideas significantly influenced Chinese architectural design, water network structure, street space, and people's way of life, attitudes, and inferior concepts. As a kind of public architecture, the ancestral hall has become the settlement's public space and spiritual sharing place. The centrality and cohesiveness of Chinese traditional villages in landscape form, with distinct internal landscape structures and orderly distribution of laneways, water channels, and buildings.

4.2 Historical and Humanistic Influence

The settlement water cultural landscape contains all kinds of human elements, the most historical and cultural landscape in the Chinese settlements. Therefore, most settlements are generally close to rivers. In the Ming and Qing dynasties, water system management was essential in constructing the traditional settlements. Water is introduced into the water system by the river, through the inlet, then through all levels of water channels, ponds, and finally through a natural circulation path into the river. In traditional settlements, water plays the role of irrigation, breeding, production, living, security, fire prevention, and geomantic omen. Most of the ancestors come from the Central Plains. The Fengshui culture of the central Plains has been introduced into Anhui Province and gradually integrated into the Chinese culture, which is most evident in the villages such as Xixinan Village, Hongcun Village, and Chengkan Village. From the current situation, the Feng Shui of Xixi South Village settlement not only meets the spiritual needs of people but also increases the water area based on protecting extensive forests. Feng Shui forest has many similarities with nature reserves in contemporary territorial planning. In the long run, the Fengshui forests and fish ponds play a significant role in the long-term stability of the complex ecosystems.

4.3 Impact of Economic Development

Among these settlements in China, land resources are related to residents' rights, crop cultivation, and economic benefits, so economic development is an indispensable factor affecting the formation of settlement mechanisms. It is mainly reflected in the following two aspects: settlement development and construction need a lot of human resources, land resources, and construction funds; rural areas with different economic development conditions have various driving forces. The financial investment in settlement construction will directly affect rural building forms' location, scale, and distribution. Economic forests and terraces have also become a significant part of the local agricultural landscape.

5 Settlement Niche Expansion and Spatial Structure Succession

Niche Expansion refers to the development status of human society and the ability to influence the environment, compared with the improvement of other species' dominance or influence ability in the ecosystem. Niche Expansion refers to the increase of states and potentials due to the infinite growth potential of biological units. The evolution of the human Ecological Niche refers to the improvement of the development state of human society and the ability to influence or dominate the environment compared with the state and potential of other species in the biosphere.

The self-organizing process in which an ecosystem uses energy to develop or change its structure is called succession. As an ecological system, the structure of traditional settlements is not fixed. With the implementation of the national economic policy in the early 1980s, the most of Chinese settlements underwent a significant spatial succession process, which was primarily seen as follows: Farmers in some settlements contracted the Fengshui forest, tea land, forest land, and cultivated land as part of a system of shared responsibility for agricultural production with the family. Thus, The spatial form of settlements is still undergoing dynamic change.

6 Conclusion and Discussion

Villages exist in the ecology of "nature, society, history, and economy". Analysing the features of the village landscape from macro, meso, and micro perspectives allows researchers to better understand its spatial characteristics and evolution process, as well as the mechanisms influencing the production of settlement landscape characteristics. Natural elements create the groundwork for settlement building, historical and human causes encourage further diversification of settlement traits, and economic development factors boost settlement spatial aspects. All sorts of villages must first meet the process of adapting to the natural environment, which is favourable to flood management and irrigation, in order to reach a state of coordinated growth of "mountain-water-house-field-forest" and people under both natural and artificial conditions. Traditional communities and homes are influenced by many natural settings, human influences, and economic situations. Ecological Niche Theory is continually evolving in economic, social, and cultural construction. It has injected new blood into relevant disciplines of study in terms of human settlement growth.

Acknowledgement. The work was supported in part by Anhui West University Campus Level Social Science Key Project "Research on the Renewal and Utilization of Residential Buildings in the Dabie Mountain Area" (No. WXSK202265).

References

1. Li, X., Lan, H.: Research on landscape zoning of traditional settlements in fujian and taiwan and its application. Fujian Forum **2014**(03), 124–129 (2014)
2. Li, X., Guo, W.: Research on landscape characteristics of traditional settlements in Hangjiahu area. Mod. Urban Res. (01), pp. 62–68 (2022). (in Chinese)

3. Chen, X.H., Xie, W.Z.: The spatial evolution process, characteristics and driving factors of traditional villages from the perspective of the cultural ecosystem: a case study of Chengkan Village. Habitat Int. **104** (2020)
4. Lu, S., Zhang, X.J.: Spatial-temporal distribution and controlling factors of traditional villages in Huizhou region. Sci. Geogr. Sinica **38**(10), 1690–1698 (2018)
5. Gao, Y., Yang, D.H.: Research on the spatial form and culture features of water adaptability of the traditional settlements on Hainan Island. New Archit. (02), 126–129 (2021)
6. Fan, J.F., Xiong, X.Y.: Ecological factors and impact in space form evolution of Dong settlement. **12**(01), 57–61+77 (2011)
7. Huang, T., Guo, Q.: H Research on the socio-economic niche of villages in the complex rural ecosystem. Acta Ecol. Sinica **42**(01), 105–115 (2022)
8. Bailey, R., Davies, K.: Cultural ecology and cultural critique. Arts (2019)
9. Porter, P.W.: Ecology, cultural. Int. Encycl. Soc. Behav. Sci. 4035–4041 (2001)

Towards a Sustainable Future: Timber Waste Management in New Zealand's Construction Industry

Dat Tien Doan(✉) and Ping Sun

Auckland University of Technology, Auckland 1010, New Zealand
dat.doan@aut.ac.nz

Abstract. This research paper provides a comprehensive overview of construction and demolition waste (C&DW) management, specifically emphasising timber waste. Firstly, it assesses timber's environmental and functional advantages in the construction industry. Despite the benefits of timber, the paper identifies significant challenges in managing chromated copper arsenate (CCA)-treated timber waste due to its potentially hazardous impacts. To address these issues, the study proposes enhanced public understanding, innovative timber treatment processes, and effective waste management strategies. The role of stakeholders and policymakers in shaping sustainable waste management practices is also explored, highlighting the importance of implementing site waste management schemes and extended producer responsibility programs. The paper culminates in the presentation of innovative approaches for waste management, encompassing statistical modelling, automation, and systemic carbon reduction strategies. The research concludes with a set of recommendations aimed at promoting sustainable material selection, managing CCA-treated timber, enhancing stakeholder engagement, and adopting lifecycle consideration in material selection.

Keywords: Sustainable Waste Management · CCA-Treated Timber · Timber Treatment

1 Introduction

The construction sector plays a significant role in environmental issues, primarily through the emission of greenhouse gases [1]. The lifecycle of building materials, from production to demolition, greatly affects global resource usage. The sector is responsible for 40% of carbon dioxide emissions, contributes to landfill waste, and creates airborne particles [2]. Furthermore, the construction industry's dependency on large-scale material extraction results in substantial energy consumption and the release of harmful environmental pollutants [3]. At the end of a building's lifecycle, the demolition process generates a vast amount of material, necessitating recycling or disposal. [2] estimated that construction and demolition (C&D) waste constitutes 13% to 60% of all landfill waste.

W. Guo et al. (Eds.): GBCESC 2023, LNCE 328, pp. 560–569, 2024.
https://doi.org/10.1007/978-981-99-9947-7_58

In residential construction within New Zealand, timber, steel, and concrete are the primary materials utilised [4]. However, the narrative surrounding timber usage extends beyond single or double-storey buildings. The global adoption of multi-storey timber construction is on the rise, attributed to its manifold advantages [5]. Innovations in technology and the advent of novel timber products have bolstered the popularity of timber as a construction material. Glued laminated timber (glulam), cross-laminated timber (CLT), and laminated veneer lumber (LVL) are particularly popular [6].

In New Zealand, untreated timber is commonly employed in construction. It is frequently repurposed for doors, windows, and structural elements such as beams and walls [7]. Additionally, when combined with other materials, it serves to create composite products such as fibreboard. However, treated wood finds significant use as landfill waste [8]. The use of chromated copper arsenate (CCA) treated substances in the domestic market produces timber that is not recyclable, thereby posing disposal challenges [8].

The evolving demands of New Zealand's building sector necessitate environmentally sustainable construction practices and materials. Reducing the carbon content in materials can reduce the associated carbon emissions during production, construction, and usage [9]. This study aims to examine the benefits of timber usage within the construction industry, focusing on its impacts on climate change and waste management in New Zealand. It advocates for sustainable timber waste management practices to alleviate climate change effects and foster an environmentally conscious built environment. It seeks to answer the following questions:

- What factors contribute to timber's preferred status as a building material in construction?
- What strategies can be implemented to address the current issue of recycling CCA-treated timber?
- How can changes in the attitudes of stakeholders and policymakers significantly affect the management of construction timber waste and facilitate waste reduction?
- How might innovative approaches and technologies revolutionise waste management practices in the construction industry to promote sustainability?

2 Research Methodology

This research employed a systematic literature review methodology involving a comprehensive review of extant publication databases to evaluate scientific publications thoroughly [10]. The adopted approach corresponds with the framework suggested by [11]. This method enables a profound comprehension of pertinent issues by synthesising and summarising previous research, thus promoting transparency for future researchers, reducing bias, and serving as a reliable information source.

The investigation into journal publications associated with timber waste in residential construction within New Zealand necessitated an extensive perusal of the Auckland University of Technology (AUT) databases, Google Scholar, and Scopus. The keywords used are presented in Table 1.

The articles were methodically categorised and filtered to facilitate a detailed comparison, thereby aiding in identifying critical factors influencing the efficiency of construction waste management. This also supported the establishment of quality control for

Table 1. Keywords

How (OR/AND)	Where (OR/AND)	What (OR/AND)	When (OR/AND)
Timber waste	New Zealand	Residential	Design
Timber recycles	Global	Housing	Production
Timber reuse		Building	Manufacturing
Timber reduces			Construction
Timber recover			End-of-life
Timber regenerate			
Timber management			

construction waste management through public policies and legislation. Initially, a total of 784 articles were identified. These underwent rigorous screening to exclude those not pertinent to the engineering field, written in languages other than English, or lacking clarity in their relevance. As a result, a total of 41 articles were meticulously selected for inclusion in the study.

3 Findings

3.1 Benefits of Using Timber in Construction

A study analysing the demand for timber-based products and their substitutes in the German construction sector, performed by [12], indicated the critical need to promote timber usage in the industry. The study calculated the elasticities of demand in relation to the price of timber and its alternatives. [12] further highlighted numerous advantages of timber, such as fire resistance, durability, stability, and sound insulation. Beyond its functional qualities, timber is also associated with health benefits, aesthetic appeal, and minimal ecological impact, making it an optimal choice for residential construction. [13] demonstrated that replacing concrete or steel with CLT could yield a carbon emission reduction of 40%. Furthermore, CLT is recognised for its high strength-to-weight ratio, ease of installation, and pleasing aesthetics. According to [14], CLT buildings offer superior overheating resistance compared to their concrete counterparts during the summer months.

[15] compared energy consumption and operation in a typical three-storey office building constructed with three different materials. The findings revealed that timber buildings possess lower embodied energy than those built with concrete and steel. Similarly, [16] suggested that CLT, as a construction material, could deliver performance at par or even superior to that of steel, brick, and concrete systems. [17] applied a material flow analysis model to forecast the material intensity in the EU-28 in the forthcoming years. The findings proposed that timber and straw are optimal for carbon storage in European edifices, particularly during the critical period of climate change and shortage of non-renewable resources [18].

[19] evaluated timber's potential as a construction material, utilising both quantitative impact assessment and participatory retrospective analysis. The findings showed an ongoing preference for timber due to its user-friendliness, rapid construction, standardisation, and cost-effectiveness, making it a popular choice among consumers. The study also predicted that technical enhancements in recycling systems could enable complete recycling and reuse of waste wood post-2050. [1] administered an Unanswered Questions and Future Challenges questionnaire to 757 panellists from the Ministry of Wood Construction and Forestry. The results revealed that a majority (57%) of Austrian millennials have a favourable attitude towards timber construction.

3.2 CCA Treated Timber

In New Zealand, CCA-treated timber is widely used in the construction industry due to its impressive resistance to insects, fungi, and water damage, achieved via the application of specific preservatives, particularly copper, chromium, and arsenic [20]. CCA timber preservatives are compounds of arsenic, copper, chromium substances, oxides, or salts, typically used for vacuum pressure treatment of timber and are commercially accessible. Their principal role is to protect timber products from pests such as timber-destroying insects, marine borers, and decay fungi [21]. CCA treatment has been prevalent in preserving timber in New Zealand since 1955. However, CCA-treated timber is recognised as potentially hazardous globally and is prohibited in countries such as Germany, Sweden, Japan, and Indonesia [22].

In New Zealand, the use of CCA preservatives is authorised in specified hazard categories. Timber is extensively employed for residential applications, including decking, playground equipment, and garden furniture. These uses are categorised as hazardous due to exposure to weather, above and below ground, or fresh water. For these applications, the arsenic content in treated timber ranges from 0.11 to 0.22% [21]. Australia and New Zealand are the two leading consumers of CCA-treated timber worldwide.

Unsustainable processing and modification techniques often undermine the green resource credentials of timber. In both New Zealand and Australia, timber is frequently treated with CCA to deter termites and mosquitoes, resulting in inseparable mixtures and composites that are non-reusable or non-recyclable. The literature analysis suggests the following three key solutions to mitigate the adverse impacts of CCA:

Public Perception
Enhancing the public's understanding of CCA-treated timber could facilitate improved categorisation of discarded materials [21]. High levels of arsenic have been identified in the ash of treated and untreated timber, highlighting the need to caution against disposing of treated wood in organic bins. Timber waste is frequently discarded in organic containers or included in demolition consignments from municipal contractors.

Timber Treatment
In the process of recycling timber waste, it is crucial to separate treated timber from untreated timber. Various extraction methods can render treated timber safe for disposal or recycling without intricate filtration systems. These methods include biological extraction, steam exposure extraction, wood liquefaction, chemical extraction, and electro-dialytic, see Table 2 [23].

Table 2. CCA extraction methods

Extraction methods	Notes
Biological	A technically viable solution with the potential to extract almost 100% of CCA components, but not effective in terms of time and cost
Steam exposure	This method is not practical from either an economic or technical standpoint, as it fails to achieve more than 90% extraction
Wood liquefaction	Up to 85% of the CCA components can be eliminated, but further research is required
Chemical	A complex and time-consuming process that requires substantial resources; this method needs further research
Electrodialytic	Similar to the steam exposure method, this method is not a viable extraction solution, either economically or technically

[23] identified that treated CCA waste timber could be recycled into wood plastic composites (WPC). They further investigated the potential use of WPC in construction and its long-term conservation in the construction industry within a circular life cycle. [24] suggested that timber residue could serve as an emerging fuel source based on assumptions and input data from timber waste landfilling. This residue could be directly burnt as a heating fuel or combined with coal and other fuels. It also generates fewer air pollutants than traditional fossil fuels. Timber pellets, in particular, hold the potential for heat provision in industries such as cement manufacturing and power generation.

Management of CCA-Treated Timber
In addition to the previously mentioned treatments and recycling methods for CCA, [23] outlined various management strategies for CCA-treated timber. These strategies include substituting treated timber with alternative materials such as brick, untreated cedar, and teak; utilising 3D scanning to collate wood parameter information for material integration, thus reducing the need for excessive CCA wood system development and design; and minimising the use of treated timber during the design process. Applying WPC could also decrease the construction industry's reliance on raw materials, including treated wood, thereby reducing the volume of treated timber waste.

Regarding waste management, while landfills need substantial space and present the risk of leaching, incineration can generate energy but sacrifices carbon storage. Thermal treatments offer a comprehensive solution but require technical support and can be time-consuming. Therefore, clear prioritisation is necessary to identify the most suitable option for the future.

3.3 Roles of Stakeholders and Policymakers in Timber Waste Management

[25] observed diverging opinions on the necessity for companies to implement C&DW management. However, it is pivotal to consider various sustainability assessments before developing a C&DW system. Several methods exist for creating C&DW systems, aiding in determining their environmental impact. Focusing on key areas such as waste materials

and metal composition is crucial to achieve tangible improvements in environmental performance. Gathering information on sectors such as wood scrap can drive progress towards meeting national C&D waste recycling targets.

[26] argued that councils across New Zealand should consider site waste management schemes as crucial for enhancing resource efficiency in the construction industry. Their research examined the global implementation of these schemes and their successful application in Auckland. Key success factors in Auckland included council-led schemes, stakeholder consultation, legal enforcement of site waste management schemes, phased implementation, template processes, minimising bureaucratic requirements, and engagement with the waste and construction industry during development.

[22] explored the most effective processes for treating timber waste and identifying the responsible producers of treated timber. It would be advantageous for New Zealand timber producers to establish a voluntary Extended Producer Responsibility (EPR) scheme. EPR is increasingly adopted worldwide. New Zealand is considering an EPR scheme for waste-treated timber, which would inevitably require legislation. Before enacting such legislation, timber producers should establish a voluntary environmental responsibility programme. Forestry companies, wood manufacturers, and chemical processors should collectively assess each other's treatment procedures to determine responsibility for treated timber more effectively. Suitable solutions such as carbon sequestration, energy production, or circular thinking can be discussed and selected accordingly.

3.4 Innovative Approaches to Timber Waste Management

Recent research presents innovative strategies for construction waste management. [27] introduced a statistical model to predict waste generation in residential construction. By enhancing the accuracy of waste forecasting, this model could improve the construction planning phase, leading to more sustainable design choices and efficient material utilisation. Future studies could expand upon this research by incorporating diverse datasets, qualitative measurements, and extending the scope to various building types and sizes.

Automation and data extraction methods also emerge as potential solutions. [28]'s script, developed for the Revit BIM database, streamlines the process of collecting precise material data. This tool can track the use of recycled or reused materials and components, offering a pathway towards a more circular economy in the construction industry.

[29] presented a systemic perspective with their Urban Equilibrium (UE) approach. By considering buildings as carbon pools to counterbalance urban greenhouse gas emissions, the UE approach underscores the role of sustainable design choices in waste management and carbon reduction. The study's suggestion of promoting timber use in urban development serves as a case in point, highlighting the environmental benefits of intelligent material selection.

[30] focused on the crucial role of material selection in managing construction waste. Their study distinguishes the divergent carbon emissions of light timber and steel framing houses, demonstrating how the recyclability and sustainability of materials can innovatively reduce the environmental impact of construction. While the recyclability of steel

offers an advantage, the potential of sustainably sourced timber to function as a carbon pool presents a novel approach to waste management.

[4] expanded upon this idea in their lifecycle analysis of building materials. This study innovatively maps out the carbon footprint of each material over its lifecycle, providing a broad perspective on their environmental impacts. It emphasises the need for construction industry stakeholders to shift their focus from merely the immediate use of materials to their long-term environmental implications. While the recyclability of steel offers clear benefits, the sustainable sourcing of timber materials can lead to a similar, if not more significant, reduction in a project's carbon footprint.

In summary, these innovative strategies highlight the intersection of construction waste management with other domains such as statistical modelling, automation, and systemic carbon reduction. By adopting a multi-faceted and forward-thinking approach to material selection, it is possible to revolutionise waste management practices in the construction industry, bringing us closer to a sustainable future in this sector.

The key findings of the research are summarised in Fig. 1.

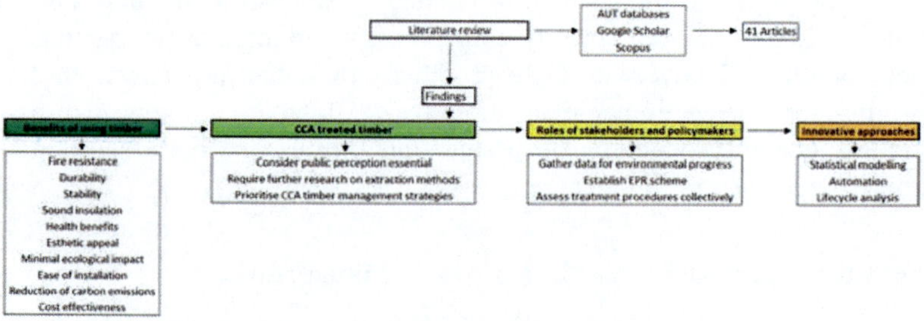

Fig. 1. Research findings

4 Conclusion

The construction industry plays a critical role in shaping the world's environmental trajectory. This research paper has addressed a lacuna in the existing literature concerning the sustainable management of construction waste, focusing primarily on timber, a much-favoured building material. Through a comprehensive literature review, the paper scrutinised the benefits of timber, the challenges posed by CCA-treated timber, the impact of stakeholders and policymakers on timber waste management, and the innovative approaches to waste management.

The findings highlighted numerous advantages of timber, including its durability, insulation properties, aesthetic appeal, and positive environmental impact. The potential of timber to act as a carbon sink emerged as a compelling solution to offset urban greenhouse gas emissions. CLT stood out for its significant reduction in carbon emissions compared to traditional materials like concrete and steel.

Nonetheless, the paper also revealed the challenges inherent in managing CCA-treated timber waste due to its potentially hazardous effects. Consequently, a deeper understanding, alternative treatment processes, and effective management strategies were proposed to alleviate these issues. Further, stakeholders and policymakers were recognised as instrumental in managing construction waste. Their roles range from implementing waste management schemes to establishing voluntary environmental responsibility programmes. Innovative strategies such as statistical modelling, automation, and systemic carbon reduction also emerged as potential solutions for managing construction waste.

Based on the findings, the following recommendations are made:

- Sustainable Material Selection: Prioritising timber in construction can dramatically lower carbon emissions whilst offering practical benefits such as fire resistance and ease of installation. Industry professionals and policymakers are urged to promote the increased use of sustainable timber in construction projects.
- Improved Management of CCA-Treated Timber: Stakeholders should investigate alternative timber treatments and advanced recycling processes to manage CCA-treated timber waste sustainably. Furthermore, public education is crucial to enhance the proper classification and disposal of treated timber.
- Policy and Stakeholder Engagement: Policymakers should introduce legislation to encourage the reduction of CCA-treated timber use, while the construction and forestry industries should actively participate in waste reduction strategies. Implementing EPR schemes could hold timber producers accountable for the waste they generate.
- Promotion of Innovative Strategies: Embracing technological advancements, such as statistical modelling and automation, can streamline waste management practices and promote efficient resource use. Construction industry stakeholders should invest in these innovative strategies to enhance waste reduction and recycling.
- Lifecycle Consideration: Stakeholders should adopt a lifecycle perspective when selecting materials to understand their long-term environmental impacts fully. Focusing on immediate use, recyclability, and end-of-life disposal can significantly reduce construction waste.

References

1. Hart, J., Pomponi, F.: More timber in construction: unanswered questions and future challenges. Sustainability **12**(8), 3473 (2020)
2. Luangcharoenrat, C., Intrachooto, S., Peansupap, V., Sutthinarakorn, W.: Factors influencing construction waste generation in building construction: Thailand's perspective. Sustainability **11**(13), 3638 (2019)
3. Labaran, Y.H., Mathur, V.S., Muhammad, S.U., Musa, A.A.: Carbon footprint management: a review of construction industry. Clean. Eng. Technol. **9**, 100531 (2022)
4. Taylor, C., Roy, K., Dani, A.A., Lim, J.B., De Silva, K., Jones, M.: Delivering sustainable housing through material choice. Sustainability **15**(4), 3331 (2023)
5. Ahmed, S., Arocho, I.: Mass timber building material in the US construction industry: Determining the existing awareness level, construction-related challenges, and recommendations to increase its current acceptance level. Clean. Eng. Technol. **1**, 100007 (2020)
6. De Araujo, V., et al.: Is cross-laminated timber (CLT) a wood panel, a building, or a construction system? A systematic review on its functions, characteristics, performances, and applications. Forests **14**(2), 264 (2023)

7. John, S., Buchanan, A.: Review of end-of-life options for structural timber buildings in New Zealand and Australia. University of Canterbury (2013). https://ir.canterbury.ac.nz/bitstream/handle/10092/10296/12653668_Review%20of%20end%20of%20life%20for%20timber%20Final%2026%20Feb%202013.pdf?sequence=1&isAllowed=y. Accessed 20 July 2023
8. Altaner, C.: Preservative treated timber products in New Zealand. Cellul. Chem. Technol. **56**(7–8), 705–716 (2022)
9. Marsono, A.K.B., Balasbaneh, A.T.: Combinations of building construction material for residential building for the global warming mitigation for Malaysia. Constr. Build. Mater. **85**, 100–108 (2015)
10. Chen, Q., Hall, D.M., Adey, B.T., Haas, C.T.: Identifying enablers for coordination across construction supply chain processes: a systematic literature review. Eng. Constr. Archit. Manag. **28**(4), 1083–1113 (2020)
11. Wolfswinkel, J.F., Furtmueller, E., Wilderom, C.P.: Using grounded theory as a method for rigorously reviewing literature. Eur. J. Inf. Syst. **22**(1), 45–55 (2013)
12. Jochem, D., Janzen, N., Weimar, H.: Estimation of own and cross price elasticities of demand for wood-based products and associated substitutes in the German construction sector. J. Clean. Prod. **137**, 1216–1227 (2016)
13. Younis, A., Dodoo, A.: Cross-laminated timber for building construction: a life-cycle-assessment overview. J. Build. Eng. **52**, 104482 (2022)
14. Dong, Y., Wang, R., Xue, J., Shao, J., Guo, H.: Assessment of summer overheating in concrete block and cross laminated timber office buildings in the severe cold and cold regions of China. Buildings **11**(8), 330 (2021)
15. Cole, R.J., Kernan, P.C.: Life-cycle energy use in office buildings. Build. Environ. **31**(4), 307–317 (1996)
16. Petruch, M., Walcher, D.: Timber for future? Attitudes towards timber construction by young millennials in Austria-marketing implications from a representative study. J. Clean. Prod. **294**, 126324 (2021)
17. Göswein, V., Reichmann, J., Habert, G., Pittau, F.: Land availability in Europe for a radical shift toward bio-based construction. Sustain. Cities Soc. **70**, 102929 (2021)
18. Kirsch, A., Ostendorf, K., Euring, M.: Improvements in the production of wood fiber insulation boards using hot-air/hot-steam process. Eur. J. Wood Wood Prod. **76**, 1233–1240 (2018)
19. Kunttu, J., et al.: Targeting net climate benefits by wood utilisation in Finland: participatory backcasting combined with quantitative scenario exploration. Futures **134**, 102833 (2021)
20. Nelson, J., Pickering, K.L., Beg, M.D.: Assessment of the potential of waste copper chromium and arsenic (CCA)-treated timber fibre reinforced polypropylene composites for construction. J. Compos. Sci. **7**(2), 48 (2023)
21. Safa, M., O'Carroll, D., Mansouri, N., Robinson, B., Curline, G.: Investigating arsenic impact of ACC treated timbers in compost production (a case study in Christchurch, New Zealand). Environ. Pollut. **262**, 114218 (2020)
22. Love, S.: Extended producer responsibility of treated timber waste. In: SB07 Sustainable Building Conference. Building Research, Auckland, New Zealand (2007)
23. Mohajerani, A., Vajna, J., Ellcock, R.: Chromated copper arsenate timber: a review of products, leachate studies and recycling. J. Clean. Prod. **179**, 292–307 (2018)
24. Hossain, M.U., Leu, S.Y., Poon, C.S.: Sustainability analysis of pelletised bio-fuel derived from recycled wood product wastes in Hong Kong. J. Clean. Prod. **113**, 400–410 (2016)
25. Dahlbo, H., et al.: Construction and demolition waste management–a holistic evaluation of environmental performance. J. Clean. Prod. **107**, 333–341 (2015)
26. Seadon, J., Aladwan, A., Golding, K.: Mandatory site waste management plans (2022). https://www.researchgate.net/profile/Jeff-Seadon-2/publication/363158378_Mandatory_Site_Waste_Management_Plans/links/630fe69d61e4553b95554bb2/Mandatory-Site-Waste-Management-Plans.pdf. Accessed 20 July 2023

27. Domingo, N., Batty, T.: Construction waste modelling for residential construction projects in New Zealand to enhance design outcomes. Waste Manage. **120**, 484–493 (2021)
28. Jayasinghe, L.B., Waldmann, D.: Development of a BIM-based web tool as a material and component bank for a sustainable construction industry. Sustainability **12**(5), 1766 (2020)
29. Stocchero, A., Seadon, J.K., Falshaw, R., Edwards, M.: Urban equilibrium for sustainable cities and the contribution of timber buildings to balance urban carbon emissions: a New Zealand case study. J. Clean. Prod. **143**, 1001–1010 (2017)
30. Dani, A.A., Roy, K., Masood, R., Fang, Z., Lim, J.B.: A comparative study on the life cycle assessment of New Zealand residential buildings. Buildings **12**(1), 50 (2022)

Quantitative Study on the Evolution of Urban Residential Spaces from the Perspective of Regionalism: A Case Study of Shanghai Lane Houses

Yaru Xu(✉)

Qingdao University of Technology, Qingdao 266520, Shandong, China
m15866170128@163.com

Abstract. Urban development in China has entered a stage of stock renewal, and regional development has become an important issue. The regional identity of buildings is crucial for promoting regional development. However, it is evident that architecture faces a crisis in cultural identity during the modernization. As one of the essential residential buildings in modern Shanghai, Lane houses are integral to urban renewal and represent an important aspect of Shanghai's culture. This study uses space syntax and M-shaped Diagram to explore the evolution of typical internal spatial structures of lane houses in different stages of development and their cultural connotations based on four variables: integration, choice, courtyard coupling, and loopiness. The results show that the central nodes (living room and traffic core) and courtyards' essential positions have remained unchanged throughout the evolution process and have been deeply influenced by traditional Chinese culture while retaining the depth of Eastern space. However, the distinctive features of traditional Chinese residential spaces have gradually lost been lost in evolution process, and the loopiness of service space, which is the main activity area for people, is gradually reduced.

Keywords: Regional Development · Quantitative Study · Interior Space

1 Introduction

With the rapid development of technology, urban development in China has generally entered a phase of stock renewal [1], and architecture faces a profound cultural identity crisis in the modernization process. Regionalism has become a theoretical consensus linking local identity and global discourse [2]. As a typical representative of the fusion of Chinese and Western cultures in Shanghai, Lane houses are one of the mainstream forms of residential buildings in modern Shanghai [3] and have become an important cultural symbol of Shanghai, highlighting its regional characteristics [4]. Therefore, it is urgent and vital to explore the profound logic of the internal spatial structure of lane houses. This study attempts to groundbreakingly use the space syntax and M-shaped Diagram to explore the evolution of the internal spatial structure of lane houses and summarize excellent experiences for their renewal.

W. Guo et al. (Eds.): GBCESC 2023, LNCE 328, pp. 570–579, 2024.
https://doi.org/10.1007/978-981-99-9947-7_59

2 Literature Review

2.1 Overview of Research on Interior Space of Shanghai Lane Buildings

Research on the changes in the interior space of Shanghai Lane buildings is primarily qualitative. For example, Xiao Xiong [5] summarizes the layout features of traditional stone-gate buildings, new-style stone-gate buildings, new-style garden-style apartments, and garden-style lane buildings, deepening the understanding of traditional lane buildings and lifestyles. Deng Huiyu [6] summarizes the evolution of the layout of stone-gate buildings, new-style lane buildings, and garden-style lane buildings and argues that the courtyard space is an essential means of creating a living atmosphere. Feng Qi [7] summarizes the layout of traditional Jiangnan courtyard houses, stone-gate buildings, and new-style lane buildings and argues that the new and local lane housing models have a continuity. Xu Ling [8] summarizes the evolution of the interior structure of stone-gate buildings.

Overall, the current research on the interior space structure of lane buildings needs a comprehensive summary of the evolutionary rules, incomplete analysis of the types of research objects, and scarce quantitative analysis. Therefore, this study will use Space Syntax and M-shaped Diagram to quantitatively analyze the interior space structure of lane buildings at various developmental stages and summarize the evolutionary rules of the interior space structure of lane buildings.

2.2 Theories of Space Syntax and M-Shaped Diagram and Current Research Status

Space syntax is a theory and method for analyzing space's topological structure and social attributes. [9] It is commonly used for the quantitative analysis of spatial structures. It can interpret the social and cultural logic hidden behind spatial structures well [10], such as Chen Chuanwen's research and optimization of the spatial form of Gannan Hakka houses based on space syntax [10]; Zhao Li et al.'s analysis of the spatial syntax of traditional homes in Songjiazhuang [11]. Compared to the entirely topological J map in space syntax theory, Wu Yihao [2] proposes M-shaped Diagram (morphological map) in *Research on the Inheritance of Courtyard-style Housing Space Quantification under the Perspective of Regionalism*, which has more excellent compatibility and coverage and is suitable for use in horizontal comparisons of a large number of samples. Space syntax theory can quantitatively study the structural form of space, but compared with M-shaped Diagram, it needs the restoration of the location and form features of building space. Therefore, this study will combine Space Syntax and M-shaped Diagram to study the interior space of lane buildings.

3 Research Method and Model Determination

The main spatial syntax theory research methods include axial analysis, convex space analysis, and visual analysis. [12] Among them, axial analysis is commonly used in large-scale urban space research [13], while visual analysis is often used to study visual spaces. Therefore, this study adopts the convex space analysis method, supplemented by M-shaped Diagram, for further analysis.

Considering that the Lane architecture incorporates the layout characteristics of traditional Jiangnan dwellings [14], the variables used in this study mainly include integration, choice, courtyard coupling, and loopiness.

(1) Integration: It represents the relationship between a space and the local or overall space. [15] The higher the integration, the higher the accessibility of the space within the entire spatial system, [16] which is easier to attract human flow.
(2) Choice: It represents the potential of attracting cross flow, and the trafficability of the space was evaluated by the number of times that any two nodes in the entire system pass through the node as the shortest path, except for that node. The higher the choice, the greater the potential of the space to attract cross-flow. In this study, the modified Choice [Norm] value is used as the data, which reflects the relationship between depth value and choice. [17]
(3) Courtyard coupling: Courtyard is the most significant feature of traditional Chinese dwellings in spatial organization and is related to the interaction between the outer space of the courtyard and the interior space at the construction level. [2] Considering the spatial status of dwellings, this study uses this indicator for the well-courtyard space in the lane houses. The calculation formula is as follows:

$$\mathrm{Yc} = \frac{s + 0.5r - 0.5k + 0.25v}{S} \tag{1}$$

Note. s is the number of directly accessible lines between all interior nodes and the courtyard node; r is the number of adjacent and inaccessible lines between all interior nodes and the courtyard node; k is the number of adjacent, inaccessible, and windowless lines between all interior nodes and the courtyard node; v is the number of lines between all interior nodes and the courtyard node that are separated by a corridor and not directly adjacent but can be seen through the window; S is the total number of solid lines in the M-shaped Diagram.
(4) Loopiness: It refers to the closed path from a node to the same node, which does not pass through repeated lines and nodes. [2] Compared with the circulation space in Chinese gardens, the higher the loopiness, the stronger the circulation experience of the space, and the richer the spatial hierarchy experience. Its calculation method is as follows:

$$\mathrm{R} = \frac{L}{2n - 5} \tag{2}$$

Note. L is the total number of basic loops in the system.

4 A Quantitative Comparison of the Development of Internal Spaces in Shanghai Lane Buildings

4.1 Sample Selection and Atlas Drawing

Chen Congzhou and Zhang Ming have proposed in *A Draft of the Modern Architectural History of Shanghai* [18] that Shanghai lane buildings have gone through six stages: old-style shikumen houses, "Guang-style" houses, new-style shikumen houses, new-style lane houses, garden lane houses, and apartment-style lane houses. The selection of research samples considers the comprehensiveness of the cross-span of Shanghai lane building stages and the typicality of the stages. This study selected the public Shun Lane as a sample for the old-style shikumen houses, Badaitou as a sample for the "Guang-style" houses, Dong Si Wen Lane as a sample for the new-style shikumen lane houses, Huai Hai Fang as a sample for the new-style lane houses, Shang Fang Hua Yuan type A as a sample for garden lane houses, and Xin Kang Garden as a sample for apartment-style lane houses.

4.2 Statistical Analysis of Sample Data

Firstly, using Depthmap to calculate the integration degree and selection degree of all the spaces in the samples, the results are shown in Table 1. This study then conducted a statistical analysis of the primary data of the M-shaped Diagram of each sample and calculated the courtyard coupling and loopiness of each sample. In order to better preserve trends and improve effectiveness, the inverse sine square root is used to process these two numerical values, and the corrected values of Yc^* and R^* are substituted for Yc and R, as shown in Table 2.

A Brief Overview of the Analysis of Integration, Selection, and Loopiness. In terms of integration, the samples generally exhibit the following standard features: the central node (guest hall, traffic core) is greater than the courtyard space; the main house is greater than the wing rooms. The central node is the highest status space in the internal space and is an important part of family life, except for the "Guang-style" house Badaitou, where the traffic nucleus was the central node due to the abolition of the front patio and the single flow line. The degree of integration of the main house is higher than that of the wing, which the traditional culture of hierarchy and inferiority should influence. The degree of integration reflects the tendency for lane buildings to converge from the periphery to the center under the influence of Confucian ritualism.

In terms of choice, the samples generally exhibit the following standard features: the central node (guest hall, traffic core) is greater than the other spaces; the main house is greater than the wing rooms. The central node has the most substantial traffic potential, highlighting its status as a pass-through public space, while the traffic potential of the main room is greater than that of the compartment, proving that the lane building has basic control over the public and private nature of the space.

In terms of loopiness, all the representative lane buildings of the period have a loopiness index, except for the "Guang-style" house Badaitou, which has a loopiness of 0. The main reason for this is that Badaitou has a single internal flow, no circulation, and a monotonous spatial hierarchy experience. The difference in the loopiness of the other samples is not too great but basically shows a trend of "rising, falling and rising again", which shows that the lane buildings have always retained the unique depth of oriental space in the process of development.

In conclusion, it can be seen that in the evolution of the interior space of lane buildings from traditional dwellings to modern residences, while ensuring the basic zoning, they have always been influenced by traditional Chinese culture, retaining the colors of oriental culture and the depth of oriental space.

Table 1. Calculation results of Convex Space Model Syntax for Lane Residences

Case Name	Plan	Integration	Choice
Old-style Shikumen Houses Gongshun Lane	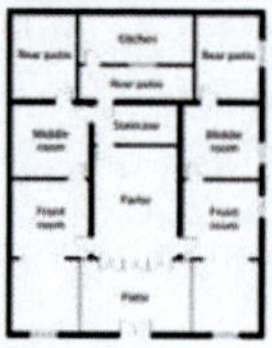	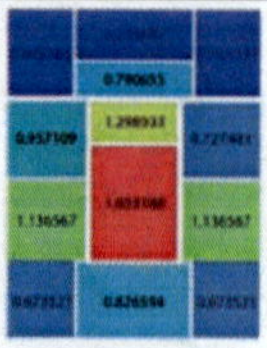	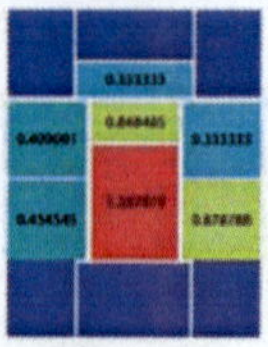
"Guang-style" Houses Bayatou	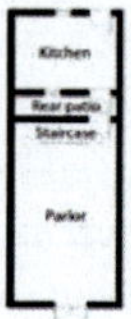	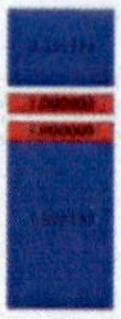	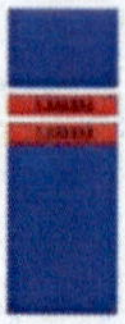
New-style Shikumen Houses Dongsiwen Lane	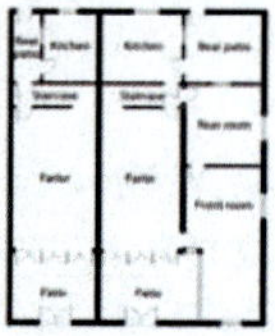	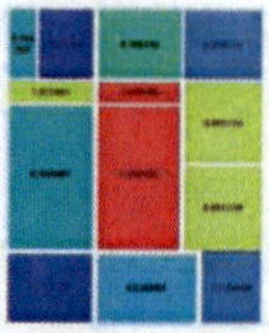	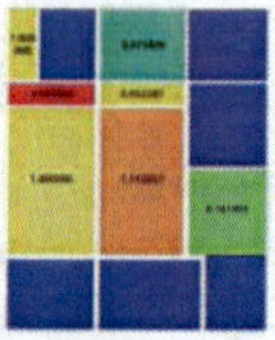
New-style Lane Houses Huaihaifang	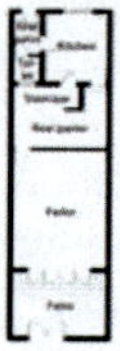	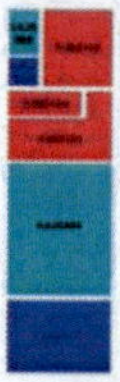	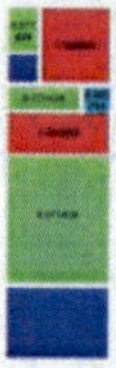
Garden Lane Houses Shangfang Huayuan Type A	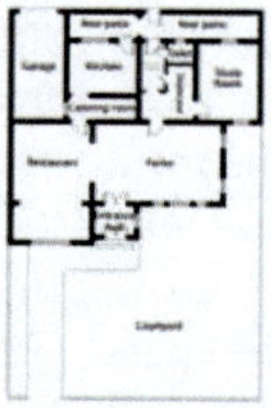	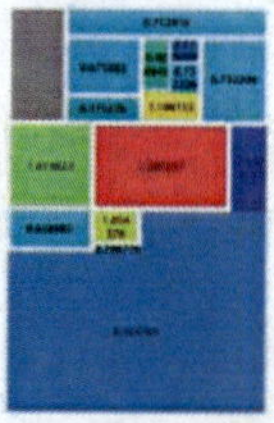	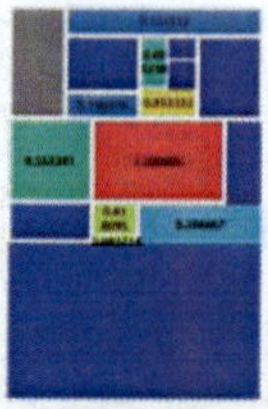
Apartment-style Lane Houses Xinkang Garden	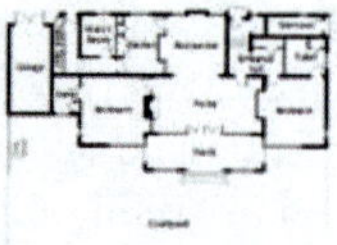	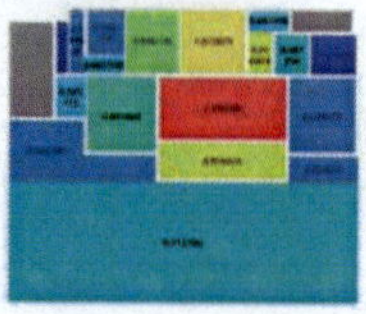	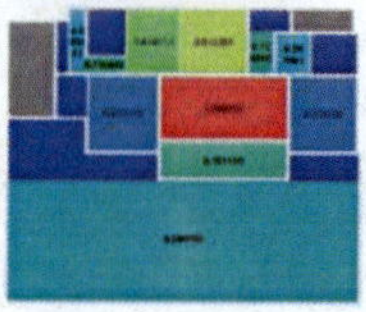

Note.The plan is sourced from books[19-21]. The integration and choice graphs were drawn by the author.

Table 2. M-shaped Diagram, Basic Data, and Attribute Indicators of Lane Residences

Case Name	M-shaped Diagram	Basic data	Courtyard coupling Yc*	Loopiness R*
Old-style Shikumen Houses Gongshun Lane	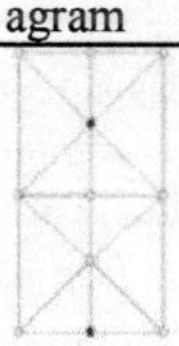	s=3 r=6 k=4 S=11 v=0 n=11 L=1	0.647284	0.244980
"Guang-style" Houses Badaitou		s=2 r=0 k=0 S=3 v=0 n=4 L=0	0.955317	0.000000
New-style Shikumen Houses Dongsiwen Lane	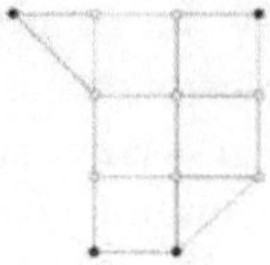	s=5 r=3 k=1 S=10 v=0 n=12 L=1	0.886077	0.231478
New-style Lane Houses Huaihaifang	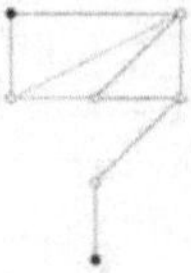	s=3 r=0 k=0 S=7 v=0 n=7 L=1	0.713724	0.339837
Garden Lane Houses Shangfang Huayuan Type A	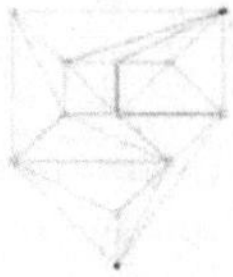	s=3 r=6 k=1 S=11 v=0 n=12 L=1	0.743683	0.231478
Apartment-style Lane Houses Xinkang Garden	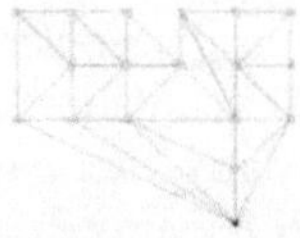	s=1 r=4 k=1 S=17 v=0 n=18 L=2	0.393564	0.256814

Note.The image was drawn by the author.

Analysis of the Calculation Results of the Central Node (Hall and Traffic Nucleus). Generally speaking, the central nodes (the hall and the traffic nucleus) are at the heart of the dwelling, taking on the functions of receiving guests, gathering families, and separating the front and back spaces.

From the point of view of plan features, the location of the hall and the traffic nucleus are close to each other. In terms of integration and choice, the guest halls and transport nuclei of each sample jointly occupy the highest and second highest positions in the system in terms of integration and choice, as well as accessibility and transport potential, and have the function of linking the front and back spaces in the spatial system.

It is easy to see that all the central nodes are connected to the courtyard space on at least one side and are a special feature of the dwellings.

In Huaihai Fang, a new-style lane house, the parlor is divided into a front, and a back parlor, with the front parlor being less integrated and less selective than the back parlor, i.e., the front parlor is the second most integrated and less selective in the system, with more accessibility than traffic, and is the 'nucleus' of the whole system.

In summary, the central node has always had high accessibility and transportability in the evolution of the lane and is complemented by the courtyard space, which provides a good landscape space. If the front and rear halls were separated, the result would be more accessibility than traffic, further emphasizing the central position of the halls as unassailable.

Analysis of the Results of the Courtyard Space Calculation. Generally speaking, the courtyard space has functions such as lighting, ventilation, rest, transition between indoor and outdoor spaces, and connection between the front and back planes. From the perspective of planar features, the courtyard space is enclosed by rooms and walls [22] instead of being enclosed by other rooms as in traditional culture, indicating a shift towards Western-style housing. In terms of integration and choice, apart from the "Guang-style" house's Badaitou, the courtyard space has reasonable integration and choice but is not at the highest level, which corresponds to its function of providing a private and comfortable place for recreation and regulating the indoor climate. From the perspective of coupling degree, the numerical value shows a trend of "rising, falling, flattening, and decreasing" in the evolution process, indicating that the interaction between the courtyard and indoor space in the organization of traditional Chinese houses gradually decreases during the transformation into Western-style housing.

The front courtyard underwent a process of "existence, disappearance, existence, and transformation into a courtyard" during the evolution, indicating the necessity of such space for people. However, due to its peripheral position and function as a transitional space, it has low integration and virtually no choice. The back courtyard showed a process of "existence, deviation from the center axis, and disappearance" during the evolution, in which it initially had functions such as lighting, ventilation, and transition between front and back spaces but lost these functions partially after deviating from the center axis, and ultimately disappeared. This is one of the main reasons for the decrease in courtyard coupling.

In summary, during the evolution of the lane-style architecture, the attributes of the courtyard space always conformed to its positioning, but it lost not only the introverted aspect of traditional culture but also the coupling with indoor space, gradually losing the characteristics of traditional Chinese houses in spatial organization.

Analysis of Results in the Circulation Space. "Circulation space" refers to a space that can be rotated back and forth. In traditional residential buildings, it mainly appears in the interaction of serviced spaces, which increases the permeability and profundity of the experience in serviced spaces and enhances the opportunities for family members to communicate unconsciously. Therefore, circulation space has meaning.

In terms of numerical values, except for the loopiness of the "Guang-style" house's Badaitou is 0, the loopiness of each sample does not differ significantly. Therefore, this study summarized the combination modules of circulation space for each sample (see Table 3) and further analyzed the circulation space.

Table 3. Analysis of loopiness spaces in lane houses

Case Name	Gongshun Lane	Dongsiwen Lane	Huaihai-fang	Shangfang Huayuan Type A	Xinkang Garden
Plan	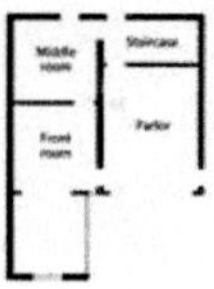	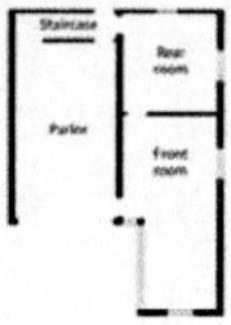	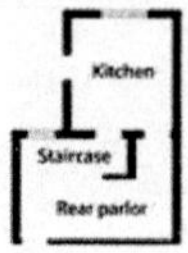	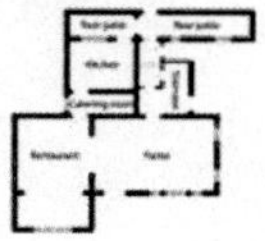	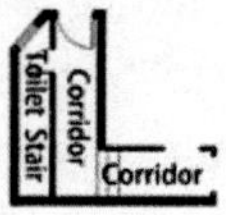
M-shaped Diagram	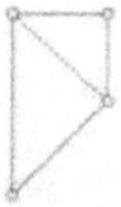	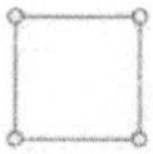	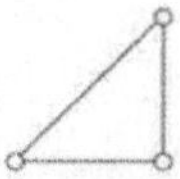	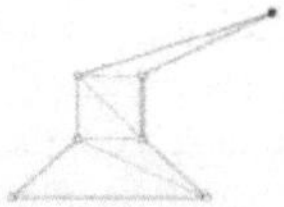	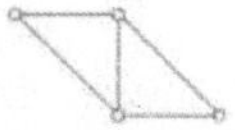

Note.The floor plan is from books [19-21]. The M-shaped Diagram is drawn by the author.

As can be seen from the table, the circulation space of the lane building basically consists of the guest room, dining room, compartment, and service space, and the well-yard space does not appear in it, and the circulation space in all stages basically has the participation of the service space. In the course of the evolution, the service space occupies an increasing proportion, and by the time the apartment-style lane Xinkang Garden is fully composed of service space, this relationship makes the circulation space lose some of its significance.

In conclusion, in the evolution of the lane architecture, the circulation space lacks the participation of the well-courtyard space and gradually loses its real meaning, the loopiness of the service space becomes weaker, and the hierarchical experience is gradually lacking.

5 Conclusion

Based on the theory of spatial syntax and the M-shaped Diagram, this study establishes a convex spatial model and an M-shaped Diagram for six representative buildings in six stages of the development of Shanghai Lane architecture. Finally, it is found that the overall analysis of the integration, choice, and loopiness shows that the internal space of the lane building has always been influenced by the traditional Chinese culture of Confucianism and hierarchy and has the depth of the basic oriental space during its evolution. The analysis of the well-yard space shows that the coupling between the well-yard space and the interior space has decreased, and the weakening and disappearance of the rear patio is one of the main reasons for this; the analysis of the circulation space shows that the loopiness of the service space, which is the main activity of people, has gradually become weaker.

In summary, this study suggests that in the future renovation and reconstruction of lane buildings, the essence of the spatial grouping of traditional Chinese dwellings should be combined with modern house types better to reshape the internal spatial structure of lane buildings. The research methodology and modeling approach of this study need to be adjusted and integrated. Due to the small scope of the study, the conclusions reached still need to be verified in future research explorations, and the laws summarised need to be further expanded and deepened.

References

1. Li, J.X.: Spatial response and institutional optimization research on urban stock industrial land renewal policies. Master's thesis, Tianjin University (2018)
2. Wu, Y.H., Wang, H.: Quantitative inheritance study of courtyard-style residential space from the perspective of regionalism. South. Architect. **11**, 31–40 (2022)
3. Tian, Q.R.: Research on interior transformation and reuse of Shanghai shikumen architecture. Master's thesis, Shanghai University (2020)
4. Wu, W.Z.H., Zhao, B., Chen, Q.: Bibliometric analysis and review of Shanghai lane house research (1980–2021). Shanghai Urban Plann.(04), 82–88 (2022)
5. Xiao, X.: Study on modern residential design based on the spatial form of traditional Shanghai lane house. Master's thesis, Xi'an University of Architecture and Technology (2020)
6. Deng, H.Y.: Research on the characteristics of modern Chinese lane house residential space and its modern interpretation. Master's thesis, Tianjin University (2020)
7. Feng, Q.: A comparative study of regionalism in modern Tianjin and Shanghai lane house residential. Master's thesis, Suzhou University of Science and Technology (2015)
8. Xu, L.: Research on interior design transformation of Shanghai shikumen lane house residential. Master's thesis, Qufu Normal University (2019)
9. Xin, L., Wei, Z.J.: Interpretation of academy ceremonial culture space based on spatial syntax: a case study of Yuelu academy. Urban Architect.(01), 127–130 (2022)
10. Yang, X.F., Chen, Y.L., Wan, Z.H.W., Zhang, A.H., Wu, L.Q.: Recognition of traditional Hakka residential space structure based on two syntactic models. Reg. Res. Dev. (02), 163–168(2019)
11. Chen, C.H.W., Fu, L., Chen,W.J.: Study and optimization of spatial form of Hakka residential houses in Southern Jiangxi based on spatial syntax. Furniture Interior Decoration (02), 122–126 (2023)
12. Zhao, L., Wang, J.P.: Analysis of spatial syntax in courtyard space of traditional Songjiazhuang residence. Industrial Architecture Magazine (eds.) Industrial Architecture Special Issue 2017 (2017)
13. Shan, Z.H.H.: Analysis of spatial view of Changtai plaza in Shanghai based on spatial syntax. Shanxi Architect.(07), 10–12 (2018)
14. Liu, Y.C.H.: Study on accessibility of comprehensive park in Shenyang based on spatial syntax. Master's thesis, Shenyang Agricultural University (2020)
15. Yu, Z.H.G.: Study on the evolution, protection and renewal of Shanghai lane house residential. Master's thesis, China Central Academy of Fine Arts (2010)

16. Cai, Y.M., Guo, Q.: Analysis of Lingnan garden space and "verticality" based on spatial syntax. Urban Architect. (01), 175–179 (2020)
17. Zhao, Y.C.H., Qin, M.L., Shi, Q.Q.: Accessibility analysis of urban park green space based on spatial syntax: a case study of the central district of Guangzhou. Geomatics World (02), 40–45 (2022)
18. Hu, X.R.: Study on the spatial form of well courtyard in southern Hunan residential houses based on spatial syntax. Master's thesis, University of South China (2020)
19. Chen, C.Z.H., Zhang, M.: Sketches of modern architecture history in Shanghai. Shanghai Sanlian Bookstore, Shanghai (1995)
20. Cao, W.: Shanghai Housing After the Opening to Foreign Trade. China Architecture & Building Press, Beijing (2004)
21. Shen, H.: Shanghai Lane Houses. China Architecture & Building Press, Beijing (1993)
22. Huang, H.: Analysis of the courtyard. Master's thesis, Chongqing University (2016)

Thermal Process Analysis in Passive Solar Dormitories in Plateau Areas: Onsite Case Study in Zoige

Diqing Wang[1], Jifan Cao[1], Yin Zhang[2](✉), and Dongsheng Huang[1]

[1] College of Architecture and Environment, Sichuan University, Chengdu 610065, China
[2] School of Architecture, Southwest Minzu University, Chengdu 610225, China
cdzhangyin@163.com

Abstract. In order to understand the thermal process of passive solar dormitories in highland areas, this paper establishes a heat transfer model for passive solar student dormitories and analyses the effects of convective heat transfer coefficient and thermal conductivity of their envelope surfaces on the heat storage, and then analyses the effects on indoor air temperature. The results show that increasing the convective heat transfer coefficient of the floor surface and the thermal conductivity of the floor can both increase the indoor air temperature. Once the thermal conductivity has been increased to 1.5 W/(m^2·K), further improvements are no longer noticeable in terms of temperature increase. Increasing the convective heat transfer coefficient of the floor surface and the thermal conductivity of the floor both enhance the heat storage of the floor. When the thermal conductivity is less than 1.5 W/(m^2·K), increasing the thermal conductivity has a greater impact on the heat storage of the floor, while increasing the convective heat transfer coefficient has a less obvious impact on the heat storage of the floor. This paper is a guide to optimizing the energy efficiency of passive solar buildings of the same type.

Keywords: Highland Areas · Passive Solar Dormitory · Energy Efficiency Optimization · Plateau Area

1 Introduction

The commercial energy consumption of rural dwellings in 2020 is 229 million tce, accounting for about 22% of China's total building energy consumption in that year, of which 344.6 billion tce is consumed by electricity [1]. With economic growth, China's building energy use will continue to grow and building energy conservation remains a challenge for China at present. Cold highland areas have the characteristics of dry and cold in winter, large daily ranges, long heating period, many strong winds, lack of fossil resources, inconvenient transportation and fragile ecological environment, but are rich in solar energy resources [2]. In order to use solar energy effectively and achieve the goal of energy saving in buildings in highland areas, solar house construction technology has come into being. Under the condition of making full use of solar energy, it can save 75%–90% of energy consumption [3].

W. Guo et al. (Eds.): GBCESC 2023, LNCE 328, pp. 580–590, 2024.
https://doi.org/10.1007/978-981-99-9947-7_60

Dengjia Wang [4] tested indoor and outdoor air temperature and humidity, solar radiation intensity and indoor wind speed in rural residential buildings and urban residential buildings in Lhasa, and investigated and analyzed the activities and clothing of indoor personnel in winter by means of a questionnaire survey. The results show that the use of passive solar technology in the Lhasa region can basically solve the problem of thermal comfort in winter. A comprehensive test and analysis of the effect of solar energy on the thermal environment of farming and herding settlements in the Qinghai-Tibet Plateau region showed that passive solar technology in the Qinghai-Tibet Plateau region has a significant improvement on the building's indoor thermal environment, and that the energy savings of a single-family solar building during the heating period can reach 76.0 × 106 kJ, with an energy saving rate of 86% [5]. Qinglong Gao [6] tested the Warm Nest Project of Zoige, and the results showed that the optimized design of passive solar buildings in solar-rich, high-altitude, cold regions can significantly improve the thermal environment of buildings in high cold regions, achieving an average daily indoor temperature of no less than 12 °C and a minimum temperature of no less than 10 °C at an average outdoor temperature of 10 °C below zero.

Most of the current studies are based on the analysis, testing and optimization of passive solar residential buildings in plateau areas, and lack the analysis and optimization of passive solar thermal processes in student dormitories in plateau areas. In this paper, a heat transfer model of the Passive Solar Building of Warm Nest 1 will be established, and given the design thermal parameters of each enclosure structure, the heat storage and discharge of each enclosure structure is calculated, and the influence of convective heat transfer coefficient and thermal conductivity of the surface of the enclosure structure on the heat storage is analyzed, and the direction of optimization of building energy efficiency is proposed. This paper has certain guiding significance for the optimization of energy saving in the same type of passive solar buildings.

2 Research Methods

2.1 Interlayer Heat Transfer Model

The air between the outer single-pane glass and the outer surface of the south wall is called the small interlayer, and the space corresponding to the double-glazed inner window is called the large interlayer, and the large interlayer and the small interlayer are connected to each other and the air flows freely (Fig. 1).

Small interlayer air heat balance equation.

$$c_a \rho_a V_1 \frac{dT_{a1}(\tau)}{d\tau} = Q_1(\tau) + Q_2(\tau) - Q_3(\tau) \tag{1}$$

where Q_1 is the convective heat exchange between the outer surface of the south wall and the air in the small interstitial layer, the expression is:

$$Q_1(\tau) = h_S A_1 (T_{w,o}(\tau) - T_{a1}(\tau)) \tag{2}$$

Q_2 is the convective heat transfer between the inner surface of a single glazed window and a small interstitial layer, expressed as:

$$Q_2(\tau) = h_1 A_1 (T_{b1}(\tau) - T_{a1}(\tau)) \tag{3}$$

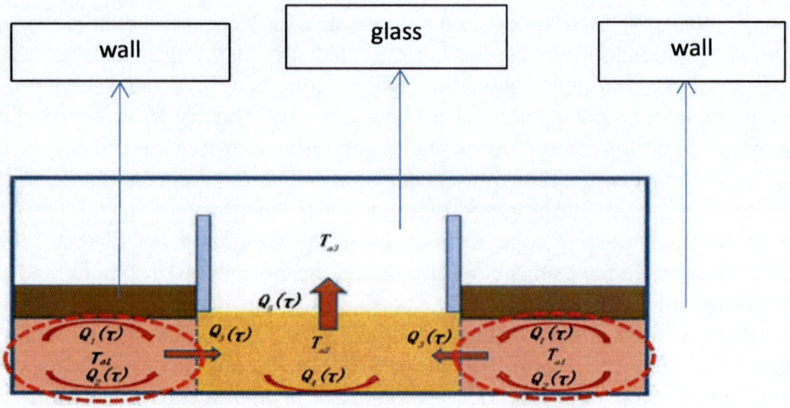

Fig. 1. Schematic of interlayer heat transfer

Q_3 is the convective heat transfer between the large interlayer and the small interlayer due to the thermal pressure, expressed as:

$$Q_3(\tau) = m_1(\tau)c_a(T_{a1}(\tau) - T_{a2}(\tau)) \tag{4}$$

where m_1 is the air flow rate for the thermal pressure convection between the large and small interstitial layers [7], expressed as:

$$m_1(\tau) = \frac{1}{3}\rho_a C_d A \sqrt{\frac{gH_1(T_{a1}(\tau) - T_{a2}(\tau))}{T_{a1}(\tau)}} \tag{5}$$

The large interlayer is the air layer between the double glazed inner window and the single glazed outer window. As the double glazed inner window opens during the day when solar radiation is strong and closes when solar radiation is weak and at night when there is no solar radiation, the heat exchange between the air in the large interlayer and the indoor air is different under different circumstances.

Large interstitial air heat balance equation:

$$c_a \rho_a V_2 \frac{dT_{a2}(\tau)}{d\tau} = 2Q_3(\tau) + Q_4(\tau) - Q_5(\tau) \tag{6}$$

where Q3 is the convective heat exchange between the small and large interstitial layers on one side due to thermal pressure. Q4 is the heat exchange between the inner surface of the single glazed outer window and the air in the large interstitial layer, expressed as follows:

$$Q_4(\tau) = h_2 A_2(T_{b2}(\tau) - T_{a2}(\tau)) \tag{7}$$

Different formulas are used to calculate the Cd values for different ratios of height H to room height, as follows.

Q5 is the convective heat transfer between the inner surface of a single glazed external window and the small interstitial layer, the expression for which varies depending on the opening and closing of the window.

Opening windows:

$$Q_5(\tau)= 2m_2(\tau)c_a(T_{a2}(\tau) - T_{in}(\tau)) \tag{8}$$

Closing windows:

$$Q_5(\tau)=h_3A_2(T_{a2}(\tau) - T_{b3}(\tau)) \tag{9}$$

where m_2 is the air flow rate for the thermal pressure convection between the large interstitial layer and the small interstitial layer, expressed as:

$$m_2(\tau) = \frac{1}{3}\rho_a C_d A_2 \sqrt{\frac{gH_2(T_{a2}(\tau) - T_{in}(\tau))}{T_{a2}(\tau)}} \tag{10}$$

In the above equations, C_a: Air specific heat capacity, kJ/(kg·K); ρ_a: Density of air, kg/m^3; T_{al}, $T_{a2}$and T_{in}: small and large interstitial and indoor air temperature, K; $T_{w,o}$: small interstitial south wall outer surface wall temperature, K; T_{bl}, T_{b2} and T_{b3}: small and large interstitial single glazing temperature and double glazing inner window outer glass temperature, K; h_1, h_2, h_3 and h_s: small and large interstitial single glazing outer surface convection heat transfer coefficient, double glazing inner window outer glass convection heat transfer coefficient and south wall outer surface convection heat transfer coefficient, W/(m^2·K); A_1, A_2 and A: Area of the external surface of the south wall of the small compartment, area of the single glazed external window of the large compartment and area of the opening between the small and large compartments, m^2; V_1, V_2: individual small and large interstices space volume, m^3; m_1, m_2: Air mass flow due to thermal pressure convection between the large and small compartments, between the large compartments and the room, kg/(m^3·s); g: acceleration of gravity, m/s^2; $H1$, $H2$: height above and below the opening, height of the double-layered inner window, m.

C_d is the flow coefficient, which is determined by the geometrical characteristics of the openings [8], for openings inside the building, different formulas are used to calculate the C_d value depending on the ratio between the height of the internal opening H_0 and the height of the room H, as follows:

$$C_d = \begin{cases} 0.0558 & \left| \dfrac{H_0}{H} \leq 0.2 \right. \\ 0.609\dfrac{H_0}{H} - 0.066 & \left| 0.2 < \dfrac{H_0}{H} < 0.9 \right. \\ 0.4821 & \left| \dfrac{H_0}{H} \geq 0.9 \right. \end{cases} \tag{11}$$

Among the large and small interstitial openings involved in Warm Nest 1 $H_0/H = 0.91 > 0.9$.

2.2 Transparent Envelop Heat Transfer Models

Due to the small heat capacity of the glass, the temperature of each layer is homogenized and treated separately as a temperature node. The equation for the heat balance of a single layer of glass corresponding to a small interlayer is as follows (Figs. 2 and 3).

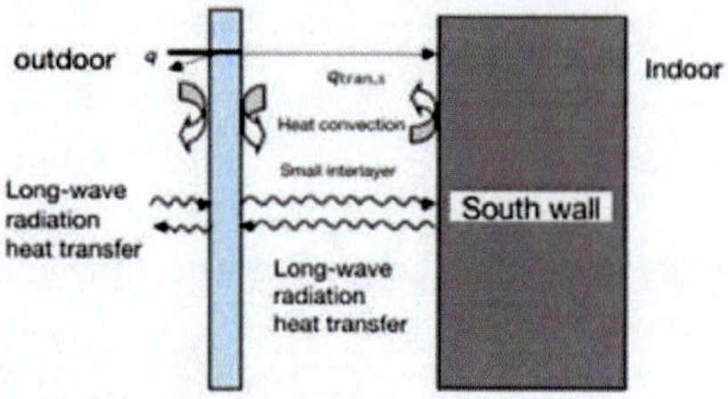

Fig. 2. Heat transfer diagram for single glazing

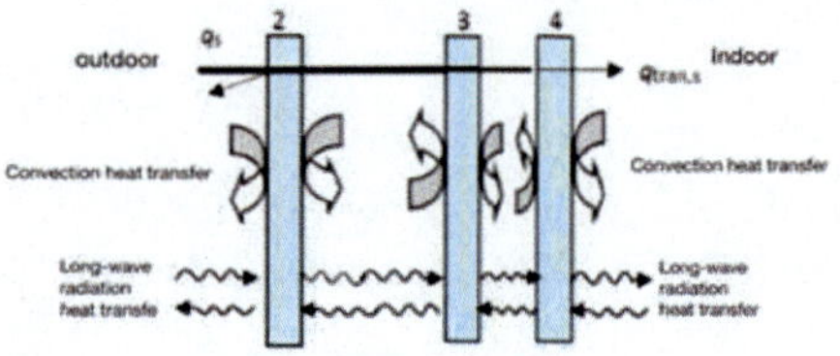

Fig. 3. Heat transfer diagram for triple glazing

Heat balance equation for external single glazed windows:

$$c_{P1}\rho_1\Delta x_1\frac{dT_{b1}}{d\tau} = h_{out}(T_{out} - T_{b1}) + h_1(T_{a1} - T_{b1}) + q_{r,out} + q_{r,w} + q_{1,s} \tag{12}$$

In which, $c_{P1}\rho_1\Delta x_1$ denotes the heat capacity of the outer, middle and inner glass layers, respectively; T_{b1} denotes the temperature of the single glass layer, respectively, K; $q_{r,w}$ indicates the indoor thermal disturbance obtained by the inner surface of the window and the long-wave radiant heat from the inner surface of the rest of the envelope, W/m^2; $q_{r,out}$ indicates the long-wave radiant heat obtained by the outer glazing, W/m^2; $q_{1,s}$ indicates the solar radiant heat absorbed by the outer and inner glazing respectively, the values of which are calculated according to Jiang Yi's transmission calculation model [3].

When the double glazed inner window is closed, the single glazed window and double glazed window together constitute a triple glazed window, again due to the smaller heat capacity of the glass, so the temperature of each layer of glass is treated as an even temperature, respectively, as a temperature node. The heat balance equation for each layer of glass in a triple glazed window is as follows.

$$c_{P2}\rho_2\Delta x_2\frac{dT_{b2}}{d\tau} = h_{out}(T_{out} - T_{b2}) + h_2(T_{a2} - T_{b2}) + h_{r,23}(T_{b3} - T_{b2}) + q_{r,out} + q_{2,s} \tag{13}$$

$$c_{P3}\rho_3\Delta x_3\frac{dT_{b3}}{d\tau} = h_3(T_{a2} - T_{b3}) + h_{r,23}(T_{b2} - T_{b3}) + h_{34}(T_{b4} - T_{b3}) + q_{3,s} \tag{14}$$

$$c_{P4}\rho_4\Delta x_4\frac{dT_{b4}}{d\tau} = h_4(T_{in} - T_{b4}) + h_{34}(T_{b3} - T_{b4}) + q_{r,in} + q_{4,s} \tag{15}$$

$$h_{ij} = h_{c,ij} + h_{r,ij} \tag{16}$$

where $c_{P2}\rho_2\Delta x_2$, $c_{P3}\rho_3\Delta x_3$, $c_{P4}\rho_4\Delta x_4$ denote the heat capacities of the outer, middle and inner glass layers respectively; T_{b2}, T_{b3}, T_{b4} indicate the temperature of the outer, middle and inner glass layers respectively, K; $q_{r,in}$ indicates the indoor thermal disturbance obtained from the inner surface of the window and the long-wave radiant heat from the inner surface of the rest of the envelope, W/m^2;

$q_{r,out}$ indicates the long-wave radiant heat gained by the outer glass, W/m^2; $q_{2,s}$, $q_{3,s}$, $q_{4,s}$ represent solar radiation heat absorbed by inner glass.

Where $h_{c,ij}$ and $h_{r,ij}$ are the air and long-wave radiation heat transfer coefficients between adjacent glazing.

$$h_{r,ij} = \sigma\left(\frac{1}{1/\varepsilon_{i,2} + 1/\varepsilon_{j,1} - 1}\right)[(T_i + 273)^2 + (T_j + 273)^2]$$
$$[(T_i + 273) + (T_j + 273)] \tag{17}$$

$\varepsilon_{i,1}$ and $\varepsilon_{i,2}$ are the long-wave emissivity of the left and right surfaces of the ith layer of glass, respectively.

2.3 Heat Transfer Model for Non-transparent Enclosures

In this paper, the heat capacity method model is used as the mathematical model and applied to all non-transparent enclosures. For non-linear enclosures, all physical properties except the volumetric specific heat capacity are considered as constants and the effect of cooling and heating processes on the variation of the volumetric specific heat capacity with temperature is neglected. To facilitate the calculations, the heat transfer of the non-transparent enclosure is treated as one-dimensional thermal conductivity, and since it satisfies a height and width of 8–10 times its thickness, the calculation error is no greater than 1% [9]. The control equation is expressed as follows (Fig. 4):

$$\rho c \frac{\partial t_w}{\partial \tau} = \lambda \frac{\partial^2 t_w}{\partial x^2} \tag{18}$$

Where t_w is the temperature of the material at moment τ. The heat capacity c, density ρ and thermal conductivity λ of the material do not vary with temperature and can be taken to the appropriate values depending on the material.

The boundary conditions are:

$$q_{\mathrm{r,in}} + q_{lw} + h_{\mathrm{in}}(t_{a,\mathrm{in}} - t_{\mathrm{w,in}}) = \lambda \left.\frac{\partial t_{\mathrm{w}}}{\partial x}\right|_{x=L} \tag{19}$$

$$q_{\mathrm{r,out}} + h_{\mathrm{out}}(t_{\mathrm{a,out}} - t_{\mathrm{w,out}}) = -\lambda \left.\frac{\partial t_{\mathrm{w}}}{\partial x}\right|_{x=0} \tag{20}$$

The initial conditions are

$$t(x, \tau)|_{\tau=0} = t_{\mathrm{init}} \tag{21}$$

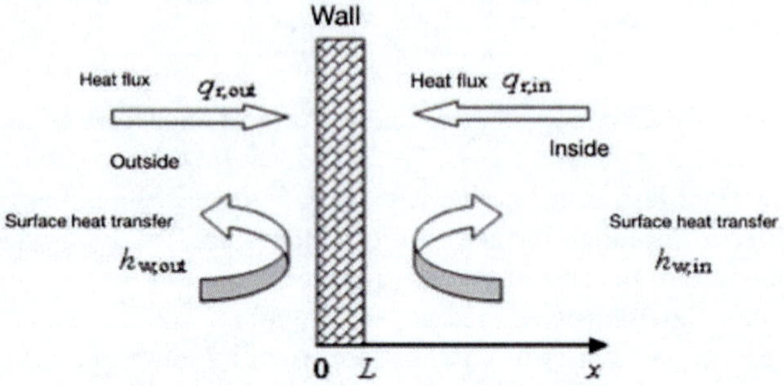

Fig. 4. Schematic diagram of heat transfer to the wall panel surface

where, t_w, $t_{a,out}$ - non-transparent envelope node temperature and outdoor air temperature, °C; λ - thermal conductivity of non-transparent enclosures, W/(m·K); x - non-transparent envelope coordinates, m; $h_{w,in}$, $h_{w,out}$ -convective heat transfer coefficients on the internal and external surfaces of non-transparent enclosures, W/(m^2·K); $q_{r,out}$, $q_{r,in}$ - radiant heat exchange heat flow from the inner and outer surfaces of non-transparent enclosures, W/m^2.

$q_{r,in}$ is mainly from internal heat disturbance (lighting, human body, equipment, etc.), mutual radiation from the surface of the envelope and solar energy entering the room through windows; $q_{r,out}$ is mainly from solar radiation, sky long-wave radiation and earth long-wave radiation; the convective heat transfer coefficient of the inner surface of the non-transparent envelope structure is calculated according to natural convection, the calculation formula is shown in ASHRAE manual [10], the convective heat transfer coefficient of the highland cold area will be different from the plain area due to the thin air density. The internal maintenance structure is considered to be insulated in its middle layer, meaning that the sum is 70% of the solar radiation entering the room is absorbed by the floor, while the remaining 30% is distributed equally by area to the other internal surfaces of the envelope [11].

2.4 Determination of Relevant Parameters

In the process of analyzing the thermal environment of a building, the convective heat transfer coefficient on the surface of the envelope is an important parameter, and the recommended values given in the heating, ventilation and air conditioning design codes [12] are all given for standard atmospheric pressure. At high altitudes, due to the low atmospheric pressure, the physical parameters of the air change significantly, and according to the theory of heat transfer, the convective heat transfer coefficient on the surface of the envelope will change significantly. The effect of altitude on the convective heat transfer coefficient has been studied by many scholars [12–14] and calculations have been done in Xiao Wei's study [15].

As the location of Warm Nest 1 is approximately 3500 m above sea level, the convective heat transfer coefficient is 2.3 W/(m^2·K) for the internal surfaces of the walls, 2.4 W/(m^2·K) for the floor and 0.65 W/(m^2·K) for the ceiling. The average convective heat transfer coefficient for the outside surface of the envelope is approximately

23 W/(m^2·K). The convective heat transfer coefficient is 1.4 W/(m^2·K) for the small interlayer, 6.9 W/(m^2·K) for the large interlayer and 2.4 W/(m^2·K) for the double glazed interlayer which is purely thermally conductive.

In this paper, we will use the above passive solar dormitory heat transfer model to calculate and study the effect of heat storage in the room envelope on the thermal comfort of the indoor environment based on the envelope design parameters shown in Table 1, and propose an optimization scheme.

Table 1. Table of building envelope design parameters

Categories	Value
shape factor	0.5
Number of floors	3
Thickness of outer glass	6 mm
Outer glass U-value	5.5 W/(m^2·K)
Solar transmission ratio of the outer glass	0.82
Internal window construction layer thickness	6Low-E + 12air + 6
Internal window U-value	2.5 W/(m^2·K)
Solar transmission ratio of inner windows	0.51
Roof U-value	0.28 W/(m^2·K)
Outside wall U-value	0.30 W/(m^2·K)

3 Results and Discussion

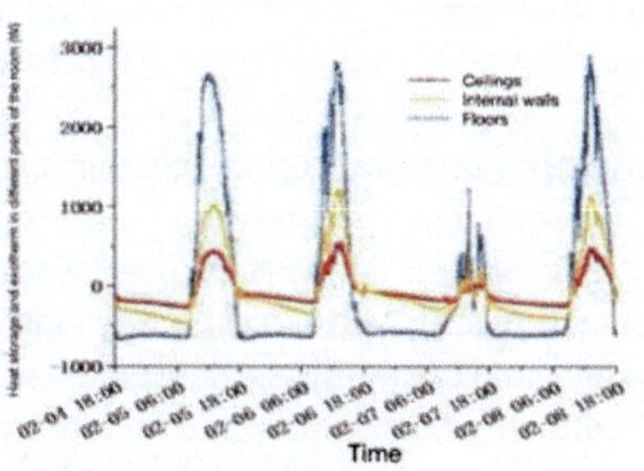

Fig. 5. Heat storage and exotherm in different parts of the room

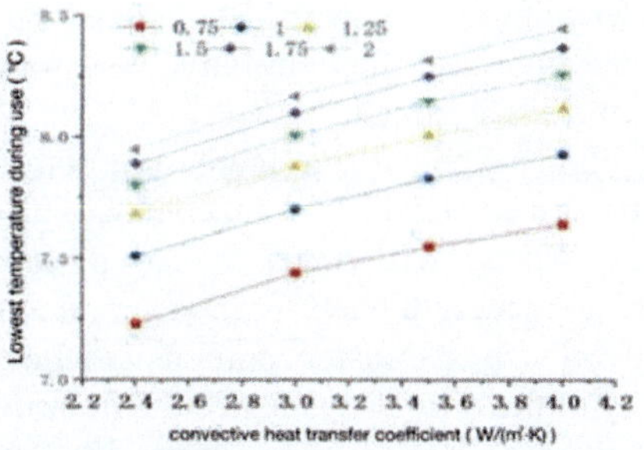

Fig. 6. The relationship between the minimum air temperature and convective heat transfer coefficient during the use period at different thermal conductivity

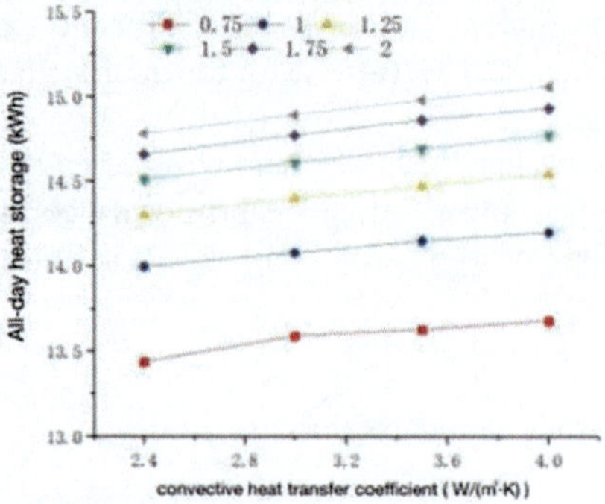

Fig. 7. Relationship between floor heat storage and convective heat transfer coefficient under different thermal conductivity

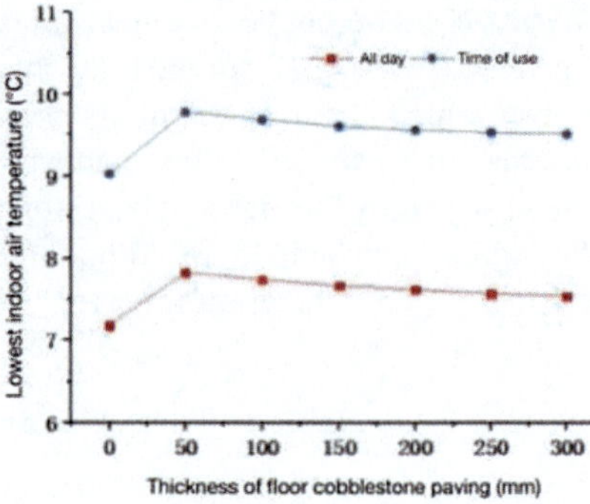

Fig. 8. Improving the effect of floor thermal storage on air temperature

Figure 5 shows that by step function controlling the south-facing transparent envelope, the interior can be effectively heated, with solar radiation first irradiating the interior walls during the day to heat the walls, and then heating the interior air in the form of convection to raise the interior air temperature, with the floor receiving the most direct solar radiation during the day, so making the greatest contribution to raising the interior air temperature.

The air density at 3500 m altitude at room temperature is about 0.82 kg/m^3. Due to the influence of air density, the convective heat transfer coefficient of the inner surface of the envelope is smaller than that of the plain area, so in order to improve the indoor thermal environment from the perspective of thermal storage, this paper continues to study in depth the effect of changing the thermal conductivity of the floor and the convective heat transfer coefficient of the surface on the indoor thermal environment. In order to improve the indoor thermal environment from the perspective of thermal storage, this paper continues to investigate in depth the impact of changing the thermal conductivity of the floor and the convective heat transfer coefficient of the surface on the indoor thermal environment.

Figure 6 shows that increasing both the convective heat transfer coefficient of the floor surface and the thermal conductivity of the floor can increase the indoor air temperature. When the thermal conductivity increases to 1.5 W/(m^2·K), further improvement is no longer significant for the temperature increase. Figure 7 shows that increasing the convective heat transfer coefficient of the floor surface and changing the thermal conductivity of the floor can enhance the floor thermal storage, when the thermal conductivity is less than 1.5 W/(m^2·K), increasing the thermal conductivity has a greater effect on the floor thermal storage, and the increase of the convective heat transfer coefficient has no significant effect on the floor thermal storage.

In order to strengthen the room heat storage, combined with the actual situation, on the basis of the original concrete floor paved with goose heating stone, the relationship between the lowest indoor air temperature and the thickness of the pebble paving is shown in Fig. 8, when the thickness of the pebble paving is 50mm, the lowest indoor air temperature can be increased to 9.8 °C during the use period.

4 Conclusions

In this paper, a heat transfer model for the Warm Nest 1 passive solar building is developed and the heat storage and discharge of each envelope is calculated given the design thermal physical parameters of each envelope. The results show that the floor contributes the most to the indoor air temperature increase as it receives the most direct solar radiation during the day. The effect of varying the convective heat transfer coefficient of the floor surface and the thermal conductivity of the floor on the amount of heat stored in the floor is then investigated, and the results show that when the thermal conductivity is increased to 1.5 W/(m^2·K), further improvements are no longer significant for the increase in temperature. Increasing the convective heat transfer coefficient of the floor surface and changing the thermal conductivity of the floor can both enhance the thermal storage of the floor, and when the thermal conductivity is less than 1.5 W/(m^2·K), increasing the thermal conductivity has a greater impact on the thermal storage of the floor. The effect of increasing convective heat transfer coefficient on floor thermal storage is not obvious. Combined with the actual situation, goose warming stones were laid on the basis of the original concrete floor. When the thickness of the pebbles laid was 50 mm, the minimum indoor air temperature could be increased to 9.8 °C during the use period. This paper can provide theoretical support and application guidance for optimizing the indoor thermal environment of solar passive buildings in the same type of plateau region.

Acknowledgement. This work is supported by National Natural Science Foundation of China (No. 52108032) and Sichuan Science and Technology Research Program (No. 2022NSFSC1148).

References

1. Hou, E.Z.: Annual development research report on building energy efficiency in China 2022 (public buildings topic) issue. Build. Energy Efficiency (04), 146 (2022). (in English and Chinese)
2. Li, K., He, F.N.: Regional analysis of terrestrial solar resource development potential in China. Adv. Geograph. Sci. **09**, 1049–1054 (2010)
3. Yan, Y.F., Zhang, Z.E., Zhang, L.: Solar energy utilization technology and its applications. J. Solar Energy **S1**, 47–56 (2012). https://doi.org/10.19912/j.0254-0096.2012.s1.001
4. Wang, D.J., Liu, Y.F., Wang, Y., Wang, Y.: Evaluation of winter indoor thermal environment testing in residential buildings in Lhasa. Build. Sci. **12**, 20–24 (2011). https://doi.org/10.13614/j.cnki.11-1962/tu.2011.12.013
5. Wang, D.J., Liu, Y.F., Liu, G.P.: Experimental study on the heating performance of passive solar buildings on the Qinghai-Tibet Plateau. Sichuan Build. Sci. Res. **02**, 269–274 (2015)
6. Gao, Q.L., Rong, X.Y., Qian, F., Li, P.Y.: Zero heating energy consumption student dormitories in high altitude cold regions - an example of Ruoerge warm nest 1 project. Constr. Sci. Technol. (08), 57–59+65 (2019). https://doi.org/10.16116/j.cnki.jskj.2019.08.010

7. Larsen, T.S., Heiselberg, P.: Single-sided natural ventilation driven by wind pressure and temperature difference. Energy Build. (6) (2006). https://doi.org/10.1016/j.enbuild.2006.07.012
8. Ruan, F.: Experimental and simulation study of natural ventilation characteristics of single openings. Doctoral dissertation, Hunan University (2016)
9. Balcomb, J.D., Hedstrom, J.C., McFarland, R.D.: Simulation analysis of passive solar heated buildings Preliminary results. Solar Energy (3) (1977). https://doi.org/10.1016/0038-092X(77)90071-8
10. Thevenard, D.J., Humphries, R.G.: The calculation of climatic design conditions in the 2005 ASHRAE handbook-fundamentals. ASHRAE Trans. (1) (2005)
11. Li, Y.Z., Di, H.F., Fang, X.D.: Principles of passive solar house and its design. Energy Press (2003). GB 50019-2003 (1989)
12. Cui, Y.: Some technical problems in the design of HVAC in high altitude areas. HVAC **06**, 39–42 (1999)
13. Liu, Y.D., Song, L.X., Zang, J.B.: Study on the heat transfer performance of plate electric heaters under low air pressure. Fluid Mech. **06**, 56–59 (2004)
14. Hu, S.T., Zhu, C., Wang, D.: Testing of natural convection heat transfer performance of electric heaters under low pressure conditions. HVAC (03), 22–24+21 (2006)
15. Xiao, W.: Research on direct benefit solar heating in remote areas of southwest Tibet. Doctoral dissertation, Tsinghua University (2010)

Functional Floor Plan Adaptation for Age-Friendly Housing in the Context of Ageing in Place

Feng Wang[1], Bo Zhang[1], Xiangyun Wang[1], and Jie Liu[2(✉)]

[1] Shandong Youth University of Political Science, No. 31699 Jingshi Dong Road, Jinan, Shandong, China
[2] Qilu Normal University, No. 2 Wenbo Road, Jinan, Shandong, China
20188528@qlnu.edu.cn

Abstract. Global ageing has further increased and, in line with national conditions, China will adopt ageing in place as the main form of retirement. The number of age-friendly residential space renovations is increasing. At the same time, China's existing residential stock is large and voluminous, mostly with wall bearing structures. At the same time, due to the current economic situation, the layout of existing homes cannot be modified too much, and the vast majority of homes have to retain their original wall and window and door openings. This requires the designer to have a clearer understanding of the floor plan characteristics of the home, in order to provide the most efficient and age-appropriate space layout for the elderly. In this paper, a common house plan is chosen. Based on the theory of spatial syntax, it analyses in depth the degree of connectivity, integration, control and depth value indicators of the individual spaces of the house plan. The characteristics of each functional space are quantified and analysed, and finally the most suitable space is selected for matching with the specific function. It is hoped that this will provide a cognitive approach to the ageing retrofitting of existing homes in China.

Keywords: Existing homes · Ageing in place · Spatial syntax · Ageing-friendly retrofit

1 Research Background

The growing size of the elderly will peak in 30 years, according to the World Population Prospects programme projections and revised estimates based on the Seventh Universal Data, the population over 60 years of age will be 256 million, reaching 299 million in 2025 rising to 441 million in 2035, and will peak in 2052 into a slow decline.

As the ageing population continues to increase, China has not adopted the community care or nursing home care of other countries, but has proposed ageing at home. On the one hand, Chinese elderly people have a strong sense of family, believing that home is the warmest place and has a greater sense of belonging than nursing homes or retirement

W. Guo et al. (Eds.): GBCESC 2023, LNCE 328, pp. 591–597, 2024.
https://doi.org/10.1007/978-981-99-9947-7_61

flats. It is also less costly for the children and does not impose an additional financial burden on the family.

As the elderly age, their physical functions continue to decline. In addition to safety, the space environment needs to be more efficient in its use. In other words, the use of space should be as efficient as possible, while keeping the space unchanged, so as to reduce the physical energy consumption of the elderly [1]. The functional areas that are frequently used are placed in the most convenient locations. Ultimately, the efficiency of space use is improved to compensate for the reduced mobility of the elderly due to the continued decline of their physical functions.

China's existing housing stock is large and voluminous, and most of the housing stock is of brick and concrete construction under the opportunity of reform and opening up. Urban housing construction is still in a stable stage of development. The renovation of existing dwellings is particularly important in light of the national measures for ageing in place. Restricted by the original building structure, the economics of renovation and other factors, the current renovation of age-friendly spaces has significant restrictions on the renovation of walls, the original building plan is less likely to be renovated, and the vast majority of design is to be carried out on top of the original plan. In other words, the designer has to go about planning the layout of an age-friendly floor plan without any changes to the flat walls. This requires a clearer and more accurate understanding of the original building plan. This requires a clearer and more accurate understanding of the original building plan, a better understanding of the spatial 'DNA' of the building plan and a more rational layout of the space functions.

The theory of spatial syntax is a formal study of the "genes" of the spatial plan, which is a theory and technique of space and society, mainly reflecting the connection and constraints between space and space without other factors as a background. This theory is applicable to all residential spaces in China. In this paper, we study the internal space types of Chinese residences, mainly through the Depthmap software, which is used in the process of spatial syntax analysis, to quantify the internal characteristics of interior spaces and finally combine them with the use of functions to achieve the most reasonable layout of floor plan functions [2].

2 Research Contents

2.1 Spatial Syntax Theory

Spatial syntax theory has been accepted by many scholars for the analysis of the internal connection of space, this paper mainly selects several concepts related to the efficiency of space use for the analysis of residential space. Connectivity value, is the number of nodes in the spatial system with the space adjacent to other spaces, this indicator to judge the index of spatial permeability, the better the permeability, the greater the degree of connectivity; Integration value, is the spatial nodes and all nodes in the whole system contact measure space accessibility indicators, the more integration the higher the degree of convenience, the better the accessibility; Control value, said a space in the system to its surrounding space to judge the degree of influence of space in the whole space, the greater the degree of influence control, the greater the influence on the rest of the space. Depth, ignore the geometric scale, only consider the connection way to judge the space convenience of indicators, the greater the depth, from a node to another node of the topographic step, the farther away from the central space, the worse the convenience [3].

Through the analysis of the above indicators, the intrinsic genetic value of each space in the residential space will be analysed in depth, and the space use efficiency of each space will be comprehensively derived, so that the data can be used to better cognise the current situation of each space in the residential house, and ultimately configure the space layout more reasonably.

2.2 The Current Situation of Existing Residential Space Form

In this paper, we have selected a 131 sq. ft. flat with three bedrooms and two bathrooms, with a square layout, a dining room, a living room running through, a large balcony on the south side, and a long and narrow corridor running through the entrance. The original corridor connects the 3 bedrooms to the bathroom, the kitchen to the dining room, and the flow from the living room to the kitchen is long.

In the original space for the elderly, the flow of activities and the use of space does not provide a comfortable living space for the elderly, we use spatial syntax software to quantitatively analyse the layout and design of the plan, which can provide a greater degree of convenience for the elderly in the home life of dynamic flow, improve the efficiency and experience of life.

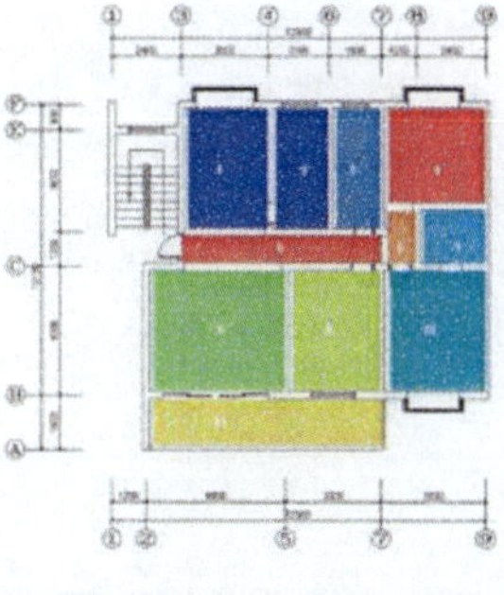

Fig. 1. Graphical spaces in different colors

Before the quantitative analysis of the interior spaces, a note on the graphical representation of the spaces [4]. Due to the limitations of the software recognition. As shown in Fig. 1, in order to meet the initial recognition by the software, a colour is given to each graphical space. But without any relation to the subsequent calculation results, setting a code to each space, used only as a way to differentiate the spaces when exporting the data [5].

3 Results

Firstly, the connectivity value for each space is analysed. Connectivity value indicates the number of spaces to which a space can be directly connected [6]. The higher the connectivity value, the better the accessibility of the space; conversely, the lower the connectivity value, the worse the accessibility of the space, which is generally less accessible to the population (Fig. 2).

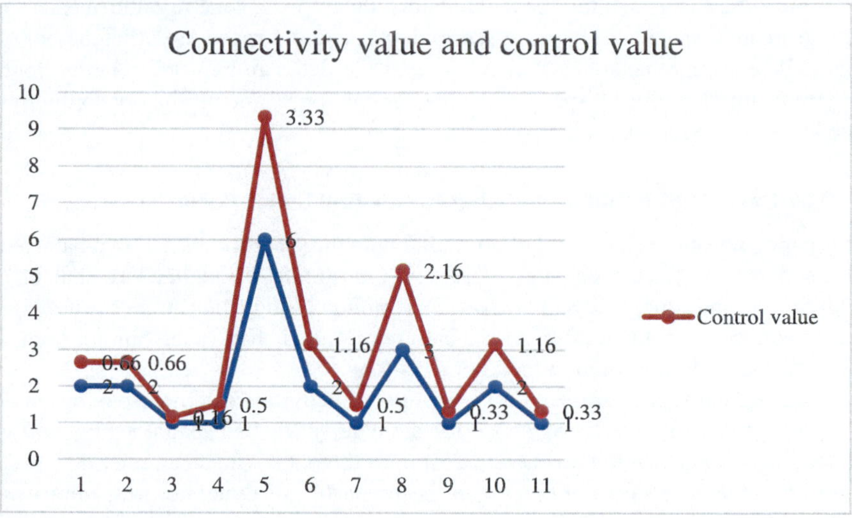

Fig. 2. Comparison of connectivity value and control value

The connectivity value graph shows the results of the analysis [7]. The space with the highest connectivity value is corridor 5, which has a connectivity value of 6, i.e. it connects 6 convex spaces, followed by space 8 (living room) with a connectivity value of 3, spaces 1, 2, 6 and 10 with a connectivity value of 2, and spaces 3, 4, 7, 9 and 11 with a connectivity value of 1 (Fig. 3).

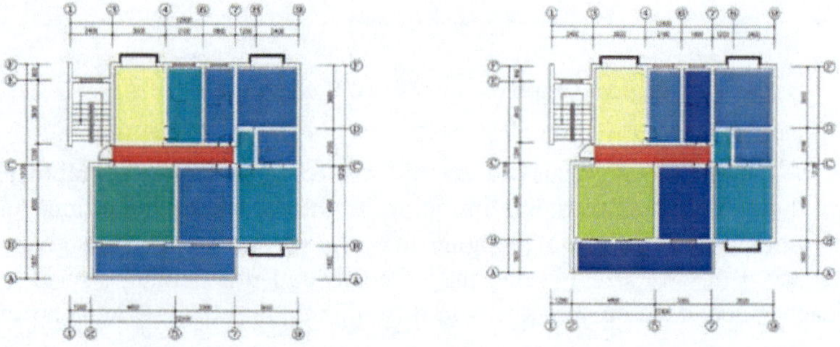

Fig. 3. Plan view of connectivity value and control value

In the analysis of the value of control of the convex spaces in the graphical representation of the degree of control, space 4 has the highest value of connectivity value and the highest value of control, with a control of 3.3. It is to be expected that the degree of control of a large degree of connectivity value is correspondingly higher, but the size of the space and the proportion of the space in the space are different, which means that the weights are not the same, leading to different results (Fig. 4).

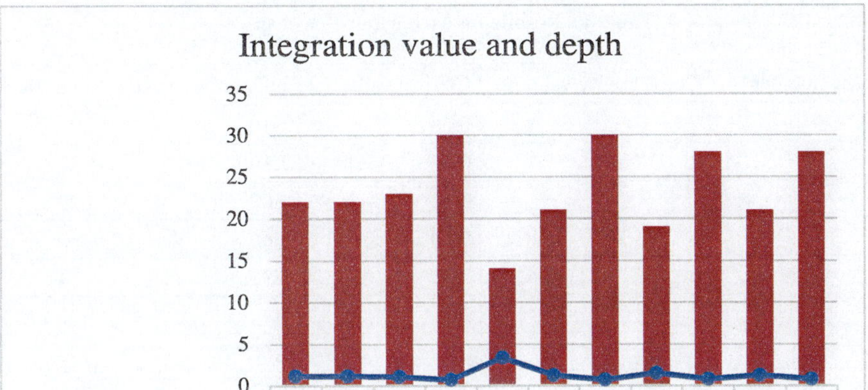

	1	2	3	4	5	6	7	8	9	10	11
Depth	22	22	23	30	14	21	30	19	28	21	28
Integration value	1.1	1.1	1.02	0.66	3.31	1.2	0.66	1.47	0.73	1.2	0.73

Fig. 4. Comparison of integration values and depth

Spatial depth is measured by the depth value, which reflects the ease and accessibility of an element in the overall system, and what is important is the number of spatial transitions, independent of the actual distance (Fig. 5).

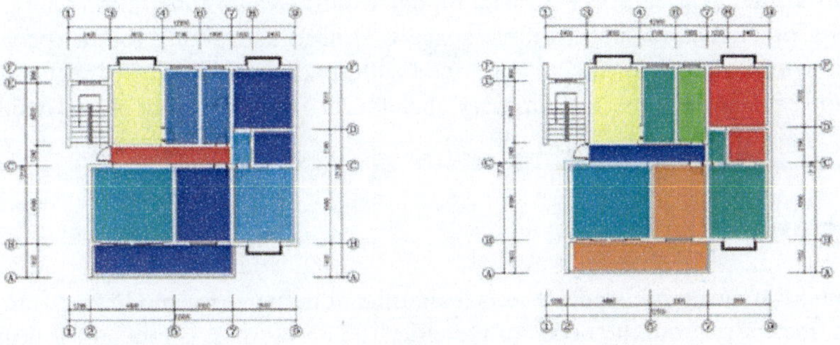

Fig. 5. Plan view of integration value and depth

Therefore, elements whose depth value is closer to blue (the smallest depth value) in colour are advantageous. This is because the shallow spatial depth means that it may take less time and effort to get from any other place in the system, to this central space. Space 5 in the spatial depth diagram has the smallest depth value of 14, which means that space 5 is the most accessible in the whole space. The red colour indicates poorly accessible spaces, which can be used for rooms with high privacy requirements such as bedrooms and study rooms.

Table 1. Results for each parameter in space.

Space number	Connectivity value	Control value	Integration value	Depth
1	2	0.66	1.10	22
2	2	0.66	1.10	22
3	1	0.16	1.02	23
4	1	0.50	0.66	30
5	6	3.33	3.31	14
6	2	1.16	1.20	21
7	1	0.50	0.66	30
8	3	2.16	1.47	19
9	1	0.33	0.73	28
10	2	1.16	1.20	21
11	1	0.33	0.73	28

As shown in Table 1, combining all the calculation results, space 4 and space 10 have the deepest space depth, while considering other factors such as area, and are more suitable as bedroom spaces; space 5 is used as a corridor connecting the living room and kitchen space, meeting the minimum distance requirement for the elderly's body displacement; space 8 has a higher Control value Integration value, while considering other factors such as area Space 8 has a higher Control value Integration value, while considering other factors such as area, space 8 is used as a living room; space 11 is connected to the balcony; space 2 is combined with space 1 to reduce the traffic flow and improve the efficiency of use. In summary, the elderly-friendly layout of this residential space has been completed.

4 Conclusion

In the context of aging in place, there is a significant demand for modifying numerous existing homes to cater to the needs of the elderly. Therefore, it is crucial for designers to gain a more precise understanding of the spatial relationships within these homes. By applying the spatial syntax theory, residential plans can be evaluated using quantifiable indicators, and subsequently reconfigured based on the functional requirements of older adults. This method demonstrates a high level of repeatability and generates relatively accurate outcomes.

However, it is important to recognize that the spatial layout requirements of the elderly are influenced by multiple factors. This method primarily generates mechanical quantitative data regarding the residential floor plan, offering designers a preliminary functional layout. Nevertheless, the ultimate floor plan should also be tailored to the specific needs of each individual elderly person. In the future, it will be essential to explore ways in which the needs of older adults can be effectively integrated with the calculated results obtained through this method.

Acknowledgment. This study is derived from the Innovation and Entrepreneurship Training Program for College Students of ShanDong Youth University of Political Science (Project No. 2022142771063).

We would like to express our gratitude to the participants of this project: students: Xinyu Zhao, Bingyan Sun, Yanan Zhou.

References

1. Kothiyal, K.: Anthropometry for design for the elderly. Int. J. Occup. Saf. Ergon. **7**(1), 15–34 (2001)
2. Yichen, Q.: Visual communication-based virtual reality design of imaging information collection and display system. Wirel. Commun. Mob. Comput. **1**, 1–12 (2022)
3. Mishra, S.A., Pandit, R.K.: Space syntax spproach for analyzing crime preventive urban design: concept review. J. Adv. Res. Constr. Urban Archit. **1**(1), 30–35 (2016)
4. Xu, Y.: Towards sustainable heritage tourism: a space syntax-based analysis method to improve tourists' spatial cognition in Chinese historic districts. Buildings **10**(2), 1–18 (2020)
5. Eirini, N.: Evaluation and experimental design in data mining and machine learning. Big Data **8**(5), 20–21 (2020)
6. WalidMahfoud, D.: Noise and spatial configuration in Biskra, Algeria—a space syntax approach to understand the built environment for visually impaired people. Sustainability **13**, 19 (2021)
7. Seungkyu, R.: Considering space syntax in bicycle traffic assignment with one or more user classes. Sustainability **13**, 21 (2021)

Identifying Architectural Forms and Evaluating the Climate Adaptability of Traditional Buildings in Southwest China

Chenpeng Xu(✉) and Keding Lu

Huzhou University, Huzhou 313000, China
02597@zjhu.edu.cn

Abstract. The unique and various forms of Miao buildings in Southwest China promote harmony with nature and are an important part of Miao culture as well. This paper firstly finds out that some architectural forms are inherited in new Miao buildings during the transformation through on-site surveying and comparison. Based on the output of the software of Climate Consultant, both the traditional and new are tested through the design strategies set out on plan, sections and building forms respectively, to find that traditional Miao buildings can adapt to the local climate well, while for the modern ones, their functions for improving thermal comfort is disappearing. The paper suggests that more attention should be paid to traditional Miao's potential efforts to adapt to the environment, rather than simply symbolic icons.

Keywords: Miao Buildings · Architectural Forms · Climate Adaptability · Climate Consultant

1 Introduction

There is adequate understanding of the influence of climate on vernacular buildings and settlement clearly identifiable characteristics that have been successful and developed through different periods. Amos Rapoport states in House Form and Culture that both socio-cultural and climatic factors have an important role in the derivative process of dwellings [1]. A Miao "Diaojiaolou", which is a stilted building, is a distinct building type resulting from a wide range of environmental factors, including other nationalities' cultures [2]. Further researches found that some specific design strategies in Miao stilted buildings may lead to a better adaptation to local climate [3, 4]. In terms of typology, these kinds of characteristics have become a significant part of local culture as well [5]. Furthermore, with more factors taken into consideration, the characteristics are not static. For instance, the plan for traditional Miao stilted buildings is believed to transform continuously as people's lifestyles change simultaneously. However, research on newly constructed Miao dwellings is still limited. This paper is divided into two sections. The first section mainly focuses on the inheritance and innovation of architectural forms of Miao dwellings in terms of typology through field measurements and comparisons. In the

W. Guo et al. (Eds.): GBCESC 2023, LNCE 328, pp. 598–606, 2024.
https://doi.org/10.1007/978-981-99-9947-7_62

second section, Climate Consultant will be used to discuss climate-responsive strategies based on the results of the first section and simulation tools. This software has been widely to analyse climate adaptability of traditional dwellings in various circumstances [6, 7].

The innovation of the paper lies in the its focus on both new and old traditional Miao buildings, aiming to investigate whether the climate-related forms are inherited and whether those forms still have a role to play today through on-site surveying and computer simulation.

2 The Inheritance of Forms of Miao Buildings

Xijiang Town is a typical historic Miao village in southeastern Guizhou Province in Southwest China, where a large number of Miao buildings exist. This section first demonstrates how the architectural forms are inherited and interpreted in terms of culture, followed by a discussion of new Miao buildings.

2.1 Traditional Stilted Building

'Banbianlou' (Half Building) and 'Sankaijian' (Three Rooms) are the two most obvious characteristics of Miao buildings. They are believed to be related to the local environment as well as to some customs.

'Banbianlou' (Half Building). A stilted building is very common in a traditional Miao village such as Xijiang Town. The prototype of this kind of building dates back to Hemudu culture, approximately 7,000 years ago. To adapt to the mountainous landscape, the original stilted building, which was entirely detached from the ground, gradually evolved and resulted in a half-stilted building known as 'Banbianlou' in Chinese. Our ancestors had to face wild animals, insects and moisture from soil and a chapter called "twelve eggs" of Miao ancient songs explains that their ancestors came from eggs and lived in a nest. In other words, the prototype of "Banbianlou" as a stilted building helps demonstrate the collective subconsciousness in Miao people's minds and serve as a solution to topography and climate.

Normally, a Miao stilted building consists of three kinds of spaces for production, living and storage. It is clear from the drawing (Fig. 1) below that those three functions occurred on different floors to adapt to the topography and make the most use of space. While the first floor serves as the space for daily life, the ground floor under it provides the place for raising livestock, and the attic is used for storage.

'Sankaijian' (Three Rooms). The influence of the Han-style architecture was profound. 'Sankaijian', which means that there are three rooms placed side by side in a single building, including one hall and two ordinary rooms, can be widely found in vernacular architecture in China. A typical 'Sankaijian' plan consists of a hall, bedrooms, a kitchen and other spaces. Normally, in a Han building, the hall is the ritual place for meeting and worshipping.

In terms of Miao buildings, some modifications have been implemented that consider the local environment. The hall tends to be in the centre with the other spaces attached to it, and a balcony occurs in the south of the building as a part of the entrance to enhance the spatial relation between the interior and exterior areas. Both the hall-centred plan and the open balcony emphasize the identifiability and importance of the hall.

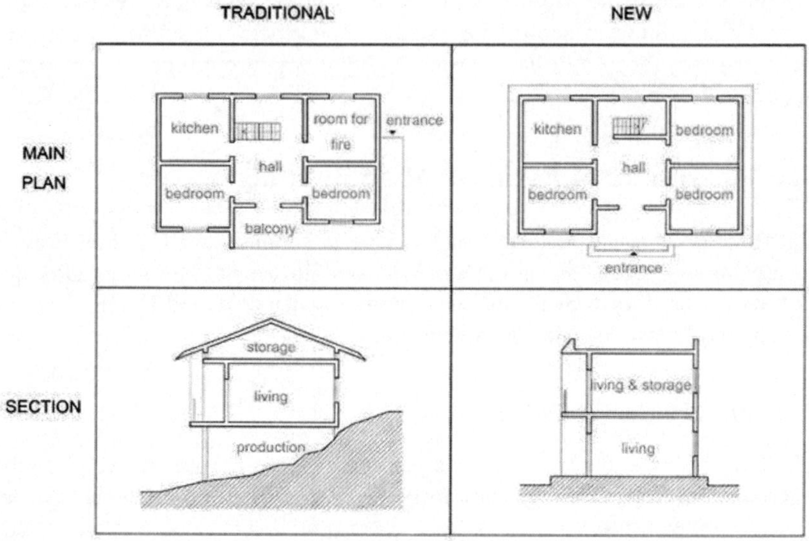

Fig. 1. A comparison between traditional and new Miao buildings

2.2 Modern Miao Building

Under the influence of social and economic development, recent decades have seen a transformation in the newly constructed Miao buildings. Since the shortage of cultivated land has been addressed, the Miao people have been encouraged to build their new dwellings on the plain in recent years. Many characteristics are borrowed from traditional buildings to show cultural coherence, while most of the buildings have changed based on new circumstances.

Ground Floor and the Hall. Placed on the plain instead of a hillside slope, the ground floor of the new Miao building does not need to adapt to the topography and is available for living space. Like Han buildings, the hall and kitchen in a new Miao building are on the first floor, with bedrooms and space for storage on the second floor. For the 'Sankaijian' plan, even though the hall is still in the middle, it is divided and not as important as before.

Entrance and Balcony. The balcony in a traditional building is on the second floor and a part of the entrance, while in a new building, the balcony is no more than a symbol on the facade rather than a space for communal activities.

2.3 Inheritance of Architectural Forms

It is clear that some architectural forms were inherited during the transformation of Miao buildings, regardless of whether they are new or traditional. However, with the development of socio-cultural and economic environments, it is hard to determine whether this kind of succession is meaningful since architectural forms are actually related to the people in the area and the adaptation to the environment in addition to decoration or style.

3 Analysis of Climate Adaptability

It is clear from the last section, for the Miao people, some architectural forms have remained a part of their culture. However, these characteristics have been partially rather than entirely inherited. This part of the paper aims to analyse the climate adaptability of architectural forms in both traditional and new Miao buildings on the basis of simulation and comparison.

3.1 Methodology

Climate Consultant and Comfort Model. Simulation and analysis software can take climate, thermal comfort and architecture design into consideration simultaneously. This paper uses Climate Consultant, which was developed by UCLA and can provide a series of design strategies to improve building climate adaptability.

The software provides 4 kinds of comfort models that are defined in the ASHRAE Handbook of Fundamentals Comfort Model through 2005, and the comfort model was chosen according to local climatic characteristics. The setting was for people dressed in normal winter clothes, and the effective temperature ranged from 20 °C to 23.3 °C (measured at 50% relative humidity), which means that the temperatures decrease slightly as humidity rises. The upper humidity limit was 17.8 °C wet bulb with a lower dew point of 2.2 °C. If people are dressed in lightweight summer clothes, then this comfort zone shifts to be 2.8 °C warmer [8].

Comparison. Based on the output of Climate Consultant, both traditional and modern Miao buildings will be tested through the design strategies set out on plan, sections and building forms respectively, to investigate whether they can adapt to the local climate well.

3.2 Climatic Characteristics

Xijiang Town, at latitude 26°40′ N and longitude 108°17 E, is in a subtropical region with a humid monsoon climate and a mountainous landscape. According to the Building Climate Zoning Standards in China, it is a in a hot summer and cold winter zone. The table below shows some key meteorological parameters of Xijiang Town based on the 'China Ground Data Set of Annual and Monthly Values' (1981–2010) provided by the China Meteorological Administration and meteorological data provided by NASA. It

Table 1. Table of meteorological parameters in Xijiang Town.

MONTH	Monthly average temperature (°C)	Monthly average maximum temperature (°C)	Monthly average lowest temperature (°C)	Monthly average relative humidity (%)	Monthly average precipitation (mm)	Maximum precipitation (mm)	Average wind speed at 10 m (m/s)	Average solar radiation on surface (kWh/(m^2·d))
JANUARY	4.7	8.7	2.0	81	34.9	70.9	2.3	1.8
FEBRUARY	6.9	11.1	4.1	80	46.1	124.3	2.5	2.2
MARCH	10.8	15.5	7.6	79	60.8	133.1	2.6	2.8
APRIL	16.1	21.1	12.6	79	115.3	231.9	2.5	3.4
MAY	19.9	24.8	16.3	80	191.0	358.3	2.3	3.7
JUNE	22.6	27.1	19.4	82	261.0	468.0	2.1	3.8
JULY	24.5	29.1	21.3	79	188.0	533.8	2.2	4.5
AUGUST	23.8	29.4	20.2	81	134.6	301.4	1.9	4.3
SEPTEMBER	20.7	26.5	16.9	81	85.6	265.0	2.2	3.7
OCTOMBER	16.3	21.2	13.1	81	87.9	249.9	2.3	2.8
NOVEMBER	11.7	17.0	8.3	80	53.8	217.3	2.3	2.5
DECEMBER	6.9	11.9	3.5	78	25.7	79.1	2.2	2.2
ANNUL	15.4	20.3	12.1	80.1	107.1	252.8	2.3	3.1

can be seen from the table (Table 1) that the monthly mean temperature through the year ranges from 4.7 °C to 24.5 °C. The hottest period is from June to August, the mean highest temperature is 29.4 °C, and the mean lowest temperature during the coldest period from December to the next February is 2 °C. The relative humidity is high year around and is approximately 80%. Accordingly, it is necessary to reduce humidity and improve natural ventilation and thermal isolation to promote thermal comfort.

3.3 Design Strategies

As shown in the psychometric chart (Fig. 2) provided by Climate Consultant, most points are out of the comfort zones, and the relative humidity tends to be quite high. Generally, this is an area with hot summers, cold winters and high humidity throughout the year, which indicates that only 6.8% of the year is comfortable. There are 15 design strategies given by the software that may improve thermal comfort, 13 of which work as shown in the table below. Generally, heating and internal heat gain are the best ways to improve thermal comfort considering the long and cold winters, while cooling and dehumidification are especially significant in summer.

3.4 Climate Adaptability of Miao Buildings

It is clear from the table (Table 2) that summer and winter provide the lowest thermal comfort of the year, with approximately 1.5% of the summer considered comfortable and no more than 3% of the winter. This section focuses on assessing both traditional and new Miao buildings in summer based on the effective passive design strategies provided in the software, while in winter, passive design strategies seem inadequate.

Dehumidification. Summer in Xijiang Town is hot and humid, with a relative humidity of approximately 80%. Therefore, among the passive design strategies, dehumidification is the most efficient in summer. According to the table, using dehumidification only will make 43.3%, 45.8% and 42.9% of month comfortable enough for people in June, July and August, respectively.

The stilted space in a traditional Miao building not only adapts to the topography but also prevents the humidity that transfers from the soil to the air from entering the home, leaving the living space dry. However, in a new Miao building, the ground floor is the main space for meeting and living, and even though water-proofing measures are properly taken, it is hard for it to avoid getting damp in summer.

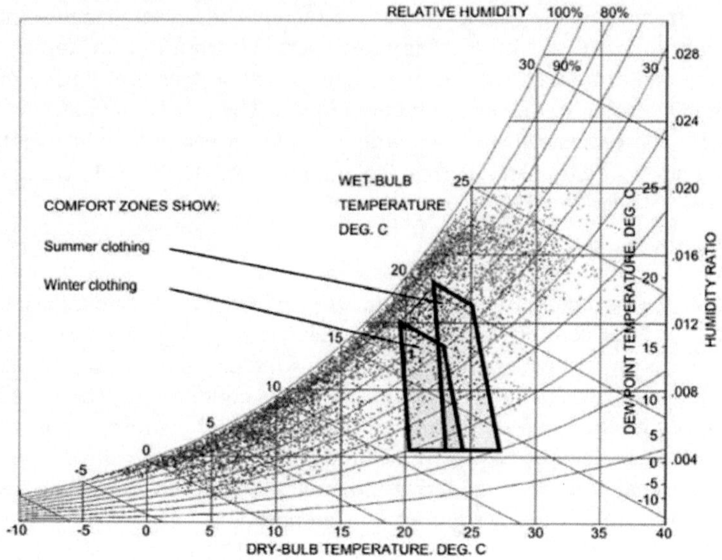

Fig. 2. Psychometric chart

Natural Ventilation. Natural ventilation is an effective passive design strategy to address hot and humid summers. The table shows that during approximately 20% of June and July, natural ventilation cooling is needed to improve thermal comfort.

A traditional 'Sankaijian' plan is suitable for natural ventilation because the wind can blow through the hall in the middle and not be blocked. This kind of cool alley, as a passive-cooling technique, is very common in Han historic buildings. In regard to new Miao buildings, the hall in the middle is separated by a wall, and the wind is blocked. As a result, the efficiency of ventilation is expected to be low, and fan-forced ventilation cooling is necessary. Additionally, the balcony in a traditional Miao building also allows this grey space to be naturally ventilated and cooled.

Sun Shading. Sun shading is also necessary in summer and approximately 15–20% of the time from June to September.

Due to the large amount of rainfall, traditional Miao buildings usually adopt a combination of a sloping roof and a broad eave, which contributes greatly to sun shading and reduces heat gain. In addition, the balcony in a traditional building also helps improve the effect of sun shading in summer. However, the sloping roof and eave have become no more than a symbol without any practical functions, as drainage systems have been well promoted. As a result, apart from the balcony on the second floor, the eave itself seems to play no role in sun shading in new Miao buildings.

Table 2. Table of various design strategies and comfortable hours using selected strategies.

DESIGN STRATEGIES	January	February	March	April	May	June	July	August	September	October	November	December	JAN to DEC
Comfort	0.0%	2.4%	3.1%	13.3%	17.2%	10.1%	1.5%	1.3%	13.8%	14.9%	3.8%	0.0%	6.8%
Sun Shading of Windows		0.1%	0.4%	5.7%	7.4%	13.2%	17.2%	22.8%	14.0%	6.2%	0.3%		7.3%
High Thermal Mass					8.3%	3.8%	1.6%	2.7%	3.2%	4.6%			2.3%
High Thermal Mass Night Flushed				4.4%	8.6%	4.6%	3.1%	3.2%	3.2%	4.6%			2.7%
Direct Evaporation Cooling				1.8%	3.6%	1.3%	0.5%	0.9%	3.1%	3.1%	0.1%		1.2%
Two-Stage Evaporative Cooling				2.5%	5.4%	2.8%	1.2%	1.9%	3.2%	3.5%	0.1%		1.7%
Natural Ventilation Cooling				7.1%	11.6%	19.3%	16.7%	6.7%	6.7%	3.2%			6.0%
Fan-Forced Ventilation Cooling				9.2%	15.9%	25.4%	22.2%	9.3%	9.3%	10.4%	6.6%		8.3%
Internal Heat Gain	2.6%	5.2%	18.8%	48.8%	56.0%	16.4%	1.5%	3.0%	43.1%	36.3%	22.4	2.8%	21.6%
Passive Solar Direct Gain Low Mass	1.9%	3.0%	3.8%	2.5%	0.8%	0.4%		0.1%	0.3%	1.2%	4.9%	2.0%	1.7%
Passive Solar Direct Gain High Mass	1.9%			5.0%	4.4%	2.5%	0.7%	2.0%	5.5%	7.3%	6.5%		3.3%
Dehumidification Only		2.1%	2.6%	4.4%	13.8%	43.3%	45.8%	42.9%	18.9%	5.5%			14.7%
Cooling, add Dehumidification if needed				0.3%	2.0%	18.1%	39.4%	46.8%	16.7%				10.4%
Heating, add Humidification if needed	96.0%	90.8%	76.5%	26.7%	1.6%				1.5%	34.3%	67.8%	97.0%	40.8%

4 Conclusion

The purpose of this research on the forms of Miao buildings and their climate adaptability was to explore the significance behind these symbolic forms and how this intrinsic relevance develops in the context of China's rapid development.

Miao people have adjusted the prototype of their buildings according to the environment, and for them, these developed architectural forms are inherited as a part of their culture. The forms tend to be related to some conditions or circumstances, while the forms also have transformed and been passed on as some obscure symbols in a building.

Through the comparison of field surveys and computer simulations, we found that identifiable forms such as 'Sankaijian' and 'Banbianlou' enable adaptation to the local climate to various degrees. Regrettably, nowadays, as the Miao people experience changes in their living environment, these symbolic architectural forms are losing their functions for improving thermal comfort. The research suggests that more attention should be paid to traditional Miao's potential efforts to adapt to the environment through various climate-related architectural forms and how they work, rather than simply symbolic icons.

Acknowledgement. This work was supported by the science and Technology Bureau Foundation of Huzhou (No. 2021GZ12).

References

1. Rapoport, A.: House Form and Culture. Prentice-Hall Inc., Upper Saddle River (1969)
2. Wang, H., Zhang, H.: Study on residential suitability evaluation of stilted houses in Southeast Guizhou Province. Guizhou Ethnic Stud. **43**(04), 132–137 (2022)
3. Shao, L., He, L.: Analysis OF climatic adaptability of Guizhou Xijiang thousand households MIao Village hanging tower. Shanxi Archit. **44**(16), 1–2 (2018)
4. Wu, Z.: Influencing factors and terrain adaptability of traditional stilted buildings in Wuling area. Huazhong Archit. **40**(06), 148–151 (2022)
5. Hu, M., Meng, Z.: The cultural connotation of the hanging footed buildings of the Tujia ethnic group in Xiangxi. Yihai **02**, 94–96 (2022)
6. Liu, J., Saierjiang, H.: Analysis on climate adaptability of dwellings in arid oasis area by climate consultant——a case of Kashi in Xinjiang Uygur autonomous region. Urban. Archit. **18**(10), 112–116 (2021)
7. Li, X., Zhang, Y.: Climatic adaptation analysis of traditional Huizhou style houses by climate consultant. New Build. Mater. **49**(01), 132–137 (2022)
8. ASHARE: ASHRAE Handbook of Fundamentals, ASHRAE Inc., Atlanta (2005)

Research on the Characteristics and Influencing Factors of Spatial Soundscape Perception in University Campuses: Guizhou University as an Example

Yonghe Ma(✉), Honggang Tang, and Xiaoheng Zhou

College of Architecture and Urban Planning, Guizhou University, Guiyang 550025, China
214519645@qq.com

Abstract. In order to investigate the characteristics of soundscape perception in different functional spaces of campus and its influencing factors, Guizhou University was used as the research object to obtain soundscape perception data based on on-site research and questionnaire survey, and four dimensions of campus functional characteristics, sound source perception, soundscape perception and soundscape satisfaction were analyzed to explore the interactions among the dimensions and the main influencing factors of campus soundscape. The results show that birdsong sound has a high Sound dominant degree and Sound harmonious degree in ecological green areas and lake water areas, and traffic sound has a high sound dominant degree in the whole campus; the soundscape pleasantness and soundscape eventfulness of different functional spaces are significantly different, with higher soundscape pleasantness in ecological green areas and lake and water areas, and more obvious soundscape eventfulness in student living areas and cultural and sports activities areas; some sound source elements have strong correlation with the soundscape perception of functional space, birdsong sound can play a role in increasing the soundscape pleasantness and soundscape eventfulness, and conversation sound can enhance the soundscape eventfulness in the lake water body area; the campus is easily accessible, so the negative impact of traffic sound on the campus environment is most obvious. The research conclusions can provide a basis for the relationship between sound source and soundscape perception in functional space. The overall sound environment can be adjusted through soundscape characteristics, which is of great significance for optimizing the quality of campus soundscape.

Keywords: Sound Source Elements · Soundscape Perception · Sound Harmonious Degree · Functional Space

1 Introduction

As a place for students to learn and live, colleges and universities are densely populated. The beautiful campus environment plays a role in enriching students 'life and alleviating physical and mental pressure [1]. In 2019, China's Assessment.

W. Guo et al. (Eds.): GBCESC 2023, LNCE 328, pp. 607–620, 2024.
https://doi.org/10.1007/978-981-99-9947-7_63

Standard for Green Building that a campus in harmony with the natural environment should be established, and a healthy, suitable and efficient teaching and living environment should be provided for teachers and students while saving energy and reducing emissions. Therefore, exploring strategies for designing and renewing campus landscapes that enhance students' life experiences is an important issue today. For the creation of the landscape, in addition to the visual landscape, researchers in recent years have found that the soundscape, as an important part of the landscape, has a great impact on the overall quality of the landscape [2]. At the same time, colleges and universities are unique places with a variety of soundscapes. The pleasant soundscape on campus helps to relieve stress and is one of the important factors affecting students 'lives.

ISO (The International Organization for Standardization) defines a soundscape as "an acoustic environment perceived, experienced, or understood by an individual or group in an environment in a particular scene" [3]. In recent years, in terms of soundscape perception, studies have been conducted mainly on public areas [6] such as plant landscape spaces, urban parks [4], and ancient towns [5], and it was found that different functional spaces have different effects on soundscape. However, fewer studies have been conducted for campus soundscape, and campus soundscape mainly focuses on campus soundscape evaluation [7, 8] and campus green space soundscape [9]. It has been found that soundscape perception varies across functional spaces and that soundscape has a strong dependency on space [10], therefore, this paper emphasizes the functional space of the campus as a factor affecting soundscape perception.

This study takes the acoustic environment of the western campus of Guizhou University as an example. By analyzing the campus soundscape and sound source perception, it explores the main characteristics and influencing factors of the soundscape of different functional spaces on campus, and provides a theoretical basis for improving the quality of campus soundscape.

2 Methods

2.1 Case Study Area

The West Campus of Guizhou University is located in Huaxi District of Guiyang City, China. Adjacent to Huaxi Park and Shilihetan Wetland Park, with a rich humanistic atmosphere and beautiful environment. The terrain of the school is undulating, and the climate is plateau monsoon humid climate. The total area of the West Campus of Guizhou University is 2.04 million square meters, with a total construction area of 1.24 million square meters, which can accommodate 19,000 teachers and students. The overall environment is relatively beautiful and the campus is rich in vegetation types, covering a variety of types of land such as grassland, woodland, farmland, and buildings. There are a variety of plants such as osmanthus, silver wattle, horsetail pine, willow fir, magnolia, bamboo, lateral cypress, red maple, willow, hammock, cherry, purple leaf plum, ginkgo and maple.

Since the perceived quality of the landscape is closely related to the functions carried by the space itself [11], a total of 15 data sampling points in five types of campus public spaces, namely, ecological green area, Lake water area, Sports activity area, student living area, and School entrance area, were selected for this study, as shown in Fig. 1.

Through field research and combining the characteristics of common sound sources on campus, five typical sound sources were selected for analysis from four categories of sound sources: biological sound, mechanical sound, human activity sound, and geophysical sound: bird song, wind sound, activity sound, conversation sound, and traffic sound.

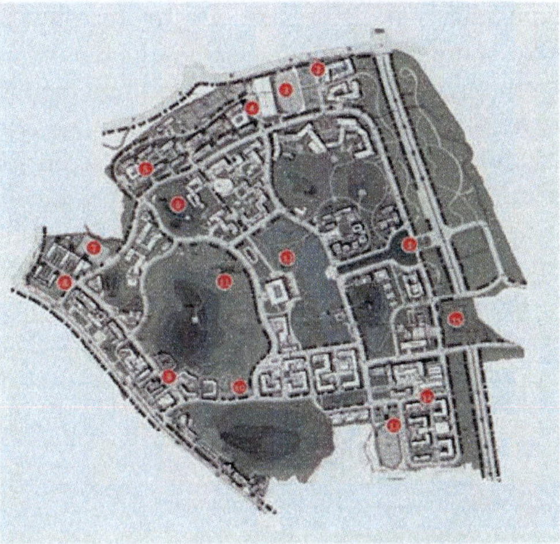

Fig. 1. Schematic diagram of measurement points

2.2 Data Source

A survey was conducted to obtain subjective information on the perception of soundscape in each functional area of the university campus. The questionnaire consists of three parts: (1) Evaluation of the perception of sound sources in different functional space sites, using a 7-level scale to evaluate the perceived frequency (1-very little, 7-very frequent), perceived intensity (1-very weak, 7-very strong) and preference (1-very disgusting, 7-very like) of sound sources. (2) The evaluation of the perception of the soundscape of the space in which it is located, the selected evaluation indexes include pleasant, harmonious, comfortable, energetic, eventful, vivid, variable, and disorderly, and is evaluated using a 7-level scale from 1 "strongly disagree" to 7 "strongly agree". (3) The overall satisfaction of each functional space soundscape was evaluated from 1 "very dissatisfied" to 7 "very satisfied".

To reduce the perception bias caused by individual differences, 30 college students with good hearing in Guizhou University were invited as subjects. The study was conducted in April-May 2023, and the survey time was 9:00–11:00 and 15:00–18:00 of the day. Clear weather with low wind speed was selected, and the questionnaire was used to collect subjective perception data at 15 sample sites.

2.3 Data Analysis

The final number of valid questionnaires recovered was 614, with an effective rate of 97.4%. The validity test showed that the KMO = 0.638 > 0.6, and the Cronbach's alpha coefficients of sound source perception, soundscape perception and soundscape satisfaction were 0.76, 0.85 and 0.79 (>0.7), respectively, so the validity of the data was good.

In the outdoor sound field, in order to more effectively reflect the subjective perceptual characteristics of the sound source elements, the sound dominant degree (SDD) and sound harmonious degree (SHD) were introduced based on the three sound source perception indicators obtained from the questionnaire [12]. Sound dominant degree (SDD) means the degree of dominance of the sound source in a given soundscape, the higher the value, the greater the degree to which the sound is perceived; Sound harmonious degree (SHD) means the degree to which the dominant position of the source in the environment is consistent with people's preference for it, the higher the value, the higher the harmony between the state and satisfaction of that sound in the soundscape. The formula is as follows:

$$SDD_{(ji)} = POS_{(ji)} \times PLS_{(ji)} \tag{1}$$

$$SHD_{ji} = \left[\frac{1}{e^{\sum_{j=1}^{n} \frac{PFS_{ji}}{n} - PFS_{ji}} + 1} - 0.5 \right] * SDD_{ji} \tag{2}$$

where j is the j-th sample; i is the i-th sound source; and n is the sample size. A mathematical model constructed with three indicators Perceived preference of sound source (PFS), Perceived occurrences (POS) and Perceived loudness (PLS) is used to represent.

3 Results

3.1 Analysis of Different Spatial Sound Source Perception

Analysis of Sound Dominant Degree. The different types of functional spaces Sound dominant degree is shown in Fig. 2. From the perspective of each functional area, the dominance of birdsong sound and wind sound in the ecological green area and the lake water body area is greater, and the dominance of activity sound is the lowest; In the sports activities area, the dominance of activity sound as the main sound element is much greater than that of other sound source elements; In the student living area, all five Sound dominant degrees are low, and the activity sound dominance degree is the smallest; in the school entrance area, the traffic sound dominance degree is larger; In terms of the overall sound source, the overall sound dominance degree of birdsong sound and wind sound is higher than that of other sound source elements. Especially in the ecological green area and lake water area, bird song has a clear dominance degree. The wind sound dominance is highest in the lake water area, and the activity sound dominance is highest in the cultural and sports activity area. The dominance of conversation sound is greater in the sports activities area and the student living area than in other areas. In the entrance area of the campus, the traffic sound dominance is the highest, and for the convenience

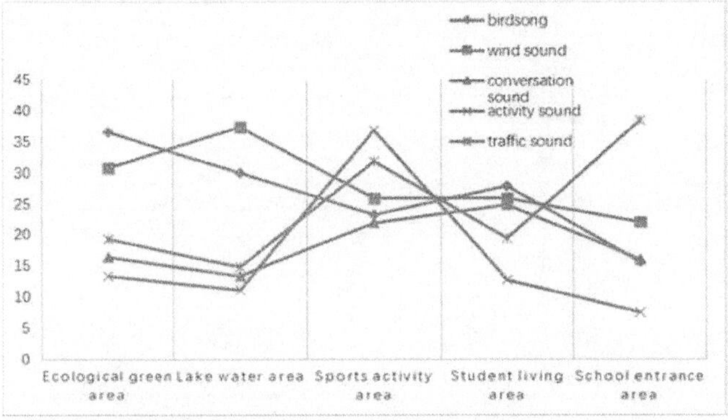

Fig. 2. Different functional space sound dominant degree

of travel inside the campus, carriageways are set up in various places, and all of the campus has more obvious traffic sound inside.

Analysis of Sound Harmonious Degree. The different types of functional spaces Sound harmonious degree is shown in Fig. 3. From the viewpoint of each functional space, the ecological green area has a rich plant landscape, and the harmony of birdsong sound and wind sound are both rated high in the natural environment. The harmony of wind sound in the lake water area is higher. The perceived preference of birdsong sound is higher in the lake water area, but the harmony value is slightly lower, indicating that enhancing birdsong sound can significantly contribute to the soundscape of the lake water area. The harmony of activity sound in the sports activity area is high. High harmony of conversation and bird song in the student living area. In the entrance area of the school gate, although the traffic sound is the most dominant sound element, its perceived preference is small, so the traffic sound harmony is the lowest. The traffic sound causes the deterioration of the overall sound environment in the entrance area of the school, and the other sound elements have lower dominance, so the overall soundscape harmony is low. In terms of sound source perception, the highest harmony is the sound of birdsong sound in the ecological forest area. The lowest harmony is the sound of traffic in the entrance area of the school gate. Due to the high perceived preference and relatively high dominance degree of birdsong sound and wind sounds, both have a high degree of harmony throughout the functional space of the campus. Traffic sound is less harmonious degree in all areas due to lower perceived preference and convenient roads in the campus area, where traffic sound is obvious. Except for the sports activities area and the student living area, the harmony degree of conversation sound did not vary much in the other three functional spaces.

Sound Source Perception Variability Analysis. Due to the differences in sound source perception in each functional space, the Kruskal-Wallis test was performed on the variability of sound source perception in different functional spaces based on the data samples obtained from the sound source perception evaluation. The results show that there are significant differences in the sound dominant degree and sound harmonious degree of

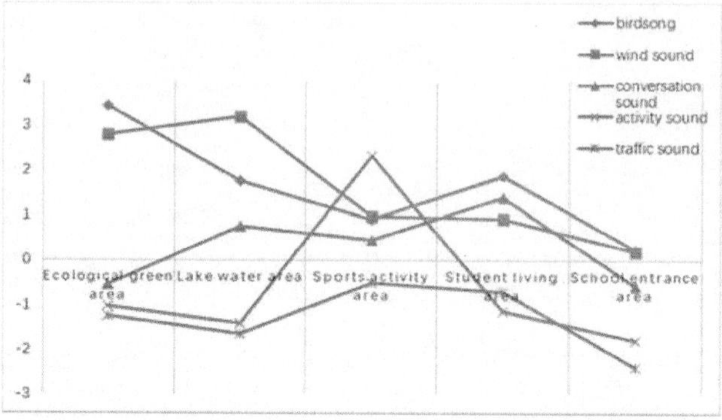

Fig. 3. Different functional space Sound harmonious degree

five sound elements: bird song, wind sound, conversation sound, traffic sound and activity sound in different functional spaces, as shown in Table 1. The differences of sound source elements in five functional spaces are further compared and analyzed in pairs. As shown in Table 2, the dominance degree of bird song, wind sound, traffic sound and the harmony degree of bird song and wind sound are significantly different between the ecological green area and the school entrance area. The dominance degree of activity sound is significantly different between the sports activity area and the school entrance area.

Table 1. Spatial sound source perception difference analysis

Sound source type	Sound dominant degree	Sound harmonious degree
Birdsong sound	86.054	9.368
wind sound	36.547	7.263
conversation sound	28.691	2.549
activity sound	41.743	2.961
traffic sound	93.671	3.215

3.2 Soundscape Perception Analysis

Soundscape Perception Dimensional Analysis. Principal component analysis (PCA) was used to extract the prominent factors of eight soundscape perception elements. The full sampling measurement value of KMO was 0.75 (> 0.7), and Bartlett's Test of Sphericity results showed $P = 0.000$ (< 0.01), which met the requirements of factor analysis. The two main factors obtained after extraction are shown in Table 3, denoted as soundscape pleasantness and soundscape eventfulness, respectively. Their variance contribution rate was 87.69%.

Table 2. Pairwise comparison analysis of different spatial sound sources

Indicators	Sound Type	EG-LW	EG-SL	EG-SA	EG-SE	LW-SL	LW-SA	LW-SE	SL-SA	SL-SE	SA-SE
SDD	BS	6.524	14.67	8.183	19.94	7.518	1.382	12.89	5.064	5.641	11.68
	WS	−6.557	5.174	5.353	8.716	11.392	11.876	15.212	0.344	3.618	3.269
	CS	2.109	−5.214	−6.812	0.954	−7.259	−8.867	−1.085	−1.695	6.179	7.695
	AS	2.574	−19.74	1.793	5.074	−22.38	−0.679	2.559	21.683	24.965	3.247
	TS	0.369	−14.39	−2.016	−15.93	−14.68	−2.294	−16.29	12.403	−1.579	−14.01
SHD	BS	7.516	10.172	6.945	12.493	2.687	−0.496	5.012	−3.228	2.276	5.485
	WS	−0.824	7.326	7.479	10.031	8.152	8.316	10.791	0.164	2.748	2.579
	CS	−5.486	−4.392	−6.975	1.324	1.083	−1.475	6.814	−2.591	5.716	8.304
	AS	1.032	−13.12	0.292	1.884	−14.03	−0.716	0.876	13.314	14.871	1.592
	TS	1.644	−2.813	−2.061	4.575	−4.447	−3.58	2.931	0.746	7.375	6.627

Note: EG:Ecological green area, LW:Lake water area, SL:student living area, SA:Sports activity area, SE:School entrance area; BS:birdsong sound, WS: wind sound, CS: conversation sound, AS: activity sound, TS: traffic sound

The soundscape perception in different functional areas is shown in Fig. 4, and the overall rating of campus soundscape perception is high. The soundscape pleasantness and soundscape eventfulness of the ecological green area are high; the soundscape pleasantness of the lake water area is high, but the soundscape eventfulness is poor; the soundscape eventfulness of the student living area and the sports activity area is high; the soundscape pleasantness and soundscape eventfulness of the school entrance area are rated relatively low.

Table 3. Principal component analysis table of soundscape perception

Principal Components	Type	Ingredients		Variance percentage/%
		Factor 1	Factor 2	
soundscape pleasantness	pleasant	0.839	–	43.537
	harmonious	0.847	–	
	comfortable	0.865	–	
	energetic	0.852	–	
soundscape eventfulness	eventful	–	0.887	44.153
	vivid	–	0.863	
	variable	–	0.874	
	disorderly	–	0.841	

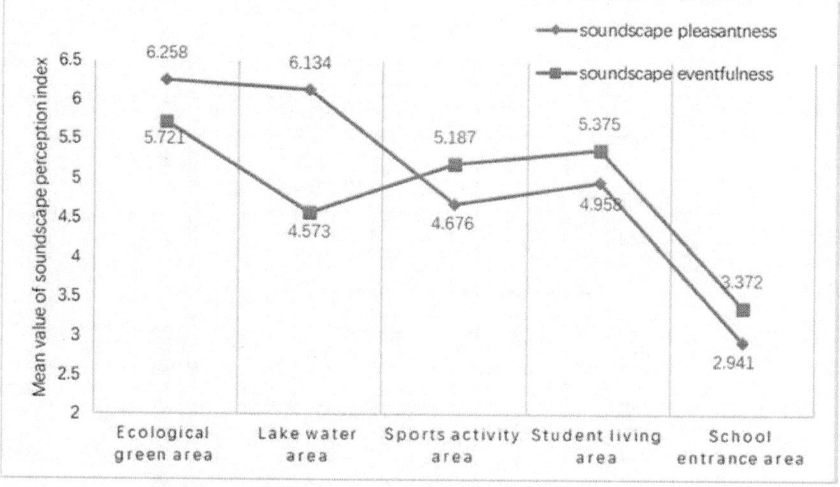

Fig. 4. Overall perception of soundscape in different functional spaces

Differential Analysis of Soundscape Perception. The Kruskal-Wallis test was used to analyze the pleasantness and eventfulness of the soundscape in different functional spaces. The results showed that both pleasantness ($Z = 22.673$, $p < 0.001$) and eventfulness ($Z = 45.436$, $p < 0.001$) were more significant. Further pairwise comparisons of the differences in the overall perception of the soundscape in each functional space were conducted, and the results are shown in Table 4. The ecological green space and lake water bodies are significantly different in soundscape pleasantness than the other three functional areas (student living area - sports activity area - school entrance area). This also validates the conclusions of Chen Wei [13]. That is, people are very fond of natural sounds, and the higher the perception of natural sounds, the higher the soundscape pleasantness. In terms of soundscape eventfulness index, except for ecological green area, student living area has more significant difference than other 4 types of functional spaces.

Table 4. Pairwise comparative analysis of spatial soundscape perception differences

Indicators	EG-LA	EG-SA	EG-SL	EG-SE	LW-SA	LW-SL	LW-SE	SA-SL	SA-SE	SL-SE
soundscape pleasantness	0.496	5.528	4.81	12.868	5.032	4.304	12.372	−0.728	7.34	8.068
soundscape eventfulness	4.192	2.536	0.984	8.996	−1.656	−3.208	4.804	−1.552	6.46	8.012

Note: EG:Ecological green area, LW:Lake water area, SL:student living area, SA:Sports activity area, SE:School entrance area;

3.3 Influence of Sound Perception on Soundscape Perception

In order to detect the contribution of the dominance degree and harmony degree of sound elements to the pleasantness and eventfulness of soundscape perception, therefore, a relativity analysis was used between sound perception and overall soundscape perception for different functional spaces. Typical sound elements with high perceptual relevance were selected so as to explore the influence on soundscape perception. We take the dominance and harmony of the selected sound elements as independent variables, and the soundscape pleasantness and the soundscape eventfulness as dependent variables, respectively。and construct a multivariate stepwise linear regression analysis model. The results of the analysis are shown in Tables 5 and 6. After the Collinearity test, none of the regression model variables had a collinearity problem ($VIF < 5$), and the model explained better.

The Influence of Sound Source Perception on Soundscape Pleasantness. In the ecological green area, birdsong sound had a positive effect on soundscape pleasantness. Wind sound, traffic sound and conversation sound have a negative effect on soundscape pleasantness, as shown in Table 5; In the lake water area, the positive effect of conversation sound on soundscape pleasantness is strong, but its dominance degree is low, so the soundscape pleasantness can be improved by increasing the conversation sound. In the sports activity area, all four sound were included in the model except for birdsong sound, and the positive effect of activity sound on the soundscape pleasantness was most obvious; the dominance degree and harmony degree of conversation sound in the student living area had opposite effects on the pleasantness of the soundscape, indicating that when Perceived preference of sound source was higher, the soundscape was more pleasant; the wind sound in the entrance area of the school had a positive effect, and the negative effect produced by traffic sound was extremely obvious.

Table 5. Results of the analysis of sound source perception and soundscape pleasantness

Space type	Arg	Beta	t	Collinearity diagnostics		R^2	F
				Tolerance	VIF		
Ecological green area	BS(a)	0.325	2.965	0.995	1.005	0.294	9.974
	WS(a)	−0.193	−2.142	0.995	1.005		
	TS(a)	−0.416	−3.579	0.983	1.017		
	CS(a)	−0.251	−2.184	0.983	1.017		
Lake water area	BS(a)	0.237	2.947	0.973	1.028	0.318	26.571
	WS(a)	0.214	3.273	0.973	1.028		
	CS(a)	0.262	2.986	0.965	1.036		
	TS(a)	−0.316	−2.454	0.965	1.036		
Sports activity area	WS(a)	0.151	1.257	0.985	1.015	0.372	19.765
	AS(a)	0.183	2.635	0.985	1.015		
	CS(b)	0.129	1.348	0.986	1.014		
	TS(a)	−0.317	−2.339	0.983	1.017		
	TS(b)	−0.225	−1.964	0.969	1.032		
student living area	BS(a)	0.129	1.876	0.944	1.059	0.347	13.161
	WS(b)	0.113	1.319	0.968	1.033		
	CS(a)	−0.142	−1.971	0.985	1.015		
	CS(b)	0.184	2.154	0.996	1.004		
	TS(b)	−0.127	−2.232	0.970	1.031		
School entrance area	WS(a)	0.113	1.173	0.963	1.038	0.221	14.905
	TS(a)	−0.576	−3.929	0.966	1.035		

Note: (a): SDD, (b): SHD; BS: birdsong sound; AS: activity sounds; WS: wind sounds; CS: conversation sound; TS: traffic sound

The Influence of Sound Source Perception on Soundscape Eventfulness. In the ecological green area, birdsong sound and wind sound have a positive effect on the soundscape eventfulness. The Sound dominant degree and Sound harmonious degree of conversation sound have opposite effects on the soundscape richness, In the ecological green area, it is suggested that the dominance degree of conversation sound should be reduced so as to enhance the soundscape eventfulness; In the lake water area, wind has a strong correlation in soundscape pleasantness, but a slightly weaker correlation in soundscape eventfulness, while conversation has a strong positive effect on soundscape eventfulness in this space, so we can enhance the eventfulness of the soundscape by appropriately increasing the sound of conversation; In the sports activities area, the dominance of activity sound has a positive effect on the eventfulness of the soundscape, and campus activities enhance the eventfulness of the campus soundscape; in the student live area, birdsong sound and conversation sound have a positive effect, and dominance of wind sound and traffic sound have a negative effect; in the school entrance area, traffic sound has the strongest negative effect, and both conversation sound and wind sound can have a positive effect.

Sound elements have different effects on soundscape eventfulness in different functional spaces. From the viewpoint of sound elements, wind sound and conversation sound harmonious degree have positive effects in each functional space, and traffic sound has a strong negative effect. Universities often set up carriageways in various places on campus, which greatly affects the campus sound environment while facilitating travel. Birdsong sound has a positive effect on the soundscape pleasantness and soundscape eventfulness of each functional space. And the correlation between the dominance of sound elements and soundscape perception is stronger.

3.4 Overall Satisfaction Analysis of Functional Space Soundscape

By investigating the overall soundscape satisfaction of different functional spaces. It is found that the ecological green area has the highest satisfaction, followed by the lake water area and sports activity area, and the school entrance area has the lowest soundscape satisfaction due to the higher traffic sound dominance degree. The difference between the sports activity area and the school entrance area is smaller for the traffic sound dominance degree, while the overall sound satisfaction of the sports activity area is much higher than that of the school entrance area. This indicates that the activity sound has a masking effect on the traffic sound, and the overall campus sound environment can be improved by using the sports activity area to mask the traffic sound at the entrance.

Table 6. Results of the analysis of sound source perception and soundscape eventfulness

Space type	Arg	Beta	t	Collinearity diagnostics		R^2	F
				Tolerance	VIF		
Ecological green area	BS(a)	0.284	2.095	0.967	1.034	0.253	8.562
	WS(b)	0.257	2.322	0.955	1.047		
	CS(a)	−0.176	−2.109	0.949	1.054		
	CS(b)	0.233	2.359	0.892	1.121		
	TS(a)	−0.296	−3.206	0.933	1.072		
Lake water area	BS(a)	0.217	2.326	0.965	1.036	0.224	22.487
	WS(a)	0.163	1.597	0.982	1.018		
	CS(a)	0.297	2.865	0.982	1.018		
	TS(a)	−0.254	−3.069	0.978	1.022		
Sports activity area	WS(a)	−0.102	−1.587	0.976	1.023	0.286	23.614
	WS(b)	0.327	3.968	0.982	1.018		
	AS(a)	0.235	2.651	0.958	1.044		
	CS(a)	0.146	2.876	0.971	1.030		
	TS(a)	−0.227	−3.534	0.993	1.007		
student living area	BS(a)	0.189	2.895	0.939	1.065	0.341	18.537
	WS(a)	−0.121	−2.103	0.951	1.052		
	WS(b)	0.184	2.657	0.968	1.033		
	CS(a)	0.254	3.019	0.940	1.064		
	TS(a)	−0.247	−2.894	0.951	1.051		
	TS(b)	−0.106	−1.865	0.981	1.019		
School entrance area	WS(a)	0.129	1.976	0.978	1.023	0.196	9.685
	CS(b)	0.166	2.036	0.994	1.006		
	TS(a)	−0.331	−3.293	0.962	1.039		

Note: (a): SDD, (b): SHD; BS: birdsong sound; AS: activity sounds; WS: wind sounds; CS: conversation sound; TS: traffic sounds

4 Conclusions

Taking the West Campus of Guizhou University as an example, this paper explores the characteristics and differences of sound source perception and soundscape perception in different functional spaces of universities。Selects five types of functional spaces and five typical sound sources for analysis, and the research results find that:

(1) Different functional spaces on campus have different sound source characteristics due to their different spatial environments and functions. The same sound source has different sound dominance degree and harmony degree, and the difference is more significant. Birdsong, as a high preference sound element, has a positive effect on the sound environment of all five types of functional spaces. For this matter, the method of planting plants that attract bird habitats can be used to isolate the noise while introducing the more popular bird sounds and thus optimizing the sound environment [14].
(2) The soundscape preferences of different functional spaces are different. With higher soundscape pleasantness in ecological green areas and lake water areas. The soundscape eventfulness is stronger in student living area and sports activity area. Different sound elements have different effects on the soundscape pleasantness and eventfulness. Birdsong sound has a positive effect on the soundscape pleasantness and eventfulness of each area, and traffic sound as the main sound source with negative effects, has a poor effect on the entrance area of the school. Conversation sound has a positive effect on improving the soundscape eventfulness, and in the lake water area, student communication can be introduced to enhance the soundscape eventfulness in this area. The soundscape dominance has a stronger correlation with the soundscape perception, and we can adjust the soundscape quality by modestly increasing or decreasing the frequency and intensity of the sound elements through internal planning.
(3) The satisfaction of the soundscape of the ecological green area, lake water area, student living area and sports activity area inside the campus is better. In the area near the entrance of the school, the dominance degree of traffic sound is the highest and the satisfaction of its soundscape is low, so we can adopt the introduction of activity sound to play the role of sound masking to optimize the sound environment of the campus.

This study. The analysis was conducted for the characteristics of soundscape in different functional spaces of the campus. Only April-May, when the weather is good, is selected, and the differences in students' behavioral activities in different seasons are not considered. And only the perceptual experience of a single soundscape was discussed, which has limitations. There is still a need for in-depth research on the influence mechanism of the soundscape of campus functional space.

References

1. LI, R., OU, D.: Acoustic landscape survey and research on university campuses. Build. Sci. **33**(08), 59–67 (2017)
2. Wu, S.: Some comments on landscape soundscape. Chin. Landscape Archit. **31**(5), 38–39 (2015)
3. InternationaI Organization for Standardization Acoustics-Soundscape—Part 1: Definition and ConceptuaI Framework.IS0 12913–1: 2014(E). Switzerland (2014)
4. Xue, Z., Han, Y., Yin, Y.: Research on evaluation of audio-visual perception quality of urban parks: a case study of Chengdu outer-ring ecological zone. Landscape Archit. **29**(9), 26–32 (2022)

5. Yang, L., Zhang, J., Xu, Y., et al.: Soundscape subjective evaluation based on tourists' preference: a case study of Dayan ancient town. Appl. Acoust. **39**(4), 625—631 (2020)
6. Xu, X., Guo, X., Han, F., et al.: Research of soundscape perception in Wulingyuan world heritage site and influence factor. Landscape Archit. **26**(6), 97–102 (2019)
7. Weng, Y., Zhu, Y., Dong, J., et al.: Effects of soundscape on emotion and attention on campus green space—a case study of Fujian agricuiture and forestry university. Chin. Landscape Archit. **37**(2), 88–93 (2021)
8. Yue, H., Yang, G., An, J., et a1.: Research and evaluation of university campus soundscape based on audiovisual effects: taking Huaxi campus of Guizhou Normal University as an example. Tech. Acoust. **41**(1), 116–123 (2022)
9. Li, Z., Lin, L.: Assessment factors extraction of campus soundscape based on semantic differential method: a south campus of sun Yat-sen university case. Planners (z1), 285–288 (2015)
10. Hong, J.Y., Jeon, J.Y.: Exploring spatial relationships among soundscape variables in urban areas: a spatial statistical modelling approach. Landsc. Urban Plan. **157**, 352–364 (2017)
11. Lian, Y., Ou, D., Pan, S., et al.: An evaluation study on soundscape of different landscape space types. Build. Sci. **36**(8), 57–63 (2020)
12. Liu, J., Xiong, Y.C., Wang, Y.J., et al.: Soundscape effects on visiting experience in city park: a case study in Fuzhou. China. Urban Forestry Urban Greening **31**, 38–47 (2018)
13. Chen, W., Daniel, W., Liu, J.: Study on soundscape regulation of urban park based on landscape elements. Chin. Landscape Archit. **38**(6), 91–96 (2022)
14. Li, G., Xiong, J., Xu, M., Dong, L.: Comprehensive study on the noise reduction ability and the visual effect on the edge of green space of Beijing urban park. J. Beijing Forestry Univ. **39**(03), 93–104 (2017)

Numerical Simulation of Indoor Air Quality and Aerosol Diffusion in Gym

Zhiqiang Kang, Baorui Hao(✉), Ning Yin, and Tong Wang

Shenyang Jianzhu University, Shenyang 110000, Liaoning, China
335445935@qq.com

Abstract. An indoor ventilation system plays an important role in the improvement of indoor air quality and the control of virus transmission. Therefore, based on Computational Fluid Dynamics (CFD), this paper establishes a full-scale gym ventilation calculation model under full-load conditions and simulates three different ventilation modes in the gym. The spatial distribution of indoor CO_2 was obtained by solving the component transport equation, and the diffusion process of aerosol generated by the coughing of virus carriers in the gym was calculated by the Lagrange method. The results show that the CO_2 con-centration in the indoor human activity area is lowest when the underfloor air distribution system is used. The top supply top return air and underfloor air distribution ventilation mode is more conducive to the removal of pollutants. The underfloor air distribution mode can minimize the mixing of air and aerosol particles in the gym and reduce the diffusion range of aerosol particles. The research results can provide a reference for the design of the ventilation system of the gym.

Keywords: CFD Analysis · Gym · Ventilation Mode · Carbon Dioxide · Aerosol

1 Introduction

Maintaining social distance is an effective way to prevent the spread of airborne pathogens [1], but in certain indoor environments, such as gyms, the required social distance is not applicable. Gyms have high population densities and enclosed spaces, and air quality is generally poor, making it very easy for respiratory infections diseases to spread [2, 3]. In a polluted environment, the body will breathe in more pollutants due to the amount of exercise and the body's high metabolic rate [4].

According to the scientific literature, the existing research results mainly focus on monitoring the concentration of pollutants in gyms, without studying the influence of air distribution on the air quality in gyms and without revealing the internal relationship between the aerosol diffusion mechanism and the mode of air supply in gyms.

Therefore, in this paper, using the spinning bicycle room in the gym as an example, the numerical simulation is carried out using ANSYS Fluent and the air distribution characteristics in the gym are obtained. The diffusion of CO_2 in gyms was studied using the mathematical model of component transport, and the diffusion of transient non-isothermal aerosol in the air in gyms was studied using the Lagrange mathematical

W. Guo et al. (Eds.): GBCESC 2023, LNCE 328, pp. 621–628, 2024.
https://doi.org/10.1007/978-981-99-9947-7_64

model. This paper evaluates the removal efficiency of CO_2 and aerosol particles in gyms by different ventilation modes, which provides guidance for improving indoor air quality in gyms and reducing the risk of aerosol transmitted infections.

Table 1. Air inlet and outlet parameter settings.

Mode		Mode 1	Mode 2	Mode 3
Inlet	Size	(0.3×0.3) m^2	(0.3×0.3) m^2	(0.3×0.3) m^2
	Number	6	6	9
	Location	Celling	Celling	Floor
Outlet	Size	(0.3×0.6) m^2	(0.4×0.7) m^2	(0.3×0.6) m^2
	Number	3	2	6
	Location	Celling	Wall	Celling

2 CFD Numerical Simulation Method

2.1 Physical Models

In this paper, the spinning bicycle room of a commercial gym in Shenyang, China is selected as the research object, as shown in Fig. 1. Its length, width and height are 7.25 m × 6 m × 3 m. The mannequin is simplified into three parts: head, torso and legs. Its mouth is 4 cm × 4 cm in length and width, which can accommodate 17 people at the same time during the business period.

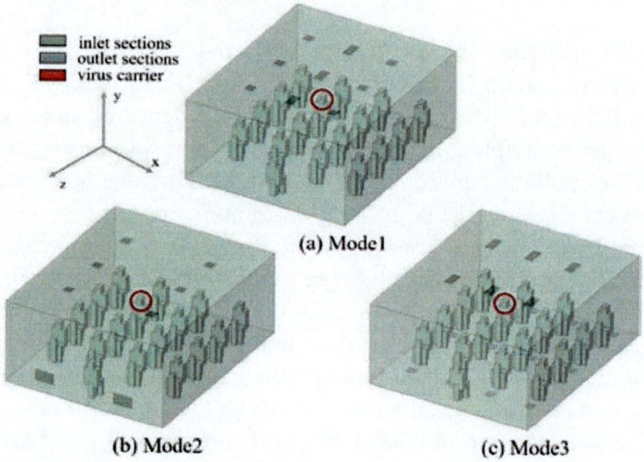

Fig. 1. Geometric characteristics of three ventilation modes.

Three modes are constructed according to the different types of air supply. Mode 1 is top supply top return air, Mode 2 is top supply bottom return air, and the air outlet is located 40cm away from the floor on the side wall, and Mode 3 is floor air supply. The setting conditions of air supply outlet and air return outlet of three ventilation systems are shown in Table 1.

2.2 Mathematical Model

The Reynolds Stress-averaged Navier-Stokes (RANS) method is used to model the turbulence. The commonly used turbulence models for simulating indoor air flow are RNG k-ε [5], Realizable k-ε [6], and Standard k-ε [7]. The research results show that the RNG k-ε model has a high accuracy in simulating the air distribution in closed space. Therefore, the author uses RNG k-ε in this paper. The governing equations of mathematical model include: continuity equation, momentum equation, energy equation, turbulent kinetic energy k equation, component transport equation and dissipation rate ε equation, and their general expressions are as follows:

$$\frac{\partial(\rho\varphi)}{\partial t} + \nabla(\rho\varphi v) = \nabla \cdot (\Gamma\nabla_{\varphi}) + S \tag{1}$$

where: ρ is the air density; t is time; v is the velocity vector; Γ is the effective diffusion coefficient of each variable; S is the generalized source term.

In this paper, the Lagrange model is used to study the motion state of aerosol in the gym and its distribution with time. The calculation of particle trajectories is based on the orbital model of Newton's second law:

$$\frac{du_p}{dt} = F_{drag} + G_g + F_a \tag{2}$$

where u_p represent the velocity of air; F_{drag} is the drag force; G_g is gravity; F_a is the additional force.

According to the literature review, the additional force terms [8] mainly include Brownian force, external force caused by pressure gradient and virtual mass, Saffman lift force [9] and thermophoresis force. In turbulent flow, turbulent diffusion plays a major role and the Brownian force can be ignored for particle diffusion. Saffman lift and thermophoresis force are mainly considered in this paper because they affect fine particles and places with large temperature gradients. Considering the influence of turbulence pulsation on particle diffusion, the random orbit model is selected to describe the particle motion [10].

2.3 Boundary Conditions

The amount of carbon dioxide exhaled by the user is calculated using the formulae given in [11], and the amount of carbon dioxide exhaled by the user is 0.76 L/min. The exerciser's sensible heat output is 145 W.

Because the droplets produced by coughing will evaporate quickly into droplet cores and aerosols [12], this paper uses droplet cores and aerosol particles with a particle size of

5 um as the research object, and the density is set at 1000 kg/m^3. In this paper, the aerosol is released by the coughing of fitness personnel 6 (red-marked personnel). According to Liu et al. [13] research achievement, the human cough speed is set at 10m/s, and a cough lasts for 0.5s. Assuming that the aerosol particles are trapped when they meet human body and the physical wall, the air supply inlet and air return outlet are set to escape. The specific boundary conditions are given in Table 2.

Table 2. Boundary Condition Settings.

Zone	Boundary condition
Floor, ceiling, surrounding walls	Adiabatic boundary
Human body surface	Heat flux is74 w/m^2
Mouth	Inlet velocity: 0.158 m/s, temperature: 35 °C
Mode1 inlet	Inlet velocity: 1.25 m/s, temperature: 20 °C
Mode2 inlet	Inlet velocity: 1.25 m/s, temperature: 20 °C
Mode3 inlet	Inlet velocity: 0.613 m/s, temperature: 20 °C
Outlet	Pressure outlet, 0Pa

2.4 Grid Sensitivity Verification

Fluent Meshing is used to discretize the geometric model and Poly-Hexcore is used to complete the global grid subdivision. Perform local grid encryption on the area where the fitness equipment, human body and tuyere are located. As the three types of ventilation use the same grid size, Mode 1 is used in this paper to verify grid independence. The results show that the average relative error in air velocity between the medium and fine grids is 1.96%, and the relative error between the medium and coarse grids is 6.89%. The independence test shows that the medium grid meets the accuracy requirements of the simulation. Therefore, the medium grid is used for the simulation calculation in this paper.

3 Analysis and Discussion of Results

3.1 Steady Flow Field Analysis

Figure 2 shows the velocity flow field distribution in the reference plane at z = 0 m for three ventilation modes. After the Mode 1 and Mode 2 incoming airflow enters the room from the supply air inlet, the jet area gradually increases and the wind speed gradually decreases from the center of the airflow, leaving the air in the non-jet area essentially undisturbed, so that the center of the room is not affected by the supply air. The air velocity around the human body is less than 0.4 m/s, and people have no sense of blowing. Due to the symmetrical arrangement of the supply air inlets, there is an ascending airflow in the aisle at the center of the room, and two vortices of the same height are formed at the bottom of the fitness area on either side of the supply air outlet, which is not conducive to the removal of pollutants. Mode 3 has low air supply velocity. There is no airflow vortex in the fitness area. The air velocity on this interface is maintained at 0.01 m/s–0.24 m/s, which makes people feel comfortable.

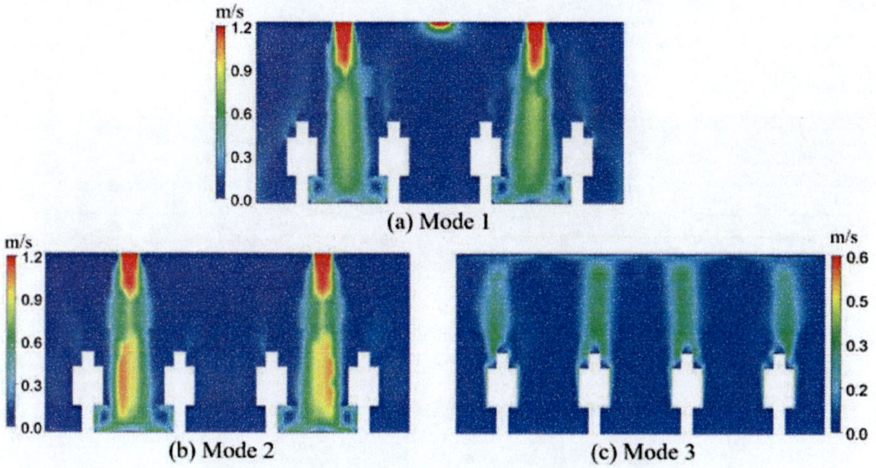

Fig. 2. Velocity distribution of plane z = 0m.

3.2 Temperature Field Analysis

Figure 3 shows the temperature field distribution on the cross section at x = 0.75 m. It can be seen from the figure that the temperature of the flow field around the human body is relatively high due to the heat dissipation and breathing of the fitness personnel, and the air heated by the human body rises under the action of buoyancy, forming a typical "hot plume" around the human body. The average temperatures at the cross sections of the three ventilation modes are 297.7 k, 298.4 k and 296.9 k respectively, which meet the thermal comfort requirements of indoor personnel.

3.3 Analysis of CO_2 Concentration Field

In this paper, the mass fraction of carbon dioxide is converted to ppm, and the minimum value of the concentration distribution is set to 0 ppm. The carbon dioxide con-centration distributions of Mode 1, Mode 2 and Mode 3 are shown in Fig. 4. The results showed that the concentration of CO_2 in the mouth and nose of people exceeded the limit value.

The indoor CO_2 concentration in Mode 2 is higher than that in Mode 1 and Mode 3. Because the air outlet is at the bottom, the carbon dioxide accumulated on the roof cannot be removed in time, so it spreads with the air supply into the area where people are active. Mode 3 is superior to Mode 2 in CO_2 removal effect. Mode 3 removes CO_2

from the release source with the airflow, reducing the intensity of CO_2 exposure in the area of human activity.

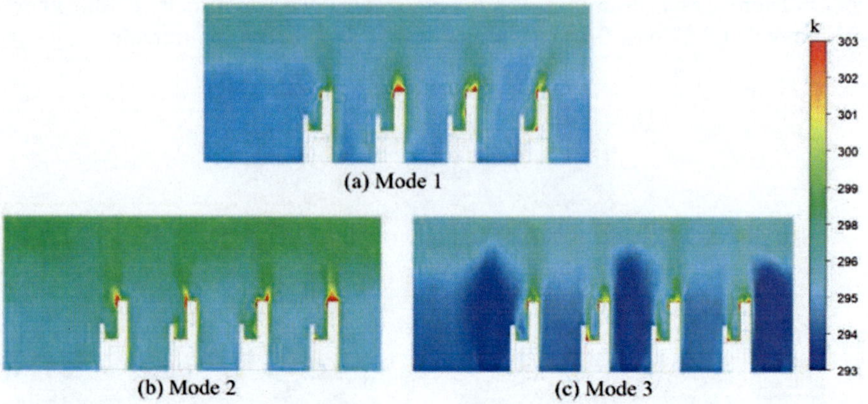

Fig. 3. Temperature contour of plane x = 0.75 m.

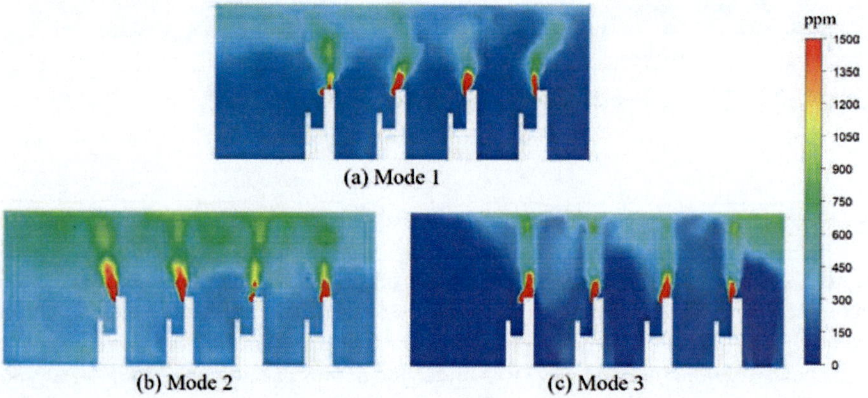

Fig. 4. CO_2 concentration distribution in plane x = 0.75 m.

3.4 Analysis of Aerosol Diffusion in Gym

The aerosol diffusion trajectory in the gym is shown in Fig. 5. The aerosol diffusion characteristic maps at 20 s, 50 s and 80 s are intercepted respectively, and the red part shows the spatial position of the 5 μm aerosol. As can be seen from the figure, under the three ventilation modes, aerosol particles spread upwards and horizontally after exhalation from the human mouth, and more aerosol particles gather in the area above the infected person's head.

In Mode 1, at t = 20 s, some aerosol particles arrive near the air outlet and start to be exhausted from the room, while the aerosol particles arriving near the air inlet move downwards with the supply air flow to mix with the indoor air. The Mode 2 air outlet is not on the roof, so the aerosol particles accumulated on the roof are not discharged, and the aerosol particles diffused near the air inlet are diffused downwards with the supply air flow. Influenced by the indoor longitudinal airflow, aerosol particles have an obvious tendency to spread to the front row. At t = 80 s, there are still a lot of aerosol particles in the room that have not been discharged. Mode 3 uses the floor air supply system and the aerosol generated by the human cough begins to spread to the air outlet. As there is no longitudinal airflow in the room, the aerosol particles spread out over a small area, concentrating mainly near the return air outlet.

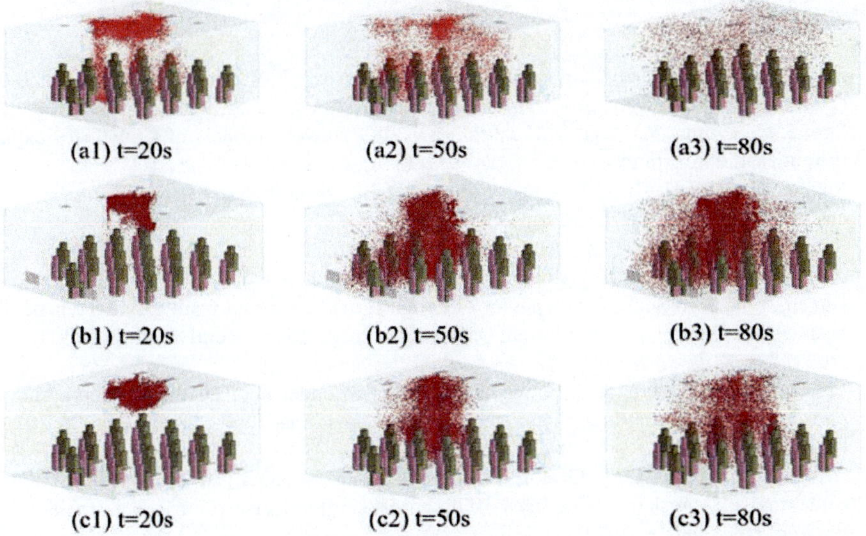

Fig. 5. Side view of aerosol diffusion in different ventilation modes. (a1) (a2) (a3) for Mode 1, (b1) (b2) (b3) for Mode 2, (c1) (c2) (c3) for Mode 3.

4 Conclusion

In this paper, the effects of different ventilation modes on indoor air quality and aerosol diffusion in gyms are studied by CFD numerical simulation, using component transport equation and discrete phase model. The main conclusions are:

(1) Three ventilation modes can obtain good thermal comfort, and indoor fitness personnel have no obvious feeling of blowing. Of the three ventilation modes, floor air supply can quickly remove CO_2 from the room in the personnel activity area, making the air in the personnel activity area the cleanest.
(2) The distribution characteristics of aerosol particles in gyms are closely related to the location of the return air outlet and air distribution. If the air return outlet is close to the infected person, aerosol particles can be quickly discharged from the air return outlet to the outside.

(3) Compared to the ventilation mode of top supply bottom return air, the top supply top return air of ventilation mode and the underfloor air distribution system are more conducive to aerosol discharge. Of the three ventilation modes, the underfloor air distribution system can minimize the mixing of air and aerosol in the gym.

Acknowledgement. This paper is supported by the scientific research project of Liaoning Provincial Department of Education: Optimal control of indoor local thermal comfort environment based on indoor personnel positioning technology (No. Z2219121).

References

1. Hua, Q., Te, M., Li, L., et al.: Indoor transmission of SARS-COV-2. Indoor Air **31**(3), 639–645 (2021)
2. Serina, C., Emma, P., Pang, W.K., et al.: Mobility network models of COVID-19 explain inequities and inform reopening. Nature **589**(7840), 82–87 (2021)
3. Sukbin, J., Si, H.H., Ji, Y.R.: Cluster of coronavirus disease associated with fitness dance classes. Emerg. Infect. Dis. **26**(8), 1917 (2020)
4. Karolina, B., Wioletta, R.K., Adrian, B.: Indoor air quality in sports center: assessment of gaseous pollutants. Build. Environ. **208**, 108589 (2022)
5. Huyen, T.D., Kyo-Seon, K.: Behavior of cough droplets emitted from Covid-19 patient in hospital isolation room with different ventilation configurations. Build. Environ. **209**, 108649 (2022)
6. Chang, H.C., Beungyong, P., Seong, R.R.: Effect of under-floor air distribution system to prevent the spread of airborne pathogens in classrooms. Case Stud. Therm. Eng. **28**, 101641 (2021)
7. Arpino, F., Cortellessa, G., Grossi, G., et al.: A Eulerian-Lagrangian approach for the non-isothermal and transient CFD analysis of the aerosol airborne dispersion in a car cabin. Build. Environ. **209**, 108648 (2022)
8. Renze, X., Fan, W., Xueliang, L., et al.: Numerical comparison of ventilation modes on the transmission of coughing droplets in a train compartment. J. Wind Eng. Ind. Aerodyn. **231**, 105240 (2022)
9. Qiqi, L., Cuiyun, O., Jian, H., et al.: Role of pathogen-laden expiratory droplet dispersion and natural ventilation explaining a COVID-19 outbreak in a coach bus. Build. Environ. **220**, 109160 (2022)
10. Sunil, K., Maria, D.M.: Numerical investigation on indoor environment decontamination after sneezing. Environ. Res. **213**, 113665 (2022)
11. Persily, A.K.: Evaluating building IAQ and ventilation with indoor carbon dioxide. ASHRAE Trans. **103**(2), 193–204 (1997)
12. Balachandar, S., Zaleski, S., Soldati, A., et al.: Host-to-host airborne transmission as a multiphase flow problem for science-based social distance guidelines. Int. J. Multiph. FlowMultiph. Flow **132**, 103439 (2020)
13. Liu, L., Wei, J., Li, Y., et al.: Evaporation and dispersion of respiratory droplets from coughing. Indoor Air **27**(1), 179–190 (2017)

Ecological Construction Strategy of Traditional Houses in Qianzhong Tunpu: A Case Study of Yunshan Tun

Fang Han[1(✉)], Jian Yue[2], Zhuo Chen[3], and Jian Liu[4]

[1] Guizhou University College of Architecture and Urban Planning, Guiyang 550025, Guizhou, China
369839837@qq.com

[2] Guizhou University Survey, Design and Research Institute Co. Ltd., Guiyang 550025, Guizhou, China

[3] Tongji Architectural Design (Group) Co. Ltd., Shanghai 200092, China

[4] Guiyang Architecture Reconnaissance Design Co. Ltd., Guiyang 550025, Guizhou, China

Abstract. The Tunpu village in the Qianzhong region of Guizhou, China, was formed as a military or commercial settlement after the Ming Dynasty army migrated into the territory of Guizhou. This form of settlement is a relatively closed cultural legacy of the Ming dynasty, which was progressively influenced to form a distinctive Tunpu village with Han Chinese culture. Traditional Chinese villages and dwellings have evolved over a long period to produce ecological construction techniques and methods that are in harmony with the natural environment and are rich in ecological wisdom and rationale. Based on the disciplinary background of architecture, this paper will discusses the typical construction concepts and ecological techniques of traditional dwellings of traditional Tunpu in the Qianzhong region from three aspects: regional characteristics, construction concepts, and construction techniques, to explore the ecological methods and concepts involved and provide a theoretical basis for the conservation and renewal of traditional dwellings, while developing sustainable perspectives for green architecture.

Keywords: Tunpu · Traditional houses · Ecological construction

1 Basic Information

1.1 History

Tunpu culture is a remnant of the livelihoods of soldiers, merchants, and their families who came to Yunnan and Guizhou from Jiangnan accompanied by the army or merchants in the Ming Dynasty. Its dwellings, costumes, diet, civil beliefs, and entertainment methods are well-preserved. Yunshan Tun remains the most intact cluster of Tunpu villages. The existing stone the Tuen Mun, tower, wall, residence, Caishen Temple, theater and other buildings are relatively well preserved. The variety of building types, which contain significant construction wisdom and an understanding of geographical and ecological conditions, is worth excavating, pondering, and learning from the ecological concept of constructing traditional villages and buildings.

W. Guo et al. (Eds.): GBCESC 2023, LNCE 328, pp. 629–637, 2024.
https://doi.org/10.1007/978-981-99-9947-7_65

Yunfeng Area is located 18 km south of Anshun City, Guizhou Province, there are eight Tunpu villages including Yunshan Tun, Ben Zhai, Lei Tun and Xiaoshan Zhai, which is the core area of the southern army garrison in the early Ming Dynasty (see Fig. 1). Yunshan Tun has 187 households, 897 people, approximately 505 acres of paddy fields, 90 acres of dry land, and 48 wells [1]. Yunvin Temple is situated on the highest mountain range in the Yunfeng area, Yunvin Mountain. It was established before the village, which winds around the Yunvin Mountain gorge from southeast to northwest, high in the middle and low on both sides, in a deep and secluded location (see Fig. 2). The buildings in front of Tun were built between the Ming and Qing dynasties, the buildings in back of Tun were built during the Republican period. The front Tuen Mun is a Gatehouse building (see Fig. 3), and the back Tuen Mun is a stone arch (see Fig. 4 and 5), which forms the defence system of the Urn City. The front and back stone walls are connected to Tuen Mun, forming a natural barrier with the natural topography. A defensive system that is easy to defend and difficult to attack has been formed. In addition, both the front and back Tuen Mun are dedicated to the Temple of the Land. The majority of the buildings, including houses and watchtowers, are located on both sides of the hillside in a stepped pattern, following the trend of the gorge to form a strip of space.

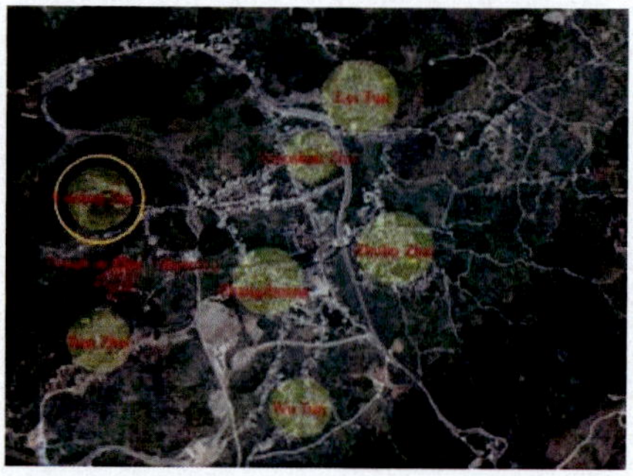

Fig. 1. Location map of Tunpu in Yunfeng area

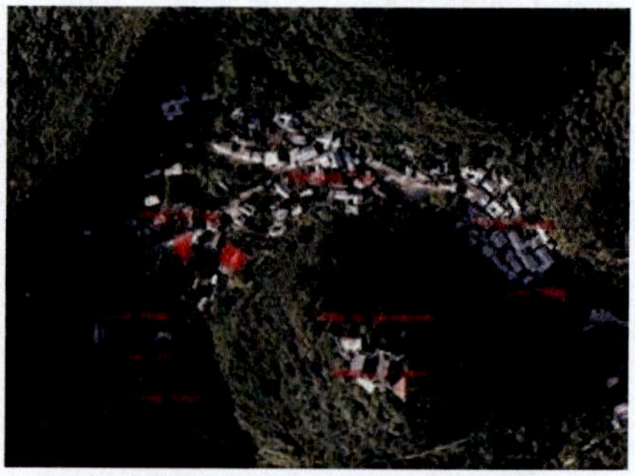

Fig. 2. Yunshan Tun Orthophotography

1.2 Regional Characteristics

Yunshan Tun is located in Pingba Yunfeng Area, Anshun City, Guizhou Province. Pingba District is located in the hinter land of Qianzhong, known as "the belly of Qian, the throat of Yunnan, the lips and teeth of Shuyue". The territory is flat and crisscrossed by rivers, with the highest altitude of 1645.6 m, the lowest altitude of 963 m, and an average altitude of 1250 m. It has a subtropical humid monsoon climate, without severe cold in winter and scorching heat in summer, four distinct seasons, and abundant precipitation. The Ultraviolet radiation intensity is low, and the climate is mild. The territory of the Pingba area belongs to the distribution zone of subtropical evergreen broad-leaved forest, and the biological resources are relatively affluent.

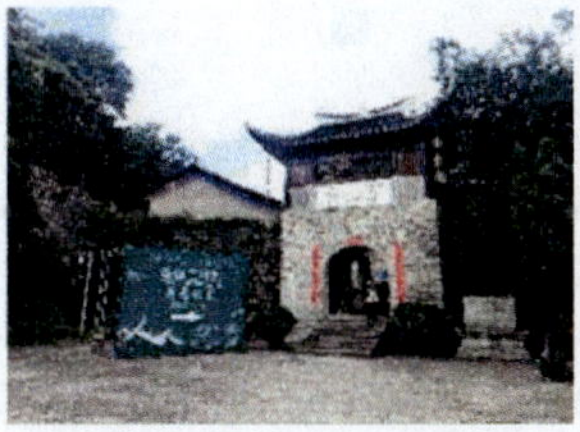

Fig. 3. Tuen Mun in front of Tun

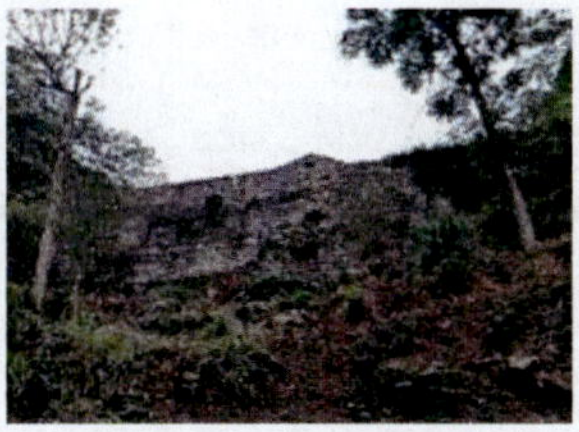

Fig. 4. Tuen Mun in back of Tun (Outside)

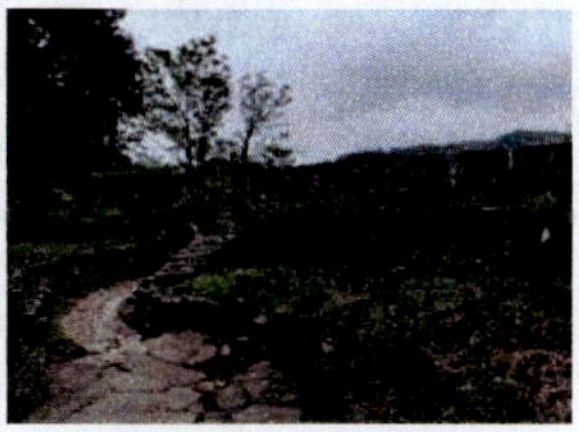

Fig. 5. Tuen Mun in back of Tun (Inside)

2 Ecological Construction Concept

The plateau and mountainous terrain of central Guizhou are remarkable. And Tunpu is mainly concentrated here, which is related to the migration route of the Ming army at that time, At the same time, the cultural system of this area was influenced by the culture of Jiangnan and changed geographically according to the production and living time, including the form of the buildings, the shape of the villages, the location, and the spiritual & cultural support, all of which incorporated natural, social and cultural factors.

2.1 Site Ecology

Tunpu people often choose flat "dams" as the main space of "tuntian", and choose sunny slopes, such as the south slope, southeast slope and southwest slope with moderate slope as residences, which greatly alleviates many problems such as dampness and drainage. The location of the village is inseparable from water resources and land resources, mainly used for agricultural production. This is because the Han Chinese have a long history of farming technology based on civilization, production, and using surface water and groundwater resources for living. The central part of Guizhou is a plateau karst geological type enriched in surface water and groundwater resources. Therefore, Yunshantun has a well-developed underground water system with 48 wells throughout the village, which addresses agricultural irrigation and domestic water use while considering water for fire-fighting. However, soil and water conservation is an essential part of the construction of the residence due to the treacherous terrain and slope of the area. For this reason, Tunpu people cultivated the feng shui forest behind their houses and built water diversion ditches. They transformed the green slopes to achieve soil consolidation and water conservation, ensuring the safety of the houses and resisting rainwater intrusion. Culturing feng shui forests also resists the cold winds from the north while creating ecological diversity. A multidimensional, ecological spatial pattern of mountains, water, forests, and fields has been formed (see Fig. 6).

2.2 Functional Needs-Oriented Space Layout

Yunshan Tun is known as the "second residence" for the prominent families of Tunpu in the Yunfeng area, as it was initially built to stockpile supplies. It was therefore located in the military stronghold of Yunhu Mountain Gorge, an easy-to-defend and difficult-to-attack site. Gradually, the function evolved from Hoard supplies to a complex place, for commercial transactions, and living. Due to its functional complexity, the layout of the village has distinctive characteristics. It is reflected in the following aspects:

As the core of the front Tuen Mun, The temporal and spatial evolution pattern of villages are growing to the back Tuen Mun.

The terrain is dangerous, easy-to-defend and difficult-to-attack;

The spatial layout differs from a military cantonment's. Although there is a defense system and fortifications everywhere, the level of defense is not as high as that of a military cantonment.

The pattern of a Strip villages parallel to the main commercial street and the residential street;

Form a functional composite structure: front commercial space and back residential space;

The spatial functions are clearly divided, forming an independent sacrificial space (Caishen Temple) and public space (theater and square), and located in the geometric center of the village;

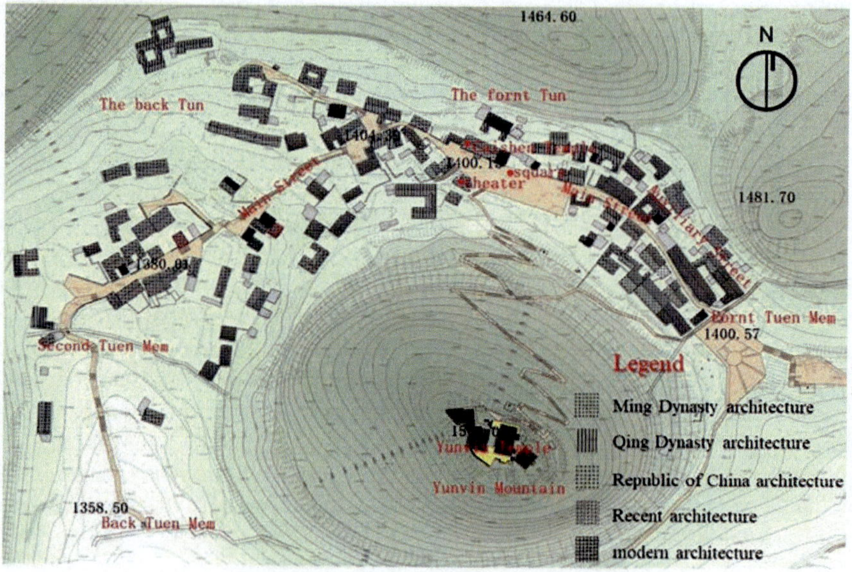

Fig. 6. Yunshan Tun general layout plan

2.3 Construction Technology

Terrain Utilization

Terrace: Guizhou is known as: "the sky is not three days clear, the ground is not three feet flat". The building conditions are very different from those of Jiangnan. The wise Tunpu people brought the building techniques of Jiangnan to the area. They adopted the construction of their houses by transforming the sloping land into a multi-stage terrace to suit the house building needs, namely by evolving from a watery village in Jiangnan to a three-dimensional village in the mountains of the Guizhou plateau.

Walking system: The whole village is organized in a strip-like space for conforming to the terrain, forming the main street along the strip space, for the commercial space, and forming auxiliary streets parallel to the main street to organize the living space. Several connecting roads are interspersed in the middle, and steps are cleverly arranged according to the height difference of the terrain to form a three-dimensional mountain traffic form.

Building layout: Due to the restrictions of terrain conditions, the form of building combination is not as regular as Jiangnan architecture. However, the Tunpu people will flexibly organize the site according to the terrain conditions, and arrange courtyards or independent buildings. These buildings are generally located along the main street and, owing to their composite function, are mainly in the form of freestanding buildings in a monolithic shape with three openings, fitting the functional needs of the shops, with the freestanding buildings forming ventilation corridors through the pedestrian path in front and behind the buildings on auxiliary streets, for the needs of simple residential functions. The courtyard building patio undertakes the lighting and ventilation of the interior of the building.

Drainage system: The building ground has a rigorous drainage system. The entire ground of the building is framed by a stone ridge wall with an exhaust channel, which is then filled with the particle size of 30 mm ~ 50 mm crushed stone, and tabia is laid on the surface that creates a complete system of waterproofing and damp-proofing of the building ground. Some are also laid with bluestone slabs, which not only protect the ground, but also play a good decorative effect. As a large amount of mountain water needs to be diverted on the side of the building near the mountain, it is generally provided with interceptor ditches or rainwater ditches that use the natural topography to guide its discharge to the road. The rainwater collected by the roof of the building, the ground, and the slope of the mountain is finally all collected on the road surface along the main commercial street, with a complete masonry drainage ditch under the road surface, which follows the topography from high to low, and then at a suitable location is discharged to the fields for soil run-off. Some are also laid with bluestone slabs, which not only protect the ground, but also play a good decorative effect.

Architectural Form. The combination of the buildings is mainly in the form of courtyards and independent buildings, with the courtyard type being based on the triple and quadruple courtyards, which are different from the courtyards of Jiangnan and have a more richer, and more multidimensional space. There are U-shaped, return-shaped, and L-shaped buildings. However, the main building still uses the three-room system, and the two side wings are used as auxiliary space for miscellaneous storage, livestock rooms, and other functions (see Fig. 7).

Building Material

Stone: The stone is stable and unchanged under natural conditions, with superior resistance to pressure, water, moisture, and ultraviolet. The Qianzhong region produces stones in abundance and with superior quality. The stone has been used in large quantities in walls, roofs, floors, and other building parts after several experiences of predecessors. At the same time, it was processed into various shapes, such as blocks, strips, and flakes, to suit the needs of the building parts as required. The stone was also used extensively in the production of household items, such as water tanks, pig troughs, and decorative stone objects.

Timber: Wood with excellent overall properties is easy to process and transport, has good tensile properties, compressive properties, bending resistance, suitable for purlins, rafters, fang, walls and other parts of wooden roof trusses. The ridge purlins, columns, and purlins through the fang are double purlins to increase the cross-sectional size of the load-bearing elements. Wood has been refined to the greatest possible extent to utilize the material's properties and maximize savings in construction materials.

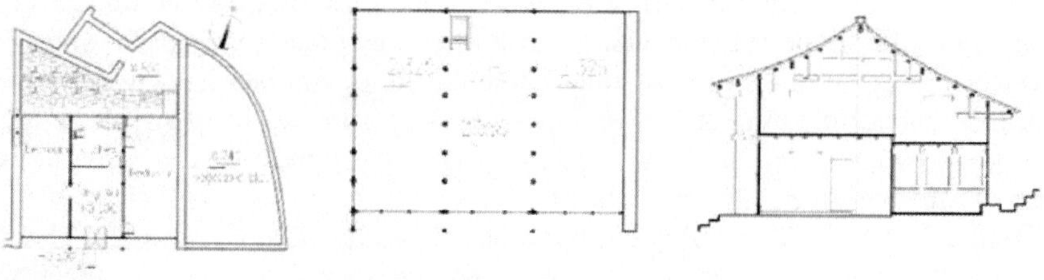

The first floor The attic floor Profile view

Fig. 7. Independent building drawings

Structural Form

Mixed Load-Bearing Structural System of Stone Outer Gable and Wooden Row Inner Transverse Walls: Most houses use masonry stone as the gable, which is convenient for take materials, and has superior mechanical properties, fire performance, and thermal insulation performance. Wise ancestors have mastered a set of design and construction principles for masonry structures. It has achieved an architectural civilization that has stood for hundreds of years. The masonry of the wall is cleverly staggered and lapped, the internal transverse wall is made of wooden structure through the row, and the paneled a wooden wall is used as a room partition, which is flexible and lightweight, and the rapid construction. The building orientation can be enlarged as much as possible, considering wooden structures can enlarge the openings of doors, windows, and shops.

Row-Lock Cavity Wall: The bottom layer of the stone wall is generally thicker, ranging from 400 mm ~ 800 mm. The second layer of the thicker wall at the bottom chooses the masonry method of the row-lock cavity wall, which not only reduces load but takes care of the beauty of the wall but also makes the air interlayer inside the wall, which plays a good role in thermal insulation and heat insulation (see Fig. 8).

Roof Truss Structural System: The building adopts a Column and Tie Construction through a timberwork system. The roof truss adopts the purlin type, the purlin on the stone gable, and the head of the column, through the connection of the purlin, forming an integral structure. The choice of the slope is mainly for the roof bluestone slab load and lap mode consideration, generally using five points of water, namely 26.6°, compared with the small green tile roof, as single piece load is greater, the size is also larger, the roof has not been draught phenomenon, so the roof forms a hard straight slope, rather than glazed tiles, the curved slope of small green tiles (see Fig. 9).

Stone-longkou: Stone-longkou is the main stressed of the cornice that can be extended outward, and the excellent "outrigger" function is obtained through the masonry method, section form, and material selection, which not only protects the wall but also obtains a good shading effect. Avoid the wall temperature increased by solar radiation in summer and transfer heat energy to the indoor room, which can effectively reduce the temperature of the indoor room in summer (see Fig. 10).

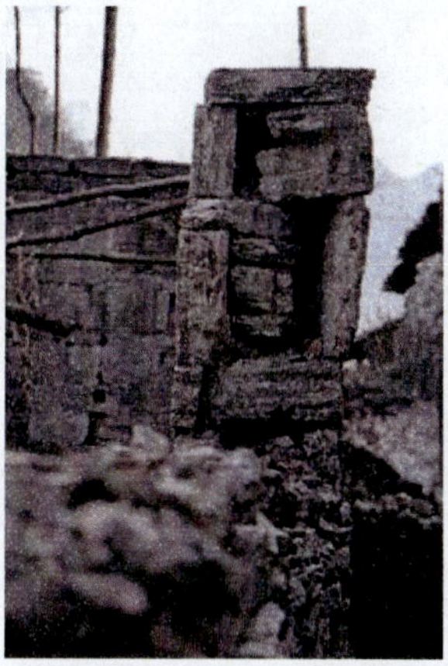

Fig. 8. Row-lock cavity wall

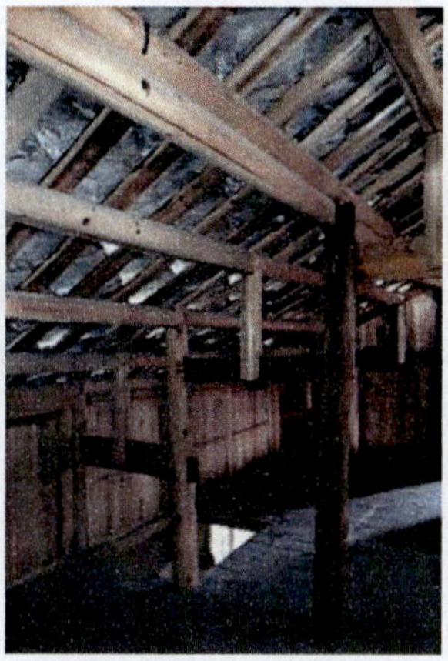

Fig. 9. Roof truss structural system

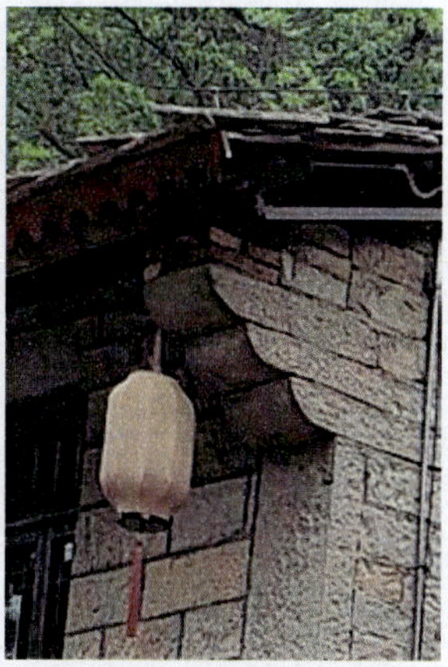

Fig. 10. Stone-longkou

Building construction: Eaves; Dripping of the eaves; Ground; Steps; Gutter; Resisting the erosion of rain and moisture has always been a concern in the construction of building houses, The form of sloped roofing is easy to organize drainage and simple to construct. At the same time, the eaves protrusion protects the walls from rainwater erosion. It consists of a 45° diagonal staggered lap of bluestones slabs, forming a jagged rhythmic shape at the drip, which serves both a functional and a very decorative architectural purpose (see Fig. 11).

Based on years of experience, it has been found that connecting the difference in height between the indoor and outdoor areas by steps is an effective way to avoid rainwater erosion (see Fig. 12). Although no concept of water dispersion was available, it was realized that drainage ditch covers sloping appropriately towards the road could act as a good diversion of rainwater from the ground and maintain the building dry. After collecting the rainwater from the building, it is organized and fed into the road storm drains (see Fig. 13).

Fireproof. For fire protection, the main measures taken are building fireproof mountain walls, leaving fire spacing, creating fireproof units, and creating water wells in public spaces. The stone has the characteristics of a non-combustible material, while there are no cavities in the gable surface. Therefore a split hill firewall is formed between the individual households. Furthermore, the spacing of the walls between households ranging from 1.5 m to 3 m creates the conditions for the formation of natural fire breaks and creates the spatial layout of the main street cascading alleyways as natural fire protection units. In other words, there is a separation between each unit in the event of a fire. Furthermore, the infrastructure, such as the roads, is made of stone masonry, which is a good barrier to the spread of fire. In addition, the large number of wells in the outdoor communal areas can be used as a source of water to control and extinguish the fire in the initial fire stage, which provides good conditions for fire prevention.

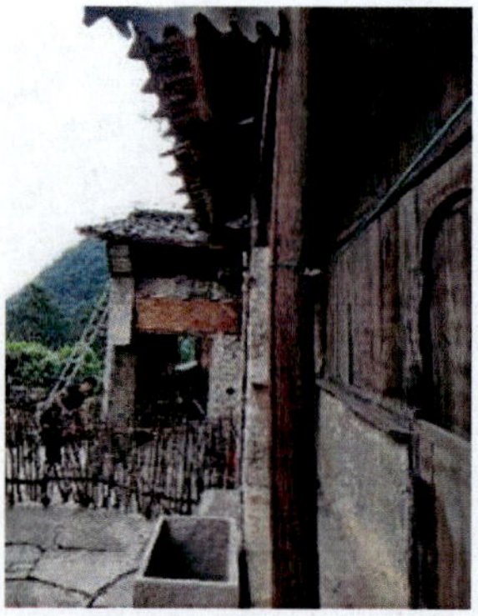

Fig. 11. Eaves and Dripping of the eaves

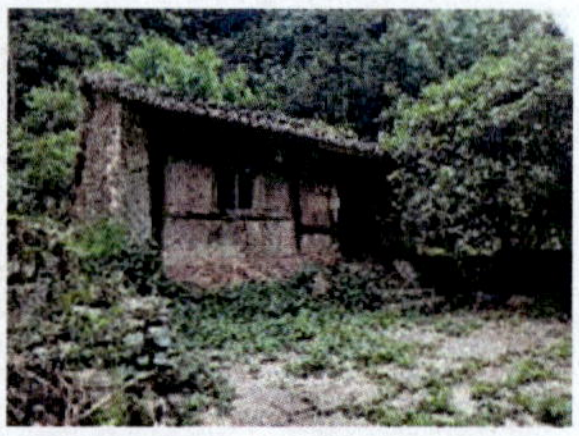

Fig. 12. Ground and Steps

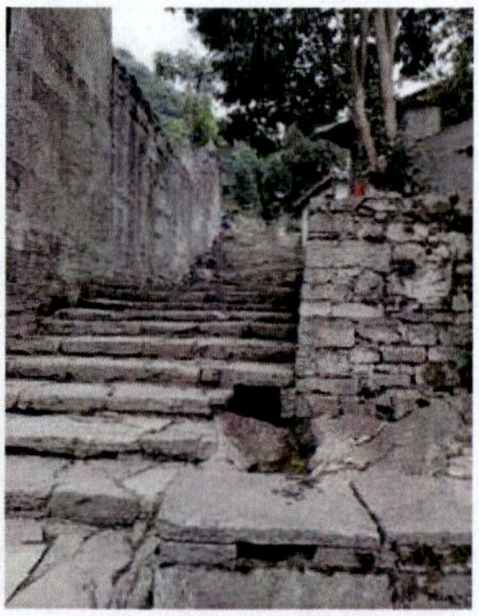

Fig. 13. Gutter

3 Conclusion

The traditional dwellings of Tunpu contain a significant element of ecological wisdom and technology, from site selection and layout to building construction. The use of passive energy-saving methods creates a comfortable living environment and achieves the purpose of saving energy and protecting the environment. From the Case Study of Yunshan Tun, we conclude the following aspects:

In general, the Yunshan Tun Traditional Houses are energy-saving architecture, The critical elements of energy-saving architecture are reasonable site selection and scientific building layout. The choice of building orientation to satisfy the conditions of proper lighting and ventilation, the control of the building size factor, and the improvement of the thermal insulation of the building maintenance structure, i.e., the building walls, roofs, and floors, are all ecological construction measures that effectively reduce building energy consumption. At the same time, the appropriate wall-to-window ratio determined by combining life experience meets the needs of light and ventilation. It takes care of heat insulation in winter and heat dissipation in summer. What's more, it's a modular building based on a deep understanding of building materials' physical and mechanical properties, forming the prototype of prefabricated building.

Acknowledgement. The work was supported in part by Guizhou Province Science and Technology Support Program (2023) ordinary313.

References

1. Yue, J.: Guizhou Workcamp -Column and Tie memorabilia. Kongxuetang publishing house. Guizhou Guiyang (2021)
2. Yue, J.: Pushingriver Stonesupthe Slopes. Guizhou Science and Technology Publishing Houst, Guizhou Guiyang (2008)
3. Gao, F.Q., Yang, Y.l., Zou, K.Z., Jiao, Y.M.: The study of the basin differentiation of traditional villages in Yunnan province. J. Yunnan Normal Univ. (Natural Sciences Edition) **42**(01), 64–72 (2022)
4. Huang, D., Dai, S.H.: Study on regionality and construction context of rocky folk house in central part of Guizhou. Architectural J. (05), 105–110 (2013)
5. Du, J., Hua, C., Yu, Y.F.: Research on spatial evolution of traditional settlements: the case study of Qianzhong Tunpu. Urban Dev. Stud. **24**(02), 47–53 (2017)
6. Qiao, X.X.: Column and tie construction technique in Tongren. Guizhou. Tradit. Archit. (00), 16–28 (2019)
7. Lu, X.S.: Technology design of ecological architecture of traditional houses in Huanggang. In: 2016 International Conference on Manufacturing Construction and Energy Engineering
8. Liu, Z.H.: Construction of green housing design methods for ancient villages in southwest minority areas. In: 3rd International Conference of Green Buildings and Environmental Management
9. Learning from Vernacular Architecture: Ecological Solutions in Traditional Erzurum Houses. Conference on Urban Planning and Architectural Design for Sustainable Development
10. Liu, M., Zhang, B.G., Ren, J.W. Lian, C.L., Yuan, J., Hao, Q.L.: Whole life-cycle ecological footprint of rural existing houses in Northern China. Buildings **8**(7), 1–19(2018)

Research on Forward Design of Green Office Building Based on BIM

Hui Cao, Ruixin Ju(✉), Zhibo Wang, and Dan Fu

CSCEC Northeast Design and Research Institute Co., Ltd., Shenyang 110000, China
ju_rita@yeah.net

Abstract. With the implementation of the dual carbon strategy, we will promote the development of office green building design and build a new mechanism for its development based on the principle of green ecological sustainability and under the guidance of BIM's multiple collaborative concept. In the full cycle design stage of the building, the digital simulation of the green performance of office buildings and the structural overrun design are mainly used to optimize the architectural design, so as to avoid the phenomenon that the defects in the architectural scheme stage lead to the later use of technology stacking to make up for its "congenital deficiency". In the Yangtze River Delta port front headquarters project of Shanghai runYou technology, the BIM positive design process is proposed from the perspective of architect responsibility system, and then the BIM optimization model is constructed with green performance as the variable, integration and environmental simulation analysis and structural overrun calculation, and energy consumption and comfort as the goal. The research results help to enhance the low-carbon design performance of buildings, It provides theoretical and practical basis for design decision.

Keywords: Office building · BIM forward design · Green building performance simulation · Multi objective optimization · Structural overrun design

1 Introduction

With the promotion of the global dual carbon strategy, the construction industry, as an important carrier of people's good life, is transforming and upgrading to green, healthy and digital. Based on the technical advantages of BIM forward design, under the exploration of the EPC mode led by the design institute and the architect responsibility system, in the whole cycle of office architecture design, Take the Shanghai Runyou Technology Yangtze River Delta Port Frontier Headquarters Project in China as an example, the process of multi-performance objectives of green digital design is established to improve the quality of green design and architecture, and realize the organic generation of architectural form and the creation of high-quality space from the starting point of environment and nature, Select the most appropriate scheme by means of digital simulation and overall balance, and give play to the leading role of architects and the collaborative promotion of various disciplines.

W. Guo et al. (Eds.): GBCESC 2023, LNCE 328, pp. 638–651, 2024.
https://doi.org/10.1007/978-981-99-9947-7_66

2 Project Overview

The runYou science and technology headquarters project is located at the intersection of Haigang Avenue and huanhunan Second Road in the development area of the first ring headquarters Bay in Lingang District, Shanghai, adjacent to the park in the South and Dishui Lake in the north. The project covers an area of 12047.80 m^2, Total construction area 56236 m^2, The design building height is 49.6 m. It is a landmark project of runYou brand with high value and milestone comprehensive use in the Yangtze River Delta region. It is a mixed function building integrating headquarters R & D and office, sports big data R & D, education and popular science exhibition base and supporting functions. The design pursues a modern and unique architectural style consistent with the temperament of the enterprise. The shape of the building is derived from the interpenetration of the blocks. On the site with limited height, the design extends the blocks with the posture of lateral growth, emphasizing the lateral force on the site. The external surface is covered with a unit type aluminum curtain wall system with a perforation rate of 40%, and partially with a glass curtain wall to form a "floating glass box" (Fig. 1).

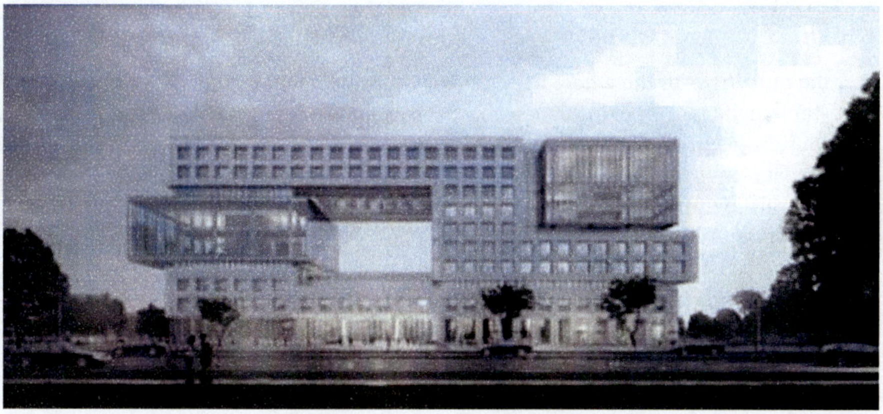

Fig. 1. Project rendering

3 BIM Forward Design

3.1 BIM Design Cycle

The digital design method is adopted in the scheme design stage, preliminary design stage and construction drawing design stage. In the later stage, the comprehensive model and pipe network comprehensive model are integrated, and finally the two-dimensional drawings, BIM model and corresponding BIM application are delivered (Fig. 2). According to the BIM design process, since the project has been designed using BIM Technology since the scheme stage, frequent large-scale adjustments are required, and the accuracy control of Revit model is very important for the project schedule. After the practice in the project, the basic parameters of Revit in the scheme stage should be locked first (such as floor height, outline, etc.). The adjustment of internal functions can be synchronized using 3D + 2D, that is, Revit controls the space, and AutoCAD controls the plane function area. In the process of Revit modeling, the function of parameter family should be fully used, and the control requirements that may occur in the design process of components must be fully considered when setting parameters.

BIM forward design is adopted in the project, and the characteristics of the model in the preliminary design, construction drawings and as built drawings are fully used in the whole BIM design cycle. After the scheme is adjusted and stabilized, the electromechanical specialty has carried out the pipeline synthesis of the main pipe, and the civil engineering quantities of the project can be obtained directly from the model. This greatly strengthens the architect's control over the overall cost. In particular, due to geographical conditions, several rounds of scheme adjustment have been carried out under the deep diving pool to avoid karst cave geology. Using dynamo programming to make small plug-ins can quickly design BIM models to extract relatively accurate quantities of specific components. This greatly improves the implementability of the adjustment scheme.

3.2 Simulation Analysis of Physical Performance of Green Building

Under the objectives of the whole life cycle of the project and the green construction two stars, establish the multi-performance objective process of green design (Fig. 2). With the goal of building a green, low-carbon, ecologically sustainable, healthy and comfortable building space environment, the planning and layout of the general plan of the building, the design of the building plane and elevation all respond to the natural ventilation and winter sunshine. At the same time, the needs of energy conservation, water conservation, building material selection and indoor comfortable environment will be met after the preliminary design of the building is completed, The preliminary results of building physical performance analysis are formed and adjusted with the overall shape design, and the indoor and outdoor analysis are fine tuned simultaneously. In this way, we can not only start from the climate analysis, but also put forward some passive design strategies for green buildings from the overall shape of the building, such as building orientation, shape, natural ventilation, facade design, sun shading design, etc., which can reduce the building's energy dependence on the active mechanical system to a great extent. Therefore, in terms of building physical performance analysis, the Revit model is used to export the DWG file of the building, and the sville green building design series software, an analysis plug-in based on AutoCAD platform, is used to identify the corresponding layers. After the auxiliary layer calculation model, a number of green building performance analysis simulations are carried out, as follows (Fig. 3):

(1) **Outdoor wind environment simulation**

The project is designed according to the green building two star standard. The outdoor wind environment assessment is based on the requirements of the items related to the outdoor wind environment in Shanghai green building assessment standard DG/TJ 08–2090-2020: the wind environment in the site is conducive to outdoor walking, comfortable activities and the natural ventilation rules of the building. Through the analysis

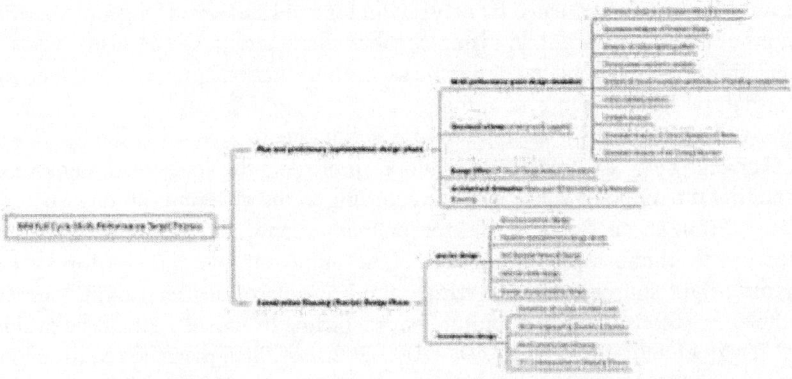

Fig. 2. BIM full cycle multi performance target flowchart

of wind speed reaching the standard, it is concluded that there is no wind speed overrun area around the building, and the building layout can be adopted; Based on the analysis of the wind pressure on the windward and leeward sides of the building, it is concluded that the building meets the requirement that the wind pressure difference between the windward and leeward surfaces of the building should not exceed 5pa, except for the first row of buildings in the windward direction.

(1) Under the conditions of typical wind speed and direction in winter, the wind speed of the pedestrian area around the building at 1.5 m above the ground is less than 5 m/s, the wind speed of the outdoor rest area and children's entertainment area is less than 2 m/s, and the outdoor wind speed amplification factor is less than 2; Except for the first row of buildings facing the wind, the wind pressure difference between the windward and leeward surfaces of buildings shall not exceed 5pa.
(2) Under the conditions of typical wind speed and direction in transition season and summer, there is no vortex or windless area in the human activity area of the site; The wind pressure difference between indoor and outdoor surfaces of more than 50% openable external windows is greater than 0.5pa.
(3) **Simulation and analysis of outdoor noise**

Considering the complexity of the surrounding noise environment after the completion of the project, this report needs to use software to simulate and calculate the daytime and nighttime noise values respectively, including the plane noise distribution of the project site, the noise sensitive building at the height of 1.5 m along the bottom contour line of the building (daytime and nighttime) and the noise sensitive building facade noise distribution (daytime and nighttime), and according to the technical code for the division of acoustic environment functional areas (GB/T 15190–2019), It is judged that the environmental noise simulation results in the site meet the relevant provisions of the environmental quality standard for noise GB3096–2008 and the evaluation standard for green buildings DG/TJ 08–2090-2020 in Shanghai.

(C) **Analysis of indoor lighting effect**

The light environment created by natural light is used and loved by people because of its economic, natural, pleasant and irreplaceable characteristics. The visual test results of various light sources show that under the same illumination condition, the recognition ability of natural light is better than that of artificial light. Natural lighting is not only conducive to lighting energy saving, but also conducive to increasing the natural information exchange between indoor and outdoor, improving the space health environment and regulating the mood of space users. According to the different lighting coefficients required by different functional rooms, reasonable window and glass curtain wall layout is carried out to obtain ideal lighting effect, so as to improve the comfort of people using natural light sources, so as to reduce the use time of artificial light sources and reduce the energy consumption of buildings. According to article 5.2.8 of the evaluation standard for green buildings (DG/TJ 08–2090-2020) of Shanghai, the requirements for the proportion of the area of the main functional rooms of public buildings with the daylighting coefficient reaching the standard are made. The area of the project for the indoor space with the natural daylighting reaching the standard is 76% (Table 1).

Table 1. Statistics of building daylighting compliance rate

Room type	Daylighting type	Standard value		Area (M2)		Compliance rate (%)
		Average daylighting coefficient (%)	Indoor natural light design illuminance (LX)	total area	Area up to standard	
Lobby	side	2.20	300	2410.06	2410.06	100
restaurant	side	2.20	150	2561.12	1735.79	68
Gym	side	2.20	300	1407.06	1182.17	84
Exhibition hall	side	3.30	450	1033.48	262.34	25
office	side	3.30	450	10342.18	8118.85	79
Activity space	side	2.20	300	1081.92	908.85	84
Conference room	side	3.30	450	401.32	0.00	0
Proportion of total qualified area (%)				76		

(D) **Driving route simulation**

Due to the requirements of Shanghai Planning Department for the line sticking rate of the facade of above ground buildings along the street, the driving space of motor vehicles around the building is too narrow. In order to ensure that large vehicles on the ground can enter the north side of the site, special software is used to simulate their driving trajectory in the preliminary design stage, and the position of the exit ramp of the underground garage is adjusted to meet the driving demand. In the process of traffic consultation, the planning department requires that two passenger car parking spaces be set on the ground. In order to confirm the feasibility of setting motor vehicle entrances and exits to urban roads in the base, the designer analyzed and simulated the passenger car driving trajectory through software, and finally through traffic consultation.

(E) **Analysis of sound insulation performance of building components**

Through the calculation and analysis of the weighted sound insulation and spectrum correction of the exterior wall, partition wall, exterior window, floor, door of the office building and the weighted standardized impact sound pressure level of the floor, it can be seen that the sum of the weighted sound insulation of the exterior wall and the spectrum correction of traffic noise is 45dB; The sum of the weighted sound insulation loss of the building partition and the spectral correction of pink noise is 50dB; The sum of the weighted sound insulation loss of the building exterior window and the traffic noise spectrum correction is 30dB; The sum of the weighted sound insulation loss of the building floor and the spectral correction of pink noise is 50dB; The project adopts a single-layer solid door, which has a good sound insulation performance after the door seam treatment, that is, the sum of weighted sound insulation and pink noise spectrum correction can reach 30 ~ 40 db; The weighted standardized impact sound pressure level of the building floor is less than 65dB. In general, it meets article 5.1.4 of Shanghai green building evaluation standard DG/TJ 08–2090-2020, "the sound insulation performance of exterior walls, partitions, floors, doors and windows of main functional rooms meets the low and medium limit requirements of the current national code code for sound insulation design of civil buildings GB 50118".

(F) **Simulation Analysis of Other Green Buildings**

The project is designed according to the green building two star standard. According to the requirements of the local government, the indoor lighting, sunshine analysis, indoor background noise, ventilation rate (Table 2) of the project are simulated and analyzed. Finally, according to the analysis of data display and scoring, the project meets the green building two-star standard.

Table 2. Statistics of air change times of public buildings under typical working conditions in transition seasons

Area ratio with air changes greater than 2 times /h		
Area with air changes greater than 2 times /h	15900.14	m^2
total area	19785.01	m^2
Area ratio RR	80.36	%

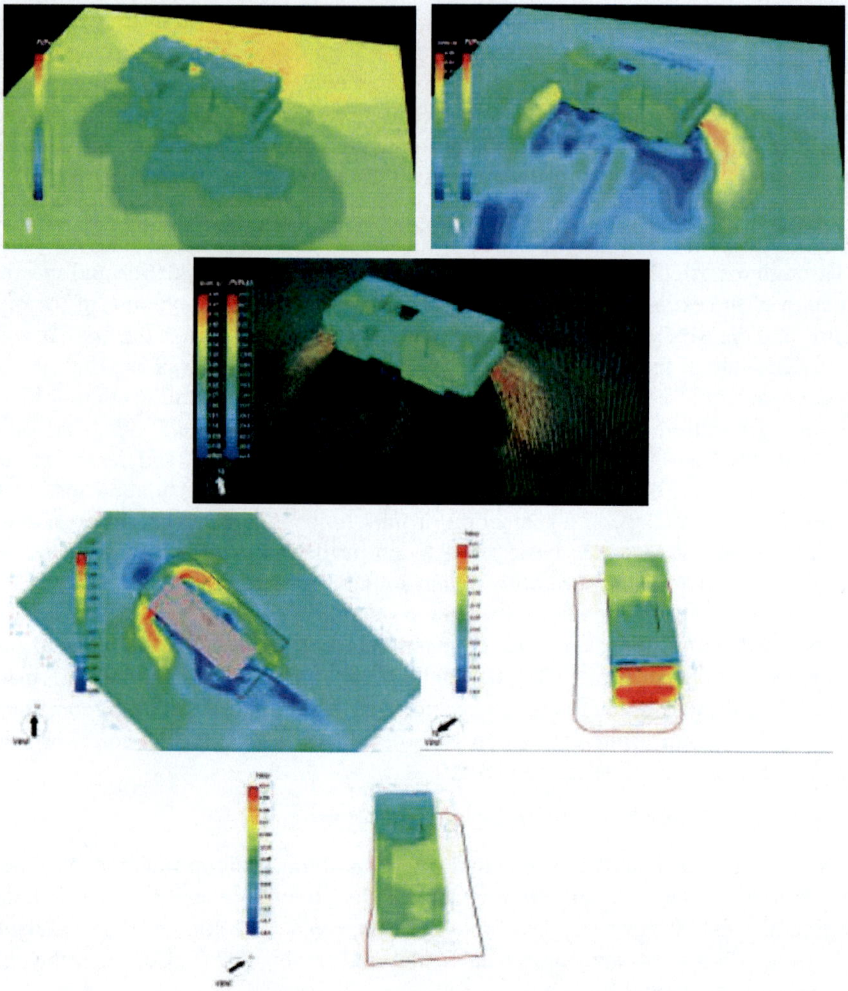

Cloud picture of wind speed magnification factor at 1.5m height;Cloud picture of wind pressure on windward side of buildings;Cloud picture of wind pressure on leeward side of buildings

(a)Simulation analysis of outdoor wind environment

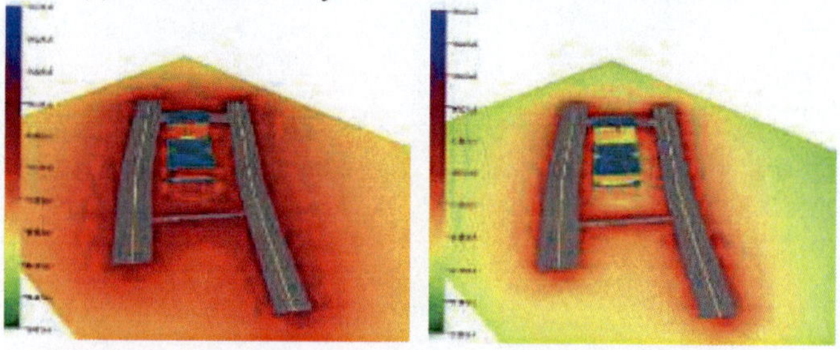

Fig. 3. Performance based simulation analysis of buildings

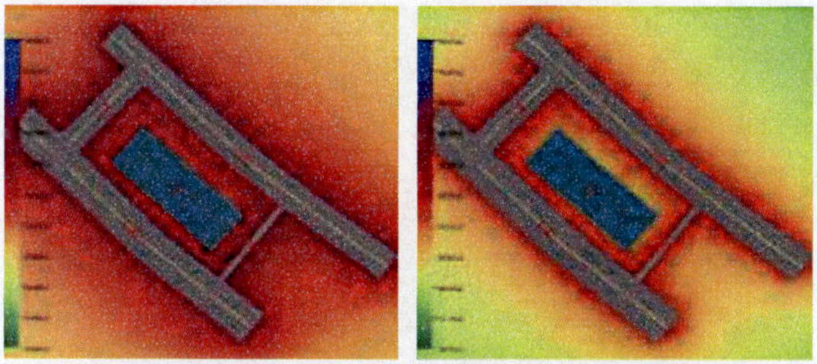

Distribution map of sound pressure level at 1.5m height of the site (day and night); aerial view of site noise distribution (day and night)

(b)Simulation and analysis of outdoor noise

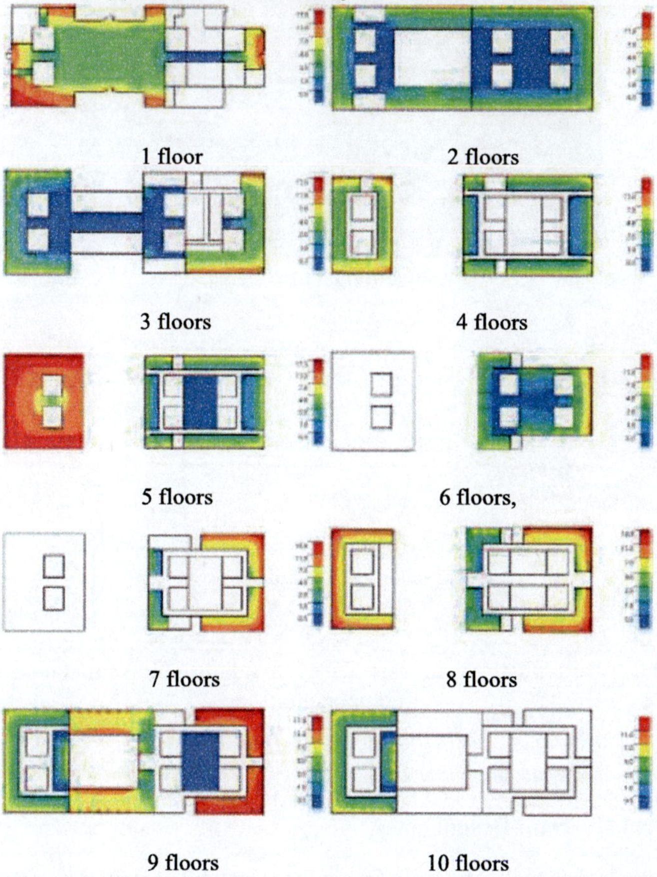

(c)Indoor lighting analysis (color map of lighting effect analysis)

Fig. 3. (*continued*)

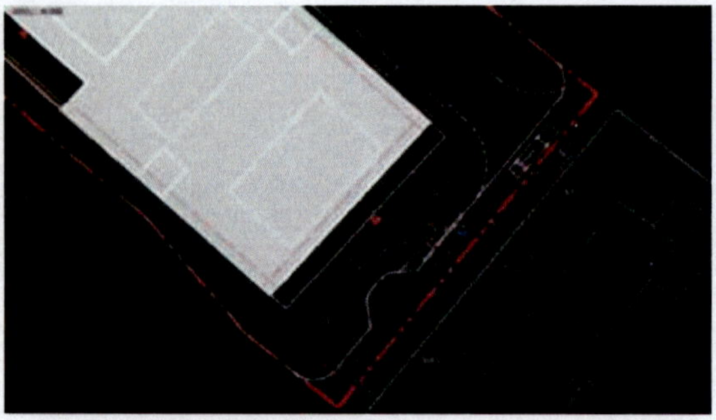

(d) Feasibility analysis of large vehicle routing

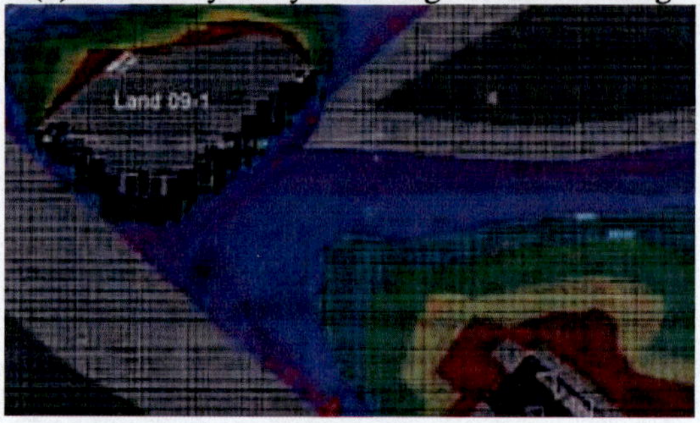

(E) sunshine analysis

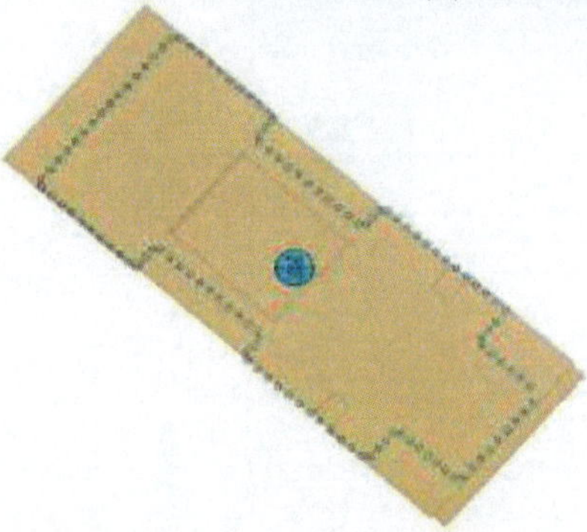

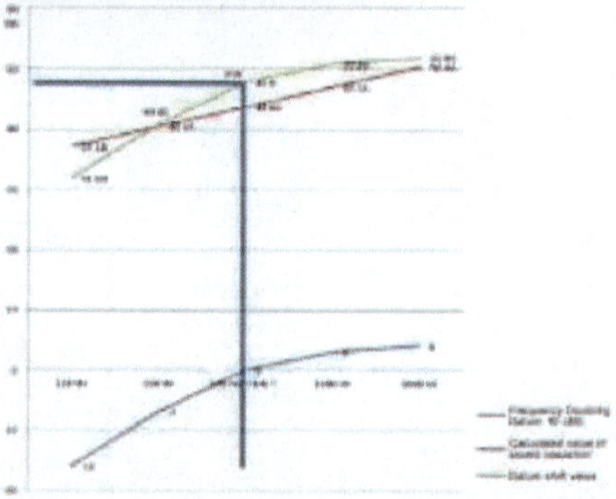

(f) background noise simulation analysis (g) sound insulation simulation analysis of components

Fig. 3. (*continued*)

3.3 Structural Overrun Design

The project adopts the steel composite frame (concrete-filled steel tubular column steel beam) - concrete core tube structure system. The embedded part is selected in the basement roof, and the bottom reinforcement part is 1–2 floors. The ratio of height to width of the structure is 1.05, mainly shear deformation. The large opening in the facade leads to the blocking of floors on floors 5–7, and most floors are connected as a whole building.

Due to the special shape of the building, the project is classified as a class a high-rise building with irregular transfinite. The transfinite items include torsional irregularity, local discontinuity of floor slab, irregular lateral stiffness (overall horizontal cantilever) and complex structure (complex connected structure formed after opening a large hole in the facade), with a total of 4 general irregularities. Therefore, the building structure needs to be designed beyond the limit. Since the initial scheme has a 50m deep indoor pool, the structural design is still very difficult.

In order to reduce the impact of the above irregularities on the structural safety, the structural design team conducted parallel adjustment and calculation of multiple structural software, and compared and analyzed the results. Through the analysis of structural small earthquake elasticity, equivalent elasticity under earthquake action of fortification intensity, equivalent elasticity under rare earthquake action, dynamic elastoplastic time history of rare earthquake and other aspects, as well as the special research on floor slab, large cantilever, connector and other aspects, the structural design is improved. After the expert argumentation, it is considered that the structural design is feasible, and then the project passed the special review of seismic fortification of high-rise building engineering beyond the limit (Fig. 4).

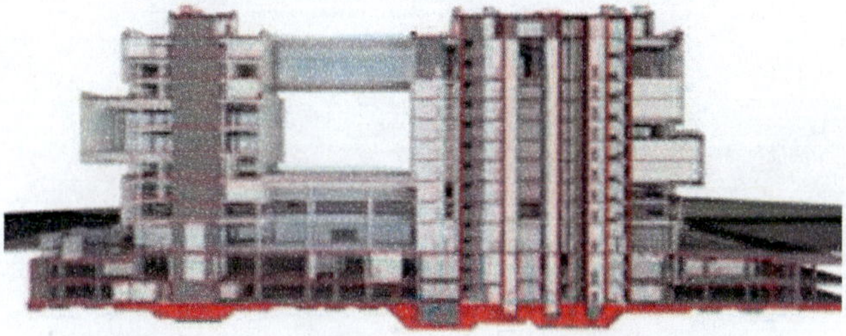

(a) Structural vertical section analysis

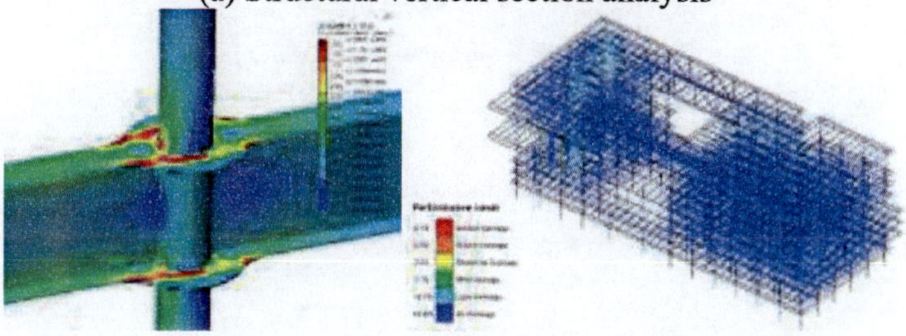

(b) component performance analysis

Fig. 4. Structural design analysis

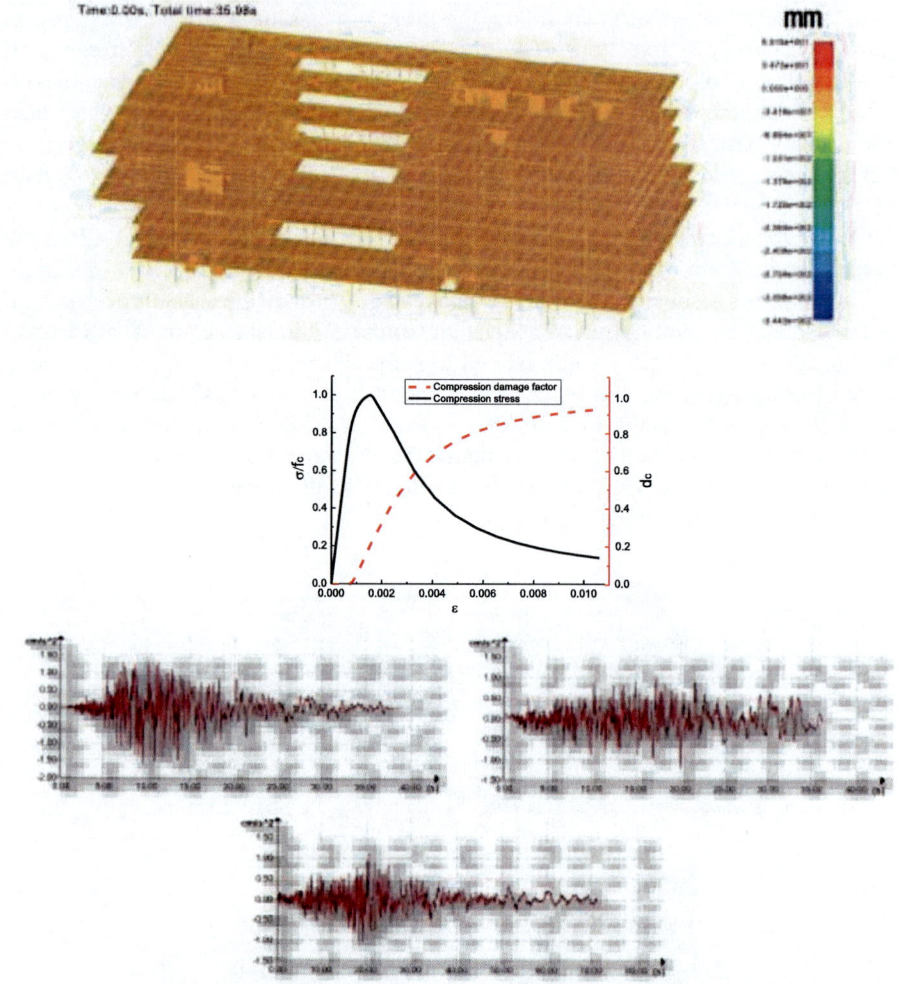

(c) Design and calculation of structure overrun

Fig. 4. *(continued)*

In the process of construction drawing design, BIM forward design solves many detail problems. For example, due to the structural design optimization, the local beam height is too high, and the pipeline cannot be laid normally. After structural optimization, the original high section beam is changed into a double-layer beam to ensure that the clear height of the service space is not affected. In the forward design process of BIM, the pipeline is integrated, and the clear height of the corridor is not less than 2.6m. The final effect is verified by the IMI roaming model. In the forward design process of BIM, each discipline links the design model, and provides the design information by providing the view and model parameters, and finally directly draws the drawing through the Revit model. In the construction drawing review stage, the BIM model is used to communicate with the drawing review unit, which is more convenient for the drawing review teacher to review the specific design content. The final construction drawings are directly exported by Revit, and the model is provided to construction, curtain wall, fine decoration, landscape lighting at night and other units, and the BIM model is transferred to the next stage of the whole life cycle of the building (Fig. 5).

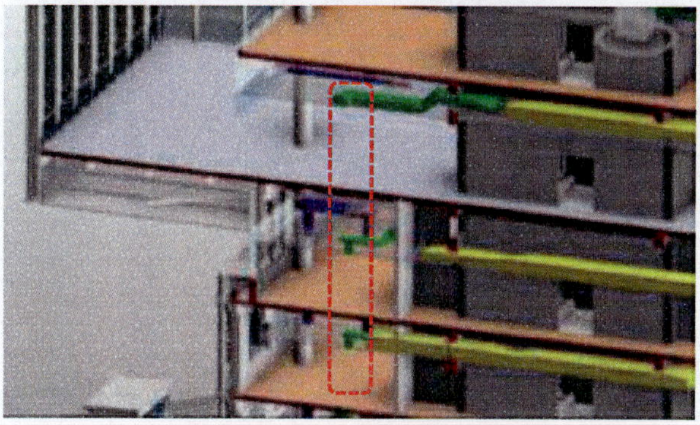

(a)Comprehensively consider the beam height and air duct, adjust the high section beam to double-sided beam

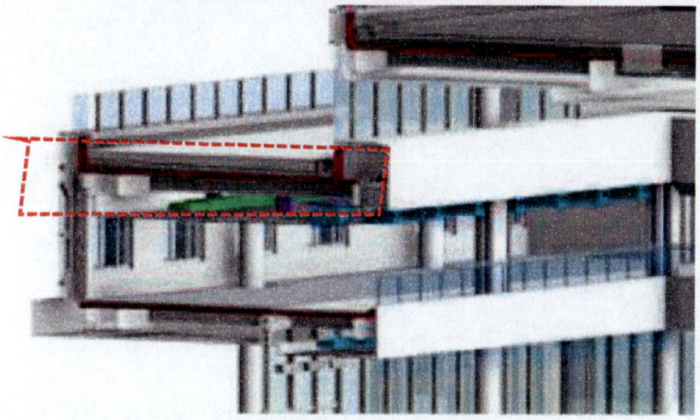

(b)Comprehensively consider the layout of air outlet, shutter and steel beam, and overhang the lower floor slab

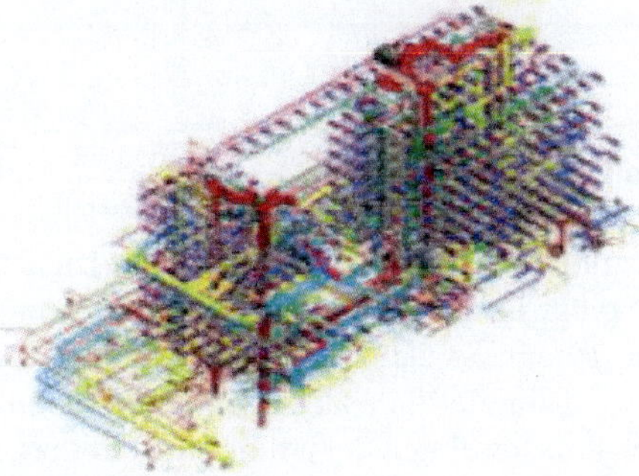

(c) Equipment discipline integrated management model

Fig. 5. BIM construction drawing design

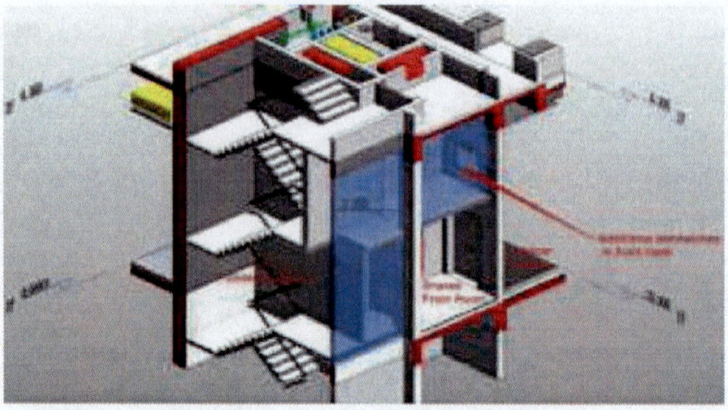

(d) 3D view of drawing review and communication

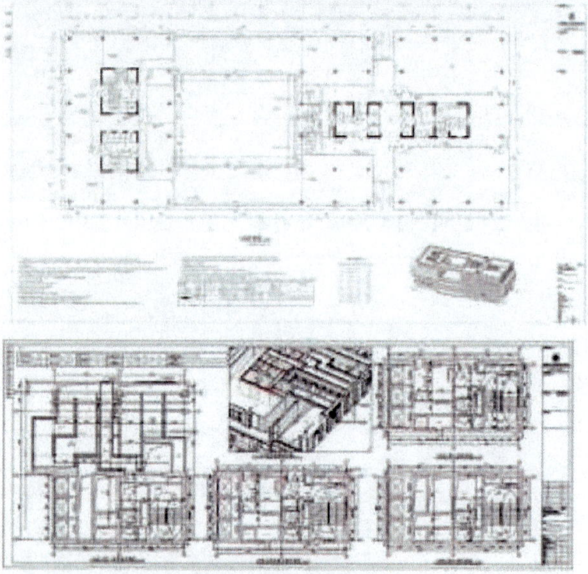

(e) Revit construction drawing design and plotting

Fig. 5. *(continued)*

4 Epilogue

In the whole life cycle of high-quality green and low-carbon digital buildings, the BIM positive design of Shanghai runYou technology Yangtze River Delta port front headquarters project is to integrate energy-saving, water-saving, material saving, indoor ventilation and lighting and other healthy and comfortable environmental factors into the preliminary design of the scheme to the construction drawing stage, and use the multi-objective performance-based simulation analysis and integration of green building BIM for structural overrun design and refined design of pipe network synthesis. At the same time, according to the ecological concept of green and low-carbon, the appropriate design scheme is comprehensively selected in the deep exploration practice of the architect responsibility system to maximize the goal of green and low-carbon energy saving, which can promote the effective implementation of the national "double carbon" strategy.

References

1. Yang, H.: Design method and prediction model of green transformation of existing buildings based on performance. Tianjin University, Tianjin (2016)
2. Yang, H.: Research on performance-based design method for green transformation of existing office buildings in cold regions. In: Proceedings of 2020 International Conference on green building and building energy efficiency. In: 16th International Conference on green building and building energy efficiency and new technologies and products Expo, pp. 327–332. China City Press, Jiangsu (2020)
3. Hu, H.: Discussion on the application of green building design concept under the background of carbon neutrality. Intell. Build. Smart City **06**, 127–129 (2022)
4. Pang, Y.: Analyzing the design application of BIM Technology in green building with a project. Value Eng. **04**, 258–259 (2020)
5. Mao, Z.: China Construction Under the "Double Carbon" Goal. China Construction Industry Press, Beijing (2022)
6. Liao, H.: Research on the application of BIM technology in green public building design. Constr. Technol. (08), 74–77 (2021)
7. Yu, W., Zao, X., Lin, Q.: Green building design practice of a headquarters office building in Shenzhen. Green Technol. (05), 202–205 (2021)
8. Zhang, L., Xia, B., Liu, W., et al.: Research on the integrated application of green technology in the library information center of Dalian University of technology Liaodongwan campus. Constr. Technol. (05), 34–36 (2020)
9. Cui, K., Liu, H., Ren, Q., et al.: Design guidelines for high quality green buildings in the new era. Constr. Technol. (07), 26–30 (2022)

Research on the Comfort of Outdoor Thermal Environment in Old Communities in Mild Climate Areas

Yiming Xing and Yan Wang(✉)

Guizhou University, Guiyang 550000, China
53783147@qq.com

Abstract. Taking four representative old communities in Guiyang City, a representative city in mild climate areas, as the research objects, the current situation of thermal environment and thermal sensation of residents in the old communities in each season were understood through field measurement and questionnaire survey, and relevant thermal comfort indexes—PET, SET* and UTCI were calculated to analyze and determine the functional relationship between each index and subjective feeling. The neutral thermal sensitivity range, thermal comfort zone and 90% thermal acceptance range of residents in different seasons were obtained. Based on the prediction accuracy of the comfort range of different indicators, the thermal comfort ranges of indicators in different seasons were respectively: SET* (20.8 °C–26.1 °C) in spring, SET* (20.2 °C–31.9 °C) in summer, PET (≤25.5 °C) in autumn, and UTCI (9.9 °C–13.6 °C) in winter.

Keywords: Old community · Mild climate area · Thermal comfort evaluation

1 Introduction

Outdoor thermal environment is related to human settlement comfort and human flow activities, which is restricted by weather, space, greening and other factors. In China's economically underdeveloped mild climate areas, densely populated residential areas, there are a large number of old communities before 2000. Poor ventilation, insufficient shading and insufficient green amount are common problems in these old communities. Therefore, understanding and mastering the outdoor thermal environment of old communities in relevant areas is helpful to improve the current situation of the thermal environment, and enhance the comfort level of residents and the utilization rate of space.

1.1 Research on Outdoor Thermal Environment

Early research on thermal environment mainly focused on the field of climatology, focusing on the urban thermal environment in large areas [1], In the later stage, it gradually shifted to small and medium scale or micro-space, and improved the quality of human settlements by studying the thermal environment [2]. At present, the research

W. Guo et al. (Eds.): GBCESC 2023, LNCE 328, pp. 652–664, 2024.
https://doi.org/10.1007/978-981-99-9947-7_67

on outdoor thermal environment mainly focuses on thermal environment evaluation, influencing factors and thermal comfort evaluation.

Outdoor thermal environment evaluation can be divided into two types: thermal safety evaluation and thermal comfort evaluation [3], For example, Wet Bulb Globe Temperature (WBGT) is often used as the basis for thermal safety evaluation of residential thermal environment. In terms of evaluation methods, observation and simulation are usually used [4], Due to its characteristics of both flow [5] and fixed [6], observation method has been widely used in the study of residential thermal environment [7].

The research on the factors affecting outdoor thermal environment is more extensive. In addition to meteorological factors, the main environmental factors include urban spatial form, vegetation cover, skin material and so on [8], and individual characteristics and differences will also affect the perception of outdoor thermal environment [9]. Therefore, the research on outdoor thermal environment gradually began to break the disciplinary barrier, involving more fields including physiology, medicine, mathematics, computer and so on, in order to seek the most authentic environmental evaluation and perception.

1.2 Research on Thermal Comfort

The research on thermal environment comfort evaluation has a history of nearly one hundred years, during which it mainly experienced three stages of development, namely the early single measurement data stage, the empirical model stage and the mechanism model stage. The most common thermal comfort index based on the mechanism model at present, Including Physiological Equivalent Temperature (PET), New Standard Effective Temperature (SET^*), Universal Thermal Climate Index (UTCI), etc.[10], because of its consideration of many factors, including physical environment parameters and human physiological behavior factors, so it is widely used in outdoor thermal comfort research. Some scholars found in the adaptive study of thermal comfort indexes in Shanghai that UTCI and PET indexes in summer and winter are more appropriate to predict subjective feelings [11]. In the study on the thermal environment of urban slow walking space in Guangzhou, some people concluded that the most comfortable voting interval of pedestrian SET^* in spring is 18.0–22.0 °C [12], Some people think that UTCI can accurately predict outdoor thermal comfort in cold regions in Tianjin [13].

To conclude, existing studies mainly focus on cold areas with significant climate characteristics or areas with hot summer and cold winter, rarely involving mild climate areas, and research objects rarely involve old urban communities. With economic development, residents have increasingly higher requirements for the comfort level of human settlements, so there is still a large space for exploration on the thermal environment characteristics of mild climate areas and the applicability of various thermal comfort indexes. Taking four old communities in Guiyang as an example, this paper analyzed the outdoor space thermal environment in different seasons and determined the applicability and value range of thermal comfort indexes PET, SET^* and UTCI through on-site meteorological parameter measurement and residents' actual thermal sensation survey, which provided theoretical basis for the research on outdoor thermal environment of old communities in mild climate areas.

2 Research Content

2.1 Research Object

Guiyang is located in the middle of the original hills of the central Guizhou mountains on the Yunnan-Guizhou Plateau. The climate type of Guiyang belongs to the subtropical monsoon humid climate. In the building thermal zone of China, it is a mild area with abundant rainfall and mild climate in all seasons. The research sites are mainly Nanming District and Huaxi District, Guiyang City. The target residential areas are mainly multi-storey residential areas built before 2000, including original unit residential areas, old-fashioned apartments, factory family areas, etc. Among them, Dianli Lane community and Memorial Tower community are selected in Nanming District, Xibei Community and Ziguangxuan Community are selected in Huaxi District. Due to their early construction, Long time, often lack of sufficient external activity space, easy to produce poor thermal environment experience.

2.2 Field Test

The test time is divided according to the season, and the weather suitable for outdoor activities is usually sunny or cloudy. The test time is two days in each season, from 8:00 to 20:00 on the same day. See Table 1 for the specific test time. Two public Spaces with frequent residents' activities were selected as measuring points in each community, with a total of eight measuring points. The specific distribution is shown in the figure (Fig. 1). The weather on the test day was clear and rainless, which was suitable for outdoor activities.

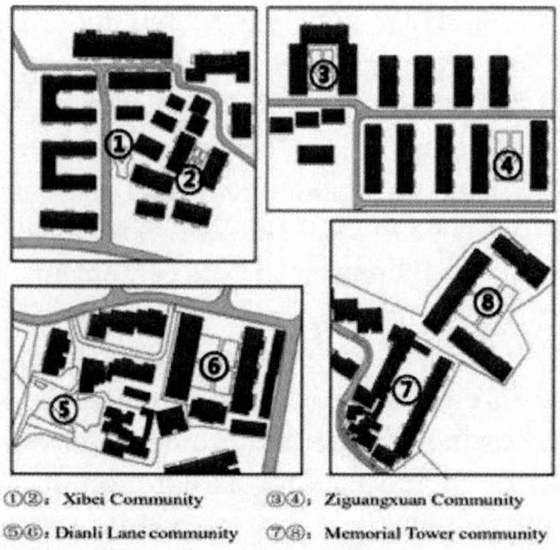

Fig. 1. Distribution of measuring points

The main outdoor thermal environment parameters collected during the measurement include air temperature (Ta), relative humidity (RH), bulb globe temperature (Tg),

Table 1. Parameter records and questionnaire survey time

Year	Season	Date	Parameter records	Questionnaire survey	Quantity
2022	Spring	5.3, 5.4	7:00–19:00	8:00–12:00&14:00–18:00	329
	Summer	7.12, 7.13	7:00–20:00	7:00–12:00&14:00–18:00	310
	Autumn	10.23, 10.24	7:00–19:00	8:00–12:00&14:00–18:00	302
	Winter	12.13, 12.14	7:00–19:00	8:00–12:00&14:00–18:00	280

wind speed (V)and solar radiation (G), and each thermal comfort index and average radiant temperature (Tmrt) are calculated accordingly. The calculation of average radiant temperature is shown in Formula (1) [14].

$$T_{mrt} = \left[(Tg + 273.15)^4 + \frac{1.10 \times 10^8 V^{0.6}}{\varepsilon D^{0.4}} (Tg - Ta) \right]^{1/4} - 273.15 \quad (1)$$

where: Tg is the bulb globe temperature (°C); ε is black sphere radiation coefficient, ε = 0.95 in this study; D is the diameter of the black sphere (m), D = 0.075 in this study; V is wind speed (m/s); Ta is the air temperature (°C).

The main instruments used in the testing process include temperature and humidity instrument, anemometer, black ball thermometer and solar radiation instrument. The specific parameters are shown in Table 2.

Table 2. Details of test instrument parameters

Parameter	Instrument	Range	Accuracy	Resolution
Ta	JTR08 Temperature and humidity meter	– 20–80 °C	± 0.2 °C	0.1 °C
RH		0–100%	± 1.5% RH	0.1%
Tg	AZ8778 Thermal index meter	– 10–80 °C	± 1.5 °C	0.1 °C
V	YGY-FSXY2 Anemometer	0–30 m/s	± 0.3 m/s	0.1 m/s
G	TES-1333 Solar radiometer	0–2000 W/m^2	± 5%	0.1 W/m^2

2.3 Questionnaire Survey

In addition to the measurement of thermal environment parameters, a questionnaire survey on the thermal environment was carried out among the residents living in the measurement area. A total of 1221 valid questionnaires were collected in four seasons, including 329 in spring, 310 in summer, 302 in autumn and 280 in winter. The questionnaire mainly consists of two parts. The first part is the basic information of the respondents, including age, height, weight, clothing (clothing heat resistance), activity status and activity duration. The second part is the respondents' thermal environment feeling, including cold and hot feeling, comfortable state, environmental acceptance, thermal environment expectation, etc. In order to improve the matching degree between the questionnaire content and the actual Thermal Sensation, the Thermal Sensation Vote (TSV) adopts the bipolar 9-level (extended) scale. TSV ranges from − 4 to 4, indicating the temperature from cold to hot. Thermal Comfort Vote (TCV) and Unaccept Rate Vote (URV) are single-level scales [15], TCV ranges from 0 to 4, indicating comfort to extreme discomfort, URV ranges from 0 to 3, indicating complete acceptance to difficult acceptance. At the same time of questionnaire survey, relevant thermal environment parameters of interviewees in the range were recorded.

2.4 Calculation of Thermal Comfort Index

The thermal comfort indexes such as PET, SET* and UTCI are calculated with RayMan-Pro software. The environmental parameters involved and residents' personal characteristics are collected through measurement and questionnaire. The individual metabolic rate was determined as sitting (60 W/m^2), standing (70 W/m^2), walking (115 W/m^2), light exercise (120 W/m^2), moderate exercise (200 W/m^2), vigorous exercise (350 W/m^2) and taking care of children (150 W/m^2) according to the main activity state of the respondents within 15 min.

3 Result Analysis

3.1 Subjective Questionnaire Analysis

As shown in Fig. 2, residents' outdoor thermal environment sensation voting in spring and summer is generally hot (+1), and 49% and 52% of residents' thermal sensation voting is slightly warm or above, respectively. This may indicate that in spring, residents still retain the memory of cold winter and the climate feature of large temperature difference after rain in Guiyang, which makes residents' clothing thicker in spring and overall thermal sensation hotter. Summer may be due to the overall warmer climate. The overall thermal sensation of residents in autumn is neutral, and 54% of them vote as not cool or hot. The reason may be that the outdoor thermal environment in autumn is relatively stable. The overall thermal perception of residents in winter is relatively cold, and 73% of them vote slightly cold or above for thermal perception, which may be due to the cold overall thermal environment in winter and the inherent cold impression of residents in winter.

As shown in Fig. 3, residents' outdoor thermal environment comfort vote shows that in autumn, residents' comfort (0) accounts for the highest proportion, about 64%, probably because the overall temperature in autumn is suitable and does not fluctuate much. The proportion of residents' comfort in spring and summer is similar, about 48%. Large temperature fluctuation in spring, high overall temperature in summer and heat energy generated by exercise may reduce residents' comfort in thermal environment. The proportion of residents' comfort in winter is the lowest, which is about 19%. The decrease of exercise in winter and the increase of instantaneous wind speed may reduce residents' thermal comfort.

As shown in Fig. 4, the outdoor thermal environment acceptance vote of residents shows that the overall outdoor thermal environment acceptance of residents is about 84.7% in spring, 83.5% in summer, 90.7% in autumn and 61.1% in winter, which reflects that residents have poorer adaptability to outdoor thermal environment in winter and prefer autumn with stable thermal environment.

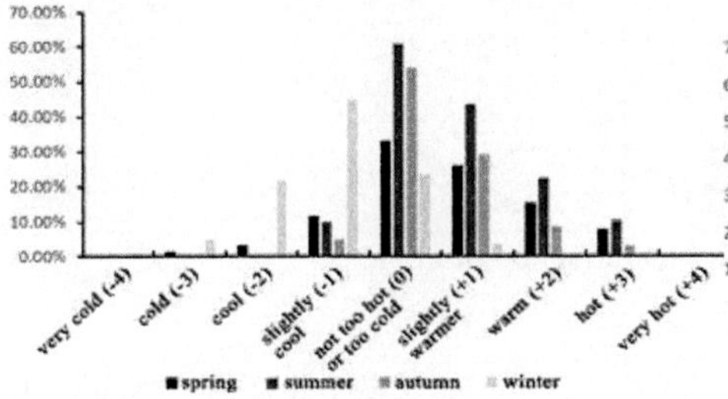

Fig. 2. Thermal Sensation Vote (TSV)

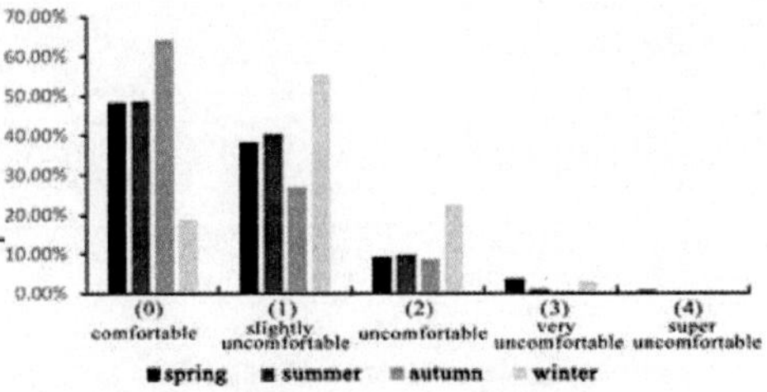

Fig. 3. Thermal Comfort Vote (TCV)

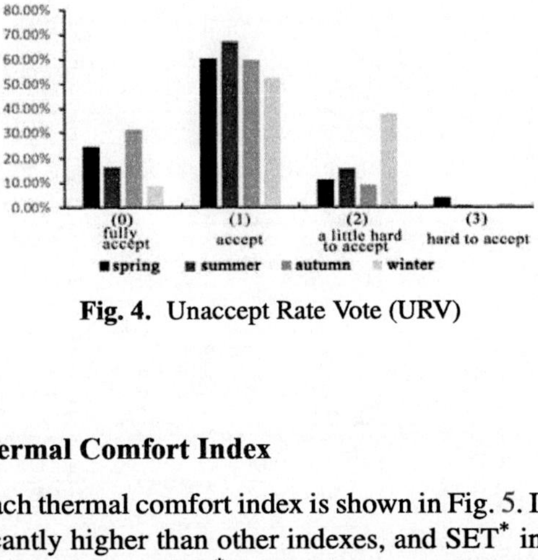

Fig. 4. Unaccept Rate Vote (URV)

3.2 Analysis of Thermal Comfort Index

The result range of each thermal comfort index is shown in Fig. 5. In spring and summer, PET index is significantly higher than other indexes, and SET* index is between other indexes on the whole. The value of SET* index in summer and autumn is lower than other indexes on the whole; The SET* index value in winter was higher than other indexes, and the UTCI index value was between the other two indexes in all four seasons.

In order to better grasp the relationship between the thermal environment comfort index and residents' subjective thermal environment perception, correlation analysis should be carried out between the index and residents' voting. Each index value should be grouped at 1 °C and the average value of TSV, TCV and URV within each group should be calculated. Linear regression analysis was conducted for each thermal comfort index through the above voting, and the analysis results were as follows:

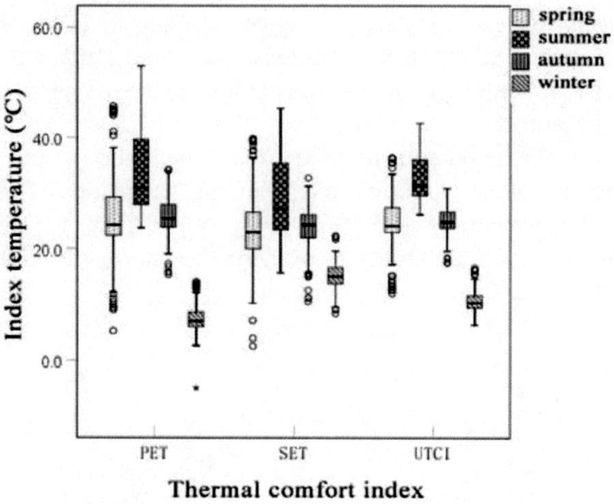

Fig. 5. Results range of each thermal comfort index

Thermal Sensation Vote (TSV). The regression equations of thermal voting and PET, SET* and UTCI of residents in different seasons are shown in Table 3:

Table 3. Regression equations between different indexes and TSV

	PET	SET*	UTCI
Spring	$TSV = 0.1PET - 2.06$ $(R^2 = 0.726)$	$TSV = -0.0022SET^{*2} + 0.2SET^* - 2.92(R^2 = 0.726)$	$TSV = 0.12UTCI - 2.36(R^2 = 0.864)$
Summer	$TSV = 0.06PET - 1.42$ $(R^2 = 0.653)$	$TSV = 0.0036SET^{*2} - 0.16SET^* + 2.22(R^2 = 0.680)$	$TSV = 0.12UTCI - 3.2$ $(R^2 = 0.778)$
Autumn	$TSV = 0.09PET - 1.72(R^2 = 0.842)$	$TSV = 0.017SET^{*2} - 0.78SET^* + 9.2(R^2 = 0.722)$	$TSV = 0.03UTCI^2 - 1.46UTCI + 18.11$ $(R^2 = 0.897)$
Winter	$TSV = 0.02PET^2 - 0.31PET + 0.14(R^2 = 0.653)$	$TSV = 0.06SET^* - 1.99(R^2 = 0.541)$	$TSV = 0.08UTCI - 2.07$ $(R^2 = 0.690)$

When TSV = [−0.5, 0.5], it is known as the neutral thermal sensation range, which is generally acceptable. Through calculation, the neutral thermal sensation range of different indexes in each season can be obtained, as shown in Table 4. In spring, summer and autumn, the neutral thermal sensation range of PET index is the largest, while in winter, There is no neutral thermal sensitivity range in SET* and UTCI, which may be related to the high proportion of cold overall thermal sensitivity in winter.

Table 4. Range of neutral heat sensation in different seasons

Season	PET	SET*	UTCI
Spring	[15.6 °C, 25.6 °C]	[14.4 °C, 22.8 °C]	[15.5 °C, 23.8 °C]
Summer	[15.3 °C, 32.0 °C]	[18.2 °C, 26.2 °C]	[22.5 °C, 30.8 °C]
Autumn	[13.6 °C, 24.7 °C]	[19.2 °C, 26.4 °C]	[22.1 °C, 26.6 °C]
Winter	[−1.0 °C, 2.5 °C] ∪ [13.1 °C, 16.5 °C]	–	–

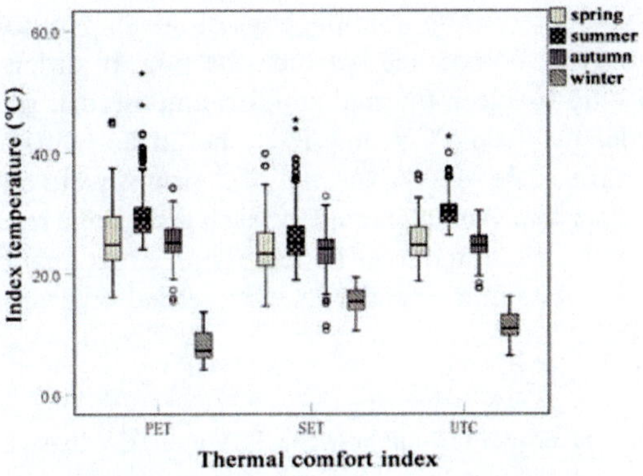

Fig. 6. Range of each index when TCV = 0

Thermal Comfort Vote (TCV). Thermal comfort zone is the range that residents deem comfortable subjectively. In this study, the value of each index when TCV = 0 is used to draw a box diagram, so as to judge the thermal comfort zone of different indexes and their data dispersion degree in different seasons. As shown in Fig. 6, UTCI index has the lowest degree of dispersion, and the 50% voting data (box type part) of each index is relatively concentrated. Therefore, it is relatively objective to take this part of data as the criterion for judging the thermal comfort zone. Through analysis, it can be obtained that the thermal comfort zone of different indexes in different seasons is shown in Table 5.

As can be seen from Table 5, the 50% voting range of the three indexes is close in spring and autumn; The index temperature range of UTCI index was narrower and

Table 5. 50% voting range when TCV = 0 for different index

Season	PET	SET*	UTCI
Spring	[21.9 °C, 27.6 °C]	[20.8 °C, 26.1 °C]	[22.5 °C, 26.7 °C]
Summer	[26.7 °C, 31.6 °C]	[23.2 °C, 28.5 °C]	[28.6 °C, 31.8 °C]
Autumn	[23.4 °C, 27.5 °C]	[21.8 °C, 25.6 °C]	[23.7 °C, 25.4 °C]
Winter	[5.8 °C, 10.0 °C]	[14.4 °C, 17.5 °C]	[9.9 °C, 13.6 °C]

higher in summer. The SET* index temperature range is significantly higher than the other two indexes in winter, which indicates that each index has different adaptability to different seasons. When the temperature is too high or too low, the difference between the indexes begins to appear.

From the analysis (1) and (2), it is not difficult to find that when the thermal sensation is voted as lukewarm state, that is, when TSV = 0, it is not the most comfortable state for residents. The most comfortable temperature in the above indexes is higher than the neutral thermal sensation (TSV = 0) temperature, so the thermal sensation and thermal comfort of residents are not equivalent. Relatively speaking, thermal sensation is only a perception of cold and hot physical environment, while thermal comfort is more complex, which is affected by multiple physiological, psychological and physical environments, and a dynamic balance between the two. In order to accurately grasp the dynamic relationship between thermal comfort and thermal sensation of residents, a mathematical model between TCV and TSV should be built, and TCV should be grouped according to the scale of TSV. A total of 9 groups were obtained. The average value of thermal comfort data was calculated for each group, and regression analysis was conducted with thermal sensation data. The regression equations for each season were shown in Table 6. TCV in summer and winter increased or decreased monotonically with the change of TSV.

Table 6. Linear relationship between TSV and TCV in each season

Spring	$TCV = 0.2TSV^2 - 0.02TSV + 0.29\ (R^2 = 0.857)$
Summer	$TCV = 0.445TSV + 0.42\ (R^2 = 0.914)$
Autumn	$TCV = 0.16TSV^2 + 0.32TSV + 0.14(R^2 = 0.888)$
Winter	$TCV = -0.45TSV + 0.59(R^2 = 0.755)$

Unaccept Rate Vote (URV). According to the internationally recognized ASHRAE. [16] standard, a comfortable thermal environment should have a thermal acceptance rate of at least 90%. Therefore, in this study, when the non-acceptance rate is less than 10%, it is considered to be the acceptable thermal comfort range for residents.

Each thermal comfort index was grouped according to 1 °C, and the proportion of acceptance voting values of 2 (slightly difficult to accept) and 3 (difficult to accept) in each group was calculated, namely, the rejection rate of each group. The regression analysis of URV and each index in each season can be obtained, as shown in Table 7. The 90% acceptance (URV $\leq$ 10%) of each index was calculated by using the correlation formula in Table 7. The results are shown in Table 8. In the presented results, PET index is always a first-order linear relation, while SET* and UTCI index are simultaneously a first-order linear relation or a quadratic polynomial relation. Therefore, the calculated 90% acceptance voting range of SET* is closer to UTCI in expression form.

Table 7. Linear relationship between seasonal indexes and URV

	PET	SET*	UTCI
Spring	URV = 0.02PET – 0.39 (R^2 = 0.803)	URV = 0.018 SET* – 0.32 (R^2 = 0.698)	URV = 0.024UTCI – 0.48 (R^2 = 0.762)
Summer	URV = 0.026PET – 0.73 (R^2 = 0.670)	URV = 0.0019 SET*2 – 0.1 SET* + 1.27 (R^2 = 0.644)	URV = 0.00571UTCI² – 0.35UTCI + 5.4 (R^2 = 0.818)
Autumn	URV = 0.03PET – 0.62 (R^2 = 0.736)	URV = 0.0038SET*2 – 0.16SET* + 1.77 (R^2 = 0.626)	URV = 0.013UTCI² – 0.655UTCI + 8.278(R^2 = 0.750)
Winter	URV = – 0.04PET + 0.7 (R^2 = 0.681)	URV = – 0.03SET* + 0.77 (R^2 = 0.773)	URV = 0.01UTCI² – 0.36UTCI + 2.78 (R^2 = 0.634)

Table 8. Voting ranges of 90% acceptance of different indexes

Season	PET	SET*	UTCI
Spring	≤ 24.5 °C	≤ 23.3 °C	≤ 24.2 °C
Summer	≤ 32.0 °C	[20.2 °C, 31.9 °C]	[27.5 °C, 33.9 °C]
Autumn	≤ 25.5 °C	[17.1 °C, 26.2 °C]	[22.9 °C, 27.4 °C]
Winter	≥ 15.0 °C	–	–

3.3 Analysis of Correct Rate of Index Prediction

Prediction Accuracy of Thermal Sensation. In order to further understand which thermal comfort index is more suitable in each season, it is necessary to carry out applicability analysis of each index. For this reason, related concepts are introduced for Predicting Accuracy (η) as a reference for seasonal applicability of indicators. In the concept of prediction accuracy, the larger the η value is, the higher the prediction accuracy and applicability is. Considering the high correlation degree between residents' thermal sensation voting (TSV) and thermal comfort, and the more detailed scale division of thermal sensation voting, the thermal sensation voting will be used as the analysis basis in the prediction accuracy analysis. The prediction accuracy formula of each index is shown in Formula 2:

$$\eta = \frac{Q_{\text{TSVP-TSV}}}{Q_{all}} \times 100\% \quad (2)$$

where: $Q_{\text{TSVP-TSV}}$ is the same number of votes as the predicted thermal sensation vote and the actual thermal sensation vote; Q_{all} is the total number of votes cast.

The calculated prediction accuracy of thermal sensation in each season is shown in Table 9. As can be seen from the results in Table 9, SET* index has the highest prediction accuracy for residents' thermal environmental sensation in spring. Therefore, SET* is the first choice for outdoor thermal environment comfort evaluation index to predict thermal sensation in mild areas in spring. In summer, autumn and winter, the thermal sensation prediction accuracy of UTCI index was higher.

Table 9. Prediction accuracy of thermal sensation in each season

	Spring	Summer	Autumn	Winter
PET	69.3%	78.4%	88.1%	72.1%
SET*	71.7%	78.1%	86.7%	65.0%
UTCI	69.3%	78.7%	89.1%	74.7%

Prediction Accuracy of Thermal Comfort Range. However, thermal sensation is not completely equivalent to thermal comfort, and the prediction accuracy rate of residents' thermal comfort range still needs to be calculated. The proportion of comfort state (TCV = 0) within different comfort definition ranges is the prediction accuracy rate of index thermal comfort range. The prediction results of different comfort definitions of different indicators in different seasons are sorted out and analyzed, and the contents are shown in Table 10. As can be seen from the analysis results in Table 10, the prediction accuracy of the average index in the thermal comfort range in autumn is the highest, while the prediction accuracy in winter is the lowest, with an average of about 15%. The main reason is that the thermal comfort condition in winter is poor and the overall comfort level of residents is not high, resulting in fewer samples in the comfort state. At the same time, the temperature difference on the test day in winter is small, resulting in a small comfort range. In view of the above situation, according to the analysis results of (1), UTCI indexes with higher accuracy of thermal sensation prediction are selected to predict the comfort range, and thus the comfort range of outdoor thermal environment in Guiyang old communities in different seasons is finally determined, as shown in Table 11.

Table 10. Prediction accuracy of different comfort ranges in different seasons

	TSV = [−0.5, 0.5]			50% voting data (TCV = 0)			90% acceptable range		
	PET	SET*	UTCI	PET	SET*	UTCI	PET	SET*	UTCI
Spring	52.51%	51.11%	55.96%	57.93%	60.30%	56.29%	51.53%	50.90%	52.56%
Summer	66.46%	67.15%	66.93%	63.25%	68.04%	61.54%	66.46%	68.10%	64.00%
Autumn	77.08%	69.05%	67.45%	67.33%	70.90%	66.36%	77.12%	69.48%	64.76%
Winter	10.00%	–	–	14.06%	18.46%	–	–	20.63%	–

Table 11. Outdoor thermal comfort range in different seasons

Season	Thermal comfort range
Spring	$SET^* \in [20.8\ °C, 26.1\ °C]$
Summer	$SET^* \in [20.2\ °C, 31.9\ °C]$
Autumn	$PET \leq 25.5\ °C$
Winter	$UTCI \in [9.9\ °C, 13.6\ °C]$

4 Conclusion

Through the questionnaire survey on the outdoor thermal environment of Guiyang residents in different seasons, it is found that the overall acceptability of autumn is the highest, about 90%, and that of spring and summer is roughly the same, respectively 84% in spring and 83% in summer. The acceptability of winter is the lowest, about 61%, which to some extent reflects that residents' adaptability to outdoor thermal environment in winter is worse. Prefer the more stable thermal environment in autumn.

The applicability of each index can be judged from the results of the accuracy of thermal sensation prediction in each season. When analyzing the overall thermal environment sensation of Guiyang residents, SET* index should be used in spring, PET index should be used in summer, and UTCI index should be used in autumn and winter.

SET* index should be adopted in Guiyang in spring and summer, PET index should be adopted in autumn, and UTCI index should be adopted in winter when calculating the prediction accuracy and judging the applicability of each index from the comfort range of different indexes in different seasons. Meanwhile, based on the prediction accuracy of different index comfort ranges, the thermal comfort ranges of each season were SET* (20.8 °C–26.1 °C) in spring, SET* (20.2 °C–31.9 °C) in summer, PET (≤25.5 °C) in autumn, and UTCI (9.9 °C–13.6 °C) in winter.

The study on outdoor thermal environment of old communities in mild climate areas can understand the overall thermal environment level and comfort level of residents in this area, and to a certain extent can provide theoretical and data reference for the improvement of outdoor space of old communities in this area. However, this paper only focuses on Guiyang, a representative of a mild climate area, and studies on the thermal environment and related thermal comfort indexes in other mild climate areas are far from enough. The research objects can be expanded in the future to provide reference and suggestions for improving the outdoor thermal environment in more old communities in mild climate areas.

Fund Information. 1. Doctoral Fund of Guizhou University [(2020) No. 11].

References

1. Zhang, J., Jiao, G., Ye, Q., Gu, X.: The impact of urban expansion on the urban thermal environment: a case study in nanchang Jiangxi China. Sustainability **14**, 24–25 (2022)
2. Li, K., Li, X, Yao, K.: Outdoor thermal environments of main types of urban areas during summer: a field study in Wuhan, China. Sustainability **14**, 952–977 (2022).
3. Xi, X.H.: Current Situation and Evaluation of Outdoor Thermal Environment in High-Rise Residential Areas in Xiamen City. Huaqiao University, Xiamen (2020)
4. Meng, Q.L., Wang, P., Li, Q.: Urban thermal environment assessment methods. Chin. Landscape Archit. **30**(12), 13–16 (2014)
5. Unger, J.: Intra-urban relationship between surface geometry and urban heat island: review and new approach. Clim. Res. **27**(3), 253–264 (2004)
6. Li, C., Li, N.P. Gao, F. Le, D.: Test methods for microthermal environment in residential areas. Build. Sci. **08**, 61–64+75 (2007)
7. Chen, Z.L., Zhao, L.H., Meng, Q.L., Wang, C.S., Zhai, Y.C., Wang, F.: Microclimate measurement and analysis of typical residential district in Guangzhou. Archit. J. **11**, 24–27 (2008)
8. Zhang, N.N., Su, X.C., Zhao, X.D., Zhang, C.H., Yang, X.B.: Effect of underground space development on the outdoor thermal environment in a residential area. IOP Conf. Ser. Earth Environ. Sci. **233**(2), 02 (2019)
9. Xi, T.Y., Wang, Q.C., Qin, H., Jin, H.: Influence of outdoor thermal environment on clothing and activity of tourists and local people in a severely cold climate city. Build. Environ. **173**, 25–28 (2020)
10. Lin, H.J., Ma, H.Y., Zhang, M.: Analysis of the variation characteristics of human thermal comfort in summer of China from 1980 to 2019 based on UTCI. Clim. Change Res. **18**, 58–69 (2022)
11. Wang, Y., Pan, C., Huang, Z.S.: Applicability comparison of PET and UTCI in different seasons in Shanghai. Build. Sci. **36**(10), 55–61 (2020)
12. Li, K.M., Zhang, H.W., Zhao, L.H., Xue, S.F.: Study on outdoor thermal comfort characteristics of residential areas in humid and hot areas in different seasons. Archit. J. **35**(06). 22–29+162 (2019)
13. Lai, D.W., Zhou, C.B., Jiang, Y., Liu, J.J.: Research and development of outdoor open space thermal comfort model and simulation program based on ASV. Eco-city Green Build. **1**, 30–33 (2012)
14. ISO. International Standard7726 (1998): Ergonomics of the thermal environment-Instruments for measuring physical quantities. International Organization for Standardization, Geneva (1998)
15. Su, Y., Wang, C.J., Wu, X.Z.: Analysis of high thermal environment of different activities in a city square in Dalian. Archit. Cult. **209**(08), 243–245 (2021)
16. ASHRAE. ANSI/ASHRAE Standard 55-1992: Thermal Environmental Conditions for Human Occupancy. Atlanta (1992)

Research on Reformation of Illuminance Uniformity in University Ladder Classroom

Saitong Li[1], Yan Wang[1], Qihua Kuang[1](✉), and Qiang Wu[2]

[1] College of Architecture and Urban Planning, Guizhou University, Guiyang 550003, China
1078503880@qq.com

[2] China Railway Seven Bureau Group, Wuhan Engineering Co., Ltd., Wuhan 430000, Hubei, China

Abstract. In order to further explore and improve the current situation of illumination uniformity in college classrooms, promote the development of green buildings and energy saving, and effectively improve the lighting comfort of college classrooms. Taking classroom 108 in the West Building of Gui-zhou University as an example, based on the research of classroom lighting in the teaching building, the illuminance of selected objects at different measuring position points was measured at different times of the day to study the influencing factors of illuminance and put forward corresponding improvement measures. By changing the arrangement of classroom lighting fixtures and adding lamps, the classroom lighting optimization program was proposed and simulated by using Dialux evo lighting design software, and the advantages and disadvantages of the optimization effect were analyzed and compared, which finally improved the average illuminance of the classroom, met the illuminance uniformity and greatly improved the classroom lighting quality, in order to provide reference significance to the lighting design transformation of the university classroom.

Keywords: Ladder Classroom · Illuminance uniformity · Lighting Quality · Lighting Retrofit

1 Introduction

In recent years, the problem of poor visual acuity among students in China has become prominent, and the country attaches great importance to the problem of myopia among youth [1]. Classrooms are the main place where students learn, and the quality of classroom lighting will directly affect the productivity and resourcefulness of students and teachers, and have a very important impact on the energy efficiency of educational facilities [2]. Ensuring the quality of classroom light environment remains an important issue that needs to be addressed urgently. Numerous evaluation studies conducted by scholars have shown that poor classroom lighting environments are a significant contributor to students' vision loss while good lighting environments enhance students' visual work and have a positive emotional impact [3–7]. Poor classroom lighting, especially if the classroom illumination is not uniform, can cause students to over-regulate their visual

W. Guo et al. (Eds.): GBCESC 2023, LNCE 328, pp. 665–676, 2024.
https://doi.org/10.1007/978-981-99-9947-7_68

organs and over-strain their eye muscles in order to adapt to the uneven lighting environment, which can cause visual fatigue [8]. University students spend more than 60% of their day in the classroom, so it is especially important to create a quality lighting environment for university classrooms [9]. Therefore, in this paper, based on the theoretical literature and practical operation, we conducted parameter measurements and Dialux evo software simulations to investigate the measures and effects of light environment optimization, and to provide a reference for light environment research and lighting renovation.

2 Analysis of the Current State of Classroom Light Environment

2.1 Selection of Test Subjects

In this paper, we selected the classroom in the West Building of Guizhou University as the test object. The West Building is a new teaching building built in recent years, located in the center of the West Campus, with six floors, a building height of 21.53 m and a floor height of 4 m. The building is oriented north-south and is surrounded by tall trees. In order to get a better understanding of the characteristics of the light environment of the student teaching building, we selected the 108 classrooms of the West Building for measurement at different time periods (as in Fig. 1) (Table 1).

Table 1. Test classroom overview

Floor	Ladder Classroom	Window Orientation	Window-to-ground ratio	Number of lamps
1	Yes	East-West orientation	0.15	34

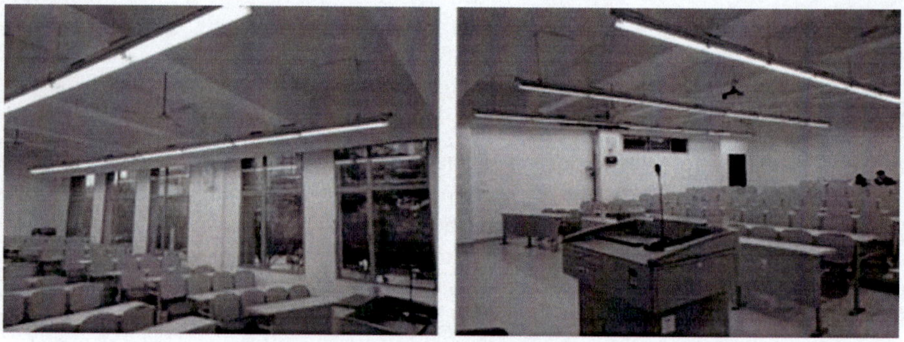

Fig. 1. Classroom lighting test site

2.2 Field Testing and Analysis

Test Method. The illuminance of the sample classroom table and chair area and blackboard area was measured using a CL-200A color illuminance meter with reference to GB/T5700–2008 "Lighting Measurement Methods" [10] and GB/T5699–2017 "Lighting Measurement Methods" [11], and the time period of relatively stable indoor and outdoor lighting was selected, so the research time of natural illuminance was from 10:00 a.m. to 12:00 p.m., and the measurement of artificial lighting illuminance was from 21:00 p.m. to 23:00 p.m.

The choice of classroom illuminance measurement is generally on the working surface at a height of 0.8 m from the ground, this measurement takes the classroom desktop height 0.75 m as the working surface, in the desktop in the region for isometric distribution of points, the quality of the classroom light environment is mainly to examine the illuminance of the classroom desktop measurement points, according to the size of the classroom selected different measurement points.

The more the number of experimental measurements sampled, the smaller the error between the final average and the actual value. However, too much sampling is neither practical nor necessary to determine the minimum number of sampling points according to the relationship between the chamber shape index and the number of sampling points in Table 2 [12]. The desktop measurement points were selected as 77 points, and the measurement points were arranged at equal distances, respectively, and the distribution of measurement points (as in Fig. 2, Fig. 3).

Table 2. The choice of chamber index and tread number

Room shape index RI	Minimum number of sampling points
<1	4
1–2	9
2–3	16
>3	25

2.3 Field Testing and Analysis

Natural light analysis.

Analysis of the current situation of average illuminance of lighting. In the classroom natural lighting, the average illuminance of the desktop are up to the specification value of 300 lx Most of the classroom natural lighting can meet the requirements, however, the illuminance of the classroom blackboard are lower than the desktop illuminance, and the blackboard illuminance are not meet the specification requirements. Students gazing at the blackboard for a long time in such a classroom will easily cause visual fatigue (Fig. 4).

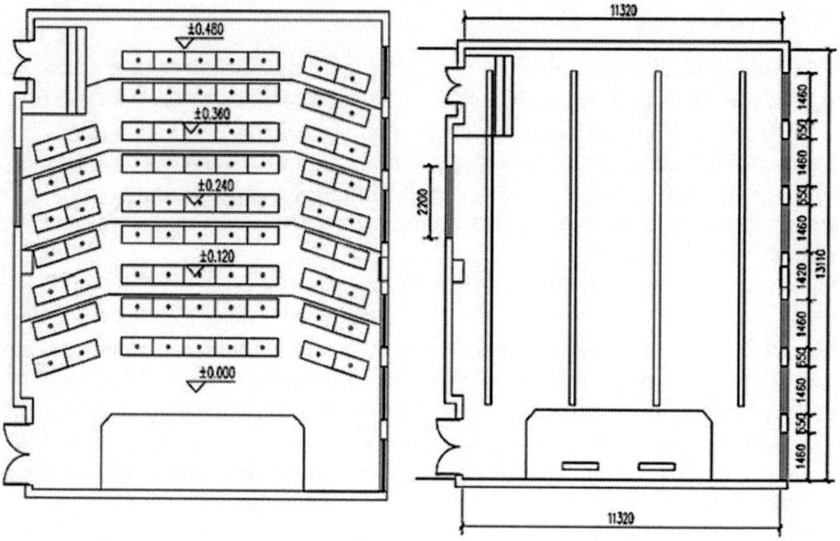

Fig. 2. Layout plan of measuring points and lamps in the classroom

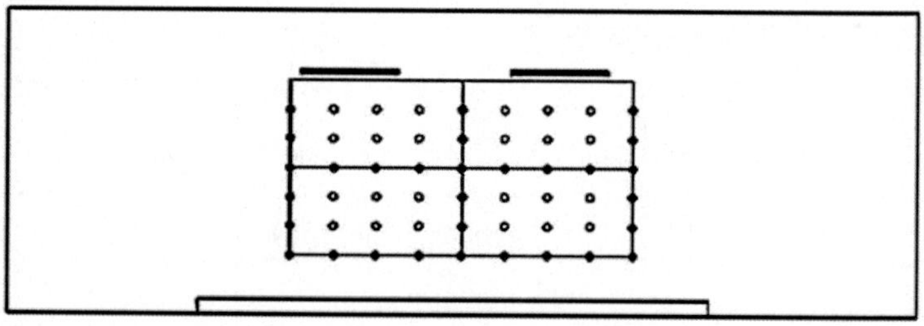

Fig. 3. Plan of blackboard measuring points

Analysis of the current situation of lighting uniformity. There are results can be seen that most of the desktop illuminance is lower than the standard value, mainly concentrated in about 0.6, only one work surface reached the specification value requirements, lighting illuminance uniformity is poor. On-site research found that the average illuminance of the desktop by the window side is higher, and the illuminance of the area by the interior is lower, so it is the phenomenon of uneven illuminance in the room, and if the illuminance uniformity is too small, students are more likely to cause visual fatigue and affect learning efficiency. The illumination uniformity of the blackboard is also not very good, and most of them are lower than the normative value. In summary, the brightness of the area near the window is too high for students' writing and reading, so the window needs to be shaded and modified (as in Fig. 5).

The natural illuminance values of the classroom measurement points and the location coordinates were analyzed, and the maximum value of desktop illuminance reached 889.72 lx, and the minimum value of illuminance was 208.16 lx in the windowless area

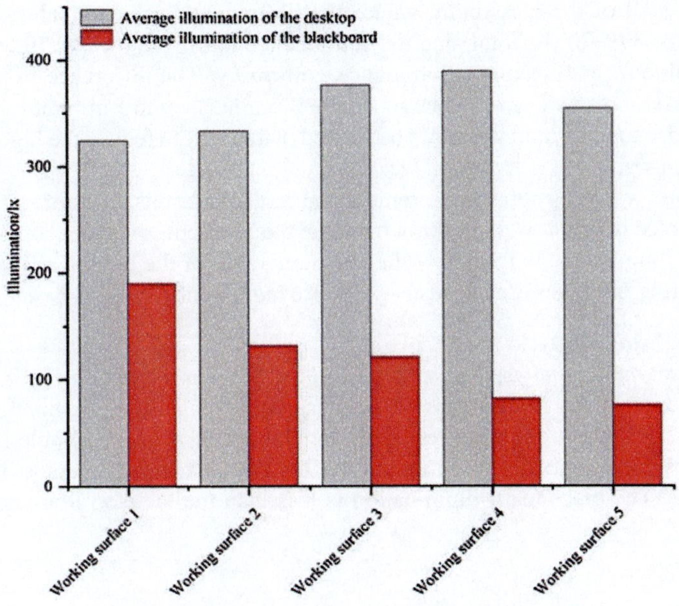

Fig. 4. Average illumination of classroom lighting

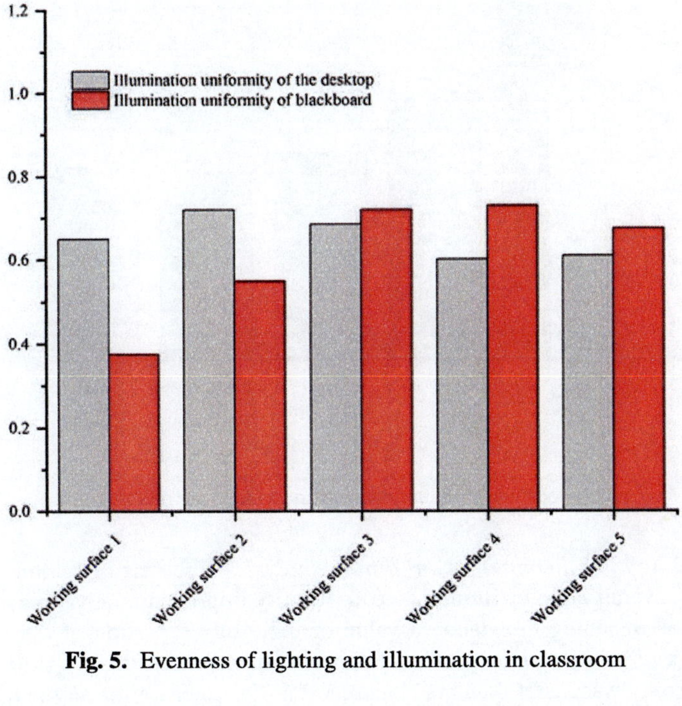

Fig. 5. Evenness of lighting and illumination in classroom

against the wall of the classroom, where the illuminance values greater than 300 lx accounted for 37% of the total, and the illuminance uniformity was 0.59 less than the standard value of architectural illuminance uniformity. The difference in illuminance between classrooms and rooms is large, so the illuminance uniformity of rooms should be improved, and the windows should be treated in this way to reduce the high brightness area of windows.

Combined with the actual measurement and research results, the uniformity of desktop illuminance is very low and cannot meet the requirements of visual comfort, and the blackboard illuminance is far below the standard value of the building. Therefore, the natural lighting of classrooms needs to be improved urgently.

Artificial lighting analysis

Average illuminance analysis of artificial lighting. When the classroom is artificially lit, none of the tabletop illumination in the classroom reaches the illumination standard of 300 lx According to the field research, the classroom is using tri-pole male lamps for T8 lamps, and the indoor illumination reaches the unstandard, average illumination of 285.59 lx. The blackboard illumination is less than the tabletop illumination (as in Fig. 6).

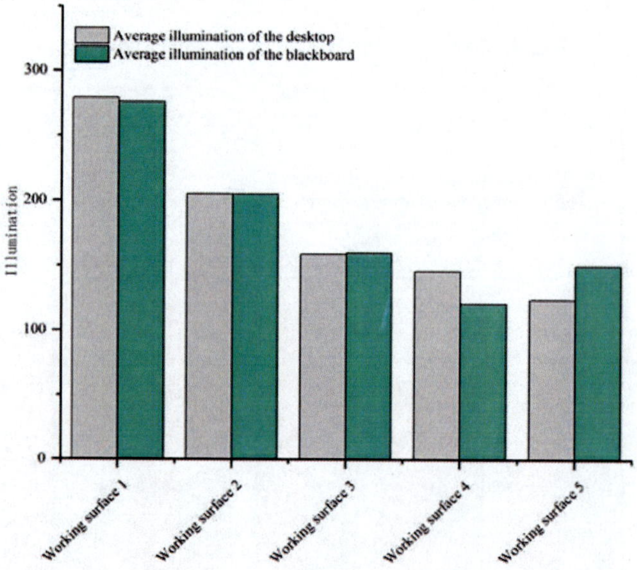

Fig. 6. Average illumination of artificial lighting in classroom

Artificial lighting illumination uniformity analysis. When the classroom was artificially lit, the overall tabletop illuminance uniformity improved relative to natural lighting, with most reaching the standard value of 0.7. Only a portion of the blackboard illuminance uniformity met the standard requirement, indicating that the tabletop illuminance uniformity was better than the blackboard illuminance uniformity in the artificial lighting case (As shown in Fig. 7).

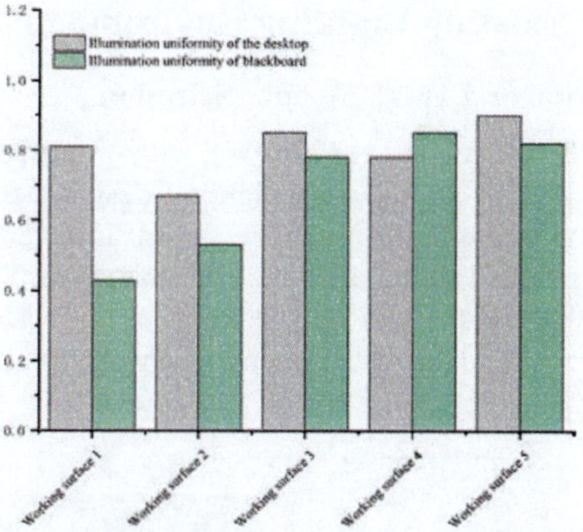

Fig. 7. Illuminance uniformity of artificial lighting in classroom

Analysis of the illuminance value of each measurement point in the case of classroom artificial lighting, the illuminance of the central area of the luminaire is higher than other areas, the overall illuminance value is mostly lower than the standard value of the plane, the average illuminance of classroom artificial lighting is lower than the normative value, the lighting conditions are poor.

Through the analysis of the existing classroom lighting arrangement situation: to measure the distribution of classroom light intensity, respectively, planning a reasonable program, the random application of lighting arrangement method, not only can not achieve the purpose, but also a waste of resources. Lighting arrangement is not the more dense the better. Different models of lamps and lanterns have different illumination ranges, according to the situation in the classroom to choose the right lamps and lanterns and layout program to achieve the purpose of reasonable energy saving. The height of the lamps should also be appropriate, if there are no special requirements, as far as possible, arranged in 2.5 m from the ground, so that the limited illumination of lamps to give full play.

In the classroom lighting layout should be noted that we should not take illumination as a single evaluation criterion, should also consider the uniform distribution of illumination, while meeting the requirements of the standard, to reasonably arrange the lamps and lanterns, appropriate control of the number of lamps and lanterns, and thus reduce the waste of electricity; in the layout should also take into account the natural lighting, wall reflection ratio and the installation height of lamps and lanterns and other types of influencing factors. Most classrooms use old fluorescent lamps and incandescent lamps, and the use of a long period of time, the luminous brightness of the lamps far from the needs of the classroom light environment. Classroom lamps and lanterns maintenance cycle is long, lamps and lanterns and more dust, more limited the brightness of the lamps and lanterns, so the average desktop illumination of the classroom and the normative value of a large difference. The average illuminance of all classroom desktops during artificial lighting is lower than the normative value, which is mainly related to the luminaire arrangement.

3 Analysis of the Existing Lighting Environment in Classrooms

3.1 Model Construction and Work Surface Selection

The classroom size is 13. 6 m (L) × 11. 42 m (W), with 4 steps, each step is 0. 2 m high. According to GB50034–2013 architectural lighting design standard [7], the working surface height is 0. 75 m horizontal surface, because it is a step classroom. Therefore, the working surface is selected as the desk above each step of 0. 75 m, the middle desk is 5.1 m (L) × 0.5 m (W), and the two side desks are 2.15 m (L) × 0.5 m (W). The 3D model of the step classroom and the distribution of the working surfaces in the classroom plan (as in Fig. 8).

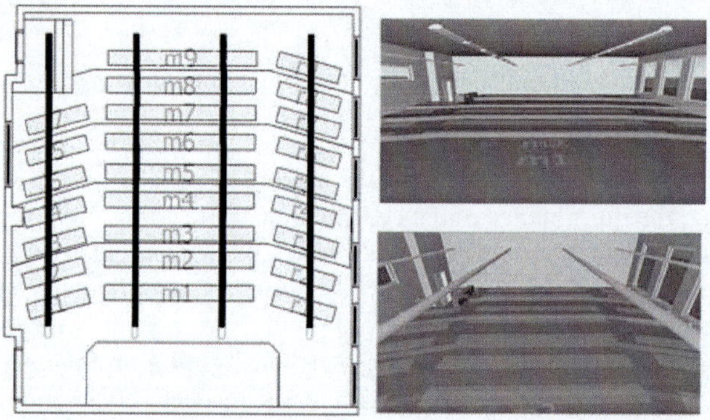

Fig. 8. Classroom working surface and classroom model

3.2 Simulation Analysis of Existing Lighting Effects in the Classroom

The classroom light source arrangement in China is usually arranged in a uniform and equally spaced manner, and the experimental classroom luminaires are distributed in 8 rows and 4 columns with a total of 32 grille luminaires, with equal spacing between each column of 3 m and a luminaire height of 3 m. The luminaires were selected from Sanxiong Aurora: PAK160816 PAK-A06-136-JS, and the luminaires were imported into the established model with the luminaire parameters shown in Fig. 2. After importing the luminaire, the calculation was carried out by Dialux software, and the calculation results of each working surface were output as shown in Table 4.

(See Table 3).

3.3 Simulation Analysis of the Optimized Lighting Effect with Reference to the Optimization Scheme of Kong Y P

According to the optimization program proposed by Kong Y P to the lamp position for reference adjustment, when the 1–3 columns of lamps and lanterns at the same time to the right 1.5 m, the right side of the lamps and lanterns to take away 4 when the entire step classroom in each work surface will obtain the maximum uniformity of illumination (as in Fig. 9), the output of each work surface calculation results are shown in Fig. 10.

Table 3. Lighting tool parameters

Quantity	Luminaires (light distribution information)	
33	PAK Sanxiong Aurora - PAK310816 PAK-A06-136-JS Eye protection series - T8 straight fluorescent lamp chalkboard light Light distribution information 1 Quantity: 1xT8 fluorescent lamp / 36W / polarized 5000K Absolute light measurement Luminous flux of the lamp: 2880 lm Power: 36.0W Luminous efficiency: 80.0 lm/W Colorimetric instructions 1xT8 fluorescent lamp/36W/polarized 5000K: CCT 5000K, CRI 82	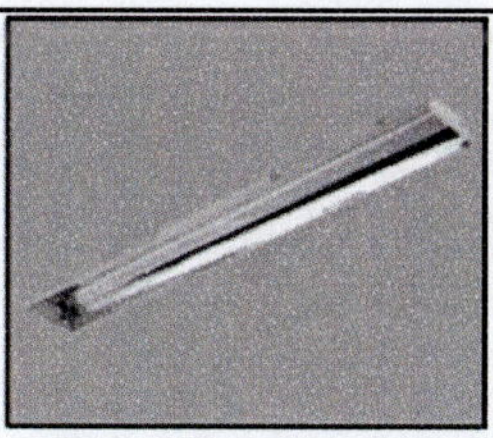 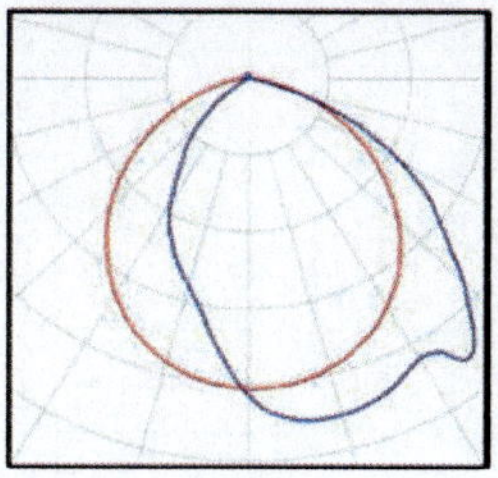

Total luminous flux of light source: 95040 lm, total luminous flux of lamps: 95040 lm, total power 1188.0W, luminous efficiency: 88.0lm/W

3.4 Discussion

From Table 4, we know that the illuminance uniformity (minimum illuminance/average illuminance) of each working surface in the classroom with conventional lighting is 0.50, 053, 0.51, 0.67, 0.48, 0.52, 0.44, 0.46, 0.40, 0.66, 0.70, 0.65, 0.67, 0.58, 0.57, 0.47, 0.44, 0.31, 0.74, 0.75, 0.68, 0.66, 0.61, 0.47, 0.36. The illuminance uniformity of 25 working surfaces, 0.44, 0.31, 0.74, 0.75, 0.68, 0.66, 0.61, 0.47, 0.36, all failed to reach 0. 7. From the comparison in Fig. 9, we found that the illuminance uniformity of all working surfaces improved, but there were still 13 working surfaces that still failed to reach the national standard of 0. 7. For this problem, the number of lamps or the power of lamps is usually increased to improve.

However, from Table 4 and Fig. 9, it is found that the average illuminance of each working surface of the classroom in both cases has reached the national standard of 300lx and is much higher than this indicator, if the number of lamps to change the uniformity of the classroom by increasing the number of lamps will certainly bring a certain waste of energy. In addition, the study conducted by Kong Y P and others when proposing the optimization plan also showed that increasing the power of lamps and lanterns did not improve the illuminance uniformity much, and increasing the power of lamps and lanterns must have the same problems as above.

Table 4. Calculation results on each work surface in lecture theatre

Number	Name	Average illuminance/lx	Minimum illuminance/lx	Maximum illumination/lx	Maximum illumination/average illumination	Minimum illumination/average illumination
1	R1	1024	683	1375	0.67	0.50
2	R2	1071	740	1395	0.67	053
3	R3	1041	737	1450	0.71	0.51
4	R4	913	720	1069	0.79	0.67
5	R5	1117	750	1562	0.67	0.48
6	R6	1163	802	1555	0.69	0.52
7	R7	1185	754	1697	0.64	0.44
8	R8	1142	725	1590	0.63	0.46
9	R9	1019	586	1450	0.58	0.40
10	M1	529	427	649	0.81	0.66
11	M2	605	501	718	0.83	0.70
12	M3	636	499	764	0.78	0.65
13	M4	651	515	774	0.79	0.67
14	M5	656	468	808	0.71	0.58
15	M6	649	460	801	0.71	0.57
16	M7	631	386	829	0.61	0.47
17	M8	590	351	790	0.59	0.44
18	M9	505	235	757	0.47	0.31
19	L1	427	371	498	0.87	0.74
20	L2	532	455	609	0.86	0.75
21	L3	601	479	707	0.8	0.68
22	L4	661	518	782	0.78	0.66
23	L5	756	557	918	0.74	0.61
24	L6	820	501	1058	0.61	0.47
25	L7	755	392	1079	0.52	0.36

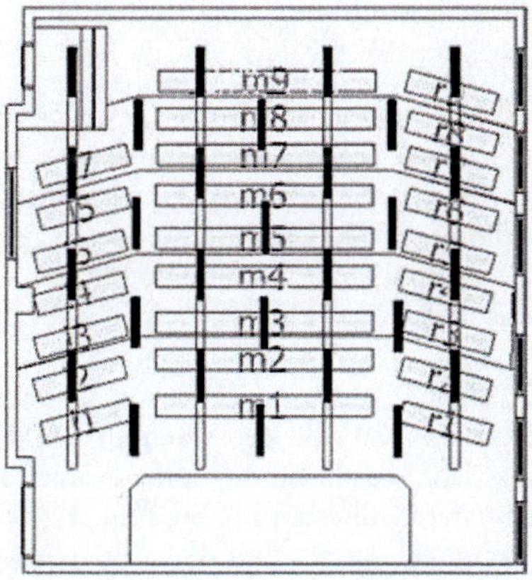

Fig. 9. Lighting optimization schemes

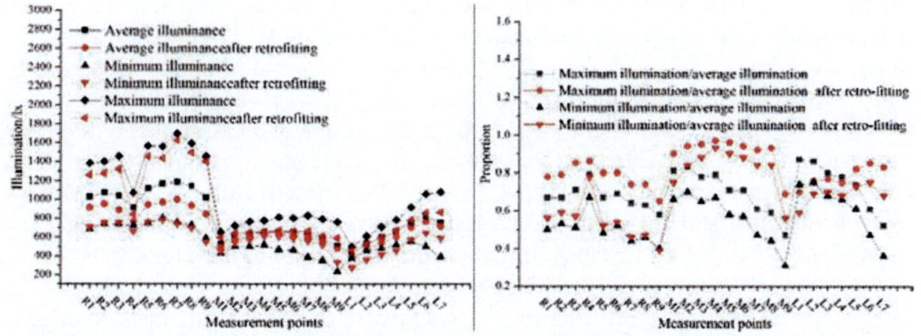

Fig. 10. Comparative calculation results for each working surface after optimization

4 Conclusion

This paper takes the university classroom as the research object, and studies the problems and shortcomings of the classroom through the field research and field measurement of the classroom. Through the analysis of the current situation and the actual measurement results of the classroom, the optimization plan pro-posed by Kong Y P and the simulation results by Dialux software, the simulation results of Dialux software are basically consistent with the actual measurement data, and then the optimization strategy for natural lighting and artificial lighting is proposed. However, it still cannot make the illumination uniformity of each working surface meet the national standard, so special lighting and lighting design should be carried out according to the specific classroom design requirements, and better performance lamps should be used to replace the traditional lamps to provide better lighting solutions.

Acknowledgement. The work was supported in part by Doctoral Fund of Guizhou University [(2020) No. 11].

References

1. Winterbottom, M., Wilkins, A.: Lighting and discomfort in the classroom. J. Environ. Psychol. **29**(1), 63–75 (2009)
2. Zhuang, J.X., Yu, R., Gao, X.Y., Xie, W.H.: Light environment analysis of university classrooms before and after lighting retrofit. Light Sources and Lighting **08**, 35–37 (2020)
3. Wang, M.L., et al.: Analysis of the current situation of teaching environment hygiene in some primary and secondary schools in Hebei province in 2020. Chin. J. Prev. Med. **23**(09), 672–675 (2022)
4. Zhang, G.M., Wang, G.H., Zhao, J.P., Luo, T.: Stroboscopic study of classroom lighting retrofit in primary and secondary schools and colleges in Beijing. Build. Sci. **37**(06), 200–205 (2021)
5. Hua, L.S., Hu, X.B., Yang, H.L., et al.: Study on the evaluation method of classroom light environment and its influencing factors. J. Lighting Eng. **31**(3), 145–150 (2020)
6. Shin, Y.B., Woo, S.H., Kim, D.H., et al.: The effect on emotions and brain activity by the direct/indirect lighting in the residential environment. Neurosci. Lett. **584**, 28–32 (2015)
7. Xiao, Y.H., Bian, G.G.: Classroom lighting light environment retrofit case sharing. China Lighting Appliances **02**, 19–24 (2022)
8. Mingyeh, H.: Effects of illuminance distribution, color temperature and illuminance level on positive and negative moods. J. Asian Archit. Build. Eng. **14**(3), 709–716 (2015)
9. Kong, Y.P., Zhao, Y.S., Lv, K., Liu, L.: Model construction and solution for optimizing the uniformity of classroom lighting. Saf. Environ. Eng. **19**(02), 61–65 (2012)
10. General Administration of Quality Supervision, Inspection and Quarantine of the People's Republic of China, China National Standardization Administration Committee. GB/T 5700-2008 lighting measurement methods [S]. Beijing: China Standards Press, 3–14 (2008)
11. General Administration of Quality Supervision, Inspection and Quarantine of the People's Republic of China, National Standardization Administration of China. GB/T 5699-2017 light measurement methods, pp. 3–10. China Standard Publishing House, Beijing (2017)
12. Li, Y.J., Peng, H.M., Xu, B.: Comparison and research of several illuminance calculation methods. J. Three Gorges Univ. (Nat. Sci. Ed.) **01**, 30–32 (2003)

Analysis and Study on Climate Adaptability of Traditional Houses in Jianghuai Region of Anhui Province, China

Ziteng Han[1,2], Shan Wu[3], and Wei Wang[1,3](✉)

[1] Anhui Provincial Institute of Architectural Science Research and Design, Anhui Key Laboratory of Green Building and Prefabricated Construction, Hefei 230031, China
Vivi.gan@126.com
[2] School of Architecture and Urban Planning, Anhui Jianzhu University, Hefei 230601, China
[3] School of Architecture and Art, Hefei University of Technology, Hefei 230601, China

Abstract. This paper takes the traditional houses in the Jianghuai area of Anhui Province of China as the research object, and selects five traditional houses as typical cases for field research and simulation. According to the climate adaptation design strategy analyzed by Climate Consultant software, the experience of jianghuai traditional houses in climate adaptability is qualitatively analyzed from three aspects: focusing on shading treatment, promoting ventilation and cooling, and strengthening heat storage, and provide ideas and guidance for contemporary design combined with climate and passive regulation methods.

Keywords: Jianghuai region of Anhui Province of China · traditional houses · climate adaptability · passive strategy

1 Introduction

When science and technology were relatively backward, traditional houses responded to the impact of climate factors on buildings with reasonable site layout, spatial form and material structure. After thousands of years of development, a set of passive construction technologies that make full use of local natural resources and adapt to local climatic conditions have been formed. With the sustainable development requirements of further promoting energy conservation and emission reduction and achieving. The goal of double carbon, the development trend of design combined with climate has gradually strengthened, and the design idea of passively regulating the building climate has gradually returned.

2 Based on Climate Condition Analysis by Climate Consultant

2.1 A Subsection Sample Climate Profile of Jianghuai Region

The Jianghuai region of Anhui Province of China, that is, the area south of the Neihuai River Basin and the north of the Yangtze River Basin in Anhui Province, China, stretches from the Dabie Mountains in the west to the eastern part of Anhui of China in

W. Guo et al. (Eds.): GBCESC 2023, LNCE 328, pp. 677–686, 2024.
https://doi.org/10.1007/978-981-99-9947-7_69

Mountains in the west to the eastern part of Anhui of China in the east. The Jianghuai region has a humid monsoon climate in the northern subtropics and has the most superior hydrothermal conditions in the province. Throughout the year, climate change is characterized by variable temperatures in spring; hot and rainy summers with obvious plum rain characteristics; high and refreshing days in autumn; and cold and dry winters with large north-south differences [1].

2.2 Simulation Object Selection

The traditional houses in Jianghuai area of Anhui of China mainly include: Jianghuai courtyard houses, Jianghuai patio houses, northwest Anhui of China polder villages, southeast Anhui of China big houses and Tongcheng family houses [2].

According to different traditional residential types, a representative residential building is selected for surveying and mapping modeling and software simulation, among which, the Jianghuai courtyardtype residential house chooses Zhang Zhizhong's former residence, the Jianghuai patiostyle residential house chooses Liu Tongxinglongzhuang, the Northwest Anhui of China Weizhai residential house chooses Liu Mingchuan's former residence, the Southwest Anhui of China House chooses Zhujiawan Ancient Residence, and the Tongcheng family mansion chooses Ni House.

2.3 Analysis Model Settings

Climate Consultant has a total of 4 thermal comfort analysis models to choose from, namely California EnergyCode, ASHRAE Standard 55, and 2005 ASHRAE Handbook of Fundamentals and ASHRAE Standard 55-2010. According to the research in this article, the 2005 ASHRAE Handbook of Fundamentals was chosen. The model assumes that there are changes in clothing to adapt to the climate, with an effective temperature of 20.0 to 23.3 °C with a relative humidity of 50%, a maximum wet bulb temperature of 17.8 °C, and a minimum dew point temperature /b121 > 2.2 °C, corresponding to the lower moisture content of about 4.6 g/kg [3].

2.4 Analysis of Climatic Conditions

The meteorological data for the Jianghuai region was obtained from Climate Consultant (CSWD) to generate an enthalpyhumidity diagram [4] (see Fig. 1). From Fig. 1, it can be observed that the Jianghuai region experiences cold winters and hot-humid summers. By excluding active design strategies such as mechanical cooling, heating, ventilation, and strategies with a time fraction of only 0.1% throughout the year, the most suitable passive design strategies for the local climate were identified using the climate adaptation design strategy combinations provided by Climate Consultant. These strategies, arranged according to their effective time duration, are as follows: (1) Internal heat gain (17.8%); (2) Dehumidification only (10.1%); (3) Window shading (9.4%); (4) Natural ventilation for cooling (7.6%); (5) Passive solar heating with low thermal mass (4.1%); (6) Passive solar heating with high thermal mass (3.4%); (7) High thermal mass with nighttime ventilation (2.8%); (8) High thermal mass (2.3%); (9) Indirect evaporative cooling (2.2%); (10) Direct evaporative cooling (1.7%).

The software suggests that the internal heat gain strategy accounts for 17.8% of the time fraction. However, based on actual research findings [5], it was discovered that traditional residences in the area have limited means of generating internal heat, typically relying only on heat from cooking stoves, which has a low heat output and short duration. Therefore, the internal heat gain strategy is not suitable for this region.

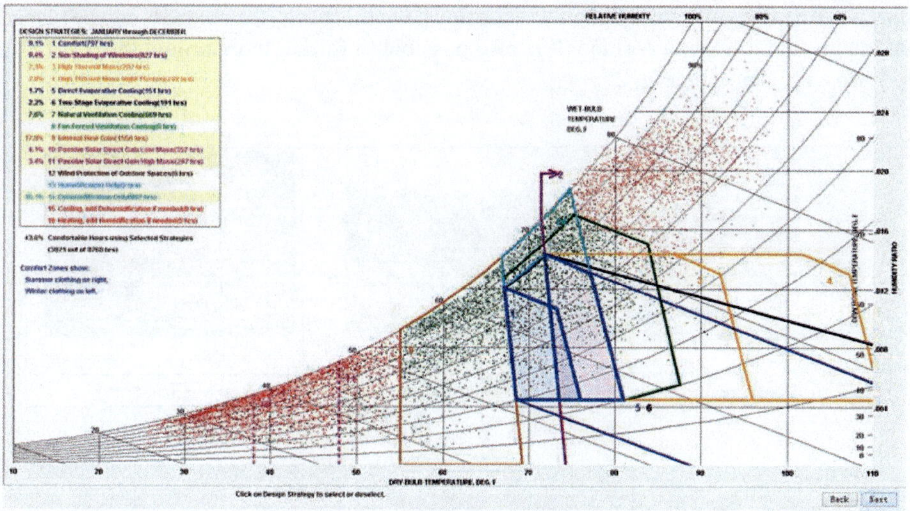

Fig. 1. Enthalpy wet diagram

The above nine passive measures are summarized and summarized, mainly divided into three categories: strengthen heat storage, promote ventilation and cooling, and pay attention to shading treatment.

According to the three types of climate adaptive design strategies proposed by Climate Consultant: focusing on shading treatment, strengthening heat storage, and promoting ventilation and cooling, the climate adaptation design methods of traditional houses in Jianghuai are specifically analyzed.

3 Climate-Adaptive Design Strategies Focusing on Shading Treatment

3.1 Deflect the Orientation of the Building Group Layout

The summer climate in the Jianghuai region is hot, and the direction of the layout of the building group should focus on preventing the impact of strong heat radiation in summer, and take into account the heat in winter. Therefore, the orientation of the building group layout generally avoids the due south direction and deflects a certain angle, so as to achieve the purpose of mutual shading and self-shading with the help of high-density group layout and internal shading components of the building [6].

3.2 Design a Variety of Forms of Shading Space

Traditional houses in the Jianghuai area design a variety of forms of shading space, including eaves, eaves, cloisters, etc. Its climatic adaptability significance, in addition to reducing the damage of rainfall to the envelope, is mainly aimed at reducing direct solar radiation on the outer surface of the wall and doors and windows in summer. Eaves picking is the most common practice in traditional houses, directly extending the slope roof and overhanging the eaves on all four sides. On this basis, the cornices can be placed on the ground to form a cornice. It is also possible to form a loop around the courtyard to form a cloister. See Fig. 2.

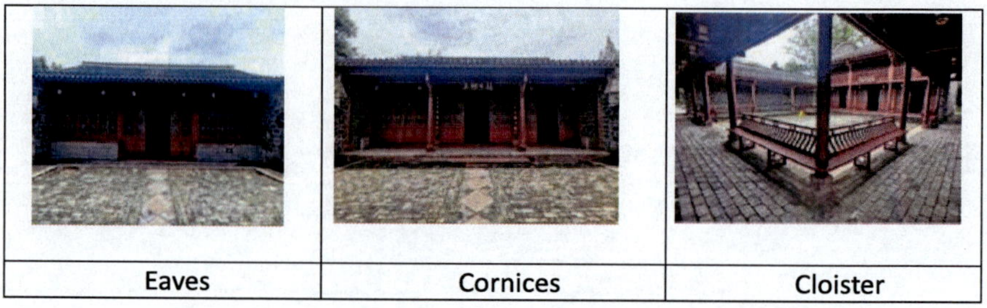

Fig. 2. Different shade forms

3.3 Recessed into the Wall of Traditional Houses

The traditional houses in the Jianghuai area are recessed at the entrance, and the wall where the entrance door is located is inclined at a certain angle, and the eaves above do not recess with it. This entrance treatment is conducive to shading and rain, which can play a role in climate buffering, and also expand the entrance space and form a space node. Tilt a certain angle in order to face the auspicious position in the feng shui concept.

4 Climate Adaptive Design Strategies to Promote Ventilation and Cooling

4.1 The Entrance is Facing the Direction of the Dominant Summer Wind

Good ventilation plays an important role in cooling, evaporation and dehumidification, and improving human comfort in summer [7]. The orientation of the layout of the building group has a great impact on the speed, strength and weakness of the natural wind flowing inside it, and the appropriate orientation can make the interior of the settlement obtain a comfortable wind environment. The direction of the traditional architectural group layout in the Jianghuai area is mostly towards the dominant wind in summer, and according to the specific situation, the entrance of the residential houses is slightly deflected to meet the dominant wind.

4.2 Layout Transforms the Micro-environment with the Help of Landscapes

Mountains and water are important environmental elements for the formation and development of settlements, and while meeting the needs of daily production and life, transportation and trade, their ecological characteristics are also conducive to creating a good settlement environment. The traditional settlements in the Jianghuai area make full use of the water environment and mountain environment to create a good internal environment of the settlement and play a certain role in ventilation and cooling. There are three main forms of layout formed by the combination of traditional settlements and landscapes in the Jianghuai area, see Fig. 3.

Kowloon Holds Pearls, and Ditches Run Vertically and Horizontally. The most representative form of village layout in the Jianghuai hilly area is "Jiulong Pearl" [8], which refers to the excavation of a half-moon-shaped pond in the middle or front of the village, and the construction of open ditches and hidden ditches leading to the pond, which are used to receive rainwater, and the underground canals are built under the building to connect with pipes to form a scientific drainage system; the residential buildings are distributed between the laneways with open ditches, forming a building group of "one pond and nine lanes".

The Back is Adjacent to the Mountain Range, and the Pond is Built in Front. In the Dabie Mountains in the western part of the Jianghuai Hills, the large houses are scattered in the flat land between the mountains, showing a collective layout form of independent ethnic groups, surrounded by mountains behind the house, and a pond is excavated in front of the house, the terrain is high and refreshing, the Mingtang is open, independently dispersed, and the cultivated land is saved. Drainage ditches are built on both sides in accordance with the slope of the mountain, and drainage ditches are set up below the residential floor to form a drainage network that combines light and dark, and the rainwater is discharged into the halfmoon pond built in front of the house.

Surrounded by Trenches, Strict and Orderly. Another representative village layout form in the Jianghuai area is the polder form, which is often selected in the pit to save cultivated land, and the polder is connected with the river and canal, taking into account farmland irrigation, flood drainage, wetland transformation and safety defense [9], the internal water body regulates the local microclimate, creating a warm winter and cool summer living environment, while also obtaining a good landscape.

4.3 The Patio Courtyard Organizes Internal Ventilation Nets

The characteristics of the spatial organization of Jianghuai traditional residential buildings are outward-facing closed and inward-facing openness. Open spaces such as patios and courtyards are set up in the internal space, which improves the ventilation and cooling effect of the summer indoors in the densely populated state.

Patio. The scale of the patio in the Jianghuai area is between the courtyard of the traditional houses in northern Anhui of China and southern Anhui, the terrain is flatter than that of southern Anhui, and the need for winter lighting is higher than that of southern Anhui, so the scale of the patio is slightly larger than that of southern Anhui,

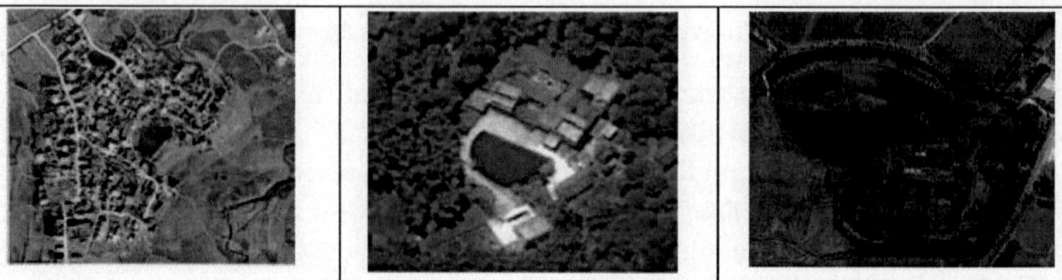

Kowloon holds pearls, and ditches run through them	The back is a mountain range, and the front is a pond	Surrounded by polders, it is strict and orderly

Fig. 3. Landscape combined with layout form

at the same time, the requirements for summer heat protection are higher than that of northern Anhui, and the scale of the patio is slightly smaller than that of northern Anhui of China to reduce solar radiation heat.

The patio plays a very important role in buffering outdoor climate change and improving the indoor space environment: forming an outdoor air circulation system with the eaves and corridors, etc., and forming an indoor air circulation system with the hall, hall, box, etc., thus building a passive indoor and outdoor climate exchange network. The plants planted in the patio and the accumulated rainwater can also regulate the air humidity and assist in fire protection. Patios are also important spaces for receiving sunlight and improving indoor lighting conditions.

Courtyard. The courtyard form is also commonly found in the traditional houses in the Jianghuai area, and there are two main ways to organize the courtyard in the houses, one is to organize the indoor space of the residence together with the patio, and the other is to organize the indoor space of the residential house in multiple courtyards. The Jianghuai region is the dividing line between the north and the south, and the spatial scale of the courtyard is between the north and south dwellings, which are built according to the terrain, which is smaller than the residential buildings in the northern Anhui of China Plain, but compared with the residential houses in the mountainous areas of southern Anhui, the area is larger, and the scale of the wide and open spaces is larger. The demand for shade in summer is higher than that in winter for lighting and warmth, so the longitudinal scale of the courtyard is narrower than that of the northern courtyard to produce a good air duct conduction. Compared with the courtyard of the southern Anhui of China houses that emphasize the role of shading and vertical wind pulling, it is more open and open, and the longitudinal series combination of multiple courtyards is conducive to forming a wind through the hall (Fig. 4).

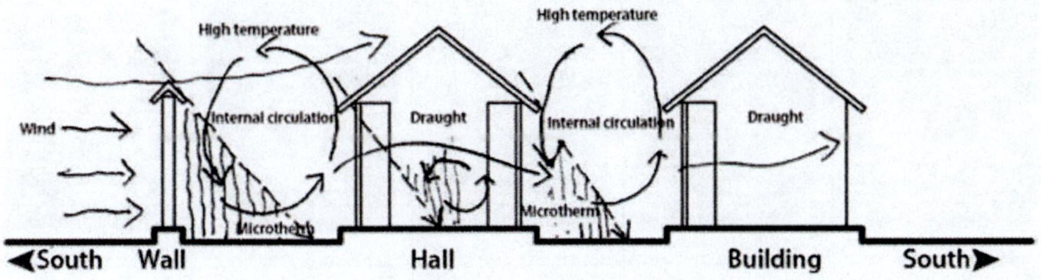

Fig. 4. Schematic diagram of internal ventilation

5 Climate-Adaptive Design Strategies for Enhanced Heat Storage

5.1 The Wall Uses Thermal Insulation and Heat Storage Materials

The wall materials of Jianghuai traditional residential buildings can be divided into three categories: one is the green brick solid wall made of clay, the other is the green brick irrigation bucket wall, the green brick empty bucket wall, and the other is the rammed earth wall and the adobe wall. See Fig. 5. Due to the high thermal resistance and heat capacity, the green brick wall is insensitive to the temperature change of the outer surface of the wall, so it is conducive to maintaining the dynamic balance of indoor temperature and improving indoor thermal comfort. Rammed earth walls and adobe walls have good thermal insulation and storage properties, and can store solar radiant heat in the wall during the day in winter and release it indoors at night to increase the temperature [10].

Green brick is mainly used for external walls and part of the interior wall, which has good air permeability and water absorption, which helps to regulate air humidity. The green brick wall adopts a variety of masonry methods such as one flat and one smooth, one smooth and one ding, two flat and one smooth, the outer wall is about 300 mm thick, and the partition wall is 200–230 mm thick. The main material of green brick masonry mortar is sand and lime, and its ratio is generally 1:4, depending on the family economic situation, glutinous rice juice, tung oil, magnetic ash and so on can also be added.

Green brick irrigation bucket wall, the two sides of the wall with brick standing, or vertical, flat alternate masonry, the central hollow part with broken bricks, mud and other filling, not only save bricks, but also has a good role in thermal insulation and sound insulation. Strong and thick, warm in winter and cool in summer, enhancing defensive function [11]. The green brick hollow wall is hollow and unfilled.

Rammed earth walls are made of a mixture of clay, straw and lime, and then reinforced with brick or stone blocks, without decoration, and harmoniously integrated with the surrounding environment. Adobe walls are generally made of adobe bricks in the upper part, with larger eaves, and the wall foundation is made of stone. The thickness of the adobe wall is generally 300–350 mm, which has good thermal insulation performance and can absorb or release water vapor according to the air humidity.

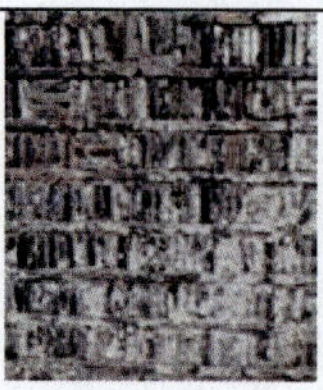	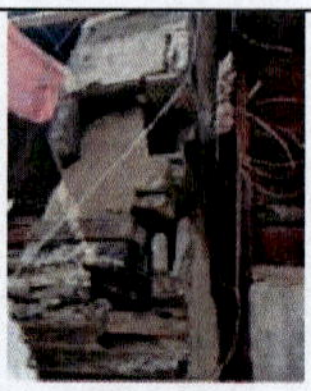	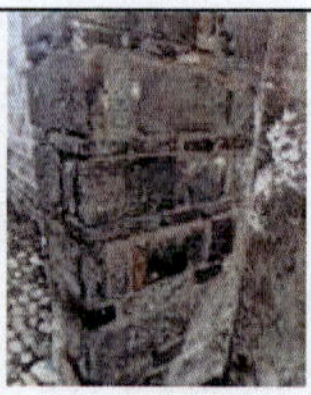	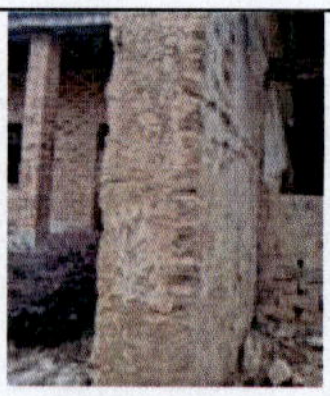	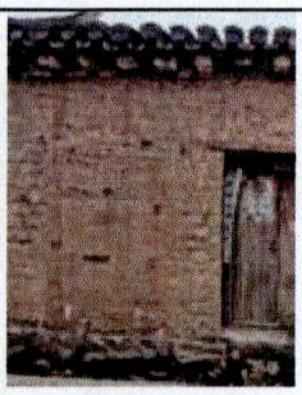
Green brick solid walls	Green brick empty bucket wall	Green brick bucket wall	Rammed earth walls	Adobe walls

Fig. 5. Schematic diagram of the wall material

5.2 Roof Tile Laying Method to Strengthen Insulation

The roof of Jianghuai traditional residential houses is mainly based on small green tile hard slope roof and small green tile hanging hillside roof, and the tiles are laid and seven dews are pressed. Traditional roofing practices are mainly divided into three types: Tile roofing without sheathing, wood flat tile roof and shedthins tile roof. The Wood flat tile roof and the shedthins tile roof are first laid on the purlin with a layer of shedthins board or shedthins tile, and then covered with tiles, and the tiles are laid in a way that is oneup and one-pitch pressure. The width of the tiles is 170 to 180 mm, the length is about 200 mm, and the spacing between the tile ridges is 200 mm. The cold tile roof is directly laid on the rafters, without reed foil, stucco, etc., and the tile bottom is exposed from the interior. See Table 1. Among them, the thermal insulation and storage capacity of the Wood flat tile roof and the brick roof is better, because the air interlayer is formed between the lookout plate, the shedthins tile and the green tile on the roof, which increases the thermal inertness of the roof and improves the thermal insulation performance of the roof.

5.3 Auxiliary Spatial Buffer Temperature Change

Distributed in the Dabie Mountains and along the River Plain, the traditional residential houses of Jianghuai adopt the spatial type of the upper partition of the bottom living. Auxiliary spaces such as pavilions and lofts are set up, which become an air interstitial layer above the living floor of the building. It has played a certain role in keeping warm and cold and ventilating the main living room.

The wooden frame of the traditional houses of Jianghuai occupies a higher space, and the pavilion and attic are erected at this height. There are various forms of treatment: one can erect the upper floor, use the area as a bedroom when the area is insufficient, the specific method is the dense beam wooden floor, with the crisscrossing primary and secondary beams as the skeleton, which is covered with 30mm thick wooden planks, and the wooden boards are coated with tung oil; the other can also be simply erected with wooden boards and bamboo poles for piling up debris, providing moisture barrier, rat prevention, antitheft and other space for storing rice [13].

Table 1. Schematic diagram of different roof constructions

Class	Construction mode	Thermal insulation characteristics	Construct diagram
Wood flat tile roof	A lookout board is nailed to the rafter, a stucco is added to the load-bearing layer as an insulation layer and binder, and finally the surface layer is laid with small green tiles, and the bottom tiles and cover tiles are arranged on an upward basis.	The gap between the lookout plate (lookout brick) and the green tile creates an air sandwich that increases the thermal inertness of the roof and improves the thermal insulation of the roof structure [12].	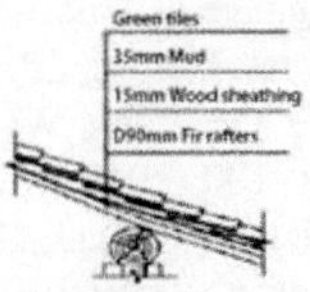
Tile roofing without sheathing	The bottom tiles are laid directly on the rafters, and the bottom tiles are arranged one positive and one negative or "one yin and one yang". There is no stucco in between.		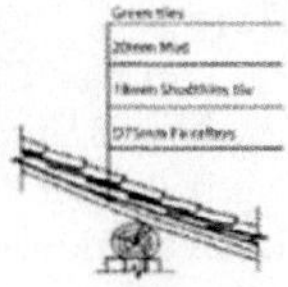
Tile roofing without sheathing	The bottom tiles are laid directly on the rafters, and the bottom tiles are arranged one positive and one negative or "one yin and one yang". There is no stucco in between.	The thermal insulation effect is not as good as the first two.	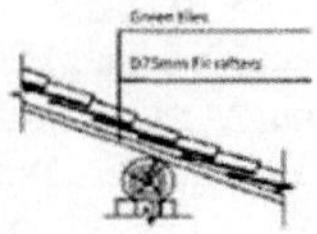

Under the conditions of humid and hot climate, the implantation of the partition is based on the layout mode of lowrise and high density under the environmental conditions of hilly terrain, which expands more building space under the condition of limited land area, and is also a response to the regional climate of hot summer and cold win-ter.

6 Conclusion

Passive building techniques in design fully utilize local natural resources, adapt to local climate conditions, and embody the principles of human-centeredness and harmonious integration between architecture and nature. Therefore, the exploration of traditional vernacular dwellings' experiences in climate adaptation and their transformation into contemporary architectural construction methods holds significant guiding significance for promoting energy efficiency, reducing emissions, achieving dualcarbon goals, and facilitating sustainable development.

7 Funding Support

Grant for Key Laboratory of Green Buildings and Prefabricated Construction in Anhui Province (Anhui Institute of Architectural Science and Design Institute), Open Fund Project (No. 2022-JKYL-004).

References

1. Zhang, L., Zhang, Q., Zhou, B.: Anhui Resources and Environment. Hefei University of Technology Press, Hefei (2010)
2. Ministry of Housing and Urban-Rural Development of the People's Republic of China, ed. Chance of Traditional Chinese Architecture Anhui Volume. China Architecture and Building Press, Beijing (2016)
3. He, Q., Wang, W., Liu, J., Yang, L.: Lhasa based on climate consultant. J. Build. Sci. **33**(04), 94–100 (2017)
4. Meteorological Data Room, Meteorological Information Center, China Meteorological Administration, Department of Building Technology Science, Tsinghua University, Hina, China Architecture & Building Press, Beijing (2005)
5. Li, T., Lei, Z.: Ings based on climate consultant analysis. World Archit. (09), 102–105+120 (2018)
6. Yang, L.: Logy. China Architecture & Building Press, Beijing (2010)
7. Li, H., Wu, B., Luo, Y., Liu, Q., Xu, S., He, Q.: Gdao. Archit. Tech. **28**(02), 62–68 (2022)
8. Wang, Y., Shen, Z.: Town. Constr. Sci. Technol. (16), 71–76 (2019)
9. Ministry of Housing and Urban-Rural Development of the People's Republic of China, ed. Types on China Architecture & Building Press, Beijing (2014)
10. Liu, X.: Face. Xi'an University of Architecture and Technology (2021)
11. Yang, X., Ning, D.: Nhui. China Architecture & Building Press, Beijing (2019).
12. Zhu, M.: Lity. Anhui Jianzhu University (2021)
13. Hao, S.: Uses. Tsinghua University (2016)

Thermal-Economic Performance Evaluation of Air Conditioning for Office Buildings

Jifan Cao[1], Diqing Wang[1], Hongli Sun[1], and Yin Zhang[2](✉)

[1] College of Architecture and Environment, Sichuan University, Chengdu 610065, China
[2] School of Architecture, Southwest Minzu University, Chengdu 610225, China
cdzhangyin@163.com

Abstract. Building energy consumption accounts for about 40% of the total energy consumption in China. Among them, for hot summer and cold winter areas, the energy consumption of air conditioning and refrigeration in summer is the main part, based on the current situation of air conditioning application in this area, it is significant to improve the energy efficiency of refrigeration equipment in summer by analyzing the energy-saving potential of operation energy consumption and comprehensively evaluating its economy. Based on the theoretical thermodynamic perfection of equipment refrigeration, taking the split type air conditioning system in Guiyang office buildings as an example, this paper selects the typical refrigeration equipment and analyzes its energy-saving potential and economy based on its energy price status, in order to clarify the office building split air conditioning equipment energy-saving capacity and its operation economy. The preliminary analysis shows that when the service life of air-conditioning exceeds 5 years, its energy-saving performance and economy can reach the agreement, that is, the best energy-saving air-conditioning economy is also the best.

Keywords: Thermal Consumption · Energy Consumption Evaluation · Energy Saving Potential · Air Conditioning

1 Introduction

For the past few years, China's economy has developed by leaps and bounds, but this process has also brought us a huge amount of energy consumption. Among the major energy consumption of our society today, building energy consumption accounts for about 40% of the total China energy consumption. For HVAC energy consumption, which is the bulk of building energy consumption, the study of its energy saving potential and energy saving economy is an important direction for the development of building energy efficiency [1]. Most studies today are based on the technical and operational evaluation of air conditioning units in large public buildings in typical cities. These studies provide an in-depth analysis of the energy efficiency of HVAC equipment operation, but rarely analyze the economic aspects of such equipment, making it impossible to make an intuitive evaluation of the energy efficiency and economy of building air conditioning. In this study, a typical city in the hot-summer and cold-winter region of Guiyang is

W. Guo et al. (Eds.): GBCESC 2023, LNCE 328, pp. 687–695, 2024.
https://doi.org/10.1007/978-981-99-9947-7_70

chosen as the target for the study. The study analyses the energy consumption of split air conditioning in office buildings, explores its energy saving potential and establishes a dynamic model of the economics of air conditioning operation on this basis, so that the energy saving and economics of building air conditioning can be weighed more intuitively.

2 Calculation of Thermodynamic Perfectness and Research Methods

2.1 Ideal Coefficient of Performance

As shown in Fig. 1, the reverse Carnot cycle is the ideal cycle of refrigeration, consisting of two isothermal processes and two adiabatic processes. Assuming that the thermodynamic temperature of the low-temperature heat source, the cooled object, is T_0, and the thermodynamic temperature of the high-temperature heat source, the ambient medium, is T_k, the temperature of the mass in the heat absorption process is T_0, and in the exothermic process is T_k, the process assumes that there is no temperature difference between the mass and the low-temperature heat source and the high-temperature heat source in the heat absorption and exothermic processes, and that the heat transfer process is carried out isothermally, and that the compression and expansion processes are carried out without any losses. The cycle process is as follows: the mass draws heat from the cold source under T_0 and carries out isothermal heat absorption process, then through the adiabatic compression process, its thermodynamic temperature rises from T_0 to the thermodynamic temperature of the ambient medium T_k, and then carries out isothermal exothermic process under the temperature of T_k and releases heat to the high temperature heat source q_k, and finally carries out adiabatic expansion process to make its temperature drop from T_k to T_0, and the mass returns to the initial state4, thus completing a cycle. The cooling coefficient of this cycle is:

$$COP = \frac{T_0}{T_k - T_0} \tag{1}$$

The COP based on the reverse Carnot cycle is an economic indicator used to characterize the corresponding refrigeration cycle, where the value of the cooling coefficient under this condition depends only on the thermodynamic temperature T_k of the heat source and the thermodynamic temperature T_0 of the cold source.

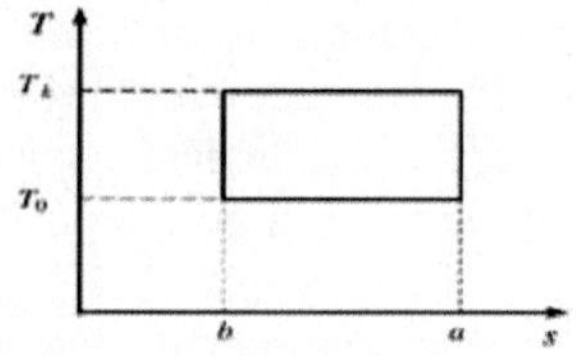

Fig. 1. The reverse Carnot cycle

2.2 Actual Coefficient of Performance

The irreversible process in the actual air conditioning operation makes the Coefficient of Performance under the actual working condition smaller than it under the reverse Carnot cycle, so the COP' under the actual working condition is defined as the ratio of the cooling capacity E_0 produced under the actual cycle working condition to the cycle power consumption E of the actual cycle.

$$COP' = \frac{E_0}{E} \tag{2}$$

2.3 Thermodynamic Perfectness

Since the actual refrigeration cycle generates irreversible losses, thermodynamic perfectness η is introduced to measure the degree of irreversibility of the actual cycle. Thermodynamic perfectness η is the ratio of the COP of the actual refrigeration cycle to the COP of the reverse Carnot cycle in the same operating temperature interval.

$$\eta = \frac{COP'}{COP} \tag{3}$$

Its amount reflects the closeness of this refrigeration cycle to the reverse Carnot cycle at the same temperature, depending on the level of the air conditioning manufacturing process, and can be regarded as a constant value within a certain range of thermal operating conditions [2].

2.4 Research Methodology

Guiyang is located in the southwest of China, in the hot summer and cold winter climate region of China, the summer temperature is always moderate and rarely hot, the average temperature is always at 16–24 °C, and it belongs to the area without central heating. Research shows that the thermal neutral temperature in hot summer and cold winter areas obtained by PMV calculation is around 18 °C, and in areas without centralized heating, people not only have lower expectations of indoor temperature, but also have some psychological comfort with lower indoor temperature [3]. Under the condition of meeting the indoor thermal comfort zone, the summer interior design temperature is taken to be 18 °C.

Under the laws of China, the workers who establish labor relations in various organizations and enterprises implement the working hour system of 8 h of work per day and 40 h of work per week, and according to the survey statistics, most of the workers in Guiyang enterprises work daily in the time period of 8–18 h [5], without considering the influence of overtime hours, so the summer day by day 9 to 17 h is selected as the office building air conditioning operation time for calculation.

The outdoor meteorological data used in the study were obtained from the meteorological data stored in the Energy Plus database, and the 24-h, day-by-day temperature of March 2005 in Guiyang was used to calculate and analyze the energy consumption of split air conditioners in office buildings based on different thermodynamic perfectness conditions.

3 Energy Consumption and Economic Analysis

3.1 Typical day Energy Consumption Analysis

The study uses four air conditioners of KFR-35GW type with different energy efficiency levels for simulations, and the performance parameters of each air conditioner are shown in Table 1. In order to analyze the performance of different air conditioners, the COP value for level 5 is higher.

Table 1. Performance parameters of air conditioners of various energy efficiency levels

Energy efficiency level	Refrigerating capacity [W]	Refrigeration power [W]	*COP*	*COP'*	η
1	3500	750	24.4	4.7	0.19
2	3500	900	24.4	3.9	0.16
3	3500	980	24.4	3.6	0.15
5	3500	1130	29.3	3.1	0.11

The highest average temperature day and the lowest average temperature day from 9 to 17 in summer as a typical day, air conditioning operation energy consumption analysis: as shown in Fig. 2, the highest average temperature day from 9 to 17 is July 31, with an average temperature of 29.4 °C; the lowest average temperature day from 9 to 17 is August 22, with an average temperature of 18.3 °C, while the typical daily power consumption of each different thermodynamic perfectness of air conditioning is shown in Fig. 3.

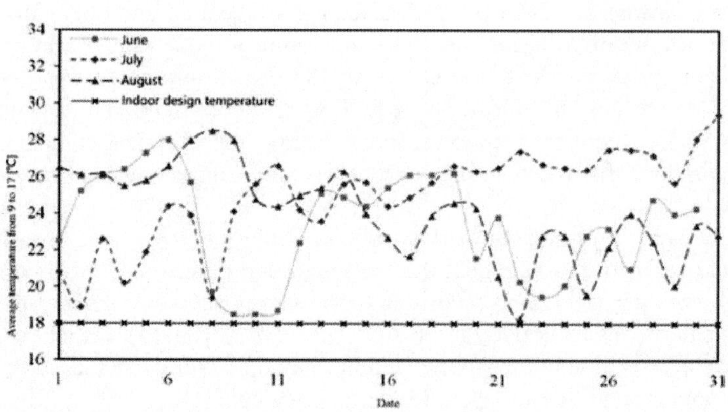

Fig. 2. Average temperature change curve for summer months from 09–17 pm

As shown in Fig. 3, the energy consumption of air conditioning decreases with the increase of thermodynamic perfectness, when the thermodynamic perfectness increases

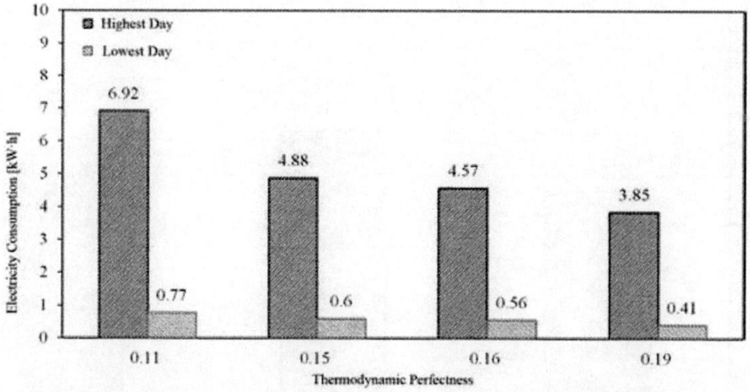

Fig. 3. Electricity consumption on a typical summer day for air conditioners with different degrees of thermodynamic perfectness

from 0.11 to 0.19, the energy consumption of air conditioning on the highest day decreases from 6.92 kW·h to 3.85 kW·h, a decrease of 44.36%. The difference in energy consumption between the same air conditioner on the highest day and the lowest day is huge, as the thermodynamic perfectness increases from 0.11 to 0.19, the difference in energy consumption between the highest day and the lowest day decreases from 6.15 kW·h to 3.44kW·h, a reduction of 44.07%. The energy saving of the same air conditioner on the lowest day compared to the highest day is nearly 90%. This shows that the higher the degree of thermodynamic perfectness under the same working conditions, the smaller the energy consumption of the air conditioner and the smaller the energy savings of the air conditioner; the outdoor temperature has a greater impact on the energy consumption of the air conditioner than the degree of thermodynamic perfectness.

3.2 Month-By-Month Energy Consumption Analysis

The analysis of the energy consumption of the air conditioner on a typical day mentioned above shows that the outdoor temperature and the difference between indoor and outdoor temperatures have a significant impact on the power consumption of the air conditioner in operation. The smaller the thermodynamic perfectness of the air conditioner, the greater the power consumption under the same operating conditions. Therefore, the energy analysis is performed with the air conditioner of energy efficiency class 5 as the operating air conditioner to find the maximum energy consumption in the selected air conditioner, and thus the maximum energy saving potential of the air conditioner can be obtained.

From the analysis of a typical day, it can be seen that the air conditioner with the smallest thermodynamic perfectness has the largest power consumption under the same temperature environment conditions, so the air conditioner with five energy efficiency levels is used as the operating air conditioner for energy consumption analysis in order to obtain the maximum energy consumption among the selected air conditioners, so as to obtain the largest air conditioner energy saving potential value, and its monthly power consumption for each month of operation in summer is shown in Fig. 5.

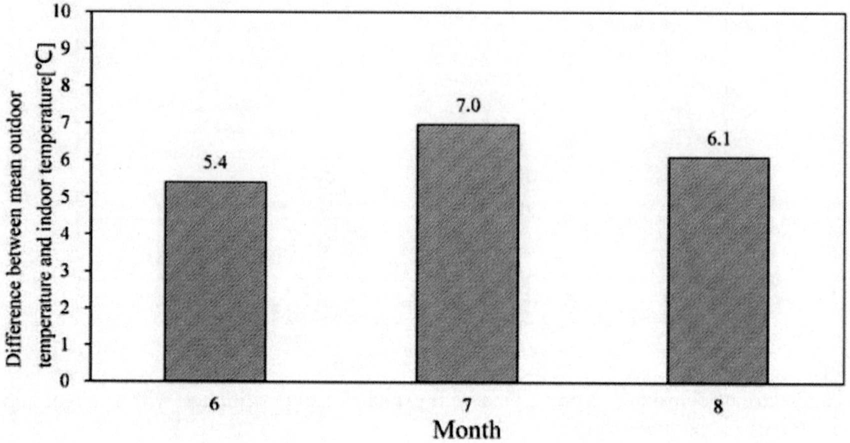

Fig. 4. Average difference between indoor and outdoor temperatures from 9 to 17 h in each summer month

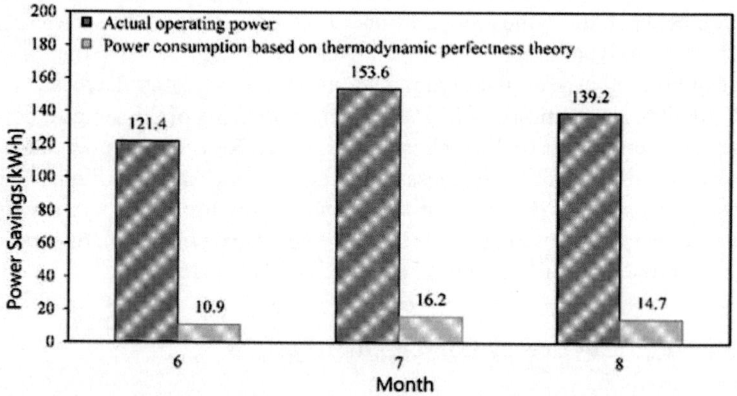

Fig. 5. Power consumption for each month of operation of class 5 energy-efficient air conditioners

Figures 4 and 5 show the difference between the average indoor and outdoor temperatures from 9 to 17 in each summer month and the power consumption of energy-efficient air conditioners in each month, respectively. The results show that the higher the average difference between indoor and outdoor temperatures from 9 to 17, the higher the power consumption of the air conditioner. 1.6°C higher than the average difference between indoor and outdoor temperatures in July than in June, the power consumption of the air conditioner is 32.2kW-h more, and the relative percentage of power consumption is 26.5%. That is, for the same model of split type air conditioner, a small change in the temperature difference between indoor and outdoor can lead to a large change in power consumption. The graph also shows that for the air conditioner with the smallest thermodynamic perfectness, the actual power consumption due to various irreversible factors is much higher than the power consumption calculated based on the theoretical thermodynamic perfection, and the actual power consumption in each month is almost 10 times of the theoretical power consumption, which means that the split air conditioner has great potential for energy saving in operation.

3.3 Energy-Saving Measures and Economic Analysis

From the analysis of the monthly operating energy consumption of the air conditioners in energy efficiency class 5, it is already known that there is a lot of room for energy saving in the actual operation of the air conditioners in summer. For the four selected split type air conditioners with different thermodynamic perfectness, their energy saving potential in summer operation is shown in Fig. 6.

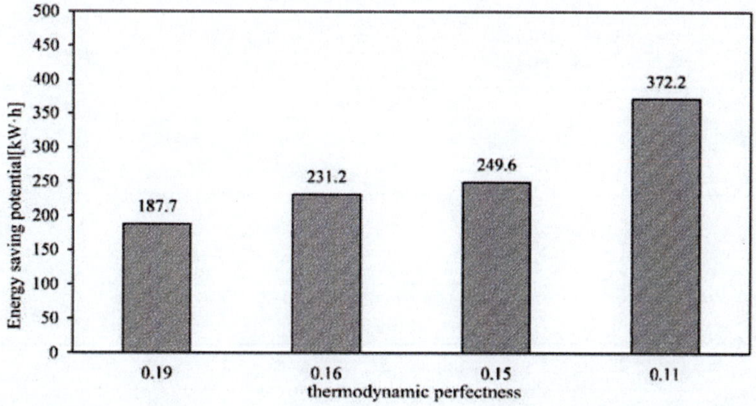

Fig. 6. Graph of air conditioning energy saving potential with thermodynamic perfectness

As shown in Fig. 6, the higher the thermodynamic perfectness degree, the smaller the energy saving potential of the air conditioner, because the increase of the thermodynamic perfectness degree means that the actual working condition of the air conditioner is closer to the theoretical working condition, and the space for energy saving will be smaller, but at the same time, it also means that the energy saving effect is better. From the graph, we can also see that when the thermodynamic perfectness degree is 0.11 and 0.19, the energy saving potential is almost double, that is, the energy saving effect is almost double, so we can start from improving the thermodynamic perfectness degree of split type air conditioners to do energy saving design, improve the efficiency of the compressor, increase the condenser and outdoor heat source side of the heat transfer capacity and effective use of condensate to take away condensation heat and other measures can be effective. The thermodynamic perfectness of split air conditioners can be improved [6], and it can be seen from Fig. 6 that a small increase in thermodynamic perfectness within a certain range can lead to a large increase in energy efficiency, which means that it is very cost-effective to improve energy efficiency by increasing thermodynamic perfectness within a certain range. However, as shown in Fig. 7, as the degree of thermodynamic perfectness increases, the rate of improvement of the energy-saving effect by increasing the degree of thermodynamic perfectness will become smaller and smaller, which will lead to a smaller and smaller cost performance by increasing the degree of thermodynamic perfectness, therefore, when increasing the rate of energy-saving by increasing the degree of thermodynamic perfectness in split type air conditioning equipment when the degree of thermodynamic perfectness is larger, it is necessary to make a certain trade-off between economy and energy-saving.

China regulations, the general use of split-type air conditioners do not exceed 10 years, so the maximum use of the air conditioner for ten years to calculate the length of air conditioning operation, Fig. 7 shows the four different thermodynamic perfectness of split-type air conditioners for ten years of operating investment costs curve.

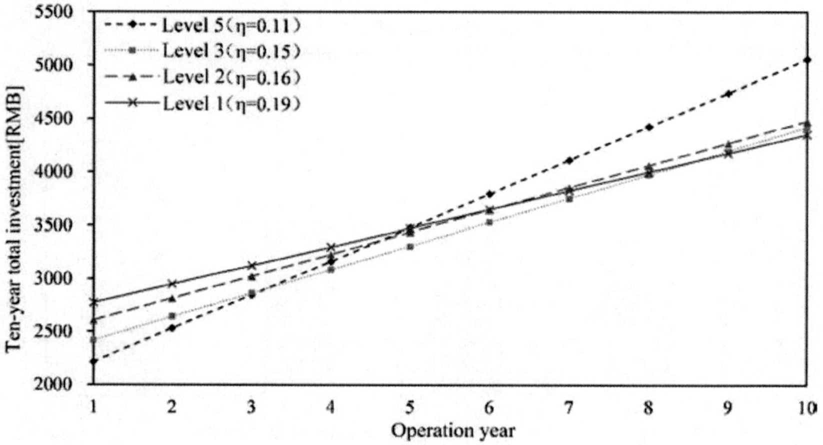

Fig. 7. Ten-year operating investment curve for each split type of air conditioner

As shown in Fig. 7, the initial investment of the split type air conditioner with five energy efficiency levels is the lowest, but the slope of its total investment curve is the largest, which means that its annual investment is the largest, and the initial investment of the split type air conditioner with one energy efficiency level is the highest, but correspondingly its total investment curve slope is the smallest and its annual investment is the smallest, and the total investment after ten years of operation under the same conditions is the smallest for the split type air conditioner with one energy efficiency level and the largest for the split type air conditioner with five energy efficiency levels, from the perspective of economy, if the long-term use of the same air conditioning equipment, the use of thermodynamic perfectness of the best economy of split-type air conditioning equipment; short-term use of the same equipment, you can choose a smaller degree of thermodynamic perfectness of split-type air conditioning equipment to seek a better economy. From the point of energy efficiency, whether long-term or short-term, the use of thermodynamic perfectness of split-type air conditioning equipment has the best energy efficiency.

4 Conclusion

Based on the theoretical thermodynamic perfection of refrigeration equipment for building air conditioning, taking the split type air conditioning system in Guiyang office buildings as an example, this paper selects the typical refrigeration equipment and analyzes its energy-saving potential and economy based on its energy price status, in order to clarify the office building split air conditioning equipment energy-saving capacity and its operation economy. From the preliminary case study, the following conclusions can be drawn.

(1) When the low temperature heat source temperature is certain, the smaller the temperature difference between the cold and hot sources, the larger the cooling coefficient of the split type of air conditioner and the higher the energy efficiency.
(2) Improving the thermodynamic perfectness of a split type of air conditioner within a certain range can significantly improve the energy efficiency of the air conditioner, but as the thermodynamic perfectness increases, the cost of improving the thermodynamic perfectness increases while the energy efficiency rate decreases, and the cost performance decreases.
(3) The greater the degree of thermodynamic perfectness, the better the economy of split air conditioners in long-term investment; and regardless of long-term investment or short-term investment, priority should be given to split air conditioners with a greater degree of thermodynamic perfectness in the case of energy-saving effects.

Acknowledgement. This work is supported by National Natural Science Foundation of China (No. 52108032) and Sichuan Science and Technology Research Program (No. 2022NSFSC1148).

References

1. Wu, J., Li, H., Yu, Z.: Optimized operation analysis of HVAC system in a near-zero energy office building. Build. Sci. **36**(06), 35–41 (2020)
2. Hou, B.S., Chen, J.H., Ge, F.: Discovery of non-standard thermal perfection laws for screw-type water source heat pump units and their application to water system energy consumption models. Refriger. Air Condition. **30**(4), 479–482 (2016)
3. Shi, H.L., Wang, Z.W., Yan, Z.F.: Study of heating interior design temperature in hot summer and cold winter areas. Sichuan Build. Sci. **41**(04), 123–127 (2015)
4. Zhang, X.Z., Jiang, F., Chen, P.Z.: A study on the effects of standard working hours on working hours and wage returns - an empirical test based on the effects of the implementation of the Labor law. Academic Res. **11**, 89–95 (2020)
5. Ding, Y., Shi, L.S.: Analysis of the energy-saving potential of split air-conditioning condensate reuse. Build. Energy Efficien. **44**(11), 97–101 (2016)

Research on Green Construction Technology of Traditional Buildings in Northern Anhui Based on Climate Adaptability

Manting Zhu[1,2], Anqi Liu[3], and Wei Wang[1,3](✉)

[1] Anhui Provincial Key Laboratory of Green Building and Prefabricated Construction, Anhui Institute of Building Science and Design, Hefei 230031, Anhui, China
Vivi.gan@126.com
[2] School of Architecture and Urban Planning, Anhui Jianzhu University, Hefei 230601, China
[3] School of Architecture and Art, Hefei University of Technology, Hefei 230601, China

Abstract. Climate is a prerequisite for traditional architecture in the process of site selection, design and construction, and also an important reason for the regionalization of traditional architecture. Therefore, the research on the green construction technology of traditional buildings adapting to climate is helpful to excavate the context of traditional architecture and inherit the construction wisdom of traditional architecture. In this paper, the spatial layout of traditional buildings in northern Anhui was studied in depth from the perspective of climate adaptability by means of field investigation, ECOTECT and PHOENICS simulation analysis, and the green construction technology for the spatial layout of traditional buildings in northern Anhui was summarized from three aspects of plane layout, external surface design and detailed structure. Therefore, it can be applied to the design of modern architecture living environment with the same climatic conditions in northern Anhui and its surrounding areas.

Keywords: North Anhui · Climate adaptation · Traditional architecture · Architectural features · Construction technology

1 Introduction

China's traditional architecture after thousands of years of cultural precipitation and artisan inheritance, has accumulated a wealth of green design and construction of wisdom strategy [1]. Recent research findings highlight a notable emphasis on studying traditional architecture within Anhui's southern region, particularly concentrated to the south of the Huaihe River [2, 3]. In contrast, scholarly attention to the northern Anhui area remains comparatively limited. However, the region's traditional architecture is a historical product of the fusion of northern and southern cultures, from the analysis of the existing traditional architecture can still be seen from the ancient people's wisdom of construction [4], which is worthwhile for us to learn from the mining.

W. Guo et al. (Eds.): GBCESC 2023, LNCE 328, pp. 696–702, 2024.
https://doi.org/10.1007/978-981-99-9947-7_71

2 Overview of the Northern Anhui Region

The scope of the northern Anhui region refers to the area north of the Huaihe River in Anhui, including the six cities of Suzhou, Huaibei, Bengbu, Fuyang, Huainan and Bozhou. The region has many plains, a warm temperate semi-arid, semi-Zijian climate, humid and rainy summers, dry and less rainy winters, and large temperature fluctuations in spring and autumn. The average temperature throughout the year is 15 ~ 17 °C. The summer is not too sultry, with an average temperature of 27 ~ 28 °C. The climate is cold in winter, with an average temperature close to 0 °C in January and a low extreme minimum temperature. Rainfall throughout the year is relatively moderate, between about 8000 ~ 1200 ml, with the characteristics of more in the south and less in the north, with significant plum rain and abundant summer. The region is rich in light resources, with long hours of sunshine and a relatively dry climate. Therefore, buildings across the region need to consider heat protection, ventilation and drainage in summer, and also cold protection in winter [5].

3 The Current Situation of Traditional Architecture in the Northern Anhui Region

The architectural style prevalent across the region harmoniously integrates elements of the northern traditional courtyard style. Green bricks and grey tiles emerge as the primary materials, imbuing the structures with a subdued yet elegant and dignified ambiance [6]. Guided by national, provincial, and municipal cultural heritage protection lists, a meticulous selection process identified well-preserved singular buildings of significant research value. These include residential homes, monetary establishments, and commercial structures which used to cater to the multifaceted needs of human production and livelihood, these 15 exemplary traditional edifices were dispersed across five cities within northern Anhui.

4 Climate-Adaptive Green Building Technology for Spatial Layout in Northern Anhui Province

4.1 Space Types

The fundamental spatial configurations of traditional buildings within the northern Anhui region encompass three primary categories: "courtyard," "com-pound," and "patio" [7]. These foundational spatial layout are further amalgamated to yield six fundamental forms, as expounded in Table 1.

4.2 Layout

Spatial Scale. The shape of patios and courtyards directly affects the daylight environment of a building. Their shape is mainly measured by the aspect ratio and height-to-width ratio, as expounded in Table 2.

Table 1. The basic shape of spatial layout in the northern Anhui region

Name	Multi-deep Sanheyuan	Multi-deep courtyard	Single-deep compound	Multi-deep compound	Single-deep patio type	Multi-deep courtyard combination
Layout	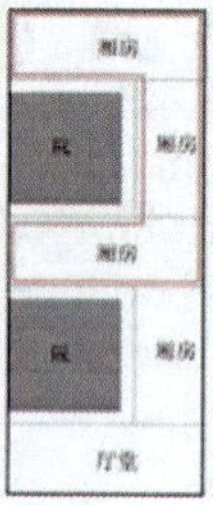	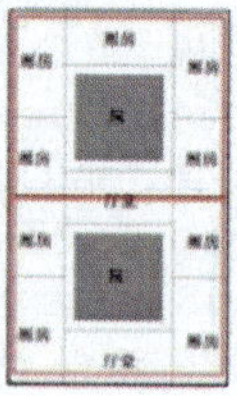	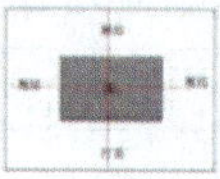	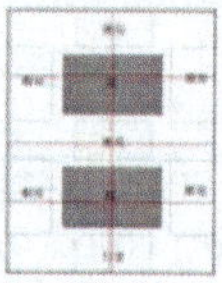	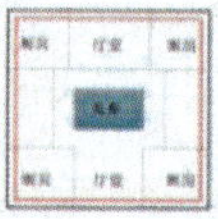	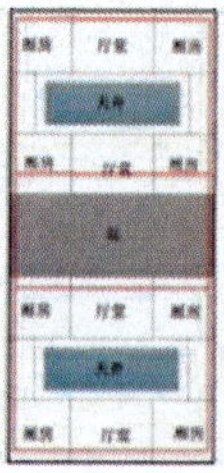
Functional order diagram	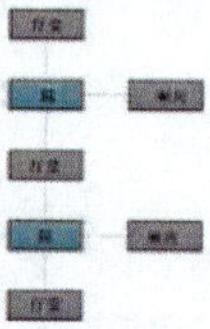	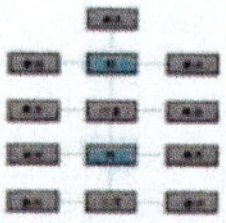	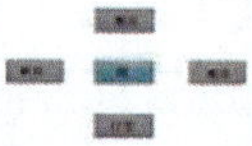	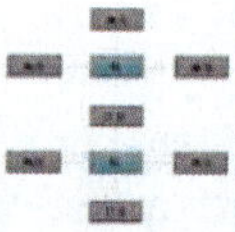	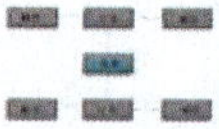	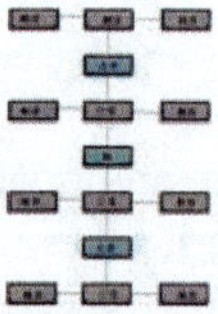
Typical case	Bozhou Nanjing Bank House	Ju Ruixing Pharmacy	Linhuan Shen's compound	Linhuan Wenchang Palace	Yingshang Jingde Hall	Yingshang Fushou Hall

Upon reviewing the Table 2, it becomes apparent that the length-to-width ratios characterizing the courtyards of traditional structures in northern Anhui predominantly fall within the range of 1.2/1 to 1.5/1, and the spatial form is similar to a square. This is a departure from the courtyards found in the northern regions, where the courtyard's depth greatly exceeds its width at the surface. In contrast, the length-to-width ratio of patios in northern Anhui spans between 1.2/1 and 1.8/1. When compared to the courtyard layout, the surface width and space ratio of patio are larger, and its form is mostly presented as a narrow rectangle that is transverse.

A substantial portion of courtyards within this region adheres to a height-to-width ratio ranging between 1/1.5 and 1/1.1. This underscores that the expansive dimension of courtyards within northern Anhui is harmonized with the solar-thermal conditions inherent to higher latitudes. This design choice aims to strive for more solar radiation. In contrast, the height-to-width ratio of patios is notably greater than that of courtyards, spanning from 1.2/1 to 1.9/1. However, it's important to note that, for the patio dimensions in southern Anhui that are much larger than 2/1, the patio in northern Anhui is

Table 2. Spatial enclosure scales of courtyards in northern Anhui region

District	Case name	Type of internal space	Spatial enclosure scale	
			Length-width ratio	Aspect ratio
Bozhou	Nanjing Bank House	Patio, Courtyard	1/1.3–1/2	1.8/1–1.2/1
	Beiguan Shop	-	-	-
	Zhang Xugu Residence	Courtyard	1.5/1	1/1.6
	Ju Xingrui Pharmacy	Courtyard	1/2	1.7/1
Linhuan	Shen's compound	Courtyard	2.9/1	1/3.3
	Yuan's House	Courtyard	1.2/1	1/1.3
	Wenchang Temple	Courtyard	1.5/1–2/1	1/1.6–1/2.1
Xiaoxian	Yang's Table	Courtyard	1.2/1–2/1	1/1.2
Yingshang	Jingde Hall	Patio	1.4/1	1.9/1
	Fushan Hall	Patio, Courtyard	1.2/1–1.8/1	1.7/1–1.2/1
	Wenchang Hall	Patio	1.4/1	2.4/1
	Wenrui Hall	Patio, courtyard	1/1	1.1/1
	You's Garden	Point	2.6/1	1.5/1
Mengcheng	Ma's Gungfu	Courtyard	1.3/1–1.5/1	1/2.5–1/2.7
Fuyang	Chen Wenbing Residence	Courtyard	2/1	1.1/1

obviously much more open. This phenomenon suggests that the internal core space within this region should not only consider the formation of shading to mitigate intense summer sunlight, but also regulate spacing to facilitate passive heating opportunities during winter. This delicate balance ensures a harmonious interplay between climatic adaptation and interior comfort.

Space Microenvironment Modelling

Wind Environment Analysis. In line with meteorological data collection and consultation specific to northern Anhui, the average wind speeds for simulation have been designated as 2.3 m/s during winter and 2.7 m/s during summer. Furthermore, the wind direction for winter is defined as the north-easterly wind (NE), while the summer wind direction is specified as the south-westerly wind (SSW).

As this simulation mainly focuses on the internal transition space, with the patio and courtyard as the main simulation object, and the height of the centre of gravity of human beings is usually considered to be at 1.5 m. Therefore, when evaluating the wind environment, the outdoor wind environment at a height of 1.5 m is used as the basis of evaluating the wind environment of the internal space [8], as shown in Fig. 1.

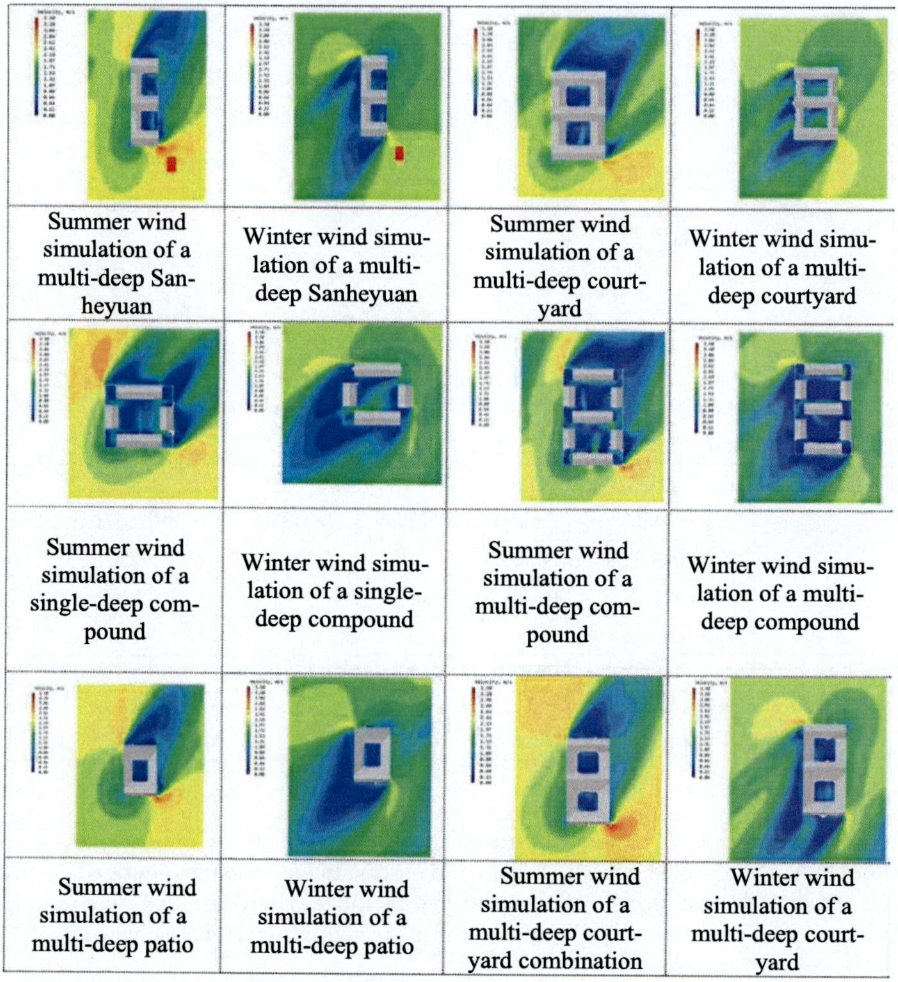

Fig. 1. List of wind environment in each space

From the Fig. 1, it is evident that the "large courtyard" layout with relatively expansive dimensions can engender a discernible wind pattern, facilitating the swift dissipation of indoor humidity and heat. During the winter months, a single-depth courtyard can harness the sheltering effect of towering walls and trees to mitigate wind impact; The "patio-style" configuration ensures internal wind stability, necessitating the influence of external wind pressure to shape the indoor air circulation. During summer, it's imperative to keep doors and windows open to induce airflow; In the case of the "compound-style" layout, it can allow the interior space to be ventilated in summer while maintaining relatively stable indoor wind conditions throughout winter.

Light Environment Analysis. Due to the relatively cold climate of the northern Anhui region, in order to reduce the heat loss of the external envelope, traditional buildings in this region feature small sized and infrequent window openings on their external walls. Therefore, light from patios and courtyards is an important source of light for residents' lives.

The lighting environment of their internal space is simulated and analsyed with EOC-TECT. The simulation environment is set up according to the provisions of "Lighting Measurement Methods" GB/T5699, using the external space of a fully cloudy day with no direct light and a light source environment dominated by diffuse light. According to the light climate zoning map in the Lighting Design Standard for Buildings (GB/T 50033–2013), the whole territory of northern Anhui belongs to the IV light climate zoning. Therefore, the critical illuminance of 4500lx of natural light outdoors in this climate zone was set, and the test height was at 750mm from the ground to evaluate the transitional spatial light environments of the six basic models, and the results are shown in Fig. 2.

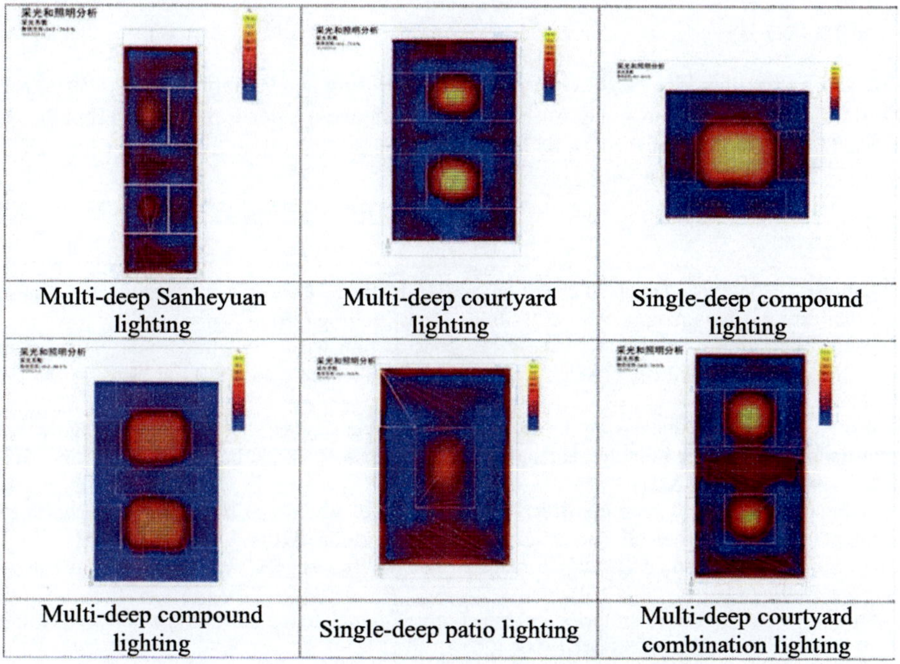

Fig. 2. Overview of lighting environment in each space

The incorporation of patios and courtyards within traditional buildings demonstrates a crucial "compensatory effect" on their lighting conditions. It can alleviate the difficulties in lighting brought about by the reduction of external window openings in order to insulation and heat storage in winter. When assessing these architectural styles from a comprehensive standpoint, the courtyard layout emerges as the most effective in optimizing lighting conditions. Following closely is the multi-courtyard arrangement, while the triad layout slightly surpasses the patio design in terms of lighting quality. Consequently, in the northern Anhui region, extended family tend to opt for the compound layout. This choice provides an open and luminous space. For settings such as production workshops or mixed-use spaces that blend commercial and residential functions, the "commercial and residential" combination, where lighting demands are comparatively lower, favors the courtyard or patio layout.

5 Conclusion

Green building technology embodies human-centred, harmonious unity of architecture, man and nature, while the climate is the most important reference index for building construction activities in various regions, so exploring the green building technology of traditional architectural spatial layout, and forming green building technology and methods suitable for the contemporary era with Chinese characteristics are of great significance in guiding the healthy and sustainable development of green building in China.

6 Funded by

The Anhui Provincial Key Laboratory of Green Building and Prefabricated Construction (Design Institute of Anhui Academy of Building Sciences) Open Fund Project (No.2022-JKYL-004); 2020 Anhui Provincial Department of Education quality project first-class teaching materials (2020yljc031).

References

1. Zhou, R., Zhao, A.: Green Technology and Humanistic Concepts in Traditional Chinese Architecture. China Architecture & Building Press, Beijing (2017)
2. Wang, W., Pan, L.: Research on the green construction wisdom of traditional buildings in southern Anhui province. J. Xi'an Univ. Architect. Technol. (Soc. Sci. Ed.) **40**(05), 23–31 (2021)
3. Wang, W., Liu, Y.: Analysis on the spatial distribution characteristics and influencing factors of traditional villages in mount Huangshan city. J. Xi'an Univ. Architect. Technol. (Soc. Sci. Ed.) **40**(02), 22–31 (2021)
4. Wang, W., Zhu, M.: Green construction technology of traditional buildings in the north of Jianghuai river based on climate adaptability. Industr. Build. **51**(04), 40–45+98 (2021)
5. GB 50352–2019: Unified Standard for Design of Civil Buildings. China Construction Industry Press, Beijing (2019)
6. Jin, N., Shen, X.: The main types and types of traditional architecture in the northern Anhui region. Industr. Build. **42**(05), 161–164 (2012)
7. Cao, W., Zheng, W.: The spatial pattern and morphological characteristics of traditional Chinese residential buildings under regional background: taking Huizhou traditional residential buildings as an example. Huazhong Architect. **39**(11), 116–119 (2021)
8. Geng, H., Zhao, D., Zhu, X.: Research on climate adaptability construction strategy of "Mulan stone masonry" residential buildings in northeast Hubei province: taking Luojiagang village, Huangpi district, Wuhan city as an example. Huazhong Architect. **38**(05), 130–134 (2020)

Simulation of Different Ventilation Methods on Indoor Air Quality and Thermal Comfort in College Classroom

Zhiqiang Kang, Ning Yin(✉), Baorui Hao, and Yunyi Wang

Shenyang Jianzhu University, Shenyang 110000, Liaoning, China
1343441760@qq.com

Abstract. Indoor air quality in university classrooms is closely related to human thermal comfort, and proper air distribution can dilute indoor air pollutants in time and create a healthy and comfortable indoor environment. This paper selects the classroom of a university in Shenyang as the research object, through numerical simulation of the ventilation modes of up-flow and up-return, up-flow and down-return, and side-flow and down-return, the indoor temperature field, velocity field, CO_2 concentration field and PMV-PPD index were obtained. The results show that the most ideal temperature field, velocity field, CO_2 concentration field and thermal comfort can be obtained by using the side-flow and down-return air supply mode in the classroom, the temperature in the classroom is in the range of 16–24 °C, the wind speed is low, and the air distribution is uniform in the area where people are concentrated; The pollutant removal efficiency is high, and CO_2 is taken away from the release source position with the air flow, which reduces the CO_2 exposure intensity of the breathing microenvironment; Students' thermal feeling is in a slightly warm state, and the percentage of dissatisfaction with the thermal environment is the lowest.

Keywords: University Classroom · Airflow Organization · Thermal Comfort · CO_2 Concentration

1 Introduction

Research into human thermal comfort and indoor air quality is often carried out in isolation, but there is now a greater willingness to create a comfortable indoor environment while improving indoor air quality. As the main place of teaching activities, college classrooms have the characteristics of concentrated personnel and closed space. The lack of fresh air in the classroom will affect students' physical and mental health and reduce their learning efficiency [1, 2]. Currently, classrooms in colleges and universities are usually dominated by natural ventilation, which saves energy, but in winter, doors and windows are mostly kept closed, because natural ventilation in winter is easy to produce serious cold airflow and it is difficult to ensure indoor thermal comfort [3, 4].

Therefore, this paper uses Fluent software to simulate the temperature field, velocity field, CO_2 concentration field and PMV-PPD index in classrooms under different air

W. Guo et al. (Eds.): GBCESC 2023, LNCE 328, pp. 703–710, 2024.
https://doi.org/10.1007/978-981-99-9947-7_72

supply modes. By analysing the results of the numerical simulation, it provides a reference for the selection of air supply modes when designing fresh air systems to improve indoor air quality and thermal comfort.

2 Setting up the Model, Boundary Conditions and Working Conditions

2.1 Mathematical Model

According to the characteristics of the classroom air flow and the actual use, the model of the proposed physical problems is simplified on the basis of the following assumptions: Indoor air is an incompressible gas; The influence of gravity is taken into account; the thermal radiation from windows and decorative structures in the room is not taken into account; The Boussinesq hypothesis is followed [5]; The flow is steady state turbulent.

The control equations describing the flow, heat transfer of indoor air and the diffusion and mass transfer of pollutants include continuity equation, momentum equation, energy equation, gas composition equation and so on [6], the unified writing can be as follows:

$$\frac{\partial(\rho\varphi)}{\partial t} + \mathrm{div}(\rho v \varphi) = \mathrm{div}(\Gamma \mathrm{grad}\varphi) + S \quad (1)$$

In the formula: φ is a general variable that can represent any solution variable; υ is a velocity in the three-dimensional direction, m/s; Γ is the diffusion coefficient, m^2/s; S is the source term.

Under the above assumptions, the RNG k-ε turbulence model in Fluent is selected for simulation, which can better solve the near wall flow problem using the wall function method [7].

2.2 Physical Model

This paper takes a classroom in a university in Shenyang as a research object. The design features of this classroom are that there is no window in the inner wall, central air conditioning is used for heating in winter, and there is no fresh air in this system. The physical model of the classroom shown in Fig. 1 is created using SpaceClaim software.

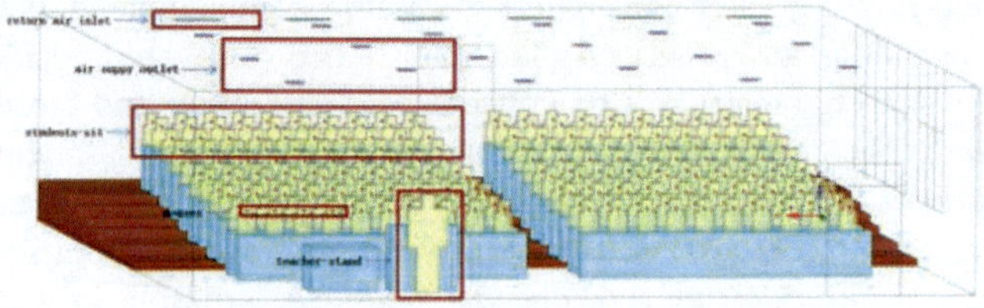

Fig. 1. Classroom physical model diagram.

2.3 Setting Up Working Conditions

In this paper, the univariate analysis method is adopted for the selected classroom, and the air distribution shape is changed only by changing the positions of the supply and return air outlets without changing the sizes and shapes of the original supply and return air outlets in the classroom, the indoor temperature field, velocity field, CO_2 concentration field and PMV-PPD are calculated respectively, and the design scheme is analysed and selected. See Table 1 for specific working conditions.

Table 1. Working condition setting table.

Working condition	Air supply mode [8]	Physical model diagram of ventilation scheme
Condition 1	Up-flow and up-return	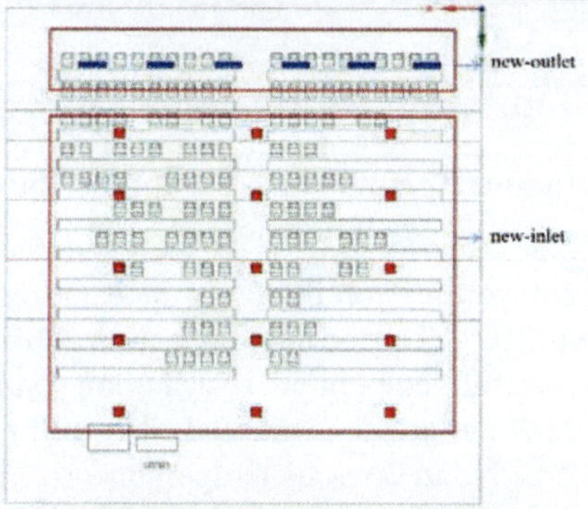
Condition 2	Up-flow and down-return	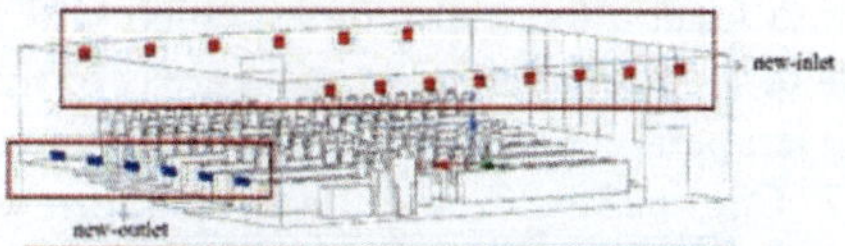
Condition 3	Side-flow and down-return	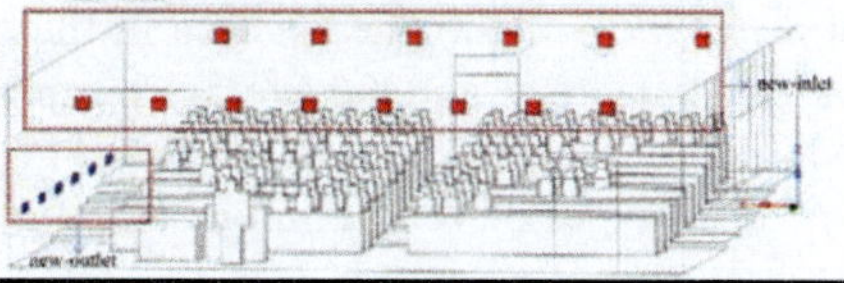

2.4 Setting Up Boundary Conditions

To simplify the problem, this paper only considers the influence of CO_2 generated by breathing in the classroom on indoor air quality [9], and the specific boundary conditions are shown in Table 2.

The hexahedral grid is used to encrypt the air supply outlet, air return outlet and human mouth. Through grid independence and grid quality inspection, the number of grids in the three working conditions is 2759600, 2919071 and 3099235 respectively, and the minimum orthogonal quality of the three working conditions is 0.41, 0.39 and 0.42 respectively, which is greater than 0.10, which meets the requirements of numerical simulation calculation.

Table 2. Boundary condition setting table.

Name	Boundary type	Parameter setting
Old and new air supply outlet	Velocity-inlet	0.75 m/s-20 °C
Return air outlet	Pressure-outlet	Pressure is ambient pressure
Students and teacher	Wall (heat flux)	45 w/m^2, 70 w/m^2; The volume fraction of exhaled CO_2 is 4.5% [10]
CO_2	Velocity-inlet	0.107 m/s-34 °C
Floor and wall	Wall (adiabatic)	-
Indoor initial value	-	The CO_2 concentration is 0.04% and the temperature is 18 °C

3 Simulation Results and Analysis

3.1 Indoor Temperature Field Distribution

As can be seen from Fig. 2, the area with the highest indoor temperature field under three different working conditions all appear around the human body, which can be maintained at about 30 °C, but the range of action is limited, which cannot have a significant impact on the overall temperature distribution of the classroom. In the cross section of the area where people are concentrated, due to the influence of human body heat dissipation, there will be an obvious temperature rise around each human body and the overhead area to form a thermal plume, which will affect the overall temperature stratification. Heat dissipation is good in most areas of the classroom in working condition 3, and the high temperature area is small. The average temperatures of the three planes are: 23.6 °C, 23.7 °C and 23.9 °C, and the temperature values are all in the range of 16–24 °C; The heat dissipation effect of each area in the classroom under working condition 2 is the most unsatisfactory, and the average temperatures of the three planes are all outside the upper limit of thermal comfort of 24 °C [11]. The degree of temperature comfort formed by the three types of ventilation is: side-flow and down-return > up-flow and up-return > up-flow and down-return.

3.2 Indoor Velocity Field Distribution

The distribution of velocity field and temperature field is closely related, so the distribution of flow field in the classroom is observed through the same three planes, as shown in Fig. 3. From the figure, we can clearly observe the indoor air flow on every plane. Part of the air flowing into the classroom from the air supply outlet flows through the crowded area and part of it flows into the room close to the wall, and finally flows out from the air supply outlet. The airflow moves around the human body, which is one of the reasons why the above heat accumulation cannot be effectively diffused. When the air encounters the surrounding hotter room occupants, it creates a more pronounced upward flow trend, which is then typical of a thermal plume. The uniformity of air distribution created by the three ventilation modes is: side-flow and down-return > up-flow and down-return > up-flow and up-return.

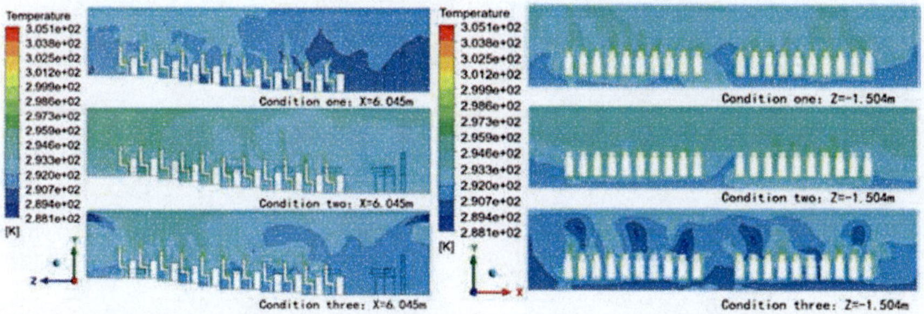

Fig. 2. Comparison of temperature fields in typical planes (X = 6.045 m and Z = −1.504 m) with three ventilation modes.

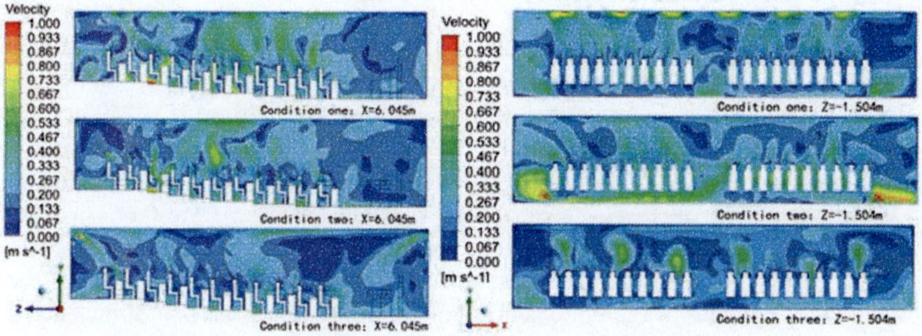

Fig. 3. Comparison of velocity fields in typical planes (X = 6.045 m and Z = −1.504 m) with three ventilation modes.

3.3 Distribution of Indoor CO_2 Concentration Field

After analysing the velocity and temperature distribution in the classroom, because the change of CO_2 concentration may be affected by them, the same analysis planes as above are taken to study the distribution law of CO_2 concentration, and the results are shown in Fig. 4, the concentration of CO_2 in the graph is in molar concentration (mole), 1 mol $= 10^6$ ppm. Observing the cloud image, it can be seen that the concentration distribution in the two planes of working condition 2 is obviously higher than that of working condition 1 and 3, the concentration in the breathing position exceeds the standard, and there is a blind area above the classroom where pollutants are difficult to be eliminated, which is not conducive to the diffusion of CO_2; Compared with the other two working conditions, the first working condition shows good uniformity at the same time under the premise of low CO_2 concentration, and the concentration is high only in the breathing position; In the third case, the overall concentration distribution is low, and the pollutant removal efficiency is high. In the vertical plane, a backflow is formed in the room and CO_2 is removed from the release source with the airflow, reducing the CO_2 exposure intensity of the breathing microenvironment. The advantages and disadvantages of the three ventilation methods for CO_2 removal are as follows: side-flow and down-return > up-flow and up-return > up-flow and down-return.

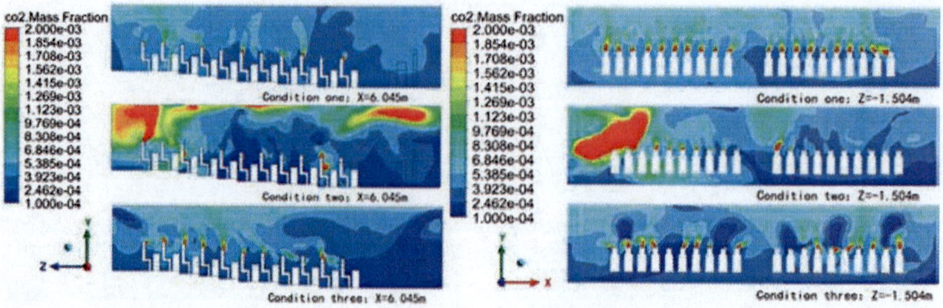

Fig. 4. Comparison of CO_2 concentration fields in typical planes (X = 6.045 m and Z = −1.504 m) with three ventilation modes.

3.4 Analysis of Thermal Comfort of Indoor People

PMV is an evaluation index of human thermal response, and its magnitude is influenced by several effects such as ambient temperature and wind speed. As shown in the Fig. 5, the average PMV of the two planes in the first working condition is 0.459 and 0.451 respectively, and the values are between −0.5 and +0.5, which meets the requirements of thermal comfort in ISO7730 standard [12], and the thermal sensation of the personnel area is moderate; The average values of the two planes in case 2 are 0.661 and 0.861 respectively, and the average values of the two planes in case 3 are 0.634 and 0.828 respectively, the typical plane is warmer under these two air supply modes. The advantages and disadvantages of the three ventilation modes for PMV are as follows: up-flow and up-return > side-flow and down-return > up-flow and down-return.

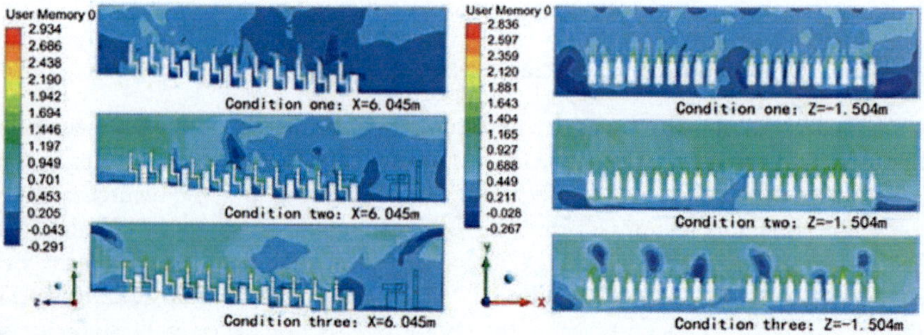

Fig. 5. Comparison of PMV in typical planes (X = 6.045 m and Z = −1.504 m) with three ventilation modes.

PPD is the percentage of people's dissatisfaction with the thermal environment reaction, the standard of PPD is between 5% and 10% [12]. As shown in Fig. 6, the average values of the two levels in case 1 are 10.67% and 12.89%, which exceeds the standard of 10%. This calculation result is contrary to that of PMV, which shows that there may be cases where the percentage of predicted dissatisfaction exceeds the standard when the thermal comfort meets the standard. The average values of the two levels in case 2 are 17.32% and 24.38%, well above the standard of 10%, which is the same as the PMV calculation. In case 3, the overall performance of the PPD value is good and the values of typical positions are within the standard range. The advantages and disadvantages of the three ventilation methods for PPD are as follows: side-flow and down-return > up-flow and up-return > up-flow and down-return.

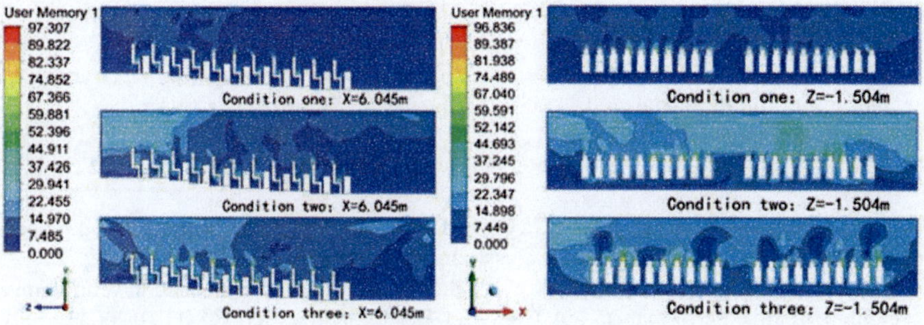

Fig. 6. Comparison of PPD in typical planes (X = 6.045 m and Z = −1.504 m) with three ventilation modes.

4 Conclusion

In this paper, under the condition that the building conditions and the sizes of air supply outlet and air return outlet are unchanged, the simulation research on improving indoor air quality and thermal comfort by different air supply outlet and air return outlet position design schemes is studied, and the following conclusions are obtained.

(1) For the temperature field and velocity field, the best temperature field and velocity field can be obtained by the side-flow and down-flow air supply mode, which meets people's needs for uniformity of room temperature distribution and air flow, the average temperatures of three sections in the classroom are: 23.69 °C, 23.79 °C and 23.96 °C, all within the upper limit of thermal comfort of 24 °C; The air distribution of the up-flow and up-return and up-flow and down-return air supply modes is uneven, and the heat dissipation effect in all areas of the room is not ideal, and the temperature is high.

(2) Overall, the up-flow and up-return air supply mode is the worst for CO_2 dilution, and there is a dead angle at the rear of the classroom. The average concentration of CO_2 in the classroom is not much different between the side-flow and down-return and up-flow and up-return air-supply modes, but for the area with relatively dense students in the classroom, the effect of removing indoor pollutants is slightly better with the side-flow and down-return air supply mode than with up-flow and up-return.

(3) Only for the thermal comfort of the students, the up-flow and up-return supply mode meets the PMV value in the classroom is between −0.5 and +0.5, which meets the standard requirements for thermal comfort, and the thermal sensation of the staff is moderate. However, the side-flow and down-return supply mode meets both indoor air quality and thermal comfort requirements, and the percentage of students' dissatisfaction with the thermal environment is the lowest.

Acknowledgement. This paper is supported by the scientific research project of Liaoning Provincial Department of Education: Optimal control of indoor local thermal comfort environment based on indoor personnel positioning technology (No. Z2219121).

References

1. Jing, J., Dengjia, W., Yanfeng, L., et al.: A study on pupils' learning performance and thermal comfort of primary schools in China. Build. Environ. **134**, 102–113 (2018)
2. Xiaojing, Z., Caixia, Z., Tianyang, Z., et al.: Association of indoor temperature and air quality in classrooms based on field and intervention measurements. Build. Environ. **229**, 109925 (2023)
3. Luca, S., Angelamaria, M., Laura, C., et al.: The effect of ventilation strategies on indoor air quality and energy consumptions in classrooms. Buildings **9**(5), 110 (2019)
4. Widiastuti, R., Hasan, M., Bramiana, C., et al.: CFD simulation on the natural ventilation and building thermal performance. IOP Conf. Ser. Earth Environ. Sci. **448**(1), 012004 (2020)
5. Stabile, L., Buonanno, G., Frattolillo, A., et al.: The effect of the ventilation retrofit in a school on CO_2, airborne particles, and energy consumptions. Build. Environ. **156**, 1–11 (2019)
6. Dongyoon, P., Seongju, C.: Effects of combined central air conditioning diffusers and window-integrated ventilation system on indoor air quality and thermal comfort in an office. Sustain. Cities Soc. **61**, 102292 (2020)
7. Xingwang, Z., Qingyan, C.: Optimal design of an indoor environment using an adjoint RNG k-ε turbulence model. E3S Web Conf. **111**(5), 04037 (2019)
8. Elang, P.W., Ardiyansyah, S.Y.: Air quality improvement in COVID-19 pandemic: numerical study of ventilation system in a classroom. CFD Lett. **15**(1), 26–38 (2023)
9. Youngtae, C., Jun-shup, S., Jinhyeon, P., et al.: Inadequacy of air purifier for indoor air quality improvement in classrooms without external ventilation. Build. Environ. **207**, 108450 (2022)
10. Yingying, F., Runping, N., Kuichao, Y., et al.: Numerical simulation of air distribution and CO_2 concentration in different air supply locations in winter classroom. IOP Conf. Ser. Earth Environ. Sci. **546**(3), 032022 (2022)
11. Carolina, B., Renata, D.V., Veridiana, S., et al.: Thermal preference and comfort assessment in air-conditioned and naturally-ventilated university classrooms under hot and humid conditions in Brazil. Energy Build. **211**, 109783 (2020)
12. Ghogare, A.G., Shobha, L.S., Tikendra, N.V., et al.: Numerical simulation for energy consumption and thermal comfort in a naturally ventilated indoor environment under different orientations of inlet diffuser. Build. Environ. **217**, 109071 (2022)

A Study on the Scenic Beauty of Huangguoshu Scenic Area Based on SD and SBE Method

Xinren Zhang and Zongsheng Huang(✉)

College of Architecture and Urban Planning, Guizhou University, Guiyang 550025, China
hzsxjh@126.com

Abstract. To promote the protection and development of tourism resources in waterfall-type scenic spots, scenic beauty estimation procedures (SBE) and semantic difference (SD) methods were used to study the scenic beauty of Huangguoshu. The results showed that the landscape beauty of the Huangguoshu Scenic Area was relatively good, and the overall beauty of Huangguoshu Waterfall was the highest. Among different scenic spots, the SBE value of the Huangguoshu Waterfall Scenic Area had the maximum value, and it was significantly different from that of the Tianxing Bridge Scenic Area and Doupotang Scenic Area. The SBE value of different landscape types, from high to low, was water system, overall landscape, mountain, vegetation, bridge, and building, and the beauty of the natural landscape was higher than that of the artificial landscape. The landscape of waterfalls showed a decreasing trend from large to small. The main factors affecting the beauty of the Huangguoshu Scenic Area were landscape identifiability, landscape coordination, naturalness of mountains, surrounding plant communities, majesty and beauty of water systems, relationship between mountains and rivers, and plant space enclosure. The scenic spots in the Huangguoshu Scenic Area that were lower than the lower limit of the scenic beauty threshold (13.73) were all artificial landscapes. The Huangguoshu Scenic Area was characterised by the beauty of harmonious symbiosis of green ecology, majestic beauty, and natural beauty of karst geology. Beauty threshold monitoring can provide technical support for the dynamic protection and optimisation of the scenic area. The results of this study provide a theoretical basis for the dynamic management and regulation of scenic spots and are of great significance for maintaining the aesthetic value of scenic spots and planning and building scenic spots with regional characteristics.

Keywords: Landscape beauty · scenic beauty estimation procedure(SBE) · aesthetic characteristics · Huangguoshu Waterfall · scenic area

1 Introduction

1.1 A Subsection Sample

Tourism resources, as the material basis for the development of tourism, have always been a hot spot in tourism research, but most of the previous studies have focused on the comprehensive evaluation of the scale, quality, grade, development prospects and

W. Guo et al. (Eds.): GBCESC 2023, LNCE 328, pp. 711–729, 2024.
https://doi.org/10.1007/978-981-99-9947-7_73

development conditions of tourism resources from the perspective of economic resource utilisation [1]. Since the aesthetic process includes the direct effects of aesthetic subject, aesthetic object, and aesthetic subject and object, and there is no clear and recognised definition of aesthetic value, the aesthetic evaluation of tourism resources cannot be concrete, especially because of the lack of aesthetic value of tourism resources, leisure and sports value, research on ecological protection value and historical and cultural value. When analysing tourism resources, people's aesthetics are often ignored, especially in terms of the value of pleasing form (beauty of form), value of leisure and value of refreshing spirit formed in the aesthetics of Chinese mountains and rivers [2]. Adopting a resource-intensive tourism economic development model in the construction of scenic spots will inevitably damage the protection and development of the aesthetic value of tourism resources [3], and development is important. At present, research on scenic spots mainly focuses on the causes of formation, overall planning, and capacity research, while there are few studies on the quantitative and threshold value of landscape aesthetics. Landscape aesthetics is a broad and interdisciplinary concept that is closely related to nature, ecology, the environment and other factors [4]. At present, landscape aesthetics studies on beauty have mainly focused on forest landscapes [5], traditional village landscapes [6], wetland landscapes [7], grassland landscapes [8], ancient town landscapes [9], urban waterfront landscapes [10], expressway landscapes [11] and other aspects, but the quantitative analysis of scenic spots, especially waterfall landscape aesthetic elements, is relatively lacking. Based on this, this paper takes the Huangguoshu Waterfall, a typical karst landform in Guizhou Province, as the research object and uses the landscape aesthetics evaluation method and the semantic difference method to evaluate its beauty to provide a theory for the protection and management of scenic tourism resources, sustainable development and utilisation, and landscape quality improvement.

2 Materials and Methods

2.1 Study Area and Sample Plot Selection

The Huangguoshu Scenic Area is located at the junction of Zhenning Buyi and Miao Autonomous County and Guanling Buyi and Miao Autonomous County, Anshun city, southwestern Guizhou Province. The geographic coordinates are 105°35′–105°42′E, 25°52′–26°02′N. It was approved as a national scenic spot by the State Council in 1982, and it has an area of 163 km2. This area belongs to the subtropical humid monsoon climate, with severe cold in winters and hot summers, with an average annual temperature of 16 °C, an annual rainfall of 130 mm, a relative humidity of 80%, and an average altitude of 900 m. It belongs to the Baishui River section of the lower reaches of the Kebu River, a tributary of the Dabang River in the main stream of the Xijiang River in the Pearl River system. The terrain is high in the north and low in the south [12]. With the Great Falls as the core, there are waterfall groups with different scenery, which belong to three scenic spots: Great Falls, Tianxing Bridge and Doupotang. The Great Falls is as famous as the three major waterfalls in the world, Niagara Falls, Victoria Falls and Iguazu Falls. It is the largest waterfall in Asia, with a height of 77.8 m and a width of 101 m. It is a karst erosion waterfall. The waterfall group with it as the core is an ideal place to study the beauty of the waterfall landscape. There is a stone village in the Huangguoshu Scenic

Area, and it is inhabited by Buyi residents. The Buyi people have a rich culture. Most of the buildings are stone houses, and the residents live in harmony with nature. In 2022, it will be listed on the IUCN Green List.

The selected samples followed the principles of typicality, randomness and completeness [13]. To ensure that the research results were more scientific and representative, we chose sunny days to shoot, did not use flashlights when shooting, and used the same camera to ensure that the shooting angles were consistent [6]. After shooting, the photos that could not represent the plot were deleted, and the photos that best represented the scenic spots were selected from the remaining photos. Finally, 23 photos were selected as the evaluation objects, and these 23 photos represented 23 scenic spots (Fig. 1).

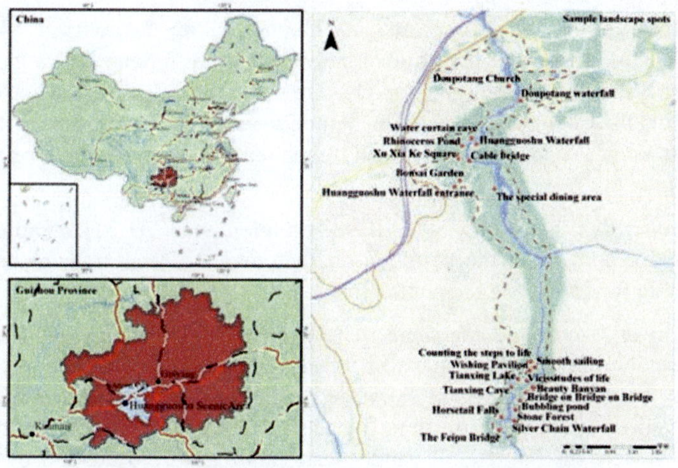

Fig. 1. Geographical location of the selected spots

2.2 Research Methods

Scenic Beauty Estimation Method (SBE). At present, the beauty evaluation method has been widely recognized [14]. At present, this method has been widely used in various landscape evaluations and has been recognised as an evaluation method with high reliability and sensitivity [15]. The main steps are as follows:

Selection of SBE Judges. The evaluators selected in this study included professionals and non-professionals [16].

Evaluation Method. The 23 photos were randomly arranged in the questionnaire survey, and each judge was asked to rate each photo. The larger the value is, the more beautiful the landscape [17].

Scenic Beauty Value Calculation. The calculation method of the average Z value was calculated according to the method of DANIEL [18], and the calculation formula is as follows:

$$MZ_i = \frac{1}{m-1}\sum_{m-1}^{m}(1 - cp_{ik}) \tag{1}$$

$$SBE_i = (MZ_i - BMMZ) \times 100 \tag{2}$$

where MZi is the average z value of the tested sample i; cpik is the frequency at which the score value of the tested sample i is k or greater than k given by the judge; m is the grade number of the evaluation; SBE_i is the SBE of the tested sample i; and BMMZ is the mean Z-score of the baseline group.

Decomposition and Quantification of Elements of Landscape Judgement by the Semantic Difference Method. The semantic difference (SD) method, also known as the sensory recording method [18]. In the study of landscape aesthetics, it is judged by using the Likert five-level scale method to judge and assign values to images to determine the judges' understanding of landscape features based on their psychological feelings [19].

Decomposition of landscape elements. Based on references [20, 21], 45 adjective pairs were screened according to the aesthetic characteristics of waterfalls to construct an evaluation table for landscape elements (Table 1).

Quantification of Landscape Elements. The qualitative evaluation of landscape elements was carried out through a questionnaire survey, and five evaluation scales were selected [22]. Played through PowerPoint, in order for judges to better understand the landscape features, the evaluation time for each photo was 90s [23]; additionally, 60 questionnaires were distributed, 54 valid questionnaires were used, and the reliability of the questionnaire was tested by the method of detecting Cronbach's α value. The test result showed that the reliability value of the overall scale was 0.804, which was greater than 0.8, indicating that the scale in this questionnaire had good reliability for analysis purposes.

Landscape Beauty Threshold. The threshold value, also known as the critical value, is used to define the critical point that can maintain a certain stable state under disturbance. According to this value, the environment can be artificially guided and repaired, effectively dredging the material and energy cycle of the ecosystem [24]. By testing the distribution of the beauty value of the Huangguoshu Scenic Area, the results showed that the SBE value was normally distributed, and combined with the principle that the threshold value of landscape beauty should be large [25], this study used the normal distribution method to evaluate landscape beauty. This method has the advantages of high precision, wide application and easy operation.

Table 1. Decomposition of landscape elements.

Landscape type	Landscape elements	Factor decomposition
Overall landscape	Landscape layering	Blurred layers - clear layers
	Landscape recognisability	Landscape common - Landscape exceptional
	Landscape orderliness	Landscape clutter - Landscape order
	Landscape naturalness	Low naturalness - High naturalness
	Landscape openness	Closed view - Open view
	Landscape coherence	Incoordination - Coordination
	Diverse topography	Flat with no variation - High and low variation
	Topographical peculiarities	Common - Peculiar
	Landscape seclusion	Clutter - Seclusion
	Landscape artistic beauty	Weak sense of ambience - Strong sense of ambience
	landscape attraction	Low attraction - High attraction
Mountain landscape	Mountain form	Worse - Better
	Mountain nature	Low naturalness - High naturalness
	Mountain flora	Monotonous - Rich
	Mountain function	Lack of function - Leisure function
	Mountain novelty	Plain - Novelty
	The openness of mountain vision	Narrow vision - Broad vision
Water landscape	Water quality	Worse - Better
	Waterfront morphology	Not graceful - Graceful
	Landscape relations	Poor landscape relationship - Good landscape relationship
	Natural extent around water body	Artificial Rules - Original Nature
	Water reflection beauty	Lack of aesthetics in reflections - High aesthetics in reflections
	Water attraction	Low attraction - High attraction
	The majesty of the water system	Trickle - Majestic grandeur

(*continued*)

Table 1. (*continued*)

Landscape type	Landscape elements	Factor decomposition
	The extent of water system seclusion	Plain - Deep and beautiful
	Water system novelty	Normal - Novelty
Plant landscape	Plant growth	Withered - Luxuriant
	Plant space hierarchy	Level fuzzy and single -Level clear and rich in change
	Plant colourfulness	Monotonous colours - Rich colours
	Plant coverage	Low vegetation coverage - High vegetation coverage
	Plant community morphology	Closed - Open
Architecture landscape	Architectural appearance texture	Relatively simple - Relatively rich
	Architectural colour features	Relatively monotonous - Relatively rich
	Architectural appearance feeling	Closed and thick - Open and transparent
	Architectural quality	Worse - Better
	Architectural and the surrounding environment coordination	Incoordination - Coordination
	Architectural style coordination	No ethnic and regional characteristics - Full of ethnic and regional characteristics
	Architectural novelty	Normal - Novelty
Bridge landscape	Bridge paving form	Single - Diverse
	Bridge wild interest degree	Plain - Full of wild fun
	Bridge vision range	Narrow - Wide
	Bridge quality	Worse - Better
	Bridges in harmony with their surroundings	Incoordination - Coordination
	Bridge novelty	Normal - Novelty
	Bridge nostalgia	Low - High

2.3 Data Processing

Data processing was carried out by Excel and SPSS 19.0 software. The average value was calculated to obtain the quantitative values of each evaluation element, and then the data were statistically analysed using variance analysis ($\alpha = 0.05$), positive state distribution analysis and correlation analysis.

3 Results

3.1 Analysis of Landscape Beauty

Overall Analysis of the Landscape Beauty of the Huangguoshu Scenic Area. As seen in Table 2 and Fig. 2, the overall landscape of Huangguoshu Falls, which has the highest value of scenic beauty, is composed of a water system, mountains and vegetation. The water quality of Huangguoshu Falls is clean, and the water is vast and in good harmony with the mountains, reflecting its characteristics as a typical waterfall with erosion cracks in the karst landscape. The SBE value was 46.02. The lowest SBE value was obtained for the wishing pavilion in the Tianxing Bridge Scenic Area, with an SBE value of -3.9. The landscape was a resting node with a single view, and the interior of the pavilion has a coin-like wishing pool, and the view is rather cluttered.

Analysis of the Scenic Beauty of Different Scenic Spots in the Huangguoshu Scenic Spot. Figure 3 shows that more scenic spots are located in the Steepotang Scenic Area. This means that the scenery beauty in the Doupotang Scenic Area is better on average, and the overall landscape, mountain landscape and water system landscape all had their own characteristics, especially in the combination of mountain and water system landscapes, highlighting the characteristics of karst landforms and giving them a vivid image, thus leaving a good impression on tourists. The Huangguoshu Falls scenic area had the highest scenic beauty value, indicating that it was the most prestigious tourist attraction in the scenic area. Huangguoshu Falls, where the natural and man-made environment coexist in harmony and the magnificence of the falls is most remarkable.

Comparison of SBE Values for Different Landscape Types. Figure 4 shows that the SBE values of each landscape type were, in descending order, water system, overall landscape, mountain, vegetation, bridge and architecture. Through one-way ANOVA, the overall landscape was significantly different from the architectural landscape, the overall landscape was different from the water system landscape but not significantly, and the overall landscape was not significantly different from the mountain landscape, vegetation landscape and bridge. There was no significant difference between mountains, vegetation and bridges, and there was a significant difference between water systems and architecture but no significant difference between mountains, vegetation and bridges. This result indicates that the Huangguoshu Scenic Area has a natural landscape as its main body and a human landscape as its auxiliary, while the water system landscape is its core landscape, reflecting the karst natural regional landscape characteristics of the Huangguoshu Scenic Area, which is famous for its waterfalls. Therefore, in daily scenic area proposal management, the protection of the natural ecological environment should be the main focus.

Analysis of the Scenic Beauty of Different Sizes of Waterfalls in the Huangguoshu Scenic Area. According to the majestic index, secluded index and peculiar index calculated by Jin Yuanhuan [26] for the Huangguoshu Waterfall Group. Table 3 shows that as the waterfall decreased, the SBE value also decreased. As the core attraction within the Huangguoshu Scenic Area, the waterfall landscape is the main tourist destination for visitors.

Table 2. Scenic beauty value of each landscape.

Landscape sample point number	SBE value	Name of attraction	Landscape type
11	−3.9	Wishing Pavilion	Architecture
6	10.55	Xu Xia Ke Square	Architecture
18	12.73	The Feipu Bridge	Bridge
23	13.58	The special dining area	Architecture
10	26.60	Smooth sailing	Overall
1	36.55	Huangguoshu Waterfall entrance	Mountain
12	40.45	Tianxing Lake	Water
9	44.66	Counting the steps to life	Water
17	46.02	Bridge on Bridge on Bridge	Bridge
16	48.84	Vicissitudes of life	Vegetation
20	53.37	Beauty Banyan	Vegetation
7	56.61	Cable bridge	Overall
19	58.23	Horsetail Falls	Water
22	61.36	Doupotang Church	Architecture
14	72.53	Bubbling pond	Water
13	76.28	Tianxing Cave	Mountain
2	84.94	Bonsai Garden	Vegetation
21	97.99	Stone Forest	Water
8	126.45	Doupotang waterfall	Water
3	136.53	Water curtain cave	Mountain
15	140.6	Silver Chain Waterfall	Water
5	141.72	Rhinoceros Pond	Water
4	155.72	Huangguoshu Waterfall	Overall

3.2 Factors Influencing Scenic Beauty

Factors Influencing Overall Scenic Beauty. Figure 5 shows according to the correlation coefficient, from large to small, the selected landscape elements had the following influence on the overall landscape beauty: landscape recognisability, landscape coordination, landscape hierarchy, topographic uniqueness, landscape attractiveness, landscape orderliness, topographic diversity, landscape naturalness, landscape seclusion, landscape artistic beauty, and landscape openness. The key factors influencing the beauty of "water" and "mountain" scenery in the Huangguoshu Scenic Area are attraction uniqueness, coordination, topography and public perception of the area.

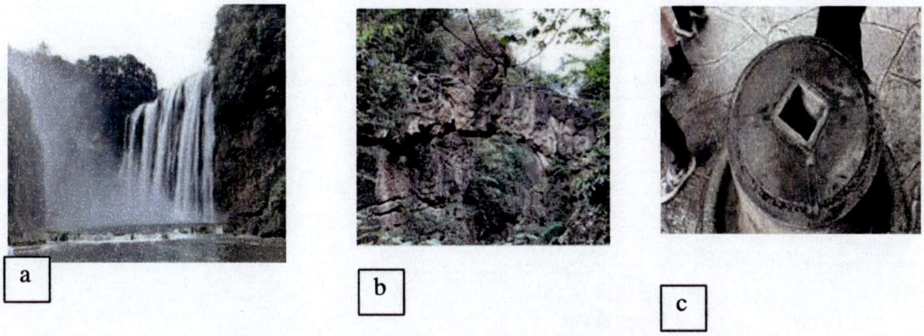

Fig. 2. Judging landscape sample points 4, 17, 11.(a):4;(b):17;(c):11.

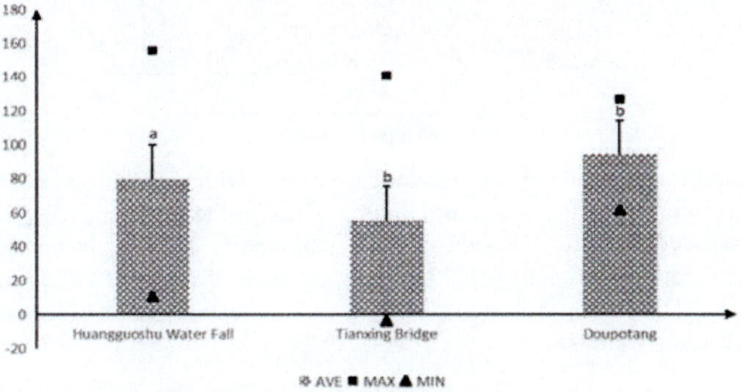

Fig. 3. SBE values for different scenic areas.

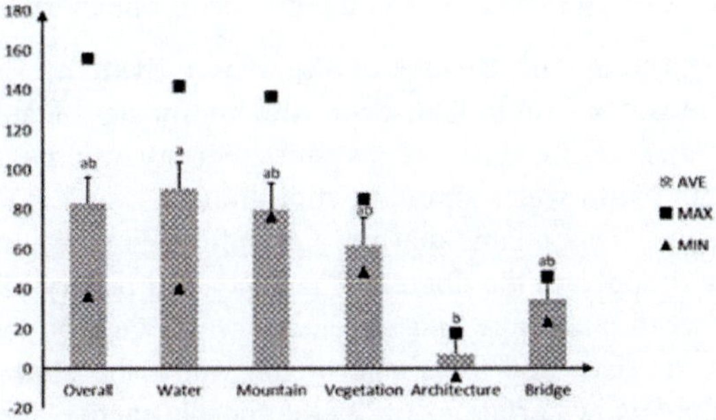

Fig. 4. SBE values for different landscapes.

Factors Influencing the Scenic Beauty of the Mountain Landscape. Figure 6 shows that the mountain landscape SBE values were highly significantly correlated with the degree of mountain naturalness, mountain flora, and mountain view openness and significantly negatively correlated with mountain function; the selected landscape elements had the greatest to lowest influence on mountain landscape scenic beauty, namely, mountain

Table 3. SBE values for different sizes of waterfall landscapes.

Waterfall type	The great waterfall	Middle waterfall	Small waterfall
Average SBE	155.72	93.91	58.23

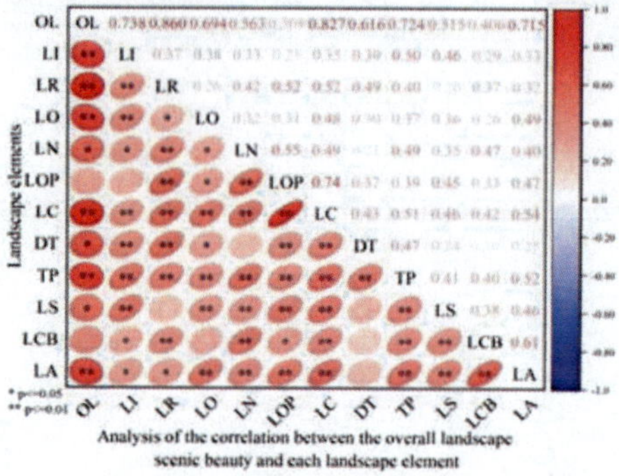

Fig. 5. Correlation analysis of the overall landscape beauty with each landscape element. "OL" is the overall landscape, "LI" is the landscape hierarchy, "LR" is the landscape recognizability, "LO" is the landscape orderliness, "LN" is the landscape naturalness, "LOP" is the landscape openness, "LC" is the landscape coherence, "DT" is the topographic diversity, "TP" is the topographic features, "LS" is the landscape seclusion, "LCB" is the Contextual beauty of the landscape, "LA" is the landscape attractiveness.

flora, mountain view openness, mountain naturalness, mountain function, mountain novelty, and mountain form, indicating that the key influencing elements were the naturalness of the mountain, the surrounding flora, and the degree of openness of the view.

Factors Influencing the Scenic Beauty of the Water System. Figure 7 shows that the SBE value of the water system landscape was highly significantly correlated with water quality, landscape relationship, water system majesty, water system seclusion and significantly correlated with water shoreline morphology, water system attractiveness, water system seclusion. This means that the key influencing factors of the beauty of the water system landscape are the degree of majesty and beauty of the water system, the relationship between landscape and water, the water quality, the attractiveness of the water system to tourists, and the shape of the water shoreline. Therefore, in the Huangguoshu Scenic Area waterfall group, as the most attractive area, we should focus on the protection of the source, and the key to protecting the water system attractions is to build a reasonable relationship between landscape and water.

Factors Influencing the Scenic Beauty of Vegetated Landscapes. Figure 8 shows that vegetation landscape SBE values were highly significantly correlated with plant spatial hierarchy and plant colour richness and significantly correlated with plant cover, plant growth, and plant community morphology. This result shows that the key factors

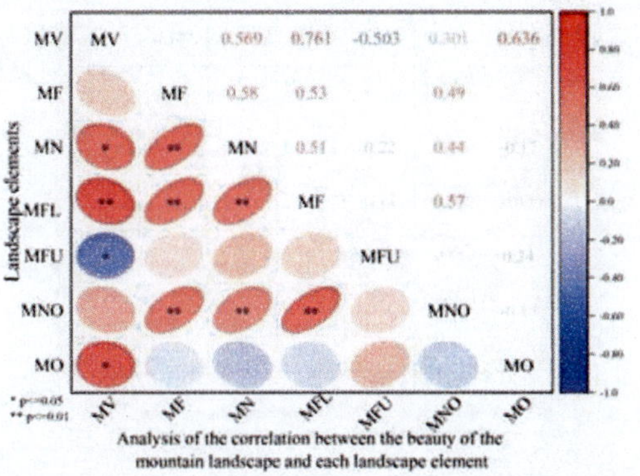

Fig. 6. Correlation analysis between the scenic beauty of the mountain landscape and each landscape element. "MV" is the mountain landscape, "MF" is the Mountain form, "MN" is the mountain naturalness, "MFL" is the mountain flora, "MFU" is the mountain function, "MNO" is the mountain novelty, "MO" is the mountain view openness.

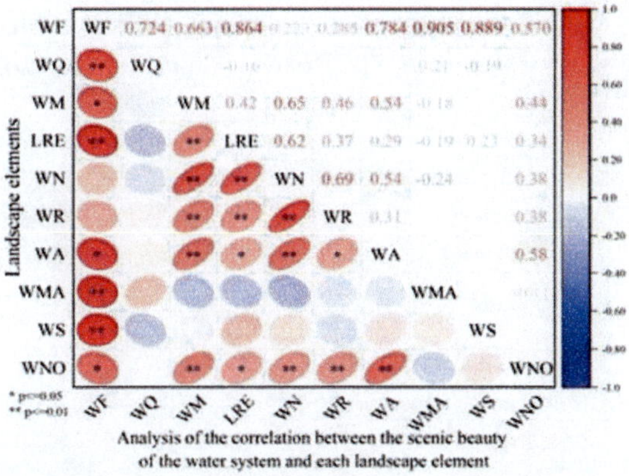

Fig. 7. Correlation analysis of the scenic beauty of the water system with each landscape element. "WF" is the water landscape, "WQ" is the water quality, "WM" is the water shoreline morphology, "LRE" is the landscape relationship, "WN" is the natural environment around water bore, "WR" is the beauty of the reflections on the water, "WA" is the water system attractiveness, "WMA" is the water system majesty, "WS" is the water system seclusion, "WNO" is the extent of water system seclusion.

influencing the scenic beauty of the vegetation landscape are plant spatial enclosure, plant community morphology, degree of dieback and colour richness.

Factors Influencing the Scenic Beauty of the Architectural Landscape. Figure 9 shows that the SBE value of the architectural landscape was highly significantly correlated with the architectural colour characteristics, architectural harmony, architectural appearance and feeling, and architectural harmony with the surrounding environment, indicating that the key influencing elements of architectural landscape scenic beauty are

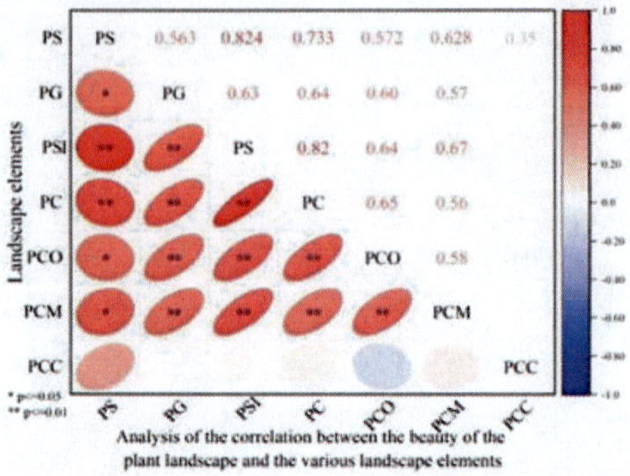

Fig. 8. Correlation analysis of plant landscape scenic beauty with each landscape element. "PS" is the plant landscape, "PG" is the plant growth, "PSI" is the plant spatial hierarchy, "PC" is the plant colour richness, "PCO" is the plant cover, "PCM" is the plant community morphology, "PCC" is the Plant community composition.

architectural colour, architectural harmony, architectural quality and architectural shape. Therefore, when protecting the original pavilions and other buildings in the Huangguoshu Scenic Area, attention should be given to not destroying the original texture, shape and colour of the buildings.

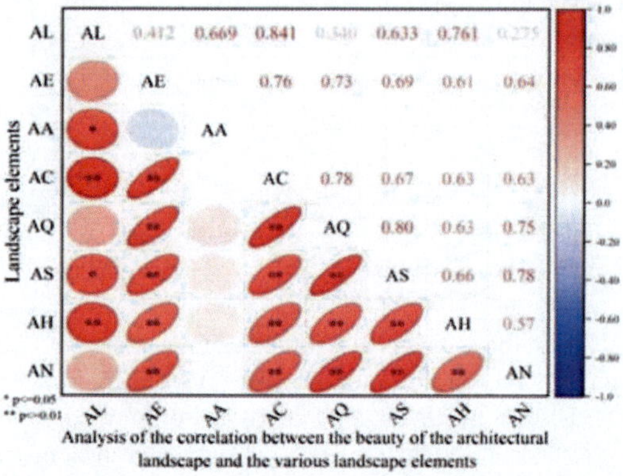

Fig. 9. Correlation analysis of architectural landscape scenic beauty with each landscape element. "AL" is the architectural landscape, "AE" is the Architectural facade texture, "AA" is the architectural appearance and feeling, "AC" is the architectural colour characteristics, "AQ" is the architectural quality, "AS" is the architectural harmony with the surrounding environment, "AH" is the architectural harmony, "AN" is the architectural novelty.

Factors Influencing the Scenic Beauty of the Bridge. Figure 10 shows that the SBE value of the bridge landscape was highly significantly correlated with the bridge paving form, bridge wildness, bridge quality, bridge harmony with the surrounding environment, and bridge novelty, indicating that the combination of landscape relationship, plant community enclosure form, and rich and scenic paving in the scenic area can improve the visitors' experiences regarding the sense of wildness and novelty.

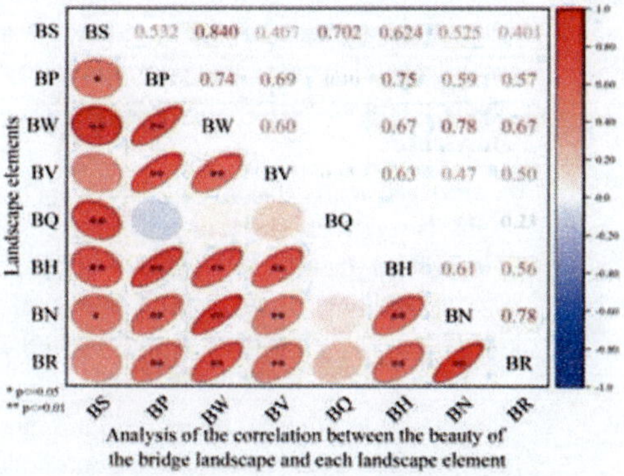

Fig. 10. Correlation analysis of bridge landscape scenic beauty with each landscape element. "BS" is the bridge landscape, "BP" is the bridge paving form, "BW" is the bridge wildness, "BV" is the range of views of the bridge, "BQ" is the bridge quality, "BH" is the bridge harmony with the surrounding environment, "BN" is the bridge novelty, "BR" is the sense of Bridge Retention.

3.3 Landscape Scenic Beauty Threshold Analysis

Table 4 and Fig. 11 show that the lower threshold of scenic beauty of the Huangguoshu landscape was 13.73, in which four scenic beauty values were lower than the threshold, namely, the Wishing Pavilion (−3.9), Xu Xiak Square (10.55), the Feipu Bridge (12.73) and the special dining area (13.60). According to the survey, these four landscapes are all artificial, presenting a single form of architecture, cluttered pedestrian flow, and simple forms of bridge paving. With the exception of these four landscapes, the beauty values of all the other sites were greater than the lower threshold of the landscape, further indicating that the sites in the Huangguoshu Scenic Area had a high aesthetic value overall.

Table 4. Sample sites in the landscape below the threshold.

Samples	Number	Reasons for being below the threshold
Wishing Pavilion	11	Compared with other architectural landscapes, the pavilion here is slightly monotonous in architectural form and colour, the surroundings are slightly cluttered and the landscape has a strong sense of closure

(continued)

Table 4. (*continued*)

Samples	Number	Reasons for being below the threshold
Xu Xia Ke Square	6	Compared to the architectural landscape in other scenic areas, this landscape sample site has a flawed and monotonous architectural appearance, poor continuity as a landscape node, general colour, variation and hierarchy, and a poor impression on visitors
Feipu Bridge	18	Compared with other bridge landscapes, this landscape sample point is relatively thick in colour, with sparse vegetation in the visual field, lacking in layered overlay, and relatively simple paving style and shape, which is generally not strong enough for the visual impact and attractive enough for visitors, lacking in novelty and contextual beauty
Special dining area	23	Compared to other architectural landscapes, this commercial building has a single texture, monotonous colours, poorly permeable surroundings and a dull and intrusive enclosed passageway

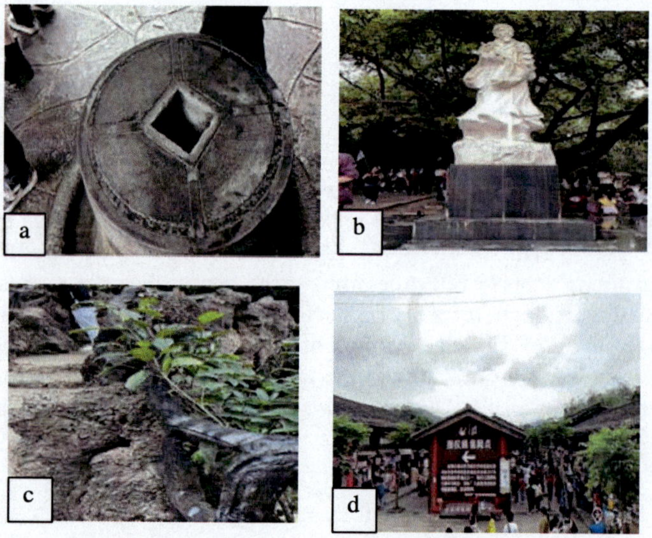

Fig. 11. Map of sample points in the landscape below the threshold. (a) is Wishing Pavilion; (b) is Xu Xia Ke Square; (c) is Feipu Bridge; (d) is Special dining area.

4 Discussion

4.1 Landscape Aesthetic Characteristics of the Huangguoshu Scenic Area

In this study, the overall landscape scenic beauty was influenced by the novelty, hierarchy, spatiality, coordination, naturalness and attractiveness of the landscape and topography, which was consistent with the findings of Liang Yuan [5]. Mountain landscape scenic beauty was influenced by the naturalness, openness and novelty of the landscape, which was consistent with the findings of Zhong Ming et al. [27]. The scenic beauty of the water system was influenced by factors such as the majesty of the water system, its beauty, the quality of the water body, the relationship between the landscape and the water, and the morphology of the water shoreline, which was basically consistent with the results of Jin Yuanhuan [26] and Hu Jiesi et al. [28]. The scenic beauty of the vegetation landscape was influenced by factors such as vegetation communities, spatial levels and colour richness, which was generally consistent with the results of Gui Guohua et al. [29] and Wang Lingling et al. [22], so the results of this study are considered reliable and general.

This study showed that the SBE value of the natural landscape was significantly higher than that of the artificial landscape and that the landscape in the scenic area was influenced by the natural landscape of mountains, water and vegetation, which had the greatest characteristics of green, ecology, harmony and symbiosis; therefore, the beauty of green ecological harmony and symbiosis in the Huangguoshu Scenic Area is one of its aesthetic characteristics. The waterfall brings inspiration while ensuring the harmony of the landscape environment and the openness of the view and reveals the harmony of the natural and man-made landscapes created by a high sense of harmony, which can bring visitors more pleasure and a sense of retention [30]. It also presents the beauty of ecological harmony and symbiosis, which is also the most desired quality of the landscape in waterfall-type scenic spots [31].

This study showed that the water system landscape, known for its major waterfalls, had the highest SBE value and was influenced by the majority of the water system and the relationship between the landscape and water. The unique geochemical processes in the karst region have shaped the unique hydrological landscape and cave landscape of the Huangguoshu Scenic Area, so the most representative aesthetic features in the Huangguoshu Scenic Area are its majestic and spectacular beauty and the natural beauty of the karst geology, which are complementary to each other. The majestic beauty of the Huangguoshu Scenic Area lies in the "inspiring" dimension of the water system, which, as the most attractive area in the entire landscape, forms a stunning waterfall landscape in accordance with the mountain. Around the famous Huangguoshu Falls in the scenic area, you can see the flying flowers and jade, water falling into deep pools, the rushing and churning of the water and a beautiful scene [32].

4.2 Insights into the Construction of Landscape Scenic Beauty Thresholds for Scenic Areas

Thresholds are used to define the threshold at which a certain steady state can be maintained under disturbance and have proven to be important in environmental management and as a basis for landscape protection, allowing for the artificial guidance and restoration of the environment and the effective unblocking of the material and energy cycles of ecosystems. The ecological environment is the basis for the survival, development and sustainable utilization of Huangguoshu Scenic Area. Based on the identification of scenic beauty thresholds and the combination of existing landscape planning and protection methods, visualisation analysis methods and big data, a dynamic monitoring system can be established, which can be applied not only in the landscape protection and optimisation planning of the Huangguoshu Scenic Area but also to other tourist-oriented scenic spots, e.g., the landscape protection and restoration of scenic areas, the protection of landscape and landscape development management, and the control of visitor capacity in scenic areas. Through the identification and measurement of thresholds, the threshold for the restoration of sample landscape spots can be judged, which improves the discovery of reasons for the reduction in the aesthetic value of the landscape and encourages us to take timely restoration and optimisation measures.

4.3 Landscape Conservation and Development Strategies for Scenic Areas Based on Scenic Beauty and its Influencing Elements

At present, there are certain problems in the scenic spot landscapes, such as natural landscapes with few plants, single landscape configurations, thin colour richness, affecting the beauty and preferences of tourists. These problems are focused on nature, economic aspects, cultural aspects and visual experience. Therefore, it is necessary to protect the landscape of the scenic area with regard to the key influencing elements and to classify the problems to take optimisation measures.

Strengthening the Protection of Natural Ecosystems and Establishing Dynamic Threshold Monitoring Systems. This study showed that a tidy environment presented a higher SBE than a cluttered environment. Scenic areas need to be maintained in a neat and orderly environment. This study also showed that the natural landscape, as the most important and attractive area in the scenic area, should be protected as the main ecological environment, with a focus on protecting the landscape vegetation, and the protection of the ecological environment, especially the protection of the quality of the water environment, should be strengthened in the sub-scenic areas where tourists are concentrated. The protection of the plant landscape lies in strengthening the control of plants in the scenic area and prohibiting random trampling.

Harmonisation of the Artificial Landscape with the Natural Environment. Taking the subject of this study as an example, for the overall landscape, it is necessary to pay attention to the protection of the karst topography, prevent soil erosion, respect the original topography and natural growth conditions, and not change or destroy the original topography, texture and plant communities. The landscape of the buildings, bridges and infrastructure in the scenic area should mainly reflect the regional Buyi characteristics of the landscape, continue the traditional architectural texture, and coordinate the natural scenery of the scenic area with the colours of the buildings and bridges. Combined with the karst terrain to create an architectural culture according to local conditions, the choice of materials should be mainly wood, wood and stone, and native soil, which should be studied and planned in a unified manner to prevent blind and disorderly construction, which would result in a chaotic architectural appearance.

Optimisation of Tour Routes and Enhancement of Ornamental Landscape Facilities. In the case of this study, attention needs to be paid to guiding the order of the tour in the area to avoid damage to environmental elements and to provide a high-quality visual experience for visitors. At the same time, to focus on the psychological feelings of tourists, a small waterfall tour route should be established so that visitors can fully experience the unique beauty of the scenic area. In addition, scenic areas can open up permeable lines, pay attention to the setting of viewing angles and select appropriate viewing points. The waterfront landscape space is also an important part of the water landscape, and attention needs to be paid to the choice of viewing position and the clearing of the viewing line, making full use of the walkway, vegetation and structures to bring out the maximum value, combining barges and walkways to set up waterfront platforms, bridges and other resting and viewing facilities to ensure the openness of the viewing line, and paying attention to the unity of style and appearance.

5 Conclusions

The Huangguoshu Scenic Area has a good overall scenic beauty, with the overall beauty of Huangguoshu Falls being the highest. The beauty of Huangguoshu Falls is the highest among different scenic spots. The scenic beauty of different landscape types are as follows: water system, overall landscape, mountain, vegetation, bridge and architecture, in descending order. The scenic beauty of the natural landscape is higher than that of the artificial landscape. The beauty of the waterfalls decreases with waterfall size. The main factors influencing the beauty of the Huangguoshu Scenic Area are landscape recognisability, landscape harmony, mountain naturalness, surrounding flora, the majesty and beauty of the water system, the relationship between mountain and water, the spatial enclosure of plants, the wildness of the bridges, and the visual and psychological feeling brought by the quality of the bridges. Sites in the Huangguoshu Scenic Area that are below the lower threshold of landscape scenic beauty are artificially landscaped. The Huangguoshu Scenic Area is characterised by the harmony and coexistence of green ecology, the majestic beauty of its grandeur and the natural beauty of its karst geology. The beauty threshold monitoring system can provide technical support for the dynamic protection and optimisation of the scenic area. The research results provide a theoretical basis for the dynamic management and regulation of scenic areas and are of great importance in maintaining the aesthetic value of scenic areas and planning the construction of scenic areas with regional characteristics.

Acknowledgement. This research was funded by the National Nature Science Foundation of China (NSFC) project (Grant number 51978187 and 51868008). Funder: National Natural Science Foundation of China.

References

1. Shao, X., Cheng, S., Chen, Y.: Study on tourism resources' evaluation. Ecol. Ind. Watch. 101–103+110 (2006). https://doi.org/10.3969/j.issn.1671-4407.2006.10.125
2. Yuan, H.: Discussion on aesthetic value evaluation index of landscape resources. Chin. Natl. Expo. 188–189 (2018). https://doi.org/CNKI:SUN:MZBL.0.2018-11-099
3. Zhang, W.: Study on the Aesthetic Characteristics and Aesthetic Value of the Yellow River Customs Line in Lanzhou City (2019). https://kns.cnki.net/kcms2/article/abstract?v=3uoqIhG8C475KOm_zrgu4lQARvep2SAkHr3ADhkADnVu66WViDP_3JYs-e79ghNRzmo2mHc-6_w2g4uGV1kni1njMix23MjP&uniplatform=NZKPT
4. Zhang, G., Qi, T.: Summary of landscape aesthetics research. In: The 31 ~ (st) Annual Meeting of Tourism Geology and Geopark Research Branch of China Geological Society and Symposium on Construction and Tourism Development of Ningde World Geopark (23 ~ (rd)). pp. 160–163. China Geological Society Tourism Geology and Geopark Research Branch, Chinese Academy of Geological Sciences, Fujian Provincial Department of Land and Resources, Fu 'an Municipal People 's Government (2016)
5. Liang, Y.: Landscape evaluation of Hainan Tropical Rain Forest National Park Based on SBE-SD method (2022). https://doi.org/10.27073/d.cnki.ghadu.2022.000198
6. Feng, W., Huang, Z., Zhang, Y., Liu, Y., Pang, M.: Landscape scenic beauty of traditional Dong Minority villages in southeast of Guizhou Province. Chin. J. Ecol. **38**, 3820–3830 (2019). https://doi.org/10.13292/j.1000-4890.201912.030
7. Liu, C.: Research on Plant Landscape Aesthetic Evaluation of the Yanghu Wetland Park in Changsha by SBE method (2015). https://kns.cnki.net/kcms2/article/abstract?v=3uoqIhG8C475KOm_zrgu4lQARvep2SAk6nr4r5tSd-_pTaPGgq4znPBn4kwS-QeeO5oCRVNqtgg1MDCppSKVwskJ1Is4H9LQ&uniplatform=NZKPT
8. Tang, G., Yu, H., Zheng, W., Wang, X., Zhu, Y.: Evaluation of grassland landscape aesthetic value and analysis of influencing factors in Kanas scenic spot. Chin. J. Agric. Resour. Reg. Plan. **38**, 64–71 (2017). https://doi.org/10.7621/cjarrp.1005-9121.20170209
9. Wang, H., Mu, J.: Aesthetic evaluation of bridges in Lijiang ancient city based on scenic beauty estimation procedures and semantic differential. J. Southwest Forestry Univ. (Soc. Sci.). **6**, 70–77 (2022)
10. Xu, C.: Beauty inquiry of Chengdu Nanhu Park waterfront landscape (2013). https://kns.cnki.net/kcms2/article/abstract?v=3uoqIhG8C475KOm_zrgu4lQARvep2SAk8URRK9V8kZLG_vkiPpTeIcNlUdPb2f85zuuHH60-UlJjCb92QwwZzDBgO7J0GBeb&uniplatform=NZKPT
11. Wang, Y.: Research on the Aesthetic Evaluation of Highway Landscape (2015). https://doi.org/10.7666/d.Y2921269
12. Zeng, J., Liu, J.: Overall planning of Huangguoshu scenic area. City Plan. Rev. 42–44 (1999). https://doi.org/CNKI:SUN:CSGH.0.1999-10-016
13. Shen, W., Li, K., Wang, X.: Comprehensive application of analytic hierarchy process and scenic beauty evaluation method in plant landscape evaluation. Northern Hortic. 110–117 (2018). https://doi.org/10.11937/bfyy.20174559
14. Daniel, T.C.: Whither scenic beauty? Visual landscape quality assessment in the 21st century. Landsc. Urban Plan. **54**, 267–281 (2001)
15. Gandy, R., Meitner, M.J.: The effects of an advanced traveler information system on scenic beauty ratings and the enjoyment of a recreational drive. Landsc. Urban Plan. **82**, 85–93 (2007)
16. Hull, R.B., IV., Buhyoff, G.J., Daniel, T.C.: Measurement of scenic beauty: the law of comparative judgment and scenic beauty estimation procedures. Forest science. **30**, 1084–1096 (1984)

17. Brown, T.C., Daniel, T.C.: Predicting scenic beauty of timber stands. Forest Sci. **32**, 471–487 (1986)
18. Daniel, T.C.: Measuring landscape esthetics: the scenic beauty estimation method. Department of Agriculture, Forest Service, Rocky Mountain Forest and Range … (1976)
19. Zhang, J.: The Diagnosis Methods in Planning and Design(16)----SD Method. Chin. Landscape Archit. 5 (2004). https://doi.org/CNKI:SUN:ZGYL.0.2004-10-012
20. Othman, J.: Assessing scenic beauty of nature-based landscapes of Fraser's Hill. Procedia Environ. Sci. **30**, 115–120 (2015)
21. Schirpke, U., Tasser, E., Tappeiner, U.: Predicting scenic beauty of mountain regions. Landsc. Urban Plan. **111**, 1–12 (2013)
22. Wang, L., Wang, H., He, C.: Evaluation and research of landscape quality of Wuhu Binjiang park based on SD technology. Anhui Agric. Sci. Bull. **28**, 60–63 (2022). https://doi.org/10.16377/j.cnki.issn1007-7731.2022.11.010
23. Hull, R.B., IV., Stewart, W.P.: Validity of photo-based scenic beauty judgments. J. Environ. Psychol. **12**, 101–114 (1992)
24. Yu, L., Lü, X., Song, S., Liang, J., Wang, X.: The preliminary structure of definition index system for river water ecological restoration threshold. Ecol. Environ. Sci. **22**(1), 170–175 (2013). https://doi.org/10.16258/j.cnki.1674-5906.2013.01.029
25. Sui, H., Liu, Y., Qiao, H., Yuan, C., Li, K.: An analysis on common statistical problems occurred in theses: expression and description of statistical data. Chin. J. Endemiol. **22**, 376–378 (2003). https://doi.org/10.3760/cma.j.issn.1000-4955.2003.04.032
26. Jin, H.: Comprehensive aesthetic evaluation of waterfall landscape : a case study of Huangguoshu waterfall group. Tourism Tribune. **5**, 38–44+73 (1990). https://doi.org/CNKI:SUN:LYXK.0.1990-04-013
27. Zhong, M., Zhao, Z., Zhang, C., Li, X.: The aesthetic semantic model of mountain scenic spot landscape: a case study of Taibai mountain. J. Zhejiang Univ. (Sci. Ed.). **48**, 368–376 (2021). https://doi.org/10.3785/j.issn.1008-9497.2021.03.013
28. Hu, J., Zhang, J.: The influencing factors of scenic beauty of rural waterfront landscape belt based on SD-SBE method: a case study of Quzhou Miaoyuan River. Chin. Agric. Sci. Bull. **38**, 68–78 (2022). https://doi.org/10.11924/j.issn.1000-6850.casb2021-0555
29. Gui, G., Yang, L., Gui, G., Li, D.: Evaluation on landscape of Waterfall Parks in Kunming City. J. Southwest Forestry Univ. (Soc. Sci.). **4**, 92–98 (2020). https://doi.org/CNKI:SUN:YNLS.0.2020-03-020
30. Ode, Å., Tveit, M.S., Fry, G.: Capturing landscape visual character using indicators: touching base with landscape aesthetic theory. Landsc. Res. **33**, 89–117 (2008)
31. Wang, Z., Li, M., Zhang, X., Song, L.: Modeling the scenic beauty of autumnal tree color at the landscape scale: a case study of Purple Mountain, Nanjing China. Urban Forestry Urban Greening **47**, 126526 (2020)
32. Zhu, Z., et al.: Mechanism of the Huangguoshu waterfall forest environment's promotive effect on human health in Guizhou, China. Ann. Transl. Med. **10**, 894 (2022)

A Study on the Characteristics and Optimization of the Soundscape of the Miao Settlement in Southeast Guizhou: A Case Study of the Thousand Miao Villages Xijiang

Qihua Kuang[1], Rui Yang[1], Yan Wang[1,2](✉), Xiaomei Li[1], Qiang Wu[3], and Saitong Li[1]

[1] College of Architecture and Urban Planning, Guizhou University, Guiyang 550025, China
1078503880@qq.com

[2] Research Center for Human Settlements and Green Smart Construction of Guizhou, Guiyang 550025, China

[3] China Railway Seven Bureau Group, Wuhan Engineering Co., Ltd., Hubei 430000, Wuhan, China

Abstract. The ethnic villages are the spatial representation of the ethnic continuation, with the protection of ethnic villages increasing, but the soundscape, as the characteristic component of ethnic villages, is not enough to excavate and protect, the characteristic soundscape can not be reflected in the planning and construction of ethnic settlements. In this study, the thousand miao villages xijiang was used to investigate the composition, temporal variation, spatial distribution and perception of thousand miao villages xijiang soundscape by means of sound walking, grid division and Patch Division, a thousand miao villages xijiang GIS soundscape map was obtained, and the features of thousand miao villages xijiang soundscape were analyzed by using SD semantic subdivision method. The spatiotemporal dynamic characteristics of thousand miao villages xijiang soundscape and the related influencing factors were studied, this paper reveals the feeling of soundscape in different thousand miao villages xijiang areas, and puts forward the countermeasures and suggestions for optimization and improvement to provide reference for the future planning and design of Miao settlement.

Keywords: Soundscape perception · Spatiotemporal variation · Soundscape · Thousand miao villages xijiang

1 Introduction

Spatial perception not only depends on vision, but also plays an important role in auditory perception [1]. There are many achievements in visual perception, but the research on sound perception is not enough, especially in the regional characteristics of the sound experience. The progress of science and technology and the process of urbanization have deepened people's perception of space, but the accelerated pace of production and life

W. Guo et al. (Eds.): GBCESC 2023, LNCE 328, pp. 730–744, 2024.
https://doi.org/10.1007/978-981-99-9947-7_74

has led to the destruction of the characteristic space environment and the emergence of man-made noise, which makes the sound environment worse and worse, and the noise has nowhere to hide, the legacy of sound is fading away. Profit-seeking capital and industrialized production make people tired of aesthetic appreciation of homogeneous sound sources, and at the same time, their thirst for distinctive sound is increasing. The sound scene with special regional characteristics is an important part of the local characteristic landscape and an important channel for users to perceive the traditional Miao culture, the study of its soundscape features has the practical significance of protecting traditional culture, increasing tourism features and improving people's quality of life. At present, the research on Miao settlement in southeast Guizhou mainly focuses on history, dress, architecture, culture and so on [2–5], but there is little research on its soundscape distribution.

Although previous studies have attempted to explore the ethnic village soundscape perception, planning, design [6–8], but compared with the ethnic village characteristics of the urgent need for sound landscape excavation and protection, there are still deficiencies. There is a lack of research on the acoustic environment of ethnic villages from a holistic and comprehensive perspective in time and space. Therefore, this paper takes the thousand miao villages xijiang as the research object and investigates the subjective and objective characteristics of the thousand miao villages xijiang, using Arc GIS 10.2 platform, spring sound data and spatial map were combined to form soundscape map, and the time and spatial characteristics of thousand miao villages xijiang soundscape were analyzed. This study provides a new perspective for the study of soundscape of Miao villages under tourism development, and provides a design basis for future thousand miao villages xijiang planning and improving design.

2 Research Methods

2.1 Study Area

The subject of this study is the thousand miao villages xijiang, which is located in the northeastern foothills of Raikou in the Qiandongnan Miao and Dong Autonomous Prefecture of Leishan County, Guizhou province, it is now the largest village inhabited by the Miao people in the world, and has completely preserved the original ecological culture of the Miao people. The landform is a valley slope, a typical minority settlement in Guizhou province. The difference in height of the mountains on both sides of the valley is hundreds of meters, the mountains in the southwest are steep, the mountains in the northeast are more comfortable, and the main settlement is located on the northeast side of the river, to reduce the harm of river erosion to settlement buildings. Standing on a high ground, one can overlook the numerous stilts, the trees protecting the village, the Baishui River crossing the valley, the bow-shaped wind-rain bridge, the water-rolling houses, the folk songs and dances performing fields and the terraces, the audio-visual landscape is continuous and rich, moving the scene. The tourism development of the Miao village is mature, the commercial activities are relatively prosperous, the smoke and fire breath is thick, is a natural element and non-natural element organic combination with the minority characteristics of the mountain settlement.

2.2 Content and Methods

Objective Investigation. In order to simulate the acoustic field characteristics of the Hmong settlement to the maximum extent, this chapter, based on the preliminary investigation, for example, traffic and soundscape resources, the typical case thousand miao villages xijiang study area is divided into 19550 m × 50 m, 550 m × 100 m and 43100 m × 100 m grids, and on May 1–15, 2022, clear weather 8:00 am–10:00 am, 12:00 pm–15:00 pm, 19:00 pm–22:00 at night to measure the sound level of each grid point (see Fig. 1).

Because of the large area of the thousand miao villages xijiang, the number of measuring points and the limited number of instruments, it is not possible to carry out measurements at the same time, the morning, middle and evening tests were carried out in the functional area. From May 1 to May 3, the test was first carried out in the area with a large number of people, while the flow of people was relatively small, low-impact areas were tested after May 3. To reduce the error after each functional area, to complete the whole measurement of one area in a relatively stable period of time, to ensure the data of spatial analysis is synchronic, according to the European Council of Graduate Students (ERC) SSID (Soundscape Indices, 740696), one-minute long sound data was used to represent the sound environment in the public space.

When measuring, the sound level is 1.5 m above the ground and the distance from the surrounding wall is more than 1 m, the data of average a level LAeq, maximum a level Lmax, were obtained, and the location, time and landscape were recorded in detail by using two-step app.

Subjective Investigation. Plaque division: In this study, based on the results of field investigation, using the boundary of landscape design, such as path, river and so on, and combining with the theme of landscape, divide the landscape into more complete areas. The region has relatively complete morphology and characteristics, and has obvious boundary division with other patches around it. The whole plaque area was fully experienced and felt by the individual in the subjective evaluation. Details of the zoning are shown in Fig. 2.

User Survey: Based on the relevant literature [10–12], combined with the thousand miao villages xijiang landscape status of the preliminary survey, the Comprehensive Evaluation Indicators were selected, the questionnaire included personal information, source preference, frequency and loudness of source perception, soundscape perception, and so on. Landscape patches were collected every two hours from 8:00 to 22:00. 300 questionnaires were distributed to ordinary tourists and residents, 300 were recovered, 289 were effective, the effective rate was 96.3% (Table 1).

GIS Cartography and Data Analysis. Firstly, based on the survey results and subjective and objective aspects, after collecting and arranging the collected subjective and objective data, using the arcgis 10.2 platform to convert the subjective and objective point data into area data by using the interpolation method, draw the map of sound environment and sound source perception. Secondly, through the SPSS 25 software processing each landscape patch each evaluation item score value carries on the cluster analysis obtains the sound scene division.

Table 1. Adjectives describing the soundscape characteristics of Miao village.

Category	Project evaluation	Adjectives
Soundscape	A sense of bustle	Lively -- quiet
	Pleasure	Joyous -- sad
	Energy	Vigorous -- dull
	Warmth	Warm -- cool
	A sense of character	Characteristic -- not characteristic
	A sense of unity	Uniform -- scattered
	Intimacy	Affectionate-estranged
	A sense of nature	Natural -- artificial
	Satisfaction	Satisfied -- not satisfied
	Coordination	Comfortable -- not comfortable

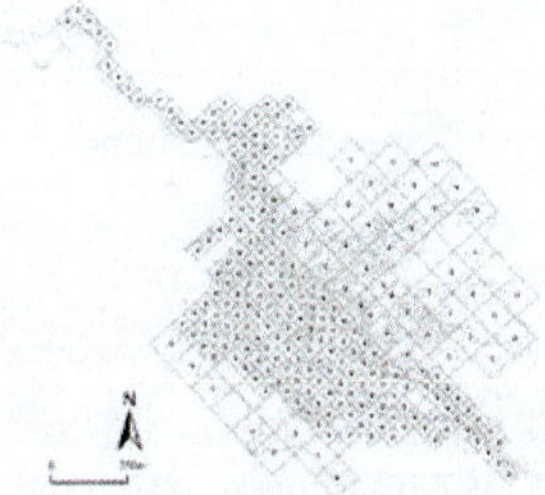

Fig. 1. Objective index measurement grid

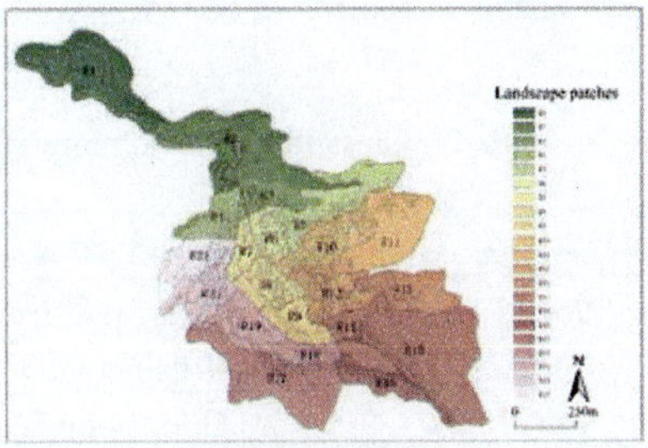

Fig. 2. Landscape patch map

3 Results Analysis

3.1 Composition of Sound and Scenery of Thousand Miao Villages Xijiang

The thousand miao villages xijiang were classified into 4 types and 40 species according to the source and nature of the soundscape. The results showed that natural sound, human sound, living sound and mechanical sound accounted for 38.65%, 17.04%, 31.49% and 10.52% respectively. The slope terrain of the valley, the high and low mountain forests and the ever-changing animals and plants have formed a natural soundscape atmosphere with vitality and charm The combination of Miao nationality's flying songs, dances, national musical instruments and Miao Village's architecture forms a humanistic soundscape atmosphere with strong local color The pastoral life of residents and the commercial activities of tourism project development constitute a complex atmosphere of life activities.

The natural sound is mainly composed of animal sound, plant sound, water sound and natural phenomenon sound, the sound of running water (1.31), birds (1.37), wind blowing leaves (1.37), Rain (1.23) and other natural sounds for users of the most favorite sound. Among them, the proportion of animal sound reaches 63.04% of natural sound, and bird sound is dominant in animal sound. The Miao village is surrounded by mountains and relatively closed natural forest land is well preserved, with a forest coverage of 58%. The Miao village ecosystem is stable, creating good conditions for birds to roost, and there are many residents who raise and walk birds in the village, it plays a very important role in the formation of the sound scene atmosphere and the molding of the characteristics of the sound scene.The humanistic sounds mainly consist of Miao language, silver ornaments and lusheng sounds. The drum sounds (1.26) and Lusheng Sounds (1.32), which are closely related to the national culture and art and the spiritual life of Miao village, are deeply loved by people, at the same time, the unique soundscape of the Miao village is created.There are many kinds of living activity sounds, mainly the sound of footsteps, conversation of tourists and living sounds of residents, people liked the sounds of children playing (0.85) and residents living (0.79), but the noise of business activities (−0.14) was annoying.The mechanical sound is mainly traffic sound and construction sound, and the construction sound (−0.75) and the machine sound (−0.5) are the most repugnant sounds, therefore, it is necessary to reasonably plan the construction operation time and noise shielding measures in the village (Table 2).

Table 2. Classification and name of thousand miao villages xijiang sound.

Sound category	Sound name
Natural Sound	Birds, poultry, dogs, cats, insects, horses, leaves, waterfalls, streams, falls, wind, rain, thunder
Human Voice	Gong sounds, drum sounds, Lusheng sounds, traditional singing and dancing sounds, silver ornaments sound, Miao language, Chung Rice Sound, water wheel sound, folk sounds
The sound of living activities	The sound of tour guides, footsteps, tourists talking, vendors peddling, children playing, cleaning the floor, residents living, playing
Mechanical Sound	Sound of vehicles (including cars, motorcycles), honking, house decoration sound, machine sound, mobile phone ring, radio sound, background music, Commercial Street Sound, bar singing, clapping sound

3.2 The Spatiotemporal Dynamic Characteristics of Soundscape Acoustic Indicators

Acoustic Environment Zoning. In order to facilitate the analysis and interpretation of the acoustic environment of 1,000 households of the Miao ethnic group in Xi River in the later period, based on the sound pressure level, the thousand miao villages xijiang sound environment is divided into three areas: Recreation Area (A), natural scenic area (B) and mixed commercial and residential area (C).

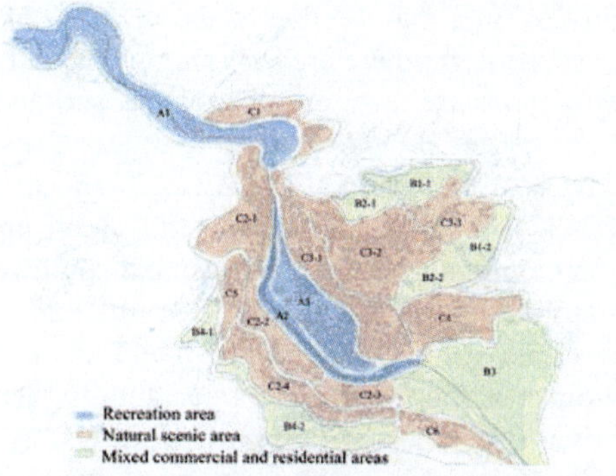

Fig. 3. Sound Environment Partition

The Recreation Area (A) comprises three areas: A1, A2, and A3. The natural scenic area (B) consists of four areas: B1, B2, B3, B4, of which B1 includes areas B1-1 and B1-2 between the high-elevation buildings and the surrounding roads to the northeast of the settlement, b 2 includes areas B1–1 and B1–2 on the north-east side of the settlement between the mid-elevation buildings and the peripheral roads, B3 is the area of paddy field between the settlement on the top of Sea Wave and the road on the south-east side of the settlement, and B 4 is the area between the road inside the settlement and the mountain on the south-west side. The mixed commercial and residential area (C) consists of six areas: C1, C2, C3, C4, C5, C6, where C1 is the area between the main road and the mountain in the northern part of the settlement, c 2 includes Area C 2–1 between the river and the mountain between the No. 5 and No. 7 storm bridges on the west side of the river, Area C2–2 between the No. 2 and No. 5 storm bridges on the west side of the river, Area C2–3 between the No.1 Storm Bridge and No. 2 Storm Bridge on the west side of the river, and Area C2–4 between the No.1 Storm Bridge and the ancient well on the west side of the river, C3 includes the area between the main road inside the settlement and the mountains on the northeast side of the settlement as C3–1 at the lower elevation, C3–2 at the middle elevation and c 3–3 at the higher elevation. C4 is the area between the main road inside the settlement and the mountains on the southeast side of the settlement, C5 is the area around the observation deck in the southwest, and C6 is the residential area in the southeast of the No. 1 Storm Bridge (see Fig. 3).

Spatiotemporal Analysis of Sound Pressure Level. (1) Time analysis. The acoustic index values of all sampling points in the three time periods were averaged to obtain the time variation of acoustic index levels in each region. In the time dimension, human activity was low in the morning (8:00–10:00), and the equivalent a sound level was low in the afternoon (12:00–15:00), at night (19:00–22:00), the night life in the village begins, and the sound around the commercial activity reaches its climax. According to the category 2 standard of "Acoustic environmental quality standard" (GB3096-2008) [13], some areas in both afternoon and evening time exceeded the national limit of 60 dba (see Fig. 4).

(2) spatial analysis. The mean sound pressure level (C2–1) in the area from No. 5 to No. 6 wind-rain Bridge is the highest in the southwest of Baishui River, the highest

sound pressure level in this area may be due to the increase of commercial activities and the number of tourists At night there are bars singing, sound equipment sound. It is found that the high sound pressure level in this area has seriously affected its daytime and nighttime rest, and measures should be taken to reduce the sound pressure level in this area. The sound levels of the river valley (A2, A3) and its adjacent mountainous areas (C2–1, C2–2, C2–3, C3–1) are higher, and the sound pressure levels decrease with the elevation of the mountains on the northeast and southwest sides, the reason of this phenomenon is mainly due to the increase of commercial activities, the decrease of human flow and the sound absorption of the high and low mountain forests. Higher on both sides of the mountain, there are some areas with higher sound levels, such as the Observation deck river (C5), the Living Road Head (C3–2) and other areas where people are active or where construction is under way. The distribution of maximum sound pressure level (MSPL) and equivalent sound pressure level (EPL) is basically the same, and the MSPL is basically distributed in the commercial area of the river valley (A2, A3). The sound pressure level in the area around the river valley is high, so it is necessary to plan and construct the sound barrier measures at the junction with the residential areas on both sides of the mountain (see. Fig. 5).

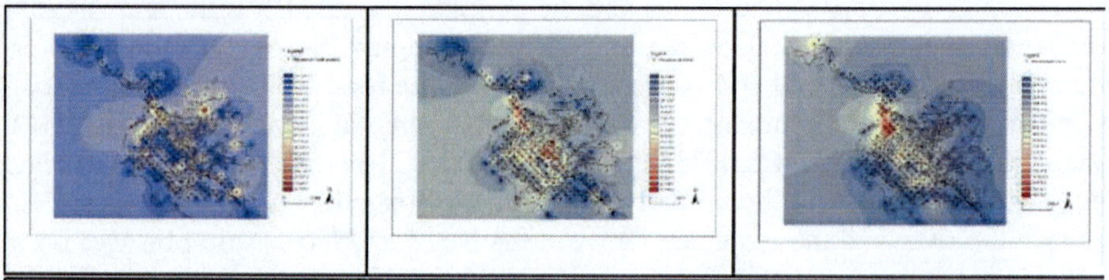

(a) A map of the sound environment in the morning	(b) Afternoon sound environment map	(c) Map of the nighttime acoustic environment

Fig. 4. The spatial distribution of sound pressure levels at different time Xijiang thousand Miao village

3.3 The Spatiotemporal Dynamic Characteristics of Sound Source Perception Indicators

Time-Varying Characteristics of Source Dominance. Perceived occurrences of sound (Pos) and perceived loudness of sound (PLS) were measured in five different time periods (1), the sound dominant degree (SDD) of each type of sound source in each region was calculated separately [14], and the time-varying trend graph of the SDD was obtained after averaging the values for each time period (see Fig. 6). 8:00–10:00,15:00–17:00: the natural sounds of the scenic spots in Miao village are mainly water sounds, bird sounds, wind sounds, etc.. 10:00–12:00: the activities of residents and tourists increase, and the artificial sound sources become more and more complex. During this period, the sounds in the Miao village are mainly the sounds of vehicles, residents' living, tourists'

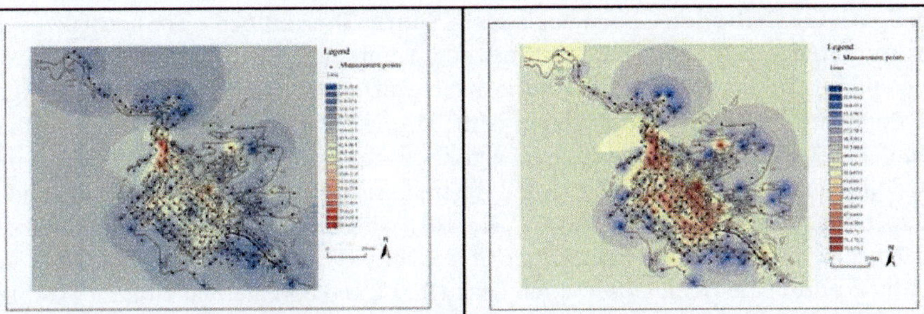

(a) LAeq mean acoustic environment map | (b) LMax mean acoustic environment map

Fig. 5. The spatial distribution of thousand miao villages xijiang mean sound pressure levels

conversation and the sounds of shopping street, all kinds of sounds are mixed, and the dominance of life activity sounds is higher. 13:00–15:00: the traditional song and dance parade begins in the village. At this time, the main attractions in the village are the human voice, such as Lusheng, traditional song and dance, silver ornaments, etc. 17:00–22:00: The Sound in the village is mixed to the climax, the number of tourists in each scenic spot increases, the tourists go to the observation deck to watch the night scene, the bar in the village starts to open, at this time, the Miao village mainly by commercial background sound, bar singing sound, sound equipment sound-based, mechanical sound of a higher degree of superiority. This phenomenon shows that the variety of thousand miao villages xijiang sounds is closely related to the natural environment, tourists' visits, traditional cultural activities and commercial activities in the village.

$$SDDji = POSji \times PLSji \tag{1}$$

Formula: j is the J sample, I is the I sound source, n is the sample size

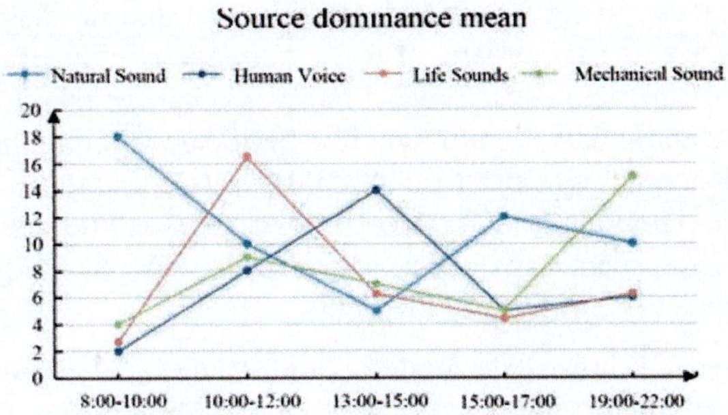

Fig. 6. The trend of source dominance with time

The Spatial Distribution Characteristics of Source Dominance. The normalized data were imported into GIS, and the full-time source dominance distribution of the thousand miao villages xijiang was obtained by the inverse distance weighting method (see Fig. 7). The dominant degree of natural sound is the strongest in Observation deck, Northeast Mountain and rural area, where human activities are relatively less and plants are more The dominance of human voice reached the highest in Observation deck (R21) and the commercial human exhibition areas (R7, R8, R9), where there were more traditional song and dance performances The dominance of live activity sound was strongest in the areas near little north gate (R2), Observation deck (R21), and commercial cultural exhibition area (R7, R8, R9), which had more tourists The dominance of mechanical sound is strongest in Observation deck (R21), commercial cultural exhibition (R7, R8, R9) and rural area (R15), and is mainly influenced by the sound of commercial activities and traffic. Compared with other sound sources, the areas with higher level of dominance of living activity sound have larger range, and the human voice is smaller, and the perception of dominance of different functional areas is different, this shows that the spatial function attributes have great influence on the type, frequency and loudness of sound source.

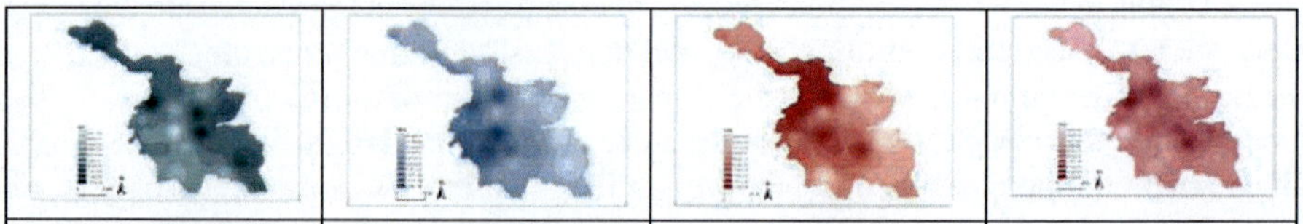

(a) Natural sound superiority	(b) Human voice superiority	(c)Life sound superiority	(d) Mechanical acoustic superiority

Fig. 7. Spatial distribution of source dominance

3.4 The Spatiotemporal Dynamic Characteristics of Soundscape Perception

Time-Varying Characteristics of Soundscape Satisfaction. The satisfaction of the thousand miao villages xijiang was calculated from the questionnaire in five different time periods. After average processing of the data in each time period, the trend graph of the satisfaction of the soundscape over time was obtained (see Fig. 8). Comparing the mean of soundscape satisfaction in five time periods, we found that the lowest level of soundscape satisfaction was from 17:00–22:00 (−1.02), and the mean of soundscape satisfaction in all time periods from 8:00 to 10:00 to 15:00–17:00 was above the middle level, the mean of soundscape satisfaction was relatively low (0.64) from 10:00–12:00, and reached the highest (1.36) from 8:00–10:00.

Spatial Distribution of Soundscape Satisfaction. The spatial distribution of the thousand miao villages xijiang soundscape satisfaction was obtained by importing the mean of the processed soundscape satisfaction score into GIS using the Kriging method (see Fig. 9). According to the figure, the soundscape satisfaction is higher in the areas near the entrance of Miao village (R1, R2), Observation deck (R21) and idyllic scenery (R15).

The main sound at the entrance of Miao village is the sound of birds and the sound of running water, in the viewing area, the sounds of conversation, silvery ornaments, motor vehicles and selling are the main sounds, while in the rural area, the sounds of birds, insects and work are the main sounds, the common features of these areas are superior landscape environment and relatively open space. The higher the elevation of the settlement, the lower the satisfaction of the soundscape (except for the observation platform area). These areas are quiet, mostly composed of wind and birdsong, it can be seen that too few people flow in quiet areas, although there is a preference for a high degree of natural sound is not very satisfactory. To sum up, the area with higher sound pressure level is higher or lower, but not the area with higher sound pressure level is lower. The level of sound satisfaction is affected by many factors, from the spatial distribution, it can be seen that the users prefer to be in the soundscape area with national culture and natural soundscape.

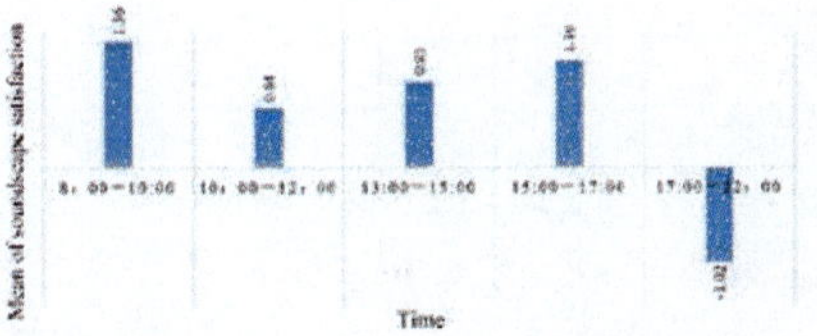

Fig. 8. The trend of soundscape satisfaction over time

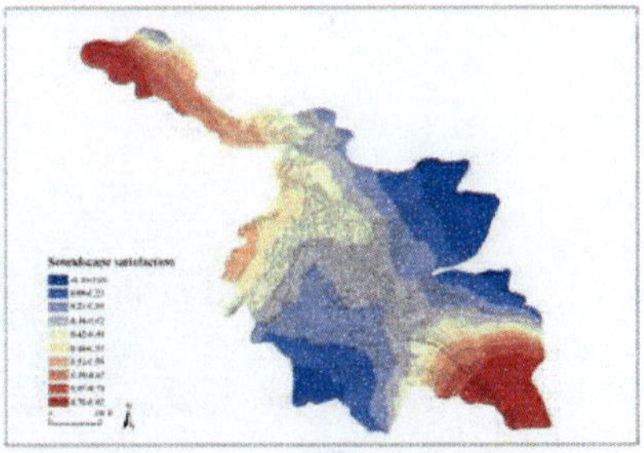

Fig. 9. Spatial distribution of soundscape satisfaction

4 Miao Village Soundscape Space Zoning

After cluster analysis calculation and mapping, the evaluation items in each landscape patch form Euclidean distance and cluster analysis dendrogram among the soundscape samples in each region, and determine the kinship of each soundscape sample according to the Euclidean distance, the Soundscape of Miao village can be divided into the following types (see Fig. 10):

Area I: Commercial Cultural Scenic Area this area is the commercial area of the Miao village and historical and cultural exhibition area, the characteristics of the soundscape features (0.812), lively (0.715), dynamic (0.548). The sound elements in this area are mainly composed of water sound, Miao language, conversation sound of tourists, peddling sound of Commercial Street, Lusheng Sound and silver decoration sound.

Area II: multi-space sound scenic area this area type for the landscape, residential and historical and cultural exhibition area, the sound scenic features for pleasant (0.713), dynamic (0.681), lively (0.660), the sound elements in this area are mainly composed of the sounds of tourists' conversation, water sound, Miao language and so on.

Area III: residential pastoral area the area type is residential and planting area, the characteristics of the soundscape are pleasant (0.577), warm (0.538), estranged (−0.192), non-characteristic (−0.154), the sound elements in this area are mainly composed of Miao language, bird calls and insect sounds.

Area IV: verdant natural sound scenic area this area type is the entrance of Miao village and Mountain Plant Landscape Area, the soundscape features are natural (0.423), scattered (−0.007) and quiet (−0.119). The sound elements in this area are mainly composed of water sound, bird sound and motor vehicle sound.

Area V: terraced landscape scenic area this area is a pastoral scenic area, the characteristics of the soundscape are natural (1.000), pleasant (0.583), characteristic (0.417), kind (0.417), quiet (−0.667), the sound elements in this area are mainly composed of Miao language, bird calls and motor vehicle sounds.

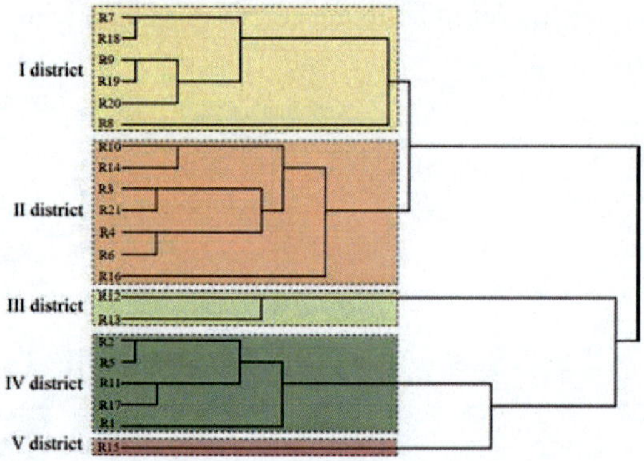

Fig. 10. Acoustic scene cluster analysis dendrogram

5 Countermeasures and Suggestions

5.1 Measures

To Improve the Environmental Quality, Create a Unique Soundscape Node. Soundscape is an important part of spatial perception, which can enhance the comfort of experience by complementing with visual scene. It was found that some areas of the thousand miao villages xijiang are dirty and unkempt, and there is insufficient environmental control. For example, the residential environment in the northeast is dirty and unkempt, the domestic sewage is directly discharged into Baishui River, and the domestic garbage is not effectively treated, which leads to the decrease of residents' well-being and the discomfort of tourists' landscape experience. Therefore, it is of great significance to strengthen the environmental sanitation, improve the sewage treatment pipeline, improve the air quality and the visual landscape experience. Moreover, there are many human landscape nodes in the Miao village, which can make full use of the existing soundscape at the nodes, through improving landscape environment design, controlling artificial soundscape, introducing artificial intelligence soundscape and configuring personalized micro-soundscape to enhance the sense of soundscape experience and enhance the interest of space environment, to attract tourists with dot-shaped characteristic soundscape, to create a characteristic soundscape environment.

To Standardize the Traffic Flow Line and Plan the Walking Route with Special Soundscape. Previous studies have shown that tourists and residents travel in a clear line consistent with the road flow, this flow line should become the Miao village characteristics of the soundscape layout, as shown in Fig. 11, Fig. 12 data from Baidu Open Data. The roads in the thousand miao villages xijiang area connect residential areas, commercial areas and scenic areas. The mixed-vehicle traffic at some junctions and along important tourist routes in the commercial areas seriously affects the quality of the tours and personal safety The traffic flow in residential area has great influence on the level space of sound pressure level, and seriously disturbs the rest of residents. Based on the previous study, the thousand miao villages xijiang road should be divided into sections as shown in Fig. 12. The station from the north gate to the small north gate should be divided into sections for pedestrians and vehicles, according to the layout of green pavement at the same time planning pedestrian, vehicular access. The section of Old Street between No. 1 Fengyu Bridge and No. 6 Fengyu Bridge is divided into a no-go section. This section is changed to the main pedestrian road. The ferry service is adjusted to the vehicle routing of the parade street, which reduces the hidden traffic danger in Old Street, the noise source area will be unified to the river bank, other traffic around the rear mountain road. The section from the No. 1 Stormwater Bridge to Observation deck street is divided into a speed-limited section, which requires noise reduction at time periods and spatial locations where the sound pressure level fluctuates significantly.

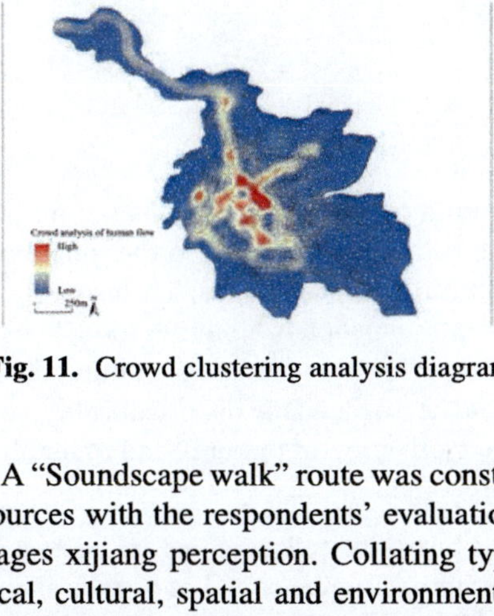

Fig. 11. Crowd clustering analysis diagram

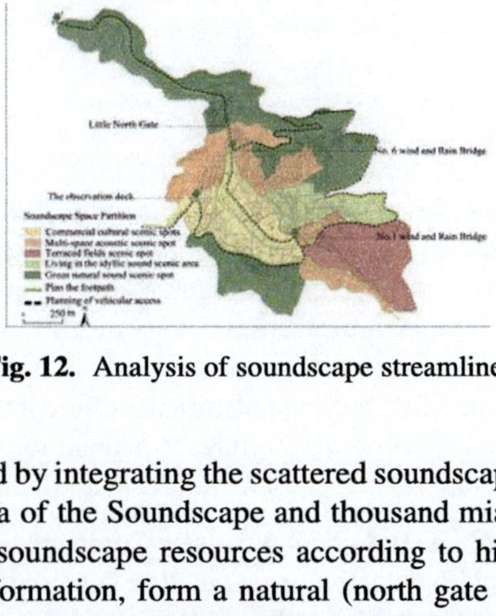

Fig. 12. Analysis of soundscape streamlines

A "Soundscape walk" route was constructed by integrating the scattered soundscape resources with the respondents' evaluation data of the Soundscape and thousand miao villages xijiang perception. Collating typical soundscape resources according to historical, cultural, spatial and environmental information, form a natural (north gate to small north gate, along the river section), humanities (along the ancient street), life (the Ancient Street North, scenic sections) composed of linear tour route.

Dealing with Noise Correctly and Planning Dynamic and Static Zones Reasonably. Within the thousand miao villages xijiang, multiple areas serve a single function. The use of the space and the time and location of the different sound sources result in the opposition between the residents' daily life and commercial activities, the high sound pressure level and the chaotic dynamic and static partition cause great disturbance to the rest of the residents. To deal with the noise correctly and to plan the dynamic and static zones rationally should become the focal point of creating the sound scene of Miao village, according to the investigation and analysis, the impact of crowd noise, bar noise, traffic noise and traditional singing and dancing thousand miao villages xijiang were found in the Recreation Area A The Natural Scenic Area B has the influence of construction noise, sound equipment noise and living noise, while the commercial and residential mixed area C has the influence of Bar Noise, crowd noise and traffic noise. Therefore, in the aspect of noise treatment: 1 it is necessary to strictly regulate the commercial activities and construction time, control the sound pressure level according to different time interval, and reduce the spatial sensing burden of sound pressure level. 2 according to the acoustics principle and the landscape, architecture, planning and design method, such as using the sound propagation characteristic to design the sound absorption screen body in different parts, using the different plant combination disposition to set up the different sound absorption, the sound insulation barrier; Measures such as using soundscape to mask noise and using existing landforms to set up natural barriers to enhance thousand miao villages xijiang environmental comfort and enhance the quality of human settlements. In the aspect of dynamic and static zoning, for the purpose of improving the design of the post-planning thousand miao villages xijiang, the soundscape can be divided into different areas according to the present soundscape perception clustering analysis (as shown in Fig. 12) and the soundscape characteristics of each area, clear the dynamic and static area, adjust the functional area reasonably, and according to the spatial positioning of the functional area, deepen the perception of soundscape and visual scene, highlight the planning theme, create multi-scale characteristic space, and form the area-like tour area.

5.2 Measures

To Explore the Cultural Connotation. This kind of national culture is not only the spiritual home and emotional sustenance of the Miao people, but also the innate condition of feeling the unique soundscape of a thousand Miao villages. Therefore, in the landscape optimization design needs to dig in-depth in people's clothing, food, housing and transportation, ceremonies, folk customs and performing arts in the cultural connotation; To inherit the historical culture in a living state, to blend the historical culture with the contemporary culture, to "Avoid the commercialization of culture" and to highlight the high-level regional culture with distinctive features.

Build a Historical Acoustic Environment. In a busy city, folk customs are visited, and KOŬP plays the new splendor: planting and strengthening the sound of historical signs according to the cultural characteristics of the settlement itself, they include: background sounds of ancient Chinese music, ethnic performances, hawkers hawking, stir-frying tea, pounding rice, water wheels, beating silver ornaments, weaving, while retaining the traditional meaning, it shows the rich historical and cultural background, typical street style and vivid Miao ethnic customs, use modern material means to design expression, ancient and modern blend, make it across the millennium time charm.

Create a Unique Soundscape Experience Space. According to the research on the resources of soundscape and the dynamic characteristics of time and space, this paper explores the best time and space position of different types of soundscape, sets up different types of soundscape experience facilities and places, the multi-dimensional and comprehensive construction of the space, the multi-sensory experience of visitors through the Miao Settlement Forest Birdsong, mountain song, festival customs, terraced fields interest and other characteristics of the Miao folk cultural soundscape, to create unique, influential and unique landscape space of the Hmong settlement, so that tourists can feel the unique cultural resources of the Hmong settlement, give play to regional advantages, and deepen the experience of visiting, stimulate interest in multiple tours.

Audio-Visual Landscape Coordination. The sound scene and the visual scene mutually cooperate, achieves the audio-visual union, can let the visitor realize from the image behind the deeper level significance. The sound scene inside the settlement should form a unity with the surrounding building environment, highlight the original environment as much as possible, and weaken the external environment noise to the sound scene environment. Just like the tourists who want to stop and rest by the White Water River, when they brew a pot of tea, they must hope to have leisure and elegant music to sweep away the noise. Artificial sound is the most likely to destroy the coordination of audio-visual landscape elements, in the commercial flavor of the pedestrian street, you can play ethnic music, scientific and reasonable placement of the audio position, adding to the settlement sound environment of the minority atmosphere, blend in with the entire Hmong settlement site.

6 Conclusion and Discussion

Soundscape is an important way for us to cognize the space environment. The study of soundscape from subjective and objective perception, quantitative and qualitative is of great significance for us to strengthen our cognition of space. This paper analyzes the composition of the thousand miao villages xijiang, the factors affecting the soundscape, and the dynamic changes in subjective and objective perception through field surveys using grid, patch partition, and sound walking, this paper, based on the characteristics of the soundscape, combines the characteristic soundscape resources with the soundscape map to explore the characteristic soundscape formed by the symbiosis of the soundscape of the Miao nationality settlement with the spatial environment and the humanistic connotation, it provides a reference for the Miao settlement to build a multi-featured space soundscape with the combination of "Points, lines and surfaces", and promotes the users' multi-sensory experience in the multi-dimensional and comprehensive space, the user perception evaluation is used to feedback the noise control and "Dynamic protective development" of the Miao settlement, to repair the ecological environment of the Miao village and improve the human settlement environment.

However, the study of this paper is only one of the ideas, Ways and means, not the final development goal of the Soundscape of Hmong settlement. The research, design and improvement of the soundscape of the Miao settlement need to be based on people's spiritual needs, social development and the development of science and technology.

The author will further explore the development of soundscape, such as the new VR technology into the Miao settlement soundscape experience platform to explore the impact of different terrain environment, building materials on the characteristics of the soundscape effect.

Acknowledgement. This work was supported in part by Study on the Theory and Supporting Technology of Applying for the Heritage of "Miaojiang Ancient Post Road" Cultural Route in Guizhou Province, Number: [2019] 2850.

References

1. Garles, J.L.: Sound influence on landscape values. Landscape Urban Plan. **43**, 191–200 (1999)
2. Zhao, X.M.: Settlement and house: Changes in settlement patterns and social clustering of Miao villages in Northeastern Guangxi Zhuang Autonomous Region. Journal **64**(3), 68–81 (2022)
3. Xiang, K.Y.: The Shouzheng innovation in the change of Miao women's dress from the perspective of non-inheritance. Journal **50**(08), 93 (2022)
4. Huang, H.J.: The regional adaptability of Basha Miao settlement and village architecture. Journal **34**(1), 111–118 (2019)
5. Chen, W.D., Guo, W.: The project conception of the World Miao Culture Exhibition Center. Journal **36**(8), 48–52 (2018)
6. Maureen, Q.: A study on the intangible cultural heritage characteristics of traditional acoustic landscape: a case study of traditional villages in Dong people Guizhou Province. Journal **1**, 108–116 (2020)
7. Lin, J M.: A study on the soundscape of traditional settlements in eastern Guizhou from the perspective of sound ecology, Harbin Institute of Technology (2014)
8. Li, J.Q.: Study on the soundscape design in the public environment of Dongzhai, Guizhou, Sichuan Fine Arts Institute (2019)
9. Hu, J.: Study on soundscape analysis and soundscape mapping based on GIS, Zhejiang University (2015)
10. Shao, Y.H., Xue, Z.Y., Hao, Y.Y., et al.: Study on the evaluation of the audio-visual perception quality of urban parks: taking the ecological zone around Chengdu as an example. Journal **29**(9), 26–32 (2022)
11. Ren, X.X., Kang, J., Liu, X.G.: Experimental study on road traffic sound evaluation under the influence of ecological water landscape vision. Journal **40**(3), 361–369 (2015)
12. Rinche, Audrey, Panson, et al.: Study on soundscape evaluation of different landscape spatial types. Journal **36**(8), 57–63 (2020)
13. Chinese Academy of Environmental Sciences.: Sound Environmental Quality Standard: GB3096–2008. China Environmental Science Press, Beijing (2008)
14. Wang, J.Y., Li, C.M., Lin, Y.L., Li, D.F.: Analysis and application of soundscape ecological data. Journal **42**(21),1–8 (2022)

Wayfinding Oriented Evidence-Based Design for Building Optimization

Qian Cao, Jingyi Li, Shuyang Li(✉), Moxuan Shen, Weiyi Liang, and Kaiyu Lu

Department of Architecture, National University of Singapore, Singapore 117565, Singapore
shuyangli@u.nus.edu

Abstract. The evidence-based design (EBD) method is deemed with significant potency in optimizing architecture designs based on qualitative and quantitative data. However, EBD methods fail to aptly collaborate with conventional design process due to the lack of practical workflow. This study proposed an EBD approach integrating spatial analysis and virtual reality and applied them to a wayfinding-oriented design optimization project. Taking the College of Design and Engineering (CDE) building group of National University of Singapore (NUS) as a study case, this research firstly identified the causes of low wayfinding efficiency via site survey and spatial analysis. Then, corresponding optimization schemes were proposed, and a VR experiment was conducted to evaluate their effectiveness. This study concluded that the lack of openness in vertical traffic space was one of the dominant causes for the low wayfinding efficiency. This issue could be optimized by improving visibility and accessibility. Besides, this study verified the feasibility and demonstrated the great potential of EBD approach, offering a practical reference to design more legible and user-friendly spaces.

Keywords: Evidence-based Design · Spatial analysis · Virtual Reality · Visibility · Accessibility · Wayfinding

1 Introduction

Evidence-based design (EBD) is a critical use of the best available evidence from existing research and practice to improve design decisions [1]. The development of design theory and computer technology has remarkably enhanced the reliability and feasibility of EBD. Some studies have adopted spatial analysis tools to reveal potential relationships between space features and human behaviors [2–5]. The methodology of spatial analysis relies on the "Isovist" theory [6] and the "Space Syntax" theory [7]. Meanwhile, other studies conducted site surveys and virtual reality (VR) experiments to explore human-centered space designs [8, 9], in which VR technology provided constructive perspectives in research and design practice.

Nowadays, computer-aided architecture design (CAAD) is an ongoing trend that has been discussed profoundly. Although as a typical CAAD method, EBD has been somehow underestimated. One reason is the disconnection between spatial analysis and

W. Guo et al. (Eds.): GBCESC 2023, LNCE 328, pp. 745–755, 2024.
https://doi.org/10.1007/978-981-99-9947-7_75

optimization assessment, which mostly fails to feed back into the original design and verifies of the validity of optimization strategies. Another reason is the insufficiency of practical experiences to demonstrate how the EBD method can fit well in architects' design process. Indeed, the EBD method surpasses other conventional design methods in design validity, feasibility, and efficiency, therefore deserves more practical adoptions.

Taking a project of wayfinding-oriented design optimization as an example, this research aims to demonstrate an EBD process with high feasibility for architects (see Fig. 1). Isovist_App was used for spatial analysis, based on which issues and corresponding optimizing strategies were proposed to facilitate the wayfinding efficiency. Hyve-3D VR was adopted to display the "before" and "after" scenarios and compare participants' wayfinding performances.

We assume that the spatial elements (scale, form, and organization method) are the most critical factors affecting users' space cognition and wayfinding efficiency, which is the primary concern for architects. Hence, this study mainly discusses spatial elements rather than other factors such as brightness, colour, material, etc. This research will test the feasibility of the spatial analysis and virtual reality-integrated EBD method. In addition, this research would provide a reference for wayfinding-oriented design optimization.

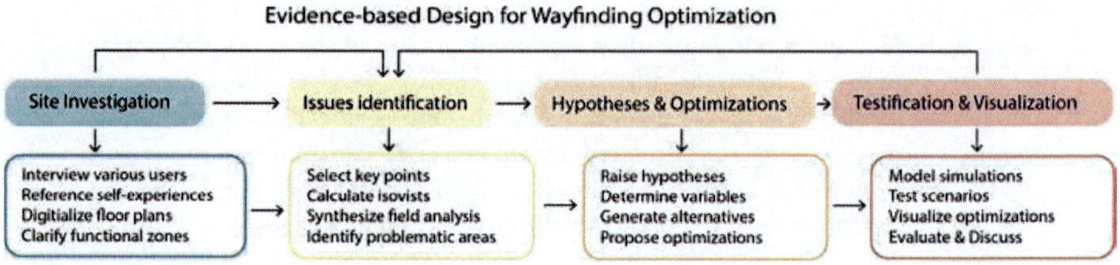

Fig. 1. Research framework.

2 Methods

This research included three phases: site survey, spatial analysis, and VR wayfinding experiment. Given that spatial analysis might only theoretically verify our optimizations in enhancing visibility and accessibility, we designed an experiment using the Hyve-3D VR platform to analyze participants' wayfinding performances and further testify the optimizations in practice.

2.1 Study Area and Site Survey

This study took the buildings of the College of Design and Engineering (CDE) at National University of Singapore (NUS) as an example. We selected nine buildings with various traffic connections vertically and horizontally (see Fig. 2). These buildings are located on an undulating slope where their ground floors have a 0 to 30-m height gap.

We interviewed students, teachers, and visitors, who all reflected on the confusion of wayfinding in the CDE and most of them were still not able to find their destinations until months' practice. According to the interview results and self-experiences, the complicated elevations of buildings' ground floors and the numerous connections between different buildings may cause low wayfinding efficiency, consequentially, discrepancies of navigation between first-time and local users.

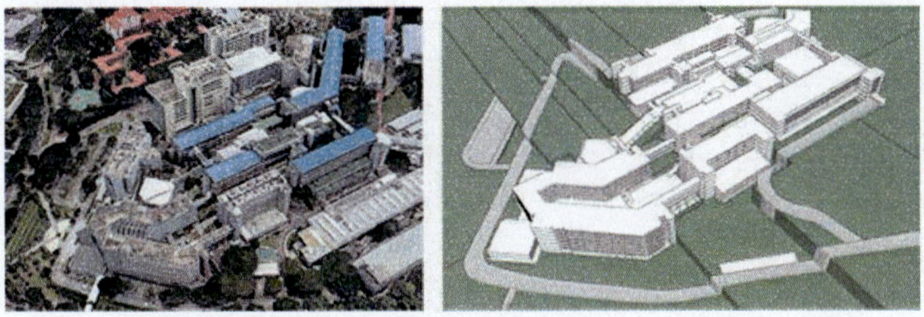

Fig. 2. CDE buildings group.

2.2 Spatial Analysis

ISOVIST. As a spatial analysis application, Isovist_App is tailored to advance the learning and calculating spatial characteristics, from fundamental understanding to advanced research and in-depth design practice. The methodology of Isovist_App relies on Benedikt's "Isovist" theory and Hillier and Hanson's "Space Syntax" theory. "Isovist" theory defined isovist as a spatial unit with a finite volume of space (size, shape and relation with others) whose properties can be quantified and mapped to create fields. "Space Syntax" aims to decipher spatial relationships on an urban scale, including behavior and configurative aspects of space.

Isovist_App can synthesize real-time analysis in point, path, region, and isovist agent analysis, which concentrates on building interiors, connections, and human aesthetic experiences and perceptions. Isovist_App augments the comprehensiveness and authenticity of spatial cognition, with various scales from a single room to urban street pattern, and provide evidence to explain specific human behaviors in the space.

Issues Identification. Fluctuant elevations complicate building connections and curate intricate vertical and horizontal traffic systems. Based on the site survey and self-experiences, the unnoticeability and illegibility of traffic spaces significantly affect wayfinding efficiency. Therefore, we used Isovist_App to calculate the visibility and accessibility of vertical traffic spaces. Directed Visibility and Co-Visibility [10] were selected as two criteria for visibility analysis. Choice, refined as how a point can be seen and visited on all shortest routes from all spaces to all other spaces [6] was identified as a metric for accessibility analysis.

According to the spatial analysis for each floor (see Fig. 3), we discovered three significant issues about the existing vertical traffic spaces:

(1) Some vertical traffic spaces are hidden inside the corner rather than directly facing the corridor, making them barely visible to users and with poor accessibility. Users need to pass multiple right-angle turns before reaching them.
(2) Some vertical traffic spaces are located at the end of a narrow corridor without an enlarged foyer space. Despite the high accessibility of such spaces, it has limited capacity to cue people their existence and encourage people's utilizations.
(3) Vertical traffic spaces have poor visibility both inside and outside, making them more difficult for people to recognize. The inclination of using them to pass through different buildings might be deprived.

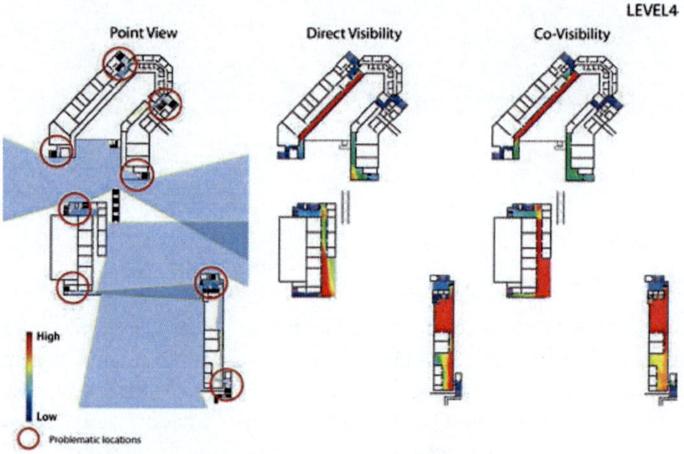

Fig. 3. Example of spatial analysis for issues identification.

Optimization Strategies. The study focused on how to adopt architectural strategies to improve wayfinding efficiency via enhancing visibility and accessibility of vertical traffic spaces. We proposed two optimizations in space form and façade material respectively.

(1) The foyers of vertical traffic spaces (stairwells and elevators) were widened by removing the redundant interior walls, then adding atriums to connect the upper and lower floors (see Fig. 4). The open spaces were designed as shared spaces for students and staffs to have meetings and discussions, which would increase the accessibility of the vertical traffic space along with the improvement of visibility.
(2) The external solid walls of vertical traffic spaces were replaced by transparent glass (see Fig. 5). This replacement can increase the visibility of vertical traffic spaces from both inside and outside. This strategy aimed to unveil the outdoor environments used as direction references for people, making them more aware of their locations in the wayfinding process.

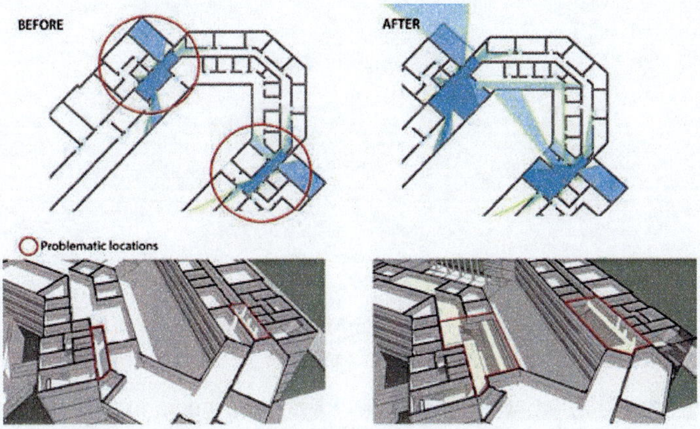

Fig. 4. Before and after change scenarios of the foyers of vertical traffic spaces.

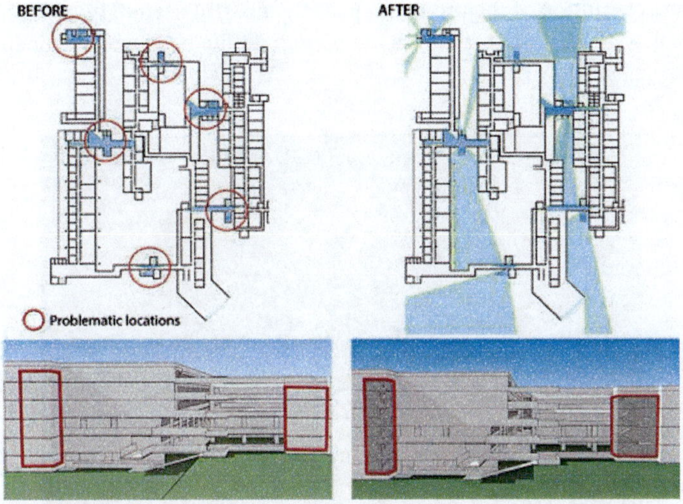

Fig. 5. Before and after change scenarios of the external walls.

2.3 VR-Based Wayfinding Experiment

Wayfinding Tasks. Based on the most used daily routes of different users, we designed three wayfinding tasks to ensure that the optimized spaces would be used in most cases. Three tasks start from significant locations (e.g., building entrances near bus stops and canteen) to different destinations (e.g., auditorium and library) (see Fig. 6). During the experiment, participants were required to walk through different buildings and go up or down stairs.

Hyve-3D VR Platform. Hyve-3D VR platform was used as a scenario presentation and wayfinding manipulation tool in this experiment. Hyve-3D is tailored as a multi-user

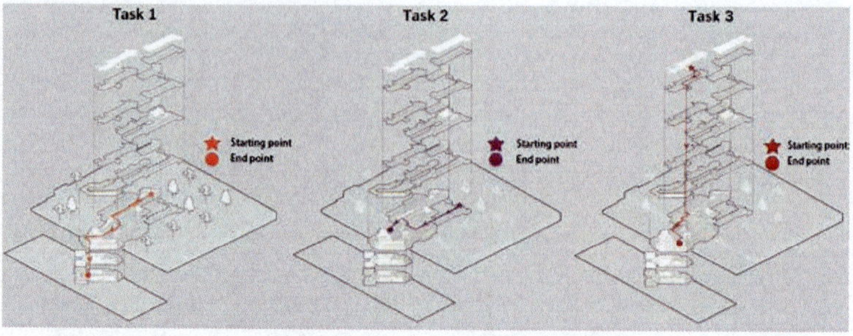

Fig. 6. Three wayfinding tasks.

VR co-design system that allows participators and designers to get immersed without headsets and to interact via 3D cursor. A spherical screen is used to display images, meanwhile a tablet PC is provided to manipulate roaming (see Fig. 7). Though the images projected by Hyve-3D VR platform might be inferior to those projected by headset VR, limited by the resolution of the projector, participants still prefer Hyve-3D VR platform for the virtual visit because it removes the burden on the head and eliminates the sense of closeness.

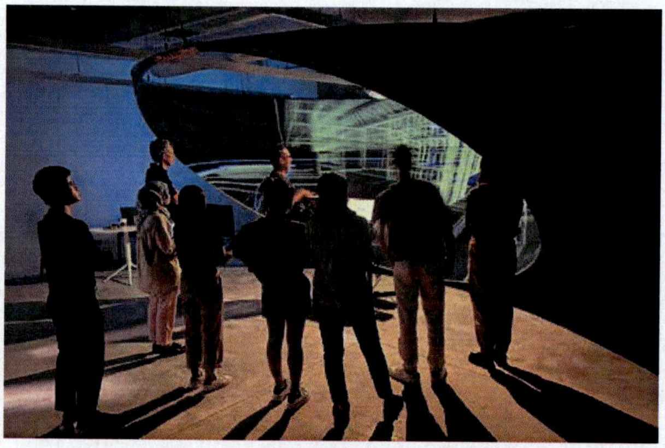

Fig. 7. Hyve-3D VR Platform.

Participants and Experiment. We invited 6 local users who have worked at CDE for more than two years. These local users are familiar with the CDE buildings group, so their wayfinding data could be used as a benchmark. Besides, we recruited 18 students who have never been to CDE, role as first-time users. We allocated 9 first-time users to the before-change group; the rest 9 were to the after-change group. Each wayfinding task was carried out by 2 local users and 3 first-time users. All the participants were randomly allocated to different groups and blinded to which group they were in during the experiment.

Videos were recorded for each participant during the VR experiment in order to retrospect the wayfinding process and observe participants' behaviors. After each participant completed the wayfinding task, we conducted a 3-min interview and inquired about what architectural components impressed them and how they helped their wayfinding decisions.

3 Results and Discussion

Combining the results of spatial analysis and VR experiment, first-time users' wayfinding efficiency was improved due to increasing space openness. However, first-time users' preferences for route selection were not fully overlap the routes with optimized spaces, which raised our attention and inspired further discussion.

3.1 Visibility and Accessibility

To verify the visibility improvement through the optimizations, we calculated the values of before and after change scenarios in points and field analysis. Since the Isovist_App treated solid and transparent surface as the same, the field analysis did not show a significant difference when the walls were replaced by glass. Hence, we compared points' Co-Visibility values for the two scenarios. Figure 8 clearly shows that replacing the solid interior wall with glass would remarkably increase the co-visibility of these selected points. Besides, using an atrium to replace some offices every two floors can significantly improve horizontal and vertical visibility (see Fig. 9).

Moreover, the accessibility increased remarkably due to the optimizations. The awkward situation of frequent-used places (vertical traffic space) but with poor visibility and accessibility has been efficaciously alleviated after the optimizations. Admittedly, the mutual improvement of visibility and accessibility after change was only proved in this research and the correlation between them might be diversified in different scenarios. It would be fruitful if future research keeps exploiting whether there will always be a co-improvement between these two components and whether their enhancement would inevitably contribute to high wayfinding efficiency in various architecture designs.

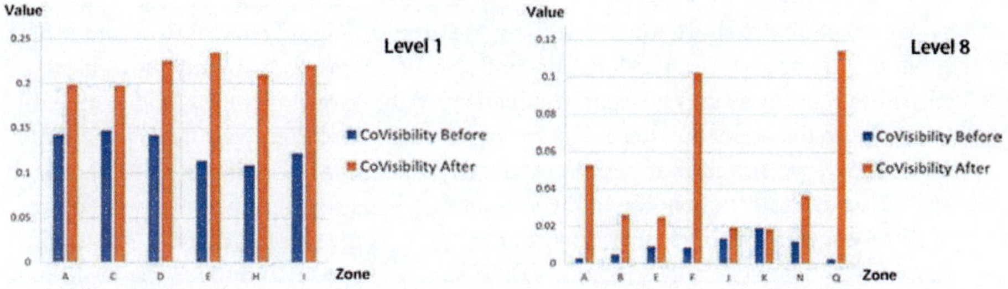

Fig. 8. Comparison of points' Co-Visibility values of before and after change scenarios.

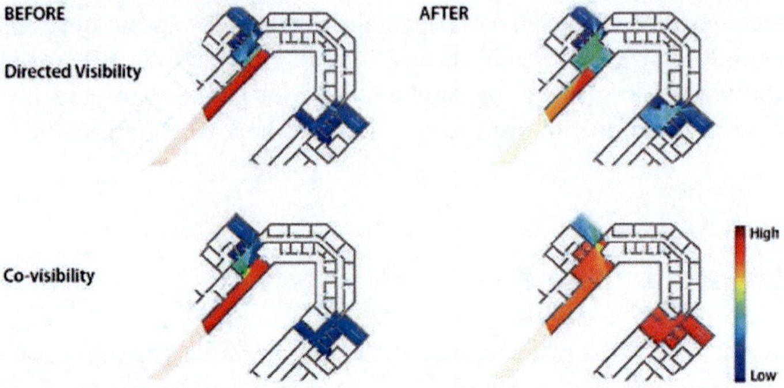

Fig. 9. Comparison of field analysis values of before and after change scenarios.

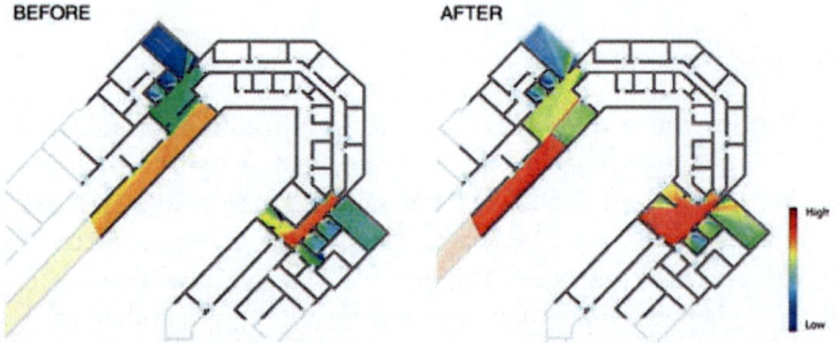

Fig. 10. Comparison of accessibility values of before and after change scenarios.

3.2 Path Length

Path length is another primary index for wayfinding efficiency. As a benchmark, the local users' average path length was obviously lower than the first-time users', regardless the before or after change scenarios (see Fig. 11). Regarding the path length of first-time users between before and after change scenarios, the latter was lower than the former, especially for wayfinding Task 1.

Most first-time users instantly decided to use these optimized vertical traffic spaces instead of detouring to look for other routes, whether intentionally or unconsciously, which could explain the shorter path length in the after-change scenario. In the interview of wayfinding Task 1, three participants indicated that they presumed the optimized vertical traffic spaces as shared spaces because of the large openness and noticeable connections with the upper and lower floors, which facilitated their wayfinding decisions.

Interestingly, we found that people especially first-time users inclined to approach outdoor spaces as their first choice in the wayfinding process. Figure 11 shows that only a slight difference between before and after change was prescented in Task 2 compared with other tasks. The start and end points of Task 2 were located outside the buildings, so outdoor spaces were subconsciously selected as wayfinding routes no matter whether the indoor traffic spaces have been optimized. While the start points of Task 1 and Task 3 were indoor spots, therefore the optimized spaces were inevitably used and then reflected an obvious difference between before and after optimizations. Undoubtedly, outdoor spaces normally own more openness than indoor spaces and provide more information as people's direction coordination, thus it is reasonable that the participants prefer to choose the route with more outdoor space.

3.3 Wayfinding Duration

Wayfinding duration is another commonly used index for wayfinding efficiency. Comparing the wayfinding duration in three experimental groups, the average time spent by the first-time users was much more than the local users, no matter before or after changes (see Fig. 12). Surprisingly, focusing on first-time users, they took noticeable longer time in the after-change scenarios than the before-change scenarios.

To exploit the reasons, we recalled the interview records. Two participants referried that the transparent glasses attracted them to spend more time observing the outdoor environment when they passed through the vertical traffic spaces, which might cause the longer duration. Another explanation could be that people tend to firstly build a comprehensive mental map (global vision) based on architectural cues and their perceptions about the spaces before the wayfinding [11]. When there are conflicts between the mental map and real-world cues during the process, people would struggle with comparisons between those information and constantly renew their maps. Majority of participants reported a deep impression of the optimized indoor spaces (see Fig. 4), because they were different from their cognitions of most educational buildings. These unusual spaces required them more efforts to update the mental map during their wayfinding process.

While the wayfinding duration have been dragged after the optimizations, it could be interpreted as beneficial due to the enhancement of spatial attraction. The modifications of indoor spaces and material of the external walls attracted participants' attention, leading them spend more time in wayfinding decisions. Meanwhile, they tended to observe the space more nuancedly, having longstanding memories of the architectural details within the routes. Inspiringly, architects can design unique spaces to retain human flows (curtail wayfinding efficiency) or make appropriate material changes of public buildings to fasten human flows (improve wayfinding efficiency).

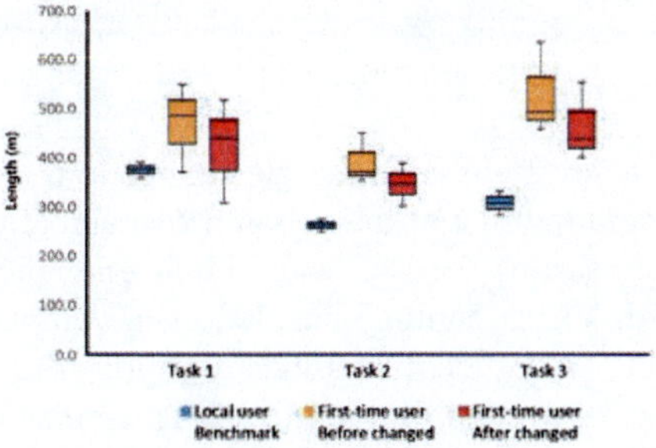

Fig. 11. Path length.

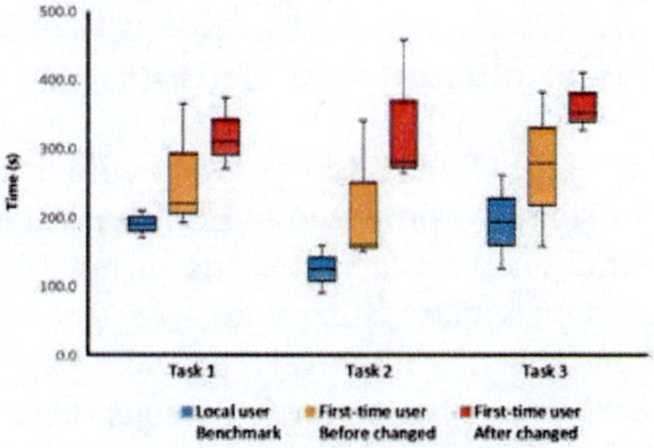

Fig. 12. Wayfinding duration.

3.4 Preferences for Route Selection

Comparing the participants' routes of local user and first-time user (see Fig. 13), we found that people especially first-time users presented two main preferences for route selection. First, they were prone to choose outdoor routes for wayfinding since the higher openness of outdoor spaces would facilitate their direction coordination system and keep them aware of their real-time location. Second, they preferred more straightforward routes with fewer turns. Most participants reflected that they were bad at sense of direction, so they could quickly forget the direction to their destination if they frequently made turns during the wayfinding process. Therefore, first-time users normally choose routes with higher simplicity in the building layouts, even if they are not the shortest routes.

The discrepancies of navigation between first-time and local users should be drawn attention and mediated in architecture design. In public buildings, the number of first-time users is normally larger than local users, which makes first-time users as a primarily focused group. Under the architecture design rules, designers should prioritize enhancement of visibility and accessibility rather than overestimate the shortest path design.

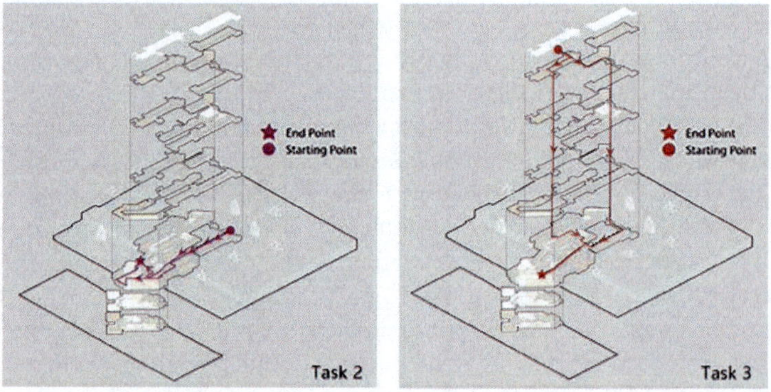

Fig. 13. First-time users' preferences for route selection.

4 Conclusion and Outlook

Taking the CDE buildings group with fluctuant elevations and intricate connections as a pilot study, we proposed an EBD approach integrating spatial analysis and virtual reality. Spatial analysis facilitated us to pinpoint the reasons for low wayfinding efficiency, based on which assumptions were put forward. VR technology enabled us to compare design schemes intuitively, providing affordable and efficient methods for optimization assessment. This research demonstrated the easy accessibility and high feasibility of EBD approach, which would encourage architects to focus on human-centered design.

This research concluded that the lack of openness in vertical traffic space was one of the dominant reasons for low wayfinding efficiency. It could be optimized by improving spatial visibility and accessibility. Admittedly, it is worth noting that other design factors (colors and materials) can also affect wayfinding efficiency, which incites in-depth research to explore more. For example, we noticed that wayfinding efficiency might also be susceptible to lightness. In terms of space organization, interior corridors (dimmer and lack of openness) and side corridors (brighter and large openness) could significantly impact people's route preference, consequentially wayfinding efficiency. Designers are expected to further ponder how to integrate various wayfinding elicitors into EBD approach and strives for more user-friendly architecture designs.

Acknowledgement. We would like to thank Dr. Panos Mavros for his advice during the research process, and we are grateful to Professor. Rudi Stouffs for providing the Hyve-3D system.

References

1. McCullough, C.S.: Evidence-based Design for Healthcare Facilities. Renee Wilmeth, Indianapolis (2010)
2. Buliung, R.N., Remmel, T.K.: Open source, spatial analysis, and activity-travel behaviour research: capabilities of the aspace package. J. Geogr. Syst. **10**(2), 191–216 (2008)
3. Hall, C.M.: Spatial analysis: a critical tool for tourism geographies. The Routledge Handbook of Tourism Geographies, pp. 178–188, London (2011)
4. Manley, E., Filomena, G., Mavros, P.: A spatial model of cognitive distance in cities. Int. J. Geogr. Inf. Sci. **35**(11), 2316–2338 (2021)
5. Ramírez, T., Hurtubia, R., Lobel, H., Rossetti, T.: Measuring heterogeneous perception of urban space with massive data and machine learning: an application to safety. Landsc. Urban Plan. **208**, 104002 (2021)
6. Benedikt, M.L.: To take hold of space: isovists and isovist fields. Environ. Plann. B. Plann. Des. **6**(1), 47–65 (1979)
7. Hillier, B., Hanson, J.: The Social Logic of Space. Cambridge University Press, Cambridge (1989)
8. Kuliga, S., Mavros, P., Brösamle, M., Hölscher, C.: Comparing human wayfinding behavior between a real, existing building, a virtual replica, and two architectural redesigns. In: Šķilters, J., Newcombe, N.S., Uttal, D. (eds.) Spatial Cognition 2020. LNCS (LNAI), vol. 12162, pp. 160–179. Springer, Cham (2020). https://doi.org/10.1007/978-3-030-57983-8_13
9. Li, S., Sun, C., Lin, Y.: A method of VR enhanced POE for way-finding efficiency in mega terminal of airport. In: Jeroen, van, A., Nicole, G., Kyung, H., Dan, L., Urvi, S. (eds.) POST-CARBON: Proceedings of the 27th CAADRIA Conference, vol. 1, pp. 79–88. Sydney (2022)
10. Turner, A., Penn, A.: Making isovists syntactic: isovist integration analysis. In: 2nd International Symposium on Space Syntax, vol. 1, pp. 103–121. Brasilia (1999)
11. Arthur, P., Passini, R.: Wayfinding: People, Signs, and Architecture. McGraw-Hill, New York (1992)

Smart City

Efficient Prediction of Indoor Airflow in Naturally Ventilated Residential Buildings Using a CFD-DNN Model Approach

Tran Van Quang[1], Nguyen Lu Phuong[2], and Dat Tien Doan[3](✉)

[1] Department of Architectural Engineering, Kyung Hee University, 1732, Deogyeong-daero, Giheung-gu, Yongin-si, Gyeonggi-do 17104, Republic of Korea

[2] Faculty of Environment, Tan Binh District, Ho Chi Minh City University of Natural Resources and Environment, 236B, Le Van Sy Street, Ward 1, Ho Chi Minh, Vietnam

[3] Department of Built Environment Engineering, School of Future Environments, Auckland University of Technology, Auckland CBD, 55 Wellesley Street East, Auckland 1010, New Zealand

dat.doan@aut.ac.nz

Abstract. Predicting indoor airflow in multi-storey residential buildings is crucial for energy-efficient natural ventilation systems. The indoor environment significantly affects human well-being due to extended indoor time and potential health risks. Precise and efficient airflow simulations are necessary to ensure thermal comfort and air quality. This study introduces a novel approach combining Computational Fluid Dynamics (CFD) simulations with machine learning techniques to predict indoor airflow. Specifically, we explore using a Deep Neural Network (DNN) model for accurate indoor airflow forecasting. The DNN effectively reproduces airflow patterns and temperature distributions. Integrating CFD simulations halves test scenario anticipation time, highlighting efficient indoor airflow prediction potential. Using a data-driven approach, this research demonstrates the feasibility of swiftly and accurately predicting indoor airflow in naturally ventilated residential buildings. Such models can optimize indoor air quality, thermal comfort, and energy efficiency, contributing to sustainable building design and operation.

Keywords: Computational fluid dynamics (CFD) · Machine learning · Deep neural network (DNN) · Indoor airflow prediction · Natural ventilation · Built environment

1 Introduction

Individuals spend around 85% of their daily lives inside buildings [1–3]. Since the early 21-century, the issues about indoor air quality, natural ventilation systems, and occupants' health in common indoor spaces have more interest in infectious airborne virus risks in buildings such as Severe Acute Respiratory Syndrome (SARS-CoV, 2003), Middle East Respiratory Syndrome (MERS-Cov, 2012), and Coronavirus Disease (COVID-19, SARS-CoV-2, 2019). Optimal indoor conditions are essential for health and productivity [4–10].

W. Guo et al. (Eds.): GBCESC 2023, LNCE 328, pp. 759–770, 2024.
https://doi.org/10.1007/978-981-99-9947-7_76

The relevance of natural ventilation lies in its potential to save energy, improve thermal comfort, and enhance indoor air quality. However, achieving this is complex, requiring an understanding of forces and their interactions [11]. While natural ventilation can prevent Sick Building Syndrome, it hinges on climate and manual control [12–15]. Computational Fluid Dynamics (CFD) has emerged as a valuable tool for predicting natural ventilation outcomes [16–25], while Machine Learning (ML), particularly DNNs, offers rapid and accurate airflow forecasts [26, 27].

However, CFD complexities and resource demands have paved the way for DNNs as a pragmatic alternative. Using CFD data, this study harnesses DNNs to predict airflow in multi-storey residential buildings. The approach delivers:

- A tailored DNN model surrogate for the CFD model.
- Holistic solutions for velocity, temperature, and pressure fields.
- Drastically reduced computation time compared to conventional CFD.

2 Related Works

Machine learning algorithms have brought significant advancements in various fields, from initial perception to deep learning models [28]. Leveraging artificial intelligence (AI), such as Deep Neural Networks (DNNs), for airflow prediction shows promise as an alternative to Computational Fluid Dynamics (CFD) analysis in the future [26, 27]. Unlike CFD, which relies on wind tunnel test data, DNNs offer the advantage of swiftly deriving wind velocity distributions based solely on input conditions of various structures within the target area. Consequently, this approach significantly reduces computation time, enabling parametric studies, short-time optimization calculations, and various other applications. In recent trends, practical research on accuracy validation has emerged in different domains, including meteorology and wind power generation prediction [29–31]. While ongoing efforts seek to enhance CFD techniques through novel algorithms and turbulence models [32, 33], there has recently been a surge of interest in new tools to replace CFD in the analysis of fluid mechanics problems, typically for faster predictions or as an aid to the CFD simulation for improved accuracy. Artificial intelligence and data-driven models are gaining popularity in applications with digitalization and large amounts of data. In particular, deep learning and DNN are particularly intriguing in terms of their skills as universal nonlinear approximates, large dimensionality fields, and computational inexpensiveness. Several studies have replaced CFD simulations with deep learning methods, using DNNs as surrogate models for numerical simulations. Depending on the situation, Surrogate modelling can enable magnitudes quicker airflow prediction [37] and even real-time forecasts addressing one of the two issues mentioned by CFD simulations [34]. As a result, Deep Neural Network (DNN) models, with their superior modelling capabilities and high-speed computing prowess, emerge as viable alternatives for airflow prediction in naturally ventilated residential buildings.

In indoor airflow prediction in naturally ventilated residential buildings, while CFD simulations can provide valuable insights into realistic wind patterns indoors and outdoors, their setup demands considerable technical expertise. Additionally, addressing numerical problems and hardware requirements for processing time can lead to substantial expenses. In the past, researchers have developed outdoor wind velocity ratio prediction models to anticipate outdoor ventilation possibilities on an urban scale, enabling faster decision-making during the early design phase. However, existing prediction models often fall short: linear regressions fail to adequately explain high-density urban airflow, and nonlinear regressions with predefined equation forms present significant barriers for urban planners and architects due to their reliance on extensive technical expertise. To address these challenges, a more sophisticated data-driven machine-learning approach emerges as a promising solution. This approach makes it possible to train nonlinear regression models without predefined equation formats, making it more accessible and adaptable for users.

3 Methodology

The methodology employed in this study is visually represented in Fig. 1, outlining the process from data generation to model training and subsequent performance evaluation. The techniques utilized to accomplish the study's designated objectives are expounded upon with greater elaboration in subsections.

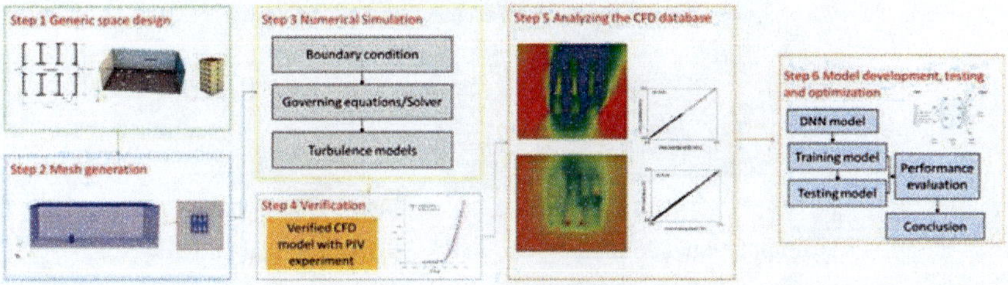

Fig. 1. Research framework of the study

3.1 Generic Space

The study utilized a generic residential building comprising six storeys, 9 m × 9 m × 17.7 m. Each floor had a corridor and six rooms, each measuring 3 m × 3.6 m × 2.9 m, with one window of size 1.5 m × 1.5 m in each room and two corridor windows of length 0.9 m × 1.5 m. As shown in Fig. 2, the residential building was generated on Rhinoceros 7 software [35]. The residential building model is located inside a computational domain with distances from their inlet, outlet, and the boundary of lateral (left and right) to their nearest surfaces of 4 H, 10 H, and 4 H, respectively, and the height of the computational domain is 5 H, shown in Fig. 3.

3.2 Governing Equations and Turbulence Models

The airflow within the computational domain is modelled for the incompressible fluid phase. The primary equations governing mass and momentum are presented in Eqs. (1) and (2).

$$\frac{\partial(\overline{u}_i)}{\partial x_i} = 0 \tag{1}$$

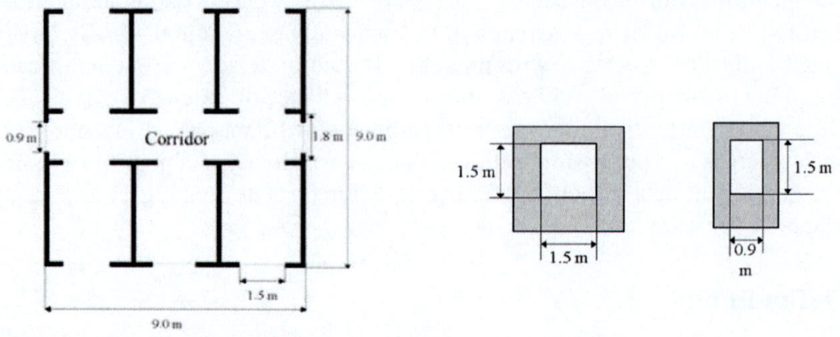

(a) (b) (c)

Fig. 2. Geometrical dimensions of the residential building; (a) cross-section of the residential building; (b) window at each room; (c) window in the corridor.

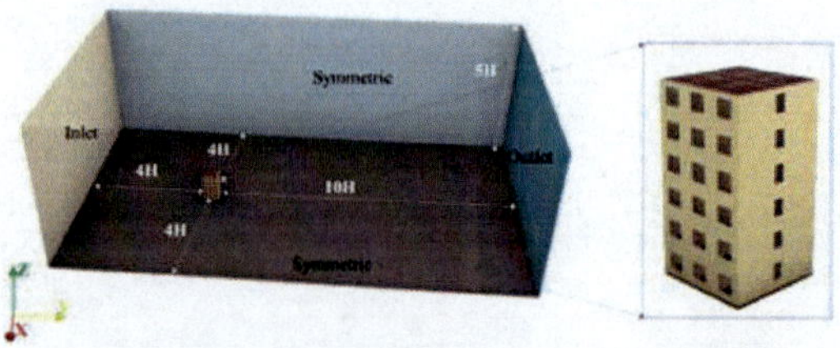

(a) Computational domain (b) Residential building

Fig. 3. The residential building 3D model.

$$\frac{\partial(\overline{u}_i)}{\partial t} + u_j\frac{\partial u_i}{\partial x_j} = -\frac{1}{\rho}\frac{\partial p}{\partial x_i} + \frac{\partial}{\partial x_j}\left(2\nu S_{ij} - \overline{u_i' u_j'}\right) \tag{2}$$

In the context of the equation discussed, the variable ν represents the kinematic viscosity, S_{ij} corresponds to the strain–rate tensor, p donates the mean pressure, ρ represents the air density, u' represents the fluctuating velocity component and $\overline{u}_i$ denotes the ensemble-average velocity.

The power law has been applied to the inlet by defining the atmosphere boundary condition.

$$U(z) = \frac{u^*}{k}\ln\left(\frac{z}{z_0}\right) \tag{3}$$

In the provided formulas, U stands for the average wind speed, y signifies the elevation, and u* corresponds to the atmospheric boundary layer (ABL) friction velocity (u* = 2.89 m/s). The von Karman constant, denoted as k, is taken as 0.415. The initial boundary conditions were measured for the simulations and presented in Table 1.

Table 1. Details of CFD simulation setting.

Computational domain	256.8 m × 150.6 m × 88.5 m (L × W × H)
Algorithm	Steady-state with a SIMPLE algorithm
Grid design	5.2 million elements (hexahedral meshes)
Turbulence model	RNG $k - \epsilon$ model
Inflow	Power law for velocity and turbulent intensity $\varepsilon(z) = C_\mu^{3/4} \frac{k^{3/2}}{l_t}$
Outflow	Fixed static pressure
Top and lateral	Symmetrical boundary
Building wall and ground surface	Power law for smooth surface wall, Standard wall function

3.3 CFD Simulation Validation

The precision of the numerical simulations was exemplified through a comparison between anticipated velocity components derived from the RANS model and empirical outcomes obtained from wind tunnel experiments conducted on a multi-story residential structure [36]. The simulation results closely matched the experimental data for velocity profiles on the upstream side of the building (Lines 1 and 2). Still, discrepancies were observed downstream (Line 3), see Fig. 4. According to previous studies, only Large Eddy Simulation (LES) can produce accurate forecasts in such cases [37]. However, the overall comparison showed that the velocity trends from the CFD model were in line with the reference research [36], indicating that CFD can effectively replicate the velocity distribution.

3.4 DNN Model and Optimization

The DNN model used in this study is an advanced version of ANN, featuring a complex architecture that enables mapping functions from input to output data. It consists of input and output layers, along with multiple hidden layers, where each layer contains one or more neurons connected to neurons in adjacent layers. Nonlinear activation functions are applied to neuron inputs during the forward pass. The model is trained by iteratively adjusting weights and biases until optimal values are attained, using error propagation and gradient descent approaches. R^2, MSE and MAE metrics are employed to evaluate the model's performance, offering a comprehensive analysis of its predictive capabilities:

$$R^2 = 1 - \left[\frac{\sum_{i=1}^{m} (y_i - \hat{y}_i)^2}{\sum_{i=1}^{m} (y_i - \overline{y})^2} \right] \tag{4}$$

$$RMSE = \sqrt{\frac{\sum_{i=1}^{m} (y_i - \hat{y}_i)^2}{m}} \tag{5}$$

$$MAE = \frac{1}{m} \sum_{i=1}^{m} |y_i - \hat{y}_i| \tag{6}$$

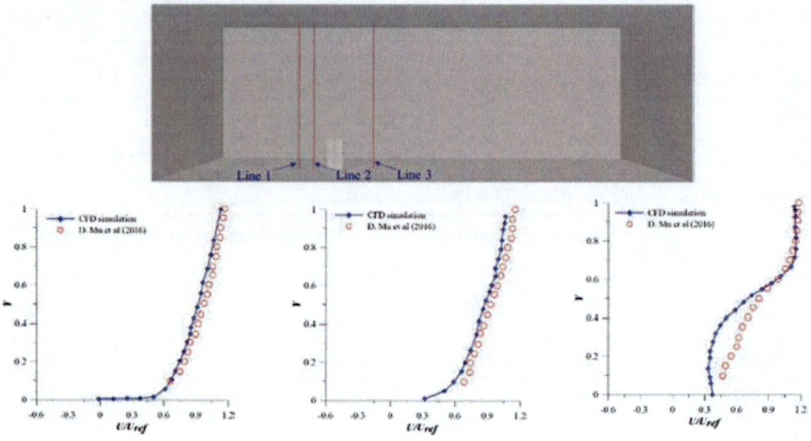

(a) Line 1 (b) Line 2 (c) Line 3

Fig. 4. A comparison of velocity profiles from experimental measurements (the dotted line) across various model sections (lines 1–3) against the CFD projections (the solid blue line)

The input layer takes in both the input vector and external environmental factors, whereas the output layer produces data related to temperature and velocity. Neurons in subsequent layers interpret signals from the preceding layer and pass them forward. The final output vector transforms into tensors for temperature and velocity, depicting the building's three-dimensional distribution of temperature and velocity. Experimental results are the basis for most boundary conditions and are consistently maintained across various scenarios. Detailed hyperparameter settings are provided in Table 2.

Table 2. Configuration of hyperparameters for the DNN

Number of hidden layers	1/3/5
Number of neurons per hidden layer	20/40/80/120/200
Activation function	Tanh
Gradient descent algorithm	Adam
Learning rate	0.001
Batch size	1

Determining when to stop training is crucial in the training process due to its non-convex and iterative nature. The training is considered complete when the monitored metric, i.e., the validation error, ceases to decrease consecutively for 5000 epochs. At this stage, the training is deemed to have reached convergence. Following termination, the parameters at the early-stopping point are retrieved. Subsequently, the DNN model undergoes testing using the testing data for performance evaluation.

4 Results and Discussion

4.1 DNN Model Performance

The CFD model for cross ventilation in multi-storey buildings was developed based on buoyancy-driven effects and adjusted to match experimental conditions (Fig. 3). Effective ventilation systems within residential structures can enhance indoor air quality and mitigate the release of pollutants. The study aimed to predict indoor airflow distribution and its impact on comfort conditions and indoor air pollution. Inadequate airflow was observed in certain areas, particularly the bedroom, highlighting the importance of well-designed ventilation systems.

Figure 5(b) shows airflow patterns under cross-natural ventilation mode. The deep learning model (Deep Neural Network-DNN) provided similar flow patterns with minor differences in recirculation details. Wind direction influenced the ventilation rate, and DNN successfully predicted its impact.

Figure 5(c) displayed the indoor temperature distribution, with static temperature fluctuations between 298^0K to 313^0K. Wind direction plays a crucial role in temperature distribution. The DNN model showed close agreement with CFD simulation results. The DNN model effectively captured indoor temperature distribution complexities.

Figure 6 illustrates the alignment of CFD and DNN predictions regarding indoor velocity and temperature for the training set. The DNN proficiently mirrors the airflow patterns observed in the training samples, with only slight deviations seen in the test set. Interestingly, the DNN displays better performance on the test set than the training set, and its consistency is on par with CFD outcomes. The relative error for predicting velocity and temperature stands below 0.5%, underscoring the DNN's precision in capturing variations.

To measure the disparity between CFD simulations and DNN forecasts, Eq. (7) establishes a metric, specifically the relative error.

$$E_{relative} = \frac{\left|\hat{y}_s - y_s\right|}{y_s} \times 100\% \tag{7}$$

where $\hat{y}_s$ represents the DNN's estimated value at a specific location, y_s denotes the actual CFD value at that identical spot, and s refers to the location.

The DNN models exhibit computational efficiency, enabling accurate predictions of indoor airflow on the test dataset. The DNN models adapt well to new design inputs for ventilation rate predictions, as indicated by their performance on the training dataset. Detailed statistical analysis of the velocity and temperature forecasting model is provided in Table 3.

4.2 Computational Time

The study compared the computational time between the DNN model and CFD simulations. GPU reference measurements were unavailable, so CPU time was used for a fair assessment. The DNN model demonstrated significantly faster computation, taking less than a second, while CFD simulations required around 20 h per case when utilizing

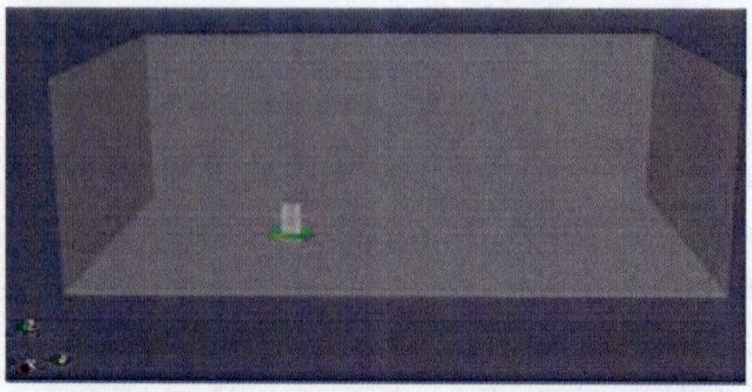

(a) Cross-section location in buildings

CFD simulation Deep learning model

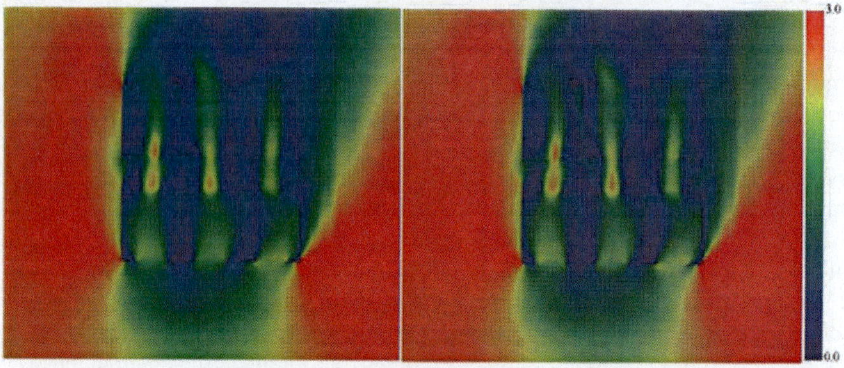

(b) Velocity distribution

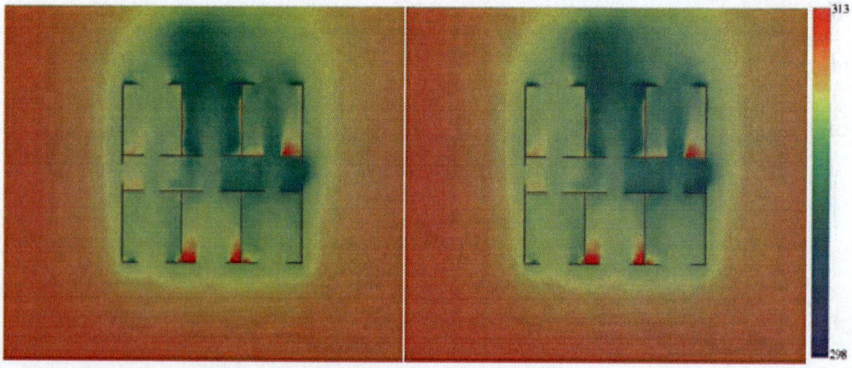

(c) Temperature distribution

Fig. 5. Contrasting CFD simulation (buoyant SimpleFoam) with Deep Learning forecasts, illustrating velocity and temperature distributions in buildings on the third floor

multiple cores. DNN installation and training took approximately 5 h. Overall, the DNN model resulted in an 80% decrease in time compared to CFD simulations, making it a promising and efficient prediction tool, see Fig. 7.

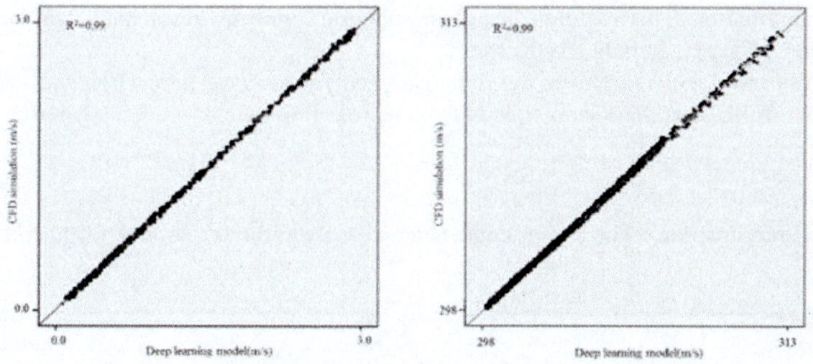

(a) Indoor air velocity (b) Indoor air temperature

Fig. 6. Correlations between the CFD simulations' and deep learning estimates' results

Table 3. Overview of the performance metrics for the created prediction models

Model	U						T					
DNN	Train			Test			Train			Test		
	R2	RMSE	MAE	R^2	RMSE	MAE	R^2	RMSE	MAE	R^2	RMSE	MAE
	0.98	0.04	0.03	0.89	0.14	0.1	0.97	0.1	0.08	0.90	0.24	0.19

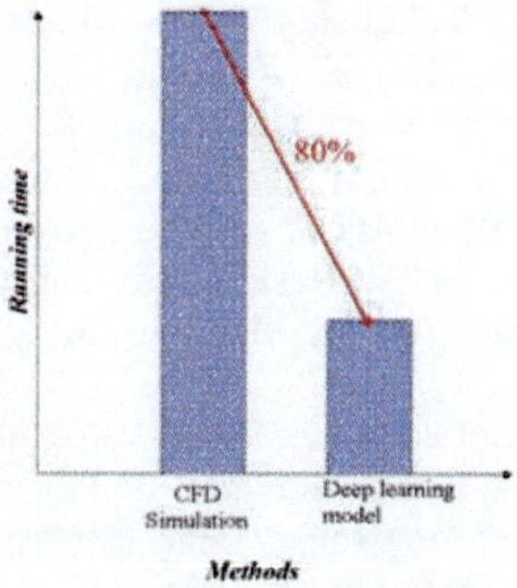

Fig. 7. Evaluating the speed of case predictions: Deep Learning model versus CFD simulation

5 Conclusions

This study presents a framework combining CFD simulation and DNN to assess wind-driven natural ventilation in buildings. The DNN model accurately predicts indoor airflow and temperature, outperforming CFD in speed. Further exploration of higher Reynolds number flows and using DNN predictions as warm-up settings for CFD simulations could enhance the approach's applicability in architectural design optimization.

Acknowledgements. This research has been supported with a Presidential Scholarship from Kyung Hee University, Republic of Korea.

The author would like to express our sincere appreciation to The Human Behaviour and Energy Laboratory (HuBEL) at the Kyung Hee University, the Republic of Korea, supported facilitates research.

Author's Contribution. The author contributed equally to the preparation of this manuscript.

References

1. Hajdukiewicz, M., Geron, M., Keane, M.M.: Calibrated CFD simulation to evaluate thermal comfort in a highly-glazed naturally ventilated room. Build. Environ. **70**, 73–89 (2013). https://doi.org/10.1016/j.buildenv.2013.08.020
2. Jamaludin, A.A., Hussein, H., Ariffin, A.R.M., Keumala, N.: A study on different natural ventilation approaches at a residential college building with the internal courtyard arrangement. Energy and Buildings **72**, 340–352 (2014). https://doi.org/10.1016/j.enbuild.2013.12.050
3. Shaikh, P.H., Bin, N., Nor, M., Nallagownden, P.: Robust stochastic control model for energy and comfort management of buildings. Aust. J. Basic Appl. Sci. **7**, 137–144 (2013)
4. Costa, A., Keane, M.M., Torrens, J.I., Corry, E.: Building operation and energy performance: monitoring, analysis and optimisation toolkit. Appl. Energy **101**, 310–316 (2013). https://doi.org/10.1016/j.apenergy.2011.10.037
5. Yang, R., Wang, L.: Multi-objective optimization for decision-making of energy and comfort management in building automation and control. Sustain. Cities Soc. **2**(1), 1–7 (2012). https://doi.org/10.1016/j.scs.2011.09.001
6. Chenari, B., Carrilho, J.D., Gameiro da Silva, M.: Towards sustainable, energy-efficient and healthy ventilation strategies in buildings: a review. Renew. Sustain. Energy Rev. **59**, 1426–1447 (2016). https://doi.org/10.1016/j.rser.2016.01.074
7. Chen, Y., Tong, Z., Zheng, Y., Samuelson, H., Norford, L.: Transfer learning with deep neural networks for model predictive control of HVAC and natural ventilation in smart buildings. J. Clean. Prod. **254**, 119866 (2020). https://doi.org/10.1016/j.jclepro.2019.119866
8. Chen, Z., Jiang, Y., Tong, Z., Tong, S.: Residual stress distribution design for gear surfaces based on genetic algorithm optimization. Materials **14**(2), 1–17 (2021). https://doi.org/10.3390/ma14020366
9. Park, D.Y., Chang, S.: Effects of combined central air conditioning diffusers and window-integrated ventilation system on indoor air quality and thermal comfort in an office. Sustain. Cities Soc. **61**(October), 2020 (2019). https://doi.org/10.1016/j.scs.2020.102292
10. Huang, M., Liao, Y.: Development of an indoor environment evaluation model for heating, ventilation and air-conditioning control system of office buildings in subtropical region considering indoor health and thermal comfort. Indoor Built Environ. **31**(3), 807–819 (2022). https://doi.org/10.1177/1420326X211035550
11. Tominaga, Y., Blocken, B.: Wind tunnel experiments on cross-ventilation flow of a generic building with contaminant dispersion in unsheltered and sheltered conditions. Build. Environ. **92**, 452–461 (2015). https://doi.org/10.1016/j.buildenv.2015.05.026
12. Common, M.: Uncertainty and the environment: implications for decision-making and environmental policy. Richard A. Young. Edward Elgar, Cheltenham, UK; Northampton, MA, 2001. ISBN 1 84064 626 8 (hardcover, 59.95 pounds sterling, $85.00). J. Econ. Psychol. **24**(3), 422–424 (2003). https://doi.org/10.1016/S0167-4870(03)00018-7
13. Al-qahtani, A.S.: Subjective assessment of indoor air quality in office buildings, Doctoral dissertation (1993)

14. Wargocki, P., et al.: Ventilation and health in non-industrial indoor environments: report from a european multidisciplinary scientific consensus meeting (EUROVEN). Indoor Air **12**(2), 113–128 (2002). https://doi.org/10.1034/j.1600-0668.2002.01145.x
15. Klepeis, N.E., et al.: The national human activity pattern survey (NHAPS): a resource for assessing exposure to environmental pollutants. J. Expo. Anal. Environ. Epidemiol. **11**(3), 231–252 (2001). https://doi.org/10.1038/sj.jea.7500165
16. Nielsen, P.V.: Fifty years of CFD for room air distribution. Build. Environ. **91**, 78–90 (2015). https://doi.org/10.1016/j.buildenv.2015.02.035
17. Liu, J., Heidarinejad, M., Gracik, S., Srebric, J.: The impact of exterior surface convective heat transfer coefficients on the building energy consumption in urban neighborhoods with different plan area densities. Energy Build **86**, 449–463 (2015). https://doi.org/10.1016/j.enbuild.2014.10.062
18. Liu, J., Heidarinejad, M., Pitchurov, G., Zhang, L., Srebric, J.: An extensive comparison of modified zero-equation, standard k-ε, and LES models in predicting urban airflow. Sustain. Cities Soc. **40**(March), 28–43 (2018). https://doi.org/10.1016/j.scs.2018.03.010
19. Liu, P.C., Te Lin, H., Chou, J.H.: Evaluation of buoyancy-driven ventilation in atrium buildings using computational fluid dynamics and reduced-scale air model. Build. Environ. **44**(9), 1970–1979 (2009). https://doi.org/10.1016/j.buildenv.2009.01.013
20. Cook, M.J., Ji, Y., Hunt, G.R.: CFD modelling of natural ventilation: combined wind and buoyancy forces. Int. J. Vent. **1**(3), 169–179 (2003). https://doi.org/10.1080/14733315.2003.11683632
21. Zhai, Z.J.: Sensitivity analysis and application guides for integrated building energy and CFD simulation 2. coupling-relevant building and environmental characteristics. Mech. Eng. **38**(9), 1060–1068 (2006)
22. Cook, M.J., Lomas, K.J.: Buoyancy-driven displacement ventilation flows: evaluation of two eddy viscosity turbulence models for prediction. Build. Serv. Eng. Res. Technol. **19**(1), 15–21 (1998). https://doi.org/10.1177/014362449801900103
23. Yazarlou, T., Barzkar, E.: Louver and window position effect on cross-ventilation in a generic isolated building: a CFD approach. Indoor Built Environ. **31**(6), 1511–1529 (2022). https://doi.org/10.1177/1420326X211061685
24. Zhang, Z.-Y., Yin, W., Wang, T.-W., O'Donovan, A.: Effect of cross-ventilation channel in classrooms with interior corridor estimated by computational fluid dynamics. Indoor Built Environ. **31**(4), 1047–1065 (2022). https://doi.org/10.1177/1420326X211054341
25. Qi, J., Wei, C.: Performance evaluation of climate-adaptive natural ventilation design: a case study of semi-open public cultural building. Indoor Built Environ. **30**(10), 1714–1724 (2020). https://doi.org/10.1177/1420326X20961495
26. Guo, X., Li, W., Iorio, F.: Convolutional neural networks for steady flow approximation. In: Proceedings of the ACM SIGKDD International Conference on Knowledge Discovery and Data Mining, vol. 13–17-August-2016, pp. 481–490 (2016). https://doi.org/10.1145/2939672.2939738
27. Umetani, N., Bickel, B.: Learning three-dimensional flow for interactive aerodynamic design. ACM Trans. Graph. **37**(4), 1–10 (2018). https://doi.org/10.1145/3197517.3201325
28. Goodfellow, I., Bengio, Y., Courville, A.: Deep learning book pdf. Nature **29**(7553), 1–73 (2016)
29. Huang, C.-J., Kuo, P.-H.: A short-term wind speed forecasting model by using artificial neural networks with stochastic optimization for renewable energy systems. Energies **11**(10), 2777 (2018). https://doi.org/10.3390/en11102777
30. Zhou, J., Liu, H., Yanhe, X., Jiang, W.: A hybrid framework for short term multi-step wind speed forecasting based on variational model decomposition and convolutional neural network. Energies **11**(9), 2292 (2018). https://doi.org/10.3390/en11092292

31. Harbola, S., Coors, V.: One dimensional convolutional neural network architectures for wind prediction. Energy Convers. Manag. **195**(February), 70–75 (2019). https://doi.org/10.1016/j.enconman.2019.05.007
32. Berger, M.J., Aftosmis, M.J., Marshall, D.D., Murman, S.M.: Performance of a new CFD flow solver using a hybrid programming paradigm. J Parallel Distrib Comput **65**(4), 414–423 (2005). https://doi.org/10.1016/j.jpdc.2004.11.010
33. Spalart, P.R., Garbaruk, A.V.: Correction to: a new "λ2" term for the spalart–allmaras turbulence model, active in axisymmetric flows. Flow Turbul. Combust. **109**(1), 253–253 (2021). https://doi.org/10.1007/s10494-021-00313-7
34. Hintea, D., Brusey, J., Gaura, E.: A study on several machine learning methods for estimating cabin occupant equivalent temperature. In: ICINCO 2015 - 12th International Conference on Informatics in Control, Automation and Robotics, Proceedings, vol. 1, pp. 629–634 (2015). https://doi.org/10.5220/0005573606290634
35. R. M. & Associates: Rhinoceros. NURBS modeling for windows, p. 256 (2008)
36. Mu, D., Gao, N., Zhu, T.: Wind tunnel tests of inter-flat pollutant transmission characteristics in a rectangular multi-storey residential building, part A: effect of wind direction. Build. Environ. **108**, 159–170 (2016). https://doi.org/10.1016/j.buildenv.2016.08.032
37. Jiang, Y., Chen, Q.: Study of natural ventilation in buildings by large eddy simulation. J. Wind Eng. Ind. Aerodyn.Aerodyn. **89**(13), 1155–1178 (2001). https://doi.org/10.1016/S0167-6105(01)00106-4

A Genetic Algorithm Using Diversity-Concern Principle to Solve Robust Influence Maximization Problem on Urban Transportation Networks

Minghao Chen and Shuai Wang(✉)

School of Intelligent Systems Engineering, Sun Yat-sen University, Shenzhen 518107, China
wangsh368@mail.sysu.edu.cn

Abstract. In recent years, the influence maximization has become a hot topic in the field of complex networks. The so-called influence maximization problem is to select a certain scale of nodes in a given network to work as seeds, and this subset can achieve the maximal information propagation effect. Networks like transportation systems are usually exposed to complex environment, its internal components are threatened by external disturbance, and its information transmission process is also at risk. The corresponding optimization problem can be modeled as the robust influence maximization problem (RIM). In cities, roads or intersections may face various failures or disruptions, such as natural disasters, traffic accidents or other emergencies. City managers need to design robust public transportation systems to minimize the impact of these situations on the overall transportation network. However, the existing work only considers the stable network structure or the damaged link, and lack the research on destructions caused by nodal failures. Also, the current optimization methods do not integrate the information of the search process well to solve the problem of robust influence maximization. To tackle these issues, the paper proposes a metric to evaluate seed robustness under node-specific attacks and a genetic algorithm called DC-GA-RIM. This method integrates the diversity concern principle and includes problem-oriented operators to improve its search ability. The effectiveness of DC-GA-RIM in solving RIM problems has been proven on a variety of networks. Stations and traffic hubs can be detected using the proposed method.

Keywords: Transportation networks · Influence maximization problem · Robustness · Optimization · Genetic algorithm

1 Introduction

In recent years, the problem of maximizing influence in the field of complex networks has been one of the hot topics of interest. Traffic congestion is a widespread issue in urban areas. The road network in a city can be conceptualized as a graph, where nodes represent road intersections, and edges represent the roads connecting them. To tackle the problem of traffic congestion, city administrators often deploy monitoring devices

W. Guo et al. (Eds.): GBCESC 2023, LNCE 328, pp. 771–781, 2024.
https://doi.org/10.1007/978-981-99-9947-7_77

at intersections to minimize its occurrence. However, given the constraints of limited budgets and resources, city administrators need to make informed decisions about which intersections to equip with surveillance equipment in order to maximize their impact. As a result, selecting the most influential "seed" intersection nodes becomes a critical issue, which can be modeled as an impact maximization problem [1]. Many methods have been developed to solve the influence maximization problem. For example, heuristic-based method [2, 3], community-based method [4, 5], and population search-based method [6, 7], all of which provide alternatives that can be referred to for seeds selection.

Networks tend to be operated with complex application environments and may experience node and link function failures due to internal or external factors. Thus, network robustness is crucial for system security in social networks, World Wide Web, and transportation networks. Network performance is closely related to its structure [8], and changes in structure can affect member behavior and influence propagation [1, 4]. Evaluating and optimizing network robustness has become a hot research topic in the field of complex networks and transportation analyses.

Challenges persist in the influence maximization problem. Network structure changes due to unpredictable attacks on nodes or links can compromise the function and cause severe damage [4, 9]. Existing work either ignores network structure change or only considers link-specific attacks, limiting its applicability. Genetic algorithms are effective but have poor population diversity and do not consider population information during the search process. Selecting seed nodes that satisfy both robustness and influence maximization is a cutting-edge research area, but its applications are limited to scenarios with uncertain parameters [10, 11]. Current research has yet to resolve the issue of designing a proven seed optimization selection scheme under node-based attacks.

To address shortcomings of current researches, this paper proposes a solution to the influence maximization problem by analyzing network impairment and node-specific attacks. It introduces a new influence performance metric based on previous research [12] and an optimization algorithm (DC-GA-RIM) that uses genetic algorithms and diversity concerns to find the most influential seed set. The algorithm incorporates various operators and effectively addresses the problem of insufficient diversity in traditional genetic algorithms. It is demonstrated that the seed set obtained under the guidance of the designed metrics can achieve better influence propagation in the process of network damage.

The paper is structured as follows, Section 2 briefly describes the influence propagation model and the associated performance evaluation methods in influence maximization problems, as well as background work on related network structure robustness metrics. Section gives an evaluation metric of seeds' robust influence performance. Section gives the flowchart of the DC-GA-RIM algorithm and the details of the operator. Section V shows the corresponding experimental results. Finally, the full-text summary is given in Section VI.

2 Background

2.1 Influence Maximization Problem and Its Related Work

There are two problems to focus on in the influence maximization problem.

The first is the information propagation model. In the process of seeds trying to spread influence outward, the propagation model defines the rules of information propagation. Considering that the IC [13] model exhibits a broader theoretical value and has received more attention in previous studies [2, 14], this work chooses the IC model as the influence propagation model to study the influence maximization problem.

The second is the influence performance evaluation method. In the paper [12], a fast approximation of the influence level based on the 2-hop neighbors (i.e. neighbors of neighbors) of the seeds is proposed, defined as follows.

$$\hat{\sigma}(S) = \sum_{s \in S} \hat{\sigma}\{s\} - \left(\sum_{s \in S} \sum_{c \in H_s \cap S} p(s, c)\left(\sigma_c^1 - p(c, s)\right) \right) - \chi \tag{1}$$

Here, H_s is the 1-hop neighbor (i.e. direct neighbors) of node s, $p(s, c)$ is the propagation probability from activated node s to inactive node c, and σ_c^1 is the 1-hop coverage area of node c. χ denotes the overlapping influence of the newly activated node on the original seed, which has been defined in the paper [12].

2.2 Network Robustness and Its Related Work

Realistic network systems are often exposed to complicated application environments, and may even suffer from malicious attacks that cause nodes or links failed. To address these issues, it is necessary to consider the robustness of the seeds in the determination process. The literature [9] proposes an evaluation factor R that considers the network robustness performance under structural attacks. The specific definition is as follows.

$$R = \frac{1}{N} \sum_{Q=1}^{N} s(Q) \tag{2}$$

Here, $s(Q)$ is the maximum connected cluster of the network after removing Q nodes, N is the total number of nodes in the network, and $1/N$ is the normalization factor to ensure that the different networks can be compared fairly with each other.

3 Robust Influence Performance Evaluation of Seeds Under Node-Specific Attacks

3.1 Evaluation Metric Design

Realistic network systems are long exposed to complex environments, for example, traffic networks are often exposed to events such as traffic accidents. Once the more influential nodes are attacked, it will inevitably cause significant disruption and damage to the function of the whole traffic network as well as the task of influence propagation. Most of previous studies are limited to the case when the network structure is not changed or only consider the case when the network structure is attacked by links, and little attention has been paid to the aforementioned case when the network node-specific attack. In this work, a new seeds robust influence evaluation metric is developed based on R from the perspective of node-specific attacks.

Next, we describe our evaluation method in detail. In the process of seeds propagating their influence outward, we re-evaluate the influence of the seed set S after each suffered attack. Since the propagation processes under suffered node attacks are uncorrelated with each other, the summation and normalization operations in R can be adopted to obtain the combined influence performance of the seed set S under node-specific attacks. According to the mechanism as above, this paper defines the robust influence performance evaluation metric of the seed set S as follows.

$$R_S = \frac{1}{N}\sum_{Q=1}^{N} \hat{\sigma}(S|Q) \tag{3}$$

Here, similar to R, N denotes the number of nodes in the current network, and $\hat{\sigma}(S|Q)$ denotes the estimated influence of the seed set S after removing Q nodes. The evaluation process of R_S is as follows, the attacker selects the node with the largest degree for each attack, evaluates the influence of the seed set on the current network, and obtains the evaluation result after normalization. In this paper, the evaluation metric R_S is designed considering the influence level of the seed set under the node-specific attack. Following the network performance evaluation in R a normalization operation is considered to ensure the comparability of the values under different network sizes. In addition, the quick evaluation method of influence level in $\hat{\sigma}(S)$ is introduced to reduce the computational cost, thus providing a more reliable guideline for the selection of the seed set.

4 DC-GA-RIM

4.1 Algorithm Framework

Algorithm 1: DC-GA-RIM

Input:

G_0: Input network;
K: Seed set size;
Mangena: Maximum evolutionary generation;
Initial Size: Initial population size;
Pop Size: Population size;
p_c: Crossover operator execution probability;
p_m: Mutation operator execution probability;
p_l: Local search operator execution probability;

Output:

$S^* = \{S_1, S_2, \ldots, S_K\}$: Seed set;

Execute the initialization operation to generate the initialized population Pop_g of size *InitialSize* based on the structure information of G_0, $Gen = 0$;
while (*Gen* < *MaxGen*) ***do***
  Execute the crossover operator at probability of p_c to expand the size of population Pop_g to *PopSize*;
  Execute the mutation operator at probability of p_m for each individual P in Pop_g;
  Execute the local search operator at probability of p_l for each individual P in Pop_g;
  Execute the selection operator to update Pop_g, find those better individuals and inherit into the next generation population with a size of *InitialSize*;
  Update S^*, $Gen = Gen + 1$;
end while;
Output S^*;

4.2 Initialization Operator

To address the shortcomings of the previous methods, the initialization operator designed in this paper follows the principle of individual diversity and combines three strategies of Degree-first, PageRank-first and Clustering-first to initialize the population. Specifically, the initialization of the first 1/3, middle 1/3, and remaining 1/3 individuals of the population performs the three strategies respectively. In detail, based on the importance information of Degree, PageRank, and Clustering as indicators for judging the importance, a given number of nodes are selected as the genes of individuals using the roulette wheel method.

4.3 Crossover Operator

In this work, a genetic diversity-based crossover ope0rator is used, utilizing three strategies (uniform, two-point, and single-point crossovers) to increase the population size to *PopSize*. Any duplicate nodes within an individual are corrected by randomly selecting a node in the network to replace the duplicate node until no duplicates exist in the individual.

4.4 Mutation Operator

This paper's mutation operator uses probability diversity, adjusting mutation probability based on an individual's fitness value. First, the fitness value of all individuals is evaluated using R_S, and the sum of their fitness is obtained. For each individual, the mutation operator executes with a probability of $p_m * \beta$ ($\beta = 1$ - this individual's fitness value/the sum of all individuals' fitness). The mutation replaces a random gene in the individual with another node while ensuring there are no duplicate nodes. If duplicate nodes exist, the correction operation is performed.

4.5 Local Search Operator

Algorithm 2: Local search operator

Input:
  p_l: Local search operator execution probability;
  Pop_g: The current population;
  Gen: Current evolutionary generation ;
  MaxGen: Maximum evolutionary generation;
Output:
  Pop_g: Updated population;

if *(rand1() <= p_l)* ***then***
  **rand1()* is a randomly-generated number in the range of [0, 1] *\
  Perform the 2-hop neighborhood search for each individual in the Pop_g;
else
  if *(rand2() <= p_l)* ***then***
    **rand2()* is a randomly-generated number in the range of [0, 1] *\
    Perform the influence value-first search for each individual in the Pop_g;
  else
    Perform the random search for each individual in the Pop_g;
  end if
end if
Output the updated population Pop_g;

DC-GA-RIM's local search operator has three parts. First is the 2-hop neighborhood search, replacing nodes randomly with neighbors. Second is the influence value first search, replacing nodes with top 10% in robust influence value. Third is random search, replacing with 10% randomly selected nodes. The operator adapts to each individual in the population and replaces nodes to improve fitness. If performance increases and there are no duplicates, replacement is accepted.

4.6 Selection Operator

At the end of each genetic iteration, DC-GA-RIM algorithm utilizes the roulette wheel method to select better individuals into the offspring population. Here some individuals with poor performance also have chances to be selected, which can enhance the diversity of the population to some extent. In addition, the first individual of the child population is saved as the global optimal solution found by the current search process, which is used to prevent the degradation of the population fitness. Such a selection mechanism also contributes to avoiding local optima.

5 DC-GA-RIM

5.1 Performance of DC-GA-RIM on Synthetic Networks

First, in order to combine the validity of the designed evaluation metric R_S and the performance of DC-GA-RIM, we implement DC-GA-RIM and other methods for seeds selection and analyze their robust influence propagation performance R_S. The experimental results are taken as the mean value of each node in the seed node, and the following experiments follow this rule. In detail, under the guidance of R_S, , we use DC-GA-RIM, genetic algorithm (GA), Degree based (Deg), PageRank based (Pag), Betweenness based (Bet), Closeness based (Clo) and Clustering based (Clu) methods to select seeds from the given network, and the performance is compared. Parameters of DC-GA-RIM are set as follows: maximum number of evolutionary generations *MaxGen* is set to 150, local search probability p_l is set to 0.7, crossover probability p_c is set to 0.5, and mutation p_m is set to 0.5, the size of the initial population *InitialSize* as 20, the size of population *PopSize* as 40; the parameter setting of GA is similar to DC-GA-RIM. Here, we conduct experiments on three synthetic networks, scale-free network [8] (SF network), Erdős–Rényi network [15] (ER network) and small-world network [16] (WS network), and set the size of seeds as $K = 5$, $K = 10$ and $K = 20$.

The experimental results are shown in Fig. 1. It can be seen that the seed set obtained by DC-GA-RIM has a higher R_S value over other methods, which indicates that these seeds can achieve better influence propagation performance in the presence of impaired network structure. The results here also show that it is often difficult to select seeds with better information dissemination performance and robustness if we rely solely on information from network structural properties (Degree, PageRank, Betweenness, etc.) in the presence of impaired network structure. Using traditional genetic algorithm (GA) is slightly improved compared with the network-structural-property-based approach, but the effect is still not marked as DC-GA-RIM.

The reasons for the improved performance of DC-GA-RIM compared to other methods can be summarized as follows. Firstly, DC-GA-RIM is based on genetic algorithm, and its population-based search strategy can make the searched seeds have better propagation ability; secondly, DC-GA-RIM is guided by R_S, and the 2-hop influence propagation estimator $\hat{\sigma}(S|Q)$ is also adopted. In addition, compared with the traditional genetic algorithm, DC-GA-RIM considers the population information while retaining the crossover operator and mutation operator, and uses a strategy based on diversity concern as well as an effective local search operator. The algorithm can effectively exploit the genetic information in the population, which greatly improves the optimization ability.

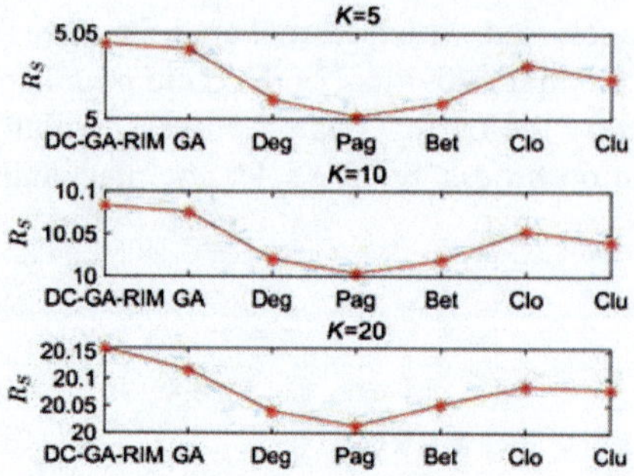

(a) SF network

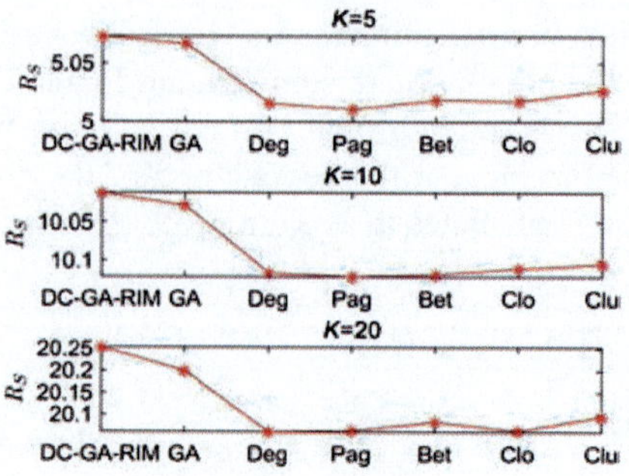

(b) ER network

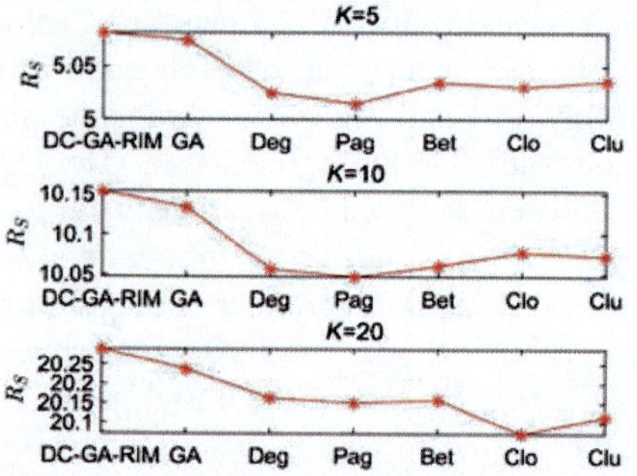

(c) WS network

Fig. 1. The robust influence performance R_S comparison of selected seeds on three synthetic networks.

5.2 Performance of DC-GA-RIM on Real Transportation Networks

In addition to the aforementioned synthetic networks, the effectiveness and practicality of R_S and the DC-GA-RIM are verified on two real networks. The first network is a regional logistics and transportation network (Frid) in Berlin given in the literature [17], which contains 224 nodes, each representing a different freight station, and 376 links, each representing a different transportation route. The second network is the traffic route network (GZTR) in Guangzhou, China, given in the database of Sun Yat-sen University, which contains 556 nodes, each representing a different traffic intersection and 761 links representing different traffic routes. The seed selection is performed for DC-GA-RIM and other algorithms guided by R_S with size of $K = 5$, $K = 10$ and $K = 20$, and their robust influence propagation performance R_S and 2-hop influence propagation level $\hat{\sigma}(S)$ are compared. The results obtained by different seeds selection schemes on these two networks are shown in Table 1 and Table 2.

It can be seen that the DC-GA-RIM algorithm still achieves satisfactory results on the two real networks compared to other algorithms, and its selected seeds are fairly robust against attacks and maintain some impact propagation performance. Meanwhile,

Table 1. Results obtained for each scenario with different K settings on the Frid network.

K	Metric	DC-GA-RIM	GA	Deg	Pag	Bet	Clo	Clu
5	R_S	**5.059**	5.057(+)	5.012(+)	5.007(+)	5.029(+)	5.027(+)	5.031(+)
	$\hat{\sigma}(S)$	5.207	5.205(+)	5.322(−)	**5.363**(−)	5.260(−)	5.250(−)	5.146(+)
10	R_S	**10.113**	10.092(+)	10.027(+)	10.020(+)	10.047(+)	10.043(+)	10.058(+)
	$\hat{\sigma}(S)$	10.404	10.343(+)	10.590(−)	**10.673**(−)	10.477(−)	10.447(−)	10.271(+)
20	R_S	**20.217**	20.183(+)	20.058(+)	20.057(+)	20.093(+)	20.075(+)	20.100(+)
	$\hat{\sigma}(S)$	20.778	20.768(+)	21.076(−)	**21.117**(−)	20.807(−)	20.734(+)	20.521(+)

Table 2. Results obtained for each scenario with different K settings on the GZTR network.

K	Metric	DC-GA-RIM	GA	Deg	Pag	Bet	Clo	Clu
5	R_S	**5.067**	5.063(+)	5.002(+)	5.002(+)	5.004(+)	5.017(+)	5.020(+)
	$\hat{\sigma}(S)$	5.154	5.154(≈)	**5.299**(−)	**5.299**(−)	5.216(−)	5.144(+)	5.092(+)
10	R_S	**10.130**	10.107(+)	10.006(+)	10.006(+)	10.013(+)	10.017(+)	10.035(+)
	$\hat{\sigma}(S)$	10.319	10.330(−)	**10.515**(−)	**10.515**(−)	10.268(+)	10.175(+)	10.186(+)
20	R_S	**20.249**	20.210(+)	20.035(+)	20.028(+)	20.059(+)	20.049(+)	20.106(+)
	$\hat{\sigma}(S)$	20.639	20.630(+)	20.876(−)	**20.916**(−)	20.474(+)	20.247(+)	20.455(+)

the selected seeds exhibit significant robustness and are capable of disseminating information even under attack. In the Frid network, these seeds are typically key freight stations in European logistics transport. Decision-makers can quickly distribute emergency medical supplies, military supplies, and other goods to other freight station nodes in the network through these stations. Similarly, in the Guangzhou transportation network, the selected seeds represent key intersections in the city's transportation road network. These critical junctions possess higher transportation efficiency than other junctions. The DC-GA-RIM algorithm employed in both the Frid network and GZTR network selects seed nodes with strong anti-interference performance. These seed nodes can maintain robust information dissemination in the face of traffic congestion, severe weather, sabotage, and other events that cause node failures. This is especially important for the military domain. In summary, the algorithm developed in this paper offers effective candidates for solving the robust influence maximization problem in both common synthetic artificial networks (SF, ER, WS), and realistic networks.

The experimental results here also allow us to note that the seed nodes obtained by DC-GA-RIM do not obtain the best influence propagation performance when R_S is used as an evaluation metric. This observation implies that there is a trade-off between seed robustness and influence propagation performance. Due to the diverse and complex environments in which networks are situated, decision-makers often prioritize selecting robust seeds over those with optimal influence propagation performance. With the guidance of R_S, DC-GA-RIM can significantly enhance the robustness of the seed nodes at the expense of some influence, yielding a seed set that is more practical and meaningful for real-world applications.

6 Conclusion

Aiming at determining influential nodes from transportation systems, this paper introduces the concept of robustness into this problem based on the deficiencies of existing research. For scenarios where the network structure is destroyed, the target is to select nodes with both robust and influential performances. Based on existing work, an evaluation metric has been proposed to comprehensively evaluate the robustness and influence performance of seeds when the network structure is under node-specific attacks. And for the shortcomings of existing algorithms, DC-GA-RIM with attention on the diversity information is proposed on the basis of the evolutionary optimization paradigm. The algorithm fully considers the information of the entire population and includes an effective local search operator. DC-GA-RIM shows good performance in experiments and provides a reliable reference solution for the design of realistic robust traffic networks.

Acknowledgement. This work was supported in part by National Natural Science Foundation of China under Grant 62203477, and in part by Guangdong Basic and Applied Basic Research Foundation under Grant 2021A1515110543.

References

1. Bond, R.M.: A 61-million-person experiment in social influence and political mobilization. Nature **489**(7415), 295–298 (2012)
2. Gong, M., Yan, J., Shen, B., Ma, L., Cai, Q.: Influence maximization in social networks based on discrete particle swarm optimization. Inf. Sci. **367**, 600–614 (2016)
3. Nannicini, G., Sartor, G., Traversi, E., Wolfler Calvo, R.: An exact algorithm for robust influence maximization. Math. Program. **2020**(183), 419–453 (2020)
4. Gong, M., Song, C., Duan, C., Ma, L., Shen, B.: An efficient memetic algorithm for influence maximization in social networks. IEEE Comput. Intell. Mag. **11**(3), 22–33 (2016)
5. Bai, M., Tan, Y., Wang, X., Zhu, B., Li, G.: Optimized algorithm for skyline community discovery in multi-valued networks. IEEE Access **9**, 37574–37589 (2021)
6. Wang, S., Liu, J.: A Memetic algorithm for solving the robust influence maximization problem towards network structural perturbances. Chin. J. Comput. **44**(6), 1153–1167 (2021)
7. Zhou, Y., Hao, J., Fu, Z., Wang, Z., Lai, X.: Variable population memetic search: a case study on the critical node problem. IEEE Trans. Evol. Comput. **25**(1), 187–200 (2021)
8. Barabási, A.-L., Albert, R.: Emergence of scaling in random networks. Science **286**(5439), 509–512 (1999)
9. Schneider, C.M., Moreira, A.A., Andrade, J.S., Havlin, S., Herrmann, H.J.: Mitigation of malicious attacks on networks. Proc. Natl. Acad. Sci. **108**(10), 3838–3841 (2011)
10. He, X., Kempe, D.: Stability and robustness in influence maximization. ACM Trans. Knowl. Discov. Data **12**(6), 1–34 (2018)
11. Chen, W., Lin, T., Tan, Z., Zhao, M., Zhou, X.: Robust influence maximization. In: Proceedings of the 22nd ACM SIGKDD International Conference on Knowledge Discovery and Data Mining, pp. 795–804. ACM, San Francisco, California, USA (2016)
12. Lee, J.-R., Chung, C.-W.: A fast approximation for influence maximization in large social networks. In: Proceedings of the 23rd International Conference on World Wide Web, pp. 1157–1162. ACM, Seoul, Korea (2014)

13. Kempe, D., Kleinberg, J., Tardos, É.: Maximizing the spread of influence through a social network. In: Proceedings of the Ninth ACM SIGKDD International Conference on Knowledge Discovery and Data Mining, pp. 137–146. ACM, DC, USA (2003)
14. Wang, S., Liu, J., Jin, Y.: Finding influential nodes in multiplex networks using a memetic algorithm. IEEE Trans. Cybern. **51**(2), 900–912 (2019)
15. Erds, P., Rényi, A.: On the evolution of random graphs. Publ. Math. Inst. Hung. Acad. Sci **5**(1), 17–60 (1961)
16. Strogatz, S.H.: Exploring complex networks. Nature **410**(6825), 268 (2001)
17. Farid, A.M.: Symmetrica: test case for transportation electrification research. Infrastruct. Complexity **2**(1), 1–10 (2015)

The Feasibility Study of the Rapid Damage Inspection Technique in Municipal Pole Structure Using UAV

Lu Zhang, Yating Liu, and Hongyu Li(✉)

College of Civil Engineering and Architecture, Guilin University of Technology, Guilin 541004, Guangxi, China
lihongyu@glut.edu.cn

Abstract. With the development of city size and modern infrastructure, various types of pole facilities were increasingly utilized in the modern cities, such as light pole, power supply tower, wind turbine and traffic sign pole. Meanwhile, such facilities may have some damage issues due to aging and other environmental factors. For ensuring the safety and municipal maintenance, it is essential to proposed some reliable and efficient inspection methods. Especially for the crowded urban area, the traditional inspection method such as manual observation and non-destructive test using the permanent sensors are not applicable. To this end, UAV inspection can be a promising technique to realize rapid evaluation of damage in the pole structure. Therefore, this paper proposed a UAV inspection method associated with computer-based visual displacement and vibration monitoring techniques to rapidly detect damage in the pole structure. Its feasibility has been discussed by methodology discussion and numerical model. Two types of common damage types of lamp poles were selected: the corrosion of lamp poles and the loosening of bolts connection were correlated to the vibration modal. The results provided a theoretical basis for UAV inspection on any municipal structure in the future.

Keywords: vibration analysis · numerical simulation · computer vision · UAV

1 Introduction

In recent years, with the development of city, all types of pole facilities play a vital role in the city. For example, light pole, power supply tower, wind turbine, traffic sign pole, etc. As the scale of the city continues to expand, the number of such facilities grows. At the same time, as the service life increases, such facilities may have some damage issues due to aging and other environmental factors. These changes lead to the accumulation of damage during the service life of such facilities, which can lead to unexpected accidents, endangering human lives and causing property damage. In order for such facilities to operate healthily, we must have some reliable and effective inspection methods.

W. Guo et al. (Eds.): GBCESC 2023, LNCE 328, pp. 782–789, 2024.
https://doi.org/10.1007/978-981-99-9947-7_78

With the development and advancement of technology, various methods and techniques of inspection have emerged. Conventional pole structure damage detection, mainly through visual inspection to determine the damage, visual inspection method mainly includes foot patrol, climbing robot and helicopter-assisted inspection [1]. Visual inspection methods are simple, but limited in accuracy and efficiency. In recent years, forms of nondestructive testing have also been gradually adopted for safety inspection; examples of non-destructive testing methods for monitoring the structural health of materials, including concrete and steel, comprise ultrasonic testing, eddy current testing, rebound hammer, stress wave methods, infrared thermography, radiographic inspection, magnetic and electrical methods, and acoustic emission detection. Furthermore, structural vibration-based inspection methods have been proposed, such as sensing displacement, velocity and acceleration [2–6]. However, such methods involve the installation of sensors, which are usually inefficient and costly, and cannot be promoted on a large scale. Several research studies have proposed vision-based measurement methods, which provide the advantage of remotely measuring structures compared to traditional sensors [7].

Therefore, this paper proposes a method for detecting damage in municipal pole structures using unmanned aircraft and combining computer vision displacement and vibration monitoring techniques. This method extracts pixel point trajectory information through overhead photography by the UAV, followed by image processing and analysis techniques [8]. Subsequently, the structural vibration is analyzed, and the response differences of the structure's vibration are compared and analyzed, ultimately leading to the calculation of the location and severity of structural damage through the corresponding algorithm. This method eliminates the need for direct contact and destructive testing of the structure, reducing the impact on structural safety and reliability. Additionally, it can capture small deformation signals, increasing the sensitivity and accuracy of structural damage detection. This approach provides a new non-contact, non-destructive, efficient, and accurate solution for structural damage detection, allowing for rapid damage detection in tall and complex urban pole structures. Furthermore, the method provides a new approach to promote and apply technologies such as UAVs, image processing, and structural vibration analysis.

2 The Principle of Dynamic Characteristic Detection Damage

The fundamental principle of the damage detection method based on dynamic characteristics is that the structure's vibration characteristics, such as frequency, vibration mode, and modal damping, are dependent on its physical properties such as mass, damping, and stiffness. The occurrence of structural damage implies changes in these physical properties, leading to alterations in the structure's vibration characteristics [9]. The principle is shown in Fig. 1. Therefore, this method can achieve the purpose of damage detection by detecting the change of dynamic response of the structure.

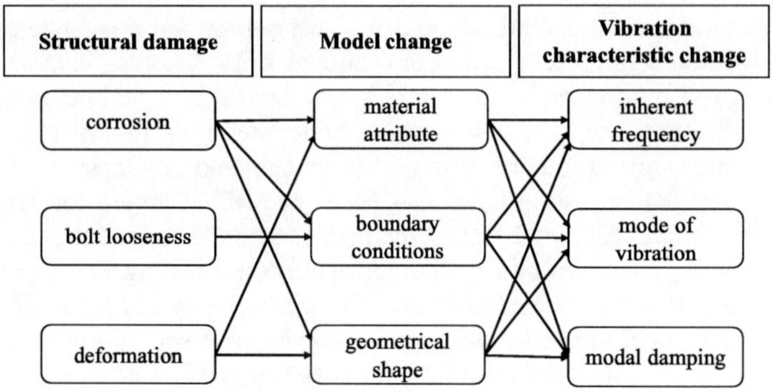

Fig. 1. The schematic of UAV measurement

3 Methodology

In this paper, a UAV inspection method combining computer vision displacement and vibration monitoring techniques is proposed to detect the damage of the pole structure. Taking the detection of light pole as an example, the method mainly includes two major parts: (i) image acquisition using UAV, and (ii); the data processing and damage identification, as shown in Fig. 2.

The UAV acquires image information by a high-resolution digital camera in any weather condition. The UAV cruises at the top of the light pole to capture the motion patterns after stimulating the free vibration on the light pole. Since the camera moves with the UAV, the acquired video cannot be automatically received by the computer, so after capturing one light pole, the UAV captures other light pole images in turn along the light poles laid on the road, and then receives all the light pole image information needed for detection by the computer in a unified way.

When processing and analyzing images on a computer, a significant amount of preprocessing required for the large number of UAV captured images. The video image information obtained includes both the motion of the structure and the motion of the camera. In order to retain only the images with structural motion, the errors caused by the camera motion are eliminated by using the planar single-strain nature [10].

The homography or perspective transformation enables images to be transformed to a different perspective [10]. To transform planar images from different locations, a homogenous coordinate system is used, which ensures that the 3 × 3 homography transformation matrix H satisfies a specific relationship:

$$[x_i, y_i, 1]^T = H \cdot [x_{i+1}, y_{i+1}, 1]^T \quad (1)$$

Assuming the structural vibration is planar motion, camera motion compensation can be achieved by using planar homography. The first of the video is taken as a reference, and subsequent frames are aligned to it by neglecting out-of-plane motion [11].

The vision-based approach can perform measurements quickly by considering each image pixel as a virtual sensor, thereby avoiding the deployment operational barriers associated with traditional sensors. By a robust computer vision algorithm (scale-invariant property) [12], which identifies the key points of imaging on a light pole at

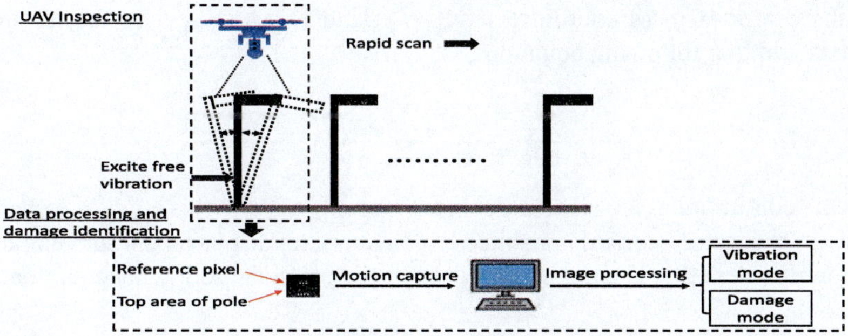

Fig. 2. Methodology

different scale spaces, we construct a Gaussian scale space $L(x, y, \sigma_i)$, To eliminate noise, the Gaussian kernel function $G(x, y, \sigma_i)$ is utilized to filter the input image $I(x, y)$ derivation of $L(x, y, \sigma_i)$ can be achieved through the following equation:

$$L(x, y, \sigma_i) = G(x, y, \sigma_i) * I(x, y) \tag{2}$$

where * is called the convolution, which is a linear operation on x and y, and

$$G(x, y, \sigma_i) = \frac{1}{2\pi\sigma_i^2} e^{-(x^2+y^2)/2\sigma_i^2} \tag{3}$$

To effectively locate stable keypoints in the scale space and identify their image characteristics, a Gaussian difference function $D(x, y, \sigma_i)$ is utilized for computation. For two adjoining filtered images $L(x, y, \sigma_i)$ and $L(x, y, \sigma_{i+1})$, $D(x, y, \sigma_i)$ can be obtained through the following equation:

$$D(x, y, \sigma_i) = L(x, y, \sigma_{i+1})L(x, y, \sigma_i) \tag{4}$$

To identify the keypoints in the scale space, we can compare the difference $D(x, y, \sigma_i)$ of the filtered image with the 26 adjacent values in the 3×3 region of the scale surrounding the sampled point. By checking if the values are the maximum or minimum ones at the location with local extreme points, we can obtain the desired keypoints. A detailed explanation and all the related equations can be found in [12].

After the SIFT method, the key point matching at the same location can be determined using the gradient amplitude $M(x, y)$ of the adjacent keypoints and the direction $\theta(x, y)$ of the adjacent pixels, which can be calculated as follows:

$$M(x, y) = \sqrt{(L(x+1, y) - L(x-1, y))^2 + (L(x, y+1) - L(x, y-1))^2} \tag{5}$$

$$\theta(x, y) = \tan^{-1}\left(\frac{L(x,y+1)-L(x,y-1)}{L(x+1,y)-L(x-1,y)}\right) \tag{6}$$

To calculate the structural displacement dk of the Kth key point based on image pixel points, we need to match the adjacent key points. The displacement can be computed using the center of mass coordinates $\left(X_k^i, Y_k^i\right)$ and $\left(X_{k+1}^i, Y_{k+1}^i\right)$ of the ith adjacent keypoints with the following equation:

$$\begin{aligned} d_k^x &= X_k^i - X_{k+1}^i; \text{ and} \\ d_k^y &= Y_k^i - Y_{k+1}^i \end{aligned} \tag{7}$$

As the conventional physical target is no longer present in this study, a camera calibration approach has been implemented to convert pixel displacement into engineering data in millimeters [8]. The conversion is achieved using the following equation:

$$R = \frac{f}{p \times Z}\left(\frac{\text{pixel}}{\text{mm}}\right) \tag{8}$$

Usually, the notation Z is used to indicate the distance between the camera and the object, wherein f represents the focal length of camera, and p denotes the unit length of the camera's sensor. The $\frac{f}{p}$ ratio is an internal camera parameter, which can be derived from the manufacturer's camera specifications. A detailed explanation and all related equations can be found in [8].

The camera is capable of capturing images at a rate of 30 frames per second, enabling displacement measurements of the structure to be acquired at the same sampling rate of 30 Hz. By analyzing the dynamic component of the displacement data, the free vibration pattern of the lamp post can be identified. To pinpoint the damage type and location with greater accuracy, the vibration pattern and damage pattern of the structure can be determined by comparing them with a pre-established numerical simulation database simulates the structure under different damage modes.

4 Numerical Simulation

Taking a light pole as an example, the following numerical simulations were conducted to establish the correlation between the intrinsic frequency decay [13], of the light pole and its corrosion damage.

A simplified model of the lamp post was established and numerical simulations were performed using COMSOL Multiphysics 6.1. This provided a numerical simulation basis for future structural health monitoring methods that combine computer vision. The height of the light pole is 10 m, the length of the pole bracket is 1.2 m, and the upper and lower calibers are φ70 mm and φ180 mm cone pole respectively, the taper ratio is 15%. The ground C20 concrete is connected to the flange using 10.9 grade M20 high strength bolts. The material constants are shown in Table 1, the division of working conditions and structural parameters are shown in Table 2.

When mesh dissection is performed, the physical field control mesh is selected and COMSOL automatically dissects the mesh according to the settings of the physical field in the model, with a cell size of refinement.

As shown in Table 2, the damage is divided into seven cases, and the simulation of corrosion is achieved by changing the geometry of the model. The geometry is removed at the location where corrosion occurs to achieve the simulation of corrosion. Case 1 is no damage to the structure. Case 2–4, by determining the corrosion location occurred in the top of the flange above 0.02 m, corrosion length of 50 mm, to achieve the numerical

Table 1. Material constant

Name	Young's modulus/Mpa	Density/(kg/m^3)	Poisson's ratio
Body of pole	200000	7850	0.3
Bolt and Nut	206000	7900	0.3
Concrete	300000	2360	0.3

Table 2. Structural parameters of the model corrosion area

Case	Corrosion rate	Corrosion area height/m	Length of corrosion area/mm
1	None	None	None
2	15%	0.02	50
3	30%	0.02	50
4	45%	0.02	50
6	30%	0.12	50
7	30%	0.22	50

simulation of the change in the corrosion rate of the cross-section. Case 3 and case-6–7, by determining the corrosion rate of 30% of the section, corrosion length of 50 mm, to achieve the numerical simulation of the change in corrosion location.

5 Validation of Method Feasibility

The numerical simulation results are shown in the Table 3 and Table 4.

The numerical simulation results are presented in Tables 3 and 4. Based on the analysis of the first 12 modes of the 10-m light pole, it is concluded that the 5th mode eigenfrequency is the most sensitive. Table 3 shows that the modal characteristic frequency decreases with an increase in the corrosion rate of the cross-section. Additionally, table 4 indicates that the eigenfrequency increases as the corrosion height approaches the flange.

Table 3. 5th mode for corrosion rate change

Case 1	Case 2	Case 3	Case 4
21.031	20.425	20.246	20.041

Based on the numerical simulation analysis results mentioned above, it can be concluded that the intrinsic frequency of the lamp pole tends to change regularly when corrosion damage happens, depending on the position and size of the corrosion. Additionally,

Table 4. 5th mode for height change of corrosion area

Case1	Case3	Case6	Case7
21.031	20.246	20.362	20.890

the structure mode undergoes varying degrees of change, indicating the occurrence of different degrees of damage in the structure. By utilizing the dynamic component of the combined computer displacement data, the free vibration law of the lamp pole can be identified, and the structural damage can be detected efficiently.

6 Conclusion

This paper proposes a method for detecting damage in municipal pole structures using unmanned aircraft and combining computer vision dis-placement and vibration monitoring techniques. Our findings indicate that this method can efficiently and accurately acquire damage information on the surface of the structure. Image processing technology can then be utilized to analyze identify image data accurately. Additionally, numerical simulation confirms that the structural modalities change regularly after different types of damage to the lamp pole structure, thus substantiating the feasibility of this proposed method.

Although the proposed method is theoretically effective in identifying damage, this paper still lacks practical laboratory and field test results. Tests are essential to establish the reliability of this method. While the UAV for rapid damage detection technique offers various benefits, it also comes with some limitations and drawbacks. Weather conditions, UAV stability, and flight altitude are several factors that can influence the detection results, and thus require consistent optimization and refinement Future studies will aim to address these limitations and optimize the method for more efficient utilization of UAVs in rapid damage detection of municipal structures.

As UAV technology continues to mature and gain wider usage in urban planning, construction, and maintenance, the utilization of UAVs for damage detection of municipal pole structures will become an increasingly important trend that can offer quick, precise, and cost-effective inspections of municipal buildings. Further research can expand the scope of application of this technology and provide more comprehensive technical support for urban planning and management.

Acknowledgement. This work was supported by National Natural Science Foundation of China (Grant Nos. 52068015 and 52168068), and Guangxi Science and Technology Base and Special Fund for Talents Program (Grant No. Guike AD20159011). Any opinions, findings and conclusions expressed in this paper are those of the writers and do not necessarily reflect the view of Natural Science Foundation of China and Guangxi Science and Technology Base and Special Fund for Talents Program.

References

1. Hosseini, M.M., Umunnakwe, A., Parvania, M., Tasdizen, T.: Intelligent damage classification and estimation in power distribution poles using unmanned aerial vehicles and convolutional neural networks. IEEE Trans. Smart Grid **11**(4), 3325–3333 (2020)
2. Schickert, M., Tümmler, U.: Initial measurement results of an automated ultrasonic scanning and imaging system. In: Proceedings of the 7th International Symposium on Nondestructive Testing in Civil Engineering, Nantes, France, 30 June–3 July 2009, pp. 359–366 (2020)
3. Szymanik, B., Frankowski, P.K., Chady, T., John Chelliah, C.R.A.: Detection and inspection of steel bars in reinforced concrete structures using active infrared thermography with microwave excitation and eddy current sensors. Sensors **16**, 234 (2016)
4. Hermanek, P., Carmignato, S.: Reference object for evaluating the accuracy of porosity measurements by X-ray computed tomography. Case Stud. Nondestruct. Test. Eval. **6**, 122–127 (2016)
5. Makar, J., Desnoyers, R.: Magnetic field techniques for the inspection of steel under concrete cover. NDT E Int. **34**, 445–456 (2001)
6. Adeli, H., Jiang, X.: Dynamic fuzzy wavelet neural network model for structural system identification. J. Struct. Eng. **132**(1), 102–111 (2006)
7. Schreier, H., Orteu, J.J., Sutton, M.A.: Image Correlation for Shape, Motion and Deformation Measurements. Springer, New York (2009)
8. Khuc, T., Necati Catbas, F.: Computer vision-based displacement and vibration monitoring without using physical target on structures. Struct. Infrastruct. Eng. **13**(4), 505–516 (2017)
9. Salawu, O.S.: Detection of structural damage through changes in frequency: a review. Eng. Struct. **19**(9), 718–723 (1997)
10. Yoneyama, S., Ueda, H.: Bridge deflection measurement using digital image correlation with camera movement correction. Mater. Trans. **53**, 285–290 (2012)
11. Weng, Y., Shan, J., Lu, Z., Lu, X.: Spencer, BF homography-based structural displacement measurement for large structures using unmanned aerial vehicles. Comput. Aided. Civ. Inf. **36**, 1114–1128 (2021)
12. Lowe, D.G.: Distinctive image features from scale-invariant keypoints. Int. J. Comput. Vision **60**, 91–110 (2004)
13. Pan, J., Zhang, Z., Wu, J., Ramakrishnan, K.R., Singh, H.K.: A novel method of vibration modes selection for improving accuracy of frequency-based damage detection. Composites Part B **159** (2018)

Identifying Key Nodes in Urban Transportation Systems Using the Information Diffusion Model

Yongbin He, Siheng Ren, Shihan Chen, and Shuai Wang(✉)

School of Intelligent Systems Engineering, Sun Yat-sen University, Shenzhen 518107, China
wangsh368@mail.sysu.edu.cn

Abstract. Urban transporting networks are a type of complex systems, widely existing in modern society. The identification of nodes with significant influential ability is crucial for scientific researches and practical applications including urban planning and city management. Through the survey of relevant literature, most of existing methods for determining influential nodes in cities ignore the topological performance or only consider an agent-based approach using structural metrics for nodal detections, which requires multi-dimensional data, resulting in insufficient robustness. To address such shortcomings, this paper first designs a metric to assess the influential ability of nodes via their structural information. Equipped with the metric, an evolution algorithm is implemented to detect a set of nodes with distinct influential ability. In the experiment, a road network data processor using Pipe4data is developed to process urban road networks from Haizhu District, Guangzhou. The effectiveness of the proposed method has been experimentally verified.

Keywords: The influence maximization problem · Urban transportation network · Complex networks

1 Introduction

Complex networks concentrate on analyzing structures of systems that are composed of a huge number of nodes and intricate relationships between them. Several properties have been discovered and summarized such as small-world effects [1], scale-free nature [2], Community structure [3]. Great attention has been paid in studying related problems. One important branch of current researches is the influence maximization problem [4], which aims to analyze the model of information dissemination processes pertaining to networks, and identify members with distinct influential ability.

As the global population size continues to grow and the urbanisation gets accelerated, more people are flocking to cities. Urban networks, as an important part of infrastructure, have an increasing impact on the efficiency of normal operations. The study of urban networks can not only reveal the internal structure and operation mechanism of cities, but also provide decision support and scientific basis for urban planning and management.

W. Guo et al. (Eds.): GBCESC 2023, LNCE 328, pp. 790–800, 2024.
https://doi.org/10.1007/978-981-99-9947-7_79

The structure and characteristics of transportation networks are an important basis for scheduling and management. Key nodes in a city are expected to combine the typological significance with the distinct inherent characteristics of nodes [5]. In urban transportation networks, it is determined that the seeds or influential nodes can support the basic transport needs of the city. Traffic paralysis can be avoided by maintaining the seeds during unexpected traffic conditions. The detection of such nodes shows practical values in both scientific results and practical applications.

To address the shortcomings of current researches, in this paper, i realistic data are collected from the Haizhu district of Guangzhou, China. An influence propagation model is employed to detect the best seed set in this urban network. A problem-specific Memetic algorithm (i.e., MA-RIM) [6] has been implemented to detect influential nodes. Possible structural perturbances are also considered to provide comparisons in the experiment. Both the topology and inherent node properties of nodes are considered to improve the efficiency. Five metrics are selected to evaluate the node performance of seeds. Results reveal that the average score of the selected seeds tends to increase as K decreases, meaning that seeds with lower scores tend to be eliminated under these five indicators, while others are retained. Such results may be helpful to provide knowledge in infrastructure planning and re-scheduling.

In this paper, the memetic algorithm is applied to urban networks, using the 2-hop influence assessment factor from the 4 and the comprehensive influence performance metric from the 24 for assessing the robustness of network structures, in an attempt to filter and analyse the important nodes of urban networks at the level of complex network analysis. Experimental results on a real urban transport network in Haizhu District, Guangzhou demonstrate that the set of seed nodes obtained by the memetic algorithm performs relatively well in both real road networks and complex network structures, and has high realistic applicability.

The paper is structured as follows, Sect. 2 briefly describes related work on the influence maximisation problem. Section 3 presents the data processor pipe4date, implements an initial simplified road network, and filters the main nodes. Section 4 gives the detailed information of the two evaluation factors and The seed determination algorithm. Section 5 shows the experimental results. Finally, concluding remarks are presented in Sect. 6.

2 Related Work

In recent years, the problem of influence maximization has attracted plenty of attention from researchers in various fields. The aim of influence maximization is to identify a small number of nodes in a network, known as seeds, that can maximise the dissemination of information or influence in the network.

One of the key challenges of influence maximisation is to develop a suitable information dissemination model for the spread of influence in the network. Many different approaches have been proposed for this purpose, including the Independent Cascade (IC) model proposed by Kempe et al. [4] and the Linear Threshold

(LT) model proposed by Rahimkhani et al. [7] The IC model assumes that each node in the network has a certain probability of adopting a new behaviour or opinion based on the behaviour of its neighbours. The LT model assumes that each node has a threshold, i.e. the number of nodes that need to adopt the behaviour or opinion before it can adopt it itself. Behaviour or opinion before it can adopt it itself. Other modelling approaches include the Trigger model (TR) [4], which is a generalisation of the IC model and LT. The TR model defines a distribution that maps a subset of a node's neighbours to a probability that indicates the likelihood that a subset of neighbours can influence the node and the weighted cascade model (WC) [8], which is a special form of the independent cascade model. In the weighted cascade model, the probability of a node v successfully activating a successor node w is the inverse of the entry degree of node w.

For solving the influence maximization problem, seeds with distinct diffusion ability are expected to be selected. Existing approaches can be broadly classified into the simulation-based, the proxy-based and the sketch-based categories. The simulation-based algorithm [4] is a method that uses native Monte Carlo simulation to estimate the influence function. This algorithm is able to handle arbitrary propagation models and network structures, has a considerable generality, and does not require any assumptions or approximations to the propagation model. However, this algorithm is difficult to apply in complex propagation models for its high complexity in the iterative simulations. The proxy-based algorithms [8] include the influence ranking agent, an algorithm that ranks each seed and generates the seed set directly from the ranking with metrics such as degree centrality, Pagerank, and the diffusion model parsimonious agent. An algorithm that improves computational efficiency by simplifying or approximating the original diffusion model is also included. Such methods do not provide any theoretical guarantees, but they offer substantial performance improvements over simulation-based methods. Sketch-based algorithms [9,10] improve theoretical efficiency by pre-computing many sketches to avoid re-running costly Monte Carlo simulations, but still requires prohibitive computational budget on large-scale networks. However, in allusion to urban transportation networks, how to apply the information diffusion model, and how to detect influential seeds considering different scenarios, these problems still remain to be solved.

3 Data Process

Not all nodes have the same utility in the structured information against urban roads. We refer to some nodes as invalid nodes, which are not priority targets in attacks against urban transportation networks.

We argue that there are two main aspects of invalid node composition. First, urban roads are divided into several classes according to the level of infrastructure and design capacity. Roads with lower grades have less traffic operation tasks. Attacking road junctions with lower grades tends to trigger trivial impact on the overall traffic operation, and these junctions are thus part of the invalid nodes. Meanwhile, some nodes do not have or have poor connectivity, i.e., their degree is extremely low in the network. Attacking this part of the node is likely to affect little on the traffic operation. We classify this part of nodes as invalid nodes as well.

A road network environment with invalid nodes would interfere with the extraction of overall structured information, and it is necessary to simplify the road network data to filter noises and trivial information. For this purpose, we design a 'Pipeline' to handle the primitive disordered road network. We named this Pipeline as Pipe4data.

Pipe4data is constructed based on GIS (Geographic Information System) and uses spatial statistical analysis to extract the structured features of urban road networks. We separate Pipe4data into two steps, the first one (step1) completes the simplification of road network, and the second one (step2) completes the nodal.

Simplifying the road network is a method to convert the complex road network into a simple connective network, which can reduce the amount of data, and improve the analysis efficiency, highlighting the main structure of the road.

Firstly, the secondary roads are deleted, due to following reasons.

1) Secondary roads tend to be service roads, residential roads, minor roads, etc., which have less impact on the connectivity and traffic flow of the whole traffic network.
2) Secondary roads are often parallel or overlapping with major roads, causing data redundancy and confusion.
3) The shape and direction of secondary roads are often irregular, which tends to be not conducive to forming a clear line network or conducting topology checks and corrections.

Secondly, the adjacent roads are merged to avoid road network clutter. For some spatial nodes with three-dimensional spatial structure, such as overpasses, tunnels, flyovers, we represent them with one node for two-dimensional processing because they are not different from ordinary roads when they are under attacked. Finally, the invalid nodes are removed including the end of the alleyway.

The obtained simplified road network is used as the input of Pipe4data step2, and nodes of the road need to be filtered subsequently. In this work, the multiple filtering degrees (MF) is defined. Defining the intersection of two or more roads as the nodes to be filtered, MF first needs to obtain the degrees of all the nodes to be filtered. The degree of a node to be filtered is defined as $deg(v)$, and the graph $G = (V_F, E_F)$ is formed with the nodes to be filtered, where E_F is the connection between the nodes to be filtered. The $deg(v)$ of a node to be filtered can be expressed as the number of edges associated with it in E_F. By filtering the $deg(v)$ of each node in V_F, the structurally important nodes can be extracted. Results obtained from this filtering step can be re-screened as V_F to achieve the competitiveness of this set. Consequently, the set of intersection points M with rich structural information can be obtained by setting the loop statement and termination condition.

In addition, considering the existence of heterogeneity in the road network, the structured information of the road network obtained by this algorithm does not achieve the completely ideal results. For this reason, this work also defines the regional screening strategy RF. RF frames a certain range of road network, marks the total road length of the region as L, and records the area of the region as S. Then the average road network density Des of the region is

$$Des = \frac{L}{S} \quad for\ L - km\ s - km^2 \tag{1}$$

If the region Des is greater than $8\,\mathrm{km/km^2}$ [11], the region is judged to be a residential area with high road network density, and the nodes need to be screened using the multiple filtering degree strategy MF separately. After the intersection set M get processed by multiple filtering degrees (MF) and regional screening RF to traverse the road network, a more desirable combination of road network intersections N can be obtained.

In order to verify the effectiveness of Pipe4data, we input the road network of Haizhu District, Guangzhou City as the primitive data for testing. The specific implementation of Pipe4data uses ArcGIS tools, and the simplified road network represents the structured urban information while highlighting the main structure of the road. The filtered nodes have good spatial characteristics, as shown in Fig. 1.

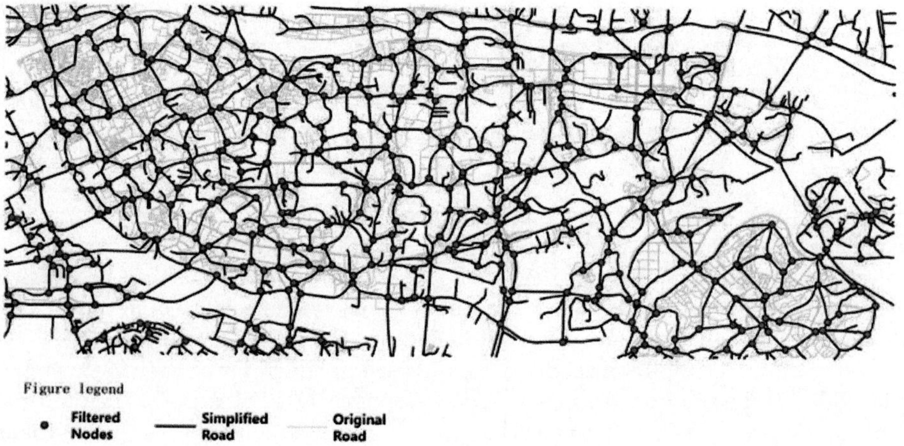

Fig. 1. Results of road network treatment in Haizhu District, Guangzhou

4 Algorithm

4.1 Two-Hop Domain Influence Assessment Based on Seed Nodes

Influence propagation is an important issue in social networks, and the choice of seed nodes has a significant impact on the effectiveness of influence propagation. The author in [12] proposed a model of influence propagation in the two-hop domain of seeds, aiming at providing an effective solution estimation of seeds' influential ability in social networks.

The 2-hop domain impact assessment based on seed nodes is a common method of influence approximation in social networks. The method considers that the influence of a seed node depends not only on its immediate neighbouring nodes, but also on the neighbours of its direct neighbours, i.e. those in the two-hop domain. Given a specific seed set, the evaluation is conducted for each seed to build its tow-hop field. Then influence of nodes is calculated its neighboring information The influence of a certain seed set is recorded and accumulated of all included seeds, and the process can be defined as follows,

$$\hat{\sigma}(S) = \sum_{s \in S} \hat{\sigma}\{s\} - (\sum_{s \in S} \sum_{c \in C_s \cap S} p(s,c) - (\sigma_c^1 - p(c,s))) - \chi \tag{2}$$

where C_s is seed s's 1-hop neighbour; $p(s,c)$ is the activation probability from activated nodes (s) to non-activated nodes (c); σ_c^1 is node c's 1-hop neighbour range; χ is new activated nodes to original seed's repeated influence.

4.2 Influence Assessment Towards Structural Perturbances

Networked systems are inevitable to be disrupted in the real world. In order to evaluate the performance of seeds under structural perturbances, a performance measure is proposed in [13] to fit the structural damage scenario, as

$$R_S = \frac{1}{M \times Per} \sum_{P-1}^{M \times Per} \hat{\sigma}(S|P) \tag{3}$$

where M is the total number of links in the network; Per is the number of links attacked as a fraction of M; $\hat{\sigma}(S|P)$ is an estimate of the selected seed set S influence in a network with P edges removed.

4.3 The Seed Determination Algorithm

The influence maximization problem aims at determining a series of influential nodes, which can be modeled as an optimization problem guided by a specific measure like in Eqs. (1) and (2). The MA-RIM algorithm proposed in [6] has shown efficacy, which is a combinatorial optimization algorithm, This solver combines an evolutionary algorithm with a local search algorithm. The basic idea of the MA-RIM algorithm is to embed a local optimization algorithm into the loop of the evolutionary algorithm to optimize the global optimal solution by combining the global search with the local optimization. Specifically, the MA-RIM algorithm typically consists of the following steps:

1. Initialize the population: a random initial solutions set with size *Popsize* is generated and used as the population for the iterative search process.

2. Evolutionary process: In each generation of evolution, new solutions are generated by performing crossover operations at a probability of p_c, variation operations at a probability of p_m, etc., and the new solutions are evaluated by means of an adaptation function, which in this paper is set to evaluate the metrics $\sigma(S)$ and R_S. Also, a local search operator is used to locally optimize some individuals at a probability of p_l.
3. Optimal search: The first individual in the next generation population is attempted to be updated using the optimal solution in the current population, and the remaining individuals are selected according to the roulette on the performance.
4. Termination conditions: When a certain number of evolutionary generations is reached or when certain termination conditions are met, evolutionary algorithm is terminated and the optimal solution in the current population is outputted.

The pseudo-code for the MA-RIM algorithm is given in Algorithm 1.

Algorithm 1: MA-RIM

input : Input network G_0, Seed set size K, Max generation $MaxGen$, Population size $Popsize$, Crossover Arithmetic execution Probability p_c, the Mutation Arithmetic execution Probability p_m, Local Search Arithmetic execution Probability p_l

output: Optimal seed set $S^* = \{S_1, S_2, ..., S_K\}$

```
Initialize an original Population P_0 with size Popsize from G_0;
for g ← 0 to MaxGen do
    Randomly select 2 unselected individuals P_i and P_j in current
    Population P_g with Probability p_c performs the Crossover Arithmetic to
    get new individual and put into temp Population P_t until all selected;
    for individuals in P_g and P_t do
        With probability p_m performs the Mutation Arithmetic;
        With probability p_l performs the Local Search Arithmetic;
    Try to update the first individual in next Population P_{g+1} with the
    optimal solution from the current population P_g, other individuals in
    P_{g+1} follow the roulette method selected from P_t and P_g;
```

5 Experimental Result

In this section, experiments are conducted to compare and verify the performance of MA-RIM on a simplified planar road network in Haizhu District, Guangzhou. Also, the plain genetic algorithm (GA) without the local search procedure is also implemented to provide comparisons. Parameters of GA are set as follows: the maximum iterations $Maxgen = 100$; the population size $Popsize = 50$, the crossover arithmetic execution probability $p_c = 0.5$, the mutation arithmetic execution probability $p_m = 0.3$. MA-RIM uses the same parameters setting while the local search arithmetic execution probability $p_l = 0.6$. The two algorithms are conducted under the condition of seed size $K = 20$ and the results are shown in Fig. 2. As it can be seen from the figure, for the same seed size K, MA-RIM converges faster than GA and the final seed set obtained has a higher score of the evaluation metrics $\sigma(S)$ and R_S, which shows that the seed set selected by MA-RIM has better propagation performance regardless of whether the network is suffering from attacks or not.

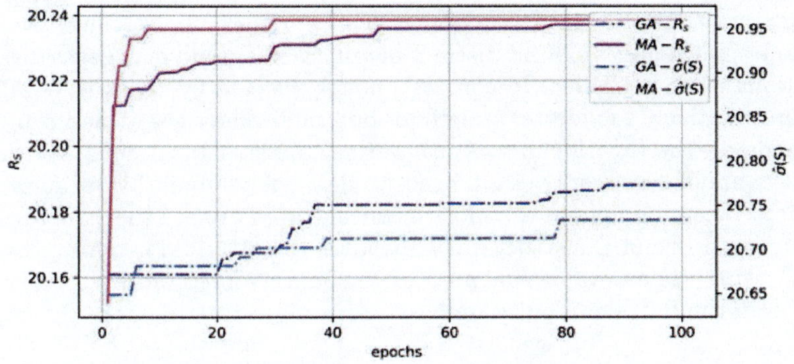

Fig. 2. Results of MA-RIM and GA based on $\sigma(S)$ and R_S

In order to verify the reliability of MA-RIM, we compare and evaluate the obtained seed set's average score under the following evaluation metrics in Fig. 3, 4, 5 and 6: degree method, betweenness-centrality, spanning tree method [14], pagerank algorithm [15] and mutual information method [16]. It can be concluded that MA-RIM selects seed nodes with generally high scores, and the importance of these nodes in the network is evident, thus proving that the MA-RIM algorithm is able to search for seeds with importance in networks.

Figure 3, 4, 5 and 6 also shows the analysis of results obtained by GA and MA-RIM when K is set as 20, 10 and 5, respectively. It can be seen that as K decreases, the selected seed's average scores shows an upward trend, which means seeds with lower score tend to be removed while the others to be keep under these five indicators. For different K, the intersection of the seed set selected by GA tends to be trivial, proving its more general selectivity for nodes with diversified importance. In contrast, MA shows superior performance in terms of key node selection and retention. This is because GA and MA-RIM exhibits different preference in the process of solving optimization problems. The former one focuses on the global search, while the latter one intends to improve the performance by combining local and global search. As a result, GA is prone to show a premature maturity, i.e., the population gradually converges to a local optimal solution during the search process and may fail in obtaining the global optimal solution. By introducing the local search operator, MA-RIM can discover and jump out of local optimum solutions with the help of improved search ability, thus avoiding premature maturity to some extent. In addition, with evaluation metrics such as $\sigma(S)$ and R_S used in this work, MA-RIM shows better performance in detecting seeds with better performance.

Based on the real road network, nodes selected by $\hat{\sigma}(S)$ as evaluation metrics are mostly clustered at the junctions of arterials of roads and bridges, s, as shown in Fig. 7. These nodes have larger intrinsic capacity and actual capacity, and the traffic flow tends to be saturated, which are hub nodes. These hub nodes are required to meet the travel needs of the transportation communities, and their connected edges often provide transportation services that meet medium and long distance connections. These nodes require higher security level over other plain nodes, Such nodes exhibit distinct invulnerability, and may cause disastrous congestions when get failed. In contrast, nodes selected by R_s as the evaluation index are relatively moderate, and those hub nodes may get attacked preferentially and loss the functionality. At this time, the structural hubs are basically paralyzed, and the network system needs to be quickly rebuilt by ordinary nodes. The selected common nodes are mostly concentrated around the structural hubs to share the traffic of the structural hubs after the attack. Therefore, the nodes selected using R_s as evaluation metrics are more robust; such strategy tends to maintain the diffusion ability under structural perturbances. Both guided by

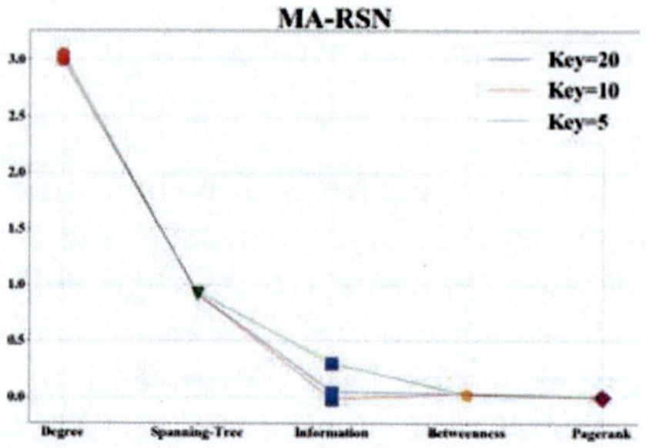

Fig. 3. Selected seed's average score of MA-RIM with Evaluation R_S

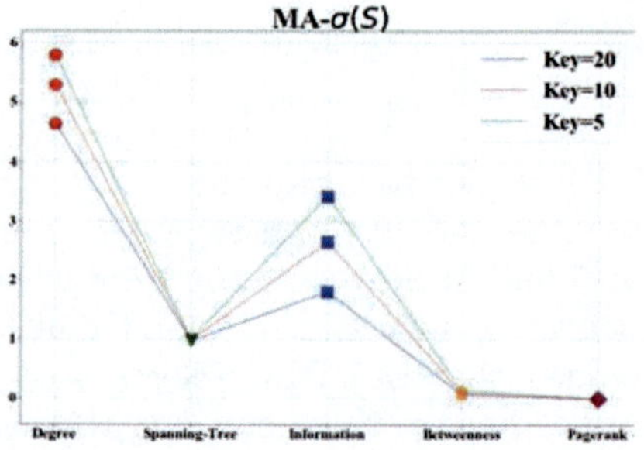

Fig. 4. Selected seed's average score of MA-RIM with Evaluation $\sigma(S)$

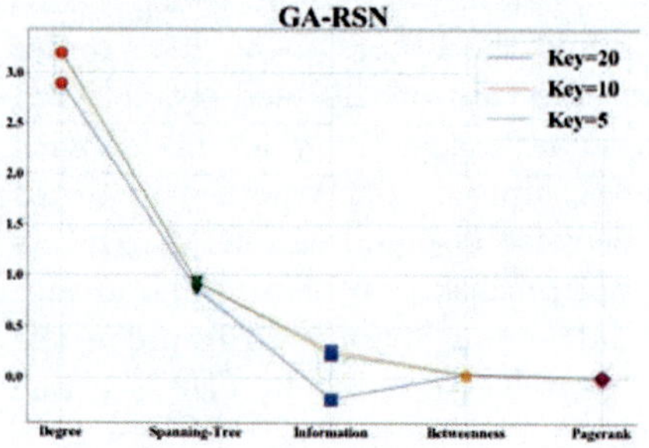

Fig. 5. Selected seed's average score of GA with Evaluation R_S

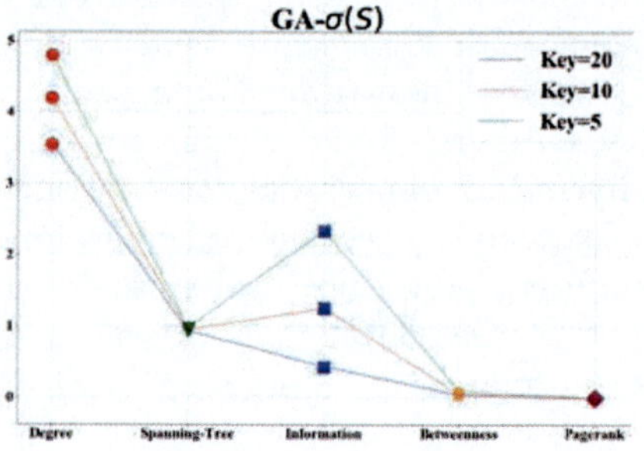

Fig. 6. Selected seed's average score of GA with Evaluation $\sigma(S)$

$\hat{\sigma}(S)$ and R_s, the selected nodes are of realistic significance, which verifies the effectiveness of the approach proposed in this work.

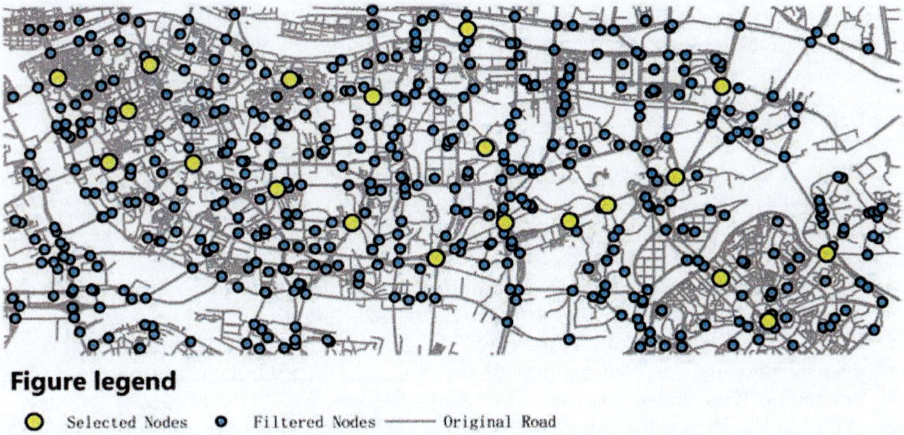

Fig. 7. Nodes obtained by filtering under the guidance of $\hat{\sigma}(S)$ using the MA-RIM

6 Conclusion

In this paper, we have investigated the influence maximization problem via a case on the urban transportation network. Experiments have been conducted on a road network in the Haizhu district of Guangzhou. Firstly based on a multiple screening degree algorithm (MF) and a regional screening algorithm to simplify This complicated road network get simplified using a multiple strategy considering the screening degree and a regional screening. Inputted by this abstract network, MA-RIM and GA are implemented to select influential nodes guided by the two evaluation metrics $\sigma(S)$ and R_S. In addition, to evaluate the reliability of the algorithm, the scores of the selected nodes are calculated based on a variety of nodal influence assessment metrics.

The experimental results show that MA-RIM performs significantly over GA in finding and retaining important nodes. We also found that the locations of influential nodes differed under various evaluation metrics. This phenomenon can be incurred by that different evaluation metrics have different concerns on the structural information. Meanwhile, the influence generated by nodes may be impacted by a variety of factors such as their connectivity and topological features. Further, the importance of selected seeds is mapped to the realistic road network, which draws a conclusion that MA-RIM has better capability of determining influential seeds from transportation networks. The obtained results provide an important reference for optimization and management of modern transportation, and the in-depth information embedded in city planning is also revealed.

Acknowledgement. This work was supported in part by National Natural Science Foundation of China under Grant 62203477, in part by Guangdong Basic and Applied Basic Research Foundation under Grant 2021A151511054, in part by Fundamental Research Funds for the Central Universities, Sun Yat-sen University under Grant 23qnpy72, and in part by Teaching Quality and Teaching Reform Project, Sun Yat-sen University under Grant 76160-12220011.

References

1. Watts, D.J., Strogatz, S.H.: Collective dynamics of 'small-world' networks. Nature **393**(6684), 440–442 (1998)
2. Barabási, A.-L., Albert, R.: Emergence of scaling in random networks. Science **286**(5439), 509–512 (1999)
3. Girvan, M., Newman, M.E.J.: Community structure in social and biological networks. Proc. Natl. Acad. Sci. **99**(12), 7821–7826 (2002)
4. Kempe, D., Kleinberg, J., Tardos, É.: Maximizing the spread of influence through a social network. In: Proceedings of the Ninth ACM SIGKDD International Conference on Knowledge Discovery and Data Mining, pp. 137–146 (2003)
5. Chen, D.-B., Xiao, R., Zeng, A., Zhang, Y.-C.: Path diversity improves the identification of influential spreaders. Europhys. Lett. **104**(6), 68006 (2014)
6. Wang, S., Liu, J., Jin, Y.: Finding influential nodes in multiplex networks using a memetic algorithm. IEEE Trans. Cybern. **51**(2), 900–912 (2019)
7. Rahimkhani, K., Aleahmad, A., Rahgozar, M., Moeini, A.: A fast algorithm for finding most influential people based on the linear threshold model. Expert Syst. Appl. **42**(3), 1353–1361 (2015)
8. Chen, W., Wang, Y., Yang, S.: Efficient influence maximization in social networks. In: Proceedings of the 15th ACM SIGKDD International Conference on Knowledge Discovery and Data Mining, pp. 199–208 (2009)
9. Tang, Y., Xiao, X., Shi, Y.: Influence maximization: near-optimal time complexity meets practical efficiency. In: Proceedings of the 2014 ACM SIGMOD International Conference on Management of Data, pp. 75–86 (2014)
10. Borgs, C., Brautbar, M., Chayes, J., Lucier, B.: Maximizing social influence in nearly optimal time. In: Proceedings of the Twenty-Fifth Annual ACM-SIAM Symposium on Discrete Algorithms, pp. 946–957. SIAM (2014)
11. Standard for urban comprehensive transportation system planning. Technical report, GB/T 51328-2018. China Standards Press (2018)
12. Lee, J.-R., Chung, C.-W.: A fast approximation for influence maximization in large social networks. In: Proceedings of the 23rd International Conference on World Wide Web, pp. 1157–1162 (2014)
13. Wang, S., Liu, J.: A memetic algorithm for solving the robust influence maximization problem towards network structural perturbances. Chinese J. Comput. **44**(6), 1153–1167 (2021)
14. Zhang, X., Tang, J., Cao, L.: Influence maximization algorithm based on reverse PageRank. J. Comput. Appl. **40**(1), 96 (2020)
15. Qiong, Q., Dongxia, W.: Evaluation method for node importance in complex networks based on eccentricity of node. In: 2016 2nd IEEE International Conference on Computer and Communications (ICCC), pp. 2499–2502. IEEE (2016)
16. Ma, R.-N., Wang, B., Wang, G., Guo, X.-C., Liu, W.-B.: Evaluation method for node importance in communication network based on mutual information. Acta Electonica Sinica **45**(3), 747 (2017)

Research on Lighting Evacuation Method Based on Visual Attention Mechanism Analysis

Ce Zheng[1,2](✉), Mingyu Zhang[1,3], and Yu Chang[2]

[1] Tianjin University, Tianjin 300072, China
763121870@qq.com
[2] Tianjin University of Commerce, Tianjin 300134, China
[3] Tianjin Key Laboratory of Architectural Physics and Environmental Technology, Tianjin University, Tianjin 300072, China

Abstract. In order to improve the evacuation efficiency of parks and scenic spots in response to unexpected situations, It integrates human visual attention mechanisms into landscape evacuation lighting design, conducts research on human oriented visual attention mechanisms, and generates an analysis model for virtual landscape evacuation lighting design. In the model, the "material field and standard solution" problem analysis and solution tool of TRIZ theory is introduced to convert the standard solution into a design basis that can be applied in landscape evacuation lighting design, and the design scheme is completed accordingly. The scheme is applied to the simulation experiment of landscape evacuation lighting design in the form of VR virtual simulation scenes to verify the feasibility of the model. It can be seen that the specific landscape evacuation lighting design can match the user's visual attention mechanism, and paying attention to the user's attention can improve the efficiency of pedestrian evacuation.

Keywords: Lighting design · Pedestrian evacuation · Visual attention mechanism · TRIZ material field · VR simulation

1 Introduction

In the face of dense crowd evacuation in emergencies, users will be surrounded by a large number of complex visual and auditory information, and cannot extract effective escape information at the first time. Especially at night, it will cause more chaotic scenes, which is not conducive to large-scale crowd evacuation and lead to malignant events. Therefore, it is urgent to carry out weight analysis of effective visual information in evacuation lighting design) and then introduce the visual attention mechanism, so as to construct the order for users to screen information at the visual level, reduce the information processing time and improve the escape efficiency. It can be seen that in order to better study users, Jin Yutong et al. [1] in order to improve the operation efficiency of virtual interactive interface, integrated human visual attention mechanism into virtual interactive interface, and proposed an optimization.

model of virtual environment interactive interface based on visual attention mechanism; Lu Liuxing et al. [2] believe that the combination of user visual attention and

W. Guo et al. (Eds.): GBCESC 2023, LNCE 328, pp. 801–812, 2024.
https://doi.org/10.1007/978-981-99-9947-7_80

image emotion is an important trend in the research and development of image emotion, and user visual attention is mostly presented in the form of capturing user eye movement; Yuan jingling et al. [3] constructed an image text emotion analysis model based on visual attention to understand users' views, attitudes and emotions. It is worth noting that many literatures have studied the theory of attention. For example, Xu Yuemei and others [4] believe that when people deal with multi-channel information, they will give priority to receiving more critical information due to physiological limitations. It follows that attention is accomplished under limited physiological conditions in which it can be mobilized in a short period of time.

In the specific practice field, Wu Si et al. [5] designed crowd counting method based on attention mechanism; Zhu Yujie et al. [6] used eye tracking technology to objectively quantify the subjects' perception, explain the environmental preference and the degree of attention recovery, and provide a research basis for exploring the interaction between visual and auditory systems in the field of attention recovery; Wang Ning et al. [7] used distance and color factors to establish a visual attention model; Bai Zhonghang et al. [8] introduced the problem analysis and solution tool of "material field and standard solution" in TRIZ theory into the interaction analysis between people and things, especially the sensory interaction questions associated with the user's attention. It can be seen that the problem analysis and solution tool of "material field and standard solution" is widely used in machinery, management and other fields. At present, it is rarely used in the field of lighting design, but it can effectively solve the problem of improving users' attention in evacuation lighting design. For the simulation analysis of the optimization model of evacuation lighting design, VR scene construction will be mainly used. Among them, Bai Han and others [9] completed the simulation research of station passenger flow evacuation by using VR technology; Wang Shuo et al. [10] applied VR simulation technology to complete data acquisition and evaluation for the problem of pedestrian traffic behavior in buildings; Xu Xiaoping et al. [11] conducted a study on the cognitive load of VR natural interaction driven by user cognition. The above research has important guiding significance, which can better calculate and simulate the visual attention mechanism of users in virtual environment.

Through the analysis of user's visual attention mechanism, the problem analysis and solution tool of "material field and standard solution" in TRIZ theory and VR simulation technology are introduced into the specific landscape evacuation lighting design application, so as to optimize the original evacuation lighting design layout, make the interface more in line with the cognitive law of user's visual attention, reduce the pressure of user identification and improve the evacuation efficiency.

2 Design Process Construction Based on Attention Analysis

2.1 Attention

Attention reflects the ability of users' senses to capture things in the external environment, and almost all the information given by external things depends on users' attention. Then, it can be understood that attention occurs through the stimulation of external things. In specific human-computer interaction situations, attention types can be divided into three types: attention to be promoted, focused attention and attenuated attention according to the degree of stimulation [12]. In the middle of the period, the attention stimulation to be improved is weak, which is not enough to attract the user's attention (the user's attention is also weak); Focused attention refers to that when the stimulus intensity matches the user's attention, the user can have focused attention; Attenuating attention means that the stimulation intensity is too high, and the user's attention cannot be focused and gradually attenuates.

2.2 Construction of Evaluation Model

In the evacuation situation, attention will be affected by the subjective feelings of users and the objective conditions of lighting design at the same time. The attention mechanism provides seev attention analysis model, which includes salience, expectation, effort and value. Therefore, the design evaluation can be completed based on seev attention model. The specific evaluation model is:

$$\mathrm{Pt} = \mathrm{aS} + \mathrm{bEx} - \mathrm{cEf} + \mathrm{dV} \tag{1}$$

Pt -- attention level;

S – salience, which refers to the salience of information interacting with vision;

Ex - anticipation, which refers to the user's expectation and is related to the user's needs;

Ef - degree of effort, which refers to the difficulty of users' cognition of information;

V – value, which refers to the user's judgment on the value of information;

a. b, c, d -- weight coefficient of each influencing factor.

The prominence, expectation and value in the model have a positive impact on attention, while the attention reflected by the degree of effort is negative. In this paper, it is convenient for design evaluation, and the degree of effort is interpreted as "identification", that is, the specific design evaluation of attention types is carried out from the four aspects of salience, expectation, identification and value.

2.3 Analysis of Attention Assessment Model and Evacuation Lighting Design

The salience and identification of lighting design itself are used to judge the stimulation degree of evacuation lighting to users' attention; From the perspective of users, expectation and value refer to users' judgment on the demand and experience value of information presented by evacuation lighting. When the stimulation degree of evacuation lighting design information is lower than the expected evacuation needs of users, and the evaluation of its experience value is low, it indicates that the stimulation degree of evacuation lighting design to users is relatively low and cannot meet the needs of users. And the type of attention problem is attention to be improved; when the information stimulation degree of evacuation lighting design is not much different from the expected needs of users and the experience value is high, the attention problem type is concentration; when the information stimulation degree of evacuation lighting design is higher than the expected demand of users and the experience value is low, it indicates that the product has relatively high stimulation to users, and the type of attention problem is attenuating attention.

Therefore, in the evacuation scenario described in this article, the user oriented sensory interaction system can use lighting as a specific external stimulus factor. Mainly through the interaction between lighting design and user sensory attention, which is defined as “material field”, that is, two substances, user (represented by the letter U) and lighting design (represented by the letter L), complete the interaction under the action of the field (represented by the letter F). The lighting design L is the emitter of the field (corresponding to S2 in the original model, the actuator); User U is the receiver of the field (corresponding to S1 in the original model, the receiver); The “field” can be understood as the evacuation lighting design function (F), which revolves around the key parameters of lighting color and light imagery in lighting design, that is, the color and light of lighting itself can be understood as the “sensory field”. Construct a hypothesis that can participate in the analysis of “matter field”. The basic model can be represented by Fig. 1, where the connection line between “L” and “U” can be represented by several symbols“ → ”Indicates “directed action”;“ ⋯→ ”Indicates “insufficient effect”;“ ⇝ ”Means “harmful effects”;“ — ”Indicates “connect.“.

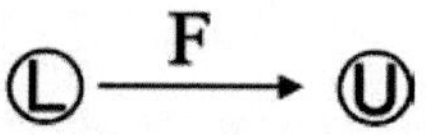

Fig. 1. The basic model of “material field” in the design of evacuation lighting

According to this, the following design flow is constructed: problem generation, problem type analysis and definition, construction of “material field” model solution for user attention analysis, VR based scheme evaluation. Specifically:

First, problem generation link. Site investigation → user analysis → attention problem extraction.

Second, the analysis and definition of problem types. Problem assessment → problem type definition. According to the type of “material field” function, four functions are summarized for evacuation lighting design.

(1) Conducive to evacuation function. The “material field” model is complete and effective, namely, the user u, lighting design L and functional field F are all complete, and can produce fully and effectively interactive functions.
(2) Incomplete evacuation function. One or more of the three elements are missing, which can not achieve the function of evacuation.
(3) Inadequate evacuation function. All three elements have, but can not achieve the required evacuation function.
(4) Functions harmful to evacuation. All three elements have the design function, but the design function is contrary to the evacuation function.

The information presented by lighting design can make the user’s attention present three types, of which, the concentration is the best level of attention, and the cognitive processing ability of information is also high, which is consistent with the function of evacuation; The attention to be improved is caused by low stimulus or even lack of stimulation, which makes it difficult to cause effective attention and effective information transmission, which is very similar to the effect of incomplete evacuation function;

Attenuating attention is caused by excessive stimulus, which leads to the decline of user attention level and cognitive ability, which is corresponding to the effect of harmful function on evacuation.

Third, the "material field" model for user attention analysis is constructed for solving. The standard solution is obtained and analyzed by using tools. Standard solution is a tool for solving TRIZ matter field problems, and it is a standard solution for invention problems summarized by Altshuller and so on through the analysis of a large number of patents [13]. Reference standard solution is a process of thinking divergence and can get effective solutions quickly. The results of the specific model are shown in Table 1.

Fourth, program evaluation. Preliminary scheme → VR scene construction → design evaluation. The VR scene is constructed and executed according to Twinmotion software and HTC-VIVE equipment. The experimental user enters the specific evacuation lighting design simulation scene through the equipment, and records the user's subjective feelings according to the level 5 scale (1–5 points). The evaluation index is "salience, expectation, identification and value" of the scheme evacuation lighting design. Here, firstly, the weight coefficient of each influencing factor should be calculated based on the assignment of focus group and experts. Specifically, five lighting design professionals with senior professional titles were selected for focus group analysis and scored (1–5 points, importance). The results are shown in Table 2:

3 Case Analysis - Evacuation Lighting Design Based on Urban Landscape Space

3.1 Problem Generation

It Conduct field investigations in Beijing, Tianjin, Shanghai and other cities, select mainstream night open spaces such as Sanlitun, Haihe square and the Bund for preliminary analysis. In addition, focus group interviews were conducted with several designers with professional landscape design background to understand the setting situation of night tour and evacuation lighting of users in the target area, and the key problems existing in evacuation lighting design of similar Spaces were summarized, as shown in Table 3.

3.2 Problem Definition

In view of the problems existing in typical users, the scale questionnaire survey is carried out from the four aspects of salience, expectation, identification and value. It is distributed to 30 users for scoring. 30 questionnaires are sent out, 30 are recovered and 29 valid questionnaires are available. After sorting out the survey data, it can be seen that:

First, question 1 in the evacuation situation, salience and recognition are more than expectation and value, which belongs to the problem of attenuating attention. The high saliency and low identifiability indicate that the lighting design as evacuation guidance is easy to attract users' attention, but the information content is not easy to be identified, and the lighting information is being identified (landscape, lighting or evacuation guidance) The process of evacuation will consume users' time and energy and reduce the efficiency of evacuation.

Table 1. Model solution for "matter field"

Classification	Serial number	Describe	Problem model	Solution model	Explain
Number 1 Class construction completion physical field model	1.1.1	The transformation from incomplete to complete physical field model is to perfect the three elements of the system, i.e. L, F, u.	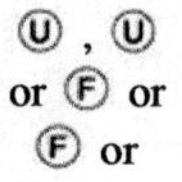	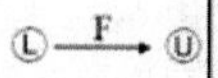	The evacuation function of lighting design fails to get the attention of users, or the lighting design lacks evacuation function, so the corresponding functional field needs to be added.
	1.1.4	The model of physical field with environment is established by using environmental resources as internal or external appendages of materials.	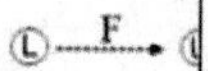	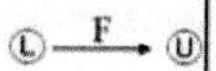	If the lighting design system cannot be changed, use environmental resources as external or internal additives to obtain users' attention and form an evacuation function.
	1.2.1	Introduce external ready-made substance S between two substances in the system to eliminate harmful functions.	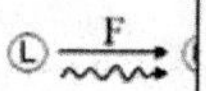	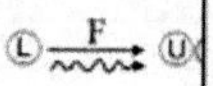	If the user does not need direct contact with the lighting in the lighting design system, substance S can be introduced to eliminate harmful effects.
	1.2.2	Introduction of variants of existing substances in the system.	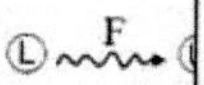	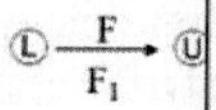	Eliminate harmful effects by changing users, lighting design or adding a field without adding new substances.
The first 2 class Enhanced field model	2.1.2	Conversion to parallel coincidence field model: establish a parallel coincidence field model by introducing the second field that is easy to control (F1).	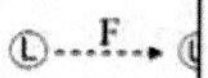	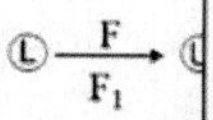	The existing lighting design evacuation function is useful but insufficient, which needs to be improved. The second field F1 can be added to enhance the role of field F.
	2.2.1	Replace with more controllable field	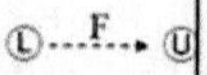	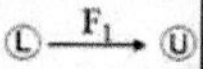	For example, sign design is introduced into the original lighting design system to better realize the evacuation function.

(*continued*)

Table 1. (*continued*)

The first 3 class Transition to multi-level system	3.1	Conversion to dual system or multi system	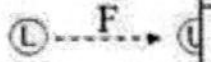	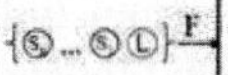	For example, the evacuation system in the lighting design is combined with other systems to form a dual system or multiple systems.
		Improve connectivity in the system		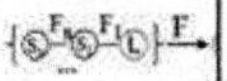	Improve the connection between lighting system and evacuation design.
The first 5 class Simplification and improvement strategy	5.1	Introduced substance	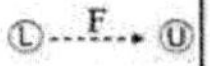	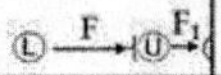	Temporarily introduce accessories to improve the evacuation function of lighting design system.
	5.2	Lead in		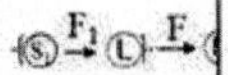	Use the field, such as using one substance to produce another field, so as to improve the evacuation function of the lighting design system.

Table 2. Evaluation index coefficient

Evaluating Indicator	Average score	Normalization Constant
Salience	4	0.25
Anticipation	3.6	0.22
Identifiability	4.4	0.27
Value	4.2	0.26

Table 3. Summary of evacuation lighting design problems in urban landscape space

Serial number	Problem description
1	The night lighting design meets the user's functions of sightseeing and walking, while it is unable to carry out targeted evacuation function conversion in response to emergencies
2	In the analysis of evacuation function in lighting design, there is no specific color light and lamp shape image analysis for users
3	In dealing with the unfamiliar night environment, users will lack access to evacuation information, and lighting information has become the key interactive information. Can they provide diversified information to match with lighting and form richer interaction with users?

Second, question 2 in the evacuation situation, salience and identification are more than expectation and value, which belongs to the problem of attenuating attention. The research on the user color light image and lamp shape image in the evacuation lighting design is conducive to capture the user's attention in a short time and improve the transmission of evacuation information.

Third, question 3 in the evacuation situation, salience and identification are less than expectation and value, which belongs to the attention problem to be improved. During night evacuation, visual interaction is the main factor, and the attention that users can allocate is limited. However, if multi-channel information can be configured in the evacuation lighting design, it is conducive to the completion of evacuation.

3.3 Determination and Analysis of Standard Solution

Through the intervention model analysis, the specific problem solutions are shown in Table 4,

Table 4. Solutions to evacuation lighting design problems

classification	Question serial number	Standard solution	Scheme model	Description of lighting scheme
attenuation attention	1、2	1.21	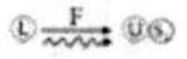	Description 1: the change of lighting information (color, illuminance, type, etc.) must be able to follow the scene for function conversion. Description 2: in case of emergency, eliminate the light information irrelevant to evacuation lighting. Description 3: the sound field is added as the auxiliary of light information. After linkage, the correct information is transmitted to improve the evacuation efficiency. Description 4: when an emergency breaks out at night, the user can download the evacuation path by QR code and applet, and the lighting information of the evacuation path fed back by the program is consistent with the actual lighting information.
		1.2.2	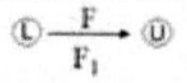	
		5.2	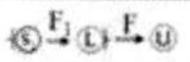	
To be promoted attention	3	2.1.2	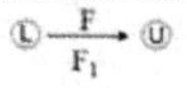	
		3.1	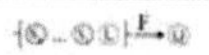	
		5.1	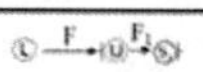	

3.4 Design Scheme Construction and VR Evaluation

First, combined with the analysis of the scheme described in the above standard solution, the evacuation lighting design is carried out for the urban landscape space. The specific scheme can be seen that the evacuation lights are arranged with linear light sources to form a targeted spatial directivity and guide people to evacuate to a safe area; Different

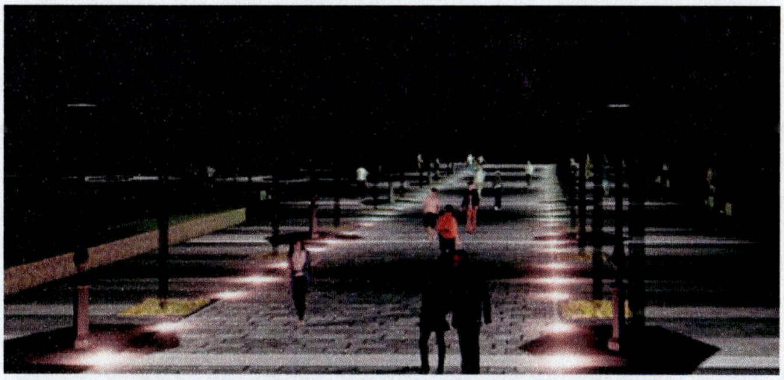

Fig. 2. Schematic diagram of evacuation lighting design scheme 1

light colors with significant differences shall be applied in evacuation area and safety area to represent safety and danger, as shown in Fig. 2.

Projection shall be carried out in the evacuation space. As shown in Fig. 3, Projection equipment shall be installed on the lamp pole, and guidance signs and dynamic information shall be put on the ground; as shown in Fig. 4, a stop sign is generated in the space to avoid the escalation of emergencies.

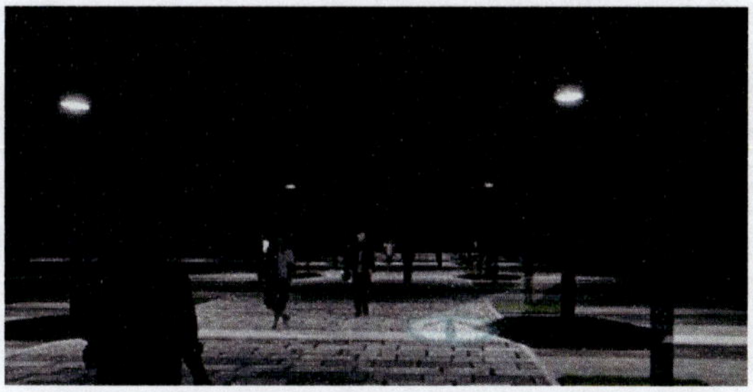

Fig. 3. Design scheme 2.

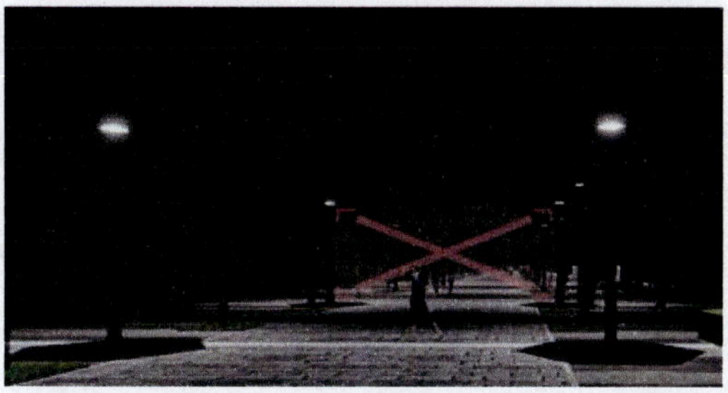

Fig. 4. Design scheme 3.

At the same time, as shown in Fig. 5, two-dimensional code is put on each lamp post, and the evacuation path is downloaded after scanning the code by smartphone.

The current position and the optimal evacuation path can be displayed, and the escape command (sound field) can be played at the same time.

Second, based on Twinmotion, the VR simulation scene is built for the evacuation lighting design scheme of urban landscape space. Specifically, in the VR experimental scene, the virtual light source is mainly white LED, and the basic parameters of the light are set according to the lighting standard as the design basis of the lighting scheme. The VR scene of Twinmotion based on BIMmotion plug-in, as shown in Fig. 6.

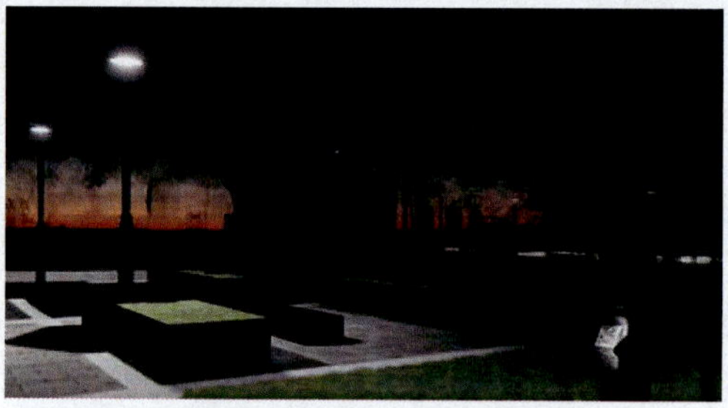

Fig. 5. Design scheme 4.

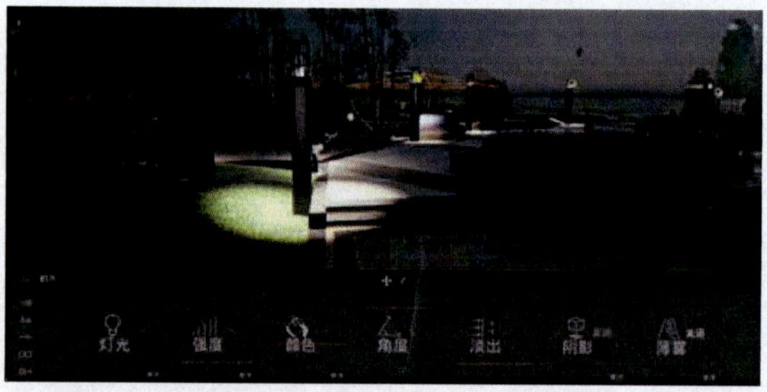

Fig. 6. VR scene

Third, program evaluation. Users can perceive and evaluate the scene through virtual reality devices. This experiment selects HTC-VIVE virtual devices. The specific experimental process is shown in Fig. 7. The subjects are expert users with senior professional titles and have passed the color blindness test (Ishihara's method), which provides a basic guarantee for the normal development of VR simulation experiment. Each scheme presents 120s, and then, the subjects evaluate and assign (90, 80, 60, 50) the "salience, expectation, identification and value" of the scheme within 60s.

Fourth, the evaluation results are divided. The evaluation results are shown in Table 5 after statistical analysis. The results can reflect the feasibility of the specific solution model and the design effect can meet the specific functional requirements.

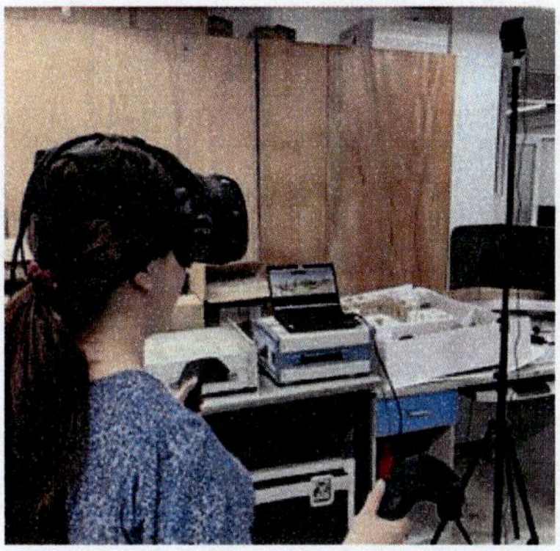

Fig. 7. VR experiment process

Table 5. Evaluation results

Evaluating indicator	Expert assignment					Evaluation results
	Expert 1	Expert 2	Expert 3	Expert 4	Expert 5	
Salience	90	90	90	80	90	81.06
Expectation	80	70	80	80	80	
Identifiability	90	80	80	90	80	
Value	80	70	80	70	70	

4 Conclusion

This paper proposes to analyze the interaction between users and evacuation lighting design from the perspective of attention, makes a specific problem analysis based on TRIZ material field model, and puts forward the innovative method and process of "material field" in evacuation lighting design. The feasibility of this method is verified by the practice of evacuation lighting design in urban landscape space. The main conclusions are as follows:

1. the user interaction situation in evacuation lighting design is explained according to the "material field", the specific "material field" model is established, and the problem is solved with reference to the standard solution, which is conducive to the generation of innovative schemes; at the same time, this paper analyzes the relationship between "people and lighting" for evacuation lighting design, which extends the analysis of TRIZ theory.
2. The interaction problem types between users and lighting design in evacuation scenes were analyzed, and the analysis results were compared with attention types, and then the attention problems generated were directly analyzed using the "matter-field" model.

3. The part of the standard solution is transformed into the standard solution of interaction problem in evacuation lighting design, which has positive effect on the solution of design problem, but the standard solution needs to be further improved according to the research.
4. In the evacuation lighting design system, the user's senses should have five kinds of sensory officialdom, among which the olfactory field and taste field are less used because of their own characteristics, which can be used as the research direction in the future.

References

1. Xu, Y.M., Shi, L.Y., Cai, L.Q.: A cross language text emotion analysis model based on emotional feature representation. Chinese J. Inf. **36**(02), 129–141 (2022)
2. Wu, S., Zhang, X.G., Fang, Y.F.: Crowd counting method based on attention mechanism. J. Saf. Sci. China. **32**(01), 127–134 (2022)
3. Wang, S., Xu, J., Li, M.P., Huang, L.: Data collection and evaluation of pedestrian traffic behavior based on VR. Southern Architect. **06**, 32–37 (2021)
4. Zhu, Y.J., Dong, J.Y., Weng, Y.X., Dong, J.W., Wang, M.H.: Audio visual interactive evaluation of forest park environment based on eye tracking technology. Chinese Gardens. **37**(11), 69–74 (2021). https://doi.org/10.19775/j.cla.2021.11.0069
5. Bai, H., Yang, S., Feng, X.J., Song, X.M., Dong, S.F.: Simulation Research on passenger flow evacuation of rail transit station based on BIM-VR. Urban Rail Transit. **34**(05), 60–65 (2021)
6. Yuan, J.L., Ding, Y.Y., Sheng, Deming, L.L.: An emotional analysis model of image text based on visual attention. Comput. Sci. **49**(01), 219–224 (2022)
7. Wang, N., Xi, M., Zhou, W.G., Li, L., Li, H.Q.: Review on the progress of depth vision target tracking. J. China Univ. Sci. Technol. **51**(04), 335–344 (2021)
8. Huang, H., Meng, Z.Q.: Multimodal emotion classification method based on two-way attention mechanism. Comput. Eng. App. **57**(11), 119–127 (2021)
9. Lu, L.X., Wu, D.: The framework of image emotion research based on visual attention. Libr. Inf. Knowl. **06**, 101–108 (2020). https://doi.org/10.13366/j.dik.2020.06.101
10. Jin, Y.T., Lv, J., Pan, W.J., Zhao, Z.J., You, Q.: Optimization of virtual interactive interface layout based on visual attention mechanism. Comput. Eng. Des. **41**(03), 763–769 (2020). 10.16208/j. issn1000-7024.2020.03.026
11. Dong, Y.F., Tan, R.H., Nie, Z.F., Yu, F., Wang, R.Q.: Research on the product redesign process model integrating the theory of material field and complexity of design process. Mech. Des. **37**(02), 47–52 (2020). https://doi.org/10.13841/j.cnki.jxsj.2020.02.006
12. Bai, Z.H., Liu, Y.A., Ding, M., Sun, H.H.: Attention oriented sensory interaction material field product innovation design method. Mech. Des. **36**(10), 133–139 (2019). https://doi.org/10.13841/j.cnki.jxsj.2019.10.022

ANN-Based High-Dimensional Multi-objective Optimal Design for Natural Lighting in Large-Span Buildings

Jinlong Zou[1(✉)], Lei Feng[1], and Zhongrong Liu[2]

[1] China Nuclear Industry 23 Construction Co., Ltd., Shunyi District, No. 58 Shunkang Road, Beijing, China
2195036026@qq.com

[2] China Nuclear Industry Huaxing Construction Co., Ltd., Jianye District, No. 79, Yunlongshan Road, Nanjing, Jiangsu Province, China

Abstract. In order to build a high-dimensional multi-objective comprehensive balance optimization system for natural lighting, and to solve the limitations of the building performance simulation (BPS) based optimization design method in terms of the number of optimization objectives and the feedback time of results when large-span buildings involve high-dimensional multi-objective optimization design. The study proposes an architect-oriented high-dimensional multi-objective optimization design method for natural lighting based on an artificial neural network (ANN) meta-model coupled with a high-dimensional multi-objective optimization algorithm, based on the Rhino + Grasshopper parametric design technology platform, using a large-span building medium-sized gymnasium as a practical research case to carry out the study. The empirical results show that: 1) the mean square error (MSE) value domain of ANN meta-model construction schemes are all less than 0.0054, and the Pearson correlation coefficient R value domain are all greater than 0.81. 2) the Pareto front solution search time for high-dimensional multi-objective optimal design is 7s. 3) the Daylight Autonomy (DA) and Useful Daylight Illuminance (UDI) values are improved by 0.82%–109.62% and 3.08%–26.71%, respectively, and the Energy Use Intensity (EUI) and Predicted Mean Vote (PMV) values are reduced by 3.81%–13.78% and 0.21%–13.77%, respectively, for the non-dominated solution optimized design schemes.

Keywords: ANN Meta-Model · High-Dimensional Multi-objective Optimization Algorithm · Large-Span Building · Structure Construction · Natural Lighting

1 Introduction

The light environment is an important part of the built environment, from the humble huts of ancient times to the complex large public buildings of modern times, all of which are inseparable from light without exception. A well-designed light environment not only creates a comfortable and efficient working and learning environment, but also significantly reduces energy consumption for lighting and air conditioning [1, 2]. However,

W. Guo et al. (Eds.): GBCESC 2023, LNCE 328, pp. 813–826, 2024.
https://doi.org/10.1007/978-981-99-9947-7_81

the existing research lacks the construction of a high-dimensional multi-objective integrated equilibrium optimization system for natural lighting, while large-span buildings of various structural forms spanning more than 30m space horizontally often involve a high-dimensional multi-objective optimization design problem with four or more optimization objectives due to the complexity of the design problem [3, 4]. The reason for this is that BPS-based multi-objective optimal design methods increase exponentially in complexity and time to search for solutions as the number of optimization objectives increases [5].

In order to address the limitations of BPS-based multi-objective optimization design method in terms of the number of optimization objectives and the feedback time of results, many scholars have explored the method of coupling the alternative model of BPS with optimization algorithm tool to improve the timeliness of optimization design problem search technology. The ANN meta-model is an effective tool for solving complex nonlinear problems, especially in cases where the classical mathematical modelling process fails [6], and is the most commonly used alternative model for objective evaluation in the BPS optimization process [7]. However, existing ANN-based multi-objective design methods for building optimization require strong computer programming skills for the development and coupling of building performance prediction and optimization algorithms [8, 9]. But architects, who have a significant influence on building design decisions, are not familiar with the development of complex building performance prediction and optimization algorithms, let alone the coupling between them through programming skills [10].

In view of the above limitations, this study proposes a coupled ANN meta-model and high-dimensional multi-objective optimization based on Rhino + Grasshopper parametric design modelling technology platform for architects with good consistency between computational cost, versatility and accuracy. While constructing a high-dimensional multi-objective comprehensive balance optimization system for natural lighting, it address the limitations of the BPS-based optimization design method in terms of the number of optimization objectives and the feedback time of results when large-span buildings involve high-dimensional multi-objective optimization design. In this study, a medium-sized gymnasium in large-span buildings is used as a practical research case.

2 Methods

2.1 Workflow

The workflow of ANN-based high-dimensional multi-objective optimal design method for natural lighting of large-span buildings consists of five steps: the first step is the selection of optimization objectives, the second step is the determination of design variables, the third step is the integration of ANN meta-model training sample data, the fourth step is the training of ANN meta-model, and the fifth step is the search of Pareto front solutions for high-dimensional multi-objective optimal design (Fig. 1).

2.2 Optimize Design Practices

Practical Research Case. This study takes a medium-sized gymnasium with a building site of 70 m × 70 m, a central competition hall site of 54 m × 40 m, and a two-way

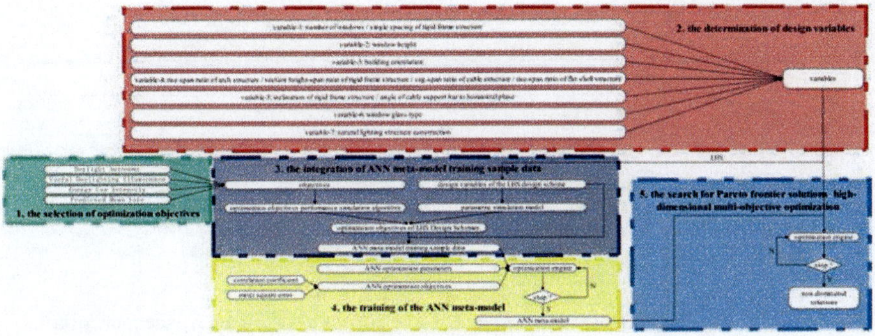

Fig. 1. Workflow

vertical natural lighting at the roof interface as a practical research case of a large-span building, and the construction address of the practical study case is Guangzhou, China.

The types of structure construction that form two-way vertical natural lighting at the roof interface vary, and can be roughly divided into four categories: the first category is mainly the use of arch structure in the thrust structure, truss structure in the bending shear structure, rigid frame structure in the bending shear structure, cable structure in the tension structure and other construction techniques designed for structure fall two-way vertical natural light type (Table 1); the second category is mainly the use of flat shell structure in the thrust structure, arch structure in the thrust structure, cable network structure in the tension structure and other construction techniques designed for structure unit combination two-way vertical natural light type (Table 2); the third category is mainly the use of arch structure in the thrust structure, flat shell structure in the thrust structure and other construction techniques designed for construction two-way vertical natural light type (Table 3); the fourth category is mainly the use of arch structure in the thrust structure, tension structure in the cable network structure and other construction techniques designed for section construction techniques for construction two-way vertical natural light type (Table 4).

Selection of Optimization Objectives. In this study, DA, UDI, EUI and PMV are selected as the optimization objectives for constructing a high-dimensional multi-objective integrated equilibrium optimization system for natural lighting of medium-sized stadiums in large-span buildings (Table 5).

Integration of ANN Meta-model Training Sample Data. The Sampler sampling module couples the Grasshopper plug-in with the Design Space Exploration component, and outputs the design variables of the Latin Hypercube Sampling (LHS) design schemes by performing LHS in the design space (Fig. 2); using the design variables of the LHS design scheme as input data, based on Rhino + Grasshopper parametric design technology platform establishes the parametric model of the practical research case of the LHS design schemes. The Ladybug & Honeybee component invokes Radiance, Daysim and Energyplus to carry on the performance simulation of the optimization objectives of the parametric model of the practical research case, and outputs the optimization objectives

Table 1. Structure fall two-way vertical natural light type

Name	Parameterized model and basic information
Natural lighting structure construction-1	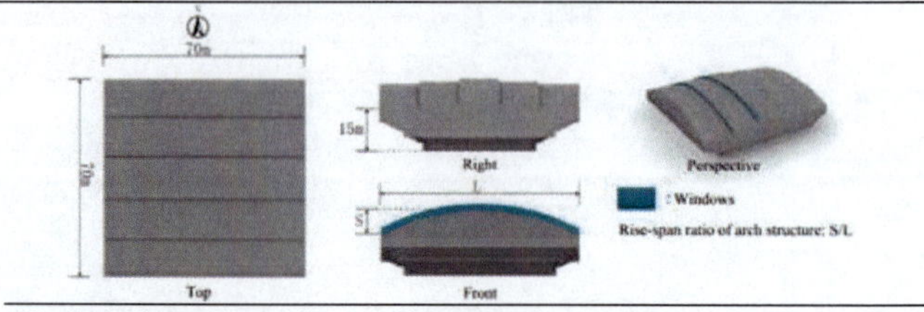 Basic information: The roof interface is constructed by the arch structure in the thrust structure, the number of windows is 4, the window spacing is 14.0m, the window height is 1.0m, the building orientation is due south, and the rise-span ratio of arch structure is 1/8.
Natural lighting structure construction-2	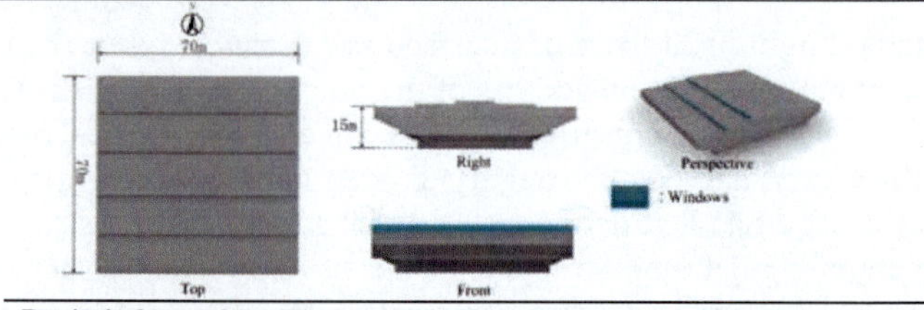 Basic information: The roof interface is constructed by the truss structure in the bending shear structure, the number of windows is 4, the window spacing is 14.0m, the window height is 1.0m, and the building orientation is due south.
Natural lighting structure construction-3	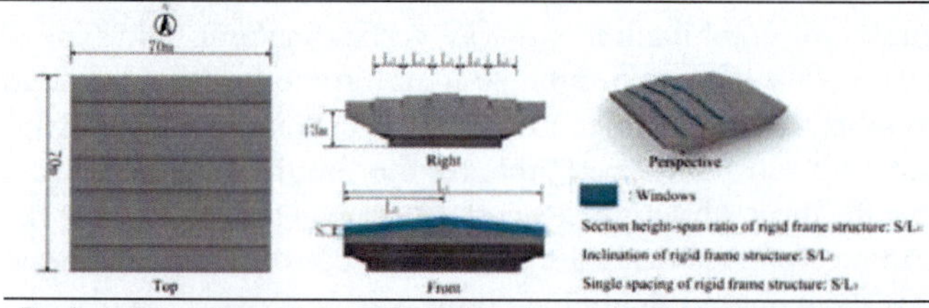 Basic information: The roof interface is constructed by the rigid frame structure in the bending shear structure, the number of windows is 6, the window height is 1.0m, the building orientation is due south, the section height-span ratio of rigid frame structure is 1/20, the inclination of rigid frame structure is 1/10, and the single spacing of rigid frame structure is 9.0m.
Natural lighting structure construction-4	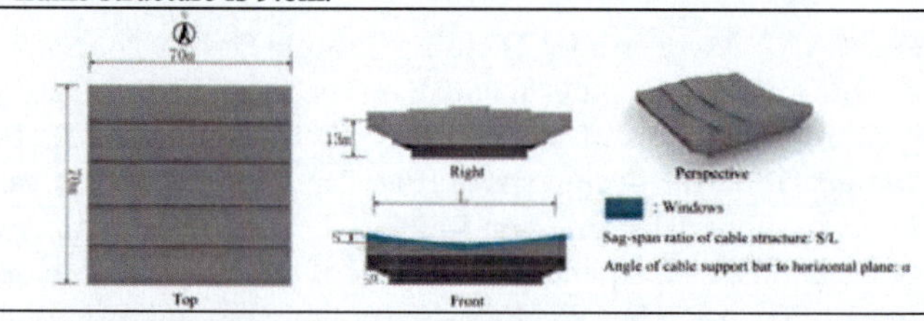 Basic information: The roof interface is constructed by the cable structure in the tension structure, the number of windows is 4, the window spacing is 14.0m, the window height is 1.0m, the building orientation is due south, the sag-span ratio of cable structure is 1/20, and the angle of cable support bar to horizontal plane is 90°.

Table 2. Structure unit combination two-way vertical natural light type

Name	Parameterized model and basic information
Natural lighting structure construction-5	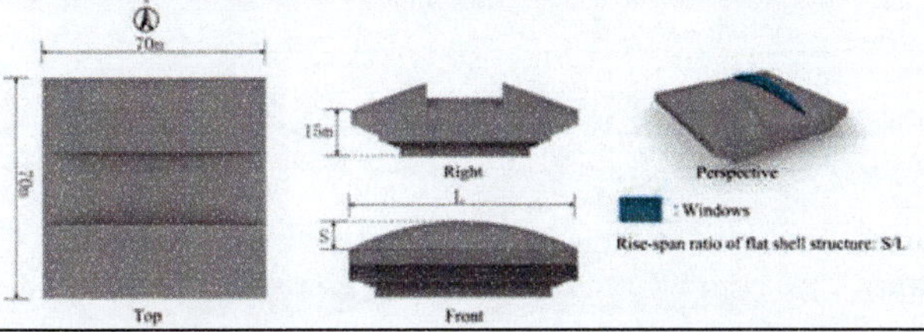 Basic information: The roof interface is constructed by the unit combination of the flat shell structure in the thrust structure and the arch structure in the thrust structure, the number of windows is 2, the window height is 5.0m, the building orientation is due south, and the rise-span ratio of flat shell structure is 1/8.
Natural lighting structure construction-6	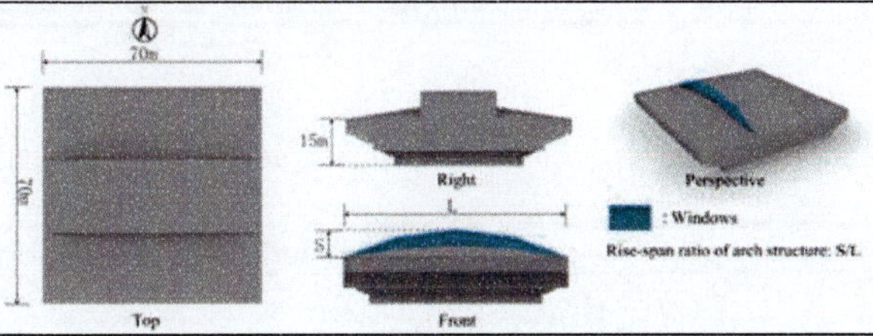 Basic information: The roof interface is constructed by the unit combination of the arch structure in the thrust structure and the cable network structure in the tension structure, the number of windows is 2, the window height is 5.0m, the building orientation is due south, and the rise-span ratio of arch structure is 1/8.

of the LHS design schemes (Fig. 3). The design variables and optimization objectives of the LHS design schemes integrate the ANN meta-model training sample data.

Training of ANN Meta-Model. The ANN meta-model training sample data is used as the input data of the module combination, through the coupling of the Crow and Octopus components, search for a multi-objective optimization Pareto front solution of the ANN meta-model with minimizing the MSE of the training and prediction dataset simulation results and maximizing the R of the validation and prediction dataset simulation results as optimization objectives (Fig. 4). The search outputs a set of non-dominated solution ANN meta-model optimization construction schemes, and the optimization construction scheme is screened by the boundary of the optimization objectives performance range to determine the better construction scheme to build the ANN meta-model (Table 6).

Search of Pareto Front Solutions for High-Dimensional Multi-objective Optimal Design. The .cnw file of the constructed ANN meta-model is loaded into the platform to calculate the optimization objectives of the natural light environment, and the Octopus component is used as a high-dimensional multi-objective optimization algorithm tool to search large-span buildings for the frontier solution design scheme of high-dimensional

Table 3. Construction two-way vertical natural light type

Name	Parameterized model and basic information
Natural lighting structure construc-tion-7	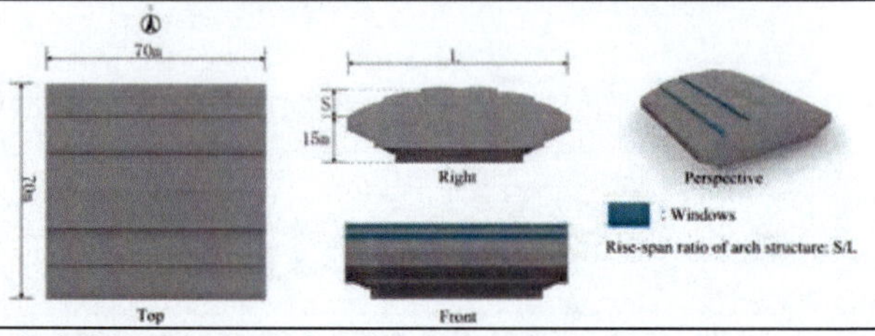 Basic information: The roof interface is constructed by the unit combination of the arch structure in the thrust structure and the flat shell structure in the thrust structure, the number of windows is 4, the window spacing is 11.7m, the window height is 1.0m, the building orientation is due south, and the rise-span ratio of arch structure is 1/8.

Table 4. Section construction techniques for construction two-way vertical natural light type

Name	Parameterized model and basic information
Natural lighting structure construction-8	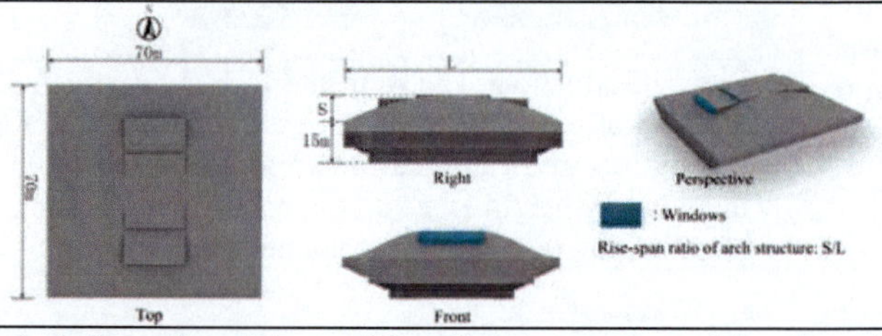 Basic information: The roof interface is constructed by the unit combination of the arch structure in the thrust structure and the cable network structure in the tension structure, the number of windows is 4, the window spacing is 11.7m, the building orientation is due south, and the rise-span ratio of arch structure is 1/8.

multi-objective optimization (Fig. 5). When the optimization search is stopped, a set of non-dominated solution optimization design schemes are output and saved as output data.

Table 5. Design variables, constraints, and modules

Name	Design variables	Constraints	Modulus
Natural lighting structure construction-1	Number of windows	4–22	1
	Window height	1.0–2.0 m	0.1 m
	Building orientation	South by east 60°–South by west 60°	6°
	Rise-span ratio of arch structure	1/8–51/280	2/350
	Window glass type	1–4	1
Natural lighting structure construction-2	Number of windows	4–22	1
	Window height	1.0–2.0 m	0.1 m
	Building orientation	South by east 60°–South by west 60°	6°
	Window glass type	1–4	1
Natural lighting structure construction-3	Window height	1.0–2.0 m	0.1 m
	Building orientation	South by east 60°–South by west 60°	6°
	Section height-span ratio of rigid frame structure	1/20–11/140	1/350
	Inclination of rigid frame structure	1/10–1/5	1/100
	Single spacing of rigid frame structure	3.0–9.0 m	1.5 m
	Window glass type	1–4	1
Natural lighting structure construction-4	Number of windows	4–22	1
	Window height	1.0–2.0 m	0.1 m
	Building orientation	South by east 60°–South by west 60°	6°
	Sag-span ratio of cable structure	1/20–11/140	1/350
	Angle of cable support bar to horizontal plane	70°–110°	4°
	Window glass type	1–4	1

(continued)

Table 5. (*continued*)

Name	Design variables	Constraints	Modulus
Natural lighting structure construction-5	Window height	1.0–5.0 m	0.4 m
	Building orientation	South by east 60°–South by west 60°	6°
	Rise-span ratio of flat shell structure	1/8–51/280	2/350
	Window glass type	1–4	1
Natural lighting structure construction-6	Window height	1.0–5.0 m	0.4 m
	Building orientation	South by east 60°–South by west 60°	6°
	Rise-span ratio of arch structure	1/8–51/280	2/350
	Window glass type	1–4	1
Natural lighting structure construction-7	Number of windows	4–22	1
	Window height	1.0–2.0 m	0.1 m
	Building orientation	South by east 60°–South by west 60°	6°
	Rise-span ratio of arch structure	1/8–51/280	2/350
	Window glass type	1–4	1
Natural lighting structure construction-8	Window height	1.0–2.0 m	0.1 m
	Building orientation	South by east 60°–South by west 60°	6°
	Rise-span ratio of arch structure	1/8–51/280	2/350
	Window glass type	1–4	1

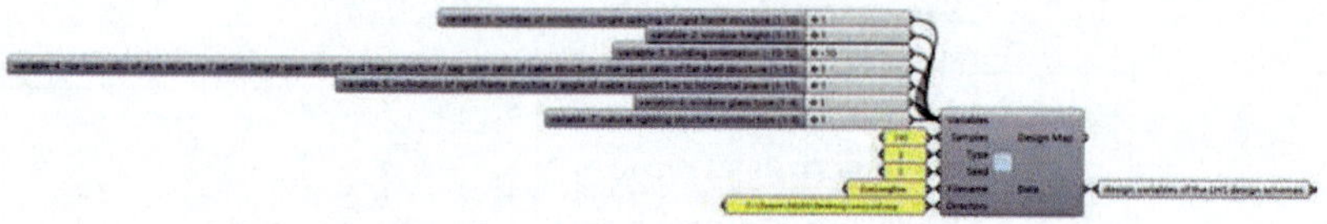

Fig. 2. Design variables of the LHS design schemes algorithm

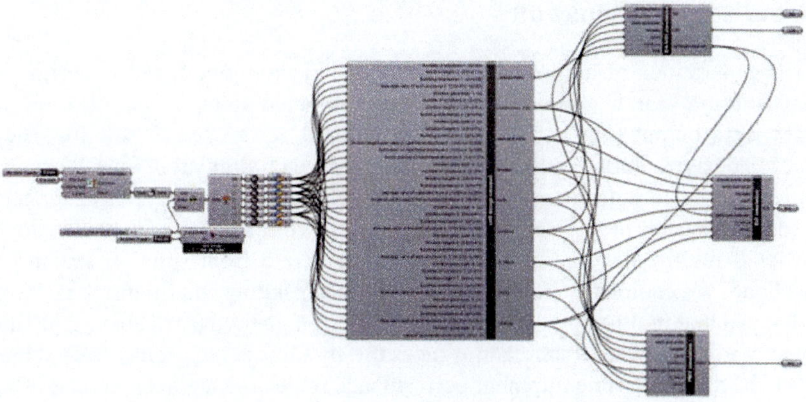

Fig. 3. Optimization objectives of the LHS design schemes algorithm

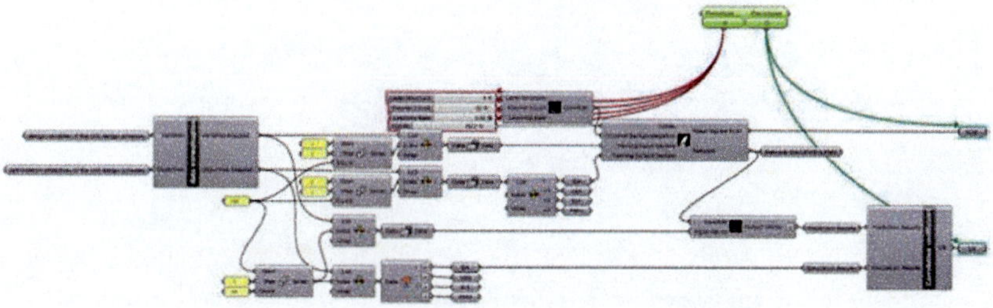

Fig. 4. Training of ANN meta-model algorithm

Table 6. ANN meta-model construction scheme for DA, UDI, EUI and PMV

Name	DA	UDI	EUI	PMV
Cycles	2176	2532	1783	2569
Layer Structure	1	0	2	4
Neuron Count	43	37	67	81
Learning Rate	0.53	0.64	0.49	0.88
MSE	0.0054	0.0019	0.0021	0.0013
R	0.83	0.86	0.85	0.81

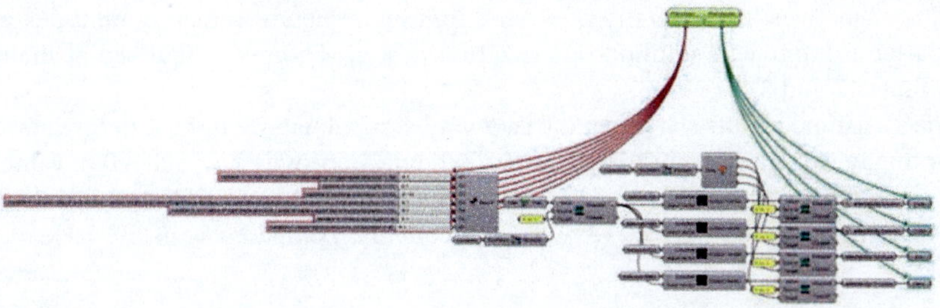

Fig. 5. Search of Pareto front solutions for high-dimensional multi-objective optimal design algorithm

3 Results and Discussion

After 50 generations of high-dimensional multi-objective optimization design Pareto frontier solution search, a total of 1363 non-dominated solution optimization design schemes were output (Fig. 6). In the entire optimal design search solution, the 140-group LHS design solution optimized target performance simulation for 10 servers took about 13 days to complete; the ANN meta-model multi-objective optimization Pareto front solution search took 1 server about 2 days to complete; the 50 generation high-dimensional multi-objective optimization design Pareto front solution search took 1 server about 7s to complete. Under the same parameter setting conditions, the BPS-based high-dimensional multi-objective optimization design method takes about 2500 days to complete with 1 server, which cannot meet the timeliness requirements of the actual project (The above server equipment performance indicators are 8G memory, 40 cores, X5472 type CPU).

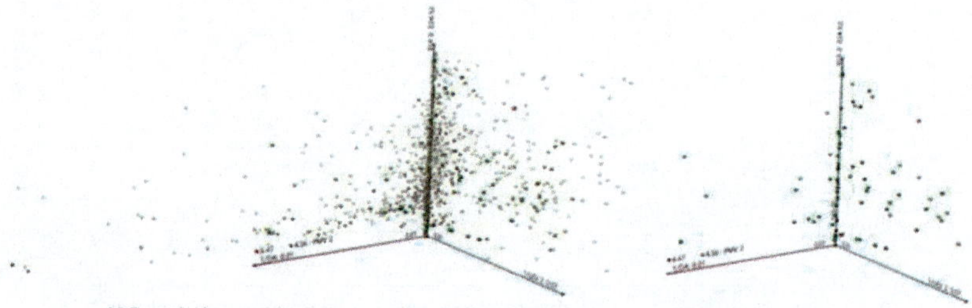

(a) Generation-by-generation convergence solves the 3D scatter plot (b) Non-dominated solution 3D scatter plot

Fig. 6. Pareto front solution for high-dimensional multi-objective optimal design

Based on the analysis and research on the non-dominated solution optimization design schemes DA, UDI, EUI and PMV optimization objectives performance value range and the distribution of natural lighting structure construction types of high-dimensional multi-objective optimization design Pareto frontier solution search output, this study proposes the optimization design schemes comparison method based on the combination of natural lighting construction type classification and optimization objectives performance average boundary limit, and then in the high-dimensional multi-objective optimization design Pareto frontier solution search output 1363 non-dominated solution optimization design schemes select the optimal design schemes for the construction type of natural lighting structure with relatively good quality. On the premise of ensuring the performance improvement of DA, UDI, EUI and PMV optimization objectives, the diversity of natural lighting structure construction types in the optional non-dominated solution optimization design schemes is realized at the same time (Tables 7 and 8).

The statistical results show that the two-way vertical natural light structure structure non-dominated solution optimized design scheme improves DA and UDI values by 0.82%–109.62% and 3.08%–26.71%, respectively, and reduces EUI and PMV values by 3.81%–13.78% and 0.21%–13.77%, respectively, compared with the LHS design scheme.

Table 7. Non-dominated solution optimization design schemes

Name	Design variables		Optimization objectives	
Natural lighting structure construction-1	Number of windows	8	DA/(%)	96.12
	Window height	1.3m	UDI/(%)	83.51
	Building orientation	South by east 6°	EUI/($kWh \cdot m^{-2}$)	162.43
	Rise-span ratio of arch structure	494/2800	PMV	4.57
	Window glass type	2		
Natural lighting structure construction-2	Number of windows	12	DA/(%)	97.48
	Window height	1.4m	UDI/(%)	81.52
	Building orientation	South by west 54°	EUI/($kWh \cdot m^{-2}$)	156.11
	Window glass type	2	PMV	4.59
Natural lighting structure construction-3	Window height	1.0m	DA/(%)	96.51
	Building orientation	South by west 60°	UDI/(%)	78.28
	Section height-span ratio of rigid frame structure	510/7000	EUI/($kWh \cdot m^{-2}$)	166.46
	Inclination of rigid frame structure	16/100	PMV	4.89
	Single spacing of rigid frame structure	4.5m		
	Window glass type	1		
Natural lighting structure construction-4	Number of windows	8	DA/(%)	96.79
	Window height	1.7m	UDI/(%)	76.24
	Building orientation	South by west 42°	EUI/($kWh \cdot m^{-2}$)	165.54
	Sag-span ratio of cable structure	350/7000	PMV	4.79
	Angle of cable support bar to horizontal plane	70°		

(continued)

Table 7. (*continued*)

Name	Design variables		Optimization objectives	
	Window glass type	2		
Natural lighting structure construction-6	Window height	1.4m	DA/(%)	96.78
	Building orientation	South by east 24°	UDI/(%)	92.17
	Rise-span ratio of arch structure	478/2800	EUI/(kWh•m^{-2})	161.7
	Window glass type	1	PMV	4.83

Table 8. LHS design schemes

Name	Design variables		Optimization objectives	
Natural lighting structure construction-1	Number of windows	8	DA/(%)	94.3
	Window height	1.3 m	UDI/(%)	78.4
	Building orientation	South by east 60°	EUI/(kWh•m^{-2})	188.4
	Rise-span ratio of arch structure	494/2800	PMV	5.3
	Window glass type	3		
Natural lighting structure construction-2	Number of windows	4	DA/(%)	95.33
	Window height	1.7 m	UDI/(%)	72.25
	Building orientation	South by west 24°	EUI/(kWh•m^{-2})	169.2
	Window glass type	3	PMV	5.05
Natural lighting structure construction-3	Window height	1.1 m	DA/(%)	95.33
	Building orientation	South by east 12°	UDI/(%)	67.5
	Section height-span ratio of rigid frame structure	47/700	EUI/(kWh•m^{-2})	173.25
	Inclination of rigid frame structure	14/100	PMV	5.03
	Single spacing of rigid frame structure	7.5 m		
	Window glass type	3		

(*continued*)

Table 8. (*continued*)

Name	Design variables		Optimization objectives	
Natural lighting structure construction-4	Number of windows	10	DA/(%)	96
	Window height	1.0 m	UDI/(%)	60.17
	Building orientation	South by west 30°	EUI/(kWh•m^{-2})	173.2
	Sag-span ratio of cable structure	47/700	PMV	5.23
	Angle of cable support bar to horizontal plane	94°		
	Window glass type	3		
Natural lighting structure construction-6	Window height	1.4 m	DA/(%)	46.17
	Building orientation	South by west 30°	UDI/(%)	89.42
	Rise-span ratio of arch structure	478/2800	EUI/(kWh•m^{-2})	168.1
	Window glass type	3	PMV	4.84

4 Conclusions

Natural light plays a vital role in human physical and mental health, and has a direct or indirect impact on the thermal environment and economic benefits of buildings. Existing research lacks the construction of a high-dimensional multi-objective comprehensive balance optimization system for natural lighting in large-span buildings; at the same time, when large-span buildings involve high-dimensional multi-objective optimization design, the BPS-based optimization design method has limitations in the number of optimization objectives and the time for result feedback, while existing ANN-based multi-objective design methods for building optimization require strong computer programming skills for the development and coupling of building performance prediction and optimization algorithms, resulting in the limitation of using large-span buildings as practical research cases. In view of the above limitations, this study proposes a coupled ANN meta-model and high-dimensional multi-objective optimization based on Rhino + Grasshopper parametric design technology platform for architects with good consistency between computational cost, versatility and accuracy. The main results of the research are as follows:

1. The high-dimensional multi-objective optimization design method based on natural lighting, the workflow and technology platform of the optimization design method are constructed. The workflow and data processing are all completed on the Rhino + Grasshopper parametric design modelling technology platform, which is highly manoeuvrable for architects.

2. The optimization design method coupled with ANN meta-model and high-dimensional multi-objective optimization algorithm can address the application bottleneck of result feedback time and optimization objectives number constraints faced by BPS optimization design method.
3. The performance of each optimization objectives of the non-dominated solution design schemes obtained from the high-dimensional multi-objective optimization design search is obviously better than that of the LHS design schemes, which verifies the effectiveness of the proposed high-dimensional multi-objective optimization design method based on natural lighting.
4. An optimization design schemes comparison and selection method was proposed based on the combination of natural lighting construction type classification and average performance boundary limitation of optimization objectives. Under the premise of ensuring the performance improvement of DA, UDI, EUI and PMV optimization objectives, the diversity of natural lighting structure construction types in the alternative non-dominated solution optimization design schemes was realized.

References

1. Asl, M.R., Zarrinmehr, S., Bergin, M., et al.: BPOpt: a framework for BIM-based performance optimization. Energy Build. **108**, 401–412 (2015)
2. Delgarm, N., Sajadi, B., Delgarm, S., et al.: A novel approach for the simulation-based optimization of the buildings energy consumption using NSGA-II: case study in Iran. Energy Build. **127**, 552–560 (2016)
3. Sysoeva, E.: Preconditions of emergence of large-span buildings in the world. MATEC web of conferences. EDP Sci. **86**, 02004 (2016)
4. Hu, J., Chen, W., Li, Y., et al.: Temperature-stress-time methodology for flat-patterning ETFE cushions in use for large-span building structures. Eng. Struct. **204**, 109607 (2020)
5. Shen, P., Braham, W., Yi, Y., et al.: Rapid multi-objective optimization with multi-year future weather condition and decision-making support for building retrofit. Energy **172**, 892–912 (2019)
6. Gupta, N.: Artificial neural network. Netw. Complex Syst. **3**(1), 24–28 (2013)
7. Roman, N.D., Bre, F., Fachinotti, V.D., et al.: Application and characterization of metamodels based on artificial neural networks for building performance simulation: a systematic review. Energy Build. **217**, 109972 (2020)
8. Gou, S., Nik, V.M., Scartezzini, J.L., et al.: Passive design optimization of newly-built residential buildings in Shanghai for improving indoor thermal comfort while reducing building energy demand. Energy Build. **169**, 484–506 (2018)
9. Si, B., Wang, J., Yao, X., et al.: Multi-objective optimization design of a complex building based on an artificial neural network and performance evaluation of algorithms. Adv. Eng. Inform. **40**, 93–109 (2019)
10. Shi, X., Tian, Z., Chen, W., et al.: A review on building energy efficient design optimization ROM the perspective of architects. Renew. Sustain. Energy Rev. **65**, 872–884 (2016)

Research on Intelligent Monitoring System of Urban Network Database

Xusheng Tang(✉) and Jie Liu

College of Landscape Architecture and Arts, Northwest A&F University, Yangling 712100, China
sunrisetang@126.com

Abstract. City network database is vulnerable to attack. Smart city IOT (Internet of Things) needs to implement intelligent monitoring systems to protect network systems. The traditional convolutional neural network needs more advantages in dealing with real-time network intrusion. This paper proposes a new detection method using a visualization method when real-time traffic data is missing. We built a real-time traffic detection platform and invented a new algorithm structure: The system uses offline data to digitize and pixelate the hacked network traffic and then carries out gray-scale mapping, a pre-processing step. The data is rotated to the right and connected to the CNN interface for training. We use real-time network data to verify the experiment and evaluate our proposed model. The results show that the scheme proposed in this paper has a good detection effect and can detect real-time network intrusion types.

Keywords: Convolutional Neural Networks · Network Security · Digitization · Data Processing · Artificial Intelligence

1 Introduction

Deep learning algorithms can learn the features of data samples adaptively under the support of mathematical principles. Convolutional Neural Network (CNN) has become a research hotspot in image processing, natural language processing, and other fields by its superior ability to extract features through autonomous learning, enabling various applications to achieve a leap in accuracy. Especially in network attack and recognition, it combines traditional algorithms' feature extraction and classification process. It can achieve arbitrary complex nonlinear expression through model self-learning, avoiding the disadvantages of manual feature extraction, solving the modeling difficulties faced by traditional target recognition, and helping to improve the quality of the field of network attack recognition [1, 2]. Therefore, deep learning algorithms have also been widely studied and applied in network attack identification.

This paper proposes an intrusion detection method based on convolutional neural networks and provides a series of visualization schemes for NLS-KDD data sets [3, 4]. Therefore, it is of great practical significance for us to explore the radar target network attack identification technology under knowledge loss in real-time, especially to solve the problem of knowledge loss in extreme offensive and defense scenarios.

W. Guo et al. (Eds.): GBCESC 2023, LNCE 328, pp. 827–834, 2024.
https://doi.org/10.1007/978-981-99-9947-7_82

2 Framework of Network Intrusion Detection System

2.1 Research Content

By studying the rapid learning ability of human beings, the researchers found that the human recognition system can quickly learn things that have not been seen before with only a few samples after a certain degree of learning. Therefore, meta-learning theory and other schemes are proposed to solve the small sample learning problem in the optical field. The core idea of meta-learning is "learning how to learn," which aims to cultivate the model's autonomous learning ability. The system learns from this experience by observing how different machine learning methods perform across a wide range of learning tasks, achieving the ability to learn new tasks [5, 6] quickly. It is hoped that the model can accumulate experience by learning the recognition process of multiple tasks so that it can learn quickly when facing new target tasks with only a few samples. Here, the target task is set as a learning target, each target has a sample, and the remaining samples of this type of target can be classified after the completion of model learning. The following figure is the processing flow of the algorithm in this paper.

2.2 Research Object

This structure Bridges the interface between the accurate data and the prediction engine by building capture, file upload, and data processing programs.

The algorithm based on optimizer learning aims to learn the optimization way of neural networks, which emphasizes the learning process of parameter updating. Figure 2 shows the detection flow of the algorithm in this paper. The core idea is that one network (meta-learner) in the model updates another network (learner) by constantly learning the optimization method so that the learner can learn the task more efficiently. Typically meta-learners use recurrent neural networks, and models are trained through reinforcement learning or supervised learning. Typical works in this field are the MAML (Model-Agnostic Meta-Learning) Model and the Reptile model, which directly optimizes the initial parameters of the network [7, 8]. This kind of method does not depend on the model and can be applied to any neural network. However, the hyperparameter search in the training process is more complicated, and the model's training is challenging (Fig. 1).

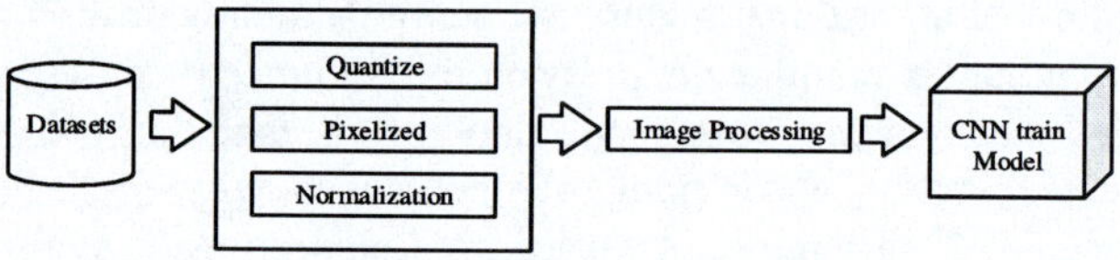

Fig. 1. Steps to establish network intrusion detection framework.

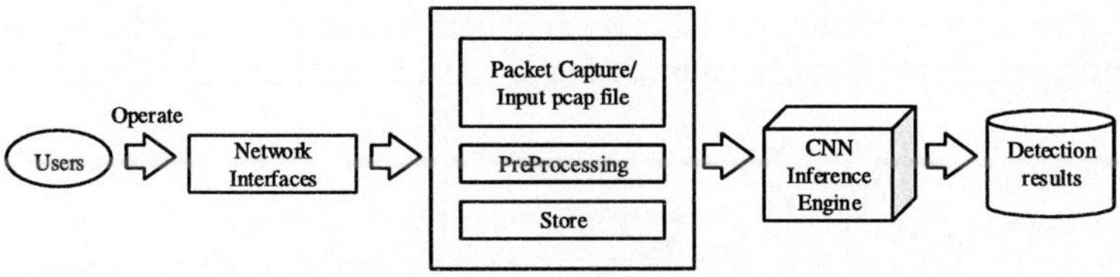

Fig. 2. Steps of detection.

3 Steps of Intrusion Detection

3.1 Data Preprocessing

The pre-processing operations include discrete digital markup, uint8 pixel gray value mapping, and normalization.

In this section, ACGAN, a typical representative of the generated countermeasure network, is used to realize the data expansion of HRRP images to increase the supervision information, expand the sample richness, and help improve the efficiency of network attack identification. The overall process of the algorithm is shown in the figure below. In the training process, the depth generation model ACGAN must first be trained to generate labeled fake samples as accurately as possible. After the model training is completed, a certain number of fake samples are generated, and the fake samples and actual samples are input into the classification model to train the model to achieve correct classification. In the testing process, the sample is directly input into the classification model to output the recognition results.

3.2 Visual Processing

When task diversity increases, meta-learning and transfer learning performance decrease significantly. However, the two are affected to different degrees, the fluctuation range of meta-learning is more extensive, and the fluctuation range of transfer learning is relatively small. Therefore, as the difference between the source task and the target task increases, the advantages of meta-learning gradually disappear, and the advantages of transfer learning become prominent. When the task difference increases to a certain extent, the transfer learning performance is better than the meta-learning performance. This performance advantage is mainly due to fine-tuning learning. Transfer learning obtains good recognition performance through adaptive learning of target tasks. The model attempts various processing methods in terms of visualizing network traffic [9].

Method 1: Detection method of linear arrangement.

In the transfer learning experiment, the amount of batch processing was 15, and a total of 200 training times were performed. The data learning method in the fine-tuning process of the target task is the same as that of meta-learning. In each experiment, three types of targets are randomly selected from the data set, and 20 samples of each type are used as the data set of this task. In this data set, five samples are randomly selected as the support set samples, and the remaining 15 samples are used as the query set samples. The training process of meta-learning requires repeated sampling training 10,000 times, and the testing process of transfer learning and meta-learning needs to be repeated 6,000 times before taking the average value of the test results as the final recognition result.

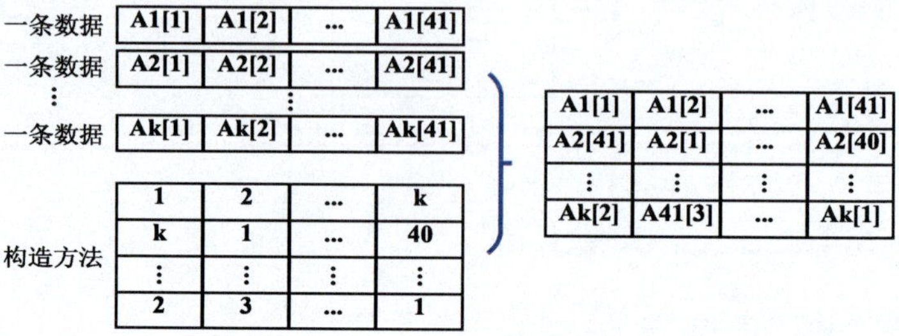

Fig. 3. Format of right-shift data.

Using the NSL-KDD dataset, Fig. 4 shows multiple grayscale effects corresponding to different network traffic types.

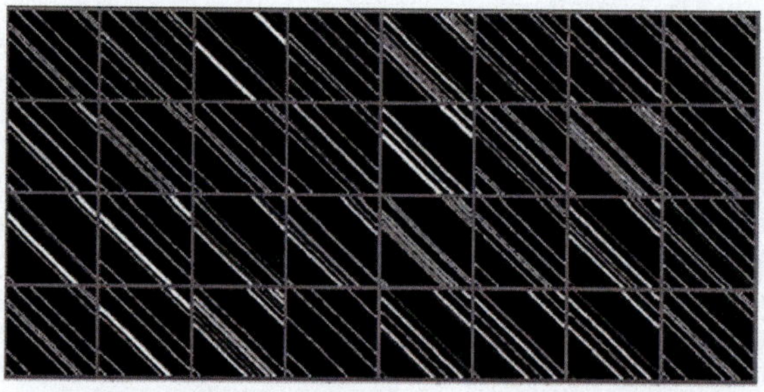

Fig. 4. Detection method of linear arrangement.

Method 2: Detection method of two-dimensional linear array.

A1 to AK in Fig. 3 are different data. Figure 5 shows multiple grayscale effects corresponding to separate network traffic types.

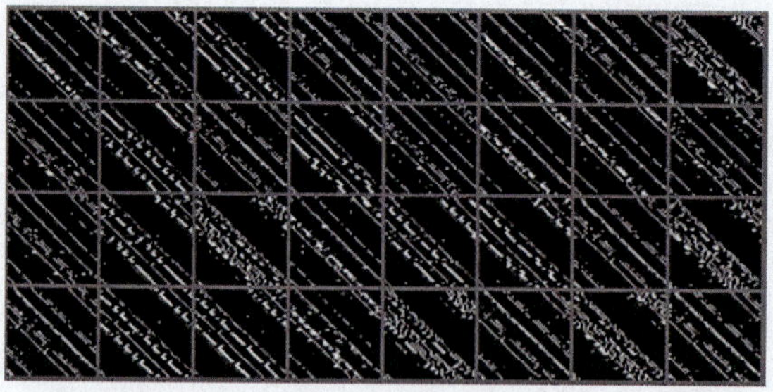

Fig. 5. Detection method of two-dimensional linear array.

3.3 The Structure of Our CNN Network

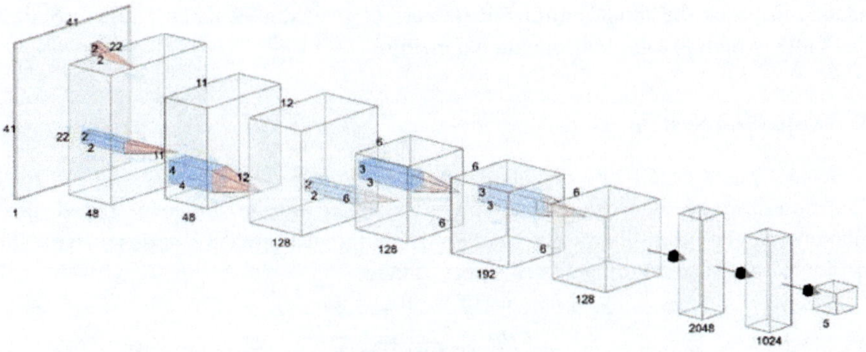

Fig. 6. CNN architecture.

Figure 6 is the CNN network structure for this article.

The one input layer, four mix-ups, and three pooling layers (after the first, second and fourth mix-ups) use an average pooling method, and the drop mechanism is installed in all connecting layers and all connecting layers [9]. And we can achieve the best overfitting prevention by adjusting its threshold to 0.5 and including the output layer.

3.4 The Process of Real-Time Detection

In today's increasingly complex international environment, how to solve the problem of real-time identification of network attacks under such extreme data loss has gradually received attention. In order to ensure the recognition effect, traditional recognition methods need to retrain the model to learn the characteristics of the new class target, but this takes a long time and cannot meet the real-time requirements. The proposal of the theory of Few Shot Learning has brought a glimmer of hope to solve this kind of problem. Small sample learning is mainly aimed at solving the problem of how to realize fast learning in the face of a small number of new class objects that have never been seen before. However, because the solution to the small sample learning problem is still developing in the field of a network attack, there are few types of research on network attack identification problems, which are currently tried to be solved by meta-learning theory.

4 Experimental Results and Data Analysis

4.1 Operating Environment

Where TP represents the number of positive class samples with correct classification, TN represents the number of negative class samples with accurate variety, FP represents the number of negative class samples with incorrect type, and FN represents the number of negative class samples with wrong type. The relationship between the variables and the evaluation method is as follows:

4.2 Data Set Introduction

Based on KDCU9999, redundant data is erased to improve a large amount of redundancy present in one of the most significant defects of the kdcu99 dataset, thus obtaining NSL-KDD, which is excellent in data balancing.

4.3 Evaluation System

The accuracy and recall rate of calculating the whole mixed matrix is related to the calculation strategy. The calculation strategy adopted by the subject is based on the weighted calculation of the number of samples in the test set. Time series visualization can be used to visualize the network's performance.

$$FPR = \frac{FP}{FP + TN};$$

$$FNR = \frac{FN}{FN + TP}$$

$$\text{Precision} = \frac{TP}{TP + FP};$$

$$\text{Recall} = \frac{TP}{TP + FN}$$

$$\text{Accuracy} = \frac{TP + TN}{\text{All Data}}$$

$$F1 - \text{score} = 2 \times \frac{\text{Precision} \times \text{Recall}}{\text{Precision} + \text{Recall}} \tag{1}$$

Table 1. Measurement result of NSL-KDD study.

Strategy	Accuacy	Precision	Recall	F_Score
CNN-First	0.786	0.812	0.765	0.743
CNN-Second	0.781	0.795	0.781	0.766
CNN-Third	0.973	0.967	0.973	0.969

Table 1 shows the measurement results of NSL-KDD study.

Although traditional CNN intrusion detection does not visualize the dataset, the relatively excellent processing method three explored in vivo has pretty good accuracy and F1 scores, and the relevant data is shown in the Table 2 compared to the appropriate strategy literature using CNN and KDD datasets. Method 3 strategies can be found in vivo higher than relevant literature.

Table 2. Compare to this paper

Strategy	Accuacy	Precision	Recall	*F1* Score
CNN-third	0.984	0.967	0.974	0.970
Ref.12	0.9437	none	0.9353	none
Ref.13	0.9556	none	0.8967	none
Ref.14	0.970	none	none	none
Ref.15	0.9218	none	0.9095	none

5 Conclusion

City network database is vulnerable to attack. Smart city IOT (Internet of Things) needs to implement intelligent monitoring systems to protect network systems.

In this paper, we study a convolutional neural network-based intrusion detection system corresponding to the inherent properties of convolutional neural networks and NCL-KDD datasets. Compared with the traditional CNN model, the system has been somewhat improved. As a result, the correct response rate of NSL-KDD for text visualization method 3 is 0.9840, the recall rate is 0.9740, and the F1 score is 0.9700.

In future work, real network traffic is also applied to train models to improve the shortcomings of the NLS-KDD dataset in the absence of current network traffic.

References

1. Priyadharshini, G., Judie Dolly, D.R.: Comparative investigations on tomato leaf disease detection and classification using CNN, R-CNN, fast R-CNN and faster R-CNN. In: 2023 9th International Conference on Advanced Computing and Communication Systems (ICACCS), Coimbatore, India, pp. 1540–1545 (2023)
2. Yanagisawa, H., Yamashita, T., Watanabe, H.: A study on object detection method from manga images using CNN. In: 2018 International Workshop on Advanced Image Technology (IWAIT), Chiang Mai, Thailand, pp. 1–4 (2018)
3. Hong, R.-F., Horng, S.-C., Lin, S.-S.: Machine learning in cyber security analytics using NSL-KDD dataset. In: 2021 International Conference on Technologies and Applications of Artificial Intelligence (TAAI), Taichung, Taiwan (2021)
4. Ingre, B., Yadav, A.: Performance analysis of NSL-KDD dataset using ANN. In: 2015 International Conference on Signal Processing and Communication Engineering Systems, Guntur, India, pp. 92–96 (2015)
5. Sheng, M., Liu, H., Yang, X., Wang, W., Huang, J., Wang, B.: Network security situation prediction in software defined networking data plane. In: 2020 IEEE International Conference on Advances in Electrical Engineering and Computer Applications (AEECA), Dalian, China, pp. 475–479 (2020)
6. Fang, L., Wang, C., Miao, Q., Liu, L.: Dynamically validate network security based on adaptive control theory. In: 2013 International Conference on Information and Network Security (ICINS 2013), Beijing, pp. 1–6 (2013)
7. Maske, S.A., Parvat, T.J.: Advanced anomaly intrusion detection technique for host based system using system call patterns. In: 2016 International Conference on Inventive Computation Technologies (ICICT), Coimbatore, India, pp. 1–4 (2016)

8. Guorui, F.: Research on the semi-supervised fuzzy clustering algorithm with pariwise constraints for intrusion detection. In: 2015 6th IEEE International Conference on Software Engineering and Service Science (ICSESS), Beijing, China, pp. 375–378 (2015)
9. Ugwoke, F.N., Okafor, K.C., Chijindu, V.C.: Security QoS profiling against cyber terrorism in airport network systems. In: 2015 International Conference on Cyberspace (CYBER-Abuja), pp.241–251 (2015)

Data Mining and Retrofit Design for Age-friendly Spaces

Jie Liu(✉)

Qilu Normal University, No. 2 Wenbo Road, Jinan, Shandong, China
20188528@qlnu.edu.cn

Abstract. Ageing in place has become the most dominant form of ageing in China. A large number of homes need to be retrofitted. Traditional retrofit designs are influenced by the personal experience of the designer, resulting in uncontrollable quality of the retrofit work. In contrast, this paper uses computer data analysis software to data mine the behaviour of users in interior spaces. The final behavioural data from the mining is used to guide the design of more rational spaces. This approach can provide a replicable methodology for all remodelling processes. In this paper, six common indoor household types for the elderly are selected for the study and the data on the home mobility behaviour of the elderly in these six representative household types is predicted with the aid of computer-related software. In the first step, the most effective path space areas are analysed. In the second step, the visual environment of the path space is analysed. Finally, a data-informed design strategy for age-friendly spaces is proposed.

Keywords: Ageing in Place · Data Mining · Ageing-friendly · Retrofit

1 Research Background

Many countries around the world are experiencing a growing population ageing problem. China is also beginning to face this problem. By the end of 2022, China's population aged 60 and over will have reached 280.04 million, accounting for 19.8% of the total population, and is likely to become the country with the largest elderly population in the world. The Chinese government has proposed a "9064" strategy for the development of the elderly. The strategy is for 90% of the elderly to be cared for at home, with the remaining 6% relying on community aged care centres to provide day care and 4% being protected by institutional care. The task of adapting home care for older people is becoming increasingly important [1]. The adaptation of residential environments for older people to ageing is also increasingly becoming a hot topic in society. We have noticed that the majority of older people live in dwellings that are already relatively outdated. Due to their early construction and relatively outdated built environment, existing dwellings do not provide adequate environmental support for older people.

Studies have shown that the choice and use of activity venues by older people is closely related to their quality of life. The presence of an interior space largely guides the behaviour of its interior users; for example, the presence of walls suggests that

W. Guo et al. (Eds.): GBCESC 2023, LNCE 328, pp. 835–843, 2024.
https://doi.org/10.1007/978-981-99-9947-7_83

pedestrians change the direction of their travel, while doorways direct people to alter the space. Among the more significant influences on the home life of older people are the behaviour of transport and the visual range of space. It is therefore necessary to document and analyse the home behaviour of older people and to propose the most logical solution for the layout of the space.

At the same time, traditional home adaptation designs are less repeatable. Most remodelling work relies on the personal experience of different designers to determine the quality of the work. This has led to a wide range of quality residential adaptations. And it is difficult to achieve a rapid, replicable model.

The conventional approach of field recording older people's home behavior faces challenges due to inefficiency and the reluctance of older individuals to be disturbed excessively. This poses a significant obstacle for the design industry in obtaining reliable research data. However, the introduction of computer data mining techniques has revolutionized the research process by enabling accurate analysis of user traffic behavior and line-of-sight analysis within a given space.

By utilizing computer data mining techniques, it becomes possible to comprehensively analyze all possible user traffic behaviors and line-of-sight scenarios under existing spatial conditions. This data provides valuable insights into the extent and patterns of user traffic behavior, serving as a direct basis for the selection and design of route spaces.

The integration of computerized data mining techniques offers a promising solution to the challenge of obtaining reliable research data. In this study, the focus is on retrofit design within living spaces, specifically targeting route spaces and sightline ranges. With the continuous advancements in computer technology and its expanding applications, the use of computer technology for mining and presenting data on older people's domestic transportation behavior and visual range becomes increasingly viable and effective.

The research utilizes computer technology to extract and analyze data related to older people's domestic transportation behavior and their visual range. Based on this data, targeted retrofitting design measures can be proposed for route spaces and age-friendly environments. By taking into account the range of vision, the retrofit design aims to create spaces that are tailored to the needs and preferences of older individuals, promoting their comfort and convenience within their living environment.

In conclusion, the integration of computer data mining techniques offers a reliable and efficient means of obtaining research data in the design industry. Moreover, the utilization of computer technology in analyzing the transportation behavior and visual range of older individuals within their living spaces allows for the development of tailored retrofitting design measures. These measures are specifically aimed at optimizing route spaces and creating age-friendly environments, ultimately resulting in an improved quality of life for older individuals.

2 Introduction to the Theory and Principle of Path Space Data Mining

This research, which is based on the theory of spatial syntax, aims to understand and analyze the spatial organization of human life at various scales, ranging from interior spaces to city-scale environments. Spatial syntax is a theory and method that quantitatively describes the spatial system and its organization. In this study, the researchers utilized Depthmap, an intelligent computer software developed based on spatial syntax theory, for data analysis.

Over an extended period of development, the theory of space syntax has significantly contributed to the detailed examination of the nature and function of architectural space. The research builds upon the theoretical foundations of space syntax and employs data mining techniques using Depthmap. By simulating the traffic trajectories and visual ranges of different types of indoor spaces, the study generates valuable insights into user movement patterns and their relationship with spatial configurations.

The analysis of traffic trajectories reveals areas of overlapping movement, indicating key spaces that efficiently support different traffic behaviors. These areas play a crucial role in facilitating smooth and efficient movement, making them essential considerations in traffic space design [3]. Conversely, areas with relatively inefficient traffic patterns are identified as potential areas for improvement. The trajectory data generated through this analysis provide designers with valuable information for selecting and optimizing traffic spaces.

Furthermore, the study utilizes visual data mining techniques to examine the visual range of specific locations within the analyzed spaces. This visual information aids in determining the functional distribution of the space, assisting designers in making informed decisions regarding the allocation of different activities within the environment. The combination of trajectory and visual data enhances the overall efficiency of age-friendly space renovation design, enabling designers to create spaces that are conducive to the needs and preferences of different age groups.

In summary, this research leverages the theory of spatial syntax and utilizes the Depthmap software for data mining and analysis. By studying traffic trajectories and visual ranges within different indoor spaces, the study provides valuable insights into spatial organization and facilitates the design of age-friendly spaces that promote efficient movement and meet the functional requirements of various user groups.

3 Basic Information

This design uses questionnaires and household interviews and semi-standardized interviews and other survey research methods to investigate the elderly. 178 questionnaires were included in this research, and the research scope of the questionnaires was within Shandong Province. The basic content of the questionnaire was categorized and summarized according to the physiological and psychological states of the elderly. At the same time, we collected the house plans of the elderly and summarized the house plans with the theory of architectural typology [4].

The household types are summarized and analyzed from the perspective of architectural typology. The hall categories of representative house types are grouped into six categories. The six common household types for elderly people are one-bedroom (45 m^2), two-bedroom (40 m^2), two-bedroom (44 m^2), three-bedroom (50 m^2), and three-bedroom (60 m^2), and four-bedroom (80.5 m^2). The house types were analyzed and designed from small to large, and the house plans were drawn linearly using the computer-aided design software AutoCAD. The floor plan was prepared for later analysis with Depthmap software. It is worth mentioning that the path space and visual range was chosen as the object of study in this paper, hiding some relatively small spaces in the larger house types that did not require computer analysis, such as bedrooms and kitchens, in order to better show the path space.

Depthmap is a software tool that begins by analyzing the basic contour of a space and subsequently examines all the internal spaces connected to it, including traffic spaces and other areas. The software simulates the ideal states of motion within the space and calculates various spatial variables. The key variables analyzed in this process include spatial depth, line of sight integration, and spatial accessibility [5].

This research paper focuses on computer data mining using two variables: traffic behavior and the line-of-sight environment within the space. The visualization presented in the paper showcases the simulation results of Depthmap, specifically the pedestrian flow simulation map and the views field simulation represented by rectangles and circles [6]. The spatial pedestrian flow simulation map provides insights into spatial hierarchy, illustrating how people move through different areas. On the other hand, the spatial view field analysis map simulates the boundaries of spatial visual occlusion, indicating areas where visibility may be obstructed. And these visualizations offer valuable information for spatial analysis. The pedestrian flow simulation map helps identify areas with high or low pedestrian movement, allowing designers to understand the dynamics of human traffic within the space. The view field analysis map provides insights into the visibility and potential obstructions within the environment, aiding in the assessment of spatial qualities and potential design interventions. By utilizing Depthmap and its visualization capabilities, the research leverages the power of spatial syntactic analysis to gain a deeper understanding of how people navigate and perceive space. This information can inform design decisions, allowing for more informed and effective design solutions that optimize spatial organization, traffic flow, and visual connectivity.

In summary, the research employs Depthmap to perform computer data mining and visual analysis of traffic behavior and the line-of-sight environment within the space. The resulting visualizations, including the pedestrian flow simulation map and view field analysis map, provide valuable insights into spatial hierarchy and visual occlusion boundaries. These findings contribute to a more comprehensive understanding of spatial dynamics and support the development of optimized design solutions (Fig. 1).

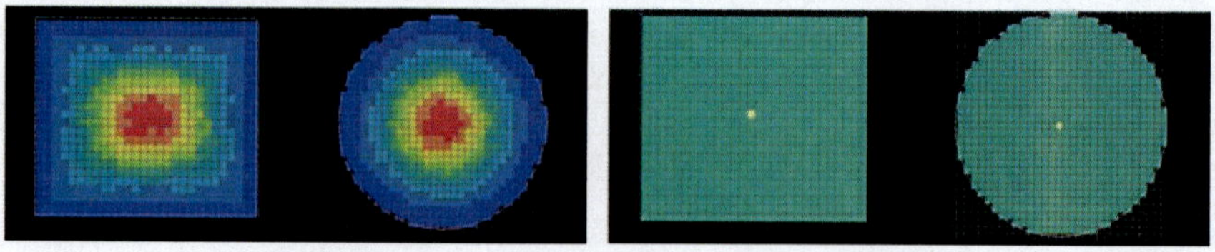

Fig. 1. Traffic and vision for software analysis

4 Application of Computer Data Mining Technology in the Spatial Design of Home Care Paths

4.1 Traffic Behavior Mining and Trajectory Space Selection

Traffic behavior mining and trajectory space selection through computer data mining software Depthmap to data mining are the six most common types of elderly living households summarized in the previous section (Fig. 2).

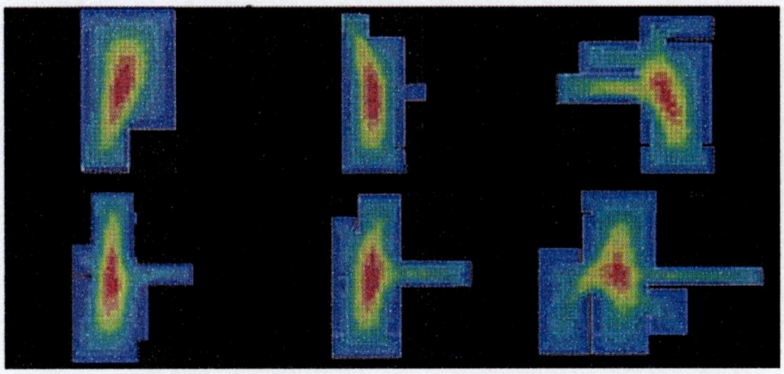

Fig. 2. Software simulation of traffic behavior data

The research involves collecting data on all possible traffic behaviors for each household type and visually presenting the results. The visual output employs a color scheme, with red indicating areas of high traffic behavior occurrence, while yellow, green, and blue indicate reduced probabilities of occurrence. This visual representation enhances the intuitiveness and clarity of the findings. The process of data analysis is swift, and the results are unambiguous, making the method highly reproducible.

The abundant data obtained from the analysis include high-frequency traffic behavior trajectories. These trajectories directly inform the selection of optimal locations for traffic spaces within the residence. Areas with a high frequency of occurrence represent regions where traffic behavior is predominantly induced, indicating higher traffic efficiency. Consequently, this provides valuable guidance for identifying suitable locations for traffic spaces specifically designed for older individuals.

By utilizing the data-driven approach, the research provides a clear and practical framework for designing traffic spaces that cater to the needs of older people. The visual representation of high-frequency traffic behavior patterns helps designers identify areas where interventions are most required to enhance traffic efficiency and improve the overall user experience. This data-driven approach offers a reliable and systematic method for decision-making in the design of age-friendly spaces.

In summary, the research employs data analysis techniques to count and visually represent all possible traffic behaviors for different household types. The visual output effectively highlights areas of high-traffic behavior occurrence, facilitating the selection of optimal locations for traffic spaces. The approach is characterized by its efficiency, clarity, and reproducibility, providing valuable insights into designing traffic spaces that prioritize the needs of older individuals (Fig. 3).

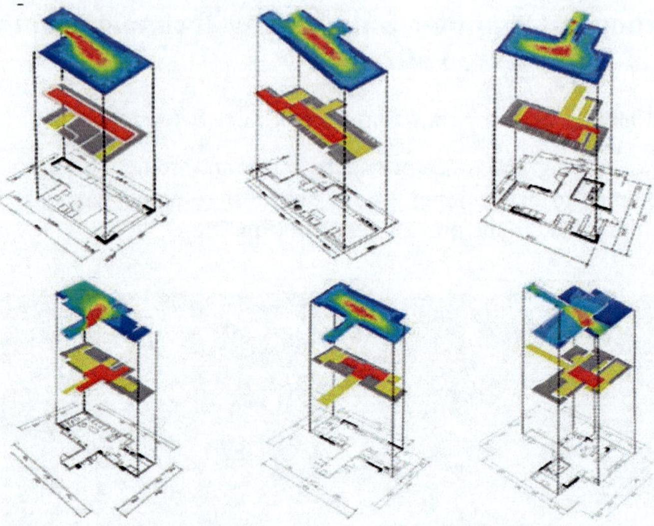

Fig. 3. Use the software to derive the trajectory space region

The Depthmap software is used to simulate the human flow in each space area and analyze all the possibilities of human pathways to a certain location in space. The human flow simulation analysis has a corresponding visualization help for the functional area analysis of the aging-friendly space. From the above figure, we can see that the range of human flow density in the space is mainly concentrated in the central area of each space, and the human flow density in the space is unevenly distributed. Regarding the traffic behavior simulation analysis, the dense human flow area should be designed from multiple angles in the aging-in-place design plan, and the red part can be designed as the most frequently used functional space in the division of functional areas, and other subsidiary functional areas can be divided on this basis. Therefore, the red area in the diagram is chosen as the general scope of the main traffic space, the yellow area is the second traffic space, and the gray area is the furniture and storage space, as well as the analysis of the daily behavior of the elderly, and finally the aging-appropriate design scheme.

4.2 Analysis of the view field environment

In the view analysis, after importing the linear space module drawn by computer-aided design software AutoCAD into Depthmap software [7]. Firstly, according to the actual area of the linear space, use the "Set Grid" command to edit the size of the module within the linear space. The larger the value of this result, the larger the grid in the linear space and the clearer the calculated data. The software will divide the space module into a network structure, and then use the "Fill" command to divide the range of the linear space operation. The final analysis of the grid structure with range boundaries is performed.

The "Make Graph Options" command converts a point into an operable model in the range of the operator space. An operable model can be understood as the number of elements that can be seen from a point in a known spatial range to other elements in the space, and the more elements seen, the warmer the color tendency of the point, and vice versa (Fig. 4).

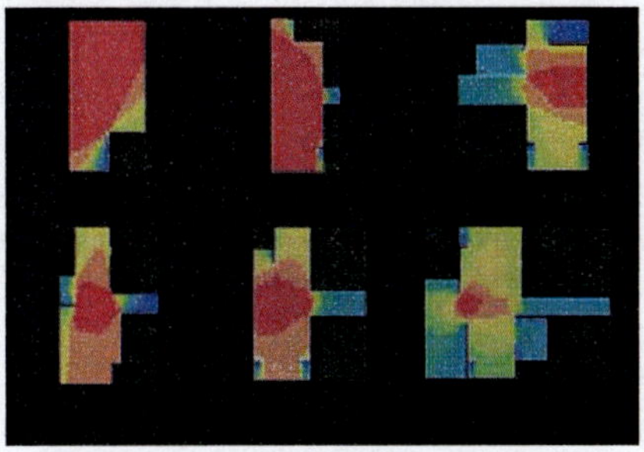

Fig. 4. Visual field Analysis

The Visual fields analysis is performed based on a computable model. The visual field is the area within reach of the line of sight. Visual field analysis can simulate the range of visual field of an elderly person at a location in space. In the analysis, multiple location points are selected to analyze the field of view in space. A computer is used to mine the visual field data of different locations in space [8]. Also, simulate the range of visual field of the elderly in different locations in the space. In the aging-friendly design of living spaces, the method uses the computer-simulated range of vision of the elderly. The designer selects the optimal space area from the obtained data. The method has a very important data reference value (Fig. 5).

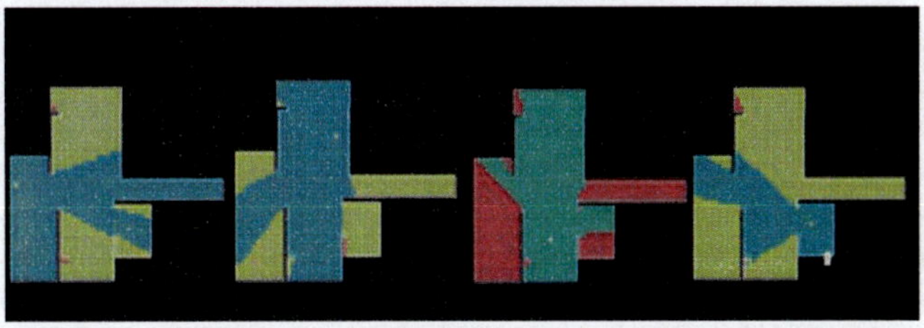

Fig. 5. Analysis of the view area of each node of house type No. 6

According to the computer for the view area fixation analysis found that in some locations, a small area of view blockage will be produced, and the area produces a hindrance to the psychological observation of the interior by the elderly, the later design can transform the area into an auxiliary class space (such as storage areas and other areas that do not require high view), or the sight environment of the area can be modified.

5 Conclusion

The computer-based comprehensive data mining approach in this study focuses on analyzing the traffic behavior and sight environment of elderly individuals within indoor spaces. By considering all possibilities within the space, the method accurately counts and displays data results, providing reliable and intuitive information for path space design. Compared to traditional design research, computer data mining enhances efficiency and accuracy, enabling designers to propose more targeted and scientifically informed design solutions for path spaces. However, it is important to acknowledge that this method primarily focuses on the influence of the physical space itself on human behavior. It currently does not incorporate a wide range of personalized parameters, such as the user's psychological preferences. Consequently, the method may not fully capture 100% of the real behavior of users within the space. In future research, it is necessary to explore and incorporate relevant means to update and improve the method. This may involve integrating additional factors that influence user behavior, refining data collection techniques, and considering user-specific preferences to provide a more comprehensive understanding of the user-space interaction.

In summary, while the computer-based data mining approach significantly improves the efficiency and accuracy of traditional design research by analyzing traffic behavior and sight environment in indoor spaces, it is essential to recognize its current limitations. Future advancements and improvements are necessary to address personalized parameters and enhance the method's ability to capture a more complete representation of user behavior in the space.

Acknowledgment. We would like to express our sincere gratitude to the participants of this project from the Innovation and Entrepreneurship Training Program for College Students of ShanDong Youth University of Political Science, namely Xinyu Zhao, Bingyan Sun, and Yanan Zhou. We appreciate their hard work and commitment to the project.

References

1. Kothiyal, K.: Anthropometry for design for the elderly. Int. J. Occup. Saf. Ergon. **7**(1), 15–34 (2001)
2. Yichen, Q.: Visual communication-based virtual reality design of imaging information collection and display system. Wirel. Commun. Mob. Comput. **1**, 1–12 (2022)
3. de Cássia Martinelli Guerreiro, T.: Data-mining, GIS and multicriteria analysis in a comprehensive method for bicycle network planning and design. Int. J. Sustain. Transp.**12**(3), 179–191 (2017)
4. Mishra, S.A., Pandit, R.K.: Space syntax approach for analyzing crime preventive urban design: concept review. J. Adv. Res. Constr. Urban Archit. **1**(1), 30–35 (2016)
5. Xu, Y.: Towards sustainable heritage tourism: a space syntax-based analysis method to improve tourists' spatial cognition in Chinese historic districts. Buildings **10**(2), 1–18 (2020)
6. Eirini, N.: Evaluation and experimental design in data mining and machine learning. Big Data **8**(5), 20–21 (2020)

7. WalidMahfoud, D.: Noise and spatial configuration in Biskra, Algeria—a space syntax approach to understand the built environment for visually impaired people. Sustainability **13**, 19 (2021)
8. Seungkyu, R.: Considering space syntax in bicycle traffic assignment with one or more user classes. Sustainability **13**, 21 (2021)

The Ecological Wisdom of Water Management in Traditional Villages in Northeast China, Hubei Province

Lilan Luo[1], Changyou Wu[1], Hui Feng[1], and Haiyu Xia[2](✉)

[1] Wuhan University of Science and Technology, Wuhan 430065, China
[2] Wuhan Natural Resources Conservation and Utilization Center, No. 55 Sanyang Road, Jiang'an District, Wuhan, China
lilanluo@wust.edu.cn

Abstract. The Northeast region of Hubei, China belongs to a low mountain and hilly terrain with abundant water systems. Villages around the river have been plagued by floods and waterlogging disasters. In the process of adapting to nature, the ancestors of the villages accumulated rich ecological wisdom for water management. This study focuses on the traditional villages in Northeast China, Hubei Province, using research methods such as literature review, field investigation, and GIS analysis. From macro to micro levels, it summarizes the ecological wisdom of water management, such as balancing the selection of sites with "hydrophilicity" and "hydrophobicity", adaptive layout based on the water environment, and spatial organization of water systems combining points, lines, and surfaces. Based on these findings, putting forward suggestions for village planning and construction.

Keywords: Traditional villages · Ecological wisdom · China · Northeast Hubei

1 Introduction

With the rapid changes and development of modern society, the living environmental is facing many water environment problems, such as flood disasters, water ecological problems, water resource shortages, etc., forcing people to actively explore sustainable water ecological solutions. The Northeastern region of Hubei Province belongs to a low mountainous and hilly landform with abundant water systems, the villages around the river channels have been plagued by floods. In the process of adapting to nature, the ancestors of the villages accumulated rich ecological wisdom for water management.Currently, research on traditional villages and water environment in China can be divided into traditional village water ecological wisdom [1, 2], water environment landscape and space [3, 4], and traditional water environment facilities [5, 6]; In terms of research areas, research on traditional villages and water environment is mostly concentrated in the southern regions with abundant freshwater resources [6, 7] and the northern regions with drought and water scarcity [8, 9]. This article takes traditional villages in Northeast China, Hubei Province as the research object, using research methods such as literature review, field research, and GIS analysis. From the aspects of site selection,

W. Guo et al. (Eds.): GBCESC 2023, LNCE 328, pp. 844–850, 2024.
https://doi.org/10.1007/978-981-99-9947-7_84

plane layout, and water system spatial organization, it condenses the ecological wisdom of water environment. And based on these findings, putting forward suggestions for village planning and construction.

2 Environment of Northeast China, Hubei Province

Northeast China, Hubei Province mainly includes Xiaogan, Suizhou, Huangpi District of Wuhan City, Luotian County, Yingshan County, Hongan County and Macheng City of Huanggang. There are 69 traditional villages in the first to sixth batches of national level. Northeast Hubei belongs to subtropical monsoon climate, with an average annual temperature of 16 °C and annual rainfall of more than 1000 mm, most of which are concentrated between 1100-1700 mm. The terrain in Northeast China, Hubei Province is high in the East and low in the west, high in the north and low in the south, high in the north and low in the northwest-southeast of the mountain range in the Shanxi section of Dabie, and plain in the south. Northeast China, Hubei Province is rich in drainage systems, including two rivers, the Yangtze River and the Hanjiang River, as well as the tributaries of the Ba River, Lift Water, Fuhe River and Rushui River. When the flood season comes, a large amount of water flows into the Yangtze River and the Hanjiang River. In addition, Northeast Hubei belongs to the landscape of low hills and hills. Other tributaries have large slopes and bending channels. When the amount of water exceeds the maximum transmission capacity of the river, flood disasters are very easy to occur. According to statistics, 90 times of flood and waterlogging disasters occurred in Northeast Hubei in the Ming Dynasty, accounting for 51% of the total number of floods in the Ming Dynasty, 115 times in the Qing Dynasty, accounting for 49% of the total number of floods in the Qing Dynasty[10, 11] (Figs. 1 and 2).

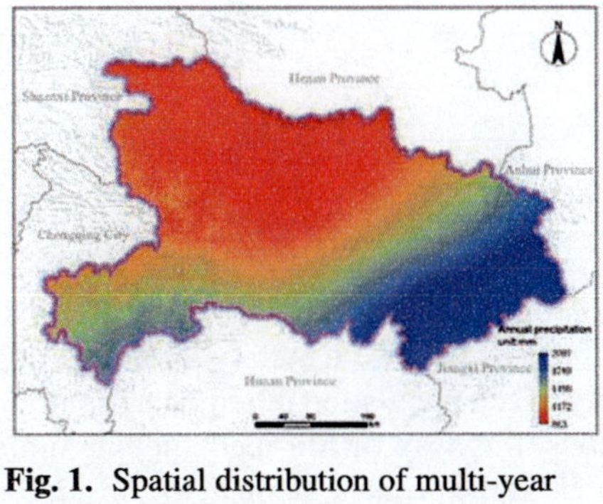

Fig. 1. Spatial distribution of multi-year average rainfallin Hubei Province, China

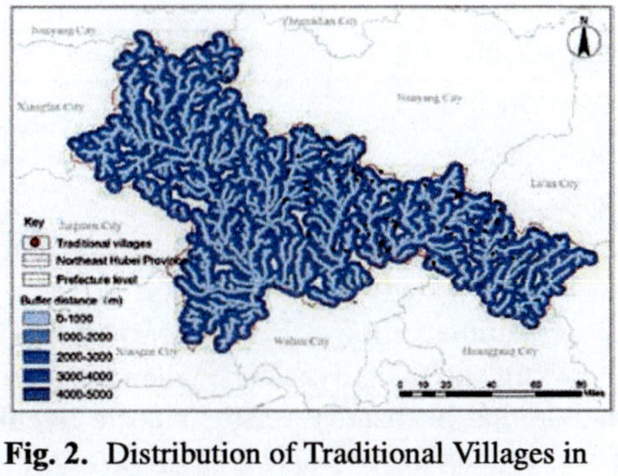

Fig. 2. Distribution of Traditional Villages in Northeast Hubei and Buffer Map of Water System Distance

3 The Ecological Wisdom of Water Management in Traditional Villages in Northeast China, Hubei Province

3.1 Water Ecological Wisdom in Village Site Selection

In this paper, the hydrological analysis in GIS spatial analysis is used to extract the water system in in Northeast China, Hubei Province. After multiple simulation calculations, 500 m is selected as the buffer radius to analyze the main stream and other main tributaries with multiple buffer zones, the relationship between village site selection and water system distance is finally obtained. According to the analysis results (Table 1), within the buffer radius range of 0–500 m and 500–1000 m of the river, there are many traditional villages, accounting for 23.19% and 20.29% of the total, respectively. In the range of 0–4000 m, the number of traditional villages is generally negatively correlated with the horizontal distance from the water system, showing a clear "hydrophilicity". In order to explore the characteristics of village site selection within a distance of 500 m from the water system, a buffer radius of 100 m was selected for multi buffer zone analysis. The results showed that villages were mainly concentrated within the range of 100–200 m from the river, accounting for 31.25%. This indicates that village site selection is more preferred in the areas of 100–200 m from the river.

Table 1. The relationship between villages and the horizontal distance from the water system

Distance/m	number of traditional villages	Ratio/%
0–500	16	23.19
500–1000	14	20.29
1000–1500	11	15.94
1500–2000	11	15.94
2000–2500	4	5.80
2500–3000	5	7.25
3000–3500	5	7.25
3500–4000	3	4.35

Traditional villages in Northeast China, Hubei Province were mainly formed during the Ming and Qing dynasties, when the majority of the population relied on agriculture and animal husbandry for their basic livelihoods, this way of life made people more dependent on the ecological environment. "Water" is the primary element to consider in the selection and construction of traditional villages. Villages show "hydrophilicity" in the selection of sites. Among the 69 traditional villages, there are 39.13% of the village names are related to "water", such as "wan" (river), "he" (river), etc. But if it is too close to the river system, it is easy to be threatened by floods. Therefore, it is necessary to avoid areas that are prone to flooding and demonstrate water resistance when selecting a site. Overall, the selection of traditional villages seeks a balance between "hydrophilicity" and "hydrophobicity" (Table 2).

Table 2. The relationship between villages and the horizontal distance from the water system (within 500 m)

Distance/m	number of traditional villages	Ratio/%
0–100	4	25.00
100–200	5	31.25
200–300	2	12.50
300–400	3	18.75
400–500	2	12.50

3.2 Water Ecological Wisdom of Village Layout

Water flow is one of the key factors in village site selection, and the location and shape of water flow also affect the layout and form of the village. Traditional villages demonstrate the ecological wisdom of adapting to water in terms of plane layout. Streamline and fishbone shapes are often used to buffer floods, and the layout of streamlined villages is consistent with the flow pattern, following the direction of the water flow, and can reduce the impact of floods. Fish bone shaped villages are generally small at both ends and large in the middle. This layout adopts biomimetic principles, and sharp corners can play a role in buffering floods. On the other hand, in order to adapt to the trend of water flow, the layout of the streets and alleys in the grid shaped village is in a grid shape. The roads in the village have a certain slope, with an overall high inside and low outside. When heavy rainfall arrives, surface runoff will follow the height difference and flow into the streams outside the village from high to low, from inside to outside, forming a complete drainage system (Fig. 3).

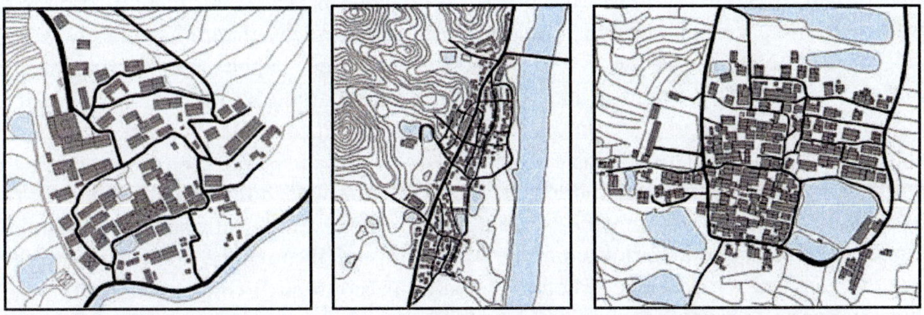

(a) Dawuyuan Village (b)Shuangqiao Village (c)Cai Guantian Village

Fig. 3. Layout of Traditional Village Form

3.3 Water Ecological Wisdom of Village Water System Spatial Organization

Traditional villages often utilize the elevation difference of natural terrain in water system organization. The peripheral water is linear, while the storage water source is point

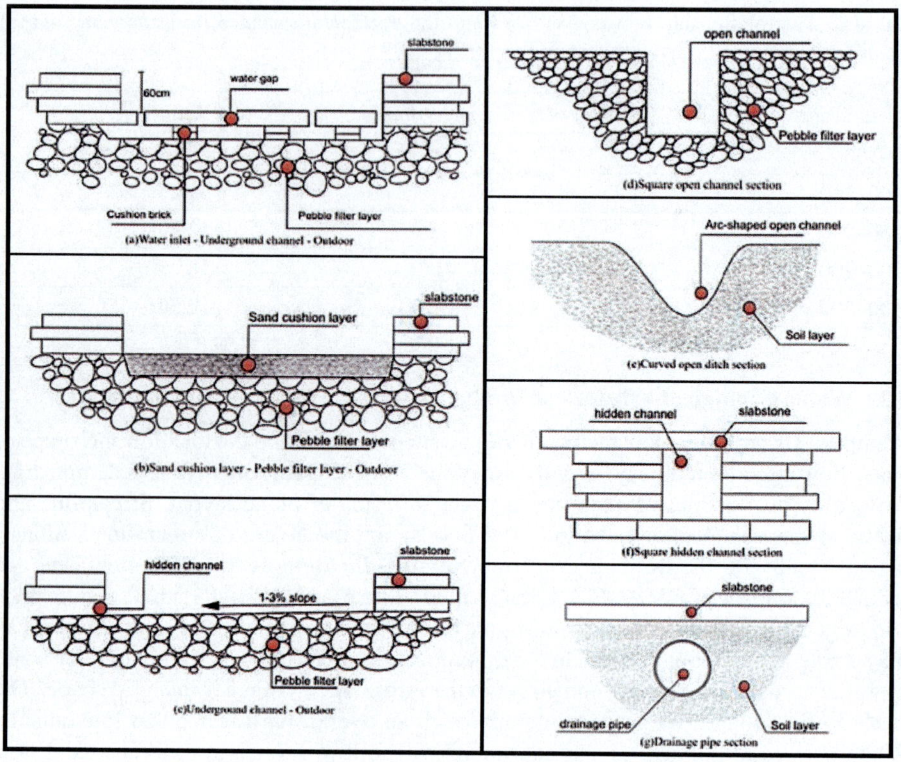

Fig. 4. Cross section of courtyard and ditch

shaped, and the combination of the inside and outside is planar, ultimately forming a water system space that combines points, lines, and surfaces. There are mainly water wells and water tanks in the dotted water system space. When encountering heavy rainfall, the rainwater stored in the wells will be discharged into the deep soil at the bottom of the wells, and water tanks are commonly used for water storage. The linear water system space mainly consists of canals and streams. Water channels are also divided into open channels and hidden channels. The rainwater collected by the courtyard inside the building flows into the hidden channel along the terrain and is discharged outside the house, and then flows into the stream or pond through the open channel. The entire village forms a drainage structure of "building (courtyard) - hidden channel - open channel - pond/stream". The surface drainage space mainly includes ponds and weirs, which can prevent flooding, regulate microclimate and breed breeding. Tangyan refers to an artificial lake built at a narrow point of a mountain or river, which can serve as a flood control, water storage, irrigation, and other functions. Some traditional villages often choose areas with tangyan upstream as their site selection (Fig. 4).

4 The Ecological Wisdom of Water Management in Traditional Villages in Northeast China, Hubei Province

Based on the analysis of various aspects of the ecological wisdom of water management in traditional villages in Northeast China, Hubei Province, and based on the actual situation, the following suggestions are proposed: (1) Plan ahead of time when selecting a site. Villages can avoid flood disasters through reasonable site selection and pre prevention methods, achieving the flood control effect of low-cost small-scale development; (2) Adopting a drainage organization based on nature. Villages can fully utilize the relationship between the surrounding environment's landscape and terrain, as well as the relationship between roads, alleys, farmland, and natural ponds, to guide the situation and unfold layer by layer, ultimately forming a complete drainage system; (3) Respect the natural environment and protect it. When planning a village, the impact of human activities on the water environment should be taken into account, and facilities such as household waste treatment and sewage treatment should be installed to reduce the negative impact of the village on the water environment. Based on the conclusions of this study, further in-depth research will be conducted on the scientific value of the ecological water management wisdom of traditional villages, and explore effective and practical methods for connecting the ecological water management wisdom of traditional villages with modern rural planning in the region.

Acknowledgement. The work was supported in part by Research on the Construction of Practical Curriculum System for Urban and Rural Planning in Universities (No. 220901375272747).

References

1. Zhao, H., Xie, W., Lu, D., Yang, B.: The ancient ecological practice wisdom and contemporary inspiration of traditional villages in northern China. Mod. Urban Res. **07**, 20–24 (2018)
2. Lu, X., Zhang, Q., Lei, Z., Jing, Y., Zhang, T.: The wisdom of water management and its planning inspiration in traditional villages in the Loess Gully Area. Mod. Urban Res. **09**, 97–104 (2022)
3. Fu, J., Xu, J., Xiao, D.: Study on the adaptability of traditional village forms and landscapes to water environment in Southern China. Chin. J. Landscape Arch. **29**(08), 120–124 (2013)
4. Xiao, H., Liu, Y., Kuang, Y., Zhang, P.: Regional landscape and ecological wisdom of spring settlements in langxi river basin of hongfan pool spring cluster in Jinan. Landscape Arch. **27**(08), 69–75 (2020)
5. Zeng, Y.: Inspiration from traditional practical wisdom: interpretation of water ecological infrastructure in hengcha village. Decoration **08**, 116–119 (2018)
6. Xiao, Y., Zhao, B., Mao, Q.: Ecological wisdom of traditional Jiangnan water management. Chin. Landscape Arch. **33**(07), 70–74 (2017)
7. Wang, B., Wang, R.: The ecological wisdom of water management in traditional villages: a case study of Diaoyuan ancient village in Jiangxi Province. J. Jinggangshan Univ. (Nat. Sci. Ed.) **37**(06), 71–77 (2016)
8. Zhao, H., Chen, Y., Xie, W., Qiu, W., Zhang, C.: Exploration of ecological water control wisdom in Yujia ancient village and its contemporary enlightenment. Mod. Urban Res. **02**, 40–44+52 (2018)

9. Zhao, B., Zhang, J., Kong, Y.: Analysis of the spatial structure of traditional settlement water systems from the perspective of flood control: a case study of spring water settlements in four northern provinces. Huazhong Arch. **32**(10), 112–116 (2014)
10. Xun, X.: A Study on the Temporal and Spatial Distribution of Flood and Drought Disasters in Hubei Province during the Qing Dynasty. Central China Normal University, Wuhan (2012)
11. Li, C.: China Fire Festival. Shanghai Science and Technology Press, Shanghai (1997)

An EV Charging Station Siting Model Based on Machine Learning

Yufang Dai(✉), Minghao Liu, and Xiangli Liao

College of Computer Science and Technology, Chongqing University of Posts and Telecommunications, Chongqing 400065, China
17316783641@163.com

Abstract. Aiming at the problems of the site selection method of EV charging station based on machine learning, such as single site selection factor, large subjective factor, insufficient sample labeling, simple labeling strategy and low location accuracy, a site selection method of EV charging station based on multi-source data fusion was proposed. This method constructed seven basic features from multi-source data, combined with PFAHP and TOPSIS methods, obtained the weight of POI in economic features, and proposed a semi-supervised learning method using mixed similarity automatic allocation markers to extend the training set. Based on the basic features and LogitBoost integration algorithm, an EV charging station siting model was constructed. In this paper, 6 districts in Shanghai were taken as the study area, and 4 sets of schemes and 8 experiments were designed to verify the superiority and applicability of the proposed method. The experimental results show that compared with the single similarity and manual labeling model, the accuracy of the recommended model is increased by 0.6%–2.6%, and the recall rate, F1 and ranking accuracy are all improved. The results show that the method proposed in this paper can better solve the problem of pre-location of EV charging stations, and has certain reference value for other public facilities.

Keywords: multi-source data · Location of electric vehicle charging stations · Machine learning

1 Introduction

The rapid development of electric vehicles can promote the green transformation of China's economic and social development to a certain extent. The number of electric vehicles in China will reach 10.45 million by 2023, and is expected to reach 16 million by 2025 and 64.2 million by 2030. China has listed the construction of charging stations as one of its key areas of "new infrastructure". Studies have shown that China's EV charging stations have many problems, such as unreasonable site selection, low utilization rate and unbalanced layout [1]. With the popularization and promotion of electric vehicles, the reasonable location of charging stations is directly related to the common interests of individuals, the society and even the country. How to solve the problem of location and layout of urban charging stations is an important content of optimizing resource allocation, and also one of the key problems in the construction of smart cities.

W. Guo et al. (Eds.): GBCESC 2023, LNCE 328, pp. 851–860, 2024.
https://doi.org/10.1007/978-981-99-9947-7_85

The location selection method based on machine learning is mainly based on multi-source data to obtain the potential relationship between resource allocation and driving factors [2]. Because machine learning localization method is rarely affected by human factors and the decision-making process is relatively objective, it has been applied in many location decisions in recent years. Most of the current studies build models based on economic indicators such as cost, expense and income. Most of them only consider the characteristics of commercial charging stations and choose a single influencing factor, with more subjective factors [3]. From the perspective of EV charging station site selection methods, some scholars have built optimization models from relatively limited dimensions and established weight relationships based on experience to analyze the site selection of charging stations. The analysis process is relatively simple, without the support of data volume, and subjective factors are relatively large [4]. These studies usually establish the corresponding mathematical model by obtaining the location of charging stations in a certain region and the relationship between various factors and the location of charging stations [5], and solve single or multiple objective functions. With the development of information technology, some more flexible data-driven methods and strategies begin to appear, mainly the application of mass data and machine learning methods combined with location [6]. Although the positioning method based on machine learning avoids the influence of human factors to a certain extent and is relatively objective, it also has many problems, such as low correlation of driver selection, simple labeling strategy when sample labels are insufficient, single selection of similar functions based on machine learning, low positioning accuracy and poor algorithm due to data imbalance [7–9].

This paper uses multi-source data and machine learning algorithm to build a model, mine the internal relationship between data and location results, propose a better location scheme within a certain target range, and improve the shortcomings of the algorithm model.

2 Methodology

The technology roadmap is shown in Fig. 1.

This research mainly includes data processing, model construction and training, model verification, model application and other modules. The data processing part includes clearing and processing the data as required, and then using relevant algorithms to extract factors such as population, vehicles and roads of a certain grid cell size. And through normalization and rasterization to form a unified resolution and coordinate data set. In the part of model construction and training, the PFAHP-TOPSIS method was used to obtain the evaluation results of the suitability of charging stations in the study area. The specific steps include: (1) Using PFAHP method to get the weight of POI factors; (2) TOPSIS method was used to obtain the comprehensive scoring index of the corresponding grid according to the weight and number of each POI; (3) Finally, the comprehensive score index is divided into two points to determine the 0 and 1 labels of the power network of the charging station. On the other hand, a machine learning model based on semi-supervised learning is constructed to obtain the suitability evaluation results of charging stations in the study area. (4) Label two samples of the corresponding grid: adaptive

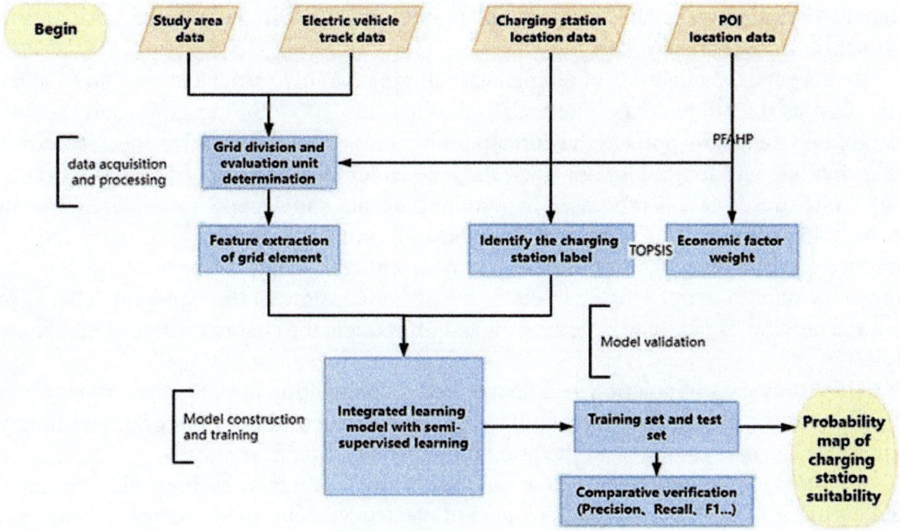

Fig. 1. Technical Roadmap

location selection and inappropriate location selection; (5) Part of labeled data and unlabeled data are randomly selected from the driving factors and the grid data set with charging stations as the training set. By analyzing various features, feature parameters are taken as the input of the location model for model training. (6) A machine learning model based on semi-supervised learning is established to transform the location problem of EV charging stations into a classification problem. Semi-supervised learning is mainly reflected in marking only a part of the data set. Through iterative learning of the model, the similarity is used to mark the unlabelled data, and finally an integrated learning model with high classification accuracy is trained. In the application part of the model, the site selection analysis and recommendation of the selected research area in Shanghai are analyzed, and the probability graph is formed.

2.1 Data Preprocessing

This paper takes into account the principles of economic, social and environmental benefits, and finally selects population, vehicles, charging demand, market competition, economic activity, road network density and construction cost as the main influencing factors of charging station location. Use the electric vehicle track data to obtain the passenger boarding and alighting positions, and then analyze the characteristics of population flow; Analyze the characteristics of population density using population data; The charging demand characteristics are extracted from the electric vehicle track data through the charging demand algorithm; The vehicle track density is extracted from the electric vehicle track data; Extract the competitive characteristics using the location data of electric vehicle charging station; Using POI(Point of Interest) data, the weights of various POI are determined based on PFAHP and TOPSIS methods, and the economic characteristics are extracted based on the number of POI. Using the house price data to extract the cost characteristics.

In this paper, six districts of Shanghai, including Xuhui District and Jing'an District, are taken as the study areas. The map is divided into 300 * 300 m grids, and the corresponding data is mapped to the corresponding range. Point data is mapped according to its latitude and longitude; For track data, consider data mapping of multiple points. The multi-source data is processed by python program and ArcGIS software. Similarly, in ArcGIS software, the fixed grid is associated with the charging pile data, and the number of charging piles in the unit grid is calculated, using the new energy vehicle trajectory data to extract the characteristics of human flow in the unit grid. That is to say, the number of passengers getting on and off reflects the characteristics of passenger flow.

This study uses population grid data to extract population density characteristics. In ArcGIS software, resample the population grid data according to the grid, and finally normalize the data to represent the population density characteristics.

Similarly, for vehicle factors, the number of track points in each grid is calculated and normalized by mapping the track data of electric vehicles to the corresponding grid area.

Characteristics of charging demand. First, the charging demand points are obtained through tracking data and charging demand algorithm, and the number of charging demand points in each grid is calculated to represent the characteristics of charging demand. The algorithm steps are as follows:

1) For each electric vehicle, select the point where the speed is 0 and the load condition is 0. 2) Filter whether the above points are within a circle with a radius of 50m from the charging station. 3) Put the matching point into the collection, and find a line segment in the track where the compliance speed of the point is always 0 and remains 40–120 min. 4) Find the nearest point in front of the corresponding clip, and the passenger load is true as the charging demand point. 5) Browse the track of each electric vehicle.

6) Collect all points that meet the requirements.

The competitive point feature is extracted from the location data of the existing EV charging station. Count and normalize the total number of 3 * 3 neighboring charging stations in each power grid.

Economic factors describe the economic environment through seven POIs: gas stations, substations, shopping centers, schools, hotels, fire stations and transportation stations. Extract economic characteristics from POI data. Calculate the number of each POI data in each grid, and use the PFAHP method mentioned below to obtain the corresponding weight of each POI. Finally, the sum of the product of each POI number and the corresponding weight is used to represent the economic characteristics, which also requires normalization.

The road length in the statistical cell grid represents the road characteristics and road density characteristics of each grid, which also requires normalization.

The average price of a region is used to represent the configuration cost of electric vehicle charging stations in that region. Draw fixed grid and house price data, calculate the average value of house price data in the grid, and use the average value to represent the cost characteristics of the grid, and normalize it.

2.2 Method and Principle

PFAHP and TOPSIS. This study uses the PFAHP method to determine the weight of each POI (Point of Interest, POI). The purpose of calculating the weight of each standardized priority is to obtain the weight of each POI driver according to the PFAHP method, provide conditions for the application of TOPSIS method, and provide basis for the determination of labels [10].

The POI weight obtained by PFAHP method is weighted and summed with the number of various POIs owned by the grid of existing electric vehicle charging stations, and the grid with charging stations is ranked according to the comprehensive score index obtained by TOPSIS method. The tag of existing charging station grid is determined by the sorting result, laying a data foundation for the following work.

LogitBoost Model and its Optimization. Because the machine learning method based on AdaBoost is too large for the penalty of misclassification points, it is sensitive to abnormal points and cannot predict the probability of classes. Therefore, the LogitBoost algorithm used can make up for these two shortcomings, it samples the forward stepwise additive model, the loss function is the logarithmic loss function, and the Newtonian descent method is considered to be applied to the error function to update the model at each step.

(1) Similarity hybrid approach building

By optimizing the LogitBoost model, that is, improving the similarity function in the LogitBoost model, the usual similarity is through a single cosine similarity or Euclidean distance, considering that the two similarities are considered in different directions, the former considers more about the difference in the directionality of the two, and the latter considers more about the difference in the number of the same direction, this paper uses the cosine similarity combined with the Euclidean distance, and maps it between 0 and 1, as shown in Eq. 1:

$$S = \frac{1 + \frac{\sum_{i=1}^{n} a_i b_i}{\sqrt{\sum_{i=1}^{n} a_i^2}\sqrt{\sum_{i=1}^{n} b_i^2}}}{2}(1-\varepsilon) + \varepsilon \frac{1}{\sqrt{\sum_{i=1}^{n} (a_i - b_i)^2} + 1} \quad (1)$$

Because cosine similarity is positively correlated with similarity and is between -1 and 1, Euclidean distance is negatively correlated with similarity, where a and b represent the two samples used to calculate similarity, i is the sample identification, and the ε is a decimal between 0 and 1, adjusting the weight between cosine similarity and Euclidean distance.

(2) Automatic labeling method based on similarity

Another optimization is the labeling method for bottom N % sample labels, the existing labeling method is to first determine the bottom N % samples through similarity, and then manually analyze them to determine the label type. Although the correct rate of labeling is guaranteed, the time cost will be greatly increased and has a certain subjectivity. In this paper, an automatic labeling method based on similarity is formed based on the similarity bottom *N* % sample labeling principle. The specific method is to sum the weighted sum of the *K* unlabeled prediction samples with the highest similarity and the labeled samples (bottom *K* %) and finally use the labels of the new prediction samples, such as Eq. 2:

$$y(x_i^v) = \beta \sum_{j=1}^{k} S\left(x_i^v, x_j^u\right)y + (1-\beta)\sum_{j=1}^{k} S\left(x_i^v, x_j^v\right)y \quad (2)$$

where β is a constant used to adjust the weight between the two, where x represents the sample, y represents the label, and u and v represent labeled and unlabeled.

3 EV Charging Station Site Selection Model

In this paper, the model is mainly divided into three parts: data acquisition, feature engineering and model training.

The data acquisition part obtains the track data of electric vehicles through the Shanghai New Energy Vehicle Public Data Acquisition and Monitoring Research Center. The track data includes taxi ID, time, longitude and latitude, included angle, instantaneous speed, passenger status and other detailed attribute information. The population data is obtained from the Shanghai Population Data Research Center, and the location data, POI data, road vector data and house price vector data of the Shanghai charging station are obtained through the application programming interface (API) crawler of Gaud Map. The feature engineering part cleans up the data in the data collection part, analyzes and combines the correlation between the electric vehicle charging station and multi-source data, and constructs the corresponding location features. In the model training part, PFAHP method is used to determine the internal weight of economic characteristics, and then TOPSIS method is used to sort the power grid with charging stations, and the label of the power grid with charging stations is determined according to the sorting results. Establish the similarity function and automatic labeling strategy, and finally use the LogitBoost algorithm model based on integrated semi-supervised learning and integrated learning to obtain the probability map of electric vehicle charging station location.

4 Analysis of Experimental Process and Results

4.1 Based on PFAHP and TOPSIS Evaluation

Based on the number and weight of POI in economic factors, parts of the sample labels are determined by PFAHP and TOPSIS methods, and the importance between each POI is evaluated by multiple experts, and finally the largest number of evaluation levels is taken as a PFAHP fuzzy number, and the weight table of POI data is shown in Table 1.

4.2 Ensemble Learning Model Building

The model used in this paper is the flow chart of the integrated learning model combined with semi-supervised learning. The steps of the model are as follows:

Table 1. Weight Table of Various POI

POI	gas station	transformer substation	mall	school
weight	0.196524	0.332918	0.049091	0.035787
POI	hotel	firehouse	transportation stops	
weight	0.013628	0.080847	0.291203	

1) Input the training set (labeled sample set U and unlabeled sample set V). The parameters involved in the model are the number of neighbors K, the first N% and the last N% of unlabeled samples, and the number of iterations T. The weak classifier selects the decision tree.
2) Give each sample a weight w = 1/n (initial sample weight);
3) Establish 40 weak classifiers (T = 40); The steps of the step 3 are as follows:

①Calculate the similarity between each sample in U and each sample in U and V in turn;②A new decision tree classifier is trained using tag data and weight distribution;

③Use a classifier to classify each label sample in U to obtain classification results;④By using a classifier to classify each label sample in V, the classification results are obtained;⑤Calculate the confidence level of each sample result in unlabeled sample set V and rank them;⑥Calculate the error rate of the decision tree classifier. If the error rate is greater than 0.5, return to step 2 to retrain; ⑦Calculate the classifier weight according to the determined error rate;⑧Select top N% and bottom N% of unlabeled samples with high confidence to add to the labeled samples, and delete them from the unlabeled samples, where bottom N% obtains its label through k nearest neighbors (N = 10); Update the original sample weight and the newly added sample weight; After obtaining T decision tree classifiers, they are fused into strong classifiers according to their weights.

This paper uses the method of combining semi-supervised learning and optimized LogitBoost to build the model. Semi-supervised learning is mainly reflected in the data set. Only part of the data is marked, that is, the power grid with charging station. In the iterative learning of the model, the similarity calculation results are used to assign labels to the data without labels, and finally the integrated learning model with high classification accuracy is trained.

4.3 Analysis of the Results

In this study, eight experiments were designed, with 80% as the training set and 20% as the test set. Accuracy, recall, F1 and ranking accuracy were used to measure the accuracy and effectiveness of the model. Through the comparison of ① and ② ③ schemes, the difference of simulation accuracy between cosine-distance hybrid similarity model and single cosine similarity model and distance similarity model is analyzed; Through the comparison of ① and ④ schemes, the influence of manual and automatic labeling on the results is analyzed; Through the comparison of ① ② ③ ④ and ⑤ ⑥ ⑦ schemes, the difference in simulation accuracy between simple semi-supervised learning model and semi-supervised learning integrated learning model is analyzed; Through the comparison of ① and ⑧ schemes, the influence of different semi-supervised learning combined with integrated learning models on the results is verified (Table 2).

Table 2. Results of The Model

Model/Metric	Precision	Recall	F1	τ
①Cosine-distance hybrid similarity automatic labeling model*	0.925	0.895	0.917	0.826
②Cosine similarity automatic labeling model	0.898	0.863	0.878	0.799
③Euclidean distance similarity automatic labeling model	0.919	0.893	0.906	0.810
④Cosine-distance hybrid similarity artificial labeling model	0.910	0.883	0.898	0.766
⑤Co-Forest	0.855	0.843	0.853	0.702
⑥Co-training	0.823	0.804	0.815	0.627
⑦SemiBoost	0.893	0.899	0.905	0.782
⑧Co-training + Integrated Learning	0.867	0.854	0.865	0.786

* The model used in this study

It can be seen from Table 2 that the results of ① and ② ③ comparison may be due to the integration of similarity functions, which makes the similarity consideration more complete, and the automatic labeling method saves time and cost. Compared with the pure semi-supervised learning model (Co-Forest and Co-training), the prediction accuracy and Kendall's hierarchical correlation coefficient of the semi-supervised learning model are greatly improved based on the cosine distance mixed similarity automatic labeling model, reflecting the advantages of the integrated learning model. Compared with SemiBoost model, the accuracy is also improved. Although F1 has little change, the ranking coefficient is greatly improved, which may be due to the impact of LogitBoost's improvement on Adaboost on accuracy; Compared with the joint training + integrated learning model, F1 has a great improvement. The key reason may be whether semi-supervised learning is better integrated into the host model.

Finally, taking the above study area as an example, the final location scheme is obtained by sorting the probability of each grid output by the model. The model output results are shown in Table 3. Figure 2 shows the probability density diagram of site selection. It can be seen from the figure that the grids with higher probability are concentrated in Xuhui and Huangpu areas in the more concentrated grids, and the experimental simulation accuracy is high.

Table 3. Sorting of Site Selection Results

Sort	The latitude and longitude position of the center of the grid	Economic characteristics	Road features	Output probability
No.1	(121.43179,31.15590)	0.63150	0.55159	0.992
No.2	(121.46344,31.23604)	0.53682	0.54154	0.922
No.3	(121.46241,31.22422)	0.48773	0.48744	0.920
No.4	(121.47494,31.24537)	0.48773	0.40897	0.917
No.5	(121.47983,31.22202)	0.48773	0.40154	0.913
....	......	......	......	......

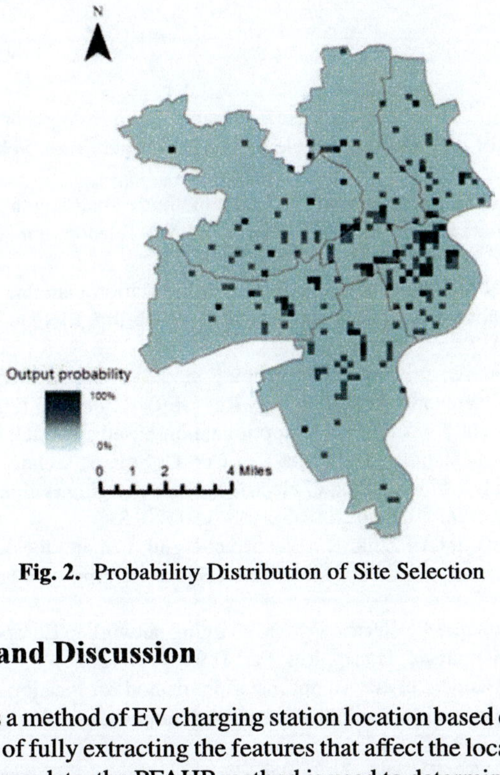

Fig. 2. Probability Distribution of Site Selection

5 Conclusion and Discussion

This paper proposes a method of EV charging station location based on multi-source data fusion. On the basis of fully extracting the features that affect the location of the charging pile using multi-source data, the PFAHP method is used to determine the weight of each POI in the economic characteristics, which solves the problem of lack of subjectivity in machine learning to a certain extent, provides a reference standard for the location label assignment, and improves the accuracy of the location to a certain extent; Finally, semi-supervised learning is combined with LogitBoost algorithm in machine learning to solve the problem of small sample data and obtain the corresponding grid output probability. In this process, the cosine distance mixed similarity automatic labeling method is constructed by reconstructing the traditional similarity measurement method, which greatly improves the simulation accuracy of the model. Comparative experiments show that the prediction accuracy of the model is improved to a certain extent compared with the model with single cosine similarity and manual labeling of bottom N % label. This algorithm is more accurate for the traditional algorithm to predict the charging station samples, and can better solve the location problem of the electric vehicle charging station.

The method proposed in this paper can better solve the location problem of electric vehicle charging station based on the type of public facilities, and also has certain reference value for the location of other public facilities. However, this study did not consider the actual landform factors and the actual capacity requirements of the charging station. Of course, there is still room for further optimization in the selection of driving factors, integration models, integration methods with semi-supervised learning, model validation methods, and charging station location ideas.

References

1. Chen, Q., Sun, J., Chen, K., et al.: Research status and development trend of battery management system for pure electric vehicle home and abroad. Mod. Veh. Power (01), 1–6+35 (2022)
2. Deng, K., Feng, H., Xu, G., et al.: Site selection method of banking facility location based on multi-source big data fusion. J. Chongqing Univ. Posts Telecommun. (Nat. Sci. Ed.) **32**(04), 664–672 (2020)
3. Liu, C., Liu, H., et al.: Multi-objective EV charging station planning with consideration of road network reliability and distribution network reliability. Electric Power Autom. Equip. **37**(08), 28–34 (2017)
4. Li, S., Sun, Q., Bai, X., et al.: Optimal planning model for electric vehicle charging stations considering users' demands. Renew. Energy Res. **36**(04), 568–573 (2018)
5. Yang, J., Dong, J., Hu, L.: A data-driven optimization-based approach for siting and sizing of electric taxi charging stations. Transport. Res. Part C: Emerg. Technol. **77**, 462–477 (2017)
6. Yang, Z., Gao, Z.: Location method of electric vehicle charging station based on data driven. J. Transp. Syst. Eng. Inf. Technol. **18**(05), 143–150 (2018)
7. Pagany, R., Marquardt, A., Zink, R.: Electric charging demand location model: a user and destination based locating approach for electric vehicle charging stations. Sustainability **11**(8), 1050–1071 (2019)
8. Falchetta, G., Noussan, M.: Electric vehicle charging network in Europe: an accessibility and deployment trends analysis. Transp. Res. Part D **94**, 1–27 (2021)
9. Yan, G., Liu, H., Han, N., et al.: An optimization method for location and capacity determination of charging stations considering spatial and temporal distribution of electric vehicles. Proc. CSEE **41**(8), 1–14 (2021)
10. Boyac, A., Iman, A.: Pandemic hospital site selection: a GIS-based MCDM approach employing Pythagorean fuzzy sets. Environ. Sci. Pollut. Res. 1–13 (2021)

Research on Influencing Factors of Smart City Construction Capability Based on DEMATEL-ISM

Xiaoheng Zhou(✉), Honggang Tang, and Wenyao Xiao

College of Architecture and Urban Planning, Guizhou University, Guiyang 550025, China
15082377610@163.com

Abstract. Currently, the construction of smart cities in China is still under exploration, and the lack of construction capacity has become an important factor restricting its construction and development. The improvement of the development capacity of smart city construction can enhance the scientificity of urban construction planning, improve the efficiency of public resource utilization, and effectively avoid construction risks, which has a significant promoting effect on the subsequent operation and maintenance of smart cities. Therefore, this article constructs a model of influencing factors on the development capacity of smart city construction through literature research, questionnaire survey, and expert interview methods. Using Decision-making Trial and Evaluation Laboratory (DEMATEL) and Interpretative Structural Modeling Method (ISM), Calculate the degree of cause and centrality of factors, identify key influencing factors, divide impact levels, and determine the specific mechanisms of action between factors. The results indicate that the ability to integrate technology, the level of network support platform construction, strategic planning ability, information collection level, and humanistic literacy level are the deep influencing factors of the construction capacity of smart cities. Strategic planning capability and information infrastructure construction level are key influencing factors. Based on the research results and key factors in each category, suggestions and strategies were proposed to promote the continuous construction and development of smart cities.

Keywords: Smart City · Influence Factor · DEMATEL · ISM

1 Introduction

At the end of the 20th century, urbanization construction in various countries around the world accelerated, and a large number of rural populations flooded into cities, providing a good opportunity for a new round of global urbanization development. However, with the continuous increase of urban population, the quality of public services in cities has sharply decreased, leading to "urban diseases" such as traffic congestion and rising prices. Widely present in international medium-sized and large cities. In addition, the population distribution in China is extremely uneven, and the phenomenon of population loss in small towns and rural areas is very serious, which has brought new and

W. Guo et al. (Eds.): GBCESC 2023, LNCE 328, pp. 861–874, 2024.
https://doi.org/10.1007/978-981-99-9947-7_86

higher requirements to social management [1]. Today, with the continuous progress of information and communication technology and network technology, smart cities have become a new development direction.

In recent years, the development and construction of smart cities have become a focus of attention both domestically and internationally. Researchers have proposed different perspectives on the current situation and development of smart cities: Li J (2021) views smart cities as innovative cities that utilize ICT and other means to improve quality of life, urban operation and service efficiency, and enhance competitiveness. Simultaneously ensuring the satisfaction of contemporary and future economic, social, and environmental needs [2]. Komninos et al. (2013) defined smart cities as investments in human and social capital, as well as traditional (transportation) and modern (information and communication technology) infrastructure, based on effective management of natural resources and participatory government. Therefore, they can promote sustainable economic development and improve the quality of life of residents [3]. Anthopoulos and Vakali (2012) argue that there is a mutually reinforcing relationship between smart cities and overall urban planning [4]. Tao Wei et al. (2021) analyzed the background and current situation of smart city construction in Hefei and suggested that Hefei smart city should adopt a comprehensive construction approach that improves the service system, promotes comprehensive construction, and breaks development barriers [5]. Zhao Qiming et al. (2021) have an impact on the construction of smart cities, including: cognitive level of smart cities, talent team construction, technology patents and intellectual property rights, ecology and energy, Internet of Things, and other big data technologies [6]. Tanaka (2021) believes that the construction of smart cities can be strengthened by improving urban planning capabilities, information perception capabilities, and increasing capital investment, thereby enhancing the comprehensive governance capabilities of smart cities [7].

Currently, there are many relevant literature on the construction capacity of smart cities in the academic community, but the analysis of the research path is not in-depth enough. Moreover, most of the existing research results evaluate the construction capacity of smart cities from a single perspective of economy, humanities, and management, and propose improvement suggestions. Smart cities are a very complex large-scale urban system, and their construction and development capabilities are influenced by multiple factors. Therefore, identifying it, clarifying its influencing factors and paths, and conducting in-depth analysis of its mechanism of action are of great practical significance for enhancing the construction capacity and quality of smart cities in China.

2 The Influencing Factors of Smart City Construction Ability

Mainly using literature research, questionnaire survey, and expert interview methods to identify influencing factors. The identification of influencing factors is mainly divided into two steps. The first step is to identify the factors that affect the development capacity of smart city construction based on literature research. The second step is to modify and simplify the identified influencing factors through expert interviews and questionnaire surveys. Finally, 18 specific influencing factor indicators are extracted from five dimensions: technology, infrastructure, government, information, and citizens.

Table 1. Model of influencing factors of smart city construction and development capacity

Third level code	Secondary code	First level code	code
Smart City Construction and Development Capability	Technical dimension	Technology integration capability	X1
		Technological innovation capability	X2
		Technical risk prevention and control capabilities	X3
	Infrastructure dimension	Information infrastructure construction level	X4
		Construction level of network support platforms	X5
		Development level of information and communication technology industry	X6
	Government dimension	Degree of capital investment	X7
		Strategic planning capabilities	X8
		Communication and coordination ability	X9
		Improvement level of management system	X10
		Improvement level of policy support system	X11
		Improvement level of public services	X12
	Information dimension	Information sharing level	X13
		Information collection level	X14
		Information protection level	X15
	Citizen dimension	Humanistic literacy level	X16
		Information literacy level	X17
		Talent cultivation and introduction ability	X18

3 DEMATEL Model Construction

3.1 DEMATEL Model Principle

The principle of Decision Experiment and Evaluation Experiment Method (DEMATEL): Decision experiment and evaluation experiment method mainly analyze the influence relationship of any two factors through the comprehensive application of graph theory and matrix, and determine the position of factors in the system by calculating the degree of cause and centrality of each factor [8].

3.2 Establish a Direct Impact Matrix

Systems science believes that a system is a w1hole formed by the organic connection of factors, and as one of the methods for modeling system structure, the DEMATEL method

	X_1	X_2	X_3	X_4	X_5	X_6	X_7	X_8	X_9	X_{10}	X_{11}	X_{12}	X_{13}	X_{14}	X_{15}	X_{16}	X_{17}	X_{18}
X_1	0	0	0	0	0	4	0	0	0	0	0	3	0	0	0	0	0	0
X_2	0	0	0	3	3	0	0	0	0	0	0	0	0	0	0	0	0	0
X_3	0	0	0	0	0	0	0	0	0	0	0	0	0	0	4	0	0	0
X_4	0	0	0	0	3	0	0	0	2	0	0	3	0	0	0	0	0	0
X_5	0	0	0	0	0	0	0	0	0	0	0	0	3	3	0	0	0	0
X_6	0	4	0	0	0	0	0	0	0	0	0	0	0	0	0	0	0	0
X_7	0	0	0	4	3	0	0	0	0	0	0	0	0	0	0	0	0	0
X_8	0	0	0	0	0	3	4	0	0	4	3	0	0	0	0	0	0	4
X_9	0	0	0	0	0	0	0	0	0	0	0	3	4	0	0	0	0	0
X_{10}	0	0	4	0	0	0	0	0	0	0	0	0	0	0	4	0	0	0
X_{11}	0	0	0	0	0	0	0	3	0	3	0	0	0	0	0	0	0	0
X_{12}	0	0	0	0	0	0	0	0	0	0	0	0	0	0	0	0	0	0
X_{13}	2	0	0	0	0	0	0	0	4	0	0	3	0	0	0	0	0	0
X_{14}	0	0	0	0	0	0	0	0	0	0	0	0	4	0	0	0	0	0
X_{15}	0	0	0	0	0	0	0	0	0	0	0	0	0	0	0	0	0	0
X_{16}	0	0	0	0	0	0	0	0	3	0	0	0	0	0	0	0	4	0
X_{17}	0	0	3	0	0	0	0	0	0	0	0	0	0	0	4	0	0	0
X_{18}	0	0	0	0	0	0	0	0	0	0	0	0	0	0	0	3	3	0

Fig. 1. Direct influence matrix X

also adheres to this idea. Firstly, identify and sort the various factors of the system to facilitate subsequent operations, and define the set of factors as $X = X_i$.

Secondly, determine the intensity of the impact between each two factors. This article uses Likert's five point scale and Delphi Method to score the impact intensity between each two factors, and defines the impact intensity from weak to strong as 0 points (no impact), 1 point (weak), 2 points (average), 3 points (strong), and 4 points (very strong), respectively. Invite expert group members to rate the intensity of the impact between the influencing factors, comprehensively process the data, and establish a direct impact matrix X.

3.3 Establish Specification Direct Influence Matrix

Process the direct impact matrix X and construct a standardized direct impact matrix D. The standard formula is as follows:

$$D = \left(1/\underset{i=1}{\overset{n}{Max}} \sum_{j=1}^{n} a_{ij} \right) * X \tag{1}$$

	X_1	X_2	X_3	X_4	X_5	X_6	X_7	X_8	X_9	X_{10}	X_{11}	X_{12}	X_{13}	X_{14}	X_{15}	X_{16}	X_{17}	X_{18}
X_1	0.00	0.00	0.00	0.00	0.00	0.22	0.00	0.00	0.00	0.00	0.00	0.22	0.00	0.00	0.00	0.00	0.00	0.00
X_2	0.00	0.00	0.00	0.17	0.17	0.00	0.00	0.00	0.00	0.00	0.00	0.00	0.00	0.00	0.00	0.00	0.00	0.00
X_3	0.00	0.00	0.00	0.00	0.00	0.00	0.00	0.00	0.00	0.00	0.00	0.00	0.00	0.00	0.22	0.00	0.00	0.00
X_4	0.00	0.00	0.00	0.00	0.17	0.00	0.00	0.00	0.17	0.00	0.00	0.17	0.00	0.00	0.00	0.00	0.00	0.00
X_5	0.00	0.00	0.00	0.00	0.00	0.00	0.00	0.00	0.00	0.00	0.00	0.00	0.17	0.17	0.00	0.00	0.00	0.00
X_6	0.00	0.22	0.00	0.00	0.00	0.00	0.00	0.00	0.00	0.00	0.00	0.00	0.00	0.00	0.00	0.00	0.00	0.00
X_7	0.00	0.00	0.00	0.22	0.17	0.00	0.00	0.00	0.00	0.00	0.00	0.00	0.00	0.00	0.00	0.00	0.00	0.00
X_8	0.00	0.00	0.00	0.00	0.00	0.17	0.22	0.00	0.00	0.22	0.17	0.00	0.00	0.00	0.00	0.00	0.00	0.22
X_9	0.00	0.00	0.00	0.00	0.00	0.00	0.00	0.00	0.00	0.00	0.00	0.17	0.22	0.00	0.00	0.00	0.00	0.00
X_{10}	0.00	0.00	0.17	0.00	0.00	0.00	0.00	0.00	0.00	0.00	0.00	0.00	0.00	0.00	0.22	0.00	0.00	0.00
X_{11}	0.00	0.00	0.00	0.00	0.00	0.00	0.00	0.17	0.00	0.17	0.00	0.00	0.00	0.00	0.00	0.00	0.00	0.00
X_{12}	0.00	0.00	0.00	0.00	0.00	0.00	0.00	0.00	0.00	0.00	0.00	0.00	0.00	0.00	0.00	0.00	0.00	0.00
X_{13}	0.11	0.00	0.00	0.00	0.00	0.00	0.00	0.00	0.22	0.00	0.00	0.17	0.00	0.00	0.00	0.00	0.00	0.00
X_{14}	0.00	0.00	0.00	0.00	0.00	0.00	0.00	0.00	0.00	0.00	0.00	0.00	0.22	0.00	0.00	0.00	0.00	0.00
X_{15}	0.00	0.00	0.22	0.00	0.00	0.00	0.00	0.00	0.17	0.00	0.00	0.00	0.00	0.00	0.00	0.00	0.00	0.00
X_{16}	0.00	0.00	0.00	0.00	0.00	0.00	0.00	0.00	0.00	0.00	0.00	0.00	0.00	0.00	0.00	0.00	0.22	0.00
X_{17}	0.00	0.00	0.01	0.00	0.00	0.00	0.00	0.00	0.00	0.00	0.00	0.00	0.00	0.00	0.22	0.00	0.00	0.00
X_{18}	0.00	0.00	0.00	0.00	0.00	0.00	0.00	0.00	0.00	0.00	0.00	0.00	0.00	0.00	0.00	0.17	0.17	0.00

Fig. 2. The specification direct influence matrix D

The direct influence matrix X (Fig. 1) is normalized to construct the direct influence matrix D, as shown in Fig. 2.

3.4 Establish a Comprehensive Impact Matrix

Calculate the direct impact matrix D of the specification and construct a comprehensive impact matrix T. The calculation formula is as follows:

$$T = D * (I - D)^{-1} \tag{2}$$

where, I is the identity matrix.

The specification direct influence matrix D (Fig. 2) is reprocessed to construct the comprehensive influence matrix T. As shown in Fig. 3.

	X_1	X_2	X_3	X_4	X_5	X_6	X_7	X_8	X_9	X_{10}	X_{11}	X_{12}	X_{13}	X_{14}	X_{15}	X_{16}	X_{17}	X_{18}
X_1	0.00	0.05	0.00	0.01	0.01	0.22	0.00	0.00	0.00	0.00	0.00	0.22	0.00	0.00	0.00	0.00	0.00	0.00
X_2	0.00	0.00	0.00	0.17	0.19	0.00	0.00	0.00	0.03	0.00	0.00	0.04	0.05	0.03	0.00	0.00	0.00	0.00
X_3	0.00	0.00	0.05	0.00	0.00	0.00	0.00	0.00	0.00	0.00	0.00	0.00	0.00	0.00	0.23	0.00	0.00	0.00
X_4	0.01	0.00	0.00	0.00	0.17	0.00	0.00	0.00	0.12	0.00	0.00	0.20	0.06	0.03	0.00	0.00	0.00	0.00
X_5	0.02	0.00	0.00	0.00	0.00	0.01	0.00	0.00	0.05	0.00	0.00	0.05	0.21	0.17	0.00	0.00	0.00	0.00
X_6	0.00	0.22	0.00	0.04	0.04	0.00	0.00	0.00	0.01	0.00	0.00	0.01	0.01	0.01	0.00	0.00	0.00	0.00
X_7	0.01	0.00	0.00	0.22	0.20	0.00	0.00	0.00	0.04	0.00	0.00	0.05	0.05	0.03	0.00	0.00	0.00	0.00
X_8	0.00	0.04	0.07	0.06	0.05	0.17	0.23	0.03	0.02	0.26	0.17	0.01	0.01	0.01	0.08	0.04	0.05	0.23
X_9	0.03	0.00	0.00	0.00	0.00	0.01	0.00	0.00	0.05	0.00	0.00	0.22	0.23	0.00	0.00	0.00	0.00	0.00
X_{10}	0.00	0.00	0.23	0.00	0.00	0.00	0.00	0.00	0.00	0.00	0.00	0.00	0.00	0.00	0.27	0.00	0.00	0.00
X_{11}	0.00	0.01	0.05	0.01	0.01	0.03	0.04	0.17	0.00	0.21	0.03	0.00	0.00	0.00	0.06	0.01	0.01	0.04
X_{12}	0.00	0.00	0.00	0.00	0.00	0.00	0.00	0.00	0.00	0.00	0.00	0.00	0.00	0.00	0.00	0.00	0.00	0.00
X_{13}	0.12	0.01	0.00	0.00	0.00	0.03	0.00	0.00	0.23	0.00	0.00	0.24	0.05	0.00	0.00	0.00	0.00	0.00
X_{14}	0.03	0.00	0.00	0.00	0.00	0.01	0.00	0.00	0.05	0.00	0.00	0.05	0.23	0.00	0.00	0.00	0.00	0.00
X_{15}	0.00	0.00	0.23	0.00	0.00	0.00	0.00	0.00	0.00	0.00	0.00	0.00	0.00	0.00	0.05	0.00	0.00	0.00
X_{16}	0.00	0.00	0.05	0.00	0.00	0.00	0.00	0.00	0.18	0.00	0.00	0.04	0.04	0.00	0.06	0.00	0.22	0.00
X_{17}	0.00	0.00	0.23	0.00	0.00	0.00	0.00	0.00	0.00	0.00	0.00	0.00	0.00	0.00	0.27	0.00	0.00	0.00
X_{18}	0.00	0.00	0.05	0.00	0.00	0.00	0.00	0.00	0.00	0.03	0.00	0.01	0.01	0.00	0.06	0.17	0.20	0.00

Fig. 3. Composite influence matrix T

3.5 Calculate the Degree of Influence, Influenced, Centrality and Cause of Each Risk Factor

Determine the degree of influence, degree of influence, centrality, and degree of cause of each factor using the comprehensive impact matrix T. The result obtained by adding the values of a certain row in the comprehensive impact matrix T is the degree of influence R of the factors corresponding to the values of that row. The sum of the values of a certain column in the comprehensive impact matrix T yields the degree of influence C of the factors corresponding to the values of that column. The calculation formula for center degree N is $N = R + C$, and the calculation formula for cause degree M is $M = R - C$.

Table 2. Calculation results of centrality and cause degree

	R	C	$M = R - C$	$N = R + C$
X1 Technology integration capability	0.52	0.22	0.30	0.72
X2 Technological innovation capability	0.52	0.33	0.17	0.84
X3 Technical risk prevention and control capabilities	0.29	0.96	−0.68	1.23
X4 Information infrastructure construction level	0.59	0.50	0.09	1.08
X5 Construction level of network support platforms	0.51	0.68	−0.17	1.17
X6 Development level of information and communication technology industry	0.34	0.47	−0.12	0.18
X7 Degree of capital investment	0.62	0.27	0.34	0.87
X8 Strategic planning capabilities	1.52	0.20	1.31	1.72
X9 Communication and coordination ability	0.54	0.81	−0.25	1.35
X10 Improvement level of management system	0.50	0.47	0.03	0.97
X11 Improvement level of policy support system	0.66	0.20	0.48	0.87
X12 Improvement level of public services	0.00	1.16	−1.14	1.15
X13 Information sharing level	0.68	0.98	−0.29	1.62
X14 Information collection level	0.37	0.28	0.09	0.65
X15 Information protection level	0.29	1.08	−0.83	1.38
X16 Humanistic literacy level	0.59	0.21	0.36	0.80
X17 Information literacy level	0.50	0.48	0.02	0.98
X18 Talent cultivation and introduction ability	0.51	0.27	0.25	0.78

Based on the above DEMATEL calculation concept, calculate the comprehensive impact matrix T to determine the degree of influence, degree of influence, centrality, and degree of cause of each factor. The calculation results are shown in Table 2.

3.6 Analysis of DEMATEL Model Results

Analysis Based on Degree of Influence and Degree of Being Influenced

Analysis Based on Impact Degree. Observing Tables 2, it was found that the influencing factors with a high degree of influence R include X_8 strategic planning capabilities (1.52), X_{13} information sharing level (0.68), X_{11} improvement level of policy support system (0.66), X_7 degree of capital investment (0.62), etc., indicating that these factors have a relatively high comprehensive impact on other factors in the influencing factor model.

Analysis Based on the Degree of Being Influenced. The degree of influence C represents the comprehensive impact of other factors on this factor. The factors with high degree of influence C mainly include X_{12} improvement level of public services (1.16), X_{15} information protection level (1.08), X_{13} information sharing level (0.98), X_3 technical risk prevention and control ability (0.96), etc. indicating that these factors are greatly influenced by other factors.

Centrality-Based Analysis. The calculation formula for centrality N is $N = R + C$, which represents the impact of this factor on other factors and the comprehensive degree of influence from other factors. Therefore, the greater the centrality of the factor, the more important it is in the entire influencing factor model system. From Tables 2, it can be seen that the factors with high centrality N mainly include X_8 strategic planning capabilities (1.72), X_{13} information sharing level (1.62), X_{15} information protection level (1.38), X_9 communication and coordination ability (1.35), X_3 technical risk prevention and control ability (1.23), etc. indicating that these factors are highly important in the influencing factor model system and have a connecting role.

The centrality of factors such as X_{14} information collection level (0.65), X_1 Technology integration capability (0.72), X_{18} talent cultivation and introduction ability (0.78), and X_{16} humanistic literacy level (0.80) is relatively low, and they are relatively independent in the influencing factor model system.

Analysis Based on the Degree of Cause

Analysis Based on Causal Factors. Factors with a positive degree of reason M belong to the category of causal factors. They are sorted according to the magnitude of the degree of reason M, and are: X_8 strategic planning capabilities (1.31), X_{11} improvement level of policy support system (0.48), X_{16} humanistic literacy level (0.36), X_7 degree of capital investment (0.34), X_1 technology integration capability (0.30), X_{18} talent cultivation and introduction ability (0.25), X_2 technology innovation capability (0.17) X_{14} information infrastructure construction level (0.09), X_{14} information collection level (0.09), X_{10} improvement level of management system (0.03) and X_{17} information literacy level (0.02).

Among them, X_8 strategic planning capabilities has the greatest impact on other factors and belongs to the root cause influencing factor; The level of X_1 technology integration capability and X_2 technology innovation capability is of great significance for the application of high-tech in smart cities; X_{16} humanistic literacy level, X_{17} information literacy level and X_{18} talent cultivation and introduction ability belong to the citizen dimension. The main body of smart city construction is urban citizens, so the improvement of citizen main body ability can effectively improve the construction ability of smart city; The X_{13} information sharing level and X_{14} information collection level refer to the ability of smart cities to process and utilize information data, which has a promoting effect on improving the construction capacity and level of smart cities; The level of investment in X_7 degree of capital investment, X_{10} improvement level of management system, and X_{11} improvement level of policy support system. At the government level, government departments should ensure sufficient funds for smart city construction, smooth funding chain, and improve the management system and policy support system.

Analysis Based on Outcome Factors. Factors with a negative degree of cause M are outcome factors, sorted by their absolute degree of cause, followed by X_{12} improvement level of public services (−1.14), X_{15} information protection level (−0.83), X_3 technical risk prevention and control capabilities (−0.68), X_{13} information sharing level (−0.29), X_9 communication and coordination ability (−0.25), X_5 construction level of network support platforms (−0.17), and X_6 development level of information and communication technology industry (−0.12).

Among them, the level of improvement of X_{12} improvement level of public services is the most important factor in the construction capacity of smart cities, so it must be given sufficient attention; X_{15} information protection level and X_3 technical risk prevention and control capabilities, by avoiding the information and technical risks faced by the construction process of smart cities, thereby enhancing the construction capacity of smart cities; X_{13}'s information sharing level and X_9's communication and coordination ability help to break the "information silo" phenomenon prevalent in the construction of smart cities; Unlike traditional urban governance, smart cities mainly rely on network platforms for comprehensive management. Therefore, the construction X_5 Construction level of network support platforms is of great significance for the management of smart cities; The development level of the X_6 development level of information and communication technology industry, application, and integration of new information and communication technologies in smart cities.

Identification of Key Influencing Factors. Centrality refers to the importance of this factor in the model system. The top 8 influencing factors for centrality N are: X_8 strategic planning capabilities (1.72), X_{13} information sharing level (1.62), X_{15} information protection level (1.38), X_9 communication and coordination ability (1.35), X_3 technical risk prevention and control capabilities (1.23), X_5 Construction level of network support platforms (1.17) X_{12} improvement level of public services (1.15) and X_4 Information infrastructure construction level (1.08).

Causality refers to the degree to which the factor contributes to the structural model. Rank the degree of cause M by numerical value, and the top 8 are: X_8 strategic planning capabilities (1.31), X_{11} improvement level of policy support system (0.48), X_{16}

humanistic literacy level (0.36), X_7 degree of capital investment (0.34), X_1 technology integration capability (0.30), X_{18} talent cultivation and introduction ability (0.25), X_2 technology innovation capability (0.17), X_4 Information infrastructure construction level (0.09), and X_{14} information collection level (0.09). According to the descending order of cause degree M, the 8th and 9th places have the same cause degree, so they are listed as 8th.

Observing the centrality and causal degree of each influencing factor, it was found that some influencing factors have both high centrality and causal degree. These factors not only have a strong impact on other elements, but also play an important role in the overall construction of the system.

4 ISM Model Construction

4.1 ISM Model Principle

Explain the principle of the structural model method (ISM): Explain the system to be analyzed by the structural model method. By sorting out and dividing the system into subsystems (factors, elements), and then analyzing the factors and the direct binary relation between them, the system structure can be decomposed into hierarchical topology without damaging the system function and structure [9].

4.2 Build Adjacency Matrix

The first and most important step in Interpretative Structural Modeling Method (ISM) is the threshold λ Set by threshold λ. The setting of simplifies the internal factor relationship of the model, and converts the comprehensive influence matrix T into the adjacency matrix K. The specific formula is as follows:

$$t_{ij} \geq \lambda(i, j = 1, 2, \cdots, n) k_{ij} = 1 \tag{3}$$

$$t_{ij} < \lambda(i, j = 1, 2, \cdots, n) k_{ij} = 0 \tag{4}$$

where, t_{ij} refers to the element of column j in row i of comprehensive influence matrix T, and k_{ij} refers to the element of column j in row i of adjacency matrix K.

The first and most important step in the ISM model is the threshold λ Because of the threshold value λ. The numerical value of is the standard for filtering subsequent impact relationships, which directly affects the subsequent hierarchical division. The specific value can be analyzed multiple times based on the dispersion of the influence degree to obtain satisfactory results. This article has taken multiple values and finally determined the threshold value $\lambda = 0.21$. Determine threshold λ Afterwards, set the threshold λ Substitute into the comprehensive impact matrix T with a threshold value λ Filter out the weak influence relationship for the standard, and construct the adjacency matrix K, as shown in Fig. 4.

	X_1	X_2	X_3	X_4	X_5	X_6	X_7	X_8	X_9	X_{10}	X_{11}	X_{12}	X_{13}	X_{14}	X_{15}	X_{16}	X_{17}	X_{18}
X_1	0	0	0	0	0	1	0	0	0	0	0	1	0	0	0	0	0	0
X_2	0	0	0	0	0	0	0	0	0	0	0	0	0	0	0	0	0	0
X_3	0	0	0	0	0	0	0	0	0	0	0	0	0	0	1	0	0	0
X_4	0	0	0	0	0	0	0	0	0	0	0	0	0	0	0	0	0	0
X_5	0	0	0	0	0	0	0	0	0	0	0	0	1	0	0	0	0	0
X_6	0	1	0	0	0	0	0	0	0	0	0	0	0	0	0	0	0	0
X_7	0	0	0	1	0	0	0	0	0	0	0	0	0	0	0	0	0	0
X_8	0	0	0	0	0	0	1	0	0	1	0	0	0	0	0	0	0	1
X_9	0	0	0	0	0	0	0	0	0	0	0	1	1	0	0	0	0	0
X_{10}	0	0	1	0	0	0	0	0	0	0	0	0	0	0	1	0	0	0
X_{11}	0	0	0	0	0	0	0	0	0	0	0	0	0	0	0	0	0	0
X_{12}	0	0	0	0	0	0	0	0	0	0	0	0	0	0	0	0	0	0
X_{13}	0	0	0	0	0	0	0	0	1	0	0	1	0	0	0	0	0	0
X_{14}	0	0	0	0	0	0	0	0	0	0	0	0	1	0	0	0	0	0
X_{15}	0	0	1	0	0	0	0	0	0	0	0	0	0	0	0	0	0	0
X_{16}	0	0	0	0	0	0	0	0	0	0	0	0	0	0	0	0	1	0
X_{17}	0	0	1	0	0	0	0	0	0	0	0	0	0	0	1	0	0	0
X_{18}	0	0	0	0	0	0	0	0	0	0	0	0	0	0	0	0	0	0

Fig. 4. Adjacency matrix K

4.3 Build Accessibility Matrix

After constructing the adjacency matrix K, iterative operations are performed on the adjacency matrix K to obtain the reachable matrix U, as shown in Fig. 5.

4.4 Modeling Multi-layer Explanatory Structures

Based on the reachability matrix U and the hierarchical division of influencing factors, an explanatory structure model is established (as in Fig. 6). The factors located in the same row in the figure belong to the same hierarchy, and the arrows indicate the direction of influence among the factors.

4.5 Analysis of ISM Model Results

From the hierarchy of influencing factors and the explanatory structural model (Fig. 5), it can be seen that the X_1 technology integration capability, the construction X_5 Construction level of network support platforms, X_8 strategic planning capabilities, X_{14} information collection level, and X_{16} humanistic literacy level are at the third level of

	X_1	X_2	X_3	X_4	X_5	X_6	X_7	X_8	X_9	X_{10}	X_{11}	X_{12}	X_{13}	X_{14}	X_{15}	X_{16}	X_{17}	X_{18}
X_1	1	1	0	0	0	1	0	0	0	0	0	1	0	0	0	0	0	0
X_2	0	1	0	0	0	0	0	0	0	0	0	0	0	0	0	0	0	0
X_3	0	0	1	0	0	0	0	0	0	0	0	0	0	0	1	0	0	0
X_4	0	0	0	1	0	0	0	0	0	0	0	0	0	0	0	0	0	0
X_5	0	0	0	0	1	0	0	0	1	0	0	1	1	0	0	0	0	0
X_6	0	1	0	0	0	1	0	0	0	0	0	0	0	0	0	0	0	0
X_7	0	0	0	1	0	0	1	0	0	0	0	0	0	0	0	0	0	0
X_8	0	0	1	1	0	0	1	1	0	1	0	0	0	0	1	0	0	1
X_9	0	0	0	0	0	0	0	0	1	0	0	0	1	1	0	0	0	0
X_{10}	0	0	1	0	0	0	0	0	0	1	0	0	0	0	1	0	0	0
X_{11}	0	0	0	0	0	0	0	0	0	0	1	0	0	0	0	0	0	0
X_{12}	0	0	0	0	0	0	0	0	0	0	0	1	0	0	0	0	0	0
X_{13}	0	0	0	0	0	0	0	0	1	0	0	1	1	0	0	0	0	0
X_{14}	0	0	0	0	0	0	0	0	1	0	0	1	1	1	0	0	0	0
X_{15}	0	0	1	0	0	0	0	0	0	0	0	0	0	0	1	0	0	0
X_{16}	0	0	1	0	0	0	0	0	0	0	0	0	0	0	1	1	1	0
X_{17}	0	0	1	0	0	0	0	0	0	0	0	0	0	0	1	0	1	0
X_{18}	0	0	0	0	0	0	0	0	0	0	0	0	0	0	0	0	0	1

Fig. 5. Reachable matrix U

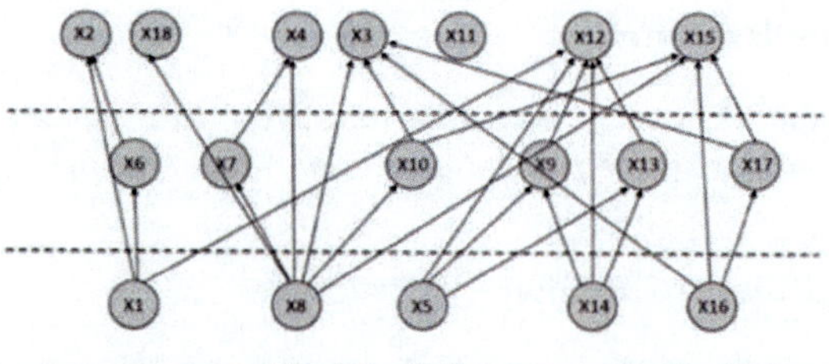

Fig. 6. The explanatory structure model of the influence mechanism of smart city construction ability

the explanatory structure model, which are the deep influencing factors of the construction ability of smart cities. X_6 development level of information and communication technology industry, X_7 degree of capital investment, X_9 communication and coordination ability, X_{10} improvement level of management system, X_{13} information sharing level and X_{17} information literacy level are at the second level in the interpretative structure model. In the interpretative structure model, it plays a connecting role between the preceding and the following.

X_2 technology innovation capability, X_3 technical risk prevention and control capabilities, X_4 Information infrastructure construction level, X_{11} improvement level of policy support system, X_{12} improvement level of public services, X_{15} information protection level, and X_{18} talent cultivation and introduction ability are located in the first layer of the explanatory structure model, which are surface influencing factors that directly affect the ability to establish smart cities.

5 Related Suggestions

Based on the centrality and causal ranking of factors affecting the construction capacity of smart cities, this article selects the top 8 factors in the centrality and causal ranking, and identifies two key influencing factors, namely X_8 strategic planning capabilities and X_4 Information infrastructure construction level.

The X_8 strategic planning capability for smart city construction mainly refers to the long-term planning of urban managers for the overall layout and development goals of future smart cities. Managers of smart cities should comprehensively consider the future layout, development direction, and construction priorities of the city. At the same time, strategic segmentation has been carried out on specific content, and management departments at all levels of government can combine intelligent technology with specific livelihood issues to improve the quality of public services. Improving the strategic planning capabilities of smart cities can clarify their development direction, enhance government regulation of public services, timely respond to residents' needs, reduce public resource waste, and improve resource utilization. At the same time, through reasonable urban planning, it can provide citizens with a comfortable and convenient living environment, while also improving people's living standards, improving people's living standards, and maintaining the natural environment. Therefore, improving the strategic planning level of smart cities can greatly promote the construction and development of smart cities.

With the further integration of information and communication technology with cities, smart cities are the development trend of future cities. Compared to traditional cities, the governance of smart cities relies more on the processing and utilization of data resources, and its governance focus has gradually shifted to the governance of virtual spaces. Therefore, the construction X_4 Information infrastructure construction level will have a direct impact on the construction capacity and achievements of smart cities. Improving the construction level of information infrastructure can in turn promote the healthy development of emerging industries such as the digital economy in smart cities, accelerate the deep integration of new generation information technology and the public sector, and significantly improve the construction effect of smart cities. Therefore, improving the construction level of information infrastructure will greatly enhance the construction and development capacity of smart cities.

6 Conclusions

This article uses the DEMATEL-ISM method to conduct in-depth research on the influencing factors of smart city construction capacity, calculating the degree of influence, degree of influence, centrality, and degree of cause of each influencing factor, and further distinguishing between causal and outcome factors based on the degree of cause of each factor. At the same time, key influencing factors are determined based on the centrality and causal ranking of the influencing factors. Among them, the strategic planning ability, the improvement level of the policy support system, the level of humanistic literacy, the level of capital investment, the ability of technology integration, talent training and introduction, the ability of technological innovation, the level of information infrastructure construction, the level of information collection, the level of management system improvement and the level of information literacy are the reasons; The improvement level of public services, information protection level, technological risk prevention and control ability, information sharing level, communication and coordination ability, network support platform construction level, and information and communication technology industry development level are outcome factors; Strategic planning capability and information infrastructure construction level are key influencing factors. And propose corresponding improvement suggestions, which provide certain reference value for improving the construction capacity of smart cities in China and improving the effectiveness of smart city construction.

References

1. Qu, Y.: Study on the Evaluation System of the Construction Level of Smart City on China. Dalian University of Technology (2017)
2. Li, J.: Characteristics and problems of smart city development in China. Smart Cities **4**(4), 1403–1419 (2021)
3. Komninos, N., Pallot, M., Schaffers, H.: Special issue on smart cities and the future internet in Europe. J. Knowl. Econ.owl. Econ. **4**(2), 119–134 (2013)
4. Anthopoulos, L.G., Vakali, A.: Urban planning and smart cities: interrelations and reciprocities. In: Álvarez, F., et al. (eds.) FIA 2012. LNCS, vol. 7281, pp. 178–189. Springer, Heidelberg (2012). https://doi.org/10.1007/978-3-642-30241-1_16
5. Tao, W., Wang, H., et al.: A study on the smart city construction of Hefei. J. Hefei Univ. **38**(01), 56–60 (2021)
6. Zhao, Q.M.: Analysis on influencing factors of new smart city based on entropy method. Project Manage. **9**, 46–48 (2021)
7. Tian, Z.: Exploring the path of strengthening the construction of smart cities and improving urban governance capabilities. China Constr. **09**, 44–45 (2021)
8. Yang, J.L., Liu, Y.N.: Research on the key influencing factors of the cross-boundary sharing of government information resource based on DEMATEL model. Libr. Inf. Serv. **62**(19), 64–76 (2018)
9. Chen, W.G., Zhang, N., Zhang, Y.S., Cheng, Z., Zhang, Y.: Study on influencing factors of urban disaster resilience based on DEMATEL-ISM. J. Catastrophology **36**(01), 1–6+17 (2021)

Research on the Renewal Design of Community Public Space Under the Smart Elderly Care Model

Lang Lyu(✉), Honggang Tang, and Biao Wang

College of Architecture and Urban Planning, Guizhou University, Guiyang 550025, China
1036378725@qq.com

Abstract. In recent years, due to changes in population structure, the number of elderly people in China has gradually increased, and the elderly population has become a hot topic and focus of attention for the whole society. At the same time, with the rapid development of information technology today, traditional elderly care models are also transitioning towards intelligence, and traditional elderly care models are no longer able to meet the elderly care requirements of our country. So, a new type of community elderly care model - smart elderly care emerged. Smart elderly care is the integration of elderly care services with information technologies such as the Internet, big data, and the Internet of Things, utilizing information service platforms to integrate resources from multiple fields such as catering, housekeeping, and healthcare, and applying these resources to community and home based elderly care services, thereby promoting the supply and demand balance of elderly care services. This article takes the physiological, psychological, and behavioral needs of the elderly as the core, and combines the Kano model analysis method to analyze and study the supply and demand relationship of community elderly care from the perspective of smart communities. It summarizes the key points of community public space design under the smart elderly care model, providing some construction references for the development of the domestic elderly care industry.

Keywords: Smart Community · Elderly care model · Kano Model

1 Introduction

According to relevant UNESCO standards, when the proportion of elderly people aged 60 and above in a country or region exceeds 10%, and the proportion of elderly people aged 65 and above exceeds 7%, it is considered an aging society [1]. The aging population is an important trend in social development and also the current social situation in China for a long period of time in the future. In response, society needs to take relevant measures to adapt to the upcoming aging population. The issue of elderly care has become a hot topic currently faced, and traditional elderly care models are clearly no longer able to meet the needs of the elderly. The elderly care industry should develop towards a higher level, so smart elderly care has become a choice that meets the needs of society and the population [2].

W. Guo et al. (Eds.): GBCESC 2023, LNCE 328, pp. 875–891, 2024.
https://doi.org/10.1007/978-981-99-9947-7_87

From the perspective of research content, domestic and foreign scholars have maintained a high level of attention to the intelligent transformation of traditional communities and related research on community public spaces. Foreign scholars mostly study smart communities from the overall perspective, constructing public spaces for smart communities that focus on user experience and are supplemented by intelligent technological means. Relatively speaking, research on the intelligent transformation of traditional community public spaces in China mainly focuses on fundamental issues such as its concept and characteristics. The research results on the intelligent transformation of traditional community public spaces are generally scattered.

Currently, in China's elderly care environment, a series of problems have emerged, such as insufficient professional level of home-based elderly care, insufficient manpower in community elderly care, and unequal resource allocation in institutional elderly care. The root cause of these problems lies in the imbalance and insufficient development of people's needs for a better life [3]. Smart elderly care has broken the traditional single service model of elderly care, utilizing new auxiliary devices such as big data, the Internet of Things, artificial intelligence, and sensors to build a diversified, rich, cooperative, and win-win operation model, thus achieving the goal of efficient integration and allocation of social resources [4].

Smart community elderly care is currently an effective way to solve the various problems of the elderly in the context of aging population in China. Smart elderly care can help optimize the elderly care service system, effectively improve the service level of the elderly, integrate supply and demand, and is of great significance to the research of smart community elderly care services.

2 Materials and Methods

2.1 Research Area

This study takes a certain place in Guiyang, Guizhou as the research object. The construction of buildings is dominated by terrain, and the overall architectural style is a mountainous city (Fig. 1).

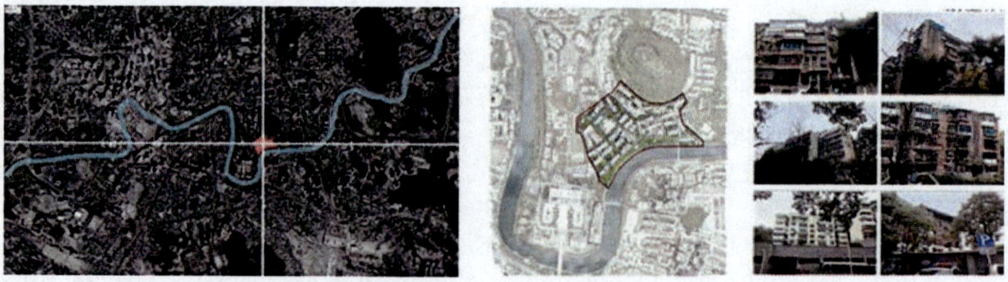

Fig. 1. Surrounding environment and current situation map of the community

In the elderly population statistics of the community, there are 1439 elderly people aged 60 and above, accounting for 45.45% of the total community population. It belongs to the community with severe aging phenomenon, and there are 439 elderly people aged 80 and above, accounting for 24.15% of the overall elderly population. Among the surveyed population, 72.72% lived with their children or spouses, 22.73% lived alone, and the proportion of elderly people living alone was relatively high. 31.82% of the population in the community suffer from basic diseases and require long-term recuperation. Most residents can take care of themselves.

2.2 Research Content

This study adopts the methods of literature review and field research, combined with a variety of data analysis, to analyze and count the public space environment and daily behavioral activities of the elderly in a community in Guiyang City from the perspective of the elderly, to explore the deficiencies in different activity environments, and to transform the community from a spatial point of view. This project aims to start from the change of spatial activity scenes to realize the change of the elderly's lifestyle in home care, to promote the construction of intelligent communities in China's cities, and to enhance the satisfaction of the elderly in home care.

2.3 Research Method

Based on the Kano model, design a survey questionnaire for the needs of users in the intelligent transformation of traditional community public spaces, conduct a quantitative sampling survey on community users, and understand the needs of community residents for community public spaces.

Based on the scope of the research, set the research base to ensure the referential nature of the data and provide materials for the data analysis in this article. This model includes two aspects: user satisfaction and product characteristics, and conducts a specific analysis of the non-linear relationship it presents. Service requirements are divided into five categories: Must-be Quality(M) , One-dimensional Quality(O), Attractive Quality(A), Indifferent Quality(I), Reverse Quality(R).

3 Research and analysis

3.1 Analysis of Activity Characteristics

The elderly in the community have a relatively long amount of free time available during the day. Apart from the time they must stay at home, most of their free time is spent outdoors, and outdoor activities have strong timeliness and regional characteristics. There are certain habits and patterns in choosing activity venues and times (Fig. 2).

In terms of travel distance, the elderly in the community mainly focus on "Home-Community-City", with static leisure activities such as eating, resting, doing household chores, watching TV, and reading newspapers accounting for the majority; The activities of the elderly in the community mainly include sports activities, family entertainment, chess and cards, neighborhood communication, etc; Activities within the city are mainly focused on specific purposes (Fig. 3).

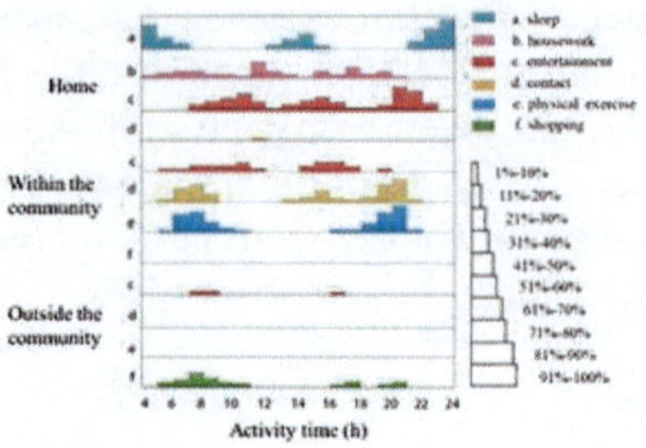

Fig. 2. Distribution of activity time for the elderly

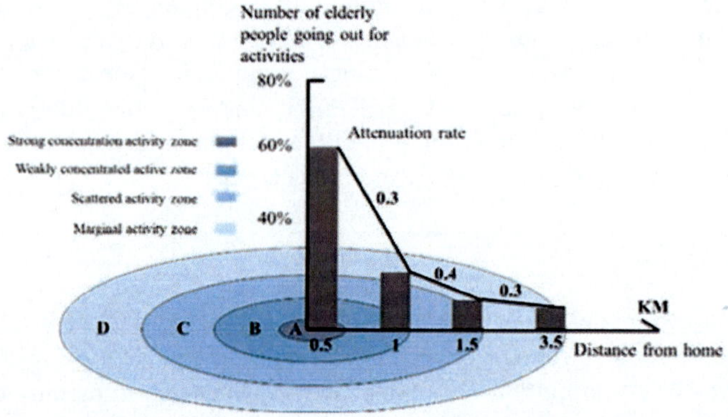

Fig. 3. The relationship between the distance of elderly people going out and the number of people traveling.

3.2 Spatial Environment Analysis

Due to the mountainous location of the internal buildings in the community, there are often height differences between them. To meet travel needs, some buildings are equipped with external staircases. This forms a special architectural spatial structure. Most of the space in this section is idle, of poor quality, and due to years of disrepair, the structure is unstable, posing certain safety hazards.

In terms of the travel mode of the elderly in the community, the pedestrian road space is narrow, discontinuous, uneven and occupied by cars. Most of the elderly choose to walk on the car road, which results in the phenomenon of mixed traffic of people and vehicles, and mixed traffic of motor vehicles and non motor vehicles, which brings great inconvenience and security risks to the elderly. The road system lacks accessibility design, and most elderly people with inconvenient travel have to choose further detours or choose not to go out directly. Due to the height difference of the site, the interconnectivity between pedestrian roads is not strong enough, the accessibility is poor, and there is a lack of systematic pedestrian transportation system planning.

Overall, due to the early construction of communities, they did not consider the rationality of venue layout after the aging phenomenon, resulting in a large number of original activity venues gradually turning into idle and abandoned spaces, which cannot meet the daily activity needs of the elderly population.

3.3 Satisfaction Analysis of Space Facilities

Through a questionnaire, satisfaction statistics were conducted on four types of public spaces within the community, namely activity space, road space, green space, and building space. The main data obtained were as follows (Fig. 4).

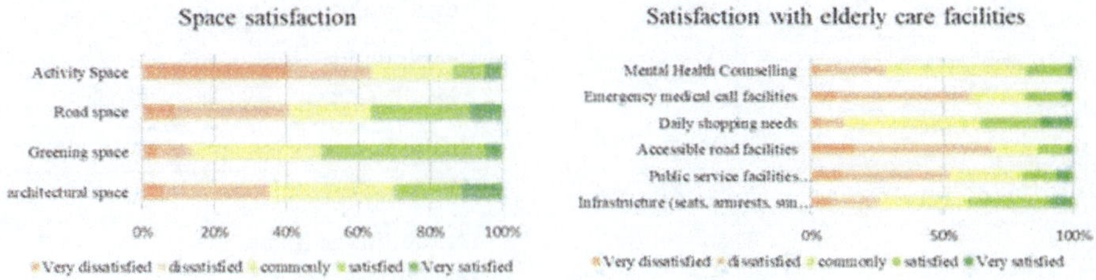

Fig. 4. Satisfaction with space and facilities

Although community facilities can ensure the most basic living functions, the outdated facilities have brought inconvenience to daily life, and the facilities have not been updated, reducing the convenience of life. When elderly people in the community go out for activities, the lack of handrail ramps to meet the needs of sun and rain protection, as well as outdated recreational seat facilities, will bring inconvenience to the elderly's travel and activities. For the elderly, whether there are self-service physical examinations and emergency channels is also an urgent issue that needs to be addressed. At the same time, smart medical service facilities are also linked to the health of the majority of people in the community, and are generally expected by residents in the community. Moreover, the configuration of smart infrastructure is incomplete, the distribution is unreasonable, and the internal guidance signs are not clear, which is not conducive to the travel of the elderly.

3.4 Analysis of Survey Results Based on Kano Model

Kano Questionnaire Design. After consulting relevant materials, this article summarizes 27 user needs and can better explain the motivation of most users to use traditional community public spaces, which has a certain representativeness (Table 1).

The respondents only need to check the options in the questionnaire based on their true feelings about traditional community public spaces (Table 2).

Each requirement item covers positive and negative sub problems, and the positive and negative options in each requirement item are corresponding to the table to obtain the attributes of that requirement item (Table 3).

Sample Reliability and Validity Analysis. This study aims to conduct a random sampling survey and select representative water observation sites, with elderly people aged

Table 1. Selection of user needs

category	Serial number	index
medical care	A1	Emergency assistance
	A2	Home care
	A3	Reminder to take medication
	A4	have an appointment with a doctor
	A5	Rehabilitation therapy
	A6	Regular physical examination
	A7	Health monitoring
	A8	Accompanying medical treatment
	A9	Health Lectures
Life care	B1	Home delivery
	B2	Haircutting at home
	B3	Housekeeping cleaning
	B4	safety protection
	B5	Help others take a bath
	B6	Help others purchase things
	B7	Home appliance maintenance
	B8	Payment and handling on behalf
	B9	Taking a taxi
Sports and entertainment	C1	Opening Classrooms for the Elderly
	C2	Listening to radio and television
	C3	Fitness Exercise
	C4	Playing cards and chess
	C5	Collective gathering
spiritual consolation	D1	Help to contact their descendants
	D2	psychological counseling
	D3	Social communication
	D4	legal aid

Table 2. Example of filling out a user demand questionnaire

Projects	like	expect	neutral	tolerate	dislike
have	O	√	O	O	O
Without	O	O	O	O	√

Table 3. Example of filling out a user demand questionnaire

demand		Negative problem(Does not have a certain function)				
		like	expect	neutral	tolerate	dislike
Forward problem(Having a certain function)	like	Q	A	A	A	O
	expect	R	I	I	I	M
	neutral	R	I	I	I	M
	tolerate	R	I	I	I	M
	dislike	R	R	R	R	Q

60 and above as the research subjects. A total of 230 questionnaires were sent out in this survey. After deleting the problematic questionnaires, 224 were ultimately obtained, with an effective rate of 97.39%. The requirements of this questionnaire survey are consistent.

In order to ensure the consistency and reliability of the survey questionnaire data, preliminary validation was conducted on the reliability and validity of the survey questionnaire data. This study used SPSS software to test the reliability and effectiveness of the survey results. According to the results, the overall Cronbach's Alpha coefficient of the questionnaire is 0.841, indicating that the internal consistency performance of this questionnaire meets the requirements and can be used for investigation and research (Table 4).

Table 4. Reliability analysis results

project	Kano questionnaire	Direct problems	Inverse problems
	Cronbach's Alpha	Cronbach's Alpha	Cronbach's Alpha
overall	0.841	0.783	0.821
medical care	0.839	0.732	0.863
Life care	0.762	0.679	0.845
Sports and entertainment	0.771	0.972	0.621
spiritual consolation	0.832	0.953	0.818

According to the results, the KMO value of the Kano questionnaire item is 0.889, and the corresponding P-values of Bartlett's spherical test are all<0.05. The questionnaire passed the validity test (Table 5).

Determination of Kano Categories for User Needs. After screening, 224 Kano questionnaire data were input into Excel. Based on the Kano category evaluation scale and Kano category analysis flowchart, the IF function was used to write calculation formulas and process the data. Finally, 27 Kano categories of user requirements were obtained (Table 6).

Table 5. Validity analysis results

KMO		0.889
Bartlett's test	Chi-square test	7896.862
	df	381
	Statistical significance	0.000

Table 6. Determination of Kano categories for user needs

category	Serial number	index	M	O	A	I	Q	R	Kano Category
medical care	A1	Emergency assistance	0.3795	0.3348	0.0714	0.2143	0.0000	0.0000	M
	A2	Home care	0.1786	0.3661	0.2411	0.2143	0.0000	0.0000	O
	A3	Reminder to take medication	0.0848	0.2188	0.5893	0.1071	0.0000	0.0000	A
	A4	have an appointment with a doctor	0.1295	0.2188	0.5402	0.1116	0.0000	0.0000	A
	A5	Rehabilitation therapy	0.2589	0.5134	0.1384	0.0893	0.0000	0.0000	O
	A6	Regular physical examination	0.2902	0.5045	0.1205	0.0714	0.0134	0.0000	O
	A7	Health monitoring	0.6071	0.2589	0.0893	0.0446	0.0000	0.0000	M
	A8	Accompanying medical treatment	0.0670	0.2500	0.5714	0.0982	0.0000	0.0134	A
	A9	Health Lectures	0.1161	0.1295	0.6339	0.1027	0.0000	0.0179	A
Life care	B1	Home delivery	0.5893	0.1384	0.1696	0.1027	0.0000	0.0000	M
	B2	Haircutting at home	0.0848	0.1295	0.4955	0.2768	0.0000	0.0134	A
	B3	Housekeeping cleaning	0.2902	0.4955	0.1205	0.0938	0.0000	0.0000	O
	B4	safety protection	0.4732	0.2098	0.1786	0.1250	0.0000	0.0134	M
	B5	Help others take a bath	0.1741	0.1607	0.2455	0.4152	0.0000	0.0045	I
	B6	Help others purchase things	0.1250	0.0580	0.6027	0.2143	0.0000	0.0000	A

(continued)

Table 6. (*continued*)

category	Serial number	index	M	O	A	I	Q	R	Kano Category
	B7	Home appliance maintenance	0.2009	0.5045	0.1830	0.1116	0.0000	0.0000	O
	B8	Payment and handling on behalf	0.2009	0.2589	0.3973	0.1429	0.0000	0.0000	A
	B9	Taking a taxi	0.1696	0.1116	0.1563	0.5402	0.0045	0.0179	I
Sports and entertainment	C1	Opening Classrooms for the Elderly	0.0893	0.0982	0.5134	0.2991	0.0000	0.0000	A
	C2	Listening to radio and television	0.1696	0.0759	0.1250	0.6295	0.0000	0.0000	I
	C3	Fitness Exercise	0.1161	0.1473	0.0938	0.6339	0.0000	0.0089	I
	C4	Playing cards and chess	0.1741	0.1384	0.1250	0.5625	0.0000	0.0000	I
	C5	Collective gathering	0.1429	0.0402	0.1295	0.6875	0.0000	0.0000	I
spiritual consolation	D1	Help to contact their descendants	0.1339	0.0848	0.2098	0.5714	0.0000	0.0000	I
	D2	psychological counseling	0.3304	0.4063	0.1696	0.0938	0.0000	0.0000	O
	D3	Social communication	0.4821	0.2589	0.0893	0.1696	0.0000	0.0000	M
	D4	legal aid	0.1250	0.1429	0.0491	0.6830	0.0000	0.0000	I

By using the principle of taking the maximum value, determine the Kano categories of all user demand items, and comprehensively grasp the user demand situation of the intelligent transformation of traditional community public spaces.

Analysis of Better-Worse Coefficient Results. Substitute the percentage values from the Kano category membership analysis table into the calculation formula for Better-Worse, and calculate the coefficient values for Better-Worse. The calculation formula is as follows:

$$Better/SI = \frac{A+O}{A+O+M+I} \tag{1}$$

$$Worse/DSI = -1\frac{O+M}{A+O+M+I} \tag{2}$$

Better represents the satisfaction coefficient when possessing the required function, and the larger the value, the greater the satisfaction value after possessing the required function, usually a positive value; Worse represents the satisfaction coefficient when eliminating a requirement item, and a smaller value indicates a lower satisfaction after not having the requirement function, usually a negative value (Table 7).

Table 7. Calculation results of coefficients for Better-Worse

category	Serial number	index	Better coefficient	Worse coefficient
medical care	A1	Emergency assistance	0.4063	−0.7143
	A2	Home care	0.6071	−0.5446
	A3	Reminder to take medication	0.8080	−0.3036
	A4	have an appointment with a doctor	0.7589	−0.3482
	A5	Rehabilitation therapy	0.6518	−0.7723
	A6	Regular physical examination	0.6335	−0.8054
	A7	Health monitoring	0.3482	−0.8661
	A8	Accompanying medical treatment	0.8326	−0.3213
	A9	Health Lectures	0.7773	−0.2500
Life care	B1	Home delivery	0.3080	−0.7277
	B2	Haircutting at home	0.6335	−0.2172
	B3	Housekeeping cleaning	0.6161	−0.7857
	B4	safety protection	0.3937	−0.6923
	B5	Help others take a bath	0.4081	−0.3363
	B6	Help others purchase things	0.6607	−0.1830
	B7	Home appliance maintenance	0.6875	−0.7054
	B8	Payment and handling on behalf	0.6563	−0.4598
	B9	Taking a taxi	0.2740	−0.2877

(*continued*)

Table 7. (*continued*)

category	Serial number	index	Better coefficient	Worse coefficient
Sports and entertainment	C1	Opening Classrooms for the Elderly	0.6116	−0.1875
	C2	Listening to radio and television	0.2009	−0.2455
	C3	Fitness Exercise	0.2432	−0.2658
	C4	Playing cards and chess	0.2634	−0.3125
	C5	Collective gathering	0.1696	−0.1830
spiritual consolation	D1	Help to contact their descendants	0.2946	−0.2188
	D2	psychological counseling	0.5759	−0.7366
	D3	Social communication	0.3482	−0.7411
	D4	legal aid	0.1920	−0.2679

Based on the results of the Better Worse coefficient, draw a smart community elderly care service demand matrix with the Better coefficient as the vertical axis and the Worse coefficient as the horizontal axis (Fig. 5).

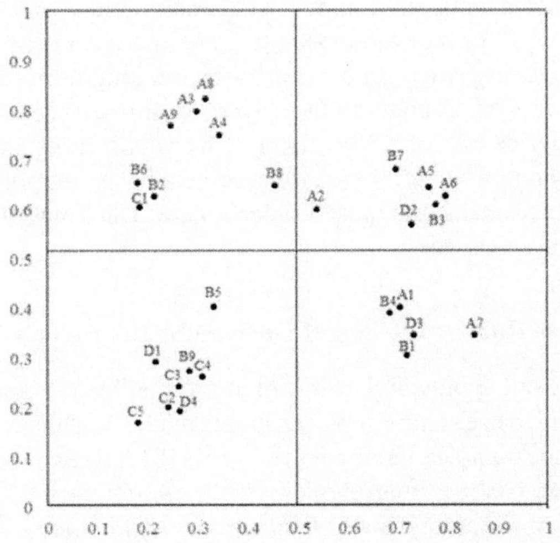

Fig. 5. Coefficient demand matrix diagram for Better-Worse

Among them, home care (A2), rehabilitation treatment (A5), regular physical examinations (A6), household cleaning (B3), home appliance maintenance (B7), and psychological counseling (D2) belong to the second quadrant area. The provision of such services plays an important role in improving the satisfaction of elderly people with smart community elderly care services, and should be given special attention.

4 Existing Problems

From the above analysis, it can be seen that the elderly have strong expectations for community elderly care needs, but there are many problems in the existing community public space [5], Covering four aspects of building space, green space, road space, and activity space within the community, the main problems in terms of elderly care needs in the context of smart community construction include the following:

4.1 Lack of Intelligent Functional Space

In the early planning and design, the construction of the smart platform center and the configuration of smart facilities were not considered, resulting in the lack of existing spatial functions [6], In addition, there is limited space available for the construction of new facilities in the community. Some service points have too few service facilities and lack multi-level information exchange, which makes their functions unable to play their due role.

4.2 Difficulty in Transforming and Updating Smart Communities

Through analysis of the data, it was found that most large communities were built in the 1970s to 1980s, and the overall environment of the community was relatively outdated, with little space available for renovation and use. The space for community elderly care facilities was also relatively small. In other aspects, the long-term lack of management in society has resulted in inefficient and time-consuming services [7]. At present, the smart elderly care service platform in Guiyang City is still in its early stages, and the spatial layout of community elderly care service centers is not yet perfect, including community elderly care and home-based elderly care. Catering and other functional venues lack community openness.

4.3 Low Acceptance Rate of Intelligent Devices for the Elderly

In terms of operation and application of information intelligent terminals, most elderly people are only willing to use some simple terminal devices, and there is a great difficulty in accepting those more complex intelligent terminals [8]. The two most frequently used devices in smart terminals are smartphones and one click pagers. In short, there are significant obstacles to the acceptance of information technology among the elderly, which poses a significant obstacle to the application and promotion of smart elderly care, and also increases the difficulty of innovation and research and development of smart elderly care products.

4.4 General Shortage of Elderly Care Personnel and Lack of Professional Teams for Smart Elderly Care Services

With the rapid development of the elderly care service industry, coupled with factors such as low social status, poor salary income, and lack of training for elderly care personnel, there is a widespread problem of manpower shortage in the elderly care service industry, which cannot be solved in a short period of time [9].

5 Transformation Strategy

5.1 Construction of an Accessible Intelligent Transportation System

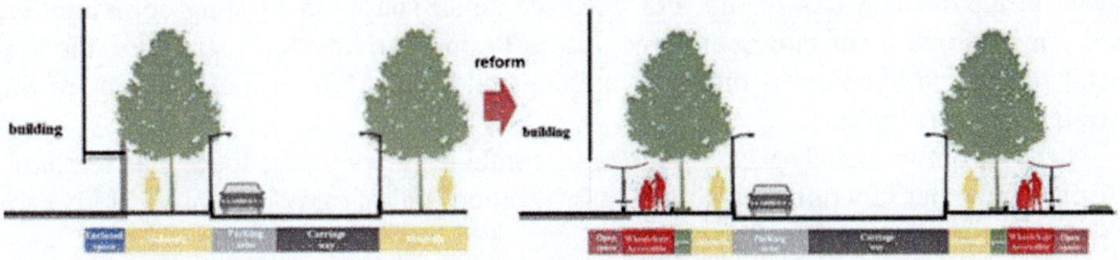

Fig. 6. Construction of an accessible intelligent transportation system

In the planning, it is necessary to clarify the level of the road network and build a safe, efficient, and intelligent road network system. Building a flow line suitable for elderly travel, pedestrian roads should connect the places that elderly people often go to, eliminate obstacles and other debris that hinder traffic, and set clear visual focus and ground signs on the road surface, as well as intelligent emergency call and safety warning systems. In addition, ground lights have been installed on both sides of the accessible passage to ensure the safety of elderly people during nighttime activities. Finally, the height difference within the road should be properly handled, and elevators should be installed in appropriate places to meet the accessibility design needs of the elderly. Attention should also be paid to the accessibility connection with outdoor venues and facilities to solve the current problem of accessibility facilities being mere formality [10] (Fig. 6).

5.2 Spatial Multifunctional Composition

In the optimization of spatial functions, based on the analysis of intelligent facilities and spatial activity modes under information technology conditions, improvement strategies for community space are proposed to guide the development of spatial functions towards sharing and intelligence [11]. In this design scheme, an intelligent control center has been added, which can monitor and respond to the daily behavior of the elderly and provide comprehensive care according to their different needs; Secondly, setting up smart companionship spaces to make the lives of the elderly more diverse and colorful;

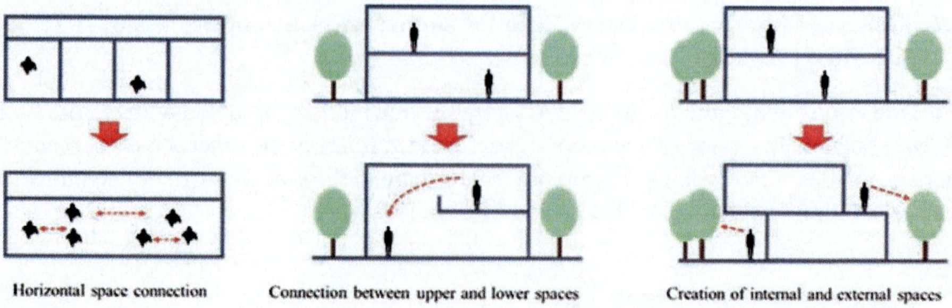

Fig. 7. Space renovation plan

Finally, establish a community volunteer service center to help the elderly realize their value in their later years of life. For enclosed small spaces in existing communities, we can integrate them into open large spaces through architectural structure, thereby achieving a combination of multiple functions and transforming the functions of the original space (Fig. 7).

In terms of functional space combination, combine indoor public space with learning, display and other functions, making it an integration station for community elderly care information resources (Fig. 8).

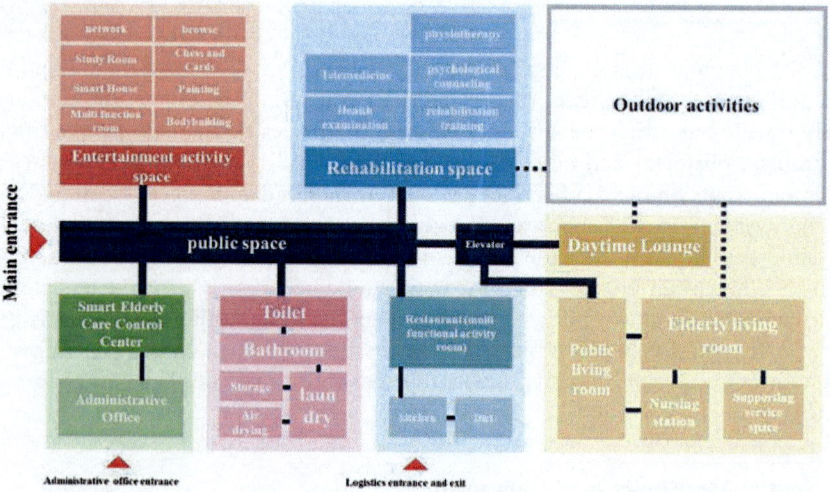

Fig. 8. Space renovation plan

The lobby space has transformed from a transportation space to a gathering space for elderly care resources, with intelligent product experience areas, learning areas, etc. set up in the space. By combining the restaurant space with a multifunctional activity space, it not only provides catering services for the elderly in the hospital, but also opens up and provides home delivery services for the elderly in the community. Expanding community care functions, setting up diversified entertainment spaces in outdoor space design, adding multi-functional rooms, outdoor communication and activity venues, rest and entertainment venues, and combining with daytime rest rooms to create an open, healthy, and intelligent indoor and outdoor activity venue.

5.3 Building a New Community Smart Elderly Care Service Structure

In the new community, based on the Internet, the Internet of Things, and intelligent platforms, information resources can be integrated to enable real-time and dynamic tracking of the living data and physical status of the elderly in the community [12]. In this case, communities can establish different types of service structures based on their actual needs. For example, communities can provide online consultation services to help residents solve problems in their daily lives. Communities can also provide offline housekeeping, maintenance and other services to provide convenience for residents; In addition, the community can also provide residents with psychological health counseling, safety knowledge popularization and other services (Fig. 9).

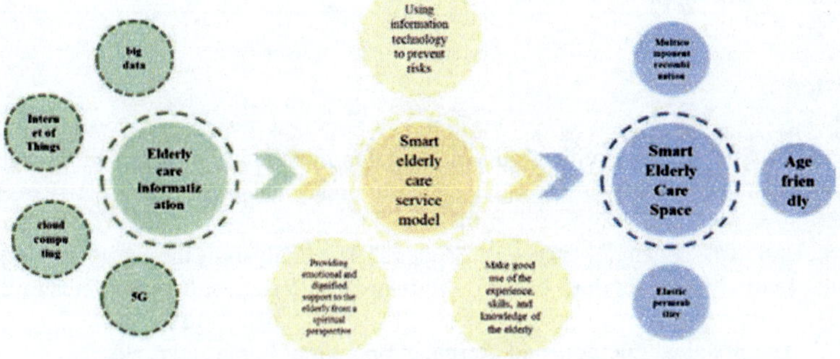

Fig. 9. Schematic diagram of smart elderly care model structure

Firstly, it is necessary to establish a hardware system that can meet the needs of different groups and provide remote medical treatment. Secondly, we can promote the complementarity of intelligent platforms and products through the Internet of Things technology, achieving the goal of information sharing. Finally, the construction of an intelligent platform allows for real-time monitoring of the activities of the elderly through the construction of an intelligent community service system. At the same time, it can also analyze the physical and psychological conditions of the elderly, providing targeted elderly care products and personalized elderly lifestyle for the elderly. In terms of IoT technology, it is necessary to enhance the perception ability and processing speed of intelligent devices to meet the different needs of users; At the same time, the intelligent platform can also achieve information sharing with other users and enterprises.

6 Conclusion

This article takes a certain area in Guiyang, Guizhou as the research object and designs a survey questionnaire using the kano model. It calculates the impact of various service items on the satisfaction of the elderly and explores the various problems faced by the elderly in the community. The specific problems are as follows: the lack of intelligent functional space; The usage rate of intelligent equipment among the elderly is relatively low; There is a general shortage of personnel and a lack of professional teams in elderly care services. And propose strategies and methods for designing community public spaces in the context of the information age in response to current issues. Although the application of information technology in Chinese society is still in its infancy, in the new social context, people's exploration of the concept of "wisdom" is also constantly deepening. In the future, the traditional community elderly care model will gradually develop towards intelligence and experiential direction. In the new community public space design, more emphasis will be placed on intelligent interaction between people and space, as well as between space and facilities. Through on-site research, we aim to gain a deeper understanding of the spatial behavior of the elderly, explore key issues in spatial design, integrate smart elderly care into their living spaces, and construct a new type of elderly care approach.

References

1. Chen, Y.H., Sun, Y.J.: Amplification and reduction: the hidden facts in China's ageing population and the implications for ageing measurement and theory. Populat. Res. **01**, 3–22 (2023)
2. Yin, S., Hao, X.N., Zhang, S., Ma, C.Y., Zheng, Y.H.: Analysis of the urban elderly's demand for smart elderly care services based on individual differences. Soft Sci. Health **37**, 92–95 (2023)
3. Ji, C.Y.: The practical dilemma and optimization path of home smart elderly care. Dongyue Tribune **43**, 182–190 (2022)
4. Liang, C.Y., Hong, W.J., Ma, Y.M.: Comprehensive elderly care: a new model for the development of smart elderly care in the new era. J. Beijing Inst. Technol. (Soc. Sci. Ed.) **24**, 116–124 (2022)
5. Zhang, D.Y., Sui, J.L., Wu, S.: Research on the suitable renovation of the old community in the 1980s and 1990s based on the model of community endowment: taking the transformation of Tiantuo community in Tianjin as an example. Constr. Econ. **43**, 392–395 (2022)
6. Zhou, X.: Development and Study of Pension Model under the Background of Intelligent Pension. Urban. Arch. **19**, 81–84 (2022)
7. Zhang, Y., Fang, J.X.: The residential public space in high level of intelligence for elderly home-based care. Sci. Technol. Rev. **39**, 52–59 (2021)
8. Wang, X.R., Liu, Y., Jia, W.Y.: Research on the construction model of community smart home-based care system for the elderly: about the construction of virtual platforms in existing communities in Tianjin. Arch. J. **S1**, 56–59 (2020)
9. Li, K.: Study on the comparison, dilemma and approach of china's main pension models from the perspective of high-quality development of pension services. Chin. Health Serv. Manag. **39**, 647–653 (2022)

10. Ying, Z.P., Sang, Y.F., Chen, L.N.: Empirical research on renovation of old residential area suitable for aging: an example for penglai-community in Zhoushan, Zhejiang Province. Constr. Econ. **42**, 97–100 (2021)
11. Chi, Y.Y., Gong, H.: Aging-adapted reform strategy for existing residential based on safety requirements. Constr. Econ. **44**, 24–30 (2023)
12. Wang, K.X., Duan, J.J., Xuan, A.Q., Li, G.F.: Construction of aging design standard system in community smart elderly care service. Soc. Welf. **11**, 17–22 (2021)

Research on the Regional Adaptability Strategies of Elderly Activity Facilities in Temperate Area

Honggang Tang(✉) and Dongyan Jiang

College of Architecture and Urban Planning, Guizhou University, Guiyang 550003, China
562283562@qq.com

Abstract. Activities for the elderly usually have high requirements for the comfort, health and safety of the place. It is necessary to combine the analysis of the local natural environment, climate characteristics and cultural environment, and adopt the design strategy suitable for the regional environment to provide the elderly with green, ecological, healthy and comfortable activity space. Taking senior activity center in a region of Guizhou as an example, through the analysis of its functional requirements, regional natural climate conditions and land use environment, the design adopts BIM technology, new structure, air layer, environmental symbiosis and other strategies. After completion, the project has been recognized by the elderly users, which has certain enlightenment for the design of the elderly activity facilities widely constructed at present.

Keywords: Elderly Activity Facilities · Regional Adaptability · BIM Technology · Air Inter-layer · Environmental Symbiosis

1 Introduction

China has entered the aging society at the beginning of this century and the public facilities for the elderly are increasing gradually, among which the elderly activity center is an important building type. As a public service facility with high energy consumption and high requirements for space comfort in terms of light environment, wind environment and thermal engineering, the senior activity center is bound to play an important role in the national strategy of energy conservation, emission reduction and the development of green buildings. In today's increasingly advanced information, science and technology, the combination of economically feasible new technologies and structures with functional layout, energy saving and emission reduction, and environmental symbiosis is used to promote the development of elderly buildings in a green and ecological direction, which will help to reduce energy consumption and improve the operating conditions of a large number of public welfare and micro-profit elderly facilities, and better provide a healthy and comfortable activity space for the elderly. Based on new technology in term of ecological suitability and new structure, the senior activity center in a region of Guizhou designed by the author is an attempt to realize green ecology and harmonious coexistence of the activity center for elderly (see Fig. 1).

W. Guo et al. (Eds.): GBCESC 2023, LNCE 328, pp. 892–902, 2024.
https://doi.org/10.1007/978-981-99-9947-7_88

Guiyang is a temperate area in the building climate zone, with high altitude and high relative humidity. The average annual temperature is 15.48 °C. The coldest month is 5.66 °C, with an extreme low of −2 °C. The average temperature of the hottest month is only 25.6 °C, with a relatively small number of days above 25 °C [1]. Even though winter temperature in Guiyang is higher than that of the north, the place still feel cold and damp due to the high humidity of the air. Guiyang does not have extreme cold and heat, which makes it easy to neglect the ecological energy-saving aspect of the design process. Under such climate conditions, if appropriate ecological technology can be adopted, the environmental comfort level can be satisfied and better ecological benefits can be obtained. Therefore, combined with the characteristics of elderly buildings and climatic conditions, the design strategy includes optimizing the layout, controlling the volume coefficient and improving the insulation performance of the envelope. Winter heating uses air conditioning combined with heat storage space; In summer, natural ventilation and lighting are fully utilized, supplemented by air conditioning in some areas. These methods can reduce building energy consumption under the premise of meeting the needs of comfort and health of the elderly activity center.

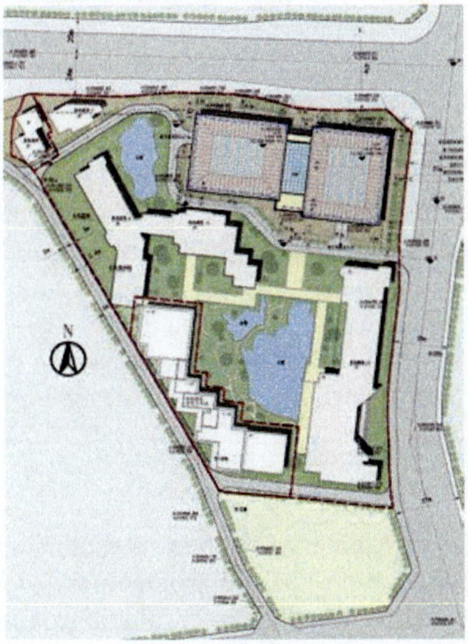

Fig. 1. General Plan

2 Optimization of Ventilation, Lighting and Thermal Engineering by Using BIM Technology in the Scheme Design Stage

BIM based simulation models can be used to test the actual effect of building design in advance, and can also focus on the impact of user behavior patterns on the actual performance of the building, which attempting to provide a forward-looking user experience evaluation criteria [2]. The parameter model can be imported into the BIM sustainable analysis software to conduct performance simulation analysis for ventilation, lighting, sunshine, thermal design, etc. Then the preliminary scheme of the building is tested, refined and optimized to meet the requirements of energy saving and comfort of green building.

2.1 Indoor Natural Ventilation Optimization

Good indoor ventilation is a prerequisite for effective diffusion of indoor heat and ensuring human comfort. The monthly charts of wind speed and wind frequency in Guiyang are obtained from the Weather Tool to determine the appropriate ventilation orientation and rationally arrange the indoor space. The auxiliary functional spaces such as toilets, dressing rooms, equipment rooms and stairwells are set on the side of the dominant wind direction in winter, and the main sports functional spaces are set on the side of the dominant wind direction in summer. Considering the larger depth of the venue, an atrium is set up in the middle of the venue as a space to organize traffic and a shared space for the elderly to communicate, and ventilation and lighting are introduced at the same time. After preliminary design and modeling, the software simulation results were used to adjust the overall layout, space combination, window area and location, so as to achieve the optimization of natural ventilation performance of the building. Finally, Phoenics software was used to simulate and analyze indoor natural ventilation performance in summer (July and August) on the third floor plane of the model's middle floor. It can be seen that indoor ventilation in the building is mainly north-south, and the flow field distribution in the large space with major functions is more uniform, and the wind speed in most areas can be kept within 1–2 m/s, which is more comfortable. The windows of the auxiliary room in the east do not face the dominant summer wind, and the indoor airflow is relatively poor. The air flow is more active in the atrium due to the effect of the patio. The result of simulation analysis shows that the overall natural ventilation condition of the building is well designed in summer (see Fig. 2).

2.2 Indoor Natural Lighting Optimization

The parameter model of the preliminary scheme was established according to the selection of the best sunshine orientation and the arrangement of the main function space, which was simulated by Ecotect. The lighting coefficient was set in the most unfavorable case and the parameter was set in the overcast day. Direct sunlight is not considered in the simulation calculation, and the simulation time is 12 noon on the winter solstice [3]. Firstly, the indoor lighting effect of the main functional space of the building is analyzed, and the percentage of the area that the main sports functional room meets the national standard of Building Lighting Design is counted. Then the overall lighting effect is analyzed to determine whether it meets the standard requirements. According to the lighting simulation results, the room with high lighting demand is placed in the southward with long sufficient sunshine, and the window opening and artificial lighting arrangement are adjusted to meet the lighting demand. The day lighting simulation analysis results of the third floor of the typical plane of Senior Cadre Activity Center show that the day lighting of the functional room is more uniform on the whole. The lighting in the middle part is not enough due to the depth of the sports space, but the top light introduced by the atrium can alleviate the lack of light. Therefore, it is necessary to control and adjust indoor lighting in groups to reduce energy consumption (see Fig. 3).

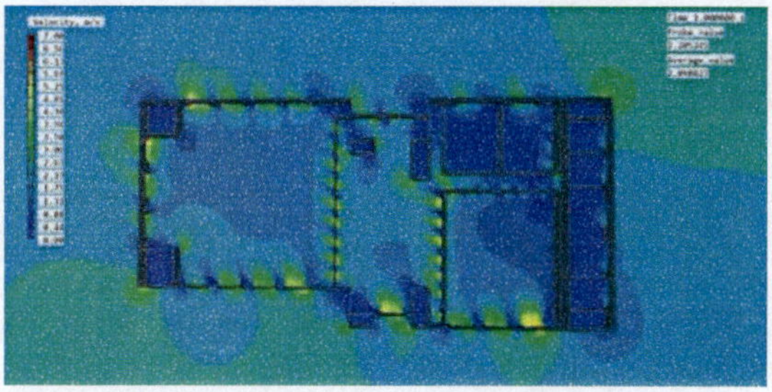

Fig. 2. Natural Ventilation Simulation Diagram

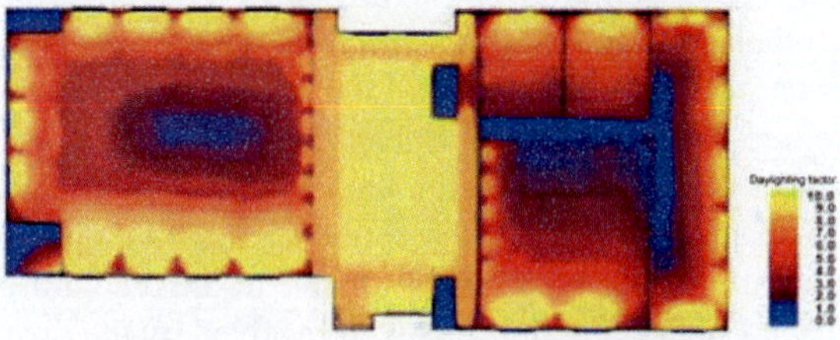

Fig. 3. Lighting Simulation Diagram

2.3 Optimization of Thermal Performance

In the process of energy saving design, good thermal performance of a room depends on the building's orientation layout, body shape coefficient, window-wall ratio, insulation materials and other factors. In winter, the peak solar radiation intensity on the south wall of Guiyang city is the largest, followed by southwest (southeast), west and east, and north. In summer, the peak solar radiation intensity of flat roof is the highest, as for wall, the west (east) is the highest, followed by southwest (southeast) wall, the northwest (northeast) wall, the south wall, and the north wall. Air conditioning heating room should avoid placement in the east-west direction and on the top floor. Effort should be made to place main function rooms to the south. Reasonable reduction in building height can maximize the utilization of solar radiation as a heating source, which can significantly reduce energy consumption. By using BIM modeling tools to conduct performance simulation analysis, which covers orientation layout, body shape coefficient control, window and wall ratio setting, parameters optimization of the envelope structure, the correlation between the heat transfer coefficient of the envelope structure and building energy consumption, so as to obtain the final thermal performance optimization scheme. In general, building performance simulation technology can provide scientific support for the design optimization process of building layout, form, facade and partial components [4].

3 The Selection of a New Structure - Hollow Sandwich Plate

Senior activity center in a region of Guizhou consists of several multistory large-span spaces, including a 500-person conference hall, table tennis court, badminton court, croquet court, tennis court and fitness rooms. The structure is combined horizontally and vertically according to the space size and net height requirements of each place. Then, according to the space characteristics, in order to maximize the use efficiency, the "Hollow Sandwich Plate" structure system developed by Academician Ma Kejian of our university is adopted after comparison of multiple structural schemes. This structure has the advantages of large span, light weight, saving material, small occupied space, using its own empty space to make equipment pipeline space, and good decorative effect of grid beam. The fourth floor adopts U-shaped steel high bolt connection hollow sandwich plate of concrete composite 39 m × 39 m and concrete hollow sandwich plate with 31.2 m × 39 m; The rest of the floors are concrete hollow sandwich plate with plane dimensions of 15.6 m × 15.6 m, 15.6 m × 39 m, 15.6 m × 23.4 m and 23.4 m × 39 m. The "hollow sandwich plate" structure hides the fire sprinkler system, cable bridge, HVAC and other equipment in the structure, without the suspended ceiling. In addition, the rhythmic grid structural beams form a neat and open visual effect, which achieving a high degree of unity of structural space, visual space and functional space. The structure is typical of ecological technology in terms of its material saving, land saving and space saving characteristics. The hollow sandwich plate on different floors are decorated with colored paint, which is conducive to the recognition and memory of the elderly. The finished space experience is well received by the owners and users (see Figs. 4, 5 and 6).

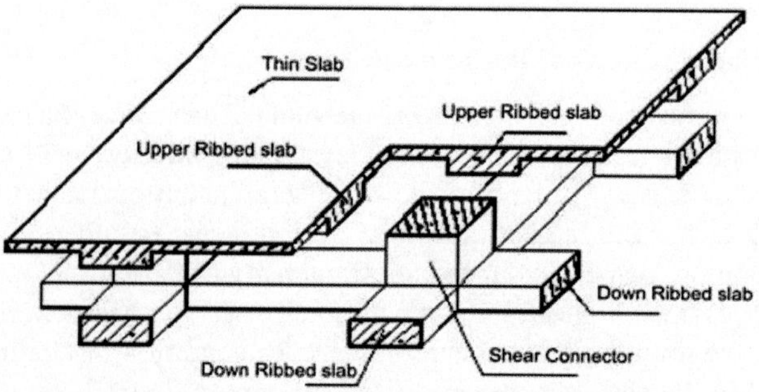

Fig. 4. Schematic Diagram of Hollow Structure Beam Plate

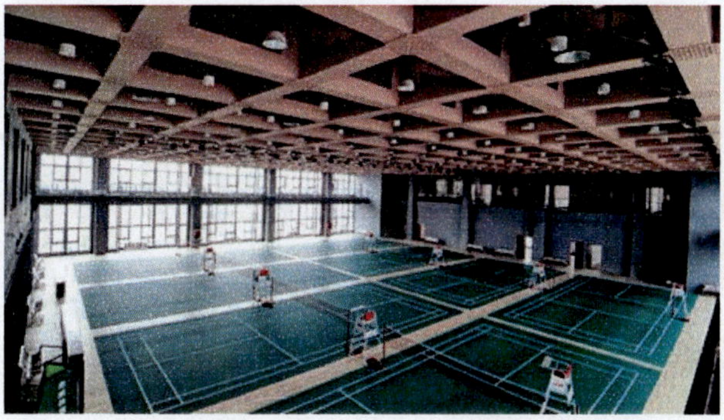

Fig. 5. Badminton Court (provided by author)

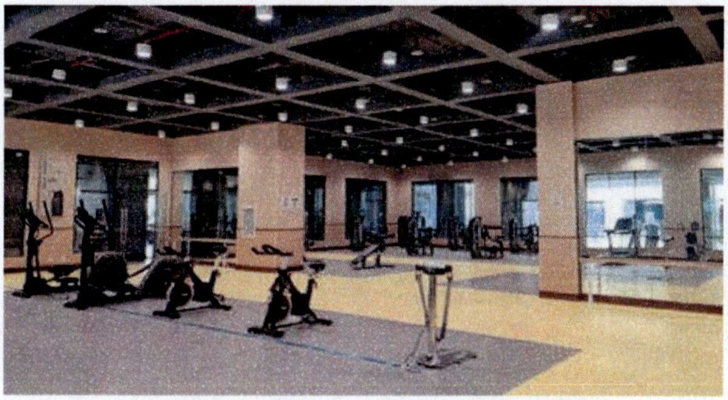

Fig. 6. Indoor Gym (provided by author)

4 Utilization of Air Inter-layer

Air inter-layer often refers to the structural facilities used for thermal insulation. When the thickness of the inter-layer is large enough to be used as a certain activity space, its role can be a heat storage space, or as a transition space for the physical conduction of heat and cold load. By controlling the ventilation organization and making good use of the wind-heat synergy effect, the air inter-layer can greatly enhance the natural ventilation of the room [5].

The east and west sides of the activity center are functional rooms, which are also the main load rooms for heating in winter. The atrium in the middle is not only a hub for traffic organization, but also a vibrant place for communication and helps to enhance air convection, which plays a regulating and buffering role in the interaction of heat and cold. Usually, the design of atrium is different in hot and cold environments. The middle atrium is suitable for hot areas, while the south-facing side atrium is more suitable for hot summer and cold winter areas, but has a negative impact on the internal environment of the building in summer. Therefore, "controlled" atrium technology must be adopted to regulate the atrium, with different control measures for summer and winter [6]. The top of the atrium of the senior cadre activity center adopts glass lighting roof to bring more sunlight to the atrium, and the windows of the north and south corridors and lounge are manually openable glass windows, which can control the opening height and window area of the floor. In winter, when the windows on both sides are closed, the whole atrium becomes a glass greenhouse to form a heat storage space, which can reduce the energy consumption of air-conditioning for indoor functional rooms; in summer, the top windows on the north side are opened to form a wind-pull effect by using heat pressure drive and wind pressure drive to increase the air flow in the atrium and also increase the natural ventilation in major functional rooms. The water feature in front of the south entrance forms a good microclimate, which can improve the temperature and humidity of indoor air (see Figs. 7, 8 and 9).

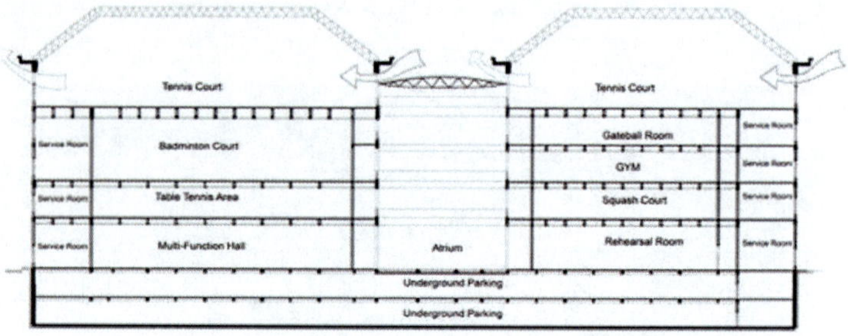

Fig. 7. Section

The roof is the component that is most influenced by outdoor heat and cold load among the building envelope, and the stuffy roof space of the Chinese traditional residential slope roof plays a certain role in heat preservation and insulation for the interior. Combined with the function of the venue, the four tennis courts are placed in two groups under two equal slope roof. The upper part is partially open in a half-overhead way, and the windows can be opened at a height of 1 to 3 m to adjust the air flow in summer. Combined with the form of the sloping roof and the trajectory of the tennis ball, the roof adopts the steel structure screen shell. The roof panels are made of pressed insulation metal panels, and the top lighting is polyurethane canopy to avoid glare. The setting of the transition air layer also eases the interaction between the fourth floor function room and the outdoor heat and cold to help reduce the heating energy consumption.

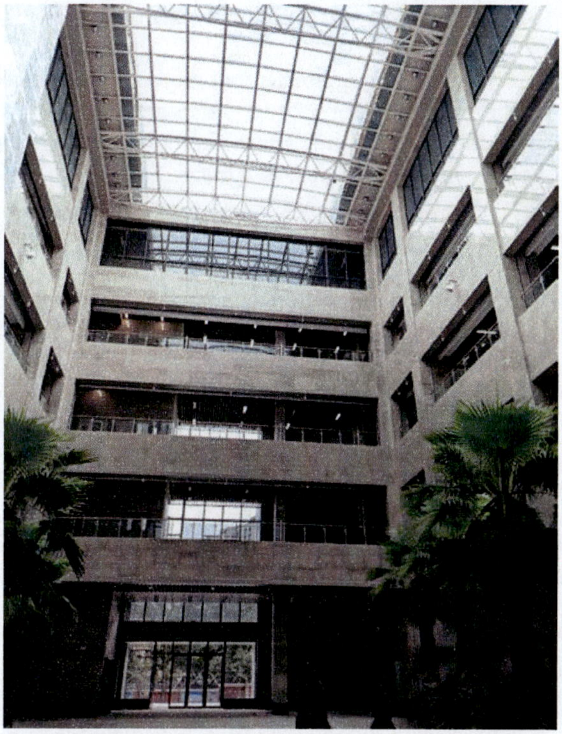

Fig. 8. Inside the Atrium (provided by author)

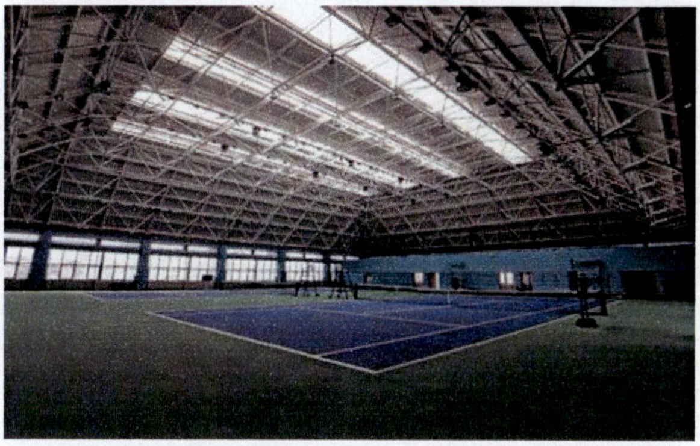

Fig. 9. Rooftop Tennis Court (provided by author)

5 Strategies for Symbiosis with the Environment

Building for elderly have higher requirements for indoor and outdoor environments, and the elderly have a clear preference for greenery. Eco-architecture pays much attention to the friendly relationship between architecture and the environment, and focuses on adapting to the environmental conditions, including climate and topography, so as to better maintain a harmonious relationship between architecture and the environment and realize the symbiosis between them. "Symbiosis" is a philosophical system, of which the basic components are: symbiosis of the local and the whole, symbiosis of the internal and the external, symbiosis of different cultures, symbiosis of different time and space, symbiosis of technology and human beings [7]. The symbiosis between different timeliness, part and whole is the focus of this design.

The former activity center for senior cadres was built in the 1980s and expanded at the beginning of this century with a garden-style space layout. The building is developed around the water feature of the atrium, with balanced relaxation and density. The architectural form is abstract and evolved from the original slope roof, which is both traditional and innovative. It is a successful example of modern architecture in Guizhou. Therefore, the general layout of the new venue is still centered on the central courtyard, forming an enclosed centripetal group with the original buildings of the first and second phase. The entrance of the building opens to the central water courtyard and establishes a relationship with the original courtyard through the axis of the foyer. The main functional space, the foyer and the atrium rest corridor are spread out facing the courtyard, forming a good borrowing and landscaping. The selection of roof, eave proportion, eave bottom decoration, vertical panel separation, and exterior wall material is consistent with the architectural style of the first phase. The original buildings, pavilions, corridors, water features, ancient trees and new buildings are in harmony with each other and have a strong recognition of the times, realizing the symbiosis between parts and whole, new and old.

The activity function of the whole center is reorganized, according to the traffic flow and the size of the building space. The original building is a low building with small space, so the space for senior music and dance is set here, which is conducive to the close integration of this part of the function with the center courtyard. The teaching room in the second phase is more square, and the original teaching space for painting and calligraphy will remain unchanged. The exercise space and mufti-functional performance hall are set in the new building, so that the whole place naturally forms three divisions: disturbance, static and movement, and the three divisions are linked by the curved corridor and paths around the courtyard and water features to form an organic whole (see Fig. 10).

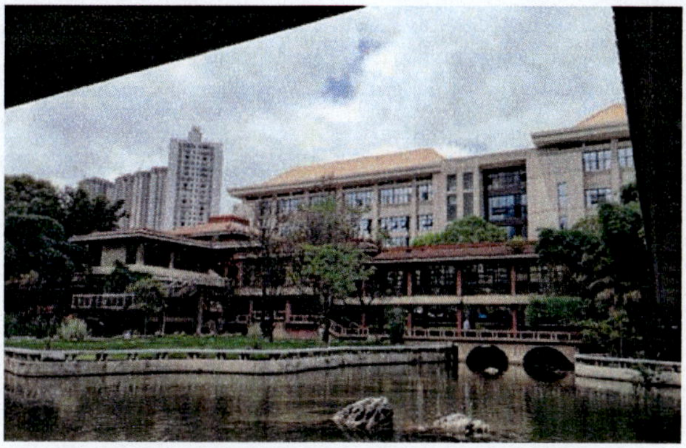

Fig. 10. Courtyard (provided by author)

6 Conclusion

According to the objective characteristics of the physiology and psychology of the elderly group, the design of the senior activity center needs to achieve compact plan, simple flow, bright space, green ecology, moderate wind speed and humidity, and also needs to create a public space conducive to the interaction of the elderly. Its demand for comfort, greenness, and publicity also provides a variety of possibilities for the application of ecological technology in architectural design. Senior activity center in a region of Guizhou combines the climatic characteristics and the basic conditions of the its construction site, by using BIM technology to optimize the ventilation, lighting and thermal design. The project chose a new structure of hollow sandwich panel, which is economically feasible and can save space, land and material. The air inter-layer was used for ventilation control, effectively reducing energy consumption for heating in winter and ventilation in summer. The whole project achieves organic harmony between the new building and the existing building with environment, and explores the use of regional adaptive design strategy to reduce building energy consumption at minimal cost while providing a green and comfortable space for elderly activities, providing inspiration for the ecological design of elderly buildings that are now widely constructed.

References

1. Yang, Z., Yang, Q., Zhan, X.Q.: Construction and demand analysis of heating model in Guiyang area. Intelligent Building and Smart City (2019)
2. Zhuang, W.M., Han, M.: Review on basic methods and advanced technologies of post-use evaluation of buildings. Times Architecture (2019)
3. Xu, L., Zhu, S.J.: Application of BIM technology in green building design. In: Proceedings of 2020 International Green Building and Building Energy Efficiency Conference, August 2020
4. Li, Z.W., Li, X.Y., Li, N.: Research on architectural design strategy of hot summer and warm winter area based on energy saving and indoor environment quality improvement: a case study of horticultural pavilion of Nanning garden expo. Architectural J. (2020)

5. Qian, F., Tang, S.N.: Study on natural ventilation design of green building – taking Jiading physical education of Tongji University. Building Science (2018)
6. Xia, B.: Research on the design of ecological building atrium space. Industrial Architecture (2013)
7. Kisho, K.: New Symbiosis Thought. China Architecture and Building Press, Beijing (2009)

Constructing a Framework for Measuring TOD Community Scenarios Based on Big Data Analysis

Suliu Chen and Yun Xiong(✉)

Shanghai Institute of Technology, Shanghai, China
18281595806@163.com

Abstract. In recent years, governments at all levels in China have advocated the integration of the TOD model into urban planning and construction, and because scene theory and TOD communities share the same core values, this paper applies scene theory to the study of the development model of TOD communities. Because of their mixed functional characteristics, TOD communities can strengthen the agglomeration effect by creating diversified characteristic scenes, attracting specific people to converge there, improving the TOD communities can attract specific groups of people to gather here and enhance the inherent vitality of TOD communities to promote sustainable development. Based on the scene analysis framework of the Chicago School, a scene measurement framework for TOD communities based on China's national conditions is constructed. Through the integration of big data such as bibliometric method and POI database, the scene measurement dimensions of TOD communities are divided into three main dimensions of functional space, public transportation space and public service space, and fifteen sub-dimensions are refined. The "UNI-CITY" TOD community is used as an empirical study, and suggestions for optimization are made based on the scenario measurement dimension framework.

Keywords: Big Data · TOD Communities · Scene Theory · Comfort Objects · Scene Measurement Dimensions

1 Introduction

TOD development model relies on high-density mixed function integrated development to attract urban people to gather, drive the growth of consumption economy and promote regional development. However, TOD development in China is still in the primary practice stage, so how to better build TOD communities and strengthen the guiding role of TOD mode construction has become an urgent problem to be solved. "Scene theory" is a new paradigm for studying urban development, and this paper applies scene theory to the study of TOD community development model. Relying on the value transformation of scene theory, different material forms form comfort facilities with different value concepts, which constitute the "scene" in a specific area, attract specific people to gather and strengthen the composite functional partition and diversified consumption scenes

W. Guo et al. (Eds.): GBCESC 2023, LNCE 328, pp. 903–915, 2024.
https://doi.org/10.1007/978-981-99-9947-7_89

in TOD communities. To apply the scenario theory to the study of TOD communities, it is necessary to build a framework for measuring the scenarios in TOD communities based on the national conditions in China because of the differences in the background, purpose and current conditions of development in foreign cities. In recent years, with the rapid development of Internet technology, the development of big data such as CNKI database, Baidu heat map data [1], Airbnb data [2], and POI (point of interest) data [3] has provided new methods and ideas for urban construction research. This paper constructs a framework of scene measurement dimensions through statistical analysis of big data to improve the scientific and rigorous construction of the framework, in order to provide reference for the construction and development of TOD communities.

2 Research Methods and Data Sources

2.1 Bibliometric Analysis

The bibliometric method is one of the main methods of bibliometric research, which reveals the academic hotspots and future development trends in the research field through quantitative or qualitative analysis of literature with different data characteristics, mainly including visual analysis of keyword co-occurrence analysis, cluster analysis, and post volume analysis [4]. The cross-correlation of the two research fields was derived, and the three main dimensions of scene construction in the TOD research field were analyzed [5]. The study selected CNKI database paper data, organized and de-duplicated to obtain a total of 1637 articles on TOD and 347 relevant literature on scene theory published between 1996 and 2022 as the data source for the analysis [3].

2.2 POI Data Analysis

The full name of POI is Geographic information points of interest data [6], which is translated as "points of interest" in Chinese. Its attribute decision can describe the entity objectively and accurately to a large extent, reflecting the behavior and activity distribution pattern of urban people [7]. More and more scholars use POI data in the research and development of urban construction, such as spatial performance of urban land and urban land development needs [8]. Through the software ArcMap with the help of location retrieval service provided by the open platform of Gaode Map, a 500 m circular buffer zone was constructed by computer programming language with the Hong Kong Kowloon MTR station as the central point to extract the Gaode Map POI data within the area one by one, as shown in Fig. 1, with a total of 1364.

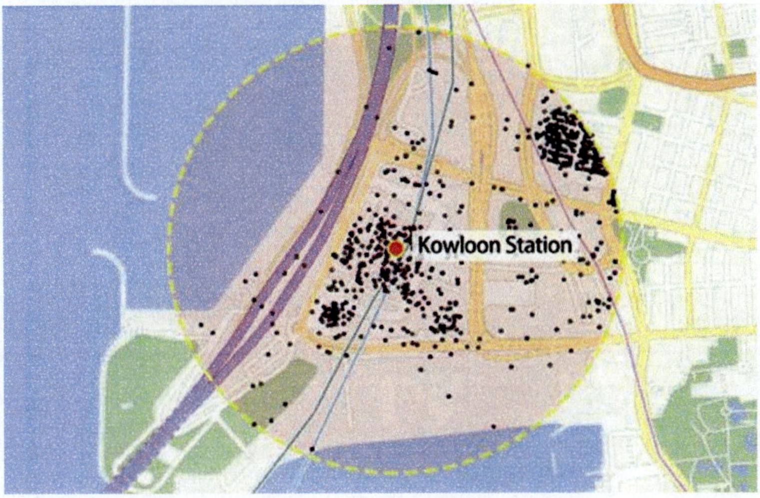

Fig. 1. Distribution of POI pickup points in the TOD community of Hong Kong Kowloon Station

3 Establishment of TOD Community Scene Measurement Dimensions Based on Bibliometric Analysis

3.1 Chicago School Scene Measurement Dimension Classification

According to the Chicago School, drama, legitimacy, and authenticity are the three dimensions for understanding "value-oriented scenes formed by different combinations of comfort facilities [9]. These three dimensions provide an initial framework for scene analysis, which presents a continuum from the physical concept of "space" to the concept of value-oriented "scenes. These three dimensions provide an initial framework for scenario analysis, which presents a range of concepts from the physical concept of "space" to the value-oriented concept of "scenario," each of which has its own specific form and purpose [10]. To make the scene analysis framework more comprehensive, The Chicago School subdivides fifteen sub-dimensions under three main dimensions: Homegrown, flashy, Traditional, Ethnic, group's, Fascinating, Leadershi, National, Good Neighbor's, Utilitarianism, Corporate, Transgressive, Egalitarianism, Rational of courtesy, Self-expression.

3.2 Scenario Measurement Master Dimension Establishment

There are 2857 keywords in TOD research in the bibliometric analysis, 45 with a frequency of 10 or more, and 895 keywords in scenario theory research, with 23 with a frequency of 5 or more. In the keyword mapping of TOD research, TOD theory is the source of research, which revolves around land use, public transportation, and integrated development. In the keyword mapping of scenario theory research, scenario theory is the research source, revolving around scenario construction, public space, and economic development (Figs. 2 and 3).

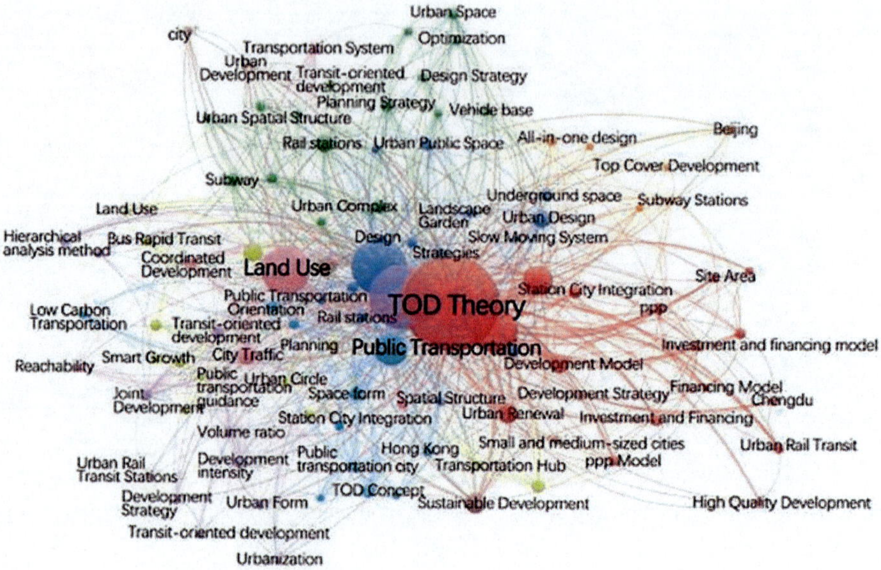

Fig. 2. TOD research keyword co-occurrence network map

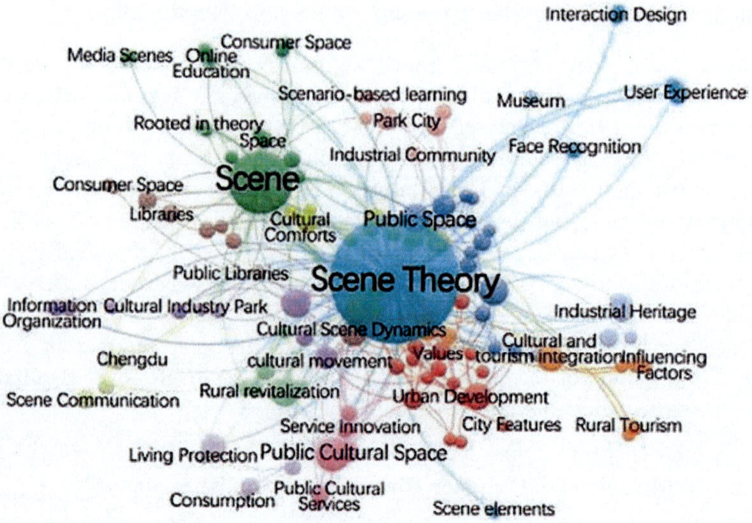

Fig. 3. Scene theory research of keyword co-occurrence network map

According to the visual analysis of bibliometric method [11], it can be seen that the public space is the main associated node in TOD theory and scenario theory, which revolves around comprehensive urban development and urban space construction to promote urban consumption practices and economic development. A TOD community is a physical space and environmental carrier for community people to produce and live, low-carbon travel and entertainment consumption, whether the physical space within a TOD community is perfect enough In addition to examining the configuration of various facilities in its internal space, it is also necessary to examine the efficiency of community people using the space. According to the research system of scene theory, combined with the real problems of TOD communities, the scene dimension in TOD communities is divided into three main dimensions: functional field space, public transportation space and public service space. Functional field space is mainly the physical space for crowd activities, which includes infrastructure, commercial facilities, entertainment facilities, sports facilities, etc. Public transportation space mainly refers to the elements of transportation facilities unique to TOD communities, which include rail stations, chronic trails, interchange spaces, etc. Public service space mainly refers to the institutional mechanism and environmental atmosphere to ensure the good development of TOD communities, which mainly includes policy guidance, functional diversity, supporting public service facilities, openness of space, etc.

4 TOD Community Scene Measurement Dimension Framework Construction

4.1 Hong Kong Kowloon Station TOD Case Overview

Hong Kong's high-density urban development model shares the same urban development trend as mainland China and, as the first city in China to become an international metropolis, has provided a very important role to learn from the development and construction of various cities in China. Hong Kong Kowloon Station is a well-known benchmark project of Hong Kong MTR Corporation's transit-oriented development, which has effectively promoted smart urban growth and intensive development in the 25-year development process based on China's national conditions. This paper selects the Hong Kong Kowloon Station TOD project as a case study of TOD community scenario dimension measurement framework, and starts the research from two perspectives: the development history of Kowloon Station, the construction background such as policy support and the POI data statistics within the Kowloon Station TOD community, and proposes fifteen sub-dimensions of scenario dimension analysis to build a TOD community scenario dimension measurement framework.

4.2 Scene Measurement Sub-dimension Building Based on POI Data Analysis

It is generally believed that GODMap POI data basically covers all facility types in the region and has a certain reasonableness and reliability as a data source [12]. Due to the common characteristics of POI and comforts, POI data points in TOD communities can also be understood as comforts facilities in the communities. The statistical classification of 1364 comfort object facilities in the Kowloon Station TOD community covers 19 categories such as food service, shopping service, living service, sports and leisure, and transportation facilities, and their data distribution is shown in Fig. 4.

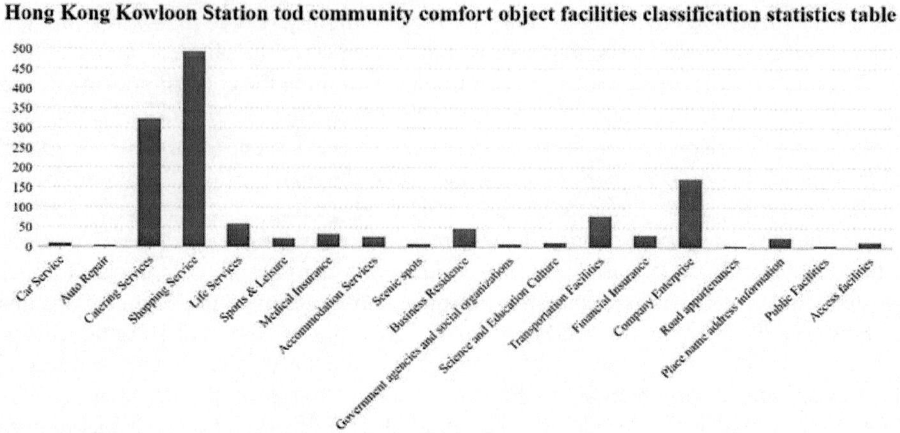

Fig. 4. Statistical chart of TOD community comfort facilities classification at Kowloon Station, Hong Kong

According to the comfort object data, it can be seen that the planning and design of Kowloon Station TOD community involves a variety of functions such as commercial, residential, office and green space mixed construction, and commercial facilities such as catering services and shopping services account for the highest proportion, followed by office facilities such as company enterprises and transportation facilities. The participation of people's activities in the community makes different combinations of comforts constitute different scenes such as consumption scenes, traffic scenes and food scenes.

4.3 Scene Measurement Dimension Framework for TOD Communities

Functional Field Space. In TOD communities, the physical practice space that community people rely on for cultural exchange and behavioral interaction generally includes the internal space of buildings and infrastructure. For example, the headquarters building of the Civil Aid Team and the Burberry store in the Kowloon Station TOD community; it also includes some cultural facilities, such as the D'Elysee Brewery and the MTR Exhibition Gallery; in addition, it includes open public activity spaces, natural landscapes and other leisure and entertainment spaces, such as the rooftop garden of the bus terminal and the Global Trade Square. The sustainable development of the physical space of a TOD community cannot be achieved without perfect infrastructure and diversified spatial functions to meet the needs of people in the community for work, consumption, living and entertainment, etc. Due to the high-density and multi-functional composite development characteristics of TOD, the construction of functional space is the key to promote the economic development of TOD communities, and diversified business forms attract more people to produce consumption behavior, which in turn The diversification of businesses attracts more people to generate consumption behavior, which in turn drives the regional economic construction. Therefore, the functional space dimension of TOD community is divided into five sub-dimensions: consumption place, business office, livable ecology, leisure and entertainment, and intelligent life.

Public Service Space. TOD communities are not only generated by the natural clustering of their instinctive living functions, but also by the guided construction mode with the intervention of government and public institutions, which is indispensable for the development of TOD communities with good resources of public service facilities and sound community system. Through the construction of functional diversity, it attracts the participation of multiple subjects to enhance the openness of the community and serve multiple types of social groups. The development body of the Kowloon Station TOD community is led by MTR, with government equity and joint developer investment to maximize the utilization and efficiency rate of the land around the MTR hub. The economic efficiency is maximized while paying more attention to the pedestrian experience by combining Kowloon Station with a man-made platform to maximize the platform; the platform has a plaza garden to create a leisure neighborhood for people. On the one hand, it realizes the effective diversion of passenger flow during peak periods; on the other hand, it provides an open and energetic social space for the surrounding residents, office workers, tourists and commercial consumers. The Kowloon Station TOD community is rich in public service resources such as ICC International Chamber of Commerce, People's Safety Service Headquarters, Café de Coral Professional Management Institute, and Pinehurst Home for the Elderly. Therefore, the public service space dimension of the TOD community is classified into five sub-dimensions: policy support, openness, crowd clustering, public service resources, and diversity.

Public Transportation Space. The core of TOD community is a community oriented by public transportation, with a good layered layout of rail transit, bus, bicycle and car, and a three-dimensional pedestrian system to link up and down, creating a three-dimensional transportation, encouraging "public transportation + walking", "public transportation + bicycle "The low-carbon mode of travel is encouraged. Kowloon Station TOD community planning takes the subway station as the core, organically integrates the subway, airport express line, bus station and high-speed rail station, and creates highly integrated transportation interchange points through the layered design of vertical space to provide convenient transportation scenarios for passengers' departure, arrival, transfer and transit to the maximum extent. Focusing on three-dimensional transportation design, it arranges cars, buses, commercial space in Elements Square, public space on the platform level of commercial space, high-grade residences, and the Global Financial Center in a reasonable hierarchy. It also provides comprehensive transportation facilities in the community, such as Jordan (West Kowloon Station), East Gate of West Kowloon Waterfront Promenade, and Station Square. Therefore, the public transportation space dimension of the TOD community is delineated into five sub-dimensions: chronic space, vertical transportation, transportation tabelfacilities, station accessibility, and interchange convenience (Table 1).

Table 1. Scene dimension analysis framework for TOD communities

Three scenario master dimensions	Functional field space	Public transportation space	Public Service Space
Fifteen sub-dimensions	Consumption places	Slow Moving Space	Policy Support
	Business Office	Vertical traffic	Degree of openness
	Livable Ecology	Transportation Facilities	Crowd Clustering
	Leisure and Entertainment	Site accessibility	Public Service Resources
	Smart Living	Convenience of interchange	Degree of Diversification

5 Shanghai Vanke "UNI-CITY" TOD Community Case Empirical Application

5.1 Scene Dimensional Analysis of the "UNI City" TOD Community

According to the framework of scene measurement dimensions, three main dimensions and fifteen sub-dimensions were used to evaluate the built scenes in the "UNI City" TOD community. The questionnaire is mainly a survey of people in the "UNI City" TOD community, trying to analyze the perception of scenes by quantitative means. In order to ensure the scientific and reliable results, seven experts in the field of TOD communities were invited to participate in the construction of the matrix weights. The scores of each scene dimension were calculated by combining the scores of the community people in the questionnaire, and then the evaluation and analysis were carried out.

Seven experts in the field of TOD communities were invited to rate the importance of the three main dimensions and fifteen sub-dimensions of the "UNI City" TOD community, and the AHP hierarchical analysis was used to construct a judgment matrix using the relative importance of the data to calculate the weights, and then a consistency test was conducted [13]. Usually the smaller the CR value, the better the consistency of the judgment matrix, and in general the CR value is less than 0.1, then the judgment matrix satisfies the consistency test. If the CR value is greater than 0.1, it means that there is no consistency, and the judgment matrix should be analyzed again after appropriate adjustment (Table 2).

Among them, the consistency test formula is:

$$CR = \frac{CI}{RI}. \tag{1}$$

CI is the consistency index, and its calculation formula is:

$$CI = \frac{\lambda \max - n}{n - 1}. \tag{2}$$

- RI is the average random consistency index.
- λmax is the maximum characteristic root.
- n is the order of the judgment torch array (Table 3).

Table 2. Results of AHP analysis of the main dimension of the scenario.

Item	Eigenvector	Weighting value	Maximum Eigenvalue	CI value
Functional field space	1.032	34.409%	3.000	0.000
Public transportation space	1.129	37.634%		
Public Service Space	0.839	27.957%		

The three main dimensions and fifteen sub-dimensions were derived from AHP hierarchical analysis, and the total weight of the fifteen sub-dimensions was derived. The formula for calculating the comprehensive evaluation score of each scene dimension of "UNI City" TOD community is:

$$D = \frac{\sum_{i=1}^{n} B_i C_i}{n} \tag{3}$$

- Bi is the total weight of a dimension,
- Ci is the score of the community population for that dimension.

In the "UNI City" TOD community, 326 questionnaires were distributed in three time slots: noon, afternoon and evening, during the rest days and weekdays, with 297 valid questionnaires. In the process of distributing and collecting the questionnaires, there were Eq. (1) strict controls for each age group and community Eq. (2), type to ensure that the research results were more objective. The scores of the 297 questionnaires were calculated by Eq. (3), and the overall scores of each dimension are shown in Table 4.

The results show that the public transportation spatial dimension in the "UNI City" TOD community has the highest score of 1.66, followed by the functional field spatial dimension of 1.50, and finally the public service spatial dimension of 1.06.

5.2 Suggestions for Optimizing the "UNI City" TOD Community Scene

Functional Field Space: Adjusting the Layout of Functional Scenes and Strengthening the High-Density Development of Communities. Among the functional spatial dimensions of the site, the highest score of 0.40 is for the consumer place dimension, while the lower score of 0.24 is for the business office dimension. As a representative project of Shanghai TOD, "UNI City" TOD community has perfect infrastructure construction and high quality physical space, which makes people in the community more satisfied with the creation of functional space dimension.

Table 3. "UNI City" TOD community scene dimension weighting value.

Main Dimension (A)	Main dimensional weights	Sub-dimensionality (B)	Sub dimensional weights	Total weighting
Functional field space (A1)	34.409%	Consumption Places (B1)	25.385%	8.735%
		Business Office (B2)	17.692%	6.088%
		Livable Ecology (B3)	17.692%	6.088%
		Leisure and Entertainment (B4)	22.308%	7.676%
		Smart Living (B5)	16.923%	5.823%
Public transportation Space (A2)	37.634%	Slow Moving Space (B6)	18.000%	6.774%
		Vertical traffic (B7)	21.333%	8.028%
		Transportation Facilities (B8)	18.000%	6.774%
		Site accessibility (B9)	23.333%	8.781%
		Convenience of Interchange (B10)	19.333%	7.276%
Public Service Space (A3)	27.957%	Policy Support (B11)	20.000%	5.591%
		Degree of Openness (B12)	20.800%	5.815%
		Crowd Clustering (B13)	18.400%	5.144%
		Public Service Resources (B14)	21.600%	6.039%
		Degree of Diversification (B15)	19.200%	5.368%

The data analysis shows that the satisfaction of the physical space construction of each functional place in the community is relatively high, but the perception and experience of some scenes do not reach the corresponding effect. For example, the construction of scenes in the consumption place dimension should not only meet the consumption needs of community residents and foreign visitors, but also focus on the weekday consumption needs of office, corporate and community management people. In

Table 4. Results of AHP analysis of the main dimension of the scenario

Main Dimension	Sub-dimensionality	Score	Total Score
Functional field space	Consumption places	0.40	1.51
	Business Office	0.24	
	Livable Ecology	0.26	
	Leisure and Entertainment	0.36	
	Smart Living	0.25	
Public transportation space	Slow Moving Space	0.30	1.66
	Vertical traffic	0.36	
	Transportation Facilities	0.25	
	Site accessibility	0.43	
	Convenience of interchange	0.32	
Public Service Space	Policy Support	0.20	1.06
	Degree of openness	0.24	
	Crowd Clustering	0.20	
	Public Service Resources	0.19	
	Degree of Diversification	0.23	

the community, we can provide all-time convenience consumption scenes, quick working meal canteen scenes and other exclusive functional scene layouts for different people in the community to strengthen the high utilization rate of TOD community space.

Public Transportation Space: Optimize the Design of Transportation Scenes and Set a Model of Low-Carbon Community. The highest score of public transportation space dimension is 0.43, followed by vertical transportation dimension of 0.36, and the lowest score of transportation facilities dimension is 0.25. Relying on the absolute location advantage, "UNI City" TOD community is closely connected with Xu Ying Road Metro Station, realizing zero transfer in real sense. In addition, various outdoor traffic lanes and walkways are equipped with multiple scene signs for each area, and the slow walking space of residents in "UNI City" TOD community is strictly separated from the traffic flow, and the walking space is integrated with greenery landscape, leisure facilities, creative corridor landscape and other scene designs, which improves the attractiveness of slow traffic scene and stimulates community walking space vitality.

Due to the strict property management system in the "UNI City" TOD community, the temporary parking of non-motorized vehicles in the community has become an inconvenience, and the traffic management scene can be optimized by adding temporary non-motorized vehicle parking points. The actual research found that the utilization rate of three-dimensional traffic corridors in residential areas is low, the ecological scene is not energetic enough, and there is a phenomenon of wasting resources. Therefore, in the design of traffic scenes, the layout of traffic and landscape resources in each functional area of TOD can be appropriately adjusted to optimize the existing space scene creation and establish a model of low-carbon community.

Public Service Space: Strengthen the Construction of Service Scenes and Promote Sustainable Urban Development. Among the public service space dimensions, the highest score is 0.24 for the openness dimension, the lowest score is 0.19 for the public service resources dimension, followed by 0.20 for the policy support dimension. The government's guidance and the intervention of public mechanism make UNICity TOD community a model for projects with diversified business forms and mixed land development. UNI City has set up several public spaces to meet the needs of daily life, and also created exclusive social spaces inside the commercial and office buildings, and cooperated with various cultural, educational and artistic institutions to create many cultural exchange activities, which enhance the diversified openness of the TOD community and create multiple types of service scenes.

In the process of research, we found that the surrounding land of "UNI City" TOD community is currently lacking in education and medical resources. Due to the comprehensive development characteristics of the TOD community, there may be uneven distribution of resources, and resources such as crowd exchange and social activities are mostly concentrated in the commercial space where people gather. This leads to a lack of interaction between neighbors in the residential area, resulting in a lack of a harmonious neighborhood atmosphere. Community service centers can create exclusive social scenes and healthy living scenes in residential areas by establishing public protection service groups to promote sustainable community development.

6 Conclusion

In this paper, we study and construct a scene measurement framework for TOD communities in the context of scene theory. Using the bibliometric method to organize and analyze the literature research on scene theory and TOD at home and abroad, three main dimensions in the scene measurement dimension framework of TOD communities are proposed on the scene analysis framework proposed by the Chicago school combined with the Chinese context. Then, by crawling the POI data of Hong Kong Kowloon Station TOD community, we classified various kinds of comfort objects in TOD community, matched the 19 types of comfort objects with the three main scene dimensions, and then subdivided 15 scene sub-dimensions to build the TOD community scene measurement dimension framework. Using Shanghai Vanke "UNI City" TOD community as the object of empirical research, the AHP hierarchical analysis method was used to derive the weight of each scene dimension in the "UNI City" TOD community, and then calculate the score of each scene dimension. Based on the scenario measurement dimension framework, we evaluated the scenario construction in the TOD community and made suggestions.

References

1. Wu, C., Ye, J.N.: Research on urban spatial structure based on Baidu map heat map - taking the central city of Shanghai as an example. Urban Plann. **4**, 8 (2016)

2. Mei, L., Jiang, H.: The process, factors and trends of spatial differentiation of B&B in Beijing based on Airbnb data. Econ. Geogr. **41**(03), 213222 (2021). https://doi.org/10.15957/j.cnki.jjdl.2021.03.022
3. Xue, B., Xiao, S., Li, J.Z., et al.: Spatial structure analysis of northeastern cities based on point-of-interest (POI) big data. Geoscience **40**(5), 10 (2020)
4. Islam, M.A., Widen, G.: Bibliometric analysis of the: 2000–2020. VINE J. Inf. Knowl. Manage Syst. **53**(3), 467–490 (2023)
5. van Eck, N.J., et al.: A comparison of two techniques for bibliometric mapping: multidimensional scaling and VOS. J. Am. Soc. Inf. Sci. Technol. **61**(12), 2405–2416 (2010)
6. Guo, J.: Evaluation and optimization of spatial characteristics of social services of urban parks in the core area of the capital function based on POI data. Beijing University of Architecture (2022)
7. Ma, C.X., Peng, F.L., Qiao, Y.K., et al.: Evaluation of spatial performance of metro-led urban underground public space: a case study in Shanghai. Tunn. Undergr. Space Technol. **124**, 104484 (2022)
8. Pozzo, D.N., Correa, K.R., Madrid, A.I.C., Campo, C.J.C., Donado, M.E.G., Biegelmeyer, U.H.: Logistics 4.0: a review of current trends using bibliometric analysis. Procedia Comput. Sci. **203**, 531–536 (2022)
9. Wu, T., Zhang, M., Yue, Z.: Progress of TOD research based on Web of Science bibliometric analysis
10. Hansen, D., Rasmuseen, C.H., Jochunsen, H.: The role of public libraries in culture-led urban regeneration. New Library World **114**, 7–19 (2013)
11. Chen, B., Yan, S.: Research on cultural value enhancement of non-heritage tourism places under scene theory-analysis based on data from 27 non-heritage tourism towns in Zhejiang Province. J. Tongji Univ. (Soc. Sci. Edn.) **33**(01), 20–32 (2022)
12. Fan, N.N.: Evaluation of urban livability in Hebei Province based on hierarchical analysis
13. Xu, L., Chen, L.: AHP-based comprehensive evaluation of parkland in Chengdu. Archit. Cult. **12**, 4 (2022)

A Computational Analysis Method to Evaluating Experiential Qualities of Historical Streets in Sustainable Townscapes

Yabing Xu[1,2], Hui Tong[1(✉)], Wenpeng Song[1], Zhao Li[1,2], John Rollo[3], Pengfei Zhao[1], Meng Chen[1], and Yolanda Esteban[3]

[1] School of Architecture and Urban Planning, Shandong Jianzhu University (Shandong University of Architecture and Engineering), Jinan 250101, Shandong, China
{xuyabing20,24249}@sdjzu.edu.cn

[2] Yellow River Basin Urban-Rural High quality Development Research Center, Shandong Jianzhu University (Shandong University of Architecture and Engineering), Jinan 250101, Shandong, China

[3] School of Architecture and Built Environment, Deakin University, Melbourne, VIC 3220, Australia

Abstract. In recent years, there have been numerous studies on the influencing factors of experiential qualities, but there is a lack of consideration for spatial cognition factors such as the perceived quality of street space by the public under a people-oriented approach. This paper uses a space syntax analysis method to evaluating experiential qualities of historical streets in sustainable townscapes. The research results show that from a global perspective, spatial quality has a certain explanatory power on the distribution of spatial vitality. From a local perspective, the impact of different dimensions of space quality on space vitality showing a more complex relationship in urban space. The difference in the impact of spatial quality on spatial vitality in different types of spaces is mainly related to location, and the improvement of spatial quality optimization on spatial vitality requires certain material prerequisites. Therefore, in actual space renovation work, it is necessary to determine whether quality optimization can be adopted to enhance vitality based on the existing material conditions of the space.

Keywords: Spatial vitality · Experiential qualities · Sustainable townscape

1 Introduction

Experiential Qualities of historical streets deserves more attention globally, because the historical streets with high quality are beneficial in promoting national unity and strengthening national image through the enrichment of cultural identity and its positive connections and connotations. Space syntax was developed as an analysis tool of architectural and urban space [1, 2].

This research aims to develop a space syntax-based analysis method to assist in evaluating the experiential qualities of historical streets in sustainable townscapes. The

W. Guo et al. (Eds.): GBCESC 2023, LNCE 328, pp. 916–925, 2024.
https://doi.org/10.1007/978-981-99-9947-7_90

case study in this research is Tongli canal town, in China, which is went through by the Grand Canal. Tongli canal town, where tourists' spatial cognitions have been analyzed in this paper. The first part is introduction which introduce the research, the second part is overview of space syntax. The third part proposes space syntax-based analysis method, and extracts spatial morphology information to analyze experiential qualities of historical streets. The fourth section presents the case study that applied space syntax and Visibility Graph Analysis (VGA) to correlate the findings of Tongli canal town in China, analysing the space of Tongli Grand canal town and the nodes that need to be strengthened for conservation. The Conclusion proposes protection suggestions and strategies for the town.

2 Overview of Space Syntax

Montello [3, 4] believed the feature definitions in the spatial data transferring process required further research. Montello [5, 6] explored the special forms of villages. He then listed "four reasons for cultural universality in spatial cognition: 1) organizational similarities in nervous systems; 2) similarities in body structures and processes; 3) similarities in learning and socialization; and 4) similarities in residential environments". Furthermore a global approach was introduced to describe the environments in a more appropriate way. At a larger scale, Franz and Wiener conducted experiments developing a novel three-pass algorithm which can identify spatial centres to formalize the behaviour. The deviations from different methods might come from minor tweaking. Their study includes several issues in architecture and spatial cognition to better design and automatically capture the spatial behaviour as well as to represent spaces sparsely and efficiently [7].

3 Methodology

The research methods base on space syntax analysis methods which introduces the methodology is the application of analysis results of experiential qualities on historical streets. Kaplan [11] mentioned the thinking in his research "Four key 'informational elements': coherence, complexity, legibility, and mystery determined the quality of a landscape setting". Kaplan [11] discussed these four factors, "mystery embodies the attraction of the bend in the road, the view partially obscured by foliage, the temptation to follow the path [11]." This research mainly focuses on the evaluation of experiential qualities. This paper provides an objective description for the strategy design with regards to promoting experiential qualities [12]. This research summarizes the factors that affect the development of the ancient canal town and its surrounding areas, the factors that affect urban development, the evolution and impact of urban spatial pattern, and the development of the canal. In terms of landscape function, regional concept, historical culture, natural environment, and socio-economic aspects, it is concluded that the canal determines the axis of Tongli canal town's development, affects urban texture, and determine the framework for town development.

4 Case Study: Tongli, China

4.1 Overview of Tongli, China

Tongli is governed by Wujiang District, Suzhou, Jiangsu Province, which is established in Song Dynasty along the banks of Tai Lake and the Grand Canal in China. Tongli is located next to Central business district of Suzhou and is located near the three largest southern cities in China, which are Shanghai, Suzhou and Hangzhou. Tongli is also part of the region referred to as the 'golden triangle' intersecting with Jiangsu, Zhejiang and Shanghai, and the central regions along China's coastline and Yangtze River delta. The total ancient town area is 8.3 square kilometres, with a population of 94,000 residents. Tongli is also gaining fame internationally due to its characteristic features of being a 'natural river town, and the re-imaging of Jiangnan' [13]. In 1992, Tongli was listed as provincial cultural relic protection town in China. The Retreat & Reflection Garden of Tongli canal town is a historic site. The town was originally constructed during the Ming and Qing Dynasty, where there are numerous gardens, temples, mansions as well as 'remarkable persons' heritage. From a geographical perspective, 15 rivers separate the town into 7 small islands connected by 9 ancient bridges, hence the theme of 'Smart bridges, falling water and heritage' becomes internationally recognised.

4.2 Interpreting the Spatial Syntax Analysis of Tongli, China

As a major method applied in this research the related theoretical system, methodology and software have become integrated for the implication of space syntax over several decades. This research applies six analysis factors of Tongli canal town and explores the reflection of these indicators. The indicators include six value. Depthmap has been widely used in the realization of space syntax in previous studies and has also been applied to specific areas in this study.

Table 1 displays the spatial characteristics of twenty major tourist streets in the Tongli ancient town area.

As the colour contours in Fig. 1 (d) and values in 1, there are four streets with the highest global integration values, namely, Xiaohe Street (217.874), Xiaochuan Street (217.693), Huaxin Street (217.610) and Xinhua Street (217.054). Xinhua Road has the highest local integration value, which is 92.775. Xiaochuan street is the greatest in connectivity of 14 among all streets in Tongli. Figure 2 presents the intelligibility value of Tongli town in China. In consideration of the high density of historic buildings which often characterise ancient canal towns [15]. Additionally, the intelligibility value (0.683) shows that the legibility in Tongli ancient town area is good.

4.3 Focused Visibility Analysis of Mingqing Historical Block in Tongli Ancient Town Area, China

Tongli possesses a much larger area that includes multiple districts and can be considered as a well-maintained canal town with much obvious characteristics. The memorial ruins with inscribed Chinese words in photo 1 and the Chinese hanging cockloft in photo 7 bring additional impressions to the ancient towns as shown in Fig. 3. The Chinese style

Table 1. Space syntax analysis variables of ancient town area in Tongli, China

Name of the roads	Connectivity value	Control value	Mean Depth value	Local integration value	Global integration value
Xiaohe Street	5	1.726	205.074	72.579	217.874
Tongli Street	5	1.726	287.469	65.291	202.908
Huanhe Street	5	1.726	336.969	82.764	204.372
Xiaochuan Street	14	4.048	185.485	76.909	217.693
Hexi Street	5	1.726	282.574	50.960	167.286
Beitu Road	5	1.726	424.733	41.695	167.219
Mingqing Road	5	1.726	225.023	48.995	167.422
Mingtu Road	5	1.726	330.902	45.720	167.452
Yuxi Street	6	2.000	191.500	57.572	170.960
Hechi Street	4	1.250	309.375	51.316	170.129
Ziru Street	6	2.000	193.918	54.709	171.202
Huaxin Road	6	2.000	237.786	60.870	217.610
Xinhua Road	5	1.726	338.135	92.775	217.054
Huayu Road	5	1.726	317.102	51.316	216.472
Zhuxing Street	4	1.250	258.402	74.771	203.111
Zhuai Street	4	1.250	259.172	54.639	213.044
Xinzhen Street	5	1.726	328.082	82.982	203.097
Yanyu Street	8	2.500	445.491	82.816	199.226
Songbei Road	3	0.841	476.020	58.588	162.517
Shenxing Street	4	1.250	498.820	50.682	142.702
Average value	5.54	1.812	315.676	64.781	188.742

wooden casements are seen in every photo. The image of Tongli is quite impressive as a successful activated example of an ancient town cultivating a healthy business profile with strong tourist visitation.

Focusing within the main tourist routes of the Mingqing historic block, both Hechi street and Mingqing road were selected to define a representative space of the ancient quarter of the town. Figure 3 displays the VGA results of the Mingqing historical block (NB the water areas of the canal are not included on the visible layer).The analysis results are from the perspective of holistic protection of the entire region, guided by the theory

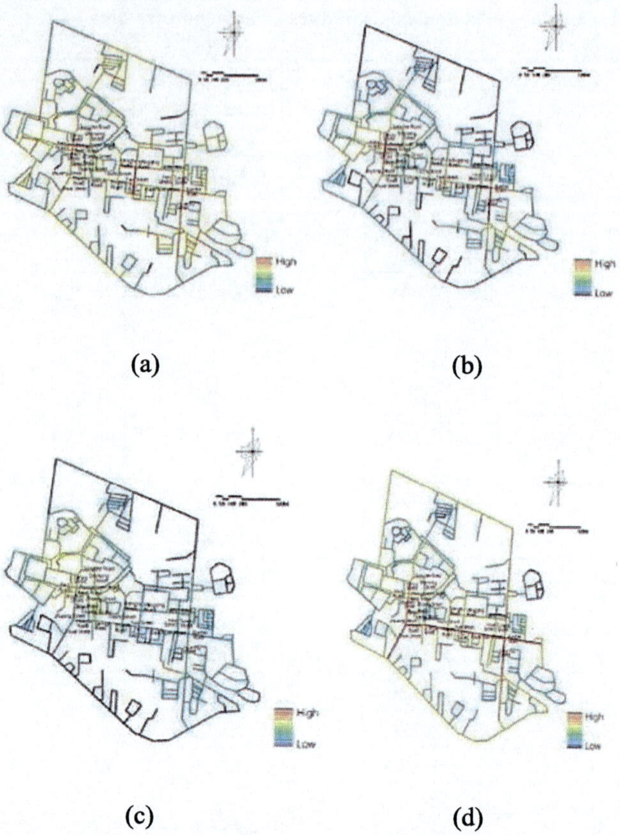

Fig. 1. The space syntax model analysis result on Tongli town: (a) Connectivity value, (b) Local integration value, (c) Mean depth value and (d) Global integration value

of cultural landscape protection, and using methods which combines environmental behavior, restoration ecology, dynamic integrity, and scientific sustainability to protect the cultural heritage of Tongli canal town. By sorting out the heritage value analysis, combined with an analysis of the current situation and problem summary of the canal cultural heritage, using a comprehensive perspective to construct the basic theory, and studying mature cases of canal renewal and conservation, in order to obtain principles and strategies suitable for the protection and optimization of Tongli canal's cultural heritage, and put them into practice.

As demonstrated in Fig. 4, the main space is distributed in the open space at the bridgehead and forms a high-intensity concentrated area. In addition, evidence has shown that streets with high syntactical accessibility attract a higher density of movements in urban areas. Additionally, this extends to the street space on the northern side, while the eastern and western sides are relatively weak. When the pedestrian flow is released in the northern side of the Mingqing historical block, the main space is distributed in the courtyard and is divided by buildings to form a triangular ring structure, which extends

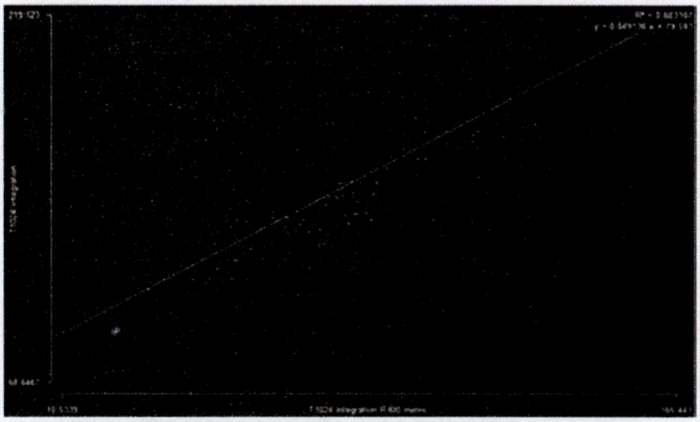

Fig. 2. The space syntax model analysis result on Tongli: Intelligibility value

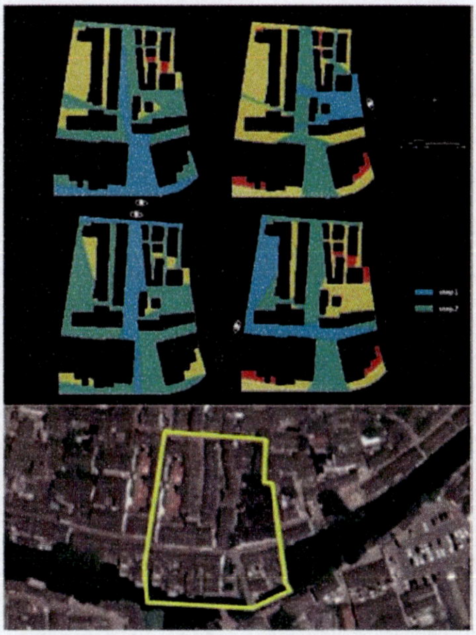

Fig. 3. Visibility Graph Analysis (VGA) of the highlighted block

to the northern and southern street spaces, but has stronger connection with the north side space. Figure 4 presents the analysis results on the VGA of the Mingqing historical block.

When the pedestrian flow is released into the northern of the Mingqing historical block, the main space is distributed in the longitudinal street space, and forms a high-intensity concentrated area in the center of the street. At the same time, the flow of people on the western side of the relatively closed courtyard decreases. When the pedestrian

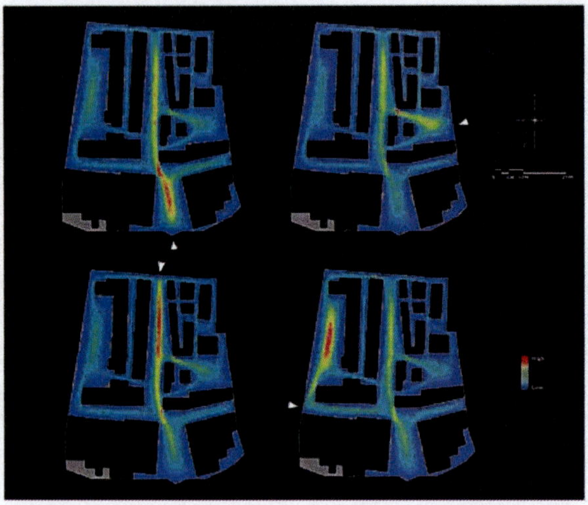

Fig. 4. Pedestrian flow simulation on VGA of Mingqing historical block in Fig. 2

flow is released into the western side of the Mingqing historical block, the main space is distributed in the open space on western side, and forms an obvious bright concentrated area. However, due to the blocking of buildings, the extension to the interior is relatively weak. Figure 5 presents the analysis results of Mingqing historical block.

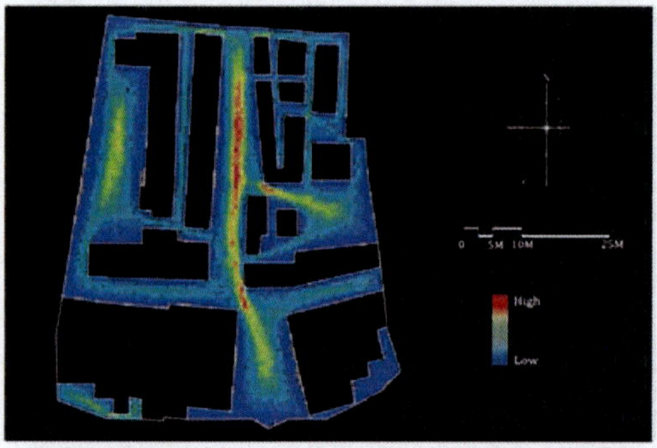

Fig. 5. Integration of pedestrian flow simulation on VGA of Mingqing historical block

As shown in Fig. 5, the human activity tracks are concentrated in the main street space in the north and the south. At the same time, the pedestrian flow in the open space and courtyard space on both sides is more obvious. The hydrophilic area on the southern side is blocked by the intersection of buildings and river banks. The main stream of people is distributed in the north-south area of the Mingqing historical block, but the pedestrian flow from different spaces forms obvious spatial distribution centers in different areas. The building directly intersecting the southern side and the river bank should be properly opened, and the outdoor space should be further integrated to realize an efficient flow of people. This could integrate the scattered buildings in the east side of the Mingqing historical block, break the closed and long continuation of the street, and introduce pedestrian flow.

4.4 Suggestions for Tongli in China and Future Research

Evaluating the spatial cognition which affected by the overlapping of various elements from the perspective of spatial heritage in Grand canal town. It is possible to systematically and clearly recognize the relationships and interactions between different elements, formulate relevant protection policies for sustainable townscapes according to local conditions and circumstances, and carry out more targeted heritage conservation practices [16].

This paper uses a computational analysis method to evaluating experiential qualities of historical streets in sustainable townscapes. The research results show that from a global perspective, spatial quality has a certain explanatory power on the distribution of spatial vitality. From a local perspective, the impact of different dimensions of space quality on space vitality showing a more complex relationship in urban space. The difference in the impact of spatial quality on spatial vitality in different types of spaces is mainly related to location, and the improvement of spatial quality optimization on spatial vitality requires certain material prerequisites. Therefore, in actual space renovation work, it is necessary to determine whether quality optimization can be adopted to enhance vitality based on the existing material conditions of the space.

5 Conclusions

The study investigated the extent of canal town. The purpose of the research is to help in evaluating the relevance experiential qualities of historical streets in sustainable townscapes. The research results show that from a global perspective, spatial quality has a certain explanatory power on the distribution of spatial vitality. From a local perspective, the impact of different dimensions of space quality on space vitality showing a more complex relationship in urban space. The difference in the impact of spatial quality on spatial vitality in different types of spaces is mainly related to location, and the improvement of spatial quality optimization on spatial vitality requires certain material prerequisites. Therefore, in actual space renovation work, it is necessary to determine whether quality optimization can be adopted to enhance vitality based on the existing material conditions of the space. Strengthening the interaction between characteristic resources and the public is a feasible approach to solve the spatial characteristic crisis of canal town resources. To determine characteristics is a common thinking among planners and architects in the past when dealing with the characteristics of the canal town attributing the crisis of characteristics to the lack of characteristic resources. The focus of guiding the spatial characteristics of the canal town is on exploring and creating characteristics. However, according to the evaluation of the spatial characteristics of the canal town, some excellent characteristic resource endowments have not been transformed into characteristics in the public's minds. It can be seen that "resources are not the determining characteristics", and are also influenced by factors such as identifiability, aesthetic attributes, meaningfulness, vitality, and activity support. In the guidance and control of urban spatial characteristics, it is necessary to deeply explore and reasonably interpret characteristic resources. However, in the case of failure of resource oriented planning and design strategies, fully utilizing existing resources and strengthening their interaction with the public is a new approach. The above mentioned factors in conjunction with computational design analysis tools such as space syntax and Depthmap. The actual application of evaluating the relevance experiential qualities of historical streets could in practice include the local residents and tourists as participants, and a survey could be conducted. For example, a set of serial vision images could be arranged in a survey and participants would be asked to select an image or sequence of images based on subjective preferences according to a favourability score, and the results could be compared with the space syntax anlysis result. This study of Tongli in China will be expanded to consider the towns of Nanyang, Wuzhen and Nanxun. In addition, Further research will also include the other factors which affect tourist experience.

Acknowledgement. The work was supported in part by Shandong Provincial Natural Science Foundation project ZR2021QE207 and Shandong Art Science Key Project (No.23QB03280021).

References

1. Tianchen, D., Xing, Z., Juan, Y.: Contradictory or aligned? The nexus between authenticity in heritage conservation and heritage tourism, and its impact on satisfaction. Habitat Int. **107**, 102307 (2021). https://doi.org/10.1016/j.habitatint.2020.102307
2. Montello, D.R.: The contribution of space syntax to a comprehensive theory of environmental psychology. In: Proceedings of the 6th International Space Syntax Symposium, Istanbul. Accessed (2007)
3. Griffiths, S., Vaughan, L.: Mapping spatial cultures: contributions of space syntax to research in the urban history of the nineteenth-century city. Urban History **47**, 488–511 (2020)
4. Penn, A.: Spaces syntax and spatial cognition: or why the axial line? In: 3rd International Space Syntax Symposium (2001)
5. ICOMOS China, Principles for the conservation of heritage sites in China (2015)
6. Hillier, B., Hanson, J.: The Social Logic of Space. Cambridge University Press, Cambridge (1984)
7. Mishra, S.A., Pandit, R.K.: Space syntax approach for analyzing crime preventive urban design: concept review. J. Adv. Res. Construct. Urban Archit. **1**(1&2), 30–35 (2016)
8. Xu, Y., Rollo, J., Jones, D.S., Esteban, Y., Tong, H., Mu, Q.: Towards sustainable heritage tourism: a space syntax-based analysis method to improve tourists' spatial cognition in Chinese historic districts. Buildings **10**(2), 1–18 (2020)
9. Lynch, K.: The image of the city, vol. 11. MIT press, Cambridge (1960)
10. Rollo, J., Barker, S.: Perceptions of place-evaluating experiential qualities of streetscapes. In: Proceedings of the 2013 6th State of Australian Cities Conference, SOAC 2013, SOAC, pp. 1–11 (2013)
11. Cullen, G.: The Concise Townscape. The Architectural Press, New York (1961)
12. Xu, Y., Rollo, J., Esteban, Y.: Evaluating experiential qualities of historical streets in nanxun canal town through a space syntax approach. Buildings **11**(544) (2021)
13. Hillier, B.: Studying cities to learn about minds: some possible implications of space syntax for spatial cognition. Environ. Plann. B. Plann. Des. **39**(1), 12–32 (2012)

14. den Hartog, H., Martanez, P.G.: Integrating heritage assets in large commercial complexes: de-contextualization and re-signification of memory in Shanghai. Habitat Int. **126**(2022), 102601 (2022). https://doi.org/10.1016/j.habitatint.102601
15. Hillier, B.: Space is the machine: a configurational theory of architecture (2007)

Research on the Construction System of Traditional Wooden Dwellings of Miao Nationality in Southeast Guizhou Based on BIM Technology

Shuang Liu, Lei Gong(✉), Li Zhou, and Chengyang Liu

College of Architecture and Urban Planning, Guizhou University, Guiyang 550025, China
gdls0531@163.com, 286009364@qq.com

Abstract. Traditional wooden dwellings of Miao nationality in southeast Guizhou show unique regional and cultural characteristics, and their main frame form is the Chuan-Dou style wooden frames. In order to analyze the component composition and construction characteristics of the traditional wooden frame system, the basic component system and three-dimensional information model were established by means of accurate design, three-dimensional visualization and quick and efficient means of BIM technology to summarize its characteristics, so as to provide strong support and guarantee for the construction, protection and inheritance of traditional dwellings.

Keywords: BIM · Traditional dwellings · Components · Wooden structure · Dwellings construction

1 Introduction

Due to China's ongoing urbanization process, traditional Chinese dwellings face many problems and challenges, such as rough construction techniques, the loss of artistic expression, and functional deficiency, while traditional dwellings have a long history and rich cultural connotation. Therefore, it is advisable to conduct in-depth research on traditional dwellings, including the sorting out of construction techniques and artistic expressions, to achieve the purpose of effective protection.

As of October 2022, the total number of Chinese traditional villages included in the list reached 8171, with a total of 757 villages distributed in Guizhou Province, and 415 villages in southeastern Guizhou, accounting for 54.8% of the total number of traditional villages in Guizhou Province. Among them, Miao villages have the highest number, with a total of 216 villages. These Miao villages have a large number of traditional wooden dwellings with ethnic and local characteristics, and are an important source of information for the study of traditional wooden dwellings. At present, in the study of Miao traditional wooden dwellings, many scholars such as Li Xiankui [1], Li Zhe [2] and Wu Zhengguang [3] have explored the expression of traditional dwelling architectural forms and styles in terms of regional and cultural characteristics. There are also a large

W. Guo et al. (Eds.): GBCESC 2023, LNCE 328, pp. 926–936, 2024.
https://doi.org/10.1007/978-981-99-9947-7_91

number of scholars who have conducted research in terms of its spatial combination [4], overall performance [5], and nodal performance [6]. As for the construction techniques, Wu Lin et al. [7] explored the oral history of ethnic artisans in Guizhou and summarized the operational methods and treatments applicable to the region from the perspective of engineering practices of building construction techniques. Zhang Xin [8] studied the settlement space, function, and cultural customs of the stilted building of the Miao nationality in southeast Guizhou.

In general, research on traditional wooden dwellings of the Miao nationality still focuses on the description of their architectural form, structure, and culture, lacking research on the entire lifecycle of their design, construction, operation, and maintenance. In recent years, some scholars have started to apply BIM technology in the field of digital construction and preservation of traditional dwellings [9, 10], which has injected new vitality into the field, but the breadth, depth and strength of its research are still lacking. Therefore, this article applies BIM technology to the study of traditional wooden dwellings of Miao nationality construction techniques in southeast Guizhou, providing a more scientific, systematic and feasible method for the inheritance and innovation of traditional dwelling construction techniques.

2 Characteristics and Construction Status

2.1 Characteristics of Traditional Dwellings

The traditional wooden dwellings of Miao nationality in southeastern Guizhou mainly use wood as the building material, and are built according to the unique terrain of the mountain. Their architectural forms are divided into two major grounding categories: flat connection type and suspension type (see Fig. 1). In terms of architectural form, it presents the characteristics of high roofs, towering foundations, and small windows. The roof form is mainly a sloping roof, which can effectively prevent the accumulation of rainwater.

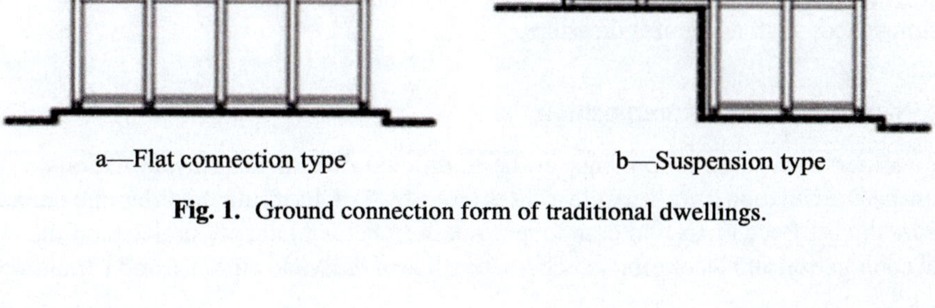

a—Flat connection type b—Suspension type

Fig. 1. Ground connection form of traditional dwellings.

2.2 Current Situation of Traditional Dwellings

In the construction of traditional dwellings of Miao nationality in southeast Guizhou, local craftsmen use the oral tradition to pass down their building skills. However, due to the lack of unified standards and systems, this inheritance method has led to the following prominent problems in the construction process of local dwellings.

① In terms of livability: Traditional dwellings in modern urban environment need to be designed and innovated to meet the needs of modern people's lives while retaining their unique architectural style and national characteristics.
② In terms of low-carbon environmental protection: The natural materials required for traditional dwellings construction have become scarce in the context of modern industrial production. It is necessary to explore new building materials and technologies to reduce waste of raw materials and environmental pollution, and achieve environmental protection.
③ In terms of construction technology: Traditional dwellings of Miao nationality construction technology relies on the experience of local carpenters, lacking scientific and theoretical basis. Therefore, production is relatively extensive, the construction steps are relatively simple, and a large amount of manpower and material resources are required. Although this method inherits ancient architectural techniques, its efficiency is relatively low and it is difficult to ensure the budget and control of construction costs.

Therefore, in response to the above issues, this article combines BIM technology to explore the construction system of traditional wooden dwellings of Miao nationality in southeastern Guizhou, seeking a new construction model for traditional wooden dwellings, in order to improve the efficiency and quality of architectural design, construction, and management, and reach the requirements of low-carbon environmental protection.

3 Component System of Traditional Wooden Dwelling based on BIM Technology

Most of the traditional wooden dwellings of Miao nationality in southeast Guizhou are the Chuan-Dou style wooden frames, whose structural feature is to vertically pass the wood through the rabbet to form a vertical Chuan-Dou structure. When bearing external loads, the loads are distributed to the whole structural system through the connectors and rabbets between the Chuan-Dou. This stress system can not only ensure the safety of the structure, but also effectively reduce the damage of the structure in extreme circumstances such as natural disasters.

3.1 Structure System Composition

The various parts and components in the traditional dwellings of Miao nationality in southeastern Guizhou have very clear titles (see Fig. 2). Clarifying the titles and connection methods of each part and component not only helps to deeply understand the cultural connotation and technological characteristics of the Miao ethnic group's traditional wooden buildings, but also helps to carry out BIM digital creation. After understanding the functions, dimensions (see Table 1), and connection methods of each component, determine the width and depth of each room in the residential area based on living needs (see Table 2).

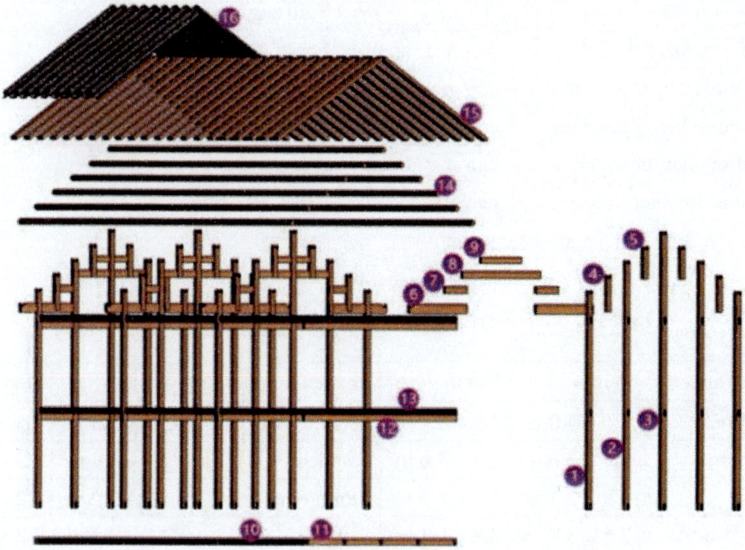

1. Side Column; 2. Second column; 3. Center column; 4. Upper short column;5. Lower short column; 6. Extension of bracket tiebeam; 7. Couple of bracket tiebeam; 8. Through three upper tiebeam; 9. Raised of bracket tiebeam; 10. Dijiao tiebeam; 11. Foot tiebeam; 12. Cross-row tiebeam; 13. Floor pillow; 14. Purlin; 15. Rafter; 16. Tile roof

Fig. 2. Typical names of various components in traditional Miao dwellings.

In traditional dwellings of Miao nationality, research has shown that the bucket rack has a prototype of modular design, and the basic module of the framework system is the step frame, which determines the size and changes of the house. One-step frame refers to the structure between each of two purlins in the framework, the commonly used step length is close to 1 m. The overall layout size of the building is determined by the number of adjustment steps for the distance between adjacent roof trusses. The spacing between the columns of the roof truss of the stilted building of the Miao nationality is between 1.6m and 2.4m. The spacing between short columns is between 0.6m and 0.8m, and its basic form is Five-columns four-melon or Five-columns four-melons with clamping columns. If the step frame is added, it can be changed to Five-columns and six-melons, while changing the number of columns can be Three-columns and four-melons, or Seven-columns and six-melons, with a maximum of Seven-columns and eight-melons (see Fig. 3).

The slope of the water surface of the roof truss of the traditional dwellings of Miao nationality in southeastern Guizhou ranges from 5 to 6 fen of water (i.e. 1:2.0 to 1:67),

Table 1. Common dimensions of components (diameter/height × width).

Category	Dimension	Category	Dimension
Column	≥ 12 cm	Floor pillow	About 3.6 cun
Melon-column	≥ 12 cm	Stopper	10 ~ 30 mm
Clamping columns	≥ 12 cm	Wall panel	7 fen
Raised of bracket tiebeam	4 cun × 1.3 cun	Foot tiebeam	5 cun × 1.5 cun
Through three upper tiebeam	4 cun × 1.4 cun	Purlin	≥ 12cm
Through three lower tiebeam	4 cun × 1.4 cun	Dijiao tiebeam	4.5 cun × 1.5 cun
Couple of bracket tiebeam	4 cun × 1.3 cun	Floor panel	6 fen
Extension of bracket tiebeam	5 cun × 1.5 cun	Cross-row tiebeam	6 cun × 1.7 cun

Note: 1 cun = 3.33 cm, 1 fen = 3.33 mm

Table 2. Plan dimensions of residential functional spaces.

Space for function	Width	Depth	Space for function	Width	Depth
Central room	3.6 – 4.0 m	3.9 – 5.1 m	Second bedroom	3.0 – 3.3 m	4.5 – 4.8 m
Dining room	2.7 – 3.6 m	2.7 – 3.6 m	Kitchen	2.1 – 2.4 m	3.3 – 3.6 m
Washroom	1.8 – 2.1 m	1.8 – 3.6 m	Storage room	1.2 – 3.0 m	1.2–3.0 m
Master bedroom	3.6 – 3.9 m	4.8 – 5.1 m	Stairwell	2.1 – 2.4 m	3.3 – 3.6 m

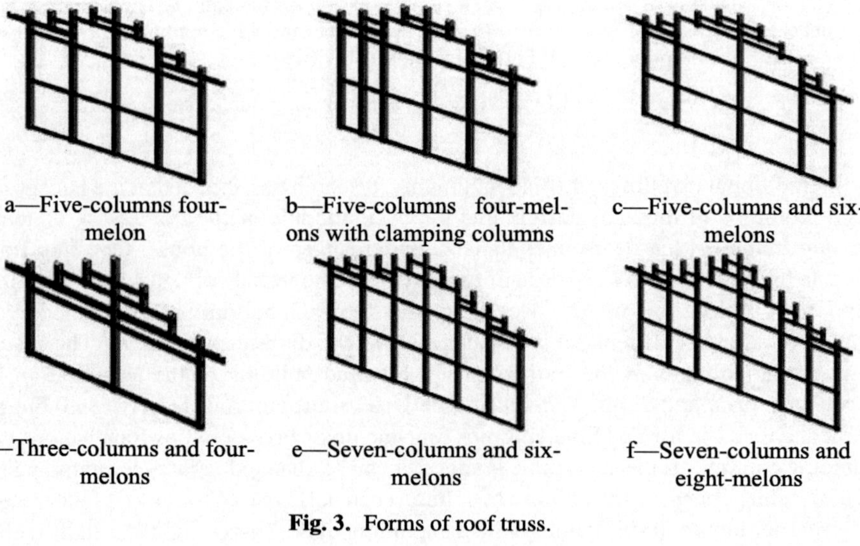

a—Five-columns four-melon

b—Five-columns four-melons with clamping columns

c—Five-columns and six-melons

d—Three-columns and four-melons

e—Seven-columns and six-melons

f—Seven-columns and eight-melons

Fig. 3. Forms of roof truss.

commonly known as 5 fen of water, with a slope of approximately 26.57°. The so-called several steps of water, is where the ratio of the height of the rack to the step length of each step frame, multiplied by 10 to obtain an integer. The treatment of slope is one of the important links in the construction of Miao wooden dwellings. If the slope is too flat or too curved, it can lead to poor drainage, and if it is too steep, the tiles will slide off. There are three specific methods for roof slope: one is the method of lifting or descending column. By raising or lowering a certain column, the conventional approach is to lower the second column by 4.5 cun or raise the eaves column by 3 cun to achieve a soft curvature and good drainage effect on the roof. The second method is the raising lifting and folding method, which involves first lifting and then folding, and treating each vertical load-bearing component such as the center column, second column, eaves column with a head and tail lifting and a middle lowering. As for the extent of folding down or up, craftsmen need to consider comprehensively based on experience, room depth, and building height. The third method is the Raising of the roof truss method, from the ridge purlin to the eaves purlin, the step water of each frame is evenly reduced, in order of 7, 6.5, 5, and 5.5. (see Fig. 4).

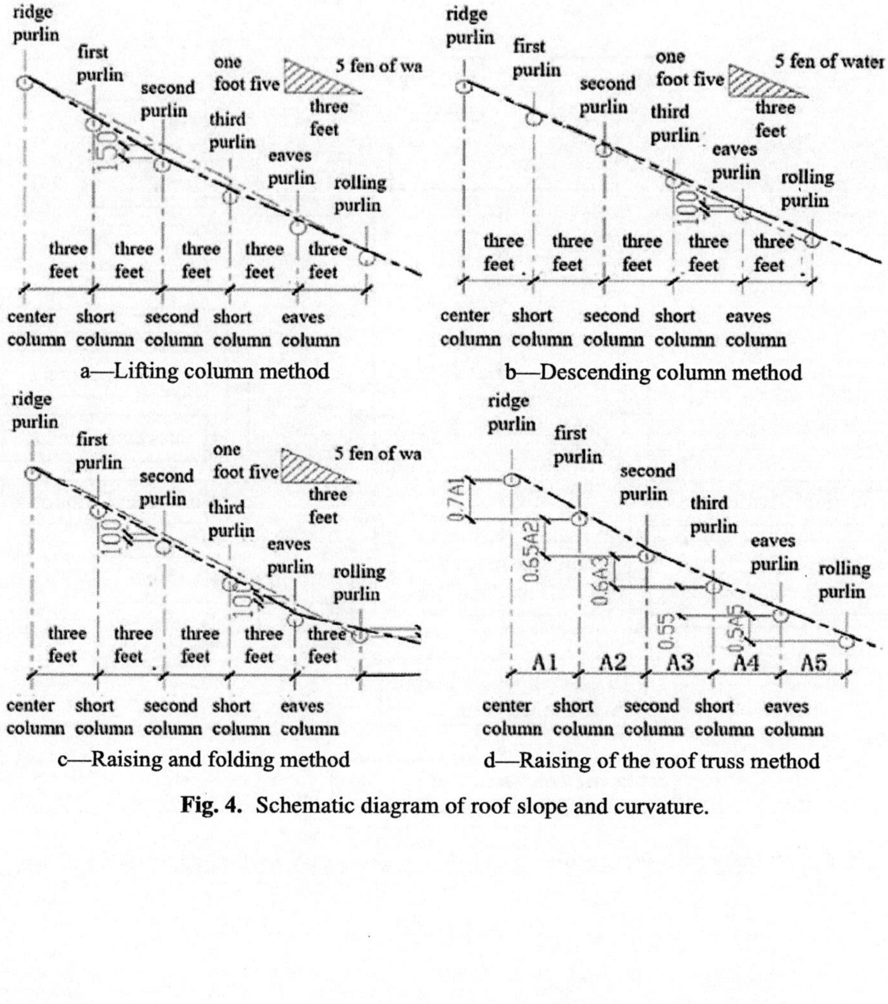

Fig. 4. Schematic diagram of roof slope and curvature.

3.2 Establishment of BIM Component Family Library

In order to build a building component library for traditional wooden dwellings of Miao nationality in southeastern Guizhou, relevant literature and materials should be searched, on-site research and mapping should be conducted, and local craftsmen should be visited. On this basis, the layout characteristics, commonly used wooden frame structural forms, and architectural elements of the traditional residential building are summarized. Based on this, the building components of the traditional wooden residential building are classified and summarized, and a corresponding building component library is constructed. (see Fig. 5).

In the process of building a traditional dwellings BIM component library, it is necessary to organize, classify, and archive the collected data, and establish data standards, classification systems, and naming conventions to ensure data consistency and maintainability. According to the characteristics of the BIM family library, it is divided into system families, loadable families, and in-place families. At the same time, it is necessary to use BIM software for 3D modeling, including components such as columns, beams, walls, roofs, doors and windows, stairs, etc. In the process of managing component libraries, it is necessary to establish a BIM component library management system,

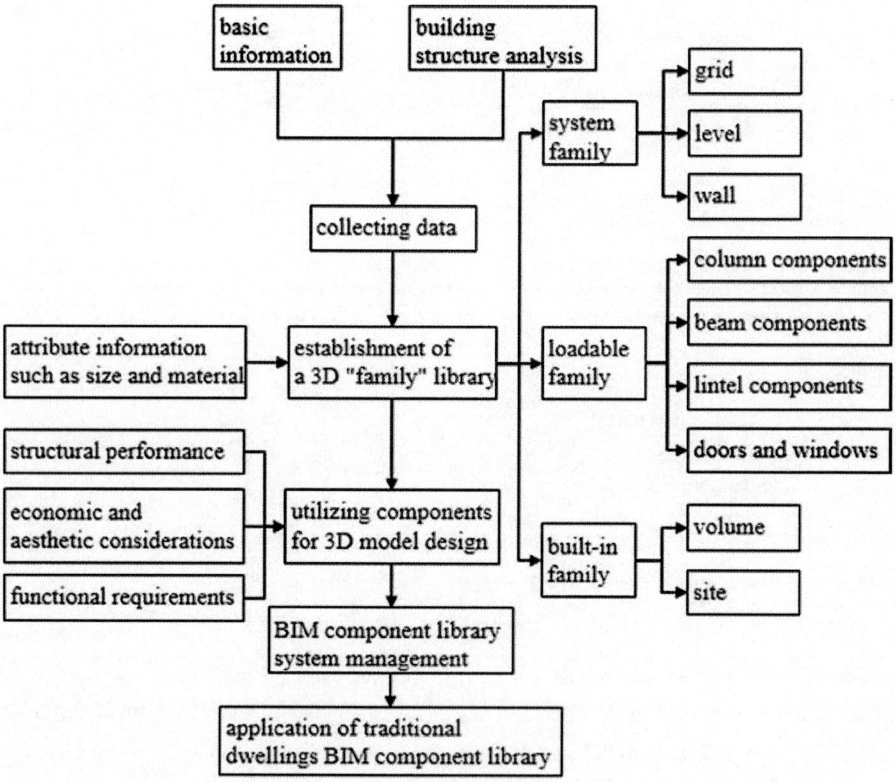

Fig. 5. The framework for constructing a BIM component library for traditional dwellings.

which includes information such as component identification, attributes, parameters. In the management system, components can be grouped, retrieved, modified, deleted, and replaced. Finally, by applying the traditional residential BIM component library, the entire process from design to construction can be digitized, improving the efficiency and quality of traditional residential design, construction, and maintenance, reducing construction risks and errors, accelerating construction progress and reducing costs, and also providing basic data and information for later maintenance and management.

4 Application of BIM Component Library

Before building a three-dimensional model of traditional wooden dwellings of Miao nationality in southeastern Guizhou, it is necessary to choose a suitable BIM software. At present, there are five mainstream BIM software, and their functional comparison is shown in Table 3. When choosing BIM software, it is necessary to consider various factors such as one's own needs, software functionality, software usability, and software compatibility. Revit software is currently one of the most popular and widely used BIM software in the market. It has powerful modeling function and rich component family library, which can facilitate the construction and editing of architectural model. In addition, Revit also has multiple plugins and extensions to meet the needs of different industries. Therefore, this article chooses Revit software for the construction and design of traditional wooden dwellings.

Table 3. Types of BIM design software.

Software	Functional focus	Functional operation	Data center	Collaborative functionality
Revit	Architecture	Easier	Based on model	Stronger
ArchiCAD	Architectural	Easier	Based on model	Weaker
Vectorworks	Architecture	Easier	Based on database	Stronger
Tekla Structures	Steel Structure	Harder	Based on model	Stronger
Bentley	Infrastructure	Harder	Based on database	Stronger

4.1 Building Three-dimensional Model Construction

After using Revit software to establish a component family library for traditional dwellings, a three-dimensional information model of the building can be constructed (see Fig. 6). Firstly, select the appropriate project template in the software, create a new project, and determine the layout and floor height of traditional dwellings based on residential needs and relevant specifications. Then levels and grids need to be created, which can be associated with other building elements such as walls, floors, beams, and columns to determine the height and position of different levels in the building.

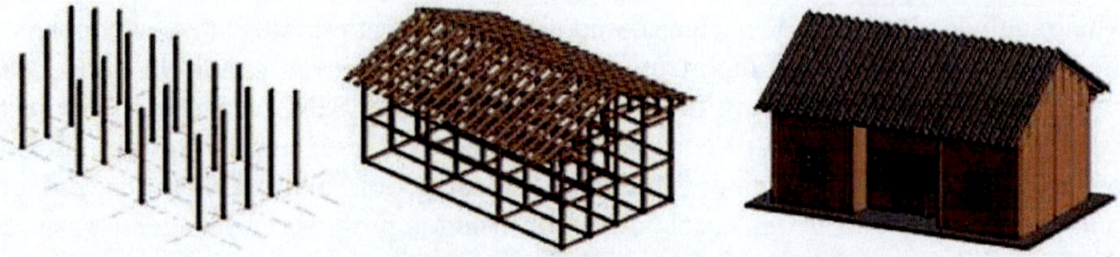

a—Column grid layout b—Horizontal element layout c—Enclosure and roof layout

Fig. 6. The construction of a 3D model for traditional dwellings buildings.

After creating levels and grids, import the previously established family library into the model, and arrange and arrange it according to the actual construction characteristics and architectural style to build an architectural model with traditional characteristics Insert a loadable family, model the main framework of the traditional dwellings, call the system family in Revit software to create a wall, place the loadable family and window family in the wall, and click the property command to modify their size, material, etc. Volume create the roof shape and load it into the project, using the metric profile - vertical stile and draw the tile profile. Load the outline into the project and find the curtain wall vertical stile - rectangular vertical stile in the browser. Create the curtain wall system, modify the settings in the type properties, and click Create System to generate the tile roof.

Finally, call the modify command in Revit to adjust the model, including adjusting component positions, dimensions, etc. Pay attention to the connection and connection between components to ensure the integrity and accuracy of the architectural model. At the same time, handle the details of the model, including doors and windows, decorations, etc. Then rendering and shading are performed to improve the quality and effect of the architectural model.

4.2 Building Information Management

The construction techniques of traditional wooden dwellings of the Miao nationality in southeastern Guizhou are mastered by craftsmen, and there is a lack of standardized drawings and specifications, which poses difficulties for the protection and inheritance of traditional dwellings. However, the emergence of BIM building information models can effectively solve this issue. By integrating and managing the design drawings, component information, and material information of traditional wooden dwellings, the accuracy and completeness of the design are ensured, while also facilitating subsequent construction and operation, ensuring the smooth progress of construction and quality control. Using software tools and technological means to digitize, visualize, interact, and collaborate various information in the process of building design, construction, and operation, achieving information sharing, transmission, and collaboration, is an efficient and sustainable building management model.

5 Conclusions

This article sorts out of construction techniques and artistic expression of traditional wooden dwellings of Miao nationality in southeastern Guizhou, analyzes the prominent problems in their construction status, and utilizes the advantages of BIM technology such as visualization, parameterization, and coordination to achieve the construction of traditional wooden dwellings, and the following are some summaries of this article.

① In terms of livability: according to human scale, relevant specifications and modern living needs, arrange reasonable space layout to achieve living comfort.
② In terms of low-carbon and environmental protection, precise design is achieved through BIM technology, and material lists and supporting construction drawings are exported to achieve standardized factory processing, achieving the goal of saving materials and shortening construction cycles.
③ In terms of construction technology: BIM technology can provide designers and owners with clearer information on building layout, material selection, and decorative effects through 3D modeling and simulation technology in the early stages of architectural design, optimizing design plans, and reducing errors and loopholes. At the same time, the BIM component library stores various component information and parameters of traditional dwellings, and designers can directly select these components to quickly build traditional dwellings, improving design efficiency.

Therefore, the significant advantages of BIM technology provide a digital platform for the creation of traditional wooden dwellings, promoting their protection and inheritance.

Acknowledgments. This research was funded by the Guizhou University Campus Level Cultivation Fund Project (Guida Survey [2022] No. 19).

References

1. Li, X.K.: Ganlan-Styled Residential Architecture of the Miao Ethnic Group. China Architecture & Building Press, Beijing (2005)
2. Li, Z., Liu, S., He, S.Y.: The research on evolution of the residential pattern of Miao ethnic group in Xiangxi. Archit. J. **S1**, 76–79 (2010)
3. Wu, Z.G.: Traditional village as museum: a case of upper Langde hamlet. China Cult. Herit. Sci. Res. **01**, 15–19 (2016)
4. Gao, P.: Study on the renovation of spatial combination model for traditional housing of Xijiang Miao village in Guizhou. Archit. J. **S2**, 66–72 (2014)
5. Wang, S.Y., Deng, H., Zhou, H.B.: Comparative study on calculation methods of lateral performance parameters for Chuan-Dou style wooden frames. Chin. J. Wood Sci. Technol. **04**, 63–69 (2022)
6. Wang, S.Y., Guo, T., Deng, H., Zhou, H.B.: Study on seismic performance of mortise-tenon reinforcement joints in Chuan-Dou style timber structure. Chin. J. Wood Sci. Technol. **01**, 36–41 (2021)
7. Wu, L., Tang, X.X., Peng, K.Q.: A study on oral history of craftsmen from a perspective of historical anthropology a case of traditional construction techniques of ethnicities in Guizhou. Archit. J. **01**, 79–85 (2020)

8. Zhang, X.: The Traditional Craftsmanship of the Miao's Pile Dwellings. Anhui Science & Technology Publishing House, Hefei (2013)
9. Du, S.H.: Research on the color protection of traditional dwellings based on BIM technology. Zhengzhou University (2021)
10. Sun, F.T., Xu, Z.J., Yang, Y.F.: Digital construction for Lingnan traditional residences based on BIM technology. S. Archit. (02), 111–114 (2021)

Analysis of the Impact Mechanism of Smart City Construction on Low Carbon Development Based on Multi Phase DID Evaluation

Wenyao Xiao(✉), Honggang Tang, and Lang Lyu

College of Architecture and Urban Planning, Guizhou University, Guiyang 550025, China
a15797868386@163.com

Abstract. With the continuous development of information technology and the continuous improvement of urban informatization application level, the construction of smart cities has emerged As a new model for the development of technology enabled cities in the digital age, smart cities play a crucial role in China's low-carbon transformation However, the current academic community mostly examines smart cities from a qualitative analysis perspective, while few scholars pay attention to the quantitative impact of smart city construction on low-carbon development, and there are even fewer studies exploring its impact mechanism. This paper uses the multi period double difference model to evaluate the implementation effect of three batches of smart city pilots on low-carbon development in China, and conducts mediating effect test and heterogeneity analysis. Smart city construction plays a very positive role in urban low-carbon construction; Smart cities promote low-carbon development of cities by enhancing their innovation capabilities and energy efficiency. In the middle and east cities and non - resource cities, the development of smart cities is more important to the low carbon development.

Keywords: smart city · carbon emissions · multi-phase double difference method (DID)

1 Introduction

Climate change is a common challenge for all mankind, and China regards addressing it as an important lever to promote ecological civilization construction and achieve high-quality development. As the world's largest energy producer and consumer, China has been committed to reducing carbon emissions. According to the 2022 Carbon Dioxide Emissions Report released by the IEA, China's carbon dioxide emissions in 2022 were 11.48 billion tons, with a per capita carbon emissions of 8.13 tons [1].

Cities are not only the major emitters of carbon dioxide in China, but also the main battlefield for energy conservation. In the context of the current "digital economy", what is the development pattern of the relationship between urbanization and carbon emissions? At present, "digital" technological innovation and "key" technological breakthroughs have provided new development paths and connotations for urbanization [2]. Among them, as the most representative product of the digital age, will smart cities,

W. Guo et al. (Eds.): GBCESC 2023, LNCE 328, pp. 937–948, 2024.
https://doi.org/10.1007/978-981-99-9947-7_92

empowered or enhanced by digital technology, have varying degrees of impact on China's carbon emissions? Through the research in this article, we can clarify the relationship between the development of smart cities in China and the ecological environment, quantitatively evaluate it, and further explore its impact mechanism, thus providing reference for the construction of smart cities and related research on low-carbon development in China.

2 Research Design

2.1 Model Settings

In this paper, the cause and effect of intelligent city building and the reduction of carbon emissions are investigated by means of multi-cycle DID. The mathematical model is shown in formula (1):

$$Y_{carbon,i,t} = \lambda + \lambda_0 X_{smartcity,i,t} + \lambda_1 X_{control,i,t} + \alpha_i + \beta_t + \varepsilon_{i,t} \quad (1)$$

In Eq. (1), $Y_{carbon,i,t}$ are the dependent variables, referring to the per capita carbon emissions; $X_{smartcity,i,t}$ are explanatory variables, referring to the construction of smart cities; I is the city, t is the year; If city i is a smart city pilot in year t, $X_{smartcity,i,t}$ is 1, otherwise it is 0; $X_{control,i,t}$ are the set of control variables; λ is a constant term, λ_0 and λ_1 is the coefficient of $X_{smartcity,i,t}$ and $X_{control,i,t}$, respectively; α_i is the fixed effect of the city; β_t is the fixed effect of the year; $\varepsilon_{i,t}$ is a random perturbation term.

Based on the above theoretical and mechanism analysis, this project refers to the "mediation test" method [3], and constructs an empirical model based on the above formula to empirically test whether smart cities can promote China's low-carbon development by improving the city's independent innovation ability and energy utilization efficiency. The regression model is shown in Eq. (2):

$$X_{mediator,i,t} = \theta + \theta_0 X_{smartcity,i,t} + \theta_1 X_{control,i,t} + \alpha_i + \beta_t + \varepsilon_{i,t} \quad (2)$$

In Eq. (2), $X_{mediator,i,t}$ are the mediating variables, referring to innovation capability and energy efficiency; θ is a constant term, θ_0 and θ_1 is the coefficient of the explanatory variable and the control variable, respectively.

The final regression model obtained after testing is shown in Eq. (3):

$$Y_{carbon,i,t} = \omega + \omega_0 X_{mediator,i,t} + \omega_1 X_{smartcity,i,t} + \omega_2 X_{control,i,t} + \alpha_i + \beta_t + \varepsilon_{i,t} \quad (3)$$

In Eq. (3): ω is a constant term, ω_0, ω_1 and ω_2 are the coefficients of the explanatory variable, intermediary variable, and control variable, respectively. In addition, the meanings of other variables are consistent with Eq. (1). Equation (3) is the regression data after adding intermediate variables.

2.2 Variable Settings

Dependent Variable (Ycarbon). Low carbon development (Y_{Carbon}) is based on per capita carbon emissions. Referring to the corresponding usage methods, measure the per capita carbon emissions of cities on a logarithmic scale. Firstly, based on the carbon emission data at the county level in China [4], the continuous dynamic distribution method was used to summarize the data from 2006 to 2019 and obtain the carbon emission data at the county level in China [5].

Core Explanatory Variables ($X_{smartcity}$). Since the start of smart city construction, there have been three groups of pilot cities. Among them, the first 90 cities (regions) were announced in January 2013; The second batch includes 103 pilot projects, which were released in August 2013; In 2014, a total of 93 provincial-level comprehensive experimental bases were officially launched nationwide, with 84 new experimental bases added, 13 expanded experimental bases, and 84 new experimental bases added. So far, there have been 290 smart city pilot projects nationwide. Due to the selection of certain districts and counties in prefecture level cities during the establishment of pilot projects, in order to prevent data errors caused by such pilot projects, such samples were excluded [6].This article selects 135 cities as the experimental group and 105 cities as the control group, with a total of 240 smart city pilot projects. Firstly, a dummy variable for a city was set, with the experimental group and control group using values of 1 and 0 as references, respectively. Next, the year dummy variable was set. After the implementation of the smart city pilot, it was set to 1, and before that, set it to 0.

Control Variable ($X_{control}$). The main contents of the control variables ($X_{control}$) include: economic growth level, population density, degree of openness to the outside world, green space rate, industrial structure, human capital, degree of government intervention, etc. Select the logarithm of per capita GDP (X_{GDP}), population density per square kilometer ($X_{populationdensity}$), proportion of FDI to GDP (X_{fdi}), green coverage rate in built-up areas (X_{green}), ratio of tertiary to secondary industries ($X_{industry}$), logarithm of per capita education years ($X_{education}$), and logarithm of local general budget expenditure to GDP ($X_{finance}$) as control variables.

Mediating Variable ($X_{mediator}$). The mediators ($X_{mediator}$) are the innovation capacity ($X_{innovation}$) and energy efficiency (X_{energy}) of the city, respectively. This study selected the urban innovation indicators from the "China Urban and Industrial Innovation Capacity Report" [7]. The smaller the logarithm of GDP/energy consumption ratio, the better the effect (Tables 1 and 2).

Table 1. Meaning of relevant variables and indicators (Source: Owner-draw)

variable	Indicator Name	symbol	Indicator Meaning
Dependent variable	Low carbon development level	Y_{carbon}	Log of per capita carbon emissions
Core explanatory variables	Smart City Construction	$X_{smartcity}$	Virtual variables (0, 1)
Mediating variable	innovation ability	$X_{innovation}$	Urban Innovation Index
	energy efficiency	X_{energy}	The logarithm of the ratio of GDP to energy consumption
control variable	Economic growth level	X_{gdp}	Per capita GDP logarithm
	Population density	$X_{populationdensity}$	Number of people per square kilometer
	Extent of openness to the outside world	X_{fdi}	FDI as a percentage of GDP
	Greening rate	X_{green}	Green coverage rate in built-up areas
	Upgrading of industrial structure	$X_{industry}$	The ratio of the tertiary industry to the secondary industry
	Human capital level	$X_{education}$	The logarithm of per capita education years
	Degree of government intervention	$X_{finance}$	The logarithm of local fiscal expenditure as a percentage of GDP

3 Quantitative Analysis and Inspection

3.1 Multiple DID Regression Results

First of all, we do the Hausman test, and we get a value of p that is 0, which means we use a fixed effect model in the panel data. Multi stage DID regression was conducted, introducing dual fixed effects of city and time based on detection, as shown in Table 3. Among them, Table (1) is the unadded control variable, while Table (2) is added, and the results show that the coefficient of smart city construction is significantly negative at the level of 1%. That is to say, smart cities can effectively promote low-carbon development, but the link with the variables is not obvious. According to the control variables in column (2), the economic growth coefficient is positive, that is, the CO_2 emissions per person will increase obviously. And did not cross the EKC curve inflection point [8], Indicating that at this stage, the majority of urban economies and carbon emissions remain tightly coupled, not decoupled. Observe the upgrading coefficient of industrial structure and find it is a positive number, which is likely due to some cities adopting

Table 2. Descriptive statistics of related variables (Source: Owner-draw)

variable	unit	Observations	mean value	standard deviation	minimum value	Maximum value
Y_{carbon}	ton/person	3120	1.7036	0.6714	−0.7536	4.2505
$X_{smartcity}$	–	3120	0.2426	0.4287	0	1
$X_{innovation}$	–	3120	10.807	53.7392	0.0051	1273.686
X_{energy}	Yuan/ton	3120	9.5042	1.1703	−0.7259	18.3363
X_{gdp}	Yuan/person	3120	10.309	0.8345	7.6697	13.1314
$X_{populationdensity}$	Person/square kilometer	3120	464.75	341.5589	10.02	2661.54
X_{fdi}	%	3120	2.0842	2.0843	0.0002	20.5654
X_{green}	%	3120	38.143	8.2152	0.36	95.25
$X_{industry}$	–	3120	0.8519	0.4372	0.1286	5.3401
$X_{education}$	year	3120	15.321	0.7624	12.7852	18.6112
$X_{finance}$	%	3120	2.6774	0.4454	0.82	5.0706

an extensive production structure, which greatly increases carbon emissions; The coefficient of government intervention is also significantly positive, which may be due to the unreasonable intervention of some city governments leading to a decrease in market resource allocation efficiency and an increase in carbon emissions. Therefore, these cities should further deepen policy system reform, optimize market resource allocation, and reduce the cost of carbon emissions, in order to achieve low-carbon development.

3.2 Mediating Effect Test

In order to further deepen the research in this paper, this paper needs to discuss how the construction of smart cities affects the path of low-carbon development. Specifically, it is to use the mediating effect to verify the hypothesis derived from the previous mechanism analysis: the low-carbon construction of smart cities can be achieved by using two paths, one is the innovation driven effect, the other is the energy upgrading effect, that is, to reduce carbon emissions by improving technological innovation and energy efficiency, Promote low-carbon development.

Innovation Ability. Analysis on the Intermediary Effect of Innovation Ability.

As shown in Table 4, Table (1) is not adding an intermediary variable and Table (2) is adding a variable. It can be seen that θ0 is significantly positive at the 1% level, ω_1 at the level of 1%, it is significantly negative, the mediating effect is significant, which means that innovation capability can effectively reduce urban carbon emissions. Because the symbols θ_0, ω_1 and ω_0 are the same, it indicates that in the process of promoting low-carbon development through smart city construction, urban innovation ability is to some extent a reasonable intermediary variable. The proportion of mediating effect

Table 3. Multiphase DID regression estimation (Source: Owner-draw)

variable	(1) Y_{carbon}	(2) Y_{carbon}
$X_{smartcity}$	−0.0469 (0.0211)	−0.0533 (0.0141)
X_{gdp}		0.7249 (0.0498)
$X_{populationdensity}$		−0.0001 (0.0001)
X_{fdi}		−0.005 (0.0265)
X_{green}		−0.0001 (0.0005)
$X_{industry}$		0.1027 (0.0265)
$X_{education}$		−0.0176 (0.024)
$X_{finance}$		0.2823 (0.0093)
Constant term	1.3419 (0.0077)	−5.935 (0.701)
Fixed time	yes	yes
Urban fixed	yes	yes
Observations	3120	3120
R^2	0.0656	0.6143

in total effect can be obtained $\theta_0\omega_1/\lambda_0 = 0.233$. The above viewpoint can be further demonstrated, that is, innovation ability plays a positive role in low-carbon construction.

Energy Efficiency. On the basis of analyzing the middle effect of improving energy efficiency, we draw the following conclusions. Table (1) shows the absence of the intermediate variable of energy efficiency, while Table (2) shows the inclusion of the intermediate variable of energy efficiency upgrading. It can be seen that θ_0 is positive, ω_1 is negative, but none is significant, therefore, the Bootstrap method is used to test the principle [9] hypothesis: $\theta_0\ \omega_1 = 0$, it is concluded that the confidence interval of direct effect and mediating effect does not contain 0, That is, the intermediary role is clear; at the level of 1%, ω_0 is negative, indicating that increasing energy quality has a greater impact on low-carbon effect; because the symbols θ_0, ω_1 and ω_0 are the same, it indicates that in the process of promoting low-carbon development through smart city construction, urban energy upgrading is also a reasonable intermediary variable to a certain extent. The proportion of mediating effect in total effect can be obtained $\theta_0\omega_1/\lambda_0 = 0.0041$. Furthermore, it can be proved that energy upgrading is an effective way to promote the development of low carbon cities (Tables 5 and 6).

Table 4. Mediating effect test of innovation driving effect (Source: Owner-draw)

variable	$X_{innovation}$	Y_{carbon}
$X_{smartcity}$	17.7383 (5.2224)	−0.0407 (0.0136)
$X_{innovation}$		−0.0007 (0.0001)
Constant term	−313.5591 (154.6704)	−6.1575 (0.675)
control variable	yes	yes
Fixed time	yes	yes
Urban fixed	yes	yes
Observations	3120	3120
R^2	0.1765	0.6158

Table 5. Mediating effect test of energy upgrading effect (Source: Owner-draw)

variable	(1) X_{energy}	(2) Y_{carbon}
$X_{smartcity}$	0.0111 (0.0547)	−0.0531 (0.0141)
Xenergy		−0.0197 (0.0126)
Constant term	5.5145 (1.6527)	−5.8262 (0.7004)
control variable	yes	yes
Fixed time	yes	yes
Urban fixed	yes	yes
Observations	3120	3120
R^2	0.1765	0.6158

3.3 Robustness Test

The conditions of applying the model are as follows: Before the policy is implemented, the difference between the test group and the control group is not significant. If the trends of carbon emissions of the two groups are not consistent before the implementation of the policy, then the results of regression analysis can not fully prove this point. In testing parallel trends, we used the event-study approach, formula (4).

$$Y_{carbon,i,t} = \eta_0 + \sum_{k \geq 4}^{4} \eta^k D_{i,t}^k + \nu X_{control,i,t} + \alpha_i + \beta_t + \varepsilon_{i,t} \tag{4}$$

In Eq. (4), $D_{i,t}^{k}$ is a dummy variable, representing the pilot of smart cities. Assuming that the year of implementing the smart city pilot in experimental group city i is mi, If $k = t - m_i \leq -4$, $D_{i,t}^{-4} = 1$, otherwise it is 0; If $k = -4, -3, \ldots, 4$, $D_{i,t}^{k} = 1$, otherwise it is 0; If $k \geq 4$, $D_{i,t}^{4} = 1$, otherwise it is 0. In order to avoid the problem of multicollinearity, this paper merges $k = -5, -6, -7$ to -4, and $k = 5$ to 4. As shown in Fig. 1, which is the result of a parallel trend test, the logarithmic differences in per capita carbon emissions between the two groups of cities were not significant prior to the implementation of the policy, i.e. the trends of the two groups were basically consistent and consistent with the parallel trend assumption.

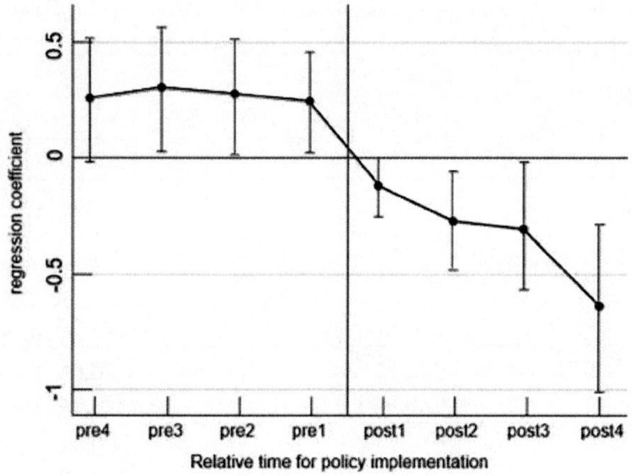

Fig. 1. Parallel trend test (Source: Owner-draw)

3.4 Placebo Test

To prevent other variables from having an impact on the results, we conducted a placebo test of the experiment [10], which randomly sampled 500 times in the experimental group and control group, and estimated the DID model multiple times to obtain the estimated values of 500 central explanatory variables. The result is shown in the figure, where the solid line is the kernel density distribution of the estimated coefficient obtained from the placebo test, and the dotted line is the actual estimated coefficient of this article - 0.0533. As can be seen from Fig. 2, most of the estimated values obtained from the placebo test are close to 0, and the conclusions obtained above are significantly different from those obtained from the placebo test. Through testing, it has been verified that the role of smart cities in promoting low-carbon development has not been affected by other uncertain factors.

3.5 Heterogeneity Analysis

Because of the great differences in economic development, innovation ability and traffic convenience, it is essential to analyse the differences and explore the effects of smart

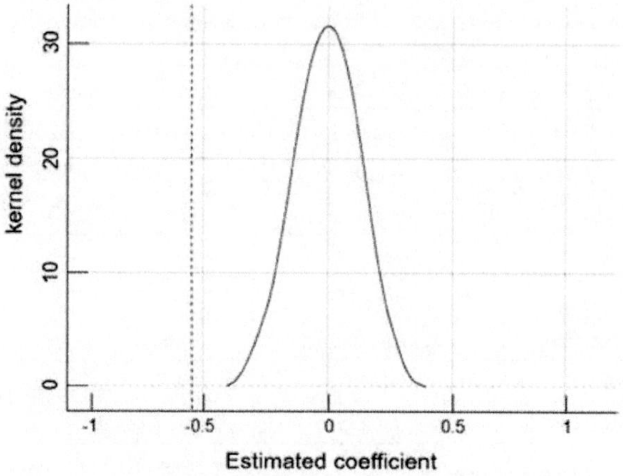

Fig. 2. Placebo test (Source: Owner-draw)

cities in different regions. Specifically, the 240 prefecture level cities in this article are divided into four regions based on their geographical locations: the eastern region (86 prefecture level cities), the central region (65 prefecture level cities), the western region (55 prefecture level cities), and the northeast region (34 prefecture level cities). And add control variables at fixed times and cities to conduct multi-stage regression analysis on DID. The results are as follows (3–4).The results show that both the east and the middle areas have a negative index, which shows that the development of intelligent cities has significantly reduced CO_2 emissions. Moreover, in the eastern region, the absolute value of the index of smart cities is larger, indicating that the low-carbon development effect of smart cities in the eastern region is better. However, in the western and northeastern areas, the "Intelligent City" index has not reached the remarkable level. This probably has something to do with the fact that the East, Middle, and West regions of China are developing more rapidly. Next-generation information technologies, such as artificial intelligence, digital twins, meta-universes, blockchain, have further developed in application and related facilities. The promotion and application of new energy and the research and development of carbon capture and storage technologies are also relatively more mature. Therefore the function is more obvious.

In addition, considering the differences in natural conditions among major cities in China, based on the "National Resource Based Urban Sustainable Development Plan (2013–2020)" [11], a sample of 240 prefecture level cities was divided into two categories. Resource-based cities and non-resource-based cities, and control variables were added for multiple period DID regression. The results are shown in Table 7. It can be seen that in both cities, the coefficient of smart city construction is significantly negative, but in resource-based cities, the correlation coefficient is even less significant. Perhaps compared to non resource-based cities, their industrial structure is more unreasonable, their dependence on natural resources is stronger, and their ecological damage is greater, resulting in less significant emission reduction.

Table 6. Analysis of Geographical Location Heterogeneity (Source: Owner-draw)

variable	(1) Y_{carbon}	(2) Y_{carbon}	(3) Y_{carbon}	(4) Y_{carbon}
$X_{smartcity}$	−0.0482 (0.0232)	−0.0413 (0.0242)	−0.0306 (0.0200)	0.0028 (0.0152)
Constant term	−3.5468 (1.4378)	−8.9600 (0.4273)	−2.7980 (1.3965)	1.1732 (0.6846)
control variable	yes	yes	yes	yes
Fixed time	yes	yes	yes	yes
Urban fixed	yes	yes	yes	yes
region	eastern	middle part	west	northeast
Observations	1118	845	715	442
R^2	0.6239	0.7135	0.2432	0.4285

Table 7. Analysis of Heterogeneity in Resource Endowment (Source: Owner-draw)

variable	(1) Y_{carbon}	(2) Y_{carbon}
$X_{smartcity}$	−0.0351 (0.0231)	−0.0600 (0.0184)
Constant term	−6.2600 (0.9734)	−6.2420 (1.0281)
control variable	yes	yes
Fixed time	yes	yes
Urban fixed	yes	yes
City category	Resource-based city	Non resource-based cities
Observations	1170	1950
R^2	0.6627	0.6527

The above conclusions are based on the theoretical and current situation analysis, supported by the panel data of 240 prefecture level cities in China from 2005 to 2019, and explore the relationship between smart city construction and low-carbon development by using the multi-phase DID model. On this basis, through mediation analysis, we verified the impact mechanism of smart cities on low-carbon development, and passed the robustness test of parallel trend test and placebo test. Finally, heterogeneity analysis was used to explore the implementation effects of smart city construction in carbon reduction for cities with different characteristics. The research results indicate that the construction of smart cities significantly reduces per capita carbon emissions and promotes low-carbon development of cities; Innovation capability and energy efficiency as mediating variables can effectively affect the low-carbon construction of smart cities; Cities in different regions and resource categories have different emission reduction effects.

4 Research Conclusion

Our country is seeking a new way of urban transformation and development, and the smart city is a feasible way. In order to scientifically evaluate the relationship between smart cities and low-carbon development in China at present, this paper regards the three pilot projects as “quasi-experiments” in terms of theory and practice, uses multi-stage dynamic difference (DID) model to measure the results, uses intermediary model to carry out empirical analysis, and combines the heterogeneity analysis of sample data to reach the following conclusions:

Firstly, the results show that the dual fixed effects of “city” and “time” are significantly negative regardless of whether there are control variables. Therefore, building a smart city plays a positive role in promoting low-carbon development.

Secondly, innovation capability and energy upgrading are two effective ways to promote low-carbon development. By improving urban innovation and energy upgrading, we can reduce urban carbon emissions and promote low-carbon urban development.

Thirdly, different types of cities, the efficiency of low-carbon development has great differences. Performance for the eastern and central cities better than the western and northeastern cities, non-resource-based cities better than resource-based cities.

5 Policy Recommendations

At present, the construction of smart cities has become the mainstream development mode of cities in China. It is recommended to adopt the following policies and measures:

Firstly, we should build a “smart city” with innovation as the driving force. Step up research and use of digital technologies, give full play to the important role of big data, cloud computing, the most advanced technologies such as the Internet of Things will be used to reduce carbon emissions in all aspects, so as to achieve meticulous, networked and smart management of the city’s carbon emissions.

Secondly, focus on optimizing the energy structure of smart cities. According to research results, optimizing energy structure is one of the important ways to achieve low-carbon development in smart cities. Therefore, it is necessary to optimize China’s urban energy structure with “carbon peak” and “carbon neutrality” as the core, and accelerate its green and low-carbon transformation, so as to achieve high-quality development of urban energy in China.

Thirdly, innovate policies based on the actual situation in various regions. According to the research results, the effect of Intelligent City on Low Carbon Development in West and Northeast China is not obvious. Therefore, we must fully understand the differences of different cities, and select the most appropriate development mode according to their developmental stages and requirements.

Fourthly, continue to push forward the development of Smart Cities and expand pilot projects. Through the summary of the first three pilot projects, we will step by step achieve the coverage of all Chinese cities, promote the development of all kinds of cities, and make management and services more intelligent, making our cities more flexible, efficient, and low-carbon.

References

1. Xu, Y.: Research on the Impact of Smart City Construction on Low Carbon Development. Zhejiang University of Science and Technology (2022)
2. Ge, L., Yu, J.: Smart city construction and urban carbon emissions: the test based on digital technology empowerment path. Sci. Technol. Progress Countermeasures **39**(23), 44–54 (2022)
3. Wen, Z., Ye, B.: Analyses of mediating effects: the development of methods and models. Progress Psychol. Sci. **22**(05), 731–745 (2014)
4. Chen, J., Gao, M., Cheng, S., et al.: County-level CO_2 emissions and sequestration in China during 1997–2017. Sci. Data **7**(1), 1–12 (2020)
5. Wu, J., Guo, Z.: Research on the convergence of carbon dioxide emissions in china: a continuous dynamic distribution approach. Stat. Res. **33**(01), 54–60 (2016)
6. Shi, D., Ding, H., Wei, P., et al.: Can smart city construction reduce environmental pollution. China Ind. Econ. **35**(6), 117–135 (2018)
7. Yu, F., Chen, J., Wang, X., Wang, Y.: Construction of China urban innovation index (CIIC) and analysis of urban industrial innovation capability. Innov. Technol. **20**(09), 1–9 (2020)
8. Zhang, B., Sun, Y.: Environmental governance of pollution emissions and high-quality economic development from the perspective of EKC. Mod. Enterprise (12), 141–142+151 (2022)
9. Ye, R., Anna. Bootstrap inference of the parameters of variance component function in the partial normal mixed effect model. Appl. Math. **36**(02), 353–363 (2023)
10. Jiang, X., Wang, L.: The upgrading effect of industrial structure of smart city policy—an empirical study based on time-varying DID. China Sci. Technol. Forum **12**, 31–40 (2021)
11. The "Action Guide" for Sustainable Development of Resource-based Cities - Interpretation of the "National Plan for Sustainable Development of Resource-based Cities (2013–2020). Land and Resources (01), 7–9 (2014)

Influence Factors Analysis of Aging Landscape Based on the Characteristics of the Elderly Human Body

Xiangyun Wang[1,2](✉), Alamah Misni[3], and Feng Wang[1]

[1] Shandong Youth University of Political Science, Jingshi Dong Lu 31699, Jinan, Shandong, China
wxiangyun1211@163.com

[2] Institute of Postgraduate Studies, College of Built Environment, Universiti Teknologi MARA (UiTM), 40450 Shah Alam, Selangor, Malaysia

[3] Centre of Studies for Landscape Architecture, College of Built Environment, Universiti Teknologi MARA (UiTM), 42300 Shah Alam, Selangor, Malaysia

Abstract. Globally, people are living longer. Today, most people may anticipate living into their 60s and beyond. The number of older people and their share of the population is increasing in every nation. With the continuous growth of age, the physiological function and psychological function characteristics of the elderly population continue to decline while they need an active lifestyle. Therefore, the demand for outdoor physical landscapes also changes. This paper aims to study and analyse the physiological and psychological function characteristics of the aged group. The characteristics of daily behaviours and activities of the elderly people and their actual needs for outdoor landscapes were obtained. The influence factors of the elderly population on the ageing outdoor landscape space are as early as in the design process of space planning, design layout, facilities, and process flow in the design and construction stages. The finding provides theoretical support and design guidance for the new construction to generate healthy, livable, ageing outdoor landscapes space.

Keywords: Design guidance · Elderly · Outdoor landscapes · Physiological and psychological characteristics · Theoretical support

1 Introduction

The older generation, who are growing in number in society, need the attention of space designers for them. The ageing problem is deepening, and the demand for the elderly to remain active is increasing, for example, for outdoor landscape use [1]. With the increasing age of the elderly group and the change of social role and social status, their physical and mental functions are constantly degraded, resulting in many physiological function and psychological function problems [2]. Outdoor landscape space in line with the actual user needs of the old people. In a certain length, it can guide and alleviate a series of problems caused by the decline of the physiological function and psychological

W. Guo et al. (Eds.): GBCESC 2023, LNCE 328, pp. 949–959, 2024.
https://doi.org/10.1007/978-981-99-9947-7_93

function of the elderly group. In the process of the design and construction of an outdoor landscape suitable for ageing, thorough study and analysis of the factors influencing the physiological function and psychological function changes of the aged group are crucial.

The study will focus on the significance and necessity of comprehensively analysing relevant literature and data. Firstly, the physiological and psychological characteristics of the elderly group were analysed, and the use demand of outdoor landscape suitable for aging is summarized. Based on the actual user needs of the elderly population, the relevant factors affecting the age-appropriate landscape design were analysed. Finally, the factors influencing the landscape design and construction suitable for ageing are summarised and analysed.

2 The Characteristics of the Aged Human Body

When people enter the ageing stage, their physiology, psychological function and behaviour will change correspondingly. In the ageing outdoor landscape design and construction process [3], it is necessary to thoroughly analyze and consider the influence of various characteristics and needs of the elderly groups on the outdoor landscape design elements [4]. When people enter the ageing stage, their physiology, psychological function and behaviour will change correspondingly. The designer needs to design and build a humanized outdoor landscape space that meets the actual user needs of the old people.

2.1 Analysis of Physiological Characteristics of the Elderly

With the increase of age, the sensory ability, motor system and nervous system gradually decline, the corresponding body structure system changes and the body gradually stopping. The decline of various physiological characteristics directly affects the user's demand for outdoor space to exercise actively, such as landscape gardens and spaces.

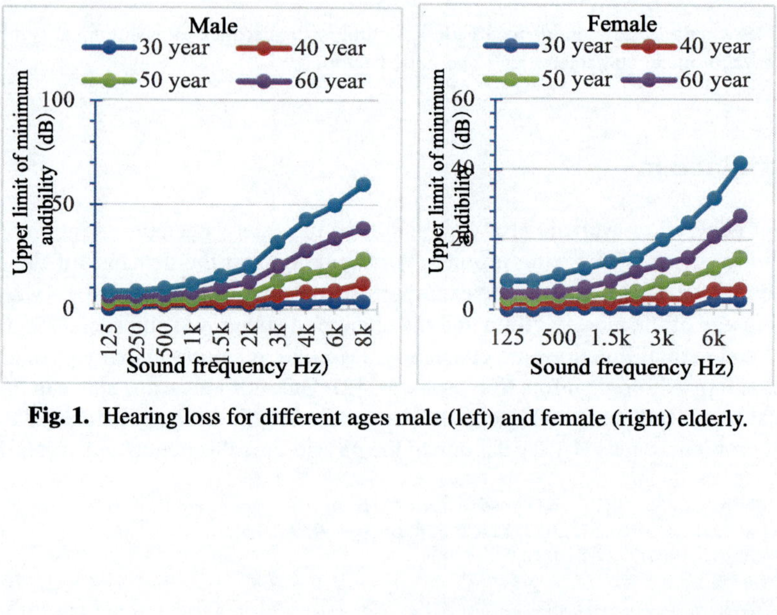

Fig. 1. Hearing loss for different ages male (left) and female (right) elderly.

The Decay of the Perceptual System. Five senses refer to the five aspects of human perception system, namely vision, hearing, taste, smell and touch [5]. When most people enter old age, their visual ability first declines, resulting in a sharp decline in environmental color recognition, color perception and color sensitivity. The ability to hear, smell, taste and other senses deteriorates (see Fig. 1) [6].

The Decline of the Motor System. Related medical research shows that in the process of ageing, older people's bones may become less elastic, less rigid and more brittle. At the same time, the amount of muscle in the body also decreases sharply with the increase of age, resulting in the decline of muscle strength in the elderly group, as the body's motor control is weakened (see Fig. 2) [6]. Due to the overall decline of motor function, the elderly group is easily injured during exercise, the recovery ability after an injury is weak, and the recovery cycle is long. In the design of park road, square, node, etc., we should fully consider the material skid resistance and appropriately reduce the ground height difference.

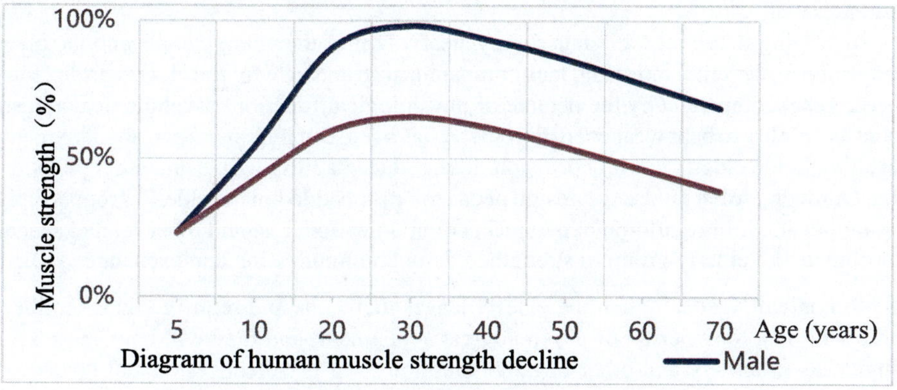

Fig. 2. Diagram of human muscle strength with age.

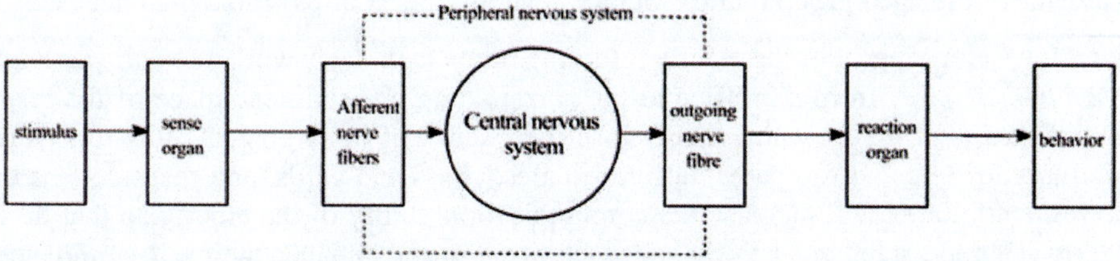

Fig. 3. The relationship between stimulus and behavior in central nervous system.

Nervous System Decline. The nervous system of old people gradually declines with age, leading to the weakening of the nervous system, which leads to the decline of respiratory, metabolic and other system functions [7]. There is a decline in memory, and a corresponding decline in behavioral responsiveness, cognitive dysfunction, comprehensive physical decline and dementia (see Fig. 3) [8]. In the process of outdoor landscape design suitable for aging, landscape space should be rationally planned according to the control factors such as the status quo of the elderly group's daily behavior and activity ability, activity characteristics and so on.

2.2 Psychological Characteristics of the Elderly

Psychological Characteristics. Due to the change of social status and social role, and the gradual decline of physical functions, the emotions such as loss, anxiety, inferiority, loneliness and attachment are the main characteristics of the psychological status of the elderly group. The sense of loss is an essential psychological feature of the elderly group. After leaving the job, out of the long-term working environment, the feeling that they cannot continue to play the value produce loneliness and depression. Loneliness is a common psychological feature of the elderly group [9]. After returning to the family from a busy job for a long time, the life circle shrinks sharply and the disposable time increases.

According to the relevant data survey, nearly half of the young people and the older people live in different cities and lack companionship and care for the elderly group. Furthermore, accompanied by the decline of physiological function, psychological adjustment and ability to bear weakened, the elderly has a sense of loneliness and abandonment, resulting in the elderly spirit depression, loss, reduced ability to live and other phenomena. Outdoor garden landscape design needs to be embedded in the elderly group's daily communication, horticultural experience, communication, rest and other related spaces to promote the elderly group to strengthen daily communication and exchanges [10].

Psychological Needs. When the elderly leaves their jobs and return to their families, there will be a long period of psychological adjustment, and they will have more sensitivity, loss, anxiety and other negative emotions. An excellent residential landscape design will help the elderly regain confidence, build a good pension mentality, and transition from psychological difficulties into a healthy ageing life. Based on the hierarchy of human needs pointed out in Maslow's hierarchy of needs theory [11], the psychological problems of the aged group mainly include several aspects of psychological needs.

Need for Belonging. For the elderly returning home from busy work, the daily activity space has changed from the office to the surrounding environment space of the residential area, and they will inevitably lose their sense of belonging. At this time, the familiar corridors and footpaths, leisure squares, chess and cards, and seating areas in the residential area will increase the communication ability of the elderly so that they elderly can understand and position themselves again and enhance their self-confidence in life. The design and construction of an excellent residential garden space environment for the elderly can accelerate their psychological adaptability and increase their sense of belonging.

Security Needs. With the improvement of age, the physical function, self-protection ability and psychological endurance of the elderly group are constantly degraded, and their awareness of security crisis is rapidly enhanced. Slippery garden paths, shuttle traffic, dimly lit garden space, unprotected waterfront acts of revenge, and other unsafe factors will increase the sense of security crisis of the elderly group [12]. In the process of design and construction, it is necessary to fully consider the safety design of the park road anti-skid, the separation of people and vehicles, lighting planning, barrier-free design and other aspects to improve the safety factor of the elderly people visiting the garden.

Demand for Neighbourhood Communication. The residential area is the primary interaction space for the elderly people in the home care group, restricted by the decline of physical function and other factors. As the activity area shrinks, the people in daily contact are mainly the elderly in the same residential area, and the neighbourhood relationship is significant. They are eager for communication and mutual respect between neighbours, so the proper planning and design of the interaction space in the residential garden landscape. Increasing leisure and entertainment facilities can help improve the daily communication of the elderly, enhance the sense of neighbourhood identity and harmony, and thus improve the quality of life of the home-based elderly care group.

Self-actualization Need. The sense of self-worth of home-based elderly care groups weakens after they leave their jobs, but they are still eager to win the affirmation and recognition of their children and the people around them through their efforts. They often participate in community painting, walking, chess and cards, photography, community service, traditional music, gardening and other activities to regain their life goals and show their self-worth [13]. In order to promote their physical and mental health to maintain a good mood.

2.3 Behaviour and Activity Characteristics of the Elderly

The Spatial Range of Behaviour and Activity of the Elderly. Affected by the decline of physical function and the need for daily activities, the research of Wang Jiangping and related scholars shows that the range of daily activities of the elderly group mainly includes three types of residential, residential and distant urban activities according to the spatial distance. The inter-house activity area is the main area for the daily activities of the elderly group, including a private courtyard, unit green space, etc. The travel distance is basically controlled within 200 areas in front of the house, and the access time is generally no more than 5 min. The second is the activity space of the residential area. Its radiation distance is within 500 m of the residential area, and the reachable time is basically controlled within 10 min. It belongs to the residential garden green landscape category and is the main space for the elderly group's daily communication and leisure activities as shown in Table 1 [14].

There are also urban activity Spaces outside the residential areas, including urban parks, theme squares, animal and botanical gardens and other large public activity areas, where the travel distance is more than 1000 m and needs to be reached employing transportation. Due to the limitations of physical functions and other conditions, the frequency of travel of the elderly is low, and most of them are concentrated in the healthy young elderly.

Table 1. Activity range of the elderly.

Area	Duration	Radius	Frequency	Range	Characteristic
Neighborhood	≤5 min	<200 m	2.3 times per day	Unit door, courtyard, green space between houses, group green space, etc.	High frequency
Region	≤10 min	≤500 m	1.5 times per day	Residential area surrounding, community landscape, street, supporting facilities	Moderate frequency
City area	20–45 min	>1000 m	0.8 times per day	City, city level parks, central squares, botanical gardens, museums, etc.	Low frequency

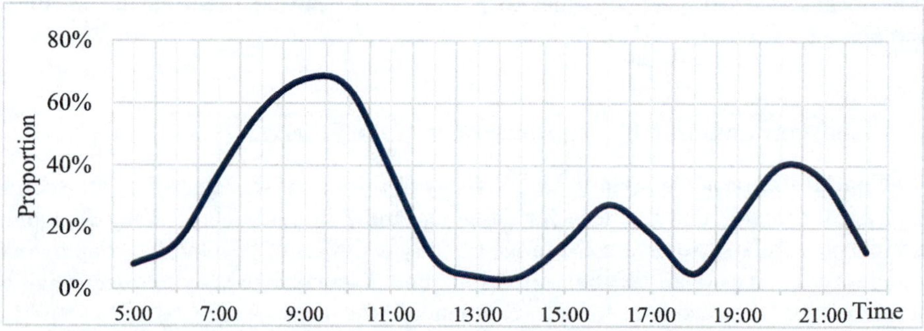

Fig. 4. Outdoor activity time pattern of the elderly.

The Behaviour and Activity Regularity of the Elderly According to the analysis of relevant scholars' research results, after the elderly leave their jobs, they will have more leisure time to develop their interests and hobbies, and the activity and travel time are relatively regular. Between 7:00 and 10:00 is the outdoor relaxation exercise time for most elderly people. 15:00–17:00 for the old people exchange leisure activities, such as chatting, sitting, chess, walking [15], etc. 19:00 to 21:00 is the bedtime fitness time for the elderly group, and the types of activities are generally group activities such as Leisure equipment fitness, walking and so on (see Fig. 4) [16]. Due to the large difference in climate between the north and the south, especially the large temperature difference between indoor and outdoor in winter in the north, the length and frequency of elderly people's trips at night are significantly reduced.

Behavioural Preferences and Types of the Elderly. With the decline of various body functions, most elderly people will increase, such as walking, Walking, jogging and other low-intensity equipment sports activities to keep the body healthy. At the same time, most elderly people lack the daily companionship of their families, so in order to relieve their loneliness, they prefer chess and cards, chatting, walking and other activities to relieve loneliness (see Fig. 5). Relaxing activities such as gardening, viewing, sitting and gardening are also important outdoor activities for the elderly. In addition, the type of outdoor activity choice of the elderly is also affected by their physical condition and gender.

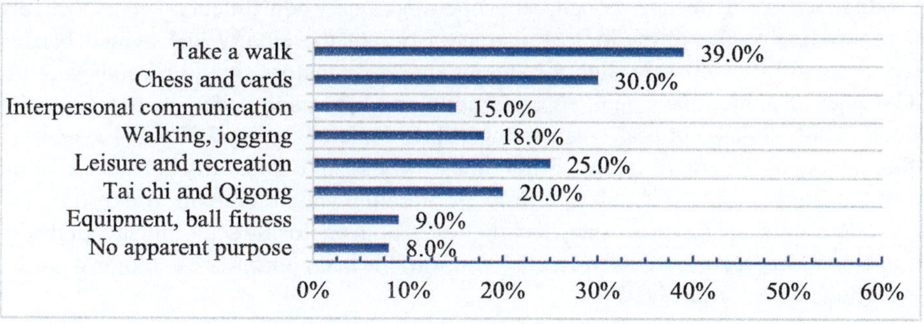

Fig. 5. Analysis of outdoor activity types of urban elderly.

Behavioral and Activity Characteristics of the Elderly. The activity characteristics of the aged group are mainly manifested in sociality and agglomeration. Regarding social activities, the groups, methods and purposes of communication are relatively fixed after the change of school colour and the decline of physical function of the elderly group. The circle of communication mainly includes family members, neighbours, old friends and like-minded people within contact distance [4]. Agglomeration is based on the demand level, which can be divided into functional and regional agglomeration.

Function mainly refers to the function of equipment facilities and site space, such as equipment fitness, fishing, gardening, chess and card activities, etc., based on the everyday use of the same activities of enthusiasts generated by the spontaneous gathering. Regional agglomeration mainly refers to spatial classification agglomeration, such as waterfront activity area, plant healing area, interaction activity area, private and semi-private retreat area and other activity categories. Due to the limitation of the outdoor activity ability of elderly groups, most outdoor activities of elderly groups are concentrated in residential areas and surrounding areas, which have spatial solid and regional characteristics.

3 The Influence of the Needs of the Elderly Behavior Activities on the Design and Layout of Garden Space

3.1 The Impact on the Overall Planning and Design Layout

Affected by the physiological function and psychological function and daily behavior needs of the elderly, the actual use needs of the elderly for landscape space should be fully considered in the overall planning and layout process of the residential age-appropriate landscape, and age-appropriate design should be carried out in the overall layout of landscape space, road traffic system, functional zoning and node design. According to the psychological needs of the elderly, the spatial layout of the garden is graded. The private space layout design of open, semi-open and lonely adjustment is set respectively.

According to the physiological characteristics of the elderly, the overall barrier-free, accessibility, anti-skid, lighting layout and other factors should be considered in the design of garden pavement, space lighting and landscape nodes. According to the actual health status, cultural level, living habits preference, psychological status and other factors of the elderly group, different user needs should be fully considered in the overall spatial layout. In the design process, attention should be paid to the static and dynamic zoning, spatial processing techniques, spatial enclosing scale and layout design of related infrastructure, so as to create a comfortable and appropriate ecological garden recreation space for the aged.

3.2 Influence on Landscape Space Design

With the overall deterioration of the body function of the elderly, in the process of garden space design, attention should be paid to the implantation of rehabilitation training elements for the elderly. For example, in garden planting, healing plants can contribute to the health of the elderly. Gorgeous colours, open spaces, quiet alone areas, etc., can enhance the sensory experience of the elderly and keep them mood happy. Relax (see Fig. 6). The shaping of landscape topography, landscape sketches, water body, plant planting and other aspects makes the overall space more suitable for recreation and daily life of the elderly group, bringing a better health experience.

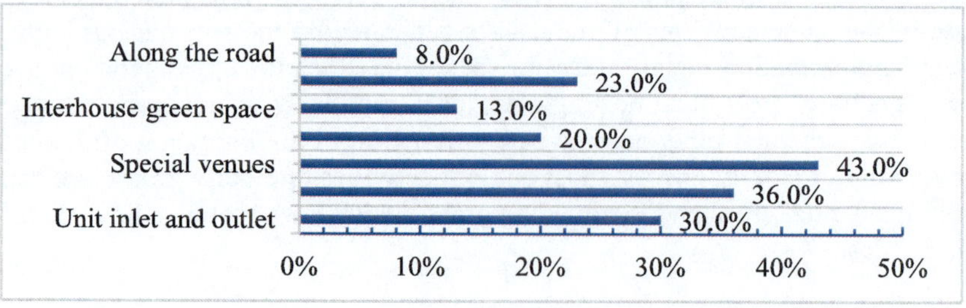

Fig. 6. Outdoor activity location distribution of the elderly in residential areas.

3.3 Influence on Construction Technology

Based on the current situation of the psychological and physiological scale of the elderly group, the layout of service facilities should be fully considered in the landscape planning and design [17], such as the material, scale and layout planning of garden seats, the material and form of the ground pavement of garden roads and squares, the ageing adaptability of recreational sports equipment facilities, the layout planning and ageing adaptability design of public toilets, and the realization of gardening and planting. Barrier-free access and the overall design of the signage system need to be designed according to the actual needs of the elderly population to enhance the overall applicability of the elderly community landscape space suitable for ageing Table 2.

Table 2. Elderly people's need for public facilities in outdoor landscape space.

Type of facilities	The main concern of the elderly	Basic requirements for facilities
Health facilities	Fitness equipment, pebble walkway, assembly square, ball square	Suitable for the elderly, strong safety, the material is warm in winter and cool in summer, smooth lines, warm colors
Recreational facilities	Chess and card seats, fishing platform, communication square, audio, exhibition platform	The space is relatively enclosed, safe, sanitary, beautiful scenery and good shade
Viewing facility	Rest viewing bench, pro - level platform, fountain, pavilion gallery flower frame, sculpture	Good shade, gentle texture, smooth lines, back and armrest
Auxiliary facilities	Restrooms, shops, signs, trash cans, community canteens, lighting	Sufficient quantity, convenient health, beautiful and practical

3.4 Behavior and Activity Characteristics of the Elderly

In the process of construction of an age-appropriate landscape, the use needs of the elderly should also be fully considered. The physical function of the elderly population is declining. The safety of landscape use is an essential consideration for age-appropriate landscape design, based on the fact that the proportion of sharp angles, steps, materials with low safety factors and vulnerable materials should be reduced as far as possible during the construction of garden space [18]. Moreover, reduce the overall slope fluctuation of the landscape space appropriately. In plant planting to achieve density and level of design, improve the overall use of safety factor.

4 Conclusions

Based on the changes in physiological function and psychological function and other characteristics of the elderly population [19], this paper analyzes the daily activity rules of the elderly population and the actual use needs of the landscape and studies the influencing factors in the overall planning and design of the outdoor landscape suitable for ageing, spatial node layout, service support facilities and construction technology. In planning and designing outdoor landscapes suitable for ageing, it is necessary to consider the actual use needs of the elderly groups, such as physical exercise, communication and daily activities. In the overall landscape planning and layout, the design and planning of related factors such as dynamic and dynamic zoning design of space types, landscape ecological health, separation of people and vehicles in park design and optimization of pavement materials, barrier-free design of various facilities and emergency rescue were carried out [20].

In landscape spatial layout, attention should be paid to the rehabilitation training elements for the elderly and the implantation of healing plants, and the shaping of landscape topography, landscape sketches, water bodies, plant planting and other aspects should fully consider the physiological characteristics of the elderly population. In terms of the layout of service facilities, it is necessary to optimize the overall design of recreational facilities, entertainment facilities, ornamental facilities and corresponding auxiliary facilities, such as the material, scale and layout planning of garden seats and the optimization design of the material and form of the ground pavement of garden roads and squares. In construction, safety is an essential consideration for the ageing landscape design. The use of acute angles, steps, and materials with low safety factors and vulnerable materials should be reduced as far as possible to improve the overall use of safety factors. Based on the above-influencing factors of the age-appropriate outdoor landscape, a comfortable, safe, appropriate scale and ecologically healthy outdoor landscape space for the aged is designed and constructed.

References

1. Dongqing, Z.: Evaluation of elderly people's satisfaction with outdoor activity space in Residential areas of Changsha City. Hunan University. Changsha (2019)
2. Lin Huaqiu, L.: Organic integration design of interior space and furnishings. Build. Struct. (07), 153–155 (2021)
3. Quanping, Q., Liuyi, Z.: Discussion on the landscape design method of community park focusing on the elderly: a case study of Huanghai Road Park in Laiyang City, Shandong Province. Design (01), 34–36 (2018)
4. Yuting, X.: Research on Aging Landscape Design of outdoor Activity Space in old communities in Nanchang City. Jiangxi University of Finance and Economics, Nanchang (2020)
5. Yunxin, O.Y.: Research on Interactive landscape design of apartment for the elderly. Soochow University, Suzhou (2017)
6. Shizhang, Y.: Urban Residential Environment Design. Chongqing University Press, Chongqing (1999)
7. Shuang, L.: Study on suitability of residential outdoor space for aging based on Structural Equation Model. Hebei University of Technology, Handan (2021)
8. Jiangping, W.: Planning and Design of Residential Environment for the Elderly. China Electric Power Press, Beijing (2009)
9. Yingxia, L.: Team work to trouble the old self-efficacy of ascension practice research. South China University of Technology, Guangzhou (2020)

10. Ruonan, R.: A Study on Absent Interaction Behavior of elderly people in the Age of Social Network. Henan University, Zhengzhou (2019)
11. Jiexia, L., Bo, Y., Zhilin, Z.: Research on design of outdoor facilities for elderly care in community based on art healing. Design **35**(07), 47–49 (2022)
12. Hong, L.: Living conditions of elderly population in big cities from the perspective of social stratification. Sun Yat-sen University Press, Guangzhou (2010)
13. Hanxue, S.: Research on Outdoor Interactive Landscape Design of nursing institutions. Wuhan Textile University, Wuhan (2020)
14. Chenguang, S.: Research Report on the Status quo and development Countermeasures of vigorous walking of the elderly in Shanghai. Shanghai University of Physical Education, Shanghai (2020)
15. Jun, Z.: Research on outdoor landscape design of suburban nursing home based on urban horticulture. Hunan Agricultural University, Changsha (2019)
16. Yudai, Q.: Architectural Design drawings for the Elderly. Science and Technology Press, Jinan (2004)
17. Haofeng, D.: Research on the Renovation of Outdoor Public Space of Elderly care Institutions under the Concept of Aging. Guangxi Normal University, Guilin (2021)
18. Yajun, C.: Research on the humanized design of park activity Space based on the behavioral characteristics of the elderly: a case study of Suzhou Park, City. **183**(06), 62–65 (2015)
19. Jingbo, Z.: Study on Outdoor space Adaptability of elderly Community. Beijing Forestry University, Beijing (2006)
20. Xian, L.: Research on outdoor leisure space environment of the elderly in Wuhan City. Wuhan University, Wuhan (2004)

Research on the Causes and Treatment Methods of Urban Flood Points

Sicheng Wang and Ruili Chang(✉)

College of Architecture and Urban Planning, Guizhou University, Guiyang 550025, China
ruilich@163.com

Abstract. Urban flooding, caused by extreme rainfall in the context of global warming, poses significant challenges to people, society, and the economy. One particular flood-prone area in Guiyang's Huaxi District serves as our research focus. Field visits revealed that inadequate infrastructure, suboptimal management systems, limited local resources, and a lack of safety awareness among residents contribute to the flooding issue. To address this, we propose implementing Low Impact Development (LID) techniques to transform the area into a sponge-like site. Using the Storm Water Management Model (SWMM), we constructed a flood simulation model to compare pre- and post-transformation scenarios. Results show that the proposed approach effectively mitigates flooding. Moreover, we emphasize improving flood prevention and control management systems by standardizing practices and enhancing drainage and flood prevention facilities. These measures aim to bolster the overall urban flood control system and ensure the city operates safely. Our study provides valuable insights for building a resilient Guiyang, acting as a reference for sponge city construction and strengthening flood control efforts.

Keywords: Guiyang city · urban flooding · sponge city · smart construction · SWMM model

1 Introduction

In recent years, urban flooding has increased due to extreme rainfall, posing threats to people's safety and property. Managing this issue is crucial for improving livelihoods and urban development. In Guiyang, rapid urbanization has led to an expansion of impervious surfaces, causing waterlogging during rainfall due to the city's topography. Despite the construction of sponge facilities, waterlogging persists due to imperfect management systems. To address this, optimizing sponge facilities and improving management systems are both necessary.

Previous research on urban flood management in mountainous areas of China has focused on the spatial distribution, risk assessment, and causes of flooding. Ding Liguo highlighted the high risk of heavy rainfall flooding in Guizhou's complex terrains, with varying risks across regions. Xueyuan Zhao et al. proposed effective prevention and control measures for inland flooding in Guiyang. Zheng Jiawei and others discovered

W. Guo et al. (Eds.): GBCESC 2023, LNCE 328, pp. 960–971, 2024.
https://doi.org/10.1007/978-981-99-9947-7_94

that topography, land use, and rainfall influence the spatial distribution of internal flooding. These studies have contributed to a scientific approach in tackling urban flooding in Guiyang. However, further research is required to deepen our understanding of renovating and managing flooded sites.

2 Study Area Overview and Data Sources

2.1 Overview of the Study Area

The flooding site on Mingzhu Avenue is in the central north of Huaxi District, Guiyang City. It covers an area of 77.7 hectares and is surrounded by three neighborhoods, a park, bus hubs, and the Huaxi District Government, serving as a transportation and administrative center. The area experiences frequent rainfall throughout the year, with abundant rainfall and rainstorms in the summer, mainly occurring at night.

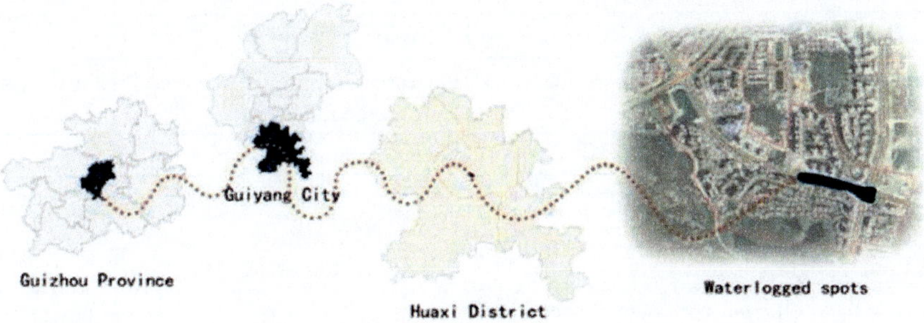

Fig. 1. Location map of the study area

The study area is predominantly residential and commercial, with limited space for transportation, green areas, and open spaces. Permeable surfaces are scarce. Drainage is primarily managed through rainwater pipes and natural infiltration methods. Rainwater pipes run underneath the roads, with inlets on both sides. Natural infiltration facilities include tree ponds, grass ditches, and barrier zones, but their coverage is relatively small in the study area.

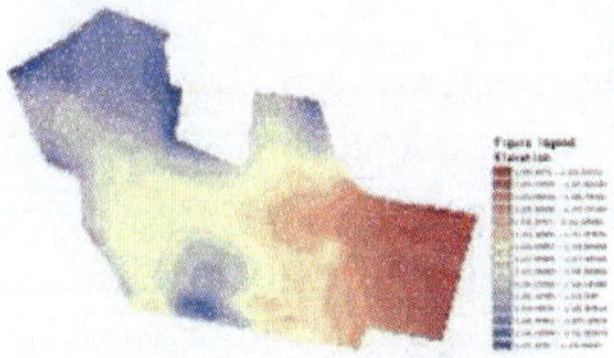

Fig. 2. Elevation analysis of study area

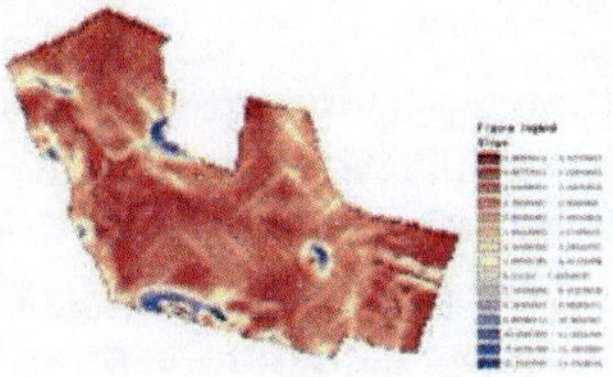

Fig. 3. Slope analysis map of the study area

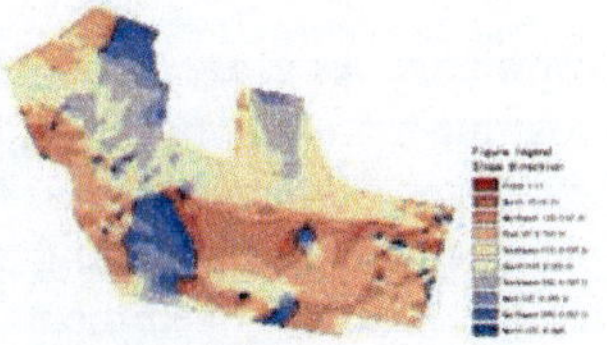

Fig. 4. Analysis of slope direction in the study area

The study area has varying elevations, with higher elevations in the northwest and lower elevations in the southeast. The intersection of Mingzhu Avenue and Huatong Road is prone to water accumulation during rainfall, leading to a high risk of waterlogging. During a heavy rainfall event on July 19, 2022, with a total of 106.3 mm of rainfall, the road section experienced three drainage wells overflowing for several days. This caused a pungent odor that significantly affected residents' daily lives. Fish were washed out of the well covers during the sewage overflow, and some residents caught and consumed them, sparking widespread social concerns.

2.2 Data Source

This study utilizes data from four primary sources, which are outlined in detail in the table below:

Table 1. Statistical table of data sources

Data Type	Access
Statistical report on road flooding in the central city of Guiyang	Guiyang Comprehensive Administrative Law Enforcement Bureau
Residents' perception of flooding	Questionnaire survey
Precipitation Data of Guiyang City	China National Meteorological Information Center Network
Elevation topographic data	Geospatial Data Cloud Platform
Land use data, pipe network data, Ground pavement, vegetation cover, building layout	Planning information, field measurements
Soil type data	Chinese Academy of Sciences Resource and Environment Data Cloud Platform

3 Analysis of the Influence of Typical Flooding Points on Pearl Avenue and the Causes of Formation

3.1 Flooding Impact Summary

Influence on Government Management. Urban flooding leads to prolonged waterlogging of municipal facilities, diminishing their lifespan and escalating maintenance expenses, placing a heavier financial burden on government entities. Floods often occur unexpectedly, challenging the emergency management capabilities and governance approaches of government agencies. Failure to address flooding in the long run can have adverse effects on social security stability and erode public trust in the government. Moreover, it can result in a decline in urban sanitation and damage the city's reputation.

Impact on Residents. Urban flooding significantly affects residents' personal lives and health. It brings inconvenience to daily travel, logistics, and express delivery, while also degrading the ecological environment in residential areas. The flooded environment becomes a breeding ground for bacteria and viruses, posing health risks and exacerbating existing illnesses. Access to medical care may become difficult for some residents. In severe cases, personal property damage and economic losses can occur.

3.2 Analysis of the Causes of Waterlogging

Hardware Facilities Do Not Meet the Demand. The drainage network layout fails to meet standards, with mixed and disorganized connections in the rain and sewage pipe network. The existing sponge facilities are spatially disconnected from the road, and can only handle the drainage demand of the community and the park. Furthermore, the sponge facilities are situated in lower terrain areas, which results in weak improvement capacity for surface runoff and inadequate relief of drainage pressure.

Management System to be Optimized. The daily management and emergency response system for flood control in Huaxi District, Guiyang City, face several challenges. The collaborative work system among flood management departments is extensive, but the transmission of disaster information is slow and difficult. Additionally, flood management in the district is mainly a top-down approach with centralized dispatching, which lacks bottom-up public participation.

Site Endowment Makes. The study area is located downstream of a reservoir, characterized by low-lying topography, abundant rainfall, and a large catchment area. The area's low permeability and infiltration rates contribute to high volumes of surface runoff. Moreover, the topographical changes in the waterlogged area hinder the flow of surface runoff, causing backflow and resulting in waterlogging.

Weak Safety Awareness Among Residents. Some residents irresponsibly dispose of garbage in drains and sewers, indicating a lack of awareness regarding the importance of maintaining unobstructed drainage flow. It is crucial to enhance residents' safety consciousness and encourage proper waste disposal practices.

4 Strategies for Managing Typical Flooding Points on Pearl Avenue

4.1 Philosophy of Governance

This study addresses flooding issues through facility and management interventions. At the facility level, the Low Impact Development concept is employed to optimize surface runoff and reduce water body pollution using decentralized source control technology. This approach maximizes rainwater utilization through source, sink, and flow stages while minimizing the impact on urban space. At the management level, the concept of a smart city is utilized, integrating communication and IT technology for flood management. This includes employing monitoring equipment, smart platforms, and network technologies to enhance management foundations, innovate methods, improve mechanisms, enhance flood risk perception, and ensure stable urban operations.

4.2 Technical Route of Treatment

This study proposes a management strategy at two levels: spongy transformation of flooded space and normalized management of flood prevention system, as shown in Fig. 5.

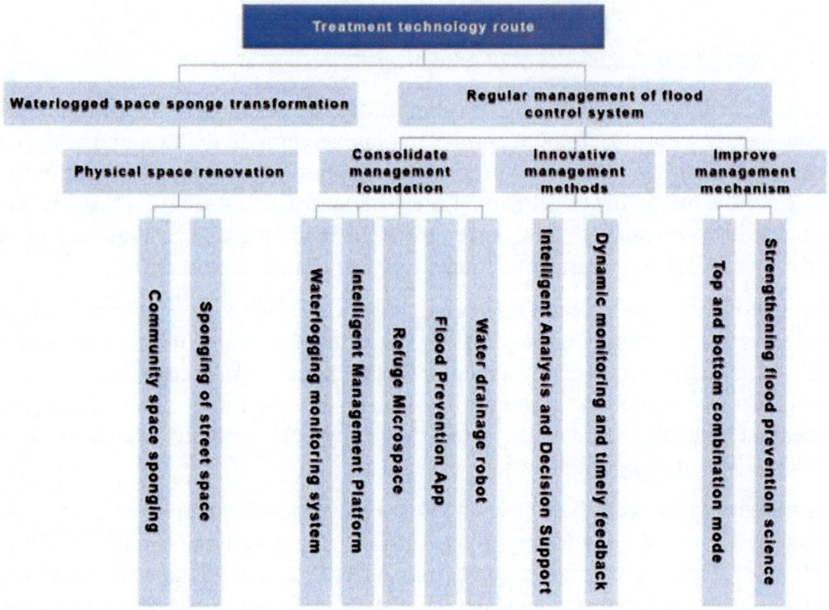

Fig. 5. Treatment technology route

5 Physical Space Renovation

5.1 SWMM Model Construction

Determination of Relevant Parameters. This simulation mainly encompasses two categories of parameters: pipe parameters and sub-catchment parameters.

The simulation focuses on two parameter categories: pipe parameters and sub-catchment parameters. Pipe parameters include rainwater pipe length, slope, diameter, and roughness. Based on site surveys and planning data, the drainage pipe network, inspection wells, and cisterns were generalized. The network consists of 73 pipe sections, 229 inspection wells, one cistern, and two drainage inlets. The drainage pipe roughness values were determined using Table 2 from the SWMM user manual.

Sub-catchment parameters include area, slope, characteristic width, imperviousness, and imperviousness Manning factor. The area is derived from generalized catchment area results, while the slope is estimated using GIS and DEM data to determine the average

Table 2. Statistical table of different types of pipe forms and roughness taking values

Pipeline Type	Shapes	Roughness
pipe canal	Rectangular pipe channel	0.035
Pipeline	Round pipe	0.013

slope. The characteristic width is approximated by treating the sub-catchment area as a rectangle and finding the square root. Imperviousness is determined using integrated runoff coefficients for different site types, as outlined in Table 3. The imperviousness Manning factor is also taken into account.

Table 3. Statistics of integrated runoff coefficients for different site types

Type of land use	Integrated runoff coefficient
Residential areas	0.6
Commercial and Service Land	0.7
Land for roads and transportation facilities	0.8
Greenfield	0.15

Sub-catchment Generalization. In this study, a comprehensive method was employed to generalize the self-catchment area. Firstly, GIS was used to divide the area based on DEM data, followed by using the Tyson polygon method based on the location of inspection wells, buildings, and roads. Finally, manual adjustments were made to divide the study area into 210 sub-catchments, as shown in Fig. 6.

Model Parameter Calibration and Validation. To validate the model, the storm intensity calculated by the model was compared with that calculated by the storm intensity formula used in Guiyang City. The storm intensity formula in Guiyang City is as follows[13]:

$$q = \frac{2146.431 \times (1 + 0.750 \lg P)}{(\mathrm{t} + 15.113)^{0.719}} \tag{1}$$

where: t - rainfall calendar time (min),P - design return period (a).

The model parameters were inputted into SWMM for rainfall runoff simulation and analysis. The results yielded a site integrated runoff coefficient of 0.87 for the Pearl Avenue section during the rainy season. Additionally, the catchment surface stormwater irrigation drains experienced an average of 2.16 h of restricted flow.

The validation results demonstrate a surface runoff error of 0.01%, indicating a high-quality simulation as per the SWMM user manual. This confirms the reliability and reasonability of the calculation results.

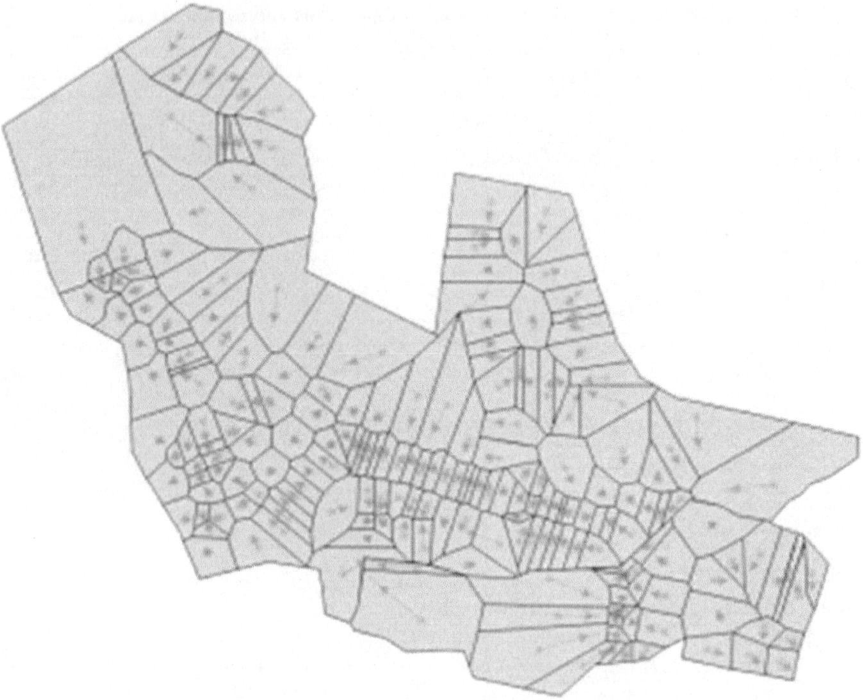

Fig. 6. Schematic diagram of the sub-catchment area

5.2 Community Space Sponging

The sponge measures for road space involve improving road surface pavement, updating drainage facilities, and increasing ecological greenery. The arrangement focuses on ensuring compatibility and coordination among facilities, maximizing their system benefits, and ensuring coordinated permeability, percolation, and water storage within the street space.

The road pavement includes both vehicle and pedestrian pavements. Semi-permeable asphalt is used as the pavement material, and permeable bricks are incorporated into the pavement. These permeable bricks, along with the underlying special planting soil, form a continuous vertical permeable system.

The drainage structure primarily consists of a new slit-type drainage ditch (shown in Fig. 7). The traditional drainage ditch is replaced with an underground linear slit drainage ditch, with only a 20 mm drainage slit remaining on the ground surface. This modification addresses the limitations of the traditional drainage ditch, ensuring pedestrian safety while allowing for water seepage during rainfall.

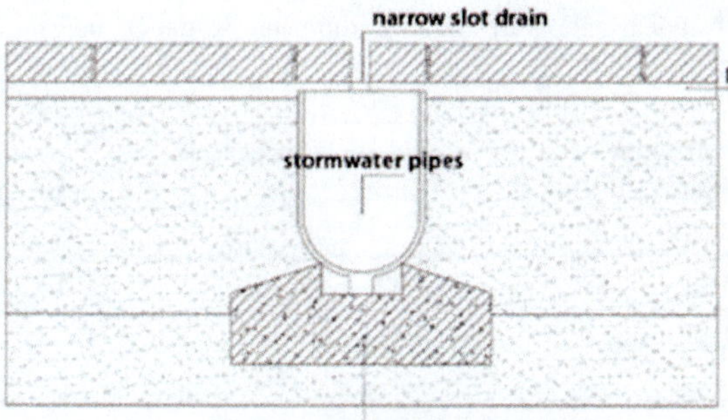

Fig. 7. Schematic diagram of slit drainage ditch

Ecological greenery planting incorporates ecological tree pits on the sidewalk and green separation zones in the middle of the road. The tree pits, permeable tiles, and special planting soil create a flat continuous permeable system, managing surface runoff on the sidewalk. The green separation zones serve as water storage and infiltration facilities for the roadway, dividing its water catchment surface and maximizing retention and infiltration capabilities.

5.3 Transformation Program Determined

Based on the literature related to sponge city construction [9], the proportion of deployment area for different types of LID facilities was determined and is presented in Table 4:

Table 4. Different LID facilities deployment area ratio statistics table

LID Facilities	Area ratio
Permeable paving	85.45% of the exposed ground and parking area
Bioretention basin	21.64% of the green area
Building cisterns	35.46% of buildings
Green Roof	19.13% of the roof area
Grass planting trench	65% of green space

After applying the renovation parameters and incorporating LID control measures, the existing model was used for fitting. The comparison of the effects before and after the renovation of Pearl Avenue is summarized in Table 5.

The data fitting analysis revealed optimization rates of 27.17% for the integrated runoff coefficient on the site carriageway, 31.98% for the sidewalk, and 63.89% for water accumulation time. These improvements align with the construction goals of a sponge city in Guiyang City(see Table 6).

Table 5. Integrated runoff index and waterlogging duration statistics before and after the renovation of Pearl Avenue

	Carriageway Integrated Runoff Index	Sidewalk Integrated Runoff Index	Pipe drainage restricted hours / hour
Before remodeling	0.92	0.82	2.16
After modification (fitted)	0.67	0.43	0.78

Table 6. Statistical table of the target requirements of each indicator in the "Planning of Sponge City Construction in Central District of Guiyang City (2018–2025)

	Integrated runoff coefficient	Permeable paving rate	Green Roof Rate	Sunken green space rate
Introductory Objectives	0.5–0.6	60%	10%	40%

The analysis of simulation results before and after optimization of the drainage system, considering design storm scenarios of 1 in 10 years and 1 in 30 years, demonstrates a substantial reduction in ponding occurrences, duration, and affected area within the study area. The optimization scheme exhibits promising improvements in drainage performance (see Fig. 8 and Fig. 9).

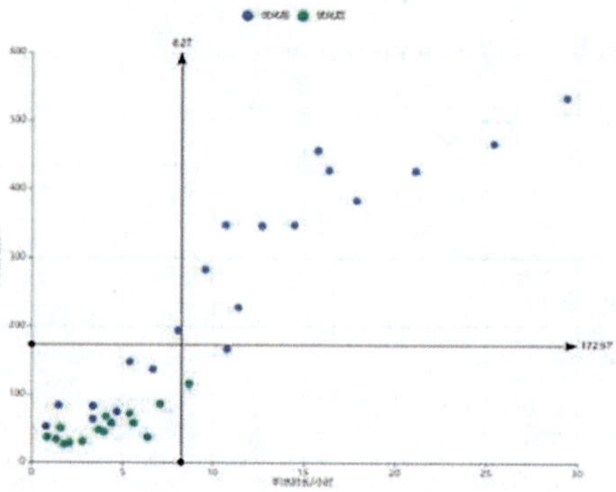

Fig. 8. Comparison of optimization results for a decade

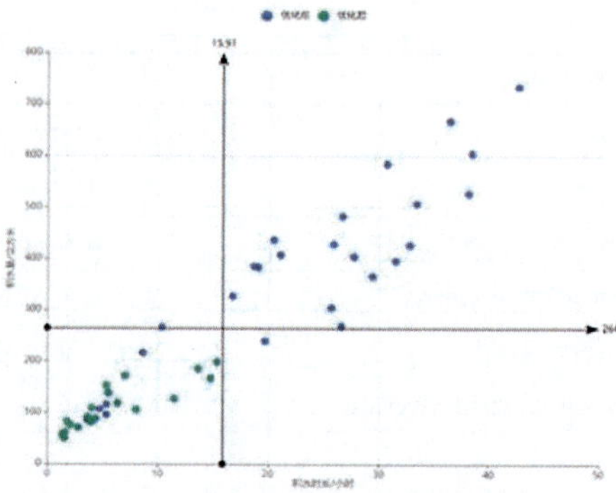

Fig. 9. Comparison of the results of the 1-in-30-year optimization

6 Regular Management of Flood Control System

6.1 Consolidate Management Foundation

Full Deployment of Flood Monitoring System. Intelligent sensing terminal devices are installed at key drainage facility nodes to monitor real-time parameters such as water flow speed, water quality, and pipe network water level. Early warning information is pushed to the intelligent drainage management platform, achieving intelligent sensing and automatic data collection of inundation data through comprehensive coverage of sensing devices.

Build a Smart Drainage Management Platform. Integrating flood information from various departments to create a flood information database, and combining feedback data from flood monitoring systems, a smart flood management platform is developed based on computer BIM technology. This platform enables real-time disaster monitoring, early warning self-triggering, rescue simulation, and evaluation of the best rescue plan as an emergency decision aid.

Creating Shelter Microspaces. Locations such as kindergartens, cafes, and restaurants are chosen on both sides of the road interface to be equipped with emergency equipment and rescue materials, and marked with flood prevention apps to create shelter micro-spaces. This ensures that people outdoors can quickly access safe locations during times of flooding.

Promote Anti-Flooding APP. The mobile flood prevention app releases a flood risk distribution map, emergency shelter locations, and various self-rescue and mutual aid methods to comprehensively enhance residents' disaster awareness and self-rescue capabilities. Additionally, the app's positioning feature can help rescuers quickly locate trapped individuals.

Introduction of Drainage Robots. Utilize the market-oriented advantages of enterprises and encourage them to actively research, develop, and introduce drainage robots to assist in drainage and rescue operations, thereby enhancing the safety of frontline personnel.

6.2 Innovative Management Methods

Intelligent Analysis and Decision Support. By combining the intelligent drainage management platform and drainage database, machine learning and other technologies can be applied to accurately identify the types of facility needs in different areas of the study area and make reasonable configurations. Simulation technology can be used to conduct 3D scenario simulations for different types of flooding, analyze the best drainage solutions, and optimize emergency plans.

Dynamic Monitoring and Timely Feedback. Real-time waterlogging data is collected via the intelligent monitoring system. The intelligent waterlogging management platform allows residents, businesses, and social organizations to provide feedback and supervise the situation. This enables prompt rectification of any issues.

6.3 Improve Management Mechanism

Combined Top and Bottom Working Mode. The management of the flood control system takes into account the needs of multiple stakeholders, creating a working model that is led by the government and involves the participation of residents, businesses, and society, with the aim of improving the overall level of regional flood control system construction.

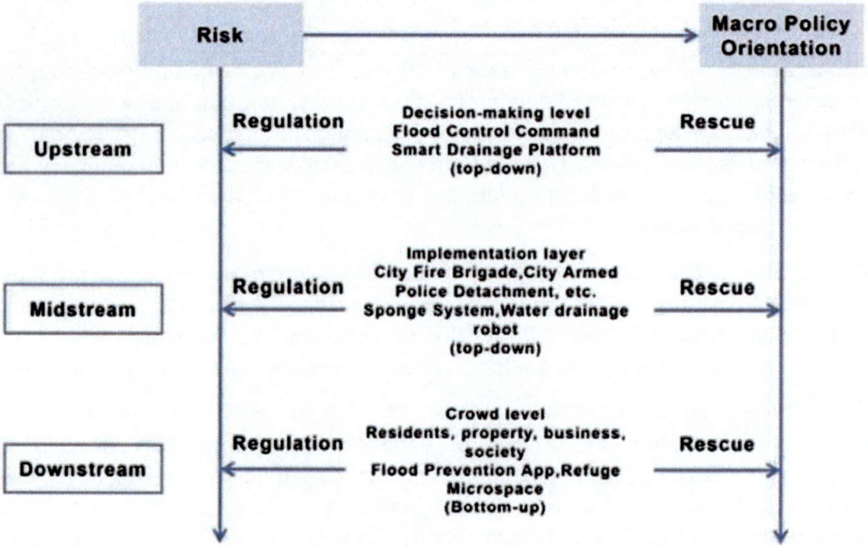

Fig. 10. Combined upper and lower working mode

Strengthen the Construction of Flood Prevention Science Using Multiple Approaches. Various methods are employed to enhance public awareness of flood prevention, including community lectures, slogans, flood simulation halls, and more. These measures aim to improve people's self-rescue and assistance capabilities during flood disasters. Additionally, visualized short videos and disaster prevention manuals are utilized on multiple platforms to increase the accessibility and engagement of science communication.

7 Conclusion

Key findings of this study indicate that the main factors contributing to flooding on Mingzhu Avenue in Huaxi District, Guiyang City, are inadequate hardware facilities, deficiencies in the management system, site conditions, and a lack of resident awareness about safety precautions. The study recommends targeted strategies to transform flooded road and community spaces, improving rainwater retention capacity. It suggests enhancing the emergency management program, developing the management mechanism, adopting innovative approaches, and establishing a normalized flood prevention system that promotes public participation and harnesses the initiative and creativity of various stakeholders. The ultimate goal is to create an effective and efficient emergency management program that meets public needs.

Funding. This research was supported by Guizhou Provincial Science and Technology Projects (Qian Ke He Ji Chu-ZK [2022] General 234).

References

1. Ding, L.G.: Study on the risk of heavy rainfall and flooding in complex terrain of Guizhou mountains (2022)
2. Fei, Y.T., Sha, Z.Z., Xue, J.B.: Analysis of the causes of waterlogging on urban roads and countermeasures to solve the problem **71**,103–106 (2016)
3. Li, M.: Impact of urban heavy rainfall and flooding on emergency medical services for highly vulnerable people (2022)
4. Liu, Y.L., Cao, C.X., Gong, H.: Research on flood prevention planning in mountainous cities coupled with "resilient city" and "sponge city". **38**, 97–103 (2022)
5. Sun, J.: Research on urban road pavement drainage design (2013)
6. Wang, X.Y.: Research on risk assessment and defensive measures for urban superstandard storm flooding disaster (2022)
7. Yin, C.Y., Dai, L., Xu, D.D., Zhao, Y.H., Zheng, J.W., Wang, P.: Risk assessment of urban flooding in mountainous areas: Guiyang City as an example, **39**, 64–69 (2022)
8. Zheng, J.W., Yin, C.Y., Dai, L., Zhou, F., Zhao, Y.H.: Spatial distribution characteristics and driving force analysis of urban flooding in southwest karst mountains at multiple scales: taking Guiyang City as an example **54**, 33–46 (2023)
9. Xing, J., Ren, H.J., Yan, Q.H., Xu, J.D.: Analysis of road waterlogging based on SWMM model.**13**, 48–50 (2020)
10. Zhang, D.H., Duan, Y., Zhou, W.Y., Li, Y., Long, L.: The derivation of storm intensity equation for Guiyang city, **40**, 95–99+11–12 (2016)

Identification the Causes of Negative Emotions in Smart Public Transportation Services Based on Social Media Data

Yanfang Shou[1(✉)] and Jianmin Xu[2]

[1] Guangzhou Institute of Modern Industrial Technology, South China University of Technology, Guangzhou 511458, China
yfshou@scut.edu.cn

[2] School of Civil Engineering and Transportation, South China University of Technology, Guangzhou 510640, China

Abstract. The negative emotional comments of smart public transportation maybe contain specific reasons behind them. If the underlying reasons can be identified, it is beneficial to timely improve public transportation (PT) services and provide a more comfortable PT experience for the public. However, the application of emotional analysis based on social media in bus service evaluation is limited, especially in the exploration of the reasons behind negative evaluations of bus services, which is still in a blank stage. Therefore, based on social media data, the paper studies the method of identifying the causes of negative emotions in PT services. Sentiment data is obtained through Sina microlog. The reasons for negative emotions are proposed and classified, which are divided into seven types: long waiting times, poor hygiene, poor driver attitude, unstable driving, overcrowding, unreasonable pricing, and other reasons. Then, a classification model for negative emotions based on feature fusion and attention mechanism, BERT-BiGRU + CNN-Attention model, is established to identify the causes of various negative emotions. Finally, the proposed method is used for analysis and validation. This model incorporates an attention model on the basis of feature fusion, which assigns higher weights to important features. Therefore, the F1 value classified on the test set reached 0.9724, achieving a good identification result.

Keywords: Smart Public Transportation · Social Media Data · Evaluation of Public Transportation Services · Text Classification · Cause Identification

1 Introduction

Smart public transportation (SPT) is an important component of smart transportation. The quality of SPT services directly affects the sharing rate of public transportation in travel, and also indirectly affects the degree of road congestion. By evaluating the services of SPT, we can identify the shortcomings in existing PT services and summarize the needs of passengers for transit, through which can improve passengers' comfort and satisfaction with bus travel services, and increase the proportion of passengers choosing PT. The

W. Guo et al. (Eds.): GBCESC 2023, LNCE 328, pp. 972–981, 2024.
https://doi.org/10.1007/978-981-99-9947-7_95

traditional evaluation survey of PT services is mainly achieved through questionnaires and interviews, and the obtained data is analyzed through modeling and other methods. Juan et al. summarized various service quality evaluation methods with the surveys of customer satisfaction [1]. Alexandre et al. proposed used a configurational method to explore the relationship between various transportation service quality parameters and overall satisfaction [2]. Although the method based on questionnaire survey can obtain more fine-grained results, it will consume a lot of time, manpower, and other costs.

With the rapid development of the Internet, people have started to publish evaluations of PT services on internet platforms. Therefore, there is a lot of comment information about PT service on social media platforms. Convenient operation makes social media data present characteristics such as large data volume, fast dissemination speed, and real-time content. Obtaining people's preferences and attitudes towards PT services from these increasingly expanding comments can help the government provide necessary decision-making basis in formulating PT policies, improving PT routes and operational services. In addition, some of the comments on social media platforms are positive and some are negative. For negative emotions comments, it is necessary to analyze certain specific reasons behind them. If the underlying reasons can be identified, it will be beneficial to timely improve PT services and provide the public with a more comfortable PT experience.

So far, sanctification analysis of public transportation services has mainly been achieved through traditional questionnaire surveys. There are few studies using deep learning models and social media data for analysis, and existing research based on deep learning has not explored the reasons behind negative emotions. In the research fields of sentiment analysis, there are many studies on serial fusion of different models, and there are few methods based on parallel feature fusion. In addition, methods combining pre-trained models and feature fusion have not yet been used. Moreover, the accuracy of existing text classification research methods needs further improvement. Therefore, in the paper, the BERT-BiGRU + CNN-Attention model is provided, with the use of Bidirectional Encoder Representation from Transformers (BERT) pre-training model and Bidirectional Gate Recurrent Unit + the Convolutional Neural Network (BiGRU + CNN) training model, which combines feature fusion with attention mechanism to enhance the accuracy of identification. The method proposed can comprehensively extract local information and contextual message, and use the attention mechanism to focus on key information. The model has a simple structure, fewer parameters, shorter training time, and can output good results with a few of data training set. It is suitable for scenarios with large data volume, fast processing speed, and high recognition requirements.

The paper is divided into sections as follows. Literature review is introduced in Sect. 2. The proposed model and solving algorithm are described in Sect. 3. In Sect. 4, the experiment testing is conducted and compared with existing text classification models. The conclusion is shown in Sect. 5.

2 Literature Review

Text sentiment analysis is the process of analyzing, processing, summarizing, and reasoning subjective texts with emotional connotations, and it is an effective approach for identifying the causes of negative emotions in PT services [3]. The methods used for text sentiment analysis mainly include emotion dictionary methods, traditional machine learning methods, and deep learning methods. The emotion dictionary methods require constructing as detailed a dictionary as possible, and often only construct for specific fields. Therefore, using the constructed dictionary in other fields often results in poor cross domain performance. Traditional machine learning methods have the advantage of high accuracy, but they rely heavily on training corpus and require a lot of time to be spent on feature engineering. Compared to the above two methods, deep learning methods can automatically learn features and are suitable for situations with high data dimensions and large data volumes, which have been widely applied in text classification recently [4]。Hochereiter et al. proposed a Long Short Term Memory (LSTM) model which allowed storage units to store information for a long time or discard previous calculation results [5]. The LSTM model has a complex structure, a lot of parameters, and needs long training time. Therefore, researchers simplified the LSTM model to a Gate Recurrent Unit (GRU) model. Bert model obtains sentence level vector representation of the entire data statement by encoding context semantic information. For different tasks, there is no need to adjust the internal structure, just add the required network layer after it, and then use the marked data to tune the model parameters [6].

The application of text analysis in traffic neighborhoods mainly includes analysis of traffic service satisfaction, traffic operation status, perception of public opinion, and traffic event prediction. Cao et al. summarized the models and methods in traffic sentiment analysis, and analyzed the disadvantages and advantages of rule based and learning based methods [7]. Cao et al. Presented a GRU model based on generative adversarial networks for Traffic status analysis, which has an accuracy improvement of 3.79% Compared to the GRU model [8]. Bondoc et al. used text analysis for road traffic state prediction and congestion identification in Metro Manila [9]. Yuan et al. proposed an effective Bert-RCNN model to classify the severity of traffic accidents in China with severe mixed traffic phenomena [10]. Social media data also has some applications in the transportation field. Lin et al. analyzed the impact of varying degrees of adverse weather on highway traffic speed using social media data [11]. Gu et al. used twitter data to obtained traffic related event information on highways and main roads, in order to get data such as event location, type, publisher, and release time [12]. Chang et al. obtained relevant comments on road traffic congestion and traffic events through Sina microblog, used to identify and predict areas of road traffic congestion and accident frequency, and verified the effectiveness of the model by comparing it with actual data [13]. Wan et al. Wan establishes an automated Natural Language Processing based model for predicting road traffic events through online traffic comment data, and sends the obtained event occurrence time, location, and status to the driver's navigation system [14].

3 Method Description

Identification the causes of negative emotions in SPT services based on social media data includes the following content: First, the microblog comments about public transit in Guangzhou are obtained. Then, the re-processed data are classified into negative and positive sentiment data. The causes of PT negative emotions are divided into seven categories. Finally, the negative sentiment data are used for the training of the cause identification model, and the trained model could recognize the causes of the negative emotions in SPT services.

3.1 Obtaining the Negative Sentiment Microblog Data

In our previous paper, we have established a BERT-BiLSTM model for obtaining negative sentiment microblog data [15]. We input the keywords to obtain data through web crawler for the Guangzhou bus service. The data pre-processing includes data cleaning, a fixed length of microblog text setting, Chinese text splitting and stop words filtering. The BERT-BiLSTM model is used for sentiment classification. Firstly, the data is input into the BERT encoding layer, where the input text is segmented in words and encoded as a vector. Embed the words into the vector and input them into the BiLSTM layer to further extract the features. Finally, we use Sigmaid activation function to output the negative sentiment microblog data.

3.2 Classification of Negative Emotions

There are many reasons for publishing negative sentiment data on social media. We conducted an unrelated analysis and summary of the existing main reasons, and divide the causes of PT negative emotions into seven types, including long waiting times, poor hygiene, poor driver attitude, unstable driving, overcrowding, unreasonable pricing, and other reasons.

3.3 Model Construction and Cause Identification

Model Framework. The paper proposes a classification model of BERT-BiGRU + CNN-Attention based on feature fusion and attention mechanism. Firstly, the BERT model segments text in words and encodes it into vectors. Then input the vectors into BiGRU and three CNN Models with different convolutional kernel sizes. BiGRU can extract association information between text before and after. The convolution kernel sizes of the three CNN models are set to 3, 4, and 5, with different convolution kernels can extract different local features. Then, the extracted features are fused, that is, the feature vectors are concatenated to obtain a new feature vector. Through feature fusion, more semantic features in the text can be captured. finally, the attention mechanism is used to weight the features, which assigns higher weights to important features, and is connected the fully connected layers for classification through the softmax activation function (see Fig. 1).

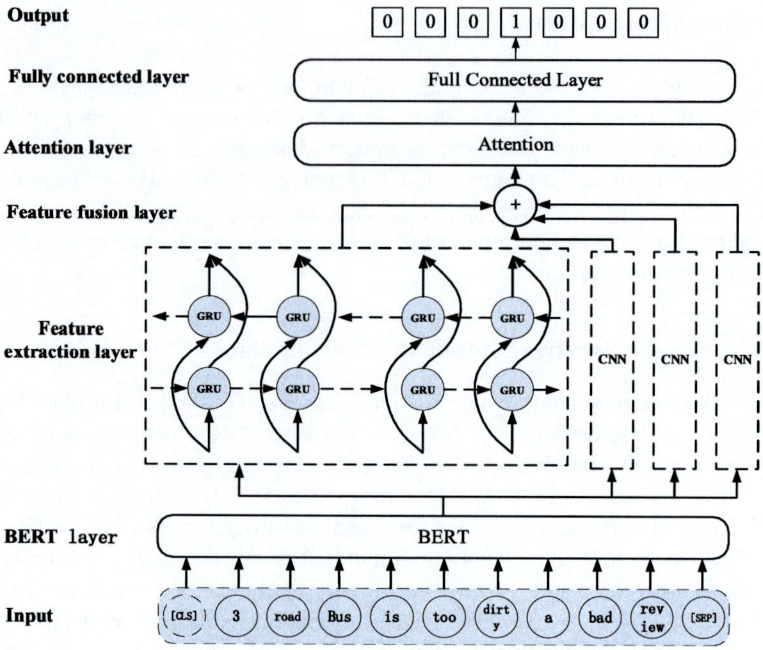

Fig. 1. Causes classification model

Attention Mechanism. The attention mechanism was inspired by the working mechanism of human vision. When humans observe things with their eyes, they often focus on key information and automatically ignore some unimportant information. Fundamentally, the attention mechanism also automatically selects most important information from a massive amount of data and focuses attention on them.

In the attention model, a new hidden state s is obtained by assigning weight coefficients α to the hidden state h, and sufficient weight is allocated on key information to highlight locally important information. The calculation process of the attention model mainly includes three steps: (1) Calculate the energy values corresponding to each hidden layer state. (2) Calculate the weight coefficients assigned to each hidden layer state based on the calculated energy values. (3) Calculate the output of the attention model based on weights and hidden layer states.

Model Algorithm Description. The algorithm description of the BERT-BiGRU + CNN-Attention model is as follows:

Input: $X = \{x_1, x_2, ..., x_n\}$, $Y = \{y_1, y_2, ..., y_m\}$, X is the preprocessed microblog text set, Y is the category corresponding to each negative sentiment microblog text.

Output: Classification results.

BERT layer: encode the input text as a vector.

CNN layer: (1)Taking the output of BERT layer as input, the height of the convolutional kernel is 3, 4, and 5. The convolutional kernel slides from top to bottom on a two-dimensional matrix in steps of 1 to extract features. (2) After the convolution operation output, GlobalAveragePooling1D pooling operation is performed.

BiGRU layer: (1)Taking the output of BERT layer as input. (2) The forward GRU goes through the following process: a. Calculate reset gating. The previous hidden layer state h_{t-1} and current input x_t are used to obtain reset gating r_t. h_{t-1} is multiplied by r_t, which splices with x_t at the current time. After splicing, the new candidate content $\tilde{h}_t$ is obtained through tanh activation function. b. Calculate updating gating. h_{t-1} and x_t are used to obtain updating gating z_t. c. Update current state. z_t and $\tilde{h}_t$ are used to output the hidden layer state h_t. (3) The input sequence direction of backward GRU is opposite to forward GRU, and the calculation process is the same. (4)BiGRU output is obtained by concatenating the outputs of forward GRU and backward GRU, and a single hidden state value is returned in the output sequence.

Feature fusion layer: The feature vector dimension of the output of three CNN layers and BiGRU layer are the same. Those outputs are spliced to obtain new features and feature fusion is achieved.

Attention layer: Taking the output of the feature fusion layer as input, the Attention model assigns corresponding weights to the input vectors. After obtaining the weight coefficients, the input vector is weighted to obtain the output of the layer of attention layer.

Fully connected layer: Taking the output of Attention layer as input and Softmax as activation function, the causes classification of negative emotions with microblog texts is obtained.

4 Experiment and Analysis

4.1 Experimental Data

We conducted experiments in Guangzhou, With the time range from January 1, 2018 to January 1, 2022. Thirteen groups of key words of PT + service + good, PT + clean, PT + like, PT + favor, PT + punctuality, PT + unreasonable, PT + emergency stop, PT + waiting + long, PT + crowded, PT + driver + temper + bad, PT + attitude + bad, PT + dirty, PT + expensive were used to conduct the Scrapy web crawler on Sina microblog, and a total of 61079 pieces of data were collected. After removing duplicates and missing values, there are still 51004 pieces of data left. Afterwards, data denoising processing were performed. The BERT-BiLSTM model with the best trained parameters was used to classify those data, and 18623 data containing user emotions were obtained, including 10692 positive sentiment data and 7931 negative sentiment data.

Experiments and tests on 7931 negative sentiment data are conducted. A total pieces of 6182 data are labeled as training sets, verification sets and test sets, with training sets accounting for 80%, verification sets accounting for 10%, and test sets accounting for 10%. Negative sentiment causes analysis of these data revealed that 824 were due to overcrowding, 601 were due to poor hygiene, 773 were due to long waiting times, 772 were due to unreasonable pricing, 1005 were due to unstable driving, 623 were due to poor driver attitude, and 1584 date were due to other reasons.

4.2 Evaluation Indicators

The paper uses precision, recall, and F1 value as evaluating indicators for the cause classification experiment [15]. Precision refers to the ratio of data predicted as positive correctly to all positive predictions. Recall refers to the ratio of data predicted as positive correctly to all positive cases. F1 value is calculated through precision and recall. The higher the value of three evaluation indicators, the better the classification performance of the model.

$$precision = \frac{TP}{TP + FP} \tag{1}$$

$$recall = \frac{TP}{TP + FN} \tag{2}$$

$$F1 = \frac{2}{\frac{1}{precision} + \frac{1}{recall}} \tag{3}$$

where TP refers to the data that is predicted as positive correctly. FP refers to the data that is predicted as positive incorrectly. FN refers to the data that is predicted as negative incorrectly.

4.3 Results and Analysis

The model parameters are set. BERT model has a network layer of 12, a hidden layer dimension of 768, and a self-attention mechanism of 12 heads, with a total of 110 M parameters. The dimension of BiGRU hidden layer is set to 128. The number of convolutional kernels in three CNN models is set to 128, and the time domain window length of convolutional kernels is set to 3, 4, and 5. The maximum input length of the text is 140 words. The Adam optimizer is selected. The learning rate is 0.00001. The batch size is 16 and the iterations number is 4.

To verify the accuracy of the proposed model, following models that have been applied in text classification are introduced for comparison:

(1) TextCNN model. Zhang et al. applied the TextCNN model to news title classification and expanded the training set using data augmentation methods. The model achieved a maximum accuracy of 95.24% [16].
(2) BiGRU model. This model is commonly used for analyzing time series data such as text. After BiGRU extracts text features, it connects the fully connected layer and uses the Softmax function to output classification results.

(3) BiGRU-CNN model. Combining the BiGRU model with the CNN model, the BiGRU model extracts features and inputs them into the CNN model for further extraction. This model is commonly used in text classification research.
(4) BiGRU-CNN Attention model. BiGRU model, CNN model, and Attention model are combined to enhance attention to important features.
(5) BERT BiGRU + CNN model. This model is used to compare the effect improvement with or without the addition of attention mechanism.

Table 1 shows the experimental results.

Table 1. Experimental result

Model	Precision (%)	recall (%)	F1(%)
TextCNN	80.51	79.03	79.76
BiGRU	78.89	77.28	78.08
BiGRU-CNN	83.47	81.98	82.72
BiGRU-CNN-Attention	85.00	75.30	79.86
BERT-BiGRU + CNN	94.51	94.14	94.32
BERT-BiGRU + CNN-Attention	97.34	97.14	97.24

In comparative experiments, the classification performance of BiGRU model is worse than that of TextCNN model, but the classification performance of BiGRU-CNN model is better than TextCNN model. This is because the addition of CNN model extracts more local features, thereby improving the accuracy of classification and the effectiveness of the model. Compared to BiGRU-CNN model, BiGRU-CNN-Attention model has 1.53% higher precision, 6.68% lower recall, and 2.86% lower F1 value. Although the addition of attention model improves the accuracy, the recall rate and F1 value decrease. This is due to the uneven amount of training data for each category, resulting in higher recognition accuracy in some categories while lower recognition accuracy in other categories. By combining BERT model with BiGRU-CNN model, the optimization effect is significantly better than other models mentioned above, with the precision of 94.51%, recall of 94.14%, and F1 of 94.32%.

BERT-BiGRU + CNN-Attention model proposed improves precision by 2.83%, recall by 3%, and F1 value by 2.92% compared to BERT-BiGRU + CNN model, indicating that the addition of Attention model improves the effectiveness. Compared with BiGRU + CNN-Attention model, it improves precision by 12.34%, recall by 21.84%, and F1 value by 17.38%. Although training data is not balanced in various classifications, there is still a significant improvement in various indicators, mainly due to the addition of the BERT model, which can achieve good model performance with a small amount of data training. Compared with BiGRU + CNN, it improves precision by 13.87%, recall by 15.16%, F1 value by 14.52%. Compared with BiGRU, it improves precision by 18.45%, recall by 19.86%, F1 value by 19.16%. Compared with TextCNN model, it improves precision by 16.83%, recall by 19.11%, and F1 value by 17.48%. Therefore, the BERT-BiGRU + CNN-Attention model for negative emotion cause classification based on feature fusion and attention mechanism proposed outperforms other models through precision, recall, and F1 value. Therefore, the BERT-BiGRU + CNN-Attention model for negative emotion cause classification based on feature fusion and attention mechanism proposed outperforms other models.

Finally, the BERT-BiGRU + CNN-Attention model with the best trained parameters was applied to 1749 unlabeled negative emotion data, and 6182 labeled data were added to obtain the classification of negative emotion reasons for all data: 873 long waiting times, 701 poor in car hygiene, 723 poor in driver attitude, 1305 unstable driving, 1024 crowded cars, 792 unreasonable charging, and 2513 other reasons.

5 Conclusion

The paper introduces the classification of negative emotions in SPT services based on social media data, and then proposes a classification model BERT-BiGRU + CNN-Attention based on feature fusion and attention mechanism. It also introduces attention mechanism, model algorithm, experimental data, evaluation indicators, and model parameter settings. Finally, the trained model was used to classify causes of negative emotions microblog data, and experiments were did and compared to other methods. The experimental results show that the classification effect of this method are better than models such as BiGRU-CNN-Attention and TextCNN. However, the training set used in model training is not large enough. In the future, larger training sets can be constructed to improve the effectiveness of the model.

Acknowledgments. This research was supported in part by the National Natural Science Foundation of China (Grant No. 51272326).

References

1. Juan, D.O., Rocio, D.O.: Quality of service in public transport based on customer satisfaction surveys: a review and assessment of methodological approaches. Transp. Sci. **49**(35), 605–622 (2015)
2. Alexandre, S., Katrin, L., Lars, E.O., Margareta, F., Satoshi, F.: Assessing travel satisfaction in public transport: a configurational approach. Transp. Res. Part D: Transp. Environ. **93**, 1–14 (2021)
3. Chen, Y.Y., Lv, Y.H., Wang, X.X., Li, L.X., Wang, F.Y.: Detecting traffic information from social media texts with deep learning approaches. IEEE Trans. Intell. Transp. Syst. **20**(8), 3049–3058 (2019)
4. Li, Y., Wang, X.T., Xu, P.J.: Chinese text classification model based on deep learning. Future Internet **10**(11), 1–12 (2018)
5. Hochreiter, S., Schmidhuber, J.: Long short-term memory. Neural Comput. **9**(8), 1735–1780 (1997)
6. Li, Q., et al.: A survey on text classification: from traditional to deep learning. ACM Trans. Intell. Syst. Technol. **13**(2), 1–41 (2022)
7. Cao, J.P., Zeng, K., Cheng, J.J., Qian, F.C., Wen, D., Gao, Y.Q.: Web-based traffic sentiment analysis: methods and applications. IEEE Trans. Intell. Transp. Syst. **15**(2), 844–853 (2014)

8. Cao, D.L., Wang, S.R., Lin, D.Z.: Chinese microblog users' sentiment-based traffic condition analysis. Soft. Comput. **22**(21), 7005–7014 (2018)
9. Bondoc, E.R.P., Caparas, F.P.M., Macias, J.E.D., Naculangga, V.T., Estrada, J.E.: An intelligent road traffic information system using text analysis in the most congested roads in metro manila. In: 2018 IEEE 10th International Conference on Humanoid, Nanotechnology, Information Technology, Communication and Control, Environment and Management (HNICEM), pp. 1–6. Philippines (2018)
10. Yuan, S.J., Wang, Q.X.: Imbalanced traffic accident text classification based on Bert-RCNN. In: 6th International Seminar on Computer Technology, Mechanical and Electrical Engineering (ISCME), pp. 1–8. Zhuhai (2021)
11. Lin, L., Ni, M., He, Q., Gao, J., Sadek, A.W.: Modeling the impacts of inclement weather on freeway traffic speed: exploratory study with social media data. Transp. Res. Rec. **2482**(1), 82–89 (2015)
12. Gu, Y.M., Qian, Z., Chen, F.: From twitter to detector: real-time traffic incident detection using social media data. Transp. Res. Part C: Emerg. Technol. **67**, 321–342 (2016)
13. Chang, H.L., Li, L.S., Huang, J.X., Zhang, Q.P., Chin, K.S.: Tracking traffic congestion and accidents using social media data: a case study of Shanghai. Accid. Anal. Prev. **169**, 1–17 (2022)
14. Wan, X.P., Lucic, M.C., Ghazzai, H., Massoud, Y.: Empowering real-time traffic reporting systems With NLP-processed social media data. IEEE Open J. Intell. Transp. Syst. **1**, 159–175 (2020)
15. Wu, Y., Xu, J.M., Lin, Y.J.: Social media-based analysis of bus service satisfaction. In: 6th International Conference on Intelligent Transportation Engineering (ICITE), pp. 540–547. Beijing (2021)
16. Zhang, T., You, F.: Research on short text classification based on textCNN. In: International Conference on Computer Big Data and Artificial Intelligence, pp.1–7. Changsha (2021)

Research on Sponge Transformation and Renewal Design of Old Residential Districts in Mountain Cities from the Perspective of Resilience

Peijun Feng(✉), Jiatan Fan, and Sicheng Wang

Guizhou University, Guiyang 550025, China
2776469129@qq.com

Abstract. Developing sponge transformation of old communities in mountainous cities from the perspective of resilience is an important means of urban waterlogging control. Based on the Xibei community site in Huaxi District, a total of 8 indicator items are constructed from three aspects: community ecological environment, community safety, and community resources. The entropy weight method and TOPSIS method are combined with SWMM software runoff simulation to construct a sponge construction evaluation system for update design research. The research shows that the two key factors to improve the sponge construction index of old communities are the average catchment area and the permeable pavement rate; Taking the elevation difference of mountainous communities as the runoff buffer area, setting linear fine seam drainage ditches, vegetation shallow ditches with small dams, and permeable pavement can absorb more than 20% of peak runoff, and reduce the comprehensive runoff coefficient of the site by more than 50%; The division of community catchment area should be flexibly adjusted according to the slope of different areas; The site pavement renewal based on LID control can reduce the comprehensive runoff coefficient of the site by more than 40%, and the peak value of road rainstorm runoff can be reduced by more than 20%; The evaluation system can provide more targeted guidance for sponge transformation design.

Keywords: mountain city · Old communities · Site assessment · Entropy weight method · SWMM simulation · Introduction to Sponge Transformation

1 Introduction

Sponge transformation in mountainous urban environments is a key focus of current urban construction. The State Council issued the "Guiding Opinions on Promoting Sponge City Construction" in 2015 to promote the construction of sponge cities; The Guiyang Municipal Government issued the "Implementation Plan for Sponge City Construction in Guiyang City" in 2019. Based on the support of such policies, it is urgent to carry out research on sponge transformation design for old residential areas.

W. Guo et al. (Eds.): GBCESC 2023, LNCE 328, pp. 982–1003, 2024.
https://doi.org/10.1007/978-981-99-9947-7_96

Sponge design has achieved success in many countries since its gradual development in the twentieth century. The United States has gradually established a comprehensive management technology for waterlogging with optimal rainwater management and low impact development as its core; New Zealand innovatively established data sources using GIS technology and RS technology, and constructed an evaluation factor scoring method for comprehensive evaluation of projects; Australia has established a unified and systematic management system for rainwater, sewage, and urban water supply, starting from the virtuous cycle of water. With the introduction of sponge construction concept into China, domestic scholars have conducted relevant research on design transformation and evaluation analysis based on foreign practical cases. Chen Jingzhen and others have constructed a construction effectiveness evaluation model for urban waterlogging prevention and control through the Analytic Hierarchy Process; Xiao Xiuli summarized the common methods of waterlogging prevention and control under the current sponge construction mode based on engineering technology; Peng Hailong took mountainous cities as the starting point and fitted the effectiveness evaluation analysis of sponge transformation technology. In terms of the current community construction in mountainous cities, the site conditions are complex and the causes of waterlogging are diverse. Therefore, in the evaluation, analysis, and transformation of sponge construction in old communities. It is advisable to conduct specific exploration and research.

2 Research Scope and Data Sources

2.1 Research Scope

The terrain of the residential area gradually decreases from No. 2812 to No. 2810 from north to south and is located in a low-lying area. Overall, it is 3.21 m lower than the adjacent municipal road surface, and the slope of the site is complex, making it a typical type of mountainous urban community. There are three buildings built on the site, with a row of above ground storage rooms uniformly built between No. 2810 and No. 2811, and a row of self-built storage rooms by residents are adjacent to the east side of the site. The internal roads of the community are one-way roads, with the only entrance and exit located in the southwest corner of the community. There is no scientifically planned green belt inside the site, only 7 trees (spaced far apart to effectively store rainwater) and brick flower beds on the ground (over 80% of the flower beds have become soil, while the rest are used by residents as vegetable beds). The internal sites of the community are all hardened and impermeable concrete floors. The drainage wells are concentrated at the lowest point of the residential area, and the wellhead layout is disorderly. When the peak season of the rainy season arrives every year, the community and the roads within 10–15 m around the community are facing severe water accumulation on the roads, causing vehicles and pedestrians to be unable to pass normally. There is an urgent need for sponge design and transformation (Fig. 1).

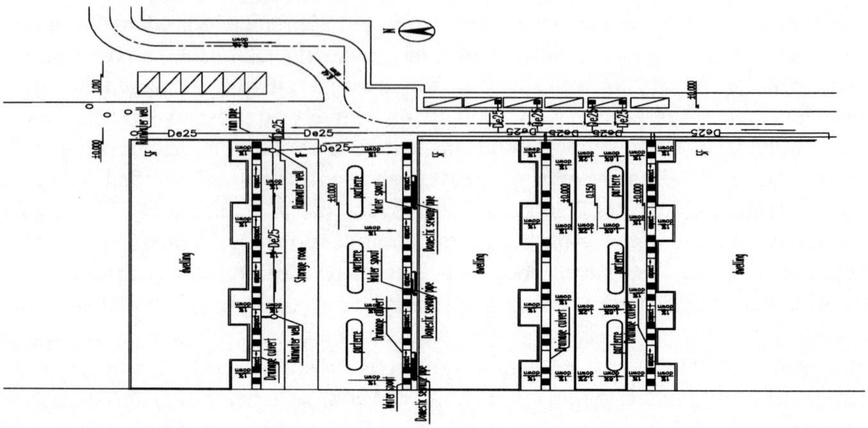

Fig. 1. Current situation diagram of site drainage facilities layout (Image source: author's self-drawing)

2.2 Data Sources

The meteorological data used in this study were obtained from Huaxi District Meteorological Bureau from 2020 to 2022, Rainfall statistics during the rainy season (March October) in the region; Site data from the author's field survey.

3 Research Methods and Technical Routes

3.1 Research Methods

Entropy Weight Method and TOPSIS Model Analysis. The influencing factors of the site's waterlogging resistance capacity are layered and summarized. The entropy weight method is used to calculate the weight. The TOPSIS method is used to calculate the comprehensive score index as the evaluation index of the site's waterlogging resistance toughness. After the design and renovation are completed, a fitting evaluation is conducted again to guide the design of the renovation with the final calculation results.

SWMM Model Analysis. Conduct parameter simulation using dynamic precipitation runoff simulation software SWMM to analyze the current situation of waterlogging on the site, and conduct further surface runoff model calculations on the site after the renovation design is completed to evaluate the renovation design results.

3.2 Technology Roadmap

Using the entropy weight method model combined with TOPSIS method analysis and SWMM software model analysis, the site of the research object is simulated to maximize the restoration of the real climate situation, and a feasible and reasonable transformation plan is obtained (Fig. 2).

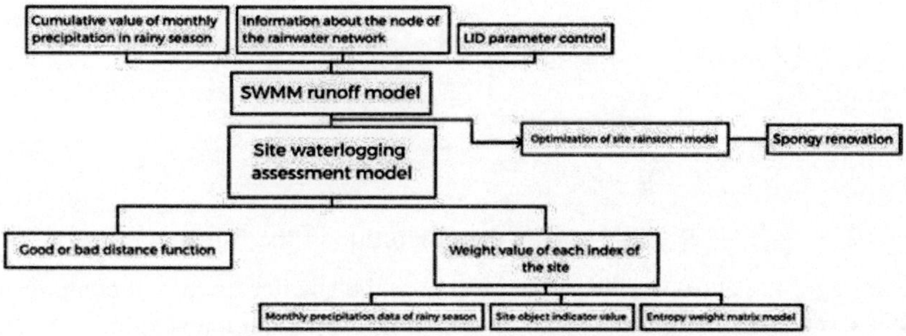

Fig. 2. Technical route for research on sponge transformation of old communities in mountainous cities (Image source: author's self-drawing)

4 Construction of Evaluation System for Sponge Construction in Old Communities

4.1 Entropy Weight Method and TOPSIS Model Calculation

The entropy weight method is an objective way to determine the weight of various information indicators through the amount of information. The entropy value can be used to evaluate the degree of change of indicators in the model. If the entropy value of a certain information is large, it indicates that its degree of change is large, and its weight in the evaluation system is low; On the contrary, it is high. The specific calculation process is as follows:

(1) By constructing an original matrix with m objects and n indicators:

$$X = [x_{ij}]_{n \times m} \tag{1}$$

(2) Use the range method to standardize the data, making it between 0 and 1 to eliminate the impact of different dimensions.

The positive indicators are treated as follows:

$$X'_{ij} = \frac{x_{ij} - min(x_{ij})}{max(x_{ij}) - min(x_{ij})} \tag{2}$$

(3) The negative indicators are treated as follows:

$$X'_{ij} = \frac{max(x_{ij}) - x_{ij}}{max(x_{ij}) - min(x_{ij})} \tag{3}$$

(4) Where: X_{ij} is the original value of the j th indicator of the i th unit, x'_{ij} is the j th indicator of the i th unit after standardization.

(5) Calculate the entropy value e_j and information utility value of indicator information d_j.

$$e_j = -k\left[\sum_{i=1}^{n} p_{ij} ln(p_{ij})\right] \tag{4}$$

$$d_j = 1 - e_j \tag{5}$$

where: $p_{ij} = \frac{x'_{ij}}{\sum_{i=1}^{n} x'_{ij}}$, $K = \frac{1}{\ln(n)}$ ·p_{ij} is the proportion of the i th year to the jth indicator. The larger the entropy value in the formula, the smaller its role in comprehensive evaluation, and the smaller its weight. Conversely, the greater the weight.

(6) Determine the weight of various indicators:

$$w_j mm = \frac{(1 - e_j)}{\sum_{i=1}^{m}(1 - e_j)} \tag{6}$$

where: $0 \leq w_j \leq 1$.

After completing the calculation using the entropy weight method, the TOPSIS model is used for target decision-making. TOPSIS is a method of determining the best solution by considering the advantages and disadvantages of each evaluation object, while also reflecting various evaluations, evaluate the gap between options. The specific calculation process is as follows:

(1) Determine the positive ideal solution X^+ and the negative ideal solution X^-:

$$\begin{aligned} X^+ &= (max\{x_{11}, x_{12}, \cdots, x_{m1}\}, max\{x_{12}, x_{22}, \cdots, x_{m2}\}, \cdots, \\ &max\{x_{1n}, x_{2n}, \cdots, x_{mn}\}) = (x_1^+, x_2^+, \cdots, x_n^+) \end{aligned} \tag{7}$$

$$\begin{aligned} X^- &= (min\{x_{11}, x_{12}, \cdots, x_{m1}\}, min\{x_{12}, x_{22}, \cdots, x_{m2}\}, \cdots, \\ &min\{x_{1n}, x_{2n}, \cdots, x_{mn}\}) = (x_1^-, x_2^-, \cdots, x_n^-) \end{aligned} \tag{8}$$

(2) Calculate the distance between each index and the ideal solution:

$$D_i^+ = \sqrt{\sum_{j=1}^{n}\left(x_j^+ - x_{ij}\right)^2} \tag{9}$$

$$D_i^- = \sqrt{\sum_{j=1}^{n}\left(x_j^- - x_{ij}\right)^2} \tag{10}$$

(3) Calculate the fit between the evaluation object and the ideal solution C_i:

$$C_i = \frac{D^-}{D^+ + D^-}, \quad 0 < C_i < 1 \tag{11}$$

Among them, C_i the larger the scale, the higher the degree of sponge construction, and vice versa.

4.2 Model Analysis Results

Based on the study of community site conditions and in combination with various indicators in the "Technical Guidelines for Sponge City Construction", the community sponge construction evaluation system is divided into three categories: community ecological environment, community safety, and community resources, with a total of 8 indicators. See Table 1 for specific indicators:

Table 1. Various indicators of the hierarchical evaluation system for community sponge construction

Target layer	Criterion layer	Indicator layer	number
Evaluation of community sponge construction	COMMUNITY ECOLOGICAL ENVIRONMENT	SUNKEN GREEN AREA/m^2	C1
		PLANT SPECIES/SPECIES	C2
		PERMEABLE PAVEMENT RATE/%	C3
	COMMUNITY SAFETY	NUMBER OF WATERLOGGING POINTS/PIECE	C4
		MAXIMUM ANNUAL SURFACE RUNOFF IN RAINY SEASON/*mm*	C5
		AVERAGE CATCHMENT AREA OF THE SITE/m^2	C6
	COMMUNITY RESOURCES	RAINWATER UTILIZATION RATE/%	C7
		NUMBER OF DRAINAGE EQUIPMENT/UNIT	C8

Based on the collected meteorological data and site survey data, the changes of various indicators in the three years from 2020 to 2022 are sorted out, and the final results are obtained by substituting them into the above formula, as shown in Table 2.

Table 2. Information entropy value and weight of site sponge construction evaluation indicators

Entropy weight method			
nape	Information entropy value e	Information utility value d	weight (%)
Sunken green area	0.766	0.234	5.643
Sunken green area	0.488	0.512	12.367
Rainwater utilization rate	0.58	0.42	10.149
Number of emergency drainage equipment	0.631	0.369	8.901
Permeable pavement rate	0.108	0.892	21.537
Waterlogging points	0.631	0.369	8.901
Maximum annual surface runoff in rainy season	0.627	0.373	9.012
Average catchment area of the site	0.027	0.973	23.490

Data sources: Calculated based on site surveying and survey parameters

On the basis of determining the weight, further calculate the current ideal solution of various indicators and 2020 through TOPSIS method, The evaluation results from 2022 to 2022 are shown in Tables 3 and 4:

Table 3. Ideal positive and negative solutions for various indicators of site sponge construction evaluation

nape	Positive ideal solution	Negative ideal solution
Rainwater utilization rate	0.89441825	0.00004472
Number of emergency drainage equipment	0.70710678	0.0000707
Sunken green area	0.70565033	0.06415003
Plant species	0.9918366	0.09016696
Permeable pavement rate	0.99987926	0.01098768
Waterlogging points	0.70710678	0.0000707
Maximum annual surface runoff in rainy season	0.7710102	0.00000126
Average catchment area of the site	0.9999956	0.00199564

Data sources: Calculated based on site survey data

Table 4. Comprehensive Scoring Index and Ranking for Sponge Construction of Buildings 2810–2812 in Xibei Community, Huaxi District from 2020 to 2022

Index value	Positive ideal solution distance (D+)	Negative ideal solution distance (D−)	Comprehensive score index	sort
2022	0.55473586	0.66608567	0.54560446	1
2021	0.83405181	0.28467875	0.25446587	3
2020	0.59636828	0.66062929	0.52556131	2

Data sources: Calculated based on site survey data

4.3 SWMM Site Runoff Simulation

The SWMM model can simulate the surface runoff generation process, the transmission process of the pipe network, and the accumulation and migration of pollutants during single or continuous rainfall events. It can also conduct low impact development simulation through the LID module to support the construction of sponge cities. The catchment area of the site at the current stage is calculated through site data collation and GIS Tyson polygon cutting, divide and draw a node network diagram in SWMM based on the layout of rainwater pipes in the community, as shown in Fig. 3:

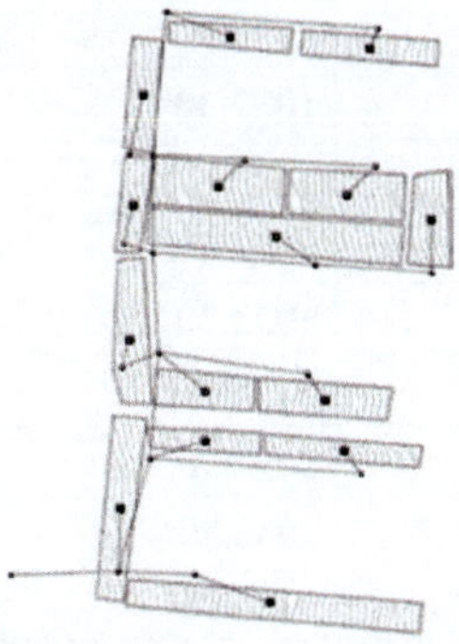

Fig. 3. Current node network diagram of the site (Image source: author's self-drawing)

Enter the rainfall data of the rainy season in Huaxi District from 2020 to 2022, set the relevant parameters of the pipeline nodes, and simulate in the form of dynamic waves to finally obtain the runoff results and the summary of the two data items of irrigation canal overload, which are summarized in Table 5–6:

Table 5. Surface Runoff Simulation Results of Building 2810–2812 in Xibei Community, Huaxi District during the Rainy Season (March October) from 2020–2021

Sub catchment area	Area/m^2	particular year	Total precipitation/mm	Peak runoff/CMS	Runoff coefficient
1	65.05	2022	625.90	22.875	0.939
		2021	692.70	23.472	0.930
		2020	1033.30	35.234	0.980
2	77.86	2022	625.90	25.575	0.925
		2021	692.20	26.246	0.922
		2020	1033.30	39.556	0.966
3	142.56	2022	625.90	36.973	0.878
		2021	692.90	37.943	0.868
		2020	1033.30	57.334	0.918
4	121.70	2022	625.90	33.561	0.890
		2021	692.90	34.445	0.882
		2020	1033.30	52.085	0.930
5	81.72	2022	625.90	26.345	0.921
		2021	692.90	27.038	0.917
		2020	1033.30	40.781	0.962
6	200.15	2022	625.90	48.851	0.842
		2021	692.90	50.114	0.829
		2020	1033.30	75.326	0.884
7	77.16	2022	625.90	25.433	0.926
		2021	692.90	26.101	0.922
		2020	1033.30	39.330	0.966
8	99.16	2022	625.90	29.634	0.906
		2021	692.90	30.415	0.900
		2020	1033.30	45.978	0.946
9	53.63	2022	625.90	20.223	0.954
		2021	692.90	20.744	0.955
		2020	1033.30	30.960	0.995
10	73.46	2022	625.90	24.674	0.930
		2021	692.90	25.321	0.927

(*continued*)

Table 5. (*continued*)

Sub catchment area	Area/m^2	particular year	Total precipitation/mm	Peak runoff/CMS	Runoff coefficient
		2020	1033.30	38.118	0.970
11	192.60	2022	625.90	44.603	0.854
		2021	692.90	45.762	0.841
		2020	1033.30	68.922	0.894
12	76.27	2022	625.90	25.252	0.927
		2021	692.90	25.916	0.924
		2020	1033.30	39.041	0.967
13	57.78	2022	625.90	21.220	0.949
		2021	692.90	21.769	0.948
		2020	1033.30	32.568	0.989
14	104.80	2022	625.90	30.646	0.902
		2021	692.90	31.454	0.895
		2020	1033.30	47.562	0.942
15	128.48	2022	625.90	34.689	0.886
		2021	692.90	35.602	0.877
		2020	1033.30	53.828	0.926

Data sources: SWMM simulation calculation based on the data provided by Huaxi District Meteorological Bureau

Table 6. Overload Results of Pipe Canals in the Rainy Season from 2020 to 2022 at the Site of Building 2810–2812, Xibei Community, Huaxi District (Outstanding Problems)

Pipe canal	particular year	Hours above normal turbulent flow	Capacity limited hours
5	2022	567.46	2.56
	2021	567.76	2.68
	2020	570.72	3.18
17	2022	792.62	0.06
	2021	793.35	0.10
	2020	792.77	0.11

(*continued*)

Table 6. (*continued*)

Pipe canal	particular year	Hours above normal turbulent flow	Capacity limited hours
20	2022	1675.13	0.01
	2021	1677.80	0.01
	2020	1678.38	0.01
25	2022	412.02	1.29
	2021	412.22	1.22
	2020	411.88	0.94

Data sources: SWMM simulation calculation based on the data provided by Huaxi District Meteorological Bureau

Based on the analysis results of site assessment data, it can be concluded that the current site is insufficient to cope with the large flow in the rainy season. The main reasons are as follows:

(1) The poor permeability of the site pavement makes the rainwater infiltration distance too small to form large surface runoff;
(2) Low vegetation coverage and no sinking transition with the ground;
(3) The unreasonable distribution of rainwater facilities in the site often leads to overloading of some pipes and canals;
(4) The recycling and utilization rate of rainwater in communities is extremely low, and water storage facilities are insufficient to form effective storage and utilization of precipitation;
(5) The drainage structure lacks adaptability and cannot adapt to local conditions to solve the drainage problems of the community with large elevation differences. The above reasons cause great pressure on the site drainage system.

5 Sponge Transformation Measures for Old Communities in Mountainous Cities

5.1 Design Strategy

According to the design requirements of sponge transformation and the site conditions, the design and transformation are carried out based on the following two strategies:

(1) Construct a community mini blue green network system and optimize the structural design. On the Deep Level of Community Greening and Rainwater System. In combination, update the community flood drainage methods and optimize the structure of the site drainage system.
(2) Strengthen and transform the foundation toughness to strengthen disaster prevention toughness. Utilizing the advantages of large longitudinal gradient and elevation difference in mountainous cities, strengthening the foundation toughness, making full use of the role of buffer zones, and conducting sponge transformation and reinforcement in the longitudinal direction of the foundation.

5.2 Community Site Renewal Design

(1) Update Site Paving

The original site pavement is made of cement concrete with poor water permeability. Replace the roadway material of the community with a permeable asphalt pavement with an average permeability coefficient of 0.0455 cm·s–1 cm·s to improve the ground's water permeability and heat dissipation, reduce rainwater evaporation, and supplement groundwater resources (Figs. 4 and 5).

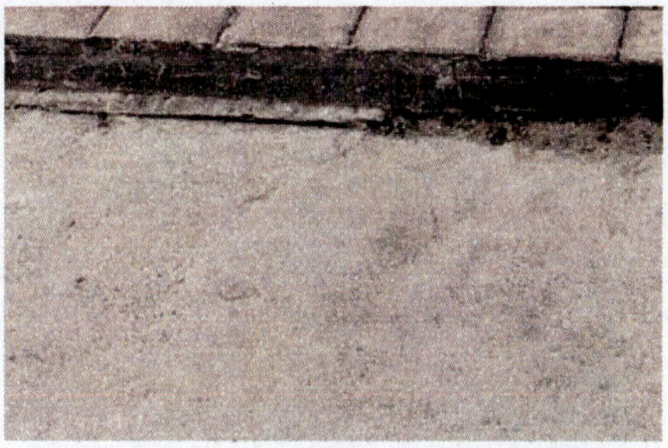

Fig. 4. Permeable asphalt pavemen material of the site (Image source: Internet)

Fig. 5. Current lane pavement (Image source: self-taken by the author)

Currently, the materials used for the site between community buildings are impermeable, and short-term heavy rainfall can only flow into the open ditch. The original foundation is replaced with permeable floor tiles in the order of water texture paving, which improves the greening rate of the community, significantly improves the absorption capacity of rainwater, effectively controls rainwater runoff, improves the degree of rainwater infiltration, alleviates surface runoff, and achieves natural infiltration and purification.

(2) Add rainwater facilities

Between 2810 and 2811, and between 2811 and 2812, there is a row of abandoned 500 mm wide brick flower beds, and some of them have been used as vegetable gardens by local residents, with minimal capacity to absorb rainwater. Dismantle the original flower pond, set up shallow vegetation ditches with several small dams on both sides of the building adjacent to the apron periphery of the building, build a blue-green net system, and use multiple vegetation covers (see Table 7) and ecological structures to collect, slow down, purify, and permeate the rainwater runoff of the site. A rainwater inlet is provided in the vegetation shallow ditch area (at the curb fracture), and 1% of the site is provided, Slope finding to promote rainwater entry; The terrain between the small dams is relatively low. After rainwater flows into the vegetation landscape area of the shallow ditch and spreads over the shallow pit with low terrain, it will be stopped by a 180 mm deep cutoff dam. The intercepting dam is composed of river pebbles and gravels, which can prevent rainwater from flowing directly through the sedimentation tank or shallow ditch area, allowing rainwater to fully collect and settle, and merge into groundwater. After flowing into the shallow ditch, rainwater will fall from one unit in the shallow ditch to another unit, achieving maximum absorption of rainwater by the shallow ditch. Some curbs far from the center of the site are not provided with openings, making the shallow trench itself a part of the dam and mitigating the flow rate. This device can enrich the ecological level of the community, improve vegetation coverage, form a sinking transition with the ground, and enhance the community's ability to resist floods and waterlogging.

(3) Inside of vegetation shallow ditch

Add a centralized water storage facility system under the open space on the south side of No. 2810, including a water storage tank, composite flow filter, rainwater waste flow device, sewage intercepting basket grille, diversion well, and a series of filtration devices to alleviate the overload phenomenon of pipe channels, improve the storage capacity and separation capacity of rainwater on the site, achieve efficient storage of rainwater, and achieve internal self-circulation of water.

(4) Enhanced drainage structure

The uncovered open ditch in the community has a small service area, poor drainage capacity, and does not have the ability to separate rain and sewage. When extreme weather comes, the uncovered open ditch has a potential safety hazard that is easy to stumble and step on. Replace the traditional open drainage ditch with a linear slot type drainage ditch (blind ditch), transfer the drainage ditch to the ground, and leave only 20 mm drainage joints on the ground. This can make up for the shortcomings of the traditional drainage ditch, meet the water seepage demand during rainfall, and ensure pedestrian safety (Figs. 6 and 7).

Table 7. Overview and zoning of planting plants in sunken green spaces

Plant name	picture	Overview of root system	Planting zoning
Ligustrum quihoui	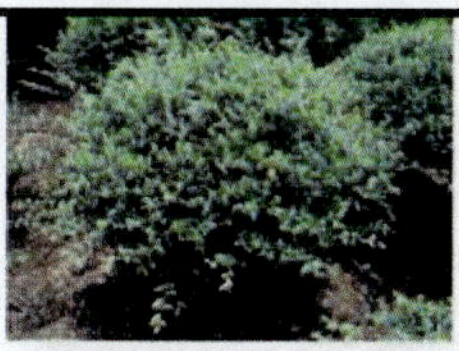	Large root system, easy to survive, can play a role in soil consolidation	Outside of vegetation shallow ditch
Fine leaved awn	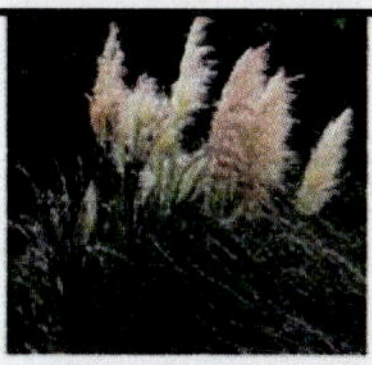	Developed root system, can consolidate soil, lodging-resistant	Near the center of the shallow vegetation trench
Chinese violet	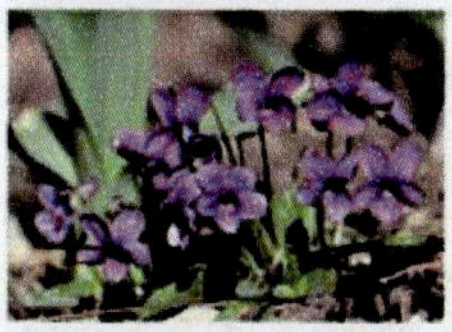	The root is a long conical shape with thin wrinkles	The outermost edge of the vegetation shallow ditch serves as a border
Bermuda grass	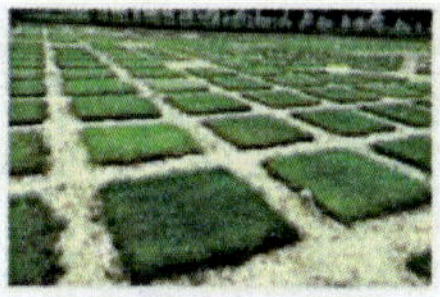	Its roots and stems have a strong spreading power and are widely spread on the ground, making it a good plant for embankment reinforcement and soil conservation	Inside of vegetation shallow ditch
Euonymus fortunei var.radicans	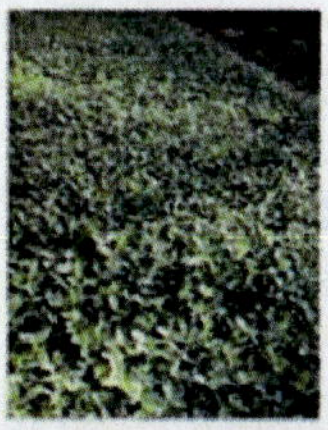	Its roots and stems have a strong spreading power and are widely spread on the ground, making it a good plant for embankment reinforcement and soil conservation	Inside of vegetation shallow ditch
Chinese pennisetum	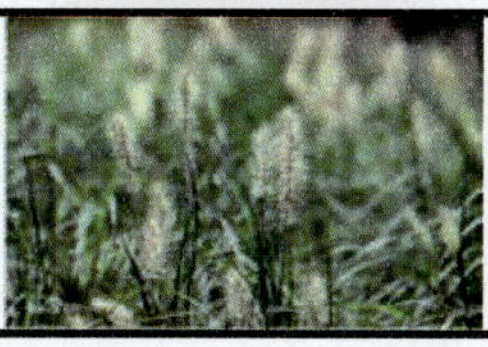	The fibrous roots are relatively strong and resistant to lodging, and can be used as a plant for consolidating embankments and preventing sand.	Inside of vegetation shallow ditch

Image source: network

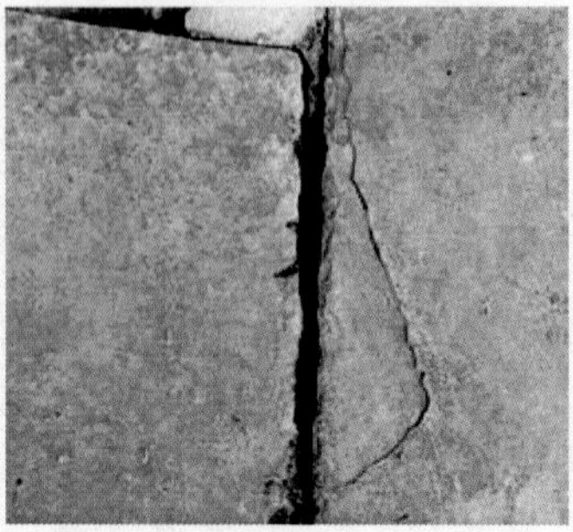

Fig. 6. Current drainage ditches on the site. Image source: author's self-photography

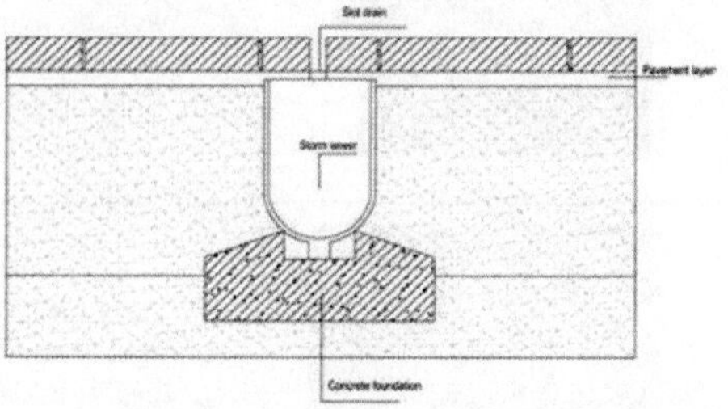

Fig. 7. Sectional Drawing of Slit Type Drainage Ditch. Image source: author's self-drawing

(5) The original trees within the site have been retained, and a circular tree pond has been added for each tree to assist the drainage facilities within the community in absorbing rainwater, eliminating the problems of soil hardening and small water seepage area that are not conducive to plant growth due to concrete pavement on the original site. At the same time, it echoes the water grain style grass planting brick field to enhance aesthetic interest; The lower part of the tree pond is connected to the municipal pipeline to timely discharge excess rainwater. In order to achieve the goal of updating community flood drainage methods and optimizing the structure of the site drainage system (Fig. 8).

Each green area within the scope of the vegetation shallow ditch is equipped with a rainwater grate 100 mm above the ground to quickly drain excess rainwater that the vegetation shallow ditch cannot handle during heavy rainfall. The drainage pipe connected below the rainwater grate is connected to the internal storage tank of the community to achieve efficient rainwater reuse.

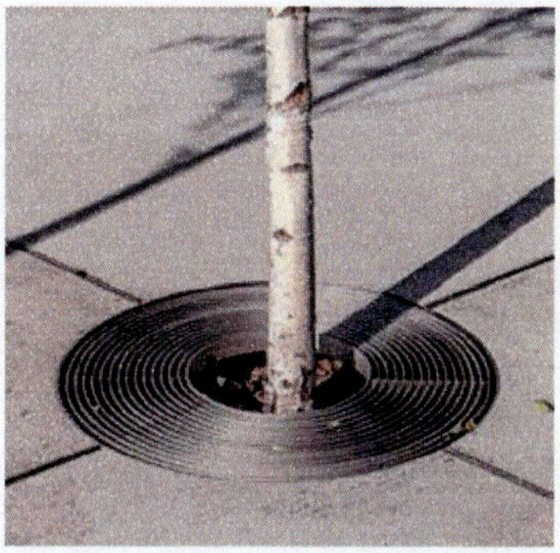

Fig. 8 Rendering of Round Tree Pool. Image source: author's self-drawing

(6) Drainage structure design based on natural longitudinal elevation difference of the site

The inner surface of the community is about 0.8 m lower than the main external transportation road. There is a height difference of about 250 mm between the inner carriageway and the squares between buildings 2811 and 2812. The terrain of the highway from the inside to the entrance of the community is in a downward trend. The lowest terrain is 19 m from the entrance of the community, with a gradient of 6.8%. Use the vertical elevation difference between the entrance of the community and the lowest point of the terrain, Set up two rows of north–south drainage ditches to divide the drainage area in two and fully absorb the rainwater entering the community from high altitudes. At the same time, the southern vacant lot of Building 2810 will be sloped by 1% based on the terrain to allow excess rainwater to flow into the drainage ditch at the lowest point of the terrain for centralized treatment. Based on the terrain trend of the roads within the community, the project places an east–west drainage ditch at the southwest corner of each building to divide the catchment area to avoid waterlogging in the community located in low terrain areas and affecting residential safety.

(7) Subdivision of catchment area

Aiming at the problem of heavy drainage pressure caused by the excessive catchment area of the community, the site catchment area is re planned: starting with increasing rainwater drainage channels, rainwater from different catchment areas is introduced into the different rainwater drainage channels set up above, so as to scientifically reduce the catchment area of the site and improve the rainwater recovery and utilization rate of the community (Fig. 9).

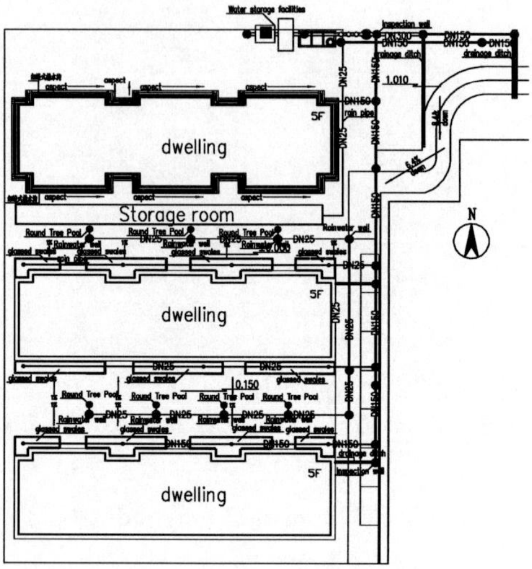

Fig. 9. Design drawing of community sponge transformation. Image source: author's self-drawing

6 Fitting Analysis of Sponge Transformation in Old Communities

6.1 Fitting Analysis of Site Parameters After Reconstruction

The parameter values of various indicators that have changed after the renovation are brought into the formula, and the comprehensive score index of the site is calculated again using the TOPSIS method. The results are compared with those before the renovation, as shown in Table 8:

Table 8. Comprehensive score index and ranking of fitting data after sponge transformation of building 2810–2812 in Xibei Community, Huaxi District

particular year	Positive ideal solution distance (D+)	Negative ideal solution distance (D−)	Comprehensive score index	sort
2022 (After transformation)	0.17935939	0.97303726	0.84435967	1
2022	0.55473586	0.66608567	0.54560446	2
2021	0.83405181	0.28467875	0.25446587	4
2020	0.59636828	0.66062929	0.52556131	3

Data sources: Calculated based on the modification design material parameters

After the transformation, the comprehensive score increased by 54.86% on average compared to the three years before the transformation, and the community's waterlogging prevention and waterlogging resistance capacity was significantly improved. The sponge transformation achieved significant results.

6.2 SWMM Runoff Simulation

According to the newly delineated catchment area, improved official network layout, and underground water storage facilities during the renovation design process, various vegetation and shallow ditch drainage facilities designed for the aboveground part are re modeled in SWMM, and LID low impact control parameters are added to fit the permeable ground conditions after the renovation. The monthly precipitation in the rainy season of Huaxi District in 2022 is taken as the rainfall meter data, and the final simulation results of rainfall and runoff on the reconstructed site are shown in Table 9 (Fig. 10):

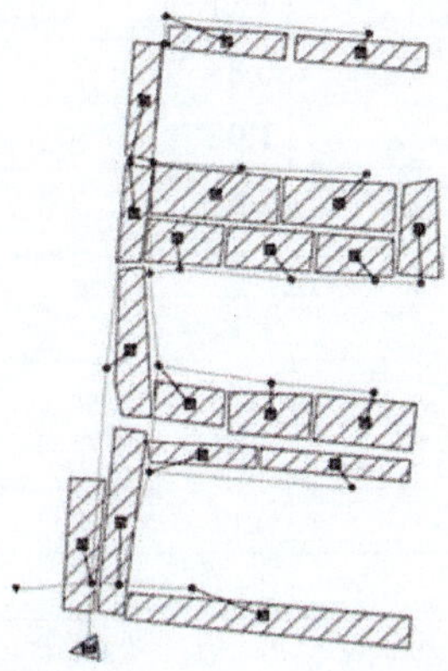

Fig. 10. Node network diagram after ground improvement. Image source: author's self-drawing

According to the comparison of the data in Table 5, based on the expected precipitation data in Huaxi District in 2022, the peak runoff value of the community has significantly decreased, with a maximum decrease of 94.396%; The surface runoff coefficient has been significantly improved, and the comprehensive runoff coefficient of the site has been reduced from 0.907 to 0.526, with an optimization rate of 42.006%. At the same time, it also meets the requirement that the runoff coefficient of the reconstruction and expansion projects in the sponge community should be controlled between 0.5 and 0.6 as stipulated in the "Sponge City Construction Plan for the Central Urban Area of Guiyang, 2018 to 2025". Due to the fact that the community entrance ramp is divided into two catchment areas and added with special slit drainage ditches, vegetation shallow ditches with small dams, and permeable pavement, surface runoff is buffered in this area.

Visual comparison between Fig. 11 and Fig. 12 shows that at the same time of simulation, the surface runoff absorbed by the buffer zone reaches an average of 25.673% of the peak flow, significantly reducing the drainage pressure on the internal sites of the community. The comprehensive runoff of the site in this area, the coefficient is reduced from 0.886 to 0.422, and the optimization rate reaches 52.370%

By comparing Figs. 13 and 14, it can be seen that after the site road pavement is updated in combination with LID control, the peak flow of road rainstorm can be reduced by 23.087%, and the maximum optimization of the site comprehensive runoff coefficient can reach 46.674%.

Table 9. Surface runoff simulation results in rainy season after reconstruction of building 2810–2812 in Xibei Community, Huaxi District

Sub catchment area	area/m^2	Peak runoff/CMS	Runoff coefficient
1	63.02	1.00	0.553
2	63.06	1.11	0.530
3	63.22	1.51	0.452
4	63.17	0.43	0.211
5	63.07	1.72	0.464
6	63.06	1.11	0.532
7	62.99	0.88	0.577
8	63.05	1.08	0.538
9	63.11	2.50	0.503
10	62.94	1.39	0.616
11	62.91	1.12	0.646
12	62.99	1.74	0.579
13	63.02	1.99	0.555
14	62.85	0.11	0.776
15	62.85	0.11	0.481
16	62.85	0.89	0.534
17	62.87	1.32	0.422
18	63.14	0.78	0.511
19	62.85	0.96	0.681
20	61.89	0.11	0.435
21	64.77	0.45	0.433

Data sources: Calculated based on relevant parameters of the renovation design

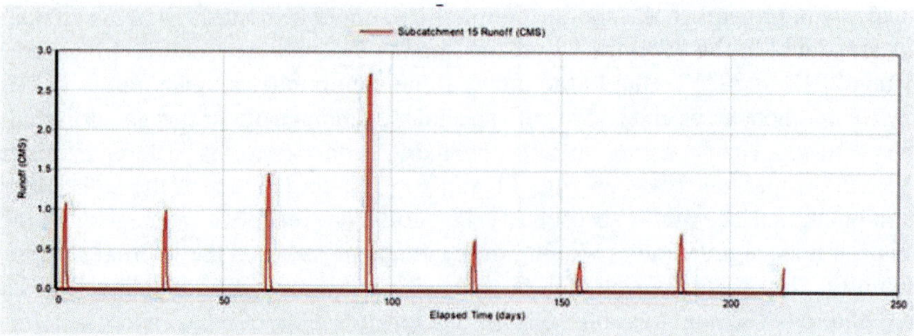

Fig. 11. Variation of runoff peak curve in rainy season (March October) before site entrance ramp reconstruction (Image source: self-drawn by the author)

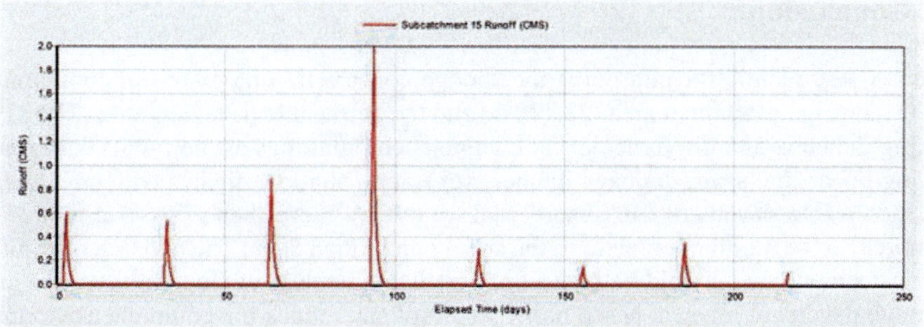

Fig. 12. Variation of peak runoff curve in rainy season (March October) after the reconstruction of site entrance ramp (Image source: self-drawn by the author)

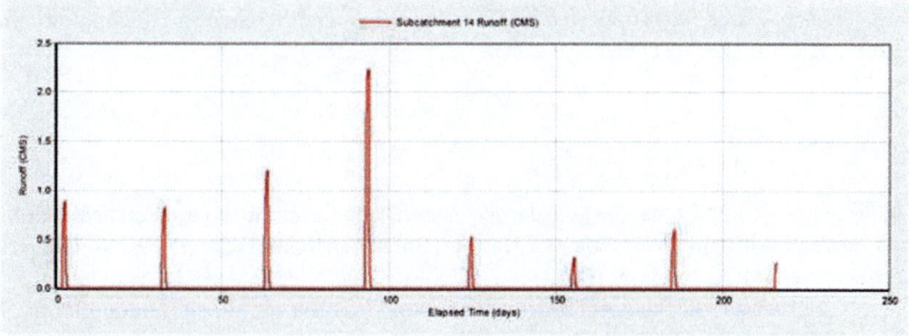

Fig. 13. Peak variation of surface runoff in rainy season (March October) before pavement pavement renewal. Image source: author's self-drawing

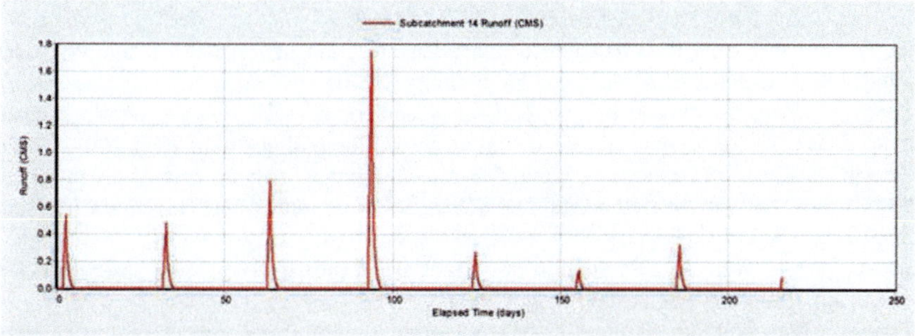

Fig. 14. Peak variation of surface runoff in rainy season (March October) based on fitting analysis after pavement renewal. Image source: author's self-drawing

7 Conclusion

The two key factors for improving the sponge construction index of old communities are the average catchment area (23.490%) and the permeable pavement rate (21.537%); Taking the elevation difference of mountainous communities as the runoff buffer area, setting linear fine seam drainage ditches, vegetation shallow ditches with small dams, and permeable pavement can absorb peak runoff by more than 20%, and reduce the comprehensive runoff coefficient of the site by more than 50%; The division of community catchment area should be based on, Elastic adjustment of slope in the same area; The site pavement renewal based on LID control can reduce the comprehensive runoff coefficient of the site by more than 40%, and the peak value of road rainstorm runoff can be reduced by more than 20%; The evaluation system can provide more targeted guidance for sponge transformation design.

Funding. This research was supported by Guizhou Provincial Science and Technology Projects (Qian Ke He Ji Chu-ZK[2022] General 234).

References

1. Xu, W.: Research on Storm Surge Adaptive Landscape Design from the Perspective of Urban Resilience: Taking the West Coast of Bao'an, Shenzhen as an Example. Xi'an University of Architecture and Technology (2012)
2. Xiao, X.: Research on the special design of sponge city in architectural community. Build. Technol. (04), 10-11 (2022)
3. Construction Plan for Sponge City in the Central Urban Area of Guiyang
4. Zhang, C.: Analysis and improvement of the current situation of mountain buildings in the Sifeng Mountain area of Jiamusi. Shanxi Archit. (36), 1–2 (2013)
5. Pan, L., Yuan, S., Liu, Y.: Current situation and research progress of sponge city construction in China. Sichuan Water Conserv. (06), 132–134 (2022)s
6. Li, P.: Research on Urban Resilience Space Construction in Response to Flood Disasters. Huazhong University of Science and Technology (2020)
7. Peng, H.: Research on waterlogging prevention and control technology under the renovation conditions of old residential areas in mountain cities. Chongqing Archit. (S1), 206–207 (2022)
8. Zhang, Z., Zhang, Y., Zhang, J., Lang, Q., Yan, J., Duan, C.: Evaluation of urban resilience based on entropy weight method and TOPSIS model: a case study of Changchun City. Disaster Sci. (01), 213–217 (2023)
9. Li, A., Liu, J., Wang, S.: Evaluation of current urban drainage networks based on SWMM model simulation. Water Purif. Technol. (11), 123–125 (2022)
10. Zhang, J., Xia, X., Dong, S., Pan, X., Ma, X.: Discussion on the sponge transformation model of existing urban residential areas. Beijing Plan. Constr. (03), 120–121 (2021)
11. Zhang, Y., Tian, F., Dong, Y., Wang, L., Wang, Y.: Discussion on the key points of sponge transformation design for old residential areas - taking Tianjin Tianshui Liyuan community as an example. Resid. Ind. (08), 88–92 (2022)
12. Zhang, H., Chang, Q., Zhu, J., Tang, G.: Research on the resilience evaluation system of sponge cities. J. Southwest For. Univ. (Soc. Sci.) (05), 57–61 (2022)
13. Shi, Y.: Enlightenment of rainwater and flood management in foreign cities on sponge city construction in China. Shanxi Archit. (09), 98–99 (2018)

14. Zuo, H.: Review of foreign research progress in sponge city construction. Ind. Technol. Forum (01), 102–103 (2019)
15. Chen, J., Pan, C., Chen, Y., Jiang, Z., Pei, Y.: Evaluation of sponge city construction effectiveness based on analytic hierarchy process. Munic. Technol. (11), 151–153 (2022)

Check for updates

Study on "Intelligent" Renewal Strategy of Block Under Stock Optimization

Yao Wu[1,2], Tianhui Li[1](✉), and António Candeias[3]

[1] Soochow University, 199 Renai Road, Suzhou, China
1098718820@qq.com
[2] International Cooperative Joint Laboratory of Cultural Heritage Conservation Science Supported by the Belt and Road Initiative, 199 Renai Road, Suzhou, China
[3] Universidade de Évora, Largo dos Colegiais 2, 7004-516 Evora, Portugal

Abstract. Blocks carry the collective memory of residents and become a rare stock space in the city, so the "intelligent" renewal of blocks is a necessary stage and has due significance in the current context of smart city development in China. Given the numerous issues with the overall appearance, street space, ancillary facilities, management, and maintenance, this paper uses Yangzhou Dongguan Block as the research object to analyze the current state of the protection and renewal of blocks in China through literature analysis and field research, we propose the "intelligent" renewal strategy based on stock optimization and intelligent renewal as an entry point, including block renewal energization, information collection visualization, block service systematization, and intelligent operation management to guide the organic renewal practice and provide a clearer renewal approach for long-term sustainable development of China's blocks.

Keywords: Stock Optimization · Block · Intelligent · Organic Renewal · Dongguan Block

1 Introduction

Promoting the intelligent renewal of blocks actively has the potential to, on the one hand, challenge wisdom about conservation and development, introduce novel concepts, technologies, and development strategies into the sustainable development of blocks, and significantly reduce conflicts between conservation and development, efficiency, and equity in the short- and long-terms [1]. It can also address issues with material, structural, and functional degradation. On the other hand, it can successfully encourage scientific advancement, effective management, and block optimization of blocks, create a modern and humanized urban stock space that better meets the needs of contemporary life, and create a situation where urban construction and block preservation promote one another. In light of this, we examine the growth of blocks and the issues they encounter against a backdrop of stock optimization. Based on the smart city and using the realization of regional cultural heritage and sustainable development as the guide, we then thoroughly explore the strategies and methods of "intelligent" renewal of blocks, finally realizing the rigorous, scientific, and intelligent renewal of blocks [2].

W. Guo et al. (Eds.): GBCESC 2023, LNCE 328, pp. 1004–1012, 2024.
https://doi.org/10.1007/978-981-99-9947-7_97

2 Stock Optimization and Organic Renewal

2.1 Concept Identification

Stock optimization refers to the renovation and optimization of the built-up city, buildings, and facilities under the current state of relative urban development saturation, as well as the sustainable development of the city by enhancing its adaptation and increasing the efficiency of the current land space. This is essentially the optimization, restructuring, and upgrading of the current stock of land resources to achieve its optimal adjustment [3]. A progressive urban development path of "high growth - efficient growth - smart growth" has been determined as a result of China's lengthy history of incremental urban growth, which has produced numerous issues like lack of style, lack of vitality, and outdated amenities. In the context of locking the urban growth boundary, stock optimization is a high-quality method for resolving a variety of practical issues. It is also a necessary condition for urban planning and development to shift from extensiveness to connotation and roughness to intensiveness in the context of sustainable development in China [4]. Integrating existing cultural resources, juggling conflicting interests, and preserving their integrity and authenticity are crucial components of the coordinated development of the city in the context of "revitalizing the stock, optimizing the quantity, and improving the quality" as an important resource of the stock economy.

Liangyong Wu introduced the idea of "organic renewal," which he defined as choosing the appropriate scale, and specifications, in line with the transformation requirements and content, and bridging the gap between the present and the future [5]. The block has several artifacts and a typical, largely unaltered landscape, but it also struggles with "protective deterioration" and "creative destruction," two serious issues. Due to the abundance of residential land and the concentration of commercial and cultural tourism industries in the area, residents, business owners, and visitors have expressed a greater need for modern production and living, infrastructure renewal, and living and recreational space renovation based on the preservation of the area's cultural heritage. Therefore, it is now necessary to investigate the conflict between protection and development, carry out research on the implementation path of organic renewal, and promote the sustainable development of the blocks rather than using the lowest common denominator of protection as the starting point for block protection. The organic renewal of blocks involves not only material regeneration but also a thorough resurrection of the blocks that addresses a wide range of issues, including social, economic, cultural, and ecological restoration as well as urban rehabilitation.

2.2 Stock Optimization Is the Value Orientation for Organic Renewal

Stock optimization theory strives to preserve the integrity and originality of urban history by achieving the full exploitation of existing resources and internal regeneration through inventory land use and excavation and efficiency enhancement [6].

The organic rejuvenation of blocks has now become a consensus among the government, scholars, and the public, while stock optimization is the current transformation stage of national urban planning and development. The purpose of organic renewal of blocks is to realize the reuse of resources and sustainable development, while actively protecting and intervening in the existing cultural heritage, and constantly renewing and renovating them in terms of their appearance, architecture, streets and alleys, production, and life, to provide residents, tourists and merchants with a more livable living environment, a richer landscape experience and more convenient commercial conditions, and finally realize the overall optimization and enhancement of blocks at different levels and with different people, this is also in keeping with the new criteria and ideals for historical and cultural heritage in the context of stock optimization in China, and encourages China's urban growth to shift from "quantitative" expansion to "qualitative" enhancement.

On the platform of organic renewal theory, we explore more rational and rigorous strategies and methods, and intervene in new technological means such as mega data, artificial intelligence, cloud computing, and VR, so that the research on organic renewal in China's blocks can be deepened and expanded to further promote the development of urban stock optimization.

2.3 Organic Renewal Is an Important Path to Promote Stock Optimization

Organic renewal advocates the exploration of urban regeneration based on the original real place, with an integrated and flexible perspective, and advocates a small-scale progressive regeneration model, which is an effective means to improve land use rate, achieve rational allocation of spatial resources, adjust industrial structure, improve human living environment and thus achieve overall optimization.

Stock optimization is mainly through urban renewal and other means to promote the functional optimization and adjustment of built-up urban areas, and the current organic renewal methods of blocks include strategies such as application-based lease surrender, restoration construction, business management, and other methods such as demolition of the old and construction of the new model, retro model, and small-scale progressive transformation model [7]. While stock optimization is a macro concept proposed by the state for urban development, their ultimate goal is to achieve sustainable urban development, "intelligent" organic renewal is more subtle and extensive in the face of new AI technology, IoT terminology, and mega data. It is suitable for implementation in urban spaces with complex property rights and heritage freeze protection, such as blocks, and it can also adapt to today's smart city development.

As an important stock of land in the city, the renewal and transformation of the block is a key and difficult task for the city to change from "incremental expansion" to "stock optimization", which is not only the result of the objective reality of resource constraint, but also a positive response to implement the national scientific development requirements and actively change the development concept.

3 The Current Situation of Conservation and Renewal of Block: Yangzhou Dongguan Block as an Example

3.1 Overview of the Block

The Dongguan Block, with a total length of 122 m and a total area of 77.54 ha, of which 32.47 ha is the core protection area, is situated in the northeastern part of Yangzhou City. It extends from the canal in the east to Guoqing Road in the west, from Wenchang

Road in the south to Yanfu Road in the north. There are 4 nationally significant cultural preservation units, including Geyuan Garden (Grand Canal World Cultural Heritage site) and Yipu, Residence of the Wangs, as well as 25 provincial and local preservation units, including Dongrong Garden, Huashi Garden, and 8 historical buildings. Since the Tang Dynasty, the Dongguan Block has served as the east-west axis of Yangzhou, the most prosperous commercial and handicraft center, the centralized embodiment of canal culture, and the axis of land and water transportation. It has preserved a more complete "river-city-street" spatial pattern, a better street system, and traditional building groups, giving it a more uniform overall appearance (Fig. 1).

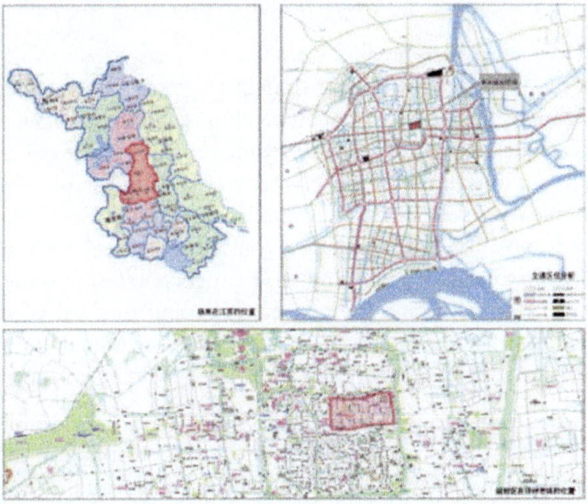

Fig. 1. Preservation Plan Location Map of Dongguan Block (Picture Source: Yangzhou Natural Resources and Planning Bureau)

3.2 The Current Situation of Conservation and Renewal

The Dongguan Block, one of Yangzhou's four existing blocks, is the oldest in terms of protective development. It has retained many of its original residents while gradually growing into a significant commercial area. Since 1997, the block has seen several conservation and enhancement initiatives, actively managing the interaction between preservation and development and conservation and regeneration. Up until 2006, researchers investigated the formation of a multiform organic renewal protection model, taking into account the individual differences of the four blocks, the Dongguan Block produced the "government-led, state-owned enterprises to operate" mode.

In response to urbanization and people's desires for use, the active preservation of the block has produced varied degrees of autonomous and passive renewal. The most notable is the commercial growth that has been sparked by the rapid growth of tourism. Currently, the Dongguan Block is home to a large number of unique culinary and historic businesses, including around 242 eateries and tourist attractions. The development of blocks is a distinctive branch of old city renewal that should consider a variety of factors including traditional culture, regional characteristics, and urban landscape, and in a local context, "preservation" should be given priority, which poses some challenges to the ongoing process of block renewal [8].

The main issue to be resolved in the current "in" situation is how to resolve the conflict between preservation and renewal, realize the integration of traditional culture and modern life, preserve the traditional appearance of the block, and show the real life of the block under the history, and at the same time allow the aborigines, merchants, and tourists in the block to enjoy the modern and convenient life brought about by intelligent technology.

3.3 Issue Analysis

On the one hand, the Dongguan Block is subject to physical, structural, and functional degradation due to complicated property ownership issues and frozen protection of cultural relics, which makes it challenging for residents to meet the needs of modern life, while traditional static maintenance and the lack of awareness of conservation among the indigenous people have resulted in the block's "protective decay"; On the other hand, because of its prime position and associated tourism benefits, it has undergone a considerable deal of "constructive destruction" to be incorporated into the city's economic growth. To successfully encourage the conservation and renewal of the block, realize the stock optimization of Yangzhou city, and advance the sustainable development of China's urban construction, it is required to delve deeper into the issues and their causes.

Overall Appearance. The Dongguan Block is Yangzhou's most emblematic old street, with its winding green brick lanes and long slate streets displaying thousands of years of urban history that led to the development of Yangzhou's unique style of living. The block's texture bears the imprint of history, and its spatial arrangement is regular and orderly. However, in recent years, urban development and construction have damaged the block's original design. On the one hand, this is because there is a lack of a thorough understanding of the block's historical context and present state, as well as a lack of knowledge about the protection measures and repair options, making it impossible to maintain the historical perspective and adopt effective protection measures. The block has gained fresh life thanks to a huge number of business influxes, but it has also lost a lot of its beauty, and several newly constructed industrial facades are no longer designed with visitors in mind due to the slow information gathering and slow popularization, this extensive landscape damage cannot be repaired. Therefore, to guide the preservation and renewal of the block style, information about the historical influence of the block, traditional architecture, material use, repair techniques, and other topics should be recorded and easily accessible.

Block Space. With streets like Guangchumen Street, Guan Alley, Qian'anjia Alley, Yaguanren Alley, and others, the Dongguan Block creates a distinctive street environment. After dredging and rectification, a complete street system can be built while essentially maintaining the block's original scale and direction. The Dongguan Block, however, was primarily a living space in ancient times, and the streets and alleys were relatively narrow. Now, however, it has to be widened in many streets and alleys to accommodate the driving of motor vehicles and the visitation of a large number of tourists, which significantly affects the scale of the vertical interface of the block space and renders the restrained and exquisite streets and alleys a victim of modern urban development. The use of intelligent transportation to direct the diversion of people and traffic flow is regarded as reconstructing an orderly block as narrow streets and large flows of people have become challenging areas in the renewal of a block.

Ancillary Facilities. Due to the long history of the Dongguan Block, its ancillary amenities are either outdated or lacking altogether, making it **impossible** for residents to meet their modern living demands and creating a significant tension between modern and traditional values. Infrastructure such as pipes, water, power, etc. has some inconveniences, sanitation facilities, and firefighting support facilities need to be improved, and visitor services and convenience services might use improvement. The block's life and production are made exceedingly "unintelligent" by the inconvenience of ancillary amenities, which also prevents the block from being renewed and developed.

Management Operations. Although it offers a glimpse into Yangzhou's millennia-old **past**, the Dongguan Block is not highly liked by locals, business owners, or tourists. The cause is that the block's management is lax, making it difficult to properly direct and control phenomena like street stalls, private residential building renovation, and traffic congestion. Operationally, it is impossible to keep up with the pace of the times and implement systematic operation strategies, and the block's entrance, the layout and direction of the tour route, and the distribution of its commercial components have not been guided effectively or scientifically. Dongguan Block has fallen into homogeneous competition due to poor management and operation, thus creating a set of intelligent management and operation strategies appropriate for its growth context is a crucial first step in fostering its distinctive development.

4 "Intelligent" Renewal Strategy of Block

Intelligent means have brought technical updates to the urban planning industry, which has led to a change in the way of thinking. The regeneration of blocks is a complicated, demanding, and tightly associated urban management task to people's production and life [9]. The concept of "people-oriented" has always been respected in the creation and development of blocks; people are the blocks' creators, users, and keepers. The "intelligent" renewal of blocks should be based on regional culture and historical heritage, with an eye toward the block's sustainable development, and use cutting-edge intelligent technology to realize the metabolism of the block and meet the needs of the community (Fig. 2).

4.1 Block Renewal Synergization

It is necessary to move beyond the technical level of "intelligent" renewal of blocks to one of systematic, integrated, coordinated development of space, economy, society, system, and management. To realize the control of block information by various groups of people and adapt to the development and requirements of smart cities more effectively

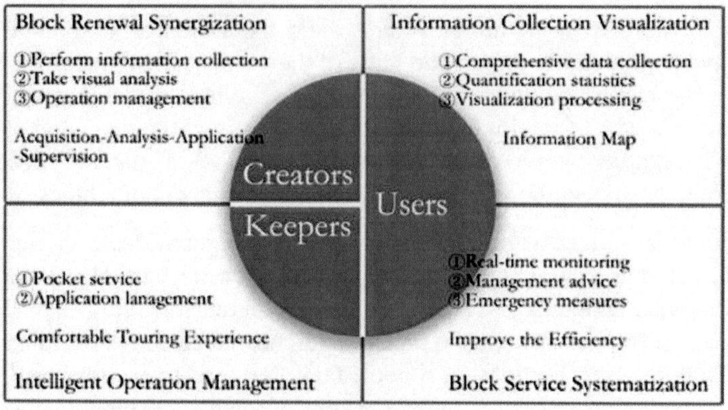

Fig. 2. Intelligent Renewal Framework Diagram

and conveniently, an intelligent update mechanism of "acquisition-analysis-application-supervision" is formed in combination with the status quo and update requirements of all levels of the block.

The tangible cultural heritage of the neighborhood is gathered using geographic information systems, global positioning systems, remote sensing technologies, etc. The intangible cultural heritage is tallied using online questionnaires and surveys, as well as the neighborhood's daily climate and the number of visitors. The information is then processed and analyzed using artificial intelligence techniques like cloud computing and mega data, which produce visual data like smart guides, electronic maps, and the suitability of the day's tour. Through smart apps and small programs, it provides services to visitors and locals. In terms of operation and management, the block's intelligent platform can also keep an eye on traffic, sanitation, flood control, and disaster prevention.

4.2 Information Collection Visualization

The block's cultural relics are complex, and as a result of the information superposition over time and the numerous upgrades and transformations carried out in the current time dimension, the block forms a network with multiple information in time and space. Additionally, some subtle information is insufficient to be written down due to its low historical value, so more traditional culture and building techniques can only be gradually lost in the block.

It is vital to picture the acquisition of information gathering to change this. First, it is important to gather information from old books, maps, texts, networks, oral histories, and other platforms about the block's location, cultural preservation units, underground caverns, former home sites, historic trees, famous trees, land use, architecture (age, floor height, quality, structure, style, and nature), public facilities, public spaces, street and lane spaces, greenery, height control, and transportation system. As a result, the block's information is more thorough and precise. The information is then broken down into a sum of data information from various logical systems, such as numbers, text, and visuals, depending on the usage requirements. The information is then quantified and tallied independently. Finally, it is visualized, and the three-dimensional model of the block is produced by digital technology, the data of buildings, streets, and alleys in the block are recorded, a complete and quantifiable information map is formed, and the heat distribution map of the protection renewal degree and progress of the block is obtained, and the current status of the block renewal is intuitively displayed. Additionally, the information gathering system is updated and corrected in real time to guarantee the accuracy and timeliness of the data and to offer the builders and maintainers the necessary references.

4.3 Block Service Systematization

The block's public service system serves as the cornerstone for "intelligence," and the development of public service platforms has an impact on the realization of various roles for "intelligent" renewal in the region [10]. Tourists want to be able to have a better, more convenient experience in the block. Afterward, based on the visualization of information collection, it can be displayed on a mobile device in other ways through the collection and analysis of data, offering tourists features like portable maps, tour suggestions, landscape planning, and tour guides and providing quick, secure, and convenient services. To increase the enjoyment of the trip and boost the effectiveness and experience of the tourists, it can also be coupled with other foods, quick movies, and other apps.

4.4 Intelligent Operation Management

The operation and management of blocks are essential to achieving the block's sustainable development, and it is required to construct a management and operation platform inside the block to increase management effectiveness and actualize intelligent management. In terms of operation, the data statistics of passenger flow are realized based on the block monitoring system and the people flow analysis system, and the real-time distribution of tourists is perceived through big data analysis and the flow of people in various areas in the future period is predicted, to improve the dredging efficiency and reasonably control the block capacity. Adopt constructive advice for merchants' façade building and sales strategies, encourage them to maintain the block style based on data gathering and analysis, and create targeted activity plans. In terms of management, we can enhance safety emergency measures in the event of some emergencies in the block, such as fire, traffic jams, etc.

5 Conclusion

The construction of smart cities is currently in full swing in China, and the country's urban development has entered a phase of stock optimization and transformation. In this context, the "intelligent" renewal of blocks is a necessary step and proper meaning of sustainable urban development. The relationship between stock optimization and organic renewal is sorted out in this paper, which also puts Yangzhou Dongguan's block through the renewal process. It also discusses the strategy for the "intelligent" renewal of blocks against the backdrop of stock optimization in China, offers theoretical support and method references for the renewal of blocks, awakens the vitality of the block, digs up the block's culture, and, on the other hand, give locals and visitors access to the smart life experiences made possible by contemporary science and technology.

Acknowledgment. The writing and research of this paper are obtained from the National Key R&D Program of China (2021YFE0200100) and the 2021 Policy Directed Program of Jiangsu Province (BZ2021015).

References

1. Summary of policies related to the construction of smart cities promulgated by the state in 2022. http://rank.chinaz.comdb.ga.net.cn/news/5232. Accessed 13 Jan 2023
2. Guangyao, S.: Study on smart city construction and urban transformation development. J. Party Sch. Fujian Prov. Comm. Communist Party China **446**(06), 19–26 (2017)
3. Bing, Z.: From "incremental expansion" to "stock optimization"-the motivation and path of transforming Shenzhen's urban master plan. Planner **29**(05), 5–10 (2013)
4. Rongzeng, L.: Urban spatial governance and reorganization based on stock optimization: the case of Zhengzhou city. Urban Dev. Res. **24**(09), 26–32 (2017)
5. Liangyong, W.: The way to improve the residential areas in the old city of Beijing: the organic renewal of urban cells and the exploration of "New Siheyuan" (07), 11–18 (1989)
6. Lisheng, Z.: Innovative thinking on planning and design methods in the context of land use inventory. In: 2015 China Urban Planning Annual Conference. Proceedings of the 2015 China Urban Planning Annual Conference, pp. 80–89. China Construction Industry Press (2015)
7. Caiyun, Q., Yang, Z.: A review of research on the optimal development of urban spatial stocks. Huazhong Archit. **36**(03), 12–14 (2018)
8. Di, Y., Guangya, Z.: Research on the normality and adaptability of conservation planning in historical and cultural districts: an example of conservation planning in Yangzhou Dongguan historical and cultural district. J. Archit. **490**(06), 40–43 (2009)
9. Qixin, G.: The characteristics and organic renewal path of Wenzhou historical and cultural district. China Ancient City (02), 73–78 (2015)
10. Haifeng, C., Xiaoling, Z., Xinmeng, D.: A study on urban historical and cultural neighborhood renewal strategies based on the concept of smart city - taking Guilin east-west lane as an example. Urban Archit. Space **29**(12), 28–31 (2022)

Evaluation of Urban Habitat Quality Based on Theory of Humanistic - A Case Study of HeFei Metropolitan Circle of China

Xingang Yang[1,2], Mengyuan Jin[1,2(✉)], and Xueyan Liu[1,2]

[1] School of Architecture Planning, Anhui Jianzhu University, Hefei 230000, China
jmy@stu.ahjzu.edu.cn

[2] Collaborative Innovation Center for Urbanization Construction of Anhui Province, Hefei 230000, China

Abstract. Under the current requirements of sustainable development, scientific and objective evaluation of the quality of urban habitat and analysis of the influencing factors of habitat quality are of great significance for the implementation of habitat improvement according to local conditions. Taking the HeFei metropolitan circle as a case study, the study uses mathematical statistics and geographic information methods to evaluate the quality of urban habitat from four dimensions: physiological need, safety need, social communication need and self-actualization need, and to analyse The study shows that the quality of urban habitat in the HeFei metropolitan circle of China is not only spatially heterogeneous but also influential. The study shows that: (1) the spatial distribution of urban habitat quality in the HeFei metropolitan circle of China varies greatly, with the quality of urban habitat in the eastern part of the city being significantly higher than that in the western part; (2) the quality of urban habitat is influenced by many factors such as the level of economic development, infrastructure arrangement and ecological civilization construction, among which the level of economic development has a significant impact. (3) The radiation effect of neighbouring cities is of great significance to the improvement of the quality of urban habitat. Through the study and analysis of the index factors affecting the improvement of urban habitat quality, we provide reference suggestions for promoting sustainable urban development.

Keywords: Theory of Humanism · Urban Human Settlement Environment · Quality Evaluation · HeFei Metropolitan Circle of China · Sustainable Development

1 Introduction

The quality of the human habitat, as a fundamental issue for human survival and development, has always been a core issue in urban research, and the urban habitat is becoming one of the hot spots for research in the fields of architecture, geography, environmental studies and planning [16]. The harmonisation of human survival and socio-economic development is of great importance for the sustainable development of society and economy and the construction of a liveable and workable human environment [2]. Since

W. Guo et al. (Eds.): GBCESC 2023, LNCE 328, pp. 1013–1029, 2024.
https://doi.org/10.1007/978-981-99-9947-7_98

China's urbanization has entered an accelerated phase, by the end of 2022, China's urbanization level has reached 65.22%, with an unprecedented scale and speed of urbanization. Urbanization is a double-edged sword. On the one hand, in these 40 years or so since the reform and opening up, China has completed the urbanization path of developed countries for more than 100 years, which is a great achievement; on the other hand, the geographical conditions, economic bases, social policies and resource allocation of different cities have not been fully considered in the new urbanization process [3]. On the contrary, the blind pursuit of high speed and over-expansion has brought about more problems, and the coordinated development of cities in terms of comprehensive carrying capacity, safety bottom line, livable quality and humanistic care is relatively insufficient [4]. The city's overdraft of resources, environmental degradation, traffic congestion, ageing before it ages, ageing before it gets rich, and challenges to sustainable development. People's living needs have also shifted from the pursuit of a pleasant physical environment to a greater focus on ecology, culture, security and equity and justice.

Since the 1950s, foreign scholars have been studying the human habitat, and the Greek scholar C. A. Doxiaodis proposed the establishment of a multidisciplinary group of human settlements, namely the "science of the human habitat". Since the Vancouver Declaration of the First Conference on Human Settlements in 1976, habitat as a discipline has received increasing attention from scholars [5]. In terms of research progress, foreign scholars have been focusing on the science of habitat as a discipline. In terms of research progress, foreign scholars have focused on the constitutive systems of the habitat environment, across the spatial scales of regions, cities and communities [6]. In addition, they have built evaluation systems and models mainly from the perspectives of ecology, human-earth relationship and urban planning, extended and deepened the evaluation content, and quantified the suitability of urban habitat environment [8]. The evaluation system and models are developed from the perspectives of ecology, human-earth relations and urban planning.

From the late 20th century, Chinese scholars began to summarise and sort out the comprehensive evaluation of urban environment at home and abroad, and explored different aspects of the evaluation of sustainable urban habitat development. Professor Wu Liangyong proposed a theoretical system of "habitat and environment science [2]. The Habitat environment, proposed by the natural system, human systems, social systems, housing systems and support systems and other five sub-systems, including global, regional, urban, community (villages and towns), buildings and other five levels of academic research on the Habitat environment [7]. It has gradually been developed to an operational level. Domestic research on habitat environment mainly includes the evaluation of the quality of urban habitat environment, the coordination and coupling of the quality of urban habitat environment with a certain development factor, and the influence of the quality of habitat environment on a region or city [8]. Li Shuai uses a combination of hierarchical analysis and entropy weighting to analyze the quality of urban habitat in Ningxia by achieving a more standardized weight distribution [9]. XIONG Ying quantitatively studied the relationship between urban habitat and economic development in Changsha. The main factors affecting the quality of urban habitat in the Shandong Peninsula urban agglomeration were investigated using principal component analysis [10]. XIONG Ying quantified the relationship between urban habitat and economic development in Changsha, and Dong Zuocheng used principal component analysis to investigate the main factors affecting the quality of habitat in the Shandong Peninsula urban cluster.

"At this stage, how to promote the optimization of urban spatial structure and quality improvement based on the concept of sustainable development has become a major concern. In order to fully implement the new requirements put forward at this stage, and to enhance the quality of urban habitat based on the concept of sustainable development, Anhui Province has issued relevant documents while accelerating the development plan for the HeFei metropolitan circle of China. In addition, the province has accelerated the construction of various economic corridors, expanded Hefei's transportation map, accelerated its integration into the Yangtze River Delta, and strengthened the Yangtze River Delta's radiation to the HeFei metropolitan circle of China to achieve a win-win situation for the region. Therefore, this paper takes the HeFei metropolitan circle of China as the research object, and analyses the urban habitat environment in various regions of the HeFei metropolitan circle of China by constructing a habitat environment evaluation system, and scientifically evaluates the quality of urban habitat environment with a view to creating a model for sustainable development of habitat environment, which is of great significance to the development of the HeFei metropolitan circle of China and Anhui Province.

2 Theoretical Foundations and Research Methods

2.1 Theoretical Connotations

Theory of humanism, a philosophical and literary movement that originated in Italy in the second half of the 14th century and spread to other European countries, is a philosophical perspective that also embraces a pragmatic and empirical approach [11]. One of the most prominent figures in humanistic theory was Maslow, who, in his study of human motivation, developed the famous "theory of needs" [12]. He believed that the hierarchy of human needs, in the shape of a ladder, gradually progresses from lower to higher needs, and consists of five major levels: physiological needs, security needs, social needs, respect needs, and self-actualisation needs. Physiological needs are the most basic, strongest, most primitive and most significant of all human needs [12]. Physiological needs after the most basic physiological needs are satisfied, people will further seek the satisfaction of security needs, including psychological security, physical security, environmental security, economic security, etc. Under this psychological need, people generally crave more order and stability. When physiological needs and security needs are satisfied, social needs will become the next level of strong needs, hoping to get support, love and understanding from others, and hoping to conduct social interactions between people. In addition, all people in society have a need or desire for self-esteem, self-respect and respect from others [13]. Finally, when all needs are met, there is a need for self-actualisation in the form of self-development [14].

The ideal urban habitat also responds to the change from low to high level. As the quality of life improves, people are no longer simply satisfied with physiological, safety and other low-level material functions, but have diversified requirements for the habitat, and increasingly need to meet the psychological perception of people pursuing high-level social activities and attention. The different levels of demand of urban residents are evident in the different stages of socio-economic development of the city in the temporal dimension, and in the spatial dimension, they depend on the coordination and interaction of the various levels of the urban territorial system, such as the background environment, infrastructure and social economy.

2.2 Selection of Indicators

The evaluation of urban habitat takes into account both the hard and soft environments. The hard environment mainly includes material factors, reflecting the physiological needs and security needs of urban residents in Maslow's hierarchy of needs theory [14]. The soft environment, on the other hand, mainly considers the cultural environment, the ease of communication and the employment situation of the residents, which reflect the higher level needs of Maslow's theory of needs.

In summary, based on Maslow's needs theory, the urban habitat environment is evaluated from different aspects and an urban habitat environment evaluation system is constructed [15] (Table 1).

First of all, physiological needs are mainly refers to the material level and can meet the basic needs of survival, such as "clothing, food, housing and transportation".

The first is livelihood security, i.e. the most basic security to live on, which is reflected by two indicators: urban residents' per capita disposable income (RMB) and urban residents' per capita consumption expenditure (RMB); the second is living conditions, reflecting the quality of urban residents' living, which is reflected by the indicator of urban per capita housing area (m^2); the second is basic medical care, which is reflected by the indicator of the number of hospital beds per 10,000 people, which is a good reflection of the level of medical care that cities can provide for their residents. Finally, infrastructure is the hard environment that best reflects the quality of life of urban residents, and is reflected by the two indicators of urban water penetration and urban gas penetration.

The second is the demand for security, that is, a human living environment that is safe, stable, dependable, free from intimidation, in need of order and the rule of law, and in which urban residents have a sense of security. 3 indicator criteria layers were selected as evaluation factors. The first is social security. A social environment with the rule of law and a structured system can well meet people's needs for security, and the number of public security cases per 10,000 people was chosen as an indicator to reflect the safety of the city. Lastly, ecological safety, with the improvement of the economic level, urban residents are paying more and more attention to the quality of the environment and ecological quality of the city, thus giving urban residents a safe ecological environment.

The third is social demand, this paper incorporates respect demand into social demand, mainly to evaluate the social interaction degree of the city's residents, whether the urban habitat environment can meet the needs of residents to obtain emotional value through social interaction, three indicator guidelines layer is selected. First, cultural needs, the number of books in public libraries per capita was selected as an indicator to reflect the cultural environment of the city; Second, social interaction, in the era of rapid development of the Internet, interpersonal interaction occurs more often on the

Table 1. Comprehensive urban habitat evaluation system.

Requirement level	Criterion layer	Indicator layer	Attribute
Physiological needs	Living security	Per capita disposable income of urban residents	+
		Per capita consumption expenditure of urban residents	+
	Living conditions	Urban per capita housing area	+
	Basic medical treatment	Hospital beds per 10000 people	+
	Infrastructure	Urban water penetration rate	+
		Urban gas penetration rate	+
Safety needs	Public security	Public security organs investigate and deal with public security cases per 10000 people	–
	Traffic safety	Traffic accidents per 10000 people	–
	Ecological safety	Per capita green area	+
		Percentage of days when air quality reaches Grade II	+
Social needs	Cultural needs	Public library collections per 10000 people	+
	Social communication	Number of computers owned by every 100 households	+
	Road traffic system	Per capita urban road area	+
Self-actualization	Job opportunities	the registered urban unemployment rate	–
	Salary	Average wage of urban non-private employees	+

Internet, mobile phones, offline environment survey is far less than the online environment can highlight social, the number of computers per 100 households was selected as an indicator, which can better reflect the degree of social contact of urban residents; Third, the road traffic system, whether the traffic is convenient or not, whether the road network density is average or not, all have a greater impact on residents' long-distance social interaction, and the urban road area per capita is chosen as an indicator to map the satisfaction of long-distance social interaction needs.

The fourth is the demand for self-actualization, which is the highest level of demand for urban habitat, and two criteria-level indicators were selected. The first is job opportunities, i.e. the ways in which residents can realize their self-worth and progress towards their ideal state. A high employment rate and low unemployment rate can provide more ways to realize oneself, which is reflected through the negative indicator of urban registered unemployment rate; the second is wage level, i.e. the economic level of urban residents, which also provides costs for self-actualization, and the positive indicator of the average wage of urban non-private sector employees is chosen to reflect this.

2.3 Research Methodology

This paper adopts the expert scoring method and AHP, according to the constructed index system, inviting experts and professors of the university and some postgraduate students to assign importance data to each different level of key index analysis, placing the data into yaahp, constituting a hierarchical index judgment assignment matrix and passing the data consistency analysis test, assigning weights to obtain the level key indexes in order.

At the same time, the data from the HeFei metropolitan circle of China in 2010, 2015 and 2020 were normalized by the polar transformation method [15]. The original data for each region falls between [0, 1], thus eliminating the effects of magnitude and order of magnitude. Assuming that the value of indicator j in year i of a city is Xij, the maximum value of indicator j is X_{max} and the minimum value is X_{min}, the normalization of the indicator can be obtained.

Drawing on relevant studies [9, 16, 17], combined with the evaluation system constructed in this paper, Eq. (1) is derived:

$$HI_{\mathrm{i}} = \sum_{\mathrm{j}=1}^{\mathrm{m}} \left(W_{\mathrm{ij}} \times N_{\mathrm{ij}}\right) \tag{1}$$

In the equation, HI_i is the Habitat Index of city i; m is the number of evaluation indicators j; W_{ij} is the weight value of the jth indicator; N_{ij} is the normalized value of the jth indicator.

3 Research Cases and Data Sources

Overview of the Study Area: HeFei metropolitan circle of China, located in the middle and lower reaches of the Yangtze River along the western end of the Yangtze River Delta, is one of the five major metropolitan areas of the Yangtze River Delta city cluster, an important hub and western gate of the Yangtze River Delta extending to the central and western parts of the region. The HeFei metropolitan circle of China includes Hefei, Huainan, Lu'an, Chuzhou, Wuhu, Ma'anshan, Bengbu and Tongcheng (county-level cities) in Anhui Province, with a regional area of 40.6% of the province and a population of 43.2% of the province [18]. With the rapid economic growth and urbanization process, the rapid development has also highlighted certain problems, which have a negative impact on the urban habitat. It is therefore of practical importance to analyse the quality of the urban habitat in the region and to give certain suggestions for its improvement (Fig. 1).

Data sources: data for the three time periods of 2010, 2015 and 2020 were selected for evaluation, statistical data for each evaluation index in the evaluation system were obtained from the Statistical Yearbook of Anhui Province (2010, 2015, 2022), other

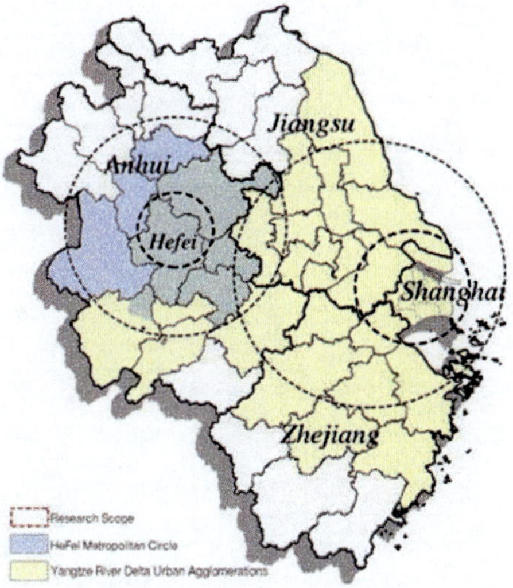

Fig. 1. Administrative map of the Yangtze River Delta city cluster.

data were obtained from the Statistical Yearbook of each city, the statistical bulletin on national economic and social development of each city (2010, 2015, 2022) and relevant statistics released by the government, and the administrative boundary data are obtained from the National Geographic Information Resources Catalogue Service System.

4 Analysis of Evaluation Results

Based on the results of expert scoring, the weights of the indicators in the urban habitat quality evaluation index system were calculated by AHP (Fig. 2), and combined with the normalized values of the indicators from, the scores of each city in the HeFei metropolitan circle of China at different levels of habitat quality and the overall scores of each city were calculated. Using ArcGIS10.3 software, the cities' habitat quality scores at different levels were graded by the natural fracture method (Table 2).

4.1 Results of Urban Habitat Measurement

In 2010, the average score of urban habitat quality in the HeFei metropolitan circle of China was 0.150551, with the highest value of 0.271154 for Maanshan and the lowest value of −0.035967 for Tongcheng, and the average score of 0.317566 in 2015, with the highest value of 0.421153 for Maanshan and 0.235138 for Tongcheng. In 2020, the average score is 0.469378, with the highest value being Maanshan City at 0.543274 and the lowest value being Huainan City at 0.394953. The increase is relatively large.

In order to analyze the spatial variation of the quality of human living environment in the HeFei metropolitan circle of China, each city was divided into five levels: lowest, lower, medium, higher and highest, using the classification criteria in Table 2 (Fig. 3).

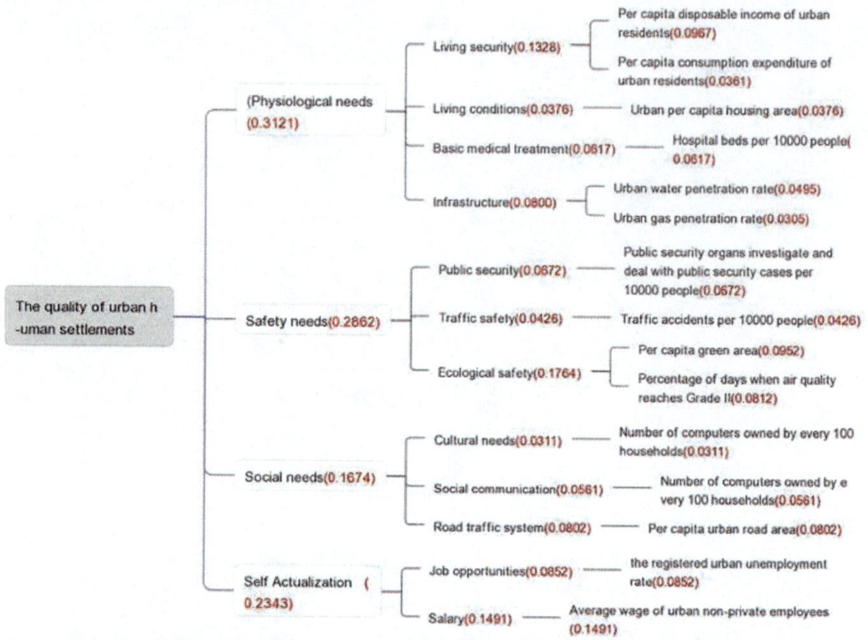

Fig. 2. Weighting of indicators for comprehensive evaluation of urban habitat quality.

The overall results of the comprehensive evaluation (Fig. 4) show that from 2010–2020, Maanshan has a greater overall growth and is leading the way. As can be seen from the figure, the southeastern cities score higher than the northwestern cities overall, especially after 2015, showing a more obvious east-west difference. Among them, Hefei, Maanshan, Chuzhou and Wuhu have been showing relatively high levels of urban habitat quality ratings in this 10-year process. Most of these cities are located in the southeastern part of the metropolitan area and belong to the Yangtze River Delta urban agglomeration, where the cities have higher levels of economic development and configuration of basic service facilities, and the quality of their habitat environment has thus been improved. However, due to the large urban population, there is still a shortage of public facilities per capita, and there is still much room for improvement in the quality of the living environment. The cities of Bengbu, Huainan, Liu'an and Tongcheng are at a lower level. These cities are mainly located in the north-western part of the metropolitan area and belong to the northern and western parts of Anhui. The lack of economic development in these regions has led to inadequate configuration of infrastructure and public service facilities, and the quality of the urban human living environment has yet to be improved.

4.2 Factors Influencing the Urban Habitat

Physical Needs. From the perspective of basic living needs, there are many factors affecting the basic physiological needs of urban habitat quality, mainly including four aspects of living security, living conditions, basic medical care and infrastructure, as

Table 2. Criteria for classifying natural break points for quality assessment in different dimensions.

	Comprehensive score	Physiological needs	Safety needs	Social needs	Self-actualization
Lowest	$-0.035967 < S \leq 0.158490$	$0.128570 < S \leq 0.025550$	$-0.017613 < S \leq 0.000000$	$0.019684 < S \leq 0.032327$	$-0.081806 < S \leq -0.045190$
Lower	$0.158491 < S \leq 0.235138$	$0.025551 < S \leq 0.103029$	$0.000001 < S \leq 0.031733$	$0.032328 < S \leq 0.061005$	$-0.045189 < S \leq -0.011884$
Middling	$0.235139 < S \leq 0.327781$	$0.103030 < S \leq 0.156849$	$0.031734 < S \leq 0.044192$	$0.061006 < S \leq 0.083793$	$-0.011883 < S \leq 0.016005$
Higher	$0.327782 < S \leq 0.465442$	$0.156850 < S \leq 0.208964$	$0.044193 < S \leq 0.063357$	$0.083794 < S \leq 0.110622$	$0.016006 < S \leq 0.055375$
Highest	$0.465443 < S \leq 0.543274$	$0.208965 < S \leq 0.278275$	$0.063358 < S \leq 0.100316$	$0.110623 < S \leq 0.140834$	$0.055376 < S \leq 0.129736$

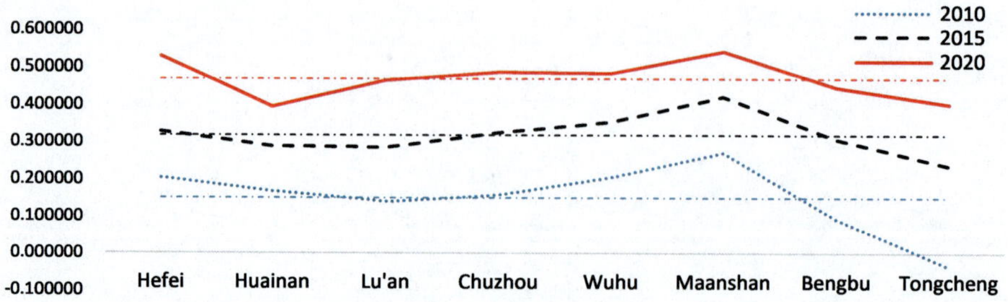

Fig. 3. Results of the comprehensive evaluation of the quality of the habitat environment of the cities in the HeFei metropolitan circle of China

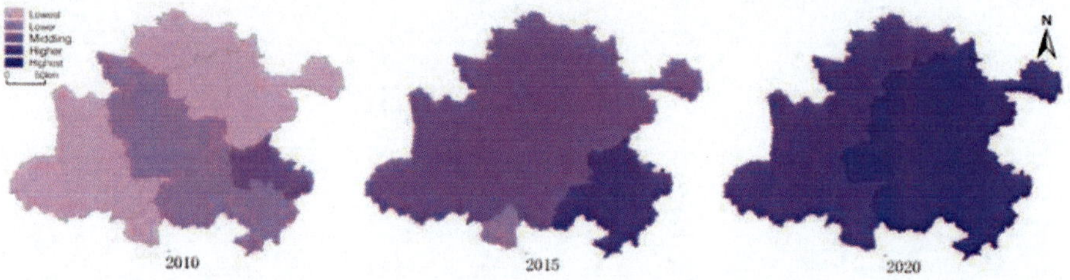

Fig. 4. Regional differences in the comprehensive evaluation of the quality of the human habitat in the HeFei metropolitan circle of China by city

well as six specific quantitative indicators such as urban residents' per capita disposable income, urban residents' per capita consumption expenditure and urban per capita housing area. The results of the physiological demand score for habitat quality show that the overall physiological demand score from 2010 to 2020 is on an upward trend, and the growth is relatively stable. Among them, Hefei has always taken the lead in the physiological needs score, which is inextricably linked to its status as the capital city of the province. In addition, as the centre of the metropolitan area, Hefei, together with Maanshan, Wuhu and Bengbu, is driven by the development of the Yangtze River Delta city cluster, and has a more developed economy and better infrastructure, which makes it more attractive to foreigners, and therefore has a higher and higher rating for the quality of the human environment in terms of physical needs. Lu'an, Huainan and Chuzhou have relatively low scores, mainly due to their large migrant populations and constraints on economic development, which have led to a lower level (Fig. 5).

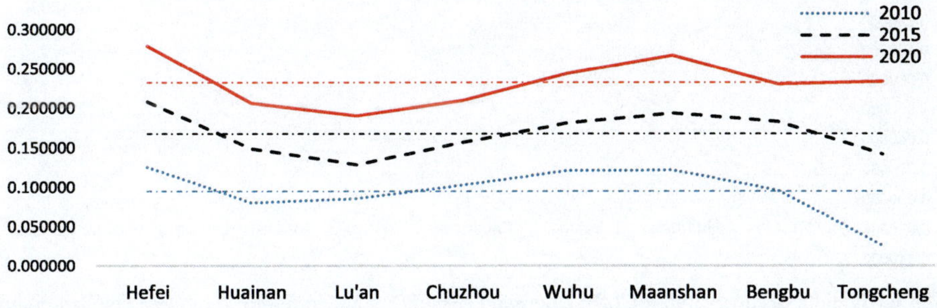

Fig. 5. Results of the evaluation of the physiological demand for habitat quality in the cities of the HeFei metropolitan circle of China

Security Requirements. From the results, we can see that the security demand level scores for urban habitat in the HeFei metropolitan circle of China are generally decreasing between 2010 and 2020, and the decrease is more obvious between 2010 and 2015.

In terms of ecological safety, the ecological quality of the city has improved to a certain extent in recent years, with the proportion of days with air quality reaching level 2 increasing significantly, and the ecological safety score is on the rise. As can be seen from the regional variance chart, the high level of urban habitat quality and safety demand scores in Lu'an is largely due to its mountainous location and good ecological environment, which makes urban ecological safety higher than that of other cities, as well as its relatively low population density, which reduces the probability of conflicts and makes social security relatively stable. Huainan has shown a more obvious downward trend in the last 10 years, due to the fact that Huainan, as a resource-based industrial city, is greatly affected by the depletion of resources and the environmental problems caused by industrial pollution. Hefei's rapid economic development and high population density has put greater pressure on ecological safety, while the economic attraction of migrant workers and the influx of a large number of migrant workers has also posed a significant threat to social security, resulting in a lower score on the safety dimension compared to other cities.

The other cities' overall scores are relatively flat, but on a declining trend. This calls for a combination of adherence to the rule of law and the rule of ethics in the subsequent development of the city, and the improvement of the quality of life of the residents, so that the social security of the city can be guaranteed, thus greatly improving the satisfaction of the city residents with the quality of their living environment (Fig. 6).

Social Needs. In the perspective of social interaction demand, the quality of urban habitat can be analysed in three aspects: cultural demand, social interaction and road traffic system.

From the results, we can see that the social demand scores of cities in the metropolitan area showed a rapid growth trend during the period 2010–2015, and a relatively slow growth during the period 2015–2020, with some cities showing a decreasing trend. As the capital city of the province, Hefei's rapid economic growth has attracted a large influx of foreigners into the city, triggering a massive gathering of people, accelerating interpersonal communication and promoting social interaction among city residents, so

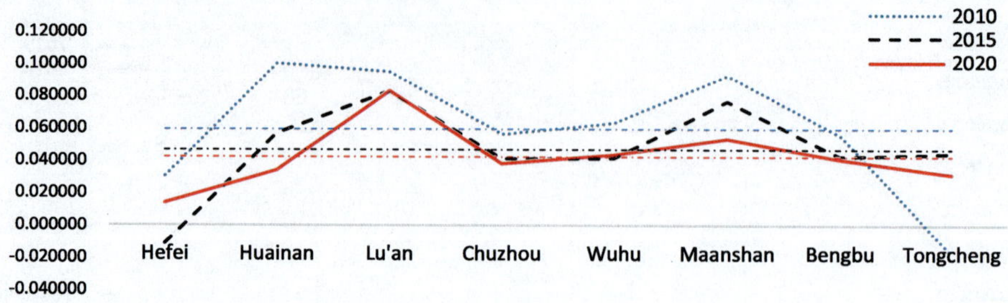

Fig. 6. Results of the evaluation of the demand for quality and safety of the human environment in the cities of the HeFei metropolitan circle of China

it ranked first in the metropolitan area in 2010. However, with the development of the metropolitan area, other cities are paying more and more attention to the development of social infrastructure and Hefei's urban infrastructure development is struggling to keep up with the rate of population growth, so the advantage of Hefei's scores is no longer obvious.

As can be seen from the figure, the social needs scores of cities in the eastern part of the metropolitan area have generally increased from low to high levels, while cities in the central and western parts of the metropolitan area have generally remained the same and are lagging behind, with significant differences in the development of social needs between the eastern and western parts. Hefei, Chuzhou, Ma'anshan and Tongcheng, which have seen rapid development of their socio-economic base over the last decade and a high degree of openness to the outside world, have also seen their social needs scores increase year on year and are above average. Huainan, as a resource-based city, is facing development challenges after the depletion of its resources, and its socio-economic development has been slow, showing a downward trend in the last five years, with solutions to be found. Bengbu and Liu'an, two cities that have experienced a significant population exodus in recent years, are experiencing slow economic development and have maintained their social interaction at the same level (Fig. 7).

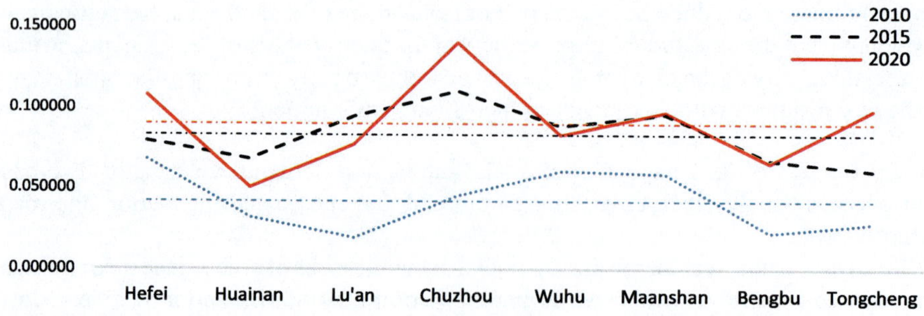

Fig. 7. Results of the evaluation of social demand for habitat quality in HeFei metropolitan circle of China cities

Self-fulfilment Needs. From the perspective of self-fulfilling demand, the metropolitan area clearly shows an upward trend during the period 2010–2020, and the growth rates are similar between the time periods 2010–2015 and 2015–2020. There are large differences between cities in 2010 and 2015, but as the development of the metropolitan area drives, the ability of each city to meet self-fulfilling demand in 2020, with the exception of Tongcheng, which is on the low side, the difference in capacity between the seven cities is not significant (Fig. 8).

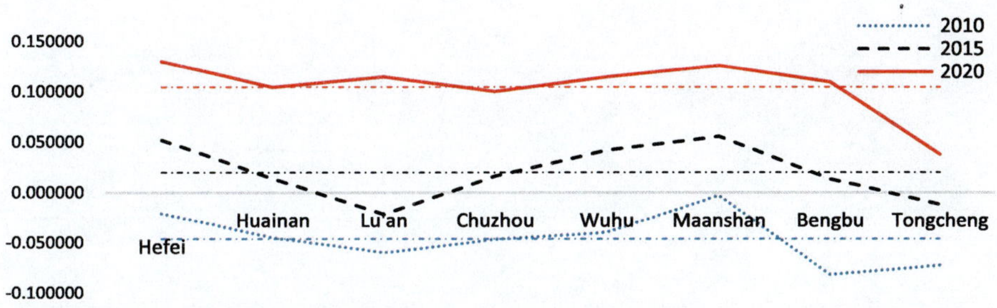

Fig. 8. Results of the self-fulfilling needs evaluation of the quality of the habitat in the HeFei metropolitan circle of China by city

From the regional divergence chart, we can see that Hefei, as the national centre of science and education, has rapid economic development, so the per capita wage is high and there are many employment opportunities, but the influx of a large number of foreign population, the urban registered unemployment rate also increased more than other cities, so although the capacity in 2020 is in the first place, but does not present an absolute advantage. The cities of Bengbu, Huainan and Maanshan have focused on for industrial development in recent years, introducing new industries and expanding the scale of industries, solving the problem of re-employment of the unemployed population and increasing per capita wages, thus driving the dimension score up from low and medium levels to higher and high levels. Chuzhou and Tongcheng, on the other hand, have more homogeneous industries, slow economic development and insufficient development opportunities for city residents, and are lagging behind in terms of urban development (Fig. 9).

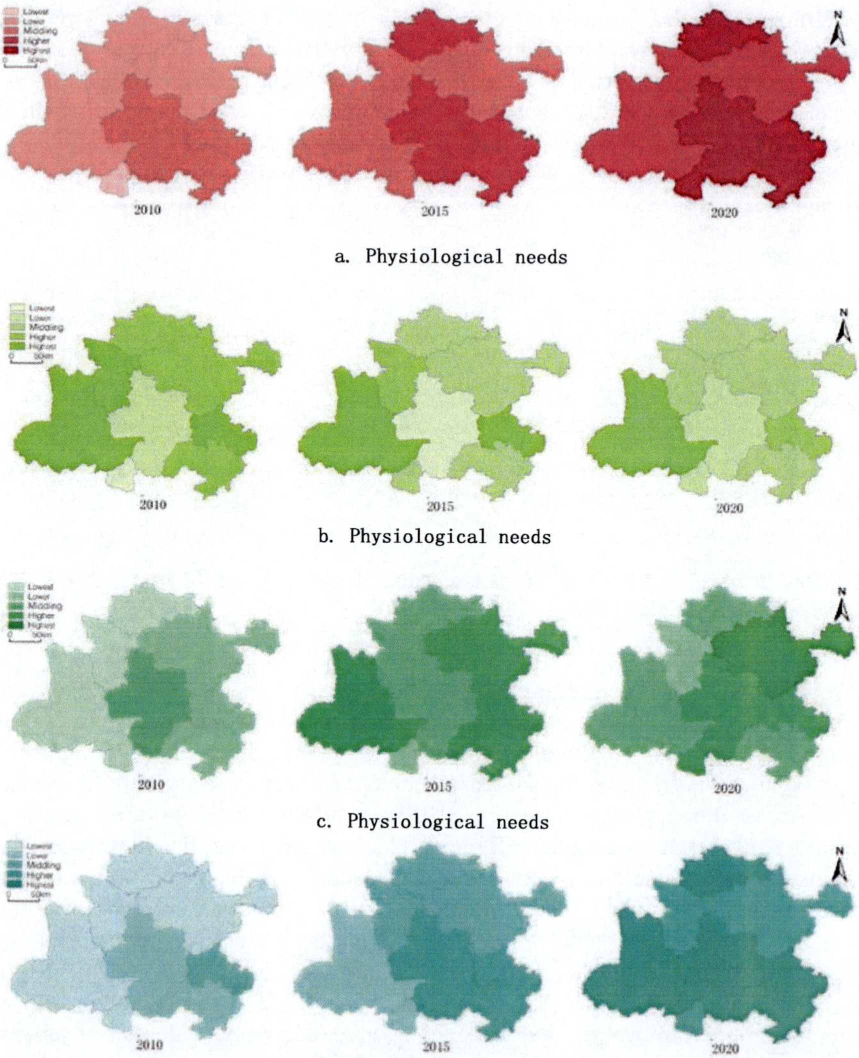

Fig. 9. Regional differences in the evaluation of the quality of the human habitat in the HeFei metropolitan circle of China by city

5 Conclusions and Recommendations

5.1 Conclusion

Based on theory of humanism, this paper constructs an evaluation system for the quality of human living environment in the HeFei metropolitan circle of China in 2010, 2015 and 2020 through AHP, and conducts an evaluation study on the quality of human living environment in each city under different dimensions. It can be seen that the overall quality of human living environment in the HeFei metropolitan circle of China has shown an increasing trend in recent years, but with the rapid development of the city, there are still some dimensions showing different degrees of decline in different cities. The main research findings are as follows:

Differences in Hefei's urban habitat quality scores are mainly in the areas of social and safety needs, including social security, traffic safety, green space per capita, air quality, library collections and road space per capita, which are most strongly associated with urban habitat quality. In addition, livelihood security, basic medical care and infrastructure construction also have a significant impact on the quality of the living environment. These factors contribute to the uneven development of cities in the HeFei metropolitan circle of China, which are important influencing factors in the development of the urban living environment.

From the results of the evaluation, we can see that the level of economic development largely affects the quality of the habitat environment. Cities with a weak socio-economic base have an obvious deficiency in the quality of their habitat environment compared to economically developed cities, reflecting that the basis for improving the quality of the habitat environment is to promote the economic development of the city.

As can be seen from the figure, cities close to the Yangtze River Delta city cluster, Jiangsu and Zhejiang, which are experiencing rapid economic development, are to a certain extent advantaged by their radiation to the surrounding cities, while cities in the west near the Dabie Mountain belt are underdeveloped in terms of habitat quality.

5.2 Recommendations

Enhancing the Economic Development of the City. An important measure to improve the quality of the urban habitat is to improve the overall economic level of the cities in the HeFei metropolitan circle of China. The cities in the western part of the metropolitan area are slow in economic development and have relatively low scores in terms of the quality of their living environment. By promoting the construction of various infrastructures, strengthening the economic ties between the cities and those with higher levels of economic development in HeFei and the eastern part of the city, and appropriately implanting new industries based on the sustainable development of the city, the city's development potential can be stimulated and the comprehensive economic level of the cities in the HeFei metropolitan circle of China can be improved.

Strengthening Ecological Civilization. In the above study, it can be seen that urban ecological civilization is an important indicator in the evaluation of the quality of human living environment. To strengthen the construction of ecological civilization control mechanisms for cities, from the perspective of sustainable urban development, attention should be paid to carbon emissions, green space system planning and other aspects to increase the area of green space per capita. In addition, for resource-based and industrial cities, it is necessary to accelerate the economic transformation of cities, scientifically co-ordinate resource development patterns and introduce green technologies, while conducting urban vulnerability analysis, establishing ecological civilization control systems and strengthening supervision and management. By strengthening the urban ecological environment, the quality of the urban habitat can be improved.

Strengthening Regional Integrated Planning. Research has found that economically developed big cities have economic, population and resource agglomeration effects, and have a strong radiation-driving ability for the surrounding cities, and the quality of the living environment of the surrounding cities will also be affected and improved to a certain extent. Therefore, it is important to strengthen regional planning at the top level, pay attention to the co-ordination between the city and the surrounding economically developed areas, stimulate the radiation-driving effect of the big cities on the surrounding cities, and drive the economic and cultural development of the city with the economic, cultural and talent resources of the big cities, so as to promote the improvement of the quality of urban living environment.

Acknowledgement. The work was supported in part by Anhui Provincial Housing and Urban-Rural Development Department Project (2020-RK04) and research on Urban and Rural Convenience Planning", Anhui Jianzhu University (HXS2019004).

References

1. Xiong, Y.: Uncertainty evaluation of the coordinated development of urban human settlement environment and economy in Changsha city. J. Geog. Sci. **21**, 1123–1137 (2011)
2. Wang, Yt.: Exploration on the application of urban habitat quality evaluation index system–based on the practice of urban physical examination pilot. Western J. Habitat Environ. **36**(06), 50–56 (2021)
3. Chen, J.: Temporal-spatial assessment of the vulnerability of human settlements in urban agglomerations in China. Environ. Sci. Pollut. Res. 1–17 (2022)
4. Xu, H.: A comprehensive and systematic evaluation of China's urban habitat construction through urban physical assessment system. Shanghai Urban Plan. **01**, 47–51 (2022)
5. Wang, Jg.: Overview: new trends and insights in urban habitat creation. J. Architect. **04**, 1–3 (2018)
6. Guan, Yy.: Comprehensive suitability evaluation of urban habitat in Liaoning Province based on geographically weighted regression model. Geoscience **42**(12), 2097–2108 (2022)
7. Wang, L.: Current situation and trend analysis of habitat research based on knowledge mapping. Green Technol. **24**(14), 164–169 (2022)
8. Cui, S.: A study on the spatial and temporal variation of urban habitat quality in Hubei Province based on entropy weight TOPSIS. J. Huazhong Normal Univ. (Nat. Sci. Ed.) **56**(04), 695–702+716 (2022)
9. Li, S.: Evaluation of urban habitat quality in Ningxia based on hierarchical analysis and entropy power method. J. Appl. Ecol. **25**(09), 2700–2708 (2014)
10. Dong, Z.: Comprehensive evaluation of habitat quality of Shandong Peninsula urban cluster. China Popul.-Resour. Environ. **27**(03), 155–162 (2017)
11. Xu, Xm.: Urban planning and community construction based on humanistic theory. Housing Real Estate **12**, 258+277 (2020)
12. Maslow, A.H.: A theory of human motivation. Psychol. Rev. **50**(4), 370 (1943)
13. Maslow, A.H.: Motivation and Personality. People's University of China Press, Beijing (2007)
14. Li, Xw.: Factors influencing rural habitat environment based on Maslow's Hierarchy of Needs theory - a green development perspective of urban and rural construction. J. Hebei Agric. Univ. (Soc. Sci. Ed.) **24**(04), 30–37 (2022)

15. Bao, J.: From survival to self-actualization: urban habitat quality evaluation from the perspective of hierarchy of needs theory–an example from Anhui Province. Urban Dev. Res. **27**(09), 88–95+140 (2020)
16. Chen, Cy.: Assessment of urban habitat quality and influencing factors in the Bohai Sea Rim. Adv. Geogr. Sci. **36**(12), 1562–1570 (2017)
17. Zhu, B.: Evaluation of rural habitat quality and spatial pattern analysis in Jiangsu Province. Econ. Geogr. **35**(03), 138–144 (2015)
18. Hefei Metropolitan Circle of China. https://baike.baidu.com/item/%E5%90%88%E8%82%A5%E9%83%BD%E5%B8%82%E5%9C%88?fromModule=lemma_search-box. Accessed 15 Mar 2023
19. Sun, Hb.: Evaluation of rural habitat environment quality and differentiated management strategies in China. J. Xi'an Jiaotong Univ. (Soc. Sci. Ed.) **39**(05), 105–113 (2019)

An Examination of the Suzhou Folk House Patio Space's Regional Characteristics Under the Concept of the Smart City

Tianhui Li[1,2](✉) and Yao Wu[1,2]

[1] Soochow University, 199 Renai Road, Suzhou, China
1098718820@qq.com

[2] International Cooperative Joint Laboratory of Cultural Heritage Conservation Science Supported By the Belt and Road Initiative, 199 Renai Road, Suzhou, China

Abstract. The idea of a "smart city" is focused on the individual, investigating more effective urban planning and design techniques from the hobbies and religious aspirations of the local populations. A space form dominates the patio and is the most recognizable space structure, which still plays a role that cannot be ignored. Suzhou residential houses are organic products adapted to the favorable natural conditions, developed water network system, long traditional culture, and thousands of years of feudal and religious system family form in the Suzhou area. This study primarily describes the spatial shape, pattern, functional elements, etc. of a Suzhou residential patio space. According to the findings, Suzhou's residential patio space prominently displays regional features in terms of its construction, spatial organization, usage of functions, and spiritual connotations, among other things. A thorough comprehension of its structure and spiritual implications is necessary for taking a more active part in the creation of the smart city.

Keywords: Smart City · Suzhou Folk House · Patio Space · Regional Culture

1 Introduction

The patio is a space environment that mixes nature, art, and mood, and Suzhou residential dwellings are renowned for being a significant feature of Chinese residential architecture [1]. Due to the various regional contexts, the patios in Suzhou homes have a vivid and distinctive spatial shape, and their design, materials, modeling, spatial creation, and building techniques all have a distinct tendency [2]. Based on the regional peculiarities of Suzhou's residential patio space for smart city renewal design, this particular expression of regional culture can strengthen residents' feelings of community and foster a sense of place. Combining the idea of a smart city with a study of regional culture is more appropriate for the future growth route of China's cities given the advancements made by the new generation of information technology, such as cloud computing and the Internet of Things.

W. Guo et al. (Eds.): GBCESC 2023, LNCE 328, pp. 1030–1036, 2024.
https://doi.org/10.1007/978-981-99-9947-7_99

2 Smart City and Regional Culture

The term "smart city" refers to the use of intelligent computer technologies in the fields of urban planning, design, construction, management, and operation to enhance the effectiveness of urban operations, citizen quality of life, and the quality of governmental services [3]. These technologies include the Internet of Things, cloud computing, big data, spatial geographic information integration, and other intelligent computer technologies. Regional culture is a crucial factor to take into account while creating a smart city since it offers a distinctive perspective and method of understanding that can be used to better comprehend and utilize the city's resources.

First, local culture can influence the design of smart cities. Every location has its distinct history, customs, and values, which can have an impact on how smart cities are built. Second, smart cities can support efforts to preserve local culture. Big data and artificial intelligence are two examples of smart city technologies that can be utilized to conserve and transmit local culture. For instance, big data can be utilized to study and comprehend local cultural patterns, or digital technology can be used to conserve and display historical structures [4]. The link between local culture and smart cities comes in third. Citizens must actively participate in the creation and maintenance of a smart city. Regional culture can affect how people act and think. For instance, people may prefer to use public transit in some locations while choosing to drive their cars in others. This necessitates that smart cities be created with these cultural variations in mind. Finally, regional culture influences how well smart cities are received. Geographical culture has an impact on how each city or town accepts new technologies. Some regions may be more accepting of new technologies than others. These geo-cultural variations must be taken into account for the development of smart cities to be successful.

3 Regional Characterization of Suzhou Folk House Patio Space

3.1 Morphological Structure

Smaller Spatial Scale. The scale of the patio space of residential homes in Huizhou, Xiangxi, Fuzhou, and other regions is often tiny, reflecting the construction concept of "small and beautiful" because the idea of "seclusion" is frequently promoted in the south [5]. The river network in Suzhou covers a significant portion of the land, which is used for construction, and the building colonies are "small in scale, to the extent that it is impossible to reduce." The total scale of Suzhou residences is typically tiny due to this high-density architecture, therefore the patio space is typically narrow and lengthy, with a width of 3–4 m, and is becoming increasingly tightly packed. Even smaller, measuring just 1.8–2 m wide, is Suzhou's one-of-a-kind "crab's eye patio".

Enclosed Interfaces are More Lushly Manicured and Open. Although the scale of Suzhou residential is small and the patio is narrow, it does not feel confined in the middle of the building because the buildings of the enclosure interface are primarily "one bright, two dark" three-room style. Additionally, the central axis centered on the patio is also an open horizontal line of sight. On the other hand, the patio of Huizhou homes is thought of as an addition to the architectural space, which is typically disregarded as a building accessory and is largely walled off [6]. In Suzhou, which is known for its picturesque surroundings, abundant vegetation, and exquisite gardening techniques, residential homes are frequently embellished, and the walls of the enclosing interfaces are typically painted white to increase spatial permeability, reflect light, and lower interior building temperatures by blocking the sun through vertical greening.

Lowering the Bottom Interface's Elevation and Positioning Drainage Structures. The majority of patio-style homes are found in regions with high rainfall, with annual precipitation typically ranging from 1,600 to 3,000 mm. Common examples are the Huizhou homes in the Jianghuai Plum Rain Region and the Suzhou homes in South China. Suzhou receives a lot of rain, so the patio area needs to be able to collect rainwater. The patio space's overall elevation is lower than the surrounding floor, and it has stone edges to reinforce the space's perimeter and prevent rainfall from spreading inside. To accommodate Suzhou's wet climate, drainage pipes, and downspouts are also positioned at the base of the interface (Fig. 1).

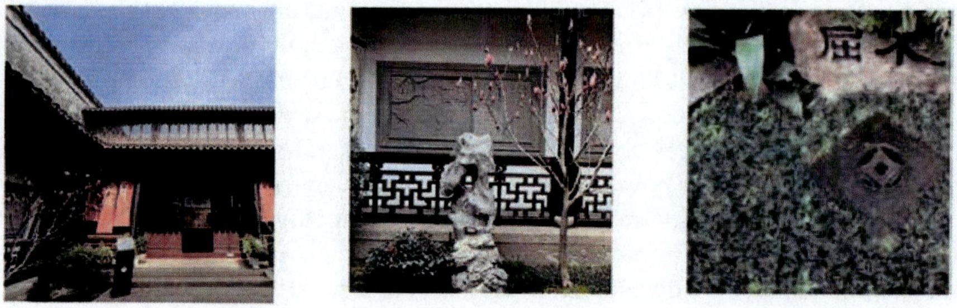

Fig. 1. Suzhou Folk House Patio Space

3.2 Space Organization

Relationships on the Bottom of the Map Centered on the Patio. All varieties of residential buildings in China are built with safety as their top priority, creating rooms that are inward-looking. However, two different relationships have developed from the identical enclosure layout. The first is the northern courtyard type of residential home, where the perimeter of the inner courtyard is surrounded by a single building, which serves as the building's core; the second is the southern patio type of residential home, where the patio space is more akin to a "dug out" within the building group. The patio area, however, resembles a "dig" in the building group more. Suzhou houses are typical southern patio homes, and their patio area has a deeper significance, i.e., the patio is the core space of Suzhou homes. On the one hand, the patio space performs the majority of the lighting, ventilation, and drainage functions of the Suzhou homes; on the other hand, the patio space is the core of the Suzhou homes, and the axial arrangement of the entire house and the arrangement of the rooms are always centered on the patio. As a result, the Suzhou residential houses have finished inverting the plan's base, creating a relationship between the base and the patio's area as the center (Figs. 2 and 3).

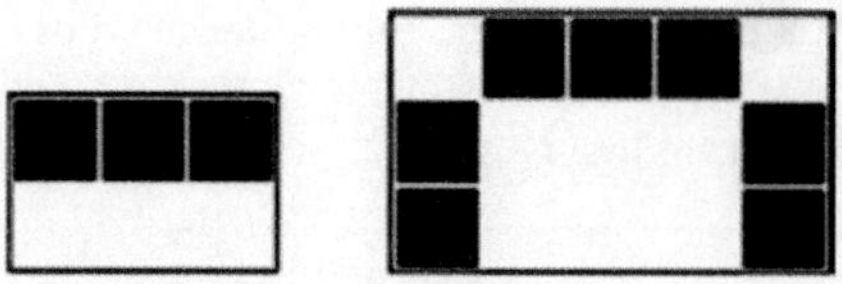

Fig. 2. The Enclosed Courtyard

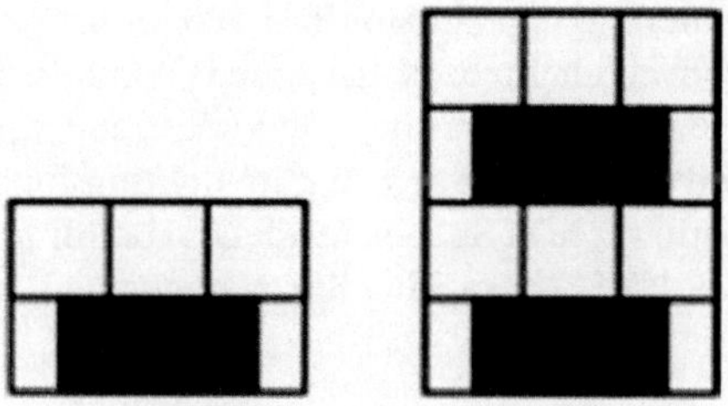

Fig. 3. The Dug-out Patio.

Axial Relationships Centered on the Patio. Suzhou homes, in contrast to other patio-style homes, employ the patio's area as their central focus, creating a distinct axial relationship that is centered on the patio. The overall development of the patio space as the core of the vertical and horizontal extension of the Suzhou is the patio, which serves as the "in" and "down" points along the axis formed by several patios. The result is a Suzhou-style home that spreads out horizontally and vertically from the patio area (Fig. 4).

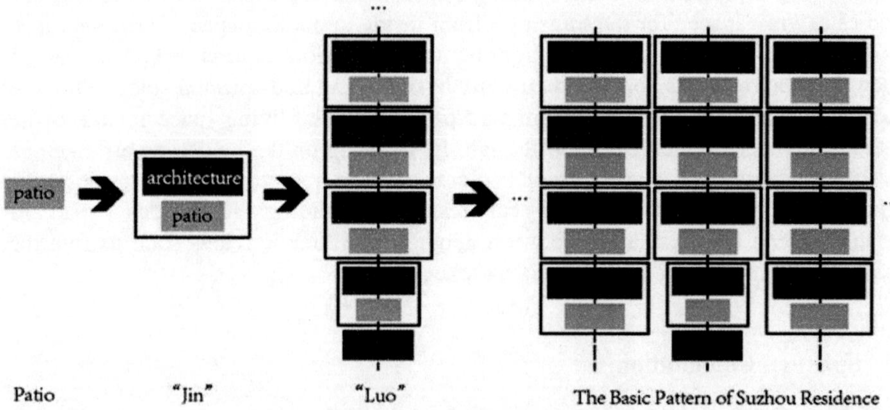

Fig. 4. The Basic Pattern of Suzhou Residence.

3.3 Usage Functions

Emphasis on Ventilation and Drainage, Light is not a Major Requirement. The patio often features several structures for lighting, ventilation, and drainage. It also plays a special role in preventing fires and enhancing the aesthetics of the space [7]. The patio space of Suzhou houses is focused on the need to take into account and satisfy the drainage of the architectural complex because Suzhou is a rainy and humid region with an average annual precipitation of 1063 mm, occasional extreme rainstorms in summer, and continuous rainfall in the rainy season, which is prone to cause extensive waterlogging. Suzhou has scorching summers, therefore to direct airflow and create convection breezes, the patio is combined with high walls and north windows. However, the use of such a small and constrained patio space does not improve the use of natural light, demonstrating that its lighting function is not the primary functional need of the patio space of Suzhou residential buildings. The average annual sunshine hours in Suzhou are 1965.0 h, making the average annual illumination low (Fig. 5).

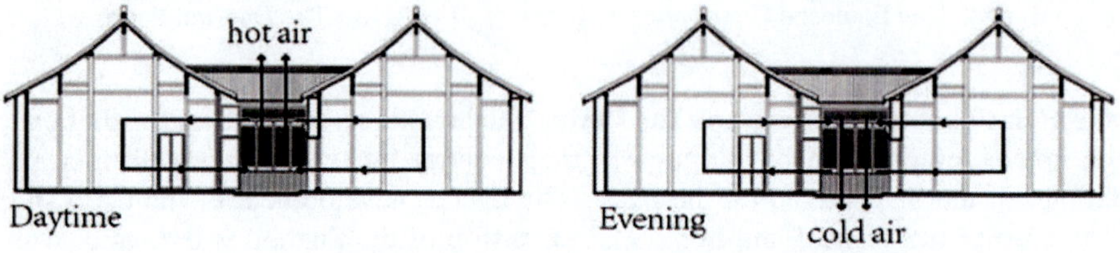

Fig. 5. Schematic Diagram of Air Circulation in the Patio.

Emphasis on Life Functions. The patio area of Suzhou Homes serves a variety of living purposes in addition to the building's specialized functions. Most patio areas are used as a "gray space" for the transition from inside to outdoor spaces, accounting for the small architectural complexes' lighting and ventilation requirements. The patio is where Suzhou residents conduct their daily business and find spiritual solace. In large-scale residences, it serves as a significant production and living space for the owner of the residence and the visitors of the family, focusing on the landscape arrangement and beautification, and plays a crucial role as a traffic node in the progressive spatial sequences of the complex; in small-scale residential buildings, it provides a space for the intersection of information between neighbors, where activities such as fighting, persuading fights, eating, and conversing take place.

3.4 Spiritual Connotation

The Ideas of "The Unity of Heaven and Man" in Philosophy and "Although Made by Man, Just Like From Heaven" in Gardening. Both Suzhou and Huizhou are ardent supporters of Confucianism and adhere to a rigid hierarchy of superiors and inferiors. The patio connects the sky, man, and the earth, reflecting the philosophical idea that "heaven and man and one," respecting the harmony between heaven and man, nature, and society as a whole, and using the patio to do so, which embodies the cultural sentiment of the city of Suzhou of not getting upset and not being shocked by insults and humiliations. The courtyard of Suzhou houses is closed and conservative. The patio space, which transitions from artificial to natural and from indoor to outdoor, reflects the gardening idea that "although made by man, it is like the sky opening up from the sky" and is an architectural space that has been long-adapted to Suzhou's climatic conditions and lifestyle [8].

Territoriality and Belonging. The patio is the inwardly enclosing space of the building complex, which is particularly centripetal and provides individuals with a sense of physical and psychological solitude and protection. Suzhou culture is introverted. The patio area in Suzhou Homes is the central area that ties the entire structure together. It allows residents to enjoy close contact and communication with nature without having to leave their homes, giving them a personal slice of heaven and earth. This meets the natural territorial needs of people for homes that date back thousands of years. The patio is the embodiment of a sense of belonging and stands in for communication, security, gratitude, and trust in the minds of the people of Suzhou.

4 Techniques for Applying Suzhou's Regional Features to Residential Patio Spaces in the Context of Smart Cities

Based on the aforementioned research, it is evident that Suzhou residential patio space has extremely unique regional characteristics in terms of shape, spatial organization, usage function, and spiritual connotation, and this regional culture is rooted in ancient gardening knowledge of China. This kind of smart space and regional features can still be developed and used under the guidance of the smart city concept.

4.1 Incorporate Environmental Sensing and Intelligent Control

The patio area of Suzhou homes is small in size, serves as a decorative area in a typical courtyard, and serves as a drainage area. To implement the monitoring of the patio space, sound, light, heat, and other important information, and to achieve the link between the patio space and the people, through the network level of access to achieve the application of information sensors, infrared sensors, laser scanners, and other devices, real-time collection of environmental data in the patio space, and analysis using big data.

4.2 Promoting Spatial Interaction and Upgrading of Services

A new smart city is an ecosystem of coordinated urban development through systematic planning, information-led reform, and innovation. It encourages the deep integration and iterative evolution of new-generation information technology and urban modernization with the primary goal of serving the people throughout the entire process, as well as urban governance in an effective and orderly manner. It is clear that the smart city is more than just a technological advancement; it also represents an improvement in urban lifestyle and a commitment to servicing the needs of its residents. The cohesion of inhabitants can be improved by creating community-shared spaces in the patio areas of certain old residential buildings and gardens, as well as by leveraging mobile internet and the Internet of Things to offer contemporary services.

4.3 Zero Energy Buildings and Green Energy Applications

Suzhou homes have well-designed patio spaces that provide good natural lighting and ventilation, reducing the need for interior lighting, air conditioning, and other equipment while also consuming less energy. To reduce the control of energy usage, it can also contain green energy equipment, such as solar panels for interior power generation in inclement weather and seasonal heating or cooling. It can also be combined with intelligent network technology. Emulate this idea in modern city architecture for passive energy-saving design driven by smart cities.

4.4 Better Historical and Cultural Preservation

It is crucial to preserve the historical and cultural aspects of Suzhou's residential buildings since the smart city idea places a strong emphasis on the peaceful cohabitation of people, the city, and the environment. To help more people understand and appreciate Suzhou's ancient culture, technologies like 3D scanning and Augmented Reality (AR) can be employed to digitally preserve and exhibit the patio spaces of Suzhou's residential structures.

5 Conclusion

Suzhou residential buildings have developed a distinctive patio space that embodies the dynamic equilibrium of harmonious symbiosis between architecture and nature after hundreds of years of climatic adaptation and historical changes. At the same time, they have formed and derived an overall similar, harmonious, and orderly architectural organization system. It strikes a harmony between people and nature, architecture and space, and fills in some of the architectural needs that the Suzhou residential buildings are too small to handle, which is also a crucial factor to take into account when building contemporary smart cities. In general, the smart city has a tight relationship to regional culture, and while developing a smart city, the influence of regional cultural traits should be completely acknowledged to create a diverse modern metropolis.

Acknowledgment. The writing and research of this paper are obtained from the 2021 Policy Directed Program of Jiangsu Province (BZ2021015).

References

1. Fangsheng, Y.: On Suzhou folk houses. Archit. J. (01), 35–43 (1990)
2. Li, L.: The discourse paradigm of traditional courtyard space-analysis of discourse components and grammatical rules. Central China Constr. **33**(09), 39–42 (2015)
3. Ya, W.: Design elements and development strategies of smart city public space based on regional culture. J. Jiangsu Second Teach. Coll. **38**(02), 9–52+75+124 (2022)
4. Dandan, Z., Xuehui, W.: The application of geographical symbols in smart city public space–the Lanzhou waterwheel as an example. Intell. Build. Smart Cities (09), 36–37+40 (2018)
5. Xiyuan, W., Xiaowen, X.: Study of patios in traditional residential architecture. J. Nanchang Univ. Hum. Soc. Sci. Ed. **41**(12), 4 (2010)
6. Biao, C.: Ecological consciousness of Jiangxi's Tianjin folk dwellings. Southern Constr. **10**, 86–88 (2006)
7. Xia, Z., Sijin, H., Yan, X.: Design combined with nature - ecological wisdom and application of traditional houses in hot summer and cold winter regions. Central China Constr. **33**(12), 4 (2015)
8. Minsu, X.: Suzhou Residential Houses. China Construction Industry Press, Beijing (1991)

Author Index

W. Guo et al. (Eds.): GBCESC 2023, LNCE 328, pp. 1037–1040, 2024.
https://doi.org/10.1007/978-981-99-9947-7